Part I –

Proceedings of the

17th International Conference on Low Temperature Physics

LT-17

Part I - Contributed Papers

Universität Karlsruhe and
Kernforschungszentrum Karlsruhe
15–22 August 1984

Editors:

U. Eckern

Fakultät für Physik, Universität Karlsruhe

A. Schmid

Fakultät für Physik, Universität Karlsruhe

W. Weber

Kernforschungszentrum Karlsruhe

H. Wühl

Kernforschungszentrum Karlsruhe

Organized under the aegis of
THE INTERNATIONAL UNION OF PURE AND APPLIED PHYSICS
by
Universität Karlsruhe and Kernforschungszentrum Karlsruhe

1984
NORTH-HOLLAND
AMSTERDAM · OXFORD · NEW YORK · TOKYO

ISBN: 0 444 86910 7

Published by:

North-Holland Physics Publishing
a division of
Elsevier Science Publishers B.V.
P.O. Box 103
1000 AC Amsterdam
The Netherlands

Sole distributors for the U.S.A. and Canada:

Elsevier Science Publishing Company, Inc.
52 Vanderbilt Avenue
New York, NY 10017
U.S.A.

Printed in The Netherlands

PREFACE

This book, in two volumes, contains the "contributed papers" of the XVIIth International Conference on Low Temperature Physics, held in Karlsruhe, Federal Republic of Germany, from August 15 to August 22, 1984. They are the first part of the conference proceedings and are made available to the participants at the beginning of LT-17. The second part of the proceedings includes the plenary lectures, the invited papers, as well as a selection of post-deadline contributions; this will appear after the conference as a special issue of Physica B. The organizers of LT-17 would like to thank the production department of North-Holland Physics Publishing, especially Ms. Mary Carpenter, for their valuable help and guidance.

We wish to mention that, for quick and convenient orientation, an abstract booklet was mailed to the conference participants in advance. It contains the conference schedule and the abstracts of all papers which are presented at the conference.

The listing of papers, which is included in these volumes, essentially agrees with the one in the abstract booklet; however, some titles were omitted in cases where the paper is not included in the proceedings.

The structure of the conference was planned to be similar to the preceding LT conferences held in Los Angeles in 1981 and in Grenoble in 1978.

In general the morning sessions begin with plenary lectures while the subsequent parallel sessions contain both invited and contributed papers. Apart from the closing day, the afternoons of each conference day are mainly devoted to the poster sessions.

The main difference between this conference and the preceding LT conferences was the plan of the organizers to put even more emphasis on posters than before. In general this plan resulted from the organizers' firm belief that the poster is a form of presentation as good as, if not superior to, a brief talk. The majority of the contributed papers (about 80%) have been assigned to poster sessions.

During the afternoons plenty of time and space is allowed for the poster presentations. This arrangement is expected to give excellent possibilities for extended discussions. If the participants feel that certain topics need further and more thorough consideration, it will be possible to arrange special discussion meetings or workshops in an ad-hoc manner. The following discussion meetings have been organized in advance:

- Hydrogen in Metals: chairman B. Stritzker
- Magnetic Monopoles: chairman B. Cabrera
- Spectroscopy with Superconductors: chairman Y.M. Rowell
- Proximity Effect at very Low Temperatures: chairman J.L. Olsen
- Neutron Scattering and Superconductivity: chairman W. Reichardt

The organizers have tried to make the conference truly international. Attempts have been made to assist scientists from developing countries to participate at the conference. The organizers wish to thank the International Union for Pure and Applied Physics for financial support. Many other institutions also supported the conference. All their help is gratefully acknowledged.

The manuscripts submitted for presentation at the conference were refereed by the Program Committee. We wish to thank all members of the Committee for their engagement. The International and National Committees were most helpful with their suggestions to the Program Committee in selecting the invited speakers.

The organizers also wish to thank various members of the Fakultät für Physik der Universität Karlsruhe who helped to prepare these volumes, in particular, Mrs. R. Schrempp, Mrs. K. Weisshaupt, Mrs. M. Sloan, Mrs. U. Bolz, Professor W. Buckel, Dr. H. Hinsch, W. Eiler, H.J. Jetter, U. Klein, W. Lehr and A. Ludviksson.

The Editors

U. Eckern
A. Schmid
W. Weber
H. Wühl

CONFERENCE CHAIRMAN

Professor Werner BUCKEL (Karlsruhe)

INTERNATIONAL COMMITTEE

G. ALBRECHT* (GDR)
K. ANDRES (FRG)
R. DE BRUYN OUBOTER* (The Netherlands)
T. CLAESON (Sweden)
F. DE LA CRUZ (Argentina)
G. DEUTSCHER (Israel)
WEI-YEN GUAN (P.R. of China)
J.P. HARRISON* (Canada)
G. KOZLOWSKI (Poland)
A.J. LEGGETT (USA)
E. LERNER* (Brazil)
O.V. LOUNASMAA (Finland)
I. MODENA* (Italy)
S. NAKAJIMA (Japan)
R. ORBACH (USA)
M. PETER (Switzerland)
F.D.M. POBELL* (FRG)
H. POSTMA (The Netherlands)
R.C. RICHARDSON* (USA)
R.S. SAFRATA* (CSSR)
Yu.V. SHARVIN* (USSR)
T. SHIGI* (Japan)
H. SMITH (Denmark)
R. SRINIVASAN (India)
M. TINKHAM (USA)
R. TOURNIER (France)
B.I. VERKIN (USSR)
W.F. VINEN* (UK)
G.V.H. WILSON (Australia)
J. WINTER* (France)

** Member of IUPAP Commission 5*

PROGRAM COMMITTEE

U. ECKERN (Karlsruhe)
R. FLÜKIGER (Karlsruhe)
R.P. HÜBENER (Tübingen)
S. HUNKLINGER (Heidelberg)
W. JUTZI (Karlsruhe)
H. KINDER (München)
P. LEIDERER (Mainz)
J. MOOIJ (Delft)
J.A. MYDOSH (Leiden)
F.D.M. POBELL (Bayreuth)
T.M. RICE (Zürich)
A. SCHMID* (Karlsruhe)
W. WEBER (Karlsruhe)
P. WÖLFLE (München)
D. WOHLLEBEN (Köln)
H. WÜHL (Karlsruhe)

** Chairman*

NATIONAL COMMITTEE

O.K. ANDERSEN (Stuttgart)
K. ANDRES (München)
E.H. BRANDT (Stuttgart)
G. EILENBERGER (Jülich)
W. EISENMENGER (Stuttgart)
H.C. FREYHARDT (Göttingen)
R. GREMMELMAIER (Erlangen)
H.D. HAHLBOHM (Berlin)
R.P. HÜBENER (Tübingen)
S. HUNKLINGER (Heidelberg)
J. JÄCKLE (Konstanz)
W. KLOSE* (Karlsruhe)
V. KOSE (Braunschweig)
K. LÜDERS (Berlin)
G. v.MINNIGERODE (Göttingen)
E. MÜLLER-HARTMANN (Köln)
D. RAINER (Bayreuth)
K.F. RENK (Regensburg)
G. SAEMANN-ISCHENKO (Erlangen)
F. STEGLICH (Darmstadt)
L. TEWORDT (Hamburg)
E.F. WASSERMANN (Duisberg)
P. WÖLFLE (München)

** Chairman*

LOCAL COMMITTEE

F. BAUMANN
P. EMMERICH
J. HASSE
H. HINSCH
A. KASTEN
H. RIETSCHEL
Mrs. E. SCHRÖDER

SPONSORS

Deutsche Forschungsgemeinschaft (DFG)
Fachausschuss Tiefe Temperaturen (Deutsche Physikalische Gesellschaft)
International Union of Pure and Applied Physics (IUPAP)
Kernforschungszentrum Karlsruhe
Ministerium für Wissenschaft und Kunst des Landes Baden-Württemberg
Universität Karlsruhe

SUPPORTERS

Brown, Boverie & Cie (BBC)
Bruker Physik
Interatom
International Business Machines (IBM Deutschland)
Linde AG
Rank Xerox GmbH
Siemens AG

EXHIBITORS

CRYOGENIC CONSULTANT
London, UK

CRYOMAGNETICS, Inc.
Oak Ridge, USA

CRYOPHYSICS GmbH
Darmstadt, FRG

DEUTSCHE L'AIR LIQUIDE
Wiesbaden, FRG

EG&G INSTRUMENTS GmbH
München, FRG

FRANCO CORRADI
Milano, Italy

ISOTEC
Centerville, Ohio, USA

KLAUS SCHÄFER GmbH
Langen, FRG

KRAUTZ GmbH
Meerbusch, FRG

MAGNEX SCIENTIFIC Ltd.
Oxford, UK

NORTH-HOLLAND PHYSICS PUBLISHING
Amsterdam, The Netherlands

OXFORD INSTRUMENTS
Wiesbaden, FRG

S.H.E. GmbH
Aachen, FRG

SPRINGER-VERLAG
Heidelberg, FRG

THOR CRYOGENICS Ltd.
Oxford, UK

CONTENTS TO PARTS I AND II - CONTRIBUTED PAPERS

PART I: CONTRIBUTED PAPERS

(first volume, pages 1–710)

AC - TEXTURES AND VORTICES IN SUPERFLUID ^{3}HELIUM I

AD - SUPERCONDUCTIVITY AND MAGNETISM I

AE - NUCLEAR MAGNETIC ORDERING

AF - MICROFABRICATION

AG - TEXTURES AND VORTICES IN SUPERFLUID ^{3}HELIUM II

AH – SOUND IN LIQUID ^{4}HELIUM AND MIXTURES

AI – SUPERCONDUCTIVITY AND MAGNETISM II: THEORY / BORIDES

AK – REFRACTORIES AND OTHER TRANSITION METAL COMPOUNDS

AL – MAGNETİSM I

AM – MAGNETISM II

AN – PROPERTIES AND TECHNOLOGY OF JOSEPHSON JUNCTIONS

BB – CHARGES AND MOLECULES AT LIQUID HELIUM SURFACES

BC – HEAVY FERMION SUPERCONDUCTORS I

BD – MULTILAYERS, SUPERLATTICES

BE – MAGNETIC PROPERTIES OF SOLID ^{3}HELIUM I

BF – HEAVY FERMION SUPERCONDUCTORS II

BG – LOW ENERGY EXCITATIONS IN GLASSES

BH – SMALL CRYOCOOLERS AND HIGH SENSITIVITY MEASURING DEVICES

BI – MAGNETIC PROPERTIES OF SOLID ^{3}HELIUM II

BK – IONS AND FILMS

BL – FLOW AND TURBULENCE

BM – HEAVY FERMION SYSTEMS

BN – SUPERCONDUCTIVITY AND MAGNETISM III: CHEVREL COMPOUNDS AND OTHERS

BO – AMORPHOUS MATERIALS, TWO-LEVEL SYSTEMS

BP – STANDARDS AND PRECISION MEASUREMENTS

CA – INTERFACES AND SURFACES

CB – NON-EQUILIBRIUM SUPERCONDUCTIVITY I

CC – SPIN GLASSES

CD – DIGITAL APPLICATIONS OF TUNNEL JUNCTIONS

CE – SPIN-POLARIZED HYDROGEN

CF – PINNING

CG – MONOCHROMATIC PHONONS, ACOUSTIC MICROSCOPY, AND KAPITZA RESISTANCE

CH – JOSEPHSON JUNCTIONS

CI – SPIN-POLARIZED ^{3}HE AND ^{3}HE-^{4}HE MIXTURES

CK – A15 COMPOUNDS I

CL – LOCALIZATION I

CM – HELIUM CRYSTAL GROWTH AND MELTING

CN – SPIN-POLARIZED SYSTEMS

CO – PINNING, VORTEX STRUCTURES

CP – A15 COMPOUNDS II

CQ – PROPERTIES OF SPIN GLASSES

CR – PHONON PHENOMENA

CS – SPATIAL STRUCTURES IN JOSEPHSON JUNCTIONS

PART II: CONTRIBUTED PAPERS

(second volume, pages 711–1394)

DC – UNUSUAL SUPERCONDUCTORS I

DD – LOCALIZATION II

DE – PERCOLATION AND SUPERCONDUCTIVITY

DF – MAGNETIC COUPLING OF 3-HE TO SOLID SURFACES I

DG – LOCALIZATION AND SUPERCONDUCTIVITY

DH – NETWORKS OF WEAK LINKS I

DI – MAGNETIC COUPLING OF ^{3}HE TO SOLID SURFACES II

DK – SOUND AND COLLECTIVE MODES IN SUPERFLUID ^{3}HE

DL – NON-EQUILIBRIUM SUPERCONDUCTIVITY II

DM – TUNNELING

DN – UNUSUAL SUPERCONDUCTORS II

DO – QUANTUM OSCILLATIONS IN 2-D ELECTRON SYSTEMS

DP – LOCALIZATION EFFECTS

DQ – NETWORKS OF WEAK LINKS II

DR – HIGH SENSITIVITY MEASURING DEVICES

EC – SUPERCONDUCTIVITY: CAPITA SELECTA

EF – CRITICAL PHENOMENA I

EH – DISSIPATIVE QUANTUM SYSTEMS

EI – CRITICAL PHENOMENA II

EK – FLOW, THERMODYNAMICS, AND SPIN DYNAMICS OF SUPERFLUID ^{3}HE

EL – MOLECULAR HYDROGEN

EM – PROXIMITY EFFECT/PENETRATION DEPTH

EN – SUPERCONDUCTIVITY: THEORY

EO – ELECTRONIC TRANSPORT PROPERTIES

EP - CHAOS AND NOISE

EQ - LOW TEMPERATURE TECHNIQUES

FA – HELIUM LIQUID-GAS INTERFACES

FC – CHARGE DENSITY WAVES

FD – MACROSCOPIC QUANTUM SYSTEMS

FE – STRUCTURE AND EXCITATIONS OF ^{4}HELIUM

FF – DEGENERATE ^{3}HE AND MIXTURES

FG – SIMPLE METALS

FH – TRANSITION METALS AND ALLOYS

FI – DISORDER, GLASSES, SUPERCONDUCTIVITY

FK – LOW DIMENSIONAL CONDUCTORS

FL – LOW TEMPERATURE PROPERTIES OF SOLIDS

LT-17 (Contributed Papers)
U. Eckern, A. Schmid, W. Weber, H. Wühl (eds)

PERSISTENT SUPERFLOW IN SUPERFLUID ^{3}HE*

P. L. GAMMEL and J. D. REPPY

Laboratory of Atomic and Solid State Physics, Clark Hall, Cornell University, Ithaca, New York 14853 U.S.A.

The a.c. gyroscope technique has been used in new studies of persistent currents in superfluid ^{3}He. The ^{3}He was in an annular channel packed with 9.5 μm α-alumina, and was studied at 15.0 bar and 29.0 bar. The critical velocity is of order 3 mm/sec, and there is much less hysteresis than for superfluid ^{4}He. At 29.0 bar there was no sign of T_{AB}, but persistent currents were found if rotation during cooling was stopped close to T_c.

The existence of persistent currents in ^{3}He-B has now been well established. (1,2) In superfluid ^{4}He the critical velocity scales as

$$v_{s,c} = (h/2md)\ln(d/2a), \tag{1}$$

where m is the bare mass, d the channel size and a the vortex core strength. Persistent currents in ^{4}He also show extreme hysteresis in the plane of $\underset{\sim}{v}_s$-v_n vs. $\underset{\sim}{v}_n$. We report experiments with a new gyroscope to investigate these phenomena in superfluid ^{3}He.

The gyroscope is shown in Fig. 1. It is nearly identical to the one used in our earlier experiments. (1) The porous media used in these experiments was 9.5 μm α-alumina, packed to 30%. SEM photographs suggest that, although the size distribution is peaked near 10 μm, there is a long tail extending to about 1 μm.

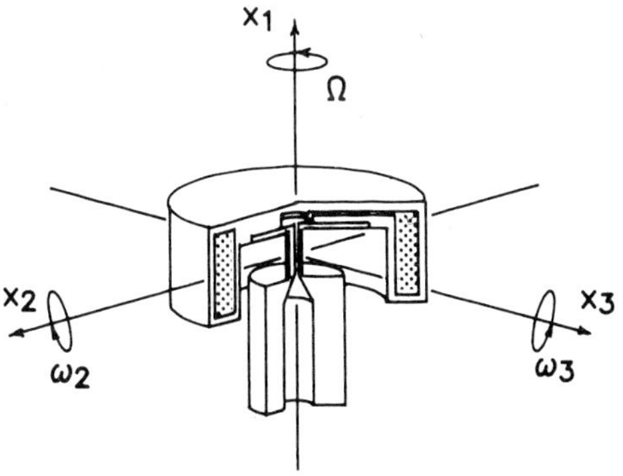

FIGURE 1
Schematic of the a.c. gyroscope. $\underset{\sim}{X}_2$ and $\underset{\sim}{X}_3$ represent the normal mode directions for tipping the cell, with resonant frequencies ω_2 and ω_3. Motion about $\underset{\sim}{X}_1$ is in the usual torsional mode.

Persistent current angular momentum is measured as before. (1) Briefly, the mode along $\underset{\sim}{X}_3$ is driven self-resonantly in a phase-locked loop. An array of capacitive electrodes detects motion in the directions $\underset{\sim}{X}_2$ and $\underset{\sim}{X}_3$. A current of angular momentum L mixes these two modes. If the mode along X_3 is initially linearly polarized, the presence of a current makes it elliptically polarized. The ratio of minor to major axes is

$$\theta_2/\theta_3 = \frac{L}{2I_2}\frac{1}{\omega_2-\omega_3}. \tag{2}$$

I_2 is the moment of inertia about the axis $\underset{\sim}{X}_2$. A frequency shift is also induced, but it is second order, and below the level of our present resolution.

Fig. 2 shows the period and dissipation of the mode along X_3 below T_c at 15.0 bar. Above T_c both curves are quite flat, as the viscous penetration depth is much larger than the interparticle spacing. The series of resonances near T_c alters the phase at which the oscillator exhibits a maximum response. They have been tentatively identified as fourth sound resonances, although they are farther from T_c than we calculate on the basis of cell size. The sharp feature near $T/T_c \sim 0.8$ is less reproducible. The data shown here are taken while slowly warming after having rotated at the lowest temperatures. At 29.0 bar the resonant structure near T_c is still present, but there are no sharp features.

Well below T_c the period and dissipation fit the hydrodynamic model used to analyze torsional oscillator response in well defined geometries. (4) Persistent current studies were limited to this regime.

* Supported by the NSF through Grants DMR-77-24221 and DMR-78-02655 and the NSF through the Cornell Materials Science Center under contract DMR-76-81083AD2, Technical Report 5141.

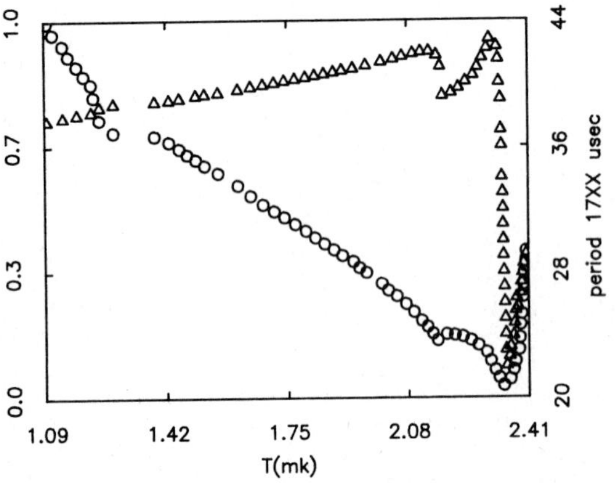

FIGURE 2
The Q (lower) and period (upper) of the X_3 mode at 20.0 bar. T_c was 2.41 mK based on LCMN thermometry.

At 15.0 bar experiments were performed by rotating the cryostat at a fixed temperature below T_c. After each rotation the cryostat was stopped and the residual angular momentum measured. In figure 3, the results of such an experiment are shown. The preparation angular velocity of the cryostat was slowly increased until the point A was reached. The direction of rotation was reversed and slowly increased to give the lower points.

The growth and saturation of the persistent current agree with earlier observations. However, saturation now sets in at about 3 mm/sec, a somewhat smaller value than our vortex-energy arguments from the 100 μm data suggested. The slow increase in angular momentum extending to the highest velocities is probably due to the pore size distribution.

The hysteretic behavior is rather different from that seen in ^4He by Kojima. (3) By comparison, the ^3He hysteresis loop is quite closed. This may imply that vortex line creation proceeds with a smaller free energy barrier in ^3He-B than in superfluid ^4He. The rapid rotational equilibrium in NMR (5) and oscillatory flow experiments (6) also suggests this.

At 29.0 bar there is no sign of T_{AB} in the hydrodynamics of this cell, rather like the fourth sound results of Kojima. (7) At this pressure, experiments were performed by pulsing the cell about T_c. Rotation was then started. The cell relaxed back to the temperature of the nuclear stage with a time constant of 30 minutes.

After the cell had cooled to some prescribed point below T_c rotation was stopped. The residual angular momentum was observed as the cell continued to cool.

The currents are seen to grow as the temperature is reduced, if we plot the angular momentum of the current vs. the superfluid density, as determined from the period of the cell. Dividing through by the superfluid density suggest the scaling is roughly in accord with conservation of superfluid velocity. At low temperatures, the angular momentum falls below what a constant superfluid velocity would imply. Similar experiments at Helsinki (2) have suggested that there is a

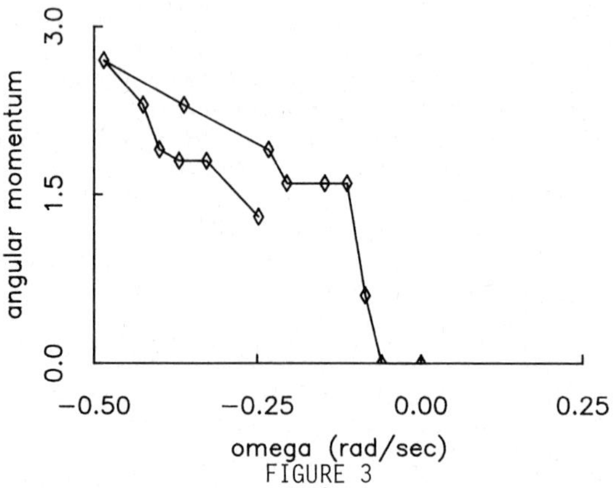

FIGURE 3
The hysteresis loop for ^3He. Data taken at 15 bar and T/T_c = 0.46.

sharp change in the critical velocity in the region of the vortex transition at T/T_c = 0.6.

One of us (JDR) would like to thank AT&T Bell Laboratories and the Massachusetts Institute of Technology for their hospitality during the period when this work was accomplished. One of us (PLG) is the recipient of an Andrew D. White Fellowship, 1980-1982.

REFERENCES
(1) P. L. Gammel, H. E. Hall and J. D. Reppy, Phys. Rev. Lett. 52, 121 (1984).
(2) Juha Simola, private communication.
(3) H. Kojima, W. Veith, E. Guyer and I. Rudnick, J. Low Temp. Phys. 25, 195 (1976).
(4) J. E. Berthold, R. W. Giannetta, E. N. Smith and J. D. Reppy, Phys. Rev. Lett. 37, 1138 (1976).
(5) P. J. Hakonen, O. T. Ikkala, S. T. Islander, O. V. Lounasmaa and G. E. Volovik, J. Low Temp. Phys. 53, 423 (1983).
(6) E. P. Whitehurst, H. E. Hall, P. L. Gammel and J. D. Reppy, to be published in this volume.
(7) T. Chainer, Y. Morai and H. Kojima, Phys. Rev. B21, 3941 (1979).

LT-17 (Contributed Papers)
U. Eckern, A. Schmid, W. Weber, H. Wühl (eds)

DISSIPATION AT LOW VELOCITIES IN OSCILLATORY FLOW OF SUPERFLUID ^{3}He-B

Ren-zhi LING, D.S. BETTS and D.F. BREWER

Physics Laboratory, University of Sussex, Brighton BN1 9QH, Sussex, U.K.

In many flow experiments in superfluid ^{3}He, including our own D.C. and oscillatory measurements through a rectangular superleak, dissipation has been higher than would be expected by comparison with superfluid ^{4}He. The important question is whether this is due to inherent dissipation mechanisms in "superfluid" ^{3}He, or to apparatus effects. We have modified our oscillation experiment to show that one such possible extraneous source of dissipation is not significant, and also, by reaching lower temperatures, shown that the dissipation is a very rapidly decreasing function of temperature in this experiment at $(T/T_c) \lesssim 0.5$.

1. INTRODUCTION

An important aim of flow experiments in superfluid ^{3}He is to determine whether it can flow with strictly zero dissipation. Recent experiments by Gammel et al (1) show the existence of persistent currents in a toroidal region filled with a fibrous material with average pore size of 10^{-2} cm at velocities around 10^{-2} cm/sec. However, several flow experiments (2)-(5) have indicated large energy losses in the flow even at very low velocity, including a fourth sound experiment (6) where observed Q factors are much lower than expected. The mechanism responsible for the losses in the flow at low velocity in some of these experiments is still not well understood.

The Sussex group diaphragm-driven oscillatory superflow experiments of ^{3}He-B, reported in an early paper (3), gave Q values in the low velocity region (around 10^{-2} cm/sec) of only 10 to 100, which is very low compared with the superfluid ^{4}He Q values of $\sim 10^4$ obtained in the same apparatus at 50 mK. We have now investigated one possible extraneous source of this dissipation, and have found a very rapid decrease of dissipation at $T/T_c \lesssim 0.5$.

2. APPARATUS

A flow cell with two chambers of ^{3}He linked by a superleak of rectangular cross section was used in these experiments, cooled by a copper nuclear refrigerator. Part of the wall between these two chambers is formed by a thin flexible plastic diaphragm, which, together with two capacitor plates situated either side of it forms the flow drive and detection system. The cross-section of the superleak is 48µm x 2.86 mm, and it is 9.16 mm long. This is narrow enough to clamp the normal fluid in the superleak itself. In the connecting channels crinkled mylar packing with spacing ~ 0.5 mm was placed in order to clamp the normal flow, but later measurements of viscosity showed that the spacing was not small enough to do so, and it might be possible for some normal flow to take place through them.

In order to investigate whether this possibility did contribute to the low Q values, the apparatus was modified by removing the mylar packing, thereby increasing the possibility of normal flow, and then repeating the experiments. A new superconducting magnet was used for nuclear cooling with larger rms field and improved field profile over the copper bundle, enabling us to reach lower temperatures than in the previous experiment.

3. RESULTS

In the low velocity region, the low frequency oscillatory superflow (around 30 to 40 HZ) observed in our experiment fits a simple damped harmonic oscillator function very well,

$$J_s(t) = A + B\, e^{-ct} \cos(\omega t + \phi) \qquad \text{(i)}$$

After collecting data from the signal averager which usually ran at an average number 64, we then processed those data which are in the form of two hundred amplitude points, using equation (i) with a five parameter least square fit employing an iterative technique with a computer. Hence the best fit for ω and c can be derived and $Q = \omega/2c$ is obtained. Figure 1 shows typical data, and Table 1 shows our results of these Q measurements in the higher Q region away from the t = 0 region where the Q is much lower.

Table 1. Q obtained under different conditions.

P (bar)	0	11.0	16.8	33.0
$(1-T/T_c)$	0.20	0.52	0.63	0.61
Q	23	24	125	118

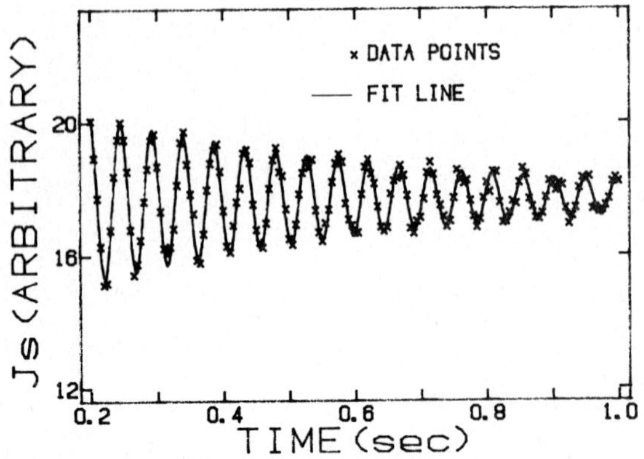

FIGURE 1
Typical data and fit line taken at $1-T/T_C = 0.317$, $P = 16.8$ bar. The fit is shown for $t > 0.2$ sec where Q is higher, but not near the beginning of the oscillations where Q is lower and the fit is not so good (see ref.3 for an explanation of this).

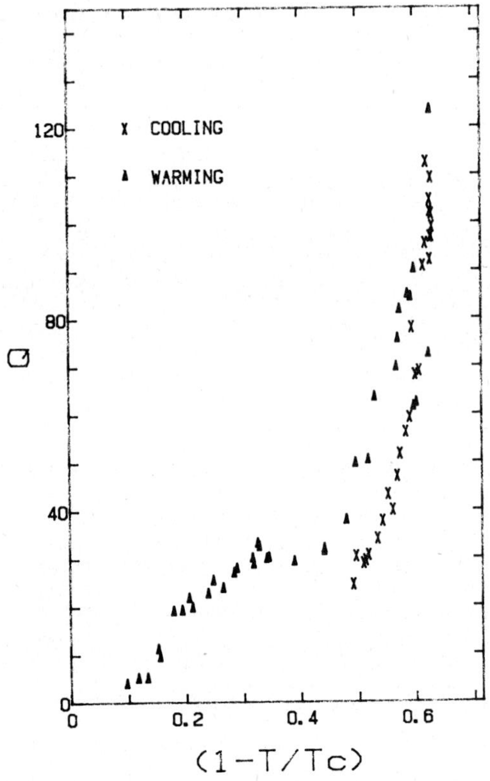

FIGURE 2
Typical values of Q at a pressure of 16.8 bar, as a function of T/T_C, for warming and cooling.

There are two features we should like to stress here. First, these measurements show that the Q has not changed very much compared with the previous data. They are only slightly higher (about 10% - 30%) which is within our error limit. This fact might suggest that the damping mechanism is likely to be occurring in the superleak rather than in the connecting channels.

Second, we found that the Q is not sensitive to pressure, but very sensitive to temperature especially at the lowest temperature. This can be seen from figure 2 which is a typical graph of Q taken at p = 16.8 bar. The rapid rise in Q as temperature is lowered may be associated with disappearance of the normal fluid fraction, and suggests that very high Q values may be observable in such experiments at lower temperatures.

Two other details of this figure are sufficiently noticeable to be commented on. First, there seems to be some hysteresis in the cooling/warming observations. Second, and perhaps more important, there is a very marked change in the temperature dependence of the dissipation on warming to temperatures above $T/T_C \sim 0.5$.

ACKNOWLEDGEMENTS

This work was supported by the SERC through Grant no. GR/B/65137 and by a Senior Fellowship (DFB); and by the Education Department of the People's Republic of China (Ren-zhi Ling).

REFERENCES

(1) P.L.Gammel, H.E.Hall and J.D.Reppy, Phys. Rev.Lett. 52 (1984) 121.
(2) J.M.Parpia and J.D.Reppy, Phys.Rev.Lett. 43 (1979) 1332.
(3) A.J.Dahm, D.S.Betts, D.F.Brewer, J.Hutchins, J.Saunders and W.S.Truscott, Phys.Rev.Lett. 45 (1980) 1411.
(4) J.P.Eisenstein and R.E.Packard, Phys.Rev. Lett. 49 (1982) 564.
(5) H.E.Hall, P.L.Gammel and J.D.Reppy (preprint).
(6) H.Kojima, D.N.Paulson and J.C.Wheatley, J. Low Temp.Phys. 21 (1975) 283.

LT-17 (Contributed Papers)
U. Eckern, A. Schmid, W. Weber, H. Wühl (eds)
©Elsevier Science Publishers B.V., 1984

EXACT SOLUTION OF THE SPIN DYNAMICS OF SUPERFLUID ^{3}He-A IN ZERO MAGNETIC FIELD

Dieter VOLLHARDT

Max-Planck-Institut für Physik und Astrophysik, - Werner-Heisenberg-Institut für Physik - Föhringer Ring 6, D 8000 München 40 (Fed.Rep.Germany)

It is shown that in the case of zero external magnetic field the equations of spin dynamics of the ABM-state can be solved exactly by elementary methods. For this we assume $\hat{\ell}$, the preferred direction of the orbital angular momentum of the Cooper-pairs, to be stationary. The solution for a typical field turn-off experiment is presented.

The spin dynamics of the superfluid phases [1-4] of ^{3}He are unusally rich due to (i) the spontaneously broken spin-orbit symmetry [5] (SBSOS), leading to a macroscopic amplification of the nuclear dipole interaction, and (ii) the intrinsic anisotropy of the superfluid. In particular, the spin dynamics of the ABM-state are governed by the dynamics of the total spin $\vec{S}$ and of $\hat{d}$, the preferred direction in spin space; $\hat{d}\cdot\vec{S}=0$. The dipole interaction, whose strength is characterized by Ω_A, the longitudinal NMR frequency, leads to a coupling of the preferred directions $\hat{d}$ and $\hat{\ell}$. Here, $\hat{\ell}$ is the direction of the orbital angular momentum of the Cooper-pairs, being a preferred direction in position space. Any motion of $\hat{d}$ therefore takes place in the potential provided by $\hat{\ell}$. The finite spin-orbit coupling implies that the total spin is not a conserved quantity.

The general, non-linear equations describing the spin dynamics of superfluid ^{3}He have first been derived by Leggett [6]. In the ABM-state they take the form

$$\dot{\vec{S}} = \gamma\vec{S}\times\vec{H} + \Omega_A^2\, \chi\gamma^{-2}\, (\hat{d}\times\hat{\ell})(\hat{d}\cdot\hat{\ell}) \qquad (1a)$$

$$\dot{\hat{d}} = -\gamma^2\chi^{-1}\, \hat{d}\times\, (\vec{S}-\frac{\chi}{\gamma}\vec{H}) \qquad (1b)$$

Owing to their complexity (due to the dipolar torque exerted on $\vec{S}$) an exact, general solution of these equations has not been possible so far. There are only a few special cases which have yielded to an analytic approach. One example is the longitudinal case [6] where the magnetic field is oriented perpendicular to $\hat{\ell}$ (and, hence, $\vec{S}\,\|\,\vec{H}\perp\hat{\ell}\,,\hat{d}$). This configuration can be solved exactly for all fields because the equation of motion of $\hat{d}$ corresponds to that of a (driven) mathematical pendulum [7]. For small external magnetic fields $\hat{d}$ oscillates around $\hat{\ell}$ with frequency Ω_A and consequently $\vec{S}$ changes its magnitude at the same frequency (longitudinal NMR). For large fields or field pulses the pendulum (i.e. $\hat{d}$) can swing around, leading to a ringing of magnetization at twice the frequency of the precession of $\hat{d}$ [7]. Another case that can be solved analytically is that of spin tipping by an arbitrary angle in a strong magnetic field, i.e. transverse NMR [6] and transverse ringing [8].

In the case of the BW-state Maki and Ebisawa [9] found that the special case of zero magnetic field can be solved exactly by quadrature, because they found the appropriate constants of motion. Brinkman and Cross [4] subsequently identified a constant vector in the problem, whose magnitude was one of these constants of motion.

The present author has recently shown [10] that the tools of differential geometry provide a particularly simple and suitable method to study the spin dynamics of superfluid ^{3}He in zero magnetic field. This special case turns out to be one which can be solved analytically. It is due to the fact that the space-curve of $\vec{S}(t)$ has zero torsion at all times. This automatically implies that there exists a vector (the "binormal" vector characterizing the orientation of the osculating plane of the space curve) which is a constant of motion. This method is quite general and is applicable to all superfluid phases. In zero magnetic field the total energy of the spin system is a constant of motion itself (if we neglect dissipation). The problem is therefore in many ways similar to that of a spinning top in classical mechanics.

In the case of the ABM-state the concepts of differential geometry actually prove to be somewhat unnecessary. This is so because one of the major results, namely that the projection of $\vec{S}$ on $\hat{\ell}$ is a constant of motion,

$$C_A \equiv \vec{S}\cdot\hat{\ell} = \text{const} \qquad (2)$$

can just as well be directly obtained from (1a) putting $\vec{H}=0$. For this to be true we have to assume $\hat{\ell}$ to be stationary. For most experimental situations this holds anyway because the spin degrees of freedom move on a much faster time scale than $\hat{\ell}$. We imagine $\hat{\ell}$ to be aligned in a fixed direction, either determined by the container geometry (e.g. parallel plates) or due to some very slow heat flow. Thus there exists a constant vector $\vec{C}_A$ along $\hat{\ell}$:

$$\vec{C}_A = (\vec{S}\cdot\hat{\ell})\hat{\ell} = \text{const} \qquad (3)$$

of magnitude C_A. This constant of motion allows for an exact solution of (1) for $\vec{H}=0$. Making use of the vector identity

$$S^2 = (\vec{S}\cdot\hat{t})^2 + (\vec{S}\times\hat{t})^2 \qquad (4)$$

where $\hat{t}=\hat{d}\times\hat{\ell}/\sin\theta$ we can express S^2 in terms of θ, the angle between $\hat{d}$ and $\hat{\ell}$, only: We take the derivative of $\hat{d}\cdot\hat{\ell} = \cos\theta$ and find $\dot{\theta} = -\gamma^2\chi^{-1}(\vec{S}\cdot\hat{t})$; furthermore $\vec{S}\times\hat{t} = C_A\hat{d}/\sin\theta$. This yields [10]

$$S^2 = (\chi/\gamma^2)^2\,\dot{\theta}^2 + C_A^2/\sin^2\theta \qquad (5)$$

The energy of the system may then be written in terms of a single variable only:

$$E_A = (\chi/2\gamma^2)\,[\dot{\theta}^2 + \tilde{C}_A^2/\sin^2\theta + \Omega_A^2\sin^2\theta] \qquad (6)$$

where $\tilde{C}_A = \gamma^2\chi^{-1}C_A$. We are then lead to the equation of motion for θ

$$\ddot{\theta} - \tilde{C}_A^2\cos\theta/\sin^3\theta + \tfrac{1}{2}\Omega_A^2\sin 2\theta = 0 \qquad (7)$$

The problem of solving the five coupled differtential equations (1) has thus been reduced to the solution of a single equation.

Once this equation has been solved, every information about the motion of $\vec{S}$ and $\hat{d}$ can be obtained by elementary techniques. In particular, $\vec{S}(t)$ is specified by (i) its magnitude (5), (ii) its projection on $\hat{\ell}$, $\vec{S}\cdot\hat{\ell} = C_A$ and (iii) its projection into the plane perpendicular to $\hat{\ell}$ where it precesses with the angular velocity

$$\dot{\varphi}_{\vec{S}} = \Omega_A^2(\chi/\gamma^2)C_A\cos^2\theta/[S^2 - C_A^2] \qquad (8)$$

It is interesting to note that (6) and (7) are very similar to the corresponding expressions for a spinning top. There C_A represents the vertical component of the angular momentum, which is a constant of motion.

Eq. (7) describes a periodic motion of $\hat{d}$ (and therefore $\vec{S}$) and can easily be solved analytically using an elliptic integral [10]. The solution depends on the magnitude of C_A. In a turn-off experiment, when an external field $\vec{H}$ is switches off, C_A can be obtained from (5). For this the initial configuration of $\hat{d}$ and $\vec{S}$ relativ to $\hat{\ell}$ and the external field $\vec{H}$, i.e.before $\vec{H}$ has been switched off, has to be determined first. (In fields $H \lesssim \Omega_A/\gamma$ the spin $\vec{S}$ is not aligned by $\vec{H}$ because the dipole coupling favors $\hat{d}\parallel\hat{\ell}$ such that $\hat{d}$ is not perpendicular to $\vec{H}$). In this case one finds that in equilibrium all vectors $(\hat{\ell},\hat{d},\vec{S},\vec{H})$ lie in a plane. It implies that initially $\vec{S}\perp\hat{d}\times\hat{\ell}$ and therefore at t=0 there is no dipolar torque on $\vec{S}$ (i.e. $\dot{\theta}=0$ at $t=0$). This is in contrast to what happens when $\vec{H}\perp\hat{\ell}$; there $\dot{\theta}\neq 0$ at $t = 0$.

Minimizing the energy with respect to $|\vec{S}|$ and Φ_0, the angle between $\vec{S}$ and $\hat{\ell}$, yields $S = (\chi/\gamma)\,H\cos(\Phi_0-\psi)$ where ψ is the angle between $\vec{H}$ and $\hat{\ell}$, and $\Phi_0 (= \frac{\pi}{2} - \theta_0)$ is determined by

$$\tan 2\Phi_0 = \sin 2\psi/[\cos 2\psi - (\Omega_A/\gamma H)^2] \qquad (9)$$

When the field $\vec{H}$ is suddenly switched off, $\tilde{C}_A$ is given by $\tilde{C}_A = \Omega_A\cos\Phi_0 z$, where $z=(\gamma H/\Omega_A)\cos(\Phi_0-\psi)$.

The resulting periodic motion of $\hat{d}$ and $\vec{S}$ is determined by (7) and depends on whether $z \lessgtr 1$. Starting from θ_0, θ initially increases in both cases. However for $z \leqslant 1$ θ cannot exceed a maximum value $\bar{\theta} = \sin^{-1}z$ while for $z \geqslant 1$ one has $\theta_0 < \theta < \pi - \theta_0$. The periods are found to be

$$T_r = \begin{cases} (4\,\Omega_A^{-1}/\sin\Phi_0)\;K(1/y) & z\leqslant 1 \\ (4\,\Omega_A^{-1}/\sin\Phi_0)\,y\;K(y) & z\geqslant 1 \end{cases} \qquad (10)$$

where $y = \sin\Phi_0/[z^2-\cos^2\Phi_0]^{1/2}$ and $K(x)$ is the complete elliptic integral.

The resulting motion of $\vec{S}$ for $z > 1$ can be described as follows: $\vec{S}$ precesses around $\hat{\ell}$ with an instantaneous angular velocity $\dot{\varphi}_{\vec{S}}$ (8) while nutating at the same time with twice the frequency of the θ-motion. This nutation is due to the finite spin-orbit coupling that wants to have $\vec{S}$ and $\hat{\ell}$ perpendicular. Note, that the precession frequency of $\vec{S}$ around $\hat{\ell}$ is quite different from that of the θ-motion. It is mainly determined by the size of $\gamma H/\Omega_A$ and the initial direction of $\vec{H}$ relative to $\hat{\ell}$. The two frequencies are generally unrelated.

Details of the calculations and numerical results for the ringing frequencies for different initial configurations can be found elsewhere [10].

I thank Peter Wölfle for discussions.

References

[1] A.J. Leggett; Rev.Mod.Phys.47, 331 (1975)
[2] P. Wölfle, Rep.Prog.Phys. 42 269 (1979).
[3] D.M. Lee and R.C. Richardson; in The Physics of Liquid and Solid Helium, eds. K.H. Bennemann and J.B. Ketterson; Wiley, New York 1978.
[4] W.F. Brinkman and M.C. Cross; in Progress in Low Temperature Physics, Vol. 7, ed. D.F. Brewer; North-Holland, Amsterdam (1978).
[5] A.J. Leggett; J.Phys. C6, 3187 (1973).
[6] A.J. Leggett; Ann.Phys. 85, 11 (1974).
[7] K. Maki and T. Tsuneto; Prog.Theor.Phys. 52, 617 (1974).
[8] W.F. Brinkman and H. Smith; Phys.Lett. 53A, 43 (1975).
[9] K. Maki and H. Ebisawa; Phys.Rev. B13, 2924 (1976).
[10] D. Vollhardt, to be published.

LT-17 (Contributed Papers)
U. Eckern, A. Schmid, W. Weber, H. Wühl (eds)

COEXISTENCE OF SUPERCONDUCTIVITY AND RANDOM MAGNETIC FREEZING IN $(Th_{0.67}Nd_{0.33})Ru_2$

D. HÜSER, G.J. Nieuwenhuys and J.A. Mydosh

Kamerlingh Onnes Laboratorium der Rijks-Universiteit Leiden, Leiden, The Netherlands

We present ac-susceptibility $\chi(T,H)$ as well as dc-resistivity $\rho(T,H)$ measurements for $(Th_{0.67}Nd_{0.33})Ru_2$. Here we find overlapping regimes of superconductivity and magnetic ordering as a function of temperature and field. Our results suggest that a cluster-glass freezing process occurs in the superconducting state without destroying the superconductivity but reducing the critical field values. The penetration depth $\lambda_L(0)$ and the coherence length $\xi(0)$ are determined by measurements of the critical fields H_{c_1} and H_{c_2} as a function of temperature.

The study of "magnetic superconductors" is of considerable interest, since the discovery of reentry behavior in several ternary compounds. (1) Another class of materials which has been investigated extensively for a possible coexistence of superconductivity and magnetism are pseudobinary laves phase compounds $(A_{1-x}B_x)Ru_2$ (2) (A, B denote two rare earth ions one which is magnetic). A susceptibility study of $Th_{1-x}Gd_x)Ru_2$ (3) indicates a spin glass type of magnetism just above the critical concentration for superconductivity. In the system $(Th_{1-x}Nd_x)$ we found recently (4) the coexistence of superconductivity and a cluster glass ordering in the concentration range $0.3<x<0.4$. Here we describe a more detailed investigation of two samples with the same nominal concentration x=0.33.

Both samples of $(Th_{0.67}Nd_{0.33})$ were prepared by repeated arc melting of the appropriate components. The sample for the susceptibility measurement had an almost spherical shape whereas the sample for the resistivity measurement was spark cut in the form of a bar (1mm x 1mm x 7mm). Both samples were annealed at 900°C for three days and then quenched into water. The ac-susceptibility measurements were performed in the mixing chamber of a He^3-He^4 dilution refrigerator by a standard mutual inductance technique and the dc resistivity was measured in a He^3 cryostat using an ultra stable current source and a nanovolt meter. External static magnetic fields could be applied and swept at any temperature.

In Fig. 1 we exhibit the behavior of the ac susceptibility as a function of temperature for various external magnetic fields. In these measurements the field was applied at 4K and the sample was field-cooled. The measured superconducting transition temperature T_c for zero

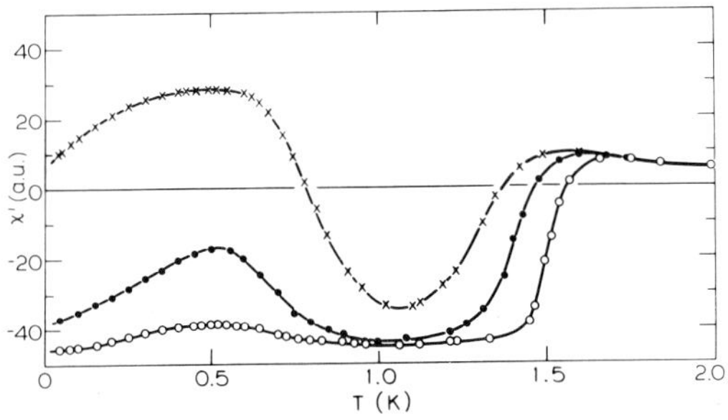

FIGURE 1
Temperature dependence of the susceptibility cooled in various magnetic fields. Note the emergence of the cluster-glass maximum as H is increased.
○ H=0 G, ● H=66 G, X H=170 G.

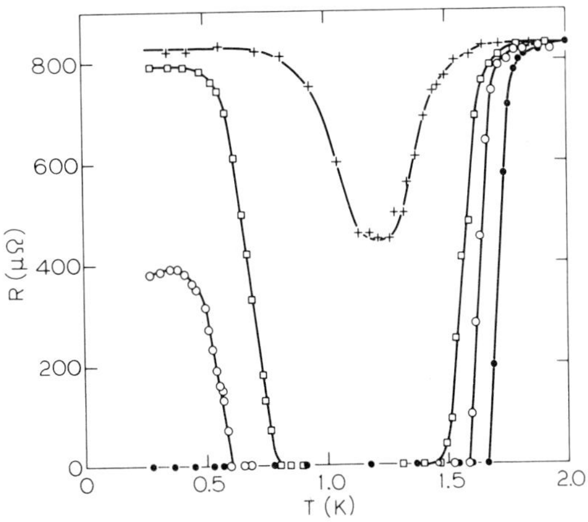

FIGURE 2
Temperature dependence of the resistivity cooled in various magnetic fields.
● H=0 G, ○ H=200 G, □ H=300 G, + H=500 G.

field is 1.50 K. At lower temperatures we find a susceptibility maximum, denoting the freezing temperature T_g=0.51K, even though the sample is in the superconducting state. This is only possible if the measuring ac-field could penetrate into the sample probably through ferromagnetic clusters or vortices. Small external fields enhance the number of vortices and drive the system from the superconducting to a magnetic state. The dc resistivity, shown in Fig.2, as a function of temperature for several external fields confirms these results. The residual resistivity is about 25μΩcm. This sample becomes superconducting at T_c=1.71 K in zero field, slightly higher than measured by ac-susceptibility probably due to a slightly different composition (see phase diagram in Ref.4). As it is clearly seen, an external field induces the reentrance from the superconducting to the normal state. Notice how the superconductivity is completely quenched below T=0.6 K for fields H=500 G, as seen by the fact, that the sample reaches again the residual resistivity. For smaller fields H≈200 G one reaches an intermediate state. Fig. 3 finally gives the critical fields for superconductivity as a function of temperature. The ac-susceptibility gives the H_{c1}- and the dc-resistivity the H_{c2}-values. The circles were obtained at increasing field and the squares at decreasing field. In the range

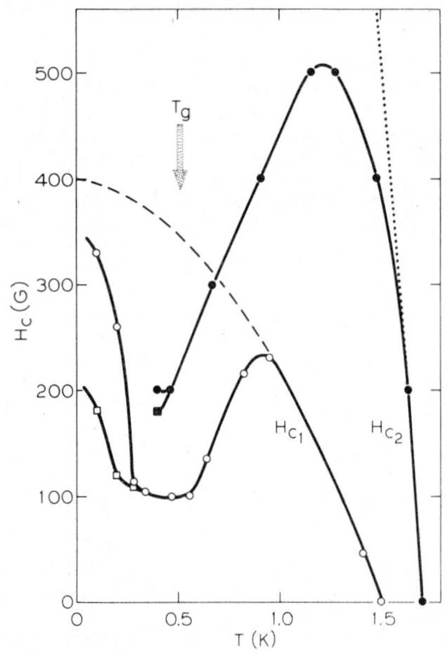

FIGURE 3
Critical fields H_{c1} and H_{c2} as a function of temperature. The lines are a visual guide. The dashed line and the dotted line are explained in the text. T_g represents the freezing temperature.

of the freezing temperature T_g the critical fields are strongly reduced followed at lower temperatures by the onset of hysteresis as a consequence of the isothermal remanent magnetization of a cluster-glass. The dashed line represents an extrapolation of $H_{c1}(T)$ according to a $1-(T/T_c)^2$ dependence, which gives a penetration depth $\lambda_L(0)\approx1300$Å. The dotted line is calculated with $h(t)=H_{c2}/(-dH_{c2}/dt)$, $t=T/T_c$ (5) leading to a coherence length $\xi(0)\approx350$Å. Both results classify the sample as a type II superconductor with $\kappa\approx4$.

In conclusion the small pairbreaking effect of Nd permits large concentrations of magnetic impurities in this system without destroying the superconductivity close to the percolation limit (x_p=0.42). Consequently, short range correlations leading to short range magnetic ordering, i.e. random freezing, can exist in the superconducting state, as long as the magnetic correlation length is smaller than the coherence length of the Cooper pairs.

ACKNOWLEDGEMENTS
The authors wish to acknowledge P.H. Kes for valuable discussions and T.T.M. Palstra for his assistance with the measurements. This work was supported by the Nederlandse Stichting voor Fundamenteel Onderzoek der Materie (FOM).

REFERENCES
(1) See, for example, M.B. Maple, J. Magn. Magn. Mater. 31-34 (1983) 479.
(2) M. Wilhelm and B. Hillenbrand, Z. Naturforsch. 2617 (1971) 141.
B.T. Matthias, H. Suhl , E. Corenzwit, Phys. Rev. Lett. 1 (1958) 449.
(3) D. Davidov, K. Baberschke, J.A. Mydosh and G.J. Nieuwenhuys, J. Phys. F7 (1977) L47.
(4) D. Hüser, M.J.F.M. Rewiersma, J.A. Mydosh and G.J. Nieuwenhuys, Phys. Rev. Lett. 51 (1983) 1290.
(5) N.R. Werthamer, E. Helfand and P.C. Hohenberg, Phys. Rev. 147. 1, (1966) 295.

LT-17 (Contributed Papers)
U. Eckern, A. Schmid, W. Weber, H. Wühl (eds)

COEXISTENCE PHASE IN FERROMAGNETIC SUPERCONDUCTORS

L.N. BULAEVSKII (a), A.I. BUZDIN (b), M.L. KULIĆ (c), S.V. PANJUKOV (a)

(a) Physical Institute "P.N. Lebedev", Moscow, USSR
(b) Physics Department, Moscow State University, Moscow, USSR
(c) Institute of Physics, P.O. Box 57, 11001 Belgrade, Yugoslavia

The theory of the coexistence phase in ferromagnetic superconductors, based on the exchange and electromagnetic interaction of electrons and localized moments gives the quantitative explanation of experimental data on $HoMo_6S_8$ and $HoMo_6Se_8$. The properties of $ErRh_4B_4$ may be explained assuming asperomagnetic ordering in the normal phase.

1. THEORETICAL PREDICTIONS

The theoretical description of an interplay between superconductivity and ferromagnetism was given in (1) where, in the framework of BCS theory, the exchange (EX) and the electromagnetic (EM) interactions between electrons and localized moments (LM), as well as the magnetic anisotropy were taken into account. It was supposed also that in the absence of Cooper pairing one realizes ferromagnetic phase below the temperature . Then in the presence of superconductivity with the critical temperature $T_{c1} \gg \theta$, the inhomogeneous transversal one-dimensional domain-like magnetic structure is realized below $T_M(\approx\theta)$ - so called DS-phase. In real compounds the EX mechanism is dominant in the formation of magnetic structure of the DS-phase ((1) and the reference therein), i.e., it determines its wave-vector Q(T) as well as competition with the ferromagnetic normal (FN) phase. The role of the EM mechanism is to renormalize the temperature θ, and to make the structure trasversal.

The value $Q(T) \approx [\Delta(T) S^2(T)/\eta(T)]^{1/2}$ and Q is of the order $(a\xi_o)^{-1/2}$, where η(T) is surface energy of domain wall, a is the magnetic stiffness of the order of atomic length, and ξ_o is the superconducting coherence length. This leads to the decrease of Q(T) on cooling, since η(T) grows and Δ(T) decreases. The ideal domain structure is characterized by the set of harmonics (2k+1)Q, whose intensities fall off as $(2k+1)^{-2}$ in a monocrystalline and as $(2k+1)^{-4}$ in polycristalline samples - k is integer. In the case of irregular domain structure the width of the (2k+1)Q peak is proportional to $(2k+1)^2$, i.e. higher peaks are much broader and weaker than the main peak Q.

On cooling the first-order transition DS→FN occurs at T_{c2} if $h(o) > h_c \approx$ $\approx T_{c1}(\xi_o/a)^{1/4}$. On heating the metastable FN-phase is preserved up to T_M, due to a large activation energy for the formation of nucleus of DS-phase.

2. INTERPRETATION OF DATA ON $HoMo_6S_8$ AND $HoMo_6Se_8$

In polycrystalline samples of $HoMo_6S_8$ Pynn et al. (2) observed the transversal magnetic structure which is characterized by peak at Q=0.03 Å^{-1} at T_{c2}=0.65 K, the value of Q being slightly larger at T_M=0.70 K (T_{c1}=1.8 K). On cooling the ferromagnetic peak is absent down to T_{c2}. On warming, from the FN-phase, this peak is present up to T_M, the Q-peak being practically absent. In polycristalline sample of $HoMo_6Se_8$ Lynn et al. (3) found the coexistence phase below T_M=0.53 K without reentrancy (T_{c1}=5.5 K). Wave-vector Q decreases from 0.087 Å^{-1} (at T_M) to the saturation value 0.062 Å^{-1} at T<0.4 K, while the magnetization is saturated at T≈0.3 K. In both compounds one needs a long time for the eqilibrium to be established on cooling below T_M(≈40h in $HoMo_6Se_8$).

These data are in accordance with the DS-phase theory. The observed values for Q agree with the estimate a≈(1-2) Å and ξ_o=1500 Å, 500 Å respectively. The exchange field h(o) is estimated from the data for $T_c(x)$ in $Ho_xY_{1-x}Mo_6X_8$ with the aid of Abrikosov-Gor'kov relation $|dTc(x)/dx|\approx\pi^2 h^2(o)N(o)/2$, where N(o) is the density of states per one LM. One gets h(o)≈25 K at N(o)≈3 $(eV)^{-1}$ in $HoMo_6S_8$, and h(o)≈15 K at N(o)≈6.5$(eV)^{-1}$ in $HoMo_6Se_8$. In former compound the EX field is large enough to destruct the DS-phase, while in $HoMo_6Se_8$ h(o) does not exceed the critical value h_c. So, the reentrant behaviour is absent in the latter case; Δ does not practically change at temperatures below T_M, and $Q(T)\sim[S^2(T)/\eta(T)]^{1/2}$, When $T\to T_M$ Q(T) grows due to the growth of $S^2(T)/\eta(T)$. For Ising ferromagnet, in the mean-field approximation, we obtain $(Q(T)/Q(o))=[1-b\, d\ln S^2/d\ln T]^{1/4}$ for $0<T<T_M$, where the fitting parameter b<<1 characterizes the coordinate dependence of interaction between LM's. This dependence agrees well with experimental data on $HoMo_6Se_8$ (3) with b≈0.15.

3. ASPEROMAGNETISM IN $ErRh_4B_4$

Sinha et al. (4) observed the ferromagnetic peaks and their satellites with Q=0.06 $Å^{-1}$ in $ErRh_4B_4$ monocrystal. So, the behaviour of $ErRh_4B_4$, at $T_{c2}<T<T_M$, differes from that of $HoMo_6X_8$, and the presence of ferromagnetic peaks cannot be explained by the above proposed theory. Besides that, $ErRh_4B_4$ is not a usual ferromagnet below T_{c2}, because a drastic disagreement was found between the value of the coherent component of magnetization and the results of Mössbauer study at T→0 (5).

To explain both anomalies we assume the magnetic ordering in real crystals of $ErRh_4B_4$ to be an irregular asperomagnetic (fan-like) one, due to the existence of chaotic easy axis in basal plane under the influence of imperfections. The origin may be explained by the weak magnetic anisotropy inside the easy basal plane of ideal crystal, together with the alternating sign of the RKKY and magneto-dipolar interactions of LM's, while in $HoMo_6X_8$ the easy-axis anisotropy stabilizes the true ferromagnetic ordering. Irregular magnetic ordering makes the formation of coherent magnetic DS-phase in whole sample impossible, except in its small parts with perfect crystal structure. Under an uniaxial stress the regular easy axis (in basal plane) may appear in $ErRh_4B_4$. The asperomagnetic ordering in $ErRh_4B_4$ may be verified by Mössbauer and NMR studies in magnetic field.

4. CONCLUSION

The theory (1) gives the quantitative explanation of the experimental data on $HoMo_6X_8$. The disagreement of the proposed theory (1) and experimental data for $ErRh_4B_4$ may be explained by the hypothesis of an irregular magnetic ordering in (low temperature) normal phase, due to the crystal imperfections.

REFERENCES

(1) L.N. Bulaevskii, A.I. Buzdin, M.L. Kulić, S.V. Panjukov, Phys. Rev. 28 (1983) 1730, J. Low Temp. Phys. 52 (1983) 137, JETP (1982) 768.
(2) P. Pynn, J.W. Lynn and J. Joffrin, Helv. Phys. Acta 56 (1983) 179.
(3) J.W. Lynn, J.A. Gotaas, R.W. Erwin, R.A. Ferrell, J.K. Bhattacharjee, R.N. Shelton and P. Klavius, Phys. Rev. Lett. 52 (1984) 133.
(4) S.K. Sinha, H.A. Mook, D.G. Hinks and G.W. Crabtree, Phys. Rev. Lett. 48 (1982) 950.
(5) F.Y. Fradin, B.D. Dinlap, G.K.Shenoy and C.W. Kimbal, Superconductivity in Ternary Compounds II, Topics in Current Physics, Vol. 34, eds. M.B. Maple and Ø. Fisher (Springer Verlag, Berlin, Heidelberg, New York, 1982) pp. 221-226.

LT-17 (Contributed Papers)
U. Eckern, A. Schmid, W. Weber, H. Wühl (eds)

ULTRASONIC ATTENUATION IN $(Er_{1-x}Ho_x)Rh_4B_4$ ($x \leq 0.15$) AT LOW TEMPERATURES

Tetsuo FUKASE, Masashi TACHIKI, Naoki TOYOTA, Yoji KOIKE, Takashi NAKANOMYO

The Research Institute for Iron, Steel and Other Metals, Tohoku University, Sendai 980, Japan

Ultrasonic attenuation of $(Er_{1-x}Ho_x)Rh_4B_4$ is measured with 15MHz longitudinal wave in the temperature range from 9 K to 0.1 K. A rather sharp peak in zero-field attenuation is observed for x = 0.05 and 0.15 near T_{c2}, the temperature at which the transition to the ferromagnetic normal-conducting phase occurs. A broad attenuation peak is also observed for x = 0.05 near 4K, and is discussed in terms of the crystal-field effects.

The reentrant superconductor $ErRh_4B_4$ has a superconducting transition at $T_{c1} \sim 8.7$ K and becomes ferromagnetic normal-conducting state at $T_{c2} \sim 0.9$ K. In the temperature range from $T_p \sim 1.3$ K to T_{c2}, a spin-sinusoidal order coexists with the ferromagnetic order. A wide variety of phenomena is expected in this pseudoternary system. The ultrasonic attenuation caused by the spin fluctuation may exhibit a characteristic dependence on temperature and magnetic field (1). The ultrasonic measurements reported here were made with 15MHz longitudinal sound wave on polycrystalline samples in the temperature range from 9 K to 0.1 K.

The experimental techniques and sample preparation are similar to previous measurements on $ErRh_4B_4$ (2). In Figs. 1 and 2 is shown the zero-field attenuation as a function of temperature for x = 0.05 and 0.15, respectively. A sharp attenuation peak is observed at T_{c2} determined by ac-susceptibility measurements for

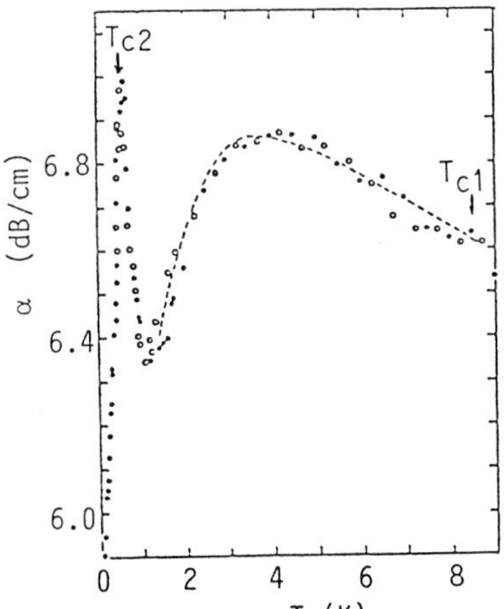

Fig. 1 Temperature dependence of the attenuation of 15MHz longitudinal wave for x = 0.05. Open and closed circles are obtained in cooling and warming runs, respectively and a broken line is obtained through eq. (4).

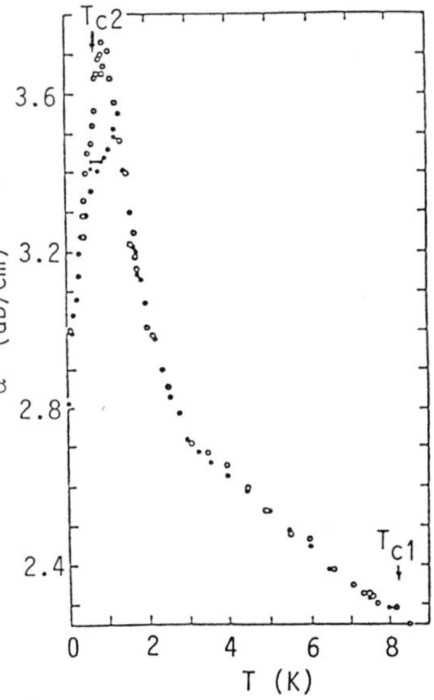

Fig. 2 Temperature dependence of the attenuation of 15MHz longitudinal wave for x = 0.15. Open and closed circles are obtained in cooling and warming runs, respectively.

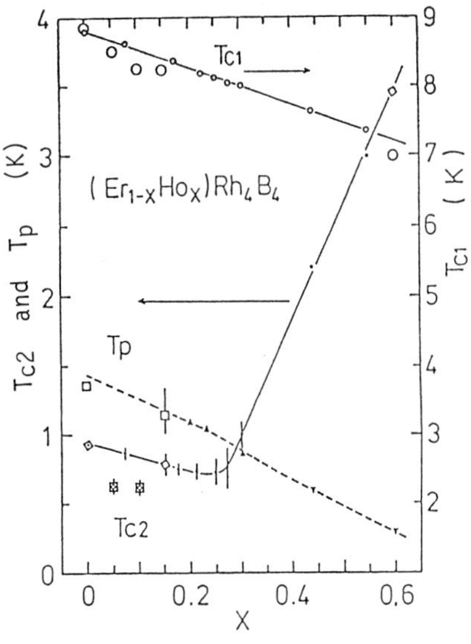

Fig. 3 Changes in the critical temperatures with Ho concentration for $(Er_{1-x}Ho_x)Rh_4B_4$. Large open circles and rhombuses represent T_{c1} and T_{c2} determined from ac-susceptibility, respectively. Squares represent peak temperatures of ultrasonic attenuation curves. Small circles and triangles are quoted from (3) for comparison.

x = 0.05, while a rather broad peak is observed for x = 0.15 at a temperature higher than T_{c2}. In the latter case T_{c2} = 0.70 K is in good agreement with the previous values (3), but in the former case T_{c2} = 0.57 K is markedly low compared with the values, notwithstanding that T_{c1} = 8.5 K is a reasonable value as shown in Fig.3. Since the residual resistance ratio of x = 0.05 is smaller than that of x = 0.15, this sample may be more nonstoichiometric or contain a lot of lattice imperfections, and T_{c2} seems to be decreased more sensitively by some kinds of lattice imperfections than T_{c1}. According to the theory (1), the attenuation is strongly enhanced by the spin-fluctuation and diverges at Tp, the temperature at which the second-order phase transition to the spin-sinusoidal phase occurs, while the attenuation may have a kink anomaly at T_{c2} and may be rapidly suppressed below T_{c2}. Therefore, the peak temperature 1.1 K for x = 0.15 in the cooling run may correspond to Tp and there is a temperature range of about 0.4 K where the two kinds of magnetic order mentioned above coexist above T_{c2}. On the other hand, in the case of x = 0.05 the peak temperature 0.58 K is very close to T_{c2} = 0.57K, so that the ferromagnetic transition seems to occur at a temperature very close to T_p or higher than Tp like the case of x = 0.10 (4). That is, Tp might be decreased by lattice imperfections more rapidly than T_{c2} in these samples.

In Fig. 1 we can also see a broad peak near 4 K. A similar peak has been observed for x = 0 (5). Although the origin of the attenuation peak has not be clarified definitely, it is understood as the energy dissipation due to the relaxation among the crystal-field levels of the Er ion. In the case of Er^{3+} (J = 15/2), the ground multiplet is split up into 8 doublets. We consider a S_0-fold degenerate ground state and S_i-fold excited states situated at E_i above the ground state. When a sound wave, $u = u_0 \exp\{-i(qr-\omega t)\}$, is present, the energy separation is modulated by the strain induced by the sound wave $\varepsilon = du/dr$ as

$$E_i = E_i^\circ + (\partial E/\partial\varepsilon)\cdot\varepsilon \qquad (1)$$

where E_i° is the energy in the absence of the sound wave. The distribution of the i-th level at equilibrium state is given by

$$n_i^\circ = \frac{S_i}{Z}\exp(-\frac{E_i}{k_BT}), \quad Z = \sum_j S_j \exp(-\frac{E_j}{k_BT}) \qquad (2)$$

and the relaxation rate of n_i is assumed to be given by

$$\frac{dn_i}{dt} = -\frac{n_i - n_i^\circ}{\tau_i} \qquad (3)$$

By substituting eqs. (1) and (2) into eq(3), we obtain the distribution n_i. The attenuation coefficient α is the power density dissipated per unit energy flux.

$$\alpha = -N < \sum_i E_i\frac{dn_i}{dt} >_{av} / (\frac{1}{2}\rho\,\omega^2 u_0^2 v_s)$$

$$= \frac{N\omega^2}{\rho v_s^3 k_B T Z^2}\sum_i \left[\frac{\tau_i S_i}{\tau_i^2\omega^2+1}\exp(-\frac{E_i}{k_BT})(\frac{\partial E_i}{\partial\varepsilon}) \times \sum_j \{\delta_{ij}Z - S_j\exp(-\frac{E_j}{k_BT})\}(\frac{\partial E_j}{\partial\varepsilon})\right] \qquad (4)$$

where N is the Er ion density, ρ the density of the sample and v_s the sound velocity. In the case of x = 0.05, ρ = 9.93, N = 0.91 x 10^{22} and v_s = 6.03 x 10^5 cm/s. An excellent fit to the data is obtained, as shown by the broken line in Fig. 1, for a doublet ground state and doublet excited states situated at E_1 = 1.4 K and E_2 = 6 K, where we assume $\tau_1 = \tau_2 = \tau$, $(\partial E_2/\partial\varepsilon)^2\tau/(\omega^2\tau^2+1)$ = 1.8 x 10^{-35} erg^2.s and $(\partial E_1/\partial\varepsilon)/(\partial E_2/\partial\varepsilon)$ = 0.2.

Although the crystal-field level schemes for $ErRh_4B_4$ have been obtained from the analysis of Schottky specific heat data (6) (7), the level schemes are not consistent with each other. The Schottky peak at 11 K observed in the specific heat measurement (6) is well fitted by assuming five doublets situated near 35 K, but a small hump due to the level separation of E_2 = 6 K appears near 2 K. If we assume a higher value for E_2 in order to eliminate this hump, the peak in the attenuation curve also shifts to a higher temperature, and the experimental data for x = 0.05 becomes in less agreement with the theoretical calculation. On the other hand, the data in the case of x = 0 (5), where the attenuation peak locates at 5 K, are well fitted by assuming E_2 ~ 12 K in agreement with the level scheme obtained by Woolf et al. (6). The substitution of Ho atoms may cause a broadening and a shift of the levels, then cause a broadening and a shift of the attenuation peak which may be realized in the case of x = 0.05. The further substitution of Ho atoms may sweep away the attenuation peak which may be realized in x = 0.15.

REFERENCES

(1) M. Tachiki, T. Koyama, H. Matsumoto and H. Umezawa, Solid State Commun. 34 (1980) 269.
(2) N. Toyota, S. B. Woods and Y. Muto, Solid State Commun. 37 (1981) 547.
(3) M. B. Maple, J. Mag. Mag. Matls. 31-34 (1983) 479.
(4) T. Fukase, S. B. Woods, N. Toyota, K. Tsunokuni, M. Ishino, M. Tachiki and Y. Muto, J. Mag. Mag. Matls. 31-34 (1983) 449.
(5) S. C. Schneider, M. Levy, D. C. Johnston and B. T. Matthias, Phys. Lett. 80A (1980) 72.
(6) L. D. Woolf, D. C. Johnston, H. B. Mackay, R. W. McCallum and M. B. Maple, J. Low Temp. Phys. 35 (1979) 651
(7) H. B. Radousky, B. D. Dunlap, G. S. Knappand, D. G. Niarchos, Phys. Rev. B27 (1983) 5526.

LT-17 (Contributed Papers)
U. Eckern, A. Schmid, W. Weber, H. Wühl (eds)

MAGNETIC ORDERING IN PRASEODYMIUM

M.S. COLCLOUGH and E.M. FORGAN

Department of Physics, Chancellors Court, University of Birmingham, Birmingham B15 2TT, U.K.

The heat capacity of a single crystal of Pr has been measured between 15 and 600mK, while the crystal was held in a way designed to minimise stress. The results are consistent with the occurrence of a cooperative transition involving the nuclei on half the atomic sites. The particular features of the ordering revealed by heat capacity measurements are discussed and the results are compared with previous measurements of heat capacity and neutron scattering.

1. INTRODUCTION

In the light rare earths, exchange and crystal field effects are similar in magnitude, and in praseodymium they are almost exactly balanced, with the exchange interaction between f electrons approximately 90% of the value which would be required to give a magnetically ordered state by the Bleaney mechanism (1). However, as calculated by Murao (2) the just subcritical exchange interaction enhances the van Vleck susceptibility of the ions and this enhanced response to the nuclear hyperfine field is sufficient for the nuclei to cause the system of nuclei and ions to order at a temperature T_c given by

$$kT_c = \frac{I(I+1)J(J+1)A^2}{3\Delta(1-\eta_{qo})} \qquad \text{----- } 1$$

where $I(={}^5/_2)$ and $J(=4)$ are the nuclear and ionic spin quantum numbers, A is the hyperfine electron-nucleus coupling constant, Δ the crystal field splitting of the two lowest states of the ion and $\eta_{qo} = 2J(J+1)\mathcal{J}(q_o)/\Delta$ where $\mathcal{J}(q_o)$ is the maximum exchange interaction. The predicted T_c is about 50mK and the ordering is antiferromagnetic.

Evidence for this ordering has been obtained by heat capacity (3) and neutron scattering measurements (4,5). These results show curious features however: the heat capacity jump is much broader than expected from fluctuation calculations and has a long high temperature tail; the neutron scattering shows possibly three different magnetic scattering wavevectors (4), the intensities of all of which vary with temperature in different ways and which extend far higher in temperature than the expected phase transition.

Presented with such complicated results, it is important to know whether these are intrinsic effects, or are caused by the experimental conditions. One possible complication is that of mechanical stress, which may be deliberately applied to modify the crystal field energies and thereby induce ordering well above 1K (6), but which may be inadvertently applied by a specimen holder.

2. WHY HEAT CAPACITY?

The observation of nuclear ordering by heat capacity measurements is of some interest because although the interactions are mediated by the f electron moments (which are observed by neutron scattering), the entropy of the transition is almost entirely nuclear in origin (the energy scale for the f electrons is too high for them to contribute any temperature-dependent entropy). After subtracting the conduction electron contribution, heat capacity measurements may be used to calculate this entropy. The heat capacity is insensitive to any electronic moment which varies more rapidly than the nuclear relaxation time (perhaps of the order of μs (8)), such as a rapidly varying ionic moment, which could give quasi-elastic neutron scattering but no nuclear ordering. The effect of the relatively long nuclear relaxation time may be seen in the exciton energy measurements in the presence of nuclear ordering (9) - the lowest energy exciton is unaltered in energy at the nuclear transition because its frequency is too high for the nuclei to follow. This is not the case for electronic ordering induced by uniaxial stress (10) - in that case the exciton mode goes "soft".

3. EXPERIMENTAL CONSIDERATIONS

Pr has an enormous heat capacity at dilution refrigerator temperatures and the design of any specimen holder must be a compromise between the need to maintain sufficient thermal contact with the specimen and the need to avoid stress due to differential thermal contraction. Doubt remained in the earlier heat capacity measurements (3) that the specimen may have been strained, so we have carried out further heat capacity measurements (extended to lower temperatures) with a new specimen (from the same source (7)) and a redesigned holder; the specimen is held within the holder only by 20 copper wires of 0.05 mm dia; conductive silver paint gives good electrical (and therefore thermal) contact with minimum stress.

The low temperature electrical resistance between specimen and holder is only about 10μΩ, but even so the 0.6g specimen has a thermal time constant of 2 hrs. at our lowest temperatures. Making the most pessimistic assumptions, our holder could apply a stress of 0.6 MPa, whereas the holder used previously could apply a non-uniform stress of up to 6 MPa. If we assume that a stress of 60 MPa is just sufficient to make the system critical ($\eta_{qq} = 1$), then according to equation 1 this would give a 10% broadening of the transition in the earlier work. In other respects, our measurements were made by the methods of ref. (3).

4. RESULTS

In the figure we show our results and the former ones (3). It is clear that there is little significant difference betwen the two, and that our worst-case analysis of stress due to the previous specimen holder was too pessimistic. It may be that the plastic binder of the silver paint used for electrical contact cushioned the specimen. We therefore have renewed confidence that the shape of the transition as revealed by heat capacity is an intrinsic effect, being independent of specimen and holder.

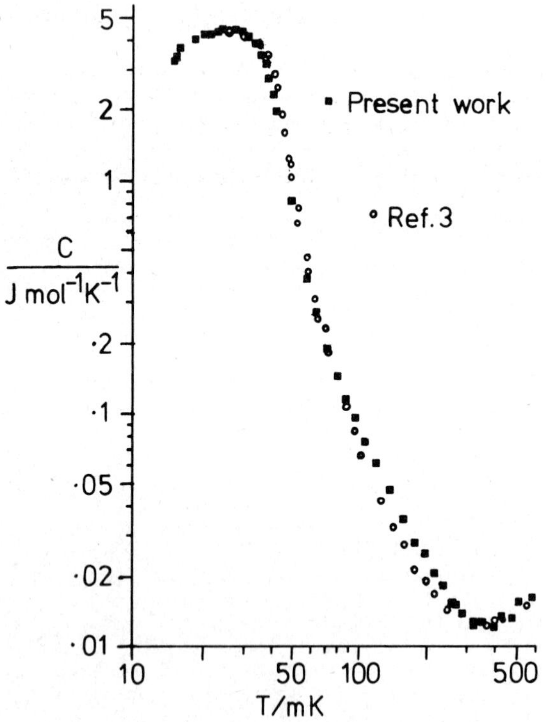

The measured heat capacity of Pr against temperature after subtraction of the (measured) holder heat capacity.

5. DISCUSSION

We comment now on the detailed form of the results: it is encouraging that on reducing our base temperature to 15mK we see the heat capacity falling below the peak near T_c. The estimated nuclear entropy removed by 15mK is 4.44 $Jmol^{-1}K^{-1}$ compared with a value of $^1/_2R\ell n6$ = 7.45 $Jmol^{-1}K^{-1}$ if the hexagonal site nuclei were completely ordered and the cubic site nuclei completely disordered.

The long tail in the heat capacity above the transition temperature was analysed in the earlier paper (3) to deduce the mean square electronic moment for comparison with the neutron scattering measurements. Reasonable agreement at high temperatures was obtained only when the very broad high temperature peak in the neutron scattering was included. This scattering is proably that calculated by Lindgård (11), which may be viewed as a sort of critical fluctuation above the Bleaney transition which would occur with a slightly stronger exchange in the absence of the nuclei. These fluctuations are, however, likely to be too rapid for the nuclei to order in their hyperfine field and contribute to the entropy change, so the apparent agreement between heat capacity and neutron scattering measurements should not be regarded as conclusive.

Nearer the transition, the broadening revealed by neutron scattering and heat capacity measurements is larger than that expected by Landau arguments (3). We conclude that the understanding of some features of the magnetic order of praseodymium is still in disarray.

REFERENCES

(1) B. Bleaney, in: Magnetic Properties of Rare Earth Metals, ed. R.J. Elliott (Plenum, New York, 1972) chap. 8.
(2) T. Murao, J. Phys. Soc. Japan 46 (1979) 40-4, and references therein.
(3) M. Eriksen et al, J. Phys. F. 13 (1983) 929-944.
(4) W.G. Stirling and K.A. McEwen in: Multicritical Phenomena eds. R. Pynn and A.T. Skjethorp (Plenum, New York, 1983)
(5) H. Bjerrum Moller et al, Phys. Rev. Lett. 49 (1982) 482-5.
(6) K.A. McEwen, W.G. Stirling and C. Vettier, Phys. Rev. Lett. 41 (1978) 343-6.
(7) Centre for Materials Science, University of Birmingham; described in(3).
(8) M. Kubota et al., Physica 108B (1981) 1093-4.
(9) K.A. McEwen and W.G. Stirling, J. Phys. C. 14 (1981) 157-165.
(10) K.A. McEwen, W.G. Stirling and C. Vettier, Physica 120B (1983) 152-5.
(11) P.A. Lindgård, Phys. Rev. Lett. 50 (1983) 690-3.

LT-17 (Contributed Papers)
U. Eckern, A. Schmid, W. Weber, H. Wühl (eds)

STUDY OF MAGNETIC ORDERING IN METALLIC Pr COMPOUNDS: IS SIMPLEST CANDIDATE PrS ?

M. KUBOTA, R.M. MUELLER and K.J. FISCHER

Institut für Festkörperforschung der KFA Jülich, D-5170 Jülich, F.R. Germany

The A.C. susceptibility of a monocrystal of PrS, a metallic hyperfine enhanced nuclear magnet (HENM) was measured as a function of temperature down to 80 μK. No evidence for magnetic ordering was observed. PrS has simple NaCl structure therefore especially suitable for comparing theory and experiment in the rapidly expanding study of nuclear magnetism.

1. INTRODUCTION

A nuclear magnet is a substance in which nuclear spin magnetic moments are located at regularly spaced lattice sites of the substance and surrounded by electronically paramagnetic media. It is a new category of magnetic materials with well localized nuclear spins together with a variety of magnetic fluctuations of the media. Since the nuclear spin moments are very small, the magnetic ordering phenomena occur at very low temperatures where the effects of other thermal excitations of the substance are neglibible though zero energy fluctuations of the media are still present.

In metallic nuclear magnets the nuclear spin-nuclear spin interaction is mediated by the conduction electrons which also provide a fast relaxation mechanism between nuclear spins and other systems in the material. It is desirable to study the ordering phenomena at equilibrium conditions of the whole substance unless one has already knowledge about various relaxation phenomena in the material. So far nuclear magnetic ordering has been studied under equilibrium conditions only in a limited number of materials, (A) the quantum solid ^3He (1) (B) metallic Hyperine Enhanced Nuclear Magnets (HENM) (2,3) and (C) nonmetallic HENM (4,5)

We have demonstrated that the HENM $PrNi_5$ orders ferromagnetically at 420 $\pm$ 20 μK and determined the parameters of the effective nuclear spin Hamilonian, for the first time for any magnetic system, purely from a thermodynamic analysis of the specific heat as a function of temperature and magnetic field. Both the enhanced nuclear saturation magnetization and the nuclear Curie constant were also determined from the analysis and they were consistent with the same enhancement factor 1+K=12.2$\pm$0.5. (3).

We have reported recently the first direct measurements of the magnetization of a metallic nuclear magnet, PrNi5, to 50 μK, well below the nuclear magnetic ordering temperature and found evidence for the long wave moment fluctuations effects in this material from the comparison of both results (6).

In order to study nuclear magnetic ordering phenomena and the related anomalous phenomena in this new kind of magnetic materials it is desirable to study the simplest systems. PrS (7) may offer such an example. It has NaCl crystal structure and Pr lies on an fcc sublattice. We have reported the demagnetization experiment to 0.7 mK (8) It was found important to improve the thermal contact in order to go to lower temperature to find a transition into an ordered state in this substance. In this letter we report on an AC susceptibility measurement of single crystal PrS to one decade lower temperature achieved by means of a new technique of thermal contact.

2. EXPERIMENT

Our PrS samples are single crystals prepared in the same manner as in ref (8). In order to characterize the quality of the sample DC susceptibility has been measured between liquid He and room temperatures (9). A temperature independent Van Vleck susceptibility χ_{VV} = 2.1 x 10^{-2} emu/mole was obtained below 20 K. The comparison with the crystal field splitting data from neutronscattering experiment suggests a very small exchange interaction which is consistent with former results (7), (8). The hyperfine enhancement factor 1+K = 5.00 $\pm$ 0.08 was obtained from the above χ_{VV} and the known hyperfine interaction constant of Pr.

To achieve thermal contact at ultralow temperatures, gold was evaporated on the surface of the single crystal sample which had been treated in vacuum at 500°C. When the sample was again heated in vaccuum, the gold reacted with PrS at the surface to form a strong bond for good thermal contact between the gold and PrS. This sample was mounted on a Cu cold finger of the Jülich two stage nuclear demagnetization cryostat (10) using a Cu clamp and a

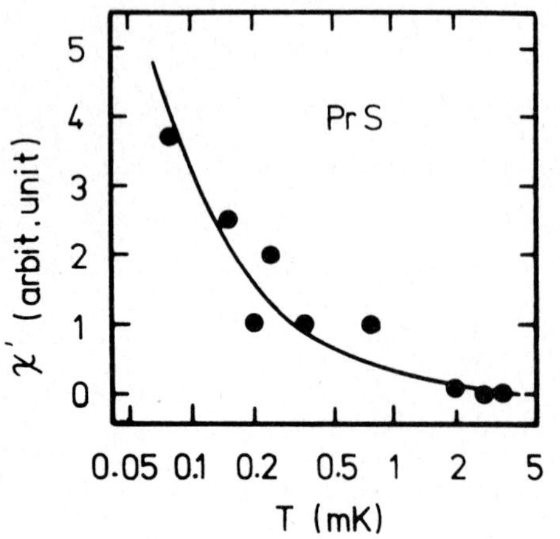

FIGURE 1
A.C. (3Hz) susceptibiltiy of single crystal Prs. The solid line through the data points represents the Curie law.

thin gold foil to cover part of the sample under the clamp.

Fig. 1 shows a preliminary result of the AC susceptibility measured in a SQUID susceptometer. Each point was measured after allowing the thermal relaxation for a day. The solid line through the measured points represents the Curie law. It is difficult to judge if there is any deviation from the Curie law because of the present poor signal to noise ratio but still we can conclude that PrS should order at a lower temperature than 100 μK.

2. DISCUSSION

Although the signal quality is very bad, PrS, the simplest metallic HENM, has not yet shown any sign of magnetic ordering to below 100 μK which is very close to the lowest lattice temperature available. There is certainly a large possibility of improving the sensitivity and there might be some possibility of reaching lower temperatures than the ordering temperature in the near future.

ACKNOWLEDGEMENTS

We gratefully acknowledge the contribution by G. Kraaij and the original suggestion of F. Pobell that PrS might be an interesting material.

REFERENCES

(1) for a Review see M. Roger, J. Magn.Mat. 31-34 727 (1983)
(2) $PrCu_6$: J. Babcock, J. Kiely, T. Manley and W. Weyhmann, Phys. Rev. Lett. 43, 380 (1979)
(3) $PrNi_5$: M. Kubota, H.R. Folle, Ch. Buchal, R.M. Mueller, and F. Pobell, Phys. Rev. Lett. 45, 1812 (1980)
(4) $HoVo_4$: H. Suzuki, N. Nambudripad, B. Bleaney, A.L. Allsop, G.J. Bowden, I.A.Campbell, and N.J. Stone, J. Phys., Paris 39, C6-800 (1978)
(5) $Cs_2NaHoCl_6$: H. Suzuki, M. Miyamoto, Y. Masuda, and T. Ohtsuka, J. Low Temp. Phys. 48, 297 (1982).
(6) M. Kubota, R.M. Mueller, Ch. Buchal, H. Chocolacs, J.R. Owers-Bradley, and F. Pobell, Phys. Rev. Lett. 51 1382 (1983)
(7) E. Bucher, K. Andres, F.J. di Salvo, J.P. Maita, A.C. Gossard, A.S. Cooper and G.W. Hull Jr. Phys. Rev. B11 500 (1975)
(8) Ch. Buchal, K.J. Fischer, M. Kubota, R.M. Mueller and F. Pobell, J. de Physique 39 L-457 (1978)
(9) G. Kraaij, Diplom thesis Jülich (1982)
(10) R.M. Mueller, Ch. Buchal, H.R. Folle, M. Kubota, and F. Pobell Cryogenics 20 395 (1980)

LT-17 (Contributed Papers)
U. Eckern, A. Schmid, W. Weber, H. Wühl (eds)

A NEUTRON DIFFRACTION STUDY OF THE HYPERFINE ENHANCED NUCLEAR ANTIFERROMAGNET $HoVO_4$*

Haruhiko SUZUKI and Taiichiro OHTSUKA

Department of Physics, Faculty of Science, Tohoku University, Sendai 980, Japan

Syuzo KAWARAZAKI and Nobuhiko KUNITOMI

Department of Physics, Faculty of Science, Osaka University, Toyonaka, Osaka 560, Japan

Ralph M. MOON and Robert M. NICKLOW

Solid State Division, Oak Ridge National Laboratory, Oak Ridge, TN 37831, U. S. A.

A neutron diffraction experiment on hfs enhanced nuclear system of $HoVo_4$ were performed following demagnetization self cooling. Magnetic diffraction peaks due to antiferromagnetic order of the hfs enhanced nuclear spin system below 4.5mK were observed.

The hyperfine enhanced nuclear antiferromagnetism of $HoVO_4$ below 4.5mK was studied by neutron diffraction experiment. This is the first observation of the hyperfine enhanced nuclear spin order by neutron diffracrion. From magnetic susceptibility (1) and nuclear orientation (2) experiments and calculation of dipole energy (3), the spin structure of $HoVO_4$ was predicted as shown in Fig.1. Our result of a neutron diffraction experiment supports this spin structure. Moreover, we also observed the spin-flop transition which is expected from this model.

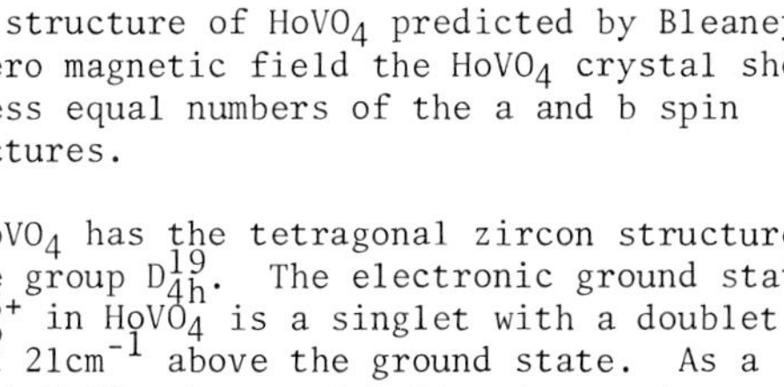

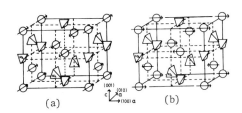

FIGURE 1
Spin structure of $HoVO_4$ predicted by Bleaney. In zero magnetic field the $HoVO_4$ crystal should possess equal numbers of the a and b spin structures.

$HoVO_4$ has the tetragonal zircon structure, space group D_{4h}^{19}. The electronic ground state of H_o^{3+} in $HoVO_4$ is a singlet with a doublet at about $21cm^{-1}$ above the ground state. As a result $HoVO_4$ shows, at helium temperatures, a large Van-Bleck temperature independent susceptibility, perpendicular to the tetragonal c-axis. This gives rise to an hyperfine enhanced nuclear magnetic moment $\mu_N = a_x g_J A_J \mu_B I_x \equiv \mu_N(e) + \mu_N(I) = 0.38\ \mu_B$, taking the crystal a axis as the x and x' axes. Here a_x stands for $2|\langle E|J_x|G\rangle|^2/\Delta$, $|G\rangle$, the ground state and $|E\rangle$, a low-lying excited state. A_J is the hyperfine interaction constant. The first term $\mu_N(e)$ is really an induced electronic moment but is proportional to the nuclear polarization and the second one, $\mu_N(I)$, is a pure nuclear moment which is negligible small in magnitude compared to the former. When this enhanced nuclear moment is fully saturated as low temperatures in this antiferromagnetic state, the internal dipole field B_x induces the Van Bleck moment $\mu_{VV}(e) = a_x g_J{}^2 \mu_B{}^2 B_x$. The magnitude of this moment was calculated as 0.08 μ_B (3). The neutron scattering length due to the electronic magnetic moment $\vec{\mu}(e)$ and the nuclear moment with a finite magnitude of the nuclear spin polarization $\langle I\rangle$ are given as follows, respectively in a quantum operation form,

$$\hat{P}_e = 0.27\ f(\vec{k})\ \vec{\mu}_{\perp}(e) \cdot \vec{\sigma} \quad (1),$$

$$\hat{P}_n = b + 0.27\ \mu_N^* \vec{I} \cdot \vec{\sigma} \quad (2).$$

Here $\vec{\mu}_{\perp}(e)$ in the compnent of $\vec{\mu}(e)$ perpendicular to the scattering vector $\vec{k}$ and $f(\vec{k})$ is the form factor of the magnetic electrons. A pseudo-magnetic moment is defined as $\mu_N^* = 1/0.27 \cdot I/(2I+1) \cdot (b^+ - b^-)$ in unit of μ_B. b^+ and b^- are the spin dependent parts of the nuclear scattering length. $\bar{b}$ is the coherent scattering length with no nuclear poralization. For $HoVO_4$, various values of μ_N^* between $-0.48\mu_B$ and $-0.75\mu_B$ are reported (4). The effective magnetic moment for neutron diffraction measurements in enhanced nuclear spin system is given as $\mu_{eff} = \mu_N^* + \mu_N(e) + \mu_{VV}(e)$. The effective moment μ_{eff} of $HoVO_4$ at low temperatures various from $-0.02\mu_B$

* Research carried out at the Oak Ridge National Laboratory under the U.S.-Japan Cooperative Program in Neutron Scattering and sponsored in part by the Division of Materials Science, U.S.Department of Energy under contract W-7405-eng-26 with the Union Carbide Corporation.

to $-0.29\mu_B$ by using the value of $\mu_N(e)=0.38\mu_B$, $\mu_{VV}(e)=0.08\mu_B$ and the various reported values of μ_N^* from $-0.48\mu_B$ to $-0.75\mu_B$. When we observe a diffraction peak for a higher angle reflection, the electronic magnetic scattering length decreases due to the magnetic form factor $f(\vec{\kappa})$ in eq.1. We thus can expect a larger magnetic scattering amplitude due to almost only the pseudonuclear moment.

A single crystal of $HoVO_4$ which was kindly supplied by Dr.H.Unoki was grown by using an image arc furnace. The size of the crystal used in our investigation was 4x4x13(mm) with the long dimension along the c-axis. The crystal was attached to the mixing chamber of the ^3He-^4He dilution refrigerator by using 300 thin copper wires. The copper wire coil was wound surrounding the crystal for rf magnetic susceptibility measurement using a low level resonant circuit. The temperature of the hyperfine enhanced nuclear spin system was determined from the magnetic susceptibility of the sample, measured simultaneously, which was normalized to the temperature dependence of the magnetic susceptibility reported by Suzuki et al (1). Starting from an initial temperature of about 30mK and an initial field of 3T along the <010> axis, the crystal was self cooled by demagnetization to 2.7mK. The experiment was carried out on the HB-1 spectrometer at HFIR of the Oak Ridge National Lab. The wavelength of the incident neutron beam was chosen to be 1.06Å to made use of a P_u filter to eliminate the $\lambda/2$ contamination. The crystal was aligned so that the scattering vector was in the <100> - <001> plane.

Below the Néel temperature we observed diffraction peaks at the (100) and (001) reciprocal positions. In Fig.2, the (001) diffraction peak at 3mK is shown together with the scan at a higher temperature $T>T_N$. The D_{4h}^{19} space group of $HoVO_4$ crystal gives the conditions limiting possible reflections of crystal structure as $hk\ell$:$h+k+\ell=2n$ and $hh\ell$:$2h+\ell=4n$. Therefore our observed (001) and (100) diffraction peaks must be superlattice reflections due to an antiferromagnetic structure of the hyperfine enhanced nuclear spin system. We also tried to observe the (002) reflection but could not observe any magnetic diffraction. These results support the spin structure shown in Fig.1. In order to observe the spin flop transition reported by previous authors (1,2), the magnetic field effects on the neutron spectra were measured. Increasing the magnetic field applied along (010) direction, the peak intensity of the (100) reflection decreased rapidly at about 80 Oe. The peak intensity of the (001) reflection showed no change arround 80 Oe but gradually decreased in higher magnetic field up to 400 Oe. The results can be understood by the spin flop transition at about 80 Oe and the completed spin poralization above 400 Oe. These results also

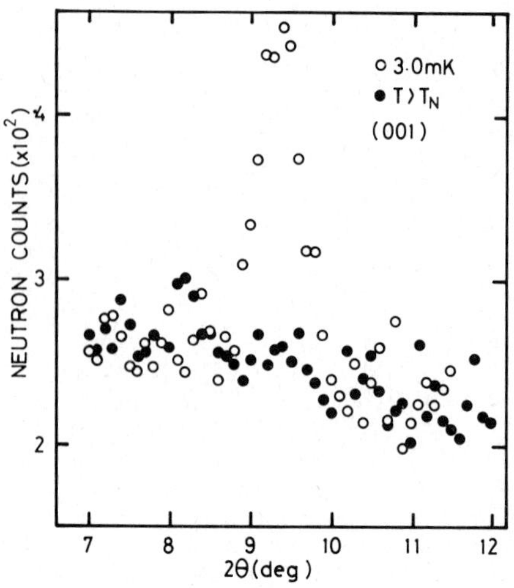

FIGURE 2
Diffraction scan through the (001) reciprocal position at about 3mK (open circle) and $T>T_N$ (solid circle).

support the spin structure shown in Fig.1.

Although it is difficult to determine the precise value of μ_N^* in our present experiment, we tried to determine its value from the ratio of the intensity of (001) reflection to its higher angle reflection (009). The ratio $I_{(009)}/I_{(001)}$ was obtained as 1.88. Taking into account the Lorentz factor, the form factor $f(\vec{\kappa})$ for Ho^{3+} calculated by Stassis et al (5) and the effect of the absorption correction, the value of μ_N^* was estimated as $\mu_N^* = 0.59 \pm 0.06\ \mu_B$. A more detailed report on our results will be published (6).

REFERENCES
(1) H. Suzuki, N. Nambudripad, B. Bleaney, A.L. Allsop, G.J. Bowden, I.A. Campbell and N.J. Stone, J. Phys. (Paris) 39 (1978) C6-800.
(2) A.L. Allsop, B. Bleaney, G.J. Bowden, N. Nambudripad, N.J. Stone and H. Suzuki, Proc. R. Soc. Lond. A372 (1980) 19.
(3) B. Bleaney, Proc. R. Soc. Lond. A370 (1980) 313.
(4) H. Glattli, G.L. Bacchella, M. Fourmond, A. Malinovski, P. Meriel, M. Pinot, P. Roubeau and A. Abragam, J. Phys. 40 (1979) 629.
(5) C. Stassis, H. W. Deckman, B.N. Harmon, J.P. Desclaux and A.J. Freemen, Phys. Rev. B15 (1977) 369.
(6) H. Suzuki, T. Ohtsuka, S. Kawarazaki, N. Kunitomi, R.M. Moon and R.M. Nicklow, Solid State Commun. in Press.

LT-17 (Contributed Papers)
U. Eckern, A. Schmid, W. Weber, H. Wühl (eds)
© Elsevier Science Publishers B.V., 1984

SERIES ARRAYS OF REFRACTORY SNS MICROBRIDGES

A. L. DE LOZANNE*, W. J. ANKLAM†, and M. R. BEASLEY*†

* Dept. of Applied Physics, Stanford University, Stanford CA 94305, USA
† Dept. of Electrical Engineering, Stanford University.

We report on the fabrication and properties of series arrays of 15 Nb/Cu/Nb SNS microbridges. The large size of the RF steps observed in single devices allows the arrays to show wide regions of constant voltage which should be useful for a practical voltage standard. These results and our previous experience suggest that it is feasible to make a practical standard operating above 10K.

An array of N Josephson devices connected in series has several advantages over a single device in a number of applications. Briefly stated these advantages are that the total resistance is proportional to N, the power generated by a coherent array is proportional to N^2 and the linewidth of this radiation is reduced as 1/N. Coherence is difficult to achieve in a practical array; several mechanisms for improving the coherence of arrays have been proposed and tested(1). There is, however, an application for which internal coherence of the array is not essential, namely the case of voltage standards. Here the coherence of the array is aided by the applied external radiation. We have made series arrays of SNS microbridges that seem quite promising for a practical voltage standard, particularly since similar single microbridges made with high-T_c Nb_3Ge have been demonstrated to operate from 3K to 19K (2,3).

Among the attractive characteristics of these microbridges is the fact that they show large and sharp steps when exposed to microwave radiation, as shown in Figure 1 at a temperature of 14.2K. The size and well-defined voltage of these steps, together with the fact that the critical currents of different microbridges are fairly reproducible, has prompted us to fabricate series arrays of these devices.

Briefly stated, these devices are fabricated by e-beam evaporation of the superconductor on a substrate that has a step. The shadowing caused by the step produces a gap in the superconductor which is bridged by in-situ e-beam evaporation of the normal metal. More details of the fabrication process have been published elsewhere (2,3,4). The resulting microbridge is shown in Figure 2a. The series connection of several microbridges is simply done by lithographically defining a serpentine pattern that goes back and forth across the step, so that a bridge is located at every crossing of the pattern with the underlying step. This is illustrated in Figure 2b. This pattern has the advantage that while the bridges are closely spaced the leads connecting them are long and wide, thus serving as a very effective heat sink. E-beam lithography and ion milling can also be used to narrow the bridge while not changing the width of the banks (4). This has not been done with the arrays presented here but will be incorporated as a future improvement.

The electrical characteristics of a 15-bridge array are shown in Figure 3. In this case we chose Nb for the superconductor to simplify the fabrication of the microbridges and deal more with the circuit problems. Curve A shows the I-V of this array at 6.70K without microwaves. On this curve one can observe different transitions out of the zero-voltage state of different bridges, indicating a moderate spread of critical currents. On a broader scale, shown in the inset of Figure 3, the I-V's look quite good

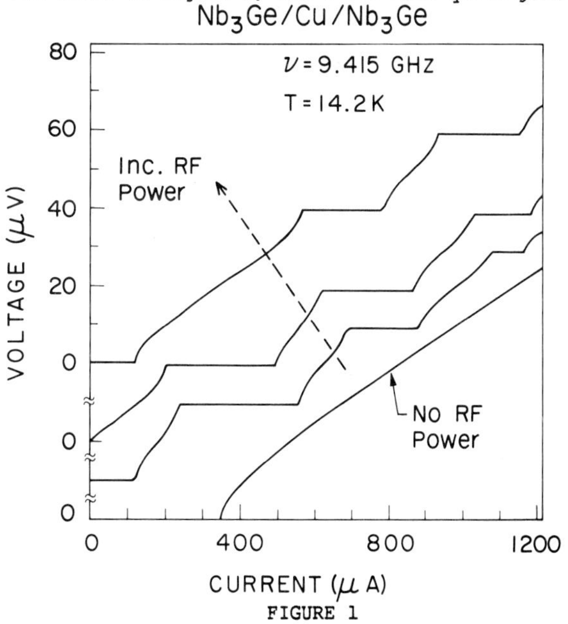

FIGURE 1
Current-voltage characteristics of a high-T_c $Nb_3Ge/Cu/Nb_3Ge$ microbridge showing sharp and wide RF steps at a temperature of 14.2 K.

and in fact are similar to those of a single device. The kink observed at 11 mA is very likely due to a short in one of the bridges which becomes normal at that point, so that for currents less than 11 mA only 14 bridges are operating in this particular array. Measurements of differential resistance (not shown) show very sharp onsets when a new critical current is reached. From the size and number of these onsets we estimate that the critical currents of the 14 operating bridges range from 2.85 to 4.80 mA with an average value of 3.60 mA and a standard deviation of 0.48 mA.

Curves B and C in Figure 3 show the constant-voltage steps induced by 1.88 GHz microwave radiation. Curve B shows a well-defined step at V=0.108 mV which corresponds to 14 bridges locked on the second constant-voltage step, effectively giving 28 times the fundamental voltage (3.88 μV) for this frequency. At higher microwave power, curve C shows a step at V= .054 mV, corresponding to 14 bridges on the first step. The measured differential resistance drops to zero (experimentally indistinguishable from the differential resistance in the zero-voltage state) over the width of these steps, typically 0.1 to 0.15 mA. These flat regions should be useful for practical voltage standards. Curves A and B also show fairly large steps at other multiples of the fundamental voltage. These steps result when all of the bridges are on some step but not necessarily all on the same step. These regions are also a useful standard since zero differential resistance for the array implies zero differential resistance for all the bridges (i.e. their voltage is a multiple of the fundamental). Better performance is expected at higher frequencies, of the order of $\hbar\omega = 2eI_cR$, as in the case of Figure 1.

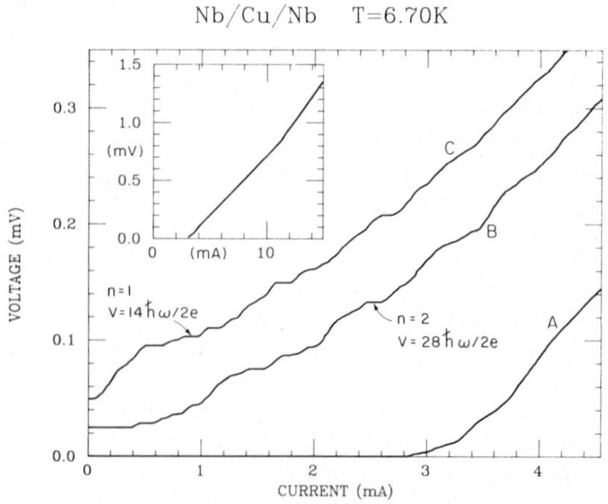

FIGURE 3
Current-voltage characteristics of an array of 15 Nb/Cu/Nb microbridges operating at 6.70K. Curve A shows the individual critical currents without microwaves, while the inset shows the same characteristics over a wider range of bias. Curves B and C are obtained with increasing microwave radiation (1.88 GHz). The ranges where all of the operating bridges are on the same RF step have been labeled with the step number n.

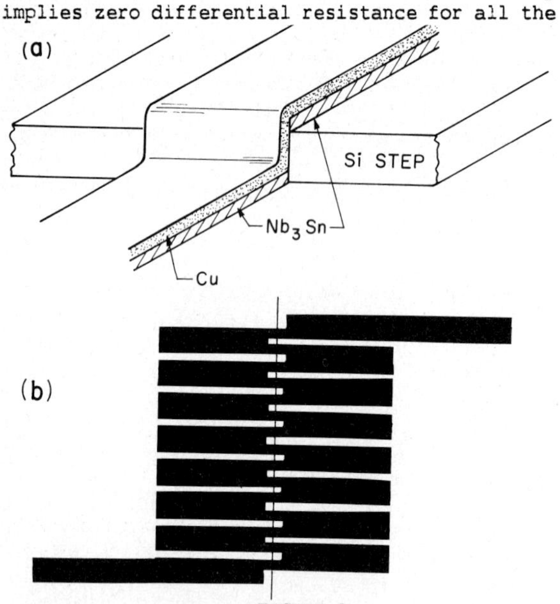

FIGURE 2
(a) Sketch of the geometry of a single microbridge. (b) Pattern used to connect 15 bridges in series. The vertical line indicates the step in the substrate.

In conclusion we note that these prototype arrays show that it is feasible to build a practical voltage standard with, say, 100 bridges in series, with a total resistance of 5-50 ohms, and frequencies in the range of 1-10 GHz that would give reference voltages up to 2 mV when operated on the first step, and 4 mV on the second step. These arrays could be operated in small closed-cycle refrigerators at temperatures above 10K.

This research is supported by the US National Science Foundation (NSF-ECS).

REFERENCES

(1) J. Bindslev Hansen, P. E. Lindelof, and T. F. Finnegan, IEEE Trans. MAG-17 (1981) 95. A. K. Jain, P.M. Makiewich, A. M. Kadin, R. H. Ono, and J. E. Lukens, ibid p 99. A.K. Jain, K.K. Likharev, J.E. Lukens, J.E. Sauvageau, Appl. Phys. Lett. 41 (1982) 566.
(2) A. de Lozanne, M. S. DiIorio, and M. R. Beasley, Physica 108B (1981) 1027.
(3) A. de Lozanne, M. S. DiIorio, and M. R. Beasley, Appl. Phys. Lett. 42 (1983) 541.
(4) M. S. DiIorio, A. de Lozanne, and M. R. Beasley, IEEE Trans. Mag. MAG-19 (1983)308.

LT-17 (Contributed Papers)
U. Eckern, A. Schmid, W. Weber, H. Wühl (eds)

On the Role of Amorphous-Silicon Barrier in Nb_3Ge Josephson Tunnel Junctions

Hajime KONISHI, *Atsushi NOYA and Shinya KURIKI

Research Institute of Applied Electricity, Hokkaido University, Sapporo, Japan
*Faculty of Engineering, Kitami Institute of Technology, Kitami, Japan

Nb_3Ge/amorphous-silicon(a-Si)/Pb Josephson tunnel junctions have good I-V characteristic when the a-Si barrier is sputter-deposited at low voltage and high Ar pressure. We have analyzed the a-Si/Nb_3Ge interface by Auger Electron Spectroscopy and found that the a-Si layer suppresses the diffusion of oxygen into the Nb_3Ge film and also changes the composition of the Ge-rich surface layer of Nb_3Ge into near-stoichiometry.

1. INTRODUCTION

We have been studying Nb_3Ge/amorphous-silicon(a-Si)/Pb tunnel junctions using high T_c and stoichiometric Nb_3Ge films. It has been found that high quality junctions are obtained with high gap voltages when the a-Si is sputter-deposited at high Ar pressure and low target voltage to reduce the pinhole density in the a-Si layer and the damage to the Nb_3Ge film. However, the role of the a-Si as an artificial barrier is not well understood. In this paper we describe the result of Auger analysis of the a-Si/Nb_3Ge interface and its relation to the tunneling characteristic. The effect of the a-Si layer on the surface chemistry of the Nb_3Ge films and the origin of Ge segregation are discussed.

2. EXPERIMENTAL METHOD

Stoichiometric Nb_3Ge films of 500 nm thickness are rf sputtered from a Nb-Ge composite target onto sapphire substrates at about 800 °C (1). Then the films are cooled for 20 minutes down below 100 °C, and an a-Si layer of 1-2 nm is sputtered in the same vacuum at rf voltage of 900 V and Ar pressure of 0.2 Torr. The a-Si layer is oxidized in air at room temperature for 1-2 hours to block pinholes. The composition depth profile of the samples are measured by Auger Electron Spectroscopy (AES) with use of the sputter-etching technique. The depth is estimated from the etching time and the etching rate which is estimated from the time required to sputter-etch the Nb_3Ge film of known thickness down to the substrate.

3. RESULTS AND DISCUSSIONS

The I-V characteristic of a Nb_3Ge/oxide/Pb junction which used a Nb_3Ge native oxide barrier is shown in Fig. 1a. Although the gap voltage of about 4.3 mV is obtained from the minimum in the dV/dI-V curve, it is hardly seen in the figure on account of the large excess current. The Josephson current is also too small to see in the figure. The multigap structure, which was reported previously (2,3) is not observed in these native-oxide junctions.

A Ge-rich segregation layer of about 7-10 nm thickness exists at the surface of as-deposited Nb_3Ge films (1,4). A possible reason of the poor tunneling characteristic of the native-oxide junctions is that the composition of the Nb_3Ge just below the oxide barrier is not stoichiometry due to the Ge segregation. Another reason is that Ge atoms in the Nb_3Ge oxide remain non-oxidized to yield the large leakage current.

On the contrary, the junctions which used an a-Si artificial barrier (5,6) show a good tunneling characteristic such as in Fig. 1b. The excess current below the gap voltage is small and appreciable amount of Josephson current is observed (not shown in Fig. 1). The AES profile for such a junction is shown in Fig. 2. The

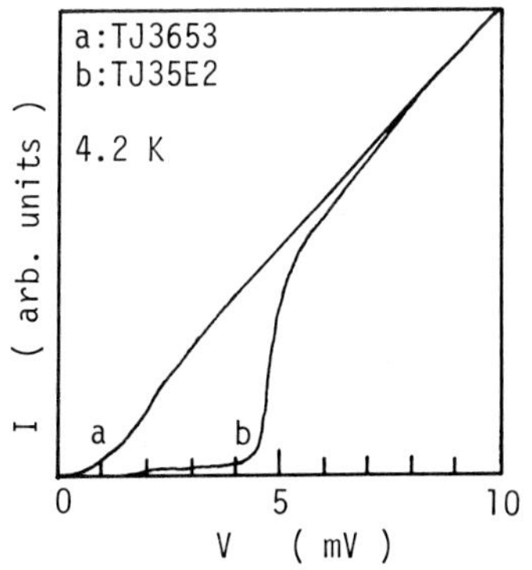

FIGURE 1
I-V characteristics of Nb_3Ge tunnel junctions which used a native oxide barrier (a) and a-Si barrier (b). Josephson current is suppressed in (b) by externally applied magnetic field.

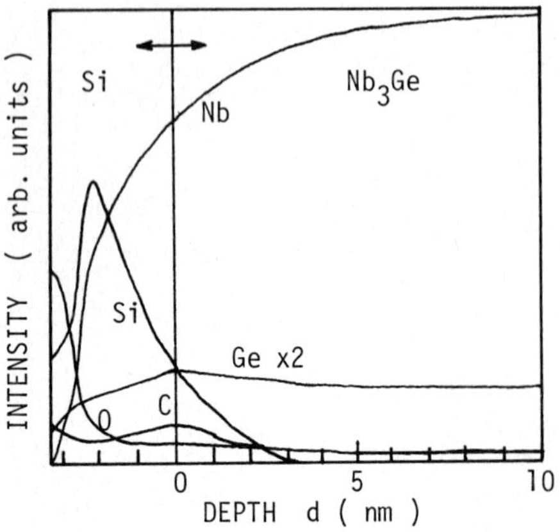

FIGURE 2
Surface AES profile of a Nb_3Ge film with a-Si barrier. The thickness of the barrier is about 3.5 nm.

a-Si/Nb_3Ge interface (the origin of the abscissa) is estimated from the maximum of carbon signal, which may result from adsorption during the cooling period before the a-Si deposition. There is no oxygen peak around this depth indicating that no oxidation occurs at the Nb_3Ge surface during the cooling period. The Ge segregation is small at the a-Si/Nb_3Ge interface.

In Table I Ge content C_{Ge} in the bulk part of the Nb_3Ge films, the increase of Ge content ΔC_{Ge} at the peak of the segregation, and the thickness of the segregation layer d_{Ge} are listed for several Nb_3Ge films prepared by different conditions. Usually, the maximum Ge content in the segregation layer reaches 35 at.%Ge, and d_{Ge} is about 8-10 nm. Even in a Nb-rich film of 21 at.%Ge the segregation layer has a Ge content 8.3 at.% larger than the bulk composition, although d_{Ge} is thinner than stoichiometric films. When the a-Si is deposited on the Nb_3Ge surface, both ΔC_{Ge} and d_{Ge} decrease gradually as the thickness of the a-Si layer d_{Si} increases. For the a-Si more than about 2 nm the composition of the segregation layer does not differ much from the stoichiometry and the segregation layer thickness becomes thinner than the coherence length of Nb_3Ge (~5 nm).

According to the binary alloy phase diagram Si and Ge make a solid solution at all compositions. On the other hand, Nb scarcely dilutes in Si. Therefore, the reason why the surface composition is compensated by the a-Si layer is considered that Ge diffuses selectively into the Si layer. If we assume that the Ge segregation is associated with oxidation of Nb_3Ge surface (7), another reason can be that the segregation is reduced by preventing oxygen from reacting with the Nb_3Ge as described below.

TABLE I Ge segregation in several Nb_3Ge films.

Sample	8908	3922	*3A31	3631	3622	3611	*3932
d_{Si} (nm)	0	0	0	1.5	2	3.5	2
C_{Ge} (at.%Ge)	24.6	21.3	26.6	23.5	23.7	23.0	24.0
ΔC_{Ge} (at.%Ge)	10.9	8.3	3.6	8.8	6.1	2.5	2.2
d_{Ge} (nm)	9.0	5.0	4.0	5.5	4.5	2.5	1.0

*:Deposited at an ambient temperature.

The a-Si layer has also the role of preventing the oxidation of the Nb_3Ge film. The Auger profile for an air-oxidized Nb_3Ge film shows that oxygen diffuses into the film more than 10 nm reaching a constant bulk level. When an a-Si layer is deposited on the Nb_3Ge film, the diffused oxygen decreases and reaches the constant level more rapidly. The a-Si layer of about 2 nm is sufficient to suppress the diffusion of oxygen into the Nb_3Ge film.

We have made some preliminary experiments to understand the origin of the Ge segregation. A Nb_3Ge film which has a usual Ge segregation is sputter-etched in the AES equipment until the ratio of Nb and Ge signal becomes constant to remove the segregation layer. Then the film is exposed to the air for one hour at an ambient temperature, put in the high vacuum again and analyzed by AES. The segregation is found to be very small compared with as-deposited films. The AES study has also been made for the Nb_3Ge films which were sputtered at an ambient temperature. The results are shown in Table I for the films with and without an a-Si layer deposited. The segregation is very small. Thus oxidation is thought not to be the main cause of the Ge segregation. Instead, it is considered to be formed during the film deposition process.

ACKNOWLEDGEMENT
We would like to thank Prof. G.Matsumoto for many helpful discussions and suggestions during the course of the work.

REFERENCES
(1) S.Kuriki and T.Ohora, J. Low Temp. Phys. 47 (1982) 111.
(2) J.M.Rowell and P.H.Schmidt, Appl. Phys. Lett. 29 (1976) 622.
(3) D.F.Moore, R.B.Zubeck, J.M.Rowell and M.R.Beasley, Phys. Rev. B20 (1979) 2721.
(4) R.H.Buitrago, L.E.Toth and A.M.Goldman, J. Appl. Phys. 50 (1979) 983.
(5) D.A.Rudman and M.R.Beasley, Appl. Phys. Lett. 36 (1980) 1010.
(6) K.E.Kihlstrom, D.Mael and T.H.Geballe, Phys. Rev. B29 (1984) 150.
(7) H.Ihara, Y.Kimura, M.Yamazaki and S.Gonda, Phys. Rev. B27 (1983) 551.

LT-17 (Contributed Papers)
U. Eckern, A. Schmid, W. Weber, H. Wühl (eds)

SMALL PLANAR SNS JOSEPHSON-JUNCTIONS IN NIOBIUM-COPPER TECHNIQUE

K.-D. KLEIN , H. LUTHER

Physikalisch-Technische Bundesanstalt, Institut Berlin, D-1000 Berlin 10, FRG

For cryoelectronic applications a type of superconductor-normal-superconductor (SNS) Josephson junction has been fabricated using Nb as the bulk material and Cu for the normal-conducting layer. Small critical current and high impedance are due to the low contact area of the sandwich (0.25 μm²). The fabrication process and some electrical characteristics are reported here. Mechanical and thermal-cycle stability of the junctions are quite well, the I-V characteristic exhibits nearly ideal behavior with respect to the resistive shunted junction (RSJ) model.

1. INTRODUCTION

Superconductor-normal-superconductor (SNS) junctions are interesting elements for Josephson devices because of their good stability and the absence of any remarkable capacitance (1). We prepared such junctions for application in resistive SQUIDs for noise thermometers (2) at 4.2 K. Here the low value $R_N I_c$ (R_N = normal resistance, I_c = critical current) is not a restrictive condition. Sufficient high resistance has been achieved reducing the contact area to 0.25 μm², patterned by electron-beam lithography. Weak dependence of the critical current on the temperature (an important property for the thermometer application) is given in the desired temperature range.

Niobium is a proper choice for the superconductors since the transition temperature is about 5 - 7 K for thin layers. Copper has been used for the normal conducting layer because of its low solubility in Nb, high long-term stability of a Nb-Cu-Nb junction can be expected.

2. PREPARATION

In Fig. 1a a schematic top-view diagram of

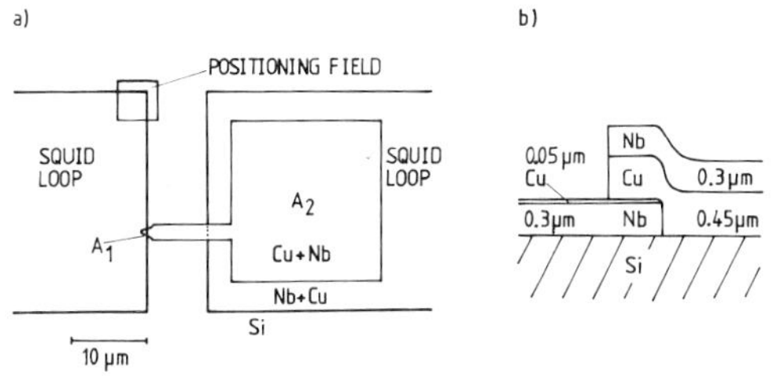

FIGURE 1
a) top view b) cross section of the junction region.

the junction configuration is shown. The area of the small SNS sandwich is marked A_1, a cross section through this area can be seen in Fig. 1b. The large area marked A_2 is a SNS junction too, but since it is so large, it will not leave the zero-voltage state during operating the device with bias currents appropriate for A_1.

In Fig. 2 a SEM picture of this "double" SNS

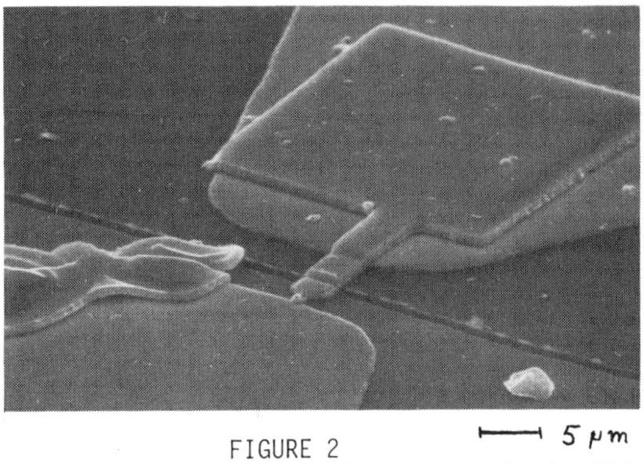

FIGURE 2
SEM picture of the SQUID-loop gap with the SNS junction.

junction is shown. Both areas A_1 and A_2 and their position at the gap of the first Nb film (which may be a SQUID loop as indicated and partially shown in Fig. 1a) can be seen. The area A_1 is shown in detail in the SEM picture Fig. 3.

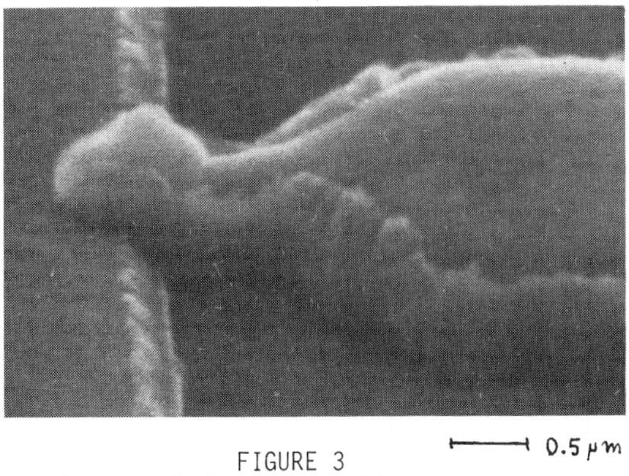

FIGURE 3
SEM picture of the SNS junction

The fabrication process is as follows: The first Nb film is patterned by optical lithography and lift-off technique on a silicon substrate. The layer is deposited by magnetron sputtering (rate 1.1 nm/s) with a thickness of 0.3 µm. During the deposition the substrate is biased with a voltage of - 70 V. For testing we measure the resistance ratio $\varrho_{295K}/\varrho_{9K}$ of samples sputtered at the same time (best value 6.5).

In the same vacuum a Cu layer of 0,05 µm thickness is rf-sputtered on the Nb. It prevents contamination of the SN interface by oxides and impurities, which may occur during the unavoidable treatment before depositing the second layer. Furthermore only soft sputter etching is required to clean the Cu surface at the beginning of the subsequent sputtering of the further layers. These films are patterned by electron-beam lithography and lift-off technique. For exact positioning of the substrate under the e-beam one corner of the first Nb-Cu layer serves as a registration mark. We took advantage in using not PMMA but Shipley Microposit 1350 J as the resist for the e-beam lithography. The minimum linewidth achieved with this resist is not so small as with PMMA but the sputterrate of 1350 J is much lower. Furthermore the redepositioned resist particles do not react with the Cu to a non-conducting layer as PMMA does. Faceting and undesired continuation of the resist development by the Ar plasma during sputter etching are smaller with 1350 J too.

The second Cu layer of 0.45 µm thickness amounts the total thickness of the normal-conduction zone to 0.5 µm. The second Nb layer is magnetron sputtered upon the Cu in the same vacuum.

3. ELECTRICAL CHARACTERISTIC

The I-V characteristic of such a junction is shown in Fig. 4 for four different temperatures. There is no hysteresis due to heating or capacitance, the shape of the curves is in good agreement with I-V characteristics based on the RSJ model. The normal resistance of this sample is 0.25 .

Fig. 5 demonstrates besides the typical dependence of the critical current on the temperature the limits of reproducibility of the preparation. The differences in I_c of the three samples shown in Fig. 5 may be explained by different crystal structures at the SN interfaces.

4. CONCLUSION

We have prepared SNS Josephson junctions of sandwich type. Using Nb and Cu we achieved good stability against mechanical treatment, thermal cycling and electrostatic discharge. Low critical current and normal-resistance values near 1 Ω are achieved using small contact area. The good agreement of the I-V characteristics with the RSJ model encourages to use this type of junction for devices as SQUID noise thermometers which demand a behavior of the Josephson element easy to survey.

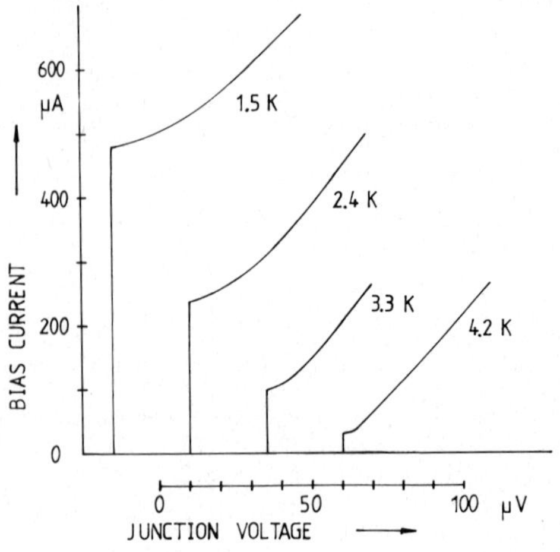

FIGURE 4
I-V characteristics of a SNS junction for different temperatures

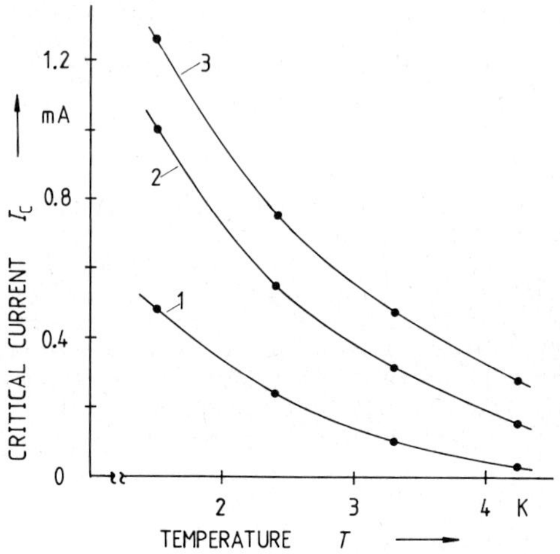

FIGURE 5
I_c - T characteristics of three SNS junctions from one charge.

ACKNOWLEDGEMENTS

We thank Dr. S.N. Erné for stimulating this work and helpfull discussions and Mrs. M. Peters for technical assistance in the fabrication.

REFERENCES

(1) R.B. Dover, A. deLozanne, M.R. Beasley, J. Appl. Phys. 52 (1981) 7327.
(2) R.A. Kamper, J.E. Zimmerman, J. Appl. Phys. 42 (1971) 132.

LT-17 (Contributed Papers)
U. Eckern, A. Schmid, W. Weber, H. Wühl (eds)

VACUUM TUNNELING APPLIED TO THE SURFACE TOPOGRAPHY OF A Pd (100) SURFACE

M. RINGGER, H.R. HIDBER, R. SCHLÖGL, P. OELHAFEN and H.-J. GÜNTHERODT

Institut für Physik, Universität Basel, Klingelbergstr. 82, CH-4056 Basel, Switzerland

K. WANDELT and G. ERTL

Institut für Physikalische Chemie, Universität München, D-8000 München, Germany

We have built a scanning tunneling microscope (STM) similar to the design published by Binnig et al. In order to test the operation of our STM, a Pd (100) surface has been studied. The mono- and double-atomic steps are clearly seen.

1. INTRODUCTION

The first report by Binnig et al. (1) on the fascinating work of vacuum tunneling was given at the last LT Conference. In the mean-time the scanning tunneling microscope (STM) has been built which is based on the phenomenon of vacuum tunneling (2,3). This instrument opens a new area of surface studies and a new field of spectroscopy.

In particular, the promises of this novel technique seem very attractive to the materials under study in our laboratory in Basel: metallic glasses, graphite and graphite intercalation compounds. The STM offers, for the first time, the possibility of direct, real space determination of surface structures, including non-periodic structures. The real space observation on an atomic scale might help to elucidate the surface topography of metallic glasses and even yield information on the short range order of these materials. Topographical problems also can be viewed of graphite and the corresponding intercalation compounds. In addition, we hope to observe the new kind of electron states in graphite (4,5) by vacuum tunneling spectroscopy.

Therefore, we have built in our laboratory a STM following very closely the published papers by Binnig and Rohrer. In our first version the STM unit is placed in a vacuum chamber reaching 10^{-8} Torr. Good candidates for testing the operation of our STM at these vacuum conditions are graphite and a Pd single crystal, which both appear to be rather inert. The next step, to operate our STM in ultra high vacuum is in progress.

2. EXPERIMENTAL

We were very much interested in the suppression of vibrations and have therefore optimized the suspension. We achieved this by calculating the different Eigenmodes of a two-stage spring system and also checking the result by measuring the movement of the platform (6). The resulting eigenfrequencies are between 1 and 3 Hz. The well-known tunnel unit consisting of piezodrive with tunnel-tip and sample mounted on the "louse" is placed on this platform. The piezodrive material, commercial piezoceramics (Philips PXE 5), is not yet calibrated. However, the length scale should be accurate within 10% according to the specifications. Moving of the piezodrive and the louse are microcomputer controlled. The tunnel current for distances in the Å range is held constant by an analog PI controller. The exponential dependence of the tunnel resistance from the displacement of the tip has been obtained between a Pt sample and a W-tip. For this experiment we have applied the described ultrasonic self-cleaning procedure (1).

3. RESULTS

The vicinal surface of the used Pd single crystal is tilted by 3.7° towards (100), leading to widespread surface steps. The height of these steps was estimated from the lattice constant of fcc Pd $a_o = 3.88$ Å to lie between 3.6 and 3.7 Å. This Pd single crystal has been characterized by the Xenon adsorption technique (7). From these results it emerges that the mono-atomic steps are not oriented parallel to [100] or [110], as observed earlier on single crystals of Pt (8). In addition, there are also macroscopic surface defects present, originating from previous sputtering procedures.

Figure 1 depicts a section of 1700 x 2000 Å from the macroscopic (100) Pd surface. The superficial roughness can be estimated to be approximately 150 Å. Resolution of details extends in this image over the range of observation of non-periodic structures currently being obtained by SEM.

Besides the rough parts in Fig.1 there are large areas seemingly flat on this scale: on higher resolution they reveal, however, the typical structures presented in Figures 2 and 3.

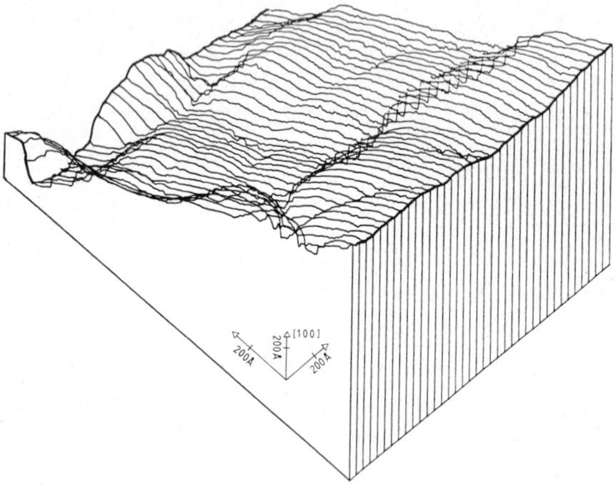

FIGURE 1
Surface topography of a large area of a Pd (100) surface.

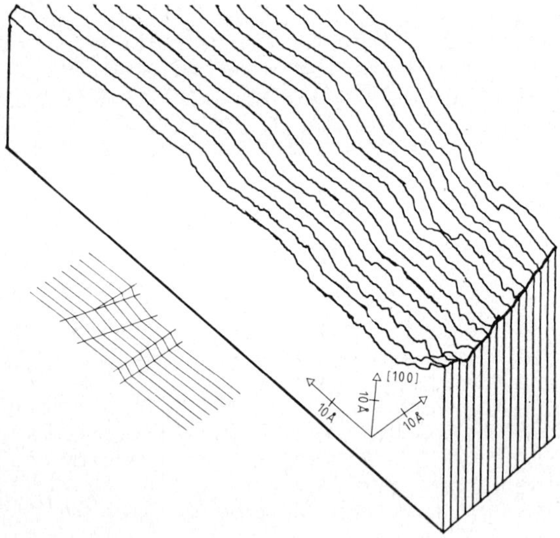

FIGURE 2
Section of a flat part with mono- and double-atomic steps. The inset models the relative orientation of the steps.

On an atomic scale, Fig.2 represents an area of 95 x 36 Å showing the co-existence of non-parallel mono- and double-atomic steps (measured height 3.5 and 7.0 Å respectively). Even non-periodic atomic structures, arising from irregular boundaries of terraces, may be imaged directly as shown in Fig.3 .

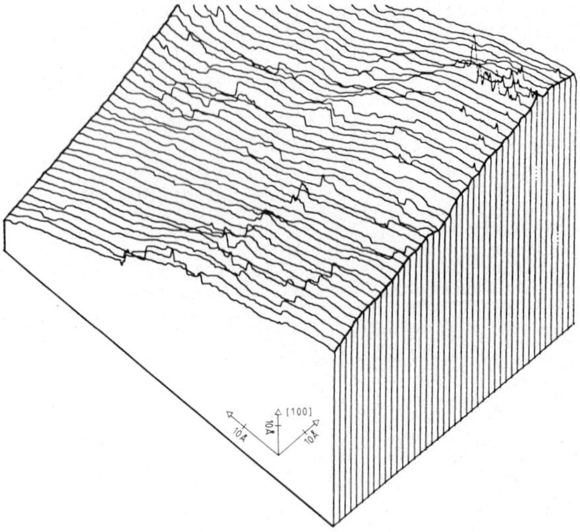

FIGURE 3
Section with non-periodic structures on an atomic scale.

ACKNOWLEDGEMENTS

We would like to thank the skillful technical support of P. Abt, H. Breitenstein, D. Holliger, E. Krattiger and A. Nassenstein. Financial support of the Swiss National Science Foundation is gratefully acknowledged.

REFERENCES

(1) G. Binnig et al., Physica 109+110B (1982) 2075.
(2) G. Binnig et al., Phys. Rev. Lett. 49 (1982) 57.
(3) G. Binnig and H. Rohrer, Helv. Phys. Acta 55 (1982) 55.
(4) M. Posternak et al., Phys. Rev. Lett. 50 (1983) 761.
(5) T. Fauster et al., Phys. Rev. Lett. 51 (1983) 430.
(6) M. Ringger, Diploma thesis, Universität Basel, 1983.
(7) R. Miranda et al., Surf. Sci. 131 (1983) 61.
(8) B. Lang et al., Surf. Sci. 30 (1972) 440.

LT-17 (Contributed Papers)
U. Eckern, A. Schmid, W. Weber, H. Wühl (eds)

DIRECT OBSERVATIONS OF THERMAL COUNTERFLOW IN SUPERFLUID ^{3}He-A; EVIDENCE FOR SUPERFLOW COLLAPSE ?

R. NEWBURY, J. SAUNDERS and D.F. BREWER

Physics Laboratory, University of Sussex, Brighton BN1 9QH, Sussex, U.K.

The thermal counterflow of superfluid ^{3}He-A has been observed directly using a Rayleigh disc, as a function of applied heat flux. At certain heat fluxes discontinuous changes in the torque experienced by the disc are seen indicative of changes in texture. No counterflow is found at temperatures above 0.95 T_c.

1. INTRODUCTION

Studies of thermal counterflow in superfluid ^{3}He-B in which the counterflow is mechanically detected by a Rayleigh disc are reported elsewhere in this volume (1). We refer to that paper for details of our apparatus and experimental technique. The B phase results may be satisfactorily analysed using phenomenological two fluid theory. This had also been found by Johnson et al (2) in their measurements of hydrodynamic heat transport in superfluid ^{3}He-B. In the A phase however their results were far more complex. One of their findings we mention in particular is a hydrodynamic thermal resistance increasing linearly with applied heat flux close to T_{AB}. Those measurements were at 29.6 bar and for applied heat fluxes varying from 0.5 nW cm^{-2} to 13 nW cm^{-2}.

2. RESULTS AND DISCUSSION

In this work the torque on a Rayleigh disc in ^{3}He-A at 29.3 bar has been measured for heat fluxes ranging from 10.9 nW cm^{-2} to 43.5 nW cm^{-2}, corresponding to a variation of applied heat from 2.13 nW to 8.53 nW. In each case the rotation of the disc is measured as the heater is turned on and subsequently turned off after a delay of order five minutes. Analysis of the B phase data up to T_{AB}/T_c = 0.851 indicates that the total torque on the disc τ is dominated by that arising from the flow of normal fluid. The resulting drag and lift forces on the disc result in a torque because of the off-axis mounting of the suspension fibre. This torque is of the form $\tau_n = A\eta_n v_n$ where η_n is the normal component viscosity, v_n the velocity of the normal component and A is a geometrical constant.

Since the boundary condition forces the $\underset{\sim}{\ell}$ vector to lie perpendicular to the surface of the disc one might expect there to be an upward jump in the torque on warming through T_{AB} of order 20% in accordance with the jump in viscosity measured by Parpia (3) for this orientation of $\underset{\sim}{\ell}$ relative to the flow. Such an effect is indeed observed at heater powers of 2.13, 3.20 and 4.27 nW. However at larger heater powers

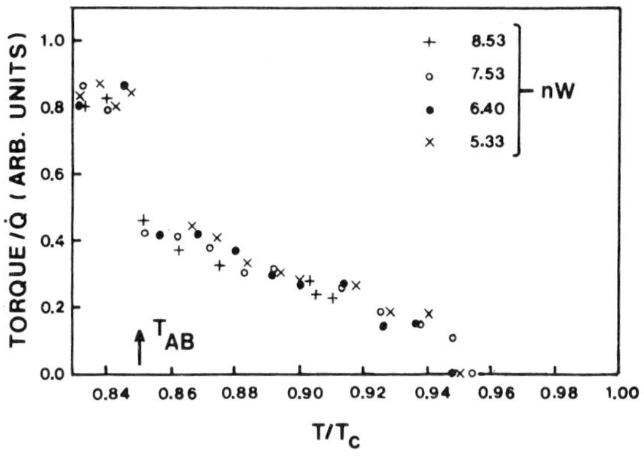

FIGURE 1
$\tau/\dot{Q}$ in the A-phase for applied powers between $\dot{Q}$ = 5.33 nW and 8.53 nW, plotted against T/T_c.

there is a downward jump in the torque at T_{AB}, indicated in figure 1. It may be seen from the figure that the values of $\tau/\dot{Q}$ lie on a single curve, indicative of the dominance of the normal fluid torque. These results are suggestive of the fact that different textures are present in the tube at the high and low heat fluxes. We note that the liquid inside the horizontal flow channel is subject to the earth's magnetic field trapped inside the horizontal niobium shield surrounding the channel However at the lowest flow velocities of v_n = 0.02 mms^{-1} at $\dot{Q}$ = 2.13 nW and T_{AB} we still expect a uniform texture with $\underset{\sim}{\ell}$ predominantly parallel to the axis of the cylinder.

Perhaps our most interesting observation is that at power levels of 3.20 nW and 4.27 nW there is a step in the torque experienced by the disc which presumably corresponds to a downward jump in v_n. An example of this effect is given in figure 2. It is tempting to interpret this in terms of a flow driven textural transition (4). Since an initially uniform texture of

$\ell || v_n || v_s$ is expected on energetic considerations it is possible that the signature may relate to the partial superflow "collapse" associated with the transition to a helical texture. We note that the torque arising from superflow, whose maximum value has been estimated using two fluid theory, is not sufficiently large that variations of v_s consequent on the anisotropy of ρ_s as $\underset{\sim}{\ell}$ is rotated are significant. The consequence of any "textural torque" that might be exerted on the disc as the equilibrium configuration of $\underset{\sim}{\ell}$ is changed is unknown.

In calculations of the stability of textures with respect to counterflow it is commonly assumed that v_n is fixed externally by $\dot{Q}$ (5). Our observations, in conjunction with the nonlinear hydrodynamic conductivity observed by Johnson et al (2), would tend to indicate that this is not the case.

For 3.20 nW, observations of the form illustrated have been made between T_{AB} and 0.93 T_c. At 4.27 nW a "transition" is sometimes observed up to 0.93 T_c otherwise the data follow the pattern of figure 1. At 2.13 nW the values of $\tau/\dot{Q}$ are consistently higher by a factor of order 2.7 than the data of figure 1, up to 0.95 T_c. However, apart from small transitory responses on first turning on a heat pulse, no torque, and hence no counterflow, has been observed (above a temperature of 0.95 T_c) at any heat flux above 10.9 nW cm^{-2}. We note that if heat were only transported diffusively at this temperature, due to a possible complete collapse of counterflow, the lowest heater power would generate a temperature difference of 0.1 mK across the column. However all the data of figure 1 extrapolate to zero torque at T_c, independent of $\dot{Q}$, so that the collapse of counterflow would appear to occur suddenly at approximately 0.95 T_c for all applied heat fluxes.

In the initial state associated with large $\tau/\dot{Q}$ there is a tendency for $\tau/\dot{Q}$ to increase with decreasing $\dot{Q}$. This is most marked at $\dot{Q}$ = 2.31 nW where at 0.941 T_c a value of $\tau/\dot{Q} = 1.05$ was twice observed on first turning on the heater. This torque subsequently decayed to zero over a time scale of a few minutes. Furthermore the value of $\tau/\dot{Q}$ after the "transition" at $\dot{Q} = 3.20$ nW is consistently larger by a factor of order 1.6 than the data of figure 2 up to 0.92 T_c.

An extension of this work would clearly be to study the influence of an axial magnetic field. An added complication here is that the disc, fabricated from niobium, rotates significantly for the fields in question (up to ten gauss).

ACKNOWLEDGEMENTS

This work was supported by SERC Grant no. GR/B/65137 and by a Senior Fellowship for one of us (DFB).

REFERENCES

(1) R. Newbury, J. Saunders and D.F. Brewer, Thermal counterflow of superfluid ^{3}HeB observed with a Rayleigh disc, this volume.
(2) R.T. Johnson, R.L. Kleinberg, R.A. Webb and J.C. Wheatley, J. Low Temp. Phys. 18 (1975) 501.
(3) J. Parpia. thesis, Cornell University (unpublished).
(4) P. Bhattacharya, T-L Ho and N.D. Mermin, Phys.Rev.Lett. 39 (1977) 1290.
(5) A.L. Fetter and M.R. Williams, Phys.Rev. B23 (1981) 2186.

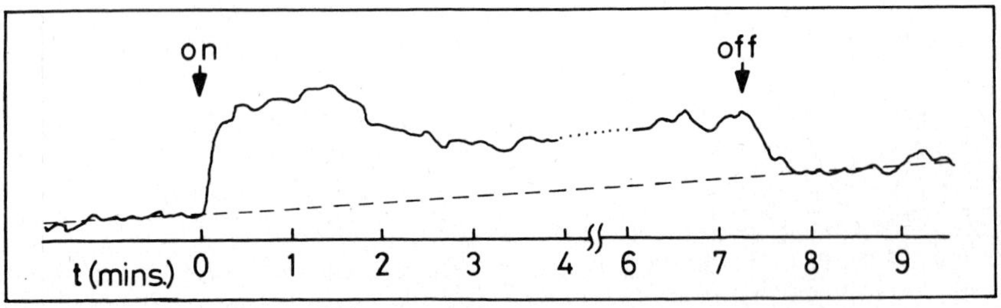

FIGURE 2
Disc response to applied heater power $\dot{Q}$ = 3.2 nW as a function of time at T/T_c = 0.92. Note that the drift in base-line (indicated by the dotted line) tends to obscure the spontaneous change in deflection.

LT-17 (Contributed Papers)
U. Eckern, A. Schmid, W. Weber, H. Wühl (eds)

SUPERFLUID DISSIPATION IN ROTATING ^{3}HE†

E. P. WHITEHURST*, H. E. HALL*, P. L. GAMMEL, and J. D. REPPY

Laboratory of Atomic and Solid State Physics, Clark Hall, Cornell University, Ithaca, New York 14853, U.S.A.

Additional dissipation of oscillatory superflow has been observed in uniformly rotating ^{3}He. The damping increases linearly with angular velocity and is interpreted in terms of vortex induced mutual friction similar to that observed in superfluid ^{4}He. In the non-rotating state the damping is amplitude dependent, and may contain a contribution from remanent vorticity.

We have observed a gross effect of superposed rotation on oscillatory flow in superfluid ^{3}He. A large extra dissipation proportional to the angular velocity Ω has been observed in both the B-phase at 20 bar and the A-phase at 29 bar. In the non-rotating state the damping of the superflow oscillations is non-linear and qualitatively similar to observations from previous experiments. (1) We concentrate this paper on the effect of rotation, which we interpret in terms of a mutual friction force proportional to vortex line density, as used to interpret related experiments in superfluid ^{4}He. (2)

The geometry of our flow cell is shown schematically in Fig. 1.

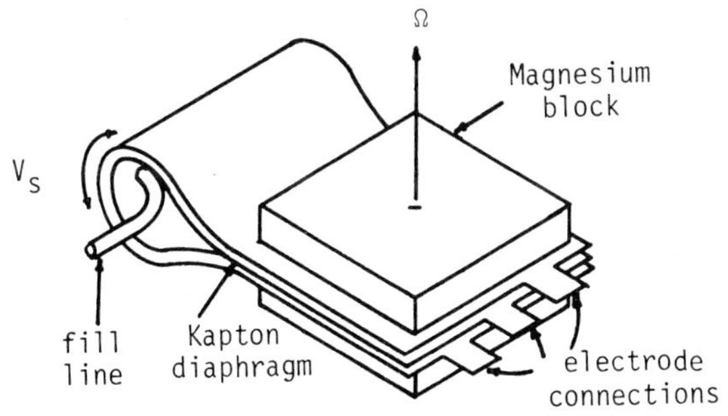

FIGURE 1
Schematic view of the flow cell, without encapsulating epoxy.

The ^{3}He sample is in the form of a thin slab approximately 9 cm long, 2.5 cm wide and 0.1 mm thick, folded over on itself so that the end regions are separated only by a thin, pressure sensitive diaphragm. The diaphragm is 12 μm Kapton, aluminised on one side, and the outer wall of the cell is formed from 75 μm Kapton sheet. The mean separation between the diaphragm and the outer wall is 100 μm. Magnesium blocks are used to hold flat the diaphragm and outer wall of the cell, and to produce tension in the diaphragm by differential thermal contraction. The entire cell as drawn is encapsulated in Stycast 2850 GT, into which is embedded a metal cage for mechanical strength and thermal contact.

Displacement of the diaphragm is used to both drive and detect flow through the central region of the slab, with a simple ratio transformer capacitance bridge and D.C. bias. Contact to bulk ^{3}He and the $PrNi_5$ nuclear refridgerant is via a 10 cm long, 1 mm I.D. copper fill line, which enters at the centre of the folded slab so that it is approximately at a pressure node of oscillatory flow.

The diaphragm has a vacuum resonance at 1.8 KHz with a Q of 25,000. When the cell is filled with ^{4}He at 20 mK the oscillatory fluid flow has a resonant frequency of $\omega_0/2\pi=6.61$ Hz with a Q of 5000; it is therefore clear that the diaphragm dissipation is very low. Since the fluid oscillation frequency is proportional to $\rho_s{}^{1/2}/\rho$ and to geometrical factors, the ^{4}He calibration enables us to deduce ρ_s/ρ in ^{3}He and hence the temperature of the experimental helium.

In ^{3}He the Q values are very much lower, in the range 200 to 40 depending on amplitude. The drastic further reduction of Q by a superposed rotation is shown in Figs. 2 and 3; even rotation at 0.5 rad/sec is sufficient to reduce the Q to about 4.

---

* Permanent address: The Schuster Laboratory, Manchester University, M13 9PL, U.K.
† Supported by the NSF through the Cornell Materials Science Center under Contract No. 76-81083 AD2, Technical Report #5221. E. P. Whitehurst is supported by an S.E.R.C. studentship.

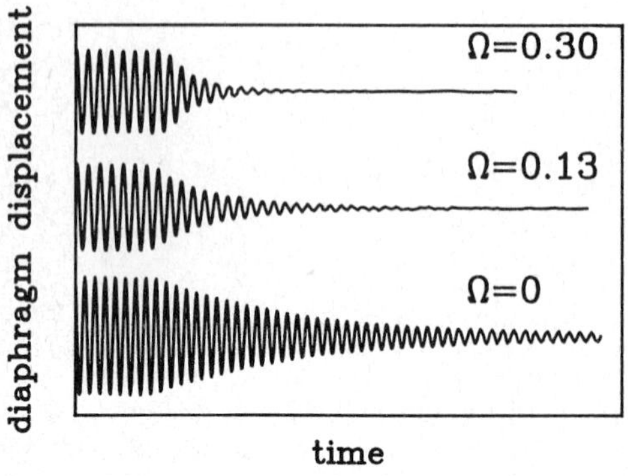

FIGURE 2
The decay of superfluid oscillations at $\omega_0/2\pi=2.2$ Hz when the drive is removed (after several cycles) for various values of angular velocity Ω. This is for the B-phase at 20 bar.

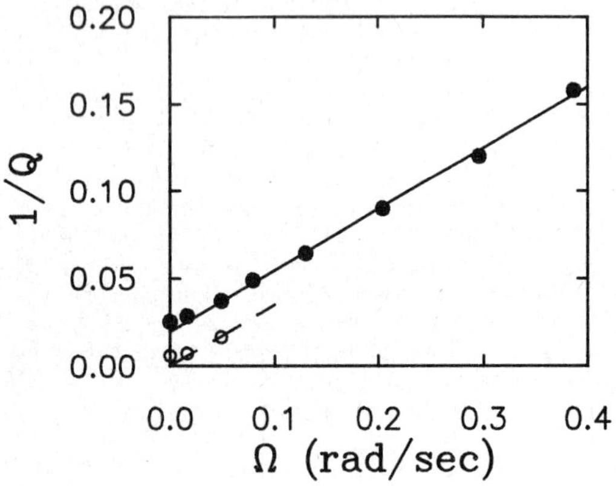

FIGURE 3
Dependence of damping on angular velocity Ω in the B-phase at 20 bar, for $\omega_0/2\pi=2.2$ Hz. Solid circles are v_s~0.1 mm/sec, open circles are v_s~10 μm/sec. The solid line is a fit to all but the lowest two high amplitude points; the broken line is a parallel line through the origin.

Since the effects of finite amplitude and of rotation on the dissipation (proportional to 1/Q) appear to be approximately additive in Fig. 3, it is simple to deduce fairly accurate values of a rotational mutual friction constant B (2), from graphs such as Fig. 3 and the formula

$$\frac{B\rho_n}{\rho} = \omega_0 \frac{d(1/Q)}{d\Omega} ; \qquad (1)$$

Equation (1) assumes all the flow is perpendicular to the rotation axis; allowance for the actual geometry of our cell would increase B values deduced from equation (1) by about 20%. Our B values are shown as a function of ρ_s/ρ (approximately proportional to Tc-T) in Fig. 4.

	$(w_0/2\pi)$Hz	ρ_s/ρ	$B(\rho_n/\rho)$	$B(\rho_s/\rho)^{1/2}$
B-phase	1.1	0.0208	8.65	1.274
20 bar	1.5	0.0386	5.19	1.061
	1.9	0.0619	5.29	1.403
	2.2	0.0830	4.85	1.524
A-phase	0.8	0.0116	5.07	0.552
29 bar	2.2	0.0875	2.74	0.888

FIGURE 4
Table of rotational mutual friction parameters for the B-phase at 20 bar and A-phase at 29 bar.

Note that the values are somewhat larger than in ^{4}He, and there is a suggestion of a similar tendency to diverge near Tc (3).

Finally we note that dissipation in the non-rotating state, which is both amplitude dependent and larger than expected theoretically in B-phase flow experiments to date, is equivalent to that produced by a rotation at a few hundredths of a rad/sec. This corresponds to a vortex density in the range 10-100 lines/cm^2, of the same order of magnitude as that found by Awschalom and Schwartz (4) in open volumes of ^{4}He. It is therefore clear that the possible presence of unintended vorticity poses problems for the interpretation of flow experiments in superfluid ^{3}He.

REFERENCES

(1) A. J. Dahm, D. S. Betts, D. F. Brewer, J. Hutchins, J. Saunders and W. S. Truscott, Phys. Rev. Lett. 45, 1411 (1980).

(2) H. E. Hall and W. F. Vinen, Proc. Roy. Soc. A 238, 215 (1956).

(3) P. Mathieu, A. Serra and Y. Simon, Phys. Rev. B 14, 3753 (1976).

(4) D. D. Awschalom and K. W. Schwartz, Phys. Rev. Lett. 52, 49 (1984).

LT-17 (Contributed Papers)
U. Eckern, A. Schmid, W. Weber, H. Wühl (eds)

PRESSURE AND TEMPERATURE DEPENDENCE OF D.C. CRITICAL CURRENTS IN ^{3}He B

Ren-zhi LING, D.S. BETTS and D.F. BREWER

Physics Laboratory, University of Sussex, Brighton BN1 9QH, Sussex, U.K.

D.C. flow experiments have been carried out in superfluid ^{3}HeB using a superleak of rectangular cross-section. In agreement with previous oscillation experiments with the same superleak, three distinct flow regimes have been identified separated by sharply defined critical currents whose temperature dependence is given by $J_c = a(1-T/T_c)^b$. As in the previous experiments we find that at pressures below the tricritical point $b \simeq 3/2$, but over a small pressure range at the tricritical pressure there is a sharp change to $b \simeq 2$, suggesting a structural change in flowing ^{3}HeB at this pressure.

1. INTRODUCTION

Several experiments have now shown that a number of different regimes exist in the D.C. and oscillatory flow of superfluid ^{3}He, over a wide temperature and pressure range and with superleaks with geometry differing both in size and shape (1). The systematics of the experiments are not well established, in the sense that there is no coherent series of measurements that determine how the critical currents and the dissipation depend on various experimental parameters such as superleak size and shape, temperature and pressure. We have concentrated our experiments in Sussex on a superleak of rectangular cross-section 2.86 mm x 48μm, and of length 9.16 mm, and first carried out oscillation experiments (2) which established the existence of three hydrodynamic flow regimes with high, medium and low dissipation, separated by sharp values of critical currents. The two sets of critical currents (between low/medium and medium/high dissipation) have both been shown to vary (at low pressure) as $(1-T/T_c)^{3/2}$, which is that predicted for the so-called "pair-breaking" critical current. The prefactors in these temperature dependences are reasonably in agreement with theory at low pressures but not at higher pressures in the weak-coupling limit. (However, the similar temperature-dependence of the lower and upper critical currents seems to be fortuitous since the lower critical current cannot be a "pair-breaking" current). Understanding of the dissipation-onset mechanisms and their critical currents, and also of the dependence of the dissipation on velocity, are still lacking in superfluid ^{3}He.

2. PRESENT EXPERIMENTS

We have now carried out additional D.C. flow experiments with the same superleak, some preliminary results of which have been described (3). Flow is initiated between two cells which are connected by the superleak and separated by a flexible gold-plated diagragm M which forms the central plate of two parallel plate capacitors; one of these has applied to it a voltage V which impresses a force proportional to V^2, the other has its capacitance to M (which measures the diaphragm displacement x) measured by a sensitive capacitance bridge. The pressure drop across the superleak is given in general, neglecting inertia effects, by

$$\Delta p = \alpha V^2 - \beta x \qquad \text{(i)}$$

where α and β are apparatus constants which must be evaluated by calibration. In static equilibrium, or with steady superflow, $\Delta p = 0$. Appli-

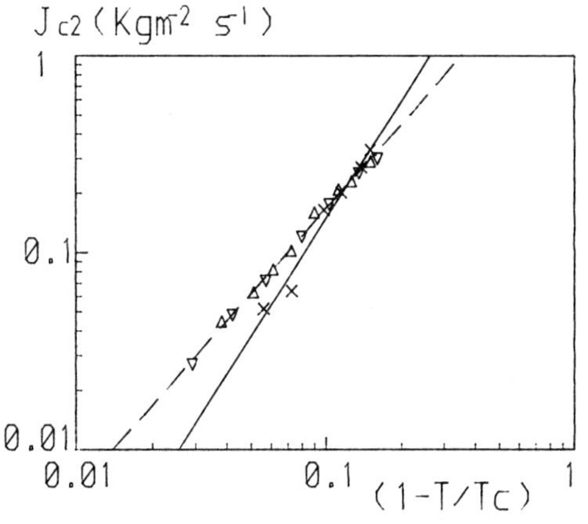

FIGURE 1
Examples of the variation of critical current J_{c2} with $(1-T/T_c)$, plotted on a log-log scale. The full line (symbols x, pressure 22.1 bar) has slope $b \simeq 1.98$; the dashed line (symbols ∇, Δ for different runs pressure 20.0 bar) has slope $b \simeq 1.42$.

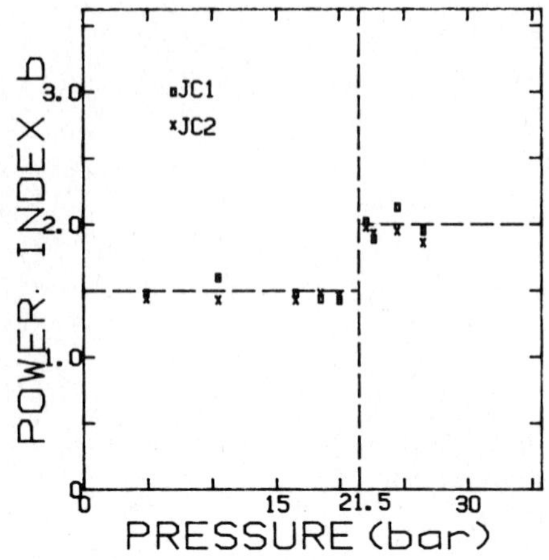

FIGURE 2
The variation of power index b (equation (ii) with pressure for both lower (□) and upper (x) critical currents in ^{3}HeB.

cation of ramp rates $V^2 = \gamma t$ where γ is constant and t is time gives, according to eqn.(i), values of subcritical velocity proportional to $\dot{x} = \alpha/\gamma$ and of Δp which is a measure of the dissipation in the superleak. These D.C. measurements confirm the main results of the oscillatory measurements: at low currents (calculated from $\dot{x}$) small or zero dissipation is observed until at a lower critical current J_{c1} a region of slowly increasing dissipation occurs up to an upper critical current J_{c2} where the dissipation increases rapidly with the current (3).

The important point which we wish to report here is a quite unexpected and abrupt change in the temperature dependence of the critical currents in ^{3}HeB *at the tricritical pressure*. At all pressures the critical currents we have observed follow with good precision the relationship

$$J_c = a(1-T/T_c)^b \qquad \text{(ii)}$$

At pressures below the pressure of the tricritical point p_{tcp}, we find as before that $b \simeq 3/2$: an example is given in figure 1. We find the striking result that at $p > p_{tcp}$ the power index changes, in an apparently discontinuous way, from $b \simeq 3/2$ to $b \simeq 2$, as again shown in figure 1. The sharpness of this b-index transition is shown in figure 2, which plots b as a function of pressure. A temperature dependence of J_c given by equation (ii) with $b \simeq 2$ had in fact been observed at $p \sim 26$ bar in our oscillation experiments (2), (4), adding support to the present more systematic investigation of the pressure dependence of the critical currents.

Such a sharp change in the temperature dependence of the critical current at the tricritical pressure, even at temperatures well below the tricritical point, must have some significance for the equilibrium conditions for the various phases of superfluid ^{3}He in dynamic states. Fetter (5) has shown that in flowing ^{3}HeB a transition to ^{3}HeA can take place at a lower temperature than in static equilibrium conditions. Our observations do not conform with this theory since our critical current transition shows a critical power exponent change from $b \simeq 3/2$ to $b \simeq 2$, whereas other experiments which we describe elsewhere (7) show that the $B \rightarrow A$ transition is accompanied by a power exponent transition in the critical current of $2 \rightarrow 1$.

This power index transition in the temperature dependence of the critical current at p_{tcp} is so clearly defined experimentally that it suggests some structural change in flowing ^{3}HeB at this pressure which needs theoretical investigation. An additional observation that needs explanation is that whereas the prefactor a at low pressures increases weakly with pressure, in a very similar way to the Helsinki results with much smaller superleaks (6), it decreases rapidly with pressure when $p > p_{tcp}$, which is in qualitative agreement with Fetter's theory (5).

ACKNOWLEDGEMENTS
We are grateful to many colleagues, especially Dr. J. Hutchins and Dr. J. Saunders for assistance and discussions of the experiments, and to Mr. C. Mills and Mr. A.J. Young for technical help. The work was supported by SERC under Grant no. GR/B/65137 and by a Senior Fellowship (DFB); and by the Education Department of the Peoples' Republic of China (Ling Ren-zhi).

REFERENCES
(1) See, for example, papers and references in: Quantum Fluids and Solids - 1983, Eds. E.D. Adams and G.G. Ihas (AIP, New York, 1983).
(2) A.J. Dahm, D.S. Betts, D.F. Brewer, J. Hutchins, J. Saunders and W.S. Truscott, Phys.Rev.Letters 45 (1980) 1411.
(3) D.F. Brewer, ref.(1) p.336
(4) J. Hutchins, D.Phil Thesis, University of Sussex (unpublished).
(5) A.L. Fetter in: Quantum Statistics and the Many Body Problem, Eds. S.B. Trickey, W. Kirk and J. Dufty (Plenum, New York, 1975) p.127.
(6) M.T. Manninen and J.P. Pekola, J.Low Temp. Phys. 52 (1983) 497.
(7) Ren-zhi Ling, D.S. Betts and D.F. Brewer this volume.

LT-17 (Contributed Papers)
U. Eckern, A. Schmid, W. Weber, H. Wühl (eds)

D.C. CRITICAL CURRENTS IN ^{3}He-A IN A RECTANGULAR CHANNEL

Ren-zhi LING, D.S. BETTS and D.F. BREWER

Physics Laboratory, University of Sussex, Brighton, Sussex BN1 9QH, U.K.

We report measurements in the A phase of the pressure and temperature dependence of D.C. critical currents through a rectangular cross-section superleak, supplementing those reported in this volume of B-phase observations (7). In contrast to the B phase results we observe lower critical currents with a different temperature dependence, and no obvious transition between different flow regimes. We discuss an analysis of the supercritical dissipation mechanisms.

1. INTRODUCTION

The flow of superfluid ^{3}He-A is a much more complex and subtle matter than in ^{3}He-B due to its anisotropic and non-irrotational nature. Several flow experiments in superfluid ^{3}He (1-5) with superleaks of different sizes and shapes show that the values of critical currents of ^{3}He-A are always lower than those of ^{3}He-B. The ^{3}He-A values vary from a few tenths to a few mm/sec.

The results we present here were obtained with a superleak of rectangular cross-section 2.98 mm x 48μm, which has been described in detail in an early paper (4). The D.C. technique we used in these experiments we have also described elsewhere (6,7). During these measurements a 98G magnetic field perpendicular to the superleak was always present. We observe a strikingly different flow behaviour in the A phase compared with the B-phase; we outline the main features below.

As explained in (6) and (7) we measure capacitively the pressure difference Δp generated by a D.C. current J_s. The latter is also generated electrostatically by means of a V^2 ramp, $V^2 = \gamma t$, where γ is constant and V is the voltage across a parallel plate capacitor formed by a flexible gold-plated membrane and a neighbouring metal plate. In general, the pressure difference is given by

$$\Delta p = \alpha V^2 - \beta x \qquad \text{(i)}$$

where α and β are apparatus constants and x is also measured capacitively.

Application of the $V^2 - \gamma t$ ramp produces, in almost all cases, a constant current which can be calculated from $\dot{x}$, and a rate of growth of dissipation $\propto d(\Delta p)/dt$ which is constant: within the possible range of our apparatus, we have never observed a steady-state condition with Δp = constant for a given flow rate.

2. RESULTS

As in the case of ^{3}HeB we observe, for low currents ($\propto \dot{x}$), that $\Delta p = 0$ within experimental precision ($\lesssim$ 1μbar). Similarly we also observe a sharp transition to supercritical flow where Δp is finite, and again as in ^{3}HeB, Δp does not reach a constant value for a given current, but increases linearly over the available ramping time, as stated above. We therefore again express our results as $d(\Delta p)/dt$, the rate of growth of dissipation, as a function of J_s, as in Figure 1.

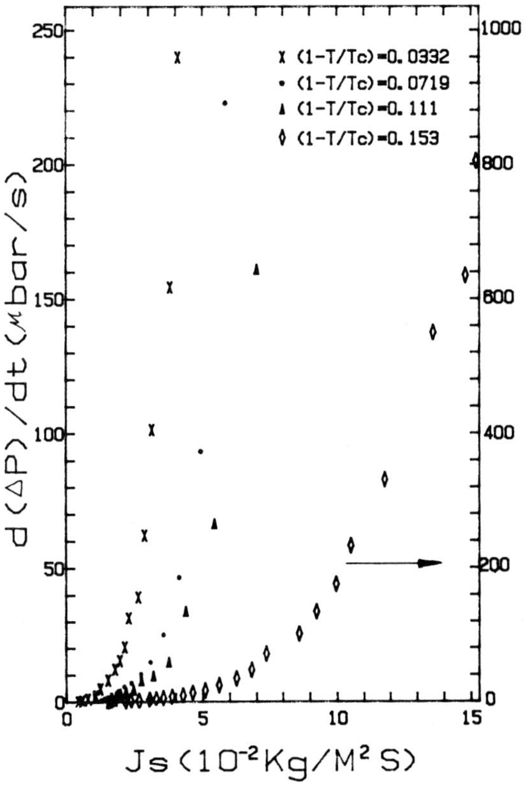

FIGURE 1
Rate of growth of dissipation vs. current p = 33 bar.

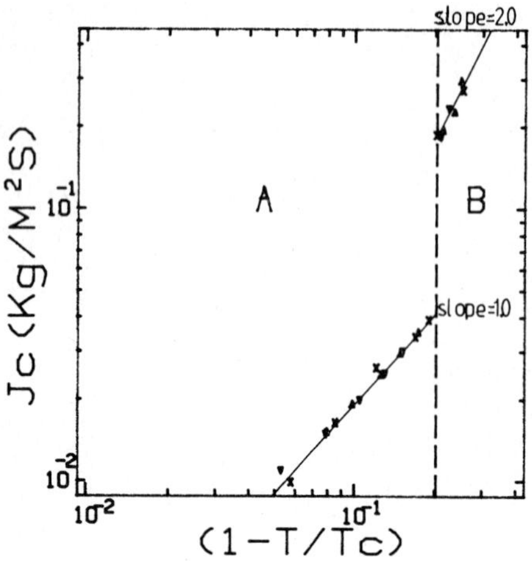

FIGURE 2
Change in critical current at the B→A transition at pressure 33 bar.

In contrast to ^{3}HeB, we now find only one critical current J_{c1}, not two, and $d(\Delta p)/dt$ does not increase linearly with J_s as it does in ^{3}HeB. It is interesting that the different types of flow regime in A and B are the exact opposite to those reported in Helsinki experiments (5) with a "Nuclepore" filter superleak consisting of many much narrower channels in a membrane: in that case, the B phase showed only one critical current, while the A phase showed two critical currents.

The values of J_c for ^{3}He are much smaller than in ^{3}HeB, and there is a sharp drop on passing from the B to the A phase at the same pressure. An example is given in Figure 2 for a pressure of 33 bar. The drop becomes smaller at lower pressures. Figure 2 also shows a sharp change in the temperature dependence at the B→A transition. Again as in ^{3}HeB, this is given by $J_c = a(1-T/T_c)^b$ where at the B → A transition the power index b changes abruptly from b ≃ 2 (7) to b ≃ 1. The prefactor a is a decreasing function of pressure. Some of the results are summarised in Table 1.

Table 1. Experimental values of J_c and v_c at $T/T_c = 0.91$, and values of power index b and prefactor a in the A phase.

p(bar)	$J_c(10^{-2}Kgm^{-2}s^{-1})$	$v_c(mms^{-1})$	b	a
26.5	2.5	4.1	0.94	0.27
33.0	1.7	2.8	1.02	0.20

3. DISCUSSION

The physical origin of these critical currents in ^{3}He has not yet been unambiguously identified. It seems clearly not to be a depairing current, for a number of reasons: (i) its values are too small, (ii) the dissipation continues to grow at higher currents, (iii) the temperature dependence is not as predicted - the power index b is ≃ 1 rather than 3/2, and (iv) its decrease with pressure disagrees with theory. Hall and Hook (8) have explained the A phase dissipation observed in our oscillation experiment (4) in terms of orbital dissipation (9), according to which the pressure difference Δp due to a constant rate of dissipation in the superleak at constant superfluid velocity v_s is given by $\Delta p = f(T/T_c)v_s^3$, where $f(T/T_c)$ is a temperature dependent prefactor. In our observations (as in the case of the Helsinki work (5)) we never reach a steady state where the dissipation rate is constant at a given velocity. If we plot the *rate of growth* of dissipation, $d(\Delta p)/dt$ (which is generally constant at constant current) as a function of J_s, we find a similar relation to the steady-state case, but with $d(\Delta p)/dt = g(T/T_c)J_s^{3.75}$ where $g(T/T_c)$ is a prefactor which increases with (T/T_c).

ACKNOWLEDGEMENTS

This work was supported by the SERC under Grant No. GR/B/65137 and by a Senior Fellowship (DFB); and by the Department of Education of the Peoples' Republic of China (Ren-zhi Ling).

REFERENCES

(1) E.B. Flint, R.M.Mueller and E.D.Adams, J. Low Temp.Phys. 33 (1978) 43.
(2) J.M.Parpia and J.D.Reppy, Phys.Rev.Lett. 43 (1979) 1332.
(3) M.A.Paalanen and D.D.Osheroff, Phys.Rev. Lett. 45 (1980) 362.
(4) A.J.Dahm, D.S.Betts, D.F.Brewer, J.Hutchins, J.Saunders and W.S.Truscott, Phys.Rev.Lett. 45 (1980) 1411.
(5) M.T.Manninen, J.P.Pekola, R.G.Sharma and M.S.Tagirov, Phys.Rev. B26 (1982) 5233.
(6) D.F.Brewer in "Quantum Fluids and Solids - 1983". Eds. E.D.Adams and G.G.Ihas (AIP, New York, 1983).
(7) Ren-zhi Ling, D.S.Betts and D.F.Brewer, Temperature and pressure dependence of D.C. critical currents in ^{3}HeB, this volume.
(8) H.E.Hall and J.R.Hook, "Hydrodynamics of Superfluid ^{3}He" (preprint).
(9) M.C.Cross in "Quantum Fluids and Solids - 1983", Eds. E.D.Adams and G.G.Ihas (AIP, New York, 1983).

LT-17 (Contributed Papers)
U. Eckern, A. Schmid, W. Weber, H. Wühl (eds)

PERSISTENT CURRENT EXPERIMENTS ON SUPERFLUID ^{3}He

J.P. Pekola, J.T. Simola, K.K. Nummila, and O.V. Lounasmaa
Low Temperature Laboratory, Helsinki University of Technology,
SF-02150 Espoo 15, Finland, and

R.E. Packard
Physics Department, University of California, Berkeley,
California 94720, USA

Persistent currents in superfluid ^{3}He have been investigated by means of ac-gyroscopic technique in a torus packed with 20 μm powder. In the B-phase there is no observable decay in the circulating current over a period of 48 h, whereas in the A-phase the current does not seem to persist at all. The dissipation-free critical velocity in the B-phase at 8 bar is 5.3 mm/s; at 29.3 bar there are two regimes with ultimate critical velocities of 5.4 and 7.8 mm/s, respectively.

1. INTRODUCTION

Dissipation in flowing superfluid ^{3}He, especially in the B-phase, has been of considerable recent interest. The first experiments on superflow utilized U-tube, torsional oscillator, and driven flow techniques.[1,2,3,4] All these tests are poor in sensitivity when they are compared with the recent ac-gyroscopic measurements by Gammel et al. [5]and, especially, with the present work.

Our experiments were carried out in a rotating nuclear refrigerator which has been described elsewhere.[6]

2. THE AC-GYROSCOPE

Our basic measuring device, is an ac-gyroscope. The torus, packed with 20 μm plastic powder, forms a path for the circulating superfluid. The ring is supported by two sets of mutually perpendicular hollow copper tubes. The ^{3}He-sample is admitted into the torus through these torsion tubes. The device is ac-driven by a small superconducting solenoid inside a Nb shield; capacitive detectors are used for monitoring the motion of the torus. Torsional modes about the two sets of tubes are determined by angles θ and ϕ.

In our experiments we observe the angular momentum L persisted in the torus. With ρ_s/ρ the relative superfluid density and v_s the superfluid velocity, $L \propto (\rho_s/\rho)v_s$ and thus gives the circulation in the ring. To detect L we use an ac-gyroscopic technique: The magnetic drive tilts the torus by θ. This vibrational motion, in turn, produces a torque $\tau = dL/dt$ in the perpendicular direction around ϕ. When the drive frequency f_o is at the resonance of the ϕ-mode we detect a signal $\phi = Q\theta L/2\pi f_o I$, where I is the moment of inertia of the torus. For our device, f_o = 69.8 Hz and the quality factor $Q = 2 \cdot 10^4$. With such a high Q, we must employ feedback electronics or a high stability frequency synthesizer to maintain the resonance.

3. RESULTS

Fig. 1 shows the increase of the persisted angular momentum in the torus as a function of Ω, the maximum angular velocity at which the cryostat was rotated. The data were obtained by speeding up the cryostat below T_c to Ω, and then by slowing it down to rest. The measurements were made with $\Omega = 0$. The entire sequence was then repeated by rotating the cryostat in the opposite direction. The inset of Fig. 1 shows a simple hysteresis loop based on the assumption that the relative velocity between superfluid and normal liquid, $(v_s - v_n)$, never exceeds the critical velocity v_c, i.e. $|\omega_s - \omega_n| < \omega_c$. This assumption separates the L vs. Ω curve into three regimes: 1) For $\Omega = \omega_n < \omega_c$ we have $L = 0$. 2) For $\omega_c < \Omega < 2\omega_c$, $L = (L_c/\omega_c)(\Omega - \omega_c)$, and 3) for $\Omega > 2\omega_c$, $L = L_c$. Our data in Fig. 1 seem to verify this behavior with v_c = 5.3 mm/s at P = 8 bar.

The resonant oscillation frequency of the gyroscope serves as a secondary thermometer in the superfluid phases. The reduction of the effective moment of inertia, and thus the increase Δf of the resonant frequency is proportional to ρ_s/ρ inside the torus. T_c and the B → A transition temperatures serve as calibration points for our ρ_s/ρ scale.

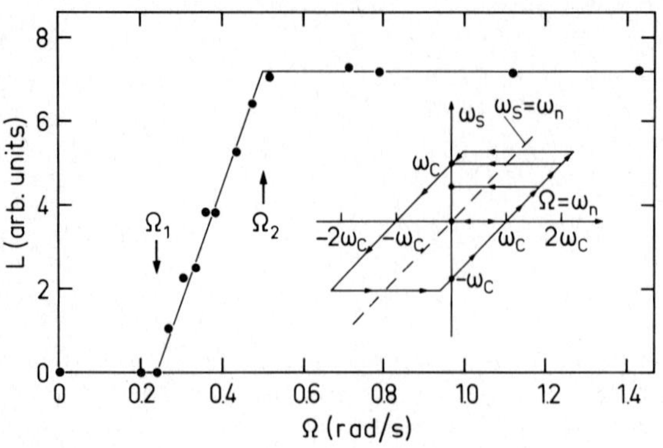

Fig. 1. Persisted angular momentum L vs. Ω at P = 8 bar; Ω is the angular velocity of the cryostat. The inset shows the idealized hysteresis loop; see text.

The persistent angular momentum L vs. ρ_s/ρ is shown in Fig. 2 at 8 bar. The data were obtained with a similar technique as for those in Fig. 1, but at a fixed angular velocity $\Omega > 2\omega_c$. The results show that, within the experimental scatter, the critical velocity v_c [$\propto L/(\rho_s/\rho)$] is independent of temperature.

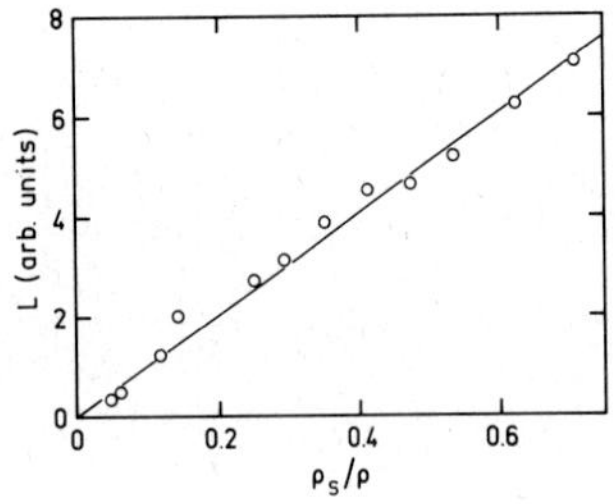

Fig. 2. L vs. ρ_s/ρ, at 8 bar and at Ω = 0.714 rad/s. The slope of the straight line corresponds to v_c = 5.3 mm/s.

At ~8 bar a check of dissipation was made by persisting the superflow for 48 h; within the experimental accuracy of 10% there was no decay of the signal. This shows that the relaxation time τ > 450 h in the B-phase.

A similar series of experiments was performed at 29.3 bar. Fig. 3 shows again an L vs. ρ_s/ρ plot, with a close-up in the A-phase. The linear regime between T_v and T_{BA} corresponds to v_c = 7.8 mm/s. Crossing the B → A phase boundary destroys the signal completely within our resolution. An experiment in which the maximum L was first created and measured in the B-phase, whereafter the torus was thermally cycled to the A-phase and back to the B-phase, agreed with this conclusion with high accuracy. The upper limit of the superflow decay time in the A-phase is approximately 1 min, which is the transient time of the signal to recover from the Coriolis shift after stopping.

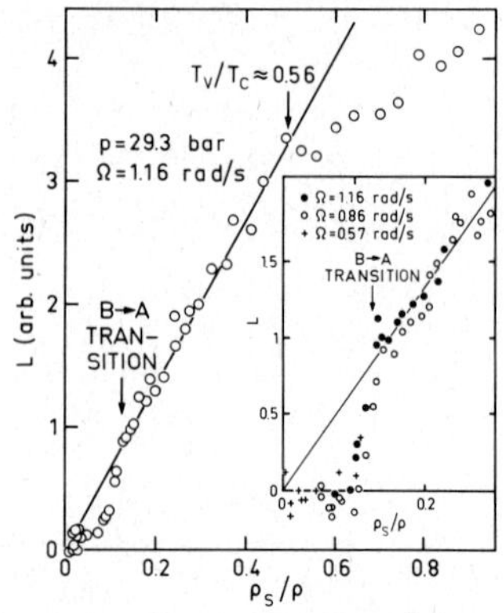

Fig. 3. L vs. ρ_s/ρ at 29.3 bar and at Ω = 1.16 rad/s. The inset shows data from additional runs around the B → A transition. The two slopes, below and above T_v, correspond the v_c = 5.4 and v_c = 7.8 mm/s, respectively.

The transition (see Fig. 3) at T/T_c = 0.56 leads to a decrease in v_c from 7.8 to 5.4 mm/s; the latter value is close to that measured for v_c at 8 bar. Experiments are in progress to test for the possible relation of the transition at T_v/T_c = 0.56 with the vortex core transition observed in the NMR-experiments on rotating ^{3}He-B at T/T_c = 0.6.[7]

REFERENCES

1) J.P. Eisenstein and R.E. Packard, Phys. Rev. Lett. 49, 564 (1982).
2) J.M. Parpia and J.D. Reppy, Phys. Rev. Lett. 43, 1332 (1979).
3) A.J. Dahm, D.S. Betts, D.F. Brewer, J. Hutchins, J. Saunders, and W.S. Truscott, Phys. Rev. Lett. 45, 1411 (1980).
4) M.T. Manninen and J.P. Pekola, Phys. Rev. Lett. 48, 812 (1983); ibid. 1969(E); J. Low Temp. Phys. 52, 497 (1983).
5) P.L. Gammel, H.E. Hall, and J.D. Reppy, Phys. Rev. Lett. 52, 121 (1984).
6) P.J. Hakonen, O.T. Ikkala, S.T. Islander, T.K. Markkula, P.M. Roubeau, K.M. Saloheimo, D.I. Garibashvili, and J.S. Tsakadze, Cryogenics 23, 243 (1983).
7) P.J. Hakonen, M. Krusius, M.M. Salomaa, J.T. Simola, Yu.M. Bunkov, V.P. Mineev, and G.E. Volovik, Phys. Rev. Lett. 51, 1362 (1983).

LT-17 (Contributed Papers)
U. Eckern, A. Schmid, W. Weber, H. Wühl (eds)

MEASUREMENT OF THE INERTIA ASSOCIATED WITH ROTATION OF $\underline{\ell}$ IN ^{3}He-A

A.D. EASTOP, H.E. HALL AND J.R. HOOK

Physics Department, Manchester University, Manchester. M13 9PL, U.K.

It is shown that the inertia associated with rotation of the orbital vector $\underline{\ell}$ in a slab of ^{3}He-A can be measured by studying the response of $\underline{\ell}$ to a precessing magnetic field.

1. INTRODUCTION

The magnitude of the intrinsic angular momentum associated with the $\underline{\ell}$ vector in superfluid ^{3}He-A has been a subject of considerable controversy(1). Recently two of us have suggested(2) that three different types of experiment which might be used to measure the intrinsic angular momentum would lead to different answers. One of the suggested categories was to measure the inertia associated with rotation of the $\underline{\ell}$ vector and in this paper we propose a specific experiment that would enable this to be done.

2. EQUATION OF MOTION OF $\underline{\ell}$

According to (2) the inertia should be characteristic of an intrinsic angular momentum density $(2C-C_o)\underline{\ell}$ and under these circumstances the equation of motion of $\underline{\ell}$ becomes

$$\mu \frac{\partial \underline{\ell}}{\partial t} - (2C-C_o)\, \underline{\ell} \times \frac{\partial \underline{\ell}}{\partial t} = \underline{\ell} \times \left[\underline{\ell} \times \left[\frac{\delta f}{\delta \underline{\ell}}\right]_\phi \right] \qquad (1)$$

where the variational derivative of the free energy density f is at constant phase ϕ. In the absence of flow but in the presence of a magnetic field $\underline{H}$ and in the dipole-locked limit

$$\left[\frac{\delta f}{\delta \underline{\ell}}\right]_\phi = -K_s \text{ grad div } \underline{\ell} + K_t \text{ curl}(\underline{\ell}.\text{curl}\underline{\ell}\ \underline{\ell})$$

$$+ K_b \left\{ \underline{\nabla}\, \ell_\alpha (\underline{\ell}.\underline{\nabla})\ell_\alpha - \text{div } \underline{\ell}\, (\underline{\ell}.\underline{\nabla})\underline{\ell} - \underline{\ell}.\underline{\nabla}[(\underline{\ell}.\underline{\nabla})\underline{\ell}] \right\}$$

$$+ C_o(\hbar/2m)\, \underline{\ell}.\text{curl}\underline{\ell}\ \underline{\ell} \times (\underline{\ell}.\underline{\nabla})\underline{\ell} + 2\epsilon\, \underline{H}(\underline{\ell}.\underline{H})$$

The coefficients μ, K_s, K_t, K_b, C and C_o have their usual meanings and the susceptibility anisotropy energy is $\epsilon(\underline{\ell}.\underline{H})^2$. Although $(2C-C_o)/\mu$ is predicted to be very small in the absence of Fermi liquid corrections(3), the existence of these corrections should lead to a value for this quantity of a few percent at the A-B transition.

3. SUGGESTED EXPERIMENT

Our initial attempts to find a suitable experiment to measure the inertial term were prompted by the assumption that if $\underline{\ell}$ were caused to lie at an angle to the z axis by applying a suitably directed magnetic field, then rotation of the field about the z axis at frequency ω_o would cause $\underline{\ell}$ to precess about z at the same rate. The effect of the inertial term would then be to cause the angle θ between $\underline{\ell}$ and z to depend on the direction of rotation. For positive $2C-C_o$ and $\theta<\pi/2$ the value θ_+ of θ for positive rotation would be less than that θ_- for negative rotation. Unfortunately equation (1) does not have this precessing solution for $\underline{\ell}$ in bulk ^{3}He-A. Instead, in the presence of such a rotating field, the polar angle ϕ of $\underline{\ell}$ does not change and θ varies periodically with frequency ω_o as the field rotates.

A solution with $\underline{\ell}$ precessing at ω_o may be obtained, however, at small rotation rates if the liquid is contained between two parallel planes perpendicular to the z axis; the boundary conditions on $\underline{\ell}$ together with the $\underline{\ell}$ bending energy then suppress the oscillatory solution. If ω_o is increased with the field magnitude kept constant then the difference in polar angles $\Delta\phi$ between the precessing $\underline{\ell}$ and $\underline{H}$ decreases from

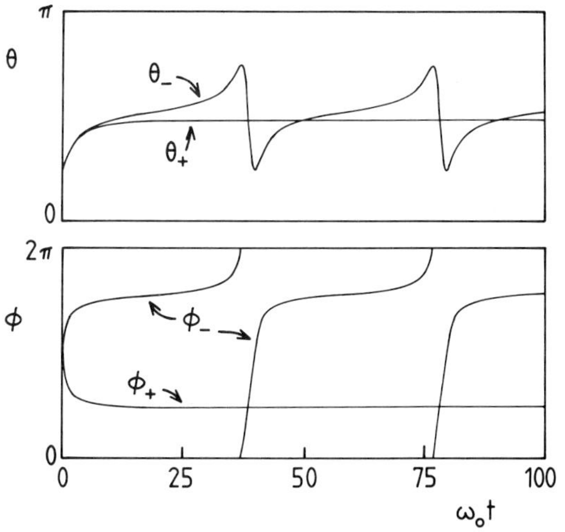

Fig. 1. Calculated motion of $\underline{\ell}$ for Ω=10.7, α=0.02, h^2=64, showing precessional motion for positive rotation and periodic motion of θ for negative rotation.

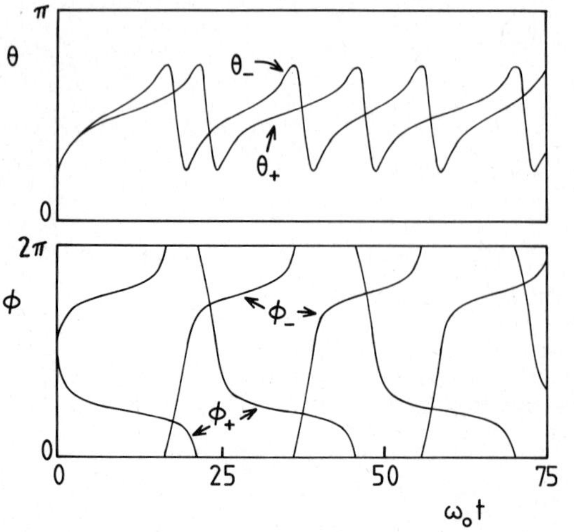

Fig. 2. Calculated motion of ℓ for Ω=11.6, α=0.02, h^2=64, showing periodic motion of θ for both directions of rotation but with different periods.

its limiting value of π at small ω_o. Eventually, ℓ is no longer able to precess at ω_o and has to occasionally slip a revolution with an associated time dependent θ during the process. For large rotation rates the solution tends to the infinite medium limit with no precession of ℓ and θ varying periodically at frequency ω_o. The inertial term causes the transition between the precessional and oscillatory solutions to occur at different values of ω_o for positive and negative rotations and this splitting provides a sensitive method for detecting the inertia.

4. SIMULATION OF EXPERIMENT

For the purpose of simulation we have assumed that the rotating magnetic field is directed at an angle $\pi/4$ to the z axis. Equation (1) has been solved numerically using an 'n' space point approximation for a slab of ^{3}He-A of thickness d (n=5 for figures 1,2 and 3). Using a finite difference approximation for the spatial derivatives integration was performed using small time steps. The sole assumption about the nature of the motion of ℓ was of spatial variation only in the z direction. In calculating $\delta f/\delta\ell$ we assumed the Ginzburg-Landau relation between the coefficients $K_b = K_s = K_t = 5\hbar C_o/4m = K$. The solutions can then be conveniently described using dimensionless units: $h^2 = \epsilon H^2 d^2/K$, $\Omega = \omega_o \mu d^2/K$ and $\alpha = (2C-C_o)/\mu$.

In figures 1-3 θ, ϕ are the polar angles of ℓ in the centre of the slab; ϕ modulo 2π is measured in a frame rotating with the field – ϕ=0 is the field direction. Subscripts + and – refer to the two directions of rotation. ℓ is parallel to z at both boundaries. All three figures are drawn for α = 0.02 and h^2 = 64, the large value of h^2 being necessary to produce the behaviour described in the previous section.

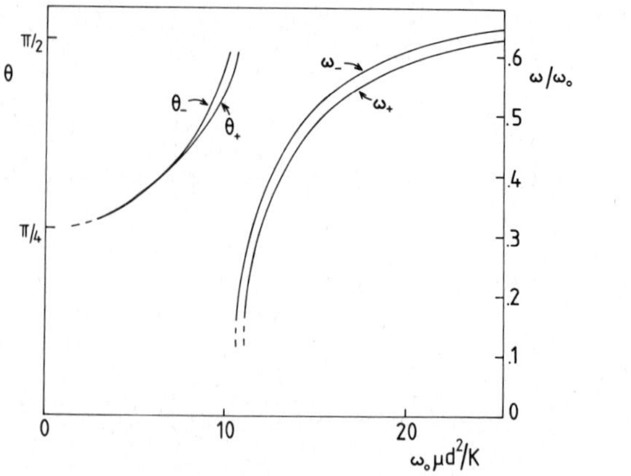

Fig. 3. Frequency of periodic motion of θ (curves on right) and values of θ in centre of slab for precessing solution (curves on left) as functions of Ω for both directions of rotation.

Figure 1. illustrates the two types of motion that can occur. For the frequency chosen (Ω=10.7), positive rotation produces the precessing solution with time independent values of θ and ϕ, whereas negative rotation produces periodic motion of θ associated with the 2π phase slips of ϕ relative to the field. At the higher value of Ω of 11.6 for which figure 2 was drawn both directions of rotation produce the solution with θ periodic in time, although the periodic frequencies ω_+ and ω_- are significantly different.

Figure 3 shows ω_+ and ω_- as functions of Ω in the vicinity of the transition and also shows that below the transition θ depends on the direction of rotation. The splitting of the transition is due to the inertia and this provides a sensitive method of detecting this quantity. Our calculations suggest that for the optimum value of h^2 it should be possible to detect values of α below 10^{-3}.

For a slab of thickness 2mm, suitable for measurements of the orientation of ℓ by ultrasonic attenuation, the field strength required is about 1 Gauss, comfortably below the dipole unlocking field. An experiment to investigate the behaviour predicted in this paper is currently in preparation at Manchester.

REFERENCES

(1) M. Liu, Physica 109 & 110B (1982) 1615.
(2) H.E. Hall and J.R. Hook, to be published in Progress in Low Temperature Physics.
(3) W.F. Brinkman and M.C. Cross, Spin and Orbital dynamics of superfluid ^{3}He, in: Progress in Low Temperature Physics, Vol. VIIA, ed. D.F. Brewer (North-Holland, Amsterdam, 1978) pp. 105-190.

LT-17 (Contributed Papers)
U. Eckern, A. Schmid, W. Weber, H. Wühl (eds)

SUPERCURRENT AND ANGULAR MOMENTUM IN ^{3}He-B INDUCED BY MAGNETIC FIELD

V.P. Mineev and G.E. Volovik

L.D. Landau Institute for Theoretical Physics, Moscow 117334, USSR

We discuss the microscopically derived supercurrent in ^{3}He-B in a magnetic field in connection with the problem of angular momentum.

The discovery of the intrinsic magnetization of vortices in ^{3}He-B[1] has converted the academic question of the existence of an orbital angular momentum in the p-wave superfluids into an urgent one. Now it has become possible to measure in NMR experiments the B-phase orbital angular momentum in a magnetic field, and this will provide a test for numerous different theories. The existence of an orbital angular momentum induced by a magnetic field in ^{3}He-B was first predicted by Leggett and Takagi.[2] Next this gyromagnetic phenomenon was discussed by Combescot,[3,4] who derived the orbital momentum from an expression for the supercurrent in the presence of a nonzero spin density, which was calculated using the method of gauge transformations. Similarly we also have worked out the same orbital momentum from the supercurrent obtained in an inhomogeneous field at constant order parameter.[5]

However, in the following paper[6] we revised the calculations on the existence of a well defined orbital momentum, as maintained in Refs. 3-5. The second theory is partially based on a general expression for the supercurrent $\vec{j}$ in a magnetic field which we obtained using the gradient expansion of Gorkov's equations.[7,8] This expression was not explicitly discussed in Ref. 6 and it is therefore the subject of the present report.

Taking into account the spacial variation of the order parameter $A_{\alpha i}$ (α is the spin and i the orbital index), the density ρ, and the magnetic field $\vec{H}$ we obtained he following terms for $\vec{j}$ linear in $\vec{H}$:

$$\vec{j} = \vec{j}^M + \vec{j}^L + \vec{j}^S \quad , \tag{1}$$

$$j_k^M = K^M e_{\lambda\mu\nu} \frac{\chi_n H_\nu}{\gamma} [A^*_{\mu i} \nabla_k A_{\lambda i} + A^*_{\mu k} \nabla_i A_{\lambda i} + A^*_{\mu i} \nabla_i A_{\lambda k} + c.c] \quad ,$$

$$j_k^L = K^L e_{\lambda\mu\nu} \frac{\chi_n H_\nu}{} \nabla_i (A_{\lambda i} A^*_{\mu k} + c.c.) \quad ,$$

$$j_k^S = K^S e_{\lambda\mu\nu} (A_{\lambda i} A^*_{\mu k} + c.c.) \nabla_i \left(\frac{\chi_n H_\nu}{\gamma}\right) \quad .$$

The last two terms in $\vec{j}$ are responsible for the orbital angular momentum in superfluid ^{3}He. If $K^L = K^S$, their sum is a pure curl manifesting the existence of a well defined angular momentum. However, we found that K^L is very small compared with K^S, it is nonzero only due to the particle hole asymmetry:

$$K^L \simeq K^S \left(\frac{T_c}{\varepsilon_F}\right)^2 \ln\frac{\varepsilon_F}{T_c} \quad , \quad K^S = \frac{5}{3} K^M = \frac{1}{3}\left(\frac{\rho_s}{\rho}\right)\frac{1}{|\Delta|^2} \quad . \tag{2}$$

This leads to the conclusion presented in Ref. 6, which was there obtained by other methods. The supercurrent $\vec{j}^L$ is related to

$$L_i = - K_L |\Delta|^2 \frac{\chi_n}{\gamma} H_\alpha R_{\alpha i} \simeq \left(\frac{T_c}{\varepsilon_F}\right)^2 \frac{\rho_s}{\rho} \frac{\chi_n}{\chi} H_\alpha R_{\alpha i} \quad . \tag{3}$$

It arises due to the orbital rotation of Cooper pairs induced by a magnetic field. Nevertheless, the total angular momentum in the vessel is large owing to the surface supercurrent $\vec{j}^S$ which is nonzero because of the existence of a density gradient at the surface, corresponding to a gradient of the normal susceptibility χ_n:

$$\vec{j}^S = \frac{2}{3} \frac{\rho_s}{\rho\gamma} \vec{H}\overleftrightarrow{R} \times \nabla\chi_n \quad .$$

Consequently, the total angular momentum is of the same order as obtained in Refs. 2-5,

$$\vec{J} = \int dV\, \vec{r}\times\vec{j} \simeq V \frac{\rho_s}{\rho} \frac{\chi_n}{\chi} \vec{H}\overleftrightarrow{R} \quad ,$$

while the local momentum is negligibly small.

We thus reach the conclusion, that the behavior of the B-phase orbital momentum is similar to that of the A-phase. Also in the

A-phase the lion's share of this momentum is caused by the superflow on the surface of the container while only a tiny part of it is of local origin from the intrinsic rotation of Cooper pairs.[9]

Such an exotic behavior of the angular momentum is intimately coupled with many other unexpected phenomena in the hydrodynamics of the p-wave superfluids. For instance, we know that the hydrodynamics of these liquids may be nonanalytic if the energy gap has a boojum-like topological singularity on the Fermi-surface.[10] Then the inconsistencies in the hydrodynamics lead to the conclusion that the density of the normal component ρ_n does not disappear at T = 0.[9] This residual ρ_n originates from the normal motion of textures both in the A- and B-phases. However, in the B-phase ρ_n at T = 0 seems to be analytic and therefore much smaller because the B-phase has no boojums on the Fermi-surface. The introduction of ρ_n at T = 0 is not a remedy for all the troubles in the hydrodynamics. One also has to allow for the striking possibility that the density of the linear momentum is not a well defined quantity. This problem already arose in the hydrodynamics of ferromagnets, where it proved to be impossible to construct a continuous single valued quantity of the linear momentum density.[11] The reason for this is analogous to that preventing the construction of a continuous single valued vector potential for Dirac's magnetic monopole.

REFERENCES

1. P.J. Hakonen, M. Krusius, M.M. Salomaa, J.T. Simola, Yu.M. Bunkov, V.P. Mineev, G.E. Volovik, Phys. Rev. Lett. 51, 1362 (1983).
2. A.J. Leggett, S. Takagi, Ann. of Phys. 110, 353 (1978).
3. R. Combescot, Phys. Lett. 78A, 85 (1980).
4. R. Combescot, J. Phys. C14, 4765 (1981).
5. G.E. Volovik, V.P. Mineev, Pis'ma ZhETF 37, 103 (1983).
6. G.E. Volovik, V.P. Mineev, ZhETF 86, no. 5 (1984) and preprint (1983).
7. N.D. Mermin, Paul Muzicar, Phys. Rev. B21, 980 (1980).
8. Paul Muzicar, Phys. Rev. B22, 3200 (1980).
9. G.E. Volovik, V.P. Mineev, ZhETF 81, 989 (1981).
10. G.E. Volovik, V.P. Mineev, ZhETF 83, 1025 (1982).
11. A. Veselov, private communication.

LT-17 (Contributed Papers)
U. Eckern, A. Schmid, W. Weber, H. Wühl (eds)

CONTINUOUS VORTICES IN ROTATING ^{3}He-A IN TRANSVERSE MAGNETIC FIELD

H.K. Seppälä

Low Temperature Laboratory, Helsinki University of Technology,
SF-02150 Espoo 15, Finland

The soft core structure and the NMR signature of the Anderson-Toulouse texture are investigated in a large transverse field. This texture is compared with experiments and other continuous 4π-vortices with a larger core energy, but a smaller overall energy at low rotation speeds.

1. INTRODUCTION

The NMR results[1] in rotating ^{3}He-A are well explained by continuous, axially asymmetric 4π-vortices. Their discrete internal symmetries may be broken in different ways, producing the "v" and "w" vortices.[2] The v-vortex remains invariant under the action of the symmetry operation $e^{i\pi}TPO^J_{y,\pi}$, while the w-vortex is symmetric under $e^{i\pi}TO^J_{y,\pi}$. Here T and P represent time and space inversions, respectively. $O^J_{y,\pi}$ rotates spin and orbital spaces about $\hat{y}$ by angle π. In this paper we report on the soft core structure of these vortices in a transverse magnetic field.

2. TEXTURES

Outside the soft core the dipole-locked $\hat{\ell}$- and $\hat{d}$-vectors are effectively clamped along $\vec{\Omega}\times\vec{H}$. Any deviation from this direction becomes costly in energy. (In an axial field, however, the texture is free to rotate around the rotation axis $\vec{\Omega} = \Omega\hat{z}$). Although the soft core of radius $\simeq 5\xi_D$ is only a small fraction of the Wigner-Seitz cell with radius $R \simeq \sqrt{\hbar/m_3\Omega}$, an energy loss, due to some misorientation outside the core, might still be compensated for by a lower core energy. This has motivated us to study the Anderson-Toulouse texture, which we call the $\hat{z}$-vortex, in a transverse magnetic field. It also has the symmetry properties of a v-vortex, its orbital order parameter can be expressed by

$$\hat{\Delta}' + i\hat{\Delta}'' = e^{-i\phi}[\hat{\phi} + i(\hat{z}\cos\eta - \hat{r}\sin\eta)] \quad . \qquad (1)$$

A texture, called the $\hat{x}$-vortex,[3] which optimizes the energy far from the core, is given by

$$\hat{\Delta}' + i\hat{\Delta}'' = e^{-i\Phi}[(\hat{z}\sin\Phi + \hat{y}\cos\Phi)\sin\eta + \hat{x}\cos\eta + i(-\hat{y}\sin\Phi + \hat{z}\cos\Phi)] \quad . \qquad (2)$$

The choice $\Phi = \phi - \pi/2$ produces a v-vortex, while $\Phi = \phi$ gives a w-vortex. In both cases $\eta(r,\phi)$ is a variational function with the boundary conditions $\eta(r=\infty) = -\eta(r=0) = \pi/2$. We have solved the Euler-Lagrange equations, corresponding to the free energy minimization, in the Ginzburg-Landau temperature region. Terms up to 2nd order in the free energy expansion have been included, except for the divergence terms. $\hat{d}$ is also allowed vary in the plane perpendicular to $\vec{H}\uparrow\uparrow\hat{y}$. Results for the sum of the dipole and gradient energies per one equivalent Wigner-Seitz cell of radius R are given for the transverse magnetic field in Table I.

TABLE I.

Vortex	Symmetry	$F_v/[\pi\rho_s(\hbar/2m_3)^2]$
$\hat{x}$	[v]	$3\ln[R/\xi_D] - 0.7$
$\hat{x}$	[w]	$3\ln[R/\xi_D] - 1.0$
$\hat{z}$	[v]	$4\ln[R/\xi_D] - 2.1$

Figs. 1a, b, and c show the components (ℓ_x, ℓ_y), $(\ell_z, -\ell_y)$, and $(\ell_z, -\ell_x)$ of the $\hat{z}$-vortex. It is interesting to note that a plot of (ℓ_x, ℓ_y) in the xy-plane for the v- and w-symmetric $\hat{x}$-vortices reproduces Figs. 1b and 1c, respectively. From the symmetry classification point of view it is essential that both the ϕ-dependence of η and the variation of $\hat{d}$, shown for the $\hat{z}$-vortex Fig. 1d, are small.

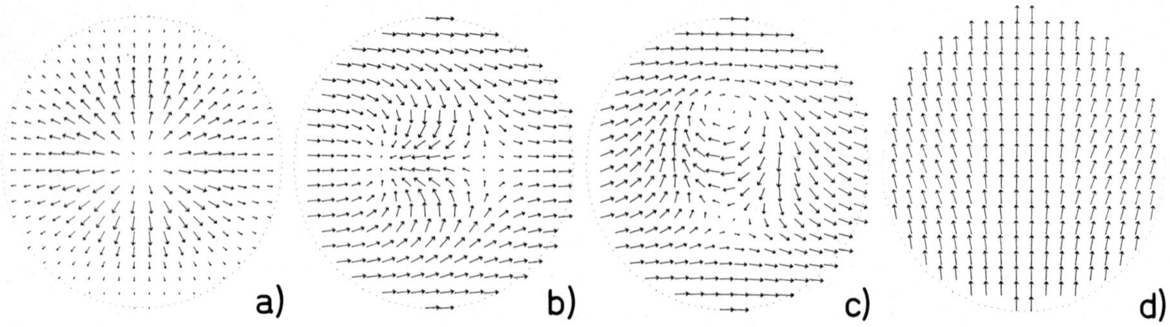

Fig. 1. Projections of the z-vortex at different locations in the xy-plane: a) (ℓ_x,ℓ_y); b) $(\ell_z, -\ell_y)$; c) $(\ell_z,-\ell_x)$; d) (d_x,d_z). The radii of the textures are 5 ξ_D.

3. SPINWAVES

The observed transverse NMR satellite peak originates from the dipole-unlocked core region. Its frequency ω_v is given by $\omega_v^2 = \omega_o^2 + R_t^2\Omega_L^2$. Here ω_o and Ω_L are the Larmor and longitudinal resonance frequencies. R_t^2 is the lowest eigenvalue of an operator Λ^g associated with the energy $\delta^2F = g\Lambda^g g$ of the fluctuating $\hat{d}$-vector. The eigenfunction $g(\vec{r})$ corresponds to the normalized amplitude of the $\hat{d}$-vector oscillations along the magnetic field $\vec{H}$. Knowing $g(\vec{r})$ the absorption intensity I_v may be calculated from

$$\frac{I_v}{I_{tot}} = n_\Omega \xi_D^2 \left\{ \frac{[\int d^2r g(\hat{d}\cdot\hat{e}_1)]^2 + [\int d^2r g(\hat{d}\cdot\hat{e}_2)]^2}{\int d^2r\, g^2} \right\} \quad (3)$$

Here n_Ω is the vortex density, ξ_D = 6 μm is the dipole coherence length, $\hat{e}_1$ is a unit vector pointing to the preferred $\hat{\ell}$-orientation and $\hat{e}_2$ is a unit vector orthogonal to $\hat{e}_1$ and $\vec{H}$. Calculated frequencies and intensities, using the boundary condition $g(r = 7\xi_D) = 0$, are compared with the experimental results for the transverse magnetic field in Table II.

TABLE II

Vortex	R_t^2	$I_v/[I_{tot}\Omega]$
$\hat{x}$ [v]	0.45	0.034
$\hat{x}$ [w]	0.51	0.038
$\hat{z}$ [v]	0.28	0.032
experiment[2]	0.70±0.10 (extrapolation to $T=T_c$)	0.090±0.010 ($T=0.76\ T_c$)

Fig. 2 illustrates $g(\vec{r})$ of the w-symmetric $\hat{x}$-vortex, which at our rotation speeds[1] is energetically the most favorable of the three vortices considered here.

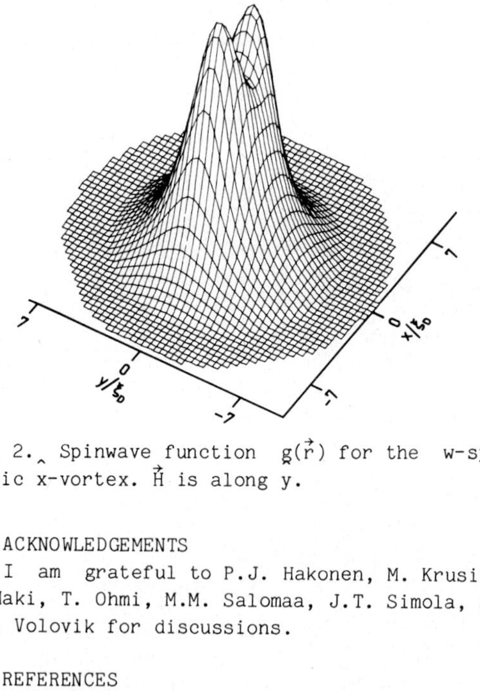

Fig. 2. Spinwave function $g(\vec{r})$ for the w-symmetric $\hat{x}$-vortex. $\vec{H}$ is along y.

4. ACKNOWLEDGEMENTS

I am grateful to P.J. Hakonen, M. Krusius, K. Maki, T. Ohmi, M.M. Salomaa, J.T. Simola, and G.E. Volovik for discussions.

5. REFERENCES

1. P.J. Hakonen, O.T. Ikkala, S.T. Islander, O.V. Lounasmaa, and G.E. Volovik, J. Low. Temp. Phys. 53, 425 (1983).
2. H.K. Seppälä, P.J. Hakonen, M. Krusius, T. Ohmi, M.M. Salomaa, J.T. Simola, and G.E. Volovik, to be published.
3. H.K. Seppälä and G.E. Volovik, J. Low Temp. Phys. 51, 273 (1983).

LT-17 (Contributed Papers)
U. Eckern, A. Schmid, W. Weber, H. Wühl (eds)

VORTEX PAIR STATE IN ROTATING SUPERFLUID ^{3}HE-A

Kazumi MAKI and Xenophon ZOTOS

Department of Physics, University of Southern California, Los Angeles, CA 90089-0484, USA

A modified version of the vortex pair state due to Seppälä and Volovik is shown to have the lowest free energy among the vortex configurations so far proposed in rotating superfluid ^{3}He-A in a magnetic field. The satellite frequency and the satellite intensity associated with the present model are compared with recent NMR experiments by the Helsinki group in rotating superfluid ^{3}He-A.

1. INTRODUCTION

As to the possible vortex state in rotating superfluid ^{3}He-A so far three candidates are considered; 1) the vortex lattice of singular vortices as in superfluid ^{4}He, 2) the rectangular lattice formed by circular and hyperbolic 2π-vortices (1,2) and 3) the lattice proposed by Seppälä and Volovik (SV) (3) formed by radial-hyperbolic vortex pairs. We have proposed recently (4) a modified version of the SV vortex pair, the lattice of circular-hyperbolic pairs which has lower free energy. We shall summarize here the main results on the circular-hyperbolic pair. First the principle advantage of a vortex pair in contrast to isolated analytic vortices lies in that the d texture describing the spin component of the ^{3}He-A condensate is free from topological singularity. Therefore in the limit of small rotation where the vortex density is dilute the vortex pair state gives the lowest free energy per vortex pair. As to the singular vortices it is believed that they appear only in the limit of high rotation say $\Omega \simeq 10\text{-}10^2 \sec^{-1}$. Furthermore far from a vortex pair the $\hat{\ell}$ vector is asymptotically unidirectional. Therefore the vortex pair state is easily accommodated even when the external magnetic field is tilted from the axis of rotation. The projection of the $\hat{\ell}$ texture on the x-y plane perpendicular to the rotation axis is schematically shown in Fig. 1a) for the original SV texture and Fig. 1b) for our modified version.

In the following we shall concentrate on three vortex pair textures; the original SV texture, our modified version and our modified version with uniform d texture. We denote these three textures I, II and III respectively.

2. FREE ENERGY

Starting from the Ginzburg Landau free energy valid near $T\text{-}T_c$, we determined variationally the free energy per vortex pair per unit length. The free energy contains a term diverging logarithmically with the radius of the container associated with the circulation around the vortex pair. We shall cut off this divergence at $r_0=(\pi n_v)^{-\frac{1}{2}}$ the intervortex distance where n_v is the vortex pair density. Since there is only one length scale $\xi_\perp=C_\perp/\Omega_A (\simeq 10\mu m)$ in the theory the free energy depends only on $r=r_0/\xi_\perp$. Our variational results are summarized as

$$f_I = \pi A \{12 \ln (\tfrac{1}{2}r) + 21.51\} \tag{1}$$

$$f_{II} = \pi A \{12 \ln (\tfrac{1}{2}r) + 16.45\} \tag{2}$$

$$f_{III} = \pi A \{12 \ln (\tfrac{1}{2}r) + 23.94\} \text{ with } r=10\xi_\perp \tag{3}$$

Here A is the coefficient of the kinetic energy and indices I, II, etc. refer to the definition given in section one. In the case of the SV III pair with uniform d the last constant depends still weakly on r due to the fact that the nuclear dipole energy depends weakly on r in this case. Among the three configurations considered the circular-hyperbolic pair with nonuniform d has the lowest free energy and therefore this should be realized in the axial magnetic field.

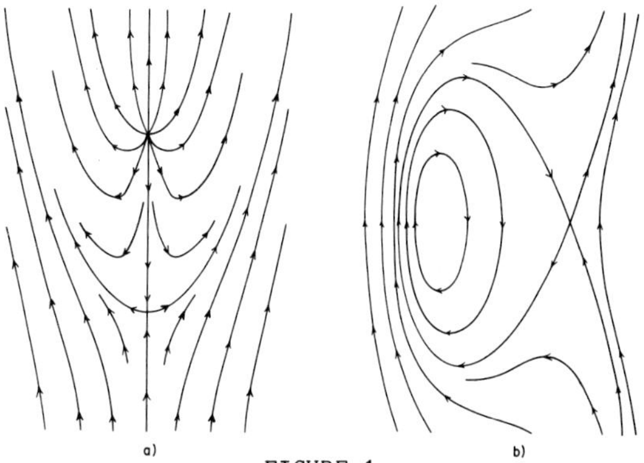

FIGURE 1
a) radial-hyperbolic pair
b) circular-hyperbolic pair

3. NUCLEAR MAGNETIC RESONANCE

In the vicinity of the vortex cores $\hat{d}$ and $\hat{\ell}$ are not parallel to each other. This gives rise to potential energy for spin waves in rotating ^{3}He-A (5). The spin wave bound states due to this potential well are seen experimentally as NMR satellites. In particular the present potential contributes one satellite each for the longitudinal and transversal resonance with satellite frequencies;

$$\omega_\ell = R_\ell \Omega_A \qquad (4)$$

$$\omega_t = \left((\gamma H)^2 + R_t^2 \Omega_A^2\right)^{\frac{1}{2}} \qquad (5)$$

where Ω_A is the Leggett frequency.

We calculate R_ℓ, R_t and the corresponding intensities variationally for three textures, which are shown together with experimental results (6) in Table 1.

TABLE 1

Textures	R_ℓ	$I_\ell/\Omega I_0$	R_t	$I_t/\Omega I_0$
I	0.69	0.072	0.69	0.062
II	0.71	0.088	0.72	0.075
III	0.77	0.094	0.87	0.116
Exp			0.90	0.058

It is somewhat suprising that the transverse satellite frequency for the circular-hyperbolic pair with uniform $\hat{d}$ is in good agreement with those observed by the Helsinki group (6) extrapolated at $T=T_c$ although the corresponding intensity appears off by a factor of two. Since extrapolation of the intensity to $T\simeq T_c$ is difficult due to large scatter in the experimental values we present the value at $T\simeq .8T_c$. Here I_0 is the intensity of the main peak in the absence of rotation. Although the circular-hyperbolic pair with nonuniform $\hat{d}$ describes not so well the observed results still this is substantially better than the radial-hyperbolic pair.

4. SOUND ATTENUATION

In the absence of rotation, both $\hat{\ell}$ and $\hat{d}$ lie in a plane perdicular to an external magnetic field. Vortex pairs in rotating ^{3}He-A induce the $\hat{\ell}$ component parallel on antiparallel to the rotation axis. Since the sound attenuation is sensitive to the angle between the sound propagation vector and the local $\hat{\ell}$ vector, the appearance of the parallel and antiparallel component is in principle detectable as changes in the attenuation coefficient.

A theoretical analysis (7) indicates that the attenuation is most sensitive to the $\hat{\ell}$ texture when the propagation vector is chosen parallel to the asymptotic $\hat{\ell}$ and $\hat{d}$ directions. In the presence of a tilted magnetic field, this direction is determined as in the x-y plane and perpendicular to the magnetic field. For example for A=C=2B where A, B, and C are coefficient of the attenuation coefficient (8)

$$\alpha(\theta) - \alpha_c = A\cos^4\theta + B\sin^4\theta + 2C\sin^2\theta\cos^2\theta$$

and θ is angle between the propagation vector and $\hat{\ell}$, the attenuation coefficient increases almost linearly with Ω in this particular goemetry. This linear slope is extremely sensitive to C the distance of pairs in the vortex pair and increases like C^2. Furthermore it is easy to discriminate the radial-hyperbolic pair from the circular-hyperbolic pair by means of ultrasonic attenuation (7).

ACKNOWLEDGEMENTS

Most of this work was done while we were staying at the Low Temperature Laboratory of Helsinki University of Technology at Otaniemi, Finland. We would like to thank the Low Temperature Laboratory for the hospitality during our stay. Particular thanks go to Martti Salomaa, Matti Krusius and Pertti Hakonen for discussions on related subjects. One of us (KM) gratefully acknowledges the support of a Nordita professorship. The present work is supported by National Science Foundation under grant number DMR-82-14525.

REFERENCES

(1) T. Fujita, M. Nakahara, T. Ohmi and T. Tsuneto, Prog. Theo. Phys. 60 671 (1978).
(2) K. Maki, Phys. Rev. B27 4173 (1983).
(3) H. K. Seppälä and G.E. Volovik, J. Low Temp. Phys. 51 279 (1983).
(4) X. Zotos and K. Maki, Phys. Rev. B (to be published).
(5) K. Maki and P. Kumar, Phys. Rev. B16 182 (1977).
(6) P. J. Hakonen, O. T. Ikkala, S. T. Islander, O. V. Lounasmaa and G. E. Volovik, J. Low Temp. Phys. 53 425 (1983).
(7) K. Maki and X. Zotos, preprint.
(8) D. N. Paulson, M. Krusius and J. C. Wheatley, J. Low Temp. Phys. 26 73 (1977).

LT-17 (Contributed Papers)
U. Eckern, A. Schmid, W. Weber, H. Wühl (eds)

INTRINSIC ANGULAR MOMENTUM IN SUPERFLUID ^{3}HE-B

Paul MUZIKAR

Physics Department, Purdue University, West Lafayette, Indiana, 47907

We discuss the term in the B phase supercurrent recently discovered by Combescot and Dombre and by Volovik and Mineev. This term arises from a spatially varying spin density, and is relevant to the question of the intrinsic angular momentum of ^{3}He-B.

The extent to which the superfluid phases of liquid ^{3}He possess intrinsic angular momentum has been investigated theoretically in recent years. One useful calculational device is to impose a spatially varying potential $U(\vec{r})$, (1) or a spatially varying magnetic field $H(\vec{r})$, (2) (3) while holding the energy gap uniform. One can then compute if any current flows in such a situation, and if so, try to interpret that current physically.

Volovik and Mineev (3) (see also Combescot and Dombre (4)) have applied these ideas to ^{3}He-B. Recall that the B phase energy gap is of the form

$$\Delta_i(\hat{K}) = \Delta_o R_{ij} \hat{K}_j e^{i\phi} \qquad [1]$$

where R_{ij} is a rotation matrix, ϕ is a real number, and Δ_o is the amplitude. Volovik and Mineev (VM) showed that in the presence of a nonuniform $\vec{H}(\vec{r})$ a supercurrent can flow even though the energy gap [1] is independent of $\vec{r}$. By using the gradient expansion of Gorkov's equations, VM showed that the current at zero temperature (we will restrict ourselves to T = o in the present discussion) is given by

$$J_i = -\frac{1}{2}\frac{\chi_B(T=o)}{\gamma}\varepsilon^{ijk}\frac{\partial H^\alpha}{\partial r_j}R^{\alpha k} . \qquad [2]$$

Since we have

$$\vec{S} = \frac{\chi_B(T=o)}{\gamma}\vec{H} \qquad [3]$$

VM rewrite equation [2] as

$$J_i = -\frac{1}{2}\varepsilon^{ijk}\frac{\partial S^\alpha}{\partial r_j}R^{\alpha k} . \qquad [4]$$

In their paper, (3) and in a following one, (5) they discuss whether it is valid to interpret [4] as meaning that ^{3}He-B possesses an intrinsic angular momentum density of

$$L_k = -R^{\alpha k}S^\alpha . \qquad [5]$$

Here, we will investigate the status of equation [4]. We can set up a spatially varying $\vec{S}(\vec{r})$ in another way: we impose a uniform $\vec{H}$ and a spatially varying $U(\vec{r})$. Then since χ_B becomes a function of $\vec{r}$ we have

$$\frac{\partial\vec{S}}{\partial r_i} = \frac{\vec{H}}{\gamma}\frac{\partial\chi_B}{\partial r_i} = \frac{\vec{H}}{\gamma}\frac{\partial\chi_B}{\partial\rho}\frac{\partial\rho}{\partial r_i} \qquad [6]$$

$$\frac{\partial\rho}{\partial r_i} = \frac{\partial\rho}{\partial U}\frac{\partial U}{\partial r_i} . \qquad [7]$$

Hence we have both a $\vec{\nabla}\rho$ and a $\vec{\nabla}\vec{S}$. We can solve the Gorkov equations to first order in $\vec{H}$ and to first order in $\vec{\nabla}U$, keeping terms of order $\vec{H}(\vec{\nabla}U)$. If we do this we find that the current is given by

$$J_i = -4\gamma\hbar^2 H^j R^{jb}\varepsilon^{kab}\frac{\partial U}{\partial r_k}\Delta_o^2 T\sum_\omega\int\frac{d\hat{K}}{4\pi}\hat{K}_i\hat{K}_a\int\frac{d\varepsilon N(\varepsilon)\varepsilon}{(\omega^2+E^2)^3} \qquad [8]$$

One might be tempted to conclude that the ε-integral gives zero by particle-hole symmetry; however, this would amount to throwing away a term of precisely the desired type. Instead, we use the following relation (6):

$$\int\frac{d\varepsilon N(\varepsilon)\varepsilon}{(\omega^2+E^2)^3} \approx \frac{N(0)}{8E_F}\int\frac{d\varepsilon}{(\omega^2+E^2)^2} \qquad [9]$$

To derive this relation we integrate by parts, and take $N(\varepsilon)$ to have the form $N(\varepsilon) = A\sqrt{\varepsilon+E_F}$. We have also ignored any ε dependence of Δ_o^2.

Using equation [9] gives

$$J_i = -\frac{1}{2}\varepsilon^{ijk}\frac{\partial S^\alpha}{\partial r_j}R^{\alpha k} \qquad [10]$$

where $\vec{\nabla}\vec{S}$ is given by [6]. Thus, the form [4] holds for a $\vec{\nabla}\vec{S}$ generated in two different ways.

Finally, we can investigate the status of terms in the current of the form $\vec{L}\times\vec{\nabla}\rho$.(5) If we assume $\vec{J}$ consists of two terms:

$$J_i = -\frac{1}{2}\varepsilon^{ijk}\frac{\partial S^\alpha}{\partial r_j}R^{\alpha k} + \lambda\varepsilon^{ijk}R^{\alpha k}\frac{S^\alpha}{\rho}\frac{\partial\rho}{\partial r_j} , \qquad [11]$$

then our results indicate that $\lambda = 0$; the effect of a density gradient is accounted for in the first term, and so no explicit $\vec{\nabla}\rho$ is needed.

It should be noted that the precise value given here for the coefficient in [10] relied on the use of the free particle form for $N(\varepsilon)$. A more general assumption, such as

$$\frac{N'(o)}{N(o)} = \frac{b}{2E_F} , \qquad [12]$$

with b not necessarily equal to one, would in general lead to a different coefficient in [10], and hence a non-zero value for λ.

ACKNOWLEDGEMENTS

I am very grateful to V.P. Mineev for pointing out an error in an earlier version of this paper.

REFERENCES

(1) N.D. Mermin and Paul Muzikar, Phys. Rev. B21, 980 (1980).
(2) Paul Muzikar, Phys. Rev. B22, 3200 (1980).
(3) G.E. Volovik and V.P. Mineev, JETP Lett. 37, 127 (1983).
(4) R. Combescot and T. Dombre, Phys. Lett. 76A, 293 (1980); R. Combescot, J. Phys. C14, 4765 (1981).
(5) G.E. Volovik and V.P. Mineev, preprint.
(6) V.P. Mineev, private communication.

LT-17 (Contributed Papers)
U. Eckern, A. Schmid, W. Weber, H. Wühl (eds)

STABILITY OF THE ZERO ENERGY EXCITATIONS IN ^{3}He A

R. COMBESCOT and T. DOMBRE

Groupe de Physique des Solides de l'Ecole Normale Supérieure+, 24 rue Lhomond, 75231 Paris Cedex 05, France

We show that zero energy excitations with finite density of states exist for a general texture.

It has been shown recently (1)(2)(3)(4) that the excitation spectrum of ^{3}He A is dependent on the local texture. The energy scale for this dependence is $\delta(\xi_0/L)^{1/2}$ where δ is the maximum of the gap, ξ_0 the coherence length and L the typical length scale of the texture. This energy scale is small and the effect is only relevant for low temperatures $T/T_c \lesssim (\xi_0/L)^{1/2}$. The origin of the effect lies in the trapping of excitations in potential wells created by the texture : a low energy excitation with fixed $\vec{k}$ has to stay in a region with $\vec{\ell} \simeq \vec{k}$ because of energy conservation. As a result the eigenstates are localized and the spectrum displays a corresponding discretization.

An important consequence of this effect (4) is the appearance of a finite density of states $N(0) = N_0 v_F|(\hat{\ell}.\vec{\nabla})\hat{\ell}|/2\delta$ for zero energy excitations. This leads at $T = 0$ to a reduction of the superfluid density $\hat{\ell}.\overleftrightarrow{\rho}^s.\hat{\ell} = \rho - k_F^2 N(0)$. We have obtained this result by keeping only the gradients of $\Delta_k(r)$ and neglecting higher order derivatives. It is important to see if the existence of a finite density of states for zero energy excitation is a general result and not an accidental feature due to our approximation.

In order to find if there are zero energy excitations, we do not need the full Green's function as in (4). We have only to find the singularities which are obtained from the homogeneous equation. This leads to the eigenvalue problem :

$$\begin{pmatrix} \xi - iv_F\partial_\rho & \Delta_k(\rho) \\ \Delta_k^*(\rho) & -\xi + iv_F\partial_\rho \end{pmatrix} \begin{pmatrix} \psi_1 \\ \psi_2 \end{pmatrix} = \varepsilon \begin{pmatrix} \psi_1 \\ \psi_2 \end{pmatrix} \qquad (1)$$

with $\rho = \vec{r}.\hat{k}$ and standard notations. We have neglected terms of order δ/E_F compared to the dominant ones. We are only interested in the solutions for $\varepsilon = 0$.

In this case, we set $\psi_1 = \exp(i\phi(\rho) - i\rho\xi/v_F)\chi_1$ $\psi_2 = \exp(-i\phi(\rho) - i\rho\xi/v_F)\chi_2$, $\Delta_k(\rho) = \delta(\rho)\exp(2i\phi(\rho))$ and make the univocal change of variable $v_F y = \int_0^\rho d\rho\ \delta(\rho)$. With pseudo-spin notation, this leads to :

$$\left(\omega(y) + \sigma_x - i\sigma_z\partial_y\right)\chi = 0 \qquad (2)$$

where $\omega(y) = \partial_y\phi = v_F\partial_\rho\phi(\rho)/\delta(\rho)$. This is transformed into :

$$\left(\omega(y) + \sigma_z + i\sigma_x\partial_y\right)\phi = 0 \qquad (3)$$

by the pseudospin rotation $\chi = \exp(i\sigma_y\pi/4)\phi$. In the generic situation $\omega(y)$ has a fixed sign. If $\omega > 0$ for example, we set $u = y + \int_0^y dy\ \omega(y)$ (we assume $\delta(\rho) \neq 0$ everywhere which gives a finite ω) and we are led to :

$$-\partial_u^2\phi_2 + \frac{1-\omega(u)}{1+\omega(u)}\phi_2 = 0 \qquad (4)$$

This is a Schrödinger equation in the potential $(1-\omega)/(1+\omega)$ and we look for an eigenstate with energy $E = 0$. For a space independent order parameter $\omega = 0$, the spectrum is $E \geqslant 1$ and we have no solution. The same result holds if ω is small everywhere. A necessary condition for a solution is that the potential gets somewhere negative which means $\mathrm{Max}\,\omega > 1$. For large ω, the potential reaches -1 and $E = 0$ eigenstates are possible. Therefore, we find as expected that zero energy excitations are only possible for large ω which implies in practice that $\delta(\rho)$ must approach zero since $\omega \sim v_F/L\delta(\rho) \sim (\xi_0/L)(\delta/\delta(\rho))$. However, it is not necessary that $\delta(\rho)$ reaches zero in order to have zero energy excitations.

In general, for a given $\Delta_k(\rho)$, we will find no $\varepsilon = 0$ excitations. But in our case we know that a good approximation of $\Delta_k(\rho)$ gives an $\varepsilon = 0$ solution. Using the exact $\Delta_k(\rho)$ will shift the solution a bit away from $\varepsilon = 0$. But we can look for a slightly different value of $\hat{k}$. In

+ Laboratoire associé au Centre National de la Recherche Scientifique.

general, the eigenvalues are analytical functions of $\hat{k}$ and therefore we will find a zero energy excitation for a nearby value of $\hat{k}$.

This can be seen explicitly by first order perturbation theory. We expand $\Delta_k(\rho)$ as :

$$\Delta_k(\rho) = \Delta_k^{(0)} + \rho\Delta_k^{(1)} + \ldots + \rho^n\Delta_k^{(n)} + \ldots \qquad (5)$$

where the overall phase of $\Delta_k(\rho)$ is chosen to make $\Delta_k^{(1)} \equiv \alpha(\hat{k})$ real. Then our approximate solution of energy $\varepsilon = \mathrm{Im}\Delta_k^{(0)}$ is given by :

$$\psi_1 e^{i\rho\xi/v_F} = \psi_2^* e^{-i\rho\xi/v_F} = \exp\left(i\,\frac{\pi}{4} - \frac{\alpha}{2v_F}\,(\rho+\rho_0)^2\right) \qquad (6)$$

where $\rho_0 = \mathrm{Re}\Delta_k^{(0)}/\alpha$. To first order the correct energy is :

$$\varepsilon - \mathrm{Im}\Delta_k^{(0)} = \mathrm{Re}\int d\rho\,(\Delta_k - \Delta_k^{(0)} - \rho\Delta_k^{(1)})\psi_1^*\psi_2 \Big/ \int d\rho\,|\psi_1|^2$$

$$= \sum_{n=2}^{\infty} \mathrm{Im}\,\Delta_k^{(n)} \left(\frac{v_F}{\alpha}\right)^{n/2} \left(\frac{i}{2}\right)^n \mathrm{Hn}\left(i\rho_0\sqrt{\frac{\alpha}{v_F}}\right)$$

$$\simeq -\frac{v_F}{4\alpha}\, H_2\left(i\rho_0\sqrt{\frac{\alpha}{v_F}}\right) \mathrm{Im}\Delta_k^{(2)} \qquad (7)$$

since the $\Delta_k^{(n)}$ term is of order $(\xi_0/L)^{n/2}$. Therefore, we obtain a zero energy excitation when :

$$\mathrm{Im}\Delta_k^{(0)} = \frac{v_F}{4\alpha}\, H_2\left(i\rho_0\sqrt{\frac{\alpha}{v_F}}\right) \mathrm{Im}\Delta_k^{(2)} \qquad (8)$$

This gives a value of $\hat{k}$ which differs from $\hat{\ell}$ by a small term of order ξ_0/L. In conclusion, we have in the general case a finite density of states N(0) which differs from our preceding result only by higher order terms in ξ_0/L.

(1) G.E. Volovik and V.P. Mineev, Zh. Eksp. Teor. Fiz. 81, 989 (1981) (Sov. Phys. JETP 54, 524 (1981)).
(2) P. Muzikar and D. Rainer, Phys. Rev. B 27, 4243 (1983).
(3) R. Combescot and T. Dombre, Phys. Rev. B 28, 5140 (1983).
(4) R. Combescot and T. Dombre, AIP Proceedings of the Sanibel Symposium on Quantum Fluids and Solids (1983). Ed. by E.D. Adams and G.G. Ihas, p.261.

LT-17 (Contributed Papers)
U. Eckern, A. Schmid, W. Weber, H. Wühl (eds)

TEXTURAL ORIENTING ENERGIES OF VORTICES IN ^{3}He-B

P.J. Hakonen, M. Krusius, G. Mamniashvili, and J.T. Simola

Low Temperature Laboratory, Helsinki University of Technology,
SF-02150 Espoo 15, Finland

In rotating ^{3}He-B two additional free energy contributions appear which affect the order parameter distribution outside the vortex core. We have measured these vortex free energy contributions as a function of temperature and pressure in the range 0.5-1.0 T_c and 10-30 bar using cw transverse NMR.

1. INTRODUCTION

During the past two years it has been realized that the vortices in ^{3}He-B are more diverse than in superfluid ^{4}He.[1] Although the asymptotic form ($r \to \infty$) of a He-B vortex is similar to a ^{4}He vortex, the core structure is different: while the B-phase components of the order parameter approach zero when $r \to 0$, other components appear with a nonzero value in the core, e.g. A-phase and β-phase as noted by Salomaa and Volovik.[2]

The deviation of the order parameter from the B-phase value leads to a nonunitary ($\langle S \rangle \neq 0$) order parameter distribution in the core region.[3] The nonunitarity and the appearance of other magnetic components[2] give rise to a spontaneous magnetic moment of the vortex core, which is observed in the measurements as the gyromagnetic free energy

$$F_{gm} = \frac{4}{5} a\kappa(\hat{\Omega}_i R_{ik} H_k). \quad (1)$$

F_{gm} is linear with respect to the axis of rotation $\vec{\Omega}$ and the magnetic field $\vec{H}$, κ is the parameter proportional to the spontaneous magnetization and the number of vortices while a is a measure of the magnetic anisotropy energy $F_M = -a(\bar{n}\cdot\bar{H})^2$. The second vortex free energy term

$$F_V = \frac{2}{5} a\lambda(\hat{\Omega}_i R_{ik} H_k)^2, \quad (2)$$

where the parameter λ has two contributions originating from the 1/r velocity field outside

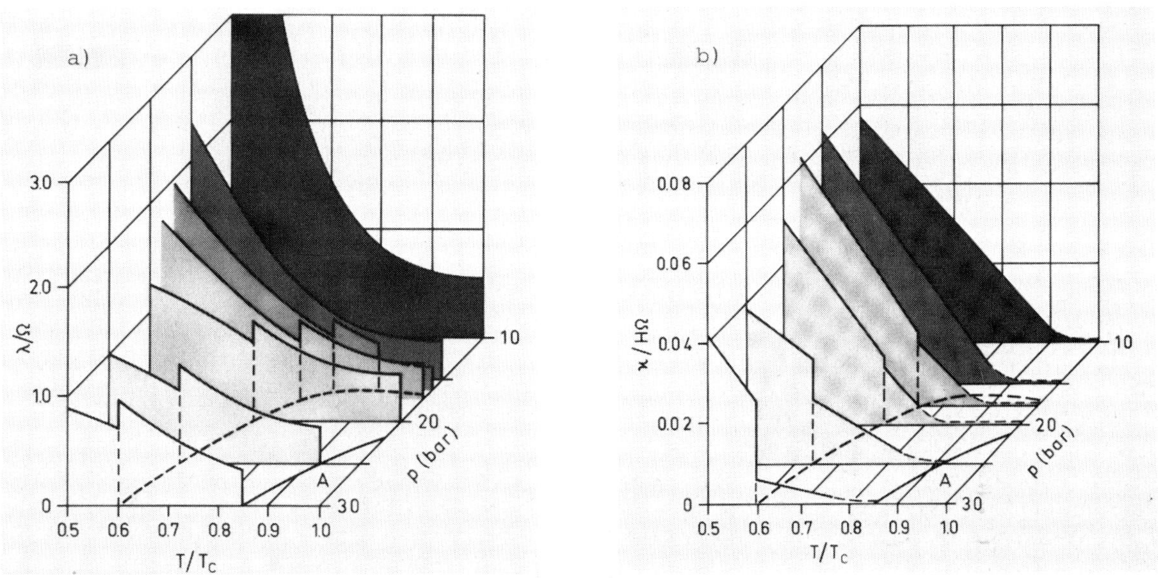

Fig. 1 λ/Ω [Fig. 1a] and $\kappa/H\Omega$ [Fig.1b] , in units 1/sec, as a function of temperature T and pressure p. The measurements have been performed in a magnetic field of H = 28.4mT.

the vortex core,[4] and from the anisotropy of the susceptibility inside the core.[1] In fact, both λ and κ correspond to two different "second moments" of the order parameter components in the core.[3] We have measured λ and κ by studying with cw-NMR techniques the changes in the equilibrium $\vec{n}$-texture, which occur during rotation.[5,6]

2. RESULTS

The results for λ/Ω and $\kappa/H\Omega$ as functions of temperature and pressure are shown in Fig 1. The λ/Ω and $\kappa/H\Omega$ curves have been measured at 29.3, 25.0, 20.5, 18.1, 17.1, 15.5, and 10.2 bar pressure; the trace at 17.1 bar is absent from the $\kappa/H\Omega$ figure. At each pressure data from many different runs are given. The data for $\kappa/H\Omega$ have been smoothed using a linear function in order to suppress a scatter of $\sim\pm 0.01$. The behavior of $\kappa/H\Omega$ near T_c cannot be reliably measured since the difference in the frequency shifts of forward and reverse rotation cannot be resolved close to T_c. There is some indication, nevertheless, that $\kappa/H\Omega$ is not approaching zero near T_c as schematically shown in Fig. 1 b.

In Fig. 2 three different runs are plotted as a function of temperature at the pressures 18.1, 17.1, and 15,5 bar. Notice that no discontinuity is present in the data at 15.5 bar. The points close to T_c have been omitted since the relative error in λ, due to the ± 25 Hz accuracy in the frequency shift measurement, diverges at T_c. The values of λ shown in Fig. 1 near T_c have been obtained by extrapolating the measured points to a constant value in the Ginzburg-Landau region.

3. DISCUSSION

At high pressures a discontinuity is observed in the λ/Ω and $\kappa/H\Omega$ curves. This is interpreted as a 1st order phase transition between two different types of vortices.[1] The data for λ/Ω at 17.1 bar show two discontinuities indicating

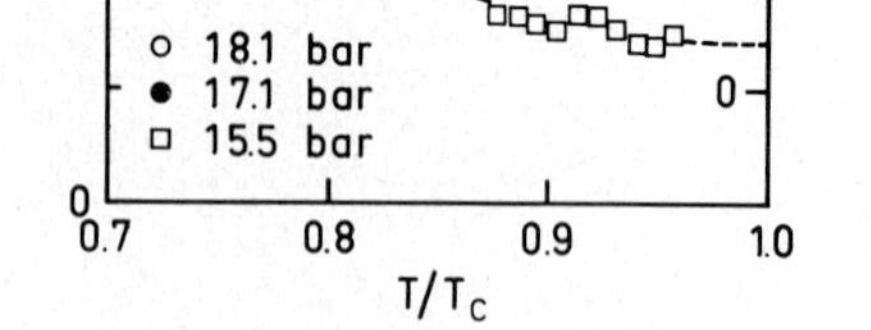

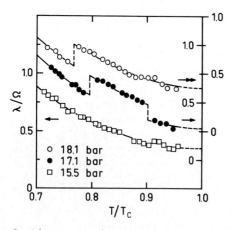

Fig. 2 λ/Ω as a function of temperature at three different pressures. Note the zero shifts of the λ/Ω scales for the different curves.

that the phase boundary in the pT-plane is curving towards higher pressures on approaching the superfluid T_c. The exact locations of the end points of the vortex phase transition line have not yet been fixed. The present data suggest that the low pressure end point is located on the T_c-curve between 17 and 18 bar.

The origin of the vortex transition is not yet fully understood. Two different alternatives have been presented: either the size of the vortex core changes roughly by a factor of two, or the core is structurally rearranged. Clearly the change in the discontinuity of $\kappa/H\Omega$ at different pressures favours the latter view; otherwise the relative importance of the discontinuity of $\kappa/H\Omega$ should be pressure independent. Salomaa and Volovik suggest that the low temperature vortex with a large spontaneous magnetic moment has a superfluid core containing A-phase and β-phase,[2] while the vortex of Ohmi et al.[3] with a normal core would have a small κ/H.

The pressure and temperature dependencies of λ and κ are qualitatively understood.[6,7] The main pressure dependence in λ originates from the cross-sectional area of the vortex core proportional to the coherence length ξ_o^2, responsible for the increase of λ towards lower pressures. In the case of $\kappa \propto (\Delta^2\chi_n)/(\varepsilon_F\gamma\hbar)$ the ξ_o^2 dependence is compensated by Δ^2 leaving a constant as a function of pressure. The quantitative understanding, of course, involves the detailed structure of the vortices which presently has primarily been discussed in the GL-region. To understand the measured temperature dependencies of λ and κ beyond the GL-region, where temperature dependencies appear to cancel out, Fermi-liquid corrections have to be taken into account.[7] Finally it should be observed that extensive numerical work is still required until a reliable identification of the B-phase vortex structures has been worked out.

REFERENCES

1) P.J. Hakonen, O.T. Ikkala, S.T. Islander, O.V. Lounasmaa, and G.E. Volovik, J. Low Temp. Phys. 53, 425 (1983).
2) M.M. Salomaa and G.E. Volovik, Phys. Rev. Lett. 51, 2040 (1983).
3) T. Ohmi, T. Tsuneto, and T. Fujita, Progr. Theor. Phys. (Kyoto) 70, 647 (1983).
4) A.D. Gongadze, G.E. Gurgenishvili, and G.A. Kharadze, Fiz. Nizk. Temp. 7, 821, (1981) [Sov. J. Low Temp. Phys. 7, 397 (1981)].
5) Yu. M. Bunkov, P.J. Hakonen, and M. Krusius, Quantum Fluids and Solids - 1983 edited by E.D. Adams and G.G. Ihas, AIP Conference Proceedings No. 103 (American Institute of Physics, New York, 1983), p. 194.
6) P.J. Hakonen, M. Krusius, M.M. Salomaa, J.T. Simola, Yu.M. Bunkov, V.P. Mineev, and G.E. Volovik, Phys. Rev. Lett. 51, 1362 (1983).
7) S. Theodorakis and A.L. Fetter, J. Low Temp. Phys. 52, 559 (1983).

LT-17 (Contributed Papers)
U. Eckern, A. Schmid, W. Weber, H. Wühl (eds)

FLOW INDUCED TEXTURES IN ^{3}He-A*

D.M. BATES, S.N. YTTERBOE, C.M. GOULD†, and H.M. BOZLER

Department of Physics, University of Southern California, Los Angeles, California 90089-0484 USA

A cell has been developed to search for changes in the texture in superfluid ^{3}He-A caused by flow. The flow is generated by a servo controlled bellows system, and texture changes are observed using the anisotropy of zero sound attenuation. At low magnetic fields we have found a phase boundary between a uniform texture region and a complex texture region in the v_s-H plane. The results are suggestive of a transition to helical textures.

We have studied induced textural changes caused by applying an increasing magnetic field parallel to a constant impressed flow. ^{3}He-A is represented by a complex order parameter (1) consisting of an orbital unit vector $\hat{\ell}$ and a spin space unit vector $\hat{d}$. The state of the superfluid is determined by the free energy (2)

$$F = F_D + F_H + F_V + F_{GRAD},$$

which is composed of a dipole energy F_D, a magnetic energy F_H, a flow energy F_V, and a gradient term F_{GRAD} involving bending of the order parameter vectors $\hat{\ell}$ and $\hat{d}$.

In the presence of flow alone the fluid should be in a uniform texture state with $\hat{\ell}||\hat{d}||\vec{v}_s$. Theoretical studies (3) predict that this uniform texture will become unstable as the magnetic field increases. Even in the absence of a magnetic field, the uniform texture is expected (4) to become unstable due to a negative term in the free energy involving both flow and gradients of $\hat{\ell}$. This zero field instability should occur for velocities on the order of 1 mm/s (3). The theory predicts a transition to a helical texture.

We have developed an apparatus (5), which creates a constant driven mass flow. A driven mass flow has the advantage that the flow can be well controlled and is independent of minor dissipation caused by textural changes in the fluid. The flow is created by using a ^{4}He filled bellows to drive a piston which forces fluid through a superleak consisting of 830,000 glass capillaries 3 mm long, with 2 μm diameters. The average orientation of the $\hat{\ell}$ vector is detected using 30.8 MHz zero sound propagating across a rectangular flow channel positioned about 3 cm from the superleak. The sound propagation is therefore perpendicular to $\vec{v}_s$ and to $\vec{H}$ which is applied along the flow channel axis.

Constant flow velocities are obtained by using a heated "bomb" to drive the ^{4}He bellows with feedback from a mutual inductance position detector. We can apply constant mass flows for several minutes at a time. All of our data is taken with the flow going towards the superleak.

Figure 1 illustrates some typical data and indicates several important features of the fluid behavior. The dashed lines represent the maximum and minimum sound levels observed during

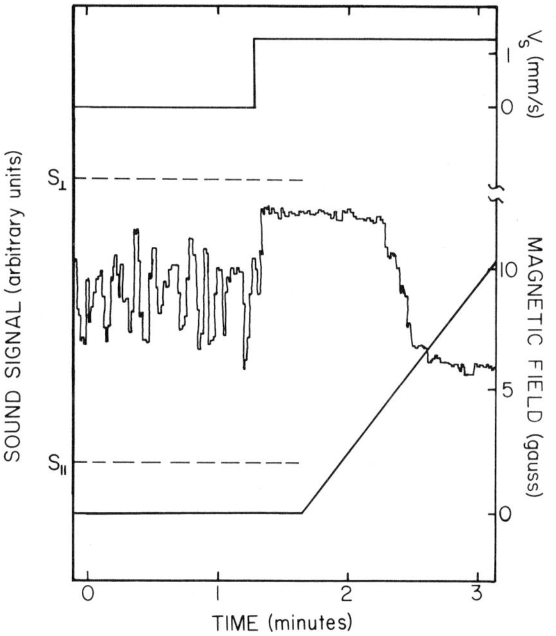

FIGURE 1.
Trace showing typical behavior of the superfluid when subjected to flow and field. $S_\perp$ and $S_{||}$ correspond to $\hat{\ell}$ normal and parallel to $\vec{q}$, respectively. The first region has both flow and field at zero. After 1.25 min. the flow is started and then the magnetic field is ramped.

* Work supported by the National Science Foundation grant numbers DMR82-00661 (HMB,DMB,SNY) and DMR82-08760 (CMG).
† Alfred P. Sloan Research Fellow

the experiment, and are associated with $\alpha_\perp$ and $\alpha_{||}$, respectively. These levels are confirmed by measurements of the sound absorption anisotropy in a rotating field (5). By rotating the magnetic field we were able to estimate the coefficients in the sound absorption, which has the form (6)

$$\alpha-\alpha_c = A\sin^4\theta + B\cos^4\theta + C\sin^2\theta\cos^2\theta, \qquad (1)$$

where θ is the angle between $\hat{\ell}$ and the sound propagation vector $\vec{q}$. $\alpha_\perp$ and $\alpha_{||}$ are given by A and B, respectively. The figure shows that the signal level for the transmission of zero sound is a sensitive indicator of the texture state of the superfluid.

An obvious feature of figure 1 is the large fluctuations of the sound signal in the first minute before either flow or field are applied. The size of the variations is due to the presence of a nearly uniform texture which fluctuates in direction on a time scale of seconds in this nearly degenerate system. The application of a constant velocity bellows stroke immediately suppresses these oscillations and brings the sound level up to near its maximum value, indicating that the $\hat{\ell}$-texture is both uniform and aligned with the flow, and hence perpendicular to $\vec{q}$. The aligned state becomes unstable as an applied magnetic field is increased.

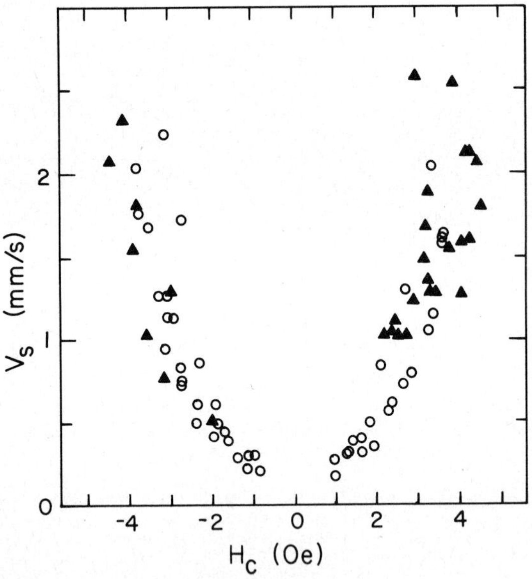

FIGURE 2.
Data at P=28.8 bars and T=.94T_c for two different flow channel cross-sectional areas: o(.054 cm^2) and Δ(.028 cm^2). H_c is determined by fitting sound traces to equation (2). Use of the smaller channel avoids exceeding the critical velocity in the superleak.

We are interested in determining the critical field H_c above which the uniform texture is no longer stable. Using $h=H-H_c$, we fit the sound signal S to a form derived from the helical texture theory:

$$S = S_0\exp\{[-{}^1/_2(C-2A)bh - {}^3/_8(A+B-C)b^2h^2]d\}, \qquad (2)$$

where A,B, and C are the coefficients in equation 1, d is the crystal spacing, and b is a measure of the rate at which the helix opening angle increases with magnetic field. S_0 is the sound signal level before the transition.

Least squares fitting to the above functional form yields values of H_c for traces taken at various velocities. At the higher velocities b in equation 2 is very small, making the determination of H_c more difficult

The results for data taken at 28.8 bars and .94T_c are plotted in figure 2. v_s is determined from the bellows velocity, known geometrical factors, and $\rho_{s||}$, the parallel component of the superfluid density tensor (7).

Qualitatively, the transitions fit the predictions of the helical texture theories. The experimentally determined boundary of the uniform texture is similar to predictions. However, the uniform texture region appears to be stable for much higher velocities than predicted. Possible explanations include: the coefficients of the relevant flow and gradient terms in the free energy may be incorrect; the actual boundary was missed because it is more difficult to detect at higher velocities; some unknown effect is stabilizing the uniform textures.

We wish to thank K. Maki for helpful discussions.

REFERENCES

(1) A.J. Leggett, Rev. Mod. Phys. 47 (1975) 331.
(2) M.C. Cross, J. Low Temp. Phys. 21 (1975) 525.
(3) D. Vollhardt, Y.R. Lin-Liu, and K. Maki, J. Low Temp. Phys. 37 (1979) 627; A.L. Fetter and M.R. Williams, Phys. Rev. B 23 (1981) 2186; and references cited therein.
(4) P. Bhattacharyya, T.-L. Ho, and N.D. Mermin, Phys. Rev. Lett. 39 (1977) 1290; and M.C. Cross and M. Liu J. Phys C 11 (1978) 1795.
(5) D.M. Bates, C.M. Gould, and H.M. Bozler, Texture transitions in flowing ^{3}He, in: Quantum Fluids and Solids, eds. E.D. Adams and G.G. Ihas (American Institute of Physics, New York, 1983) pp. 282-287.
(6) J. Serene, Thesis, Cornell University, (1973), unpublished; and P. Wölfle, Phys. Rev. Lett. 30 (1973) 1169.
(7) J.E. Berthold, R.W. Giannetta, E.N. Smith and J.D. Reppy, Phys. Rev. Lett. 37 (1976) 1138.

LT-17 (Contributed Papers)
U. Eckern, A. Schmid, W. Weber, H. Wühl (eds)

STABILITY OF THE UNIFORM TEXTURE FOR ^{3}He-A IN FINITE GEOMETRIES.

R.C.M.DOW and J.R. HOOK

Department of Physics, University of Lancaster, LA1 4YB, United Kingdom, and
Department of Physics, University of Manchester, M13 9PL, United Kingdom.

We examine the stability of the uniform texture for ^{3}He-A with increasing parallel superflow and magnetic field. The effects of two finite geometrical constraints are examined and found to produce gross modification of the unconstrained stability region.

1. INTRODUCTION

We examine the stability of the uniform texture $\underline{l} \parallel \underline{d} \parallel \hat{\underline{z}}$ in the presence of imposed parallel superflow and magnetic field, for two systems. (i) Thermally driven counterflow between two parallel plates. And, (ii) Bulk liquid superflow constrained by periodic boundary conditions over a finite length in the flow direction. The uniform texture stability can be markedly affected by the finite aspects of these systems, and we derive the stability region for each case.

2. UNITS AND ASSUMPTIONS

We use dimensionless units, with field in units of $H^* = (g/\varepsilon)^{\frac{1}{2}}$ superflow in units of $V_s^* = (2g/\rho_0)$, and length in units of $z^* = \hbar/(2^{3/2} m V_s^*)$. These units are derived from comparison of alignment energies - $g(\underline{d}.\underline{l})^2$, $\varepsilon(\underline{d}.\underline{H})^2$, and $-\rho_0 (V_s.\underline{l})^2/2$, terms in the free energy. For $(T \sim T_c)$ the coefficients give values of order 20 Oe, 1 mm/s and 15μm respectively.

We work in the Ginzberg-Landau regime $(T \sim T_c)$, assuming $V_n = 0$ as $\rho_n \gg \rho_s \rightarrow V_n \ll V_s$ for the thermal counterflow system. We consider one-dimensional flow, with variation only in the flow direction (transverse fluctuations may be neglected (1,2), and assume that the uniform texture becomes unstable first with respect to helical perturbations.

3. PARALLEL PLATE SYSTEM

The effects of finite plate separation on a dipole locked thermal counterflow have been studied by Hook and Hall (3). We do not restrict ourselves to this limit, so the dimensionless channel width 'a' can take an arbitrarily small value. The boundary conditions at the plates require $\underline{l} = \hat{\underline{z}}$ and $\partial \underline{d}/\partial z = 0$. A suitable helical form for $\underline{l}$, in Cartesian co-ordinates, is

$$\underline{l} = (\alpha s \sin kz, \alpha s \cos kz, 1 - \alpha^2 s^2/2)$$

where $s = \sin(\pi z/a)$.

This describes a helix with small opening angle 'α' and pitch $2\pi/k$; winding about 'z' in a negative sense for positive k. The term 's' incorporates the boundary condition on $\underline{l}$ and assumes that the amplitude of a small distortion of the uniform texture varies as 'sin', to leading order. The uniform texture may be characterised by $\alpha = 0$.

To consider the stability in small fields, we expect $\underline{d}$ of order $\underline{l}$, and write

$$\underline{d} = (\sin(\beta)\sin kz, \sin(\beta)\cos kz, \cos(\beta))$$

$\underline{d}$ winds in the same way as $\underline{l}$, and we can relate the small opening angle of the $\underline{d}$ helix to that of the $\underline{l}$ helix. For $\alpha, \beta \ll 1$ and $H \gg 1$,

$$\beta = \alpha s/(1 + k^2/z - H^2)$$

For small 'a', there exists a uniform state for $H \gg 1$, with $\underline{d}$ perpendicular to H. The existence of this texture with $\underline{l} \parallel V_s \parallel \underline{H} \perp \underline{d}$ is confirmed by taking the form for $\underline{d}$, appropriate to the $H \gg 1$, $a \ll 1$ limit (a form also appropriate to the $V_s = 0$ limit), where

$$\underline{d} = (0,1,0)$$

The relevant free energy for a system with externally fixed mass flow 'j' is the Gibbs energy, which will take on two forms with $\underline{l}$ and $\underline{d}$ as above. For $H \ll 1$ and $H \gg 1$ respectively, and expanding to second order in the small angle 'α', we get

$$G_1 = - j^2 + 2kj + H^2 + \alpha^2 s^2 [j^2 - 3kj + 3k^2/2 + 3\pi^2/2a^2 + (k^2 - H^2)/(1 + k^2 - H^2) + \pi^2/(a^2(1 + k^2 - H^2)^2)]$$

$$G_2 = - j^2 + 2kj + \alpha^2 s^2 [j^2 - 3kj + 3\pi^2/2a^2 + 3k^2/2 - 1]$$

where

$$V_s = j - j\alpha^2 s^2 + k\alpha^2 s^2$$

The uniform textures $\underline{l} \parallel \underline{d}$ and $\underline{l} \perp \underline{d}$ become

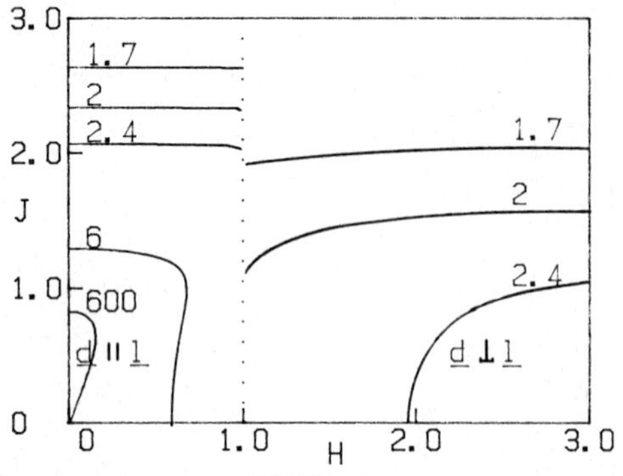

FIGURE 1

The curves of figure 1 indicate the parallel plate uniform texture stability regions for the values of 'a' indicated. From near the origin we have an expanding region of $\underline{l} \parallel \underline{d}$. For small 'a' the region $\underline{l} \perp \underline{d}$ develops.

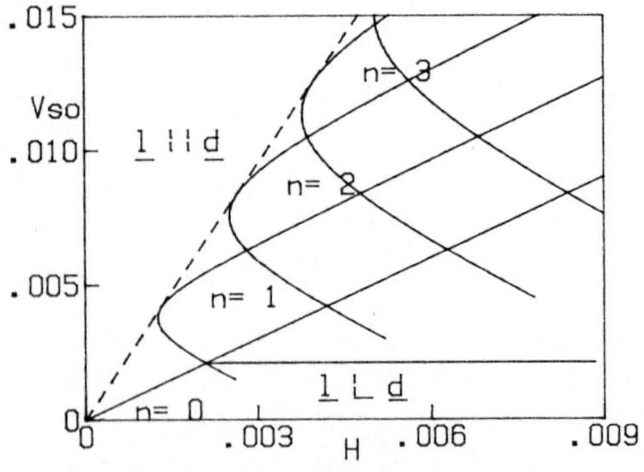

FIGURE 2

The uniform texture stability in the dipole-locked region for periodic flow with periodic repeat length 'a'=2500. The numbers 'n' indicate the winding of the helical instability.

unstable if the square bracketed terms in the relevant free energy become negative. The phase diagram suggested by these stability criteria has been confirmed by numerical solution of the dynamical equations for $\underline{l}$ and $\underline{d}$ (to be published). The results are shown in fig.1. Note (i) For 'a' large the stability region tends toward the unconstrained bulk fluid result. (ii) As 'a' decreases, the stability region increases until, (ii) for 'a' < 2.75 the $\underline{d} \perp \underline{l}$ uniform texture becomes stable. (iv) For 'a'<<3 the stable region ultimately becomes step like.

4. PERIODIC CONDITIONS IN BULK FLUID

Finite periodicity can be used to simulated a multiply connected geometry. We treat the dipole-locked case and forget about boundaries.

The mass flow in this system is not externally fixed, and the relevant free energy is the free energy F, with coefficients applicable to dipole locking. The helical form for $\underline{l}$ can now be written

$$\underline{l} = (\alpha \sin(kz), \alpha \cos(kz), 1 - \alpha^2/2)$$

With 'z' the flow direction, 'α' the small opening angle, and pitch $=2\pi/k$. We expand to second order in 'α' as before, deriving the following stability criterion for the uniform, $\underline{l} \parallel V_S \parallel H$, texture:

$$H^2/V_{SO}^2 + 3n/x - 5n^2/2x^2 - 1 > 0$$

where V_{SO} is the original superflow in the uniform state, and

$$x = a\,V_{SO}/2\pi \;;\; k = 2\pi n/x \;,\; n = 0,1\ldots$$

with 'n' the number of turns of the helix over the periodic repeat length 'a'.

The stability condition provides the phase diagram of fig.2. We have chosen a specific value for 'a', but note that altering this value by a factor x, merely results in a scaling of the V_{SO},H values by a factor 1/x. We see that a region exists near the origin, where the uniform texture remains stable to a field much larger than the bulk value- dashed line, and undergoes a first order phase transition to a perpendicular uniform $\underline{l}$ texture. Away from this region, the phase boundary is still severely modified, but the resulting transition is of second order to a nearby helical state. For larger V_{SO} we return to the unconstrained bulk liquid result. As before, these results have been checked against the behaviour of the dynamical equation for $\underline{l}$.

5. CONCLUSION

Dramatic effects arising from finite geometry have been shown to occur in one region of the unconstrained bulk fluid phase diagram. Vastly different effects occur in two possible geometries, with differing physical constraints.

We expect all of the effects generated by small separation in a parallel plate flow system to be observable. However, the effects arising from periodic boundary conditions may be more difficult to see; the effects of distant wall boundaries, or the physical bending of the texture around a real torus may render the distorted region of the phase boundary inaccessible.

REFERENCES

(1) A.L.Fetter and M.R. Williams, Phys. Rev. B23, (1981) 2186.
(2) D.Vollhardt, Y.R.Lin-Liu, and K. Maki, J. Low Temp. Phys. 37, (1979) 627.
(3) J.R.Hook and H.E.Hall, J. Phys. C12, (1979) 783.
(4) C.G.Harris, J. Phys. C13, (1980) L1061.

LT-17 (Contributed Papers)
U. Eckern, A. Schmid, W. Weber, H. Wühl (eds)
© Elsevier Science Publishers B.V., 1984

SOUND PROPAGATION IN SUPERFLUID HE II SATURATED POROUS MEDIA*

F. PASIERB†, David SINGER, R. RUEL and H. KOJIMA

Serin Physics Laboratory, Rutgers University, Piscataway, NJ 08854

Observations on the propagation of first and second sound in superfluid He II saturated porous media (fused glass beads and a rock) are reported. The measured speed of sound gives the tortuosity parameter for the porous medium. The measured attenuation of sound is interpreted in terms of viscous losses in the normal fluid component motion.

When fourth sound was first observed in He II filled packed powder [1], it was observed that the measured speed of fourth sound was reduced by a constant factor (index of refraction) from the value expected from the two fluid model. Recently there has been much interest in the problem of the acoustics of fluid-filled porous media [2]. We present measurements on the propagation of first and second sound in He II-filled fused glass beads and a naturally occurring rock. The size of the pores present in these porous materials was sufficiently large not to lock the normal fluid component of He II.

The fused glass bead sample investigated here was prepared by sintering at ∿700°C packed glass beads whose diameter is in the range 180∿210 μm [3]. A naturally occurring porous sample, Masillon sandstone, was obtained from a rock core obtained from a quarry in Ohio [4]. The Masillon sandstone sample was not treated in any way other than machining. A cylindrical plug was machined out of cores of porous materials. The plug was inserted into a cylindrical cavity whose diameter (∿2.0 cm) and length (∿1.0 cm) were adjusted for a snug fit. The ends of the cavity were closed either with first sound (plain Mylar diaphragm) or second sound (Nuclepore filter membrane) capacitance transducers.

The cylindrical cavity filled with He II constitutes a first or second sound resonator. The speed of sound c_j^* (j = 1 and 2 for first and second sound, respectively), in the presence of a porous medium was determined from

$$c_j^* = \frac{2f_m L}{m} \qquad (1)$$

where f_m is the measured resonant frequency of m-th length mode of j-th sound and L is the length of the cavity. The measured speed of sound in the presence of a porous sample was always reduced from that in the bulk [5]. The resonant frequencies were observed in the range 1 kHz ∿ 50 kHz and 500 Hz ∿ 12 kHz for first and second sound, respectively.

In Fig. 1, we show the measured speed of first sound (open circles) and second sound (dots) in an identical fused glass bead sample whose porosity ϕ = 0.345. The solid lines in Fig. 1 represent the speed of sound in bulk [5] divided by an index of refraction equal to 1.32 and 1.29 for first and second sound in the fused glass bead sample, respectively. There is good agreement in the indices of refraction of two sounds in the same fused glass bead sample. The measured speeds of

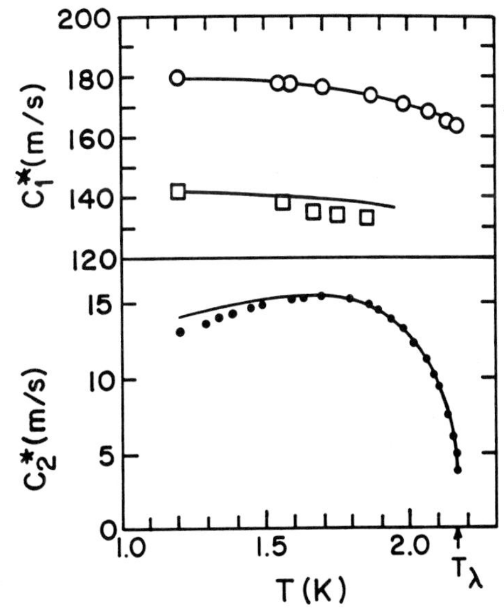

Figure 1.

*Research supported by Schlumberger-Doll Research.
†Present address: Materials Research Corporation, Pearl River, NY 10962

sound can be qualitatively described by a temperature independent index of refraction. A closer inspection of data reveals systematic deviations between the lines and the data near T_λ and below 1.5 K in first and second sound results, respectively. These small deviations may be accounted for by the inertial part of the dynamic flow resistance [6]. The speed of first sound measured in Masillon sandstone (ϕ = 0.27) is shown as squares in Fig. 1. The signal level was much weaker in the Masillon sandstone than in the fused glass beads. The index of refraction for the sandstone is about 1.67. The deviation of data from the temperature dependence of the bulk first sound may be caused in part by errors owing to low levels of signal.

Based on a self-similar model for the pore space, Johnson and Şen [7] derived the power law relationship, $n^2 = \phi^q$, where n is the index of refraction, and q is a constant dependent on a pore structure. If we let q = -1/2 [7] for the structure of a randomly placed spheres, this formula gives n = 1.30 when ϕ = 0.345. The predicted n is in good agreement with the fused glass bead result in Fig. 1. A comparison of the theory with the Masillon sandstone is difficult since its pore structure is not well characterized.

The measured quality factor of first sound (m = 2, open circles) and second sound (m = 3, dots) resonances are shown in Fig. 2. The Q of the first sound resonance decreases, whereas the Q of the second sound resonance increases as the temperature increases towards T_λ. The observed temperature dependence of Q is characteristic of an attenuation mechanism

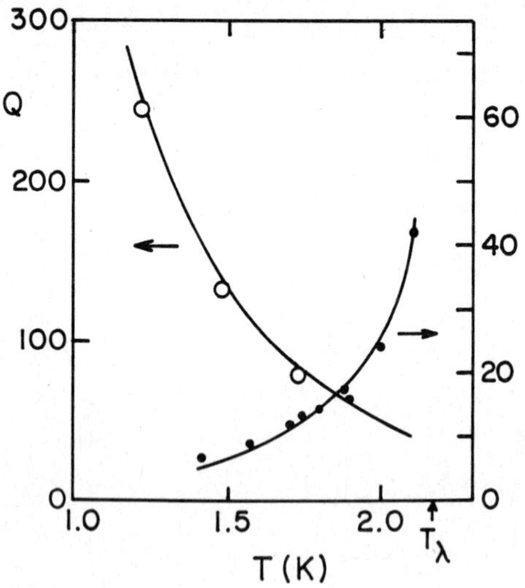

Figure 2.

dominated by viscous losses on the normal fluid component at the solid walls.

Viscous effects may be included into the equation of motion of the normal component (its velocity is $\vec{v}_n$) by adding a drag force, $-R\vec{v}_n$, where R represents a frequency and geometry dependent dynamic flow resistance [8]. We carried out a calculation by assuming our sample can be represented by a collection of tubes with an average radius r [6]. The solid lines in Fig. 2 represent the calculated Q where the value of r was adjusted to 17 μm and 19 μm for first sound and second sound, respectively. A rough estimate gives an average pore radius equal to 34 μm for our fused glass bead sample. This estimate is fairly close to the adjusted values of r used in Fig. 2.

It is our great pleasure to acknowledge very useful conversations with David Johnson, Izzy Rudnick and Tom Plona.

REFERENCES

(1) K.A. Shapiro and I. Rudnick, Phys. Rev. 137A (1965) 1383.
(2) AIP Conference Proceeding No. 107, Physics and Chemistry of Porous Media, eds. D.L. Johnson and P.N. Sen (American Institute of Physics, 1984).
(3) D.L. Johnson, T.J. Plona, C. Scala, F. Pasierb, and H. Kojima, Phys. Rev. Lett. 49 (1982) 1840.
(4) The Masillon sandstone sample was provided to us by David Johnson.
(5) J. Maynard, Phys. Rev. 14B (1976) 3868.
(6) D. Singer, F. Pasierb, R. Ruel and H. Kojima (to be published).
(7) D.L. Johnson and P.N. Sen, Phys. Rev. 24B (1981) 2486.
(8) S. Baker, D. Rudnick, and I. Rudnick, J. Acoust. Soc. Am. Suppl. 1 (1983) S59.

LT-17 (Contributed Papers)
U. Eckern, A. Schmid, W. Weber, H. Wühl (eds)

PULSED THIRD SOUND ON SUBSTRATES OF KNOWN CONTAMINATION

D.T. SMITH, M. LIEBL, M.D. BUMMER, AND R.B. HALLOCK

Laboratory for Low Temperature Physics, Department of Physics and Astronomy, University of Massachusetts, Amherst, MA USA 01003

The propagation of pulsed third sound is studied as a function of ^4He film thickness d_4 simultaneously on glass substrates both clean and dusted with $A\ell_2O_3$ powder of 0.3μm characteristic size. The time of flight ratio shows a clear signature of capillary condensation above d_4 = 7 atomic layers. The effect of this condensation on the third sound pulse width is documented.

1. INTRODUCTION

Liquid helium represents a realization of a Kosterlitz-Thouless phase transition when it has the configuration of a thin film. Detailed predictions of the properties of the dynamic helium film due to Ambegaokar et al. (1) have been confirmed through the use of a number of techniques including third sound (2), vibrating substrates (3) and thermally induced mass flow (4). Although apparently unimportant, the effects of surface roughness have never been studied (5) in detail. In an effort to determine the effects of capillary condensation on the propagation of third sound and the possible role of surface structure on the nature of the helium phase transition, we are studying mass flow and the time of flight of third sound pulses on surfaces dusted with known coverages of $A\ell_2O_3$ powder. We report here observations of third sound as a function of film thickness.

2. TECHNIQUE

In a typical experiment two substrates are prepared identically with thin film $A\ell$ bolometers deposited by vacuum evaporation techniques. Suitably biased, the bolometers serve as generators and detectors of third sound. In a preliminary experiment the two clean substrates were used at T = 1.54, 1.64, and 1.74 K to establish that the 5kHz pulsed third sound (single clamped sine pulse) time of flight was the same for each substrate to within 3% over the range of ^4He film thickness, d_4, studied: $4 \lesssim d_4 \lesssim 20$ atomic layers. All measurements made in the cell which housed the substrates were equilibrium measurements with stability established after each change in d_4.

3. RESULTS AND DISCUSSION

Subsequent to the control measurements one substrate was removed from the experimental chamber and exposed to an air suspension of 0.3μm $A\ell_2O_3$ powder. The substrate was housed in a cylindrical chamber and protected by a shutter following the creation of the suspension by an air blast to about 0.25 cm^3 of $A\ell_2O_3$. Under such conditions it is possible to produce structures which appear dendritic (see Fig. 1) and contain many hundreds of particles. These presumably grow as they fall in the cylindrical chamber under the influence of gravity; we observe a collapsed two dimensional projection. These structures deserve study in their own right, but we developed techniques to avoid them for the present work. Under proper conditions of shutter closure and $A\ell_2O_3$ drying quite homogeneous coverages of $A\ell_2O_3$ can be routinely achieved. Microscope studies of the prepared

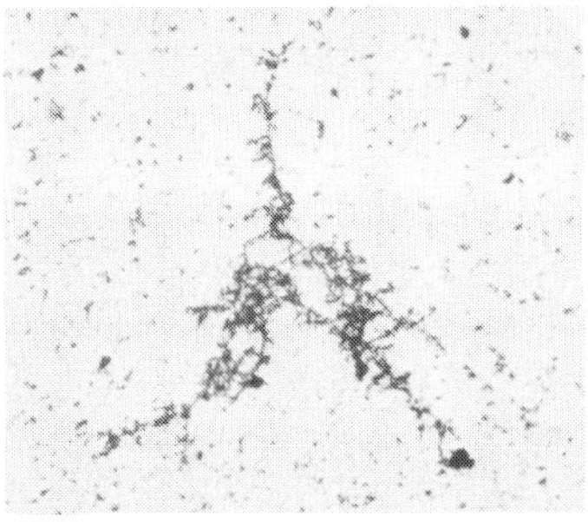

FIGURE 1
Example of an unusual structure obtained as a result of exposing our substrate to an air suspension of $A\ell_2O_3$ particles of characteristic size 0.3μm. The dendritic appearance is reminiscent of aggregates seen in diffusion dominated growth experiments.

substrates indicate single particles and a distribution of small clusters ($\lesssim$ 20 particles) are present. The shutter could be opened for a selectable time so as to allow various surface coverages to be created.

We report here third sound results at T = 1.64 K for a substrate dusted with $2.3 \times 10^4/(mm)^2$ particles of nominal size 0.3μm as compared with results from the nearly identical substrate which remained clean. As a function of film thickness the time of flight ratio $n = \tau_d/\tau_c$ has the value 1.24 ± 0.02 for $d_4 \lesssim 7$ layers and increases dramatically but smoothly for $d_4 > 7$ layers (Fig. 2). We conclude that $d_4 > 7$ represents the region of capillary condensation and below this thickness the $A\ell_2O_3$ merely affects the third sound propagation path length. A plot of d^4/τ^2 is linear at low coverages and is consistent with this interpretation since $C_3^2 \cong (\rho_s/\rho)(3\alpha d^{-4})(d-D)(1+TS/L)^2$. Here ρ_s/ρ is the bulk superfluid fraction, α the vander Waals constant, D an empirical constant, S the entropy and L the latent heat. In Fig. 3 the linear behavior of d^4/τ^2 as a function of d for $d \lesssim 7$ is apparent with D = 2.81 ± 0.10 for both clean and dusted substrates. This is in excellent agreement with the empirical value 2.82 deduced from Rudnick's measurements on clean substrates. A simple model calculation which presumes the capillary condensed region for d > 7 both slow down the third sound as they thicken and shift the path length as they grow with film thickness agrees with the time of flight data to < 5% over the full thickness range $4 < d_4 < 20$ layers.

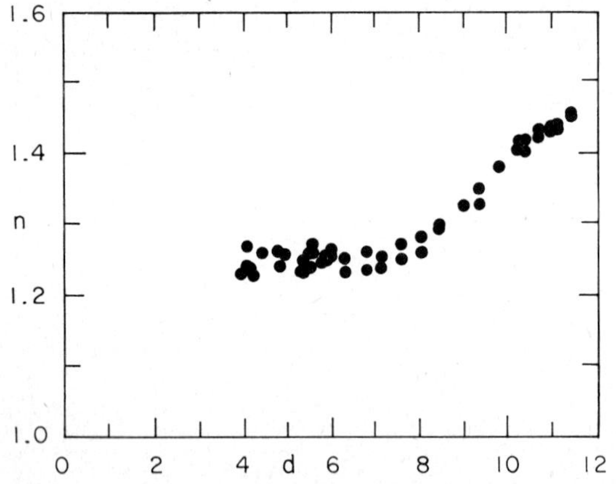

FIGURE 2
The time of flight ratio $n = \tau_d/\tau_c$ as a function of film thickness. The ratio is essentially constant for $d_4 < 7$ atomic layers.

Preliminary results for the third sound pulse width have also been examined. The presence of the capillary condensation for $d_4 > 7$ layers appears to broaden the pulses substantially. We presume this to be an effect related to pulse propagation although we cannot rule out a capillarity effect at the driver; to-date the entire substrate has always been dusted. No pulse broadening effects of this sort have been observed on the clean substrate. We have not yet looked into a model for the pulse broadening.

ACKNOWLEDGEMENTS
This work was supported by the National Science Foundation (DMR79-09248).

REFERENCES
(1) V. Ambegaokar, B.I. Halperin, D.R. Nelson, and E.D. Siggia, Phys. Rev. B21, 1806 (1980).
(2) I. Rudnick, Phys. Rev. Lett. 40, 1454 (1978).
(3) M. Chester and L.C. Yang, Phys. Rev. Lett. 31, 1377 (1973); D.J. Bishop and J.D. Reppy, Phys. Rev. Lett. 40, 1727 (1978).
(4) J. Maps and R.B. Hallock, Phys. Rev. Lett. 47, 1533 (1981) and Phys. Rev. B26, 3979 (1982); G. Agnolet, S.L. Teitel, and J.D. Reppy, Phys. Rev. Lett. 47, 1537 (1981).
(5) See, however, the mass flow measurements on rough Alumina by M.Z. Shoushtari and K.L. Telschow, Phys. Rev. B26, 4917 (1982).

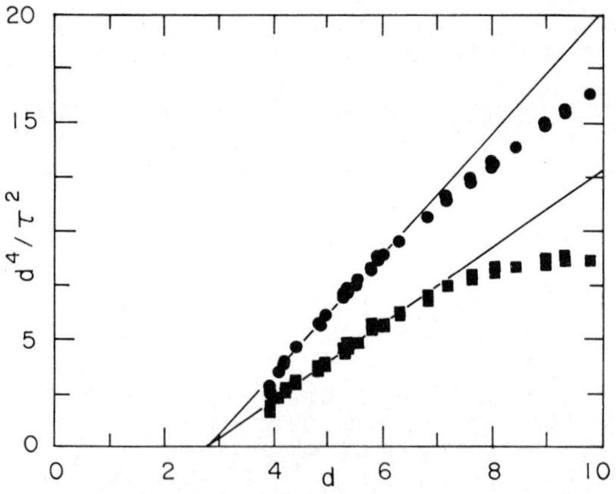

FIGURE 3
The quantity d^4/τ^2 as a function of the film thickness, d. The presence of $A\ell_2O_3$ powder does not appear to affect the thickness of the immobile layer since the intercept $d^4/\tau^2 = 0$ occurs at the same value d = D in both the clean and dusted cases. Clean = ● · ($d \equiv d_4$).

LT-17 (Contributed Papers)
U. Eckern, A. Schmid, W. Weber, H. Wühl (eds)

STUDIES OF ^3He-^4He MIXTURE FILMS BY THIRD SOUND RESONANCE

R.M. HEINRICHS and R.B. HALLOCK

Laboratory for Low Temperature Physics, Department of Physics and Astronomy, University of Massachusetts, Amherst, MA, USA 01003

^3He-^4He mixture films are studied by the technique of third sound resonance at $T \leq 0.3$ K for coverages $0 < d_3 < 2$ layers at $d_4 = 5.3$ layers. Results for the resonance frequency are consistent with a bilayer model for the film with the ^4He closest to the substrate; the resonance Q shows interesting and unexplained structure, some of which is presented here.

1. INTRODUCTION

We are investigating the properties of ^3He-^4He mixture films by means of measurements of the frequency and damping of third sound using a resonance technique. The system is of interest due to the apparent bilayer nature (1,2) of the films, the potential for two dimensional phase transitions (3) in the ^3He layer (e.g., phase separation, a 2D liquid-gas transition or a surface reconfiguration transition) and the unexplored damping in such situations. Such films seem to represent a two dimensional Fermi fluid with properties which are apparently tunable by changes in the temperature, ^3He coverage (where $\xi = 1$ atomic layer presumes a surface density $\sigma = 6.45 \times 10^{14}$ ^3He atoms/cm^2) and ^4He thickness, d_4.

2. TECHNIQUE

Our measurements are conducted in a pancake shaped resonator (4) where the third sound is driven thermally (0.1-10 nW) and detected capacitively. The detection sensitivity expressed in terms of ^4He film thickness is 0.001 Å; extremely small changes in surface coverage can be seen. Typical third sound amplitudes on resonance are $\delta d_4 \lesssim 0.02$ Å. To collect data the resonator is driven and the frequency slowly stepped (typ 1-50 x 10^{-5} Hz/sec) through the resonance. The capacitive film thickness monitor forms part of an LC circuit which operates near 21 Mhz. The LC frequency is mixed with a fixed frequency reference signal and phase lock loop techniques allow measurement of both the in-phase and quadrature amplitudes. Computer fits to the data then reveal the center frequency and Q of the resonance.

3. RESULTS AND DISCUSSION

Early time of flight measurements on ^3He-^4He mixture films (1) conducted in this laboratory suggest that the ^3He resides primarily on top of the ^4He for 0.3 K < T < 0.6 K and the data were in reasonable agreement with a simple bilayer model for the film. More recent work conducted in the resonator at d_4 = 5.6 atomic layers and T = 0.1 K shows, perhaps surprisingly, even more impressive agreement with the simple model (2). These later measurements revealed substantial structure in the resonance Q as a function of ^3He coverage. The present report explores this structure in more detail as a function of both ^3He coverage ($0 < \xi < 2$) and temperature (0.06 < T < 0.3 K) for d_4 = 5.3 atomic layers. Further work at d_4 = 4.0 atomic layers is underway.

Figure 1 illustrates the resonance Q as a function of ^3He coverage for T = 0.10 and

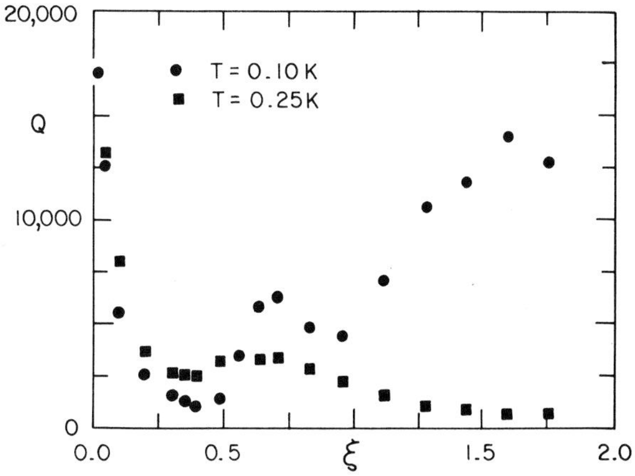

FIGURE 1
The resonance Q as a function of the ^3He coverage ξ for two temperatures.

T = 0.25 K. The most remarkable feature in the data is the dramatic rise in the dissipation in the system near a ^3He coverage of 0.35 layer. The data at T = 0.1 K are in general agreement with earlier work at d_4 = 5.6 layers. The reason for this enhanced damping is still not clear to us. The temperature dependence of the dissipation changes qualitatively in the vicinity of ξ = 0.35. This is illustrated in Fig. 2 where the dissipation is shown as a function of 1/T. At intermediate coverages, $\xi \simeq 0.35$, the dissipation falls with an increase in temperature. At very low coverages the data tends in the opposite direction and at high coverages and temperatures appears to obey an empirical rule (5) of the general form $Q^{-1} = A+B\exp(-C/T)$ and for $\xi \gtrsim 1.2$ we observe $C \simeq 1.4$ K, weakly dependent on coverage ξ. The parameter C appears to grow smaller as the ^4He coverage increases.

It is not yet clear what relationship these results have to the recent heat capacity measurements on ^3He-^4He mixture films (6) or the dramatic shifts in velocity reported on such films by Laheurte and his co-workers (7). We do observe anamalous velocity changes in the vicinity of T = 0.15 K but only for relatively thick ^3He coverages (5). It is interesting to compare the conditions explored by the various experiments and we illustrate this in Fig. 3 When completed, our measurements at d_4 = 4 atomic layers should provide interesting overlap with these other experiments.

ACKNOWLEDGEMENTS

This work has benefited from numerous conversations with R.A. Guyer. Financial support has been received from the National Science Foundation through DMR79-09248.

REFERENCES

(1) F.M. Ellis, R.B. Hallock, M.D. Miller, and R.A. Guyer, Phys. Rev. Lett. 46, 1461 (1981).
(2) F.M. Ellis and R.B. Hallock, Phys. Rev. B29, 497 (1984).
(3) A.N. Berker and D.R. Nelson, Phys. Rev., B19, 2488 (1979); J.L. Cardy and D.J. Scalapino, Phys. Rev. B19, 1428 (1979).
(4) F.M. Ellis and R.B. Hallock, Rev. Sci. Instrum. 54, 751 (1983).
(5) R.M. Heinrichs and R.B. Hallock, 75th Jubilee Conference on ^4He, ed. J.M. Armitage (North Holland, 1983), in press.
(6) B. Bhattacharyza and F. Gasparini, Phys. Rev. Lett. 49, 919 (1982).
(7) J.P. Laheurte, J.C. Noiray, and J.R. Romagnan, J. Phys. (Paris) Lett. 42, L197 (1981).

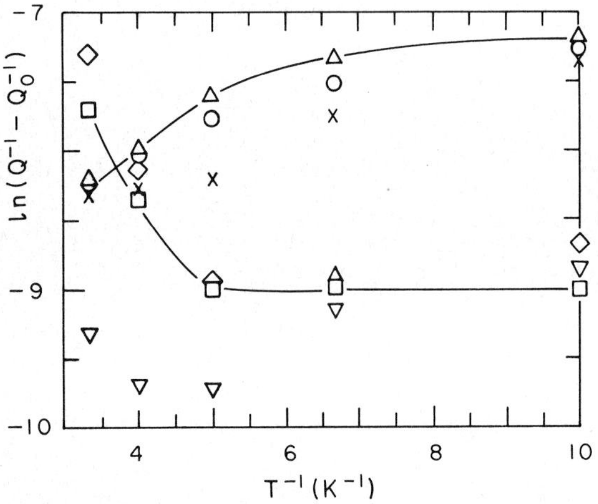

FIGURE 2
The excess dissipation $\ln(1/Q-1/Q_0)$ as a function of T^{-1}. Here Q_0 is the temperature dependent Q for pure ^4He with no ^3He added to the apparatus. Here ▽, X, △, 0, □ and ◇ refer to ξ = 0.102, 0.300, 0.350, 0.487, 0.708, and 0.830 respectively.

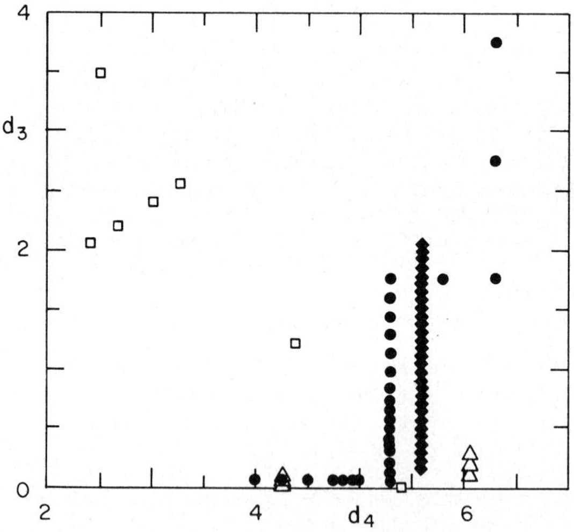

FIGURE 3
A guide to the $d_3 \equiv \xi$, d_4 coordinates studied in this work ● and in refs. (2) ◆, (6) △ and (7) □.

LT-17 (Contributed Papers)
U. Eckern, A. Schmid, W. Weber, H. Wühl (eds)

TWO-PHASE SOUND IN ^{3}He-^{4}He MIXTURES AT LOW TEMPERATURES

J.P. DESIDERI and J.P. LAHEURTE

Laboratoire de Physique de la Matière Condensée*, Université de Nice, Parc Valrose, Nice, France

H. DANDACHE

Université Libanese, Faculté des Sciences, Hadath-Beyrouth, Lebanon

G.A. WILLIAMS†

Physics Department, University of California, Los Angeles, California, USA

The velocity and attenuation of the two-phase sound modes in ^{3}He-^{4}He mixtures are studied for the case of low temperatures, T<1K. As the temperature is lowered the mode initially having the velocity of second sound drops toward zero velocity, while the upper mode velocity drops from the vapor velocity to a value intermediate between the second sound and vapor sound.

The hydrodynamic boundary conditions at the liquid-vapor interface of ^{3}He-^{4}He mixtures were recently formulated (1) and applied to the propagation of low-frequency two-phase sound. Including the diffusive current of ^{3}He resulting from the evaporation-condensation process at the free surface gives a good description of the experimental results (2) at 1.2K. The theory was able to describe the crossover from the single mode in ^{4}He to the two modes in ^{3}He-^{4}He mixtures as the ^{3}He concentration increased. In this paper we show that a similar crossover behavior should occur for a fixed ^{3}He concentration as the temperature is lowered below 1K. It appears that at least a portion of the crossover region may be accessible to experiments, and this could provide a measurement of the thermodynamic and diffusive parameters characterizing the two-phase sound.

We start with the dispersion relation obtained previously (1) (3)

$$k_{2x}^2+k_{vx}^2\,\frac{u_6^2}{u_2^2}\left(\frac{H-L}{L}\right) = -\,\frac{\beta(1-i)H-L}{\omega^{3/2}}\,k_{2x}^2k_{vx}^2 \qquad [1]$$

with $k_{2x}^2=\dfrac{\omega^2}{u_2^2}-k_z^2$ and $k_{vx}^2=\dfrac{\omega^2}{u_v^2}-k_z^2$

where u_2 is the second sound velocity, u_v the vapor sound velocity, and L the height of the liquid phase in the acoustic waveguide of total height H. u_6 and β are quantities depending on the thermodynamic and transport properties of the ^{3}He-^{4}He mixture, and are given in Eqs. 13 and 15 of Ref. (1).

As in Ref. (1) we set $k_z=k+i\alpha$, where the phase velocity is $u=\omega/k$, and α is the attenuation per unit length. Eq. 1 can then be written in a more convenient dimensionless form if we set $x=u/u_v$, $R=u_v/u_2$, $a=\beta(H-L)k^{1/2}/u_v^{3/2}$, and $b=u_6^2(H-L)/u_2^2L$, yielding

$$x^2R^2-(1+i\tfrac{\alpha}{k})^2 + b[x^2-(1+i\tfrac{\alpha}{k})^2] + \frac{a(1-i)}{x^{3/2}}\,[(x^2R^2-(1+i\tfrac{\alpha}{k})^2)(x^2-(1+i\tfrac{\alpha}{k})^2)]=0 \qquad [2]$$

This equation simplifies considerably for the case b=0, which would be the limit at very low temperatures where u_6 becomes small. One solution is then undamped second sound $u^2=u_2^2$, while the other solution is obtained by solving

$$x^3+16a^2(x^2-1)+8a\,x^{3/2}=0 \qquad [3]$$

and $$\frac{\alpha}{k}=\frac{x^{3/2}}{4a} \qquad [4]$$

The numerical solutions of these equations for u/u_v and α/k are shown as the dashed curves in Figs. 1 and 2. For large values of a it is easily seen that Eq. (3) gives $u/u_v\to 1$ and $\alpha/k\to 0$. In the opposite limit of small a the phase velocity approaches zero as $u/u_v\to[4(\sqrt{2}-1)a]^{2/3}$, and the attenuation approaches the constant value $\alpha/k=\sqrt{2}-1=0.41$.

Also shown as the solid lines in Figs. 1 and 2 are the solutions of the full Eq. 2 using the value $R=u_v/u_2=2.26$ appropriate to a 2% molar concentration in the liquid. These solutions are similar to those of Eqs. 3 and 4 except in the region $a\approx 0.2$ where there is strong "mode repulsion" as the modes try to cross. The lower mode then varies from zero to u_2 as the quantity a is increased. Since u_6 is finite for finite b, the upper mode at small a starts at the velocity given by

$$u^2 = u_2^2 \frac{\left[1 + \left(\frac{H-L}{L}\right)\frac{u_6^2}{u_2^2}\right]}{\left[1 + \left(\frac{H-L}{L}\right)\frac{u_6^2}{u_v^2}\right]}$$

which is just the same as for pure ^{4}He (4) except that now u_2 and u_6 are those appropriate to the mixture.

The attenuation shown in Fig. 2 arises from the ^{3}He diffusion, and does not include the attenuation from the evaporation at the surface (4), which can be minimized if the mode frequency is quite low. In such a case the upper mode should propagate over nearly the entire range of a, as shown in Fig. 2 (a). The quality factor of the mode is $Q=k/2\alpha$, and for b=0.3 the Q is always greater than its minimum value of $Q\approx3$ at $a\approx0.2$. The lower mode, however, will only propagate for $a>0.2$, since the Q of the mode becomes of order 1 at this point.

The two experimental points in Fig. 1 are from Ref. (2), at a molar ^{3}He concentration of 2%, and correspond to the values a=8.9 and b=0.93 at T=1.21K. Since a and b both decrease with decreasing temperature (and decreasing H-L) some portion of the crossover region should be accessible to experiments below 1K, limited only when the atomic mean free path in the vapor exceeds the waveguide dimension H-L and the vapor sound no longer propagates.

REFERENCES

*Laboratoire Associé au C.N.R.S.

†Supported in part by the NSF, Grant DMR 81-00218

(1) J.P. LaHeurte, G.A. Williams, H. Dandache, and M. Zoaeter, Phys. Rev. B28, 6585 (1983).
(2) G.A. Williams, R. Rosenbaum, H. Eaton, and S. Putterman, Phys. Lett. 72A, 356 (1979).
(3) Eq. 1 corrects an error in Eq. 12 of Ref. (1), where the minus sign on the right-hand side of Eq. 1 was inadvertently omitted.
(4) G.A. Williams, J. Low Temp. Phys. 50, 455 (1983).

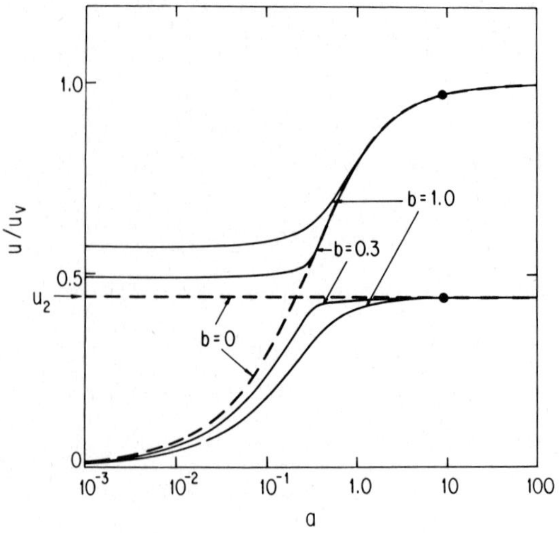

FIGURE 1
Velocities of the two phase modes as a function of the parameter a, for several different values of b.

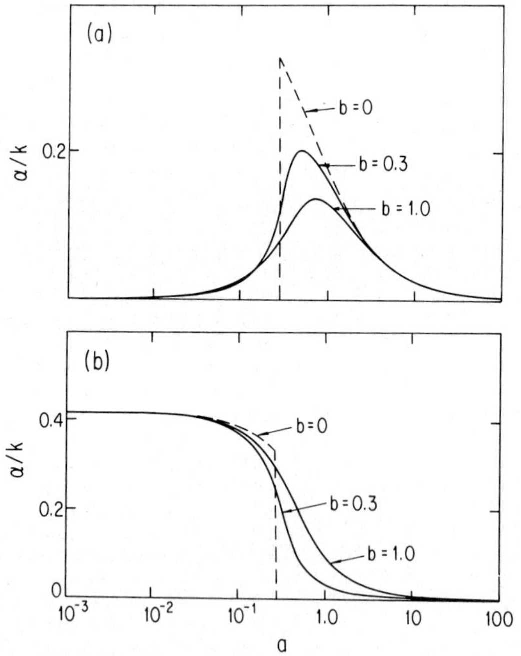

FIGURE 2
The diffusive attenuation coefficient α/k for the (a) upper mode and (b) lower mode for the same parameters as in Fig. 1.

LT-17 (Contributed Papers)
U. Eckern, A. Schmid, W. Weber, H. Wühl (eds)

THE TRANSVERSE ACOUSTIC IMPEDANCE OF ^{4}He ABOVE THE CRITICAL TEMPERATURE

M.J. LEA, D.S. SPENCER and P. FOZOONI

Department of Physics, Bedford College, University of London, Regent's Park, London NW1 4NS, U.K.

Measurements of the transverse acoustic impedance of ^{4}He at 20.5 MHz were used to obtain the viscosity from 0 to 2.7 bar at 5.239 K ($\varepsilon = (T-T_c)/T_c \cong 10^{-2}$). The results show the transition to non-hydrodynamic behaviour at low pressures and the effects of van der Waals forces in producing density gradients in the critical region.

We present here measurements of the transverse acoustic impedance $Z = R - iX$ of fluid ^{4}He just above the critical temperature T_c at pressures P up to 2.7 bar. The impedance was measured using a quartz crystal resonator as in previous work (1,2). The helium reduces the resonant frequency $f(=\omega/2\pi)$ and the quality factor Q of the loaded resonator from their values at P=0. R and X can then be found from

$$R = \tfrac{1}{4}n\pi\, R_q\, [Q^{-1}(P) - Q^{-1}(0)] \qquad (1)$$
$$X = \tfrac{1}{2}n\pi\, R_q\, [f(0) - f(P)]/f(0)$$

where n is the harmonic number of the resonance and R_q is the acoustic impedance of the quartz. The resonant frequencies were corrected for the small linear pressure dependence of the parameters of the quartz crystal itself. Measurements were made at the third harmonic at a frequency of 20.5 MHz. The temperature was stabilised to ±0.001 K at 5.239 K, 49 mK above the critical temperature at 5.190 K (on the T_{58} scale). The results for $R(P)$ and $X(P)$ are shown in Fig. 1. Note the rapid increase in R and X near the critical region and that $X > R$ at all pressures.

The transverse acoustic impedance of a hydrodynamic viscous fluid is

$$Z = R - iX = (1-i)(\pi f\eta/\rho)^{\frac{1}{2}} \qquad (2)$$

where η is the viscosity and ρ is the density. We have used the equation of state for $\rho(T,P)$ given by McCarty (3) to derive an effective viscosity

$$\eta' = R^2/\pi\rho f \qquad (3)$$

from the data in Fig. 1, and η' is plotted as a function of density in Fig. 2. For $\rho > 20$ kg m^{-3} $\eta'(\rho)$ increases almost linearly with density and an extrapolation back to ρ=0 gives a value $\eta'(0) = 13.5 \pm 1.0$ μP. This is in good agreement with the low density limit from Betts (4) of 13.55 μP at 5.239 K.. As $\rho \to 0$ the effective viscosity at 20.5 MHz decreases rapidly to zero, rather than a finite value, because of the transition to the non-hydrodynamic regime ($\omega\tau > 1$) as the collision time τ increases. From the kinetic theory of an ideal gas $\tau = K\eta/P$ where K is a factor close to unity. Thus $\omega\tau = 0.02$ at 20.5 MHz at a pressure of 0.1 bar. We have found previously (1) that the equation

$$Z = \frac{(1-i)(\eta\rho\pi f)^{\frac{1}{2}}}{1 + (1-i)\beta\ell/\delta} \qquad (4)$$

gives a good account of the transition from hydrodynamic to non-hydrodynamic behaviour, where $\ell = v\tau$ is the mean free path of the atoms with mean speed v, $\delta = (\eta/\rho\pi f)^{\frac{1}{2}}$ is the viscous penetration depth and η is the hydrodynamic viscosity. For an ideal gas $\ell/\delta = (3\omega\tau/2)^{\frac{1}{2}}$. If $\beta = 4/3\alpha$ then eq.(4) gives the correct limits for Z in both the hydrodynamic ($\omega\tau \ll 1$) and ballistic ($\omega\tau \gg 1$) regimes. The fraction of atoms which are diffusely scattered at the surface of the resonator, α, is the only free parameter. We have calculated the effective

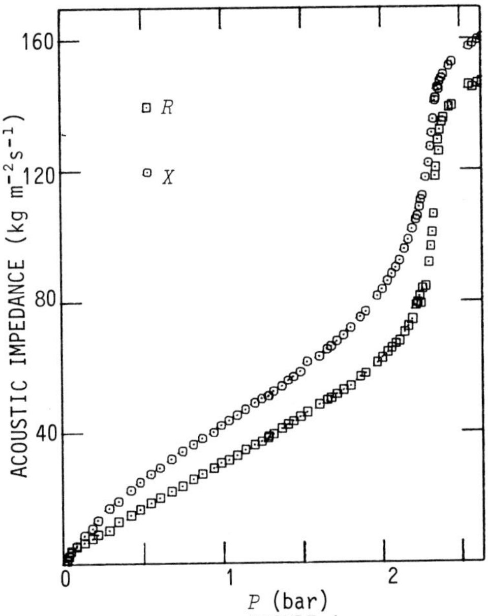

FIGURE 1
The transverse acoustic impedance $Z = R - iX$ of ^{4}He at 5.239 K

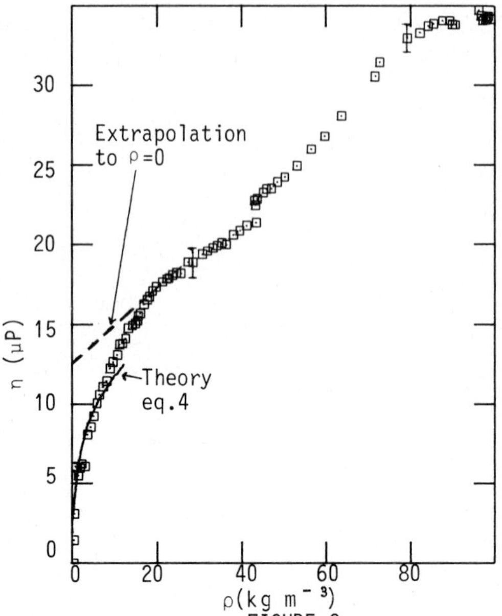

FIGURE 2
The effective viscosity, η', of ^{4}He

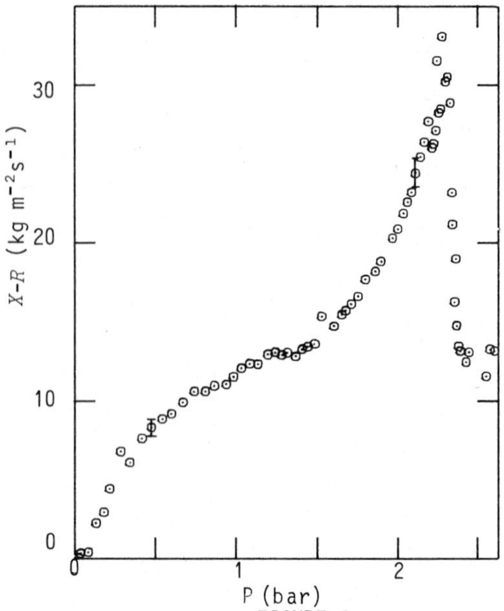

FIGURE 3
X-R for the data in Figure 1

viscosity η' at 20.5 MHz as shown in Fig. 2 for $\alpha = 0.2$ which gives the best fit. This value is similar to that found for ^{3}He quasiparticles in superfluid ^{4}He below 0.1 K (2) and this suggests that the diffuse scattering is due to surface roughness.

Another region of interest is near the critical point (P_c = 2.274 bar). Goodwin et al. (5) measured η with a vibrating wire viscometer at 4.5 kHz and obtained a value of 24.8 μP for ρ=67 kg m^{-3}at $\varepsilon = (T-T_c)/T_c = 10^{-2}$. For $\varepsilon < 10^{-2}$ they found a small critical anomaly in η along a near-critical isochore. Our value for η' at this density is 29±2 μP at $\varepsilon = 0.94\times10^{-2}$where the increase due to the critical point should be small. We are now investigating the critical anomaly in η at 20.5 MHz for $\varepsilon < 10^{-2}$. Non-local effects may be important since Lo and Kawasaki (6) have shown that for a wavenumber q the effective viscosity is

$$\eta_q = \eta(T)(1-F(q\xi)) \qquad (5)$$

where ξ is the correlation length. For our experiment $q \cong 2\pi/\delta$ where $\delta \cong$ 24 nm near the critical point. At $\varepsilon = 10^{-2}$, $q\xi \cong 1$ but $F(q\xi)$ is only 0.02 and the correction is negligible.

Also van der Waals forces will increase the density and viscosity of the fluid close to the crystal resonator. If the fluid at a distance x from the crystal is locally hydrodynamic then the change in acoustic impedance ΔZ is given by (1)

$$\Delta Z = -i\omega \int_0^\infty [\rho(x) - \rho(\infty)\eta(\infty)/\eta(x)]dx \qquad (6)$$

if $\Delta Z \ll Z$ and if the range of the forces $\ll\delta$. Thus van der Waals forces increase X, but not R, and X-R, as plotted in Fig.3, is a measure of the excess density and viscosity close to the crystal. As the pressure is increased, X-R exhibits a large critical anomaly where $(\partial\rho/\partial P)$ is large. This effect shows again that weak forces, such as gravity or van der Waals forces, can produce significant inhomogeneities near the critical point of a fluid.

ACKNOWLEDGEMENTS

We would like to thank Professor E.R. Dobbs for his encouragement and support; A.K. Betts, F. Greenough, F.W. Grimes and A. King for technical assistance and the SERC (UK) for a Research Grant and a Studentship (for D.S.S.).

REFERENCES

(1) M.J. Lea, P. Fozooni and P.W. Retz, J.Low Temp.Phys. 54 (1984) 303.
(2) M.J. Lea and P.W. Retz, Physica 107B (1981) 225.
(3) R.D. McCarty, J.Phys.Chem. Ref. Data 2 (1973) 923.
(4) D.S. Betts, Cryogenics 16 (1976) 3.
(5) J.M. Goodwin, A. Harai, T. Mizusaki, K. Tsuboi and T.K. Waldschmidt, Physica 107B (1981) 351.
(6) S-M. Lo and K. Kawasaki, Phys.Rev.Lett. 29 (1972) 48.

LT-17 (Contributed Papers)
U. Eckern, A. Schmid, W. Weber, H. Wühl (eds)

SPHERICAL SECOND SOUND SHOCK WAVES

John R. TORCZYNSKI

Graduate Aeronautical Laboratories, California Institute of Technology, Pasadena, CA 91125

Spherically converging second sound shock waves are produced in liquid helium II. Experimental measurements of the time of flight and temperature jump at the sensor agree well with weak shock theory. The relative velocities are the largest ever produced in bulk fluid (around 10 m/sec). We show that phenomena limiting the shock strength occur at the heater, not in the bulk fluid.

1. INTRODUCTION

Unique to liquid helium II is second sound[1], a temperature-relative velocity ($\vec{w} = \vec{v}_n - \vec{v}_s$) wave with acoustic velocity

$$a_2{}^2 = (\rho_s\, s^2 T)/(\rho_n c_p). \tag{1}$$

When the amplitude of a second sound wave is large, the characteristic velocity is a function of the local temperature perturbation[2].

$$u(\Delta T) = a_2\,[1 + B(\Delta T/T)] \tag{2}$$

$$B = T\left(\frac{\partial}{\partial T}\right)_p \ln\,(a_2{}^3\, c_p/T)\,. \tag{3}$$

Osborne first observed shock formation caused by the nonlinear wave speed[3]. More recently, Turner used second sound shocks to study the fundamental critical velocity[4]. The relative velocity jump is given by

$$\Delta w/a_2 = (\rho c_p/\rho_s s)(\Delta T/T), \tag{4}$$

and the Mach number is found by averaging the characteristic velocities on both sides of the shock[5].

$$M = u_s/a_2 = 1 + \tfrac{1}{2}\,B\,(\Delta T/T) \tag{5}$$

Turner found that weak shocks obey Eq. (5) very well (see Fig. 3). However, for large input heat fluxes, the data diverge significantly from the predicted values. Turner attributes this to exceeding the fundamental critical velocity (superfluid breakdown).

2. WEAK SPHERICAL SECOND SOUND SHOCKS

Spherically converging second sound shocks may be used to study the fundamental critical velocity. A converging shock tube causes a shock to strengthen as it propagates, so a given input heat flux produces a larger $\vec{w}$ than in a straight channel. Moreover, converging experiments would show whether nonsteady heat transport is limited by conditions at the heater or out in the bulk fluid where the shock is stronger.

A weak wave in a channel has a strength inversely proportional to the square root of the area . For a conical channel this becomes

$$\frac{M(r) - 1}{M(r_o) - 1} = \left(\frac{A(r_o)}{A(r)}\right)^{1/2} = \frac{r_o}{r}\,, \tag{6}$$

where r is measured from the cone's vertex. If a shock wave travels a distance L down a converging conical channel, Eq. (6) may be integrated to yield

$$\langle M\rangle = \frac{L}{a_2 t_a} = 1 - (M_o - 1)\,\frac{r_o}{L}\,\ln\left(1 - \frac{L}{r_o}\right), \tag{7}$$

where t_a is the arrival time and M_o and r_o refer to the initial conditions. Eqs.(5) and (6) may be used to recast Eq. (7) into a form containing the experimentally measured quantities, t_a and the final temperature jump ΔT_L.

$$\langle M\rangle = \frac{L}{a_2 t_a} = 1 + \tfrac{1}{2}\,B\left(1 - \frac{r_o}{L}\right)\ln\left(1 - \frac{L}{r_o}\right)\frac{\Delta T_L}{T} \tag{8}$$

It should be emphasized that these relations retain validity only so long as $(M(r) - 1) \ll 1$.

3. APPARATUS

The converging channel (see Fig. 1) is contained within a brass housing. The shock "tube" is a conical teflon channel with half-angle 15°, with an area contaction of 3:1 and a length L = 3.14 cm. The edges of the channel are normal to the heater, a Nichrome thin film vacuum-deposited atop a plano-concave lens with radius of curvature r_o = 2.70 cm. Shock pulses are produced by sending a rectangular current pulse

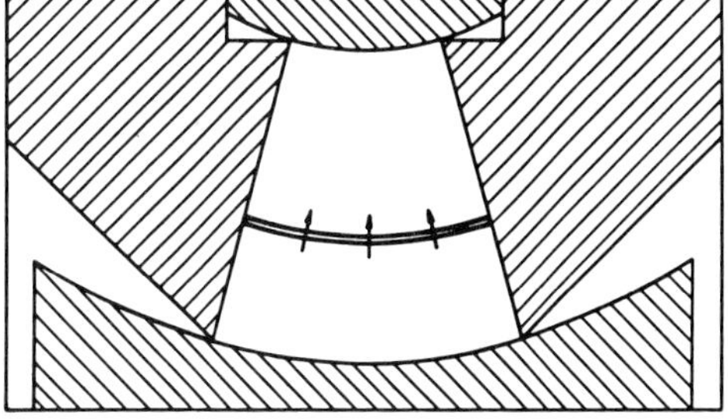

Fig. 1. The spherically converging shock tube.

through the thin film (Joule heating). Detection of these waves is accomplished by a thin film tin-on-gold superconducting sensor[6] vacuum-deposited on a plano-convex lens. Spring loading keeps the parts in contact.

4. EXPERIMENTAL METHOD AND RESULTS

Converging shock experiments are run as follows. A thermodynamic point (p_0, T_0) was selected, and a series of shock pulses are fired. For each pulse t_a and ΔT_L are measured. Three experiments were performed at 1.605 K, 1.571 K, and 1.463 K. In all cases the pressure was the saturated vapor pressure. Figure 2 shows the results of these experiments with the right side of Eq. (8) plotted against the left side. These curves are typical of those in straight channel experiments except that the Mach numbers, temperature jumps, and relative velocities here are much higher. Eq. (8) is remarkably well obeyed until there is an abrupt divergence from the prediction (breakdown). Shown in Table 1 are values associated with the maximum ΔT_L in each run. Mach numbers and relative velocities (in m/sec) from the Khalatnikov solution[2] and from calculations by Sturtevant and Moody[7] (SM) are given for comparison.

Table 1. Strong Converging Shocks

T_0	ΔT_L	M_L(Khal)	M_L(SM)	w_L(Khal)	w_L(SM)
1.605	0.071	1.098	1.148	6.0	7.6
1.571	0.076	1.115	1.182	6.5	8.3
1.463	0.087	1.161	1.302	7.6	10.7

One purpose of these experiments was to see where breakdown would first occur. To study this, it is instructive to compare values of M_0 and ΔT_0 in the above experiments (calculated from Eqs. (5) and (6)) with values from straight channel experiments at the same thermodynamic conditions. If breakdown occurs at the heater, these two data sets will lie almost atop one another. Figure 3 shows that this is indeed the case (note the different scales in Figs. 2 and 3). Hence, breakdown occurs principally at the heater.

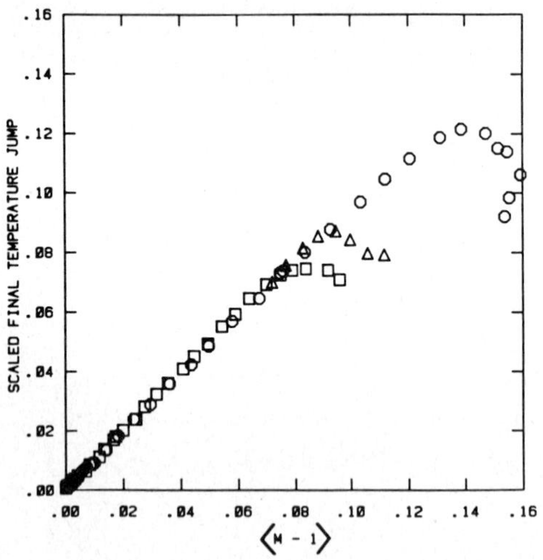

Fig. 2. The scaled final temperature jump is plotted against the average Mach number for experiments at 1.605 K_0(□), 1.571 K (**o**), and 1.463 K (Δ). The 45° line is the prediction.

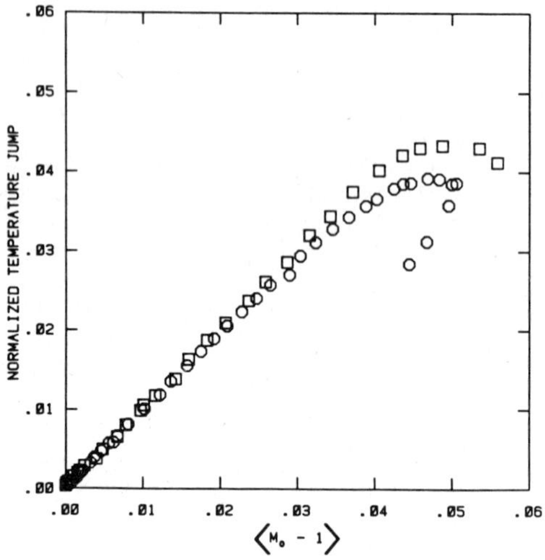

Fig. 3. Calculated conditions at the heater in a converging experiment at 1.605 K (□) are compared with results of an earlier straight channel experiment at 1.609 K (**o**).

5. CONCLUSIONS

Second sound shocks strengthen as they propagate into a conical channel. Relative velocities produced are the largest ever observed in bulk superfluid (around 10 m/sec). Calculation of conditions at the heater shows that the breakdown of superfluidity occurs in this region. After the temperature profile has been produced, it travels without further anomalous decay. Thus, the difficulty is getting the heat into the helium rather than transporting it by counterflow.

REFERENCES

(1) L. D. Landau and E. M. Lifshitz, Fluid Mechanics (Pergamon Press, Oxford, 1959).
(2) I.M. Khalatnikov, An Introduction to the Theory of Superfluidity (W.A. Benjamin, Inc. New York, 1965).
(3) D.V. Osborne, Proc. Phys. Soc. (London), 64 (1951) 114.
(4) T.N. Turner, Phys. Fluids, 26 (1983) 3227.
(5) G.B. Whitham, Linear and Nonlinear Waves (John Wiley & Sons, New York, 1974).
(6) G. Laguna, Cryogenics, 16 (1976) 241.
(7) B. Sturtevant and D.M. Moody, Phys. Fluids, in print.

LT-17 (Contributed Papers)
U. Eckern, A. Schmid, W. Weber, H. Wühl (eds)

FLOW VISUALIZATION OF SHOCK WAVES IN LIQUID HELIUM II

John R. TORCZYNSKI, Dagmar GERTHSEN, Thomas ROESGEN

Graduate Aeronautical Laboratories, California Institute of Technology, Pasadena, CA 91125

A schlieren system is used to study second sound shocks. Reflection of double-shock pulses from boundaries is observed. Second sound shocks obliquely incident on the liquid-vapor interface produce transmitted gasdynamic shocks and reflected second sound shocks. Firing strong shocks in rapid succession generates disturbances of undetermined nature.

1. INTRODUCTION

Second sound has been used by many investigators to study superfluid helium. Most studies have emphasized local (point) or average measurements. Schlieren photography, however, yields global information useful in problems involving curved nonsteady waves and wave interactions with boundaries or other waves. In such situations, schlieren reveals the flow geometry and in some cases can uncover new phenomena.

2. APPARATUS

Schlieren is a standard technique to map the gradient of the refractive index n. Since variations in the density ρ are proportional to changes in n,

$$\nabla n = [(n-1)/\rho]\nabla\rho = -(n-1)\ \beta\nabla T$$

where β is the coefficient of thermal expansion. Hence, second sound shocks (∇T is large) are seen with schlieren[1].

A shock tube with optical windows, developed by Turner[2] generates second sound shock waves (see Fig. 1). A heat flux is produced by sending a top-hat voltage profile through a planar Nichrome thin film heater located at the bottom of the rectangular test section. A plane temperature wave emerges from the heater, traverses the test section heading upward, reflects from the top end wall, and travels back down the channel. The temperature pulse bounces back and forth until it has decayed below observable levels. A 1 μsec spark source provides the illumination.

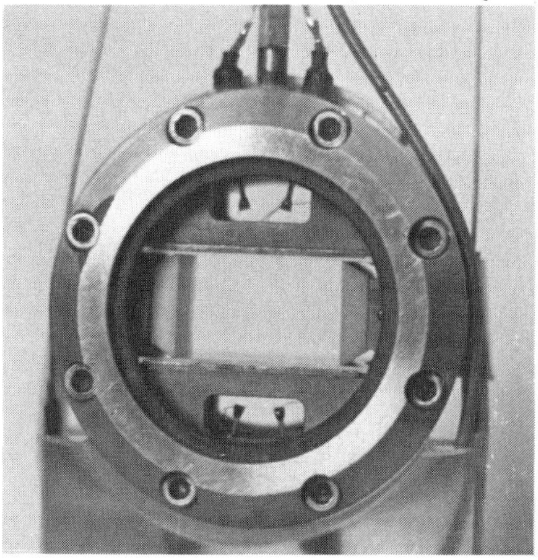

Figure 1. The optical shock tube.

3. EXPERIMENTAL RESULTS

3.1 Double-shock profiles

At temperatures slightly below 1.884 K, a positive temperature pulse steepens at both the front and the back, producing a *double-shock* pulse, studied in detail by Turner[3]. Shown in Figure 2(a-f) is a double-shock pulse at successively later times. Ambient conditions are 1.877 K and 16.22 Torr (SVP). The pulses are 300 μsec long, and the temperature jumps are about 10 mK. In the first photograph, the leading shock front has emerged from the heater (the trailing shock will be produced 100 μsec later). The next two photographs show the temperature pulse propagating upward. After 900 μsec, the leading shock front (the upper *dark* line) has reflected from the top end wall. The change from light to dark underscores the fact that schlieren reveals gradients of n. After reflection, the shock is traveling in the opposite direction, so ∇n is also oppositely directed. In the last two pictures, the trailing shock has reflected from the upper end wall and appears as a bright line. Note that both fronts remain steep and planar, even after reflection.

3.2. Shock-interface problems

The interface separating superfluid helium from its vapor is interesting since not all of the boundary conditions holding across the interface are known when evaporation is taking place. Thus, when a known second sound shock is incident upon a liquid-vapor interface, the strengths of the transmitted and the reflected shocks *cannot be calculated*[4]. Shock strength measurements may shed some light on the boundary conditions.

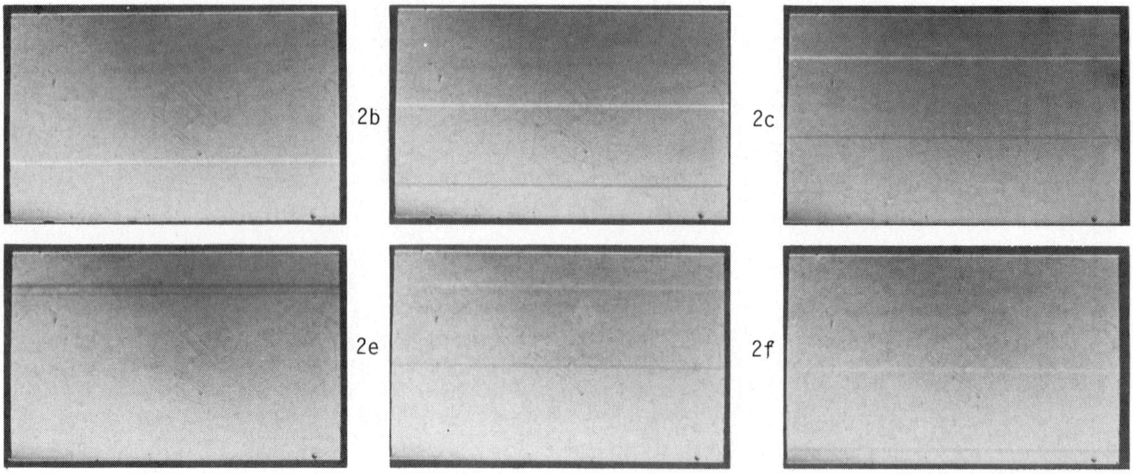

Figure 2. A double-shock pulse. Delay times are: (a) 200 μsec; (b) 400 μsec; (c) 600 μsec; (d) 900 μsec; (e) 1200 μsec; (f) 1500 μsec.

In Figure 3(a-b), the optical shock tube is rotated 12.5° about the optical axis and partially filled with superfluid helium, forming an interface (the dark band). Ambient conditions are 1.656 K and 7.21 Torr (SVP), and the picture was taken 200 μsec after the initial shock (M = 1.039) was fired (it is no longer in the field of view). A transmitted gasdynamic shock (M = 1.063) and a reflected second sound shock (M = 1.026) are produced. The Mach numbers are calculated from angle measurements and Snell's Law although convection of the reflected shock by the relative velocity set up by the initial shock must be considered - the reflected shock moves subsonically in the lab frame.

3.3. Fluctuations produced by shocks.

If several strong second sound shocks are fired, disturbances with vortical nature are generated[5]. Figure 4 is a schlieren picture of an experiment at 1.877 K and 16.22 Torr (SVP). Again, the heat pulses are of 300 μsec duration, producing shocks with 10 mk temperature jumps at 15 second intervals. The spark source is delayed 500 μsec, so the shock observed in Figure 4 is propagating upward in the cavity.

Note the appearance of mottling in place of the uniform gray background previously observed. This mottling appears before the shock (produced by previous shocks) as well as after it. It is likely that these fluctuations involve quantized vortices although the actual explanation is uncertain.

REFERENCES

(1) A.I. Gulyaev, Sov. Phys. JETP 30 (1970) 34.
(2) T.N. Turner, Ph.D. Thesis, California Institute of Technology (1979).
(3) T.N. Turner, Proc. of 16th Intl. Conf. on Low Temp. Physics, LT16B (1981) 701.
(4) T.N. Turner and D.M. Moody, Bull. Am. Phys. Soc. 26 (1981) 1255.
(5) J.R. Torczynski, Phys. Fluids, in print.

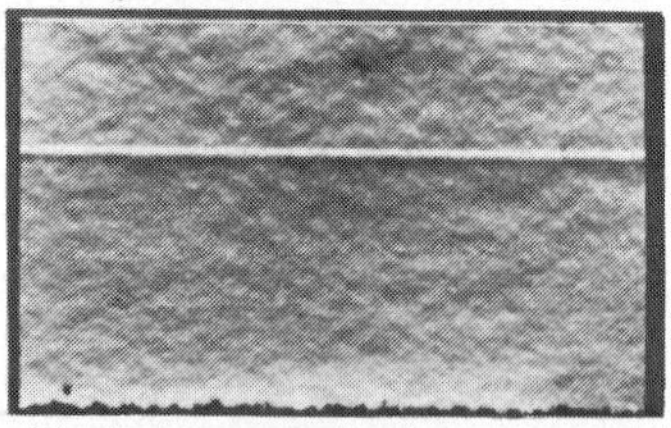

Figure 4. Firing several strong second sound shocks produces fluctuations.

3a

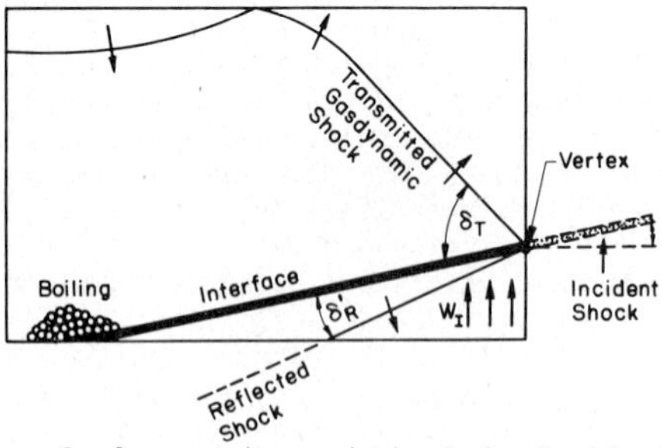

3b

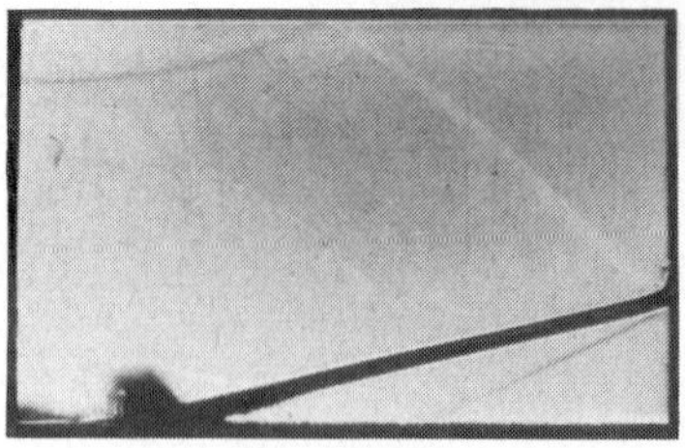

Figure 3. A second sound shock is incident upon the liquid-vapor interface.

LT-17 (Contributed Papers)
U. Eckern, A. Schmid, W. Weber, H. Wühl (eds)
© Elsevier Science Publishers B.V., 1984

SECOND SOUND SCATTERING USING FOCUSING CAVITIES

Thomas ROESGEN

Graduate Aeronautical Laboratories, California Institute of Technology, Pasadena, CA 91125

Two experiments are described which use a focusing geometry to study the scattering of second sound in liquid helium II. A double elliptical cavity was used for scattering from a small wire target, whereas a parabolic/cylindrical setup was designed to observe scattering from purely fluid mechanical disturbances, for example a counterflow jet.

1. INTRODUCTION

Scattering of second sound can be used to investigate flow phenomena in liquid helium II. Compared to normal acoustic scattering, this technique has certain advantages. In many cases, scattering amplitudes are proportional to some positive power of the wave number (1). Thus, for a given frequency, the small speed of second sound means larger scattering amplitudes. Also, the available detectors are very sensitive and have a wide band frequency response. We describe two experiments that make extensive use of focused second sound. In this way, scattering can be used nonintrusively to probe small regions of fluid away from walls.

2. SCATTERING FROM A SOLID TARGET

The scattering of second sound from a cylinder (12.7 μm diameter copper wire) is observed in a double elliptical cavity as shown in Figure 1. The configuration is a 1.25 cm deep channel whose side walls are two confocal ellipses, each with a half axis ratio of 0.9. The piece was machined out of brass on a numerically controlled milling machine with 12.7 μm step size. Surfaces generated this way are smooth on the scale of the second sound wavelengths used here. A thin film Nichrome heater deposited on a quartz cylinder at the left focus produces the incident wave. A granular aluminum transition-edge bolometer is mounted in a similar fashion at the right focus. The target is inserted in the center, where the focal areas of the sensor and heater overlap. Due to the arrangement of the thin films (see Fig. 1), only the side-scattered component from the target is observed.

The waveform used in the experiments is a sinusoidal burst of second sound with 200 μsec duration, which is much shorter than the time of flight for a round trip (6.59 msec). The quality of the cavity is high enough to observe several round trips of the incident pulse. Figure 2 shows a typical waveform received at the sensor, here signal-averaged for noise reduction. With a wavelength of 112 μm, this is an example of long wavelength scattering ($ka = .356$, with a: cylinder radius, k: wave number). No coherent signal similar to the one in Figure 2 could be seen when the experiment was repeated with the target removed.

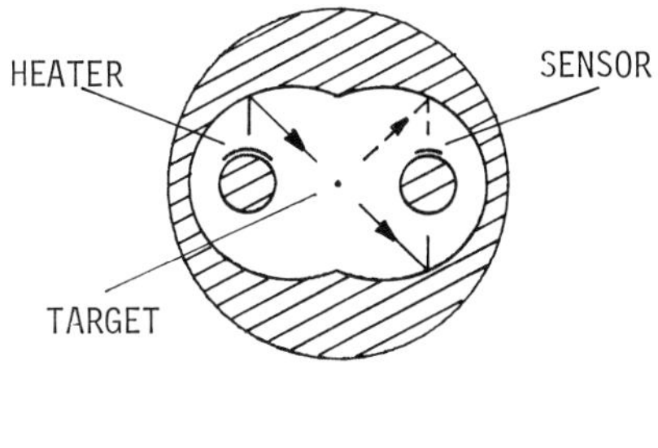

Figure 1. Schematic of the elliptical cavity; the arrows indicate the propagation direction of the incident (—) and scattered (--) wave.

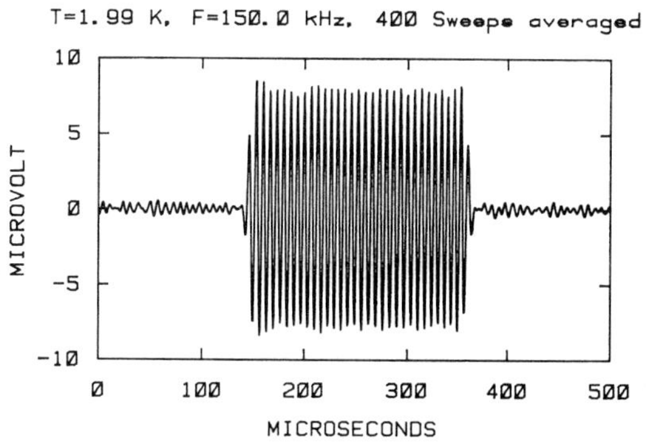

Figure 2. Scattering signature from a wire target. Signal averaging (400 sweeps) and band-pass-filtering were used to reduce the noise.

3. SCATTERING FROM FLUID DISTURBANCES

A similar philosophy is used in the cavity design depicted in Figure 3. Here, three sensors are located in the parabolic wings of the cavity, whereas the heater is formed by a cylindrical section in order to preserve uniformity of the wave amplitude in the focus. The cavity is 2.54 mm deep and was machined the same way as its elliptical predecessor.

Several advantages of this design should be pointed out: both forward and side scattering can be observed, allowing cross-correlation of the two side-scattered signals. The spatial selectivity is very high and restricts the region of maximum sensitivity to approximately two second sound wavelengths. The sensor areas are planar and allow the use of large area (large sensitivity) phase-matched detectors.

In the present experiment, aluminum bolometers were chosen, with resistances of several Megohms and sensitivities of up to 1500 volts/Kelvin. Photolithography was used to produce sensors with a 4 μm feature size and an active area of 2.3 mm x 1.65 cm. Specially designed cryogenic buffer amplifiers immersed in the liquid are incorporated to match the sensors and the output line impedances.

Preliminary observations have been made of the forward scattering from a counterflow jet created by a second thin film heater (0.1 mm x 0.1 mm) placed in the central focal area.

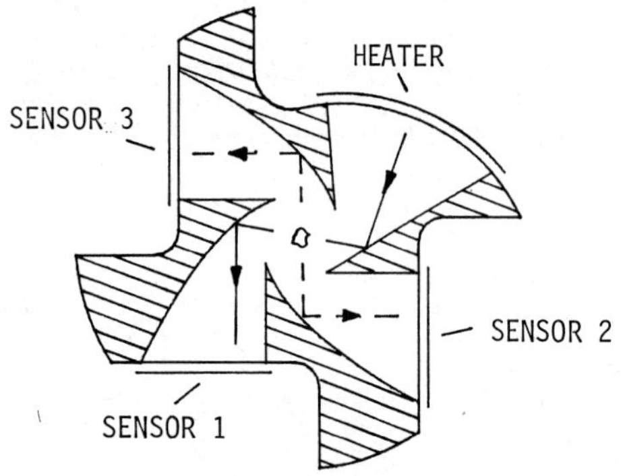

Figure 3. Schematic of the parabolic cavity; the arrows indicate the propagation direction of the incident (—) and side-scattered (--) waves.

ACKNOWLEDGMENTS

The author wishes to thank Prof. Dr. H.W. Liepmann and Dr. J.R. Torczynski for many helpful discussions. This research was supported by NSF Grant No. MEA82-00027 and Caltech funds.

REFERENCES

(1) P.M. Morse and K.U. Ingard, Theoretical Acoustics (McGraw-Hill, New York, 1968).

LT-17 (Contributed Papers)
U. Eckern, A. Schmid, W. Weber, H. Wühl (eds)

SHOCK WAVES OF STRONG SECOND SOUND IN He II

A.Yu.Iznankin, L.P.Mezhov-Deglin

Institute of Solid State Physics USSR Academy of Sciences
Chernogolovka, Moscow district USSR

A thermal bolometer has been employed to investigate the pulse shapes of first and second sounds, emitted by a heater in He II. At great powers ($W \gtrsim 20\ W/cm^2$, $\tau_0=10\mu ks$) the second sound wave shape was not coincident with the theoretical one. In this case an intensive first sound was emitted. Its appearance can be related to the formation of a gas film on the heater.

We have studied the shape of thermal pulses of first and second pulses emitted by a heater in superfluid helium. Helium is known to be a strongly nonlinear medium, therefore at sufficiently great powers one can observe shock waves of second sound, fig.1. The temperature in the second sound shock wave varies, according to the theory, like $\Delta T=\rho_s u_{20} v_n/\rho_n S$, before the formation of the velocity triangle profile

$$v_{n_0} = W/\rho S T \sim W \qquad (1)$$

and after the formation of the velocity triangle profile:

$$v_n = (2 v_{n_0} \tau_0/\alpha t_0)^{1/2} \sim W^{1/2} \qquad (2)$$

where α is nonlinearity coefficient t_0 is heater-receiver flight time, u_{20} is velocity of linear second sound, ρ_n and ρ_s are densities of normal and superfluid components, ρ , S, T are density, entropy and temperature of the liquid.

Fig.1 shows that starting with a certain power the shape of the observed pulse deviates from that predicted by the theory. It is to be noted that such deviation was observed in a number of works, published recently [1,2,3]. Normally, these attenuations are observed for a "trial" pulse propagating in the trace of a strong "exciting" pulse. In this work we present the pulse shapes recorded in the stroboscopic repetition regime, i.e., each one pulse is, simultaneously, a "trial" for the foregoing and "exciting" for the subsequent start. By varying the delay time between the "exciting" and "trial" pulses in the single start regime one can determine for how long the "trial" pulse feels the "exciting" one. It appears that this time is about 20s [3]. Concurrent with the experiments by Vinen [4] for constant heat fluxes it is assumed that a strong heat pulse, passing in helium, gives rise to vortices which are effective scattering centers for the second-sound waves. However, like in Vinen's experiments, the mechanism of the vortices generation remains obscure.

Our thermal bolometer, together with the second-sound waves, could register the first-sound waves, fig.2. The starting weak heating wave, $\Delta T>0$, (since in helium the compressibility coefficient is less than zero, corresponding to the heating wave is the rarefaction pulse $\Delta P<0$) was followed by a strong cooling ($\Delta T<0$) - compression ($\Delta P>0$) wave. It is known that a heater placed in a superfluid helium emits, together with an intensive heat wave, a very weak density wave (because of a small difference of the compressibility coefficient from zero). In this wave $\Delta P<0$. We measured the time t_1 needed for the beginning of the compression wave formation $\Delta P>0$, fig.2. It appears that this time is described well by the expression

$$W t_1^{0.5\pm0.1} = C \qquad (3)$$

Here $C \simeq 4\cdot10^{-2} W\ s^{1/2}/cm^2$ for $T<1.8$ K and it decreases as T_λ is approached. These dependences are obtained for the times of helium boiling on the heater in normal helium [5]. In these experiments $C=1.7\cdot10^{-2}\ W\ s^{1/2}/cm^2$ since simultaneously with a decrease of the second sound pulse amplitude its trailing edge is hanging-over in time, fig.1. This might be attributed to a decline of the heat exchange between the helium and the heater. In a simplest model the heat exchange between the heater and the helium is cut off completely in time τ_1 after the pulse has been fed to the heater, (for instance a gas film or a dense vortex tangle is formed). So, an effective pulse of duration τ_1 rather than τ_0 is emitted into helium. According to (2)

$$A_1/A_0=(\tau_1/\tau_0)^{1/2} \qquad (4)$$

Fig.2 presents thus calculated τ_1 is dependent on W like $\tau_1 W^{1.6\pm0.1}$=const Fig.2 shows that at first helium starts boiling on the heater (curve t_1), and, consequently, the second-sound pulse shape distorts (curve τ_1). It is, however, unlikely that the gas bubbles can exist in the superfluid helium longer than about 20s. Probably, the most effective mechanism of the vortex formation is the liquid disturbance at the moment of the bubble formation on the heater. The vortices formed worsen the heat exchange between the helium and the heater. By comparing the dependences between the boiling times $W t_1^{0.5} = 4\cdot10^{-2}\ W\cdot s^{1/2}/cm^2$ and the empiric equations for the turbulization, times [4] $W^{1.5} t = 5\cdot10^{-2}\ W^{3/2} s/cm^3$ one can assume that in Vinen's experiments the main vortex generator was the heater surface for $W \sim 1\ W/cm^2$.

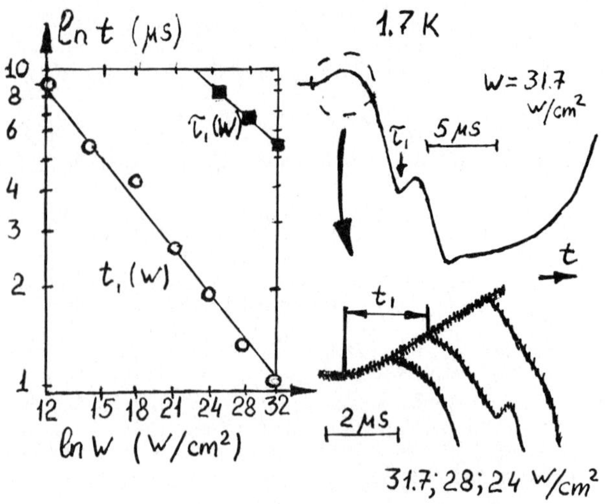

Fig.2 Shape of the observed first-sound pulses. The dependence of the time of the compression wave formation t_1 - (o) and the effective duration of the emitted second-sound pulse τ_1 - (■) on the power.

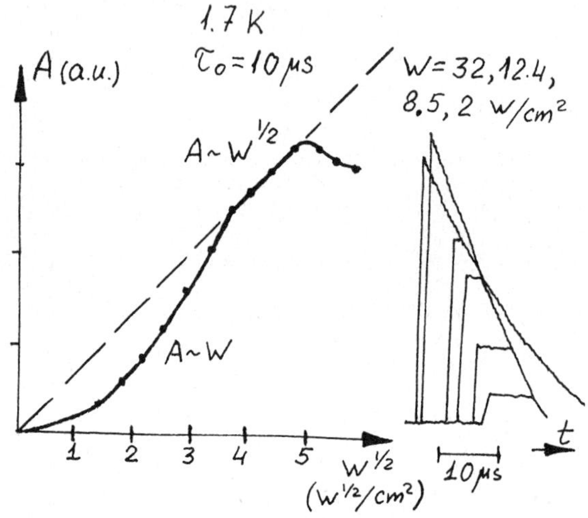

Fig.I Shape of the observed second-sound pulses and their dependence on the power. The heature measures I I cm^2 The sourse to receiver distance is I.Icm. The repetition frequency is 40 Hz.

REFERENCES

(1) J.S.Cummings, D.W. Schmidt, W. Wagner. Phys. Fluids 21 (1978) 713.
(2) S.K. Nemirovsky, A.N. Zoy. JETPh Letters (Russian) 35 (1982) 229.
(3) A.Yu.Isnankin, L.P. Mezhov-Deglin. JETPh (Russian) 84 (1983) 1378.
(4) W.F. Vinen. Proc.Roy.Soc., A240 (1957) 114.
(5) B.A.Danilchenko, V.N. Poroshin. Cryogenics 23 (1983) 546.

LT-17 (Contributed Papers)
U. Eckern, A. Schmid, W. Weber, H. Wühl (eds)

A SILICON FOURTH SOUND RESONATOR

Ken DALY and Richard PACKARD

Physics Department, University of California, Berkeley, CA 94720, USA

We have built a powderless fourth sound resonator for superfluid He-3 from silicon (Si) wafers using integrated circuit technology. The transducers are boron doped Si membranes with a displacement sensitivity of 5Å/μbar in vacuum. The resonator will be used to measure normal mode frequencies and Q's in the limit where the quasiparticle mean free path is long.

Fourth sound is a pressure-density oscillation which can be excited in a superfluid when the normal fluid component is kept stationary (1). The typical fourth sound resonator is packed with powder to viscously lock the normal fluid (2). The presence of the powder complicates the interpretation of the observed sound. In superfluid He-3, since the normal fluid viscosity is much larger than He-4, it is possible to construct a resonator of sufficiently small dimensions to lock the normal fluid without utilizing a powder. This was done by Kojima, Paulson and Wheatley (3) who constructed a resonator from a stack of closely spaced epoxy plates. Although these measurements were the first detection of fourth sound in superfluid He-3, they suffered from complex geometry; it was not possible to fully understand the mode structure or measure the dissipation.

Recently fourth sound Q's and normal mode frequencies have been calculated (4) for superfluid He-3. At the low temperatures required for superfluidity in He-3 the quasiparticle mean free path can be large compared to the geometrical dimensions of a resonator. We intend to measure fourth sound Q's and normal mode frequencies in this limit.

We have observed fourth sound in a resonator using an epoxy body and transducers made from aluminized teflon. This design lacked the necessary geometrical accuracy to understand the normal modes seen.

We believe the silicon (Si) resonator illustrated in the figure is characterized by a simple and well defined geometry. It is built using micro-machining techniques (5). The resonator is made from two Si wafers clamped together along their polished surfaces.

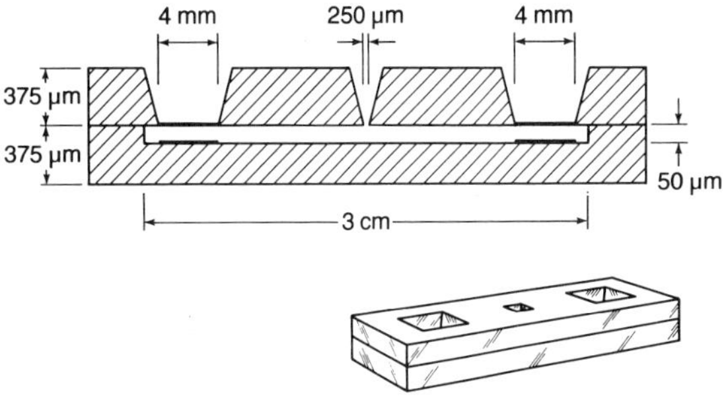

Figure 1. Cross sectional and perspective views of the Si resonator (not to scale).

A cavity is chemically etched into one wafer to a depth of 50 microns. The etch depth varies by 3% over the 3.0 cm by 0.55 cm lateral dimensions of the cavity. The roughness of the etched surface is about 1000Å.

The transducers for creating and detecting sound in the cavity are made from Si membranes approximately one micron thick etched into the second Si wafer. The membrane thickness is controlled by the depth of boron diffused into the polished surface of the wafer. Due to the presence of the smaller boron atoms in the Si lattice, the membrane is in tension; the measured resonant frequency of 0.4 cm by 0.4 cm membranes is 41 kHz in vacuum.

A pair of parallel plate capacitors is formed by evaporating about 1000Å of aluminum on each membrane and on the opposite surface at the bottom of the cavity. Each capacitor is biased at about 100V. An AC signal applied to one capacitor causes the membrane to oscillate, producing a pressure oscillation in the fluid filling the cavity. The second capacitor is configured as a standard capacitive microphone; the pressure oscillations in the fluid displace the membrane causing a capacitance change which can be measured.

A fill hole is etched through the upper Si piece. He-3 in the fill hole is the thermal link to the 700Å silver powder heat exchanger of a nuclear adiabatic demagnetization refrigerator of conventional design. We estimate the thermal time constant for cooling of normal He-3 in the cavity through the fill hole to be $0.6T^2$ s/mK^2 and the time constant for thermal equilibration within the cavity itself to be $2.8T^2$ s/mK^2.

These silicon resonators have been tested in He-4. The signal to noise observed for the fundamental mode of first sound at 1.45K is about 200 for a drive of 7V p-p, bandwidth of 10Hz and bias of 100V on both transducers. The Q of this mode was about 50. Conventional commercial electronics (PAR 113 preamplifier and HP 3580 spectrum analyzer) were used for this measurement. An estimate of the microphone sensitivity is 10μV/μbar.

An alternate detection scheme which promises greater signal to noise is based on an inductive pickup and a SQUID preamplifier (6). A coil is deposited on the membrane and a superconducting ground plane is deposited on the opposing surface at the bottom of the cavity. Oscillation of the membrane with respect to the superconducting surface modulates a persistent current in the coil which is detected by the SQUID. With a persistent current of 100mA we estimate the SQUID will enhance the signal to noise by a factor of 100 relative to the capacitive system.

The apparently ideal geometry of the resonator is complicated from an acoustical perspective by several factors. Fluid in the centrally located fill hole provides a mass-like load on the fluid in the cavity. This shifts the even harmonics of an ideal closed resonator to higher frequency and has little effect on the odd numbered modes. A simple low frequency model gives the fractional frequency shift as,

$$\frac{\Delta f_n}{f_n} = \frac{A^2 \ell}{\pi^2 S V n^2} = \frac{0.55}{n^2}$$

where A is the fill hold cross sectional area, l the length of the resonator, S the cross sectional area of the resonator, V the fill hole volume, and n (an even integer) the mode number. For the small resonator cross sectional area necessary here, these shifts are substantial.

The acoustical impedance of the membranes is small; this means that in spite of appearances the resonator is not closed. The membranes yield sufficiently to pressure oscillations in the fluid that the volume velocity is not zero at the ends of the cavity. We find that the acoustical impedance of the membrane must be large compared to the acoustical impedance of the cavity to have a closed resonator. This condition is expressed by

$$\frac{2\pi m}{f A^2} (f_0^2 - f^2) >> \frac{\rho c}{S}$$

where the additional variables are: m, the membrane mass; f, the frequency; f, the membrane resonant frequency; ρ, the density of the fluid; and c, the sound speed in the fluid. Here again the small cross sectional area of the cavity makes this condition difficult to satisfy.

Another difficulty for the proposed use of the resonator is the relatively large volume behind the membranes in which the normal fluid will not be as strongly clamped by its viscosity. Any normal fluid motion will reduce the Q's of the fourth sound resonances which we plan to measure.

Partial solutions to these problems are available. One can increase the effective stiffness of the membranes by decreasing the membrane dimensions or by making the volume behind the membrane small and relying on the incompressibility of the fluid itself. It may also be possible to eliminate the fill hole entirely, relying on the inevitable small gap between the two wafers for filling and thermal contact.

We wish to acknowledge John Clarke for the suggestion of the SQUID system. We are indebted to Dietrich Einzel for his calculation of fourth sound Q's and normal mode frequencies. Also the instruction and cooperation of all users and staff of the "Micro lab," especially Roger Howe, Kiyo Uozumi, Raif Hijab, Greg Lee and Dave Haas is gratefully acknowledged.

References

(1) K.R. Atkins, Phys. Rev. 113, (1959), 962.
(2) K.A. Shapiro and I. Rudnick, Phys. Rev. 137, (1965), A1383.
(3) H. Kojima, D.N. Paulson, and J.C. Wheatley J. of Low Temperature Physics, 21, (1975), 283.
(4) H.H. Jensen, H. Smith, and P. Wolfle, J. of Low Temperature Physics, 51, (1983), 81; D. Einzel, private communication.
(5) Kurt E. Petersen, Proceedings of the IEEE, 70, (1982), 420.
(6) H.J. Paik, J. of Applied Physics, 47, (1976), 1168.

LT-17 (Contributed Papers)
U. Eckern, A. Schmid, W. Weber, H. Wühl (eds)

NEW MECHANISM OF RELAXATION IN SUPERFLUID HELIUM KINETICS

V.I.SOBOLEV and L.A.POGORELOV

Institute for Low Temperature Physics and Engineering, UkrSSR Academy of Sciences, Kharkov, USSR

The analysis of data on liquid ^{4}He viscosity shows that the kinetics of superfluid helium involves a new relaxation mechanism related to phonon absorption (emission) by rotons. The contribution of this mechanism to kinetic coefficients increases with pressure. Investigation of the temperature dependence of phonon mean free paths taking into account the above process revealed an anomalous roton dispersion at saturated vapor pressure.

Specific kinetics of superfluid ^{4}He is connected with the features of the dispersion relation and the nature of interactions between elementary excitations. These are conveniently examined through an analysis of experimental data on viscosity η. Here the measurement results on η in ^{4}He are reported for a wide range of pressures and temperatures at constant total density of the liquid. The method of an oscillating sphere was used (1). Experimental results are shown in Fig. 1: curve 1 for saturated vapor pressure, curves 2-5 for densities 0.1540 g/cm , 0.1614 g/cm , 0.1677 g/cm and 0.1721 g/cm , respectively. For visualization, the data in the right top corner are magnified, each curve being shifted upwards with respect to the previous one by 5 poise. These data were used to find the phonon mean free path l_{ph} from the relation (2)

$$\eta_{ph} = \eta - \eta_r = \frac{1}{5}\rho_{nph} c\, l_{ph} \qquad (1)$$

where η_{ph} and η_r are the phonon and roton parts of viscosity, respectively; ρ_{nph} is the normal phonon density, c is the sound velocity. η_r was assumed to be independent of temperature T and was found from η data at T from 1.1 K up to the η(T) minima. η_r increases monotonically with density ρ from 12.7 poise at saturated vapor pressure to 18.6 poise at ρ =0.1721 g/cm .

Processing of the l_{ph}(T) dependence calculated from Eq. (1) for the temperatures, at which phonon-roton processes are essential, shows a deviation

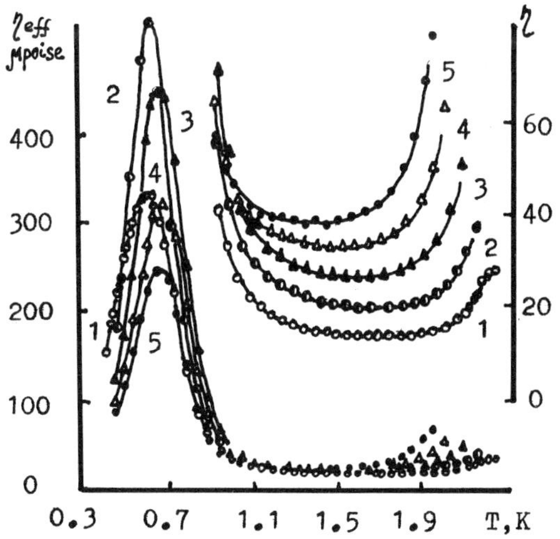

FIGURE 1

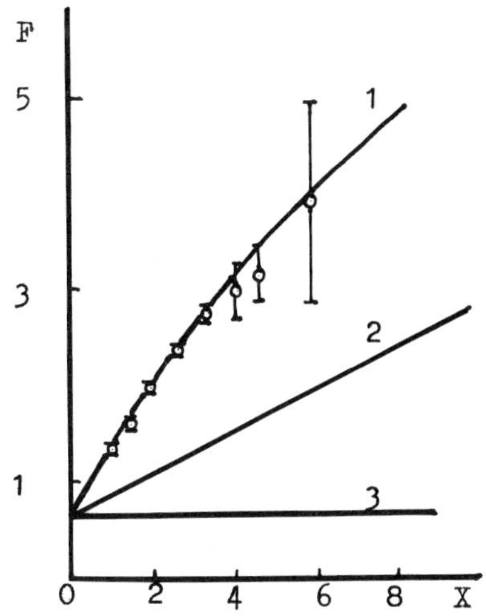

FIGURE 2

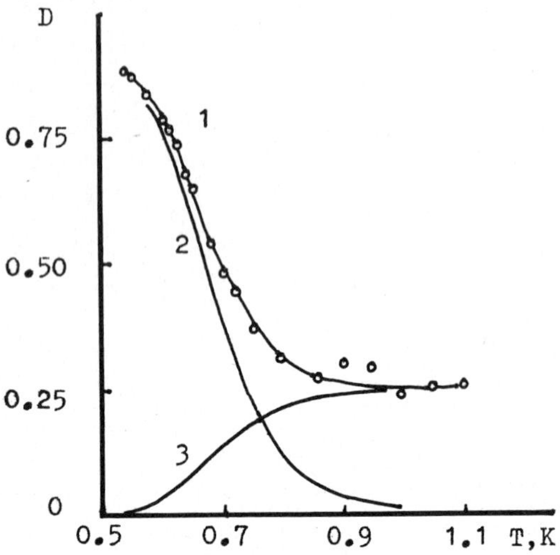

FIGURE 3

from theory (2) which considers only elastic four-particle processes leading to $l_{ph}^{-1} \sim \rho_{nr} T^5$, where ρ_{nr} is the normal roton density. This dependence, however, does not agree with experiment. The most essential discrepancy is observed at high densities. This is seen in Fig. 2, with $F=l_{ph}^{-1}/\rho_{nr} T^5$ on the ordinate and $x=\rho_{nr}/T^3$ on the abscissa. The points are found using Eq. (1) for the density 0.1721 g/cm, and curve 3 corresponds to theory (2).

It is thus necessary to take into account nonelastic processes, the three-particle one is the simplest of them. These processes related to roton energy smearing were considered earlier (3) and the formulae obtained were used to plot curve 2 (Fig. 2). It is seen that theory (3) does not describe experiment quantitatively. This is explained in Ref. (4) taking into account the maximum-velocity rotons, which are important to the process considered. The calculation (4) is in good agreement with experiment (curve 1), and the dispersion relation of rotons E(p) to the right of the minimum is

$$E(P)=E_m+b\,\mathrm{th}\, V_m(P-P_m)/b \quad , \qquad (2)$$

where E_m and P_m are the energy and momentum of a roton with the maximum velocity V_m, and $b=2\Delta-E_m$ (Δ is the roton gap). The parameters are as follows: $E_m = 10.3$ K, $P_m = 2.30$ Å^{-1}, $V_m = 0.77$ C.

A similar analysis under saturated vapor pressure shows that three-particle phonon - roton processes are also important in kinetics, though not as decisive as under elevated pressures. This is seen in Fig. 3 showing the dependence of the relative contribution of the three-particle process to l_{ph}^{-1}. Curves 2 and 3 describe phonon-phonon and phonon-roton collisions, curve 1 the total contribution of three-particle processes. These are essential at T=0.8-1.1 K, when their contribution is 25 to 30%. The following parameters of the dispersion relation, Eq. (2), were obtained: $E_m = 12.35$ K, $P_m = 2.25$ Å^{-1}, $V_m = 1.04$ C. E(P) containing the region with V > C permits an important conclusion: under saturated vapor pressure He II must experience a spontaneous decay of rotons emitting phonons (5). It is interesting to note that the observed anomaly and the anomaly of the phonon spectrum (6) seem to be related: the anomaly in the roton part leads to renormalization of the phonon spectrum; as a result, the phonon energy is increased. Complete renormalization must naturally include the phonon - phonon contribution.

REFERENCES

(1) B.N.Eselson, O.S. Nosovitskaya, L.A. Pogorelov and V.I. Sobolev, Physics B (1981) 41.

(2) L.D. Landau and I.M. Khalatnikov, Zh. Eksp. Teor. Fiz. 19 (1949) 637.

(3) I.N. Adamenko and V.I. Tsyganok, Zh. Eksp. Teor. Fiz. 82 (1982) 1491.

(4) L.A. Pogorelov and V.I. Sobolev, Fiz. Nizk. Temp. 9 (1983) 1222.

(5) L.P. Pitaevsky, Zh. Eksp. Teor. Fiz. 36 (1959) 1168.

(6) H.J. Maris, Rev. Mod. Phys. 49 (1977) 341.

LT-17 (Contributed Papers)
U. Eckern, A. Schmid, W. Weber, H. Wühl (eds)

UPPER CRITICAL FIELD OF FERROMAGNETIC SUPERCONDUCTORS

L.N. BULAEVSKII (a), A.I. BUZDIN (b), M.L. KULIĆ (c)

(a) P.N. Lebedev Physical Institute, Moscow, USSR
(b) Physics Department, Moscow State University, Moscow, USSR
(c) Institute of Physics, P.O.Box 57, Belgrade, Yugoslavia

Type of transition to the superconducting state in magnetic field (H) and near magnetic transition point T_M is studied. In $ErRh_4B_4$ monocrystalline sample the type of transition depends on demagnetizing factors. In spherical sample below 3.5K and down to 1K the transition into the intermediate state occurs. For samples in the form of disc and in perpendicular field, the inhomogeneous Larkin-Ovchinikov-Fulde-Ferrel state (LOFF) realizes at $H<H_{C2}$, and for T<5K.

1. EXPERIMENTAL RESULTS IN $ErRh_4B_4$

Crabtree et al. (1) have measured the upper critical field of monocrystalline ferromagnetic superconductors (FS) $ErRh_4B_4$ vs. $T_M \approx 1K$. For $\vec{H}$ along a-axis one gets bell-like curve with H_{max} near 5K, and critical field tends to zero as $T \to T_M$ or $T \to T_{C1} = 8.7K$. The measurements of magnetization (M) vs. T and internal magnetic field H shows that $ErRh_4B_4$ is type-II superconductor near T_{c1}, while the first-order transition from normal (N) to the Meissner (MS) state takes place near T_M. We calculate internal critical field $H_{C2}(T)$ for the formation of very small superconducting nuclei, and thermodynamical internal field $H_c(T)$ of the N-MS transition at $T>T_M$.

2. CRITICAL FIELD $H_{C2}(T)$

To calculate $H_{C2}(T)$ we take into account the orbital effect of magnetic induction (B) and paramagnetic effect of the exchange field (h (T)) of localized moments (LM's) acting on electrons. We consider clean system and Gruenberg-Günther theory (3), the exchange scattering of electrons being included by renormalization of superconducting order parameter. As $T \to T_M$ the susceptibility χ (of LM's), for $\vec{H}$ along a-axis, grows like $(T-T_M)^{-1}$ causing the increase of B and h. Thus $H_{C2}(T)$ diminishes as $T \to T_M$. In the compounds with $(h(o)/T_{C1})^2 >> 1$, $h(o) \equiv h(T=0)$, we get the bell-like curve $H_{C2}(T)$, and $H_{C2}(T) \sim (T-T_M)/h(o)$ near T_M. Below $T_o \approx 0.55\, T_{C1}$ the line $H_{C2}(T)$ separates the N- and LOFF-states.

3. THERMODYNAMICAL FIELD $H_C(T)$

The field $H_C(T)$ of N-MS transition falls off as $T \to T_M$, due to the increase of LM's polarization in the N-phase and diamagnetic character of the MS-state. $H_C(T)$ depends essentially on the demagnetizing factor N_Z. At T where M (of LM's) of the N-phase depends linearly on H at $H<H_C$, we get $H_C(T) \sim (T-T_M)[N_Z(1+N_Z)]^{-1/2}$. As a result, the type of superconducting transition near T_M depends on $a = 1.12 N_Z(1+N_Z)\theta_{em}/\theta_{ex}$, where $\theta_{ex} = h^2(o)N(o)$, $\theta_{em} = 2\pi n \mu^2$, N(o) is the density of states per one LM, μ is the ionic magnetic moment, and n is the concentration of LM's. θ_{em} and θ_{ex} are long-range contributions of the exchange and magnetic-dipolar interaction to the energy of ferromagnetic ground state (per one LM), respectively.

4. PHASE DIAGRAM IN H-T PLANE

At a<1 we get $H_C > H_{C2}$ and the first-order N-MS transition takes place, near T_M, in the internal field $H_C(T)$. For samples with $N_Z \neq 0$, the transition in an applied field (H_a) is actually continuous one, due to the formation of the intermediate state in the region $H_{a1}<H_a<H_{a2}$, where $H_{a1}=H_c(1-N_z)$ and $H_{a2}=H_c(1+4\pi N_z n\chi)$. Assuming that M~H, at $H<H_c(T)$, we obtain $H_{c2}(T)>H_c(T)$ for a>1. In this case N-LOFF transition occurs near T_M.

The shape of $ErRh_4B_4$ sample studied in (1), (2) was spherical with $\theta_{ex}\approx 1K$, $\theta_{em}\approx 1.5K$, and with M~H for $H<H_c$ at T>3K. The corresponding H-T phase diagram, coming out of our theory, is shown in Fig. 1, where VS is the vortex phase, and H_{c1} is lower critical field.

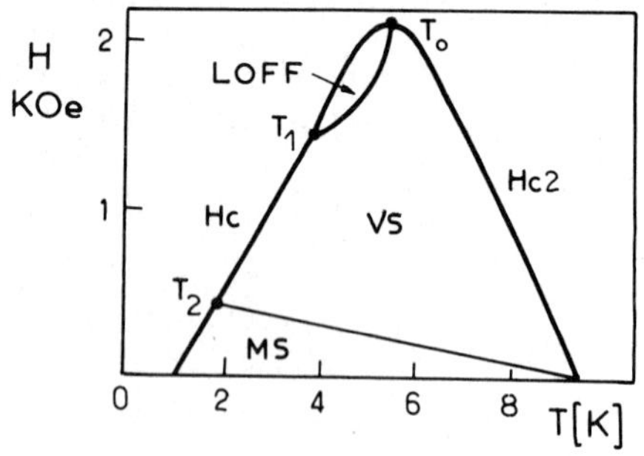

FIGURE 1
Phase diagram of spherical $ErRh_4B_4$ sample. The field is along a-axis.

Below $T_1=3.5K$ and $T_2=1.5K$ first-order transitions N-VS and N-MS occur respectively, while the intermediate state may be observed by magnetooptical and magnetic powder methods.

For the $ErRh_4B_4$ sample with $N_z=1$ (disc in perpendicular field) the phase diagram, coming from our theory, is quite different. Here the N-LOFF transition occurs at all temperatures below $T_o\approx 5K$, see Fig. 2.

Thus the FS monocrystalline sample gives the opportunity to study experimentally the LOFF state; it seems that it has never been observed up to now. The behaviour of polycristalline $ErRh_4B_4$ samples in $\vec{H}_a$ is more complicated, due to chaotic orientation of crystalites, with respect to applied field, and to strong magnetic anisotropy. Here, the superconducting transition in the field is a percolating one. $H_{c2}(T)$, in this case, for dirty systems has been calculated in (4).

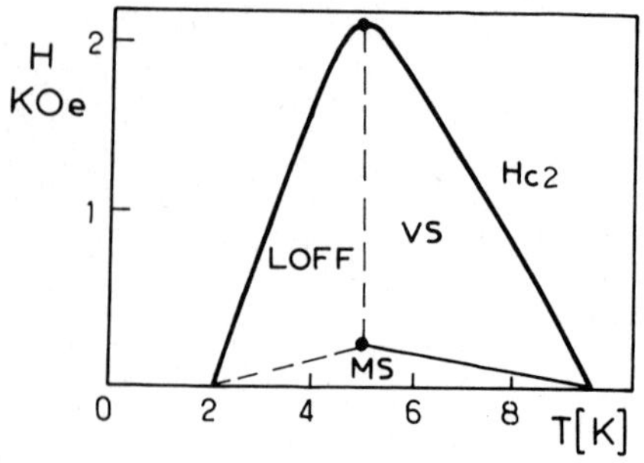

FIGURE 2
Phase diagram of $ErRh_4B_4$ in form of disc in perpendicular magnetic field. The dashed lines are unknown. The field is along a-axis.

5. CONCLUSION

So, the theory for H_{c2} and H_c explains the properties of spherical $ErRh_4B_4$ sample studied experimentally in (1) and (2). We predict quite different behaviour of disc sample in perpendicular magnetic field. Here, the LOFF state should be realized near T_M.

REFERENCES

(1) G.W. Crabtree, F. Behroozi, S.A. Campbell and D.G. Hinks, Phys. Rev. Lett. 49 (1982) 1342.
(2) F. Behroozi, G.W. Crabtree, S.A. Campbell and D.G. Hinks, Phys. Rev. B27 (1983) 6849.
(3) L.W. Gruenberg and L. Günther, Phys. Rev. Lett. 16 (1966) 996.
(4) L.N. Bulaevskii, A.I. Buzdin and M.L. Kulić, Sol. St. Comm. 41 (1982) 309, Phys. Lett. 85A (1981) 161.

LT-17 (Contributed Papers)
U. Eckern, A. Schmid, W. Weber, H. Wühl (eds)
© Elsevier Science Publishers B.V., 1984

A NEW STRUCTURE FOR FERROMAGNETIC SUPERCONDUCTORS

P. Stampfli and T. M. Rice

Theoretische Physik, ETH-Hönggerberg, 8093 Zürich, Switzerland

A new structure is proposed for the intermediate region between pure superconducting and pure ferromagnetic. It consists of planar Bloch walls and planar singularities in the superconducting order parameter which combine to lower the electromagnetic interference energies between the two states.

Experiments on $ErRh_4B_4$ have found a sinusoidal magnetic structure (1) in the transition region between pure superconductivity and pure ferromagnetism in agreement with theoretical predictions (2,3) but they also observed magnetic scattering at, or very near to, $Q=0$ (1) and in the thermal conductivity a reduction in the fraction of superconducting electrons in this region (4). It has been suggested that these results are due to macroscopic inhomogenities, stresses perhaps, which give coexisting purely ferromagnetic and purely superconducting domains (5,6). We would like to propose a new structure consisting of superconducting Bloch domain walls separated by planar singularities in the superconductivity. The electromagnetic interaction between the superconductivity and ferromagnetism is reduced in the new structure relative to the sinusoidal structure. The period of the magnetism is longer and the fraction of superconducting electrons is reduced. This new structure is a possible candidate to explain the experiments quoted above.

We will examine the free energy configurations of a ferromagnetic superconductor without an external magnetic field in a Ginzburg-Landau mean field model. The free energy density in appropriate units is

$$F(\psi,\vec{M},\vec{A}) = -|\psi|^2/2 + |\psi|^4/4 + |(\kappa^{-1}\vec{\nabla} - i\vec{A})\psi|^2/2 - \varepsilon M^2/2 + \varepsilon M^4/4 + \delta\sum_{i=1}^{3}|\vec{\nabla}M_i|^2/2 + \eta M^2|\psi|^2/2 + |\vec{B} - \mu\vec{M}|^2/2$$

with $\psi(\vec{r})$ as the complex superconducting order parameter, $\vec{M}(\vec{r})$ the magnetization and $\vec{A}(\vec{r})$ the vector potential corresponding to the magnetic field $\vec{B}(\vec{r})$. The ratio of the free energy of a purely ferromagnetic state to a purely superconducting one is given by the parameter, ε. Typically (e.g. $ErRh_4B_4$) the characteristic lengths of the superconductivity and magnetization are much smaller than that of the magnetic field so we can set $\vec{B} = \mu\vec{M}$. (This approximation will always lead to a free energy in a mixed structure which is too high.) In this way the vector potential $\vec{A}(\vec{r})$ is determined, modulo a gauge transformation, from the magnetization $\vec{M}(\vec{r})$.

It is convenient to write the complex order parameter $\psi(\vec{r})$ in terms of its amplitude $|\psi(\vec{r})|$ and phase $\phi(\vec{r})$ leading to gradient terms

$$|(\kappa^{-1}\vec{\nabla} - i\vec{A}(\vec{r}))\psi(\vec{r})|^2 = \kappa^{-2}|\vec{\nabla}|\psi(\vec{r})||^2 + |\psi(\vec{r})|^2|\kappa^{-1}\vec{\nabla}\phi - \vec{A}|^2.$$

The magnetic field $\vec{B}$ couples to ψ through the curl of a gauge-invariant effective potential, $A_{eff} = \kappa^{-1}\vec{\nabla}\phi - \vec{A}$. At vortices and planar singularities with $|\psi(\vec{r})| = 0$, $\vec{\nabla}\times\vec{\nabla}\phi$ is singular. By these means the magnetic field $\vec{B}$ is cancelled over lengths longer than the distances between singularities. Bloch walls in the magnetization and singularities in ψ reduce the electromagnetic interference term in a similar way.

Theories of structures with spiral or linearly polarized sinusoidal magnetism and weakly varying forms of $\psi(\vec{r})$ without singularities, have been proposed by several groups (2,3,5,7). Also self-induced vortex structures and structures with planar singularities in $\psi(\vec{r})$ and weakly varying forms of $\vec{\mathbf{M}}(\vec{r})$ polarized in a given direction were examined. However if the characteristic lengths of the magnetization and the superconductivity are not too different then new structures with both Bloch walls in the magnetization and singularities in $\psi(\vec{r})$ can have lower free energies. Consider a planar structure in which the measurable quantities are functions only of x and the magnetization is polarized in the z-direction. Taking a divergence-free gauge for A, we can write

$$\vec{M} = (0,0,M(x));\ \psi(\vec{r}) = |\psi(x)|\ \exp(iky)$$

$$\vec{A} = (0,A(x),0);\ dA(x)/dx = \mu M(x)$$

and the phase ky is constant inside each superconducting domain bounded by planes at which $|\psi| = 0$. Minimizing the free energy inside such a domain leads to the condition

$$\int_{ScD} |\psi(x)|^2 \, A_{eff}(x)dx = 0 \text{ with } A_{eff}(x) = \kappa^{-1}k - A(x)$$

and at the boundaries $A_{eff}(x)$ is discontinuous.

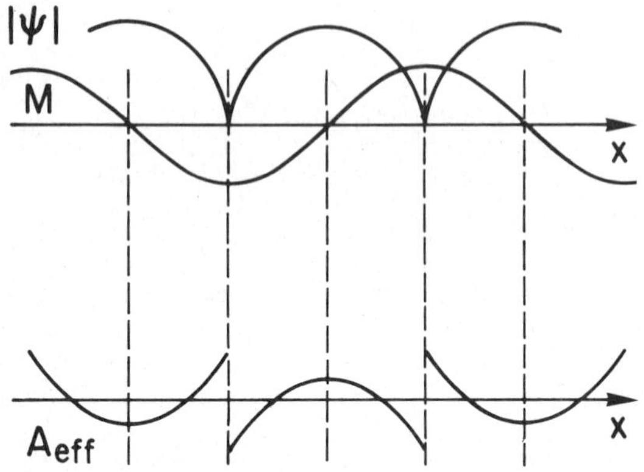

Fig. I

Fig. 1: A schematic representation of the new structure with planar Bloch domain walls in the magnetization, M(x) and planar singularities in the superconducting order parameter, $\psi(x)$. Underneath the effective vector potential $A_{eff}(x)$ (see text) is drawn.

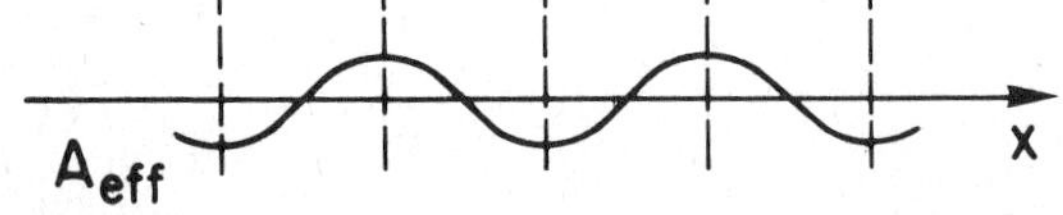

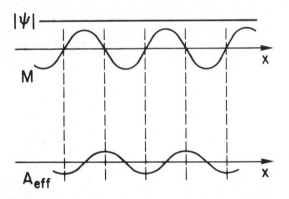

Fig. 2

Fig. 2: For comparison purposes we show a structure with twice as many Bloch walls and a constant value of $|\psi(x)|$. The corresponding effective potential $A_{eff}(x)$ is shown below.

In Fig. 1 and 2 the behavior of $|\psi(x)|$, M(x) and the electromagnetic interference term, $A_{eff}(x)$ are shown schematically. A comparison shows that by introducing the planar singularities we can make A^2_{eff} smaller simultaneously in all the Bloch walls where $|\psi(x)|^2$ is large. In the sinusoidal structure however one can only achieve a similar result by compressing the separation between Bloch walls. In the new structure A^2_{eff} is large near the planar singularities in the superconductivity but here $|\psi(x)|^2$ is small and therefore so is the contribution to the free energy. The singularities in the superconductivity cost approximately the same amount of energy as the Bloch walls but the increased spacing between the Bloch walls and between the superconducting singularities lowers the repulsive interactions and so the new structure can be stabilized.

We have made preliminary estimates of the free energy of the new structure with a simple variational ansatz. We have taken linear rises for $|\psi(x)|$ and M(x) separated by constant plateaus. Because of the discontinuous derivatives the estimated free energy is clearly too high. Nonetheless a comparison with the exact calculations for the spiral magnetic phase without singularities in $\psi(\vec{x})$, gives a lower free energy of the new structure in the case the characteristic superconducting length is smaller than that of the magnetism. Strong anisotropy forces the spiral into a linearly polarized sinusoidal structure and raises its free energy. The free energy of the new structure is unchanged making the comparison more favorable for the new structure. Refined estimates of the free energy with a view towards the application to $ErRh_4B_4$ are being made at present.

ACKNOWLEDGEMENTS

We are grateful to H. R. Ott, S. Sinha and M. Tachiki for useful conversations and to the Schweizer Nationalfonds for financial support.

REFERENCES

(1) S.K. Sinha, G.W. Crabtree, D.G. Hinks, H.A. Mook, Phys. Rev. Lett. 48, 14 (1982)
(2) E.I. Blount and C.M. Varma, Phys. Rev. Lett. 42, 1079 (1979)
(3) H. Matsumoto, H. Umezawa and M. Tachiki, Sol. State Comm. 31, 157 (1979)
(4) H.R. Ott, W. Odoni, H.C. Hamaker, M.B. Maple and L.D. Woolf in Superconductivity in d- and f-Band Metals 1982 (ed. W. Buckel, W. Weber; Kernforschungszentrum Karlsruhe) p.223
(5) L.N. Bulaevskii, A.I. Budzin, S.V. Panjukov and M.L. Kulic, Phys. Rev. B28, 1370 (1983)
(6) M. Tachiki, B.D. Dunlap and G.W. Crabtree, Phys. Rev. B28, 5342 (1983)
(7) M. Tachiki, Physica 109 & 110 B, 1699 (1982) and references therein.

LT-17 (Contributed Papers)
U. Eckern, A. Schmid, W. Weber, H. Wühl (eds)

NEUTRON-SCATTERING SIGNATURE OF A NEW SAVL PHASE FOR THE COEXISTENCE OF SUPERCONDUCTIVITY AND FERROMAGNETISM*

Chia-Ren HU

Department of Physics, Texas A&M University, College Station, Texas 77843, USA

Previously the author and T.E. Ham have predicted a new "square-antiferromagnetic vortex lattice" phase for the coexistence of superconductivity and ferromagnetism in ferromagnetic superconductors. Variational calculation and circular-cell numerical solutions within a Ginzburg-Landau model revealed oscillations of magnetization within each vortex cell. Model studies reported here give clear neutron scattering signatures of this new phase which are consistent with the present observations on $ErRh_4B_4$ but predict new off-axes, off-diagonal, satellite peaks of strong intensity, which can be used to test this model.

Previously, Hu and Ham (1) have proposed a new "square-antiferromagnetic vortex lattice (SAVL)" phase for the coexistence of superconductivity and ferromagnetism in ferromagneic superconductors, particularly with the aim to explain the coexistence state observed with neutron scattering in a single crystal of $ErRh_4B_4$ (2). This reference revealed that the magnetization $\vec{M}$ is linearly polarized along either of the two a-axes of a tetragonal symmetry, and reported the observation of four satellite peaks around each Bragg peak (for each polarization), located at $\pm(1,1)q_0$ and $\pm(1,-1)q_0$ in an ac plane perpendicular to $\vec{M}$ with $2\pi/q_0 \approx 92$Å. Theoretical attempts to explain these observations are mostly based on a generalized Ginzburg-Landau approach, which have led to the proposal of a spontaneous vortex phase (3) and a linearly polarized phase (4). Since these proposed states cannot account for the positions of the satellite peaks, as well as some other aspects, (cf. Ref. 1) we have proposed the SAVL state as an alternative model. This state is characterized by an antiferromagnetic arrangement of superconducting vortex lines in a two-dimensional square lattice. When we first envisioned this state, we pictured the magnetization to decrease monotonically to zero from the cell center to the cell boundary. This would give primary neutron-scattering satellite peaks at $\pm(1,1)\pi/a$ and $\pm(1,-1)\pi/a$ with "a" the square cell dimension. Our variational study, and numerical solutions in the circular-cell approximation, however, revealed that it is energetically more favorable for the magnetization to oscillate several times about zero between the cell center and the cell boundary (cf. Fig. 1 for a typical example), but we cannot determine the optimum number of oscillations, nor the exact profile within a square cell. In order to understand how such a magnetization oscillation affects our predictions about the neutron-scattering satellite-peak distribution and whether it is still consistent with the present observations, we have carried out the following model study. For qualitative understanding only, we assumed the magnetization within each vortex cell to be given by

$$\vec{M}(x,y) = \hat{z}\begin{cases} M_0 & (|\xi| < \delta\pi/2) \\ M_0 \cos\left[\dfrac{(2\ell+1)(|\xi|-\delta\pi/2)}{1-\delta}\right] & (|\xi| \geq \delta\pi/2) \end{cases}$$

$$\text{for } |\xi| \geq |\eta| \qquad (1)$$

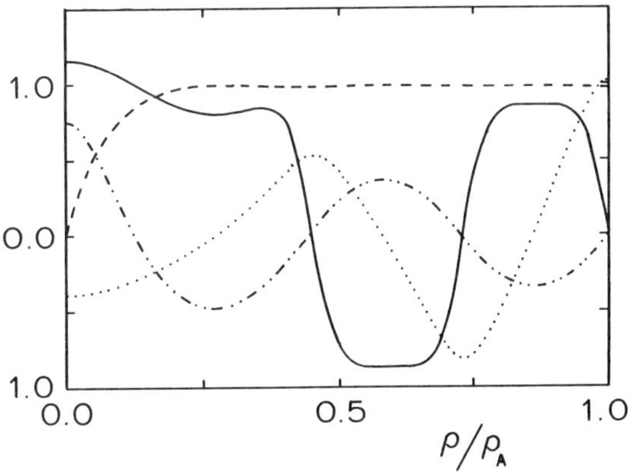

FIGURE 1
A sample vortex solution in the circular-cell approximation, satisfying the boundary conditions to form an antiferromagnetic vortex lattice. Solid line: magnetization m. Dashed line: superconducting order parameter f. For the other curves, see Ref. 1.

* Supported by the U.S. National Science Foundation, DMR 82-05697.

and with ξ replaced by η for $|\eta| \geq |\xi|$, where $\xi \equiv \pi x/a$ and $\eta \equiv \pi y/a$ are confined between $-\pi/2$ and $\pi/2$, and $M_0(T)$ is the equilibrium magnetization for the pure ferromagnetic state. Note that Eq. (1) gives a flat square plateau of dimension δa at the cell center with $\delta<1$, and then $\vec{M}$ oscillates sinusoidally outward with square-node curves, until it vanishes at the cell boundary, so that the nearest-neighbor vortex cells can have the opposite polarity. This choice is motivated by our exact circular-cell solutions, such as the one shown in Fig. 1, but we expect the exact oscillations to be not sinusoidal, and the exact node lines to be not square-shaped. To obtain the satellite-peak intensities corresponding to this M(x,y), we calculate the absolute square of its Fourier transform:

$$\frac{I(m,n)}{M_0^2} = \frac{2}{mn} p^2 \left\{ \frac{\cos(m+n)\delta\pi/2}{(m+n)^2-p^2} - \frac{\cos(m-n)\delta\pi/2}{(m-n)^2-p^2} \right\} \quad (2)$$

where $p \equiv (2\ell+1)/(1-\delta)$, and $k_x \equiv m\pi/a$, $k_y \equiv n\pi/a$, with m and n being both odd integers. In Fig. 2, we have schematically presented this satellite peak intensity distribution by a set of dots of various sizes, for $\ell=18$ and $\delta=28/139$, corresponding to a central plateau dimension d=(28/139)a, and an oscillation wavelength Λ=(6/139)a. Fig. 2 is qualitatively consistent with the observations made to date. Because m and n are both odd, no satellite peaks appear on either axis, as is reported in Ref. 3. Along the (1,1) direction, there is one isolated strong peak at (23,23), which may be identified with the observed peak at $(1,1)q_0$; and a group of strong peaks for m=n<7, which may have given rise to the observed wide tail of the central Bragg peak. The intensities of the remaining peaks along this direction are less than 5% of that of the peak at (23,23), and are probably not observable. Fig. 2 also shows many more strong peaks away from the axes and diagonal, particularly the whole chain of them along a line of slope -1 passing through the (23,23) peak, the existence of which we believe has not yet been looked for or ruled out. Thus, searching for these off-axes, off-diagonal strong peaks should provide a definitive test of this model. Some features in Fig. 2, however, should not be taken as general predictions. For example, the strong peaks at (1,45) and (1,47) are more than a factor 100 stronger than the peak at (23,23). This is due to the square node-curves assumed, which make the principal oscillations of $\vec{M}$ along the x and y axes. We believe that there should be a spontaneously broken 90°-rotation symmetry in the actual solution, so that the node curves are closer to ellipses with their principal axes aligned along the diagonals of the square cell, which should move the positions of the strongest peaks closer to the (1,1) direction. It should also make the peak at (m,n) of different intensity as the ones at ±(m,-n) but keeps the pair at ±(m,n) of equal intensity, as is observed in Ref. 3.

Finally, we note that the plane-wave like "linearly polarized state" (4) cannot give rise to such off-axes, off-diagonal peaks, and the "spontaneous vortex state" (3) should give strong peaks along the two axes, so the three models can be distinguished experimentally.

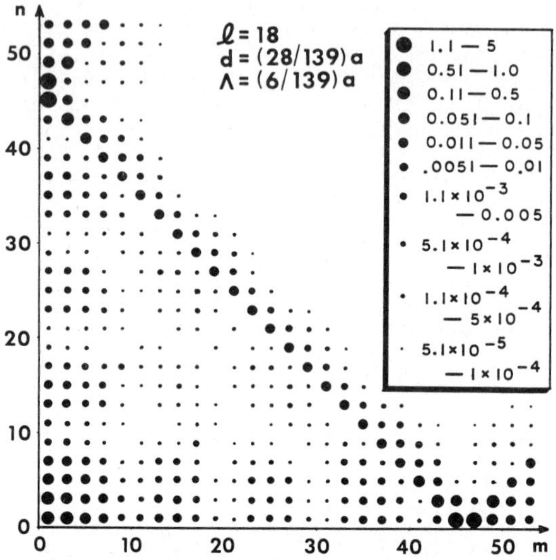

FIGURE 2
A schematic representation of the neutron-scattering satellite-peak intensity distribution for the model magnetization in Eq. (1), with ℓ=18, and δ=28/139. The exact intensity values may be computed from Eq. (2).

REFERENCES

(1) C.-R. Hu and T.E. Ham, Physica (Utrecht) 108 B+C, 1041 (1981); Bull. Amer. Phys. Soc. 28, 342 (1983); and to be published.
(2) S.K. Sinha, G.W. Crabtree, D.G. Hinks and H. Mook, Phys. Rev. Lett. 48, 950 (1982).
(3) C.G. Kuper, M. Revzen and A. Ron, Phys. Rev. Lett. 44, 1545 (1980); Solid State Comm. 36, 533 (1980).
(4) H.S. Greenside, E.I. Bount and C.M. Varma, Phys. Rev. Lett. 46, 49 (1981); and H.S. Greenside, Ph.D. Dissertation, Princeton University, 1981.

LT-17 (Contributed Papers)
U. Eckern, A. Schmid, W. Weber, H. Wühl (eds)

COEXISTENCE OF SUPERCONDUCTIVITY AND FERROMAGNETISM IN THIN FILMS

U. KLEIN

Institute for Theoretical Physics, University of Linz, 4040 Linz, Austria

A study of the coexistence of superconducting and ferromagnetic order in thin films of rare-earth ternary compounds is presented. The possibility of observing a self-induced-vortex structure in a film of a thickness of several coherence lengths is discussed.

As a result of the imperfect screening properties of a superconductor the conditions for coexistence of superconductivity and ferromagnetism are more favorable near the surface of a specimen than in the bulk. The study of these effects started in 1957 with a classical paper by Ginzburg (1) and was continued by several authors (2,3,4) after the discovery of a class of ternary compounds showing superconducting reentrant behavior (5). In the case of a single surface the main conclusion of these theoretical studies was the appearance of a surface coexistence region in a temperature range $T_s<T<T_{c3}$ which is larger than the corresponding range $T_p<T<T_{c2}$ where the bulk coexistence state of superconductivity and modulated ferromagnetic order (MOD-state) exists. T_s and T_{c3} depend on a magnetic surface parameter μ. While T_s is known for arbitrary μ [3] an exact result for T_{c3} is only available for $\mu=\infty$ (μ as defined in Ref. 4) corresponding to a vanishing slope M'=0 of the magnetization at the surface. An approximate form of $T_{c3}(\mu)$ has been calculated recently (6). The surface coexistence state, although compatible with recent tunnelling measurements (7), has not yet been veryfied experimentally. In this note results of a study of the coexistence region in thin films are reported using the method (and notation) of Ref. 4.

The upper boundary of the surface coexistence state is obtained from the condition that eqs. (4), (5) of Ref. 4 (with $F_o=1$, $M_o=Q_o=0$) possess a nontrivial solution for the infinitesimal fluctuations of the magnetisation m, vector potential q and order parameter f. The boundary conditions are now: $f'=0$, $q'=m$, $m'=\pm m/\mu$ at $z=0$ and $z=d$, taking the +/- sign at z=0/d. By an analysis similar to that outlined in Ref. 4 one can determine T_s as the highest zero of an equation whose form depends on the value of μ and on the symmetry properties (symmetric or antisymmetric) of the magnetization with respect to the middle plane of the film. As an example, for values of μ corresponding to $T_o>T_s>T_p$, T_s is determined from

$$\mu_{\pm}=\frac{2p_1p_2(\mathrm{ch}\,p_1d\pm\cos p_2d)}{p_2(1+p_2^2-3p_1^2)\mathrm{sh}p_1d\pm p_1(1-p_1^2+3p_2^2)\sin p_2d}$$

Here +/- refers to symmetric / antisymmetric magnetization, T_o is the temperature where an oscillating part of m appears near a surface or fluxline. The wave vectors p_1, p_2 are given by $p_{1,2}=(a_1^{1/2}\pm a_2/2)^{1/2}$ where a_1, a_2 may be expressed in terms of the parameters of Ref. 4 as follows: $a_1=2(a/|a|+\nu/\delta)$, $a_2=-1-2a/\zeta_M^2|a|$. Other expressions have been obtained for μ-values corresponding to $T_s>T_o$ and $T_s<T_p$ and for the corresponding spontaneous magnetizations m at T_s. The d-dependence of T_s is shown in Fig.1 for three different values of the surface parameter. The GL parameters of Ref.1, corresponding roughly to the situation in $H_oM_{o6}S_8$, are used throughout this paper. Fig. 1 shows an oscillating behavior of the transition temperature and alternating regions of symmetric and antisymmetric magnetization with increasing film thickness. Qualitatively similar results have been obtained by Takahashi et al (8) using an integral equation approach. For $\mu=\infty$, the critical thicknesses d_n where symmetric and antisymmetric transition curves coincide are given by $d_n=n\pi/p_2(t_{s\infty})$, n=1,2,... where $t_{s\infty}$ is to be determined ($t=T/T_m^o$) from $1+p_2^2-2p_1^2=0$. This analytical result holds approximately for a wide range of μ-values. For small d, $T_s(\mu,d)$ reduces to the transition temperature of a nonsuperconducting magnetic film (9) since the superconducting screening property gets lost for $d<<\lambda$.

For the stability limit $T_{c3}(d,\mu)$ of the normal-conducting ferromagnetic state (NF) an exact result exists only for $\mu=\infty$. As before (4) $T_{c3}(d)=T_{c3}(d,\infty)$ is determined by an eigenvalue problem (Eq.(7) of Ref. 1, with boundary conditions f'=0 et z=0,d) which is completely equivalent to the determination of the critical temperature of a film in a longitudinal external field H (10).

Consequently we have

$$4\pi M_o(T_{c3}) = H_{c3}(d) , \qquad (3)$$

where $H_{c3}(d)$ is the third critical field of a film calculated numerically by Saint-James and DeGennes (10). From (3) one finds

$$t_{c3}(d)=t_c^o(t_c^o\Delta-1)\{-1+[1+(2\Delta-1)/(t_c^o\Delta-1)^2]^{1/2}\}, \qquad (4)$$

where $\Delta=\nu(o)/2\bar{\kappa}^2$, $\bar{\kappa}=\kappa H_{c3}(d)/H_{c2}$.

It is still unclear whether $T_{c3}(d)$ represents a second order transition or a superheating limit of the homogeneous ferromagnetic state. To discuss this point in connection with the possibility of a self-induced-vortex (SV) state, consider Fig.1 where $T_s(d)$ and $T_{c3}(d)$ are shown together with the bulk transition temperatures T_p, $T_{c2}^{(w)}$ and T_{c2}. Here $T_{c2}^{(w)}$ denotes the upper stability limit of the NF state and T_{c2} is an estimate of the temperature where the free energies of the MOD and NF states coincide. Consider first the situation in the bulk. The SV state would be realized for $T_{c2}^{(w)}<T_{c2}$. Then the critical fluctuations of the NF state would appear with infinitesimal amplitude at the second order transition point $T_{c2}^{(w)}$. For the present parameters $T_{c2}^{(w)}>T_{c2}$ and the MOD state is realized.

In a film the fluctuations f limiting the NF state are also well known (10). For $d<d_c=1.84\xi$ f has no zero, for intermediate thickness $d>d_c$ a flux-line structure develops while for $d\to\infty$ f vanishes everywhere except near the two surfaces. These fluctuations possess physical significance provided $T_{c3}(d)<T_{c2}(d)$. As regards $T_{c2}(d)$ only its (estimated) limiting value T_{c2} for large d is shown in Fig.1. If we compare it with $T_{c3}(d)$ we see that no definite conclusion regarding the nature of the phase transition at $T_{c3}(d)$ can be

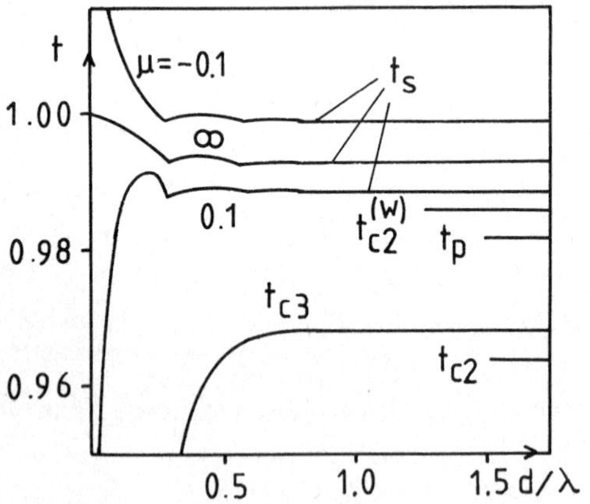

Fig.1: Phase diagram showing T_s, T_{c3} as a function of d and the approximate position of the bulk transition temperatures T_p, $T_{c2}^{(w)}$, T_{c2}.

drawn, this being in contrast to Fig.1 of Ref.4 where $T_{c3}(\infty)<T_{c2}$ was assumed. Something probably more important in view of the great uncertainty in parameters of real systems is the fact that $T_{c3}(d)$ is much closer to $T_{c2}(d)$ than $T_{c2}^{(w)}$ is. Therefore a self-induced-vortex-like structure in a film of a thickness of some ξ may be realized much easier than a SV state in the bulk of the same material. In this respect a thickness reduction has roughly the same effect as an enhancement of κ by a factor of 1.7. Strictly speaking this conclusion only holds for $\mu=\infty$ and the considered anisotropy.

We conclude with the two comments regarding the situation in $ErRh_4B_4$. Recent measurements of the GL-coefficients (11) imply the need for a sixth-order term in the free energy expansion. Consideration of this term does not change the results for bulk and surface stability limits of the paramagnetic superconducting state. Likewise the NF-state stability limits remain unchanged when expressed in ordinary units (eq.(3)). Secondly, one finds that a consistent fit of all experimental data is impossible in the framework of the present purely electrodynamical model. Nevertheless, the above remarks on the possibility of observing a self-induced-vortex structure in thin films apply here too.

REFERENCES

(1) V.L. Ginzburg, Sov. Phys. JETP (1957) 153.
(2) A. Kotani, M. Tachiki, H. Matsumoto, H. Umezawa and S. Takahashi, Phys. Rev. B23 (1981) 5960.
(3) U. Klein, Phys. Rev. B27 (1983) 1925.
(4) U. Klein, Phys. Lett. 96A (1983) 409.
(5) Superconductivity in Ternary Compounds, Vol. 2, eds. M.B. Maple and O. Fischer (Springer, Berlin, 1982).
(6) J. Kasperczyk, private communication.
(7) L.J. Lin, A.M. Goldman, A.M. Kadin and C.P. Umbach, Phys. Rev. Lett. 51 (1983) 2151.
(8) S. Takahashi, A. Kotani, M. Tachiki, H. Matsumoto and H. Umezawa, J. Phys. Soc. Japan, 52 (1983) 989.
(9) M.I. Kaganov and A.N. Omelyanchuk, Sov. Phys. JETP, 34 (1972) 895.
(10) D. Saint-James, P.G. DeGennes, Phys. Lett. 7 (1963) 306.
(11) F. Behroozi, G.W. Crabtree, S.A. Campbell, M. Levy, D.R. Snider, D.C. Johnston, B.T. Matthias, Sol. State Comm. 39 (1981) 1041.

LT-17 (Contributed Papers)
U. Eckern, A. Schmid, W. Weber, H. Wühl (eds)

THE PHASE DIAGRAM OF A FERROMAGNETIC SUPERCONDUCTOR IN THE MIXED STATE

Mircea CRISAN, Miklos GULACSI and Zsolt GULACSI

Department of Physics, University of Cluj, 3400 Cluj, Romania

The phase diagram of a ferromagnetic superconductor in the mixed state has been obtained taking into consideration an expression for the susceptibility used for the explanation of the neutron scattering experiments.

1. INTRODUCTION

Recently there has been much interest in the theory of magnetic ordering in ternary alloys (1).

Neutron scattering experiments have shown the occurence of the ferromagnetic transition below the critical temperature $T_M < T_c$, where T_c is the critical temperature for the superconductor. The mixed state of a ferromagnetic superconductor has been also studied (2), (3) but until now the theory neglected the existence of an inhomogeneous magnetic field in the vortex state.

2. THE PHASE DIAGRAM

In this paper we will obtain a new equation for the phase diagram using the results obtained by Ferrel and Batacharjee (4) about the effect of the applied field on the spin correlation in a ferromagnetic superconductor. In (4) the occurence of the shift of the neutron scattering and the other suplementary effects were explained considering the mixed state of a superconductor. In fact the ferromagnetic superconductor in the mixed phase has been discussed first by Krey (5) but in a phenomenological model. We will apply the method proposed by Machida and Younger (6) but using the susceptibility

$$\chi(q) = \chi_0 \frac{1}{q^4 + q^2 k_0^2 t_0 + k_0^2 q_s^2} \qquad (1)$$

where $t_0 = (T - T_c)/T_c$, is a constant, k_0 is of the order of reciprocal lattice vector, and q_s is the reciprocal of the penetration depth.
Using (1) we get for $\rho = 1/2\pi T_c \tau (T_c)$ the expression

$$\rho = C \frac{a}{b} \frac{1}{t^2 - 2a} \ln \frac{(bt+a)(t^2-a)}{a(t^2-a+b)} \qquad (2)$$

where

$$a = q_s^2/k_0^2, \quad b = 4k_F^4/k_0^2, \quad t = (t_c - T_M)/t_m$$

and $t_c = T_c/T_{co}$, $t_m = t_M/T_{Mo}$.
The constant C from (2) is given by $C = N(o)(I/2)^2(g_J - 1)^2/4 \; nT_{co}$, where I is the "s-d" interaction, n the number of vortices, T_M the critical temperature for the ferromagnetic superconductor.
The equation

$$\ln t_c = \Psi(z) - \Psi(1/2) \qquad (3)$$

where Ψ is di-gama function and $z = 1/2 + \rho$, will give the phase diagram given in Fig.1.

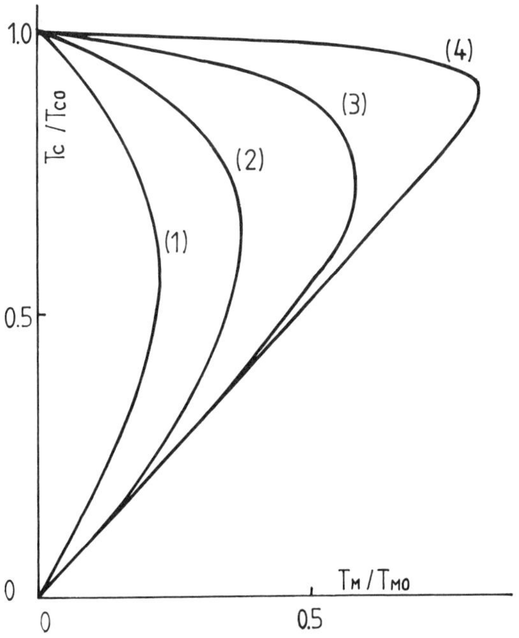

FIGURE 1

where we used the parameters $2k_F/k_0^2=0.03$, C=1 and q_s^2/k_0^2 equal to (1) $1x10^{-3}$ (2) $3.5x10^{-4}$ (3) $1x10^{-4}$ and (4) $1.5x10^{-5}$.

Recently, Buzdin et al.(7) considered a class of materials whose symmetry is consistent with the existence of weak ferromagnetism in the normal state These materials in the superconducting state can be in an inhomogeneous antiferromagnetic or weak ferromagnetic states and a transition from domain structure to a Meissner phase is also possible. At the low temperatures the phases may be described by a self-induced vortex structure. The results contained in (7) have been obtained using the Ginsburg-Landau theory (combined with some results of the classical microscopic theory) but it seems to be similar to the results obtained by Appel et al.(8) concerning a possible coexistence between itinerant-electron magnetism and superconductivity.

Recently Machida (9) reconsidered the problem of coexistence between ferromagnetism and superconductivity and showed the importance of the first order effects in the exchange interaction, which stabilizes the spatially inhomogeneous superconducting state. The Fulde-Ferrel state in this case becomes very important and in the phenomenological treatment there ate two order parameters describing the magnetic state: the magnetization and the stagered magnetization. In this model (9) the temperature dependence of the satellite intensity and hysteresis behaviour in neutron diffraction have been explained.

2. DISCUSSIONS

We obtained a phase diagram for a ferromagnetic superconductor in the mixed state which seems to be similar to that from (6). However, we have to remark that in our phase diagram $T_c=T_{co}$ if $T_m=0$ but in (6) $T_c<T_{co}$ a result which cannot be correct. We also obtained the reentrance phenomenon which is a general characteristic of the coexistence between magnetic order and superconductivity. This phase diagram has been obtained using an expression for susceptibility given by (1) obtained form the magnetic induction B(r) used by Ferrel et al.(4) in order to explain the neutron scattering experiments. In order to have more arguments for such a model, proposed in (4) we will perform in a future paper, $H_{c2}(T)$ and the specific heat calculations.

ACKNOWLEDGEMENTS

The authors thank Professors J.Appel U.Krey and J.Keller for useful correspondence.

REFERENCES

(1) P.Fulde and J.Keller, "Theory of magnetic superconductors", in Superconductivity in Ternary Compounds II, Vol.34, eds. MB Maple and Ø.Fischer (Springer Verlag Berlin, Heidelberg 1982) pp. 249-294.
(2) M.Tachiki, H.Matsumoto, T.Koyama and H.Umezava, Solid State Commun. 34 (1980) 19.
(3) H.S.Greenside, E.I.Blount and C.M. Varma, Phys. Rev. Lett. 46 (1981) 49
(4) R.A.Ferrel and J.K.Battacharjee Physica 108 B (1981) 1039
(5) U.Krey, Int. J. Magn. 3 (1972) 65; 4 (1973) 153.
(6) K.Machida and D.Younger, J. Low Temp. Phys. 35 (1979) 449.
(7) A.I.Buzdin, L.N.Bulayevskii and S.S.Krotov, Zh. Eksperim. i Teor. Fiz. 85 (1983) 678.
(8) J.Appel and D.Fay, Solid State Commun. 28 (1978) 157, Phys. Rev. 22 (1980) 1980
(9) K.Machida, J. Phys. Soc. Japan 51 (1982) 3462.

LT-17 (Contributed Papers)
U. Eckern, A. Schmid, W. Weber, H. Wühl (eds)

NEW TYPE METAMAGNETIC TRANSITION IN A BCT $ErRh_4B_4$ SINGLE CRYSTAL

Hideo IWASAKI, Manabu IKEBE and Yoshio MUTO

The Research Institute for Iron, Steel and Other Metals, Tohoku University, Sendai 980, Japan

The magnetization has been measured on a bct $ErRh_4B_4$ single crystal which is an antiferromagnetic superconductor. The magnetization process below T_N shows unique metamagnetic behavior in the tetragonal c-plane. This metamagnetism can be explained by a four sublattice model taking account of the magnetic anisotropy with fourfold symmetry in the c-plane.

In international conference on magnetism 1982, we pointed out that $ErRh_4B_4$ with body centered tetragonal (bct) structure is an antiferromagnetic superconductor [1]. Recently we have obtained a bct $ErRh_4B_4$ single crystal. The details of sample preparation are to be reported in a separate paper. The superconducting transition temperature T_c of the bct $ErRh_4B_4$ single crystal is 7.80 K. In this paper we discuss about the magnetization process below the Néel temperature $T_N(\simeq 0.65\ K)$.

Magnetization was measured by a vibrating sample magnetometer above 1.7 K and by a conventional integral technique below 1.3 K. From the magnetization measurements in the paramagnetic superconducting state we got the following results. (1) In the c-plane the high field magnetization along the [100] direction is considerably larger than that of the [110] direction, while the magnetizations in both directions coincide in low fields. This indicates that the large fourfold magnetic anisotropy exists in the c-plane and that the axis of easy magnetization is [100]. (2) The magnetization along the [001] axis is very small compared with those along the c-plane. The [001] axis is the axis of hard magnetization. At 1.7 K the values of the magnetization in 56 kOe are $5.3\mu_B$/Er([100]), $4.4\mu_B$/Er([110]) and $2.2\mu_B$/Er ([001]), respectively.

Figs. 1 show the low field magnetization at 0.1 K, at which the sample is in the antiferromagnetic (AF) superconducting state. The magnetization along the [100] direction (H// [100]) shows the two jumps with the large hysteresis at $H = H_1$ and H_3 for increasing and decreasing fields. The magnetization value in the intermediate ferrimagnetic (I) state which exists in a field between H_1 and H_3 is half of the saturation value (M_0) of the induced ferromagnetic (F) state. The magnetization with the two jumps has also been observed in the case of H// [010]. However, corresponding H_1 and H_3 in the [010] direction are a little higher than those in the case of H// [100]. This fact suggests that there is a small difference of the magnetic anisotropy between crystallographically equivalent [100] and [010] and that the [100] axis is the easiest magnetization axis. The magnetization curve along the [110] direction show only one jump at H_2. With further increasing fields the magnetization increases gradually. The magnetization after the jump takes the value between $M_0/2$ and M_0.

These metamagnetic transitions can be explained by a four sublattice model considering the magnetic anisotropy with the fourfold symmetry. In fig. 2 we show schematically the proposed magnetic structures under a magnetic field along the [100], [010] and [110] directions. The different structures apper as the I states in the cases of H//[100] and H// [010]. This is due

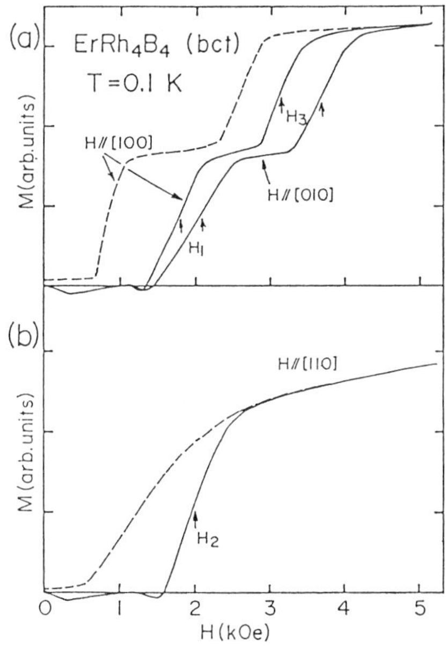

Figure 1 Low field magnetization along (a) the [100] and [010] directions and (b) [110] direction of bct $ErRh_4B_4$ at 0.1 K. Solid lines and dashed lines represent the magnetization for increasing and decreasing fields, respectively.

to the small difference of the magnetic anisotropy between the [100] and [010] axis. The Er spins belonging to the two sublattices along the [100] direction turn to the [010] direction at H_2 in the case of H// [110]. Such a state is possible only under the magnetic anisotropy with the four-fold symmetry. We name this state 90°canted ferromagnetic (CF) state, in which the magnetic moments make a right angle one another. The magnetic free energy is represented as a sum of the exchange energy and the Zeeman energy. Lorentz local field must be included because of the large magnetic moment of Er spin. The dipole field is cancelled because Er spins form almost the fcc lattice. Then we calculate the free energy by the following expression,

$$E = -\sum_{i,j} \mathcal{J}_{ij}(g_j-1)^2 \mathcal{J}_i \cdot \mathcal{J}_j - \sum_i \mu_i \cdot \left(H + \frac{4\pi}{3}\right) \quad (1)$$

where μ is the magnetic moment of Er spin. Exchange constants have been considered for the nearest J_1 and the second nearest J_2 neighbours.

Magnetic structure can not be determined uniquely from the magnetization measurements. As a plausible model we have assumed the magnetic structure in which Er spins order antiferromagnetically in all of the c-planes. Then both ferromagnetically and antiferromagnetically ordered c-planes arrange alternately in the I state. In the CF state all of the c-plane contain the Er spins along the [100] and [010] directions.

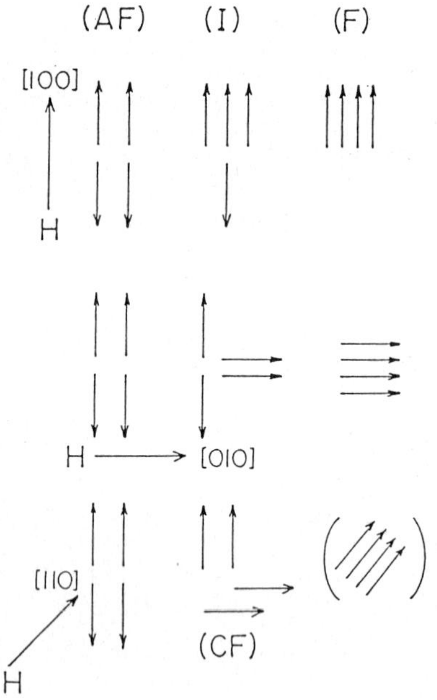

Figure 2 Schematic structures under a field along the [100], [010] and [110] directions. See text about the terms AF, I, F and CF.

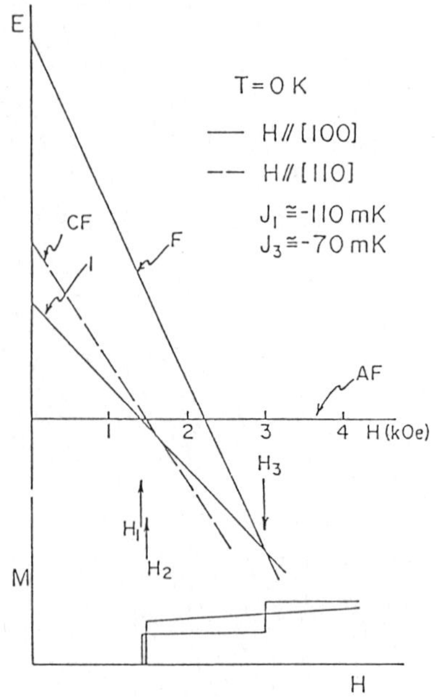

Figure 3 Field dependence of the magnetic free energy for each state shown in figure 2 in the cases of H// [100] and H// [110]. The lower part is the corresponding magnetizations.

Fig. 3 shows the field dependence of the free energy for each state. The corresponding magnetization is schematically shown in the lower part of this figure. In the case of H// [100] the magnetic structure changes from the AF state to the I state at H_1 and from the I state to the F state at H_3. In the case of H// [110] the magnetic structure changes from the AF state to the CF state at H_2, and then gradually reaches the F state. Based on this model, we obtain the values of the exchange constants $J_1 \simeq -110$mK, and $J_2 \simeq -70$ mK.

In conclusion we have observed unique metamagnetic behavior in the tetragonal c-plane of bct $ErRh_4B_4$ below T_N. The magnetization process under the fourfold magnetic anisotropy can be explained by the four sublattice model. In order to explain the magnetization process in the [100], [010] and [110] directions, it is necessary to consider the magnetic structures in which the magnetic moments make a right angle one another as intermediate states.

Acknowledgment- We would like to thank Profs. M. Tachki and S. Maekawa for valuable discussions.

REFERENCE

[1] H. Iwasaki, M. Isino, K. Tsunokuni and Y. Muto, J. Magn. Magn. Mat., 31-34 (1983) 521.

LT-17 (Contributed Papers)
U. Eckern, A. Schmid, W. Weber, H. Wühl (eds)

ON LARGE ANISOTROPY OF H_{c2} IN A BCT $ErRh_4B_4$ SINGLE CRYSTAL

Hideo IWASAKI, Manabu IKEBE and Yoshio MUTO

The Research Institute for Iron, Steel and Other Metals, Tohoku University, Sendai 980, Japan

In an antiferromagnetic superconductor bct $ErRh_4B_4$, the large anisotropy of H_{c2} has been observed between the [100] and [001] directions. Remarkable fourfold anisotropy of H_{c2} in the tetragonal c-plane has also been found above T_N. The anisotropic H_{c2} is closely related with the anisotropic behavior of the magnetization. It is concluded that the spin polarization effect due to the s-f exchange interaction is mainly responsible for the anisotropic H_{c2} of bct $ErRh_4B_4$.

The magnetic superconductor has attracted wide attention of researchers because of close interplay between superconductivity and magnetism. In order to obtain unambiguous conclusions, it has highly desired to make study on single crystals. We have succeeded in prepareing a bct $ErRh_4B_4$ single crystal, which is an antiferromagnetic superconductor with $T_c = 7.80$ K and $T_N \simeq 0.65$ K. We have reported strongly anisotropic behavior of the magnetization of bct $ErRh_4B_4$ in a separate paper [1]. In that paper, very important role of the fourfold magnetic anisotropy in the c-plane has been pointed out. In the present paper, we report on the results of the anisotropic upper critical field H_{c2} and discuss how the anisotropy of H_{c2} is affected by the magnetic anisotropy .

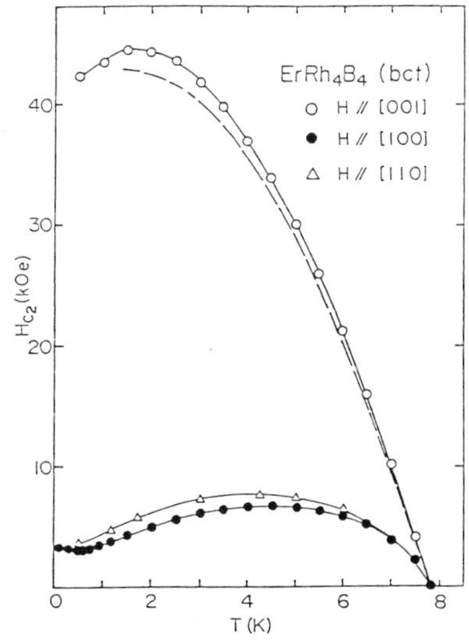

Figure 1 Temperature dependence of H_{c2} along the [100], [110] and [001] directions. Dashed curve shows H_{c2} along the [001] direction after the demagnetization.

H_{c2} was measured by an ac resistive method, sweeping magnetic fields at a constant temperature T. Definition of H_{c2} was given by the cross point of the extrapolated R-H transition curve with the zero resistance base line. As a common feature of bct $RERh_4B_4$ (RE is rare earth metal) the residual resistivity is very large and the sample is inferred to be in the dirty limit. The magnetization was measured by a vibrating sample magnetometer.

In fig. 1 we show H_{c2} along the [100], [110] and [001] directions as a function of the temperature. H_{c2} along the [001] direction which is the axis of hard magnetization [1], is very large compared with those in the c-plane. $H_{c2}(T)$ curve along the [001] direction increases down to 1.5 K, and decreases a little at lower temperatures. H_{c2} curves in the c-plane take maxima at 4-5 K, and decrease gradually at lower temperatures. H_{c2} along the [100] direction takes a

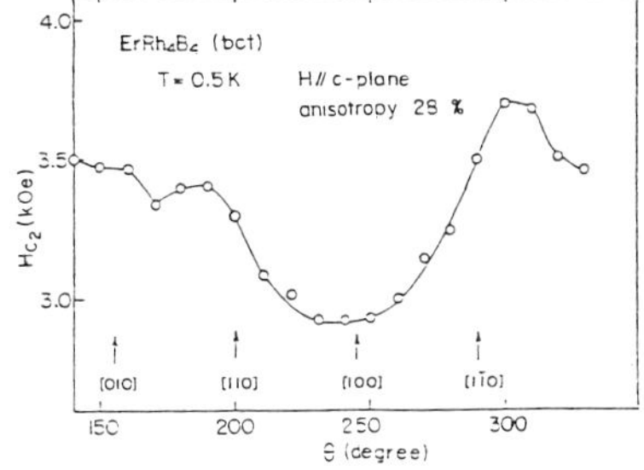

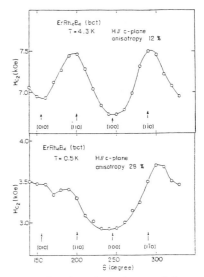

Figure 2 Angular dependence of H_{c2} at (a) T = 4.3 K and at (b) T = 0.5 K.

minimum at 0.6 K. This temperature is nearly equal to the Néel temperature T_N of this sample. H_{c2} has the fourfold anisotropy in the c-plane above T_N, as shown in fig. 2 (a). The minima of H_{c2} are in the [100] and [010] directions which are the axis of easy magnetization, and the maxima of H_{c2} are in the [110] and [1$\bar{1}$0] directions which are the axis of hard magnetization in the c-plane. Below T_N the fourfold anisotropy of H_{c2} disappears and the complex twofold anisotropy is observed in the c-plane (fig. 2 (b)): H_{c2} takes an only minimum along the [100] direction below T_N. The large anisotropy of H_{c2} between parallel and perpendicular to the c-axis and the fourfold anisotropy in the c-plane can never been expected if we do not take account of the effect of the anisotropic magnetization because the crystal structure of bct $ErRh_4B_4$ is nearly cubic as a whole.

Figure 3 shows the temperature dependence of the magnetization value at H_{c2} along the [100], [110] and [001] directions. The magnetization at H_{c2} along the [001] direction which is the axis of hard magnetization is very small. The magnetizations along the [100] and [110] directions are nearly equal at all temperatures, but the small difference should not be neglected: The magnetization at H_{c2} along the [110] direction is always smaller than that along the [100] direction.

We analyze the behavior of H_{c2}, considering the the effect of the magnetization through the electromagnetic interaction [2] and through the spin polarization due to the s-f exchange interaction [3], [4]. In the analysis we use the following relations which hold for the dirty limit.

$$H_{c2}(T) = H_{c2}^{*}(T) - 4\pi M(H_{c2},T) - A(H_{c2}(T) - I'M(H_{c2},T))^2 \quad (1)$$

$$A = 0.022\alpha/\lambda_{so}T_{co} \quad (2)$$

$$I' = 4\pi + I'' \ ; \qquad I'' = (g_J - 1)I/Ngg_J\mu_B^2 \quad (3)$$

The term $4\pi M$ represents the effect of the electromagnetic interaction and $I''M$ represents the effect of the spin polalization due to the s-f exchange interaction ($I'' \propto$ s-f exchange constant I). H_{c2}^* means orbital upper critical field without magnetic ions. λ_{so} is the parameter which characterizes the strength of the spin-orbit interaction and α is Maki parameter.

Within experimental error, the resistivity at 10 K (60~70 $\mu\Omega\cdot$cm) and the residual resistance ratio (RRR = 1.6) were found to be common in bct $ErRh_4B_4$ and $LuRh_4B_4$. For $H_{c2}^*(T)$ we used $H_{c2}(T)$ of bct $LuRh_4B_4$ which we synthesized. The bct $LuRh_4B_4$ sample has T_c = 8.61 K and $(-dH_{c2}/dT)_{T_c}$ = 20.8 kOe/K, and has no anisotropy. Experimental values of the magnetization were used for the analysis. From the conditions $H_{c2}^* = H_{c2}^*$[100] and $H_{c2}^* = H_{c2}^*$[001] we can determine the values of the parameters I' and A at each temperature. The mean values of I' and A at several temperatures are ($I' = 1112$, $A = 1.1 \times 10^{-6}$) and ($I' = -140$, $A = 1.05 \times 10^{-4}$). As the solution of I' with a minus sign does not explain the temperature dependence of H_{c2} along the [100] and [110] directions and the angular dependence on H_{c2} in the c-plane, this was discarded. Estimation of the s-f exchange constant I and the parameter λ_{so} can be made by use of the relations (2) and (3). The results are $I \simeq 74$ K and $\lambda_{so} = 2.81$. The large value of λ_{so} indicates that the spin-orbit scattering is very strong in this crystal. In the expression (1) the last term in the right hand side, in which the effect of the spin polarization due to the s-f exchange interaction is main, is very large compared with the second term which represents the effect of the electromagnetic interaction.

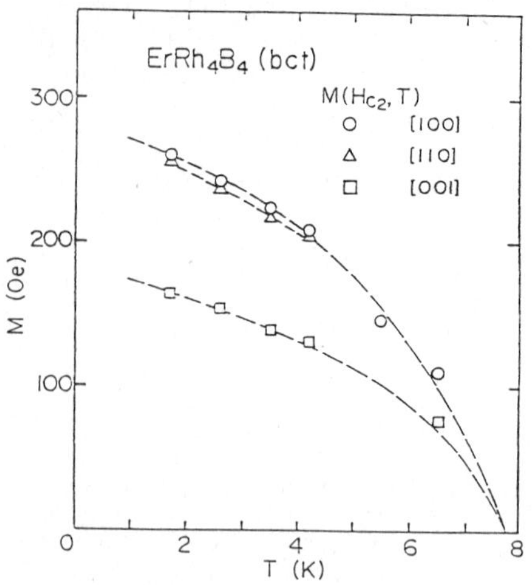

Figure 3 Temperature dependence of the magnetization at H_{c2} along the [100], [110] and [001] directions.

In summary, we have found the large anisotropy of H_{c2} on the bct $ErRh_4B_4$ single crystal. From the analysis in which the electromagnetic interaction and the spin-polarization due to the s-f exchange interaction are taken into account, we have made clear that the effect of the spin polarization is very important in this crystal. The coefficient of s-f exchange effective field I' and the spin-orbit scattering parameter λ_{so} have been determined independently.

Acknowldgment- We would like to thank Prof. M. Tachiki for valuable discussions and Mr. M. Isino for a cooperation in experiments of this work.

REFERENCES

[1] H. Iwasaki, M. Ikebe, Y. Muto, New type metamagnetic transition in a bct ErRh4B4 single crystal, this volume.

[2] O. Sakai, M. Tachiki, T. Koyoma, H. Matsumoto, H. Umezawa, Phys. Rev. B24 (1981) 3830.

[3] G. Zwicknagl, P. Fulde, Z. Physik 43 (1981) 23.

[4] K. Machida, J. Low Temp. Phys. 37 (1979) 583.

LT-17 (Contributed Papers)
U. Eckern, A. Schmid, W. Weber, H. Wühl (eds)
© Elsevier Science Publishers B.V., 1984

NMR STUDY ON LOCAL DENSITY OF STATE OF B IN BCT $RE(Rh_{1-x}Ru_x)_4B_4$

Ken-ichi KUMAGAI, Yutaka HONDA and Frank Y. FRADIN[+]

Department of Physics, Faculty of Science, Hokkaido University, Sapporo, 060, Japan,
[+] MST, Argonne National Laboratory, Argonne, ILL. 60439, U.S.A.

Nuclear spin-lattice relaxation time T_1 of ^{11}B in body centeerd tetragonal $RE(Rh_{1-x}Ru_x)_4B_4$ compounds have been investigated with the pulsed NMR. The local density of state (DOS) of B site have minimum near the critical concentration $x \sim 0.5$ where superconducting transition temperature T_s changes abruptly. The variation of DOS at B site seems to be sensitive to the atomic distance of the clustered TM and RE atoms.

The rare earth rhodium borides $RERh_4B_4$ compounds have provided an interesting study of their unusual superconducting and magnetic properties (1). After the discovery of the primitive tetragonal crystal structure of $CeRh_4B_4$ type (2), body center tetragonal (bct) $LuRu_4B_4$ (3) and orthorhombic $LuRh_4B_4$ (4) type structure have been found. All those type compounds have the similar building blocks of B_2 dimers and clusters of transition metal (TM) tetrahedra. The clustering of TM atoms is probably contributing factor to high superconducting transition temperature T_s. Detailed studies of composition dependence of T_s and lattice parameters have been carried out for both magnetic $Er(Rh_{1-x}Ru_x)_4B_4$ (5), and nonmagnetic $Y(Rh_{1-x}Ru_4)_4B_4$ compound(6).

In each of these systems, a very abrupt decreases in T_s around x=o.5 has been found to be common properties. The magnetic interaction between RE atoms also changes from ferromagnetic to antiferromagnetic with changing x. Several models are interestingly proposed to interpret the abrupt change of superconducting and magnetic properties (6,7,8). The postulated models can be tested via composition dependence of nuclear magnetic resonance at each atom site, which gives us information about the local properties of electronic structure. In this paper, we report studies of nuclear spin-lattice relaxation time of ^{11}B in bct $RE(Rh_{1-x}Ru_x)_4B_4$, where RE of the non-ragnetic Y and Lu are chosen because of the simplicity due to the lack of magnetic effect on the NMR properties.

The samples were prepared by arc melting under argon atmosphere. The samples were then wrapped in tantalum foil and sealed with an argon in a quartz tube for annealing at 1200°c for 5 days, 100°c for 4 days and 800°c for 10 days. The samples for X-ray analysis and NMR measurement were crushed and passed through a 325 mesh sieve. The X-ray diffraction investigation indicated that the samples contained the body centered tetragonal(bct) structure and were close to

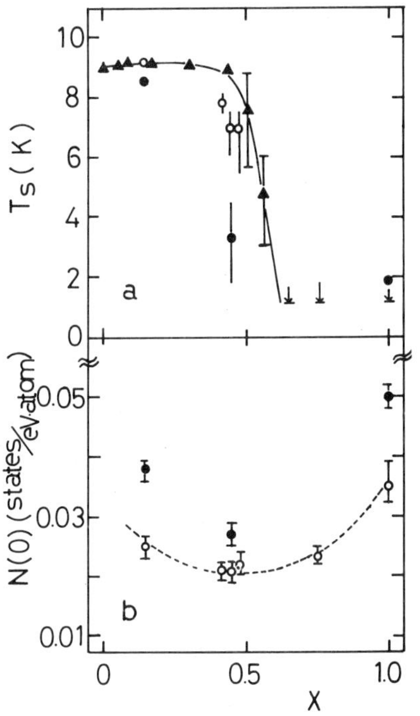

Fig.1. a) Superconducting transition temperature T_s of $Y(Rh_{1-x}Ru_x)_4B_4$; ○, and $Lu(Rh_{1-x}Ru_x)_4B_4$.●. T_s obtained by Johnston(6);▲, are also plotted. b) Local density of state of s electron of B as a function of x,: ○ for $Y(Rh_{1-x}Ru_x)_4B_4$: ● for $Lu(Rh_{1-x}Ru_x)_4B_4$.

single phase. T_s measurement was done with ac susceptibility measurement on annealed powder samples. The results of T_s are shown in Fig. 1. The non-linear variation of T_s with composition in the present samples is almost similar to those reported previously (6) i.e. T_s remains nearly constant with increasing x up to o.4~o.5. With further increasing x, T_s drops below 1.5K.

NMR measurement was made with a conventional pulesed phase coherent spectrometer. The nuclear spin-lattice relaxation time T_1 was obtained with "comb pulse" method. T_1 of ^{11}B in $RE(Rh_{1-x}Ru_x)_4B_4$ is proportional to temperature as expected in the normal metal. The Korringa relaxation constant T_1T obtained from the temperature dependence of T_1 are shown as a function of TM composition in Fig.2. In the bct compounds with RE=Y and Lu, T_1T shows the non-linear variation with x. The dominant contribution to T_1 in boron is considered to arise from the Fermi contact interaction via the 2s electron of boron atoms. The Korringa relaxation rate T_1^{-1} is given by

$$T_1^{-1} = 2h\gamma_n^2 H_{hf}^2 N_s(0)^2 k_B T,$$

where γ_n is the nuclear gyromagnetic ratio, H_{hf} is the hyperfine constant and $N_s(0)$ is the local density of state of s electron at Fermi level. Using the free atom value of $H_{hf} = 1.0\times10^{-6}$ (atom Oe/μ_B) (9), we obtain the local density of state of 2s electron at B site for each composition. The $N_s(0)$ of B are shown in Fig. 1. $N_s(0)$ decreases with increasing x at Rh-rich side. With further increasing x, $N_s(0)$ increases at Ru-rich side. The local DOS of B has minimum near the critical concentration x~0.5. Apparently, no abrupt change of N(0) are observed, which is in contrast to the model by D.C. Johnston (6). The continuous variation of DOS is revealed by the result of Auger experiment (8).

Although there are no band calculation available for $LuRu_4B_4$ type compounds, we compare the experimental results with the calculation performed for $CeRh_4B_4$ type compounds which provides us with qualitative understanding on non-linear variation of DOS, because of the similarity of the local boron environment in each crystal structure. According to the band calculation for primitive tetragonal YRh_4B_4 (10), the E_F is situated in an isolated narrow peak in DOS and that this peak structure arises from the structure in the Rh-4d contribution. The substitution of Ru for Rh reduces the total number of d electron and therefore, shifts E_F to lower energy. THe shifts of E_F causes the large variation of DOS of TM-4d electron and also DOS of B-2s electron. The gradual change of the charge transfer from B site to a high DOS peak at E_F associated with TM site may attribute to the rather large variation of DOS of B with x, although there are no evidence of an abrupt change of the charge transfer proposed by Johnston(6). The smooth variation of E_F relative to the d band structure has been demonstrated by synchrotron radiation experiment (11).

Another important result of the NMR experiment is the rather large difference of DOS of B in Y and Lu system. This may arise from the difference of the atomic radius between Y and Lu, which

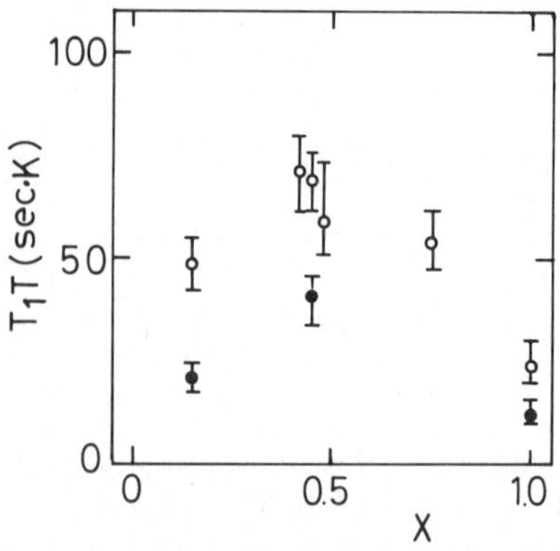

Fig.2. T_1T of ^{11}B in $Y(Rh_{1-x}Ru_x)_4B_4$:○, and $Lu(Rh_{1-x}Ru_x)_4B_4$:●, as a function of x.

suggests that the band structure in bct compounds is sensitive to the atomic distance of the TM cluster and RE atoms.

The authors would like to thank J.W. Downey for preparing the samples.

REFERENCES

(1) M.B. Maple and O. Fischer (eds.) Superconductivity in ternary compounds II. (Springer-Verlag,1982).
(2) B.T. Matthias, E. Corenzwit, J.M. Vandenberg and H. Barz, Pro. Nat. Acd. U.S.A., 74, (1977) 1334
(3) D.C. Jonhston, Solid State Commun. 24 (1977) 699.
(4) K. Yvon and D.C. Johnston, Acta. Crys. B38 (1980) 247.
(5) H.E. Horng, and R.N. Shelton,in Ref. 1,p213
(6) D.C. Jonhston, Solid State Commun. 42 (1982) 453.
(7) K. Ku, F. Acker and B.T. Matthias, Phys. Lett. 76A (1980) 399.
(8) H.C. Hamarker, G. Zajac and S.D. Bader, Phys. Rev. B27,(1983) 6713.
(9) G.C. Carter, L.H. Bennett and D.J. Kahan, Metallic Shifts in NMR (Pergamon, New York 1977) Vol I, p9.
(10) A.J. Freeman and T. Jarlborg, in Ref.1 p167-200.
(11) R. Knauf, A.Thoma, H. Adrian and R.L. Johnson, Phys. Rev. 29B (1984) 2477.

LT-17 (Contributed Papers)
U. Eckern, A. Schmid, W. Weber, H. Wühl (eds)

EFFECT OF CRYSTALLINE ELECTRIC FIELDS ON SPIN DYNAMICS IN $(Y_{1-x}RE_x)Rh_4B_4$ (RE=Tb,Dy,Ho)

Ken-ichi KUMAGAI, Yutaka HONDA and Frank Y. FRADIN+

Department of Physics, Faculty of Science, Hokkaido University, Sapporo, 060, Japan, +MST, Argonne National Laboratory, Argonne, ILL., U.S.A.

Pulsed NMR measurement of ^{11}B has been performed to study spin dynamics of RE ions in $(Y_{1-x}RE_x)Rh_4B_4$. We find the fluctuation rate of RE=Tb,Dy and Ho moment to be exponential in $\Delta/_T$. This is consistent with a model of the crystalline electric field (CEF) in which thermal activation across a crystalline field gap Δ must take place in order $\Delta m_J = \pm 1$ transition to occur. Anomalous temperature dependence of transverse relaxation time T_2 around 15K in $(Y_{1-x}Dy_x)Rh_4B_4$ is explained in connection with the CEF effect.

Many investigations on the ternary $RERh_4B_4$ compounds have been performed to study their unusual magnetic and superconducting properties (1). In previous paper(2), we reported the ^{11}B nuclear relaxation time in $(Y_{1-x}RE_x)Rh_4B_4$ (RE= Gd, Er) where the nuclear relaxation was found to be dominated by the fluctuations of the local dipole field due to rare earth ions. The most important experimental observation was the significant reduction of the electronic spin relaxation due to superconductivity in the boride compounds.

In the compounds with dilute Tb, Dy and Ho, the temperature dependence of nuclear relaxation is distinctly different from those in the compounds with dilute Gd and Er(3). This paper discusses studies of anomalous temperature dependence of nuclear relaxation and hence of spin dynamics of rare earth moments of Dy, Tb and Ho doped in YRh_4B_4. The importance of crystalline electric fields (CEF) effect on the spin dynamics of rare earth ions are discussed.

The $(Y\text{-}RE)Rh_4B_4$ samples were prepared by arc melt and homogenized at 1100°c under an argon atmosphere for 5 days. NMR measurement was made with a conventional pulsed spectrometer. The nuclear longitudinal spin relaxation time was measured with the field cycled method. The nuclear spin-lattice relaxation time due to the magnetic moment τ_1 was obtained by the same procedure described previously(2).

An unusual temperature dependence of τ_1 is obtained in the compounds doped dilute Tb, Dy and Ho with various concentrations. Important features of the results are as follows: 1) τ_1 decreases with decreasing temperature below superconducting transition temperature T_s (T_s=10.3 K at H=0 kOe), which is considered to arise from the same origin as observed in $(Y\text{-}Gd)Rh_4B_4$ (effect of superconductivity)(2). 2) τ_1 drops drastically with increasing temperature above 10K where any magnetic order or superconducting transition are not detected. Such a behavior of τ_1 is much different from that in $(Y\text{-}Gd)Rh_4B_4$ where τ_1 is independent of temperature in the normal state (2). 3) Applied field suppresses the relaxation rate.

An anomalous temperature dependence of spin echo decay time (transverse relaxation time) T_2 of ^{11}B in $(Y\text{-}Dy)Rh_4B_4$ has been also observed at the temperature region where τ_1 changes drastically. As shown in Fig. 1, T_2 drops steeply with decreasing temperature, and then recovers to the value of 400 μsec at low temperature. With increasing the concentration of Dy, enhancement of T_2^{-1} become more pronounce. T_2 becomes so small that the signal can not be detected around 15K in $(Y_{0.99}Dy_{0.01})Rh_4B_4$. Such anomalies of T_2 are also observed in the compounds with Tb and Ho at low temperature, whereas no anomalies of T_2 are observed in $(Y_{1-x}RE_x)Rh_4B_4$, RE=Gd and Er. The temperature dependence of T_2 is not due to the contribution from T_1, as the order of magnitude are still much different with each other ($T_2 \sim$ μ400 sec, $T_1 \sim 0.01$ sec at concerned temperature). The

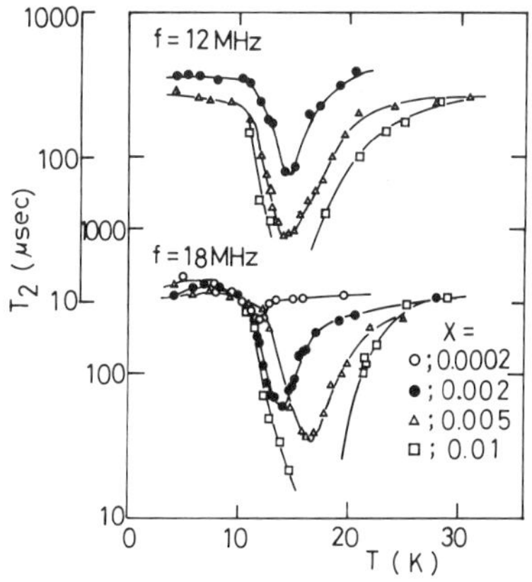

Fig.1. Spin echo decay time T_2 of ^{11}B as a function of temperature in $(Y_{1-x}Dy_x)Rh_4B_4$.

critical slowing down due to magnetic order is immediately ruled out, since the susceptibility measurement shows no sign of anomalies at the temperature around 15K.

The most plausible interpretation of these relaxation anomalies is that the fluctuation time of the localized moments τ_m are slowing down with decreasing temperature. Slow-fluctuation regime($\omega_n\tau_m \gg 1$) has been reached below 15K where τ_0 has the form of $\tau_0^{-1} \propto (\omega_n^2 \tau_m)^{-1}$ changing from the form of $\tau_0^{-1} \propto \tau_m$ in the fast-relaxation regime($\omega_n\tau_m \ll 1$) at high temperature. This slowing down is likely to the CEF effect as pointed out previously(3).

In order to take into account the temperature and field dependence of the fluctuations of magnetization of rare earth ions in highly anisotropic compounds, we use the numerical value of $k_TB(\partial\langle J\rangle/\partial M)$ obtained by Dunlap(4) instead of the derivative of Brillouin function with free spin model to calculate τ_0 as following relation (5), $\tau_1)_{LD}^{-1} = \tau_0)_{LD}^{-1} k_BT(\frac{\partial\langle J\rangle}{\partial M})_{CEF}$,

$$\tau_0)_{LD}^{-1} = (^{16}/_9)\,\pi^3(\gamma_n g_J \mu_B J N(0) x)^2 \frac{\tau_m}{1+(\omega_n\tau_m)^2},$$

where γ_n is the nuclear gyromagnetic ratio, g_J is the Lande factor, J is the RE angular momentum, x is the concentration of the magnetic ions, N_0 is the density of RE site, τ_m is the longitudinal relaxation time of paramagnetic ions and ω_n is the nuclear Larmor frequency. Figure 2 shows the τ_0 as a function of reciprocal temperature in $(Y_{0.998}RE_{0.002})Rh_4B_4$ with RE=Dy,Tb and Ho. τ_0 has the form of $\exp(-^{\Delta}/_{kT})$ at concerned temperature. The values of Δ are calculated to be 180, 26 and 36 for Dy, Tb and Ho, respectively. The exponential temperature dependence of τ_0 in the compounds with Dy,Tb and Ho are in sharp contrast to the result in (Y-Gd),(Y-Er)Rh_4B_4 (2).

The CEF parameters for $RERh_4B_4$ systems have been derived by Dunlap et al (6). For Dy case, the $^6H_{15/2}$ ground state splits into 8 Kramers doublets, with the lowest doublet separated from the next level at Δ=131K. For Ho which is non-Kramers doublets ions, the $^5I_8$ ground state splits into 5 doublets and 7 singlet levels. The lowest level is a doublet separated at Δ= 26K and 56K for Tb and Ho, respectively. There are no matrix element for the $m \leftrightarrow m\pm1$ transition within the ground multiplet for Dy,Tb and Ho, whereas the Gd and Er has the non-vanishing matrix element for spin-flip scattering within a degenerate ground state. Therefore, spin-flip scattering in Dy,Tb and Ho become very hard at low temperature, since thermal activation across a CEF gap must take place inorder for a $\Delta m = \pm1$ transition to occur. In this reason, τ_m has the form of $\exp(-\Delta/_{kT})$ where Δ is the energy between the ground and next level. The experimental values of are in good agreement with the calculated ones obtained by Dunlap(6). This fact suggests that the CEF effect is importantly

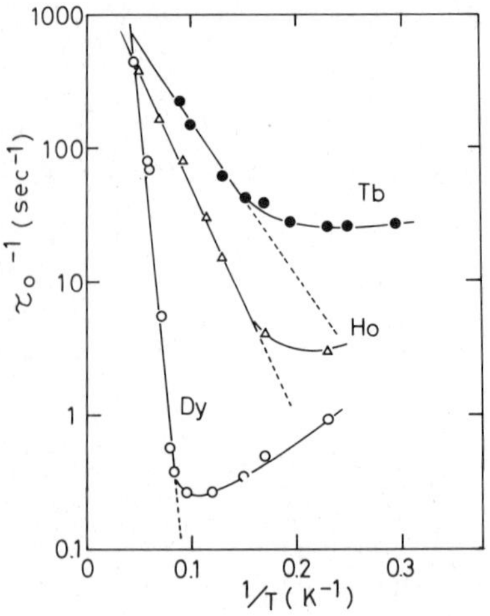

Fig.2. Nuclear spin-lattice relaxation rate τ_0 due to the longitudinal dipolar fluctuations as a function of reciprocal temperature in $(Y_{0.998}RE_{0.002})Rh_4B_4$ with RE = Tb,Dy and Ho.

responsible for the spin dynamics of rare earth ions in the boride compounds. It should be noted that Ho ground state doublet has no transverse magnetic moment, which appears to explain the Ising -like behavior of $HoRh_4B_4$ (7), and that the anomalous temperature dependence of muon relaxation is observed in $(Ho_{1-x}Lu_x)Rh_4B_4$ (8).

In conclusion, the unusual behavior of nuclear relaxation time of T_1 and T_2 must be related to the slowing down of τ_m due to the CEF effect. Further investigations are now in progress to obtain detailed analyses of the CEF effect on the spin dynamics of the local moments in the ternary boride compounds.

The authors would like to thank J. W. Downey for experimental assistance.

REFERENCES

(1) M.B. Maple and O. Fischer (eds.), Superconductivity in ternary compounds II (Springer-Verlag, 1982).
(2) K. Kumagai and F.Y. Fradin, Phys. Rev. 27B (1983) 2770.
(3) K. Kumagai and F.Y.Fradin, Bull. Am. Phys. Soc., 27 (1983) 343.
(4) B.D. Dunlap, private communication.
(5) M.R. McHenry, B.G. Silbernagel and J.H. Wernick, Phys. Rev. B5 (1972) 2958.
(6) B.D. Dunlap and D. Niarchos, Solid State Commun. 44 (1982) 1577.
(7) H.R. Ott, L.D. Woolf, M.B. Maple and D.C. Johnston, J. Low Temp. Phys. 39 (1980) 383.
(8) D.E. MacLaughlin, S.A. Dodds, C. Boekema, R.H. Heffner, R.L. Hutson, M. Leon, M.E. Schillaci and J.L.Smith, J. Magn. Magn. Mater. 31 (1983) 497.

LT-17 (Contributed Papers)
U. Eckern, A. Schmid, W. Weber, H. Wühl (eds)

PHASE-DIAGRAM, MAGNETIZATION AND CRITICAL FIELDS OF THE PSEUDOTERNARY bct SYSTEM $Ho(Rh_{1-x}Ru_x)_4B_4$

H. ADRIAN, A. THOMÄ, B. KANDOLF, and G. SAEMANN-ISCHENKO

Physikalisches Institut, Universität Erlangen-Nürnberg, Erwin-Rommel-Str. 1, D-8520 Erlangen, FRG

Measurements of the ac-susceptibility χ_{ac}, dc-magnetization and upper critical field $H_{c2}(T)$ of the pseudoternary bct system $Ho(Rh_{1-x}Ru_x)_4B_4$ are reported. In the phase-diagram a steep drop of T_c from about 6.3K to below 1.3K in a small composition range above x=0.30 is observed. For $x \leq 0.35$ and $T \leq 1.5K$ superconductivity coexists with antiferromagnetic order as deduced from magnetization and $H_{c2}(T)$ data. Besides a shallow minimum at about 1.5K the $H_{c2}(T)$ curves exhibit a hysteretic behavior below about 0.7K. For $x \leq 1$ $\chi_{ac}(T)$ indicates a ferromagnetic-type ordering.

1. INTRODUCTION AND EXPERIMENTAL

Ternary compound families are of basic interest, because of their high structural flexibility and the occurence of superconductivity in compounds with magnetic ions on a regular sublattice. This offers unique possibilities to study the influence of stoichiometric composition on superconductivity and magnetism and the interplay between these two cooperative phenomena (1). In this paper we report ac-susceptibility, dc-magnetization and resistive critical field measurements of the pseudoternary system $Ho(Rh_{1-x}Ru_x)_4B_4$. Sample preparation and X-ray-analysis were reported earlier (2). For temperatures above 1.3K transitions to superconductivity or magnetically ordered states ($x \geq 0.9$) were determined by ac-susceptibility χ_{ac}. Dc-magnetization measurements were carried out using small cylinders spark-cut from the original ingots (3).

For measurements of $H_{c2}(T)$ the superconductive transition was monitored at constant temperature in a slowly varying magnetic field oriented perpendicular as well as parallel to the cylinder axis. For this type of experiment the temperature range was extended to 5mK by the use of a ^3He-^4He dilution refrigerator.

2. RESULTS AND DISCUSSION

The phase-diagram, shown in fig.1, reveals a steep drop of T_c in a narrow range of composition above x=0.30, which is often observed in pseudoternary transition-metal borides (4). In case of $Ho(Rh_{1-x}Ru_x)_4B_4$ photoemission studies showed that the reduction of the number of valence electrons shifts E_F to lower energies which explains the T_c behavior in conjunction with the d-electron density of states (2). For $x \leq 0.37$ a transition to an antiferromagnetic ordered state at about 1.5K is deduced from $H_{c2}(T)$ data as discussed below. For $x \geq 0.90$ a strong peak in χ_{ac} indicates a transition to a ferromagnetic-type state. More results about this composition range are presented in a separate paper (7). The arrows in fig.1 indicate that no superconductive or magnetic transition was observed above 1.3K. It is interesting to note that the character of magnetic order is different for the regions with $x \leq 0.37$ and $x \geq 0.90$.

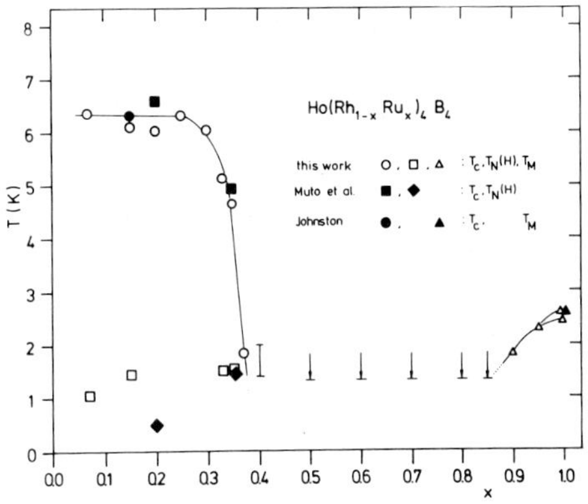

FIGURE 1
Phase-diagram including data of ref. 5 and 6.

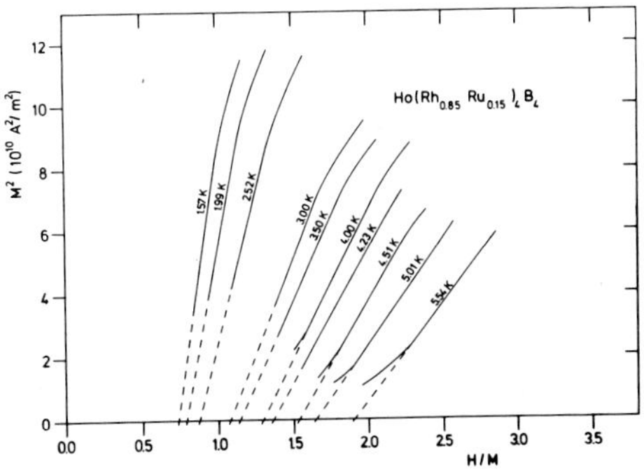

FIGURE 2
Representative Arrott-plots M^2 versus H/M.

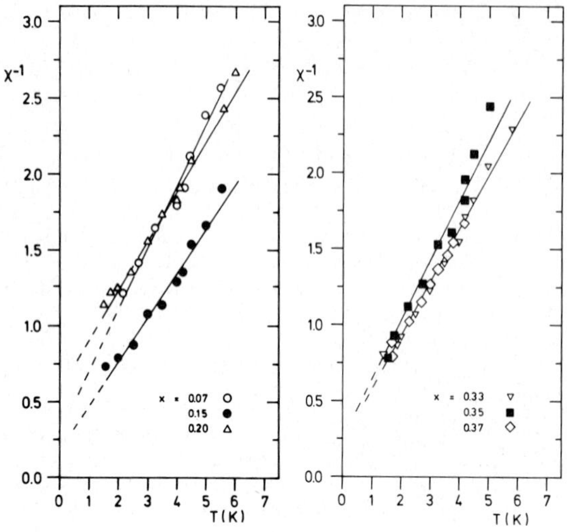

FIGURE 3
χ^{-1} versus T for several samples.

From dc-magnetization curves Arrott-plots were calculated. An example is shown in fig.2. Small deviations at the lower end of the lines are due to the onset of superconductivity. The extrapolation of the very regular curves to $M^2=0$ gives reliable values of the inverse susceptibility χ^{-1}. The results, shown for several samples in fig.3, clearly indicate for this composition range antiferromagnetic behavior with Curie temperatures of about -1K. A consequence of this ordering can also be seen in a shallow minimum at about 1.5K of the $H_{c2}(T)$ curves (see fig.4a) which is typical for antiferromagnetic order in coexistence with superconductivity. The temperature of the minimum in $H_{c2}(T)$ has been taken as ordering temperature in fig.1. Although the zero field Nêel-temperature T_N is expected to be slightly higher. The shift of the position of the minimum probably reflects the field dependence of T_N. At lower temperatures ($T<0.7K$) a hysteresis between transitions taken in increasing and decreasing fields has been observed. The reason for this interesting phenomenon is not clear. It could be a field induced change in magnetic order which is stable over a small region of magnetic fields. As this material has a strong positive magnetization at H_{c2}, demagnetization effects are important. Therefore we began to repeat the measurements with the cylinder axis oriented parallel to the field. First results are shown in fig.4b. Although the critical field is significantly reduced, the qualitatively behavior remains unchanged.

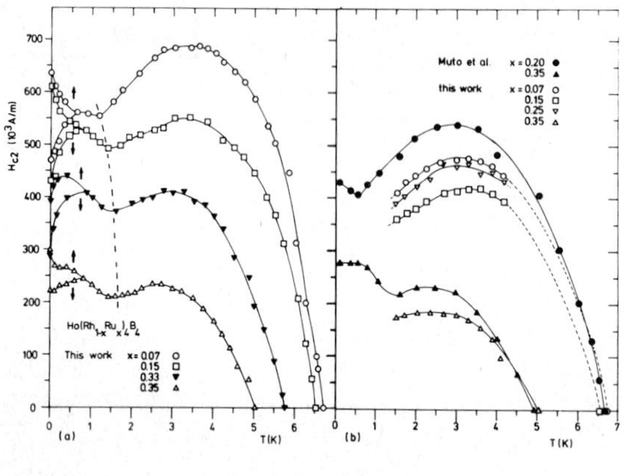

FIGURE 4
$H_{c2}(T)$ for orientation of the cylinder axis perpendicular (a) and parallel (b) to the external field.

3. CONCLUSIONS

The established phase-diagram of the pseudoternary bct system $Ho(Rh_{1-x}Ru_x)_4B_4$ exhibits the typical sharp drop in T_c in a narrow composition range above x=0.30. For $x\leq 0.35$ superconductivity in coexistence with antiferromagnetic order below 1.5K is observed. Contrary to this at $x\lesssim 1$ a different type of magnetic order containing a ferromagnetic component exists. Further experiments to clarify the change in magnetic order are in preparation.

ACKNOWLEDGEMENTS

The authors acknowledge technical assistance by the Siemens Research Laboratories in Erlangen and financial support by BMFT.

REFERENCES

(1) for a review see: Superconductivity in Ternary Compounds I+II, eds. Ø Fischer and M.B. Maple (Springer, Berlin, 1982)
(2) R. Knauf, A. Thomä, H. Adrian, and R.L. Johnson, Phys.Rev. B29 (1984), in press
(3) H. Adrian, R. Müller, and R. Behrle, Phys.Rev. B26 (1982) 2450
(4) H.E. Horng and R.N. Shelton in Ternary Superconductors, eds. G.K. Shenoy, B.D. Dunlap and F.Y. Fradin (North-Holland, Amsterdam, 1981) p. 213
(5) D.C. Johnston, Solid State Commun. 24 (1977) 699
(6) Y. Muto, H. Iwasaki, T. Sasaki, N. Kobayashi, M. Ikebe and M. Isino, in ref. 4, p. 197
(7) R. Müller, A. Thomä, T. Theiler, H. Adrian, G. Saemann-Ischenko, and M. Steiner, this volume

LT-17 (Contributed Papers)
U. Eckern, A. Schmid, W. Weber, H. Wühl (eds)

HIGH RESOLUTION PHOTOEMISSION STUDIES ON THE PSEUDOTERNARY pt $Ho(Rh_{1-x}Ir_x)_4B_4$ SYSTEM

R. KNAUF[+], R. MÜLLER[+], H. ADRIAN[+], G. SAEMANN-ISCHENKO[+], and R.L. JOHNSON[++]

+ Physikalisches Institut, Universität Erlangen-Nürnberg, Erwin-Rommel-Str. 1, D-8520 Erlangen; FRG
++Max-Planck-Institut für Festkörperforschung, Heisenbergstr. 1, D-7000 Stuttgart 80; FRG

Photoemission measurements using synchrotron radiation are presented on the pseudoternary pt system $Ho(Rh_{1-x}Ir_x)_4B_4$ which exhibits superconductivity as well as magnetic order covering the whole range of composition ($0 \leq x \leq 0.80$). The valence band spectra show a broadening of the structure of the transition metal d-states with increasing x. The results are discussed in the framework of existing band-structure calculations and provide a qualitative explanation of the observed pronounced decrease of the superconducting transition temperature T_c in the vicinity of a critical composition $x_{cr}=0.50$. A comparison with earlier work on the $Ho(Rh_{1-x}Ru_x)_4B_4$ system shows that the T_c reductions are caused by different mechanisms.

1. INTRODUCTION AND EXPERIMENTAL

The pseudoternary systems $RE(Rh_{1-x}T_x)_4B_4$ with T=Ru,Ir and RE=Rare Earth or Y show a pronounced drop in T_c as function of the composition parameter x within a small interval of typical $\Delta x=0.10$ in the vicinity of a critical value x_{cr} which may vary between $0.30 \leq x_{cr} \leq 0.50$. This feature is remarkable, because it occurs with and without magnetic RE ions, in two different crystal structures and in cases where the number of valence electrons changes (T=Ru) as well as where it is constant (T=Ir). Therefore it was speculated that a common mechanism should exist. A recent photoemission study of the $Ho(Rh_{1-x}Ru_x)_4B_4$ system revealed conclusive evidence that a reduction of the number of d-electrons causing a shift in E_F to lower energies and this way reducing $N(E_F)$ is the underlying mechanism for the observed T_c-decrease in that system (2). As this mechanism can be clearly excluded in case of the $Ho(Rh_{1-x}Ir_x)_4B_4$ system, we undertook another photoemission study which is presented here to clarify the apparent controversy.

The samples were prepared by arc-melting as reported earlier (2) and subsequent annealing. The X-ray diffraction patterns could be indexed with the primitive-tetragonal structure. In some cases small amounts of impurity phases (RhB, $HoRh_3B_2$) were present. The phase-diagram was determined by measuring the ac-susceptibility for temperatures down to 1.3K.

The photoemission spectra were obtained using synchrotron radiation from the DORIS storage ring in Hamburg. The total experimental resolution at $h\nu$=60eV was 0.33eV. Before each run the samples were cleaned by scraping in situ with a diamond file.

+ Financial support by BMFT

2. RESULTS AND DISCUSSION

The phase-diagram shown in fig.1 closely resembles the data published earlier (3). For $0.10 \leq x \leq x_{cr}=0.50$ a superconducting T_c slightly increasing with x is observed. Above x_{cr}, T_c drops to about 1.5K and antiferromagnetic order below about 3K is found. This is in contrast to the bct $Ho(Rh_{1-x}Ru_x)_4B_4$ system where for large x T_c is presumably zero and ferromagnetic order occurs.

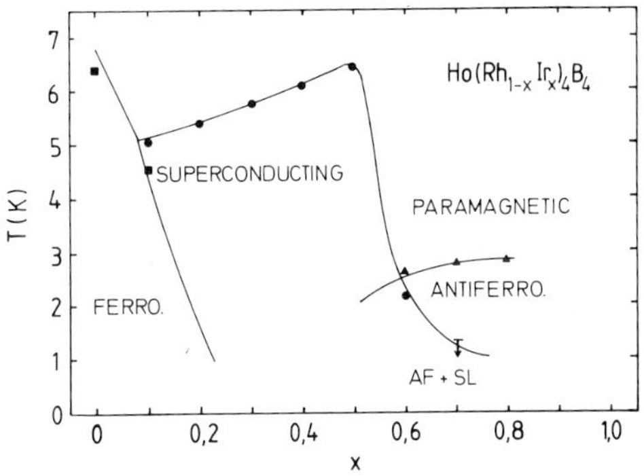

FIGURE 1
Phase-diagram of pt $Ho(Rh_{1-x}Ir_x)_4B_4$.

Similar, however, is the dependence of the valence band spectra on photon energy (2) shown in fig.2 from which the peak near E_F can be attributed to the transition-metal d-electrons. The part of a representative selection of spectra near E_F which is relevant for superconductivity is displayed in detail in fig.3a. As ex-

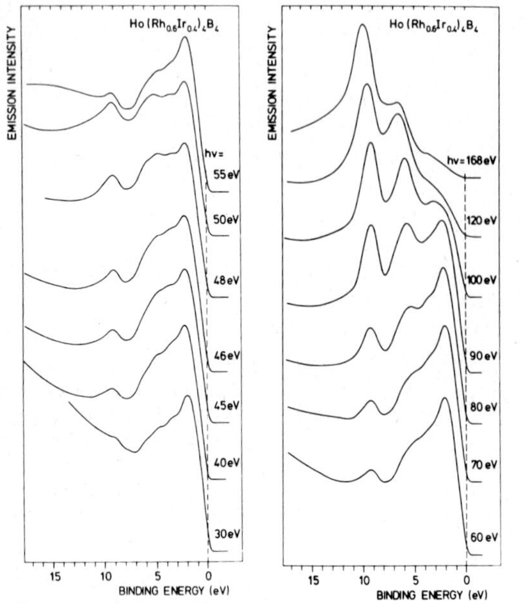

FIGURE 2
Set of valence-band spectra.

pected no shift of E_F can be detected. However, from the observed changes in the shapes of the spectra one can deduce a broadening of the structure in N(E) which is discernible only for $x > 0.50$. This broadening should arise from the more extended 5-d wave functions of Ir and is obvious by comparing the spectra of pure Rh and Ir metal.

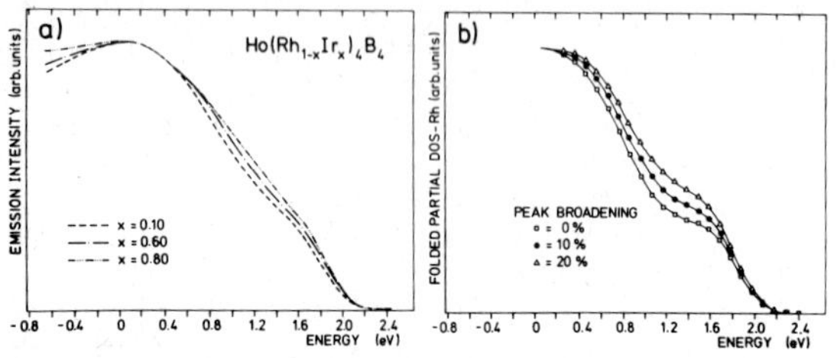

FIGURE 3
a) Experimental and b) calculated spectra at E_F.

In order to demonstrate the effect of broadening the calculated partial density of states for pt $HoRh_4B_4$ (4) was approximated by the superposition of 3 Gauss-peaks as illustrated in fig.4. The results of subsequent broadening with constant peak integrals and finally folding with a gaussian analyzer function (σ =0.20eV) are given in fig.3b. Although the experimental structures

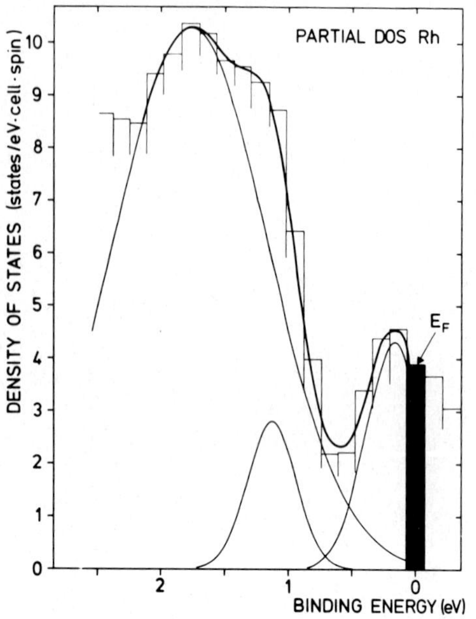

FIGURE 4
Calculated N(E) from ref.4 and approximation of N(E) by 3 Gauss-peaks used for broadening.

are less pronounced there is an excellent qualitative agreement between figs. 3a and 3b. The increasing overlap between the Gauss-peaks which becomes significant for $x > x_{cr}$ leads to a slight shift of E_F to higher energies in order to keep the total number of electrons constant. The resulting decrease of $N(E_F)$ qualitatively explains the observed T_c behavior.

3. CONCLUSIONS

A broadening of the d-electron structure by substituting Ir for Rh in $Ho(Rh_{1-x}Ru_x)_4B_4$ was observed by photoemission. This effect is presumably caused by the more extended wave functions of 5-d-electrons compared to 4-d-electrons. The experimental result was simulated by broadening the calculated N(E) which revealed that a shift of E_F is a necessary consequence. The resulting change in $N(E_F)$ qualitatively explains $T_c(x)$.

REFERENCES

(1) H.E. Horng and R.N. Shelton, in Ternary Superconductors, eds. G.K. Shenoy, B.D. Dunlap and F.Y. Fradin (North Holland, Amsterdam, 1981) pp.213-216
(2) R. Knauf, A. Thomä, H. Adrian, and R.L. Johnson, Phys.Rev. B29 (1984) in print
(3) H.C. Ku, F. Acker, and B.T. Matthias, Phys.Lett. 76A (1980) 399
(4) T. Jarlborg, A.J. Freeman, and T.J. Watson-Yang, Phys.Rev.Lett. 39 (1977) 1032

LT-17 (Contributed Papers)
U. Eckern, A. Schmid, W. Weber, H. Wühl (eds)

NATURE OF THE MAGNETIC ORDER IN $HoRu_4B_4$ BELOW 4.2K

R. MÜLLER+, A. THOMÄ+, T. THEILER+, H. ADRIAN+, G. SAEMANN-ISCHENKO+, and M. STEINER++

+Physikalisches Institut, Universität Erlangen-Nürnberg, Erwin-Rommel-Str. 1, D-8520 Erlangen, FRG
++Hahn-Meitner-Institut, Glienickerstr. 100, D-1000 Berlin 39, FRG

The ternary transition metal boride $HoRu_4B_4$ exhibits two magnetic phase transitions at 2.6K and 2.3K, which were clearly distinguished by ac-susceptibility, specific heat, and neutron-diffraction measurements. The magnetic structure of the intermediate phase could not be solved definitely, but seems to consist either of short range order or an incommensurate phase. In the low temperature phase the magnetic moments of the Ho^{3+}-ions are oriented in the basal plane of the tetragonal unit cell forming ferromagnetic chains in x- and y-direction. The magnetic moment is consistent with the moment of the free Ho^{3+}-ion.

Pseudoternary transition metal borides $RE(Rh_{1-x}Ru_x)_4B_4$ (RE=Dy,Ho,Er) have shown interesting superconducting and magnetic properties (1-3); for small Ru-concentrations ($x \leqslant 0.40$) superconductivity appears to coexist with antiferromagnetism and for $x \geqslant 0.90$ another magnetic order without superconductivity exists. In the intermediate range neither superconductivity nor magnetism is observed above ~1K.

The samples were prepared by arc-melting the stoichiometric amounts of Ho,Rh,Ru, and B. For the sample used in the neutron-scattering-experiment we used ^{11}B, because of the high absorption cross-section of ^{10}B. Powder X-ray diffraction patterns indicate all samples to be single phase bct-$LuRu_4B_4$-type. Measurements of the specific heat and ac-susceptibility show that $HoRu_4B_4$ exhibits two magnetic phase transitions at 2.6K and 2.3K as shown in fig.1. The heights of the susceptibility-peaks reveal that both phases are not simple antiferromagnets and presumably exhibit a spontaneous magnetization. The shapes of the peaks in the specific heat indicate that the transition at 2.6K is of second order whereas the second transition at 2.3K is possibly of first order. With decreasing Ru-concentration (x=0.95 and 0.90) only one magnetic transition at decreasing temperature is observed. The qualitatively different shapes of the magnetic peaks for x=0.90 and x=0.95 may indicate a change of the magnetic structure in this composition range. From Arrott-plots determined by dc-magnetization measurements an anomaly at ~2.5K is present as shown in fig.2. For temperatures above 2.5K $HoRu_4B_4$ behaves like a normal ferromagnet in the paramagnetic state with a Curie-temperature of ~2.5K. Below 2.5K the slope of the straight lines increases with decreasing temperature and the intersection with the H/M-axis is temperature-independent. This might be understood for polycristalline samples, where the magnetic moments are fixed to a crystallographic axis by crystal field effects and therefore the dependence of the net-magnetization on the magnetic field is weakened.

In order to determine the magnetic structure of $HoRu_4B_4$ neutron-diffraction was performed at the Hahn-Meitner-Institut in Berlin using a mul-

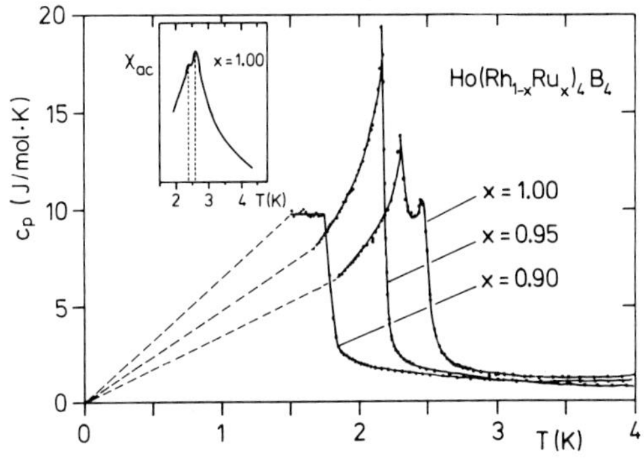

FIGURE 1
Specific heat and ac-susceptibility.

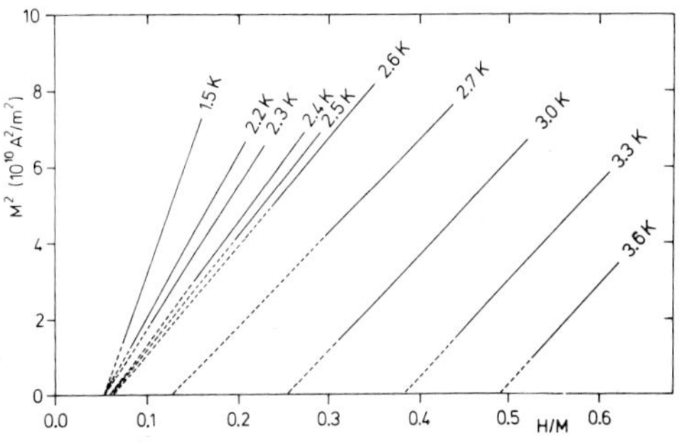

FIGURE 2
Arrott-plot of $HoRh_4B_4$.

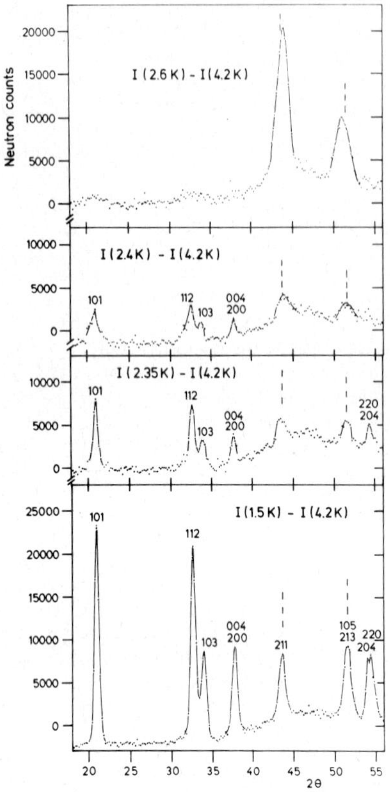

FIGURE 3
Magnetic intensities of $HoRu_4B_4$ at 2.6K, 2.4K, 2.35K, and 1.5K.

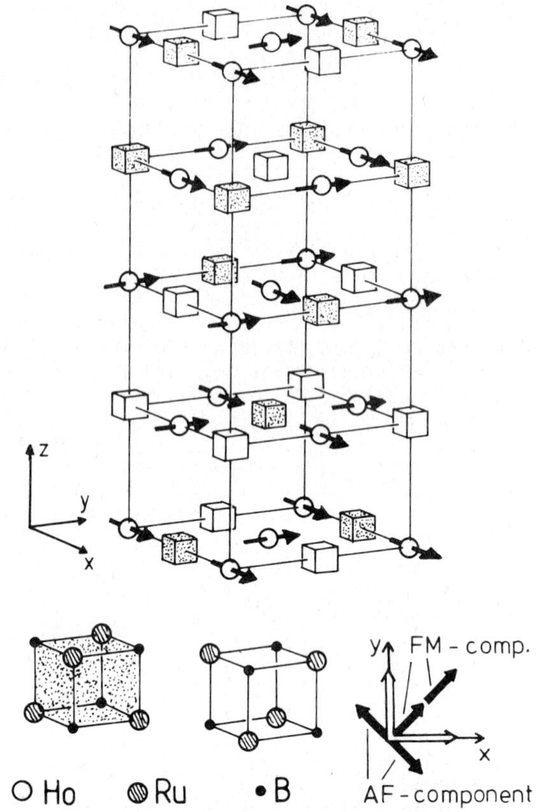

FIGURE 4
Proposed magnetic unit cell of $HoRu_4B_4$ at 1.5K.

tidetector powder-diffractometer. The neutron wavelength was 2.395Å. Fig.3 shows the magnetic intensities at various temperatures, the nuclear intensities, the He- and the Al-background are subtracted. The magnetic phase at 2.6K exhibits two broad peaks which cannot be indexed with the chemical unit cell. Because the resolution was not high enough, it cannot be ruled out whether these peaks consist of two unresolved ones due to a possible incommensurate magnetic order or show short range magnetic order. With decreasing temperature these peaks are vanishing and new peaks appear, which can be indexed with the chemical unit cell (bct, a=7.44Å, c=14.97Å). They divide into two groups, one with a ferromagnetic component (hkl=112, 200, 004,220,204) and the other with an antiferromagnetic (hkl=101,103,211,213,105). The measured intensities can be explained as a canted antiferromagnet with the magnetic moments oriented along two perpendicular axes in the basal plane. A possible structure with a reliability-factor $R=\sum(I_{obs}-I_{calc})^2/\sum I_{obs}^2=0.3\%$ is shown in fig.4. The magnetic moment of the Ho in $HoRu_4B_4$ at 1.5K is consistent with the moment of the free Ho^{3+}-ion ($\mu=10.6\mu_{Bohr}$). The Ho-moments are oriented along the x- and y-axis, respectively, and ferromagnetic zig-zag chains in these directions are formed. The two different orientations along the x- and y-axis are consistent with the chemical surrounding of the Ho^{3+}-ions; the moments are aligned parallel to the connection line of the dotted $(RuB)_4$ cubes and perpendicular to the connection line of the transparent ones. In this way a spontaneous magnetization in (x+y)-direction exists in the low temperature phase of $HoRu_4B_4$.

The authors acknowledge financial support by the Bundesministerium für Forschung und Technologie.

REFERENCES
(1) H.C. Hamaker and M.B. Maple, J. Low Temp. Phys. 51 (1983) 633
(2) H. Adrian, A. Thomä, B. Kandolf, and G. Saemann-Ischenko, this volume
(3) H.E. Horng and R.N. Shelton, in Ternary Superconductors, eds. G.K. Shenoy, B.D. Dunlap, and F.Y. Fradin (North-Holland, Amsterdam, 1981) p. 213

LT-17 (Contributed Papers)
U. Eckern, A. Schmid, W. Weber, H. Wühl (eds)

SPECIFIC HEAT-MEASUREMENTS OF AN $ErRh_4B_4$ SINGLE CRYSTAL

J.M. DEPUYDT and E.D. DAHLBERG

School of Physics and Astronomy, University of Minnesota, Minneapolis, Minnesota, 55455, U.S.A.

D.G. HINKS

Argonne National Laboratory, Argonne, Illinois 60439, U.S.A.

Measurements of the heat capacity of a single crystal of $ErRh_4B_4$ have been made in the temperature range of 0.4 K to 1.5 K. This temperature range includes the temperature for the superconducting to ferromagnetic transition. The measurements include data taken on both heating and cooling of the sample. The transition as measured with the heat capacity is hysteretic in temperature and has a sharp maximum in both the heating and cooling cycles which occurs at the same temperature as the occurrence and disappearance of the satellite lines observed in the neutron scattering.

1. INTRODUCTION

Since the discovery of the rhodium borides and the Chevrel phase compounds, our understanding of the competition between long range magnetic order and superconductivity has been greatly increased. Of these compounds, one of the most studied systems is $ErRh_4B_4$ which was found to be paramagnetic at high temperatures, superconducting for temperatures between roughly 9 K and 1 K and ferromagnetic for lower temperatures. Because of the complications of the crystal field splitting of the erbium moment, the resulting anisotropic magnetic moment(1), the sensitivity of the magnetic and superconducting properties to structural damage(2) and residual strains(3), much of the detailed understanding of the interaction between superconductivity and ferromagnetism has been the result of studies on single crystals of this compound. Studies performed thus far include resistivity, field dependent magnetization and neutron scattering. In order to provide a more complete understanding of the complicated phenomena which occur in this compound we have made measurements of the heat capacity of the same single crystal which was used in the neutron scattering measurements.

2. EXPERIMENTAL DETAILS

The present specific heat measurements were performed using a thermal relaxation technique. The basic calorimeter consists of a sapphire plate of dimensions 2.5 x 1.5 x 0.0025 cm with a thin gold heater evaporated on the sapphire. The temperature of the calorimeter is monitored with an unencapsulated miniature Ge resistance thermometer element. This temperature sensing element is attached at one end with GE varnish to the sapphire. The entire calorimeter assembly is attached to a small dilution refrigerator by a thin Delrin rod. The thermal contact from the calorimeter to the mixing chamber of the refrigerator is accomplished with two no. 40 gauge copper wires which also supply the electrical current to the heater on the sapphire plate. The single crystal $ErRh_4B_4$ sample is roughly semicylindrical with the large area surface resulting from the original crystal being sawed. The large area surface has dimensions of .50 x .35 cm. This surface of the sample was attached to the calorimeter with a thin layer of GE varnish. The maximum distance of any part of the sample from the sapphire plate was .2 cm. For the experimental results described here, a second thermometer similar to the one attached to the sapphire was attached to the free surface of the sample as far from the sapphire as possible, i.e., .2 cm. For the specific heat measurements the calorimeter heat was given a step function change in power and the resistance of both the calorimeter and sample thermometers were monitored as a function of time until after the calorimeter temperature equilibrated. The temporal evolution of the calorimeter temperature can be fitted to

$$T(t) = T_o + \Delta T(1-\exp(-t/\tau)) \qquad (1)$$

where T_o is the initial equilibrium temperature and $T_o + \Delta T$ is the final equillibrium temperature. The time constant τ is given by the ratio of the thermal conductance of the copper wires connected to the mixing chamber and the heat capacity of the calorimeter and sample. In these experiments ΔT was always less than 0.040 K and in the region of T_{c2} was typically 0.010 K. To insure the proper time constant was being measured in the region of T_{c2}, the calorimeter temperature was monitored for

10^4 sec before the next change in heater power was applied. Both the thermometer on the calorimeter and the sample produced equivalent measurements within the thermometer accuracies. The ΔT given in Eq. (1) may be either positive or negative depending on an increase or decrease in the heater power. Thus this technique allows measurements of the heat capacity with both increasing and decreasing temperature. The data shown here were acquired from several heating and cooling cycles. Prior to a cooling cycle, the sample was always heated to at least 2°K. The heating cycles always started from temperatures less than 0.4 K.

3. RESULTS AND CONCLUSIONS

Figures 1a and 1b exhibit the results of the heat capacity measurements on heating and cooling respectively. In the heating cycle, the largest observed heat capacity occurs at 0.75 K which roughly coincides with the appearance of the the satellite intensity in the neutron scattering and the disappearance of the resistance upon heating(4). The maximum heat capacity measured cooling the sample occurs at 0.715 K which coincides with the disappearance of the satellite intensity in the neutron scattering and the appearance of resistance while cooling. Thus all three measurement techniques are consistent with a hysteretic transition of roughly 35 mK at T_{c2}. This is of course consistent with the transition at T_{c2} being a first order phase transition. Within the limitations of the measurement technique described here there is no observable latent heat.

ACKNOWLEDGEMENTS

This work was supported by the Graduate School of the University of Minnesota and the Department of Energy. One of the authors gratefully acknowledges receipt of an A.P. Sloan Fellowship. The authors would also like to thank Drs. A. Goldman, W. Weyhmann, I. Schuller, B.Maple, B. Dunlap and G. Crabtree for numerous discussions.

REFERENCES

(1) H.B. Radousky, B.D. Dunlap, G.S. Knapp, and D.G. Niarchos, Phys Rev. B27, 5526 (1983).
(2) J.M. Rowell, R.C.Dynes, and P.H. Schmidt, in Superconductivity in d- and f- Band Metals, edited by Harry Suhl and M. Brian Maple (Academic Press, New York, 1980), pp. 408-418.
(3) M. Tachiki, F.D. Dunlap and G.W. Crabtree, Phys Rev. B 20, 5342 (1983).
(4) S.K. Sinha, G.W. Crabtree, D.G. Hinks, and M. Mook, Phys. Rev. Lett. 48, 950 (1982).

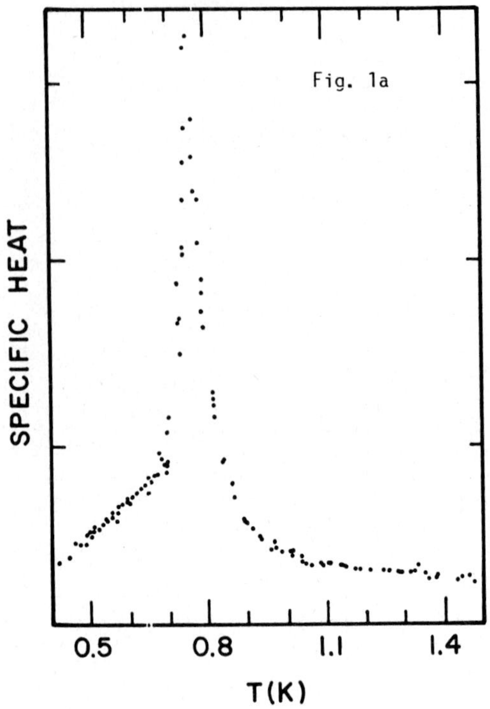

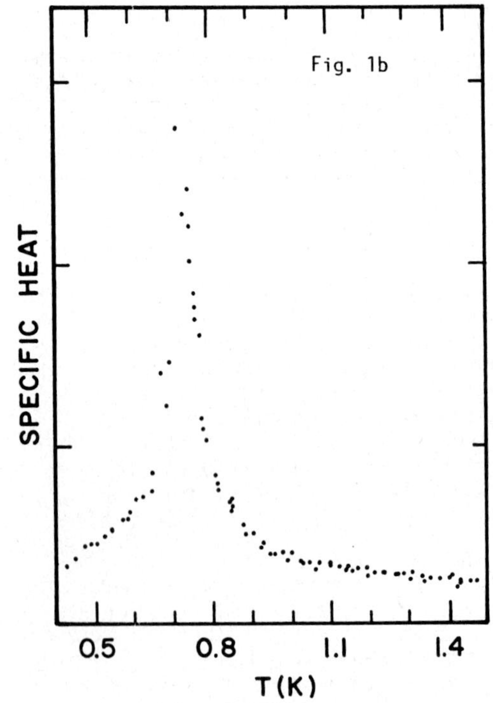

Fig. 1a. Specific heat while heating.

Fig. 1b. Specific heat while cooling.

LT-17 (Contributed Papers)
U. Eckern, A. Schmid, W. Weber, H. Wühl (eds)

DYNAMIC SUSCEPTIBILITY STUDIES ON $ErRh_4B_4$

*P. Svedlindh, *L. Sandlund, D.H.A. Blank and J. Flokstra

Twente University of Technology, Department of Applied Physics, P.O.B. 217, 7500 AE ENSCHEDE, The Netherlands
*Teknikum Uppsala University, Uppsala, Sweden

Dynamic susceptibility measurements have been performed on a spherical sample of $ErRh_4B_4$. The frequency dependent behaviour in the paramagnetic state has been satisfactorily described with the theory for the screening effect of eddy currents. The relaxation time of the Er^{3+} spins appears to be about 0.1 s at T = 4.2 K and is almost independent of the external field.

1. INTRODUCTION

Magnetic superconductors have been extensively studied by several techniques in recent years (1). The measurements concern for instance the electrical conductivity, specific heat, D.C. magnetization and susceptibility as a function of the temperature T and applied magnetic field H_a. However, almost no attention has been paid to the dynamical behaviour of magnetic superconductors in a harmonically oscillating magnetic field superimposed on a constant background field. Magnetic isolators have often been studied in this way in order to obtain information about the spin-phonon interaction. In the case of magnetic superconductors the interactions between spins, phonons and conduction electrons can be investigated and further, screening effects due to the electrical conductive behaviour of the material can be studied. In this paper we present some results of dynamic susceptibility measurements on the magnetic superconductor $ErRh_4B_4$.

2. DYNAMIC SUSCEPTIBILITY MEASUREMENTS ON $ErRh_4B_4$

The re-entrant superconductor $ErRh_4B_4$ is characterized by the critical temperatures T_{c1} = 8.7 K and T_{c2} = 0.9 K. The polycrystalline sample was made by arc melting of a pressed powder of ErB_4 and Rh and afterwards annealed during 100 hours at 1100 °C and 1 week at 950 °C. The spherical sample had a diameter of 4 mm. We used two techniques for studying the dynamic susceptibility $\chi = \chi' - i\chi''$. The field dependence of χ was studied using a standard mutual inductance bridge and the frequency dependence of χ at fixed fields by means of the SQUID susceptometer (2).

FIELD-SWEEP MEASUREMENTS

Field-sweep measurements have been performed up to about 4 Tesla at liquid-helium temperatures in the frequency range from 30 Hz to 40 kHz. Typical results for $\chi'(H_a)$ and $\chi''(H_a)$ at T = 4.2 K and ν = 34 Hz are presented in Fig. 1. $\chi'(H_a)$ is negative at weak fields indicating the superconductive behaviour, becomes positive at about 0.3 T (hysteresis phenomena, $\chi'' \neq 0$) and slowly decreases for H_a > 0.7 T according to the saturation of paramagnetic material. At large magnetic fields χ' becomes equal to the constant value $1.9 \times 10^{-2}\chi(H_a = 0)$, calculated as 3.5×10^{-6} m^3/mol. This results in a linear part of the M versus H_a curve, which is also obtained from D.C. magnetization measurements (3). This behaviour is ascribed to the level splitting of the ground-state multiplet of the Er^{3+}.

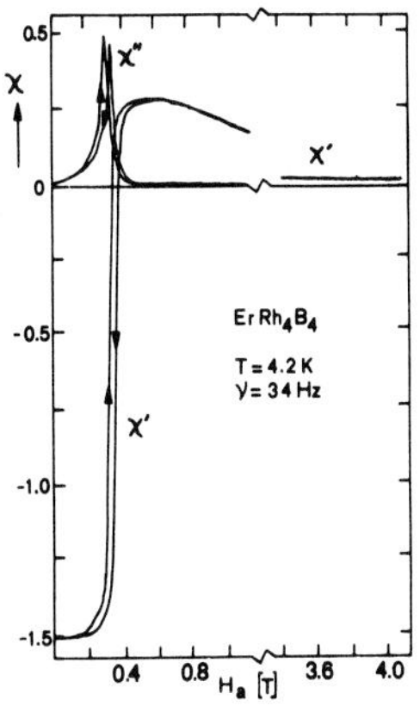

FIGURE 1
In-phase (χ') and out-of-phase (χ'') component of χ as a function of the applied field

It follows from Fig. 1 that the whole sample becomes paramagnetic at fields of about 0.7 T. Although the values of H_{c2} are reported to be considerably less, our results appear to be in agreement with the resistance measurements of Ott et al. (4).

The frequency dependence of χ in the range from 30 Hz to 40 kHz is obtained from a series of field-sweep measurements at various frequencies. In Fig. 2, a $\chi' - \chi''$ plot is presented for 1.15 T (paramagnetic region) and T = 4.2 K. The high-frequency final value is taken from the zero-field susceptibility being equal to -1.5 due to the demagnetization in the case of a spherically shaped superconductor. The shape of the

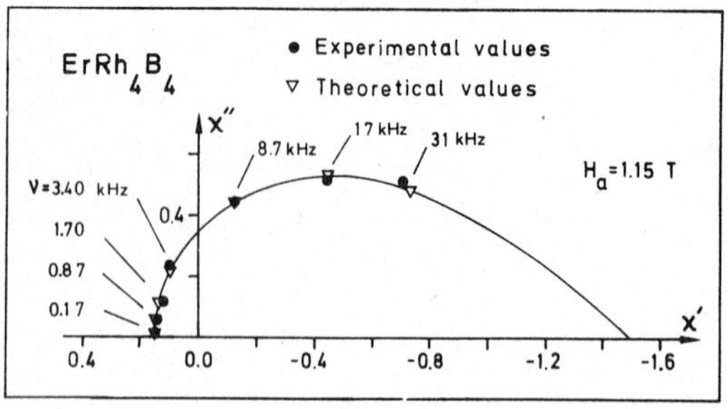

FIGURE 2
The effect of eddy currents

$\chi' - \chi''$ curve is characteristic for the effect of eddy currents in metals, however the low-frequency final value of χ is not equal to zero but to the paramagnetic susceptibility denoted by χ_{ad}. The complex susceptibility can be calculated in this case as (5)

$$\chi = -\frac{3}{2}\left(1 + 3\left(\frac{1}{1+\chi_{ad}} - 1 + \frac{1}{1+\chi_{ad}}\cdot\frac{1}{A}\right)^{-1}\right),$$

$$A = \frac{\cot ka}{ka} - \frac{1}{k^2a^2},$$

$$k = (1+i)/\delta,$$

with a = 2 mm being the radius of the sample and $\delta = (\pi\mu_0\mu_r\sigma\nu)^{-\frac{1}{2}}$ being the penetration depth. A good fit was obtained with $\sigma = 1.86 \times 10^7\ (\Omega m)^{-1}$ and $\mu_r = 1 + \chi_{ad} = 1.15$. The value of σ which can be calculated from the resistance measurements of Ott et al. (4) is equal to $9.3 \times 10^7\ (\Omega m)^{-1}$.

FREQUENCY-SWEEP MEASUREMENTS

It is not obvious whether the susceptibility measured at low frequencies with the field-sweep equipment is the isothermal or the adiabatic susceptibility. In order to decide this we performed measurements with our frequency-sweeping SQUID-susceptometer, which operates in the frequency range from 2 mHz to about 3 kHz. In Fig. 3 the measured Argand diagrams are plotted for some fields in the paramagnetic region at T = 4.2 K.

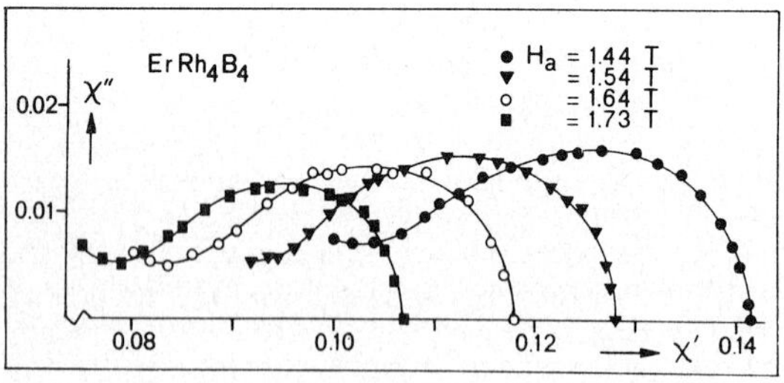

FIGURE 3
Argand diagrams measured using the frequency-sweeping technique in the range from 0.1 Hz to 100 Hz at T = 4.2 K. The curves represent the relaxation of the Er^{3+} spins.

It is found that the low-frequency final value of χ in Fig. 2 is equal to the adiabatic susceptibility $_{ad}$. A relaxation process occurs in the frequency range from 0.1 to 100 Hz, leading to semi-circular Argand diagrams at the low-frequency side. The relaxation can be described by a time constant of 0.1 s, which value is almost independent of the magnetic field.

A remarkable result of the frequency-sweep measurements is that $\chi_{ad}/\chi_T \approx 0.7$, independent of the field. In the case of a simple paramagnet a dependence according to $H_{int}^2/(H_{int}^2 + H^2)$ would have been expected where H_{int} represents the interactions between spins and environment. It has already been mentioned that the high-field behaviour of the magnetization of $ErRh_4B_4$ is influenced by the multiplet splitting and this also affects the field dependence of χ_{ad}/χ_T. It can be shown for the linear part in the M versus H curve that χ_{ad}/χ_T is indeed constant assuming that $(\delta M/\delta T)_H$ is not very dependent on the field.

We have recently performed similar measurements on a single crystal of the re-entrant superconductor $ErRh_{1.1}Sn_{3.6}$. The relaxation time appears to be of the same magnitude as that of $ErRh_4B_4$ and also the field-independent value for the ratio χ_{ad}/χ_T was found. This indicates that the energy splitting of the lowest doublets is quite similar in both compounds.

A microscopic interpretation of the relaxation of the Er^{3+} spins can not be given from these preliminary measurements. Additional experiments are necessary in order to derive the rate determining factor in the energy exchange between spins, phonons and conduction electrons. With respect to this we will perform measurements on a thermally isolated sample in order to obtain the specific heat to which the spin system is relaxing. Our frequency-sweeping technique is especially suitable for these kind of measurements because magneto-thermal effects are avoided.

REFERENCES

(1) Ø. Fischer and M.B. Maple, Superconductivity in Ternary Compounds I + II (Springer-Verlag, Berlin 1982).
(2) J.A. Overweg, H.J.M. ter Brake, J. Flokstra and G.J. Gerritsma, J. Phys. E 16 (1983) 1247.
(3) H.R. Ott, W.A. Fertig, D.C. Johnston, M.B. Maple and B.T. Matthias, J. de Phys. 39 (1978) 375.
(4) H.R. Ott, W.A. Fertig, D.C. Johnston, M.B. Maple and B.T. Mathias, J. Low Temp. Phys. 33 (1978) 159.
(5) L.D. Landau and E.M. Lifshitz, Electrodynamics of continuous media (Pergamon Press, Oxford 1975).

LT-17 (Contributed Papers)
U. Eckern, A. Schmid, W. Weber, H. Wühl (eds)

LOW TEMPERATURE PHASE DIAGRAM OF TERNARY MAGNETIC SUPERCONDUCTORS

H. C. KU*

Department of Physics, National Tsing Hua University, Hsinchu, Taiwan 300, Republic of China

Superconducting transition temperature (T_c) contour map for the ternary and pseudoternary rare earth borides with the $CeCo_4B_4$-type structure $RE(Rh_{1-x}Ir_x)_4B_4$ (RE = Gd, Tb, Dy, Ho, Er, Tm or Lu) is presented with maximum transition of 11.7 K at $LuRh_4B_4$ and a local minimum of 0.9 K at $Er(Rh_{0.2}Ir_{0.8})_4B_4$. A ground state (T = 0 K) phase diagram is also presented which show delicate interplay between superconductivity, ferromagnetism and antiferromagnetism.

1. INTRODUCTION

Ternary rare earth (RE) compounds which are magnetic superconductors have been observed in only a few crystal structure types. One of the most important is the $CeCo_4B_4$-type crystal structure (space group $P4_2/nmc$). Of the twenty six members which exhibit this crystalline structure, only seven compounds show both superconductivity and long range magnetic order (ferromagnetic or antiferromagnetic): $RERh_4B_4$ (RE = Nd, Sm, Er or Tm) and $REIr_4B_4$ (RE = Ho, Er or Tm)[1]. However, superconductivity occurs in many pseudoternary compounds $RE(Rh_{1-x}Ir_x)_4B_4$ even when one or both end members are not superconducting, which show a delicate balance between superconductivity and long range magnetic order due to the relatively weak exchange interaction between the conduction electron spins and the RE magnetic moments. This in turn, associated with transitioin metal clusters which, along with the RE ions, are the basic building blocks of theses ternary and pseudoternary compounds. This paper presents a summary report on the general occurrence of superconductivity in the pseudoternary systems $RE(Rh_{1-x}Ir_x)_4B_4$ and the interplay between superconductivity and long range magnetic order in these magnetic rare earth compounds.

2. RESULTS AND DISCUSSION

2.1. Superconductivity

The RE^{3+} ions occupy a regular sublattice 2(b) of the space group $P4_2/nmc$ and the exchange interaction between these magnetic ions cause long range magnetic order (with various magnetic structures) for all compositions. The superconductivity is believed to be primarily associated with the transition metal d-electrons that are relatively confined within the clusters and thereby interact only weakly with the RE ions. With these weak magnetic background, it was found that by changing the transition metal or rare earth metal, T_c can be varied over a wide range without disturbing the integrity of the RE sublattice.

In order to explain these peculiar features, a comprehansive picture of the T_c variation in shown in Fig. 1, where a T_c contour map with

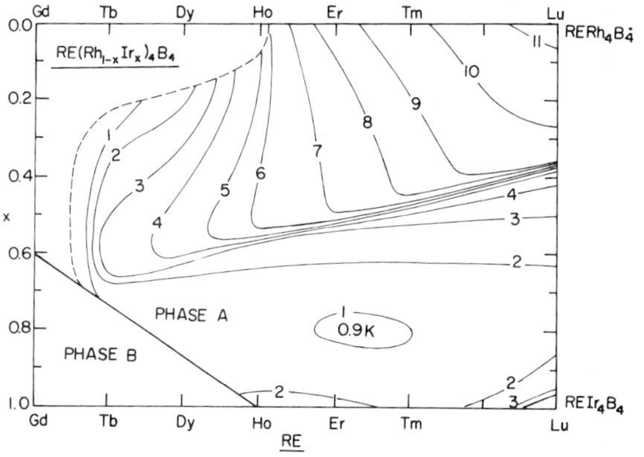

FIGURE 1
Superconducting transition temperature contour map for pseudoternary compounds $RE(Rh_{1-x}Ir_x)_4B_4$ with the $CeCo_4B_4$-type phase (phase A). Phase B ($NdCo_4B_4$-type) show no superconductivity.

constant T_c contour curves of 1 K interval is plotted for the pseudoternary compounds $RE(Rh_{1-x}Ir_x)_4B_4$ (RE = Gd, Tb, Dy, Ho, Er, Tm or Lu). The disapperance of T_c of Dy and Tb systems on the Ir-rich side is due to the disappearance of the $CeCo_4B_4$-type phase (A) and the onset of the $NdCo_4B_4$-type phase (B) which are not superconducting. The disappearance of T_c at $T \neq 0$ K for Dy and Ho systems on the Rh-rich side is due to onset of ferromagnetic order which, with large spontaneous internal magnetic field below Curie temperature, completely quench superconductivity[3,4]. In Er system where Curie temperature lower than T_c for all compositions, re-

* Research supported by the National Science Council, ROC.

entrant superconductivity was observed.[1]

The sharp drop of T_c around x $\cong$ 0.52 for Er system, x $\cong$ 0.55 for Ho system, x $\cong$ 0.62 for Dy system and x $\cong$ 0.68 for Tb system as shown in Fig. 1 is apparantly of nonmagnetic origin since drop of T_c occurred in the nonmagnetic Lu system around x $\cong$ 0.36,[5] as can be easily seen from the dense T_c contour lines in Fig. 1. This sharp variation of T_c should be contributed to the combined effects of the variation of electron-phonon interaction strength, effective phonon frequency and Coulomb interaction as the compounds with lighter 3d rhodiam metal are replaced by heavier 4d iridium metal which is in the same column of the periodic table.

At fixed x on the Rh-rich side, the decrease of T_c from Lu to Tb is largely due to the stronger magnetic depression as less magnetic RE^{3+} ions are replaced by more magnetic ions.

Maximum T_c of the $RE(Rh\text{-}Ir)_4B_4$ compounds was observed in $LuRh_4B_4$ (T_c = 12.7 K)[1], minimum T_c was observed in $Tb(Rh_{0.75}Ir_{0.25})_4B_4$ (T_c = 0.3 K) and is extrapolated to 0 K at $Tb(Rh_{0.78}Ir_{0.22})_4B_4$. The dashed line in Fig. 1 indicates the destruction of T_c by the ferromagnetic order and/or the decrease of T_c to near 0 K. Note that in $Er(Rh_{0.2}Ir_{0.8})_4B_4$, there is a local minimum about T_c = 0.9 K and T_c's on the lower half are almost constant which show the weakness of the interaction between d-electrons and RE ions.

2.2. Long range magnetic order

Coexistence of superconductivity with long range antiferromagnetic order was observed in $RERh_4B_4$ (RE = Nd, Sm and Tm), in $REIr_4B_4$ (RE = Ho, Er and Tm) and in $RE(Rh_{1-x}Ir_x)_4B_4$ (RE = Tb, Dy, Ho, Er and Tm)[1-7]. Ferromagnetic superconductors with re-entrant superconductivity was observed in $ErRh_4B_4$, $(Ho_{1-x}Er_x)Rh_4B_4$ and in $RE(Rh_{1-x}Ir_x)_4B_4$ (RE = Dy, Ho and Er)[1-4]. Ferromagnetic compounds with no superconductivity was observed in $RERh_4B_4$ (RE = Gd, Tb, Dy and Ho) and in $RE(Rh_{1-x}Ir_x)_4B_4$ (RE = Gd, Tb, Dy and Ho).[1-6]

Experiments on pseudoternary compounds provide an alternate method for studying the interaction between superconductivity and long range magnetic order and a detailed investigation of the $Ho(Rh_{1-x}Ir_x)_4B_4$ was carried out,[6] which shows the complexity of magnetic order and magnetic structure.

A ground state (0 K) phase diagram for $RE(Rh_{1-x}Ir_x)_4B_4$ compounds is proposed from all update available data which is shown in Fig. 2. Most of the magnetic compounds show antiferromagnetic order (AFM) which coexist with superconductivity (S). Antiferromagnetic Neel temperature (T_N) of 0.4 K in $TmRh_4B_4$ is expected to decrease to zero with about 50% of nonmagnetic Lu ions. Superconducting transition temperature (T_c) is expected to decrease to zero at the dashed line between AFM and AMF + S. Ferromagnetic order exists only in a small region on the Rh-rich side where no superconductivity can exist at 0 K.

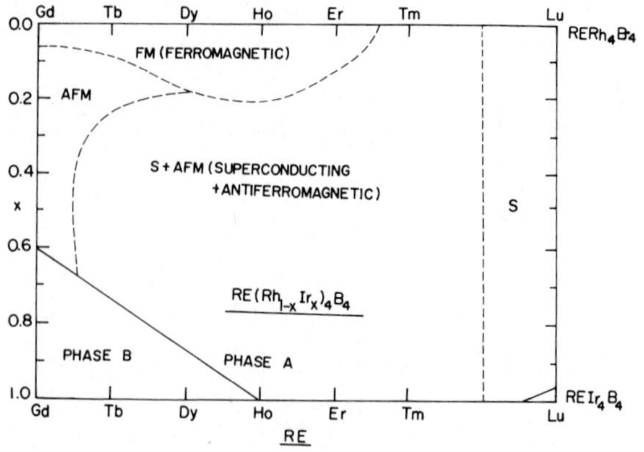

FIGURE 2
Proposed T = 0 K ground state phase diagram for pseudoternary compounds $RE(Rh_{1-x}Ir_x)_4B_4$. Superconductivity can only coexist with antiferromagnetism.

3. Conclusions

The pseudoternary rare earth boride compounds show interest interplay between superconductivity and long range magnetic order. However, more measurements on magnetic properties, especially the neutron diffraction studies, are needed to solve the complexity of the magnetic structures.

REFERENCES

(1) See Superconductivity in Ternary Compounds II, Topics in Current Physics, Vol. 34, eds. M.B. Maple and Ø. Fischer (Springer, 1982), and references cited therein.
(2) H.C. Ku, B.T. Matthias and H. Barz, Solid State Commun. 32 (1979) 937.
(3) H.C. Ku, F. Acker and B.T. Matthias, Phys. Lett. 76A (1980) 937.
(4) H.C. Ku and F. Acker, Solid State Commun. 35 (1980) 937.
(5) H.C. Ku, S.E. Lambert and M.B. Maple, Superconductivity in d and f-Band Metals 1982, eds. W. Buckel and W. Weber (Kernforschungszentrum Karlsruhe), (1982) 231.
(6) F. Acker, L. Schellenberg and H.C. Ku, Superconductivity in d- and f-Band Metals 1982, (1982) 237.
(7) K.N. Yang, S.E. Lambert, H.C. Hamaker, M. B. Maple, H.A. Mook and H.C. Ku, Superconductivity in d- and f-Band Metals 1982, (1982) 217.

LT-17 (Contributed Papers)
U. Eckern, A. Schmid, W. Weber, H. Wühl (eds)
© Elsevier Science Publishers B.V., 1984

SUPERCONDUCTING TRANSITION TEMPERATURE AND STRUCTURE OF ION IRRADIATED NbC THIN FILMS

N. KOBAYASHI*, R. KAUFMANN** and G. LINKER

Kernforschungszentrum Karlsruhe, Institut f. Nukleare Festkörperphysik, P.O.B.3640, Karlsruhe, FRG

Single phase NbC thin films have been irradiated with 600 keV Ar^{++} ions. The superconducting transition temperature decreased continuously from 11 K to 4 K for fluences up to 10^{16} ions/cm² and saturated for higher fluences. With the preservation of the B1 structure, the lattice parameter first increased, then decreased below its initial value and saturated in the T_c saturation range. The lattice parameter increase coincided with the formation of a defect structure consisting of small static displacements of the lattice atoms. The changes of the lattice parameter are discussed with respect to the agglomeration of point defects.

1. INTRODUCTION

Refractory compounds have interesting physical properties - high melting points, hardness and high superconducting temperatures(T_c). NbC is one of these compounds with a T_c beyond 11 K and a stable B1 structure over a wide range of vacancy concentrations. T_c of transition metal carbides and nitrides with B1 structure undergo less depression by neutron irradiation in comparison with A15 compounds (1) but few investigations of the irradiation effects for these compounds have been reported (2,3). Therefore, irradiation studies of NbC over a wide range of fluences and different ion species are of interest. In the following, the results of T_c changes of NbC by ion irradiation as well as the structural distortions and variations of the normal conducting properties are presented.

2. EXPERIMENTAL

Stoichiometric NbC thin films with thicknesses of about 200 nm have been deposited onto sapphire substrates by reactive RF sputtering using CH_4 gas. The partial pressure of CH_4 was 1.7 to 2.0 x 10^{-3} Torr and the substrate temperature was maintained from 830 to 1000°C during the sputtering process. Film thickness, homogeneity and composition were analyzed by RBS measurements with 2 MeV He ions. The films have been irradiated with Ar^{++} ions (600 keV) at R.T. from fluences of 10^{13} to 10^{17} ions/cm². T_c was measured resistively and inductively. Structural information was obtained by X-ray analysis employing a Guinier thin film diffractometer.

3. RESULTS

T_c decreases continuously with increasing irradiation fluence and reaches saturation values. This behaviour is shown in Fig. 1 where the T_c values are plotted as a function of the irradiation fluence and corresponding d.p.a. (displacements per atom) values. T_c shows a depression from 11 K to 4 K, saturation at a fluence of 10^{16} ions/cm² (20 d.p.a.) and a slight recovery at 10^{17} ions/cm². T_c for the maximum fluence is the result of an additional irradiation at 77 K.

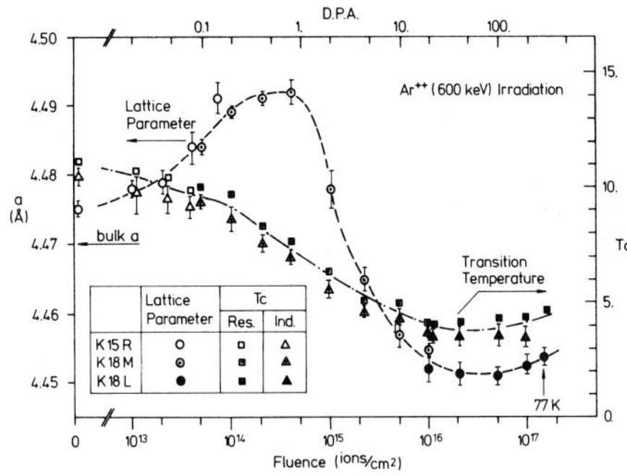

FIGURE 1
The superconducting transition temperature T_c and the lattice parameter a of argon irradiated NbC films as a function of the fluence and corresponding d.p.a. The lattice parameter of the bulk material is indicated by an arrow.

The X-ray diffraction investigation revealed appreciable changes of the lattice parameter with the preservation of the B1 structure after irradiation. It also shows some line broadening and intensity degradation of the Bragg peaks. In Fig. 1 the relation of the lattice parameter a to the fluence and d.p.a. is also shown. It increases and reaches a maximum value which is greater than the initial value by 0.4% at 4 x 10^{14} ions/cm² (0.8 d.p.a.) and then decreases down to a saturated value which is 0.5% smaller than a of undistorted NbC, in the fluence range where the T_c saturation was observed. Fig. 2 shows the variation of the static atomic displacements u as a function of the irradia-

* On leave from: Electrotechnical Laboratory, Sakura-mura, Ibaraki, Japan
**Present address: Dornier System GmbH, Friedrichshafen, FRG

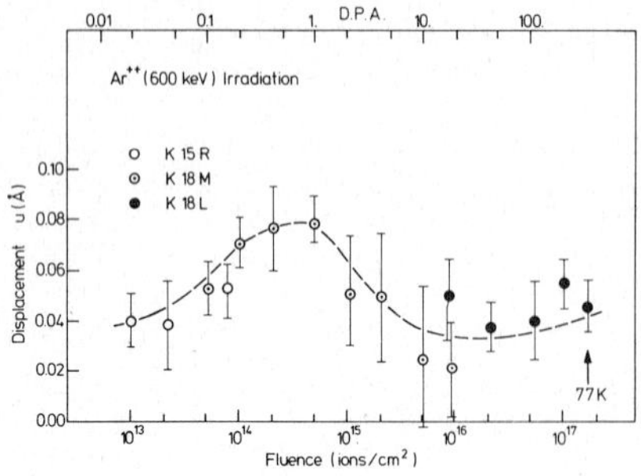

FIGURE 2
Static atomic displacements u obtained from X-ray analysis as a function of the irradiation fluence.

tion fluence. These values are calculated from the Debye-Waller factor obtained from the peak intensity change before and after the irradiation (4). u has a maximum value again at $4x10^{14}$ ions/cm², decreases and saturates for above 10^{16} ions/cm². This behaviour coincides well with that of the lattice parameter.

The monotonous decrease of T_c could be due to a continuous increase of the vacancy concentration . The increase of the lattice parameter in the first stage of irradiation, however, is thought to be due to the increase of the concentration of interstitial atoms which overcompensate the influence of vacancies, whereas after higher fluence irradiations the interstitials agglomerate or precipitate and the decrease of the lattice parameter can be, therefore, caused only by the vacancies.

Fig. 3 shows the relation of the resistivities (at R.T. ρ_{295}, residual resistivity at 12 K ρ_{12} and thermal part of the resistivity ρ_{th}) and the residual resistivity ratio r

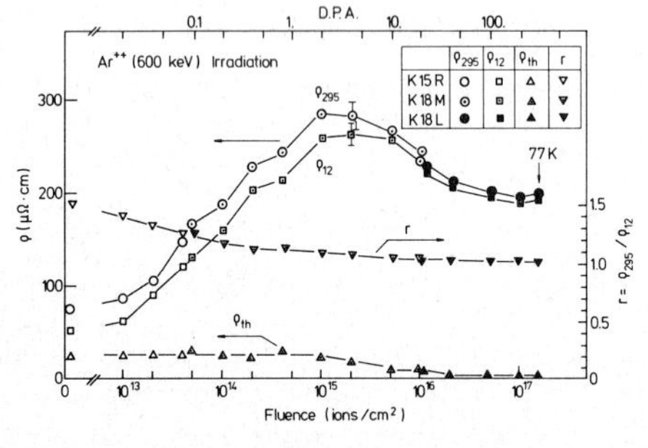

FIGURE 3
Resistivities (ρ_{295}, ρ_{12} and ρ_{th}) and residual resistivity ratio r of argon irradiated NbC films vs. irradiation fluence and d.p.a.

($r=\rho_{295}/\rho_{12}$) to the irradiation fluence and d.p.a. The tendency of a continuous decrease of ρ_{th} is in agreement with the decrease of T_c. r also decreases continously and approaches 1. ρ_{12}, which can be taken as an integral measure of disorder, has a maximum value at $2x10^{15}$ ions/cm² (4 d.p.a.) and then decreases. It thus reveals a similar behaviour like the lattice parameter however with a shift of the maximum effects on the fluence scale. This could be due to a different influence of the discussed defect agglomeration process on the lattice parameter and the residual resistivity. Definite conclusions, however, could only be drawn on the basis of a quantitative analysis which cannot be performed at the present state of the knowledge of defects in irradiated refractory compounds.

It is well known that T_c of NbC decreases strongly with increasing vacancy concentration (5,6). This is attributed to an electronic charge transfer towards the vacancy and a resulting change of the partial local density of states (7). The T_c depressions by the irradiation could be explained by the same reason. The relation of T_c versus the lattice parameter is, however, quite different in non-stoichiometric NbC_x ($x < 1$) and in irradiated NbC. This is because the T_c degradation of the irradiated samples could be due not only to carbon vacancies, but also to metal vacancies, whereas the lattice parameter is strongly affected by the metal interstitials.

4. CONCLUSIONS

T_c degradation and the structural distortions of NbC have been observed by Ar^{++} ions (600 keV) irradiation. The change of the lattice parameter may be understood by the production and agglomeration of point defects. The decrease of T_c is thought to be due mainly to the growing vacancy concentration which will affect the electronic density of states at the Fermi energy.

ACKNOWLEDGEMENT
Many helpful discussions with Dr. O. Meyer as well as the ion irradiation by Mr. M.Kraatz are gratefully acknowledged.

REFERENCES
(1) D. Dew-Hughes and R. Jones, Appl. Phys. Lett. 36 (1980) 856.
(2) V. Jung, this volume.
(3) J. Ruzicka, E.L. Haase and O. Meyer, this volume.
(4) G. Linker, Nucl. Instr. Meth. 182/183 (1981) 501.
(5) A.L. Giorgi, E.G. Szklarz, E.K. Storms, A.L. Bowman and B.T. Matthias,Phys. Rev. 125 (1962) 837.
(6) L.E. Toth, M. Ishikawa, Y.A. Chang, Acta Met. 16 (1968) 1183.
(7) G. Ries and H. Winter, J. Phys. F: Metal Phys., 10 (1980) 1.

LT-17 (Contributed Papers)
U. Eckern, A. Schmid, W. Weber, H. Wühl (eds)

Ar^{++} IRRADIATION EFFECTS IN SUPERCONDUCTING NbN SPUTTERED LAYERS

Volkhard JUNG

Kernforschungszentrum Karlsruhe, Institut für Nukleare Festkörperphysik, P.O. Box 3640, D-7500 Karlsruhe 1, Federal Republic of Germany

Irradiation of sputtered NbN layers with Ar^{++} ions at small fluences causes an decrease of the lattice parameter a. With increasing dose, a decreases continuously and reaches saturation at a fluence of 10^{17} Ar^{++}/cm^2. The superconducting temperature T_c decreases monotonously with increasing irradiation. The residual resistance ρ_o at T_c + 2 K increases to a maximum and drops again with higher irradiation fluence. Specimens containing a second phase were transformed into pure B_1 phase by irradiation with $6 \cdot 10^{16}$ Ar^{++}/cm^2.

1. INTRODUCTION

Dew-Hughes and Jones have reported that irradiation with fast neutron causes only a small drop of T_c in NbN, i.e. to T_c = 14 K (1). However, the irradiation doses used in this investigation were too small to reach the saturation of the T_c decrease. In the present work NbN has been irradiated with 600 keV Ar^{++} ions, which allows to achieve much larger doses than with fast neutrons and thus to explore the saturation limit of T_c after irradiation with very large doses.

2. EXPERIMENTAL

The sputtering conditions to prepare NbN have been investigated by Pan et al. over a wide range of parameters (2). Based on these results we have sputtered NbN layers on Al_2O_3 substrates at 750°C with an Ar partial pressure of $2 \cdot 10^{-2}$ Torr. Layers sputtered with different Ar and N partial pressures contained a second phase. The sputtered layers were irradiated at liquid nitrogen temperature and at room temperature with Ar^{++} ions of 600 keV. As the range of Ar^{++} ions of this energy (2400±550 Å) slightly exceeds the thickness of the NbN layers (2000 Å), only a few ions stopped in the NbN film. After each irradiation step the critical temperature T_c, the lattice parameter a and the resistivity ρ at RT and at T_c+2K were measured. The measurement of the lattice parameter a was carried out with a X-ray diffractometer with Seemann-Bohlin focussing. To avoid texture effects each irradiated sample was measured twice, first at 0° and then at 180° with respect to an axis normal to the layer. Thus, for the calculation of the lattice parameter up to 20 reflection positions were used.

3. RESULTS

In Fig. 1 the lattice parameter a and the critical temperature T_c are shown as function of the irradiation dose. The lattice parameter increases up to a dose of 2 to $3 \cdot 10^{14}$ Ar^{++}/cm^2. This is attributed to distortion of the lattice by displaced atoms, which after irradiation are on interstitial places. By further irradiation the lattice parameter decreases steeply, until beyond 10^{17} Ar^{++}/cm^2, it remains nearly constant. In this case there may be a balance between knock-out and diffusion back of atoms out of the lattice and from possible depositions at the grain boundaries, respectively. Generation and

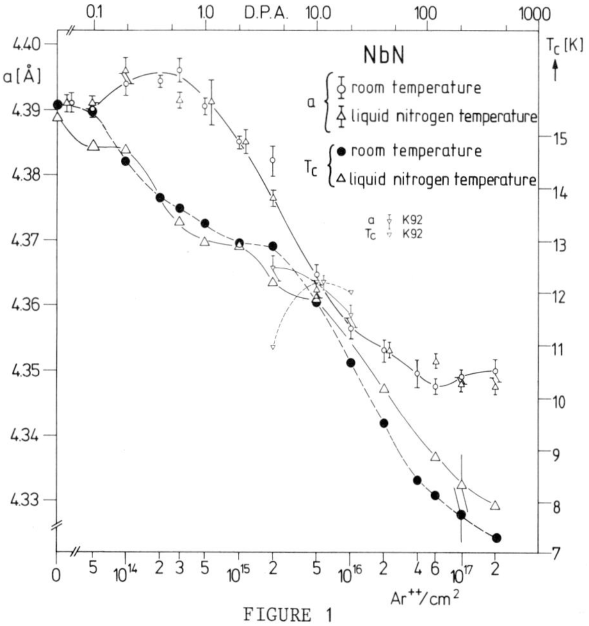

FIGURE 1

Dependence of the lattice parameter a and the superconducting temperature T_c on the Ar^{++} irradiation dose. Two parts of the same sample were irradiated at room temperature and at liquid nitrogen temperature, respectively, in steps. Another specimen (K 92, inverted triangles) was irradiated in one step with $2 \cdot 10^{15}$ Ar^{++}/cm^2.

recombination of defects are in balance. The depositions at the grain boundaries could have grown during irradiation. The decrease of $\underline{a}$ and T_c is attributed to an increasing number of vacancies in both the Nb and N sublattices.

If the plot of Horn and Saur (3) (T_c as function of the N/Nb ratio) and the plot of Brauer and Kirner (4) ($\underline{a}$ as function of the N/Nb ratio) are combined, one can draw a plot of T_c as function of the lattice parameter $\underline{a}$. The N/Nb ratio varies from 0.85 to 1.00 (maximum of lattice parameter) and then to 1.10. A nearly parallel descend of the measured T_c has been found in the present investigation for the region of the literature values N/Nb=1.0 to 1.1 (see Refs. (5) and (6)).

In Fig. 2 the specific resistance is shown as a function of the dose. The value of the unirradiated specimen (350 μΩ·cm) is about a factor 7 higher than that of bulk material. This may be due to the fact that the crystallites grow in columns normal to the layer plane, and in the direction parallel to the layer there are much more grain boundaries than normal to it. Also intrinsic defects may increase ρ. It is surprising that ρ_{RT} is smaller than ρ_0 at high fluence.

In Fig. 3 the results are shown for an irradiated specimen containing a second phase. A dose of

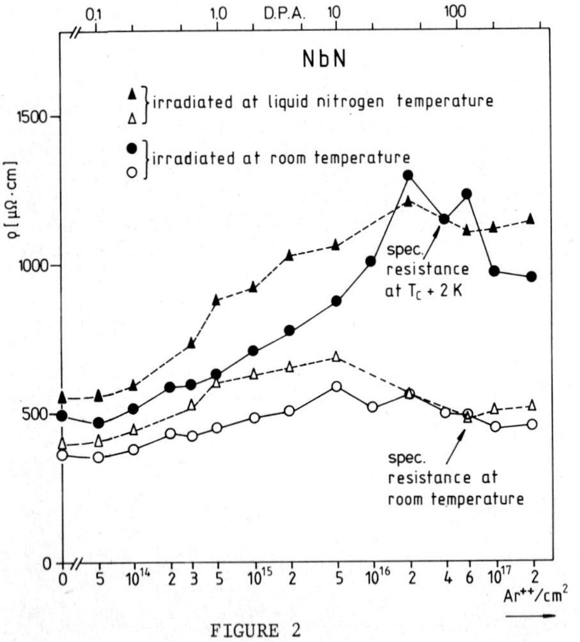

FIGURE 2
Specific resistance ρ of two halves of the same sample with pure B_1 phase as function of Ar^{++} irradiation, irradiated at liquid nitrogen temperature and at room temperature, respectively. The values of ρ have to be corrected slightly due to the reduction of layer thickness by sputtering during irradiation at high dose.

$6 \cdot 10^{16}$ Ar^{++}/cm^2 has transformed the second phase into B_1. T_c does not decrease as rapidly as for pure B_1 phase specimens and is higher by 2.7 K at $2 \cdot 10^{17}$ Ar^{++}/cm^2. The lattice parameter does increase again with higher irradiation. This result may be understood as follows: The phase fractions are not constant during irradiation, mixed B_1 and second phase at first, and relatively undamaged B_1 material forms by phase transformation at the end of the irradiation, as is indicated by the narrow superconducting transition range (ΔT_c=0.7K) for the specimen irradiated with the highest dose.

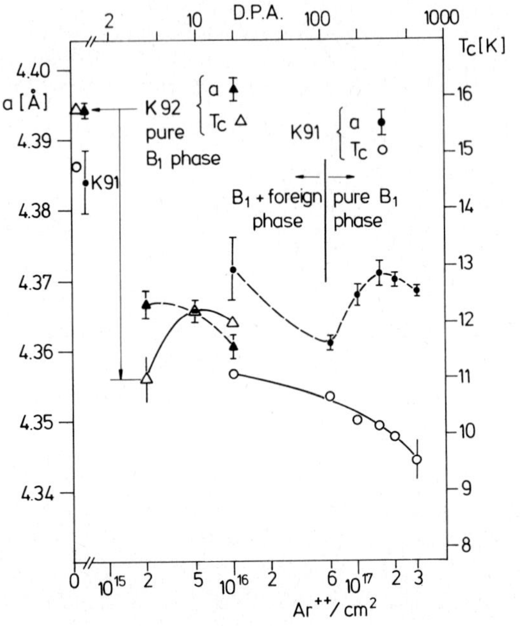

FIGURE 3
Dependence of the lattice parameter $\underline{a}$ and the superconducting temperature T_c on the Ar^{++} fluence for a specimen with pure B_1 phase (K 92) and for a specimen with B_1+second phase (K 91). After irradiation with $6 \cdot 10^{16}$ Ar^{++}/cm^2 the second phase is completely transformed.

ACKNOWLEDGEMENT

The author would like to express his gratitude to Dr. O. Meyer for stimulating discussions, to M. Kraatz for the irradiation of the specimens and to R. Smithey for the T_c measurements.

REFERENCES

(1) D. Dew-Hughes and R. Jones, Appl.Phys.Lett. 36 (10) 856 (1980).
(2) V.M. Pan et al. Cryogenics 23, 258 (1983).
(3) G. Horn and E. Saur, Z. Phys. 210, 70 (1968).
(4) G. Brauer and H. Kirner, Z. Anorg. Allg. Chem. 328, 34 (1964)
(5) N. Kobayashi et al. this volume.
(6) J. Ruzicka et al. this volume.

LT-17 (Contributed Papers)
U. Eckern, A. Schmid, W. Weber, H. Wühl (eds)

TUNNELING AND NEUTRON SCATTERING EXPERIMENTS ON SUPERCONDUCTING TiN

J. GEERK, F. GOMPF and G. LINKER
Kernforschungszentrum Karlsruhe, Inst. f. Nukl. Festkörperphysik, P.O.B.3640, Karlsruhe, FRG
K. BISCHOFF and W. BUCKEL
Physikalisches Institut, University of Karlsruhe, FRG

Tunnel junctions on reactively sputtered TiN films were prepared using artificial tunnel barriers of AlZr oxide. The junctions showed an energy gap of 0.8 meV for TiN and $2\Delta_o/kT_c = 3.7$. The peaks found in the second derivative traces in the energy region of the phonons agree well with the main features of the acoustic phonon density of states, which has been determined by inelastic neutron scattering.

1. INTRODUCTION

Compared to the extensive tunneling work done so far on the group of high T_c A15 superconductors there has been very little effort spent onto the transition metal carbides and nitrides with B1 structure which are very interesting for applications such as superconducting interferometers and high field superconducting coils. Early work on this group of superconductors are the second derivative studies of Zeller (1) on TaC and the work of Geerk et al. on NbC (2) who extracted an Eliashberg function probably strongly affected by proximity effects and compared it to the phonon density of states. These studies mainly suffered from the very poor quality of the natural oxide of these materials. The tunneling barrier height of the natural oxide of NbC turned out to be much lower than that of Nb the latter already giving serious problems in this respect. The recent progress of the tunneling experimental technique due to artificial barriers stimulated us to apply the overlayer technique to the B1 superconductor TiN. We chose TiN because its phase diagram is comparably uncomplicated in the vicinity of the cubic phase which should make the preparation of thin films easy.

2. EXPERIMENTAL

Thin film tunnel junctions of the configuration TiN-AlZr oxide-In have been prepared using the magnetron sputtering apparatus as described in Ref. (3). The TiN thin films were sputtered reactively in an Ar, N_2 mixture with $P_{Ar}=6\cdot10^{-2}$ Torr and $P_{N_2} = 5\cdot10^{-3}$ Torr and a substrate temperature of 940°C. X-Ray measurements were carried out to confirm the B1 structure of the sputtered material. The neutron scattering experiments were performed on a TiN powder sample (T_c=5 K) using a multidetektor time-of-flight spectrometer with an incident energy of 65 meV. The generalized phonon density of states $G(\hbar\omega)$ was corrected for multiphonon contributions, spectrometer resolution and intensity errors due to the difference in scattering cross section and mass of the Ti and N atoms in order to obtain a true density of states $F(\hbar\omega)$. For the latter correction we calculated a shell model fit to phonon dispersion data of Ref. (4).

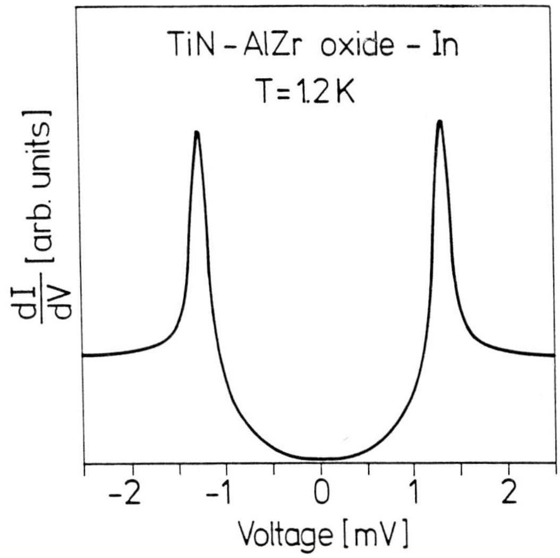

FIGURE 1
dI/dV versus V for a TiN-AlZr oxide-In tunnel junction at 1.2 K. The In counter electrode (Δ_o=0.53 meV) is in the sc state.

3. RESULTS AND DISCUSSION

The main experience from the experimental point of view we can report from these studies is that the preparation of the junctions turned out to be comparably easy, the number of shorts being restricted to about 30% of the junctions prepared. The main difficulty was the preparation of TiN films with high T_c and acceptable homogeneity. Fig. 1 shows the dI/dV versus V curve of a TiN-AlZr oxide-In junction with the

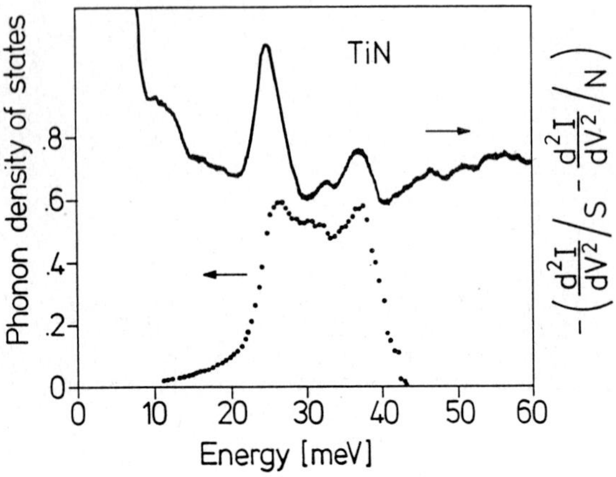

FIGURE 2
$-d^2I/dV^2$ versus energy of a TiN-AlZr oxide-In tunnel junction. In is in the nc state. The dots show the phonon density of states.

In counter electrode in the superconducting state. We see well defined peaks at the sum of the gaps at $\pm$ 1.30 mV. At zero voltage we observe a small leakage conduction and between 0.5 and 1.2 mV there is an increasing conduction probably due to tunneling into nitrogen deficient material. From Fig. 1 we deduce an energy gap Δ_o=0.8 meV for our thin film TiN sample. T_c in the vicinity of the junction has been determined inductively yielding T_c=5.0$\pm$0.2 K. As the maximum T_c of TiN is about 6 K (5) and in cause of its high sensitivity to nitrogen deficiency we conclude that our thin film sample as well as the neutron scattering sample are within 2 at% of nitrogen close to stoichiometry. Concerning the coupling strength we get $2\Delta_o/kT_c$=3.7 suggesting weak coupling for TiN.

In cause of the high excess currents and normalizing difficulties caused by the high residual resistivity of our samples we did not extract quantitative data (Eliashberg function) from our tunneling measurements but instead present second derivative data ($-d^2I/dV^2$) which show peaks where the Eliashberg function shows peaks or shoulders. In Fig. 2 we show the second derivative data along with the phonon density of states (PDOS) of the acoustic phonons of TiN. We see an excellent agreement of the peaks near 26 and 37 meV. Also the small hump in the middle of the PDOS appears in the tunnel data. The cut-off energy is marked in the tunnel data by the minimum at 41 meV in agreement with the upper flank of the PDOS. At low energies we see some structure near 12 meV in the tunnel data where we find no counterpart in the PDOS. As in the tunneling experiment the intensity of the Eliashberg function is scaled with $1/\omega^2$ we expect these low energy structures to be caused by very small structures in the Eliashberg function.

In the energy region of the optical phonons near 70 meV we could not locate any significant structure in the tunneling experiment. From theoretical calculations (6) we can expect for the optical part of the Eliashberg function a coupling strength almost equal to the acoustic one. In that case the structures in the tunnel data due to optical phonons would be about a factor of 10 smaller than those in the acoustic part, which makes it tough to separate them from multiphonon effects and also from noise.

In conclusion we have prepared tunnel junctions on TiN with AlZr oxide barriers and found excellent agreement of the tunnel data with the neutron scattering data in the acoustic part of the phonon spectrum. Better junctions with respect to homogeneity are needed for quantitative studies.

REFERENCES

(1) H.R. Zeller, Brown Boveri Research Report KLR-71-18 (1971).
(2) J. Geerk, W. Gläser, F. Gompf, W. Reichardt, E. Schneider (1975), in "Low Temperature Physics-LT 14", Vol. 2, (ed. S. Krusius).
(3) U. Schneider, J. Geerk and H. Rietschel, this report.
(4) W. Kress, P. Roedhammer, H. Bilz, W.D. Teuchert and A.N. Christensen, Phys. Rev. B 17 (1978) 111.
(5) T. Wolf, Thesis 1982, Univ. Karlsruhe.
(6) H. Rietschel, H. Winter and W. Reichardt, Phys. Rev. B 22 (1980) 4284.

LT-17 (Contributed Papers)
U. Eckern, A. Schmid, W. Weber, H. Wühl (eds)
© Elsevier Science Publishers B.V., 1984

ANNEALING AND IRRADIATION STUDIES OF MoC LAYERS

J. RUZICKA*, E.L. HAASE and O. MEYER

Kernforschungszentrum Karlsruhe, Institut für Nukleare Festkörperphysik, P.O.B. 3640, D-7500 Karlsruhe, FRG

MoC thin films were prepared by sputtering using different deposition conditions. The B1 structure is formed phase pure for substrate temperatures T_S between -100 and 900°C with T_c-values up to 12 K. The annealing behaviour is different for samples grown at different T_S and a phase transformation in Mo_2C is observed for $T_S \geq 600$°C. In contrast to NbC and NbN only minor changes in T_c and lattice parameter a_o are observed upon irradiation with 300 keV He-ions.

1. INTRODUCTION

By rapid quenching from the melt B1-MoC_x was synthesized with maximum values of x = 0.75, a_o=0.4278 nm and T_c = 14.3 K (1-3). The objective of this work was to prepare thin films of B1-MoC_x by sputtering in order to investigate whether the carbon-vacancy concentration could be further reduced by applying this nonequilibrium preparation technique. Similar to other refractory compounds, T_c and a_o are expected to increase with decreasing carbon-vacancy concentration. For compound films deposited by the sputtering technique at low substrate temperatures the composition can often be varied with ease. However, these films may be defective or even amorphous. Thus the influence of annealing procedures on the film properties was also investigated. A further knowledge on defect structures and their influence on film properties is expected by performing irradiation studies and comparing with results obtained for NbC (4) and NbN (5).

2. RESULTS

Fig. 1 shows a compilation of data of T_c vs. lattice parameter a_o. The four points with T_c=9-14 K and a_o=0.427-0.428 nm stem from rapidly quenched samples (1-3). Our results from samples sputtered reactively in an Ar-CH_4 plasma are shown as +. Samples sputtered in an Ar-plasma with a Mo-graphite composite target are shown as solid circles. For low a_o-values T_c increases linearly with increasing a_o, whereas for large a_o-values T_c saturates. Samples with large a_o-values could only be grown at T_s below 400°C. The carbon content of 40 and 43 at% is indicated in Fig. 1 We did not succeed to prepare samples with T_c-values above 12 K. This corresponds to a C-content of 41.5 at% for optimally prepared samples.

The dashed line represents an asymptotic limit for optimally prepared MoC and has a slope of 2000 K/nm. For the samples with large a_o-values T_c is either too small for the given a_o or a_o is too large for the measured T_c. Either result indicates the presence of defects for samples prepared at low T_S.

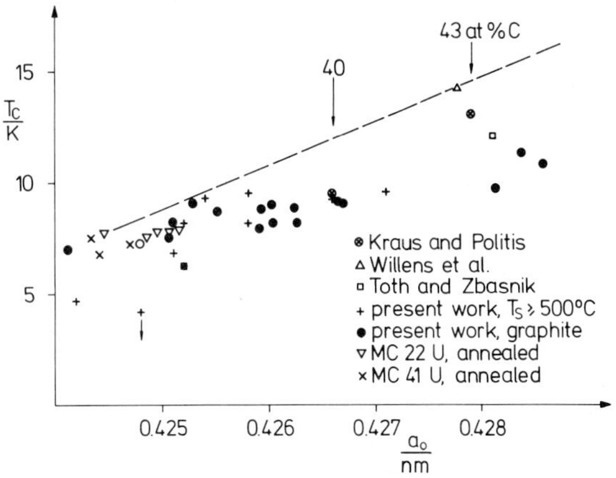

FIGURE 1
Compilation of T_c vs. lattice parameter a_o of rapidly quenched, sputtered and annealed MoC samples. T_c increases with increasing a_o as stoichiometry is approached.

Isochronous annealing experiments were performed using an annealing time of 1 h at each temperature step for films grown at low (<200°C) and high substrate temperatures. Fig. 2 shows T_c, ao and ρ(295 K) vs. annealing temperature T_A for a sample deposited at 700°C. For the T_c-values measured resistively the onset, midpoint and downset are shown. T_c drops noticably in the temperature range below 500°C, then shows an oscillating behaviour. In this region ao and the specific resistivity at RT ρ change only very little. Between 600 and 800°C the B1 phase transforms to Mo_2C. The T_c of 4.7 K at 1200°C points to α-Mo_2C. ρ has a substantially

*On leave from the Institute of Physics, Czechoslovak Academy of Sciences, Prague 8, Czechoslovakia

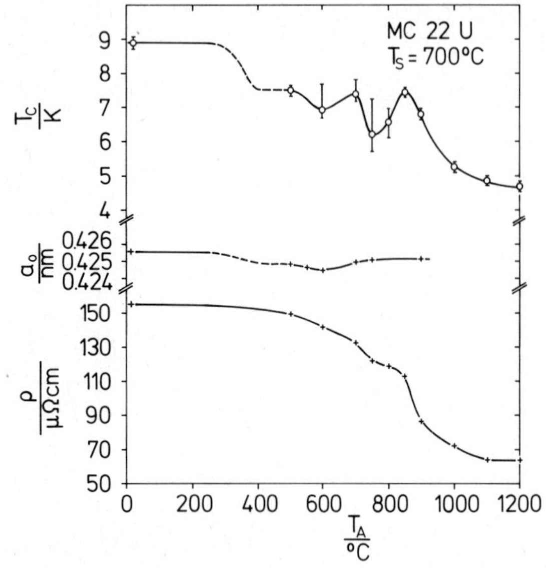

FIGURE 2
T_c, a_o and ρ(295 K) are shown as a function of the annealing temperature T_A. Between 600 and 800°C the phase transformation from B1 to Mo_2C takes place.

smaller starting value (154 μΩcm) than that for samples prepared at lower substrate temperatures (500-700 μΩcm). ρ drops noticably between 600-800°C where the phase transformation takes place. The starting residual resistivity ratio was 1.092.

Altogether five samples were annealed. For all T_c and a_o dropped noticably between 250 and 500°C, the drop of T_c and a_o being larger for the samples prepared at lower T_S. The observed $\Delta T_c/\Delta a_o$ does not correlate with the value from Fig. 1. This result indicates that the T_c- and a_o-decrease is due to the annealing of defects. The higher T_c-values up to 12 K observed for layers produced at low T_S seems also to be correlated with additional defects. The T_c-increase observed for T_A>500°C is probably caused by carbon indiffusion from carbon precipitates.

Irradiation studies were performed on a pure B1-phase MoC_x sample grown at 800°C with a grain size of about 40 nm. Fig. 3 shows the behaviour under irradiation with 300 keV He^+ at room and at liquid nitrogen temperatures. The top axis shows the number of displacements per atom (dpa). T_c rises slightly to 8.3 K for dpa-values up to 0.2 and then reaches a saturation at 7.5 K. For the irradiations at liquid nitrogen temperature a similar behaviour is observed except that the T_c-drop is slightly smaller. The T_c-increase at low dpa-values is probably correlated with an increase of defects. a_o only

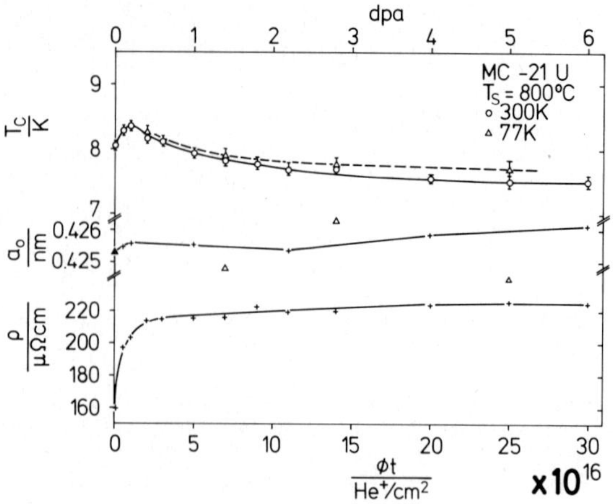

FIGURE 3
T_c, a_o and ρ(295 K) are shown as a function of the irradiation fluence.

changes minimally by 2°/oo. Both the T_c and a_o changes are much smaller than for NbC (1) and NbN (2), which is probably due to a large number of defects in the starting material. ρ(295 K) rises steeply from 160 to about 205 μΩcm for dpa-values up to 0.2 and approaches 225 μΩcm asymptotically. The residual resistance ratio starts at 1.12 and approaches 1.

3. SUMMARY AND CONCLUSIONS

The observation of Mo-segregation and of Mo_2C formation already at 600°C means that diffusion takes place over distances of the order of the grain size of 10-40 nm. This seems to prevent the formation of MoC with more than about 41.5 at% C in the B1 phase. No T_c-values above 12 K were reached. MoC behaves in the preparation as well as during annealing and irradiation differently than typical refractory materials. The steep increase of ρ and T_c in the low dpa-region is a further hint that the larger T_c-values for low T_S samples are caused by a special defect structure which can either be produced by sputtering at low substrate temperatures or by irradiating films grown at high substrate temperatures.

REFERENCES

(1) W. Krauss and C. Politis, in: IV. Conf. on Superconductivity in d- and f-Band Metals, eds. W. Buckel and W. Weber (Kernforschungszentrum Karlsruhe, 1982), 439 and private communication.
(2) R.H. Willens and E. Buehler, Appl. Phys. Lett. 7 (1965) 25.
(3) L.E. Toth and J. Zbasnik, Acta Met. 16 (1968) 1177.
(4) N.Kobayashi,R.Kaufmann,G.Linker,this volume.
(5) V. Jung, this volume.

LT-17 (Contributed Papers)
U. Eckern, A. Schmid, W. Weber, H. Wühl (eds)

SOLID STATE EPITAXIAL GROWTH OF SINGLE CRYSTAL NbN ON SAPPHIRE*

J. R. GAVALER, J. GREGGI, and J. SCHREURS

Westinghouse R&D Center, 1310 Beulah Road, Pittsburgh, Pennsylvania 15235

Single crystal superconducting B1 structure films with a (113) surface orientation have been prepared on $(2\bar{1}\bar{1}3)$ sapphire substrates by a solid state epitaxial process. These films have critical temperatures, $T_c \sim 16K$, and lattice parameters, $a_0 = 4.39$ Å, which are typical for NbN. The single crystals were formed by annealing amorphous films which had been prepared by a sputtering process.

1. INTRODUCTION

In a previous publication it was reported that superconducting B1 structure NbN films can be prepared by annealing sputtered, x-ray amorphous Nb-N. (1) Among other variables, the microstructure was found to be strongly dependent on the substrate material. Data have now been obtained which show that single crystal NbN can be grown by this method on sapphire substrates of the proper orientation. In this paper we report these data.

2. EXPERIMENTAL PROCEDURE

The method used to prepare the NbN films for this work has been described in detail in previous publications. (1,2) Briefly, it involves dc reactive sputtering of niobium in pure nitrogen or in various argon-nitrogen atmospheres. For the present experiments pure nitrogen was used as the sputtering gas. The sputtering target was made by pressing and sintering niobium and carbon powders mixed in ratio of 4 at. % Nb/1 at. % C. The use of this target rather than a pure Nb target was based upon a previously stated conclusion (3) that B1 structure NbN prepared at < 1300°C is an impurity stabilized phase with the stabilizing impurity being carbon and/or oxygen. Sputtering voltages typically were 1600 volts, current density, 10 mA/cm², and target-substrate separation, 1.3 cm. The deposition rate under these conditions was of the order of 100 Å/min. Typical sputtering times were between ten and twenty minutes. The substrates were either polycrystalline alumina or sapphire. The sapphire substrates had three different surface orientations, namely, $(2\bar{1}\bar{1}3)$, $(1\bar{1}02)$ and a third orientation which was made by a random cut through the sapphire and which was not close to any major plane. The alumina which was made by pressing and sintering Al_2O_3 powder was used as-fabricated, while the sapphire had a polished surface similar to that used in optical windows. All of the substrates were cut into 0.03 x 1.3 x 0.5 cm pieces and rinsed in acetone before being placed into the sputtering chamber. No outside heating of the substrates was used, however during the sputtering process, due to the small target-substrate separation, the temperature rose to a maximum of ~ 400°C after about 30 minutes. After deposition the films were annealed at 700°C for 30 minutes under a vacuum of ~ 10^{-7} Torr. Analyses were by Auger, TEM, RHEED, X-ray, and electron diffraction. Superconducting critical temperatures were measured by a standard four-point resistive technique using germanium thermometry.

3. RESULTS AND DISCUSSION

Films sputtered on both alumina and sapphire, were found to be amorphous, based on X-ray analyses. In these films, crystallinity first appears in the form of the B1 structure after annealing at 500°C or higher. It should be emphasized that this result is specific to our particular set of experimental conditions. Other workers have reported that NbN films sputtered onto lower-temperature substrates were already crystallized. (4) Probably the most significant parameter in the sputtering process which can account for our high crystallization temperature is the small target-substrate separation of 1.3 cm. This small separation was chosen because of the experimental observation that as the spacing between the target and substrate became smaller, the criticality of having an exact argon/nitrogen ratio to achieve optimum T_c's became less. In fact, films sputtered in pure nitrogen using the spacing, of 1.3 cm had optimum critical temperatures of $\gtrsim$ 16K. An additional and

*Supported in part by the AFOSR Contract No. F49620-83-C-0035.

perhaps a more important benefit may be a homogenization of the deposited film resulting from mixing by the energetic particles striking the substrate surface. This is indicated from chemical analyses which show flat depth profiles of niobium, nitrogen, carbon, and oxygen in the sputtered films. (The oxygen found in the film was introduced unintentionally during the sputtering process.)

Using the above procedure, B1 structure superconducting films approximately 1500 Å thick all having T_c's of ~ 16K and ΔT_c's of < 0.5K were prepared on both alumina and single crystal sapphire substrates. The films on alumina were similar to those reported previously. (1) They were found to consist of ultrafine, equiaxed, randomly oriented grains. The same type of microstructure was seen in the films crystallized on the sapphire substrates with odd orientations not close to any major plane. The niobium nitride grown on the $1\bar{1}02$ sapphire was also polycrystalline however in this case the film were highly textured with a (111) orientation. Finally the films on the $(2\bar{1}\bar{1}3)$ sapphire were found to be single crystal. Figure 1 shows a TEM and an electron diffraction photographs of one of the single crystals. This film is approximately 1200 Å thick. It has a T_c of 16.3K and a transition width of 0.4K. Its surface orientation is (113) and its lattice parameter is 4.39 Å. As can be noted, the film also contains a large number of somewhat oriented voids. Based on the T_c and lattice parameter which are typical for NbN it would appear that very little of the carbon in the film was actually incorporated into the BI structure.

We have analyzed the structure of the sapphire and its NbN overlayer shown in Figure 1

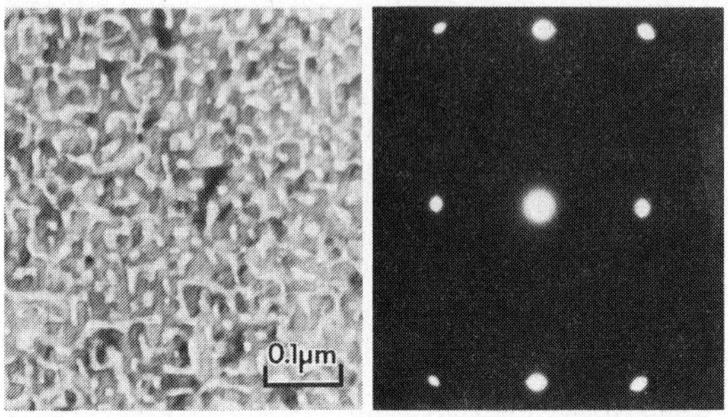

(a) (b)

FIGURE 1
Transmission electron micrograph of a single crystal NbN film (a) and its corresponding single crystal electron diffraction pattern showing a (112) B1 orientation (b).

and find that there are many major planes in the two structures which are parallel or misoriented by only a few degrees. These include the sapphire (0001) and NbN $(\bar{1}\bar{1}2)$, the $(2\bar{1}\bar{1}0)$ and (111), the $(0\bar{1}10)$ and $(1\bar{1}0)$, among several others. From these data we conclude that the NbN was nucleated at the sapphire surface and grew epitaxially into a single crystal. Since the crystallization occurred during the annealing step this is in fact a solid state epitaxial process. It was found that the three sets of major planes of sapphire and of NbN which are parallel and normal to the interface have a quite high (~ 10%) lattice mismatch. This implies that some intermediate structure at the sapphire-NbN interface may be present which would allow epitaxial growth to occur.

Whether this technique is capable of epitaxially growing NbN on sapphire orientations other than $(2\bar{1}\bar{1}3)$ is not known at this time. Based on the results of Noskov et al. (5) who grew single crystal NbN on other sapphire orientations by a vapor phase epitaxial technique, this should be possible. A more detailed discussion of the structural analyses of the single crystal films and the epitaxial growth process will be reported in a subsequent publication.

ACKNOWLEDGEMENTS

We are grateful to A. I. Braginski for many valuable discussions. We thank R. Wilmer for his dedicated technical assistance and Marilyn B. Cross for carefully preparing the manuscript.

REFERENCES

(1) J. R. Gavaler, J. Greggi, R. Wilmer, and J. W. Ekin, IEEE Trans. Magn. MAG-19, 418 (1983).
(2) J. R. Gavaler, A. T. Santhanam, A. I. Braginski, M. Ashkin, and M. A. Janocko, IEEE Trans. Magn. MAG-17, 573 (1981).
(3) J. R. Gavaler, A. I. Braginski, M. Ashkin, and A. T. Santhanam, Thin Films and Metastable Phases, in: Superconductivity in d- and f-Bend Metals, eds. Harry Suhl and M. Brian Maple (Academic Press, New York, 1980) pp. 25-36.
(4) R. T. Kampwirth and K. E. Gray, IEEE Trans. Magn. MAG-17, 565 (1981).
(5) V. L. Noskov, Yu. V. Titenko, F. I. Korzhisky, R. L. Zelenkevich, and V. A. Komashko, Kristallografiya 25, 878 (1980).

LT-17 (Contributed Papers)
U. Eckern, A. Schmid, W. Weber, H. Wühl (eds)

HIGH PRESSURE METHOD FOR OBTAINING NbN SUPERCONDUCTOR

Andrzej MORAWSKI

High Pressure Research Center "UNIPRESS" Polish Academy of Sciences,
Warszawa, Sokolowska 29/37, Poland

The high pressure-high temperature method of manufacture of B-1 NbN samples using Nb /3N/ powder or /4N/ foil was studied. The high pressure of pure N_2, up to 1GPa and temperature from 1273 K up to 1873 K was applied. The highest critical temperature /T_c/ equal to 16.72 K for the NbN foil was obtained. The range of the superconductivity transition for the foil samples was smaller than 0.5 K. The critical field at temperature of 4.2 K /H_{c2}/ was equal to 11 T. Preliminary examinations showed that high pressure-high temperature method seems to be useful in the manufacture of superconducting electronic devices.

1. INTRODUCTION

Niobium nitride was subjected to intensive studies over the last decade because of its possible applications in some electronic computer devices /1/, or in a superconducting power switches /2/. Most of the good superconductors of the /Metal-Nitrogen/ MeN type have the cubic / NaCl /, B-1 type structure. The structure is determined by the metal atoms larger than the nitrogen atoms. The structure of these superconducting compounds is characterised by the non-stoichiometry in the two sublattices /metal, nonmetal/. In the normal conditions /low pressure of N_2/ of manufacture the nonmetal sublattice vacancies are generally present. But when the sublattice of the nonmetal was nearly stoichiometric, metal vacancies have been observed. It's true that nonmetal vacancy concentration decreases with the increasing pressure of N_2 during the isothermal process. On the other hand decrease of the metal vacancy fraction was observed at increasing temperature during the isobaric process. The superconductivity properties depend strongly on both types of vacancies /3/. Thus the application of high pressure and high temperature simultaneously seems to be the proper way to minimize the concentrations of both vacancy types. This paper show what is the influence of the parameters of the synthesis on the critical temperature /T_c/, resistivity /ρ/ and critical field /H_{c2}/ of superconductors obtained. To obtain pure NbN, special systems of gas purging were applied. The analysis of the phase composition, stoichiometry and lattice constant was carried out using X - ray diffractometer. Most of the samples have been analysed by the chemical Kjeldahl's method,weight analysis and scanning electron microscopy. For the case of N_2 gas, the activity of nitrogen rapidly increases with increasing pressure /Fig.1/,/4,5/. This effect can be used in technology. We have tried to obtain high quality bulk B-1 NbN and the B-1 type MoN with the good superconducting properties/6,7/.

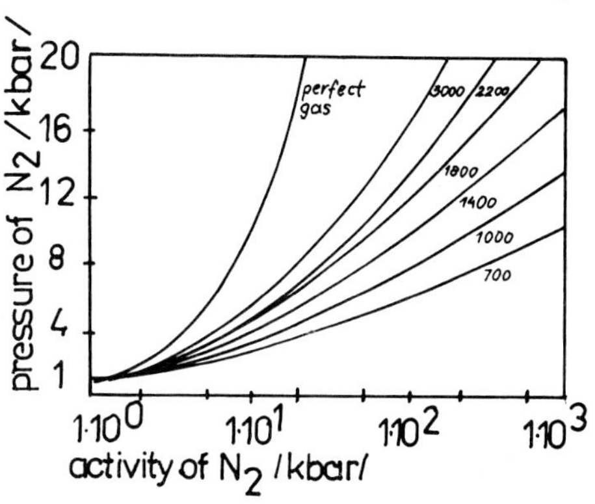

FIGURE 1

2. EXPERIMENTAL APPARATUS AND PROCEDURES

The NbN - cubic type samples were prepared by the high pressure method using 1.5 GPa "UNIPRESS" gas compressor /GC-30 type /. The precise pressure measurement was realized by the Manganin gauge inserted in the compressor. The powder or foil ofniobium was put into the alumina container and heated up to 1900 K in the high pressure up to 1 GPa. The experimental procedure was as the following ones.

The temperature was stabilized with the accuracy of ± 2 degree. The applied nitrogen gas /3 ppm purity/ was purified in two stages. The purging of the low pressure gas /12MPa/ was carried out at 673 K using copper and magnesium purifiers. Next the gas compressed to 0.35 GPa was purified at 823 K using titanium. NbN was synthetized in the Al_2O_3 crucible using the graphite or tungsten heater. The temperature was measured using Pt-Rh 18 termocouples. The conditions of the process and the results have been shown in Table 1. Kjeldahl's method /±3 % accuracy/ for the quantitative chemical analysis was used. Resistivity and T_c measurements were made in the SESI, CEN

TABLE 1.

NbN N^o	Lattice parameter /Å/	Pressure of N_2 /kbar/	Temperature /°C/	Activity of N_2 /kbar/	Time of reaction / h /	Quenching time / s /	Phase	Critical temperature T_c /K/	ΔT_c /K/	N/Nb /%at/	/RRR/ $\rho\ \frac{300-\rho 25}{\rho\ 25}$
6	4.358	8	1383	43.7	1	65	δ ε	14.0	3.6	109.1	-0.63
7	4.381	8	1385	43.6	3	55	δ	13.0	4.0	103.1	-
8	4.363	4	1371	9.7	5	60	δ	-	-	98.6	-
9	4.368	5	1371	14.9	3/4	58	δ	13.2	1.5	97.6	-0.5
10	4.363	10	1412	78.1	4	40	δ	9.7	0.5	102.1	-0.69
11	4.366	4	1482	8.7	2	52	δ	11.2	1.0	103.2	-0.56
12	4.364	4	1600	7.3	4	70	δ ε	10.5	0.5	107.5	-0.62
13	4.379	2	1600	2.9	3	60	δ	12.3	1.5	90.8	-
14	4.373	3	1250	6.2	4	38	δ	-	-	98.8	-
15	4.384	3	1150	6.5	1	85	δ	12.4	3.0	-	-
13F	-	2	1600	2.9	3	60	δ	16.28	0.4	-	-0.43
14F	-	3	1250	6.2	4	38	δ	16.72	0.5	-	-0.49

Fontenay-aux- Roses by the four point method using foil NbN samples and samples cut from sintered NbN ignots. The carbon-glass calibrated gauge was used to estimate the T_c temperature.

3. RESULTS

By the change of the pressure, temperature, time and the cooling rate during the high pressure synthesis we have obtained samples with different superconductivity properties. At the low cooling rates the samples containing hexagonal phase were obtained. Thus we have found that the time of the quenching was the most important parameter during the high pressure-high temperature synthesis of the cubic NbN.

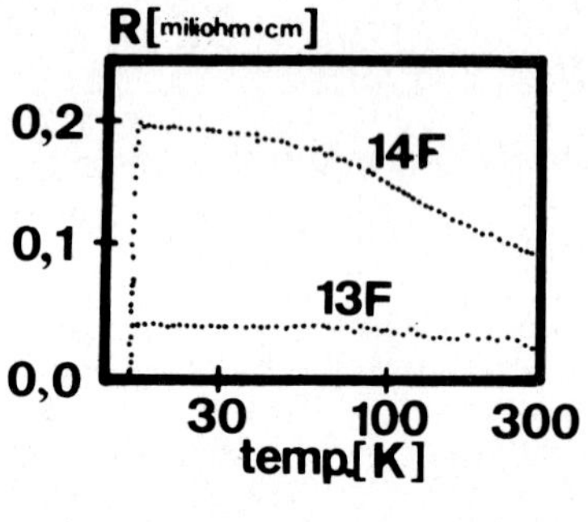

FIGURE 2

There are differences between properties of the powder and foil samples. For the powder ones range of the superconducting transition is larger than for the foil like samples. /ρn / at 25 K is about ten times larger than for a foil. The RRR coefficient depends on the time of the synthesis and on the activity of the gas. / See Table 1 and Fig. 2./ For the foil samples RRR coeficient is much smaller than for the powder ones. The T_c temperature for the powder NbN samples ranged from 10.3 to 15.5 K and was much smaller than for the foil ones. The T_c for the foil samples was equal to 16.4-16.7 K. /See Fig. 3./. H_{c2} measured at 4.2 K for a 30 um thick foilwas equal to 11 T. Many NbN samples were analysed by scanning microscopy. Thegrains of 2-12 um diameter were revealed.

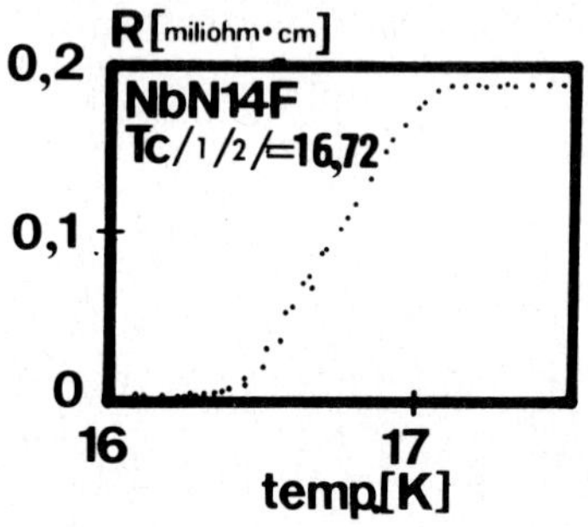

FIGURE 3

REFERENCES

1/ S. Kasaka et al. IEEE Tran. Magnetics Vol. MAG.-17,N°1, January 1981

2/ K.E. Gray and D.E. Fowler, J.Appl. Phys. 49/4/2546/1977

3/ J. Karpinski, S Majorowski, Procedings of the VIII AIRAPT Conference, Uppsala 1981

4/ J. Karpinski et al. Journal of Crystal Growth. 66/1984/ JCG00941 /NC/

5/ E. K. Storms, A.L. Giorgi, E.G. Szklarz, J.Phys. Chem. Solids 1975,36

6/ Zhao You-xiang, He Shou-an, Solid State Communications,Vol.45,No.3, pp 281-283,1983

7/ N. Pessall, C.K. Jones, H.A. Johansen, and J. K. Hulm, Appl. Phys. Lett. 7, 38/1965/.

LT-17 (Contributed Papers)
U. Eckern, A. Schmid, W. Weber, H. Wühl (eds)

ALLOYING EFFECTS ON LOW TEMPERATURE PROPERTIES IN $V_xNb_{(1-x)}$ AND $V_xNb_{(1-x)}N$

Thomas JARLBORG, Olivier PICTET, Michel DACOROGNA and Martin PETER

Département de physique de la matière condensée, Université de Genève, CH-1211 Genève 4, Switzerland

Self-consistent LMTO bandresults for $V_xNb_{(1-x)}$ and $V_xNb_{(1-x)}N$ supercells are used to study the variation of superconducting T_c, specific heat γ, and magnetic susceptibility enhancement, S, with concentration x. Spin fluctuations strongly reduces T_c in alloys containing V. Part of the T_c reduction for intermediate x, is associated with a localization of the wave functions and decreased λ.

1. INTRODUCTION

Vanadium and niobium based compounds are among the best high-T_c superconductors. These compounds include the carbides and nitrides, C15 and A15 compounds, as well as vanadium metal, are believed to show spin-fluctuations which limit their T_c's (1-4). The superconducting T_c's are in general higher for Nb based compounds than the corresponding V based compounds. In Nb and its compounds T_c decrease with applied pressure while the behaviour is opposite in V and its compounds (5). For alloys one finds reduced T_c's for intermediate compositions. This is found for the VNb system (6) as well as for the nitride system VNbN (7,8).

In this paper we study theoretically the effects of alloying on the low temperature properties, such as superconductivity, electronic specific heat, γ, and spin enhancement S, in the VNb alloys and their nitrides. As basis for our study serves self-consistent Linear Muffin-Tin Orbital band structure calculations for various supercell structures corresponding to different compositions x.

2. DETAILS OF THE CALCULATIONS

Details and references concerning the band calculations can be found in ref.4. The density-of-states (DOS) values at E_F, and their l-decompositions are used to calculate the electronic contribution, η, to the electron phonon coupling, λ, using the rigid Wigner-Seitz approximation (4,9). The phononic contributions are taken from experimental data. The obtained λ's are used to calculate the electronic specific heat coefficients :

$$\gamma = \frac{2\pi}{3} k_B^2 N(E_F) (1 + \lambda) \tag{1}$$

The superconducting transition temperature is calculated using the McMillan formula (10):

$$T_c = \frac{\omega_{log}}{1.2} \exp - \frac{1.04 (1 + \lambda)}{\lambda - \mu^*(1 + 0.62\lambda)} \tag{2}$$

where ω_{log} is from experiment and μ^* is in the range 0.08-0.18. In order to include effects from spin-fluctuations we calculate the spin enhancement or Stoner factor $S = 1/(1-\bar{S})$ from the band results (11). The contribution to λ from spinfluctuations is calculated from (12).

$$\lambda_{sf} = 4.5\ \bar{S}\ \ell n(1 + 0.047 \cdot S \cdot \bar{S}) \tag{3}$$

This additional contribution to λ is included in eq. 1 to obtain the effect on γ while on T_c, λ and μ^* are renormalized in eq. 2 as follows (13)

$$\lambda_{eff} = \frac{\lambda}{1 + \lambda_{sf}} \qquad \mu^*_{eff} = \frac{\mu^* + \lambda_{sf}}{1 + \lambda_{sf}} \tag{4}$$

3. RESULTS

a) Nitrides

The DOS-variations with x is almost linear between the pure VN and NbN values. However, λ shows a reduction relative to a linear interpolation for intermediate x, and as a consequence of this also the T_c's are reduced. The Stoner factor $\bar{S}$, is almost constant (0.55) for all compositions containing V (x>0.25), while for NbN it is considerably lower (0.25). Thus for x>0.25 T will be reduced further by spin fluctuations as seen in Fig 1. for x<0.5 the agreement with experiment is good, while for large V-concentrations (x=0.75 and 1.0) additional reduction of T_c probably from spin fluctuations is required. However, it is interesting to see that not all of the T_c reductions for intermediate compositions are due to spin-fluctuations. The reason for the λ-reduction despite the linear total DOS variation, is traced to a relative localization of the d-wave functions in the alloys. The effects on γ and susceptibility are smaller and the behaviour follows experiment fairly well for all x. More details and discussion of the nitride results are given in ref. 8.

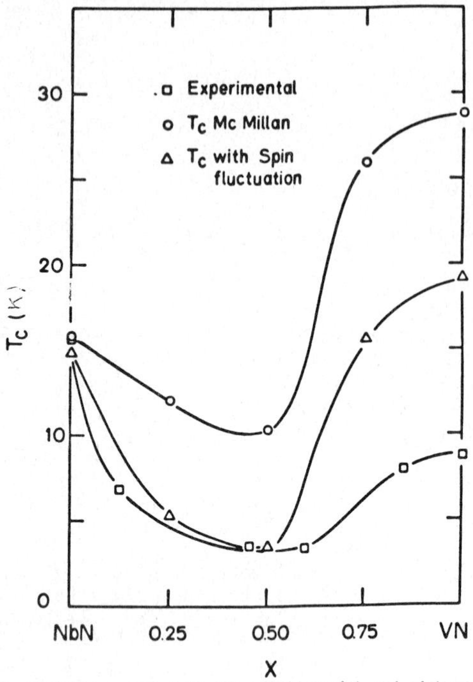

T_c variations with x in the nitrides with and without spin fluctuations. Experimental values from ref. 7 and 8.

FIGURE 1

b) VNb alloys

In order to study the question about relative localization of the d-functions due to alloying, we performed calculations of VNb for a lattice constant between that of Nb and V, and compared the results with those of pure Nb and V at the same lattice constant. Comparing the shape of the wave-functions at E_F one finds only minor differences. The value for η is indeed smaller in VNb than an average between V and Nb, although the total DOS is almost in between. The partial 4d DOS is reduced compared to the pure Nb case, while in V the 3d DOS is increased by the alloying. The net effect is an increase of the V contribution to and a larger decrease of the contribution from Nb.

When the real changes in lattice constants are taken into account one can observe a relative localization of the V-d function after alloying although smaller than for the nitrides. The lattice constants have been chosen to give similar pressure in the 3 calculations. The η is smallest for the alloy, while the T_c excluding spin fluctuations is slightly larger than for Nb, cf Table I. The large T_c reduction comes via spin fluctuations in particular for V and NbV, and the calculated T_c follows experiment. For V the effect of spin fluctuations is so large, that T_c is very sensitive to changes in lattice constants, and T_c goes up rapidly with pressure.

Table. Properties calculated from the electronic structure, with and without spin fluctuations. Experimental values from ref. 6.

	DOS	λ_{ep}	$\bar{S}$	γ	γ^{sf}	γ^{exp}	T_c	T_c^{sf}	T_c^{exp}
	$[at \cdot Ry]^{-1}$			$mJ \cdot [g \cdot atK^2]^{-1}$				[K]	
V	22.5	1.22	0.58	8.6	9.3	9.6	22.3	5.4	5.5
NbV	23.7	1.15	0.63	8.8	9.7	8.2	17.2	1.4	3.8
Nb	18.5	1.19	0.43	7.0	7.2	7.7	16.4	10.7	9.2

REFERENCES

(1) H. Rietschel and H. Winter, Phys. Rev. Lett. 43 (1979) 1256.
(2) H. Rietschel, H. Winter and W. Reichardt, Phys. Rev. B 22 (1980) 4284.
(3) O. Rapp and C. Crafoord, Phys. Stat. Sol. 64 (1974) 139.
(4) T. Jarlborg, A. Junod and M. Peter, Phys. Rev. B 27 (1983) 1558.
(5) H.L. Luo, S.A. Wolf, W.W. Fuller, A.S. Edelstein and C.Y. Huang, Phys. Rev. B 29 (1984) 1443.
(6) M. Ishikawa and L.E. Toth, Phys. Rev. B 3 (1971) 1856.
(7) C. Geibel, H. Rietschel, A. Junod, M. Pelizzone and J. Muller (to be published) M. Pelizzone Thesis, Univ. of Geneva (1982) (unpubl.)
(8) M. Dacorogna, T. Jarlborg, A. Junod, M. Pelizzone and M. Peter (to be published).
(9) G.D. Gaspari and B.L. Gyorffy, Phys. Rev. Lett. 28 (1972) 801.
(10) W.L. McMillan, Phys. Rev. 167 (1968) 331.
(11) T. Jarlborg and A.J. Freeman, Phys. Rev. B 22 (1980) 2332.
(12) S. Doniach and S. Engelsberg, Phys. Rev. Lett. 17 (1966) 750.
(13) J.M. Daams, B. Mitrovic and J.P Carbotte, Phys. Rev. Lett. 46 (1981) 65.

LT-17 (Contributed Papers)
U. Eckern, A. Schmid, W. Weber, H. Wühl (eds)

ELECTRONIC DENSITY OF STATES AND SUPERCONDUCTIVITY IN THE C15 COMPOUND V_2Zr

C. GEIBEL, H. KEIBER, B. RENKER, H. RIETSCHEL, H. SCHMIDT, H. WÜHL, and G.R. STEWART[+]

Kernforschungszentrum Karlsruhe and Universität Karlsruhe,
Postfach 3640, D-7500 Karlsruhe, Federal Republic of Germany,
[+] Los Alamos National Laboratory, Los Alamos, N. M. 87545.

We present low temperature data on the specific heat and magnetic susceptibility in V_2Zr, for both the cubic and rhombohedral phases. Whereas in the rhombohedral phase N(o) is strongly reduced, T_c remains almost unaltered. From these findings we infer that V_2Zr is one further example of T_c limitation by spin fluctuations.

1. INTRODUCTION

A high electronic density of states at ε_F, N(o), favors superconductivity in two respects. First, it is a direct measure of the number of electrons sharing the superconducting quantum state, and second, it leads to strong phonon renormalization and thus to phonon softening. On the other hand, an increasing N(o) implies a growing tendency towards itinerant ferromagnetism which manifests itself in spin fluctuations and reduces T_c (1).

In this paper, we present experimental evidence that the Laves phase V_2Zr is a further demonstration of the limitation of high-T_c superconductivity by spin fluctuations in materials with high N(o).

At about 100 K, V_2Zr undergoes a martensitic transformation from the cubic into a rhombohedral phase. In our experiments, we exploited the fact that this transformation can be hindered by small amounts of residual strain or concentration gradients, thus allowing the simultaneous observation of both phases at low temperature. Performing neutron scattering experiments and measurements of γ (coefficient of the electronic specific heat) and χ (magnetic susceptibility) on samples which only partly underwent the martensitic transformation, we were able to assign distinct values for T_c, γ and χ_{spin} to both phases.

2. EXPERIMENTS AND RESULTS

The experiments were performed on four different samples S1 to S4 which proved to be mostly single-phased V_2Zr with small traces (< 3 %) of $V_3Zr_3O_x$. S1, S2 and S4 were taken from different sites of the same ingot, S3 from a different ingot.

Specific heat data were taken on S1 to S4 in a heat pulse calorimeter and are shown in Fig.1. In the normal state, the experimental results were analysed by fitting them to $C_N(T) = \gamma t+\beta T^3+\alpha T^5$ with the usual requirement that the entropy above $T_{c\,cub}$ be independent of the superconducting phase. A further check of the extrapo-

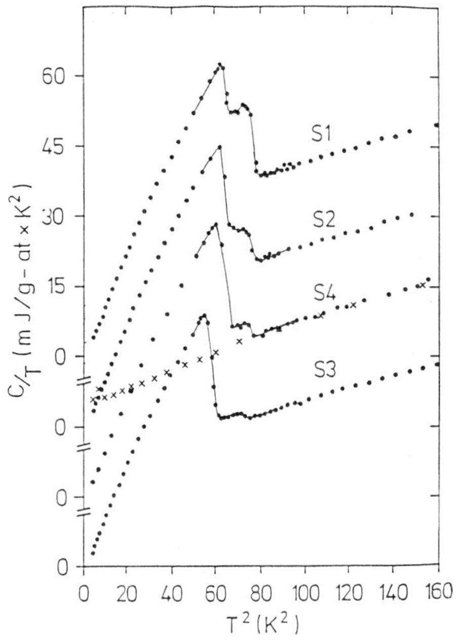

FIGURE 1
Specific heat of V_2Zr with different amounts of cubic and rhombohedral phase. The crosses represent the measurement in 13 Tesla for sample S4.

lation was done for S4 by depressing T_c below 5 K in a magnetic field of 13 T. The γ values for sample S1, S2, S3, and S4 are 18.9, 17.8, 16.2, and 17.0 mJ/g-at·K². All curves display two distinct superconducting transitions. We identify the lower (higher) transition as that of the rhombohedral (cubic) phase. In order to extract the specific heat coefficient γ_{rh} (γ_{cub}) for the individual rhombohedral (cubic) phase, we made use of the equation $\gamma = p_{rh}\cdot\gamma_{rh}+(1-p_{rh})\cdot\gamma_{cub}$, where p_{rh} is the bulk percentage of the rhombohedral phase. The values for p_{rh} in the different samples were determined through an evaluation of the specific heat discontinuities and agree with the results of neutron scattering and magnetic susceptibility measurements. This identification is of central importance for our conclusions. We found $\gamma_{rh} = 16.1 \pm 0.5$ and $\gamma_{cub} = 23 \pm 1.0$ mJ/g-at K^2. The larger error bar for γ_{cub} than for γ_{rh} is a result of the p_{rh} evaluation for sample S1. Our result for γ_{rh} compares well with those by Rapp and Vieland (2) who also found two discontinuities in their specific heat data, but were unable to identify the minor phase (7 %) as the cubic one.

Neutron powder diffraction experiments were performed on S2 and S3 at the ORPHEE reactor in Saclay. We observed the splitting of both the (111) and (220) reflex at low temperature. From these data, the quantity p_{rh} could be determined through the ratio (intensity in the split lines) /(total intensity) and was found to be in good agreement with the specific heat results.

The magnetic susceptibility was measured on S1 to S3 using a Faraday balance. The absolute error was about 3 %. The corresponding χ versus T curves are shown in Fig. 2. They all exhibit a steep decrease around $T_M \sim 100$ K. For S3 our results are very similar to those by Marchenkov and Polovov (3). Following these authors we assign this decrease to a lowering of the spin susceptibility $\chi^s(T)$ at the cubic-rhombohedral phase transition. In order to determine the low temperature susceptibility of the cubic phase, we made use of the theoretical calculation of

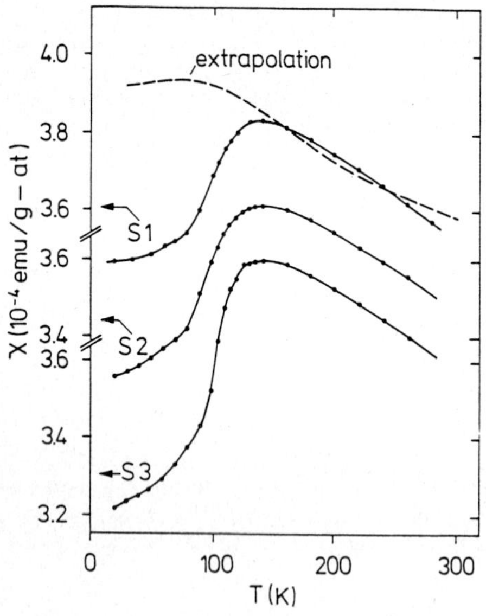

FIGURE 2
Susceptibility $\chi(T)$ of V_2Zr samples S1 - S3. The dashed line is a theoretical calculation for the cubic phase after ref. 4.

$\chi^s{}_{cub}(T)$ by Klein et al. (4). They fitted the susceptibility data for T > 100 K and extrapolated this curve down to 20 K by assuming an orbital contribution of the susceptibility of $\chi_{orb} \sim 1.8 \cdot 10^{-4}$ emu/g-at. for both phases. This fit is shown in Fig. 2 as a dashed line. We obtain $\chi^s{}_{cub}$ (20 K) $\sim (2.1 \pm 0.3) \cdot 10^{-4}$ emu/g-at. for all three samples. Since S3 is the sample with the highest amount of rhombohedral phase $\chi^s{}_{rh}$ (20 K) is extracted from this data and is found to be $(1.4 \pm 0.3) \cdot 10^{-4}$ emu/g-at. With these two values and the drop of the susceptibility we can calculate p_{rh} for sample S1 and S2. These values are in good agreement with the specific heat results.

3. CONCLUSIONS

Our central result is that in the rhombohedral phase T_c is only slightly lower (8.0 K) than in the cubic phase (8.7 K), whereas both γ and χ_{spin} are reduced by about 30 %.

Let us analyze the results in terms of the quasi-particle DOS $N(\gamma) = 3\gamma/2\pi^2 k_B^2$ and $N(\chi) = \chi^s/2\mu_B^2$. With N(o) = 110 states/Spin Ry uc in the cubic phase (4, 5), for the mass enhancement we find $(1+\lambda)_{cub} = N(\gamma)/N(o) \sim 3.6$. If this large $\lambda \sim 2.6$ were only due to electron phonon coupling, then, since $\theta_D \sim 200$ K, T_c would be much higher than the actual 8.7 K. On the other hand, the Stoner enhancement $S = N(\chi)/N(o) \sim 2.4$ (Klein et al. (4) find even $S \sim 3.8$!) indicates a strong pairbreaking influence of spin fluctuations. This is further corroborated by the fact that the ratio $N(\chi)/N(\gamma) \sim 0.7$ for both phases, i.e., that the decrease in electron phonon coupling is just compensated by a decrease in S which explains the almost unaltered T_c. This makes V_2Zr a further example for T_c limitation by spin fluctuations, joining V and VN (1).

REFERENCES

(1) H. Rietschel and H. Winter, Phys. Rev. Lett. 43 (1979) 1256.
H. Rietschel, H. Winter, and W. Reichardt, Phys. Rev. B 22 (1980) 4284.
(2) Ö. Rapp and L.J. Vieland, Phys. Lett. 36A (1971) 369.
(3) V.A. Marchenkov and V.M. Polovov, Sov. Phys. JETP 51 (1980) 535.
(4) B.M. Klein, W.E. Pickett, D.A. Papaconstantopoulos, and L.L. Boyer, Phys. Rev. B 27 (1983) 6721.
(5) T. Jarlborg and A.J. Freeman, Phys. Rev. B22 (1980) 2332.

LT-17 (Contributed Papers)
U. Eckern, A. Schmid, W. Weber, H. Wühl (eds)

SUPERCONDUCTIVITY AND NMR INVESTIGATIONS ON CUBIC LAVES PHASE HYDRIDES OF $V_2Hf_{0.5}Zr_{0.5}H_x$ ($0 \lesssim x \lesssim 2$)

W. DÄUMER, H.R. KHAN* and K. LÜDERS

Freie Universität Berlin, Fachbereich Physik, Arnimallee 14, D-1000 Berlin 33, FRG
* Forschungsinst. f. Edelmetalle u. Metallchemie, Katharinenstr. 17, D-7070 Schwäbisch Gmünd, FRG

Measurements of the lattice parameter a, superconducting transition temperature T_c, Knight shift K and relaxation rate R of ^{51}V nucleus as a function of hydrogen concentration x are made on cubic Laves phase (C-15) compounds $V_2Hf_{0.5}Zr_{0.5}H_x$ ($0 \lesssim x \lesssim 2$). T_c as well as R decrease slowly with increasing x whereas K increases up to x=1. After this concentration, hydrides do not exhibit superconductivity down to 1.5 K and K as well as R variation is roughly constant.

1. INTRODUCTION

The V-based cubic Laves phase (C-15) alloys of Hf and Zr exhibit T_c values of $\simeq$ 10 K and high critical magnetic fields of $\simeq$ 23 T (1) as well as resistance to the neutron radiation damage (2). They seem to be interesting superconducting materials for practical applications. Recently Rao et al. (3) observed that the superconducting transition temperature T_c of the cubic Laves phase (C-15) compound of composition $V_2Hf_{0.5}Zr_{0.5}$ increases upon hydriding. For example T_c values are 10.1 K and 11.8 K for $V_2Hf_{0.5}Zr_{0.5}$ and $V_2Hf_{0.5}Zr_{0.5}H_{0.5}$, respectively. The electronic structure of V_2HfH_x (4) and V_2ZrH_x (5) has been investigated by NMR of ^{51}V and ^1H nucleus. In this paper, the measurements of T_c, Knight shift K and relaxation rate, $R=(1/T_1T)$, of ^{51}V nucleus on a series of the hydrides of $V_2Hf_{0.5}Zr_{0.5}H_x$ ($0 \lesssim x \lesssim 2$) are reported. The hydriding of $V_2Hf_{0.5}Zr_{0.5}$ was performed in different ways but we could not repeat the results of Rao et al. where an increase in T_c value for the hydrogen concentrations up to x=1 is observed.

2. EXPERIMENTAL DETAILS

The sample of composition $V_2Hf_{0.5}Zr_{0.5}$ was prepared by arc melting in Ar-atmosphere on a water cooled copper hearth. The hydriding of this compound was accomplished by two different methods. In the first method, the sample was heated between 450 and 650°C in H_2-gas atmosphere using a maximum pressure of 300 mbar. The hydrogen concentration in the compound was determined from the change in hydrogen gas pressure. In the second method the hydriding of $V_2Hf_{0.5}Zr_{0.5}$ was first done in hydrogen gas at a pressure of 45 bar. This hydride of $V_2Hf_{0.5}Zr_{0.5}H_{3.5}$ together with the pure $V_2Hf_{0.5}Zr_{0.5}$ sample was annealed at 460°C for a period of 48-72 h in He-gas at a pressure of 300 mbar. In this way a homogeneous distribution of hydrogen in $V_2Hf_{0.5}Zr_{0.5}$ was accomplished. The lattice parameter were determined at room temperature by X-ray diffraction of powders using a Guinier camera and CuK_α radiation. The superconducting transition temperatures were measured inductively using Lock-in-technique in a magnetic field of $\simeq$ 3 G using a carbon glass thermometer. The accuracy is $\pm$ 0.1 K. The onset of superconductivity is used as the T_c value. The Knight shift measurements as a function of H-concentration on ^{51}V were made by cw method at a temperature of 20 K and in a magnetic field of 2.5 T. The spin-lattice relaxation time at room temperature of ^{51}V was measured using a 180°-τ-90° pulse sequence in a pulse spectrometer in a magnetic field of 2 T. For all samples the spin magnetization shows an exponential decay for room-temperature measurements, whereas for low temperatures the decay is not simple exponential.

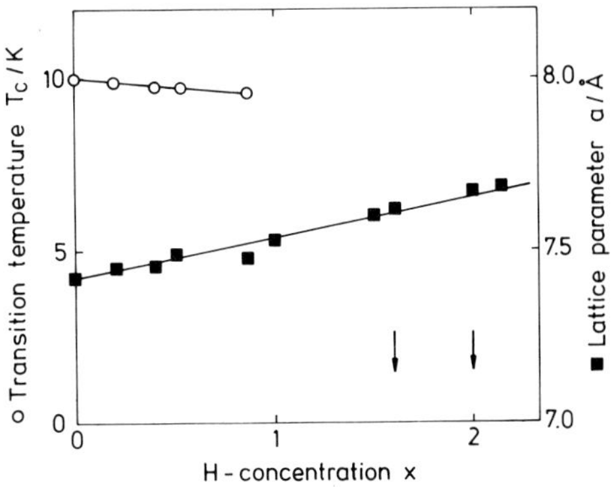

FIGURE 1
Superconducting transition temperature T_c and lattice parameter a at room temperature.

3. RESULTS AND DISCUSSION

The lattice parameter of the C-15 phase in $V_2Hf_{0.5}Zr_{0.5}H_x$ increases with the hydrogen concentration from 7.42 Å for $V_2Hf_{0.5}Zr_{0.5}$ to 7.68 Å for $V_2Hf_{0.5}Zr_{0.5}H_{2.15}$ as shown in Fig. 1. All the samples have small parts of other phases besides the cubic Laves phase. The superconducting transition temperature decreases slowly from 10 K to 9.65 K with increasing hydrogen concentration up to x=.85. An increase of T_c with the hydrogen concentration in this concentration range is not observed. Such an initial increase in T_c has been reported by Rao et al. (3). The hydriding of $V_2Hf_{0.5}Zr_{0.5}$ was done in different ways but we were not able to verify the results of Rao et al. For the H-concentration of x>1, T_c drops to a temperature lower than 1.5 K. This sudden drop in the value of T_c indicates that these hydrides transform to a low temperature phase which has a very low T_c value. The superconducting transitions are shown in Fig. 2.

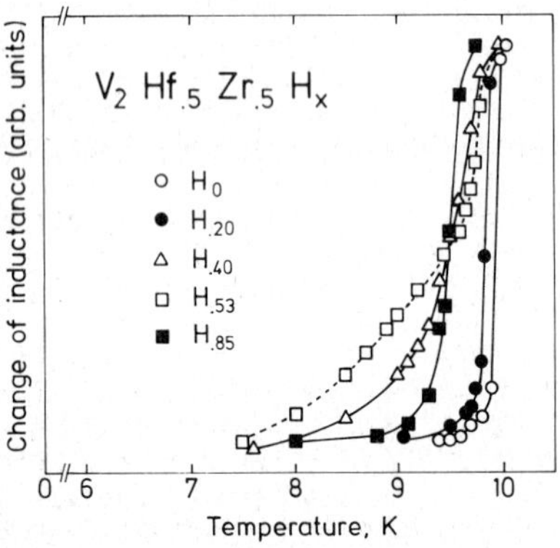

FIGURE 2
Superconducting transitions for different hydrogen concentrations

The Knight shift K of ^{51}V increases from 0.57% ($V_2Hf_{0.5}Zr_{0.5}$) to 0.604% ($V_2Hf_{0.5}Zr_{0.5}H_{0.85}$) but for higher concentrations of hydrogen it remains practically constant as shown in Fig. 3. In the same figure, the variation of the relaxation rate, $R=(1/T_1T)$, of ^{51}V is also plotted. For the initial H-concentrations, R decreases up to x=.85 and again becomes constant. For x=2.15 a second decrease in relaxation rate is observed.The contribution of the d-electrons to the Knight shift is negative. Therefore when K_d decreases, an increase in the total Knight shift is observed. K_d is proportional to the d-electron density of states at the Fermi surface, $N_d(E_F)$, whereas R_d is pro-

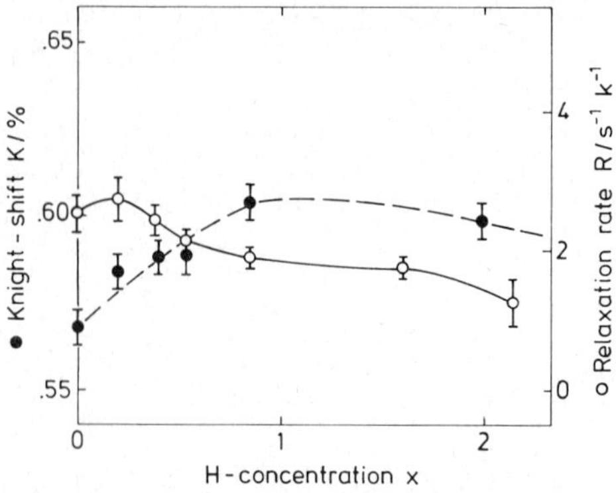

FIGURE 3
^{51}V Knight shift K at 20 K and spin lattice relaxation rate $R=(1/T_1T)$ at room temperature

portional to $N_d^2(E_F)$. Thus from these data a decrease in T_c is explained by the reduction of the d-electron density of states.

ACKNOWLEDGEMENTS
This work was supported by the Deutsche Forschungsgemeinschaft.

REFERENCES

(1) K. Inoue and K. Tachikawa, Proceedings LT 12 (Kyoto 1970) (Academic Press of Japan, 1971), p. 483.
(2) B.S. Brown, I.W. Hafstrom, and T.E. Klippert, J. Appl. Phys. 48 (1977) 1759.
(3) V.U.S. Rao, D.M. Gualtieri, S. Krishnamurthy, A. Patkin, and P. Duffer, Phys. Lett., Vol. 67A, 3 (1978) 223.
(4) D.T. Ding, J.I. de Lange, T.O. Klaassen, J. Poulis, D. Davidov, and J. Shinar, Solid State Commun., 42 (1982) 137.
(5) M. Peretz, J. Barak, D. Zamir, and J. Shinar, Phys. Rev. B., 23 (1981) 1031.

LT-17 (Contributed Papers)
U. Eckern, A. Schmid, W. Weber, H. Wühl (eds)
© Elsevier Science Publishers B.V., 1984

KNIGHTSHIFT AND NUCLEAR SPIN RELAXATION IN THE $HfV_2.(H,D)_x$ SYSTEM

A.P. SEDEE, D.T. DING*, T.O. KLAASSEN, N.J. POULIS

Kamerlingh Onnes Laboratorium der Rijksuniversiteit Leiden, Nieuwsteeg 18, 2311 SB Leiden, The Netherlands.

The ^{51}V Knightshift and nuclear spin relaxation rate in the $hfV_2.(H,D)_x$ system has been determined in the low temperature phase as a function of the hydrogen (deuterium) concentration x. For $x \leqslant 1.5$ the results show a decrease of the vanadium 3d electron density of states at E , in accordance with T_c results. For $x \geqslant 2$ the NMR results do not yet yield reliable information concerning the concentration dependence of $N_d(E_F)$.

1. INTRODUCTION

The C-15 type intermetallic compound HfV_2 is known as a superconductor with a relatively high superconducting transition temperature (T_c=9 K). It can easily absorb large quantities of hydrogen, forming stable hydrides $HfV_2.H_x$ ($x < 5$). Upon hydrogenation T_c decreases steeply (1); the T values of the deuterides are consistently lower than those of the hydrides. We investigate the influence of hydrogen absorption on the metallic and superconducting properties of HfV_2 by means of X-ray, susceptibility-, specific heat- and magnetic resonance experiments, in order to understand the basic mechanisms that determine the strong cocnentration and isotope dependence of T_c and the density of states at the Fermi level in the $HfV_2.(H,D)_x$ system.

In this paper we report on part of our low temperature ^{51}V knightshift and nuclear relaxation experiments, performed on finely powdered samples. In the high temperature C-15 phase of these hydrides the H atoms exhibit a fast hopping between various interstitial sites. Towards lower temperature the hydrogen diffusion disappears and a phase transition to a tetragonal structure takes place. Also for x=0 a slihgt tetragonal distortion occurs at T=120 K.

The presence of hydrogen in the lattice does change the local surrounding of the ^{51}V nuclei drastically. In an earlier paper we reported already on the x-dependence of the nuclear electric quadrupole interaction in the low temperature phase (2). We concluded that for the $Hfv_2.H_x$ system no crystallographic order of the H atoms exists for $0<x<4$; also no separation into low- and high H concentration phases occurs. Only for x=4 an ordered hydrogen lattice was found to exist, as evidenced by the observation of a well defined single-site ^{51}V NMR spectrum, split up by quadrupole interaction.

* On leave from Nan-Kai University, Tian-jin, China.

2. EXPERIMENTAL RESULTS AND DISCUSSION

In the figure present results on the low T Knightshift K and nuclear relaxation rate R are given. It should be realised that for $0<x<4$, due to the absence of a regular lattice, a number of crystallographically inequivalent V sites exists. As the resonance spectra of the different sites strongly overlap, the values for K and R should be considered as weighted averages over the various sites. Only for x=0,4 they reflect accurate single-site values. The V Knightshift consists of a number of contributions: $K = K_s + K_d + K_{orb}$. The positive s-electron term is proportional to the s-electron density of states at the Fermi energy and to the s hyperfine field, $K_s \propto N_s(E_F)H_s$, the negative d core-plarisation term $K_d \propto N_d(E_F)H_d$. The positive orbital Knighshift does not depend on N (E). The relaxation rate is given by $R=(T_1T)^{-1} = R_s + R_d + R_{orb}$, with $R_s \propto (N_s(E_F)H_s)^2$, $R_d \propto (N_d(E_F)H_d)^2$ and $R_{orb} \propto (N_d(E_F)H_{orb})^2 p_{orb}$. The reduction factors P_d and P_{orb} depend on the site symmetry and the relative contribution of the various d bands to $N_d(E_F)$.

Bandstructure calculations (3) show that, in the isomorphous compound ZrV_2, $N_s(E)$ is small and not very dependent on the energy, whereas $N_d(E)$ is very large at E_F and stronly energy dependent. So the observed increase of K and decrease of R when x increases from 0 to 1.5 can be understood as being mainly due to a decrease of $N_d(E_F)$. This conclusion is supported by the observed strong decrease of T_c with increasing x for $x \leqslant 1.5$. The results for $x \geqslant 2$ are more difficult to explain straightforwardly. The relaxation rate for x=1 and x=4 is about equal. When the increase of R for $x \geqslant 2$ is due to an increase of $N_d(E_F)$ one expects the T_c values to be about the same. However $T_c(1)$= 5K whereas $T_c(4)<1$K, which proves that such an increase of $N_d(E_F)$ is very unlikely (Bandstructure calculations by de Groot et al. (4) on $ZrV_2.H_4$ do show that $N_d(E_F)$ for x=4 is smaller than for x=0 by about a factor 3 to 4). Most probably the increase of R is

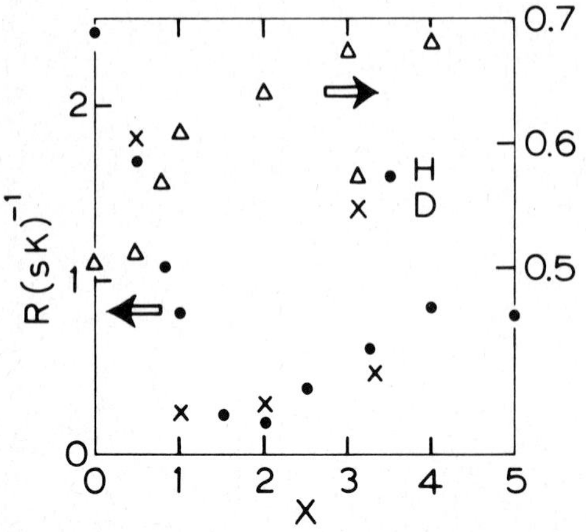

Knightshift (Δ) and relaxation rate (x,0) as a function of the concentration x.

related to an increase of the reduction factor p_{orb} as a result of the drastic change in the vanadium site symmetry and bonding upon hydrogenation. The increase of K in the same concentration range is likely to be caused by the filling up of the vanadium 3d band and possibly a reduction of the average band splitting.

The experimental results on the deuterides and hydrides are rather similar. Apparently the density of states in both series shows about the same composition dependence: the lower T values of the deuterides have probably to be attributed to differences in the phonon spectrum.

We also performed Knighshift and relaxation experiments on a series of $ZrV_2.(H,D)_x$ compounds. The results however are not reliable at all, due to the presence of a second, hexagonal, phase in the C-15 material.

Recently Belyaev et al. (5) published results of spin-echo experiments on $HfV_2.H_x$ (x=3.2 and 3.9). From the temperature dependence of the spin-echo spectrum in a magnetic field of 2T they concluded that in the x=3.9 compound two inequivalent V sites exist in the low temperature phase, with different quadrupole interaction and Knightshift. As this conclusion contradicts both our former NMR results in low magnetic fields, and neutron diffraction results (6), we performed additional NMR experiments in high field on $HfV_2.D_4$. Using a marginal oscillator we recorded the first derivative of the absorption spectrum at T=1K in various fields between 1T and 4T. In most cases we could observe all the θ=90° quadrupole satelite discontinuities of the powder spectrum; sometimes also the θ=0° discontinuities. A prelimenary analysis of the spectra shows convincingly that in $HfV_2.D_4$ only one V site is present, characterised by the nuclear interaction parameters:
$\nu_q \approx$ 500kHz, $\zeta \approx 0.12$, $K_x \approx 0.84\%$, $K_y \approx 0.66\%$ and $K_z \approx 0.50\%$. (The x, y, z axis are choosen according to the conventional definition of principal axes of an electric field gradient tensor). It must be noted that these interaction parameters can describe satisfactorily the spectrum given by Belyaev et al., i.e. the materials are alike, only the interpretation of the experimental results differs. The interaction parameters given by Belyaev et al. however do not describe our resonance spectra. A more accurate analysis, based on computersimulations of the overall lineshape will be published elsewhere.

3. CONCLUDING REMARKS

The results of $HfV_2.D_4$ show that the data on K and R in the $HfV_2(H,D)_x$ system must be interpreted with care. Because of the possibility of a large anisotropy in K - and thus likely in R - for a detailed analysis more experiments in high fields are necessary, in combination with susceptibility and specific heat results. Moreover, knowledge of the reduction factors, that strongly determine the relaxation rate, have to be obtained from detailed band structure calculations on HfV_2 and $HfV_2.H_4$ (for 0<x<4 detailed calculations seem to be not well possible because of the disoredered H lattice).

ACKNOWLEDGEMENTS

This investigation is part of the research program of the "Stichting voor Fundamenteel Onderzoek van de Materie (FOM)" which is financially supported by the "Nederlandse Organisatie voor Zuiver Wetenschappelijk Onderzoek (ZWO)".

REFERENCES

(1) P. Duffer, D.M. Gualtieri, V.U.S. Rao, Phys. Rev. Lett. 37(1976)1410
(2) D.T. Ding, J.I. de Lange, T.O. Klaassen, N.J. Poulis, Solid State Commun. 42(1982)137.
(3) B.M. Klein, W.E. Pickett, D.A. Papaconstantopoulos, L.L. Boyer, Phys. Rev. B27(1983)6721.
(4) R.A. de Groot, to be published.
(5) M.Yu. Belyaev, A.V. Skripov, V.N. Kozhanov, A.P. Stepanov, E.V. Galoshina, Solid State Commun. 48(1983)1049.
(6) A.V. Irodova, V.P. Glazkov, V.A. Somenkov, S.SH. Shilstein, J. Less- Common Metals 77(1981)89.

LT-17 (Contributed Papers)
U. Eckern, A. Schmid, W. Weber, H. Wühl (eds)

EXPERIMENTAL STUDY OF THE ELECTRON-PHONON INTERACTION IN AgD_x PREPARED BY IMPLANTATION

A. TRAVERSE, J. CHAUMONT, A. BENYAGOUB, H. BERNAS, P. NEDELLEC*, J.P. BURGER*

CSNSM, BP N° 1, 91406 ORSAY, France
*Laboratoire de Physique des Solides, Paris XI, 91405 ORSAY CEDEX, France

AgD_x is prepared by low temperature deuterium implantation in Ag, and its resistivity is measured between 1.8 K and 80 K. The D content is monitored in situ by Rutherford backscattering. No superconducting transition is measured down to 1.8 K. Two main results are obtained : i) an increase of the acoustic phonon resistivity ; ii) the existence of an optic phonon resistivity. The coupling of the electrons with the acoustic and the optic phonons are of the same order of magnitude.

1. INTRODUCTION

It is known that the electron-phonon coupling responsible for superconductivity is dominated in PdH(D) by the optic or H vibration contributions (1). We analyze here the case of AgD_x in order to see if this is a general phenomenon. Pd and Ag are both fcc and band calculations have been performed both on Pd-Ag and Ag hydrides (2).

AgD being unstable at room temperature, we have prepared it by low T implantation and checked its behaviour regarding the electron-phonon interaction by measuring the electrical resistivity.

2. EXPERIMENTAL

Silver films (thickness $\sim$ 1000 Å) were evaporated on quartz substrates in a vacuum of 10^{-7} torr, then implanted at 6 K with deuterium of 7 or 8 keV, up to 9.1 ^{17} D.cm^{-2}, the choosen incident energy depending on the thickness. The distribution profile of implanted deuterium was wide enough to obtain a rather homogeneoulsy charged sample. The deuterium content in the film was monitored by in situ Rutherford backscattering experiments.

Resistivity measurements were made during implantation at different D concentrations, during thermal cycling between 6 K and 80 K. Search for superconductivity was done by pumping down to 1.8 K.

3. RESULTS AND ANALYSIS

This section presents the resistivity temperature dependence of pure Ag and of an Ag film implanted with 4.10^{17} D.cm^{-2} ($x \sim 0,6$).

1) Stability vs. temperature :

After a continuous increase from 6 K upwards the slope of the $\rho(T)$ curve of a D implanted silver film changes irreversibly between 80 and 85 K, indicating either desorption of precipitation of deuterium in the sample : for this reason, we have never exceeded this temperature.

2) Electron-Phonon interaction

Two typical $\rho(T)$ curves are shown in fig. 1. The analysis of these curves was performed by assuming a simple additivity :

$$\rho = \rho_0 + \rho(T) \qquad (1)$$

where ρ_0 is the residual resistivity and $\rho(T)$ the temperature dependent resistivity. In pure silver, a good fit is obtained on the temperature range 10 K to 80 K (see Fig. 1) with an expression of $\rho(T)$ in terms of an electron-acoustic phonon interaction given by :

$$\rho(T)=\rho_{ac}(T)=(n-1)A\lambda_{ac}\theta_{ac}\left(\frac{T}{\theta_{ac}}\right)^n J_n\left(\frac{\theta_{ac}}{T}\right) \qquad (2)$$

$$\text{where } J_n(x) = \int_0^x \frac{z^n\,dz}{(e^z-1)(1-e^{-z})} \qquad (3)$$

λ is the electron phonon coupling parameter and A is related to the el ectronic structure of the material. With the usual definitions :

$$A = 2\pi\,\frac{m^*\,k_B}{Ne^2\,\hbar} \qquad (4)$$

where N = number of conduction electrons per unit volume.

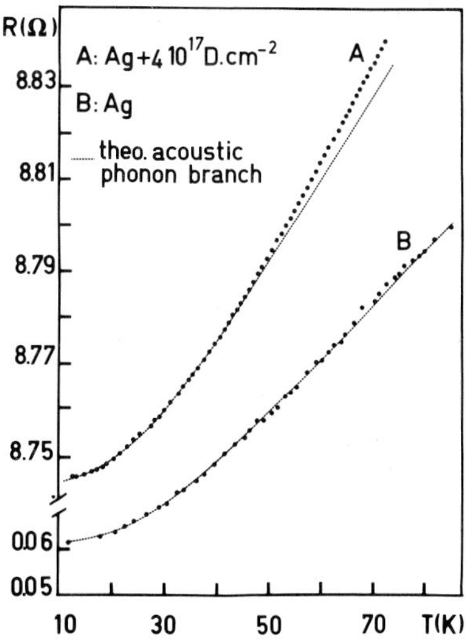

Figure 1 : Resistivity temperature dependence for pure Ag and Ag + 4 .10^{17} D.cm^{-2}

The exponent n was found to be nearer to 3 than to 5 (which is the Bloch-Grüneisen value (3)), the latter being the value found by White and Woods (4). Our curves were fitted with both exponents, and both gave the same characteristic temperature θ_{ac} within experimental uncertainties :

$$\theta_{ac}(n=3) = 210 \pm 10 \text{ K} \; ; \; \theta_{ac}(n=5) = 195 \pm 10 \text{ K}$$

This may be due to the fact that the temperature range over which it is possible to determine precisely the exponent value is between $\theta_{ac}/20$ and $\theta_{ac}/10$ (4) ; this range is reduced in our case to 10 K - 20 K, i.e. over an interval in which the temperature variation is too weak to determine the exponent value very accurately.

In the deuterium implanted silver, the experimental data used for the fit were taken during sample cooling down after annealing at 80 K, in order to eliminate any damage recombination effects. Equation 2 (with n = 5 or n = 3) provides a good fit in the low temperature range (10 to 45 K), but in the higher temperature range (45 to 80 K) the resistivity curve increases faster than Equation 2 (see Fig 1.). We account for the difference between the experimental resistivity curve and the calculated electron-acoustic phonon interaction by adding a supplementary interaction i.e. :

$$\rho(T) = \rho_0 + \rho_{ac}(T) + \rho_{op}(T) \qquad (5)$$

where $\rho_{op}(T)$ is due to electron scattering on optic phonons ; we assume an Einstein spectrum for the optical phonons :

$$\rho_{op}(T) = A\lambda_{op}\theta_{op} \; f(\theta_{op}/T) \qquad (6)$$

$$\text{where } f(x) = xe^{x} \, (e^{x} - 1)^{-2} \qquad (7)$$

Numeracal values of these fits are given in Table 1 :

TABLE I

	θ_{ac}(K)	θ_{op}(K)	$(A\lambda)_{ac}$ (Ω.m.K^{-1})	$(A\lambda)_{op}$ (Ω.m.K^{-1})
Ag	210±10		3.21 10^{-11}	0
Ag+4.10^{17}D.cm^{-2}	180±10	350±50	4.67 10^{-11}	3.69 10^{-11}

Note that, at 80 K, the difference (Fig. 1) between the experimental value of the total deuterium-implanted silver resistivity and the acoustic phonon contribution alone is 5 times larger than the uncertainty (due to the determination of θ_{ac}) on the latter.

3) Superconductivity :

During pumping down to 1.8 K, no superconducting transition were found whatever the deuterium content up to 9.10^{17} D.cm^{-2}.

3. DISCUSSION AND CONCLUSION

Our characteristic acoustic phonon temperature θ_{ac} in pure silver compares well with the value (220 K) given by White and Woods (4). It decreases slightly after deuterium implantation.

In deuterium implanted silver, the main result is that we have to introduce an optic phonon mode to fit the resistivity temperature dependence. The size of the electron-optic mode coupling is of the same order as the coupling to the acoustic phonons, with $\lambda_{op}/\lambda_{ac} \cong 0{,}8$. This phonon mode, attributed to the deuterium vibration suggests that D occupies a well defined site in the silver lattice and is not in form of precipitates.

On the other hand, the value of $A\lambda_{ac}$ increases when going from pure Ag to Ag + 4.10^{17} D.cm^{-2}. From Papaconstantopoulos'band structure calculations (2), an in sulating state is predicted for the stoechiometric Ag D system. Although this is not experimentally observed (5), we may surmise that the parameter A, which contains the electronic properties (see Eq. 4), should increase and account for the increase of $A\lambda_{ac}$. From this, we may tentatively conclude that λ_{ac} itself does not increase with the D content and may even decrease. This may also explain the absence of superconductivity in Ag D, despite the appreciable optic phonon mode contribution.

REFERENCES

(1) Hydrogen in Metals, Topics in Applied Physics, Vol. 29, Ed G. Alefeld and J. Völkl, 1978.

(2) D.A. Papaconstantopoulos et al, Journal de Physique 39 C6, (1978) 435

(3) J.M. Ziman, Electrons and Phonons, Oxford University Press, 1979

(4) G.K. White and S.B. Woods, Trans. Roy. Soc. of London, Series A, 251, (1958/59) 995

(5) A. Traverse et al., to be published.

LT-17 (Contributed Papers)
U. Eckern, A. Schmid, W. Weber, H. Wühl (eds)

ELECTRON-PHONON INTERACTION IN TRANSITION METAL DIHYDRIDES

Dimitrios A. PAPACONSTANTOPOULOS

Naval Research Laboratory, Washington, DC 20375 USA and Dept. of Physics, University of Crete, Iraklion, Crete, Greece

In a study(1) of the superconducting properties of PdH we have demonstrated that the mechanism for superconductivity in this system is the large value of the electron-optical phonon contributions to the electron-phonon coupling. This is due to the relatively high value of the s-like density of electronic states (DOS) on the hydrogen site, n_s^H, at the Fermi level, E_F, and to the soft optic modes of the phonon spectrum in this system.

Recently Gupta(2) calculated the electron-phonon (EP) interaction in ZrH_2 and NbH_2 using the results of her non-self-consistent band structure calculations. The conclusion from this work was that the above conditions (high hydrogen site DOS and soft optic modes) that make the EP interaction strong for PdH are not satisfied for these dihydrides. Indeed Gupta has shown that n_s^H has a very small value for ZrH_2 and NbH_2, and experiments(3) for ZrH_x suggest much harder optic modes than in PdH.

We have now performed self-consistent band structure calculations for TiH_2, VH_2, ZrH_2, NbH_2, and HfH_2 and used the results to calculate the EP interaction η for these compounds. Our results for ZrH_2 and NbH_2 show clear quantitative differences from those of Gupta, but our conclusions remain the same.

Our band calculations were done self-consistently using the augmented plane wave (APW) method, and included relativistic effects except spin-orbit. As an example we show in Fig. 1 the energy bands of VH_2. The lowest band that originates at the point Γ_1 represents the vanadium-hydrogen bonding states. The next set of bands centered at the levels $\Gamma_{25'}$ and Γ_{12} correspond to the d-levels of vanadium. The band connecting to the $\Gamma_{2'}$ level arises from the antibonding combination of the two hydrogen atoms in the unit cell. The position of the $\Gamma_{2'}$ state found in our calculation above the d-states and above the E_F is a point of controversy among previous calculations. Gupta and Chatterjee(4) found the $\Gamma_{2'}$ level below the d-states while Kulikov(5) has found it above. For TiH_2 the non-self-consistent calculations except that of Kulikov locate the $\Gamma_{2'}$ state below E_F. Our self-consistent calculation for TiH_2 has the same ordering of levels as that of VH_2, i.e. $\Gamma_{2'}$ lies above the d-levels and above E_F. On the contrary our calculations for ZrH_2, NbH_2, and HfH_2 place the $\Gamma_{2'}$ level below the d-states in agreement with the non-self-consistent calculations.

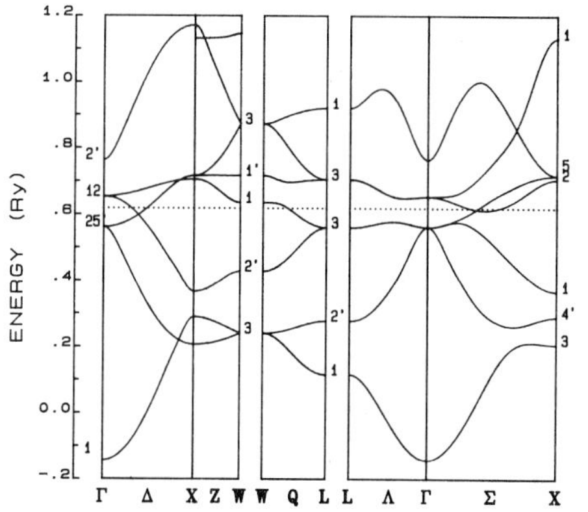

Figure 1. Energy Bands of VH_2.

In order to find the DOS we used Boyer's(6) interpolation scheme in conjunction with the tetrahedron method.(7) Our DOS for VH_2 are shown in Fig. 2. They are characterized by a double peak at an energy about 3 eV which is a mixture of vanadium p- and d-like states and hydrogen s-like states. Near the Fermi level the DOS is dominated by the vanadium d-states.

To calculate the EP interaction η we have applied the rigid muffin-tin theory of Gaspari and Gyorffy.(8) Table I shows the total DOS, n_t, the components n_s, n_p, n_d and the EP, η, for both the metal (first row) and the hydrogen sites (second row). Our results for the five dihydrides are compared with PdH. From this table we observe that for all dihydrides the η of the two hydrogen sites is smaller than the η of PdH by at least a factor of three. On the other hand the η of the metal site is at least a factor of two smaller than the corresponding pure element (quantity in parenthesis).(9) According to our previous work(1) in PdH, one needs an enhanced value of the

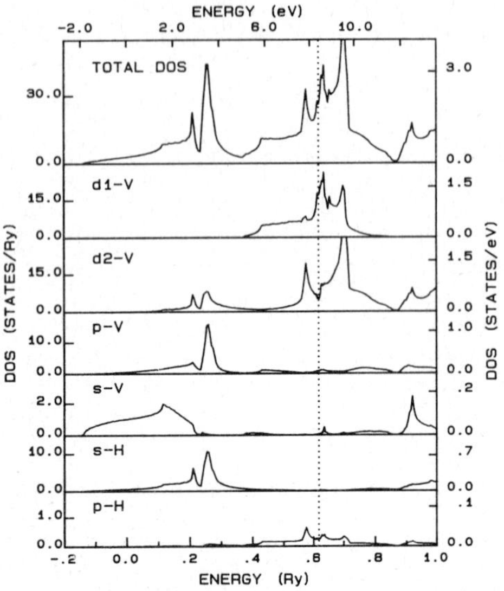

Figure 2. Densities of states for VH_2.

electron-optical phonon interaction for superconductivity to occur. In view of our results listed in Table I the only way for this to happen is to have optic phonon modes which are even softer than those of PdH. This is not likely since the experimental(3) evidence at least for ZrH_x is in the opposite direction. It should be mentioned here that Gupta(2) gives for ZrH_2 $\eta_H = 0.176$ eV/Å^2 and for NbH_2 $\eta_H = 0.204$ eV/Å^2. These are larger than ours by factors of 2.5 and 1.5 respectively, but undoubtedly lead to the same conclusions as ours. On the other hand the results of Kulikov(5) for TiH_2 and VH_2 give η_H about a factor of 15 larger than ours. We believe that this author is in error.

In conclusion we have presented calculations of the EP interaction in metal dihydrides based on self-consistent APW calculations. These calculations, in agreement with previous predictions,(2) do not anticipate superconducting temperatures of any significance in these materials.

ACKNOWLEDGEMENT - I am grateful to Drs. A.C. Switendick and B.M. Klein for helpful discussions.

Table I

	n_t	n_s	n_p	n_d	η
		states/Ry/spin			eV/Å^2
TiH_2	15.633	0.0065	0.1872	12.8436	1.3739 (4.840)[a]
		0.0038	0.3493	0.0174	0.0409
VH_2	13.049	0.0240	0.4000	11.0554	2.8587 (6.894)[a]
		0.0209	0.1265	0.0148	0.0587
ZrH_2	9.631	0.0050	0.1414	6.7053	1.5179 (4.658)[a]
		0.0078	0.3743	0.0177	0.0690
NbH_2	8.540	0.0321	0.3745	6.1029	3.9545 (7.627)
		0.0381	0.1650	0.0177	0.1400
HfH_2	7.929	0.0071	0.1436	5.2039	1.7357
		0.0083	0.3446	0.0158	0.0825
PdH	3.301	0.0590	0.1370	2.5770	0.8650 (3.590)[a]
		0.2160	0.0400	0.0030	0.3920

[a]Ref. 9

References

1. D.A. Papaconstantopoulos, B.M. Klein, E.N. Economou, and L.L. Boyer, Phys. Rev. B17, 141 (1978).
2. M. Gupta, Phys. Rev. B25, 1027 (1982).
3. W.L. Whittemore, in Symposium on Inelastic Scattering of Neutrons in Solids and Liquids (International Atomic Energy Agency, Vienna 1965) p. 305.
4. R. Sen Gupta and S. Chatterjee, J. Phys. F12 1923 (1982).
5. N.I. Kulikov, phys. stat. sol. (b) 91, 753 (1979).
6. L.L. Boyer, B19, 282 (1979).
7. G. Lehmann and M. Taut, phys. stat. sol. 54, 469 (1972).
8. G.D. Gaspari and B.L. Gyorffy, Phys. Rev. Lett. 29, 801 (1972).
9. D.A. Papaconstantopoulos, L.L. Boyer. B.M. Klein, A.R. Williams, V.L. Moruzzi, and J.F. Janak, Phys. Rev. B15, 4221 (1977).

LT-17 (Contributed Papers)
U. Eckern, A. Schmid, W. Weber, H. Wühl (eds)

THEORETICAL LATTICE CONSTANT OF THE PREDICTED HIGH-T_c COMPOUND: B1-STRUCTURE MoN

Barry M. KLEIN Larry L. BOYER, Henry KRAKAUER* and Ching-Ping S. WANG+

Naval Research Laboratory, Condensed Matter Physics Branch, Washington, D.C. 20375 USA.

First-principles LAPW calculations of the electronic structure and total energy of the predicted high-T_c compound B1-structure MoN yield values of 4.17 Å for the cubic lattice constant and 5.4 Mbar for the bulk modulus.

1. INTRODUCTION

In a paper presented at LT16, Pickett, et al. (1) predicted that B1-structure MoN would be a high-T_c material with $T_c \sim 30K$ [further discussion may be found in (2)]. This prediction was based on augmented-plane-wave (APW) energy band calculations and estimates of the phonon spectral moments for this compound as input into the theory of superconductivity [see (3) for a recent review]. Although the equilibrium phase diagram reveals a hexagonal compound δ-MoN at a strict one-to-one ratio which is stable below 800°C (4), with current methods of materials preparation (e.g. sputtering, electron beam evaporation, shock compression, vapor deposition, ion implantation), it may be possible to form metastable MoN in the B1-structure. Given the possibility of such a dramatic increase in T_c, a number of experimental efforts, so far unsuccessful, to form stoichiometric B1-structure MoN have been undertaken.

Since stoichiometric B1-structure MoN has not as yet been prepared, the lattice constant used in (1) was based on an estimate using experimental data for MoC_x and NbC_xN_{1-x} as a guide. The value of the cubic lattice constant chosen was 4.25 Å. In this paper we present results of an ab initio calculation of the lattice constant and bulk modulus of B1-structure MoN.

2. METHOD OF CALCULATION

The self-consistent general-potential linearized APW (LAPW) method has been used to determine the electronic band structures and total energies as a function of the cubic lattice constant of MoN. Details regarding the specific computational techniques used may be found in (5). The major differences between the present calculations and those in (1) are our utilization of fully general forms for the charge density and crystal potential in the whole unit cell. In (1) spherical approximations to the muffin-tin quantities were made, and constant averages in the interstitial region were used. We have used the same local density form for the exchange-correlation potential as was used in (1).

Self-consistent calculations were performed for four different lattice constants: 4.25 Å, the value chosen in (1), and 2% and 4% compressions and a 4% expansion about this value. The total electronic energies for each lattice constant were used to determine polynomial fits (to the lattice constant or volume) and also fits to Murnaghan's (6) equation of state. From these fits the equilibrium volume and lattice constant, and the bulk modulus, were determined.

3. RESULTS AND DISCUSSION

3.1 Lattice Constant and Bulk Modulus

Our total energy fits yield values of 4.17 ± 0.01 Å and 5.4 ± 0.5 Mbar for the B1-structure MoN lattice constant and bulk modulus, respectively. The uncertainties shown are estimates of the accuracy of our fits due to the limited number of points used.

Our value of the lattice constant is approximately 2% smaller than the value used in (1). Experience with ab initio calculations on systems with experimentally known lattice constants have generally shown accuracies of 1-2 % or better for semiconducting (7) or metallic (8) systems. With this in mind, our agreement with the estimate used in (1), and with other empirical estimates, is gratifying. The only experimental report of B1-structure MoN is that of Sauer and collaborators (9) who identified it as a second phase in a sample which was primarily hexagonal ("WC type") MoN.

---

*Permanent address: College of William and Mary, Williamsburg, Virginia 23185 USA. Supported by NSF Grant DMR-81-20550 at the College of William and Mary. Also at Sachs/Freeman Associates, Inc., Bowie, MD 20715 USA.

+Permanent address: University of Maryland, College Park, Maryland 20792 USA.

Their value of the cubic (stoichiometry uncertain) lattice constant is 4.16 Å in good agreement with the present value.

The value of 5.4 Mbar for the bulk modulus is large compared to values for transition metals, but consistent with values for known carbo-nitride systems (10). Typically it is found that bulk modului for the carbo-nitrides are approximately twice as large as the value for the transition-metal component alone. We find that this is the case for MoN as well.

3.2 Band Structure and Density of States

Table 1 compares some of the present LAPW eigenvalues with those from the APW calculations used in (1). The differences, a consequence of the non-muffin-tin-terms included in the LAPW calculation, are fairly modest, with the biggest shifts occurring for the Mo d states with e_g (Γ_{12})-type symmetry. These eigenvalues correspond to wave functions which bond with N p states along the x, y and z axes, and are the most sensitive to the non-muffin-tin terms.

Table 1. Comparison of eigenvalues at Γ and X from an LAPW (general potential) and an APW (muffin-tin potential) calculation. Energies are in Ry with respect to the bottom of the valence bands in each calculation.

	E(LAPW)	E(APW)	ΔE
Γ_{15}	1.079	1.083	-0.004
$\Gamma_{25'}$	1.112	1.121	0.009
Γ_{12}	1.231	1.265	-0.034
X_1	0.066	0.069	-0.003
$X_{4'}$	0.850	0.861	-0.011
X_3	0.903	0.886	0.017
$X_{5'}$	1.016	1.024	-0.008

In (1) it was shown that the density of states at the Fermi level, $N(E_F)$, for MoN was more than twice the value of NbN which has a known T_c of 15.4K. This was the fundamental reason that T_c of B1-structure MoN was calculated to be close to 30K, or possibly higher. Although our current LAPW estimates of $N(E_F)$ for MoN are somewhat smaller than the APW values (by ∿20% for the same lattice constant), they are still nearly twice the NbN value, and the theoretical T_c is well over 20K [keep in mind the uncertainties in the estimates of the MoN phonon spectral moments discussed in (1)]. The discovery of high-T_c B1-structure MoN still remains an exciting prospect.

4. CONCLUSIONS

Since any major advances in raising the maximum T_c are likely to come from stabilizing some metastable structures in systems with good prospects for high-T_c, theory such as that described here can play an important role in pinning-down some of the structural information. Of particular interest to experimentalists would be predictions of the transition pressures needed to lock-in the desired structures. Hexagonal MoN has a unit cell volume some 10% larger (11) than our predicted B1-structure value. We are currently performing LAPW calculations for this phase which, together with the present results, will yield a theoretical value for the hexagonal to cubic transition pressure. These results will be reported in a subsequent paper.

ACKNOWLEDGMENTS

We are grateful to Drs. D. A. Papaconstantopoulos, W. E. Pickett, S. B. Qadri, E. F. Skelton, and W. Temmerman for helpful conversations.

REFERENCES

(1) W.E. Pickett, B.M. Klein, and D.A. Papaconstantopoulos, Physica 107B (1981) 667.

(2) D.A. Papaconstantopoulos, W.E. Pickett, B.M. Klein and L.L. Boyer, Nature, in print 1984.

(3) B.M. Klein and W.E. Pickett, Rigid Muffin-Tin Calculations of Superconducting Parameters in d- and f-Band Metals, in: Superconductivity in d- and f-Band Metals, 1982, eds. W. Buckel and W. Weber (Kernforshungszentrum Karlsruhe GmbH, Karlsruhe, 1982) pp. 477-485.

(4) H. Jehn and P. Ettmayer, J. Less-Common Metals 58 (1978) 85.

(5) H. Krakauer and S.-H. Wei, unpublished.

(6) F.D. Murnaghan, Proc. Nat. Acd. Sci. U.S.A. 30 (1944) 244.

(7) See e.g. M.T. Yin and M.L. Cohen, Phys. Rev. B26 (1982) 5668.

(8) See e.g. V.L. Moruzzi, J.F. Janak and A.R. Williams, Calculated Electronic Properties of Metals (Pergamon, New York, 1978).

(9) E.J. Sauer, H.D. Schechinger and L. Rinderer, IEEE Trans. MAG-17 (1981) 1029; H. Bauer, E. Sauer and D. Schechinger, in Proceedings of the 14th International Conference on Low Temperature Physics, Vol. 2, eds. M. Krusius and M. Vuorio (North-Holland, Amsterdam, 1975) pp. 55-58).

(10) L.E. Toth, Transition Metal Carbides and Nitrides (Academic, New York, 1971).

(11) N. Schönberg, Acta Scandinavica 8 (1954) 204.

LT-17 (Contributed Papers)
U. Eckern, A. Schmid, W. Weber, H. Wühl (eds)

IMPURITIES AND THE ANOMALOUS SPIN-LATTICE RELAXATION IN COPPER AT SUBMILLIKELVIN TEMPERATURES

M.T. Huiku, M.T. Loponen, T.A. Jyrkkiö, J.M. Kyynäräinen, A.S. Oja, and J.K. Soini

Low Temperature Laboratory, Helsinki University of Technology, SF-02150 Espoo 15, Finland

The spin-lattice relaxation time τ_1 of copper has been studied as a function of the external magnetic field at constant electronic temperatures between 50 and 300 µK. The ratio $r = \tau_1(15\text{ mT})/\tau_1(0)$ for a high purity copper sample was found to be as high as 100. By selective oxidation r could be reduced to 2.6 in the same specimen. Present theories seem unable to explain the anomalously rapid relaxation in zero field.

1. INTRODUCTION

At low temperatures the nuclear spin system is connected to the spin-lattice relaxation process from the conduction electrons. The nuclei approach exponentially T, the temperature of the electron system, with a time constant τ_1', which in metals is given by the Korringa law $\tau_1' = \kappa/T$. In magnetic fields higher than the local field B_{loc}, the Korringa constant κ should be independent of the external field B. However, when $B < B_{loc}$, spin-spin interactions reduce the spin-lattice relaxation time through the faster relaxation of the dipolar energy storage (1). In the absence of quadrupolar interactions

$$\tau_1' = \tau_Z \frac{B^2 + B_{loc}^2}{B^2 + (\tau_Z/\tau_D)B_{loc}^2} \qquad (1),$$

where τ_Z and τ_D are the relaxation times in the Zeeman and dipolar systems, respectively. In high fields the energy of the spins is in the Zeeman interaction, whereas when B = 0 the energy is in the spin-spin interactions. In pure metals, τ_Z/τ_D should vary between 2 and 3, depending on the extent of correlations in the spin system.

In low fields (B < 15 mT) magnetic impurities, such as Fe, Cr, Ni, and Mn, can strongly influence the spin-lattice relaxation process. In the presence of impurities τ_1 can be written as $1/\tau_1 = 1/\tau_1' + 1/\tau_1^a$, where τ_1^a is the anomalous relaxation time induced by the impurities.

2. EXPERIMENTAL

Our measurements of τ_1 were carried out in a two stage nuclear demagnetization refrigerator where the second stage acted also as the sample (2). During the experiments the electron temperature of the specimen (in the range 0.05 - 0.3 mK) stayed constant for a few hours.

We always measured τ_1 at small polarizations ($p \leq 0.3$) from the relaxation of the susceptibility after demagnetization to a low external field. Two different methods were employed: In the first set of experiments the susceptibility was measured using a Robinson oscillator circuit. In the second set the static susceptibility $\chi'(0)$ was obtained in different fields using a SQUID in the feedback mode and a low frequency (usually 10 Hz) ac-technique.

The average frequency of the Robinson circuit was 153 kHz corresponding to a field of 13.5 mT. In these experiments the susceptibility was recorded by sweeping the field slowly across the resonance to obtain the absorption curve and thence, the polarization p. By sweeping the field off resonance for a certain time (and then back), the change in p during the stay gave the relaxation time τ_1.

Several specimens were used in the experiments. Some of them were heat treated at 950°C under a pressure of 10^{-4} torr of dry air in a quartz tube oven. Under these conditions copper oxides are unstable, whereas most of the other oxides are stable (3). This makes it possible to selectively oxidize impurities in copper and thereby reduce the scattering of the conduction electrons from the impurity moments. The oxidation time varied from a few hours to several days, depending on the thickness of the samples.

3. RESULTS

Anomalously rapid spin-lattice relaxation was observed in nonoxidized samples. The specimens, their magnetic impurity concentrations, their residual resistivity ratio RRR, and the ratio of the high field relaxation time to the corresponding zero field value, $r = \tau_1(15\text{ mT})/\tau_1(0)$, are presented in Table 1. In the high field re-

Table 1. Properties of the samples. Wx, Sx, and Fx denote wires, single crystals, and foils with thickness x in μm, respectively. O_2 stands for selective oxidation. Symbols refer to Fig.1. See text.

Sample	Impurity/ppm Mn	Cr	Fe	Ni	purity	RRR	r	O_2
1● W40	.18	<1.2	19		4N	200	4	no
2■ W40	<1	1	30	30	3N		7	no
3△ W125	.2	<.3	.8		5N	550	100	no
4 - S500	<.1	2	3	<.4	4N	400	10-100	no
5 - S500	"	"	"	"	"	2000	2.8	yes
6 - F125	<.1	.3	.8	<.1	5N	260	100	no
7○ F125	"	"	"	"	"	8500	2.6	yes

gion the longest τ_1(15 mT) was over 7 hours.

In Fig.1 we show τ_1 as a function of B. Data for Samples 1 and 2, measured by the SQUID-NMR-technique, have been reported earlier (2). Those results showed a decrease in τ_1 between 15 and 2 mT, which was attributed to impurities. The further decrease by a factor of 2.3 below 2 mT fits Eq.1, if the local field for copper, B_{loc} = 0.34 mT, is employed. Since the anomalous relaxation occurs above 2 mT, which is much higher than the internal fields in copper, it cannot be due to quadrupolar interactions. Therefore, it was concluded (2) that quadrupolar interactions, due to the possibly slightly distorted FCC-lattice, are neglible in our copper samples.

Specimen 3, consisting of 0.125 mm thick high purity wires, showed a rapid relaxation in the Robinson oscillator measurements. Although magnetic impurity concentrations were very low (< 1.4 ppm), r was as large as 100. In this sample, in addition to the anomalous relaxation, the NMR-lines deviated significantly from the Lorentz shape at high polarization. Both in ^{63}Cu and in ^{65}Cu a broad satellite peak was observed on the high frequency side of the NMR-line.

Sample 4, a single crystal, studied by means of the static susceptibility method, showed two clearly different relaxation times, corresponding roughly to r = 10 and r = 100 (respectively C and D in Fig. 1). About 90 % of the relaxation in zero field occured fast. By comparing results at various frequencies we concluded that the slow relaxation occurs on a surface layer about 50 μm thick. The diffusion constant of oxygen in copper suggests that oxidization in ambient air during a few years had inactivated the impurities near the surface. When the same sample was annealed, but not oxidized, only one relaxation time was observed, corresponding to r = 25 (B in Fig.1).

The oxidization hypothesis proved right when another single crystal (Sample 5) from the same batch was selectively oxidized. The RRR increased from about 400 to 2000 and r decreased to 2.8 Which is in the theoretically prdicted range. Even more dramatic effects of the heat

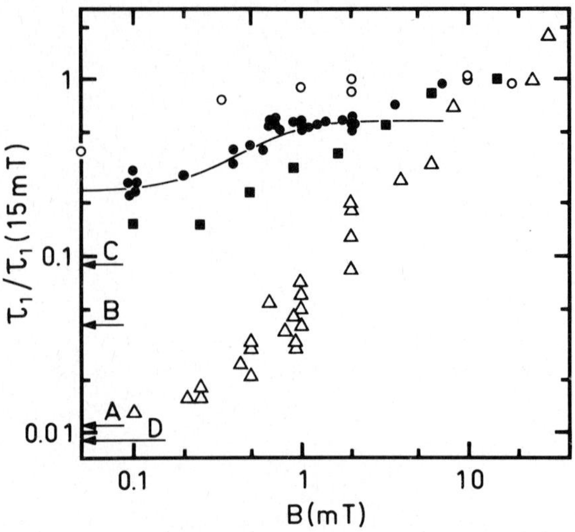

Fig.1. The relative spin-lattice relaxation time as a function of the external magnetic field. Filled dots and squares are for Samples 1 and 2, respectively (See Table 1). Triangles are for Sample 3 and open circles for Sample 7. A indicates τ_1(0) for Sample 6 and B,C and D for Sample 4. See text.

treatment were observed in high purity copper foils (Samples 6 and 7), where RRR increased from 260 to 8500 and r decreased from 100 to 2.6; this small r-value indicates that there are no extra relaxation processes left.

There are two approaches to explain the effect of impurities. However, neither of them seem to account for the observed anomalous relaxation. The Kondo-model (4) and the spin-diffusion theory (5) were developed to describe relatively small r-values.

We conclude that at submillikelvin temperatures magnetic impurities influence dramatically the spin-lattice relaxation process, but the effect can be quenched by selective oxidation. For observing the nuclear ordering in copper a long τ_1, obtained by selective oxidation, proved to be crucial (6).

REFERENCES

1. M. Goldman, Spin Temperature and Nuclear Magnetic Resonance in Solids (Claredon Press, Oxford 1970), p. 55.
2. G.J. Ehnholm, J.P. Ekström, J.F. Jacquinot, M.T. Loponen, O.V. Lounasmaa, and J.K. Soini, J. Low Temp. Phys. 39, 417 (1980).
3. J. Peterseim, G. Thummes, and H.H. Mende, Z. Metallkde 70, 266 (1979).
4. W.A. Roshen and W.F. Saam, Phys. Rev. B22, 5495 (1980).
5. P. Bernier and H. Alloul, J. Phys. F6, 1193 (1976).
6. M.T. Huiku and M.T. Loponen, Phys. Rev. Lett. 49, 1288 (1982).

LT-17 (Contributed Papers)
U. Eckern, A. Schmid, W. Weber, H. Wühl (eds)

PARAMAGNON CONTRIBUTION TO NUCLEAR SPIN RELAXATION IN NEARLY MAGNETIC FERMI SYSTEMS*

P.J. HIRSCHFELD and D.L. STEIN

Joseph Henry Laboratories of Physics, Princeton University, Princeton, N.J. 08544, USA.

The temperature dependence of the nuclear spin-lattice relaxation rate T_1^{-1} for a nearly ferromagnetic metal is calculated in a simple paramagnon model. One and two spin-fluctuation processes can account for the high-temperature behavior of the momentum-dependent susceptibility $\chi(\vec{q},T)$, and hence the observed T_1^{-1}, giving reasonable quantitative agreement with recent experiments on Palladium metal.

1. INTRODUCTION

Recently, Takigawa and Yasuoka (1) measured the nuclear spin-lattice relaxation time T_1 for the nearly ferromagnetic transition metal Pd. Below 100K they found the usual Korringa-like relation $(T_1T)^{-1}$ = constant. At higher temperatures, however, $(T_1T)^{-1}$ was found to decrease substantially. It is tempting to attribute this departure to spin fluctuations, which are expected to be important to the magnetic properties of the transition metals (2).

2. RELAXATION MECHANISMS

The Pd nucleus may relax via i) the Fermi contact interaction with conduction band electrons; ii) the interaction of the nucleus with the orbital moments of the s and p-band electrons; iii) the spin dipolar interactions with these bands; and iv) the core polarization effect, in which the d-band electrons electrostatically polarize the cloud of tightly bound electrons surrounding the nucleus. The measured relaxation rate will be the sum of the rates corresponding to each mechanism:

$$T_1^{-1}=(T_1^{-1})_{cont}+(T_1^{-1})_{orb}+(T_1^{-1})_{dip}+(T_1^{-1})_{cp} \qquad (1)$$

Each contribution is related (3) to an associated dynamical susceptibility $\chi(\vec{q},\omega,T)$:

$$(T_1T)^{-1} = H^2\gamma_N^2\sum_{\vec{q}}\frac{\mathrm{Im}\,\chi(\vec{q},\omega_0,T)}{\omega_0}\ , \qquad (2)$$

where H is a coupling constant for each process, γ_N is the nuclear gyromagnetic ratio, and ω_0 is the Larmour frequency.

Good results for the spin-lattice relaxation rates for most of the transition metals have been obtained by Asada, et al. (4) using a self-consistent band structure calculation of each of the above contributions. For Pd, however, this approach fails dramatically, presumably because it neglects the electron-electron correlations which give rise to the well known Stoner enhancement of the Pd susceptibility.

*Research supported in part by NSF DMR802063

As strong interactions among the d-band electrons should primarily enhance the core polarization effect, we adopt the rates associated with the other mechanisms as calculated by Asada, et al., and attempt to describe the enhancement of $(T_1^{-1})_{cp}$ within the well-known paramagnon model (2).

3. CALCULATION OF DYNAMICAL SUSCEPTIBILITY

Since the number of d-holes in Pd metal is small compared to the total number of d-states, we may be justified in taking as a model a single parabolic band of holes which we allow to interact via a contact repulsion between opposite spins. In the RPA, the susceptibility at T=0 is then given by

$$\chi(\vec{q},\omega,0) = \frac{\chi^0(\vec{q},\omega,0)}{1 - I\chi^0(\vec{q},\omega,0)}\ , \qquad (3)$$

where I is the energy parameter describing the interaction, and χ^0 is the Lindhard function. As emphasized by a number of authors (5-6), however, the RPA is not sufficient to account for spin fluctuation effects which occur at finite temperatures. This point is discussed formally by Mishra and Ramakrishnan (6), who develop a perturbation theory based on a functional integral from which the fermion degrees of freedom have been eliminated. Diagrams with one and two correlated internal spin fluctuations are shown to provide the leading order temperature corrections to the self-energy; in the "paramagnon" regime $(T<(1-I\chi^0(0))T_F \equiv T_{SF})$ these are of order $(T/T_{SF})^2$.

We have calculated the analogous susceptibility diagrams to find the leading order temperature corrections to $\chi(\vec{q},\omega,0)$. Some of these are shown in Figure 1. As emphasized in reference (7), these are a factor $(1-I\chi^0(0))^{-1}$ greater than the finite temperature corrections to the Lindhard function.

In order to calculate T_1^{-1} and compare with experiment, it is sufficient to take $T<<T_F$ and $\omega\sim\omega_0<<\epsilon_F$, since the Larmour frequency is small. For a quantitative comparison, however, we may not neglect the q-dependence of $\mathrm{Im}\chi$ since the

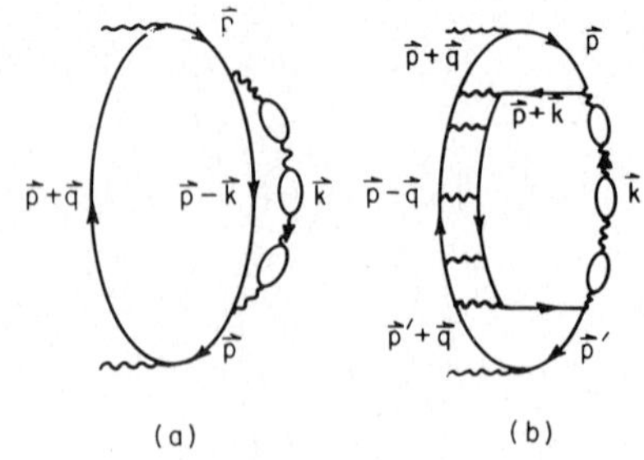

FIGURE 1
Diagrams contributing to leading-order temperature correction $\delta\chi_{fl}$ to $X^0(\vec{q},\omega,0)$: a) single spin-fluctuation self-energy correction; b) two spin-fluctuation diagram.

integrand in equation (2) assigns a larger weight to higher external momenta and peaks around $q \sim q_F$.

The full interacting dynamical susceptibility which enters equation (2) is now

$$\chi(\vec{q},\omega,T) = \frac{\tilde{X}^0(\vec{q},\omega,T)}{1 - I\tilde{\chi}^0(\vec{q},\omega,T)}, \qquad (4)$$

where the renormalized bare bubble $\tilde{\chi}^0$ is simply

$$\tilde{\chi}^0(\vec{q},\omega,T) = \chi^0(\vec{q},\omega,T) + \delta\chi_{fl}(\vec{q},\omega,T). \qquad (5)$$

After a calculation parallelling that of reference (5) for the q=0 case, the fluctuation part $\delta\chi_{fl}$ is found to be

$$\delta\chi_{fl}(\zeta,0,T) \simeq \frac{I^3N(0)}{16\varepsilon_F^2}\,\alpha(\bar{I},T)\left\{\frac{2(\bar{I}+1) - 2(2+\bar{I})\zeta^2}{(1+\bar{I})\zeta^2(1-\zeta^2)} + \frac{1}{4\zeta^3}\log\left|\frac{1-\zeta}{1+\zeta}\right| - \frac{1}{2\zeta^2(1+I\chi^0(\zeta))}\left(\log\left|\frac{1-\zeta}{1+\zeta}\right|\right)^2\right\} \qquad (6)$$

where N(0) is the density of states at the Fermi surface, $\bar{I} \equiv IN(0)$, $\zeta \equiv q/2q_F$, and $\alpha(\bar{I},T)$ is defined as

$$\alpha(\bar{I},T) = \sum_{\vec{k},} \frac{\chi^0(\vec{k}\omega)}{1-\bar{I}\chi^0(\vec{k},\omega)} \qquad (7)$$

The sum over Matsubara frequencies gives $\alpha \sim (T/T_{SF})^2$.

4. COMPARISON WITH EXPERIMENT

Using the density of d-states and hyperfine core polarization coupling constants from reference (4), it is straightforward to calculate the nuclear spin-lattice relaxation time numerically via equations (2) and (4)-(7). There are two free parameters determined by a fit to the data of reference (1): i) the Stoner enhancement factor $(1-\bar{I})^{-1}$, which from the T=0 limit of $(T_1T)^{-1}$ is found to be 10.3; and ii) the degeneracy temperature of the d-holes, T_{Fd}, which from the finite-temperature data is estimated to be 4000K (see Figure 2).

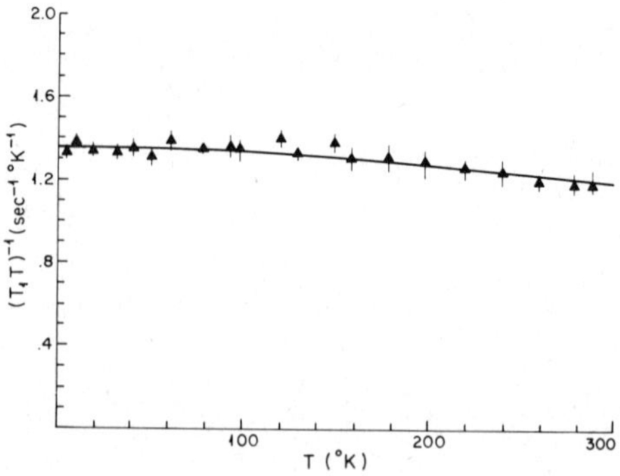

FIGURE 2
Pd nuclear spin relations rate obtained by Takigawa and Yasuoka (1) and prediction of finite-temperature paramagnon theory, with $(1-\bar{I})^{-1} = 10.3$ and $T_{Fd} = 4000K$.

The enhancement is similar to that found for Pd by a number of authors (8), including Takigawa and Yasuoka for their own susceptibility data. The d-band Fermi temperature is also consistent with the results of a number of band structure calculations (9), but due to the complexity of the Pd Fermi surface these results are rather uncertain, and T_{Fd} is probably not known to much better than a factor of two. Nevertheless, our results lend further support to the contention that the simple paramagnon model is a good description of metals like Pd near a ferromagnetic instability.

ACKNOWLEDGEMENTS
The authors gratefully acknowledge enlightening discussions with J. Sauls and P. Wölfle.

REFERENCES

(1) M. Takigawa and H. Yasuoka, J. Phys. Soc. Japan 51 (1982) 787.
(2) N.F. Berk and J.R. Schrieffer, Phys. Rev. Lett. 17 (1966) 433.
(3) T. Moriya, J. Phys. Soc. Japan 18 (1963)516.
(4) T. Asada, K. Terakura, and T. Jarlborg, J. Phys. F: Metal Phys. 11 (1981) 1847.
(5) S.K. Ma, M.T. Beal-Monod, and D.R. Fredkin, Phys. Rev. 174 (1968) 227.
(6) S.G. Mishra and T.V. Ramakrishnan, Phys. Rev.B 18 (1978) 2308.
(7) M.T. Beal-Monod, Phys. Rev. B 28 (1983) 1630.
(8) S.Doniach, Proc. Phys. Soc.(London) 91 (1967) 86.
(9) F.M. Mueller, A.J. Freeman, J.O. Dimmock and A.M. Furdyna, Phys. Rev. B 1 (1970) 4617.

LT-17 (Contributed Papers)
U. Eckern, A. Schmid, W. Weber, H. Wühl (eds)

SPECIFIC HEAT AND THERMOELECTRIC POWER OF (La,Ce)B_6

H.J. Ernst, H. Gruhl, T. Krug and K. Winzer

I. Physikalisches Institut der Universität Göttingen
und Sonderforschungsbereich 126
Bunsenstraße 9, D 3400 Göttingen, Germany

We report on measurements of specific heat and thermoelectric power of the dilute Kondo system (La,Ce)B_6. The contribution to the specific heat per Ce-impurity has a maximum $c_i/k_B = 0.40$ at T = 0.35 K, which is twice the value of the recent theoretical calculations for Kondo impurities with spin 1/2. The thermoelectric power has a maximum value of S = 66 μV/K at T = 0.65 K, which is the largest value of S observed so far for all dilute magnetic alloys. Both results are discussed with respect to a possible Γ_8 ground state of the Ce^{3+} ions in LaB_6.

1. INTRODUCTION

In some former papers we have shown that the local moment behavior of Cerium impurities in LaB_6 leads to a very strong Kondo anomaly in the resistivity with a Kondo temperature $T_K = 1.05$ K (1). The Kondo ground state is partially removed by a magnetic field, leading to a large negative magnetoresistivity. From the field dependence of the magnetoresistivity at very low temperatures we obtained a Kondo magnetic field $B_K = 1.1$ T (2). Using a scaling relation of the form $k_BT_K = g_{eff}\mu_BB_K$ a determination of the effective g-value leads to $g_{eff} = 1.42$.

The ground state of the Ce-ion in a crystal electric field of cubic symmetry splits into a Γ_7 doublet and a Γ_8 quartet. According to Murao et al. (3) the Γ_7 doublet may be described by a modified g-factor $g' = -(5/3)(6/7) = -1.43$. The excellent agreement with the amount of the experimental value encouraged us to decide that the doublet forms the ground state of the 4f electrons of Ce^{3+} in LaB_6. Measurements of the susceptibility and of the field dependence of the Hall coefficient of dilute (La,Ce)B_6 can also be successfully interpreted assuming a Γ_7 ground state (4,5).

On the other hand ESR results on dilute Er and Dy in LaB_6 show that the crystal field parameter A_4 is negative (6). The sign of A_4 should not change for different RE ions in the same matrix since it is mostly affected by the host ligands. A negative sign of A_4 is only consistent with a Γ_8 ground state for Ce^{3+}. Recent results of Raman scattering on the concentrated Kondo lattice system CeB_6 also demand a Γ_8 ground state.

2. EXPERIMENTAL DETAILS

For the specific heat measurements flat disks of 300 mg were cut from arc melted samples of 4N pure La and Ce and 4N5 pure B. The measurements were performed in an adiabatic demagnetization cryostat using a heat pulse technique and in a dilution refrigerator using a semi-adiabatic method. Both measurements gave nearly the same results.

For the measurements of the thermoelectric power (TEP) single crystals of (La,Ce)B_6 were grown by an arc floating zone technique. Samples of typically 30×0.6×0.6 mm^3 were cut by sparc erosion and measured in a dilution refrigerator down to 30 mK. The temperature differences were measured with two pairs of precalibrated Ge resistors.

3. RESULTS AND DISCUSSION

Fig.1 shows the excess specific heat per formula unit CeB_6 for two samples with 0.5 and 1 at% Ce respectively as a function of temperature. The specific heat curves show a maximum at $T_m = 0.35$ K, which is nearly proportional to the Ce concentration. From the temperature T_m a Kondo temperature of $T_K = 1$ K results, in good agreement with T_K from other quantities. For the lower concentrated sample a maximum value of $c_i/k_B = 0.4$ is obtained. This is more than twice the value 0.18 of theoretical calculations for a spin 1/2 Kondo model (7,8) (dashed-dotted curve in Fig.1).

Recently Rajan (9) has calculated the specific heat in the Coqblin-Schrieffer model. This model takes into account the impurity-mediated hopping of the electrons between various total angular momentum eigenstates of the impurity. In contrast to the results of the s-d model where the maximum of c_i is largely reduced for $S > 1/2$, in the Coqblin-Schrieffer model the specific heat at low temperatures scales with degeneracy. The result of Rajans calculation for $J = 3/2$ is given in Fig.1 (full curve).

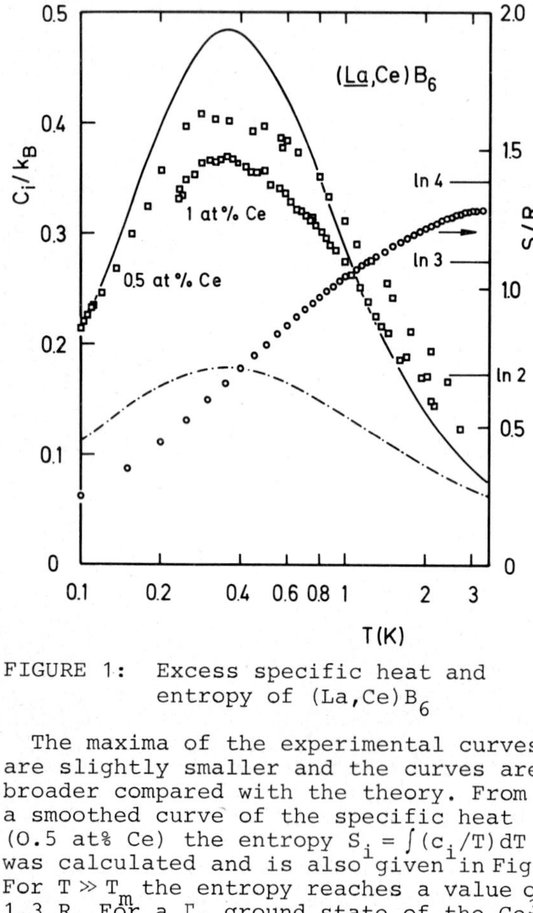

FIGURE 1: Excess specific heat and entropy of (La,Ce)B_6

The maxima of the experimental curves are slightly smaller and the curves are broader compared with the theory. From a smoothed curve of the specific heat (0.5 at% Ce) the entropy $S_i = \int (c_i/T)\,dT$ was calculated and is also given in Fig.1 For $T \gg T_m$ the entropy reaches a value of 1.3 R. For a Γ_8 ground state of the Ce^{3+} we have $J = 3/2$ and hence $S_i/R = \ln 4 = 1.38$. This high-temperature limit of the entropy strongly supports the recent experimental results which can be interpreted by a Γ_8 quartet rather than by a Γ_7 doublet.

The thermoelectric power of four (La,Ce)B_6 samples is given in Fig.2 as a function of temperature. The TEP curves show pronounced maxima around 1 K of up to 66 μV/K, which is by far the largest TEP of all dilute magnetic systems. The "single impurity" TEP S_m can be obtained using the Nordheim Gorter rule for multiple scattering, which leads to values of up to $S_m = 90$ μV/K for the lowest con-

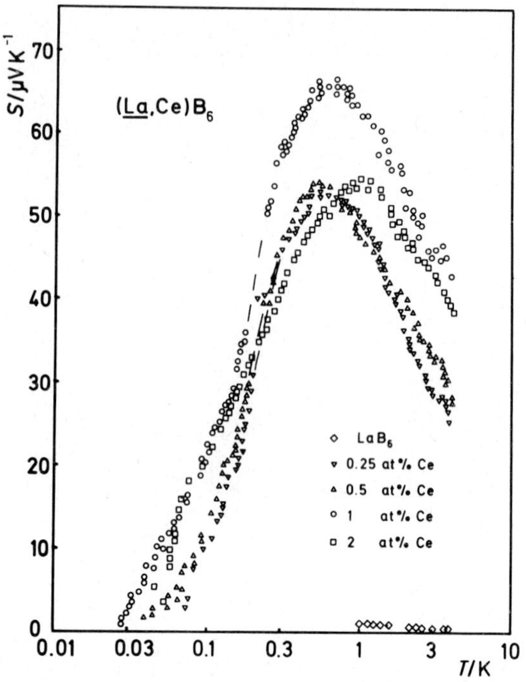

FIGURE 2: Thermoelectric power of dilute (La,Ce)B_6 single crystals

centrated sample. From Maki's theoretical calculation of the TEP (10) a maximum value of $S_m = 50$ μV/K can be obtained for $S = 1/2$, which is considerably smaller compared with the experimental result.

In conclusion, the impurity specific heat and the TEP of (La,Ce)B_6 is too large, to be reproduced by theories based on the spin 1/2 Kondo model. The specific heat data strongly support the Γ_8 quartet as ground state of Ce^{3+} in LaB_6.

REFERENCES

(1) K.Winzer, Solid State Commun. 16, 521 (1975)
(2) K.Samwer and K.Winzer, Z.Physik B 25, 269 (1976)
(3) T.Murao and T.Matsubara, Progr. Theor.Phys. 18, 215 (1957)
(4) W.Felsch, Z.Physik B29, 211 (1978)
(5) R.Dreyer and K. Winzer, Solid State Commun. 46, 71 (1983)
(6) H.Luft, K.Baberschke and K.Winzer Phys.Lett. 95A, 186 (1983)
(7) H.-U.Desgranges and K.D.Schotte Phys.Lett. 91A, 240 (1982)
(8) V.T.Rajan, J.H.Lowenstein and N. Andrei, Phys.Rev.Lett. 49, 497 (82)
(9) V.T.Rajan, Phys.Rev.Lett. 51,308(83)
(10) K.Maki, Progr.Theor.Phys. 41,586(69)

LT-17 (Contributed Papers)
U. Eckern, A. Schmid, W. Weber, H. Wühl (eds)

ANISOTROPIC BEHAVIOR OF THE DENSE KONDO STATE IN Ce-Si SINGLE CRYSTALS

Nobuya SATO, Hiroshi MORI, Hideo YASHIMA and Takeo SATOH

Department of Physics, Faculty of Science, Tohoku University, Sendai 980, Japan

Hidetoshi HIROYOSHI and Humihiko TAKEI

The Research Institute for Iron, Steel and Other Metals, Tohoku University, Sendai 980, Japan

In order to study anisotropic effects in a noncubic dense Kondo system, single crystals of the α-$ThSi_2$ type compound $CeSi_x$ (x=1.86 and 1.70) were prepared. The electrical resistivity, the paramagnetic susceptibility and the high-field magnetization were measured along the a- and c-axis directions. These quantities show the appreciable anisotropy, which we discuss in terms of the crystal field effect.

1. INTRODUCTION

The α-$ThSi_2$ type compound $CeSi_2$ was found to exhibit various anomalies associate with intermediate valence or the Kondo effect of the Ce ions (1,2). The investigation was extended to the Si-deficit system, $CeSi_x$($1.55 \leq x \leq 2.00$), retaining the same α-$ThSi_2$ structure (3). There are several important aspects of the system to be mentioned. First, in the composition range $1.85 \leq x \leq 2.00$ the system is nonmagnetic at low temperatures and for $x \leq 1.80$ the system undergoes a ferromagnetic transition around 10K. Kondo anomalies still exist in the ferromagnetic regime. Second, the spin fluctuation temperature, T_K, estimated from various physical quantities keeps on decreasing upon decreasing x (4). Third, the system is a noncubic system. Therefore the system $CeSi_x$ seems to be of particular insterest in which to explore anisotropic effects in a noncubic dense Kondo system, studies of which have started to appear only recently. Direction dependent susceptibilities and resistivities have been reported for orthorhombic CeNi (5) and tetragonal $CeCu_2Si_2$ (6, 7).

2. EXPERIMENTS

Single-crystalline $CeSi_x$ of x=1.86 and 1.70 were prepared by the floating-zone method. These belong respectively to the nonmagnetic and the magnetic regime. Typical dimensions of these crystals are 8mm in diameter and 30mm in length. Specific heats, paramagnetic susceptibilities, electrical resistivities and magnetizations were measured.

3. RESULTS AND DISCUSSION

The susceptibility results of $CeSi_{1.86}$ are shown in Fig.1, where χ_a and χ_c are the susceptibilities along the a- and c-axis, respectively. As expected from the tetragonal symmetry around the Ce ions, the susceptibility is anisotropic. At room temperature, $\chi_c > \chi_a$, and then the anisotropy reverses at 67K. The low temperature susceptibility can be fitted well with a T^2-law, $\chi(T)=\chi(0)(1+AT^2)$, over the region 4.2K to 15K.

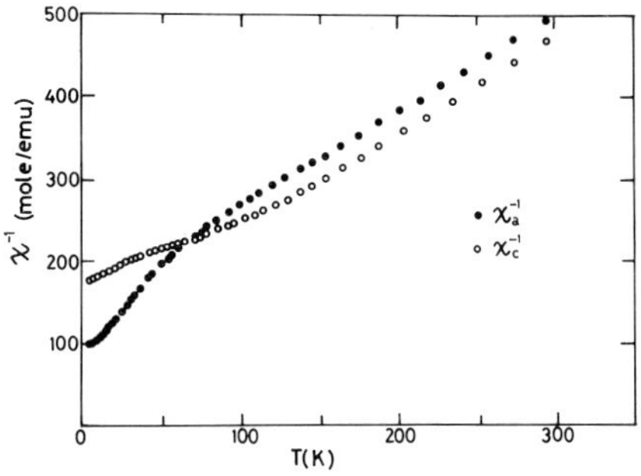

Fig.1 The reciprocal susceptibility, χ_a^{-1} and χ_c^{-1} of $CeSi_{1.86}$.

The resistivity results of $CeSi_{1.86}$ are shown in Fig.2. Both the resistivity along the a-axis, ρ_a, and that along the c-axis, ρ_c, show a small temperature dependence around room temperature. At lower temperatures, the behavior for two directions show a clear contrast; ρ_c starts dropping suddenly around 50K whereas ρ_a decreases more gradually. Below 5K, both ρ_a and ρ_c follows a T^2-dependence, $\rho(T)=\rho(0)+BT^2$.

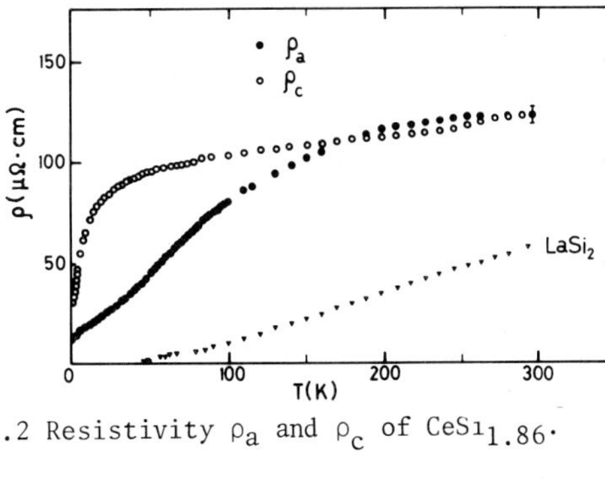

Fig.2 Resistivity ρ_a and ρ_c of $CeSi_{1.86}$.

The T^2 dependence of χ and ρ at low temperature suggests the system behaves as a Fermi liquid as found in other "anomalous" Ce compounds (8). A new feature revealed by the present measurements is the anisotropy in resistivity and susceptibility in the Fermi-liquid region of a nonmagnetic Kondo compound.

In order to understand the anisotropy of the susceptibility, we may invoke the crystal field (C.F.) acting on trivalent Ce ions. The tetragonal C.F. potential can be expressed using Stevens' operators $H_{CF}=B_2^0\,O_2^0+B_4^0\,O_4^0+B_4^4\,O_4^4$. A calculation of the C.F. susceptibilities, χ_a^o and χ_c^o, was performed, and it was found that an appropriate choice of parameters B_2^o, B_4^o and B_4^4 reproduces the above mentioned anisotropy. We impose three conditions; (i) $\chi_c^o > \chi_a^o$ at T>67K and $\chi_c^o = \chi_a^o$ at 67K, (ii) $(\chi_a^o)^{-1}-(\chi_c^o)^{-1}$ equals the observed value at room temperature, (iii) and the ratio $(g_a/g_c)^2$ is equal to the observed ratio $\chi_a(0)/\chi_c(0)=1.79$, where g_a and g_c are the calculated effective g-values for the ground doublet. Here we are assuming that $(\chi_{a,c})^{-1} = (\chi^o{}_{a,c})^{-1}+\lambda+T_K$ (the molecular field constant λ, is isotropic) and that the anisotropy in the Fermi-liquid region can also be expressed by the C.F. effect. The above conditions (i) to (iii) give as the level pattern E_2 = 0K, E_1 = 289K and E_3 = 330K (the latter two are degenerate in a cubic field). Further, if we use Yoshimori's expression (9) $\chi(0)=N_A(g\mu_B)^2S(S+1)/3k_BT_K(S=1/2)$, and substitute g with the calculated g_a=1.81 and g_c=1.35, then we get T_K=26.9K. The analysis presented here is tentative and the C.F. level pattern must be checked by other measuring methods. We stress, however, that the general anisotropic behavior of susceptibility is reproduced by a small amount of tetragonality added to a cubic C.F. potential.

To explain the anisotropy of ρ, a more complicated mechanism seems to be needed. In the present $CeSi_{1.86}$, the anisotropy is most pronounced around 50K, where ρ_c starts to drop sharply whereas ρ_a has already fallen to a much lower value. Among obvious energy scales such as the C.F. splitting and T_K, 50K is the order of our estimated T_K. Therefore the sudden drop of ρ_c may suggest a coherent Kondo-lattice formation, but still, the question of why ρ_a behaves quite differently remains unanswered. This anisotropic behavior is of a new type qualitatively different from that reported for CeNi (5) or $CeCu_2Si_2$ (6, 7).

For the $CeSi_{1.70}$ single crystal, the magnetization results are shown in Fig.3. We observe that a broad step around 40KOe, previously obseved on polycrystalline $CeSi_{1.70}$ (10), is now clearly seen in the a-axis magnetization. Results of the paramagnetic susceptibility and the electrical resistivity measurements will be discussed in comparison with those of $CeSi_{1.86}$ mentioned above.

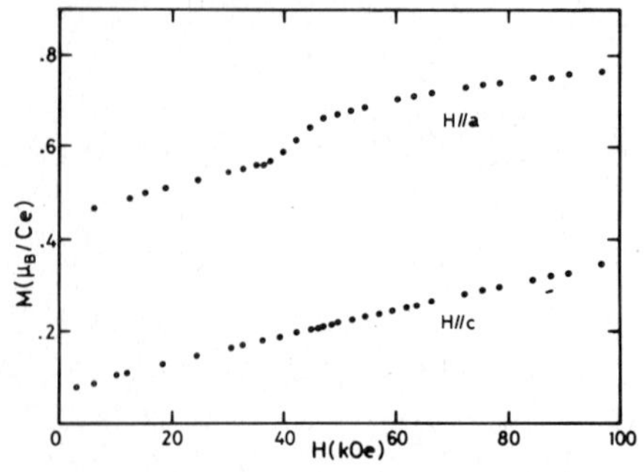

Fig.3 Magnetization of $CeSi_{1.70}$ along a- and c-axis at 4.2K.

ACKNOWLEDGEMENTS

We thank Mr. T. Miura for his help in sample preparation and Prof. T. Ohtsuka for his continuous encouragement during the study.

REFERENCES

(1) H. Yashima, T. Satoh, H. Mori, D. Watanabe and T. Ohtsuka; Solid State Commun. 41, 1 (1982).
(2) W. H. Dijkman, A. C. Moleman, E. Kesseler, F. R. de Boer and P. F. de Châtel, in Valence Instabilities, edited by P. Wachter and H. Boppart (North-Holland, 1982) 515.
(3) H. Yashima and T. Satoh, Solid State Commun. 41, 723 (1982).
(4) H. Yashima, H. Mori, T. Satoh and K. Kohn, Solid State Commun, 43, 193 (1982).
(5) D. Gignoux, F. Givord and R. Lemaire, J. Less-Common Met. 94, 165 (1983).
(6) W. Assmus, M. Herrman, U. Rauchschwalbe, S. Riegel, W. Lieke, H. Spille, S. Horn, G. Weber, F. Steglich and G. Cordier, Phys. Rev. Letters 52, 469 (1984).
(7) H. Schneider, Z. Kletowski, F. Oster and D. Wohlleben, Solid State Commun. 48, 1093 (1983)
(8) J. M. Lawrence, P. S. Riseborough and R. D. Parks, Rep. Prog. Phys. 44, 1 (o981).
(9) A. Yoshimori, Prog. Theor. Phys. 55, 67 (1976).
(10) H. Yashima, T. Satoh and H. Hiroyoshi, in High Field Magnetism, edited by M. Date (North-Holland, 1983) 179.

LT-17 (Contributed Papers)
U. Eckern, A. Schmid, W. Weber, H. Wühl (eds)

THE MAGNETORESISTIVITY OF DILUTE RhFe ALLOYS

A.Hamzić*+ and V.Zlatić*

* Institute of Physics of the University,POB 304, 41001 Zagreb,Yugoslavia
\+ Physics Department,Faculty of Natural Sciences,POB 162, 41001 Zagreb,Yugoslavia

We have measured the magnetoresistivity of dilute RhFe alloys (40 ppm,0,35% and 0,5%) for T = 1,5 K - 35 K and H ≤ 70 kG.After subtracting the normal contribution, the effect of magnetic field on LSF is obtained:the magnetoresistivity is positive at low temperatures,decreases with increasing temperature and becomes negative above T_0 (which is field dependent).The results are compared with the theoretical calculations for one-band Wolff model.

The RhFe dilute alloy system exibits very peculiar low temperature properties.Detailed measurements (1)-(4) revealed that the impurity part of the electrical resistivity can be written as $\rho_{imp}/c = \rho_0(T/T_K)$,where $T_K \simeq 15$ K is the characteristic temperature of the system; that the impurity part of the magnetic susceptibility changes from Pauli to Curie-Weiss like as T increases from $T \ll T_K$ to $T \gg T_K$, with negative $\Theta_{CW} \simeq T_K$, and that the magnetoresistance (for T up to 4,2 K) is positive for T≃0, and it decreases with increasing temperature.In order to obtain further information about this system, we have extended the magnetoresistivity measurements to higher temperatures (up to 35K) and higher fields (up to 70 kG).

The effect of magnetic field on the resistivity is well accounted for by the quantity $\rho_i(T,H) = (\rho_{\parallel} + 2\rho_{\perp})/3$, where $\rho_{\parallel}$ and $\rho_{\perp}$ are the longitudinal and transverse magnetoresistivity obtained separately by the a.c.method.The zero field resistivity is compensated at each temperature, and the field dependence is obtained by recording the signal from the sample during the sweep up and down of the field.Thus, $\rho_i(T,H)$ measures directly the difference $\rho(T,H) - \rho(T)$.The samples have nominal concentration 40 ppm, 0,35% and 0,5%, and are in the form of thin wires.

Figure 1 shows the results for $\rho_i(T,H)/\rho(T)$ for Rh-0,35% Fe sample,as a function of temperature and for several field values.The measured ρ(4.2 K) value was 310 nΩcm (which is somewhat higher than the published result (2)) and our $\rho(T)$ data showed the same relative changes as Rusby´s data (2).

Trying to understand the observed behaviour, we first note that de Haas-van Alphen results (5) have shown that the sheets of the Fermi surface in Rhodium, which contribute about 80% of the conductivity, are d-like.This means that both the conductivity and the screening are due to d-electrons.

Since the susceptibility data suggest that the peculiarities of RhFe are due to the fluctuations of the local magnetization at Fe sites, we belive that it is appropriate to discuss the properties of this system in terms of the one-band Wolff model (6),(7). Indeed, this model treated in the localised spin fluctuation (LSF) approximation has been successfully used (6),(7) to describe the zero-field resistivity and the changeover from the Pauli to Curie-Weiss behaviour of the susceptibility.Furthermore, the calculations for finite magnetic fields (8) have shown that the

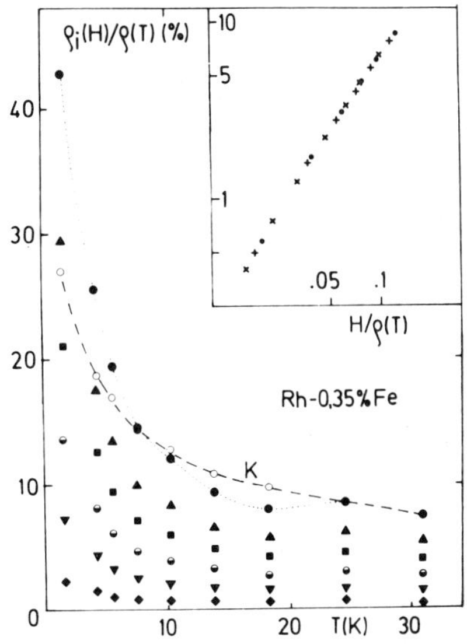

Figure 1. $\rho_i(H)/\rho(T)$ (in %) vs temperature,for field values: ● 69 kG, ▲ 53 kG, ■ 42,5 kG, ◒ 32 kG, ▼ 21 kG, ◆ 10,5 kG.Dashed line represents the Kohler term for 69 kG.Inset: $\rho_i(H)/\rho(T)$ (in %) vs $H/\rho(T)$ (in kG/nΩcm) for 0,35% (● 24,5 K, + 31 K) and 0,5% (✖ 30 K).

magnetoresistance at T = 0 is given by

$$\Delta\rho_{LSF}/\rho_0^\infty = (\mu_B H/k_B T_K)^2$$

where ρ_0^∞ is a constant. For a given field the magnetoresistance $\Delta\rho_{LSF}(T,H)$ decreases with increasing temperature, and for large enough temperature it becomes negative. The crossing temperature T_0 increases linearly with increasing field for $\mu_B H \ll k_B T$, and saturates for $\mu_B H \gg k_B T$.

In order to compare the experimental data with the theoretical results, we have assumed that the measured resistivity could be written as $\rho(T,H) = \rho_{LSF}(T,H) + \rho_m(T,H)$, where ρ_m corresponds to the "normal" field dependent resistivity one would obtain for the Rhodium alloy of the same residual resistivity but without the LSF. The magnetoresistivity due to the LSF is defined as $\Delta\rho_{LSF}(T,H) = (\rho(T,H) - \rho(T)) - \rho_K(T,H)$, where $\rho_K(T,H) = \rho_m(T,H) - \rho_m(T)$ and denotes the "normal" Kohler part of the magnetoresistivity. The quantity $(\rho(T,H) - \rho(T))$ is just the measured quantity $\rho_i(T,H)$, while ρ_K is estimated in the following way.

We notice first that for $T > T_K$ the LSF contribution to $\rho(T)$ is saturated and that, as it is usually the case in dilute alloys, the LSF part of the total magnetoresistance is negligable. Thus for $T > T_K$ we take $\rho_i = \rho_K$ and assume that the plot of $\rho_i/\rho(T)$ vs $H/\rho(T)$ defines an universal function which is characteristic of the "non-magnetic" matrix. Indeed, for T > 25 K, the functional form of $\rho_i/\rho(T)$ is identical for all the samples and temperatures (see inset of Fig.1).

To obtain the LSF contribution for $T < T_K$, we read off, for given $H/\rho(T)$, $\rho_K/\rho(T)$ from the Kohler plot, and assume that the values thus obtained correspond to the normal magnetoresistance of the matrix. These values (for H = 69 kG) are shown as a dashed line in Fig.1. Finally, the LSF part of the magnetoresistance is defined (at 69 kG) as the difference between the experimental data (dotted line) and the Kohler part for this field value. The same procedure is repeated for other field values.

The temperature variation of $\Delta\rho_{LSF}(T,H)/\rho(T)$ obtained with the above analysis shows that $\Delta\rho_{LSF}$, which is positive for lowest temperatures decreases with the increasing temperature and finally becomes negative for $T > T_0(H)$. High temperature negative magnetoresistance is just what one would expect to observe in a dilute magnetic alloy above T_K. The $\Delta\rho_{LSF}$ values extrapolated to $T \to 0$ are, within the error of the procedure, field independent, allowing the estimation of T_K. One obtains $T_K = (20 \pm 5)$K, which agrees well with the T_K value previously determined (2).

The field dependence of T_0 for 0,35% alloy shows that T_0 tends to saturate above 50 kG; however the experimentally determined T_0 are (Fig.2) about three times higher then it is predicted by the theory. Finally, we point out that similar behaviour (the change of sign of $\Delta\rho_{LSF}$ and T_K of the order of 20 K) was found for other two concentrations as well.

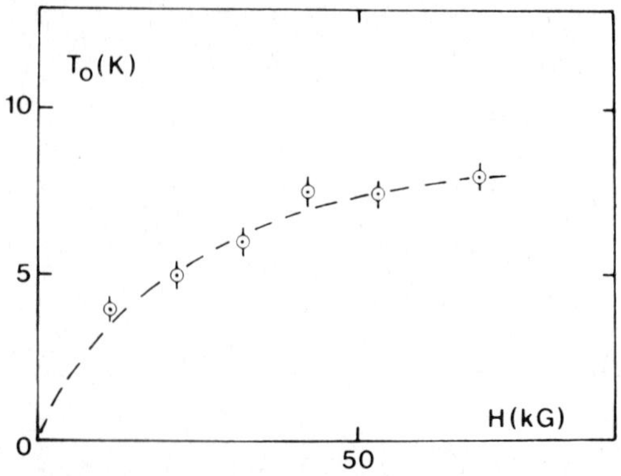

Figure 2. T_0 vs H for Rh-0,35% Fe

To conclude, although we are aware of the fact that the performed analysis could be open to some critisism, we believe that the presented experimental results demonstrate that the low temperature magnetoresistivity of dilute RhFe is determined by the fluctuations of the local magnetization at Fe sites.

ACKNOWLEDGEMENTS

We would like to thank Dr R.Rusby for providing the samples, Prof.B.R.Coles for the fruitfull discussions and Đ.Drobac for help during some stage of the experimental work. The financial support from the Humboldt foundation is gratefully acknowledged.

REFERENCES

(1) B.R.Coles, Phys.Lett. 8 (1964) 243
(2) R.L.Rusby, J.Phys. F4 (1974) 1265
(3) G.S.Knapp, J.Appl.Phys. 38 (1967) 1267
(4) A.P.Murani and B.R.Coles, J.Phys. F2 (1972) 1137
(5) L.S.Cheng, R.J.Higgins, J.E.Graebner and J.J.Rubin, Phys.Rev. B19 (1979) 3722
(6) N.Rivier and V.Zlatić, J.Phys. F2 (1972) L99
(7) K.Fisher, J.Low Temp.Phys. 17 (1974) 87
(8) V.Zlatić, J.Phys. F8 (1978) 489

LT-17 (Contributed Papers)
U. Eckern, A. Schmid, W. Weber, H. Wühl (eds)

EFFECT OF Au AND Rh IMPURITIES IN SPIN FLUCTUATING PtFe ALLOYS

J. RAY and GIRISH CHANDRA

Tata Institute of Fundamental Research, Homi Bhabha Road, Bombay-400005, India

This paper reports recent measurements on the effect of Au and Rh impurities on the long range magnetic interaction between Fe atoms diluted in a Platinum matrix. For this purpose, d.c electrical resistivity of ternary alloys $PtAuFe_{0.2}$ and $PtRhFe_{0.2}$ are studied between 25 mK to 40 K. It is found that the characteristic spin fluctuation temperature is either depressed or enhanced depending on the presence of Au or Rh in the alloy systems.

1. INTRODUCTION

Our present understanding of dilute magnetic alloys of exchange enhanced metals like Pd, Pt and Rh with small amounts of 3d transition metal impurities is still not well understood because of peculiar band structure effects (1,2). PtFe alloys provides an interesting study as it shows characteristics of 'local moment' behaviour as revealed through magnetic data (3), whereas transport measurements (4) show effects of local spin fluctuation (LSF) behaviour. The impurity atom manifests as spin and charge densities which fluctuate in space and time, rather than appear as a moment with fixed magnitude. Resistivity of several PtAuFe and PtRhFe alloys are carried out to study whether LSF effects are also present when Au and Rh atoms are diluted in a Pt host.

2. EXPERIMENTAL DETAILS

The ternary alloys were all prepared keeping the Fe content constant at 0.2 at.%. The Au and Rh content was varied between 2 to 8 at.% approx. All the alloys were prepared by dilution of a master alloy of PtFe melted in a water cooled hearth under flowing argon condition. The buttons were melted several times to ensure homogeneity and the compositions were analysed by atomic absorption spectroscopy. The starting materials used were all from Johnson-Matthey of 5N purity. The buttons were cold rolled to an approx. size of 7 cms x 0.2 cm x 0.01 cm, then etched and finally annealed in vacuo at 850°C for five days, to remove mechanical strains. Measurements of dc electrical resistivity were carried out in ^3He-^4He dilution refrigerator (supplied by Oxford Instr. U.K.) using a standard four probe technique. Voltages were measured using a Keithley dc nanovoltmeter (Model No.181) with a precision of 1 part in 10^5. A constant current source of the same order of precision (Model No.5752 of Tinsley Instr. U.K.) was used. Temperatures were measured with ^{60}Co$\underline{Co}$ nuclear orientation thermometer between 25 - 60 mK, and higher temperatures using a calibrated Ge sensor from Lake Shore Cryotronics, USA.

3. RESULTS AND DISCUSSIONS

The impurity magnetic resistivity of Fe atoms in the ternary alloys have been analysed by using the following eqn.

$$\Delta\rho = \rho_{PtMFe} - \rho_{PtM} \qquad (1)$$

where M refers to Au or Rh atoms and other symbols have their usual meanings. Fig.1 shows the variation of $\Delta\rho/c$ against temperature plotted on a logarithmic scale in the region of interest between 0.2 - 40 K. The magnetic behaviour of the alloys are well explained on the basis of LSF picture. The temperature dependences of $\Delta\rho$ are fitted using the expression

$$\Delta\rho = A+B\ln[(T^2+T_s{}^2)^{\frac{1}{2}}] \qquad (2)$$

where T_s is the LSF temperature. A and B are constants depending on the nature of the magnetic impurity and its concentration. This function, so chosen, predicts the low temperature T^2 behaviour for $T<<T_s$ (5). The fitting is shown in Fig.1 and values of the parameters are depicted in Table 1.

All the curves exhibit a characteristic feature of rapid increase of $\Delta\rho$ above 10 K which is invariably associated with deviations from Matthiessen's

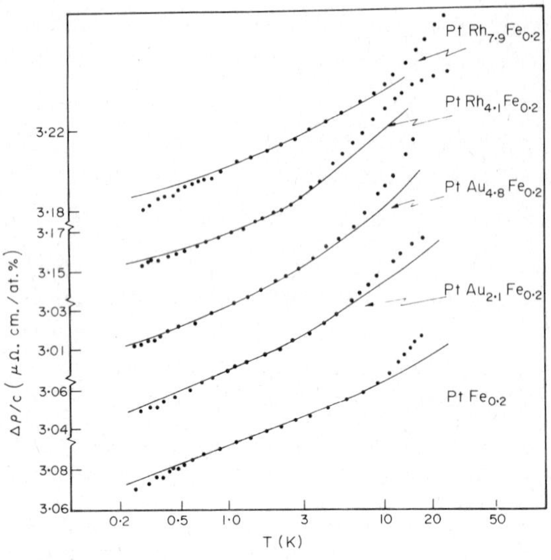

Fig.1. $\Delta\rho/c$ vs. log T plots of Pt ternary alloys using eqn.2

rule (DMR). This is established by the fact that the Fe impurity scattering term has a much weaker temperature dependence than that arising from DMR, as is observed in other alloys also (4,6). Therefore, only the low temperature points below 10 K where eqn.(1) holds good, are used in the fitting analyses. Our study reveals that at low temperatures (T<0.4 K) the experimental curves deviate from the fitted one. This departure from T^2 behaviour is most probably due to the onset of impurity-impurity interactions observed also in PtFe and many other systems (4).Another important finding is the variation of T_s in our systems. The addition of 4.8 at.% Au depresses the T_s value over the pure PtFe value down to 193 mK, whereas 7.9 at.% Rh enhances it to 1.75 K. This is a novel feature when we compare this with the pure PtFe alloys (4), where T_s is found independent of Fe concentration below 0.2 at.%.

The implication of T_s variation is quite difficult to interpret on the basis of transport data alone. The variation of T_s is most probably related to the variation of host susceptibility (χ) of Pt alloyed with Au or Rh. Rh, itself being an exchange enhanced metal, is quite likely to increase the χ of PtRh alloys, which may then also be responsible for the rise of T_s in the corresponding ternary alloys. On the other hand, Au being

Table I

Alloys (at.%)	A ($\mu\Omega$.cm)	B ($\mu\Omega$.cm)	T_s (K)
$PtFe_{0.2}$	0.612	0.0062	0.732
$PtAu_{2.1}Fe_{0.2}$	0.608	0.0058	0.431
$PtAu_{4.8}Fe_{0.2}$	0.602	0.0056	0.193
$PtRh_{4.1}Fe_{0.2}$	0.630	0.0068	0.900
$PtRh_{7.9}Fe_{0.2}$	0.636	0.0060	1.75

a normal metal, could reduce the χ of PtAu and consequently a reduction of T_s values in the corresponding ternary alloys. In fact, such a correlation has indeed been established in a similar LSF system PdRhNi where enhancement of T_s with increase of Rh concentration is linked to the increase of χ of the correspondng PdRh alloy(7). However, the confirmation of such a correlation holding true also for the Pt based alloys must await magnetic measurements. Also, the influence of electron mean free path on the addition of Au or Rh, needs to be considered in a detailed study.

REFERENCES

(1) K.H. Fischer, Phys. Repts. 47 (1978) 225.

(2) G.J. Nieuwenhuys, Adv. Phys. 24 (1975) 515.

(3) J. Crangle and W.R. Scott, J.Appl. Phys. 36 (1965) 921.

(4) J.W. Loram, R.J. White and A.D.C. Grassie, Phys. Rev. B 5 (1972) 3659.

(5) A.B. Kaiser and S. Doniach, Int. J. Magn. 1(1970) 11.

(6) J.S. Dugdale and Z.S. Baszinski, Phys. Rev. 157 (1967) 552.

(7) H.G. Purwins, Y. Talmor, J. Sierro and F.T. Hedgcick, Sol. St. Comm. 11 (1972) 361.

LT-17 (Contributed Papers)
U. Eckern, A. Schmid, W. Weber, H. Wühl (eds)

SPINFLUCTUATION SUPPRESSION IN THE MAGNETORESISTIVITY OF $TiBe_2$

J.M. van Ruitenbeek [a], H.W. Myron [b], R.W. van der Heijden [b] and J.L. Smith [c]

(a) Physics Laboratory and (b) High Field Magnet Laboratory, Katholieke Universiteit Nijmegen, The Netherlands
(c) Los Alamos National Laboratory, Los Alamos, New Mexico, USA.

Here we present magnetoresistance and magnetization measurements on $TiBe_2$ and $TiBe_{1.8}Cu_{0.2}$ in fields up to 200 kOe. Using these results the apparent controversy of the suppression of spin fluctuations only above 50 kOe is explained.

The strongly exchange enhanced paramagnet $TiBe_2$ has been studied intensively during the last few years (1) and has now become one of the most well defined metallic systems for studying spin fluctuations (SFs). The suppression of SFs only above 50 kOe, as observed in specific heat experimets (2), has raised some controversy. This suppression is also observed in magnetoresistance (MR) (3). Here we extended the MR measurements along with magnetization measurements and present a consistent interpretation of these results.

For the MR experiment a polycrystalline $TiBe_2$ sample was used, having an extrapolated resistivity ratio R(300 K)/R(0 K) = 95 and a resistivity at 0 K of 0.8 μΩcm. The MR data presented in this paper are longitudinal MR results using a 4-probe low frequency AC technique. Transverse MR measurements at low temperatures showed a larger positive effect as well as less structure. The magnetization of $TiBe_2$ was measured on material from the same batch, using a moving sample magnetometer.

The low field (H<50 kOe) magnetization of $TiBe_2$ is given in the Arrot plot in figure 1a. The data are well described by the solid line, which corresponds to $H/M = \chi_o^{-1} + BM^2$, where $\chi_o = 9.86\cdot10^{-3}$ emu/mole and $B = -4.63\cdot10^{-5}$ $(\text{emu/mole})^{-3}$. In figure 1b the susceptibility $\chi = \Delta M/\Delta H$ is plotted as a function of H. The broken line corresponds to the fit to the Arrot plot. This curve diverges at H_m = 60 kOe. This observation was first made by Acker et al. (4) and our results agree with theirs. At higher T the maximum in χ gradually disappears as well as being shifted to lower fields.

Figure 2 shows the MR $\Delta\rho(H,T) = \rho(H,T) - \rho(0,T)$ for $TiBe_2$. At the lowest temperatures $\Delta\rho(H)$ is positive and uniformly increasing. At higher T $\Delta\rho(H)$ initially follows the same curve and then, at about 50 kOe, suddenly changes to a smaller slope. This effect is more pronounced and the deviation from the low T curve starts at a lower field as T increases. At about 10 K $\Delta\rho$ is negative for all field strengths and uniformly decreases with increasing field. For T>25 K, $\Delta\rho(H)$ once again increases but remains negative. For T > 6 K we find no saturation of the MR up to 15 T. The MR of $TiBe_{1.8}Cu_{0.2}$ at 4.2 K (fig. 3) is negative and no T dependence is measured below 4.2 K.

For the interpretation of these results we assume (a) nearly meta-magnetism and (b) spin fluctuations. We start with a Stoner description of $\chi(H)$ in figure 1b.

(a) In the Stoner model the density of states, N(E), near the Fermi level $E_o = E_F(H=0)$ is approximated by a quadratic function in E $N(E) = N_o + N'E + \frac{1}{2}N''E^2$ where we have put E_o = 0. As long as this approximation is correct the Stoner model gives straight Arrot plots. For $TiBe_2$ the Arrot plot is straight in two separate regions: between 0 and 40 kOe with a negative slope and between 100 and at least 200 kOe with

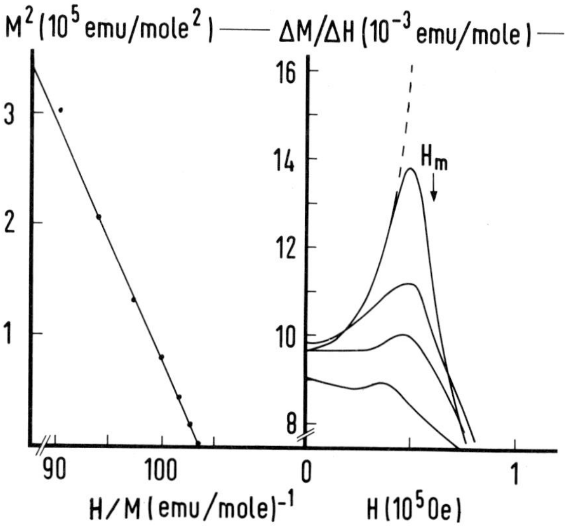

FIGURE 1
1a. (left) Arrot plot of the magnetization of $TiBe_2$ for H<50 kOe and T = 3.5 K.
1b. (right) Susceptibility $\chi = \Delta M/\Delta H$ as a function of H for T=3.5 K, 10 K, 15 K and 20 K.

a positive slope. In figure 1a only the low field part is shown. A large negative slope requires a large positive value for N". This is indeed found in bandstructure calculations on $TiBe_2$ (5). With an applied field N(E) and χ(H) increase with H. At some H, N(H) will reach its critical value for spontaneous magnetism and χ(H) diverges. This is the field H_m in fig. 1b. However χ goes through a maximum for $H<H_m$ and then strongly decreases with H. This can be interpreted as a break-down of the Stoner approximation or in other words, the Fermi level crosses a peak in N(E) before the metamagnetic state is reached. Once the peak is crossed a new set of constants N_o, N′ and N" are needed to describe the high field magnetization.

(b) The effects of SFs on the low temperature properties of $TiBe_2$ have been described by Béal-Monod (7). Ref. (7) predicts no, or small effects on the low temperature field dependence of the magnetization, so that the observations under (a) are unaffected by the introduction of SFs. The specific heat results of Stewart et al. (2) show convincing evidence for SFs in $TiBe_2$. In a magnetic field they observed a suppression of SFs, only for H > 50 kOe. This is also observed in the low temperature MR (fig. 2).

The following simple picture may explain this effect. At T = 0 no SF contribution to the MR is observed, as predicted by (7), and the positive MR is caused by ordinary Fermi surface effects, fig. 2. As the temperature increases

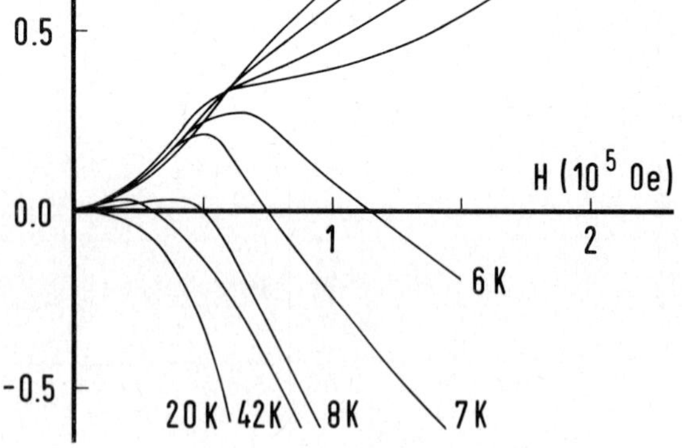

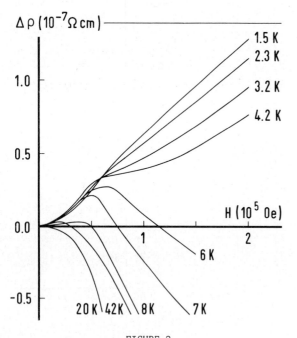

FIGURE 2
The MR of $TiBe_2$ for several values of T.

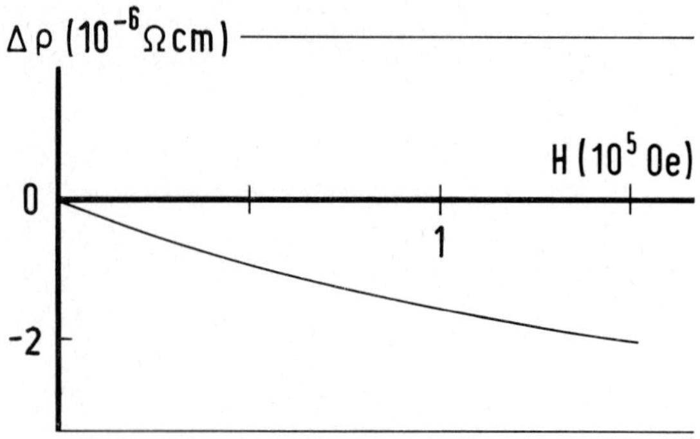

FIGURE 3
The MR of $TiBe_{1.8}Cu_{0.2}$ at T = 4.2 K.

ρ(T) increases with a large T^2 term (2), characteristic of a Fermi liquid behaviour (7) when we place $TiBe_2$ in a strong magnetic field at a temperature 0<T<10 K the SFs are indeed supressed by the field, as is evident for H>50 kOe. However for H<50 kOe the strong exchange splitting and the positive curvature of N(E) near E_o causes N to be enhanced with respect to N(0). The higher N(E) enhances SFs. Thus for H<50 kOe we have a competition between a suppression by the field and an enhancement by the incrasing N(H). Once E_F has reached the maximum in N(E), N(H) decreases and now both the N(H) and the field itself suppress the SFs.

At higher temperatures the structure in N(E) is smeared out by the temperature and by SFs. Above about 25 K, $-\Delta\rho(H,T)$ decreases for fixed H and increasing T, suggesting $T_{SF} \approx 25$ K, in agreement with (2). In ferromagnetic Cu-alloyed $TiBe_2$ the specific heat shows an equally strong SF contribution (2). In $TiBe_{1.8}Cu_{0.2}$ the intrinsic exchange splitting is such that E_F is beyond the maximum in N(E), thus the SFs are suppressed at all fields (fig. 3).

The results and the model presented explain in terms of a competition between suppression and enhancement, why a suppression of SFs in $TiBe_2$ is apparent only above 50 kOe.

REFERENCES

(1) J.M. van Ruitenbeek et al. J. Phys F, in print and references therein.
(2) G.R. Stewart et al. Phys. Rev. B 25 (1982) 5907 and Phys. Rev. B 26 (1982) 3783
(3) O. Laborde and G. Chouteau, to be published and I.A.Campbell and G. Creuzet, Proc. of the Workshop on 3d Metallic Magnetism, eds. D. Givord and K.R.A. Ziebeck (ILL-Grenoble, August 1983) 29.
(4) F. Acker et al. Phys. Rev. B 24 (1981) 5404
(5) R.A. de Groot et al. J. Phys. F 10 (1980) L235 and T. Jarlborg and A.J.Freeman, Phys Rev. B 22 (1980) 2332
(6) E.P.Wohlfarth, J. Phys. Lettr. 41 (1980) L563
(7) M.T. Béal-Monod Physica 109&110B (1982) 1837

LT-17 (Contributed Papers)
U. Eckern, A. Schmid, W. Weber, H. Wühl (eds)
© Elsevier Science Publishers B.V., 1984

De HAAS-Van ALPHEN EFFECT IN Pd(Ni) ALLOYS

M SPRINGFORD* and P WISE

School of Mathematical and Physical Sciences, University of Sussex, Brighton, East Sussex, UK

L W ROELAND, A TAL and J C WOLFRAT

Natuurkundig Laboratorium der Universiteit van Amsterdam, 1018 XE Amsterdam, Nederland

Anomalies in the dHvA effect in dilute Pd(Ni) alloys are interpreted as direct evidence of spin fluctuations. Additionally, the exchange enhancement factor varies with Ni content, position on the Fermi surface and with magnetic field.

1 INTRODUCTION

Pd(Ni) alloys have intriguing magnetic properties. For dilute alloys ($c < 1$ at.%) it is customary to have in mind a local enhancement model in which, although the Ni does not carry a local moment, the magnetization of the host is locally enhanced in the vicinity of the impurity cell (1). At higher concentrations a measure of association between impurities occurs (2) giving rise to some complex magnetic centres and, for $c \simeq 2.25$ at.%, the system becomes ferromagnetic (3).

Of particular interest to us is that electron scattering is weak in the alloys, making it possible to investigate the dHvA effect throughout the region of interest. Electron-electron interactions are expected to influence the dHvA effect by (a) renormalisation of effective masses, (b) enhancement by a factor S of the g-factor and consequently affecting the relative amplitudes of the dHvA harmonics r through the spin factor $\mathcal{S}_r$,

$$\mathcal{S}_r = \cos\left(\frac{r\pi m g S}{2 m_0}\right) \qquad (1)$$

and (c) leading to amplitude anomalies when $\hbar\omega_c > kT_{s\ell}$ where $T_{s\ell}$ is the local spin fluctuation temperature, estimated to be 20 K (15 T) in Pd(Ni) (4).

2 EXPERIMENTAL

We report measurements for alloy concentrations between 0.1 and 1.0 at.% in magnetic fields up to 40 T. Quantum oscillations are observed from all sheets of the Pd Fermi surface, but we discuss here only the X-(hole)-ellipsoid for a magnetic field directed along <100> for which in pure Pd, $F = 572$ T and $m^* = 0.628\ m_0$ (5). The temperature range is 1 - 4 K. Single crystals were grown by the strain/anneal technique and Ni concentrations deduced from resistivity measurements (6).

3 RESULTS

The X-ellipsoid effective mass increases with Ni concentration initially by 13% per at.% with no indication of any temperature dependence of m^*. We note that this figure is comparable to the change in electronic specific heat (7), although of course the latter samples the whole Fermi surface. At low concentrations of Ni the electron lifetime, as measured by the Dingle temperature is 1.4 K per at.%.

Measurements of absolute amplitude reveal changes in $\mathcal{S}_r$ with Ni concentration and an objective of this research is to investigate the variation with c of $g^* = gS$. Such a measurement is complicated, however, by another effect. In the more dilute alloys ($c < 0.35$ at.%) the relative amplitudes of the dHvA harmonics find a consistent interpretation if it is assumed that $\mathcal{S}_r$ and hence S are also field dependent. The experimental result for an 0.35 at.% alloy is shown in Fig 1. In general such results yield the argument of $\mathcal{S}_r$ modulo 2π (or π if the infinite-field-phase is determined). In the present case, however, an approximate value of S is known from another source. The spin-splitting is sufficiently great ($g^* \sim 20$) that the dHvA frequency is significantly field dependent, in order to satisfy the requirement of a constant number of electrons. The frequency, or equivalently phase, dependence on field may be analysed (5) to yield that $S \sim 8$ at low fields in this alloy.

Experimental results in the more concentrated alloys cannot be interpreted consistently in the

* Work supported by the SERC

same way. Whilst in this case we cannot exclude the possibility that several mechanisms may be operative simultaneously, it would seem that the semiclassical one-electron theory of dHvA breaks down in the general sense indicated by Engelsberg and Simpson (8). We believe that this breakdown arises from the influence of spin-fluctuations and that the dHvA effect provides, therefore, a rather direct way of observing these ellusive excitations.

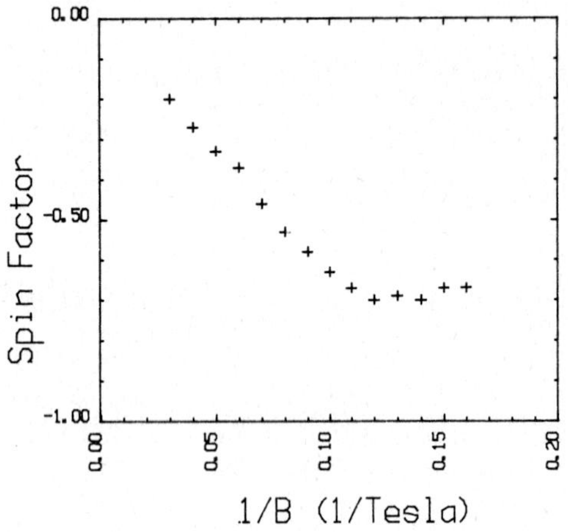

FIGURE 1
Variation of the spin factor $\mathcal{S}_1$(r = 1) with magnetic field in Pd(.35 at.% Ni) for the X-ellipsoid at <100>.

REFERENCES

(1) P Lederer and D L Mills, Phys Rev 165 (1968) 867; Phys Rev Lett 20 (1968) 1036
(2) T D Cheung, J S Kouvel and J W Garland, Phys Rev B23 (1981) 1245
(3) R M Bozarth, D D Davis and J H Wernick, J Phys Soc Japan suppl B1 (1962) 112
(4) G Chouteaux, R Fourneaux, R Tournier and P Lederer, Phys Rev Lett 21 (1968) 1082
(5) L W Roeland, J C Wolfrat, D K Mak and M Springford, J Phys F 12 (1982) L267
(6) R A Beyerlein and D Lazarus, Phys Rev B7 (1973) 511
(7) K Ikeda, K A Gschneidner and A I Schindler, Phys Rev B28 (1983) 1457
(8) S Engelsberg and G Simpson, Phys Rev B2 (1970) 1657

LT-17 (Contributed Papers)
U. Eckern, A. Schmid, W. Weber, H. Wühl (eds)

ELECTRICAL RESISTIVITY AND SUSCEPTIBILITY OF $GdRh_{1.1}Sn_{4.2}$ COMPOUND

A. Rojek, C. Sulkowski, A. Zygmunt and G. Kozlowski

Institute for Low Temperature and Structure Research Polish Academy of Sciences, 53-529 Wroclaw, Prochnika 95, Poland.

Single crystals of compound in the system $GdRh_{1.1}Sn_{4.2}$ have been grown from a tin solvent in vacuo. The results of electrical resistivity and magnetic susceptibility measurements on the compound are given. The compound $GdRh_{1.1}Sn_{4.2}$ exhibits antiferromagnetic order below T_N = 11.6 K. The temperature dependence of the electrical resitivity has been determined.

In 1980 Remeika et al |1,2| have reported the superconducting and magnetic properties of new series of rare earths-rhodium-tin compounds. The compounds containing heavy rare earths/Tm, Yb, Lu/ are superconducting and those with light ones /Eu, Gd, Tb, Dy, Ho/ are magnetic. The compound $ErRh_{1.1}Sn_{3.6}$ exhibits a reentrant superconductivity with T_c = 1.22 K and T_m = 0.57 K. In this note we report results of the electrical and magnetic measurements performed on the single crystals of $GdRh_{1.1}Sn_{4.2}$ compound. The crystals were prepared by reacting Gd and Rh with a surplus Sn in a sealed evacuated quartz tube. The grown crystals had dimensions along the octahedral edges 2 - 4 mm. X-ray microanalyzer JXA - 5A JEOL was used to determine the chemical composition. The composition for three samples /obtained with mole ration Gd to Rh 2 : 1, 1 : 1, 1 : 2/ was always $GdRh_{1.1}Sn_{4.2}$. Powder X-ray diffraction indicated that the compound was single phase and had primitive cubic type structure with lattice constant a = 9.647 A. The magnetic susceptibility χ were measured on the powder sample in magnetic field 10 kOe. The magnetic properties are presented in Figure 1.

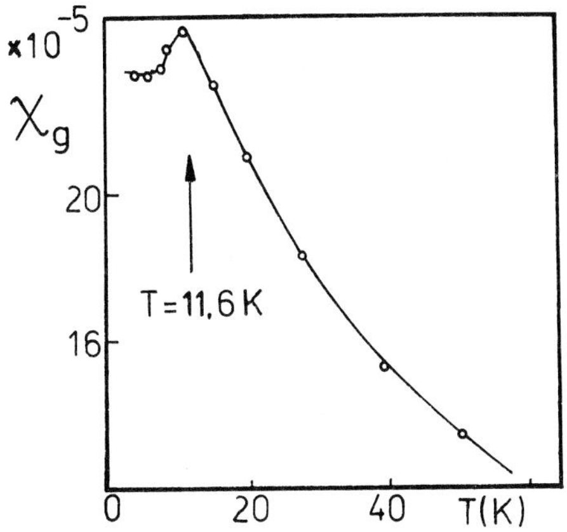

FIGURE 1

The relation of magnetic susceptibility with respect to temperature have a peak at Neél temperature T_N = 11.6 K. From the dependence χ^{-1} on T the paramagnetic Curie temperature Θ_p = -28.8 K and the effective magnetic moment μ_{eff}= 7.96μ_B have been obtained. The experimental value of magnetic moment of gadolinum ion equals the above value. Resistivity measurements of single crystal $GdRh_{1.1}Sn_{4.2}$ were carried out in temperature range from 2.15 K to 300 K. Specific resistivity ρ our compound equals 614 ± 40 $\mu\Omega \cdot$ cm at 295 K. The temperature dependence of resistivity is presented in Figure 2. One can observe four temperature ranges where resistivity follows different relations: $\rho \sim T^{1.5}$, $\rho \sim T^2$ $\rho \sim T$ and $\rho \sim \ln T$. At lower temperatures $T^{1.5}$- and T^2-dependent resistivity is connected with magnetic ordering of Gd atoms sublattice. The apperance of T^2-, T-, and lnT-dependent contributions to resistivity seems to be originating from spin fluctuations of locatizaed Rh 4d-electrons. The characteristic spin-fluctuation temperature T_{SF} = 44.7 K obtained from T and lnT dependeces is four times the Neél temperature T_N = 11.6 K.

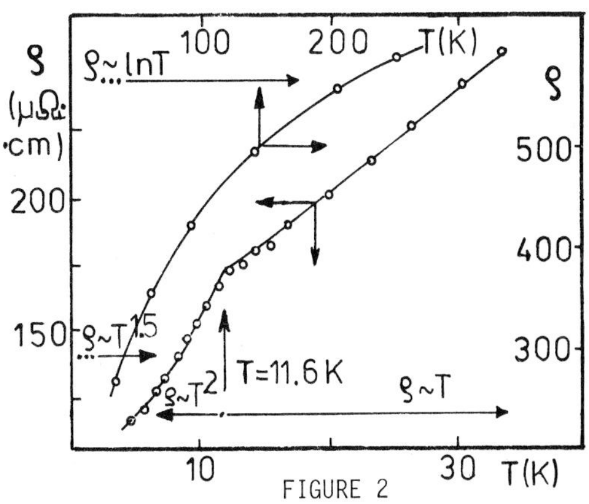

FIGURE 2

REFERENCES

(1) J.P. Remeika et al, Solid State Commun. 34, 923 (1980)

(2) G.P. Espinoza, Mat. Res. Bull. 15, 791 (1980)

LT-17 (Contributed Papers)
U. Eckern, A. Schmid, W. Weber, H. Wühl (eds)

REORIENTATION IN FERROMAGNETIC $TmAl_2$

S. HORN[+], W. LIEKE[+], M. LOEWENHAUPT[§] and F. STEGLICH[+]

[+] Institut für Festkörperphysik, Technische Hochschule Darmstadt, D-6100 Darmstadt, F.R.G.

[§] Institut für Festkörperforschung, Kernforschungsanlage Jülich, D-5170 Jülich, F.R.G.

The low temperature behavior of $TmAl_2$ polycrystals and one single crystal was studied by neutron diffraction and specific heat measurements. Our results reveal that the collinear-to-canted ferromagnetic phase transition at T ≃ 3K previously discovered is connected with a change in the easy direction of magnetization.

Neutron diffraction experiments showed that $TmAl_2$ undergoes two phase transitions: a collinear ferromagnetic phase (I) forms at $T_{c1} \simeq 4K$ and a canted ferromagnetic one (II) at $T_{c2} \simeq 3K$ (1,2). Since these experiments were done with powdered samples, no information about the direction of magnetic moments in phase I could be obtained. For phase II it was concluded that the moments are oriented within (111)-planes and include an angle of about 56° with those of adjacent planes.

In this note we report neutron-diffraction measurements between 1.6 and 40K on several polycrystalline and one single crystal samples of $TmAl_2$ and specific-heat measurements on a $TmAl_2$ polycrystal between 0.5 and 5K, both at zero magnetic field and B = 0.5T.

The neutron diffraction experiments were performed at T = 2K, 2.6K, 3K, 3.8K, 4.2K, 8K, 15K and 40K, respectively, using the multidetector instrument D1B of the Institut Laue-Langevin with neutrons of incident wavelength of 2.54Å. Figs. 1a-d show the patterns at T = 2K, 3K, 3.8K and 15K, after subtraction of the 40K pattern. The latter shows only nuclear Bragg peaks and diffuse magnetic scattering, but no indications of magnetic correlations. Subtraction of the diffuse magnetic scattering from those patterns displaying magnetic correlations is the reason for the negative counting rates seen in figs. 1a-d (note the different vertical scales used in these figures).

At T = 15K, a maximum appears in the diffuse scattering intensity around Q = (1/2 1/2 1/2), and there is increase of intensity as Q → 0. This modulation of diffuse scattering points to the presence of both antiferromagnetic and ferromagnetic correlations. At T = 8K, the modulation has increased and, at 4.2K, weak magnetic Bragg reflections at the positions of the nuclear ones indicate the onset of long-range ferromagnetic order. Still, there is a steep increase of magnetic scattering (critical scattering) as Q → 0. Lowering the temperature further (to 3.8K and 3.4K) we find both a decrease of critical scattering around Q = 0 and an increase of magnetic Bragg scattering. At T = 3K, a sharp Bragg peak has emerged from the maximum in the diffuse scattering at Q = (1/2 1/2 1/2) and, in addition, the other superlattice reflections with propagation vector τ = (1/2 1/2 1/2) have shown up. There is still a modulated diffuse magnetic background as most clearly seen by the wings of the (1/2 1/2 1/2) peak. At even lower temperatures, diffuse scattering intensity appears further reduced.

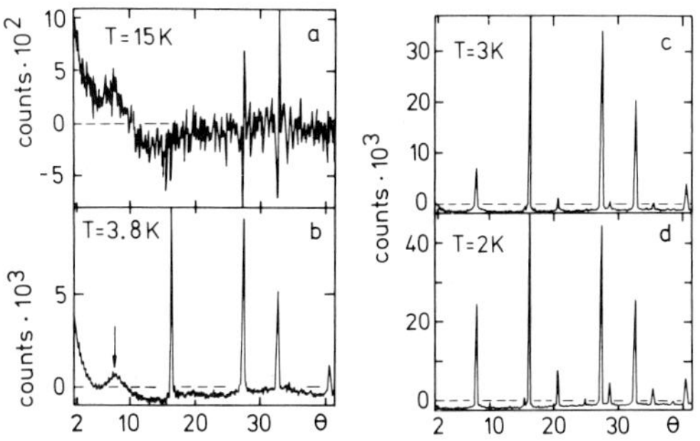

FIGURE 1
Neutron diffraction patterns for $TmAl_2$ at differing temperatures. The patterns are formed from the difference between the pattern for the respective temperature and the one for T = 40K. The arrow in (b) marks Q = (1/2 1/2 1/2).

The transition from the collinear to the canted ferromagnetic structure has been confirmed by neutron-diffraction measurements on two other powdered samples and also by the specific-heat experiment. The temperature dependence of the specific heat, C(T), at B = 0 and B = 0.5T is shown in fig. 2. For B = 0, two distinct maxima occur at the transition temperatures T_{c1} = 3.87K and T_{c2} = 3.26K, determined from a C/T vs. T plot in the usual way. Upon application of B=0.5T, T_{c2} is seen to shift to 3.46K, whereas the maximum at T_{c1} is suppressed almost completely, which is always found for ferromagnets in sufficiently high fields.

An attempt to determine the directions of magnetic moments relative to the crystal axes was made using a $TmAl_2$ single crystal, which was grown by the Czochralski method. It was characterized by neutron- and γ-diffractometry and found to be of good quality; in particular, the width of its rocking curve was smaller than the resolution of the neutron-three-axes spectrometer used (< 1°). Surprisingly, for temperatures T $\geq$ 2K the single crystal did not show the transition into the canted ferromagnetic structure. Rather, we found only the collinear ferromagnetic structure and, in addition, antiferromagnetic correlations. A similar behavior was observed for $TmAl_2$ polycrystals containing either 10% La or Y on Tm sites. Already a La concentration of 1% caused a drop of T_{c2} from 3.2K to 2.5K. It is interesting to note that a similarly low T_{c2} ($\simeq$ 2.2K) was detected, by neutron diffraction, for the polycrystalline charge, left in the crucible, from which the single crystal had been grown. In all of these samples, antiferromagnetic correlations were present and, within the experimental error, T_{c1} was the same as in $TmAl_2$ polycrystalline material.

In the following, we wish to discuss especially the nature of the collinear-to-canted ferromagnetic transition at T_{c2}. At first glance, the orientation of the moments within (111)-planes below T_{c2} appears surprising, because from the crystal-field (CF) parameters (3), <111> directions are expected to be the easy axes of magnetization (at B = 0). Using the same CF parameters, however, it can be calculated that at a finite field of ~ 3T, <110> directions become easy axes. This was demonstrated by measurements of the magnetic anisotropy (4) for a $TmAl_2$ single crystal (5). Therefore, we propose that the moments in the collinear ferromagnetic phase point along <111> and that at T_{c2} the internal field causes a reorientation of the moments from <111> to <110>. Furtheron, owing to the antiferromagnetic component of the magnetic interaction, the moments on two adjacent <111> planes are oriented along different <110> directions, both of which perpendicular to the same <111> direction. This interpretation is supported by our specific-heat results showing that addition of an external field to the internal one shifts the reorientation transition towards higher temperatures (fig. 2).

Finally, the observed sensitivity of this canted ferromagnetic structure to lattice defects, which is much higher than that of the collinear structure, should be ascribed to the more complex arrangement of moments in the former one. This offers a plausible explanation of the fact that the $TmAl_2$ single crystal did not show the onset of the canted phase above T $\simeq$ 2K: Because of the rather high vapor pressure of Tm at the melting temperature of $TmAl_2$ (~ 1500K), it is likely that this single crystal was grown slightly off stoichiometric. This is in accordance with the depression of T_{c2} in the polycrystalline charge (left behind in the crucible), which is of the same order than the T_{c2} depression in the $Tm_{0.99}La_{0.01}Al_2$ alloy.

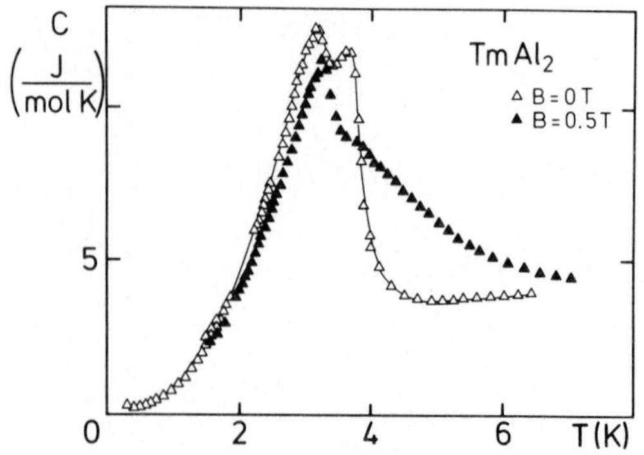

FIGURE 2
Specific heat as function of temperature for a $TmAl_2$ polycrystal at B = 0 and B = 0.5T.

We gratefully acknowledge useful conversations with Dr. H.E. Hoenig. The work was supported by SFB 65 and SFB 125.

REFERENCES

(1) M. Loewenhaupt, S. Horn, H. Scheuer, W. Schäfer and F. Steglich, in: The Rare Earth in Modern Science and Technology, Vol. 2, eds. G.J. MacCarthy, J.J. Rhyne and H.B. Silber (Plenum 1980), p. 339.
(2) S. Horn, M. Loewenhaupt, H. Scheuer and F. Steglich, J. Magn. Magn. 14 (1979) 239.
(3) M. Loewenhaupt, S. Horn and B. Frick, in: Crystalline Electric Field Effects in f-Electron Magnetism, eds. R.P. Guertin, W. Suski and Z. Zolnierek (Plenum 1982) p. 125.
(4) H.E. Hoenig, Habilitationsschrift Frankfurt (1979), unpublished.
(5) H.E. Hoenig, private communication.

LT-17 (Contributed Papers)
U. Eckern, A. Schmid, W. Weber, H. Wühl (eds)

CANTED FERROMAGNETISM IN DILUTE PdMn ALLOYS

J. L. THOLENCE*, I. A. CAMPBELL+

*C. R. T. B. T., 166 X, 38042 Grenoble Cédex, France
+Physique des Solides, Bât. 510, Université Paris-Sud, 91405 Orsay, France

Measurements on PdMn alloys with 2 and 3.2 at % Mn show a drop in initial susceptibility χ', peak in χ'', and a sudden increase in coercivity at temperatures of 0.55 K and 0.2 K respectively.

The magnetic phase diagram of dilute PdMn alloys shows spin glass (SG) ordering for Mn concentrations $x > 5$ % and a "ferromagnetic" region for $x < 5$ % (1). The change of behaviour can be understood as due to competition between long range ferromagnetic interactions through the polarization of the Pd matrix and near neighbour antiferromagnetic Mn-Mn interactions, which become important at higher concentrations. Although the magnetization of samples in the "ferromagnetic" range appear to saturate very easily the individual Mn moments remain randomly canted with respect to the overall magnetization direction and true saturation needs enormous fields (2, 3).

We have measured the ac susceptibilities χ', χ'' and the dc magnetization of PdMn samples with 3.2 at % and 2 at % Mn ; we will only discuss data for the 3.2 % Mn sample, similar results being observed for the 2 % Mn sample.

Figure 1 shows χ' and χ'' measured at 112 Hz in an ac field of about 0.4 oe and zero nominal external field. Below $T_c \simeq 5.5$ K, χ' reaches a plateau value of about 2.5 emu/cm^3, well below the demagnetization factor limit for this sample which is about 40 emu/cm^3. χ'' shows a peak just below T_c. Arrott plots of the dc magnetization data confirm the T_c value of 5.5 K but at lower temperatures show almost zero spontaneous magnetization in the ordered state.

The apparent saturation magnetization in moderate applied fields corresponds to only 1.9 μ_b/Mn atom at 0.11K. As the local moment on each Mn atom is about $7\mu_B$ (2) this apparent saturation is only a pseudo-saturation and the local spins are highly canted in the "ferromagnetic" state (2, 3).

At temperatures below 1 K there is a drop in $\chi'(T)$ with a peak in $\chi''(T)$ at 0.55 K. The temperature at which $d\chi'/dT$ is maximum corresponds to the maximum in $\chi''(T)$; neither of these temperatures is very sensitive to small static applied fields up to 35 oe. There is a dramatic increase in the coercive field around 0.55 K, figure 2.

PdMn in this concentration range near the SG regime is thus a canted ferromagnet at low temperature. From the magnetization measurements we cannot specify at what temperature the local transverse magnetization goes to zero as in AuFe (4). However we find the low temperature drop off in $\chi'(T)$ which has labelled the "reentrant spin glass transition" in other similar systems (5). In our sample, this drop off is centered at 0.55 K and is accompanied by a sharp peak in $\chi''(T)$ and a sudden rise in coercivity. These properties would seem easier to understand in terms of an intrinsic domain wall blocking setting in at this temperature, as the degree of ferromagnetic order does not vary in this temperature range. If this is a "transition", then it ressembles a roughening transition where domain wall movements become restricted.

The fact that the Arrott plots indicate that the spontaneous magnetization remains essentially zero below T_c may be related to the canting ; the behaviour ressembles that seen in "random field" systems.

REFERENCES

(1) J. Rault and J. P. Burger, C. R. A. S. 269 (1969) 1085
S. C. Ho et al., J. Phys. F 11 (1981) 1107
(2) W. M. Star et al., Phys. Rev. B 12 (1975) 2690
J. J. Smit et al., Sol. St. Comm. 30 (1979) 243
(3) A. Kettschau et al., J. M. M. M. 37 (1983) L 1
(4) F. Varret et al., Phys. Rev. B 26 (1982) 5195
(5) A. P. Murani et al., J. Phys. F 6 (1976) 425
Y. Yeshrun et al., Phys. Rev. Lett. 45 (1980) 1366.

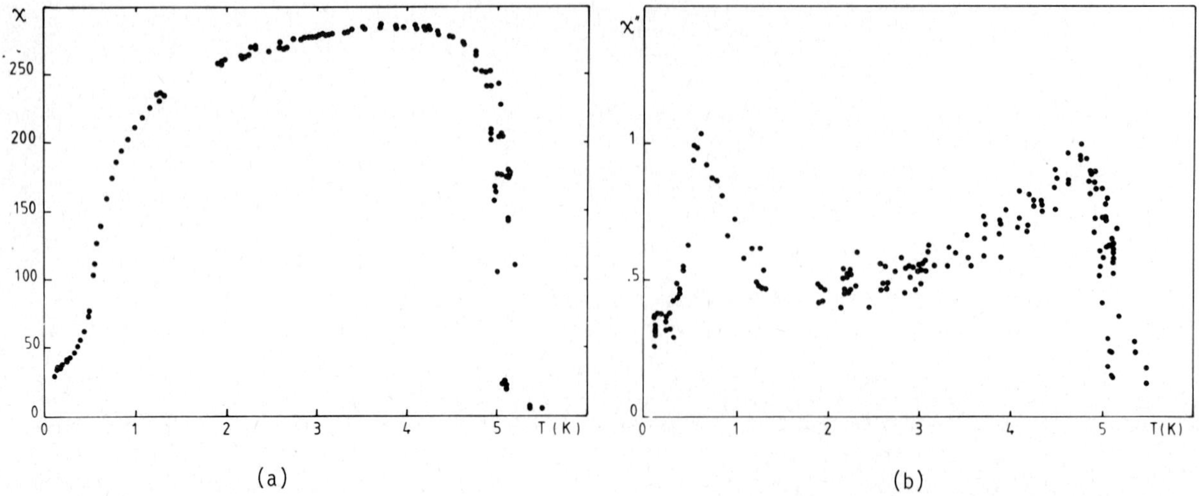

Figure 1 : Susceptibility of PdMn 3.2 % as a function of temperature (arbitrary units)
(a) χ'(T) (b) χ"(T)

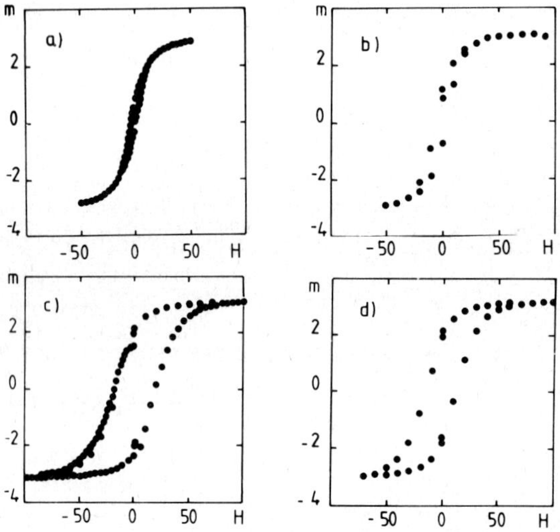

Figure 2 : Hysteresis curves of PdMn 3.2 %. Fields in oersteds
Magnetisation in arbitrary units.
Temperatures : (a) 1.25 K, (b) 0.45 K, (c) 0.11 K,
(d) 0.3 K

LT-17 (Contributed Papers)
U. Eckern, A. Schmid, W. Weber, H. Wühl (eds)

FERROMAGNETIC TRANSITION AND NEGATIVE MAGNETORESISTIVITY IN DILUTE Pt-Co ALLOYS

Franco PAVESE

CNR - Istituto di Metrologia "G.Colonnetti", strada delle Cacce 73, 10135 Torino, Italy

Resistivity measurements have been performed on dilute ferromagnetic alloys of Pt with 0.45, 0.50, 0.75, 1.06 and 2.15 at% Co, in the temperature range 2 - 28 K and in magnetic fields up to 6 T. Curie temperature have been found to occur at 6.0 K, 14.2 K and 32 K (independent of magnetic field value) for alloys with 0.75, 1.06, and 2.15 at% Co respectively; no transition was found for lower concentrations. Negative magnetoresistivity has been observed to occur in both phases, in a temperature and magnetic field range increasing with cobalt concentration. The transition resulted in a second order discontinuity of the R-T characteristics at B=0 and at B = 3.8 ± 0.2 T with 0.75 at%Co and 8.0 ± 0.5 T with 1.06 at%Co, while for other field values the discontinuity was first-order.

1. INTRODUCTION

Alloys of platinum and palladium with transition elements such as cobalt, iron, nickel are known since early 1960's (1) to exhibit ferromagnetic behaviour persisting at very strong dilution of the transition element, with Curie temperatures in the criogenic range.

Recently, another property of the dilute alloy of cobalt in platinum, the enhancement of resistivity at low temperatures by Kondo effect, has been investigated for application in thermometry, in order to enhance the sensitivity of platinum thermometers below 20 K (2). For the same purpose, specimens of five different Pt-Co alloys were measured in zero magnetic field and up to 6 T between 2 K and 28 K, in order to understand the change of magnetoresistivity with cobalt concentration, after preliminary checks had shown a region with negative values on a Pt-0.5at%Co alloy (3).

The aim of this paper is to present this set of new measurements in connection with magnetic properties of dilute Pt-Co alloys.

2. EXPERIMENTAL METHOD AND SPECIMEN DATA

The experimental apparatus is a vacuum calorimeter with automatic data acquisition, described elsewhere (3); it allows an overall temperature accuracy of ± 5 mK and a resistance accuracy of ± 0.01 %. Resolution was ± 0.1 mK for temperature and ± 100 µΩ for resistance. Eleven specimens have been measured, as indicated in Table 1.

A basic feature of the specimens is that thermometers are actually made of them, with a wire-wound wire dia 0.05 mm, mounted strain-free and well annealed. Samples 4,5 are in fact commercial thermometers; the others are prototypes obtained on loan from dr. H.Sakurai (2). Only termometers 10,11 of more simple fabrication, gave rise to some stability problems, at a few 0.01 K level.

Measurements have been taken at 25 temperatures between 2 K and 28 K and at 10 magnetic field values between 0 and 6 T, with a total of 250 experimental points per sample.

Table 1: Pt-Co alloy specimens

Sample n°	Cobalt (at%)	R(4.2)/R(273.15)
1,2	0.45	0.070
3,4,5,6	0.50	0.075
7,8	0.75	0.130
9	1.06	0.180
10,11	2.15	0.280

3. FERROMAGNETIC TRANSITION IN Pt-Co ALLOYS

As the resistance of a ferromagnet contains a contribution which is constant at temperatures above the Curie point and falls to zero as temperature goes to 0 K (4), in zero field the Curie point is marked by a second-order discontinuity on the R-T characteristics of the specimen. The measured temperature value of this discontinuity for the different alloys is: 6.0 ± 0.05 K with 0.75 at%Co, 14.2 ± 0.05 K, with 1.06 at%Co and, by extrapolation, 32 ± 0.2 K with 2.15 at%Co.

Fig.1 shows these results obtained in zero field, compared with those obtained by previous authors with different methods. The agreement is good, except for the data below 1 at%Co. No transition has been found above 2 K for cobalt concentrations of 0.45 and 0.50 at%; this has been confirmed by our measurements in non-zero fields. Extrapolation of our data to 0 K, would set above 0.5 at%Co the limit composition for the transition to occur .

4. NEGATIVE MAGNETORESISTIVITY IN Pt-Co ALLOYS

The resistance change due to the effect of magnetic field has quite a complicated behaviour, which is different for the two phases

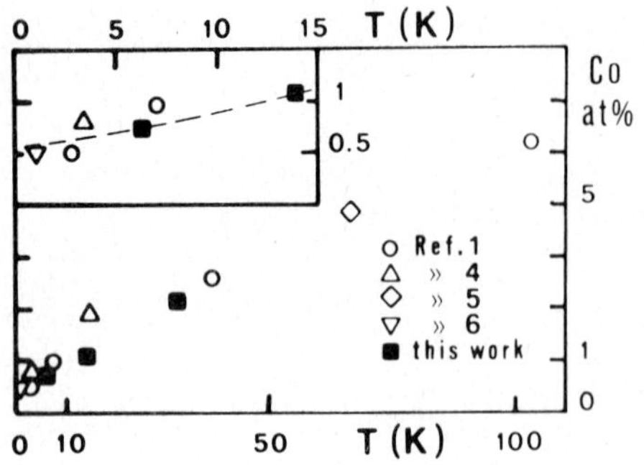

Fig. 1 : Ferromagnetic transition temperatures.

across the transition, as shown in Fig.2 for the 0.75 at%Co alloy: in the vertical scale of the surface, resistance changes have been transformed in equivalent temperature changes using the local dR/dT values. Figs. 3a) to d) show the constant-level maps of these surfaces for four alloys. Changes appear to be quite high at temperatures below 10 K and rising the magnetic field, since dR/dT too is quite affected (lowered) by magnetic field: for example the 2.15 at%Co alloy undergoes a slope reversal below 4 K at 6 T.

The surfaces for the two phases join at the transition temperature - which has been found constant with B within ± 0.2 K - with a first order discontinuity. It becomes a second-order one (ΔR=0) at two points: B = 0, and B = 3.8 ± 0.2 T with 0.75 at%Co and 8.0 ± 0.5 T (extrapolated) with 1.06 at%Co.

5. CONCLUSIONS

Results on ferromagnetic transition temperatures were found in good agreement with published data above 1 at%Co, while increasing discrepancy occurred for lower cobalt concentrations: in fact no transition has been found with Co ≤ 0.50 at%. The transition has been

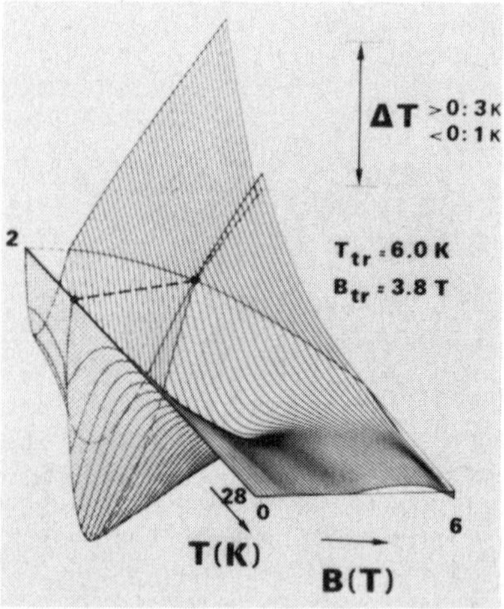

Fig.2 : Magnetoresistivity of 0.75 at%Co alloy (shown as temperature-equivalent error).

found to persist in magnetic fields up to 6 T, as a first order discontinuity and at a constant temperature, crossing the zero-change plane in another point besides B=0. This was due to a broad region where negative magnetoresistivity has been found for all concentrations extending in both phases across the transition.

REFERENCES

(1) J.Crangle and W.R.Scott, J.Appl.Phys. 36 (1965) 921.
(2) T.Shiratori, K.Mitsui, K.Yanagisawa and S. Kobayashi, Temperature, AIP, New York, 5 (1982) 839.
(3) F.Pavese, L.M.Besley and M.P.Sassi, IMGC Int.Report (1981), paper MD-5; full results to be published on Cryogenics.
(4) J.C.G.Wheeler, J.Phys.C 2 (1969) 135.
(5) D.M.S.Bagguley and J.A.Robertson, Phys. Lett. 27A (1968) 516.
(6) B.M.Boerstoel and C.VanBaarle, J.Appl.Phys 41 (1970) 1079.

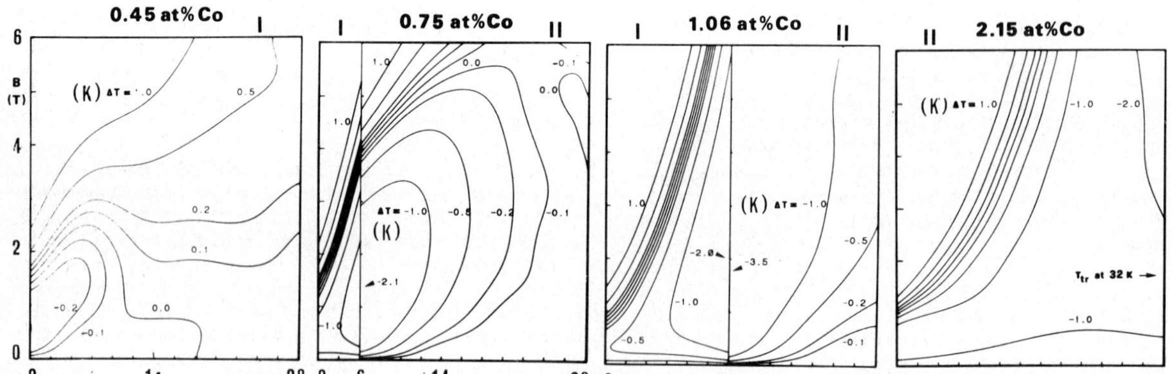

Fig.3 : Maps of temperature error due to magnetoresistivity in thermometers made of four alloys.

LT-17 (Contributed Papers)
U. Eckern, A. Schmid, W. Weber, H. Wühl (eds)

THERMOMAGNETIC PROPERTIES OF GdPd*

J. A. BARCLAY, W. C. OVERTON, Jr., and C. B. ZIMM

Group P-10, MS K764, Los Alamos National Laboratory, Los Alamos, NM 87545, U.S.A.

The magnetic susceptibility, magnetization, and field-dependent heat capacity of ferromagnetic GdPd have been measured. The magnetic moment of the Gd spin obtained from the saturation magnetization at 4.2 K is enhanced over the free Gd value. The heat capacity shows a large lambda-type magnetic anomaly at the Curie temperature of 40.0 + 0.5 K.

1. INTRODUCTION

Ferromagnetic alloys of gadolinium are excellent choices as refrigerants in magnetic refrigerators because they exhibit a large magnetocaloric effect near their Curie temperatures. One potential refrigerant is GdPd, which is an intrinsically interesting ferromagnetic system because palladium is well known for its polarizability by other magnetic moments. The polarization may enhance the thermomagnetic properties.

This alloy has not previously been well studied. The orthorhombic CrB crystal structure (1) and phase diagram (2) have been determined. The congruent-melting PdGd alloy has a melting point of 1380°C. It can be prepared by induction melting or argon-arc melting. The magnetic susceptibility has been previously measured (1), (3) indicating a Curie temperature of 39.5 + 1 K and an effective magnetic moment of 8.31 β and 8.2 β, respectively in the references. The only other investigations of GdPd systems are studies of amorphous alloys, which have distinctly different ferromagnetic properties (4).

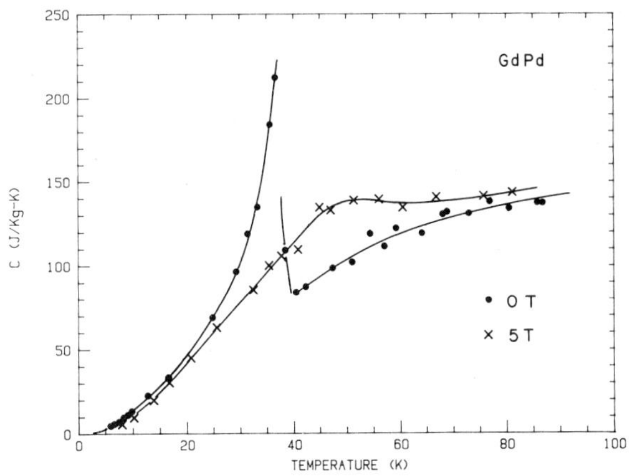

FIGURE 1.

2. EXPERIMENTAL METHOD AND MEASUREMENTS

The sample was obtained from Ames Materials Preparation Center and was prepared by argon-arc melting of 99.99% pure Gd and 99.99% Pd. After several remelts in the arc furnace, no further heat treatment was done. Metallography indicated a single-phase material. The a.c. magnetic susceptibility, measured from 4 K to 300 K in zero applied magnetic field, indicates a strong, albeit conventional ferromagnetic transition. The maximum slope in the susceptibility signal gives a Curie temperature of 40.0 + 0.5 K.

The magnetization at 3.95 K as a function applied magnetic field up to 7.5 T was measured in a vibrating sample magnetometer. The saturation magnetization at 3.95 K is 1.61 $\pm$ 0.03 x 10^6 A/m, which yields a magnetic moment μ_{eff} of 8.78 $\pm$ 0.19 Bohr magnetons. Without correcting for demagnetizing effects, the residual induction after removal of the applied field was less than 10^{-3} T.

The heat capacity was measured from ∿ 4 K to ∿ 80 K in fields up to 9 T. The apparatus used for these measurements has been previously described (5). The zero field and 5 T results are shown in Fig. 1.

3. DISCUSSION AND CONCLUSIONS

The measured Curie temperature is in agreement with the previous results, but the magnetic moment is larger than earlier reported. A moment of 8.78 β is significantly higher than 7.94 β

*Funding for this project has been provided by the U.S. Department of Energy, the Defense Advanced Research Projects Agency, and the National Aeronautics and Space Administration.

for a free gadolinium spin. The probable explanation of this enhanced magnetic moment is polarization of the conduction electrons in palladium by the presence of the gadolinium. The conduction electrons may in turn may polarize the d-electrons in the palladium contributing to the enhanced moment. Whatever the mechanism, the extra magnetic moment must come from the palladium because the gadolinium moments have a maximum value at 7.94 β. The interesting question is, does this extra moment contribute to the magnetic entropy at the ferromagnetic transition? We tried to answer this question by analysis of the heat capacity data.

Because the experimental data on the heat capacity of LaPd or LuPd are not available, the lattice contribution to the heat capacity was approximated by the Debye model. The only free parameter is Θ_D, which was used to fit the experimental zero-field heat capacity from >40 K to 80 K. The average value for ten points over this temperature range is Θ_D = 200 + 10 K. The calculated lattice heat capacity was subtracted from the total experimental heat capacity to obtain the magnetic contribution. The resultant magnetic heat capacity was integrated to obtain the magnetic entropy. For the zero-field results, we obtained a total magnetic entropy at 40 K of 64.5 ± 3.2 J/kg K. The expected value from a free Gd spin in GdPd would be 65.5 J/kg K. If the enhanced magnetic moment fully ordered, the zero-field magnetic entropy for GdPd would be 68.5 J/kg K. Unfortunately, the precision of the data is insufficient to stringently test whether or not the enhanced moment contributes to the magnetic entropy. The approximation of the Debye model for the lattice heat capacity also makes any definite conclusion very speculative. The magnetic entropy results do suggest, however, that the enhanced moment does not contribute.

Another measure of the effect of the enhanced magnetic moment on the thermomagnetic properties of GdPd is the adiabatic temperature change upon magnetization or demagnetization. The expression for the adiabatic temperature change is given by

$$\ln\left(\frac{T_f}{T_i}\right) = \int_{B_i}^{B_f} \frac{1}{C_B}\left(\frac{\partial M}{\partial T}\right)_B dB \qquad (1)$$

where T_i, T_f are the initial and final temperatures, C_B is the total field-dependent heat capacity, M is the magnetization, and B_i, B_f are the initial and final applied magnetic fields.

The total experimental heat capacity data for 0 and 5 T were integrated to obtain the entropy at these two fields. The maximum adiabatic temperature change obtained from the entropy curves was 9.0 + 1.0 K for a 0 to 5 T field change starting at 37.5 K. The molecular field model plus the Debye model for the lattice predicts 7.2 K for μ_{eff} = 7.94 β and 7.7 K for μ_{eff} = 8.78 β. Both predicted values are below the experimental value, which suggests that the enhanced moment may contribute to the adiabatic temperature change or that the models are just inadequate.

The small residual induction of the GdPd alloy indicates that the sample is magnetically soft, even at 3.95 K. Because hysteretic heating can be a serious source of inefficiency in a magnetic cycle, magnetic softness is an important requirement, especially near the Curie temperature. The large magnetic entropy near the Curie temperature and relatively low lattice heat capacity yields a larger adiabatic temperature change than found in GdNi (5). Because the efficiency of magnetic cycles is proportional to the adiabatic temperature change, a large change is desirable.

In summary, we have measured some thermomagnetic properties of GdPd and shown that, on the basis of the data available, this alloy is an excellent potential refrigerant.

4. ACKNOWLEDGEMENTS

It is a pleasure to thank W. Stewart, W. Johanson, C. Olson, R. Chesebrough, and M. McCray for their assistance with the preliminary measurements on GdPd.

5. REFERENCES

(1) J. Pierre and E. Siand, C. R. Acad. Sc. Paris 266 (1968) 1482.
(2) O. Loebich, Jr. and E. Raub, J. Less-Comm. Metals 30 (1973) 47.
(3) O. Loebich, Jr. and E. Raub, J. Less-Comm. Metals 31 (1973) 111.
(4) K. H. J. Buschow, H. A. Algra, and R. A. Henskens, J. Appl. Phys. 51 (1980) 561.
(5) J. A. Barclay, W. F. Stewart, W. C. Overton, Jr., R. Chesebrough, M. McCray, and D. McMillan, Adv. in Cryog. Eng., to be published 1984 (Proc. of Int'l. Cryog. Mat'ls. Conf., Colorado Springs, CO, Aug. 15-19, 1983.

LT-17 (Contributed Papers)
U. Eckern, A. Schmid, W. Weber, H. Wühl (eds)

ELECTRICAL RESISTANCE OF THIN DYSPROSIUM FILMS AT LOW TEMPERATURES

J.DUDÁŠ

Department of Theoretical and Experimental Electrotechnics, Technical University, 041 38 Košice, Czechoslovakia

A.FEHÉR

Faculty of Sciences, Šafarik University, 041 54 Košice, Czechoslovakia

The temperature dependence of eletrical resistance of thin dysprosium films was measured in the temperature range from 4.2 K to 300 K. A decrease of Néel temperature with decreasing thickness was observed. An increase of the spin disorder resistivity with decreasing thickness below 100 nm was also observed.

1. INTRODUCTION

Considerable interest has been devoted to the study of electrical and magnetic properties of the rare earth thin films in the last decade (1). The size effect of heavy and some light rare earth thin films prepared by vacuum evaporation in the high vacuum was studied and experimental results were compared with the theories. Relatively small number of papers appeared on the study of low temperature properties of these films. Only three papers were devoted to the study of spin disorder resistivity of this films.

Dysprosium is very interesting rare earth metal having two magnetic transition temperatures (2). Being paramagnetic at room temperature, it orders antiferromagnetically below the Néel temperature T_N=179 K and ferromagnetically below the Curie temperature T_C=85 K. Although Kaul te al. (3) and Cadim and Al Bassam (4) have studied size effect on thin dysprosium films, the work of Lodge (5) devoted to the low temperature study of some thin dysprosium films stimulated us to study spin disorded resistivity and low temperature behaviour of eletrical resistance of thin dysprosium films.

2. EXPERIMENTAL

Dysprosium thin films of various thicknesses were prepared by evaporating the bulk material (99,95 % pure) onto pre-cleaned glass substrates in the high vacuum coating unit. Prior to the film deposition the silver electric contacts were deposited. Thin dysprosium films were deposited at a pressure better than 1.3×10^{-4} Pa.

The liquid nitrogen trap was used to obtain better vacuum and we didn´t allow condensation on the substrate until the gettering action has finished- -several outgassings were made in order to get only weakly rising pressure at evaporation-and films were coated with sufficient layer of SiO. Electric contacts were made by cementing thin silver wires at appropriate positions by silver paint.

The temperature of thin film in the helium cryostat was measured by Lake Shore Cryotronics, Inc. germanium (4.2 K to 80 K) and platinum (80 K to 300 K) thermometers.

A conventional d.c.arrangement was used to measure the electrical resistance with the accuracy of 0,05 %.

The film thickness was measured with the accuracy of ±5 nm by optical Tolansky method using monochromatic light.

3. RESULTS

Prior to the thin film study we have measured the resistance R versus temperature T behaviour of the bulk material from which the films were prepared. The result of this measurement is illustrated on fig.1 as resistance ratio $R/R_{4.2}$ vs. T. The change of the slope corresponding to T_C is clearly seen on this figure at 92.0±0.1 K. The Néel temperature obtained by the method (5) is 180.5±0.1 K. Spin disorder resistivity estimated by the method (2) provides the value of 47.5 $\mu\Omega$cm.

We have studied thin dysprosium films in the thickness range from 30 nm to 300 nm. Four typical temperature dependences of the electrical resistance of these films are illustrated on fig.2. As we see from this figure the onset of the antiferromagnetic spin ordering is less pronounced on thin films as on the bulk sample. The Néel temperature estimated by the method (5) decreases with decreasing thickness. From R vs. T dependences shown on fig. 2 the following values of T_N were obtained: sample 1 - 178.7 K, sample 2 - 178.9 K, sample 3 - 175.3 K and sample 4 - 176.2 K. This decrease of T_N with decreasing thickness

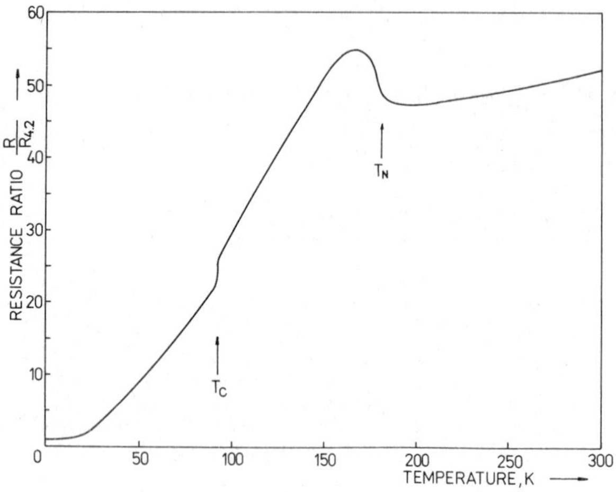

FIGURE 1
Resistance ration $R/R_{4.2}$ as a function of temperature of the bulk dysprosium

was also observed by Lodge (6) who found T_N=171 K for thickness of 27.7 nm. However, the quality of thin dysprosium films prepared by Lodge and also by us doesn't allow to study this decrease of T_N vs. thickness in more detail in order to compare this dependence with the theory of magnetic phose transitions in thin films.

We also have estimated spin disorder resistivities from R vs. T dependences of our films. The values obtained by the method (2) for dependences shown on fig.2 are following: sample 1 - 39.3 μΩcm, sample 2 - 46.4 μΩcm, sample 3 - 50.7 μΩcm and sample 4 - 85.5 μΩcm. In a disagreement with the results of Lodge (6) we observed an increase of spin disorder resistivity with decreasing thickness of the films below 100 nm. We suppose, that this discrepancy is probably caused by various defects in the structure of thin films.

Moreover, we have found, that size effect of our thin dysprosium films doesn't exhibit an anomalous behaviour observed by Kaul et al. (3) or Cadim and Al Bassam (4) in their films.

4. CONCLUSIONS

We have measured the temperature dependence of electrical resistance of thin dysprosium films as well as of bulk dysprosium in the temperature range from 4.2 K to 300 K. Detailed R vs. T study was done in the vicinity

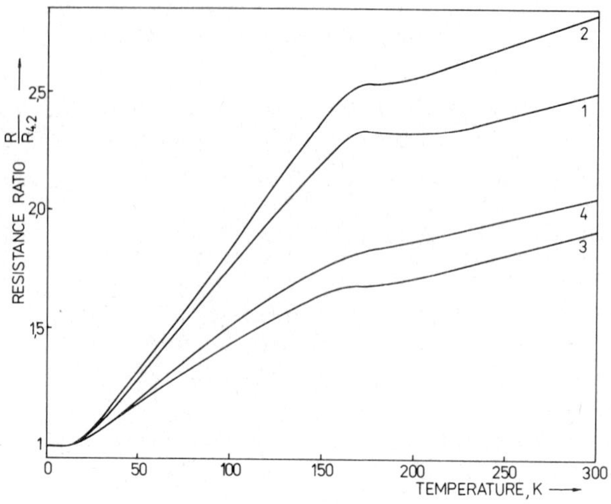

FIGURE 2
Resistance ratio $R/R_{4.2}$ as a function of temperature of four thin dysprosium films: 1-250 nm, 2-163 nm, 3-114 nm, 4-45 nm

of magnetic phase transitions. We have observed a decrease of the Néel temperature with the decreasing film thickness and an increase of spin disorder resistivity below 100 nm.

ACKNOWLEDGEMENT
We are grateful to Dr.Š.Jánoš for several helpful discussions.

REFERENCES
(1) M.Gasgnier, phys.stat.sol. (a) 57(1980), 11.
(2) K.N.R.Taylor and M.I.Darby, Physics of Rare Earth Solids (Chapman and Hall, London, 1982).
(3) V.K.Kaul et al., Thin Solid Films 30(1975), 65.
(4) M.Cadim and T.S.Al Bassam, Thin Solid Films, 41(1977), L9.
(5) K.V.Rao et al., J. Phys. C 6(1973), L231.
(6) F.M.K.Lodge, Thesis, University of Durham, Durham, Great Britain.

LT-17 (Contributed Papers)
U. Eckern, A. Schmid, W. Weber, H. Wühl (eds)

1-D SYSTEMS WITH LOCAL MAGNETIC MOMENT

Fu-sui LIU, Qi-du JIANG

Physics Department of Beijing University, Beijing, China

The exact 1-D RKKY interacton is given. Meantime, three forms of 1-D RKKY interaction is discussed. In this paper the effect of RKKY coupling between magnetic impurities to 1-D susceptibility and resistivity are also discussed.

1. 1-D RKKY INTERACTION

1-D magnetic systems become a new research subject. (ref. 1,2) A lot of authors hava discussed the form of 1-D RKKY interaction in magnetic alloy,(ref. 3,4,5) but an better form of 1-D RKKY interaction is still lacking.

To find the form of 1-D RKKY, we should calculate the correction of the second order of exchange Hamiltonian for the magnetic systems.

$$H_{sd} = -(J/2N)\underset{(kk'j)}{SUM}\exp[i(k'-k)R_j] \cdot [(a^+_{ku}a_{k'u} - a^+_{kd}a_{k'd})S^z_j + a^+_{ku}a_{k'd}S^-_j + a^+_{kd}a_{k'u}S^+_j] \quad (1)$$

SUM means the summation over k, k' and j. u (d) implies spin up (down), J is the s-d exchange integral, N the numbers of unit cell, $\vec{S}$ the d-electron's spin vector. Similar to Kondo's (ref.6) treatment, we get the energy correction:

$$E = \underset{(ji)}{SUM}(\vec{S}_j \cdot \vec{S}_l)J(|R_j - R_l|) \quad (2)$$

Where

$$J(|R_j - R_l|) = -(J_2/25.12)N(O)\int_{+\infty}^{+\infty}\exp(iq|R_j - R_l|) \cdot [\ln|(q+2k_F)/(q-2k_F)|]/qdq \quad (3)$$

put $t = q|R_j - R_l|$, $x = 2k_F|R_j - R_l|$ and define

$$I = \oint_c \exp(it)/t[\ln|(t+x)/(t-x)|]dt \quad (4)$$

It is easy to show that

$$I = \oint_c [\exp(it)/t]\ln|(t+x)/(t-x)|dt \quad (4')$$

$$\oint_c [\exp(it)/t]\ln[(t+x)/(t-x)]dt = 0 \quad (4'')$$

Subtracting (4″) from (4′), we get

$$I = 3.14\int_X^{\infty} 2\sin t/tdt \quad (4''')$$

The integral contour C is a upper half circle in t complex plane without points −x, O and x.

At last we obtain the 1-D RKKY interaction:

$$J(x) = -(J^2/4)N(O)\int_X^{\infty}\sin t/tdt \quad (5)$$

For large x, eq. (5) becomes

$$J(x) = -(J^2/4)N(O)(\cos x/x + \sin x/x^2) \quad (5')$$

By the Kittel's approch (ref.4)

$$J(x) = (J^2/8)N(O)[3.14 - \int_X^{\infty} 2\sin t(2t)/tdt] \quad (6)$$

It is easy to see that the above expression is not correct Becanse the J(x) in eq.(3) is equal to zero at the limit of infinite x, but the J(x) in eq. (6) is not.

By Roth's (ref.3) and Abrikosov's (ref.5), the J(x) is proportional to cosx/x. In comparision with our result (eq.5′) it can be seen that the above form of J(x) is only an approximate expresson at large x. Even in large x condition the term of $\sin x/x^2$ in eq.(5′) will be important for the numerical calculation in spin-glass.

2. 1-D SUSCEPTBILITY

One author of this paper has gotten the formula of susceptibility for 3-D dilute magnetic alloy. By the similar derivation ot ref. 7, we get the 1-D form as

$$X = X_0 + X_1 + X_2 + \cdots \quad (7)$$

where X is 1-D magnetic susceptibility, and

$$X_0 = (2B)^2WS(S+1)/(3T) \quad (8)$$

where B is Bohr magneton, W is the number of magnetic atoms. The first order X_1 is

$$X_1 = 3.14(2B)^2S(S+1)/(48T^2)J^2N(O)2 \cdot [\underset{(n)}{SUM}\int_{2k_F}^{\infty} dq\sin(q|R_n|)/q] \quad (9)$$

In the condition of large $k_F|R_n|$, We have

$$X_1 = 3.14(2B)^2 2S(S+1)/(48T^2)J^2N(O)\underset{(n)}{SUM}[\cos(2k_F|R_n|)/(2k_F|R_n|) + \sin(2k_F|R_n|)/(2k_F|R_n|)^2] \quad (10)$$

In the nearest neighbor magnetic atom pair approximation, we have

$$X_1 = 3.14 \, (2B)^2 2S(S+1)/48 \, T^2) \, J^2 N(0)c$$

$$[\cos(2k_F a)/(2k_F a) + \sin(2k_F a)/(2k_F a)^2] \quad (11)$$

where the a is the nearest neighbor distance, c is magnetic impurity concentration.

$$X = (2B)^2 S(S+1)/(3T)W\{1 + [3.14 \, J^2 N(0)/(8T)$$

$$\underset{(n)}{\text{SUM}} \int_{2k_F}^{\infty} dq \, \text{siu}(q|R_n|)/q/[1 - 3.14 J^2 N(0)$$

$$\underset{(n)}{\text{SUM}} \int_{2k_F}^{\infty} dq \sin(q|R_n|)/(4qT)]\} \quad (13)$$

Because $\int_{2k_F}^{\infty} dq \sin(q|R_n|)/q$ is great than zero for small enough $2k_F \, |R_n|$, the quasi 1-D phase transition in this case is ferromagnetic.

3.   1-D RKKY COUPLING RESISTANCE

The 3-D RKKY coupling resistance is given in ref. 8 and 9. By the similar derivation we get the 1-D form. Under the condition of $2k_F a$ is greater than unit, the 1-D RKKY coupling resistance is

$$R_{RKKY} = R'(T_0/T) \, c^2 \cos^2(k_F a)[\cos(2k_F a)$$

$$+ \sin(2k_F a)/(2k_F a)] \quad (14)$$

where $\qquad R' = 3mJ^2 3.14^2/(32e^2 E_F h)$

and $\qquad T_0 = J^2 N(0)/16$

It is easy to get the 1-D Kondo resistance formula as

$$R_{kondo} = R'c[1 - 2JN(0)\ln(D/T)] \quad (15)$$

It is the same as in 3-D case.

REFERENCES

(1) A.A.Abrikosov, J. LOW Temp. Phys. 37(1979) 111

(2) U.Larsen, K.Carneiro, Chem. Scr. 17(1981)71.

(3) L.Roth et al, Phys. Rev. 149(1966)519.

(4) C. Kittel, Solid State Physics, V. 22, (Academic Press, New York, 1967).

(5) A.A.Abrikosov, J. Low Temp. Phys. 34(1979) 595.

(6) J.Kondo, Solid State Physics, V. 23, (Academic Press, New York, 1969).

(7) Fu-sui Liu, Commun. in Theor. Phys. (Beijing, China) 1 (1982) 779.

(8) Fu-sui Liu, J.Ruvalds, Physica,107B(1981)623.

(9) J. Ruvalds, "Advances in Superconductivity", (North-Holland, Amsterdam), pp. 475—514.

LT-17 (Contributed Papers)
U. Eckern, A. Schmid, W. Weber, H. Wühl (eds)

ON THE SUSCEPTIBILITY OF RARE EARTH IONS IN RANDOM CRYSTAL FIELDS

A. GUESSOUS and K. MATHO

Centre de Recherches sur les Très Basses Températures, C.N.R.S., BP 166 X, 38042 Grenoble-Cédex, France

We discuss the initial susceptibility $\chi(T)$ for rare earth ions submitted to a second order crystal field Hamiltonian including non-axial terms. The exact results are compared with a simple algebraic approximation.

Purely quadrupolar crystalline electric fields (CEF), including non-axial terms (1), are an important limiting case for rare earth ions in disordered structures. If there is no symmetry in the local charge distribution, they can dominate the higher order multipoles. The CEF-Hamiltonian, depending on two parameters C(energy) and η, $|\eta| \leq 1$ (2), can be written as

$$H_2 = -C(\eta/3\ O_2^\circ - \frac{1}{2}\sqrt{1-\eta^2}\ \{J_zJ_x+J_xJ_z\}) \qquad [1]$$

We report results on the initial susceptibility $\chi(T)$ for a general (Hund's rule) ground multiplet J, assuming that excited J-states are higher than the total splitting of H_2, $E_{max}-E_\circ = \Delta = C\delta_J(\eta)$. With $\delta_J(1) = J^2$ or $J^2-1/4$, for integer or half-integer J, respectively. The function $\delta_J(\eta)$ is closely reproduced by

$$\tilde{\delta}_J^2(\eta) = \delta_J^2(1)-\delta_J(1)(1-\eta^2)(5J/6 - 3/4) \qquad [2]$$

We also define the reduced effective moment function $r(T) = \chi(T)/\chi_c(T)$ with the Curie function for free multiplets in the denominator. $\chi(T)$ is evaluated at fixed C and η, but for an isotropic "powder average" over the orientations of principal axes.

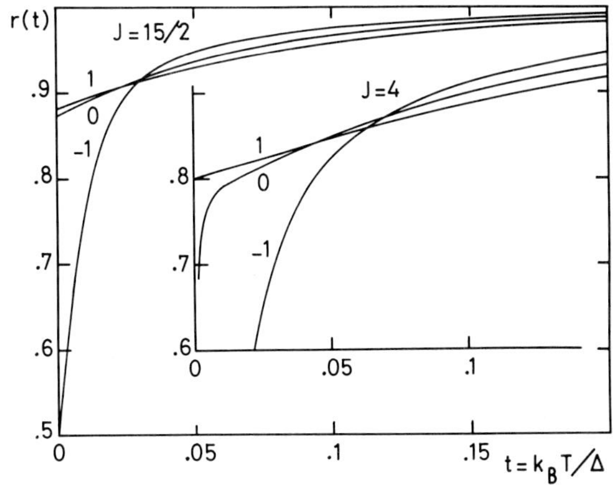

FIGURE 1
Fixed point property of the reduced effective moment function r(t) for J = 15/2 and J = 4. Curves are labeled by their η-values.

In order to carry out further averages over model distributions (1,3) of C and η, one needs a rapid access to $\chi(T)$ without diagonalizing each time the Hamiltonian.

The main purpose of this short note is to point out that the exact $\chi(T)$ has a simple scaling property if a reduced temperature scale $t = k_BT/\Delta$ is used. In particular, at a "fixed point" $t_f(J)$ the effective moment $r(t_f) \equiv r_f(J)$ is very nearly independent of η.

Figure 1 shows two examples of r(t), for half integer J = 15/2 and integer J = 4. All curves for values $|\eta| \neq 1$ pass very nearly through the crossing point $\eta = \pm 1$ curves, which is our definition for $r_f(J)$ and $t_f(J)$. At t_f, maximal deviations for half integer J are about 2 % for J = 5/2, less for increasing J. For J = 3/2, r(t) is independent of η. For small integer J, the fixed point is less well defined. The values of the fixed point are numerically given by

$$t_f = (J+1)/(5J^2),\ r_f = \frac{5J/6 + 1/3}{J(J+1)}, \qquad [3]$$

showing that the characteristic temperature, $T_f = t_f.\Delta$, and the deviation from the free moment, $J(J+1)(1-r_f)$, are both or order J.

Except for $k_BT < k_BT_s = C[(1-\eta)/2]^J$, the half integer and integer cases are "reunified". The smallest scale, T_s, enters the problem for integer J, giving the separation of two low lying singlet states (2). The corresponding regime $T \lesssim T_s$ of the free energy function, which can be treated in the approximation of a two-level system (4), gives $r(t) \to 0$.

In the half integer case, $\mathcal{H}_2$ has well separated Kramers doublets. Then, low temperature effective moments and Van-Vleck terms are rigorously defined for $T \to 0$, as $r(0) \equiv r_\circ(\eta)$ and $(dr/dt)_\circ = r'_\circ(\eta)$.

Making use of the fixed point, we need only these two η-dependent quantities to construct a simple approximation

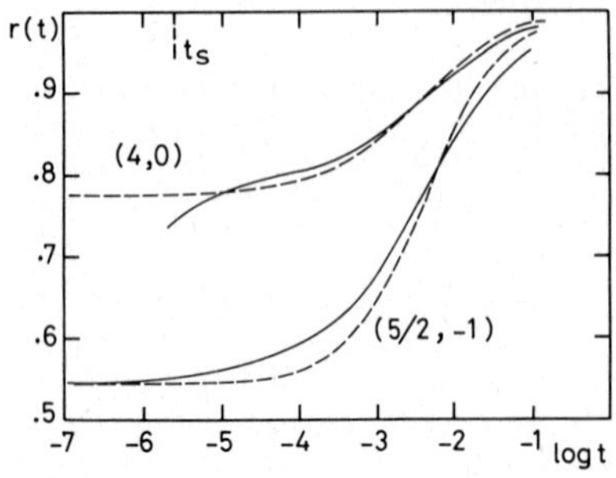

FIGURE 2
Comparison between the exact effective moment (—) and approximation $\tilde{r}(t)$ (---), defined by equations [3-8]. The label is (J,η). For $t_s = k_B T_s/\Delta$, see text.

$$r(t) = (r_o + X(t)/(1+X(t)) \qquad [4]$$

$$X(t) = r'_o t + \left(\frac{r_o - r_f}{1-r_f} - t_f r'_o\right)(t/t_f)^2 \qquad [5]$$

The quality of this approximation is shown in figure 2. The "reunification" of half integer and integer J means that, above T_S, interpolated values for r_o and r'_o can be used to define $\tilde{r}(t)$ for integer J. Below T_S, it would be sufficient to replace the constant r_o apparent in [4] by the two-level approximation (essentially a strong Van-Vleck term) and keep r_o in [5].

The functions $r_o(J,\eta)$ and $r'_o(J,\eta)$ are given approximately by the expressions

$$J(J+1)(1-r_o(\eta)) \cong J + \frac{1}{2}(J+1/2)(J-3/2)G_{2J-1}(\eta) \qquad [6]$$

$$J(J+1)r'_o(\eta) \cong (J+1/2)\left[J + \frac{1}{2}(J^2+J-1/4)(J-3/2)G_n(\eta)\right] \qquad [7]$$

$$n = 5(2J-3)/J \; ; \; G_n(\eta) = \left(\frac{1-\eta}{2}\right)^n \left(1+n\,\frac{1+\eta}{2}\right) \qquad [8]$$

These are exact for $|\eta| = 1$ and sufficient in accuracy for all J and η, to be used in the approximation $\tilde{r}(t)$.

The main characteristic of the quadrupolar CEF's is an anomalously weak reduction of the effective moment. The weak dependence on η, except at very low temperatures, is also striking. During an investigation of Ce^{3+} (J = 5/2) in amorphous $CeAl_3$ we were led to consider H_2 in competition with a higher order CEF of hexagonal symmetry (5). Such effects of competition are probably the general rule and the plain dominance of H_2 will be an exception.

REFERENCES

(1) R.W. Cochrane et al., J. Phys. F : Met. Phys. 5 (1975) 763.
(2) A. Fert and I.A. Campbell, J. Phys. F : Met. Phys. 8 (1978) L57.
(3) G. Czjzek et al., Phys. Rev. B 23 (1981) 2513.
(4) E. Borchi and S. De Gennaro, J. Phys. F : Met. Phys. 11 (1981) L47.
(5) A. Guessous et al., J. Non Crist. Solids (in print).

LT-17 (Contributed Papers)
U. Eckern, A. Schmid, W. Weber, H. Wühl (eds)

MONTE CARLO MOLECULAR DYNAMICS STUDY OF IRON

B. Sriram Shastry

Tata Institute of Fundamental Research, Colaba, Bombay 400 005, India

The spin dynamics of the classical Heisenberg model on the BCC lattice is computed using a Monte Carlo Molecular Dynamics approach in the paramagnetic phase at T=1.025Tc and 1.275Tc. The resulting scattering function S(q,w) and its integral over restricted energy windows are computed and compared with the recent neutron scattering experiments on Iron at Brookhaven and Grenoble. We find a reasonable overall agreement with much of the experimental data.

Recent experiments at Brookhaven [1] and Grenoble [2] on paramagnetic Iron using spin polarized neutron scattering provide an important and exciting challenge to the theories of the magnetism of Iron. These measurements yield the scattering function S(q,w) and its convolution with resolution functions of a given energy width *on an absolute scale.* These measurements are of considerable importance since they provide stringent tests of possible theories. The existing theories are broadly in two classes. One class of theories assume that vast short ranged order (SRO) is present in the paramagnetic phase of Iron [3] as suggested by the early work using unpolarized neutron scattering [4]. Another class of theories [5,6,7] view Iron as a disordered local moment situation wherein fluctuating Anderson-like local moments exist in the paramagnetic phase. These theories have been rather successful in explaining the thermodynamic data on Iron.

The present work addresses the problem of a reliable calculation of the scattering function S(q,w) for the Heisenberg model in the paramagnetic phase. We avoid analytical approximations and seek to estimate S(q,w) from a Monte Carlo Molecular Dynamics approach [8,9]. This is a problem which is made difficult by the large value of the effective exchange coupling relative to the typical experimental resolution widths, and our computation is, we believe, the longest attempted for arrays of the size considered. We set up 2.16^3 sized classical spin arrays following the usual Monte Carlo (MC) algorithm. This generates typical sample arrays representing the equilibrium states of a system described by the Hamiltonian,

$$H = -Js(s+1) \sum_{<ij>} \vec{\sigma}_i \cdot \vec{\sigma}_j \qquad (1)$$

Here the variables $\vec{\sigma}_i = \vec{s}_i/(s(s+1))^{1/2}$ (s=1) are unit vectors, and the exchange constant J(=22.42 mev), the single parameter in the theory, is determined by fitting the room temp spinwave spectrum. This leads to Tc ≃ 1069 °K in good agreement with experiments (∿ 1000 °K). The semiclassical approach to the dynamics [8,9] assumes that the dynamics of $\vec{\sigma}_i$ is obtained from the equation of motion

$$t_0 \frac{d}{dt} \vec{\sigma}_i = \vec{\sigma}_i \times \sum_{j \varepsilon n i} \vec{\sigma}_j \qquad (2)$$

where $t_0 = h/j(s(s+1))^{1/2}$ is the natural time scale in the problem. The initial arrays are used as initial conditions for the evolution equations Eq.2. The maximum time to which we integrate these equations is $24t_0$ and the maximum time lag for the correlations is $12t_0$. Stable integration algorithms for this length of integration are non-trivial, and after considerable experimentation we finally used the Gill fourth order Runge Kutta method [10,11]. Each dynamical evolution requires 6 hours CPU time on a CYBER 170:730 computer. The spin fluctuations $\vec{\sigma}_q(t)$ were monitored for t=.3,.6,...,24. for the eight nonzero q vectors in the <110> direction and estimates of autocorrelations were obtained by the usual procedure. We averaged the correlation function over ten independent samples and found reasonably stable results. The static correlations were estimated from the MC procedure and were used to normalize the correlation functions in time. From our computation of S(q,w), we obtained the pseudo-static correlations defined as

$$M_q^2(w_0) \equiv 4 \int_{-\infty}^{+\infty} S(q,w)R(w/w_0)dw \qquad (3)$$

Here the instrumental resolution function $R(w/w_0)$ is assumed to be a Gaussian with width w_0 (FWHM) which was picked as 20 mev and 43 mev to compare with the Brookhaven and Grenoble experiments respectively. The resulting functions M_q are compared with the two experiments in Figs.1 and 2. Fig.1 is at T=1.025Tc and the circles are from the Brookhaven data. The largest deviations from theory occur at very samll q, which is the region of largest theoretical uncertainty. Fig.2 is at T=1.275Tc. The triangles are the quoted Grenoble data on an absolute scale. The circles are also from the Grenoble data normalized to the theory at q=0 but without form factor corrections. In both figures the dashed line is from the earlier approximate results of Shastry, Edwards and Young (Ref.6). This theory has very different predictions for the functional form of S(q,w) but the integrated M_q is remarkably similar to the present work. We emphasize that ours is essentially a one parameter theory and we believe that the agreement between theory and experiment as seen the two figures demonstrates that the Heisenberg model provides a reasonable description of the dynamics of Iron. The calculation also shows that in order to extract the true equilibrium correlations, the energy windows have to be much larger, of order 200 mev or so. We propose that meaningful results can be extracted from experiments by varying the window widths at a fixed temperature and comparing the differences with theoretical predictions such as those implicit in the present work.

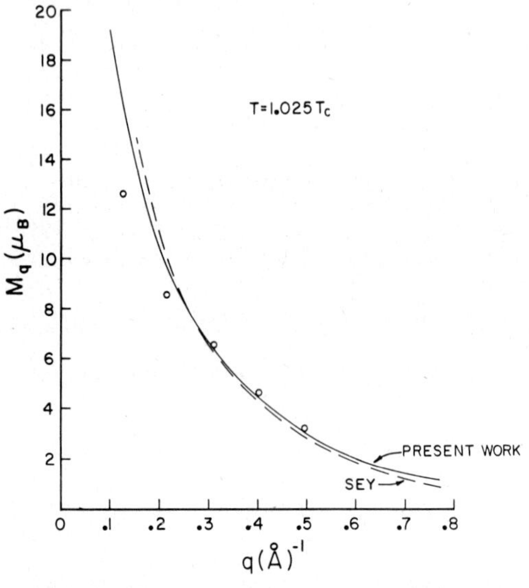

Fig.1. M_q vs q at 1.025 Tc compared with Brookhaven data. Dashed line from Ref.6.

ACKNOWLEDGEMENTS

I thank Profs. G. Shirane, J. Wicksted and D.M. Edwards for useful discussions and a valuable correspondence.

REFERENCES

(1) J.P. Wicksted, G.Shirane and O. Steinsvoll, Phys. Rev. B29, 488 (1984).
(2) P.J. Brown, J. Deportes, D. Givord and K. Ziebeck, in Proceedings of the I.L.L. Workshop on 3-d Metallic Magnetism 1983 (Ed. D. Givord and K. Ziebeck).
(3) R. Prange and V. Korenman, Phys. Rev. B19, 4691 (1978); H. Capellmann, Zeit, Physik B34, 29 (1979).
(4) J. Lynn, Phys. Rev. B11, 2624 (1975).
(5) J. Hubbard, Phys. Rev. B20, 4584 (1979); Phys. Rev. B23, 5974 (1891).
(6) B. Sriram Shastry, D.M. Edwards and A.P. Young, J. Phys. C14, L665 (1981).
(7) H. Hasegawa, J. Phys. Soc. Japan, 46, 1504 (1979); T. Moriya, J. Phys. Soc. Japan, 51, 420 (1982).
(8) M. Evans and C. Windsor, J. Phys. C6, 495 (1973).
(9) R.E. Watson, M. Blume and G. Vineyard, Phys. Rev. 181, 811 (1969).
(10) B. Sriram Shastry, to be published.
(11) S. Gill, Proc. Cambr. Phil. Soc. 47, 96 (1951).

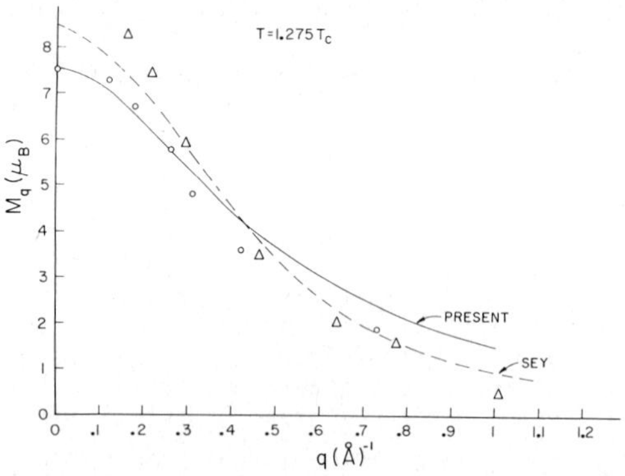

Fig.2. M_q vs q at 1.275 Tc compared with Grenoble data. Triangles are quoted absolute scale results and circles are raw data normalized at q=0 without form factors.

LT-17 (Contributed Papers)
U. Eckern, A. Schmid, W. Weber, H. Wühl (eds)

HYPERFINE MAGNETIC FIELDS IN UP, UAs, USb, USe and UTe FROM LOW TEMPERATURE SPECIFIC HEAT MEASUREMENTS

H. RUDIGIER, H.R. OTT

Laboratorium für Festkörperphysik, ETH-Hönggerberg, 8093 Zürich, Switzerland

Specific heat measurements on U-monopnictides and chalcogenides between 0.1 K and 1 K reveal the onset of a nuclear Schottky anomaly from which the nuclear hyperfine (hf) field of the pnictogen nuclei, and in the case of the chalcogenides of the uranium nuclei, can be extracted. Our results of published values for the hf-fields at the pnictogen-nuclei show a remarkable correlation with the paramagnetic Curie temperatures.

1. INTRODUCTION

Measurements of hyperfine (hf) magnetic fields at the nuclei of both the cation and the anion in magnetically ordered systems provide information about the nature of magnetic interactions. The most common techniques to obtain such local fields include Mössbauer and nuclear-magnetic-resonance (NMR) experiments. In this paper we want to demonstrate that hf magnetic fields in the NaCl-type uranium monopnictides and chalcogenides can also be obtained from specific heat (C_p) experiments. At temperatures below 1 K a Schottky-type contribution C_N to the heat capacity is observed, which is due to the change in the thermal population of the Zeemann-split nuclear ground state. As the maximum of the Schottky anomaly occurs below 0.1 K in these materials, the full expression for C_N can be expanded in inverse powers of T to give $C_N = a_2 \cdot T^{-2} + \ldots$(1). Neglecting quadrupole interaction, the coefficient a_2 is given by(1)

$$a_2/R = 1/3\cdot(\mu_N\cdot H_{eff}/k_B)^2\cdot I\cdot(I+1)/I^2 \qquad (1)$$

In Eq. (1) R denotes the gas constant, μ_N the nuclear magnetic moment, k_B Boltzmann's constant, and I the nuclear spin. In the UX compounds both the cation (^{235}U-isotope) and the anion contribute to C_N:

$$a_2/R = \alpha_U\cdot\nu_U\cdot H^2_{eff}(U) + (\Sigma_i\alpha_x^{\,i}\cdot\nu_x^{\,i})\cdot H^2_{eff}(X) \quad (2)$$

where $\nu_x{}^i$ is the percentage of the natural abundance of the isotope i of the element X and $\alpha_x{}^i$ is defined by Eq. (1).

2. EXPERIMENT AND RESULTS

The experiments reported here were performed on single crystals (80-200 mg) in the temperature range between 0.1 K and 1 K using a thermal relaxation method. Because of the internal radioactive heating, the specimens could not be cooled below 0.1 K. From the experimental value of C_p, C_N was obtained by subtracting the lattice- and electronic contributions as derived from experiments on the same samples in the temperature range between 1.5 K and 12 K (2,3). The T^{-2} behaviour of C_N is shown in Fig. 1 for UP, UAs, USe and UTe. For the data of USb see Ref. 2).

For UP and USb the agreement between the calculated values of a_2 (Eq. 2) inserting published values for H_{eff} of U (4,5), P (6) and Sb (4) and the experimental value is within 2% for UP, 4.3% for USb. Taking H_{eff} = 3700 kOe from the linear relation between H_{eff}(U) and the ordered moment μ_n of the 5f-electrons (5) we obtain H_{eff}(As) = (68 ± 5)kOe.

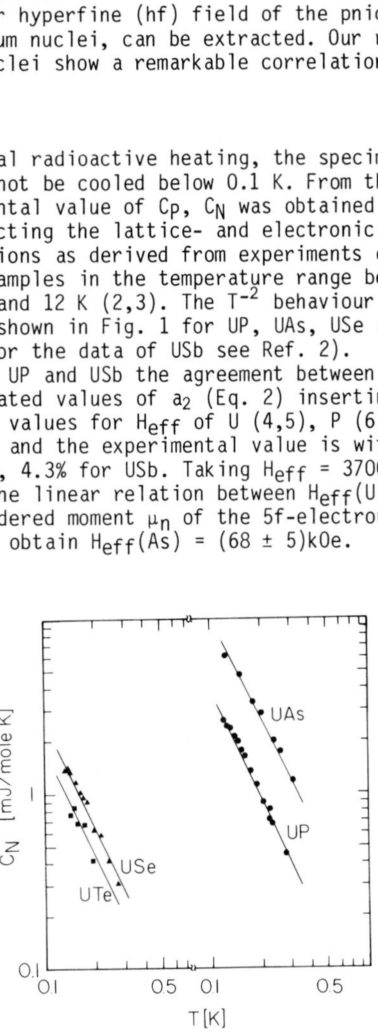

FIGURE 1
T^{-2} behaviour of the nuclear specific heat of UP, UAs, USe and UTe.

The total natural abundance of ^{123}Te and ^{125}Te isotopes of only 7.87% (^{77}Se 7.58%) make the coefficient a_2 more sensitive to the U-hf field $H_{eff}(U)$ = (1850 ± 600) kOe for UTe. As no values for neither $H_{eff}(Se)$ nor $H_{eff}(U)$ (USe) could be found in literature we have assumed that a similar relation to $H_{eff}(Sb)/\mu_n$ = = $H_{eff}(Te)/\mu_n$ = 60 kOe/μ_B also exists between $H_{eff}(As)$ and $H_{eff}(Se)$ we get $H_{eff}(Se)$ = 61 kOe and from a_2 $H_{eff}(U)$ = 3200 ± 400)kOe for USe.

3. DISCUSSION

Two mechanisms, both of which are proportional to the moment of the 5f-electrons, lead to a net spin density at the anion nuclei: i) overlap and covalency effects, and ii) conduction-electron polarization. The importance of mechanism i) was demonstrated by Perscheid et al. (8) in the case of UTe.

The polarization of the conduction electrons leads to a finite spin density at the anion site, giving raise to a hf-field via contact interaction. This polarization couples the magnetic ions and in terms of a molecular field theory the paramagnetic Curie temperature Θ_p is proportional to the polarization.

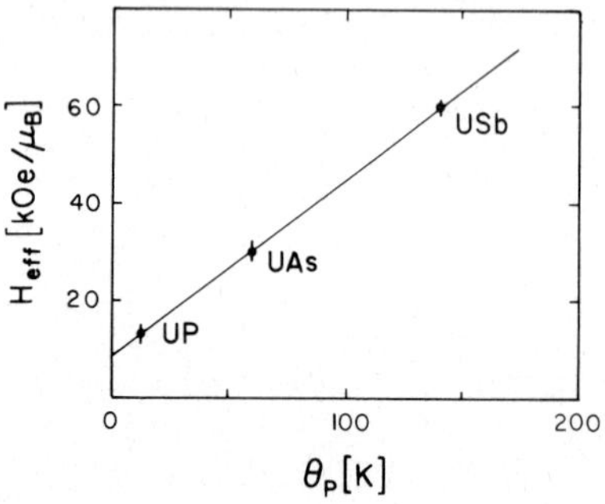

FIGURE 2
Anion hf-fields as a function of the paramagnetic Curie temperature.

In Fig. 2 we have plotted $H_{eff}(X)/\mu_n$ vs Θ_p, showing that $H_{eff}(X)$ is directly related to the strength of the interaction between the magnetic ions.

Whereas for the pnictides the uranium hf-fields increase with increasing moment of the 5f-electrons (Ref. 5) $H_{eff}(U)$ decreases with increasing μ_n in the chalcogenides.

We relate this to the difference in the magnetic moment as deduced from neutron scattering-experiments (μ_M) and bulk magnetization measurements μ_n suggesting antipolarized d-electrons. The decrease of $H_{eff}(U)$ is due to the s-d exchange polarized s-shells which lead to a partial compensation of the h-f-field produced by the 5f-moments. This is shown in Fig. 3.

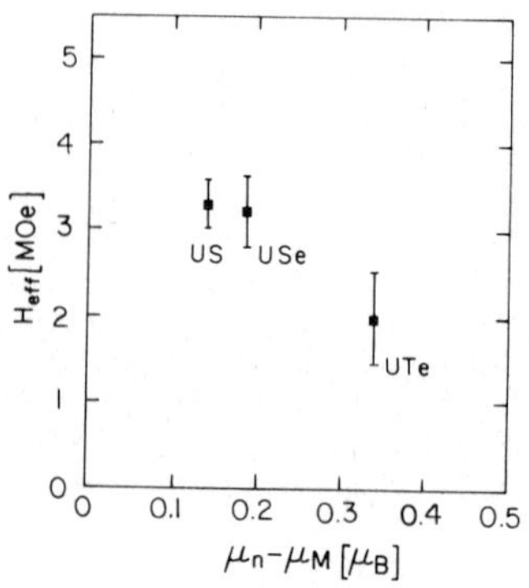

FIGURE 3
Uranium hf-fields as a function of the moment difference μ_n-μ_M as defined in the text.

ACKNOWLEDGEMENTS
The authors are very grateful to Dr. O. Vogt and K. Mattenberger for the generous supply of the samples. Part of this work was supported by the Schweizerische Nationalfonds zur Förderung der wissenschaftlichen Forschung.

REFERENCES

(1) O.V. Lounasmaa, Nuclear Specific Heats in Metals and Alloys, in "Hyperfine Interactions, eds. A.J. Freeman and R.B. Frankel (Academic Press, New York, London 1967) pp. 467-496.
(2) H. Rudigier, Ch. Fierz, H.R. Ott and O. Vogt, Solid State Commun. 47 (1983) 803.
(3) to be published.
(4) G.K. Shenoy, G.M. Kalvius, S.L. Ruby, B.D. Dunlap, M. Kuznietz and F.P. Campos, Intern. J. Magn. 1 (1970) 23.
(5) G.K. Shenoy, M. Kuznietz, B.D. Dunlap and G.M. Kalvius, Phys. Lett. 42A (1972) 61.
(6) C.L. Carr, C. Long, W.G. Moulton and M. Kuznietz, Phys. Rev. Lett. 23 (1969) 786.
(7) G. Longworth, F.A. Wedgewood and M. Kuznietz, J. Phys. C.: Solid State Phys. 6 (1973) 1652.
(8) B. Perscheid and G. Kaindl, Actinides-1981 (Lawrence Berkeley Laboratory, University of California, 1981) 163.

LT-17 (Contributed Papers)
U. Eckern, A. Schmid, W. Weber, H. Wühl (eds)

A STUDY OF THE THERMAL CONDUCTIVITY OF FERROMAGNETIC EuS

C. ARZOUMANIAN, A.M. DE GOËR, B. SALCE and F. HOLTZBERG*

SBT/Laboratoire de Cryophysique, Centre d'Etudes Nucléaires, 85 X, 38041 GRENOBLE Cedex, France.
*I.B.M. Yorktown, Po Box 218, N.Y. 10598, U.S.A.

Thermal conductivity measurements K of an EuS single crystal have been carried out in the temperature range 60 mK to 100 K and in magnetic fields up to 7 Teslas. The observed increase of K in H = 7 T in the whole temperature range rules out the possibility of a pure magnon contribution to the heat transport. The results, in the low temperature range, are tentatively explained by considering the coupled phonon-magnon system.

1. INTRODUCTION

Recent studies of the low temperature thermal conductivity K of $Eu_xSr_{1-x}S$ crystals have shown a magnetic field dependence of K in spin-glass samples : x = 0.44 (1), x = 0.25 (2), and in a frustated ferromagnet : x = 0.54 (2). In all cases, the conductivity was increased by magnetic field. We have interpreted our results as giving evidence of the presence of "magnetic two level systems" which scatter phonons in the spin-glass crystal (1). On the other hand, Lecomte et al (2) have suggested that spin wave-like excitations induced by the applied field could contribute to the heat transport. In order to test the possible importance of such a magnon contribution, we have measured the thermal conductivity of a ferromagnetic EuS single crystal from the same origin as the mixed crystals studied before, in a large range of temperatures (60 mK-100 K) and magnetic fields (0-7 Teslas). EuS has a NaCl structure and is an Heisenberg ferromagnet with $T_C \simeq 16$ K. Previous measurements on this system have been made only above 1.5 K (3) (4). The thermal conductivity was observed to *decrease* with applied field below 4 K, and this result has been interpreted as an evidence of a magnon contribution to the heat transport (3)(4).

2. EXPERIMENTAL

The EuS crystal was approximately parallelepipedic, with a length of 4.9 mm along (100) and a cross-section of ~ 1 mm^2. The quality of the crystal has been tested in two ways. The magnetic susceptibility measured as a function of temperature has given a Curie-Weiss temperature $\theta = 21$ K, and a Curie temperature $T_C = 16.5 \pm 0.5$ K, in good agreement with the values calculated from the exchange constants (5). This is a sensitive test as θ and Tc largely increase in EuS lightly doped with different impurities (Gd, Nd, Cl...) and in non stoechiometric EuS (6). γ diffractometry has shown the existence of small misorientations between parts of the crystal with a maximum value of 40'.

3. RESULTS

The experimental results are plotted in fig.1 as a function of T, with zero field and with H = 7 Teslas. The effect of this *large magnetic field* is to *increase* K in the whole temperature range. The accuracy of measurements with applied fields above 10 K is poor, but the increase of K certainly persists up to Tc and above. Measurements versus field have been carried out at fixed temperatures below 1.5 K. Some of the re-

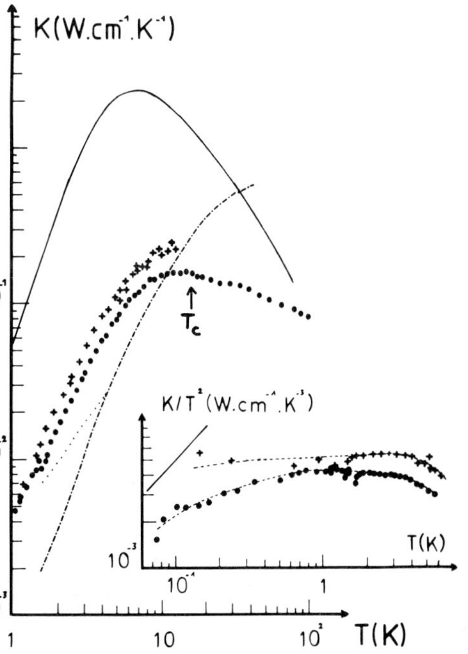

FIGURE 1 - K versus T. This work : •, H = 0 ; +, H = 7 T ; — theoretical curves ; ---- to guide the eye. From ref.(4): ... , H = 0 ; -·-, H = 6.45 T

sults are given in fig.2. A *decrease* of K at *low fields* is observed at T = 0.63 K (and also at 1 and 1.2 K) but is barely detectable at 0.24 K, a temperature where the *relative increase at high fields* is larger. (The behavior at 0.15 K is qualitatively similar to that at 0.24 K). In all cases, the saturation is apparently not achieved at H = 7 Teslas.

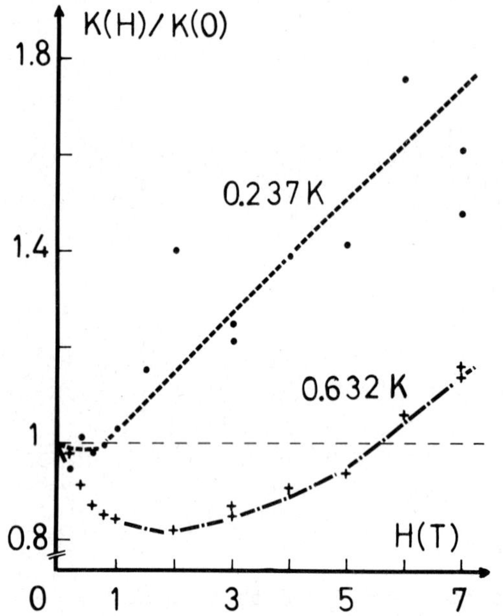

FIGURE 2

4. DISCUSSION

The difference between our results and those of the previous studies (3)(4) is striking. We have reported in fig.1 the results of ref. (4), obtained with a single crystal. In the low temperature range, our absolute values of K are noticeably larger than those of (4), and the effect of a large magnetic field is opposite. Also the magnetic field effect extends to higher temperatures in our case. We have calculated the theoretical phonon conductivity integral in zero field, using a total relaxation rate $\tau^{-1}=V/L + A\omega^4 + B\omega^3T^2$ (the phonon-phonon term $B\omega^3T^2$ has been obtained from the fit of the curve of the dilute $Eu_{0.017}Sr_{0.983}S$ crystal). This is also shown in fig.1 and it is clear that our experimental results are still largely smaller despite the good quality of the crystal. (The addition of a scattering term by dislocations in τ^{-1} does not change noticeably the curve, up to $N_d \simeq 10^8/cm^2$). This fact together with the effect of the magnetic field, gives evidence of phonon scattering processes which are magnetic in origin. On the other hand, the estimated magnon conductivity in zero field (the mean free path being supposed to be limited by the size of the crystal, i.e. $\simeq$ 1 mm) would be very large ($\simeq$ 10 Watt/cm.K at 1.5 K). Therefore an interpretation of the results by independent contributions of phonons and magnons is ruled out, and will need to consider the coupled phonon-magnon system.

Qualitatively, the low temperature results ($T/T_c < \simeq 0.1$) could be explained from the fact that phonon and magnon dispersion curves actually cross at a non-zero energy and small wavevector, as there is an energy gap for the magnon curve. This gap $\hbar\omega_0$ is related to the existence of an internal field of about 0.2 Tesla - the origin of which is not clear - (5) and is therefore of the order of 1 K, that is the energy of the dominant phonon $\hbar\omega_{max}$. at $\simeq$ 0.25 K. The effect of the magnetic field is to increase the gap so that, at 0.25 K, the phonons involved in the resonant coupling are less important and the conductivity increases. At higher temperature, $\hbar\omega_0$ (in zero field) is smaller than $\hbar\omega_{max}$ so that the conductivity firstly decreases with applied field and then increases again, as observed in the range 0.6 - 1.5 K. For temperatures larger than $\simeq$ 5 K, T/Tc is not small, and the situation is more complicated : the magnon energies start to decrease with temperature (5) and the crude description given above probably does not apply. Moreover, critical fluctuations, which have been involved to explain the increase of K with magnetic field around Tc in other magnetic systems (7) could play a role. More experimental work is planned as well as more quantitative analysis of the results.

ACKNOWLEDGEMENTS

We are very grateful to Dr. Chouteau for susceptibility measurements and Mr.Arnaud and Favre for technical assistance in the K(T) measurements.

REFERENCES

(1) C.Arzoumanian et al, J. Phys. Lettres 44 (1983) 2229, and in "International Conference on Phonon Scattering in Condensed Matter" (Stuttgart, 1983) (in print).
(2) G.V.Lecomte et al, J. Magn. Magn. Mater. 38 (1983) 235.
(3) D.C. Mc Collum et al, Phys. Rev. 136 A (1964) 426.
(4) G.V.Lecomte et al, in "International Conference on Phonon Scattering in Condensed Matter" (Stuttgart, 1983) (in print).
(5) H.G.Bohn et al, Phys. Rev. B22 (1980) 5447.
(6) S.Methfessel et al, Magnetic, electric and optical properties of rare earth chalcogenides, in : Colloques Internationaux du CNRS vol.II eds C.N.R.S. (Paris, 1970)pp.565-577.
(7) G.S.Dixon and D.Walton, Phys. Rev. 185 (1969) 735.

LT-17 (Contributed Papers)
U. Eckern, A. Schmid, W. Weber, H. Wühl (eds)

THE DELOCALIZATION OF IMPURITY MAGNETIC EXCITATIONS AT EXTERNAL FIELDS

V.V.EREMENKO, V.M.NAUMENKO and V.V.PISHKO

Institute for Low Temperature Physics and Engineering, UkrSSR Academy of Sciences, Kharkov, USSR

The spectrum rearrangement of a weakly doped crystal caused by the delocalization of impurity excitations is studied on CoF_2 crystals doped with MnF_2, and on $CoCO_3$ doped with $FeCO_3$, $MnCO_3$, $NiCO_3$ for the impurity concentrations of 10^{-5} to 10^{-2}. The radiospectroscopy measurements are carried out in the frequency range of 1 to 40 cm^{-1} at magnetic fields up to 16 T.

For weakly doped crystals with the impurity concentration $C \ll 1$, the impurity excitations are considered to be localized. These excitations however may be delocalized through an indirect (crystal-matrix) interaction between separate impurity centers. The interaction value is dependent on the energy gap between an impurity excitation and a host excitation band edge which can be easily changed by an applied magnetic field.

For the frequency range $\nu = 1 \div 30 cm^{-1}$, the measurements were carried out with a pulse submillimeter spectrometer(I), and for the range $\nu = 22 \div 42 cm^{-1}$ with a diffraction long-wave IR spectrometer (2).

The measurements were carried out on the single crystals of CoF_2 with the MnF_2 impurity concentration of $2 \cdot 10^{-5}$ to 10^{-2} and on $CoCO_3$ with the $FeCO_3$, $MnCO_3$ and $NiCO_3$ impurity concentrations of 10^{-3}, 10^{-4} and 10^{-4}, respectively. The impurity concentration was controlled by the spectrum analysis method so that the error of the impurity concentration estimate was not more than ±30% for $C=10^{-4}$ and ±5% for $C=10^{-2}$. The samples were plane-parallel or slightly tapered plates of 3×3 mm^2. The sample thickness varied between 0.06 and 0.1mm. And the absorption intensity in the region of interaction was 50 to 70%.

The delocalization of impurity magnetic excitations and the related rearrangement of weakly doped crystals spectra were observed in a number of experiments. These effects were studied in detail on a tetragonal antiferromagnet CoF_2 doped with MnF_2 (3). As the frequencies of host and impurity excitations are brought closer together, the localization regions of impurity excitations grow larger and are overlapped producing a collective rearrangement of the crystal spectrum. The single impurity approximation typical of the weakly doped crystal description becomes then invalid because the interaction between separate impurity centers bound through the host excitations should be taken into account (4).

It has been found that in CoF , despite a low impurity concentration (between $2 \cdot 10^{-5}$ and 10^{-2}), there exists the critical concentration $C_{cr} \sim 10^{-4}$ above or below which the behaviour of spectrum rearrangement changes. For $C < C_{cr}$, an incoherent rearrangement of spectrum is realized. The average energy of impurity-impurity interaction does not exceed the dispersion energy due to random distribution of impurity centers in the bulk crystal. The concentration-induced broadening of impurity levels is larger than the dispersion of impurity magnons, and the impurity magnon band is not formally realized. At a varying applied magnetic field the narrow impurity absorption line is intensified, when coming closer to a broad AFMR line, and merges into the latter without a considerable broadening. The impurity excitation does not manifest itself as a resonance in the crystal spin-wave band. For $C > C_{cr}$, there occurs a coherent rearrangement of spectrum. The average energy of interaction between impurity centers is higher than the dispersion energy, the impurity excitations show the coherence in the bulk crystal and form the impurity magnon band. The frequency-field dependences of host and impurity absorption lines are splitted in this case which is followed by the change of their characteristics. As the impurity excitation penetrates the spin-wave band, it no longer manifests itself as a resonance.

In a trigonal $CoCO_3$ with three different impurities ($FeCO_3 \sim 10^{-3}$, $MnCO_3 \sim 10^{-4}$ and $NiCO_3 \sim 10^{-4}$), the delocalization of impurity excitations is also observed; it is followed by a coherent spectrum rearrangement due to the interaction between the impurity excitations and a low-frequency AFMR mode. The frequency-field dependences of host and impurity absorption lines are shown in the figure.

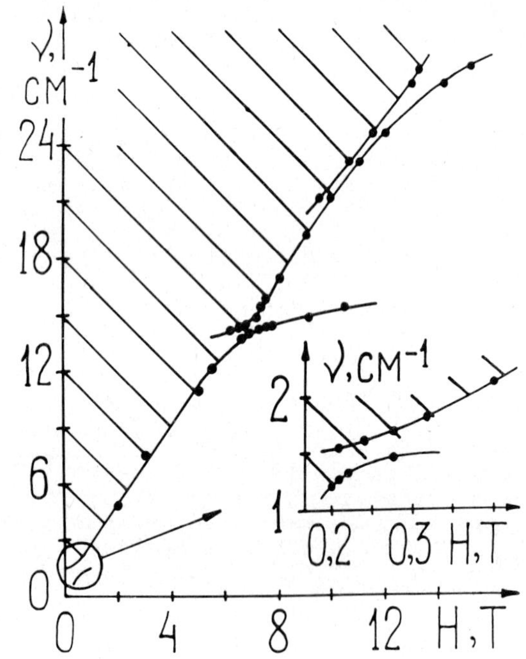

The magnetic field is applied perpendicular to the crystal trigonal axis. The spin-wave band is shaded. The excitation near the frequency 1.5cm^{-1} is due to the $FeCO_3$ impurity. An anomalously low excitation frequency can be accounted for by the orbital degeneration of the ground state of Fe^{2+} ions in the $CoCO_3$ matrix. The Fe^{2+} ions spins are directed along the crystal trigonal axis and are orthogonal to the matrix spins (the orthogonal-type impurity (5)). The excitations of the $MnCO_3$ and $NiCO_3$ impurities are observed near the frequencies 14cm^{-1} and 24cm^{-1}, respectively. The fact that these frequencies are higher than the excitation frequency of the $FeCO_3$ impurity suggests that $MnCO_3$ and $NiCO_3$ are impurities of a collinear type. A weak dependence of the $MnCO_3$ impurity excitation frequency on external magnetic field corresponds to a ferromagnetic exchange interaction of Mn^{2+} ions with the $CoCO_3$-matrix, and a strong dependence for $NiCO_3$ to an antiferromagnetic interaction of Ni^{2+} ions with the $CoCO_3$ matrix (5).

REFERENCES

(1) V.M. Naumenko, V.V. Eremenko and A.V. Klochko, Prib.i Tekhn.Exper. 4 (1981) 159.
(2) V.M. Naumenko, V.I. Fomin and V.V. Eremenko, Prib.i Tekhn.Exper. 5 (1967) 223.
(3) V.V. Eremenko, V.M. Naumenko, S.V. Petrov and V.V. Pishko, ZhETF 82 (1982) vyp.3 813.
(4) M.A. Ivanov, V.M. Loktev and Yu.G. Pogorelov, Fiz.Nizk.Temp. 7 (1981) 1401.
(5) M.A. Ivanov, V.M. Loktev and Yu.G. Pogorelov, Fiz.Tv.Tela 25 (1983) vyp.6 1644

LT-17 (Contributed Papers)
U. Eckern, A. Schmid, W. Weber, H. Wühl (eds)

QUANTUM CROSSOVER AT LOW TEMPERATURE IN THE TRANSVERSE ISING MODEL

K. LUKIERSKA-WALASEK

Institute of Physics, Technical University of Wrocław, Wybrzeże Wyspiańskiego 27, 50-370 Wrocław, Poland

The quantum to classical crossover at low temperature in the transverse Ising model is investigated using the field theoretic renormalization group method. The crossover scaling forms of the inverse correlation lenght, susceptibility and equation of state are given for $d<3$ and $3<d<4$, where d is the spatial dimension.

1. INTRODUCTION

The Ising ferromagnet in a transverse magnetic field Γ is the simplest example of a class of systems for which the critical line $\Gamma_c(T)$ terminates at the multicritical point $[\Gamma_c(0),\ T = 0]$ characterized by new, specifically quantum-mechanical exponents (1), (2). At sufficiently low temperature the region in which the quantum fluctuations are irrelevant becomes small and one expects to see quantum-classical crossover associated with the zero-temperature multicritical point. In this report we present crossover scaling form of the correlation length, susceptibility and equation of state.

The spin-operator Hamiltonian of our system is the following:

$$H = -\tfrac{1}{2}\sum_{ij} I_{ij}\, S_i^z S_j^z - \Gamma\sum_i S_i^x - B\sum_i S_i^z$$

where the spines are located on the sites of d-dimensional simple hypercubic lattice; I_{ij}, Γ and B denote the exchange integral, transverse and longitudinal magnetic fields, respectively.

In order to get the appropriate crossover scaling form for the thermodynamic functions we use the field theoretic renormalization group method discussed in detail in (2).

2. CROSSOVER SCALING FORM OF THE CORRELATION LENGHT AND SUSCEPTIBILITY

We consider the system in the zero longitudinal magnetic field (B = 0) approaching the critical line from above $\Gamma \to \Gamma_c^{+}(T)$. The inverse correlation lenght ξ^{-1} in the crossover region can be represented as follows:

$$\xi^{-1} = h^{\nu_q} X(z) \equiv h^{\nu_{eff}(z)} X(0)$$

where ν_q is the multicritical correlation lenght exponent, $z \sim Th^{-\varphi}$ is crossover scaling variable, $h = (\Gamma - \Gamma_c)/\Gamma_c$, φ denotes the crossover exponent, X(z) is the scaling function having different forms for $d<3$, $d=3$ and $3<d<4$ cf. (2).

The results for the correlation lenght are illustrated by figures 1, 2 and 3, where the plot of the effective correlation lenght exponent ν_{eff} vs $z/(1+z)$ are given for $d<3$, $d=3$ and $3<d<4$, respectively.

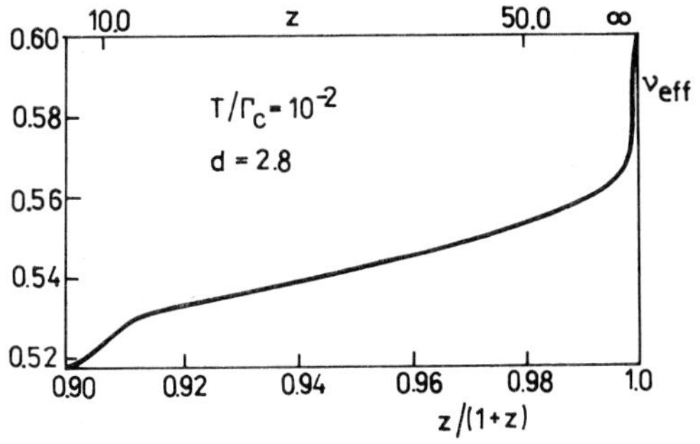

FIGURE 1
Exponent ν_{eff} vs $z/(1+z)$ for $d = 2.8$.

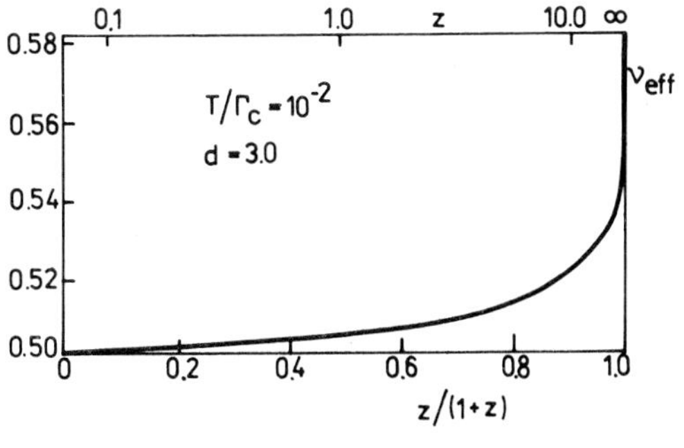

FIGURE 2
Exponent ν_{eff} vs $z(1+z)$ for $d = 3$.

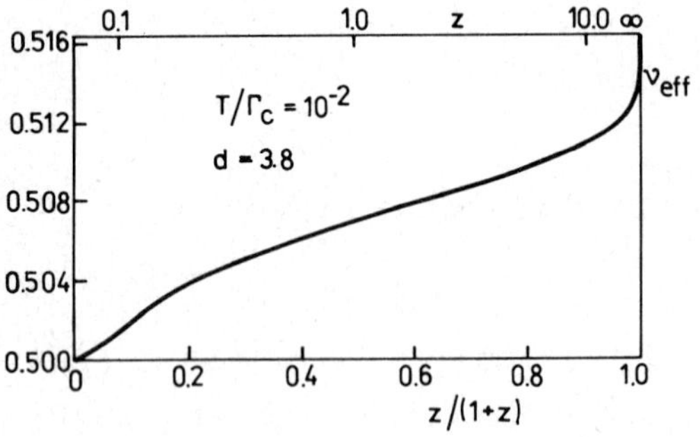

FIGURE 3
Exponent ν_{eff} vs $z(1+z)$ for $d = 3.8$.

The crossover scaling form of the susceptibility χ can be easily obtained taking into account that in our approximation $\chi^{-1} \sim \xi^{-2}$.

3. THE EQUATION OF STATE

The crossover scaling form of the equation of state we calculate for $\Gamma = \Gamma_c(T)$ and $H \to 0$.

It has the following form

$$H = M^{\delta_q}\Psi(z) = M^{\delta_{eff}(z)}\Psi(0) \quad ,$$

where M is the longitudinal magnetization, $z \sim TM^{-\varphi/\beta_q}$, β_q is the multicritical exponent for magnetization, $\Psi(z)$ is scaling function given in explicite form in (3).

The plots illustrating the variation of δ_{eff} with $z/(1+z)$ for $d<3$, $d=3$ and $3<d<4$ cf.(3) are similar respectively to those shown above for ν_{eff} .

4. CONCLUSIONS

Comparing the plots of effective exponents given in figures 1 - 3 it is seen that the range of classical asymptotic behaviour is very small.

This shows that at low temperatures the effect of quantum fluctuations on the critical behaviour is ruther significant.

REFERENCES

(1) I.D. Lawrie, J. Phys. C : Solid State Phys. 11 (1978) 3857.
(2) K. Lukierska-Walasek and K. Walasek, J. Phys. C : Solid State Phys. 16 (1983) 3149.
(3) K. Lukierska-Walasek, Phys. Lett. 98A (1983) 346.

LT-17 (Contributed Papers)
U. Eckern, A. Schmid, W. Weber, H. Wühl (eds)

EFFECT OF CONCENTRATION AND PRESSURE ON THE SPIN-HAMILTONIAN PARAMETERS OF $NiSiF_6 \cdot 6H_2O$

V.P.DYAKONOV, G.G.LEVCHENKO, I.M.FITA

Donetsk Physico-Technical Institute of the Ukrainian Academy of Sciences, 340114 Donetsk, USSR

The effect of pressure and concentration on the magnetic susceptibility of $Ni_xZn_{1-x}SiF_6 \cdot 6H_2O$ is studied under pressure up to 10 kbar, in the temperature range 4.2-0.05K, at Ni-concentration from 0 to 1. It is stated that the effect of concentration on the spin-spin interactions is nonlinear and that at x=0.15 the sign reversal takes place due to exchange couplings variation and nonequiprobable distribution of interacting ions with concentration. The Curie temperature - concentration relationship is derived, the critical concentration value, at which the long-range magnetic order disappears, is estimated. The results are interpreted in terms of the mean-field theory.

1. INTRODUCTION

The magnetodielectric $NiSiF_6 \cdot 6H_2O$ has special place among fluosilicates as its isotropic exchange interaction and single-ion anisotropy parameters are of comparable order of magnetude. Below T_c=0.15 K exchange interactions provide ferromagnetic ordering in this crystal, the magnetic moment is oriented along the trigonal axis. The sign reversal of the initial split D=-0.16K under pressure is a result of the level inversion for the ion basic state Ni^{2+}. The complex nonmonotonous Curie temperature - pressure relationship (in the pressure range up to 10 kbar the derivative dT_c/dP reverses the sign twice) is determined by the value-and-sign corelation of the exchange interaction parameter $T_o=-2/3\sum z_iJ_i$ and single-ion anisotropy D /1/. The control of the latter makes it possible to suitable variate the magnetic properties of crystals with the aid of high hydrostatic pressures and magnetic dilution (substitute diamagnetic in place of magnetic ions). Compounds $Ni_xZn_{1-x}F_6 \cdot 6H_2O$ are convenient for studies in extrema conditions due to their high compressibility and inclination to dilution.

2. EXPERIMENTAL

A convenient and reliable method to determine the exchange parameter T_o under pressure involves estimation of the paramagnetic Curie temperature $\Theta(P)$. Our experimental setup is of original construction with the following advantages: estimation of absolute magnetic susceptibility values in the temperature range 10-0.4 K under multilateral compression up to 10 kbar; replacement of specimens in the cooled state of cryostat. The susceptibility values were taken with the aid of induction technique at 32 Hz, a modulation field amplitude was 0.4-25 Oe. Pressure was produced with the aid of a conventional cylider-piston technique in a high-pressure cell of beryllium bronze(outer and inner diameters 4 and 1,3 mm, respectively). Specimens were monocrystalline, of cylindrical shape. For tests in a magnetoordered phase a ^3He-^4He refrigerator providing temperature 50 mK was used.

3. RESULTS AND COMMENTS

In the molecular-field approximations expressions for susceptibility along C_3 and in the basic plane of zero magnetic field at high temperature are as follows /1/

$$X_{\parallel}=C/(T-T_o-D/3)$$
$$X_{\perp}=C/(T-T_o+D/6) \qquad (1)$$

hence,

$$\Theta_{\parallel}=T_o-D/3 \; ; \quad \Theta_{\perp}=T_o+D/6 \qquad (2)$$

Pressure vs. effective spin-spin interaction, anisotropy, $\Theta_{\parallel}$ and $\Theta_{\perp}$ relationships were determined (see Table 1).

Parameter T_o monotonously grows with pressure. According to EPR data /2/ the effective exchange interaction

Table 1

Pressure vs. Θ, T_o and D for Ni- fluosilicate

P,kbar	Θ ,mK	Θ ,mK	T_o ,K	D, K
0	165	85	0.111	-0.16
2.5	117	194	0.162	+0.15
5.0	71	245	0.186	+0.35
7.0	41	260	0.188	+0.46
9.0	17	275	0.190	+0.52

parameter varies under pressure differently in comparison to $T_o(P)$ variation. It might be due to difference of exchange coupling between ions in pure and diluted Ni-fluosilicate.

Alike concentration vs. temperatures $\Theta_{\parallel}$ and $\Theta_{\perp}$ relationships for $Ni_xZn_{1-x}SiF_6 \cdot 6H_2O$ were derived. They exhibit explicit nonlinear character.

Using relationships $\Theta_{\parallel}(x)$, $\Theta_{\perp}(x)$ and paramagnetic Curie temperature vs. spin-Hamiltonian parameters we derived curves $T_o(x)$ and $D(x)$ (see Fig.1).

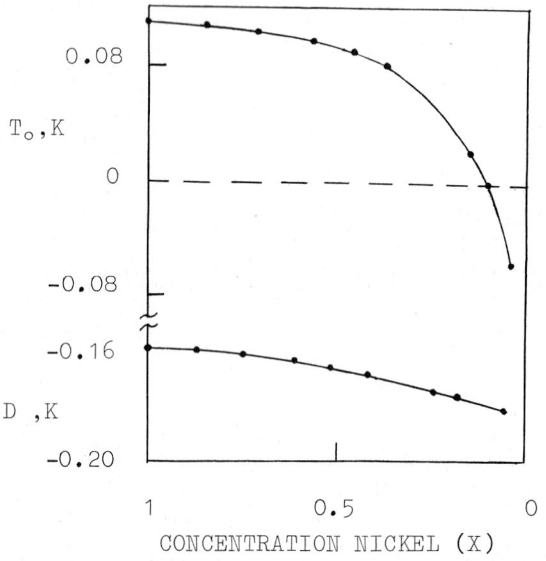

FIGURE 1

Magnetic ion concentration vs. spin-Hamiltonian parameter for $Ni_xZn_{1-x}SiF_6 \cdot 6H_2O$

Ni-concentration decrease results in essentially nonlinear reduction of spin-spin interactions and even sign reversal at x=0.15. Nonlinear character of $T_o(x)$ might be due to superexchange paths variation and contribution of nonequiprobable distribution of magnetic ions over coordinational spheres and contesting interactions of different signs between nearest and next nearest Ni-ions.

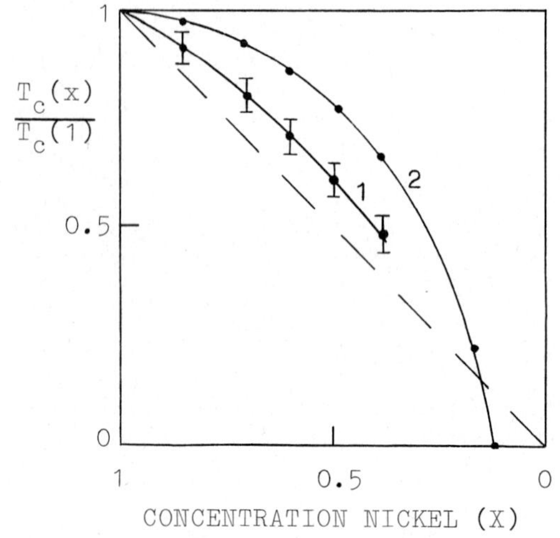

FIGURE 2

Normalized Curie temperature vs. concentration for $Ni_xZn_{1-x}SiF_6 \cdot 6H_2O$

1-experimental, 2-estimation with the aid of data for T_o and D (see Fig.1).

Variation of phase transition temperature with concentration is shown in Fig.2, also calculated curves of $T_c(x)$ values is due to nonlinear pattern of spin-spin interactions at Ni-concentration.

REFERENCES

(1) I.M.Vitebskii, V.P.Dyakonov, G.G.Levchenko, I.M.Fita, G.A.Tsinsadze, Fiz. Tverd. Tela 25(1983) 1546.

(2) A.A.Galkin, A.Yu.Kozhukhar', G.A.Tsintsadze, Fiz. Tverd. Tela 70(1976) 248.

LT-17 (Contributed Papers)
U. Eckern, A. Schmid, W. Weber, H. Wühl (eds)

VERY HIGH FIELD MAGNETIZATION AT LOW TEMPERATURES IN THE SPIN SYSTEM OF THE FREE RADICAL DPPH

Willy BOON and Lieven VAN GERVEN

Laboratorium voor Vaste Stof-Fysika en Magnetisme, Katholieke Universiteit Leuven, Leuven, Belgium

Magnetization measurements of the pure free radical DPPH in high magnetic fields up to $\mu_o H$ = 20 T, in the temperature range 1.8 K to 4.3 K indicate the existence of a linear alternating antiferromagnetic spin chain in the system.

1. INTRODUCTION

The magnetic properties of the free radical 1,1'-diphenyl 2-picryl hydrazyl (DPPH) have been studied for a long time. Proton magnetic resonance measurements by Fujito (1) and by Verlinden *et al.* (2) and the static field susceptibility measurements by Grobet *et al.* (3), (4) have shown that the unpaired electron spins of two molecules of DPPH are associated in pairs. By overlap of both electron spin wave functions in one spin pair one finds a ground state singlet (S = 0) and an excited state triplet (S = 1), separated from the ground state by an energy gap Δ. This model is called the isolated pair model. Using this model, Grobet *et al.* (3), (4) have shown that Δ/k = 17.5 K and that the intrapair interaction energy J is antiferromagnetic.

2. THEORETICAL CONSIDERATIONS

So far all measurements by (1), (2), (3) and (4) were performed in low magnetic fields ($\mu_o H$ < 2 T). An applied external magnetic field splits the degenerate triplet level and can even bring the lowest triplet level below the singlet level. At T = 0 this happens at a certain critical field called the level crossing field H_{lc}: $\mu_o H_{lc} = \Delta/g\mu_B$. In pure DPPH $\mu_o H_{lc}$ is about 13 T. At T = 0 the magnetization is zero for $H < H_{lc}$. For $H > H_{lc}$ the magnetization is saturated and equals 0.5 $gN\mu_B$. In the isolated pair model an exact expression for the magnetization versus the magnetic field exists for all temperatures.

In a real system of radical molecules, the spin pairs are linked into linear chains by a weaker interaction J' called the interpair exchange interaction energy. Such a system can be described by a linear alternating Heisenberg model with alternation parameter $a = J'/J$. The Hamiltonian describing our system may then be represented in the following way:

$$H = -2J\sum_i^{n/2}(\bar{S}_{2i}\cdot\bar{S}_{2i+1} + a\bar{S}_{2i}\cdot\bar{S}_{2i-1}) - g\mu_B\sum_i^n \bar{B}\cdot\bar{S}_i$$

where n is the number of electron spins in a chain.

The eigenvalues of this Hamiltonian have been computed for finite chains (n up to 10 spins) and for different values of a by Duffy *et al.* (5). They also derive some thermodynamical properties for H = 0 at finite temperatures. Magnetization versus magnetic field is only calculated at T = 0. Since no analytic expressions for magnetization at finite temperatures in high magnetic fields are available for linear alternating Heisenberg chains, we decided to write the magnetization M in a first order perturbation calculation assuming $|J'/J| \ll 1$.

$$M = m\, y_1\, (1 + y_2)$$

where $m = 0.5\, N\, g\mu_B$

$$y_1 = (\mathrm{sh}x)(\beta + \mathrm{ch}x)^{-1} \qquad (1)$$

$$y_2 = (J'/kT)(\beta\mathrm{ch}x + 1)(\beta + \mathrm{ch}x)^{-2}$$

$$x = g\mu_B\mu_o H/kT$$

$$\beta = 0.5\,\{1 + \exp(-2J/kT)\}$$

where N is the number of spins per unit volume, H is the magnetic field and g, μ_B, μ_o and k are physical constants. With $J' = 0$ this expression gives exactly the magnetization for the isolated pair model.

3. EXPERIMENTAL SET-UP

The magnetic field is generated by the discharge of a capacitor bank (6500 µF, 5 kV) through a wire wound coil precooled by liquid nitrogen. The magnetic field can be as high as $\mu_o H$ = 30 T with a rise time of 13 ms. A liquid helium dewar, inserted in the magnet, contains the measuring probe. This probe consists of pick-up coils which measure time derivative of magnetic field and magnetization. The sample with nearly spherical shape is mounted in the center of the pick-up coils. The signals from the pick-up coils are stored in an transient recorder. Later on they are displayed on a paper chart recorder. Magnetic field H and magnetization M are obtained by numerical integration.

The temperature of the sample can be changed by pumping off the liquid helium. The temperature ranges from 4.3 K to 1.8 K. Details of experimental set-up are explained in reference (6).

4. SAMPLE

The pure, solvent free DPPH was prepared in our Laboratory (3), (4) and has a very high concentration of unpaired electrons ($C \simeq 0.98$). Since the electron spin-lattice relaxation time is very short ($\simeq 5.10^{-8}$ s) compared to the rise time of the field (13 ms), the magnetization is measured under almost isothermal conditions.

5. EXPERIMENTAL RESULTS

Expressions (1) for the magnetization are fitted by a least squares method to the data of our measurements at different temperatures, using a digital computer. From fittings the quantities J' and J can be determined. A typical series of experimental points, taken at $T = 1.9$ K, is shown in fig. 1. On the vertical axis is the relative magnetization $\sigma = M/m$; the horizontal axis is the magnetic field $\mu_0 H$.

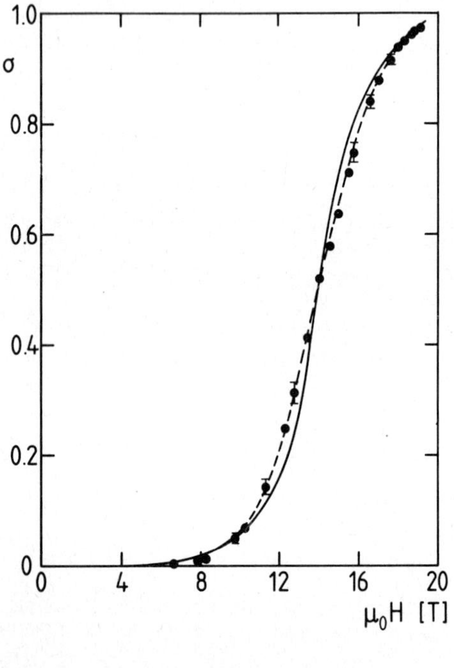

Figure 1

The dashed curve represents the least squares fitting between the data points and the interacting pair model. The full curve represents a similar fitting with the isolated pair model ($J' = 0$). Fittings at all temperatures ($T = 4.24$ K, $T = 2.57$ K, $T = 2.11$ K and $T = 1.9$ K) yield finally the most probable values for J'/k and J/k: the intrapair exchange energy $J/k = (-9 \pm 0.1)$ K and the interpair exchange energy $J'/k = (-1 \pm 0.1)$ K. All interactions are clearly antiferromagnetic and the alternation parameter $a = J'/J = 0.11$.

Brooks Harris (7) gives à theoretical expression for the energy gap in a linear alternating Heisenberg chain provided $|J'/J| << 1$: $\Delta \simeq (2 - a)|J|$. Putting in this expression our values for J and a we find $\Delta/k = 17.0$ K, which is in good agreement with the value of 17.5 K given by Grobet *et al.*, but we disagree with their model and their J-value.

6. CONCLUSIONS

A simple model with intrapair interaction and a weak interpair interaction can describe the magnetization of the free radical DPPH. This model is equivalent to a linear alternating Heisenberg chain. We have shown that both the intrapair interaction and the interpair interaction are antiferromagnetic: $J/k = -9.0$ K; $J'/k = -1.0$ K; $a = 0.11$. Comparison with earlier results shows clearly the necessity of very high fields for a detailed study of magnetic chains.

ACKNOWLEDGEMENTS

The authors acknowledge financial support to this work by the Belgian *Interuniversitair Instituut voor Kernwetenschappen* in the framework of the project "Research on electron-magnetic systems at low temperatures and very high magnetic fields (Malthov)".

REFERENCES

(1) T. Fujito, Chem. Letters, 7 (1972) 557
(2) R. Verlinden, P. Grobet, L. Van Gerven, Chem. Phys. Lett. 27 (1974) 535
(3) P. Grobet, L. Van Gerven, A. Van den Bosch, J. Van Summeren, Physica 86-88B (1977) 1132
(4) P. Grobet, L. Van Gerven, A. Van den Bosch, J. Chem. Phys. 68 (1978) 5225
(5) W. Duffy, K. Barr, Phys. Rev. 165 (1968) 647
(6) W. Boon, L. Van Gerven, submitted to J. Magn. Magn. Mat.
(7) A. Brooks Harris, Phys. Rev. B7 (1973) 3166

LT-17 (Contributed Papers)
U. Eckern, A. Schmid, W. Weber, H. Wühl (eds)

INDIRECT NUCLEAR SPIN-SPIN INTERACTION IN $CUSO_4.5H_2O$

W.G. BOS, T.O. KLAASSEN and N.J. POULIS

Kamerlingh Onnes Laboratorium der Rijksuniversiteit Leiden, Nieuwsteeg 18, 2311 SB Leiden, The Netherlands

The anomalous finestructure and transverse relaxation rate of proton resonance lines in $CuSO_4.5H_2O$ have been studied. The effects are found to be caused by an indirect nuclear spin-spin coupling between the proton spins and copper nuclear spins, via the paramagnetic electron spins.

1. INTRODUCTION

In the unit cell of $CuSO_4.5H_2O$ two crystallographically inequivalent copper ions are present. The ions at the (0,0,0) sites form a 1D antiferromagnetic Heisenberg system, whereas those at (½,½,0) experience only weak exchange interactions and behave as a 3D paramagnetic system. The final aim of this research is the further investigation of the dynamic properties of the 1D antiferromagnetic system, through the study of the proton NMR lineshapes (1,2). However, also the paramagnetic spin system is found to have a profound influence on the proton lineshapes. In order to sort out both effects, first the latter influence has been studied. Part of earlier CW lineshape experiments by Wittekoek et al.(1) has been reproduced and complemented with transient NMR experiments, yielding transverse nuclear relaxation rates T_2^{-1}.

For magnetic fields below 1T and temperatures below 1K, the resonance lines of those proton spins that strongly interact with the paramagnetic system show a peculiar, often asymmetric, finestructure, that depends strongly on magnitude and direction of the applied field and on temperature. Also the transverse relaxation rate shows an anomalous behaviour. We will present in this paper only a (representative) part of our results; a more detailed presentation will be given elsewhere.

2. EXPERIMENTAL RESULTS AND DISCUSSION

The angular dependence of the overall splitting of the finestructure is depicted in fig. 1 for two resonance lines. (H_0=0.3 T, T=0.3 K). It should be noted that for ϕ=35°, i.e. the direction for which the field is parallel to the (electronic) $g_\perp$ - direction of the paramagnetic system, the linesplitting disappear. The results of the T_2^{-1} experiments for one of the lines are given in fig. 2.

2.1 Static effects

We introduce an indirect interaction between the proton spins (I=½) and the copper nuclear spins (J=3/2) of the paramagnetic system as the explanation for the observed phenomena. This indirect nuclear spin-spin interaction can be visualised as follows: The copper nuclear spin has a strong hyperfine interaction with its own electron spin. As a result of this, the electron spin experiences a small, but not negligible internal field $\vec{h}_j$ from the copper nuclear spin. This internal field influences the time averaged magnetisation of the electron spin, both in magnitude and in direction, which in turn influences the total field on the proton spins. Thus, the copper nuclear spin influences the resonance frequency of the proton spins in an indirect way.

The copper nuclear spin (J=3/2) has four distinct eigenstates, leading to four different values for the field $\vec{h}_j$ experienced by the electron spin. Consequently a fourfold splitting of the resonance line would be expected. In fact, in most of the observed proton spectra only twofold splittings have been found. One of the reasons for this is the fact that when the value of the frequency splittings are comparable to the line width of each separate component, the fourfold structure on the resulting resonance line shape is lost. Using the copper nuclear interaction parameters, A_{zz}=-400MHz, A_{xx}=-100MHz, A_{yy}=0 and ν_q=40MHz (2), the fields h_j experienced by the electron spin can be calculated. Assuming that the average lifetime of the copper nuclear spin states is long, the four resonance frequencies of proton i are given by:

$$\nu_{ij} = |\gamma_I \vec{H}_0 + \tfrac{1}{2} p(\vec{H}_0+\vec{h}_j, T) \frac{\vec{H}_0+\vec{h}_j}{|\vec{H}_0+\vec{h}_j|} \cdot \vec{A}_i|$$

where p(H,T) denotes the local electron spin polarization, and $\vec{A}_i$ the proton hyperfine interaction tensor. The so calculated differences between maximum and minimum frequencies $\Delta\nu$ for both lines are shown in fig. 1 as drawn curves. The field directions for which the maximal line splittings occur are well reproduced by the calculations. The calculated values, although of the right order of magnitude, are however always larger than the experimental values. The origin of this discrepancy is the, contrary to the assumption, relatively short lifetime of the copper nuclear spin states, leading to a reduction of the splittings.

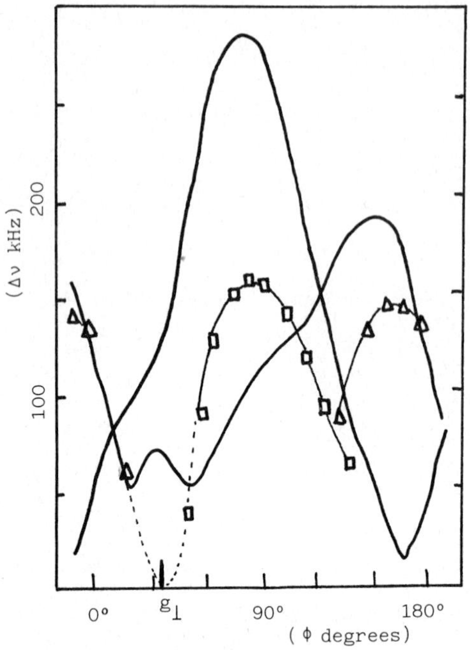

Fig. 1 The angular dependence of the experimental and theoretical value of the line splitting.

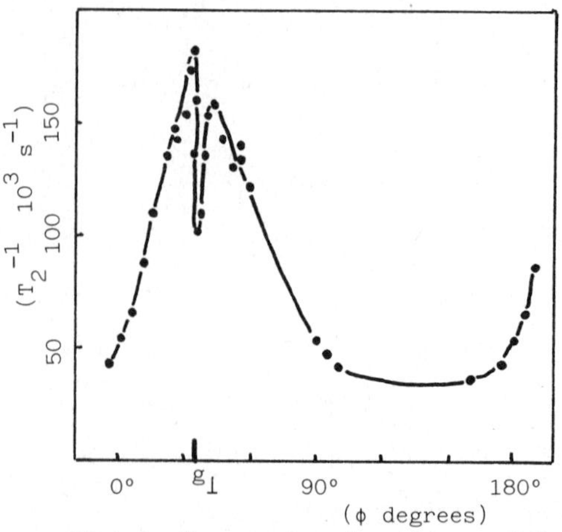

Fig. 2 The angular dependence of the transverse relaxation rate T_2^{-1}.

2.2 Dynamic effects

In principle it is possible to calculate the autocorrelation function of the time dependent field at the proton site, which determines both the CW and transient lineshape (3). An evaluation of the proton lineshape is however far to complex for the qualitative description we want to give here. Hence, for the sake of simplicity we will treat the effects of these complicated time dependent fields on the proton spins as if the effects originate from a fluctuating field with a single effective interaction strength Δ^{eff} and a single fluctuation rate R^{eff}. The influence of the dynamic behaviour of the copper nuclear spin on both the finestructure and T_2^{-1} of the proton spins can be described in a relatively simple way, using Δ^{eff} and R^{eff} (3). The value of R^{eff}, defined as the inverse average lifetime of the four copper nuclear spin states, can be calculated from the known dynamic behaviour of the paramagnetic system (4). The anisotropy of the copper hyperfine interaction is found to lead to a strong angular dependence of R^{eff}:R^{eff} is maximal in the g direction (ϕ=35°) and minimal for ϕ=125°. For $\phi \approx$ 125° the proton spins therefore feel a quasi-static field from the copper spins, leading to a well defined finestructure; the slowly fluctuating field does not contribute much to the transverse relaxation rate of the proton spins. Approaching the $g_\perp$ direction R^{eff} increases, which leads to a smaller line splitting than calculated for the "static" case, due to partial motional narrowing. The contribution of the fluctuating field to T_2^{-1} increases. Very near, and at, the $g_\perp$ direction R^{eff} becomes so large that complete motional narrowing of the finestructure occurs. This results in the disappearence of the line splitting and a decrease of the contribution to T_2^{-1} (note the dip in the T_2^{-1} -ϕ curve at $g_\perp$).

ACKNOWLEDGEMENTS

This work is part of the research program of the "Stichting voor Fundamenteel Onderzoek der Materie (F.O.M.)".

REFERENCES

(1) S. Wittekoek, N.J. Poulis, Physica 32(1966)2051.
(2) J.P. Groen, W. Broeksma, T.O. Klaassen, N.J. Poulis, Physica 113B(1982)1.
(3) M.W.L. Bovy, M. Glasbeek, J. Chem. Phys. 76(1982)1676.
(4) L.S.J.M. Henkens, T.O. Klaassen, N.J. Poulis, Physica 94B(1978)27.

LT-17 (Contributed Papers)
U. Eckern, A. Schmid, W. Weber, H. Wühl (eds)

RELAXATION AT THE SPIN-FLOP TRANSITION OF THE LINEAR-CHAIN ANTIFERROMAGNET $CsMnCl_3.2H_2O$

*M. Chirwa and J. Flokstra

Department of Applied Physics, Twente University of Technology, P.O. Box 217, 7500 AE ENSCHEDE, The Netherlands
*Physics Department, University of Zambia, Zambia.

Low-frequency relaxation behaviour at the spin-flop transition in the linear-chain antiferromagnet $CsMnCl_3.2H_2O$ has been investigated experimentally between 1.4 and 4.2 K. The relaxation rate τ^{-1} shows an exponential temperature dependence, $\tau^{-1} = \omega_0 \exp(-E/kT)$, where the activation energy $E/k = 3.19 \pm 0.04$ K is approximately equal to the magnitude of the intrachain exchange interaction J_a/k.

1. INTRODUCTION

$CsMnCl_3.2H_2O$ is a typical quasi one-dimensional Heisenberg antiferromagnet with an orthorhombic crystallographic structure of space group Pcca. The dominant superexchange interaction between the magnetic Mn^{2+} ions occurs along $Mn^{2+} - Cl^- - Mn^{2+}$ chains in the direction of the crystallographic a-axis. Below $T_N = 4.89$ K its magnetic space group is $P_{2b}c'ca'$ (1). The exchange interactions along the hard (a-), easy (b-) and intermediate (c-) axes of magnetization are $J_a/k \gtrsim -3.2$ K, $J_b \sim 0.01\, J_a$ and $J_c \sim 0.01\, J_a$ respectively (2).

In a magnetic field H applied parallel to the b-axis, the magnetic phase diagram of $CsMnCl_3.2H_2O$ exhibits the disordered paramagnetic (PM) and the ordered antiferromagnetic (AF) and spin-flop (SF) phases, with the bicritical point at $T_b = 4.3541$ K, $H_b = 1.6030 \times 10^6$ Am^{-1} (3). At the first-order spin-flop (AF-SF) phase transition, re-orientation of the magnetic moments from parallel to the b-axis in the low-field AF to parallel to the c-axis in the high-field SF phases takes place. This transition occurs over a narrow field interval in which the AF and SF phases coexist as domains (4).

Our investigation of the spin-flop transition is concerned with the relaxation behaviour of the domain walls between the AF and SF phases. In a previous paper (5) on the isostructural compound $CsMnBr_3.2H_2O$, the spin-flop relaxation was found to exhibit an exponential temperature dependence of the relaxation rate:

$$\tau^{-1} = \omega_0 \exp(-E/kT) \qquad (1)$$

with the value of $E/k \gtrsim 3.6$ K nearly equal to the magnitude of its intrachain exchange interaction constant $J_a/k \gtrsim -3$ K. Here we present the results of relaxation experiments on $CsMnCl_3.2H_2O$ which were performed in order to explore further the relationship between E/k and J_a/k in these one-dimensional antiferromagnets.

2. EXPERIMENTAL

The magnetic field was aligned parallel to the b-axis of a 10.9 mg single crystal placed in direct contact with liquid helium. Parallel dynamic susceptibility, $\chi(\omega) = \chi'(\omega) - i\chi''(\omega)$, measurements were performed between 1.4 and 4.2 K in the frequency range 0.1 Hz - 3.0 kHz with the frequency-sweeping SQUID susceptometer (6). The spin-flop transition was identified with the peak in the temperature- and field-dependences of $\chi(\omega)$ for $T < T_b$ and $H < H_b$. The frequency dependence of $\chi(\omega)$ was measured at constant temperature in a magnetic field with the amplitude $h_p < 10$ Am^{-1} of the oscillatory component less than 0.4% of the half-width ΔH at half height of the $\chi(H)$ peak. A relatively small h_p is necessary because $\chi(H)$ is very sensitive to changes in H near the spin-flop transition peak at $H = H_{SF}$, where $\Delta H \gtrsim 10^{-3} H_{SF}$ for $CsMnCl_3.2H_2O$.

3. RESULTS AND DISCUSSION

At a constant temperature T the relaxation

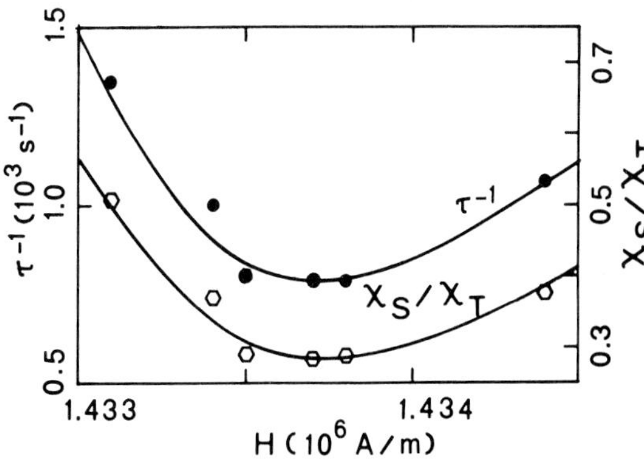

FIGURE 1
Field dependence of τ^{-1} and χ_S/χ_T at T = 1.955 K near $H = H_{SF}$.

rate τ^{-1} at the spin-flop transition depends on the magnetic field. Fig. 1 shows an example of the magnetic field dependence of τ^{-1} and the ratio χ_S/χ_T of the adiabatic to the isothermal susceptibilities in a small field-interval $\sim 0.5\ \Delta H$ around the spin-flop transition field H_{SF} at T = 1.955 K. It is clear from the figure that τ^{-1} and χ_S/χ_T have similar field dependences and in particular τ^{-1} has one value for a fixed χ_S/χ_T value irrespective the applied field within the transition region. For example $\tau^{-1} \sim 950\ s^{-1}$ for $\chi_S/\chi_T = 0.35$ at both $H = 1.4334 \times 10^6\ Am^{-1}$ and $H = 1.4342 \times 10^6\ Am^{-1}$. Using the thermodynamic identity $\chi_S/\chi_T = C_M/C_H$ and the relation $\tau^{-1} = \alpha/C_H$, where α is a constant and C_M and C_H are the magnetic specific heat capacities at constant magnetization and constant field respectively, we obtain the equation:

$$\tau^{-1} = (\alpha/C_M)\ \chi_S/\chi_T. \tag{2}$$

Then a plot of τ^{-1} against χ_S/χ_T at constant T should yield a straight line with slope α/C_M and intercept at origin of co-ordinates, provided α/C_M is independent of H. A linear dependence of τ^{-1} on χ_S/χ_T is indeed found as shown in Fig. 2 at several temperatures. These results show that α/C_M is independent of H at constant T.

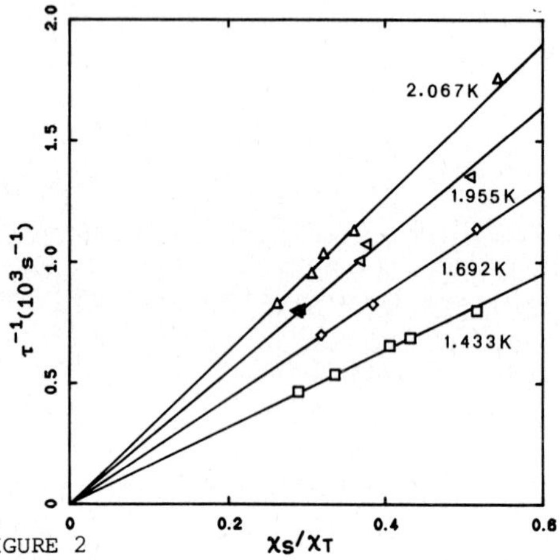

FIGURE 2
Variation of τ^{-1} with χ_S/χ_T at spin-flop transition

In order to determine the temperature dependence of τ^{-1}, we used the above results (eq. (2)) in conjunction with the fact that at the very top of the spin-flop susceptibility peak, χ_S/χ_T was observed to have a temperature-independent value. Thus τ^{-1} was evaluated at each temperature for a given value of χ_S/χ_T and the data was fitted to eq. (1). The results obtained are summarized in Fig. 3 where the activation energy E/k for the relaxation process and the coefficient ω_0 are plotted against χ_S/χ_T. Clearly E/k is independent of χ_S/χ_T and has the value 3.19 ± 0.04 K, approximately equal to the magnitude of

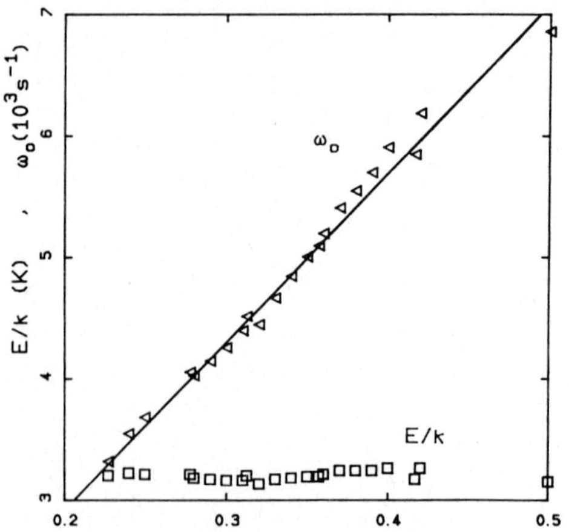

FIGURE 3
E/k and ω_0 as functions of χ_S/χ_T

$J_a/k \approx -3.2$ K for $CsMnCl_3.2H_2O$. This relationship between E/k and J_a/k supports the results found earlier for $CsMnBr_3.2H_2O$ (5). The parameter ω_0 increases linearly with χ_S/χ_T according to $\omega_0 = d.\chi_S/\chi_T$ with $d = 1.41 \times 10^4\ s^{-1}$. By combining eqs. (1) and (2) and re-arranging terms it can be shown that $d = (\alpha/C_M)\exp(E/kT)$.

In this paper we have shown that the activation energy for domain wall motion at the spin-flop transition of $CsMnCl_3.2H_2O$ is of the order of $|J_a/k|$. Further work aimed at giving a theoretical explanation to this behaviour is in progress.

REFERENCES

(1) R.D. Spence, W.J.M. de Jonge and K.V.S. Rama Rao, J. Chem. Phys. 51 (1969) 4694 and J. Skalyo Jr., G. Shirane, S.A. Friedberg and H, Kobayashi, Phys. Rev. B2 (1970) 1310.
(2) K. Kopinga, Phys. Rev. B16 (1977) 427 and I.R. Jahn, J.B. Merkel, H. Ott and J. Herrman, Solid State Comm. 19 (1976) 151.
(3) E. Vélu, R. Mégy and J. Seiden, Phys. Rev. B27 (1983) 4429.
(4) A.R. King and D. Paquette, Phys. Rev. Lett. 30 (1973) 662 and V.P. Novikov, V.V. Eremenko and I.S. Kachur, Sov. Phys. JETP 55 (1982) 327.
(5) M. Chirwa, J. Top and J. Flokstra, Physica 123B (1983) 53.
(6) J.A. Overweg, H.J.M. ter Brake, J. Flokstra and G.J. Gerritsma, J. Phys. E 16 (1983) 1247.

LT-17 (Contributed Papers)
U. Eckern, A. Schmid, W. Weber, H. Wühl (eds)

CRITICAL DYNAMICS BELOW T_c OF THE UNIAXIAL FERROMAGNET $LiTbF_4$

Andreas FROESE, Jürgen KÖTZLER

Institut für Angewandte Physik, Universität Hamburg, 2ooo Hamburg 36, FRG

Measurements of the dynamical susceptibility between 5 Hz and 6oo MHz reveal below T_c=2.871 K two relaxation steps between the frequency-independent plateaus 1/N, χ_1 and χ_2. The domain relaxation rate Γ_d turns out to be by a factor of $1o^3$ smaller than the spin-lattice rate Γ_{sl} and is about the same factor larger than the spin-spin relaxation. χ_1 agrees quantitatively with the adiabatic susceptibility χ_s, calculated from specific heat and magnetization data, but is at variance with a renormalization-group prediction. The lowest plateau, χ_2, most likely corresponds to the isolated (van Vleck) susceptibility, χ_v, since for $T<<T_c$, the data approach the mean-field limit. These are the first unambiguous observations of the critical behaviours of χ_s and χ_v in a ferromagnet.

1. INTRODUCTION

In contrast to the presently well established knowledge on the critical (q=o) magnetization dynamics above the Curie temperature of ferromagnets (see e.g. Ref. 1), the understanding is much less advanced below T_c. Measurements of the dynamical (q=o) susceptibility on uniaxial, low-temperature ferromagnets (2,3) have been analyzed assuming a sequence of two Debye-relaxation processes:

$$\chi(\omega) = \frac{1/N-\chi_1}{1+i\omega/\Gamma_d} + \frac{\chi_1-\chi_2}{1+i\omega/\Gamma_s} + \chi_2 \qquad (1)$$

The rate $\Gamma_d \sim \exp(-E_A/k_BT)$ leading to the maximum value 1/N (N=demagnetization coefficient) has been associated with thermal activated domain wall hopping across the anisotropy barrier E_A (2,3). For $LiTbF_4$, the effect of magnetic field and impurities on Γ_d has been reported recently (4).

Among the other quantities in Equ. 1, only χ_1 could be determined as isothermal susceptibility χ_T for the 3d-ferromagnets $CuRb_2Br_4 \cdot 2H_2O$ and $CrBr_3$ (3), while otherwise χ_1, χ_2 and Γ_s remained unexplained (2). Here we make a new attempt towards a better understanding of the critical behaviour of the order-parameter susceptibility below T_c of uniaxial ferromagnets. We believe $LiTbF_4$ to be a favourable example for such a purpose because its static behaviour belongs to one of the best understood critical phenomena: $LiTbF_4$ represents one of archetypal systems having the attainable upper critical dimension d*=3. Here mean-field (MF) laws enhanced by logarithmic corrections, e.g. in the specific heat (5), correlation length (6), and magnetic equation of state (7) have been detected as a direct proof of Wilson's renormalization-group (RG) equations (8).

2. RESULTS

We have measured the complex susceptibility using a Hartshorn- and Schering-Bridge up to o.2 and 3o MHz, respectively, and by a UHF-susceptometer at the highest frequencies. Fig. 1 shows the results for the dispersion part, which as

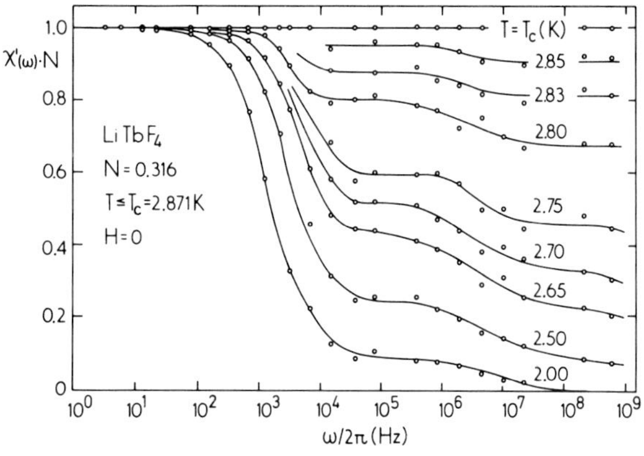

FIGURE 1
Real part of the dynamic susceptibility vs. frequency. Full lines are guides to the eye.

essential feature exhibits the two-plateau structure following from Equ. 1 for $\Gamma_d<<\Gamma_s$. According to Fig. 1, Γ_d varies between $5 \cdot 1o^3$ and $5 \cdot 1o^4$ Hz while the spin-relaxation rate Γ_s increases from 12 MHz to 6o MHz. In order to identify the faster process, we have to explain the measured values of χ_1 and χ_2.

3. DISCUSSION

From Fig. 2 we defer that the internal values of χ_1 and χ_2 can be well parameterized by power laws, $\chi_{1i}=0.10(1)\cdot(-t)^{-1.28(2)}$ and $\chi_{2i}=0.04(1)\cdot(-t)^{-1.41(3)}$. They are significantly smaller than χ_T deduced from magnetization data (7). This is consistent with the observation of a very small spin-lattice rate, $\Gamma_{S1}\simeq 1$ Hz, measured just above T_c (4). Consequently, the magnetization cannot reach thermal equilibrium at frequencies $\omega\simeq\Gamma_d$ and falls below χ_T.

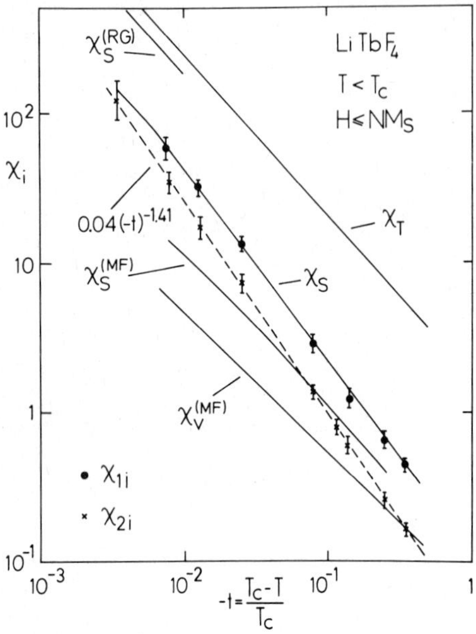

FIGURE 2
Plateau-susceptibilities corrected for demagnetization, $\chi_i=(\chi^{-1}-N)^{-1}$ compared to various calculations explained in the text.

In a concentrated magnet one expects the spin-spin relaxation rate, Γ_{SS}, to be very fast: using an estimate by Finger (9) for $T\gtrsim T_c$ we find Γ_{SS}=50 MHz, which is just in the order of our result for Γ_S. Thus a comparison of χ_1 with the adiabatic susceptibility

$$\chi_S = \chi_T - \frac{T}{c_H}\cdot\mu_0\left(\frac{\delta M}{\delta T}\right)_H^2 \qquad (2)$$

is indicated. Inserting c_H (5) and $(\delta M/\delta T)_H = dM_S/dT$ with $M_S/M = 1.9\cdot(-t)^{0.41}$ for the normalized spontaneous magnetization (7) we obtain excellent agreement with χ_1. On the other hand, our results are clearly at variance with two theoretical predictions: (i) $\chi_S=(5/6)\cdot\chi_T\cdot|\ln(-t)|^{-1}$ based on phenomenological arguments (10) and (ii) $\chi_S=(3/4)\cdot\chi_T$ from RG-considerations (11). Also shown in Fig. 2 is a MF-calculation starting from

$$\chi_S^{-1} = \chi_T^{-1} + \frac{T}{c_M}\cdot\mu_0\left(\frac{\delta H}{\delta T}\right)_M^2 \qquad (3)$$

c_M has been approximated by its zero-field value (12) noting that this quantity will not be very sensitive to magnetic field (13). Apparently at $T\ll T_c$ χ_S approaches the data fairly well, thus demonstrating the strong effect of the critical fluctuations near T_c.

Similar is true for the van-Vleck susceptibility, χ_V, arising from the zero-field splitting of the Tb^{3+} groundstate-doublet, 2Δ=1.4 K (14), which in MF-approximation reads:

$$\chi_V^{(MF)} = \frac{\Delta}{g}\cdot\left(\frac{\Delta}{g}\right)^2\left(\frac{M_0}{M_S}\right)^2 \qquad (4)$$

with λ= Curie constant, $g=\Delta/\tanh(\Delta/T_c)$ = MF constant. To our knowledge, the behaviour of χ_V near T_c has not yet been investigated and it would be interesting whether a theory, taking the critical fluctuations into account, would reproduce the observed, possibly effective critical exponent γ=1.41 .

REFERENCES

(1) J. Kötzler, J.M.M.M. 15-18 (1980) 393 and Phys. Rev. Lett. 51 (1983) 833.
(2) P.M. Richards, Phys. Rev. 187 (1969) 690; A. Schlachetzki and J. Eckert, Phys. Stat. Sol. (a) 11 (1972) 611; P.H. Müller et al., Phys. Stat. Sol. (b) 119 (1983) 239.
(3) W.C.L. Rutten and J.C. Verstelle, Physica 86-88 B (1977) 564; J. Kötzler and W. Scheithe, Phys. Rev. B 18 (1978) 1306.
(4) J. Kötzler and D. Sellmann, subm. to J.M.M.M.
(5) G. Ahlers et al., Phys. Rev. Lett. 34 (1975) 1227.
(6) J. Als-Nielsen, Phys. Rev. Lett. 37 (1976) 1161.
(7) R. Frowein et al., Phys. Rev. Lett. 42 (1979) 739 and Phys. Rev. B 25 (1982) 3292.
(8) see e.g. K.G. Wilson and J. Kogut, Phys. Rep. C 12 (1974) 75.
(9) W. Finger, Phys. Lett. 60 A (1977) 165.
(10) D. Stauffer, Ferroelectrics 18 (1978) 199.
(11) A.D. Bruce, Phys. Rev. Lett. 44 (1980) 1.
(12) L.M. Holmes et al., Phys. Lett. 50 A (1974) 163.
(13) A.T. Skjeltorp and W.P. Wolf, Phys. Rev. B 8 (1973) 215.
(14) J. Magariño, Physica 86-88 B (1977) 1233.

LT-17 (Contributed Papers)
U. Eckern, A. Schmid, W. Weber, H. Wühl (eds)

PRESSURE AND MAGNETIC FIELD EFFECTS ON PHASE TRANSITIONS IN HEISENBERG FERROMAGNETS

B.R. GERMAN, V.P. DJAKONOV, V.I.MARKOVICH

Donetsk Physico-Technical Institute of the Ukrainian Academy of Sciences, Donetsk, USSR

The results of the investigation of uniform compression and magnetic field effects on phase transitions in Heisenberg ferromagnets $CuRb_2Br_4 \cdot 2H_2O$ (T_c=1.87K) and $CuRb_2Cl_4 \cdot 2H_2O$ (T_c=1.02K) are given.Magnetic susceptibility and magnetization measurements were carried out using modulation method on single crystals at pressure up to 10 kbar in the temperature range 4.2 - 0.4 K and the fields up to 500 Oe. For $CuRb_2Br_4 \cdot 2H_2O$ the baric coefficient of T_c shift is equal to dT_c/dP= +0.05 deg.kbar^{-1}. One observes nonlinear T_c increase for $CuRb_2Cl_4 \cdot 2H_2O$. The magnetic phasediagram is built. In a field applied along an easy magnetization direction there is a phase transition from a multidomain to a singlemagnetized phase but in a perpendicular field - from an angular to the collinear one. The results agree with predictions of the phase transition theory.

1.INTRODUCTION

The Heisenberg theoretical model which is a base for an investigation of magnetic collective phenomena in dielectric crystals with localized spins and isotropic exchange interactions got an experimental confirmation after discovery and study of a number of magnetic dielectrics: CrBr, EuO, EuS, $CuM_2X_4 \cdot 2H_2O$, where M- NH_4, Rb, Cs, K; X-Cl, Br. In this group of crystals of divalent copper compositions with a spin S=1/2 take a special place.

In a homological set of copper complex compounds $CuM_2X_4 \cdot 2H_2O$ the salt $CuRb_2Br_4 \cdot 2H_2O$ has the largest lattice parameters (c=8.435A, a=7.985A) and the highest Curie temperature (T_c= 1.87K). $CuRb_2Cl_4 \cdot 2H_2O$ has magnetic ordering temperature T_c=1.02 K and lattice sizes c=8.042A and a=7.609A. Magnetic dielectrics $CuRb_2Br_4 \cdot 2H_2O$ and $CuRb_2Cl_4 \cdot 2H_2O$ have rather high comcompressibility, this allows to study pressure effect on their magnetic properties.

2. EXPERIMENTAL

Magnetic susceptibility measurements were carried out with a low-frequency differential magnetometer up to 10 kbar in the temperature range 4.2 - 0.4 k and the fields up to 500 Oe.

Amplitude of a variable magnetic field at 30 Hz does not exceed 0.5 Oe. A measuring coil being interior of a high pressure vessel consists of three coaxial sections winded in an opposite direction. Single crystal samples having a cylinder form (diameter-1mm, lengh-5mm) oriented along and perpendicular to an easy magnetization direction which for $CuRb_2Br_4 \cdot 2H_2O$ coinsides with an axis C_4 and for $CuRb_2Cl_4 \cdot 2H_2O$ is in a basic plane (direction 110 and 1$\bar{1}$0) were studied.

3.RESULTS AND COMMENTS

3.1. The compound $CuRb_2Br_4 \cdot 2H_2O$.

The results of susceptibility and magnetization measurements show that an easy axis character of $CuRb_2Br_4 \cdot 2H_2O$ is preserved under pressure. With pressure increasing the curves X(T,P) are shifted to the range of higher temperatures. Figure 1 shows that Curie temperature increases linearly with pressure increasing (dT_c/dP = +0.05 deg.kbar^{-1}).

The results of magnetization measurements of $CuRb_2Br_4 \cdot 2H_2O$ under uniform compression show that the saturation magnetization under pressure is reached in higher magnetic fields. The anisotropy field is equal to 290 Oe at T=0.37K and P=0 and is increased up to 370 Oe under pressure of 9.6 kbar.

3.2. The compound $CuRb_2Cl_4 \cdot 2H_2O$.

The substitution of a diamagnetic anion Br for Cl doesn't only decreases the critical temperature by a factor of 1.8, but changes an orientation of spin moments in a magnetoordered state from an "easy" axis direction to the "easy" plane. In ferromagnetic $CuRb_2Cl_4 \cdot 2H_2O$

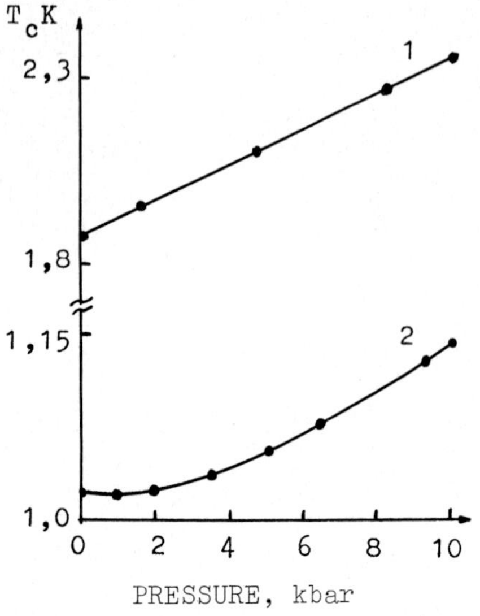

FIGURE 1

Dependence of Curie temperature T_c on pressure: 1- $CuRb_2Br_4 \cdot 2H_2O$, 2- $CuRb_2Cl_4 \cdot 2H_2O$.

one discovered nonlinear behaviour $T_c(P)$ in the range of low pressures (Fig.1). At pressure from 5 to 10 kbar the Curie temperature changes linearly, the baric coefficient $dT_c/dP=+0.012$ deg.kbar^{-1}.

The character of an isotherm magnetization change shows that the spontaneous magnetization $CuRb_2Cl_4 \cdot 2H_2O$ decreases under pressure and the anisotropy field at P=9.5 kbar, T=0.4K is equal to 125 Oe in comparison with 95 Oe at P=0.

An analysis carried out an approximation of a molecular field allows to suppose that the observed character of Curie temperature change is due to the change of the parameter tetragonality of local surrounding of Cu^{2+} ion leading to the change of an exchange bond paths.

It is established that the critical behaviour of investigated ferromagnets near T_c under high pressures satisfies to the scaling theory.

3.3. Phase transitions in a magnetic field.

In contrast to the usual susceptibility curve in a zero field singularities in the form of a distinct maximum in a paramagnetic phase at T_h and a diffuse maximum bn an ordered phase at T_H are exhibited when a constant magnetic field is superimposed. Peruliarities related to the effect of the magnetization process are observed both in an easy and hard magnetization directions. The form of the curves is conserved under pressure. With the field increasing T_h is increased and T_H is decreased.

Figure 2 shows H-T phase diagram for $CuRb_2Cl_4 \cdot 2H_2O$.

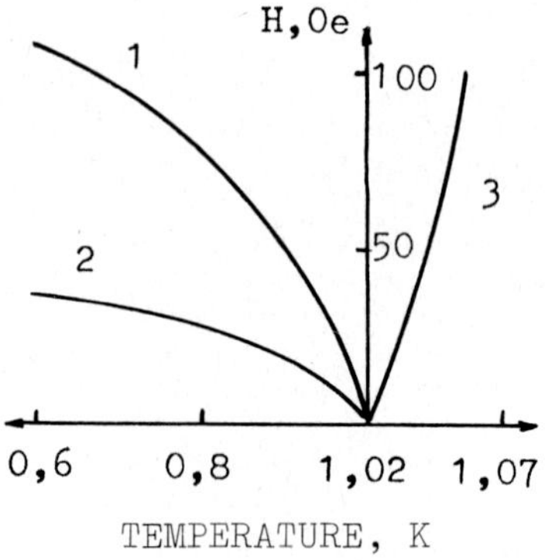

FIGURE 2

H-T phase diagram of $CuRb_2Cl_4 \cdot 2H_2O$.

The line 1 (the field H II C_4) divides angular and colinear phases and is determined by an anisotropy value. The line 2 is a phase boundary of multidomain and homogeneously magnetized phases (in a field which is parallel to an easy magnetization direction). The dependence $T_H(H)$ is approximated by the power function: $1-T_H/T_c=AH^w$. For $CuRb_2Cl_4 \cdot 2H_2O$ w=2.7, for $CuRb_2Br_4 \cdot 2H_2O$ w=2.5. The line 3 corresponds to a change of the temperature of the maximum X(T,H) in a paramagnetic phase.

The power dependence of T_h on the field: $T_h-T_c = BH^p$ with the exponent p= 0.7 for $CuRb_2Br_4 \cdot 2H_2O$ and p= 0.8 for $CuRb_2Cl_4 \cdot 2H_2O$.

The experimental results are consistent with the theories of phase transitions in magnetic field.

LT-17 (Contributed Papers)
U. Eckern, A. Schmid, W. Weber, H. Wühl (eds)

MAGNETIC ORDERING IN $Dy(BrO_3)_3 \cdot 9H_2O$ BELOW 170 mK*

Satoru SIMIZU, G. H. BELLESIS, and S. A. FRIEDBERG

Physics Department, Carnegie-Mellon University, Pittsburgh, PA 15213, U.S.A.

We have measured the magnetic susceptibilities of hexagonal crystals of $Dy(BrO_3)_3 \cdot 9H_2O$, DyBR, parallel and perpendicular to c-axis between ∿ 0.06 and 4.2 K. The data indicate the onset of dipolar ferromagnetism below $T_c \cong 0.17$ K with the c-direction an easy axis. Above ∿ 1K, $\chi_{||}$ and $\chi_{\perp}$ obey Curie-Weiss laws with $g_{||}$ ∿ 8.5 and $g_{\perp}$ ∿ 12.1. This is in striking contrast to the behavior of the structurally similar salt $Dy(C_2H_5SO_4)_3 \cdot 9H_2O$, DyES, for which $g_{||}$ ∿ 10.8 and $g_{\perp} = 0$. We believe that the two lowest Kramers doublets in DyBR are inverted with respect to those in DyES and much closer together.

1. INTRODUCTION

We have been studying (1) the low temperature magnetic behavior of a number of rare earth salts having the formula $R(BrO_3)_3 \cdot 9H_2O$ or RBR where R = Pr, Tb, Dy, Er and Tm. All of these compounds are now believed to have a hexagonal structure (2,3) belonging to the space group $P6_3/mmc\text{-}D_{6h}^4$. The metal ions are magnetically equivalent and arranged in essentially the same way as in the rare earth ethylsulfates [$R(C_2H_5SO_4)_3 \cdot 9H_2O$ or RES] (4,5) except that for the RBR the unit cell volume is about 30% smaller while the c/a ratio is about 14% larger. The symmetry of the nine-water coordination structure about the R^{3+} ion in RBR is D_{3h} with six H_2O's at the corners of a trigonal prism and three H_2O's out from the centers of the prism faces in the equatorial plane. In RES the equatorial triangle of H_2O's is rotated by 5° from the symmetric position reducing the site symmetry to C_{3h}. However, the crystal potential perturbing the free R^{3+} ion ground state has the same number and types of terms in RBR and RES. Since the dimensions of the polyhedron of H_2O ligands differ by only a few percent between corresponding compounds of the same rare earth, one might expect level splittings in the RBR qualitatively similar but perhaps quantitatively different from those in the RES. Magnetic interionic coupling in RBR is expected to be predominantly or perhaps purely dipolar as it is in RES.

In a recent paper (1), we have reported the first results of AC and DC magnetic susceptibility measurements between ∿ 0.06 and 4.2 K for the RBR salts listed above. The data for the Tb, Tm, and Er bromates indicate that the ground states are similar in character to those known to occur in the corresponding RES and that low-lying excited states are at comparable energies. In the Dy and Pr salts, however, the situation is strikingly different. In the present paper, we report more complete susceptibility data on DyBR and their interpretation.

2. EXPERIMENTS

The DyBR compound was obtained by adding hot barium bromate solution to that of dysprosium sulfate. After insoluble barium sulfate was filtered out, crystals of DyBR were obtained by slow evaporation at room temperature. Very often only a glassy material was obtained due to rapid nucleation and several recrystallization processes were necessary to get single crystals with hexagonal prism shape.

A SQUID magnetometer was used to measure DC susceptibilities in a He^3-He^4 dilution refrigerator. The susceptibility along the c-axis was measured with a thin needle-shaped crystal. For the perpendicular direction, a crystal shaped into a sphere of about 1.6 mm in diameter was employed. AC susceptibilities were also measured between 0.5 K and 4.2 K at 80 HZ to calibrate the SQUID magnetometer.

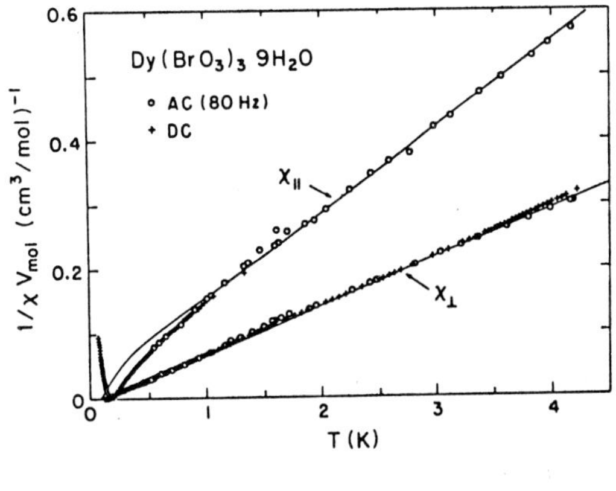

FIGURE 1
Inverse susceptibilities of $Dy(BrO_3)_3 \cdot 9H_2O$.

*Work supported by NSF Grant DMR 81-06491

3. RESULTS AND DISCUSSION

The inverse susceptibilities of DyBR are shown in Fig. 1. The data are corrected to infinite needle geometry. These results differ sharply from those for DyES (6). $\chi_\perp$ of DyBR follows a Curie-Weiss law with $g_\perp \sim 12.1$, while DyES shows a small temperature-independent susceptibility below 4 K. $\chi_\parallel$ of DyBR can be approximated by a Curie-Weiss law with $g_\parallel \sim 8.5$, but this is much smaller than $g_\parallel \sim 10.8$ for DyES. These results apparently contradict our preliminary assumption of very similar crystal field splittings in the two compounds. One notes, however, that in DyES, the first excited doublet, $\cos\theta|\pm7/2\rangle - \sin\theta|\mp5/2\rangle$, with $g_\parallel \sim 5.4$ and $g_\perp \sim 8.5$ is separated in energy, from the ground doublet $\cos\varepsilon \cos\eta|\pm9/2\rangle - \cos\varepsilon \sin\eta|\mp3/2\rangle + \sin\varepsilon|\mp15/2\rangle$, for which $g_\parallel = 10.8$ and $g_\perp = 0$, by only 23 k. (6,7) We find that the order of the two doublets can be interchanged by a small change in crystalline field parameters, without substantially altering the g-values. The exceptionally high value of $g_\perp$ in DyBR can be explained by a second-order Zeeman contribution if the two doublets are close.

The molar susceptibilities were calculated for the two lowest doublets, neglecting interactions. They are

$$\chi_\parallel^\circ V_{mol} = \frac{N\mu_B^2}{4kT}\,\frac{g_{1\parallel}^2 + g_{2\parallel}^2 \exp(-\Delta/kT)}{1 + \exp(-\Delta kT)}$$

$$\chi_\perp^\circ V_{mol} = \frac{N\mu_B^2}{4kT}\,\frac{g_{1\perp}^2 + 8g_J^2\alpha^2(1-\exp(-\Delta/kT))kT/\Delta}{1 + \exp(-\Delta/kT)} + c,$$

where the subscripts 1 and 2 refer to the ground and excited doublets, respectively, and $\Delta(>0)$ is their energy separation; g_J is the Landé g-factor and α is the matrix element of J_x between the two doublets. We added a temperature-independent term c to $\chi_\perp$, because the contribution of higher levels, due to the second order perturbation effect, is estimated to be rather large ($\sim$3% at 4 K). The molecular field formula,

$$1/\chi = 1/\chi^\circ + [4\pi D - (p + 4\pi/3)]$$

was then employed, assuming purely dipolar interactions. In the above formula, D is a demagnetizing factor and p is a direct lattice sum of the dipolar field inside a virtual sphere. (6) It is calculated to be 2.64 and -1.32 for the $\parallel$ and $\perp$ directions, respectively, in DyBR. As shown in Fig. 1, a good fit with the data was obtained with $g_{1\parallel} = 5.80$, $g_{1\perp} = 8.72$, $g_{2\parallel} = 11.12$, $\alpha = 3.54$, $\Delta/k = 1$ K and $c = 0.20$ cm^3/mole (solid lines). To obtain the best fit of the data, it proved necessary to relax constraints on the g-values and α slightly by multiplying the best set of theoretical values by 1.04. This may be due to the neglect of higher excited states as well as small systematic errors in calibration of the susceptibility data. Although Δ/k and c were determined as adjustable parameters, the values obtained appear to be consistent with crystal field calculations within experimental error. We hope that heat capacity and EPR measurements now in progress will give us more concrete evidence for the scheme of low-lying levels inferred from the above analysis.

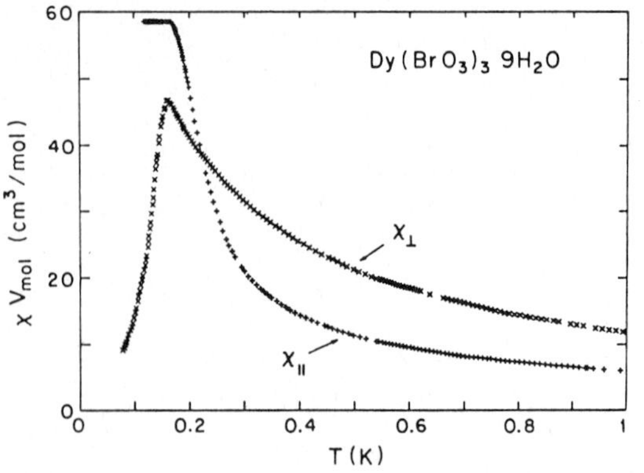

FIGURE 2
DC susceptibilities of $Dy(BrO_3)_3 \cdot 9H_2O$.

The DC susceptibilities below 1 K are shown in Fig. 2. The $\chi_\perp$ data are for a spherical sample while the $\chi_\parallel$ results are corrected to that shape. $\chi_\parallel$ exhibits a kink at Tc $\sim$ 0.17 K below which it assumes a nearly constant value determined by the sample shape. Such behavior is typical of an easy axis ferromagnet in the absence of significant remanence. On the other hand, $\chi_\perp$ reaches a maximum at 0.165 K and then falls sharply with temperature. When the ground doublet is split due to spontaneous ordering along the c-axis, it can be shown in a molecular field approximation that the perpendicular field has no effect on it regardless of the magnitude of $g_\perp$. Therefore the only contribution to the susceptibility comes from the second order perturbation energy due to excited doublets. $\chi_\perp \times V_{mol}$ at 0 K is given by $2N(g_J\mu_B\alpha)^2/\Delta$ and this amounts to 13 cm^3/mole after demagnetizing field correction. Although this exceeds the lowest value observed, the discrepancy is not very serious given the uncertainty in the constants α and Δ.

REFERENCES

(1) S. Simizu, G. H. Bellesis, and S. A. Friedberg, J. Appl. Phys. 55 (1984) 2333.
(2) S. K. Sikka, Acta Cryst. A25 (1969) 621.
(3) J. Albertsson and I. Elding, Acta Cryst. B33 (1977) 1460.
(4) J. A. Ketelaar, Physica 4(1937) 619.
(5) D. R. Fitzwater and R. E. Rundle, Z. Krist. 112 (1959) 362.
(6) A. H. Cooke, et al. Proc. Roy Soc. (London) A306 (1968) 313 and earlier papers cited therein.
(7) S. Hüfner, Zeits. Physik 169 (1962) 417.

LT-17 (Contributed Papers)
U. Eckern, A. Schmid, W. Weber, H. Wühl (eds)

NMR OF ORIENTED ^{54}Mn NUCLEI IN ANTIFERROMAGNETIC $MnBr_2.4H_2O$.*

B.G. TURRELL

Department of Physics, University of British Columbia, Vancouver, B.C., V6T 2A6, Canada.

NMRON lines at 499.8 MHz and 501.6 MHz have been observed for ^{54}Mn doped into $MnBr_2.4H_2O$. A value for the hyperfine interaction strength A/h = -202.0 ± 0.1 MHz is deduced. The 1.8 MHz splitting of the lines is due to second order magnetic "pseudoquadrupole" interaction (1.4 MHz) and pure quadrupole interaction (0.4 MHz).

1. INTRODUCTION

The first observation of nuclear magnetic resonance of oriented nuclei (NMRON) in an antiferromagnet was reported by Kotlicki and Turrell (1) who studied ^{54}Mn-$MnC\ell_2.4H_2O$. Pursuing a program to investigate various manganese compounds, we have initiated an investigation of ^{54}Mn doped into $MnBr_2.4H_2O$. This salt was chosen because, although the magnetic and crystallographic structures are the same as the chloride, the molecular fields are quite different (2). In $MnC\ell_2.4H_2O$, the exchange and anisotropy fields are B_E = 1.04 T and B_A = 0.22 T respectively, whereas in $MnBr_2.4H_2O$ the exchange field is B_E = 1.26 T and the anisotropy field is much stronger being B_A = 0.87 T. The Néel temperature of the bromide salt is 2.12 K.

2. EXPERIMENT AND RESULTS

A single crystal was grown from a solution containing ^{54}Mn ions and a specimen containing an activity ~ 5μCi was obtained. This was attached with a paste of Apiezon N grease and conducting paint to a copper fin connected to the mixing chamber of a dilution refrigerator. A Ge-Li detector was mounted to measure the intensity of γ-rays emitted along the crystalline c-axis which is close to the easy axis of magnetization. The crystal was cooled to a temperature $\lesssim$ 100 mK and the γ-ray intensity monitored while the frequency of an RF field applied to the sample was swept in the vicinity of 500 MHz. A section of a run in which the frequency was stepped in approximately 0.11 MHz intervals with a modulation of 0.12 MHz through the range 495-505 MHz is shown in Figure 1 and resonances can be seen at 499.8 ± 0.1 MHz and 501.6 ± 0.1 MHz.

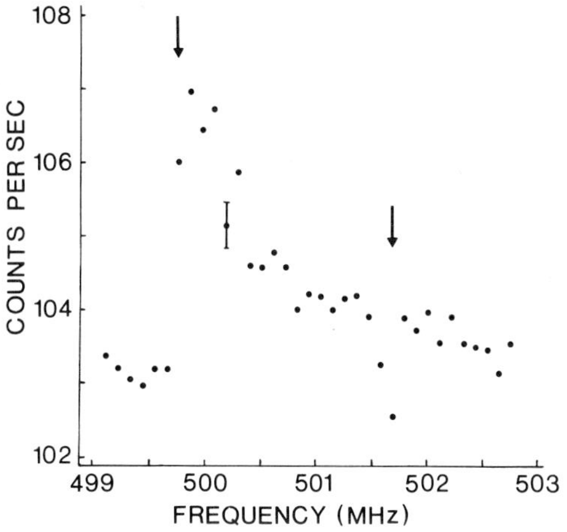

FIGURE 1
An experimental run in which the RF frequency is swept in 0.11 MHz steps with 0.12 MHz modulation. Resonances can be seen at 499.8 MHz and 501.6 MHz.

3. DISCUSSION

The spin Hamiltonian for the ^{54}Mn ion in $MnBr_2.4H_2O$ can be written

$$H = -g\mu_B B_E S_z - \frac{2}{5} g\mu_B B_A S_z^2 + A\vec{I}.\vec{S} + P[I_z^2 - \frac{1}{3}I(I+1)] \qquad (1)$$

where biaxial terms in S_x^2, S_y^2 have been ignored. Second order perturbation theory then yields for the strongest transition

* Work supported by grants from the Natural Sciences and Engineering Research Council of Canada.

($S_z = -\frac{5}{2}$, M = -3 → $-\frac{5}{2}$,-2) and the next-strongest transition ($-\frac{5}{2}$,- 2 → $-\frac{5}{2}$,-1) respectively

$$h\nu_{-3\,-2} = -\frac{5}{2}A - 3P' - 5P \qquad (2a)$$

$$h\nu_{-2\,-1} = -\frac{5}{2}A - 2P' - 3P \qquad (2b)$$

where

$$P' = \frac{5}{2}A^2[g\mu_B(B_E+1.6B_A)]^{-1} \qquad (3)$$

The term in P' results from a second order magnetic "pseudoquadrupole" interaction, whereas the term in P results from the pure quadrupole interaction.

For ^{54}Mn nuclei with spin I=3 oriented dominantly by an $\vec{A}\vec{I}.\vec{S}$ interaction, the normalized γ-ray intensity measured along the axis of magnetization is

$$W = 1 - 0.495\,B_2 - 0.447\,B_4 \qquad (4)$$

where the 'orientation parameters' are given by

$$B_k = \sum_{M=-3}^{3} (2k+1)^{1/2}\,C(3k3;M0)p_M \qquad (5)$$

Here C(3k3;M0) is a Clebsch-Gordan coefficient and p_M is the population of the Mth Zeeman energy level.

The change in γ-ray intensity (signal) produced when the populations p_{-3} and p_{-2} are equalized in the M=-3→-2 transition can be calculated from Equation 4 and is

$$\Delta W_{-3\,-2} = 0.833(p_{-3} - p_{-2}) \qquad (6a)$$

The signal for the M=-2→-1 transition is

$$\Delta W_{-2\,-1} = -0.167\,(p_{-2} - p_{-1}) \qquad (6b)$$

Note that for thermal equilibrium when $p_M > p_{M+1}$, the signal $\Delta W_{-3\,-2}$ is positive whereas $\Delta W_{-2\,-1}$ is negative, i.e., in the latter case, the γ-ray anisotropy 1-W is enhanced. These changes reflect the different contributions of B_2 and B_4.

The magnitudes and signs of ΔW for the transition allow the identification $\nu_{-3\,-2}$ =499.8 ± 0.1 MHz and $\nu_{-2\,-1}$ = 501.6 ± 0.1 MHz. Also, it has been observed that the relaxation following the M = -2→-1 transition is relatively fast because of the B_4 contribution (3). We calculate that A/h = -202.0 ± 0.1 MHz and P'/h = 1.4 ± 0.1 MHz leaving a pure quadrupole contribution P/h = 0.4 ± 0.1 MHz. These resonances and the inferred hyperfine parameters can be compared with those for ^{54}Mn-$MnCl_2.4H_2O$ for which $\nu_{-3\,-2}$ = 500.4 ± 0.1 MHz, $\nu_{-2\,-1}$ = 503.5 ± 0.1 MHz, A/h = -203.8 ± 0.1 MHz, P'/h = 2.7 ± 0.1 MHz and P/h = 0.4 ± 0.1 MHz (4,5). It is not surprising that the values of P in the two salts are very similar because they do have the same structure. Also one expects the A values to be closely the same for the S-state manganous ion in the two similar environments.

We intend to investigate the dependence of the NMRON lines on an applied magnetic field. It will be interesting to compare the results with the $Mn.Cl_2.4H_2O$ data (6,7) and, in particular, to observe whether or not similar cooling effects occur near the magnetic phase transitions (7).

REFERENCES

(1) A. Kotlicki and B.G. Turrell, Hyp. Int. 11 (1981) 197.
(2) C.H. Westphal and C.C. Becerra, J. Phys. C13 (1980) L527.
(3) A. Kotlicki, B.A. McLeod, M. Shott and B.G. Turrell, Hyp. Int. 15/16 (1983) 747.
(4) A. Kotlicki, B.A. McLeod, M. Shott and B.G. Turrell, Phys. Rev. B29 (1984) 26.
(5) A. Kotlicki and B.G. Turrell, Phys. Lett. A in print.
(6) A. Kotlicki, M. Shott and B.G. Turrell, Hyp. Int. 15/16 (1983) 751.
(7) M. de Araujo, G.J. Bowden, R.G. Clark and N.J. Stone, Hyp. Int. 15/16 (1983) 969.

LT-17 (Contributed Papers)
U. Eckern, A. Schmid, W. Weber, H. Wühl (eds)

A NUCLEAR ORIENTATION STUDY OF A SINGLE CRYSTAL OF THE β-PHASE OF IRON-GERMANIUM §

J.M. DANIELS*, A.M. SABBAS+, M.E. CHEN, P.S. KRAVITZ, J. GAVILANO, J.L. GROVES and C.S. WU

Physics Dept., Columbia University, New York, NY 10027, USA

1. INTRODUCTION

The β-phase of iron-germanium has the unit cell composition $Fe_{2+2y}\,Ge_2\square_{2-2y}$ $(0.5\leq y\leq 1)$ and has the $B8_2$ (Ni_2In) structure (1). There are 3 unit cell sites: 2a at (0,0,0) and (0,0,½), 2c at (2/3, 1/3, 1/4) and (1/3, 2/3, 3/4), and 2d at (1/3, 2/3, 1/4) and (2/3, 1/3, 3/4). It is canted ferrimagnet, the hexagonal c-axis is a hard direction of magnetization, and the a-axis is an easy direction (2)(3).

The Mössbauer spectrum shows three partially resolved 6-line patterns which have been identified as arising from Fe in the a,d and c sites respectively (4)(5). The strongest pattern arises from 2 Fe in the unit cell, whose location could only be the a-sites (6).

2. PREPARATION OF SPECIMENS

Single crystals of composition $Fe_{1.67}$ Ge were grown by the Bridgman method, as described by Daniels et al. (5). Pieces 10.8 mm x 1.2 mm x 1.2 mm, were cut with the long dimension along an a-axis and one short dimension along a c-axis. About 10-20 μCi of ^{54}Mn and of ^{60}Co, in the form of carrier free chloride in 0.5 M HCl were placed on an a-c face of separate crystal slices, and dried. The activity was diffused into the crystal in a furnace in a hydrogen atmosphere, and then the crystal was lightly etched to remove any activity left on the surface.

^{60}Co was diffused for 3½ hr at 900 °C, followed by 6½ hr at 950 °C; the procedure used by Amersham Ltd. to produce the single crystal source used by Daniels et al. (5), who showed that 85% of the cobalt goes into the a-sites. The diffused layer is about 2 μm thick.

Manganese, however, goes almost entirely into the d and c sites (7). The thickness of the diffused layer was measured by measuring the relative intensities of the 835 keV γ-ray from ^{54}Mn, and the 5.4 keV Cr Kα x-ray which accompanies this decay. Knowing the absorption length of the x-ray in this alloy, the thickness of the diffused layer is easily calculated. We found that the diffusion coefficient can be represented by

$$D = D_0 \exp(-Q/RT)$$

where $D_0 = 9.15 \times 10^3$ cm^{2}s^{-1}, and $Q/R = 3.3 \times 10^4$K. Diffusion for 3 hr at 626 °C followed by 1 hr at 657 °C was both predicted and subsequently verified to produce a diffused layer 2 μm thick.

3. EXPERIMENT

The single crystal specimens were soldered on to the bottom end of a cold finger which was cooled by a ^{3}He-^{4}He dilution refrigerator to temperatures in the range 14 mK to 200 mK. This equipment has been described in greater detail by Chirovsky et al. (8). The crystals were orientated with the long direction, an a-axis, horizontal, and a c-axis vertical. The a-c face containing the activity faced downwards. A magnetic field of 0.26 T was applied along the a-axis by means of a quasi-Helmholtz pair of superconducting coils. This was more than sufficient to saturate the magnetization of the crystal. The γ-rays were counted by NaI/Tl counters placed in the same horizontal plane as the specimen; several counters were used, but most of the data came from the counter on the a-axis. The specimen could be irradiated with R.F. power from a loop antenna positioned underneath it. The R.F. source covered the range 1 MHz to 1400 MHz in three bands, and could be frequency modulated in the manner usual for NMR/ON.

4. RESULTS

An anisotropy of both the 835 keV γ-ray of ^{54}Mn, and the 1173 keV γ-ray of ^{60}Co, was seen as the temperature of the specimens was lowered. The decrease in counting rate along the a-axis was twice the increase perpendicular to this axis, showing that both ^{54}Mn and ^{60}Co are aligned along the axis. The counting rate for the ^{54}Mn γ-ray along the a-axis, as a function of temperature, was fitted to the theoretical expression (9) giving a value of 21.7 ± 1.3 T for the hyperfine field at the ^{54}Mn nucleus with

---

+Present address: Mail Stop D456, P3, Los Alamos National Laboratory, Los Alamos, NM 87545, USA.
*Present address: Dept. of Physics, University of Toronto, Toronto, Ontario, Canada, M5S 1A7.
§Supported by the National Science Foundation (USA), the Natural Sciences and Engineering Research Council (Canada), and the Guggenheim Foundation.

a reduced χ^2 = 4.3 (A value of χ^2 = 1 was obtained for the fit to the measurements for T > 40 mK.) This rather high value of χ^2 can be attributed to the fact that ^{54}Mn occupies more than one crystal site and the value of H is an R.M.S. average.

In another experiment, two crystals, one containing ^{54}Mn and the other containing ^{60}Co, were mounted together on the cold finger. The counting rate of the ^{60}Co γ-ray was measured simultaneously with that of the ^{54}Mn γ-ray, and the latter was used to determine the temperature. From the anisotropy of the ^{60}Co γ-ray, an R.M.S. value of 11.2 ± 1.1 T was deduced for the cobalt hyperfine field.

NMR/ON was attempted on both ^{54}Mn and ^{60}Co without success, although there appeared to be some suspicion of NMR/ON signals buried in the noise. Large broad peaks were seen, however, in the counting rate, which were attributed to eddy current heating. This effect was used to measure the nuclear spin-lattice relaxation time in the following way.

The radio frequency was switched back and forth every 40 sec. between a value which caused heating and one which did not, and the counts along the a-axis were accumulated in a 400 channel multiscaler synchronized to the RF switching cycle. Thus, when the RF heating was switched off, the multiscaler displayed the decay of the nuclear spin temperature. A satisfactory fit was obtained to an exponential decay and the relaxation time was deduced from the fit. These relaxation times are given in the table.

Table.

Nuclear spin relaxation time of ^{54}Mn and ^{60}Co in β-Fe-Ge

	T (mK)	τ (sec)	χ^2
^{54}Mn	28.17	3.78 ± 0.14	0.95
^{54}Mn	26.06	4.13 ± 0.21	1.0
^{54}Mn	24.96	4.22 ± 0.25	1.1
^{60}Co	24.67	10.9 ± 3.6	1.1

5. DISCUSSION

The anisotropy of the γ-rays is consistent with the hypothesis that the magnetization is saturated along the a-axis, and that this is the direction of alignment of the nuclei. It is not surprising that ^{60}Co is aligned along the a-axis, since it goes mainly into the a-sites (5), but ^{54}Mn goes almost exclusively in the d and c-sites (7), and Fe spins in these sites are canted relative to the a-axis (4)(5). This might be explained by the model proposed by Daniels et al. (4) where Fe in the d and c-sites replaces Ge, and where the 3d-shell contains only 4 electrons giving rise to a localized spin-only moment of 4 Bohr magnetons. Mn in this site would have 3 3d electrons, and a different wavefunction with different directional properties.

The relaxation time for ^{60}Co is a nuclear spin-lattice relaxation time, but the relaxation time for ^{54}Mn is, strictly, an upper limit of the nuclear spin-lattice relaxation time, but it is the right order of magnitude for it to be a nuclear spin-lattice relaxation time.

ACKNOWLEDGEMENTS

We are grateful for the financial support of the National Science Foundation, the Natural Sciences and Engineering Research Council of Canada, and the Guggenheim Foundation.

REFERENCES

(1) L. Castellitz, Z. Metalk. 46 (1955) 198.
(2) J.J. Becker and A.E. Austin, J. Phys. Chem. Solids 26 (1965) 1795.
(3) Y. Tawara, J. Phys. Soc. Japan 20 (1965) 237.
(4) J.M. Daniels, F.E. Moore and S.K. Panda, Can. J. Phys. 53 (1975) 2428.
(5) J.M. Daniels, H-Y. Lam and P.L. Li, Can. J. Phys. 60 (1982) 1564.
(6) E. Adelson and A.E. Austin, J. Phys. Chem. Solids 36 (1965) 1000.
(7) S.K. Panda, Ph.D. Thesis, University of Toronto (1974).
(8) L.M. Chirovsky, W.P. Lee, A.M. Sabbas, A.J. Becker, J.L. Groves and C.S. Wu. Nucl. Instrum. and Meth. 219 (1984) 103.
(9) R.M. Steffen, Angular distributions and correlations of radiation emitted from oriented nuclei. Los Alamos report LA-4565 (1971).

LT-17 (Contributed Papers)
U. Eckern, A. Schmid, W. Weber, H. Wühl (eds)

EXPERIMENTAL STUDIES ON THE ENHANCED NUCLEAR SPIN SYSTEM IN $Cs_2NaTbCl_6$ AND $Cs_2NaTmCl_6$

Yumiko MASUDA, Haruhiko SUZUKI and Taiichiro OHTSUKA

Department of Physics, Faculty of Science, Tohoku University, Sendai 980, Japan

Peter J. WALKER

Clarendon Laboratory, Parks Road, Oxford OX1 3PU, U. K.

Hyperfine enhanced nuclear spin systems of elpasolite compounds $Cs_2NaTbCl_6$ and $Cs_2NaTmCl_6$ were investigated between about 10 mK and 4.2 K by magnetic susceptibility, specific heat and SQUID·NMR experiments. The magnetic susceptibility of $Cs_2NaTbCl_6$ showed an anomalous peak at about 280 mK. In the specific heat measurement of this compound we could not find any anomaly in zero magnetic field at about 280 mK. The Schottky type heat capacity due to the Zeeman splitting of the enhanced nuclear spin levels of Tb was observed in magnetic fields of 3T and 5T.

1. INTRODUCTION

Most of the elpasolite compounds, $Cs_2NaLnCl_6$ have the cubic structure in the space group $Fm3m(O_h^5)$, with the lattice constant of about 1nm, at room temperature. They undergo a distortion to tetragonal symmetry above 4.2K. $Cs_2NaTbCl_6$ and $Cs_2NaTmCl_6$ compounds are also reported to have a small distortion resulting in an anisotropic signal of enhanced NMR around helium temperature (1). The ground state of Tb^{3+} and Tm^{3+} ions in $Cs_2NaTbCl_6$ and $Cs_2NaTmCl_6$ is Γ_1 singlet with a low-lying Γ_4 triplet at about 36 cm^{-1} and 58 cm^{-1}, respectively. At low temperatures the first excited triplet is split by crystal distortion. It is also reported that these van Vleck paramagnetic compounds show hyperfine enhanced nuclear moments with enhanced factors of 38 for $Cs_2NaTbCl_6$ and 29 for $Cs_2NaTmCl_6$. We carried out magnetic susceptibility, heat capacity and SQUID NMR measurements between about 10 mK and 4.2 K. An anomalous temperature dependence of the magnetic susceptibility was observed in $Cs_2NaTbCl_6$. The Schottky heat capacity of enhanced nuclear spin of ^{159}Tb was observed in magnetic fields of 3T and 5T. We have already reported observation of the ordered state of the enhanced nuclear spin system in $Cs_2NaHoCl_6$ (2). We also tried magnetic cooling experiments on these compounds, but have not been alble to realize nuclear spin order at present.

2. EXPERIMENTS

Single crystals used in our experiments were grown at the Clarendon Laboratory of Oxford University. Experiments were mainly conducted by using a ^3He-^4He dilution refrigerator.

2.1. AC magnetic susceptibility and demagnetization cooling

The temperature dependence of the ac magnetic susceptibility was measured by using a SQUID based mutual inductance bridge below 4.2 K. As shown in Fig.1 the magnetic susceptibility of $Cs_2NaTbCl_6$ showed a maximum at about 280 mK.

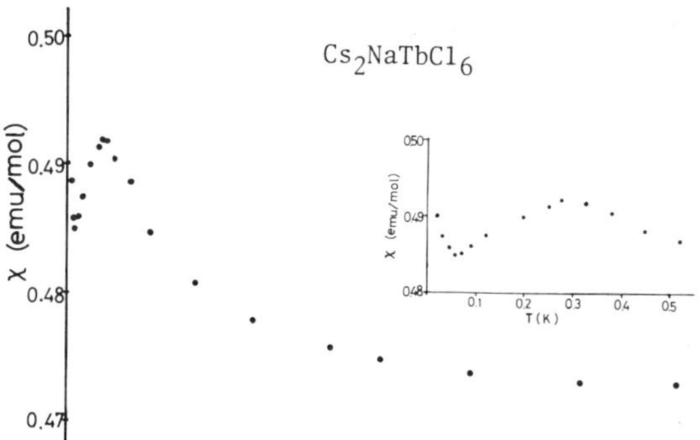

Fig.1 Magnetic susceptibility of $Cs_2NaTbCl_6$ against temperature. The inset shows the low temperature part on an expanded scale.

The peak of this magnetic susceptibility appears to depend slightly on the measuring frequency. From the susceptibility below about 60 mK the effective moment was obtained as 0.02 μ_B by assuming the Curie law, which roughly agrees with the calculated value 0.04 μ_B, by using the value of $\gamma/2\pi$=295MHz/T obtained in our present experiment at 1.7 K. At temperature higher than 280 mK, the susceptibility fits the Curie-Weiss law, $\chi=C/(T+\Theta)$, with an effective moment of 0.2 μ_B and Weiss temperature Θ=120mK. The value of 0.2 μ_B is much larger than the enhanced nuclear moment of 0.04 μ_B. A magnetic susceptibility of $Cs_2NaTmCl_6$ showed enhanced nuclear paramagnetic behaviour down to about 10 mK. By demagnetization cooling from various initial fields up to 0.5 T, the enhanced nuclear spin systems of $Cs_2NaTbCl_6$ and $Cs_2NaTmCl_6$ cooled slightly below the initial temperature of about 14 mK. The lowest temperatures were only 6.3 mK for $Cs_2NaTbCl_6$ and 8.2 mK for $Cs_2NaTmCl_6$, which were estimated from the measured susceptibilities.

2.2. Heat capacity

The heat capacity of the Tb compound was measured by the conventional adiabatic heat pulse method between 100 mK and 600 mK. No anomaly of the heat capacity was found around 280 mK in zero magnetic field as shown in Fig.2, where heat capacities in magnetic fields of 3T and 5T are also shown. The results can be understood by the Schottky specific heat due to Zeeman splitting of the enhanced nuclear spin levels. Experimental data fit well calculation (solid line in Fig.2) using the value of $\gamma/2\pi$=295 MHz/T which was obtained by our present SQUID NMR experiment.

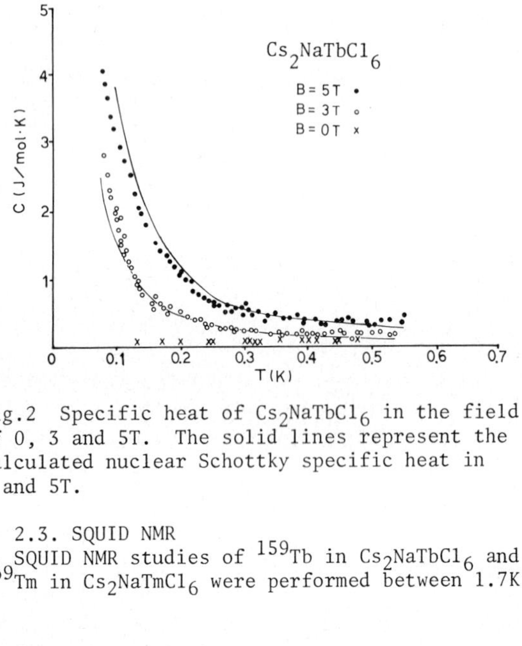

Fig.2 Specific heat of $Cs_2NaTbCl_6$ in the field of 0, 3 and 5T. The solid lines represent the calculated nuclear Schottky specific heat in 3 and 5T.

2.3. SQUID NMR

SQUID NMR studies of ^{159}Tb in $Cs_2NaTbCl_6$ and ^{169}Tm in $Cs_2NaTmCl_6$ were performed between 1.7K and 4.2 K. Our measured values of the gyromagnetic ratio $\gamma/2\pi$ was 295 MHz/T for the (001) direction of the $Cs_2NaTbCl_6$ crystal and 105 MHz/T for the (111) direction of the $Cs_2NaTmCl_6$ crystal, which agreement with the conventional NMR results was very well (1). The spin-lattice relaxation time T_1 of the enhanced nuclear spin system was measured to be 0.2 sec in $Cs_2NaTbCl_6$ and 0.4 sec in $Cs_2NaTmCl_6$ at 1.7 K. The temperature dependence and the magnetic field dependence of T_1 were also measured. T_1 became shorter with increasing the temperature and longer with increasing the magnetic field in both compounds. Similar field dependence of T_1 has already observed in the enhanced nuclear spin system of $TmPO_4$ (3).

3. DISCUSSIONS

The origin of the maximum at 280 mK in the magnetic susceptibility of $Cs_2NaTbCl_6$ is not clear at present. One possibility is an impurity effect. However if we attribute this anomalous susceptibility to the impurity, a few percent of impurities must be contained in the specimen, which is improbable. Another possibility is some type of phase transition. Our result of the heat capacity measurements suggest that below 280 mK the enhanced nuclear spin system still remains in the paramagnetic state. Further investigations on this problem are in progress.

REFERENCES

(1) B. Bleaney, F. R. S., A. G. Stephen, P. J. Walker and M. R. Wells, Proc. R. Soc. Lond., A381 (1982) 1.
(2) H. Suzuki M. Miyamoto, Yu.Masuda and T. Ohtsuka, J. Low Temp. Phys., 48 (1982) 297.
H. Suzuki, Yu. Masuda, M. Miyamoto, S. Sakatsume, P. J. Walker and T. Ohtsuka, J. Mag. Mag. Met., 31-34 (1983) 741.
(3) H. Suzuki, Y. Higashino and T. Ohtsuka, J. Low Temp. Phys., 41 (1980) 449.

LT-17 (Contributed Papers)
U. Eckern, A. Schmid, W. Weber, H. Wühl (eds)

A STEP EDGE SUPERCONDUCTOR-SEMICONDUCTOR-SUPERCONDUCTOR JOSEPHSON JUNCTION

Alicia SERFATY M. and Miguel OCTAVIO

Fundación Instituto de Ingeniería and IVIC, Apartado 40200, Caracas 1040 A, Venezuela.

We describe a simple technique for the fabrication of superconductor-semiconductor-superconductor weak-links. We have fabricated Pb-Si-Pb step edge junctions where the Si is doped to the solubility limit. The junctions exhibit large normal state resistances which should be useful in applications and they show little or no hysteresis. The critical currents do not agree with theoretical models.

Among superconducting weak-links, those in which the barrier is formed by a normal metal or doped semiconductor present several advantages over other types of weak-links. In particular, they are expected to exhibit ideal Josephson behavior and should have the higher resistance values which are needed in many applications. The use of a doped semiconductor should provide even greater flexibility as the level of doping modifies the coherence length and resistivity of the weak-links allowing for the desing of the properties of the junction.

We report on the fabrication of Pb-Si-Pb step-edge Josephson junctions which can be fabricated with relatively simple techniques and which allow the controlled variation of the weak-link length. The technique is a step-edge technique similar to that used in the fabrication of other types of weak-links (1,2), except that the substrate is an active part of the final device. Fig. 1 shows the fabrication sequence used and final device geometry. First, a Cr film is e-beam evaporated onto a Si wafer partially covered by a photoresist strip. The photoresist and Cr on top of it are removed by lift-off leaving a Cr film covering part of the wafer. The surface is then ion-milled using Ar or ClF_4 ions incident on the surface at a normal angle. Due to the slower milling rate of the Cr film, a step is created in the Si wafer. While the removal rate of the ClF_4 was found to be almost 3 times that of Ar, Ar ion milling was found to yield smoother steps with significantly less redeposition than ClF_4. After this step, the Cr film is removed and phosphorus doping is carried out at 1100 C in order to achieve the solubility limit of $\sim 10^{20}$ cm^{-3}. Then Pb is evaporated at an angle, at 77 K, leaving two lead films separated by a step. We note that prior to the Pb evaporation the surface is once again ion milled for a short time in order to remove any possible oxides on the Si wafer. Inmediately after this step Pb is evaporated assuring a good electrical contact. It should be pointed our that this step was found to be crucial for the observation of supercurrents in these devices. After this step, a Pb strip (2-4 μm in width) is patterned.

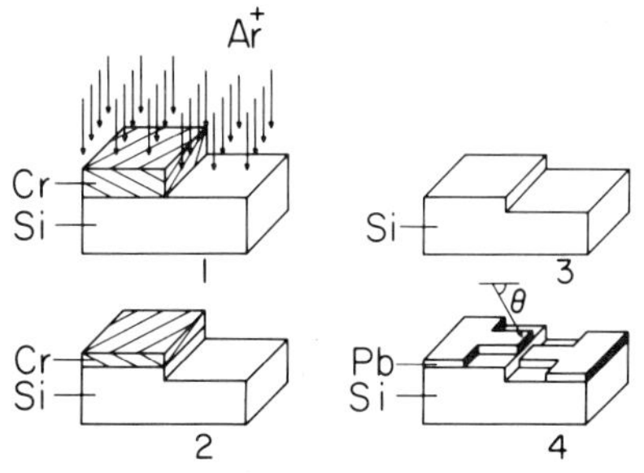

FIGURE 1
Fabrication sequence for step-edge Pb-Si-Pb junction.

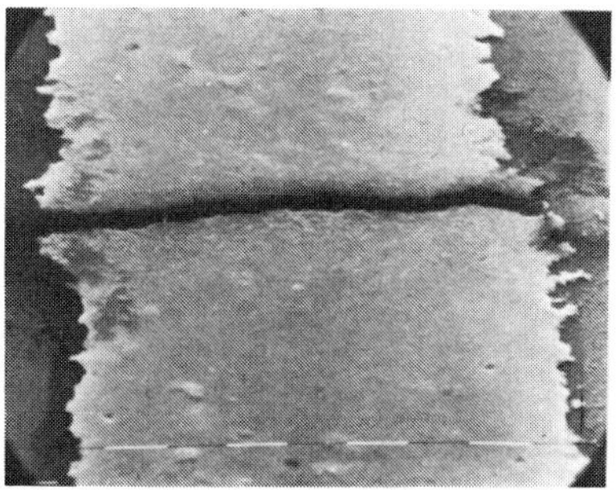

FIGURE 2
SEM micrograph of a step edge Pb-Si-Pb device (each bar is 1 μm).

The length of the device is defined by the difference between the step height and the lead thickness and in theory one should be able to vary it down to very small lengths. In practice, one wants to avoid small microshorts across the step, thus in this initial phase of our work we have used step heights of the order of .2-.4 μm and Pb films from .06 to .3 μm. The device thus obtained has negligible capacitance, variable resistance through doping and, since the coupling occurs in the massive substrate, it should not be affected by heating. The technique should be applicable to higher T_c materials except for A-15"s since the contact to the Si would depress the T_c of the banks (2). Fig. 2 shows a SEM picture of a finished device.

Josephson junctions with Si barriers have been made in sandwich-type structures (3) and coplanar configurations (4,5). In the latter, application of the theory of SNS junctions (6) does not explain the relatively large I_cR_n products observed, furthermore, low yields have been reported. In our case, the observation of Josephson coupling depends on the fabrication sequence, the critical parameters being the step height and the sharp separation of the films which form the banks. All of our devices show high normal state resistances ranging from 1.26 -10 Ohms. This is a significant advantage over most weak-link devices, and points to the possibility of even higher values with reduced doping. The I_cR_n products vary widely depending on the length of the sample ranging from 0.24-3mV. In order to understand these values one can use models for SNS junctions (6) or semiconductors (7). In both models the temperature dependence is expected to be of the form exp $(-\ell/\xi)$, since typical lengths used are of the order of 0.2 μm and $\xi(T_c)$= 150-200 Å for a mobility of 200 cm^2/V-sec. Thus, even at the lowest temperatures reached, ℓ/ξ= 5-10 and one would not expect to achieve the full I_cR product of 2.10 mV for the short bridge limit in the SNS case, (6). This is in contrast with some of the observed values above 3 mV. We do notice that devices with similar critical currents have dissimilar resistances suggesting a difficulty with the definition of the normal state resistance. In fact, our devices exhibit a nonlinear I-V characteristic at high voltage (~10 mV), presumably due to impact ionization at high current densities. Evidence for the existence of Schottky barriers comes from comparing the expected resistance of the device with the actual resistance. For typical dimensions used, one expects an upper limit of ~0.25 ohm for R_n in contrast to the observed values of up to 10 ohms and typical values of the order of 3 ohms.

Fig. 3 shows the temperature dependence of $V= I_cR_n$ normalized to $V_o= \pi\Delta(0)/2e$ for a device with R_n= 2.24 ohm. For a wide range near T_c the critical current can be fit to the SNS model (6). For $\ell/\xi\ (T_c) \cong 3$, somewhat smaller than expected.

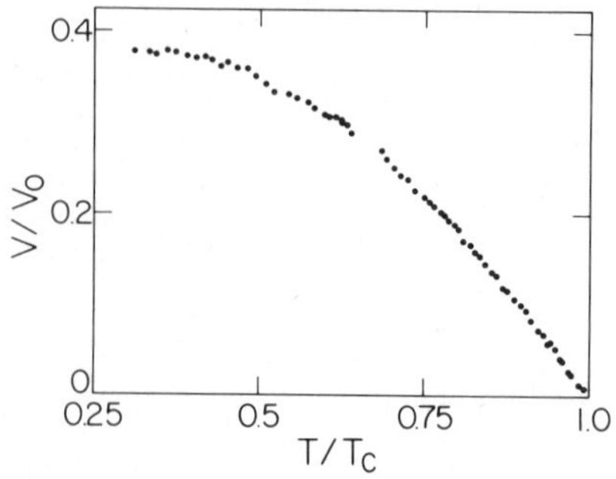

FIGURE 3
Temperature dependence of $V= I_cR_n$ normalized to $V_o= \pi\Delta(0)/2e$. The device has R_n= 2.24 ohm.

However, at low temperatures the critical current does not approach the expected value for the same normalized length. In fact, from the model one expects the saturation of I_c to occur at lower reduced temperatures than those for which it is observed.

The I-V characteristics of the devices show no or little hysteresis down to the lowest temperatures with subharmonic gap structure at 2 Δ/ne. In contrast to microbridges, the structure at 2 Δ/e is quite weak. They exhibit Josephson steps when irradiated with microwaves, however, we have been unable to couple enough power to observe the first zero of I_c. This is quite surprising given the high resistance of the junctions.

ACKNOWLEDGEMENTS
This work supported in part by CONICIT and the Office of Naval Research Contract N00014-G-006.

REFERENCES
(1) M.D. Feuer and D.E. Prober, IEEE Trans. Mag. MAG-15, 578 (1979).
(2) A. De Lozanne, M.S. Di Iorio, and M.R. Beasley, Appl. Phys.Lett. 42, 541 (1983).
(3) M. Shyfter, J. Maah-Sango, N. Raley, R. Ruby, B.T. Ulrich and T. Van Duzer, IEEE Trans. Mag., MAG-13, 862 (1977).
(4) R.C. Ruby and T.Van Duzer, IEEE Trans. El. Dev, ED-28, 1394 (1981).
(5) N.V. Alfeev, Sov. Phys. Solid State, 22, 1951 (1980).
(6) K.K. Likharev, Sov. Phys. Tech. Phys. Lett. 2, 12 (1976).
(7) L.G.Aslamazov and M.V. Fistul, Sov. Phys. JETP 54, 206 (1981).

LT-17 (Contributed Papers)
U. Eckern, A. Schmid, W. Weber, H. Wühl (eds)

PLANAR THIN-FILM DC SQUID, FABRICATION AND MEASURING SYSTEM

K.-H. BERTHEL, F. DETTMANN

Friedrich-Schiller-Universität Jena, Sektion Physik, DDR-6900 Jena, Max-Wien-Platz 1, GDR

The design of a thin film dc SQUID and its fabrication are described. The experimental apparatus, especially the readout scheme, is discussed. The measuring system works with good resolution for applications and high reliability.

1. INTRODUCTION

Recent advances made in the fabrication of Josephson junction devices permit the production of sensitive thin-film SQUIDs for normal use with external superconducting circuitry such as coils. In this paper the design of a thin-film SQUID and the experimental apparatus for use in biomagnetic measurements is described. The aim of our investigation was to achieve an increase of the yield of the fabrication method of the shielded two-hole SQUID with two Nb-NbO_x-Pb tunnel junctions (1) and to improve the reliability of the experimental apparatus. The design of the SQUID and its fabrication are described in the first section of the paper. The experimental apparatus, especially the readout scheme, are discussed in the second section. The last part presents the characteristic values of operation of the complete system.

2. SQUID DESIGN AND FABRICATION

A schematic of the left part of the double hole device is shown in fig.1(The right part of the device is symmetrical to the left). Figure 1 does not show insulation films.

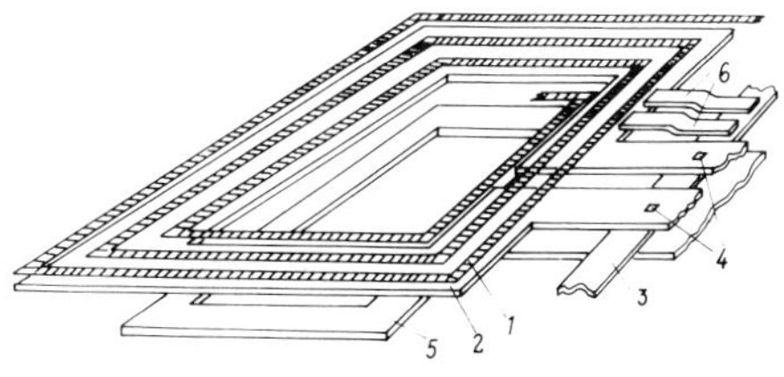

FIGURE 1
Schematic of the left and middle part of the SQUID.

For simplicity in the figure the thin-film input coil 1 has only 3 turns in each hole and not 15 as in the actual device. The input coil consists of a niobium film, 200 nm thick. The double hole loop 2 of the SQUID consists of lead, 300 nm thick, and acts as a gradiometer. In the middle part the double hole loop is closed by the junctions 4 and the niobium film 3. The SQUID is shunted by two (in reality, 5) normal conducting stripes 6 of silver/indium. The arrangement is completed by the superconducting screening film 5. The width of the screening stripe b_s is smaller than that of the SQUID loop 6.

The inductances of the double loop SQUID L and the two input spiral coils connected in series L_E are estimated to be

$$L = (d_2+p+p_s)\mu_o l/(2b_s) \qquad (1)$$

and

$$L_E=((d_2+p+p_s)/b_s+(d_1+p+p_1)/(b-a))\mu_o ln^2/2. \qquad (2)$$

μ_o is the free space permeability, l is the circumference of one SQUID loop, a is the sum of the distances between the turns of one spiral. d_1 and d_2 are the thicknesses of the insulation between spiral and SQUID loop and SQUID loop and screening loop, respectively. p, p_1 and p_s are the London penetration depths of the SQUID film, the spiral film and the screening film. n is the sum of the turns of the two input coils. The input coupling constant is given by

$$k = (1+b_s(d_1+p+p_1)/(b-a)(d_2+p+p_s))^{-1}. \qquad (3)$$

From equations (2) and (3) we obtain sufficiently high input inductances and coupling constants approaching unity with reliable insulation thicknesses d_1 and d_2 if the width of the screening film is much smaller than that of the SQUID loop. With such high input inductance the device presented here was successfully coupled to a gradiometer to get good sensitivity (3).

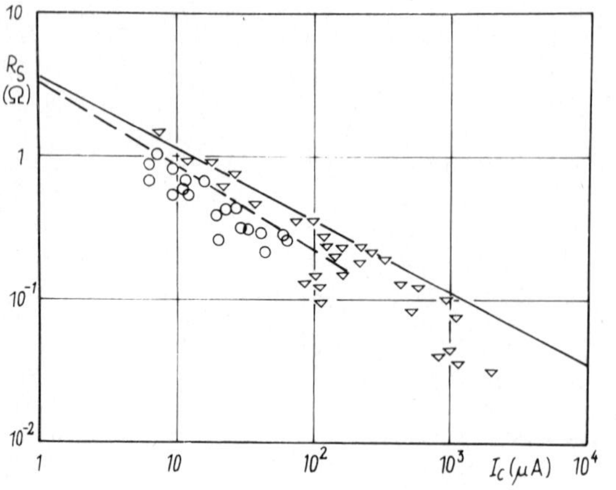

FIGURE 2
Adjusted shunt resistance of different SQUIDs as a function of critical current. (Circles - without hysteresis, triangles - with hysteresis, solid line - theoretical limit of hysteresis).

After simultaneously fabricating of nine devices with total dimensions of 8 • 12 mm^2 on one silicon substrate the critical currents and the resistances of the shunt stripes could be determined. Then by scratching one or more shunt stripes the maximum output voltage of the SQUIDs was adjusted. In fig. 2 the adjusted shunt resistance of different devices is shown versus the critical current. Current-voltage characteristics without hysteresis are represented by circles. From fig. 2 it is apparent that useful devices can be manufactured over a wide range of critical current (up to about 60 μA).

3. EXPERIMENTAL APPARATUS

For measurements in application the conventional phase locked loop dc SQUID readout scheme is used (2). The matching circuit between the SQUID and the amplifier is a resonant transformer at room temperature as shown in fig. 3 . The resistance of the current leads R_z is not larger than the shunt resistance R_s. The bias current is coupled to the SQUID circuit by a small resistor with a resistance of 0,01 Ohm. Therefore the voltage of the SQUID is more strongly adjusted than the current.

The cryogenic probe contains the resonant transformer at room temperature and the sample holder at helium temperature. All leads between the two ends of the probe are twisted and doubly screened by an aluminium foil and by German silver tubes with small wall thickness. The SQUID is enclosed in a lead box. The leads to the SQUID output consist of bonded aluminium wires,

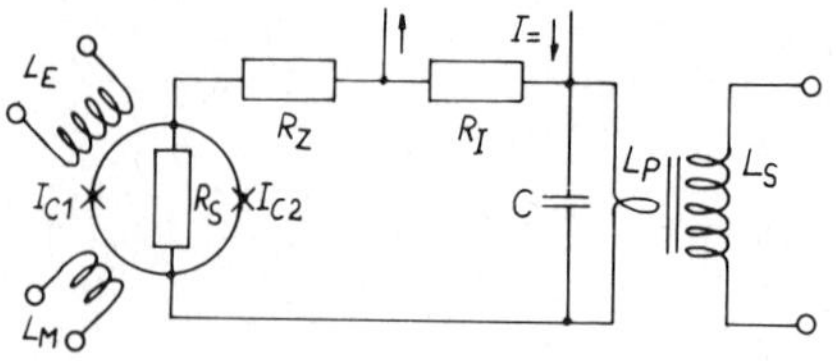

FIGURE 3
Matching circuit between SQUID and amplifier.

those to the input of bonded niobium wire. The lead box and all the connections are enclosed in a lead cylinder.

4. RESULTS

We succeded in fabricating SQUIDs with critical currents between 5 and 60 μA. Input current periods from 0.2 to 1.5 μA and input inductances from 600 nH to 70 nH were reached. A coupling constant of 0.77 was measured; this is in agreement with the calculation according to equation (3). 198 devices were fabricated and 17 % of them were found to exhibit the desired parameters. In measurements with the flux locked loop regime a flux resolution of 10^{-4} $\Phi_o/\sqrt{Hz}$ was achieved. This corresponds to an energy sensitivity of $3 \cdot 10^{-28}$ J Hz^{-1}. Measurements performed with the same SQUIDs on a special electronic equipment yielded a resolution of $2 \cdot 10^{-5}$ $\Phi_o/\sqrt{Hz}$. Long-duration measurements with the described apparatus revealed that perturbations caused by switching pulses such as flux jumps or jumps of the critical current could not be observed.

ACKNOWLEDGEMENTS

The authors are indebted to the cryoelectronic and technological groups of the Sektion Physik of the Friedrich-Schiller-University and, in particular, to Dr. K. Blüthner, Dr. W. Richter, Dr. W. Vodel and Dr. P. Weber.

REFERENCES

(1) F. Dettmann, W. Richter, G. Albrecht, W. Zahn, phys. stat. sol. (a) 51 (1979) K 185
(2) M.B. Ketchen, IEEE Trans. Magn. Mag-17 (1981) 387
(3) G. Albrecht, W. Haberkorn, G. Kirsch, H. Nowak, H.-G. Zach, Recent results of biomagnetic measurements with a dc SQUID system, this volume.

LT-17 (Contributed Papers)
U. Eckern, A. Schmid, W. Weber, H. Wühl (eds)

THIN-FILM DC SQUIDs FOR SUPERCONDUCTING MAGNETOMETER DEVICES

S.I. BONDARENKO, V.V. KRAVCHENKO, E.A. GOLOVANEV and N.M. LEMESHKO

Institute for Low Temperature Physics and Engineering, UkrSSR Academy of Sciences, Kharkov, USSR

Thin-film dc SQUIDs intended for connection to the antennae of superconducting magnetometer devices are described. As to the type of magnetic coupling with the antenna the SQUIDs are divided into those with autotransformer, inductive-autotransformer and inductive coupling. The electrophysical parameters of the SQUIDs are presented.

1. INTRODUCTION

Superconducting magnetometer devices incorporate, as a rule, a superconducting antenna loop to receive the magnetic flux being measured and to convert it into electric current, and a SQUID to detect this current. Thin-film dc SQUIDs are now used in a variety of instrument applications. To reach an optimum match between a SQUID and an antenna, the equation $L_a = L_i$ should be held, where L_a is the antenna inductance, L_i is the SQUID input inductance. The coupling between the SQUID loop and its input inductance is characterized by the coupling constant k given by:

$$k = \sqrt{\frac{E_o}{E_i}} \quad . \qquad (1)$$

Here

$$\Delta E_o = \frac{\phi_o^2}{2L_o} \quad , \qquad (2)$$

which is the energy required to change the flux in a SQUID loop with self-inductance L_o by a flux quantum,

$$\Delta E_i = (\Delta I_H)^2 \, L_i/2 \qquad (3)$$

is the corresponding change of energy in the input inductance (ΔI_H is the input current modulation period of the SQUID voltage). Substitute Eqs.(2) and (3) into Eq.(1) and obtain

$$k = \frac{\phi_o}{\Delta I_H \; L_i L_o} \quad . \qquad (4)$$

As to the type of connection with the antenna, SQUIDs can be divided into those with autotransformer, inductive-autotransformer and inductive coupling. We have developed three SQUID designs whose descriptions and electro-physical characteristics are given below.

2. AUTOTRANSFORMER COUPLING SQUID (SQUID-1)

SQUID-1 (1) comprises a film loop formed by superconducting films 1, 2 (Fig. 1) separated by insulation 3 and closed at edges, and variable-thickness bridge junctions 4 shunted by film shunts 5. SQUID-1 is connected to the antenna by means of superconducting leads I_H attached to control film 2 (see Fig. 1) which is part of the SQUID loop. The control film is narrowed in the SQUID loop region. For SQUID-1 $L_o = (2\text{-}3) \times 10^{-11}$ H, $L_i \simeq L_a$, $\phi_o = \Delta I_H L_i \simeq \Delta I_H L_o$. According to Eq.(4), $k \simeq 1$.

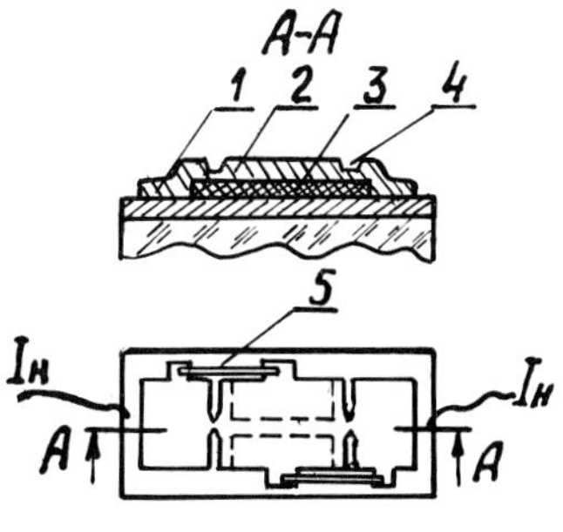

FIGURE 1
Diagram of SQUID-1

TABLE

SQUID type	I_c, μA	ΔI_c, μA	ΔI_H, μA	ΔV, μV	L_o, H	L_i, H	dV/dI_H, V/A	k
SQUID-1	200-500	50	130	50	$(2-3)\times10^{-11}$	2×10^{-11}	0.8	∿ 1
SQUID-2	200-500	20-30	50	20	$(2-5)\times10^{-11}$	5×10^{-8}	0.8	0.04
SQUID-3	200-500	10-20	1.7	10	8×10^{-11}	5×10^{-8}	12	0.57

2. INDUCTIVE-AUTOTRANSFORMER COUPLING SQUID (SQUID-2)

This structure (2) comprises a SQUID similar to that described above which is inserted into the film loop of a superconducting autotransformer. The latter is inductively coupled to a wire-wound superconducting solenoid connected to an antenna. The solenoid inductance L_i is selected equal to that of the antenna. The coupling constant of the SQUID loop with the solenoid calculated by Eq. (4) at $L_i = 5\times10^{-8}$ H, $L_o = 2\times10^{-11}$ H, $\Delta I_H = 50$ μA was 0.04.

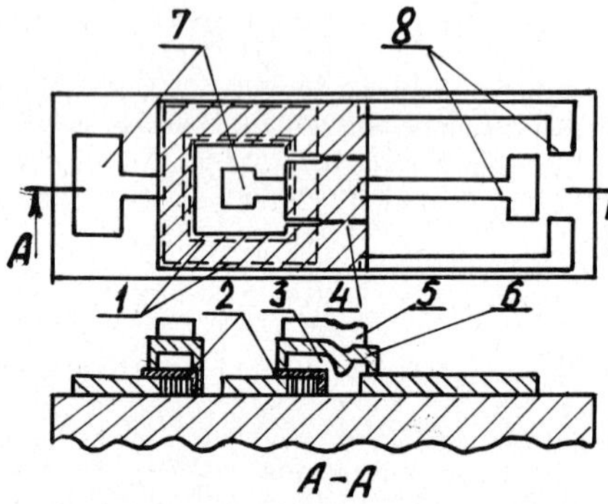

FIGURE 2

Diagram of SQUID-3 (the groundplane not shown in the plan of Fig. 2)

3. INDUCTIVE COUPLING SQUID (SQUID-3)

SQUID-3 (3) is an all-film planar structure (Fig. 2). The input inductance is flat spiral 1 with 15 turns 10 μm wide with the same turn-to-turn distance. The spiral is separated by insulation layer 2 from SQUID loop 3 with bridges 4 arranged above the spiral. Superconducting groundplane 5 separated from the SQUID loop by insulation film 6 reduces the loop inductance. SQUID-3 is coupled to the antenna by means of superconducting leads I_H connected to termination pads 7. Terminals 8 supply transport current to the SQUID and register voltage. The input inductance L_i, both calculated and found experimentally (3) was $(5.3 \pm 0.1) \times 10^{-8}$ H. At $L_i = 5.3\times10^{-8}$ H, $L_o = 8\times10^{-11}$ H, $\Delta I_H = 1.7$ μA, k was 0.57.

Basic electrophysical parameters of the SQUIDs described above at a temperature of 4.2 K are summarized in the Table (I_c is critical current, ΔI_c is the amplitude of magnetic modulation of the critical current, ΔV is the amplitude of the SQUID voltage modulation, dV/dI_H is the input current-to-voltage transfer function).

4. CONCLUSION

SQUIDs with different types of coupling with the antenna are described. It is expedient to connect SQUID-1 to a low-inductance antenna. SQUID-2 can be matched with antennae of various inductances; however, low k is a drawback of this structure. SQUID-3 actually matches the antennae with the inductance 10^{-8} to 10^{-7} H. Its advantage over SQUID-2 is substanially higher k, as well as greater dV/dI_H when compared with other SQUIDs.

REFERENCES

(1) S.I.Bondarenko, B.I. Verkin, E.A.Golovanev et al., Geofiz. Apparatura 69 (1979) 9.
(2) S.I.Bondarenko, V.V. Kravchenko, E.A. Golovanev, M.F. Salun, and E.N.Til'chenko, Geofiz. Apparatura 76 (1982) 3.
(3) S.I.Bonarenko, V.V.Kravchenko, M.M. Lemeshko, and E.A.Golovanev, Cryogenics 23 (1983) 263.

LT-17 (Contributed Papers)
U. Eckern, A. Schmid, W. Weber, H. Wühl (eds)

FABRICATION OF NIOBIUM VARIABLE - THICKNESS BRIDGES FOR DC-SQUID APPLICATIONS

K.-D. KLEIN, H. KOCH, H. LUTHER

Physikalisch-Technische Bundesanstalt, Institut Berlin, Abbestraße 2-12,
D-1000 Berlin 10, FRG

The fabrication of Nb Josephson-junctions with a variable-thickness bridge-structure is described. The electrical characteristics are presented. A DC-SQUID based on such junctions has been operated successfully.

1. INTRODUCTION

Superconducting quantum interference devices (SQUIDs) have become an important mean in fields like biomagnetic measurements etc. For high resolution of the often used DC-SQUID the following characteristic data have to be fullfilled: modulation parameter $\beta = 2\ I_c\ L/\phi_0 \approx 1$ (I_c = critical current of the junction, L = SQUID-loop inductance, ϕ_0 = flux quantum), high value of the product R_qI_c (R_q = quasiparticle resistance of the junction), low junction capacitance and good coupling to the input coil (1). Practical reasons require a working temperature of 4.2 K and good mechanical and thermal-cycle stability. Niobium is a proper choice for the bulk material. A suitable junction configuration is a microbridge configuration in order to obtain a neglectable capacitance. In a bridge of variable thickness heating effects are reduced. In this paper we report about the fabrication process and some electrical characteristics of microbridges prepared by depositing the bridge over an edge, following a suggestion of FEUER and PROBER (2).

2. FABRICATION

In fig. 1a a top view of the junction region

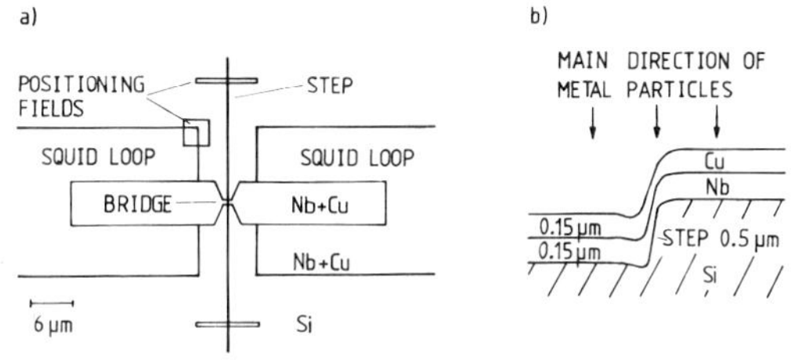

FIGURE 1
a) top view b) cross section of the junction region.

of the SQUID loop and the bridge structure are shown. The SQUID loop (only partially shown in fig. 1a) is fabricated by optical lithography and the bridge structure by electron-beam lithography.

The fabrication starts with the etching of a 0.5 µm high step into a silicon substrate by rf sputtering in an argon atmosphere. The step etched in the homogeneous material seems to be more stable than the edge of a deposited film as used in ref. (2).

The thin film loop is prepared by Nb magnetron-sputtering and lift-off technique. Its inner diameter is 1 mm (L ≈ 1 nH) and the film thickness 0.3 µm. A layer of 0.05 µm Cu is sputtered immediately after Nb deposition in order to prevent an oxydation of the Nb film.

The bridge structure is formed by electron-beam lithography. While PMMA is the common choice for high resolution e-beam lithography we used the photoresist Shipley Microposit 1350 J because it has proved to be more resistent against the subsequent sputtering process. The disadvantage of enlarged linewidth is tolerable.With reduced scan area ("positioning fields" in fig. 1a) the electron beam is focussed and the substrate is exactly positioned. Then the bridge structure is exposed with a distance between the dots of 0.4 µm and a dosis of $3\ .\ 10^{-5}$ C/cm² for a resist layer of 0.7 µm thickness. Before depositing the second Nb film by magnetron sputtering the Cu film in the overlapping area is cleaned by soft rf sputter-etching. As can be seen from fig. 1b the main deposition direction is nearly parallel to the vertical part of the step. This makes the second layer of Nb thinner in the vertical region than in the horizontal parts. Thus a variable-thickness bridge is formed. The final Cu layer decreases I_c by the proximity effect and serves as a heat conductor.

In fig. 2 a SEM picture of the bridge region of a sample is shown. Two features typical for non-reactive sputter-etching can be seen: the trench in front of the step and the redepositioned material at the step edge. The latter can be reduced by mechanical wiping.

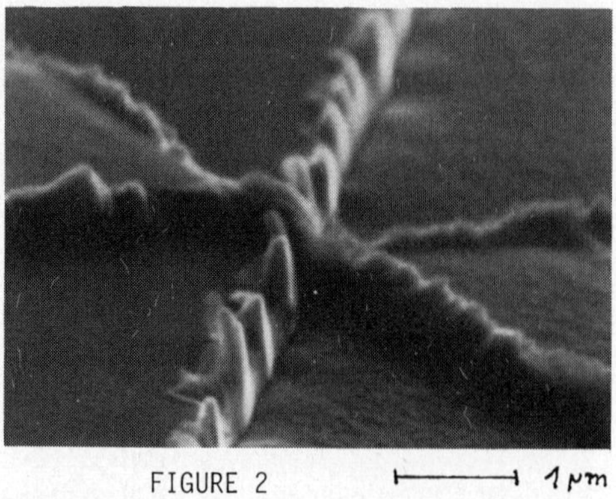

FIGURE 2

SEM-picture of the bridge region.

3. I-V-CHARACTERISTIC

In fig. 3 the I-V-characteristic of one of our microbridges is shown for different temperatures. The quasiparticle resistance is about

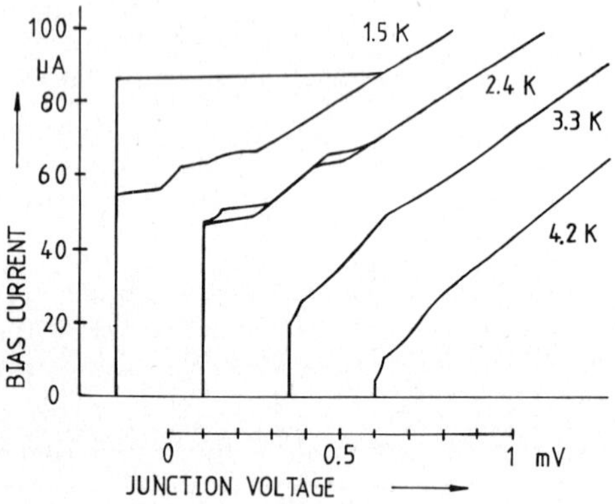

FIGURE 3

I-V characteristics of a microbridge for different temperatures.

17 Ω, at 4.2 K the critical current amounts to 4 μA. For temperatures lower than 2.5 K hysteresis occurs. The increased critical current causes increased power dissipation that cannot be removed effectively by heat conduction, due to the small dimensions of the bridge (width 0.7 μm, thickness 0.15 μm). The parameters of this microbridge are quite suitable but the nonsufficient reproducibility of the product $R_q I_c$ indicates that (independent of the geometrical parameters) the crystal structure of the superconducting material varies from sample to sample. Nevertheless with the method of trimming by electric pulses (3) we were able to adjust the critical current to appropriate values.

4. CONCLUSION

We have developed a fabrication process for stable variable-thickness bridges with low critical currents. The $R_q I_c$ products are of suitable magnitude for SQUID applications, some work has to be done to make the process more reproducible. Bridges of this type have been mounted into a thin-film DC-SQUID loop. After critical-current adjustment of each junction the loop has been closed by a Nb connection. The parameters of the microbridges did not change significantly during the closing procedure. With the non-optimized large loop inductance of 1 nH we achieved at a critical current of 5 μA a SQUID-voltage modulation of 1 $\mu V/\phi_0$, which is a reasonable value for a microbridge DC-SQUID.

ACKNOWLEDGEMENTS

The authors wish to thank Dr. S.N. Erné for stimulating discussions and Mrs. M. Peters for technical assistance in the preparations.

REFERENCES

(1) J. Clarke, C.D. Tesche and R.P. Giffard, J. Low. Temp. Phys. 37 (1979) 405.
(2) M.D. Feuer and D.E. Prober, Appl. Phys. Lett. 36 (1980) 226.
(3) D. Duret, P. Bernard and D. Zenatti, Rev. Sci. Instr. 46 (1975) 474.

LT-17 (Contributed Papers)
U. Eckern, A. Schmid, W. Weber, H. Wühl (eds)

SOURCE IMPEDANCE EFFECTS ON DC SQUID PERFORMANCE

R. S. GERMAIN, M. L. ROUKES, M. R. FREEMAN, R. C. RICHARDSON and M. B. KETCHEN*

Laboratory of Atomic and Solid State Physics, Cornell University, Ithaca, New York 14853, U.S.A.
*IBM Thomas J. Watson Research Center, Yorktown Heights, New York 10598, U.S.A.

Preliminary results concerning the effect of source impedance upon thin-film dc SQUID performance are presented. We observe significant changes in the flux voltage curve with decreasing values of shunt impedance. This suggests that the frequency scale at which these effects occur is set by the fundamental self-resonance of the input coil.

It is now possible to fabricate dc SQUIDs with noise performance far superior to that of rf SQUIDS. (1,2) Although rf SQUIDS are widely used, to date there have been relatively few applications of dc SQUIDS to laboratory experiments. Despite the similarities between the two, different practical considerations enter into their use. As part of our development of dc SQUID based experiments (3,4), we are studying how device performance is changed by coupling to external circuits. Below we present data showing that the internal impedance of a signal source connected to a dc SQUID's input coil can affect the device characteristics.

These studies are motivated by the transformation shown in figure 1. A dc SQUID is coupled inductively through L_{in} to a signal source impedance $Z_s(\omega)$. This picture is equivalent to one in which an additional impedance, $Z_{eff}(\omega)$ appears in the superconducting loop of an isolated SQUID. The effective impedance appearing in the loop is

$$Z_{eff}(\omega) = \frac{\omega^2 k_{sq}{}^2 L_{sq} L_{in}}{Z_s(\omega) + j\omega L_{in}}$$

where L_{sq} is the SQUID loop inductance. Both theoretical modelling (5) and the observed characteristics of different SQUIDS lead one to expect that the device performance may be significantly altered by this extra dissipation and reactance added to the SQUID ring.

Simple pictures such as that of Fig. 1 however, must be viewed with caution. At typical operating voltages of order 100 μV the frequency scale of Josephson radiation in the SQUID ring is roughly 100Ghz. Digital simulation shows that many harmonics of the fundamental component contribute to the dc device characteristic in this extremely non-linear system. (6) One cannot expect to model the precise nature of $Z_{eff}(\omega)$ over all relevant frequency scales by a simple lumped circuit model. It is for this reason that we have undertaken an experimental investigation of the problem.

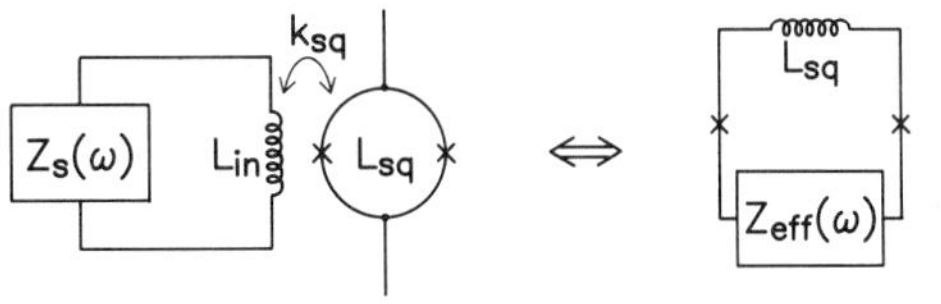

FIGURE 1. An impedance shunting the input coil places an effective impedance in the SQUID loop.

In our study we use IBM dc SQUIDs, fabricated by J. Greiner and coworkers, which are similar in design to those of Ketchen and Jaycox. (7) The input coils, however, are larger, having one hundred turns and a self inductance of 0.8 microhenries. The input coil is tightly coupled to the SQUID loop ($k_{sq}{}^2$=0.9). At 4K we typically measure flux noise of 5×10^{-7} $\phi_0/Hz^{1/2}$ giving coupled energy sensitivity of 10h measured in the KHz region. (6) The experiments carried out thus far involve shunting the thin-film input coil with various impedances. Both the slowly swept flux bias and the small ac signal, used to measure $V(\phi)$ and $dV/d\phi$ respectively, are applied to an additional 1/4-turn control coil on the chip.

For each value of source impedance we record the flux-voltage characteristic at many different bias currents. A representative family of these curves taken at bias current I_b=100 μA for a variety of shunt impedances is pictured in figure 2. Sharp resonant features on the flux voltage curve occur for large values of shunt impedance. As this impedance is decreased these are dramatically smoothed and the characteristic becomes more sinusoidal in form. This behavior raises the possibility of "tuning" a dc SQUID's performance to suit a particular application after it has been fabricated. Jaycox and Ketchen (8) have observed qualitative changes in the I-V characteristic occuring when a short circuit across the input coil (actually fabricated on the chip) was opened. We perform our experiments with discrete components attached to Nb screw terminals which are connected electrically to the SQUID chip by superconducting wire bonds. Low loss rf capacitors with very short superconducting leads are employed in order to minimize the effects of dissipation and parasitic inductance.

It is perhaps surprising that relatively large values of shunt capacitance, >300pF, are required to appreciably affect the SQUID's flux-voltage curve. For $C_s \sim 300$ pF, the frequency scale set by $(L_{in}C_s)^{-1/2}$ is of order 10 MHz. Closer inspection of the SQUID design reveals, however, that a large capacitance per unit line length, 10 pF/cm, is associated with the SQUID input inductor. The large input inductance of the SQUID is obtained through roughly 25 cm of this line--yielding an effective input coil self-capacitance of order 250 pF. This explains why shunt capacitances smaller than 300 pF are ineffectual. It is remarkable that resonant behavior at 10 MHz

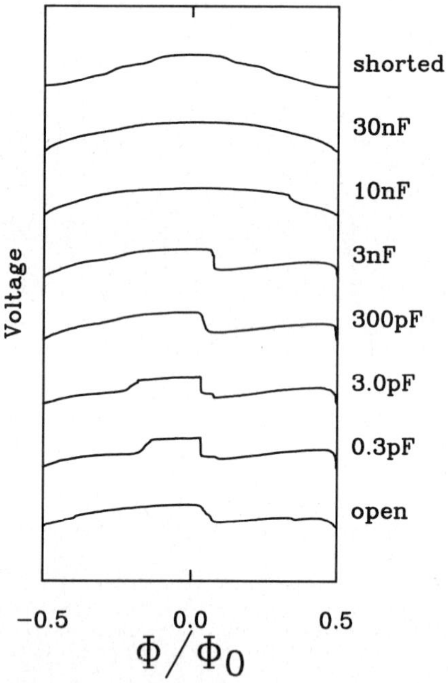

FIGURE 2. Flux voltage characteristics for different shunting impedances. The traces are displaced for clarity.

affects the characteristics of a SQUID operating at 200 μV, since 2eV/h=100GHz. We are currently exploring how this behavior changes for SQUIDs with smaller input inductors and, therefore, much higher self-resonant frequencies.

We are evaluating the dependence of the dc SQUID's intrinsic flux noise upon shunting impedance using an rf SQUID voltmeter to read out the voltage noise from the dc SQUID. Qualitatively we find that there is marked improvement as the shunt capacitance is increased, particularly for C_s=30 nF. For such a value of C_s the effective input loop self-resonance is high enough (1 MHz) that the device remains an extremely sensitive small signal amplifier over a useful frequency range.

Our results imply that one should measure SQUID characteristics in the actual circuit in which it will be used for an experiment. This is necessary because of the apparently strong effects of source impedance.

ACKNOWLEDGEMENTS

Two of us (MLR and RSG) would like to thank the staff at the IBM T. J. Watson Research Center for use of their facilities. This work was supported by the ONR and the Semiconductor Research Corporation.

REFERENCES

(1) J. Clarke. IEEE Trans. on Elec. Dev. ED-27 (1980) 1896.
(2) M. B. Ketchen. IEEE MAG-17 (1981) 387.
(3) M. L. Roukes, R. S. Germain, M. R. Freeman and R. C. Richardson, this volume.
(4) M. R. Freeman, M. L. Roukes, R. S. Germain and R. C. Richardson, this volume.
(5) C. D. Tesche, Appl. Phys. Lett. 41 (1982) 490.
(6) M. L. Roukes, unpublished.
(7) M. B. Ketchen and J. M. Jaycox, Appl. Phys. Lett. 40 (1982) 736.
(8) J. M. Jaycox, Master's Thesis, M.I.T. (1981).

LT-17 (Contributed Papers)
U. Eckern, A. Schmid, W. Weber, H. Wühl (eds)
© Elsevier Science Publishers B.V., 1984

MAGNETIC HYSTERESIS OF CRITICAL CURRENTS IN SUPERCONDUCTING GRANULAR ALUMINUM BRIDGES

T. AOMINE* and A. YONEKURA**

* Research Institute of Fundamental Information Science, Kyushu University, Fukuoka 812, Japan
** Department of Information Systems, Kyushu University, Kasuga 816, Japan

Hysteresis of critical currents vs. perpendicular magnetic fields in granular Al uniform-thickness bridges has been studied as a function of temperature. The temperature below which the hysteresis occurs differs with bridge width. By adding pulse magnetic fields to bias dc magnetic fields, samples are switched from voltage v=0 to v≠0 and vice versa.

1. INTRODUCTION

It has been observed (1) that critical currents Ic of bridges with geometrically symmetrical constriction and uniform thickness, which were made of type II superconducting granular aluminum films, show hysteresis in the limited range of a perpendicular magnetic field H. The hysteresis is different from that due to a localized normal hotspot maintained by Joule heating (2).

In this paper we describe experimental results on temperature dependence of hysteresis curve of Ic(H), the variation of Ic with H, and a voltage response for pulse magnetic fields, which were not examined in a previous work.

2. EXPERIMENTAL METHOD

The experimental technique used in this work is essentially the same as used in the previous work (1). Samples have such a geometry as in Fig. 1. A waist of constriction is geometrically symmetrical. Bridge width w ranges from 5 μm down to 2 μm and the bridge length about 1 μm.

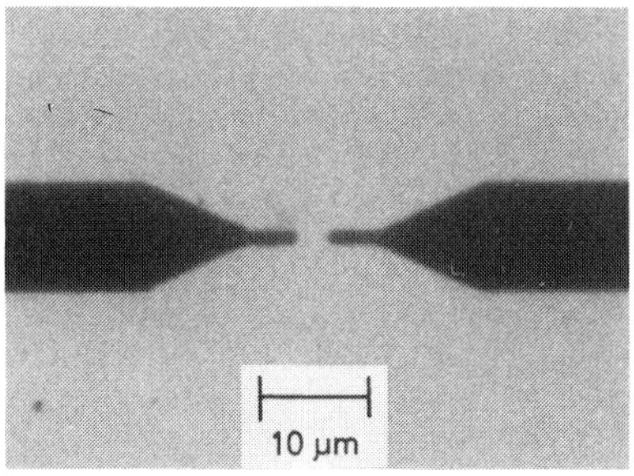

FIGURE 1
Photograph of the sample with w=3 μm taken with an optical microscope.

All the samples have the same bank width, 190 μm. Their thickness d is about 40 nm. The values of $\lambda_{\perp}=2\lambda^2/d$ in the entire investigated temperature region are larger than w of the samples used in this work. Here λ is the bulk penetration depth of the material. We used the apparatus similar to that in (1), allowing recording of the curve Ic(H) on an X-Y recorder.

3. RESULTS AND DISCUSSION

The observed behaviour in Ic(H) is similar to that in the previous work. The hysteresis occurs within the limited values of H. As a typical hysteresis curve of Ic(H), the case for the sample with w=2 μm at T=1.45 K is shown in Fig. 2. Here curves ①, ②, ③ correspond to those obtained by the following ways: ① increasing H from zero to the maximum value, Hm, in the figure, ② decreasing H from Hm to zero, ③ increasing H, after the operation

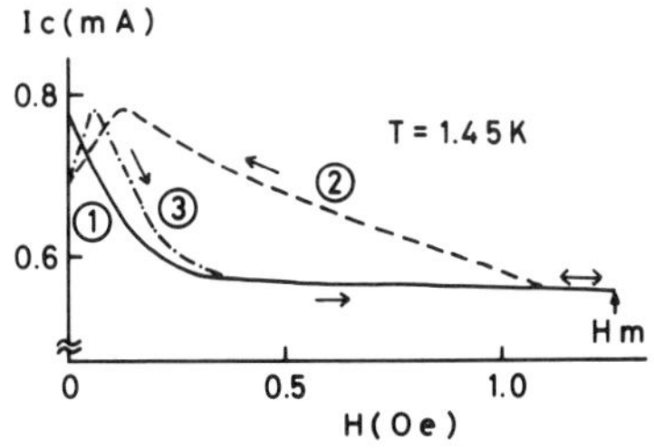

FIGURE 2
Ic(H) curve for the sample with w=2 μm at T=1.45 K. ① increasing H from zero to the maximum value, Hm, in the figure, ② decreasing H from Hm to zero, ③ increasing H, after the operation corresponding to curve ②, from zero to Hm.

corresponding to curve ②, from zero to Hm.

The curve obtained when H is decreased from Hm to zero is always the same as curve ②. The value of Ic at H=0 usually has two values. The state with the higher value of Ic is more stable than with the lower value, which is always realized when H is decreased from Hm to zero (see curve ②). It is not rare that Ic at H=0 changes from the lower value to the higher value due to an electric shock. There are cases that Ic's at H=0 are the other values between the higher and lower values, although they are not frequently seen and not so stable as the higher value. The higher value of Ic at H=0, called Ic(0), is almost consistent with the maximum of Ic in curve ②. The Ic(H) curves with hysteresis are the same for the reversal of sign of both the sample current I and H, but when the sign of either I or H is changed, the curves are not the same.

There are samples in which the hysteresis is not observed near the critical temperature Tc. The temperature below which the hysteresis occurs differs with w. The difference between Ic(0) and the value of Ic in the high H portion of Ic(H) with no hysteresis, δIc, changes linearly with temperature. Ic(0) has also linear temperature dependence. The value of δIc normalized by Ic(0), δIc/Ic(0), is plotted against T/Tc in Fig. 3, where Tc is determined from the extrapolation of Ic(0) vs. T to Ic(0)=0. The

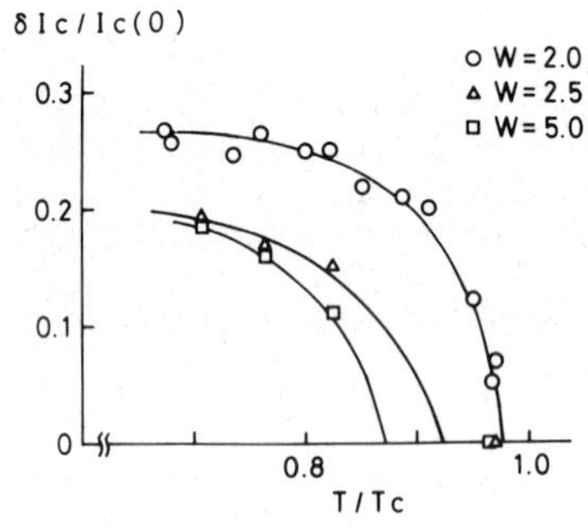

FIGURE 3
δIc/Ic(0) as a function of T/Tc for three samples with different bridge width.

temperature dependence of δIc/Ic(0) is reflected by the fact that δIc and Ic(0) have the nearly same temperature dependence, but the temperatures extrapolated to δIc=0 and Ic(0)=0 are different.

As for the magnetic hysteresis thus obtained, it is difficult to find out its cause in the bridge; owing to the method of Ic measurement, the bridge becomes normal at each value of H and has nothing which remembers the value of H. Considering that in addition to the method of Ic measurement, the temperature at which the hysteresis begins in reducing temperature depends on w, we think that the joint between the bank and the bridge is associated with the hysteresis.

Next let us study a sample voltage response v for pulse magnetic fields h, in applying the sample current of triangular waveform and the bias dc magnetic field, which correspond to "x" in Ic(H) of inset of Fig. 4. The observed result is shown in Fig. 4. Thus it is confirmed

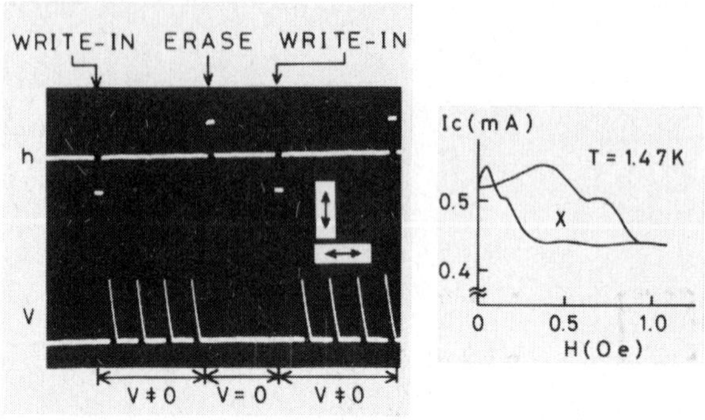

FIGURE 4
Sample voltage V (sample with w=3 μm) and pulse fields h for write-in and erase vs. time. The vertical scales for V and h are 10 mV/div and 0.49 Oe/div, respectively. The horizontal scale is 10 ms/div. One division for each scale is given in the photograph. The used sample current and bias dc magnetic field correspond to "x" in Ic(H) of inset.

that the sample has the two states with v=0 and with v≠0 under the bias dc magnetic field, depending on write-in and erase pulses of h.

4. SUMMARY

The hysteresis of Ic(H) curve in granular Al uniform-thickness bridges is found to have the temperature dependence. The temperature below which the hysteresis occurs depends on bridge width. By using the hysteresis switching from v=0 to v≠0 or from v≠0 to v=0 is accomplished under pulse magnetic fields in addition to bias dc magnetic fields.

ACKNOWLEDGEMENTS

We would like to thank Dr. K. Mizuno for helpful discussions.

REFERENCES

(1) K. Mizuno and T. Aomine, J. Phys. Soc. Jpn. 52 (1983) 4311.
(2) W.J. Skocpol, M.R. Beasley and M. Tinkham, J. Appl. Phys. 45 (1974) 4054.

LT-17 (Contributed Papers)
U. Eckern, A. Schmid, W. Weber, H. Wühl (eds)

THERMAL COEFFICIENTS OF SQUID MAGNETOMETERS*

J. LUU, T.C.P. CHUI and J. A. LIPA

Physics Department, Stanford University, Stanford, CA 94305

We report measurements of the temperature and temperature gradient coefficients of a commercially available (1) SQUID magnetometer system. We find that the gradient coefficient is sufficiently large to explain observations of excess noise sometimes seen when the magnetometers are located in a helium bath and operating at their maximum sensitivity. Measurements with the input circuit open and with the sensor in vacuum indicate the excess noise may be due to thermo-electric effects.

1. INTRODUCTION

In the application of SQUID magnetometers to experiments which demand the maximum possible signal to noise ratio, it is essential to minimize all possible sources of interfering signals. One noise source that appears to be intrinsic to the SQUID is the white noise spectrum above about 0.1 Hz that has been extensively studied by Clarke and others (2). Less well understood is the 1/f portion of the spectrum extending to much lower frequencies. However both these noise contributions exist as a "minimal background" present even in the most carefully controlled circumstances. In this paper we describe some results from an investigation of a low frequency excess noise source, which we found to exist to varying degrees in a number of commercially available magnetometers (1) when operated in typical experimental helium dewars. This noise source appears to be related to thermal phenomena occurring in the probe assembly that forms an intimate part of the magnetometer system, normally housing the SQUID sensor and the rf transmission line, and is probably due to low level thermoacoustic oscillations. As part of our investigation of this phenomenon we measured both the temperature and temperature gradient coefficients of a number of sensors, and report these results here. The gradient coefficient was surprisingly high and adequate to generate noticable pick-up in the SQUID input circuit from thermo-electric currents flowing in the normal metals which form integral parts of the probe and SQUID assemblies.

2. APPARATUS

Since preliminary observations indicated that the excess noise phenomena were too complex to be explained in terms of the SQUID temperature coefficient alone, we instrumented four independent SQUID probe and sensor units to measure both their temperature and temperature gradient coefficients. The devices were equipped with two internally mounted thermometers and two external heaters as shown in Figure 1. The measurements were made with the low temperature portion of the SQUID probe assembly contained in a vacuum-tight metal cylinder filled with helium exchange gas to a pressure of about 50 mm Hg. The lower portion of the cylinder was made from 0.1 cm thick OHFC copper to promote thermal equilibrium in the neighborhood of the SQUID with the heaters off. To generate the temperature offsets and gradients heater power levels in the range 0.1 to 10 mW were used. The resulting d.c. offsets of the SQUID output were of the order 10^{-3} Φ_o/mW, independent of the polarity of the heater currents. Because of the difficulty in establishing good thermal contact between the thermometers and the SQUID sensor body, the exact meaning of the temperature gradients measured within the probe is somewhat unclear. Consideration of the thermal impedances involved indicates that the gradients are probably more representative of the brass SQUID holder assembly than of the sensor body itself.

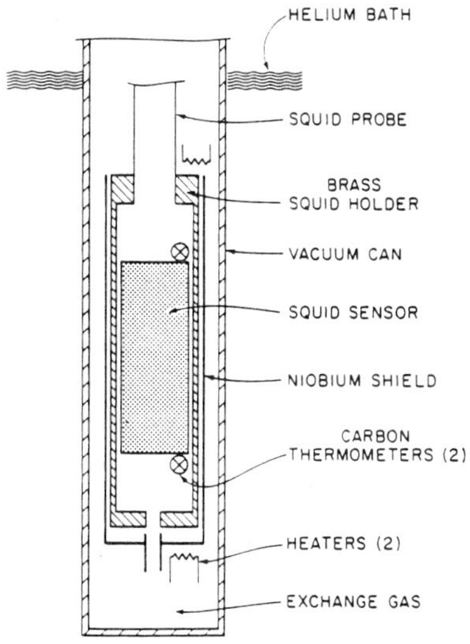

Figure 1: Schematic view of test assembly.

* Work supported by JPL Contract #955057

3. OBSERVATIONS

A typical example of the SQUID output data under various conditions is shown in Figure 2. The curve between A and B shows the initial cool-down characteristic for the sensor, with a thermal time constant of 10-15 sec. Between C and D 4 mW was dissipated in the upper heater, and between E and F, 2 mW was dissipated in the lower heater. This last section is interesting in that it shows the SQUID output falling below the zero power base-line even though both thermometers showed a rise in temperature. Such an effect is difficult to explain on the basis of an equilibrium temperature coefficient alone. Indeed, the major change between the portions CD and EF was a reversal of sign of the temperature gradient. Initially we attempted to fit the data from the two thermometers and the SQUID with a model assuming a simple temperature coefficient was the only source of the SQUID offsets, but had no success. With the inclusion of an offset term proportional to ∇T, we obtained a greatly improved fit, except for a few departures which appeared to be related to small changes in the SQUID baseline.

The values we obtained for the temperature and gradient coefficients of the four sensors are listed in Table 1. All sensors were studied with a superconducting short across the input terminals to represent a minimal pickup experimental situation. Measurements were also made on two sensors with the input terminals open circuit to examine the extent to which the coefficients could be attributed to phenomena associated with the input circuit. We found that the gradient coefficient was considerably reduced in the open circuit case, indicating that significant magnetic fields were being generated in the vicinity of the SQUID. This observation is not surprising in view of the existence of large quantities of normal metals within the SQUID probe assembly. It is difficult to eliminate these materials entirely since the hermetic SQUID sensor case itself is fabricated from a

SQUID #	$A_T(\phi_o/K)$	$A_G(\phi_o/K/cm)$
1	.071	.208
1 (o/c)	.081	.060
2	.028	.584
3	.007	.292
3 (o/c)	.006	.020
4	.003	.336

Table 1: Temperature coefficients (A_T) and gradient coefficients (A_G) of four SQUID sensors. All measurements except those indicated were made with shorted input terminals. Sensors were of the toroidal point contact type (1), except for #4 which had a thin film junction (S.H.E. Model TSQ).

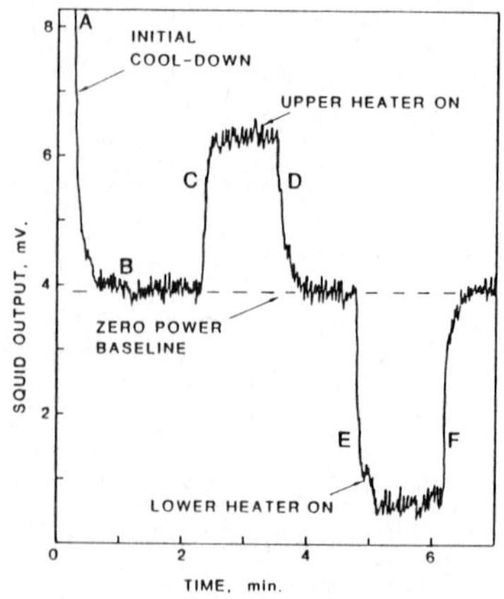

Figure 2: SQUID output under conditions described in the text.

beryllium/copper alloy. Selzer (3) has reported easily detectable thermomagnetic fields associated with temperature gradients in slabs of normal metals, even without the presence of junctions between dissimilar materials. Scaling his results to the case of beryllium/copper, we estimate a temperature gradient of 10^{-5} deg/cm along the SQUID case would be sufficient to generate offsets of the same order as those seen in the present experiment.

4. CONCLUSION

The existence of a significant temperature gradient coefficient can be important in experiments concerned with the low frequency noise and d.c. stability of SQUID magnetometers. For example, in an experiment to study the lambda-transition of helium using a high resolution thermometer with a SQUID readout system (4), a dramatic reduction in spurious noise was obtained by relocating the sensors in a vacuum space where they were housed in superconducting shields containing no normal metal. In general we would expect noticeable additional noise to be present in the low frequency regime in any experiment involving a sensor connected to a superconducting input ciruit, unless special precautions are taken to avoid the effects of the gradient coefficient.

REFERENCES

(1) Model SP with Model 330 electronics, S.H.E. Corporation, Sorrento Valley Road, San Diego, California 92121
(2) J. Clarke in SQUID Devices and their Applications (de Gruyter, Berlin, 1977) eds: H. D. Hahlbohm and H. Lubbig, p. 213.
(3) P. Selzer, Thesis, Stanford University, 1974 (unpublished)
(4) J. A. Lipa and T.C.P. Chui, Phys. Rev. Letts., 51, 2291 (1983).

LT-17 (Contributed Papers)
U. Eckern, A. Schmid, W. Weber, H. Wühl (eds)
© Elsevier Science Publishers B.V., 1984

SATURATION THICKNESS FOR JOSEPHSON TUNNEL OXIDE ON NIOBIUM BY ION BEAM OXIDATION.

Ralf HERWIG

Institut für Elektrotechnische Grundlagen der Informatik, University Karlsruhe,
P.O. Box 6380, D 7500 Karlsruhe+

Nb-Nb O -PbAu Josephson junctions have been fabricated by ion beam oxidation. The maximum Josephson current is found to reach a saturation limit with no further dependence on oxidation time.

In the fabrication of Josephson junctions the plasma oxidation process (1) is the common method for preparing the tunneling barrier. On Pb alloy base electrodes during oxidation a saturation oxide thickness can be reached. On Nb, however, the high oxidation rate cannot be compensated by the sputter rate that way and the current density remains time dependent (2). Time dependencies up to proportional t^{-5} have been reported (3) being detrimental to the reproduction of current densities.

A fabrication method for Josephson junctions using an ion beam oxidized barrier on niobium or niobiumnitride has already been published (4,5,6). Since in contrast to plasma processing the oxidation rate by the oxygen atoms in the background gas and the sputter rate of the argon ion beam can be controlled separately over a wide range there might be a chance to find a steady state process between oxidation and sputter remove of niobiumoxide.

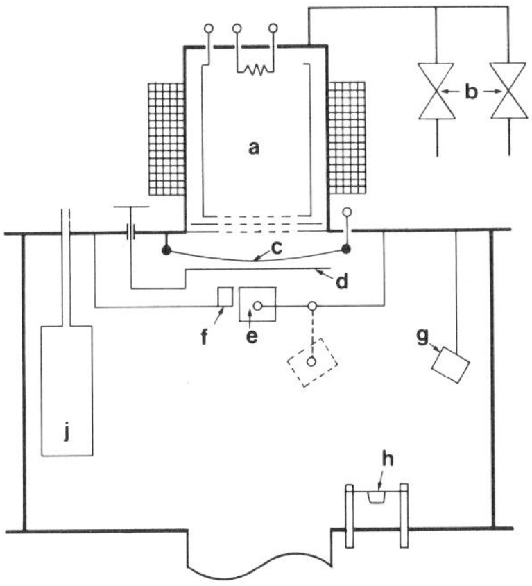

FIGURE 1
Schematic diagramm of the experimental arrangement. The indicated components are: a. Plasma chamber, b. Gas inlet, c. Neutralization filament, d. Beam shutter, e. Substrate holder, f. Beam analyzer, g. Quartz filmthickness monitor, h. Evaporators, j. Liquid nitrogen baffle.

The tunnel junction used in the present experiments were fabricated on thermically oxidized Si wafers. A niobium layer was evaporated from an electron beam heated source with high purity Nb to a thickness of 200 nm at a substrate temperature of about 350 °C. It was structured by wet chemical etching and a photoresist mask of AZ 1350 J. Then, an isolating window layer was applied by evaporation of SiO with the window areas screened by a lift off mask made from AZ 1350 J photoresist in chlorobenzene technology. After SiO window formation the samples were covered with another lift off mask for the top electrode produced in the same way and mounted in the oxidation equipment shown in Fig. 1. As ion beam source a Kaufmann type plasma chamber of a Veeco Microetch machine with three extraction grids was used. The pressure was controlled by piezoelectric actuated valves and by ionizing gauge measurement combined with a quadrupol mass analyzer and a rotating ball viscosity vacuum meter for calibration. A five grid analyzer attached close to the sample was used for monitoring beam energy and neutralization conditions. After pump down below 10^{-4} Pa during a 300 s cleaning procedure about 20 nm of niobium was removed by an 600 eV Ar beam. For oxidation O_2 background pressures from 10^{-4} up to 10^{-3} Pa were applied in mixing Ar and O_2 in the gas input flow. The ion beam was space charge neutralized and its energy was varied from 100 eV up to 600 eV. After oxidation the substrat holder was revolved to a position opposite the evaporation sources and the top electrode consisting of Pb with 1.7 wght.% Au and a thickness of 300 nm was evaporated. Additional layers for isolation and control currents can be deposited in the usual manner.

Fig. 2 shows the measured dependence of the current density of two series of samples on oxidation time. A saturation current density corresponding to a saturation thickness of the

+ Supported in part by the German Minister of Research and Technology under grant 423-7291-NT 2515 6 tunnel oxide is clearly achieved. The saturation time constant decreases with increasing ion beam energy. If the oxidation rate is described with $dX/dt = K \exp(-X/X_0)$ according to (7) and the sputter rate with $dX/dt = R$ where X is the oxide thickness the measured points in Fig. 2 are matched by the following parameters; a) R = 0.05 nm/s, Xo = 2 nm and K = 0.12 nm/s; b) R = 0.01 nm/s, Xo = 2 nm and K = 0.03 nm/s. K depends mainly on the oxygen partial pressure and the structural disorder in the niobium oxide owing to the energy of the ion beam.

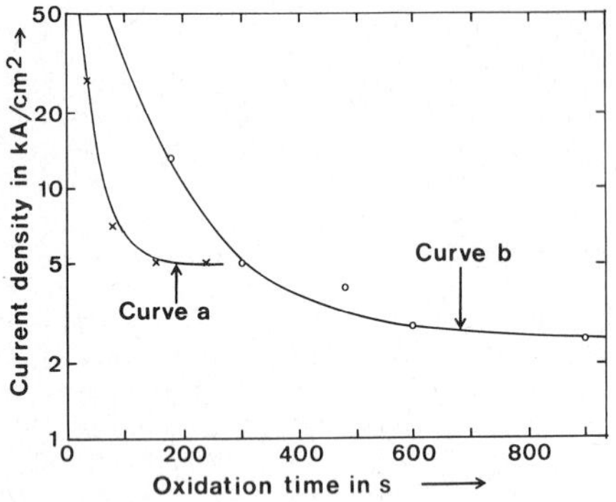

FIGURE 2
Oxidation time dependence of the Josephson current density. Curve a: Beam energy = 600 eV, beam current density = 0.15 mA/cm^2, oxygen partial pressure = 0.9 mPa.
Curve b: 300 eV, 0.12 mA/cm^2, 0.7 mPa.

Fig. 3 shows the I-U characteristic of 31 Josephson junctions (8μm diam.) connected in series with a current density of 6 kA/cm^2 at 4.2 K. The standard deviation s=5% is mainly due to geometrical errors on the mask produced by microprojection. The relatively high leakage current is consistent with published data from processes in clean environment (8). Because of the absence of any backsputtering effect during the ion beam oxidation process there is only very small contamination. Without application of an additional diffusion barrier a heterogeneous oxide layer with suboxides is formed, which can carry single particle currents below the gap voltage (9).

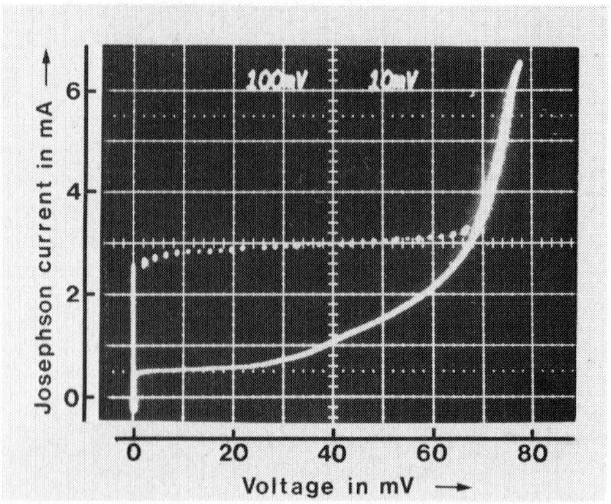

FIGURE 3
I-U characteristic of 31 Josephson junctions connected in series.

To sum up, we have fabricated Nb-Nb_xO_y-PbAu Josephson junctions by an ion beam oxidation technique which is applicable to integrated devices and potentially allows to overcome the problem of current density tolerances. Further work is in progress to lower the subgap currents and to obtain more information about the oxidation process.

The author would like to thank Prof. W. Jutzi for supporting this work and for many useful discussions and Dr. M. Neuhaus for fabricating the niobium films.

(1) J.H. Greiner, J. Appl. Phys. 42 (1971) 5151
(2) P.C. Karulkar and J.E. Nordmann, J. Appl. Phys. 50 (1979) 7051
(3) R.F. Broom et al., IEEE Trans. on Electr. Dev. 27 (1980) 1998
(4) R. Herwig, Electr. Lett. 16 (1980) 850
(5) A.W. Kleinsasser and R.A. Buhrmann, Appl. Phys. Lett. 37 (1980) 841
(6) R.B. van Dover and D.D. Bacon, IEEE Trans. Magn. 19 (1983) 951
(7) J.H. Greiner, J. Appl. Phys. 45 (1974) 32
(8) W.J. Gallagher, S.I. Raider and R.E. Drake, IEEE Trans. Magn. 19 (1983) 807
(9) J. Halbritter, Surf. Sci. 122 (1982) 80

LT-17 (Contributed Papers)
U. Eckern, A. Schmid, W. Weber, H. Wühl (eds)

CYLINDRICAL RF-SQUIDS MADE OF Nb_3Ge FILMS

Shinya KURIKI, Masayuki SUEHIRO and Hajime KONISHI

Research Institute of Applied Electricity, Hokkaido University, sapporo, Japan

Nb_3Ge thin-film microbridges are fabricated on a cylindrical substrate to form rf-SQUIDs. High temperature operations well above 10 K (up to 17.5 K) are obtained. An intrinsic flux noise of the order of 10^{-5} $\phi_0/\sqrt{Hz}$ is estimated from the step rise parameter. Comparison with a theoretical noise indicates a non-sinusoidal current-phase relation of the microbridge.

1. INTRODUCTION

Several reports have described (1 - 4) dc- and rf-SQUIDs which are made of high T_c films of NbN, Nb_3Sn and Nb_3Ge and operate at temperatures more than 10 K. It is thought that weak links made on a cylindrical substrate are suited for the element of practical rf-SQUID because of low inductance and good coupling to the input coil. In the present work we have fabricated microbridges using Nb_3Ge films deposited on a cylindrical substrate and measured the rf-SQUID characteristics.

2. FABRICATION

2.1. Nb_3Ge films

Quartz and sapphire rods (1 mmϕ x 15 mm) are used as substrates. Nb_3Ge films have been rf-sputtered from a composite target of Nb and Ge plates at an Ar pressure of 0.2 Torr. The quartz or sapphire rod is placed in a box-heater, which is joule-heated by current, made of a Ta sheet. The temperature of the rod is raised to about 800 °C by radiation of the three walls of the heater. The Nb_3Ge is deposited on a half side of the rod without the wall. The rod is then rotated by 180 deg in the same sputtering run, so that the Nb_3Ge is deposited on the other half side. The film thickness is 1000 - 1500 Å on one side where the microbridge is fabricated and 1500 - 2000 Å on the other.

2.2. Microbridges

Electron-beam exposure of PMMA resist is used to form a microbridge pattern on the Nb_3Ge film. The long focal length of the electron beam is favorable when a curved curface such as cylindrical substrate is exposed. The Nb_3Ge film is then plasma-etched with CF_4 gas. A low rf voltage of 400 V and high gas pressure of 0.5 Torr are used to minimize the bombardment which may reduce the T_c of the Nb_3Ge film. SQUID samples having a microbridge with lengths l = 0.6 - 0.8 μm and widths w = 0.3 - 0.6 μm are fabricated.

3. MEASUREMENTS

The cryostat consists of a vacuum tight μ-metal can filled with He exchange gas and an aluminum block held in the can. The temperature of the block is controlled above 4.2 K with a temperature sensor and a heater cemented to the block. The SQUID sample around which a coil of thin Cu wire is wound is inserted in a hole of the block. The coil forms the tank circuit with a capacitor in parallel. The self inductance L of the SQUID is estimated to be about 0.5 nH from the values reported for cylindrical SQUIDs of 1 - 3 mm in diameter. Commercial electronics (SHE Corp., San Diego, Calif., USA, Model 330X) with an operating frequency of 19 MHz are used in the rf measurements.

3.1. I-V characteristic and critical current

An example of rf I-V characteristic showing a step pattern is shown in Fig.1 for different temperatures. The steps have a slope which originates in the uncertainty of the external flux at which the SQUID ring jumps to adjacent fluxoid states. The slope reflects the width of the transition distribution due to thermal fluctuations. Another source of the slope is non-sinusoidal current-phase relation (CPR) of the weak link in the SQUID as Jackel and Buhrman have shown (5).

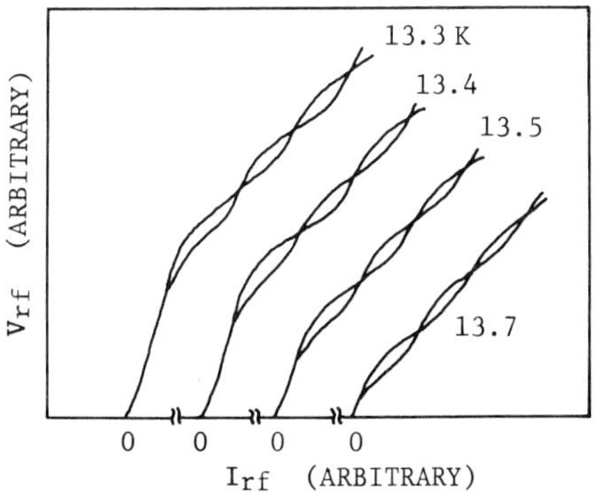

FIGURE 1
I-V characteristic at different temperatures.

TABLE I. Characteristics of the SQUIDs

Sample No.	l μm	w μm	T_{op} K	α	$\delta\phi_\alpha$ $10^{-5}\phi_o/\sqrt{Hz}$	$\delta\phi_i$ $10^{-5}\phi_o/\sqrt{Hz}$
771	0.7	0.6	6 - 13	0.35 - 0.5	5.6 - 8	2.5 - 3.8
752	0.6	0.3	16.5 - 17.5	0.5 - 0.55	8.5 - 9	3.4 - 5.3
OAB	0.8	0.4	14 - 15	0.5 - 0.6	8 - 9.6	5.0 - 8.2
OBE	0.8	0.3	10 - 13	0.4 - 0.5	6.4 - 8	4.0 - 4.4

The value of LI_c/ϕ_o is estimated from the rf I-V characteristic measured at various temperatures. It shows a gradual and then steep rise as the temperature decreases from T_c. The temperature variation of the critical current is approximated by a form of $I_c \propto (T_c - T)^n$, which gives the n values between about 1.5 and 2 for different samples. These values are slightly lower than the values of 2 - 2.5 which are obtained from the dc I-V characteristic of Nb_3Ge variable-thickness bridges in our previous work (6). In Table I the range of temperatures T_{op} for which the SQUID operation is observed is shown together with dimensions of the microbridge. The samples 771 and 752 are made on the quartz substrates and OAB and OBE are on the sapphire substrates. The lowest T_{op} is limited by the steep rise of I_c with temperature decrease, while the highest T_{op} is given by T_c of the microbridge. Though small in range, a high temperature operation above 17 K is obtained in the sample 752. The sample 771 has a wide range of T_{op} between 6 and 13 K.

3.3. Noise

As mentioned before the slope of the steps in the rf I-V characteristic reflects the flux noise of the SQUID. In terms of the fractional step rise parameter α, defined as the ratio of the voltage rise along the step to the voltage difference between the steps, the mean square value of the flux noise is given by (5),

$$\delta\phi_\alpha \simeq 0.7\alpha f^{-1/2}\phi_o , \qquad [1]$$

where f is the operating frequency of the SQUID. This equation gives the intrinsic noise of the SQUID with any CPR. The measured values of α in the operating temperature range and the calculated $\delta\phi_\alpha$ are shown in Table I. The samples have $\delta\phi_\alpha$ values between 5 and 10×10^{-5} $\phi_o/\sqrt{Hz}$.

Kurkijarvi has calculated (7) the width of the transition for an ideal rf-SQUID having a sinusoidal CPR, from which the intrinsic flux noise $\delta\phi_i$ is calculated as follows:

$$\delta\phi_i \simeq 0.7(3\pi/2\sqrt{2})^{2/3} f^{-1/2}(LI_c)^{1/3} \times (k_BTL/\phi_o)^{2/3} . \qquad [2]$$

The range of $\delta\phi_i$ calculated from the $LI_c(T)$ for each sample is compared with $\delta\phi_\alpha$ in Table I. The result that $\delta\phi_\alpha$ is 1.5 - 2 times larger than $\delta\phi_i$ indicates that the intrinsic flux noise is affected by the CPR which deviates from the sinusoidal one. From a short coherence length (~50 Å) of the Nb_3Ge it is expected that the microbridges are in the vortex flow regime having a skewed CPR.

The flux noise $\delta\phi_n$ of the total SQUID system has been measured by closing a flux-locked loop. The noise voltage at the output of the SQUID electronics is measured with a 100 Hz low pass filter, and converted to a flux noise per $\sqrt{Hz}$ to be $\delta\phi_n = 2 - 4 \times 10^{-4}$ $\phi_o/\sqrt{Hz}$. Since $\delta\phi_n$ includes the tank circuit noise and the preamplifier noise, it usually exceeds the intrinsic noise $\delta\phi_\alpha$. It is noted that one of the best rf-SQUID systems commercially available has a $\delta\phi_n$ value of the order of 1×10^{-4} $\phi_o/\sqrt{Hz}$. In view of high temperature operation well above 10 K the noise level of our SQUIDs is in a reasonable range.

ACKNOWLEDGEMENT

The authors would like to thank Professor G. Matsumoto for many helpful discussions.

REFERENCES

(1) C.T. Wu and C.M. Falco, J. Appl. Phys. 49 (1978) 361.
(2) S.A. Wolf, E.J. Cukauskas, F.J. Rachford and M. Nisenoff, IEEE Trans. Magn. MAG-15 (1979) 595.
(3) T. Fujita, M. Suzuki, S. Ikegawa, T. Ohtsuka and T. Anayama, ICEC 9th Conf. Proc. (Kobe, 1982) 369.
(4) M.S. DiIorio, A. de Lozanne and M.R. Beasley, IEEE Trans. Magn. MAG-19 (1983) 308.
(5) L.D. Jackel and R.A. Buhrman, J. Low Temp. Phys. 19 (1975) 201.
(6) S. Kuriki, A. Yoshida and H. Konishi, J. Low Temp. Phys. 51 (1983) 149.
(7) J. Kurkijarvi, Phys. Rev. 6B (1972) 832.

LT-17 (Contributed Papers)
U. Eckern, A. Schmid, W. Weber, H. Wühl (eds)

HIGH FREQUENCY RESPONSE OF THE SELF-RESONANT CURRENT STEP OF A JOSEPHSON TUNNEL JUNCTION*

Håkan OLSSON and Tord CLAESON

Physics Department, Chalmers University of Technology, S-412 96 Göteborg, Sweden

Superconducting Pb tunnel junctions were irradiated with 10 GHz microwaves. The first self-resonant current step of the I-V curve decreased, and higher order satellite steps appeared as the microwave power increased. The behaviour speaks in favour of a model where the rf signal mixes with the internal Josephson oscillation producing a voltage component at the resonant frequency.

1. INTRODUCTION

When a Josephson oscillator (with a frequency of $f_0=2eV_0/h$) is coupled to a resonator, a current step (in the I-V curve) occurs at a voltage where the frequency equals the resonance frequency, f_r. The resonator can be an external waveguide resonator (1), a microstrip LC-circuit (2), or a tunnel junction cavity (3,4). The dc power fed into the oscillator equals the resonator losses. Hence resonance step heights and widths depend upon the resonator Q-value (4). We have studied the response of the first self resonant, or Fiske (3) step to an applied rf signal. Satellite steps occur when the sum of the Josephson frequency and a multiple of the rf frequency equals the resonant frequency. We argue that the different current steps roughly show squared Bessel function dependences on the rf voltage.

2. EXPERIMENTS

Pb tunnel barriers were grown by thermal oxidation. The junction was connected to a $\lambda/4$ long microstrip transforming the 50 Ω impedance of the coaxial cable to 14 Ω at the junction. The critical Josephson current, I_0, could be tuned magnetically. Measurements were performed at 4.2 K.

Both in-line and crossed strip junctions were tested. Depending upon junction length and critical current, first order resonance frequencies, f_r, varied between 100 and 550 GHz. The results reported here were taken on an in-line junction with I_0=485 μA, L/λ_j=0.4, and f_r=100 GHz. The applied frequency was f_1=10.7 GHz. To observe the self-resonant step, the zero voltage current was depressed by a magnetic field.

3. RESULTS

Slowly swept I-V curves are shown in Fig. 1, taken with and without rf power. The Fiske step decreased and successively higher order satellite peaks appeared as the power was increased.

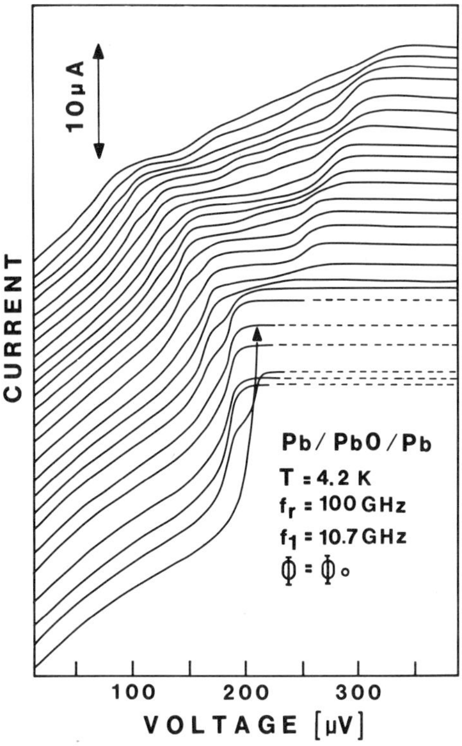

FIGURE 1
I-V curves in a region around the first self resonant step. The lower curve was taken without rf radiation, while the rf power increased by 17 dB going from the next lowest curve to the one at the top. Satellite steps occur approximately 22 μV apart. For clarity, the curves were shifted vertically.

The voltage separation of the peaks was $\Delta V=hf_1/2e$. The new current steps had widths reflecting the width of the first resonant step. Small Q junctions showed satellite bumps rather than peaks.

*Supported by the Swedish Natural Science Research Council and the Swedish Board for Technical Development.

4. DISCUSSION

To make a quantitative analysis we use a voltage biased junction model and assume that only current and voltage components at frequencies f_o, f_1, and f_r have to be considered, i.e.:

$$V = V_o + v_1 \cos(2\pi f_1 t+\phi_1) + v_r \cos(2\pi f_r t+\phi_r).$$

The Josephson current is expanded in Bessel functions:

$$I = I_o \sum_{m=-\infty}^{\infty} \sum_{n=-\infty}^{\infty} J_m(\alpha_1)J_n(\alpha_r) \times$$

$$\sin\{2\pi(f_o+mf_1+nf_r)t + \phi_o+m\phi_1+n\phi_r\}$$

where $\alpha_1 = 2ev_1/hf_1$, and $\alpha_r = 2ev_r/hf_r$.

The dc step height and the rf amplitude at f_r when biased at V_o, given by $f_o+mf_1=f_r$,

$$I_m = I_o\, J_m(\alpha_1)\, J_1(\alpha_r)$$

$$i_r = I_o\, J_m(\alpha_1)\, J_o(\alpha_r)$$

At the top of an induced step, the admittance is real. With $Y_r = i_r/v_r$ we get

$$\alpha_r = (2eI_o/hf_rY_r)\, J_m(\alpha_1)\, J_o(\alpha_r)$$

If α_r is small, $J_o(\alpha_r) = 1$, $J_1(\alpha_r) = \alpha_r/2$, and

$$I_m = (eI_o^2/hf_rY_r)\, J_m^2(\alpha_1).$$

The steps occur at voltages $m\Delta V$ from the Fiske step and their heights are squared Bessel functions of the rf voltage.

In Fig. 2 we give step heights as functions of α_1. They were determined by subtracting oscilloscope I-V curves traced with and without rf power. The full curves of Fig. 2 are drawn using the relation for I_m derived above. In order to normalize the calculated curves, we used Y_r=11 mho and related α_1 to the square root of the applied rf power at one point such that a resonable fit was obtained for the first m=0 lobe. Note that this calibration of α_1, also gave a correct α_1 location of the maxima of the satellite steps.

(To check the calibration of α_1, we compared with the rf power dependence of the normal rf induced Josephson steps when no magnetic field was applied. The regular variation of the first 5 of these steps with power gave an α_1 vs square root of the power relationship that agreed within ±5% with the one obtained from the self resonance step variation, surprisingly good accord. Josephson steps of higher order than 5 needed a considerably higher power to give a certain α_1.)

While the locations of the higher order peaks in Fig. 2 are correct, their heights are smaller than expected from the m=0 peak. This is not surprising. We have used a voltage bias model and ignored at least two complications. Y_r is not independent of biasing and power dissipated at harmonic frequencies contribute to step amplitudes. Both contributions may depend on α_r which in turn depends on $J_m(\alpha_1)$. Hence both Y_r and the harmonic contribution may depend on m.

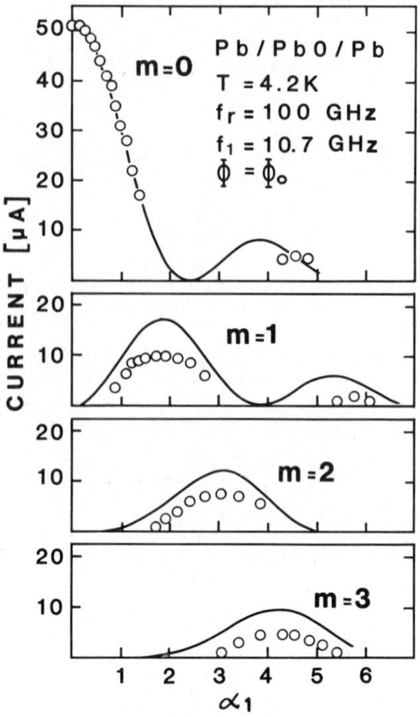

FIGURE 2
The rf voltage dependence of step amplitudes for m=0,1,2,3. Experimental points are compared with squared Bessel function curves.

5. CONCLUSION

The resonant current step due to a tunnel junction cavity varies regularly with applied rf power. Extra, induced steps appear. These may be explained using a mixer model - the rf signal combines with the internal Josephson radiation to give a component at the resonant frequency where power is dissipated. Rf voltages determined both from satellite and rf induced steps agreed well.

REFERENCES

(1) A. Longacre Jr, Proc 5th Appl. SuperCo Conf, Pap, Annapolis, Md, May 1-3 1972, p 712.
(2) H.H. Zappe and B.S. Landman, J. Appl. Phys. 49 (1978) 344.
(3) M.D. Fiske, Rev. Mod. Phys. 36 (1964) 221.
(4) I.O. Kulik, Zh. Tekh. Fiz. 37 (1967) 157. (Sov. Phys. Tech. Phys. 12 (1967) 111)

LT-17 (Contributed Papers)
U. Eckern, A. Schmid, W. Weber, H. Wühl (eds)

RESONANT MODES FOR A JOSEPHSON JUNCTION COUPLED TO A STRIPLINE RESONATOR

Gianfranco PATERNO'

Associazione EURATOM-ENEA sulla Fusione, Centro Ricerche Energia Frascati, C.P. 65, 00044 Frascati, Rome, (Italy)

Anna Maria CUCOLO, Giuseppina MODESTINO
Dipartimento di Fisica, University of Salerno, 84100 Salerno (Italy)

Resonant modes in small Nb-NbO_x-Pb Josephson junctions coupled to a superconducting stripline resonator have been observed. The dependence of the resonance amplitude on the magnetic field has been investigated. The experimental data have been compared with a simple theoretical expression derived for small values of the coupling parameter Γ.

1. INTRODUCTION

Resonant modes due to the coupling with an external cavity have been observed for point contact (1,2), and for bridges (3,4). As for as we know the only observations for tunneling junctions have been reported by Finnegan and coworkers (5) for Pb-PbO_x-Pb devices. The problem has been theoretically investigated by Werthamer and Shapiro (6) and by Smith (7).

In this paper we report on the observation of resonant modes in small Nb-NbO_x-Pb Josephson junctions coupled to a superconducting stripline resonator. The dependence of the amplitude of the resonances by the applied magnetic field has been investigated. The experimental data have been compared with the theoretical results valid for small values of Γ the coupling parameter. A simple expression for the magnetic field dependence is derived. A good agreement has been found between exprerimental and theoretical results.

2. THEORETICAL CONSIDERATIONS

Let us briefly summarize the main results of the theory of resonances in a junction coupled to an external resonant circuit. The maximum amplitude I_n of the nth step is given by (6):

$$\frac{I_n}{I_o} = \frac{\delta^2}{2n\Gamma} \qquad (1)$$

where I_o is the maximum Josephson current. $\Gamma = \omega_j^2/\gamma\omega_c$ is a coupling parameter defined in terms of the plasma frequency $\omega_j = (2e/\hbar)(I_o/C)$, and the frequency of resonance of the circuit ω_c. C is the junction capacitance and γ is a frequency independent loss factor. The variable δ has to be determined by solving the non-linear equation:

$$|J_{n-1}(\delta) + J_{n+1}(\delta)| = \delta/\Gamma \qquad (2)$$

where $J_n(x)$ are Bessel functions of order integer. In the limit of $\Gamma \to 0$ only the first resonance n = 1 exists (6). By using the asymptotic expressions for the Bessel functions from the 1 and 2 we get for the amplitude of the first step (7):

$$I_1/I_o = \Gamma/2 \qquad (3)$$

By inserting in Γ the magnetic field dependent expression for ω_j (8), the amplitude of the first resonance as a function of the magnetic flux applied to the junction becomes:

$$\frac{I_1(\phi)}{I_o} = \frac{\Gamma_o}{2} \left| \frac{2J_1(\pi\phi)}{\pi\phi} \right| \qquad (4)$$

$\phi = \phi_e/\phi_o$ is the flux normalized to the flux quantum ϕ_o. Γ_o is the value of the coupling parameter for ω_j (0) the plasma frequency in zero field; J_1 (x) is the Bessel function of order one. The magnetic field dependence valid for a junction of circular geometry (9) has been used.

3. EXPERIMENTS

A schematic of the geometrical configuration of the samples investigated is shown in the inset of Fig. 1. A superconducting strip resonator is formed between a Nb and Pb films separated by an insultating layer of SiO. A Josephson junction is defined in the center of this structure through a "window" opened in the insulating layer. The fabrication procedure was the following. A Nb film 4000 Å thick is deposited by r.f. sputtering onto a 7059 corning glass substrate and its geometry is defined by photolithography and chemical etching. An SiO layer about 3000 Å thick is then deposited by vacuum evaporation. By a liftoff technique a window is opened in the insulat-

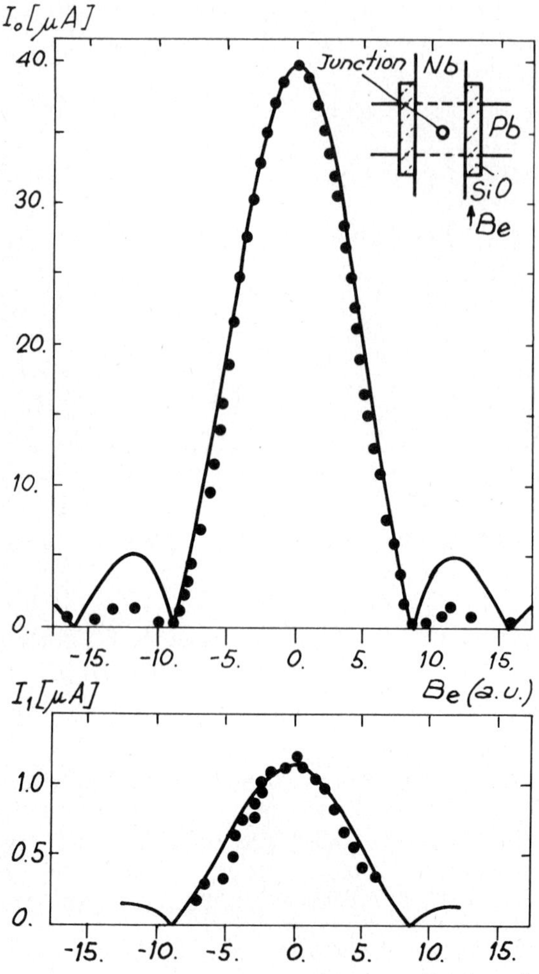

Fig. 1

a) Josephson current vs magnetic field
b) Resonance amplitude at V_1= 380 μV as a function of the magnetic field

ing layer to define the junction dimension. After a sputteretching process the samples are exposed to the atmosphere at ambient temperature for few minutes to obtain the Nb oxide barrier. Finally a Pb layer is deposited and patterned by photolithography and chemical etching. The junction diameters ranged from 13 to 30 μm and the width of both superconducting films was 100 μm. The samples have been tested at 4.2 K. The resonances amplitude was measured by using a dc bias superimposed to a low frequency ac signal and observing the current voltage characteristics on an oscilloscope (10). The magnetic field coupled to the junction was provided by a solenoid mounted around the samples and was parallel to the Nb first electrode.

In Figure 1 typical experimental data are shown. For this sample the junction diameter was 13 μm and the maximum Josephson current was 40 μA. Only one resonance was observed at a voltage V_1= 380 μV, and of maximum amplitude I_1 = 1.15 μA. From the ratio I_1/I_o= 0.03 the value $\Gamma \cong 0.06$ is deduced. In figure 1a the magnetic field dependence for the maximum Josephson current is shown. The solid line is the theoretical expression valid for a circular junction. The magnetic pattern for the first resonance is reported in Fig. 1b. The experimental data are compared with the theoretical dependence (solid line) computed from expression 4.

REFERENCES

(1) J.E. Zimmerman, J. Appl. Phys. 42 (1971) 30.
(2) S.I. Bondarenko, I.M. Dmitrienko and T.P. Narbut, Sov. Phys. Solid State, 14 (1972) 295.
(3) T. Ganz and J.E. Mercereau, J. Appl. Phys. 46 (1975) 4986.
(4) N.F. Pedersen, O.H. Soerensen, J. Mygind, P.E. Lindelof, M.T. Levinsen and T.D. Clark, Appl. Phys. Lett. 28 (1976) 562.
(5) T.F. Finnegan, J. Toots, L.B. Holdeman, S. Wahlsten, P. Mukhopadhyay and T. Claeson, Bull. Am. Phys. Soc. 21 (1976) 340.
(6) N.R. Werthamer and S. Shapiro, Phys. Rev., 164 (1967) 523.
(7) T.I. Smith, J. Appl. Phys., 45 (1974) 1875.
(8) N.F. Pedersen, T.F. Finnegan and D.N. Langenberg, Phys. Rev. B11 (1972) 4151.
(9) A. Barone and G. Paternò, Physiçs and Applications of the Josephon effect (Wiley New York 1982) chapter 9.
(10) G. Paternò and J. Nordman, J. Appl. Phys., 49 (1978) 2456.

LT-17 (Contributed Papers)
U. Eckern, A. Schmid, W. Weber, H. Wühl (eds)

THE INVERSE AC JOSEPHSON EFFECT IN JOSEPHSON TUNNEL JUNCTIONS AT ENERGY GAP FREQUENCIES

M. RIEDEL, H.-G. MEYER, W. KRECH, P. SEIDEL, K.-H. BERTHEL

Friedrich-Schiller-Universität Jena, Sektion Physik, DDR-6900 Jena, Max-Wien-Platz 1, GDR

On the basis of the Werthamer equation it is shown that the inverse ac Josephson effect is possible in tunnel junctions at much higher frequencies than predicted by the RSJ model.

Since Levinsen (1) has proposed a voltage standard realization on the basis of the inverse ac Josephson effect, there are extensive investigations dealing with this subject (2-4). Within the frame of the RSJ model (5), Kautz (2) showed that the inverse ac effect is possible only for hysteretic junctions (McCumber parameter $\beta_c \gg 1$) and at low external frequencies ($h\cdot f \lesssim 1.16 eR_N I_c$). The second limitation is mainly caused by the simple linear Ohmic current component in the resistively shunted junction. Unfortunately, the low-frequency region in hysteretic junctions shows significant indications of chaotic behaviour. Consequently, there is a great sensitivity of the effect, e.g. to small variations of the intensity of the external radiation.

High-capacitance tunnel junctions, however, reveal a highly nonlinear quasiparticle current component and low excess current for voltages within the gap region. Hence, the inverse ac Josephson effect should be observed at higher frequencies, too. For quantitative examination of this assumption we considered the Werthamer equation (6,7)

$$i(s) = \frac{g^2}{4}\beta_c \ddot{x}(s) - g\int_0^\infty ds'(p(s')\sin\tfrac{1}{2}(x(s-s')+x(s)) + q(s')\sin\tfrac{1}{2}(x(s-s')-x(s))) = i_o + i_1\sin(ws) \,. \quad (1)$$

The real retardation functions $p(s')$, $q(s')$ give the solid-state properties of the tunnel junction, $x(s)$ is the phase difference depending on the normalized time $s = t(\Delta_1+\Delta_2)/\hbar$, and $w = hf/(\Delta_1+\Delta_2)$ is the dimensionless frequency of the external ac bias current. Furthermore, we have introduced a constant $g = (\Delta_1+\Delta_2)/eR_N I_c$ depending on the energy gaps at zero temperature of both superconducting electrodes.

We anticipate a periodic solution which obeys the relation

$$x(s+\frac{1}{w}2\pi m) = x(s)+2\pi l \quad , \quad (2)$$

i.e. we make the ansatz

$$x(s) = x_o + \frac{l}{m}ws + \sum_{k=1}^{\infty} a_k \sin(\frac{k}{m}ws+b_k) \quad (3)$$

with the unknown constants x_o, a_k, b_k. An approximate solution of the corresponding nonlinear algebraic equation system in the Fourier space for small values of the parameter

$$p_o = 1/g\beta_c w^2 \ll 1 \quad (4)$$

can be found (8). In this case, the dc current through the junction at the Josephson step with the mean voltage $(l/m)(w/2)$ consists of three terms:

$$i_o^{(l,m)} = i_{o_{aut}} + i_{o_{pa}}(i_1) + i_{o_{step}}(i_1, x_o) \,. \quad (5)$$

Here we only represent the results up to the first order in p_o and for $l = m = 1$:

$$i_{o_{aut}} = g(\ \mathrm{Im}\ i_q(\tfrac{w}{2}) + 2p_o \mathrm{Re}\ i_p(\tfrac{w}{2}) \mathrm{Im}\ i_p(\tfrac{w}{2})\ )$$

$$i_{o_{step}} = 2p_o \mathrm{Re}\ i_p(\tfrac{w}{2})\ i_1\ \sin(x_o) \,. \quad (6)$$

$i_p(w)$ and $i_q(w)$ are the Fourier amplitudes of $p(s)$ and $q(s)$, respectively, using the sign convention of Poulson (9). The first term of eq. (5) in general contains the autonomous excess current contributions ($i_1=0$) and is identical with Schlup's result (10). The second term, arising from the non-linearity of the current amplitude $i_q(w)$, describes a photon-assisted tunneling process and remains zero in the lowest order of p_o. The third term depends on the absolute phase x_o and produces the current step, which reaches the axis $i_o=0$ if

$$\frac{2\, i_1}{g\beta_c}\frac{1}{w^2}\ \mathrm{Re}\ i_p\left(\frac{w}{2}\right) \geq i_{o_{aut}} + i_{o_{pa}}(i_1) \quad . \tag{7}$$

Contrary to the RSJ model this condition can be satisfied at high frequencies, too. In fig. 1 we compare the numerical solution of the eqs. (1) and (3) with the corresponding RSJ result.

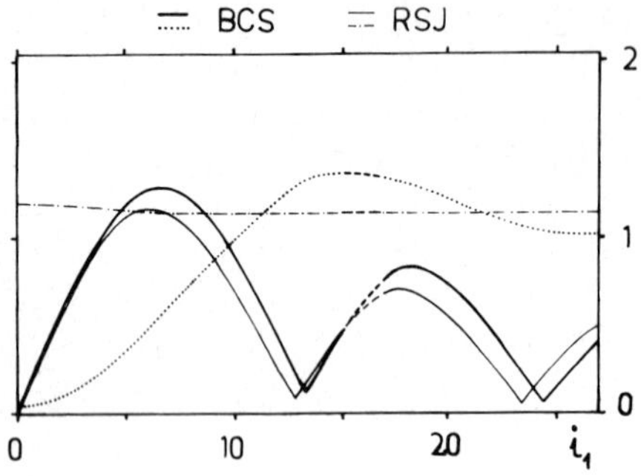

FIGURE 1
Numerically calculated height of the first Josephson step (l=m=1) as a function of i_1. The thick line represents the solution of the Werthamer equation with identical BCS electrodes at zero temperature, the thin line gives the corresponding RSJ result. The broken lines show the i_1-dependence of the step centre (enlarged by factor 2) in both models. At the value $i_1 \approx 16$ the periodic solution does not exist for all x_o (8). Junction parameters: $\beta_c = 10$, $w \triangleq 0.9$ and $g = 4/\pi$.

In the case of BCS electrodes at zero temperature the inverse ac Josephson effect is possible for parameters

$$0.2 \lesssim i_1 \lesssim 9.7 \quad , \tag{8}$$

whereas in the RSJ model the effect only can occur for $i_1 \approx 6.5$.

An additional advantage is that the effect remains stable within a great range of the interval (8). The optimum value for the normalized frequency w is determined by eq. (7). The right hand side should not be too large and the height of the first Josephson step must be as large as possible. Thus we get the conditions $w < 2$ and $w \approx 1$. Recent experimental results of Danchi, Habbal and Tinkham (11) seem to confirm our conclusions, though the authors did not mention the inverse ac Josephson effect in their paper.

REFERENCES

(1) M.T. Levinsen, R.Y. Chiao, M.J. Feldmann and B.A. Tucker, Appl. Phys. Lett. 31 (1977) 776.
(2) R.L. Kautz, J. Appl. Phys. 52 (1981) 3528.
(3) N.F. Pedersen and A. Davidson, Appl. Phys. Lett. 39 (1981) 830.
(4) D. D'Humieres, M.R. Beasley, B. A. Huberman and A. Libchaber, Phys. Rev. A 26 (1982) 3483.
(5) W.C. Stewart, Appl. Phys. Lett. 12 (1968) 452.
(6) N.R. Werthamer, Phys. Rev. 147 (1966) 255.
(7) R.E. Harris, Phys. Rev. B 13 (1976) 3818
(8) M. Riedel, to be published
(9) U.K. Poulson, Report No. 121 Lyngby 1973.
(10) W.A. Schlup, Phys. Rev. B 18 (1978) 6132.
(11) W.C. Danchi, F. Habbal and M. Tinkham, Appl. Phys. Lett. 41 (1982) 883.

LT-17 (Contributed Papers)
U. Eckern, A. Schmid, W. Weber, H. Wühl (eds)

SELF-CAPACITANCE EFFECT ON PLASMA OSCILLATIONS IN A JOSEPHSON SIS TUNNEL JUNCTION

G. BRUNK[+] H. LÜBBIG[++] Ch. ZURBRÜGG[+]

+ Technische Universität Berlin, Berlin, West Germany
++ Phys.-Technische Bundesanstalt, Berlin, West Germany

The influence of the high-frequency resistivity of ideal Josephson tunnel junctions on plasma oscillations is studied in terms of a small signal analysis in the complex frequency domain. The plasma frequency is shown to be directly related to the reactive pair- and quasiparticle current components involving logarithmic singularities of the Riedel peak type.

1. INTRODUCTION

Superconducting Josephson junctions whether DC-current biased subcritically ($I_{DC} < I_c$) or flux controlled such as in a hysteretic rf-SQUID performance have been shown to be promising systems to observe quantum phenomena on a macroscopic scale at low temperatures, (1). The equation of motion $(\Phi_0/2\pi)C\ddot{\varphi} + I_d(\varphi,\dot{\varphi},t) = dU/d\varphi$, where φ is the quantum phase, contains a macroscopic potential U exhibiting a set of metastable states and walls inbetween.

The system can jump from one to another metastable state via thermal fluctuation or quantum mechanical tunneling. The influence of the junctions dissipation by which the quantum system is coupled to the environment is of fundamental interest in this type of process, (2). One of the macroscopic parameters which determine the dynamical properties of the transition is the plasma frequency that describes the eigenoscillations of the "particle" around the bottom of the valley, φ_0.

It is the purpose of the present note to study eigenoscillations of an ideal tunnel junction on the base of the microscopic Werthamer tunnel model (WTM), and to indicate deviations and supplementations with respect to the restrictive "resistively shunted junction model" (RSJM) used widely.

As starting point we choose the small signal admittance due to the WTM:

$$\begin{aligned} Y = {} & (e/\hbar\omega)(I_{Ji}(\omega)\cos\varphi_0 + I_{QPi}(\omega)) \\ & + i\omega C \\ & - (e/i\hbar\omega)((I_{Jr}(\omega) + I_{Jr}(0))\cos\varphi_0 \\ & \quad + I_{QPr}(\omega) - I_{QPr}(0)) . \end{aligned}$$

Therein the complex amplitudes I_J and I_{QP} describe the tunneling of pairs and quasiparticles resp. and the real (r) and imaginary (i) parts are correlated by an Hilbert transform, (3). The RSJM is the following borderline case of the WTM: $I_{Jr}(\omega)\to I_{Jr}(0)=-I_c$, $I_{QPr}(\omega)\to I_{QPr}(0)$; $I_{Ji}(\omega)\to I_{Ji}(0)=0$, $I_{QPi}(\omega)\to(\hbar\omega/eR_N)$. Here R_N is the normal state resistance of the junction. In the present context we are mainly interested in the plasma resonant state of the junction as defined by Im Y = 0, i.e. $\omega_p^2 = 1/\mathcal{L}C$. Here $\mathcal{L}$ denotes the effective induction coefficient

$$\begin{aligned} \mathcal{L}^{-1} = {} & \mathcal{L}_0^{-1}(1/2)(1 + I_{Jr}(\omega_p)/I_{Jr}(0)) \\ & - (e/\hbar)(I_{QPr}(\omega_p) - I_{QPr}(0)). \end{aligned}$$

Therein $\mathcal{L}_0 = (\hbar/e2I_c\cos\varphi_0)$ is the quantum phase dependent induction coefficient due to the RSJM. The amplitudes I_{Jr} and I_{QPr} which modify $\mathcal{L}_0$ are even and continuous functions of the frequency except at $\omega_\Delta = 2\Delta/\hbar$ (for the case of a symmetric junction) where they exhibit logarithmic singularities. The singularity of I_{Jr} is known as Riedel peak. At $\omega\to\infty$ one has $I_{Jr}\to\omega^{-1}$ and $I_{QPr}\to\omega^{-3}$.

2. METHOD

To analyse the appearance and the stability of small plasma oscillations $\varphi = \varphi_0 + \mathrm{Re}\,\psi\exp(-iut)$, $|\psi|\ll 1$, the complex frequency domain $u = \omega(1-i\delta)$ is introduced. The corresponding complex amplitudes $I_J(u)$ and $I_{QP}(u)$ can be constructed from the $I_J(\omega)$ and $I_{QP}(\omega)$ resp. by means of analytic continuation. Then the characteristic equation follows

$$(\hbar/e)C\,u^2 + (I_J(u) + I_J(0))\cos\varphi_0 + I_{QP}(u) - I_{QP}(0) = \begin{cases} 0 \\ 2I_J(0)/\beta \end{cases}$$

for the case of current- and of flux control resp., where $\beta = 2\pi L I_c/\Phi_0$ with L the inductance coefficient of the SQUID.

3. RESULTS

The characteristic equation has been analysed for the current biased junction. The main result is that there exist plasma oscillations even if one assumes vanishing (geometric) susceptance, $C = 0$. This self-capacitance effect is caused by the intrinsic susceptance which is related directly to the real parts of the pair- and the quasiparticle tunnel current.

In figure 1 the eigenvalue condition is demonstrated schematically on the real frequency axis: Assuming first $C = 0$ one finds the plasma frequency corresponding to the intercept of the displaced real parts of the quasiparticle- (dashed line) and the pair tunnel current amplitude (dotted line), the latter modulated by $\cos\varphi_0$. An increase of the biasing current I_{DC} ($d\cos\varphi_0 < 0$) entails a decrease of the plasma frequency as indicated in figure 1. Since no intercept is possible above the energy gap $\omega_\Delta = 2\Delta(T)/\hbar$ is limiting frequency for plasma oscillations. - The addition of a finite capacitance (dash-dotted line) lowers the plasma frequency and - under certain conditions - the limiting frequency as well.

For $T = 0$ the conductance Re Y vanishes in the sub-gap region $\omega < \omega_\Delta$ and, consequently, there exist undamped oscillations.

At finite temperatures ($T/T_c \lesssim 0.6$) a small sub-gap conductance arises as indicated in figure 1 for $T/T_c = 0.43$ and $I_{DC}/I_c = 0.91$ (solid line). Consequently one finds very weakly damped plasma oscillations as shown in table 1.

In the low frequency regime the conductance increases logarithmically. Therefore, if the biasing current comes near to the critical current the plasma frequency decreases rapidly ($\varphi_0 \simeq \pi/2$) and, consequently, the system reaches the aperiodic limit.

Approaching the critical temperature the subgap conductance is raised and therefore only aperiodic solutions are expected in this regime.

TABLE 1

I_{DC}/I_c	1	.998	.983	.910	.828	0
ω_p/ω_Δ	0	.438	.688	.925	.988	1
δ		.059	.019	.005	.002	0

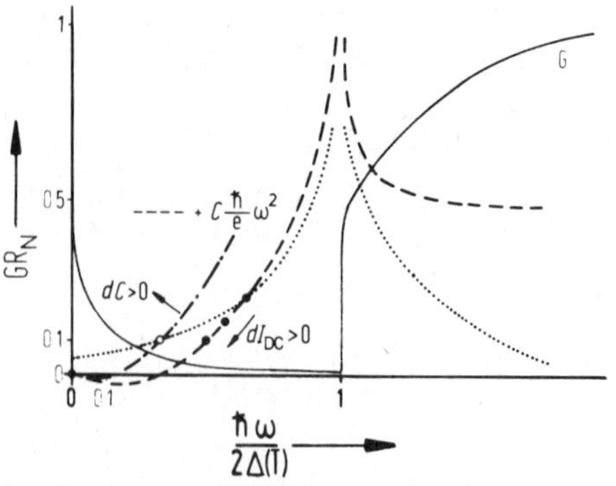

FIGURE 1

Schematic graph of $-(I_{Jr}(\omega) + I_{Jr}(0))\cos\varphi_0$ (dotted line) and $I_{QPr}(\omega) - I_{QPr}(0)$ (dashed line) as function of the normalized frequency. The dash-dotted line represent the influence of the capacitance. In addition, the solid line shows the conductance $G = \text{Re } Y$ for $\varphi_0 = 66^0$ and $T/T_c = 0.43$.

In summary, the reactive tunnel components due to pairs and quasiparticles influence the (phase depending) inductance coefficient significantly in the sub-gap region. This is related directly to the temperature dependent conductivity which is sensitive in the low frequency regime and near the energy gap.

(1) R. de Bruyn Ouboter et al., Physica 112 B (1982) 15: (1983) 1.
R.F. Voss, R.A. Webb, Phys.Rev.Lett. 47 (1981) 265.
(2) A.L. Leggett, Prog.Theoret.Phys.Suppl. 69 (1980) 80.
(3) D.G. Jablonski, J.Low Temp.Phys. 51 (1983) 433.

LT-17 (Contributed Papers)
U. Eckern, A. Schmid, W. Weber, H. Wühl (eds)

ELECTRICALLY CHARGED SUPERFLUID FILMS

L. Wilen and R. Giannetta

Dept. of Physics, Princeton University, Princeton, NJ 08544+

Electrons are deposited on thick (400-750Å) superfluid helium films above a glass substrate. Design of the experimental cell, mobility measurements, and charging methods are discussed.

1. INTRODUCTION

It is well established that a two dimensional gas of electrons may be confined near to the free surface of liquid helium. By reducing the thickness of the helium layer, d, from macroscopic dimensions ∿1 mm, to values ∿100-1000Å, several new physical phenomena are likely to arise. The Van der Waals attraction of the helium film to its underlying substrate should render the charged film stable against hydrodynamic instability up to areal charge densities of order (1)

$$n_c = (\rho\Lambda\tau)^{1/4} / d\sqrt{2\pi e^2 \varepsilon}$$

In this expression, ρ = liquid helium density, τ = surface tension, d = film thickness, e = electronic charge, ε = dielectric constant of substrate, and Λ/d^4 is the Van der Waals potential. For d = 100Å, ε = 10, charge densities up to ∿10"/cm^2 should be stable, allowing one to study the electron gas in a degenerate regime. For thin films, the substrate screens the Coulomb interactions between surface-state electrons, leading to a dipolar interaction potential and a phase diagram in which quantum zero point effects become important at accessible areal densities (2). In addition, strong electron-ripplon coupling may lead to polaron formation (3,4,5) and ripplon mediated attraction between electrons.

2. EXPERIMENTAL CELL, MOBILITY MEASUREMENTS

We are presently studying charged helium films using the cell shown in Fig. 1. A saturated film of thickness d = 400-750Å covers a clean glass substrate onto the back of which is evaporated a pair of concentric ring electrodes. The impedance of the electrode structure is measured at audio frequencies using a GR-1615A bridge at drive levels from 10-100 mV. A mobile layer of surface state electrons changes the capacitance and conductance between electrodes in a predictable manner (6) allowing a measurement of the low frequency ($\omega \ll 1/\tau$ where τ is the scattering time) conductivity of the charge sheet. Assuming a Drude model for the electron mobility, μ, applying Poisson's equation and the continuity equation for charge in the electron layer, we obtain a simplified expression for the conductivity change due to surface state electrons above the film,

$$G = \omega g(k)(\frac{ek}{\omega} n_s\mu)/1+f(k)(\frac{ek}{\omega} n_s\mu)^2$$

where k is the fundamental wavenumber defined by the cell geometry, g(k) and f(k) are geometric factors, and $\omega/2\pi$ is the drive frequency (typically 1 kHz < $\frac{\omega}{2\pi}$ < 100 kHz). As the conductivity, $en_s\mu$, is increased (by decreasing the temperature), G goes through a maximum. The peak height scales linearly with frequency and the peak shifts to lower temperatures as ω is increased. Just this behavior is observed for electrons on a 400Å film as shown in Fig. 2, indicating that the electron layer is indeed mobile.

In Fig. 3 we plot the surface state electron mobility above a 750Å helium film. The mobility is close to that obtained for gas-atom dominated scattering above a deep liquid layer (7). We find that for film thickness less than 400Å it is difficult to obtain reproducible data. Indeed, even at 750Å we find considera-

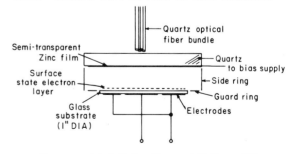

Expermental Cell for Charged Helium Films

FIGURE 1

+ Research supported by NSF Grant DMR-8211548 and Research Corporation Grant #9823.

ble hysteresis ($\sim$0.1 K) upon warming and cooling. Data near to T_λ is difficult to obtain since we suffer abrupt signal loss upon warming to this temperature. The signal due to mobile charge can be fully recovered by cooling sufficiently far below T_λ.

3. CHARGING METHODS, MAGNETIC EFFECTS

We have not, as yet obtained charge densities in excess of $5\times10^8/cm^2$, for any film thickness. Our cell has been designed with a photoemissive layer of zinc as the top plate electrode. Low energy electrons may be produced by pulsing a room temperature Hg-Xe arc lamp which is focussed onto a bundle of 6 quartz optical

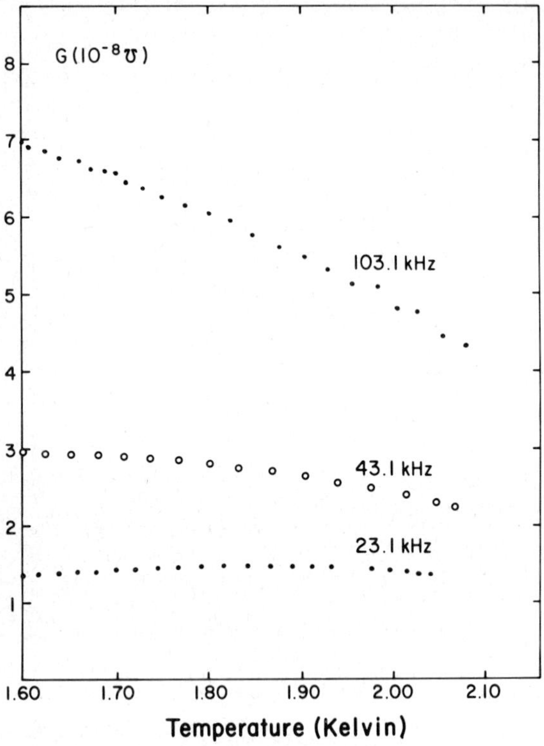

FIGURE 2
Conductance versus temperature

fibers that enter the experimental cell. Typical currents are 10 pA with small heating. However, this scheme does not charge the superfluid films. Instead we resort to a glow discharge within the cell. We do not, as yet, understand why the glow suceeds where the photo-emitter fails.

In addition to the measurements discussed earlier, we have applied DC magnetic fields of up to 600 Gauss in order to obtain the magnetoresistance of the charge layer. Using the relation

$$\mu(H) = \mu(H{=}o)/1+\left(\frac{\mu(H{=}o)H}{c}\right)^2$$

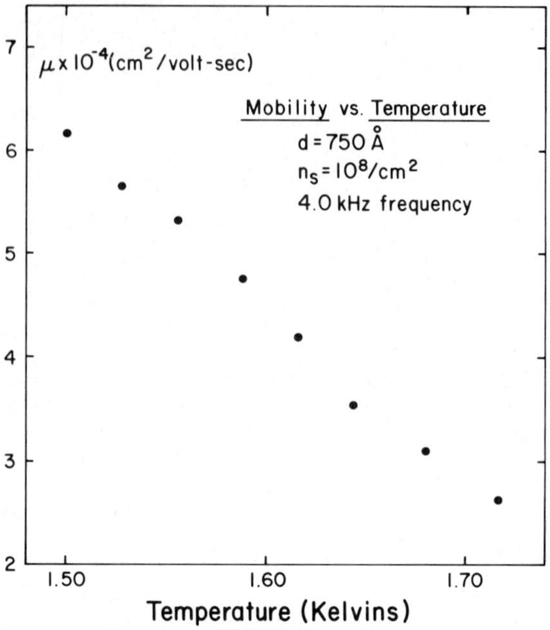

FIGURE 3
Mobility of electrons on a 750Å helium film.

we obtain mobilities in close agreement with those of Fig. 3. At 1.6 K and H = 600 Gauss, $\omega_c\tau \sim 0.4$ where ω_c is the cyclotron frequency. As the temperature is lowered we expect the conditions to be favorable for cyclotron resonance. Using a transmission type RF spectrometer operating at 2 GHz we plan to search for large effective mass shifts associated with the formation of electron-helium film polarons (3).

ACKNOWLEDGEMENTS

We wish to thank G.C. Grimes, P.M. Platzman, S. Jackson, E. Andrei, G. Williams, and H. Ikezi for numerous useful discussions.

REFERENCES

(1) H. Ikezi and P.M. Platzmann, Phys. Rev. B, 23, no. 3 (1981) 1145.
(2) F.M. Peeters and P.M. Platzman, Phys. Rev. Lett., 50, (1983) 2021.
(3) E. Andrei, preprint.
(4) S. Jackson and P.M. Platzman, Proc. of 4th International Conference on Electronic Properties of Two-Dimensional Systems, ed. F. Stern (North-Holland, Amsterdam, 1981) pp. 401-404.
(5) O. Hipolito, G. Farias, and N. Studart, ibid, pp. 394-400.
(6) J. Theobald, M.L. Ott-Rowland, and Gary A. Williams, Proceedings of LT-16, Part II, Physica B+C, 108, (1981) p. 957.
(7) Yasuhiro Iye, Jour. Low Temp. Phys., 40, nos. 5/6, (1980) 441.

LT-17 (Contributed Papers)
U. Eckern, A. Schmid, W. Weber, H. Wühl (eds)

POSSIBLE CORRELATION EFFECTS IN THE DENSITY DEPENDENT MOBILITY OF A TWO-DIMENSIONAL ELECTRON FLUID

R. MEHROTRA,* C.J. GUO,+ Y.Z. RUAN,++ D.B. MAST, and A.J. DAHM

Department of Physics, Case Western Reserve University, Cleveland, Ohio 44106

Measurements of the mobility of electrons supported by a liquid helium substrate in the temperature range of 150 mK to 1K are discussed. Excellent agreement with one-electron theories is found for the lowest density sample, $n = 5.3 \times 10^7 cm^{-2}$. At higher densities the mobility is less than the predictions of these theories by an amount which increases with density. The effect of correlations is examined and shown to be unimportant within the framework of the one-electron theories.

Presented herein are measurements of the low frequency mobility of a two-dimensional electron fluid for motion parallel to the underlying bulk helium substrate. Regimes in which the mobility is limited by ripplon scattering (low temperatures) and helium vapor atom scattering (high temperatures) are investigated. Excellent agreement with one-electron theories (1) is found at low areal densities. However, large density dependent deviations from these theories are observed at higher areal densities. The effect of correlations in single electron scattering processes is shown to be negligible at our densities, and high temperatures and we suggest that multi-electron or plasmon mode processes may have to be included in a more accurate theory.

The Sommer-Tanner technique (2) is used to measure the mobility. Our adaption of this technique to include the inductive response has been discussed (3) and a detailed description of our cell will be published elsewhere. This technique involves the application of an oscillating voltage at angular frequency ω to an electrode located near one end of the sample and measuring a signal capacitively induced on an electrode located near the opposite end of the sample when the electrons move in response to the applied voltage. The mobility μ and effective mass m^* were obtained respectively from the in and out of phase components of the signal (4).

Measurements were carried out in the small signal limit at a frequency of 1 MHz and at the saturated density $n = E_\perp/2\pi e$, where $E_\perp$ is the holding field. The effective mass was measured to be independent of temperature and less than ten electron masses. Measured values of the inverse mobility plotted versus temperature for six densities are shown in Fig. 1. Absolute errors are indicated by the error bars. There is a possible $\cong 4\%$ relative error due to uncertainties in C. The sharp rise in μ^{-1} at low temperatures indicates the formation of a two dimensional electron crystal at a temperature Tm.

The curves in Fig. 1 represent the theoretical prediction of the inverse mobility due to a combination of scattering mechanisms. For ripplon scattering we extended the one-electron theory of Monarkha (1) to include a finite $E_\perp$, and for helium vapor atom scattering we used the one-electron theory of Saitoh (1).

A comparison of the data and the theoretical curves show a nearly temperature independent and strongly density dependent discrepancy for densities $\gtrsim 10^8$ cm^{-2}. The validity of the one-electron theories in the limit of low density is demonstrated by the excellent agreement between experiment and theory for the lowest density sample. However, these theories are clearly not adequate at higher densities.

The effect of electron correlations on the high frequency ($\omega\tau > 1$) mobility (5), (6) has been examined and shown to be unimportant (6). We show by inspection that correlations are also not important within the framework of the present one-electron theories for the low-frequency, ripplon-limited mobility.

The electron-ripplon interaction is given by (1)

$$H_{e-R} = \sum_q A^{-\frac{1}{2}}\rho(q)V(q)(b^+_{-q} + b_q) \qquad [1]$$

where A is the surface area, $\rho(q) = \Sigma_j \exp(i\vec{q}\cdot\vec{r}_j)$ is electron density operator for electrons (at coordinates $\vec{r}_j$), $V(q) = Q(q)(eE_\perp + \gamma(q))$, $Q(q) = (\hbar q/2d\omega(q))^{\frac{1}{2}}$, and $\gamma(q) = -(\hbar^2 q^2/2ma_0)[1+\ln(qa_0/4)]$ results from the polarization interaction of electrons with the substrate. Here $A^{-\frac{1}{2}}Q(q)$ is the amplitude of a ripplon of wave-vector q and energy $\hbar\omega(q)$, b_q is a ripplon annihilation operator, d is the density of

*National Physical Laboratory, Hillside Road, New Delhi, 110012 India.
+National Taiwan College of Education, Changua, Taiwan.
++University of Science and Technology of China, Hefei, Anhui, People's Republic of China.

liquid helium, m is the electron mass, and $a_0 \cong 76$ Å is the effective Bohr radius for the electrons in surface states. The average of square of $\rho(q)$ is the structure factor $S(q)$. The use of plane wave states to describe the electrons leads to the following expression for the momentum scattering frequency for an electron of momentum p

$$\nu(p) = (m/2\pi\hbar^3) \int_0^{2\pi} d\phi |\langle V(q)\rangle|^2 S(q)[2N(q)+1](1-\cos\phi) \qquad [2]$$

where $N(q)$ is the Bose occupation factor and $\phi = 2\sin^{-1}(\hbar q/2p)$. In a Boltzman transport calculation μ^{-1} is proportional to an appropriate average over $\nu^{-1}(p)$.

For the highly correlated electron fluid (7) $S(q)$ is approximately zero for small q, has an absolute maximum at $q = q_0 \cong 6.7\, n^{\frac{1}{2}}$, and oscillates with decreasing amplitude about the large q limiting value of unity. In the one-electron theories $S(q)$ was set equal to unity. An examination of Eq. [2] shows that this assumption is justified. The integrand has a broad maximum for thermal q values [$q_{th} \cong 8.9\times 10^4 (T/n)^{\frac{1}{2}} q_0 \cong 10^6$ cm^{-1}] for $S(q) \cong 1$. For an areal density of $n < 10^9$ cm^{-2}, the suppression of $S(q)$ at small q values (Debye screening) and the enhancement near $q = q_0$ occur in the region where the rest of the integrand is small. Thus, corrections to the one-electron theory resulting from correlations cannot account for our observed discrepancy.

A sharp peak occurs (7) in $S(q)$ at $q = q_0$ for $T \cong 2$ Tm and the amplitude of this peak increases rapidly as $T \to$ Tm. In the presence of these strong correlations, the use of plane wave states in deriving Eq. [2] may not be appropriate. Computer simulations of a two-dimensional electron fluid show oscillations in the velocity auto-correlation function (8),(9). Kalia et al. (9) suggested that this is a result of coupling between single particle motion and plasma modes with wave-vector $q \sim n^{\frac{1}{2}}$. We suggest that similar effects may be important in momentum loss scattering processes, and that multi-electron processes must be considered in the derivation of the mobility.

Finally we comment that a density dependent mobility has been reported by Kajita (10) for electrons supported by a neon substrate coated with a thin helium film.

We wish to thank L. Foldy, P. Platzman and B. Segall for helpful conversations and the NSF for support under grant #DMR-82-13581.

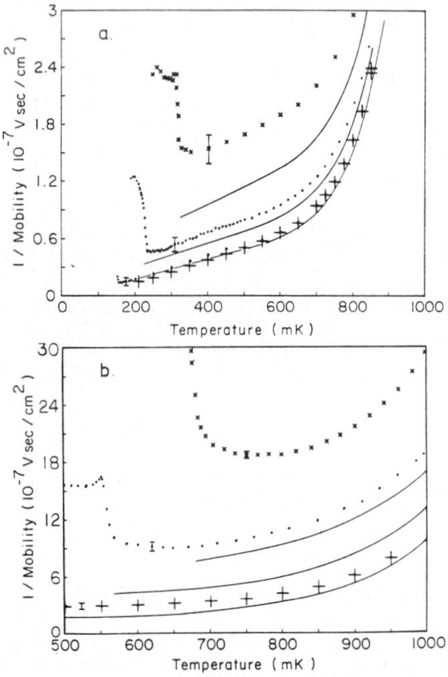

FIGURE 1
Inverse mobility vs. temperature for various densities. The solid curves are the theory for scattering from ripplons and helium gas atoms. The density in units of $n_8 = 10^8$ cm^{-2} are: Fig. 1(a) - (+)$n_8 = .53$; (•)$n_8 = 1.05$; (x)$n_8 = 2.1$; Fig. 1(b) - (+)$n_8 = 3.2$; (•)$n_8 = 6.3$; (x)$n_8 = 9.5$. The theory curves are shown for sequentially increasing densities.

1. V.B. Shikin and Y.P. Monarkha, J. Low Temp. Phys. 16, 193 (1974); P.M. Platzman and G. Beni, Phys. Rev. Lett. 36, 626 (1976); 36, 1350E (1976); M. Saitoh, J. Phys. Soc. Japan 42, 201 (1977); Yu.P. Monarkha, Fiz. Nizk. Temp. 2, 1332 (1976) [Sov. J. Low. Temp. Phys. 2, 600 (1976)].
2. W.T. Sommer and D.J. Tanner, Phys. Rev. Lett. 27, 1345 (1971).
3. R. Mehrotra, B.M. Guenin, and A.J. Dahm, Phys. Rev. 48, 641 (1982).
4. R. Mehrotra et al., to be published, Phys. Rev. B Rapid Comm.
5. P.M. Platzman, A.L. Simons, and N. Tzoar, Phys. Rev. B 16, 2023 (1977).
6. H. Totsuji, Phys. Rev. B 22, 187 (1980).
7. R.C. Gann, S. Chakravarty, and G.V. Chester, Phys. Rev. B 20, 326 (1979); F. Lado, Phys. Rev. B 17, 2827 (1978).
8. J.P. Hansen, D. Levesque, and J.J. Weis, Phys. Rev. Lett. 43, 979 (1979).
9. R.K. Kalia, P. Vashishta, S.W. deLeeuw, and A. Rahman, J. Phys. C 14, L991 (1981).
10. K. Kajita, to be published in Surface Science in the Proceedings of the Fifth Int. Conf. on the Electronic Properties in Two-Dimensional Systems.

LT-17 (Contributed Papers)
U. Eckern, A. Schmid, W. Weber, H. Wühl (eds)

OBSERVATION OF A PROPAGATING SHEAR MODE IN THE 2D ELECTRON SOLID ON LIQUID HELIUM

G. DEVILLE, A. VALDES, E.Y. ANDREI and F.I.B. WILLIAMS

Service de Physique des Solides et de Résonance Magnétique, CEN-SACLAY, 91191 Gif sur Yvette Cedex, France

We have observed the propagation of transverse sound in the two-dimensional electron solid in the domain of the phase diagram from the melting temperature T_m to about 1/3 T_m. Propagation velocities and damping have been measured. The shear modulus is seen to decrease linearly with temperature at low temperature and its value just prior to melting agrees with that derived from a model of melting by dislocation unpairing.

1. INTRODUCTION

Shear waves are a fundamental attribute of a solid but, in a two-dimensional system, no direct experimental evidence has yet been given that such waves propagate. Electrons constrained to the surface of liquid helium make a nearly ideal two-dimensional system of particles with scalar interactions. It has been amply demonstrated that this system undergoes a phase transition to a solid phase at $\Gamma=\Gamma_m=135\pm10$ [1,2,3,4] where the reduced thermodynamic parameter $\Gamma = 2e^2(n\pi)^{1/2}/(1+\varepsilon)T$ is the ratio of the potential energy to the kinetic energy ; e is the electronic charge, n the particle density, T the temperature in energy units and ε the dielectric constant of the helium substrate. The experiment described here shows explicitly that transverse phonons (shear waves) propagate in this system and it gives the manner in which propagation ceases as melting is approached [5].

2. EXPERIMENT

The principle of the experiment is to detect the resonance of the coupling between electrons and an electromagnetic probe field when the polarization, wavelength and frequency of this field match those of the phonons. The measurements of the transverse velocity v_t gives very directly the shear modulus μ through $v_t^2 = \mu/mn$ where m is the electron mass. The radio frequency (R.F.) power is transmitted from a frequency swept source to a detector through a constant impedance meander transmission line which interacts with the electrons. The meander periodicity of $\lambda_L = 120$ µm produces a *longitudinal* electric field of wave vector $k_L = 2\pi/\lambda_L = 520\text{cm}^{-1}$, in the plane of the electrons. This field is coupled to the transverse motion when a uniform and static vertical magnetic field H is applied. As the electrons have a *longitudinal* velocity induced by the longitudinal electric field, they are subject to a *transverse* force with the same vector k_L. The observed signal frequency ω is shifted from the bare transverse frequency $\omega_t = v_t k_L$ by two ways :

i) the effect of the applied vertical magnetic field H ;

ii) the effect of the interaction of the electrons with the helium surface, which depends on the vertical electric field $E_\perp$ felt by the electrons [4,6].

We can eliminate these two contributions by measuring, at each temperature, the asymptotic value of the resonant frequency for $H\to0$ and $E_\perp\to0$. Thus, we obtain the transverse velocity of the solid as a function of temperature.

We want to emphasize the fundamental difference between this kind of measurement of v_t at $k=k_L=520\ \text{cm}^{-1}\simeq10^{-2}k_{Debye}$, in the region of the propagative regime where $\partial\omega/\partial k \simeq \omega/k$, and a previous experiment [4] in which the measurement of the electron localization related to the thermal fluctuation of all vibrational modes led to a thermally averaged value of the shear modulus μ. The experiment described in this paper provides a direct measurement of v_t and thus represents a non ambiguous determination of $\mu(k_L,\omega)$. Fig.(1) gives an example of experimental signal for various temperatures.

3. RESULTS AND COMMENTS

The variations of the shear modulus with temperature are directly deduced from the experimental values of $v_t(T)$. Figure (2) shows the thermal behaviour of μ for a given sample. The low temperature behaviour is described by $\mu(T)/\mu(T=0) = 1 - \alpha(T/T_m)$ where $\alpha = 0.25 \pm 0.1$. This value of α is in good agreement with values obtained either from analytic calculation of the effect of anharmonicity (α=0.18)[7,8] or from molecular dynamic calculation (α=0.24)[9]. The absolute value of $\mu(T_m^-)$ supports the idea that the melting proceeds by unpairing of dislocation pairs for the last measured value of the ratio $B = \mu(T_m^-)a_0^2/4\pi T_m = 0.9 \pm 0.2$ is compatible with the Kosterlitz-Thouless stability criterion $B \geq 1$ [10].

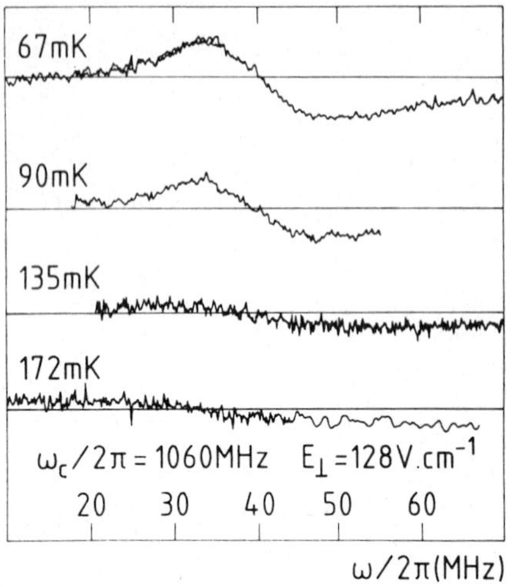

FIGURE 1
Transverse sound resonance observed at fixed magnetic and electric fields for different temperatures. The magnetic field amplitude H is represented by the electron cyclotron frequency $\omega_c = eH/mc$. The electron density is $n = 6\ 10^7\ cm^{-2}$.

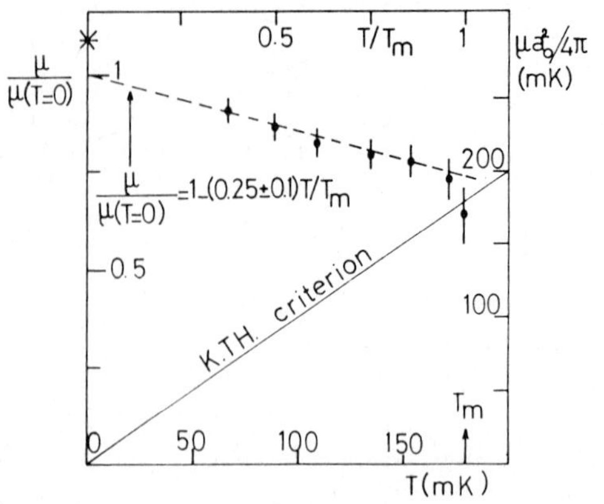

FIGURE 2
The shear modulus, normalized to its extrapolated zero temperature value $\mu(T=0)$, plotted as a function of the temperature T for $n = 6\ 10^7\ cm^{-2}$. The right hand scale $\mu a_0^2/4\pi$ enables a comparaison to be made with the Kosterlitz-Thouless stability criterion : $\mu a_0^2/4\pi \geq 1$. The asterisk * situates the calculated classical zero temperature value [12].

We claim that this experiment clearly demonstrates the existence of propagating transverse sound in the 2D-electron solid. We checked the transverse nature of the observed mode by following the magnetic field dependence of the signal intensity and its frequency shift. This mode becomes progressively more damped as the temperature increases and it disappears at $T \to T_m$. The linewidth is seen to increase linearly with temperature for $T/T_m \leq 0.7$ and to increase more strongly as T_m is approached. As the above measurements strongly suggest that melting arises due to dissociation of dislocation pairs so favoring the appearance of an intermediate oriented liquid phase (hexatic) [11], we expect that it will be possible to measure with this method the caracteristic viscosity signature of this intermediate phase.

ACKNOWLEDGEMENTS
We acknowledge J. Poitrenaud and C. Glattli for suggestions and critical readings of the manuscript.

REFERENCES

[1] C.C. Grimes and G. Adams, Phys. Rev. Lett. 42, 795 (1979)
[2] D. Marty, J. Poitrenaud and F.I.B. Williams J. Physique Lettres 41, L311 (1980)
[3] R. Mehrotra, B.M. Guenin and A.J. Dahm, Phys. Rev. Lett. 48, 641 (1982)
[4] F. Gallet, G. Deville, A. Valdes and F.I.B. Williams, Phys. Rev. Lett. 49, 212 (1982)
[5] G. Deville, A. Valdes, E.Y. Andrei and F.I.B. Williams, to be published
[6] D. Fisher, B.I. Halperin and P.M. Platzman Phys. Rev. Lett. 42, 798 (1979)
[7] D.S. Fisher, Phys. Rev. B26, 5009 (1982)
[8] M. Chang and K. Maki, Phys. Rev. B27, 1646 (1983)
[9] R. Morf, Phys. Rev. Lett. 43, 931 (1979)
[10] J.M. Kosterlitz and D.J. Thouless, J. Phys. C6, 1187 (1973)
[11] A. Zippelius, B.I. Halperin and D.R. Nelson Phys. Rev. B22, 2514 (1980)
[12] L. Bonsall and A.A. Maradudin, Phys. Rev. B15, 1959 (1977)

LT-17 (Contributed Papers)
U. Eckern, A. Schmid, W. Weber, H. Wühl (eds)

NUCLEAR SPIN RELAXATION AND DYNAMIC MAGNETIC BEHAVIOR OF THE HEAVY-FERMION SUPERCONDUCTOR UBe_{13}

W.G. CLARK (a), Z. FISK (b), K. GLOVER (a), M.D. LAN (a), D.E. MACLAUGHLIN (c), J.L. SMITH (b), and Cheng TIEN (c)

(a) Physics Department, University of California, Los Angeles, CA 90066, U.S.A.
(b) Los Alamos National Laboratory, Los Alamos, NM 87545, U.S.A.
(c) Physics Department, University of California, Riverside, CA 92521, U.S.A.

Measurements of the ^9Be nuclear spin-lattice relaxation time (T_1) and spin-phase memory time (T_2) are reported for the superconducting and normal states of the heavy-fermion superconductor UBe_{13}. Below the transition temperature (T_c), there is a rise in $1/T_1$ that is anomalously large for a type-II superconductor, and a sharp increase in $1/T_2$ that is interpreted as evidence for slow magnetic fluctuations in the superconducting state. Above T_c, $1/T_1$ behaves as expected for conduction electrons moving in a very narrow band.

1. INTRODUCTION

One of the important questions regarding heavy-fermion superconductors is the nature of magnetic fluctuations of the conduction electrons in both the normal and the superconducting states. In addition to the conventional aspects of this problem (1), there is evidence for features in heavy-fermion superconductors associated with the unusually narrow f-band (2), large spin fluctuations (3), and the possibility of unusual Cooper pairing in the superconducting state (4). In this paper, we present preliminary results for the nuclear spin-lattice relaxation time (T_1) and spin phase memory time (T_2) of the ^9Be nuclei in the heavy-fermion superconductor UBe_{13} (transition temperature T_c approximately 0.85 K) over the temperature range 0.55-300 K.

It is well known that in metals the rate $1/T_1$ reflects the power spectrum of fluctuations in the electron nuclear interaction (5) or, alternatively, the electronic density of states. Our measurements confirm the narrow bandwidth reported (6) for UBe_{13}, and show an unusually large increase in the relaxation rate upon entering the superconducting state. Furthermore, the behavior of T_2 is opposite to that usually observed in type-II superconductors, and may indicate the presence of very slow magnetic fluctuations.

2. EXPERIMENTAL DETAILS

The samples were grown from an Al flux (6). This yielded chunks having several well developed faces a few mm on an edge. A gently powdered sample was prepared by breaking the chunks into small pieces under acetone. Particles in the size range 0.1-0.5 mm diameter were selected for the NMR sample. This sample had a superconducting transition at 0.8 K. The NMR measurements were made at a frequency of 9.000 MHz in an applied field (H_o) of 15.56 kOe.

3. EXPERIMENTAL RESULTS AND INTERPRETATION

3.1. Spin-lattice relaxation rate

The temperature dependence of $1/T_1$ is shown in Fig. 1. The increase in $1/T_1$ below T_c is qualitatively similar to that often seen in conventional superconductors (1), where an observed maxmimum in $1/T_1(T)$ is attributed to the increased density of quasiparticle excitations near the edge of the BCS gap. The peak value of $1/T_1$ has not been attained above 0.55 K, but must be greater than twice $1/T_1$. This suggests that pair breaking in the mixed (vortex) state is weak in UBe_{13}. This, in turn, is consistent with the conclusion that the system is in the extreme dirty limit (6), for which orbital pair breaking by an applied field $H_o \ll H_{c2}(T=0)$ is weak. Our result for UBe_{13} differs from conventional behavior, however, in that the maximum in $1/T_1$ must occur below a reduced temperature $T/T_c = 0.65$, compared to approximately 0.9 in ordinary superconductors.

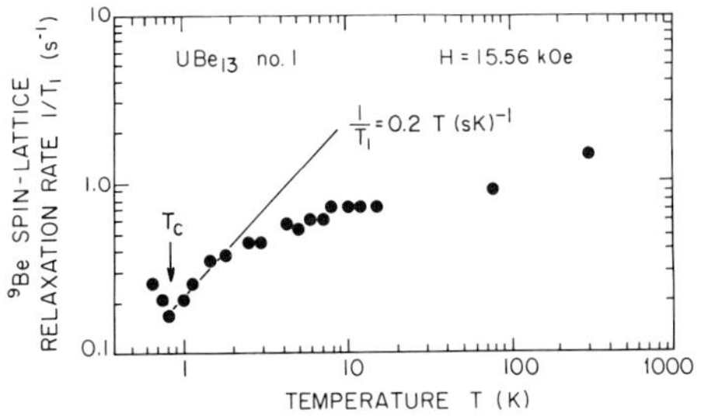

FIGURE 1
Temperature dependence of ^9Be $1/T_1$ in UBe_{13}.

For T somewhat above T_c, $1/T_1$ varies linearly with T. This appears to be the well known Korringa relaxation of nuclei by a degenerate electron gas (5). Above about 10 K $1/T_1$ rises sublinearly with T. It can be shown that this deviation can arise either from the electrons becoming nondegenerate in a very narrow band (bandwidth about 10 K), or from a temperature dependent reduction in the density of states above 10 K.

3.2. Spin-phase relaxation rate

The behavior of $1/T_2$ in UBe_{13} is unusual in two respects. First, the best fit to the decay of the spin-echo envelope is much closer to a gaussian than to an exponential. Our measured values of $1/T_2$ in the vicinity if T_c, given in Fig. 2, are based upon a gaussian fit to all but the earliest part of the echo decay envelope. The second unusual feature is that $1/T_2$ increases upon entering the superconducting state. Let us first consider this increase.

By using a standard calculation (7), we estimate that the second moment of the ^{9}Be resonance line due to the nuclear dipolar interaction in UBe_{13} would yield $1/T_2 = 1.3 \times 10^4\ s^{-1}$ if quadrupolar broadening, magnetic relaxation, and anisotopic magnetic field shifts were ignored. This is an order of magnitude larger than the observed value. The standard explanation for the decrease of $1/T_2$ is that mutual spin flips are inhibited because the interacting nuclei are "detuned" by static inhomogeneous interactions. The quadrupolar interaction in UBe_{13} is available to play this role. Usually $1/T_2$ decreases further in the superconducting mixed state, as penetration of the flux lattice develops further detuning of the nuclei on a microscopic distance scale (8). For this reason the increase of $1/T_1$ below T_c in UBe_{13} is anomalous. Either there is some property of the superconducting state that reduces the detuning effect, or an additional relaxation process affects $1/T_2$ below T_c. It is difficult to construct a reasonable mechanism for the former; we speculate, therefore, that an additional relaxation mechanism becomes operative in the superconducting state.

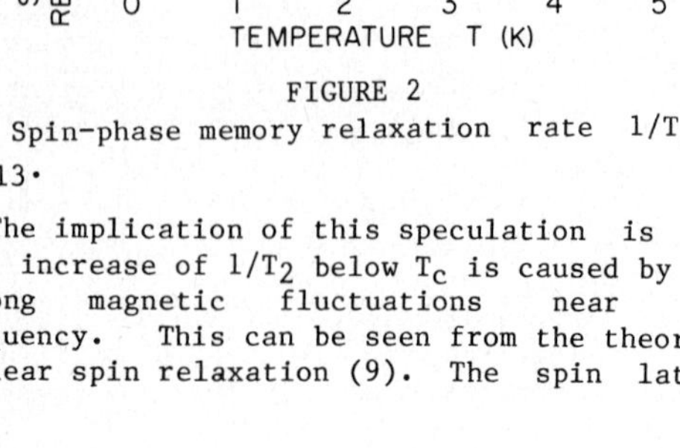

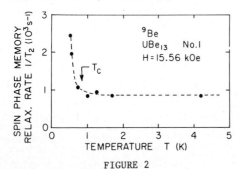

FIGURE 2

^{9}Be Spin-phase memory relaxation rate $1/T_2$ in UBe_{13}.

The implication of this speculation is that the increase of $1/T_2$ below T_c is caused by very strong magnetic fluctuations near zero frequency. This can be seen from the theory of nuclear spin relaxation (9). The spin lattice relaxation rate, which is on the order of $0.2\ sec^{-1}$ in the superconducting state, is proportional the power spectrum of fluctuations of the electron-nuclear interaction at the nuclear Larmor frequency. The spin-phase memory relaxation rate ($2 \times 10^3\ s^{-1}$) is, in addition to the above, proportional to zero frequency power spectrum. Experimentally, $1/T_2 \gg 1/T_1$, so that this model indicates the presence of very slow magnetic fluctuations associated with the electrons in the superconducting state.

The gaussian shape of the spin echo decay envelope is probably explained along the lines of the stochastic model of Klauder and Anderson (10) for an inhomogeneously broadened line. In UBe_{13} the inhomogeneous broadening is provided by the ^{9}Be quadrupolar interaction, which our experiments show is spread over about 200 Oe in both the normal and superconducting states. A similar result was found for A-15 superconductors (8).

4. CONCLUSIONS

Measurements of $1/T_1$ and $1/T_2$ for ^{9}Be in UBe_{13} in the superconducting and normal states are reported. The data indicate an anomalous increase of $1/T_1$ below T_c, and benavior indicative of very narrow bands above T_c. A dramatic increase in $1/T_2$ is seen below T_c. It is argued that this may be evidence for very slow magnetic fluctuations in the superconducting state.

ACKNOWLEDGEMENTS

This work is based upon research supported by the U.S. National Science Foundation (U.S. NSF) Division of Materials Research-Solid State Chemistry Program Grant DMR-8103085, (W.G.C., K.G., M.D.L), U.S. NSF Grant DMR-8115543 (D.E.M., C.T.), and the U.S. Department of Energy (Z.F., L.S.).

REFERENCES

(1) For a review, see D.E. MacLaughlin, Solid State Phys. 31 (1976) 1.
(2) F. Steglich et al., Phys. Rev. Lett. 43 (1979) 1892; H.R. Ott et al., Phys. Rev. Lett. 50 (1983) 1595; G.R. Stewart et al., Phys. Rev. Lett. 52 (1984) 679.
(3) M.T. Beal-Monod, Phys. Rev. B 24 (1981) 261.
(4) C.M. Varma, Proc. NATO Adv. Summer Inst. on the Formation of Local Moments in Metals, ed. by W. Buyers, to be published.
(5) See A. Abragam, Principles of Nuclear Magnetism (Clarendon Press, Oxford, 1961) pp. 355-375.
(6) H.R. Ott el al., ref. 2.
(7) A. Abragam, op. cit., Ch. IV.
(8) B.G. Silbernagel et al., Phys. Rev. 153 (1967) 535.
(9) A. Abragam, op. cit., Ch. VIII.
(10) J.R. Klauder and P.W. Anderson, Phys. Rev. 125 (1962) 912.

LT-17 (Contributed Papers)
U. Eckern, A. Schmid, W. Weber, H. Wühl (eds)

SUBSTITUTION OF Ga, Cu and B IN Be SUBLATTICE OF HEAVY FERMION SUPERCONDUCTOR UBe_{13}*

A.L. Giorgi, Z. Fisk, J.O. Willis, G.R. Stewart and J.L.Smith

Los Alamos National Laboratory, Los Alamos, New Mexico 87545

A series of compositions was prepared in which Ga, Cu or B was substituted for part of the Be in the Be sublattice of the heavy fermion superconductor UBe_{13} and the effect on the superconductivity determined. A linear decrease in T_c with increasing Ga content contrasted sharply with a dramatic suppression of superconductivity (T_c less than 15 mK) by low concentrations of Cu or B. Low temperature specific heat measurements have been made on the composition with low Cu ($UBe_{12.94}Cu_{0.06}$) and the γ value compared with the corresponding value for pure UBe_{13}.

1. INTRODUCTION

The three compounds $CeCu_2Si_2$, UBe_{13} and UPt_3 representing a new class of superconductors referred to as heavy fermion superconductors, are attracting considerable interest because of their very unusual properties. They have highly correlated electrons which interact so strongly that these electrons have an effective mass almost 200 times the mass of a free electron. Their electronic heat capacity coefficients have values above 1000 mJ/mole K^2 and although their T_c is below 1.0 K their dH_{c2}/dT values are extremely high. In an earlier investigation (1) small amounts of various metals were substituted for U in UBe_{13} and the effect on the superconducting and normal state properties determined. The present investigation is an extension of that study to determine the effect of substitution of small amounts of Ga, Cu or B for the Be in the Be sublattice.

2. EXPERIMENTAL

All of the compositions were prepared from ^{238}U, high purity Ga, Cu, B(99.99+%) and Be (99.6%) by arc melting under a purified argon atmosphere. Proper weights of the constituents were melted together to give the desired composition and the melting repeated four times with the button turned after each melting to insure homogeneity. A portion of each sample was wrapped in tantalum foil, sealed under helium in a quartz tube, annealed for 5 days at 950°C and then quenched in water. Room temperature x-ray diffraction patterns were taken of the finely powdered samples as arc cast and after the annealing process and the lattice parameter determined. The T_c values were measured using standard ^3He and dilution refrigerator techniques. The low temperature specific heat measurements were performed using a small sample calorimeter described in reference (2).

*This work was performed under the auspices of the U.S. Department of Energy.

3. RESULTS AND DISCUSSION

A tabulation of the compositions, T_c values and the lattice parameters of the arc cast and annealed preparations is given in Table I.

TABLE I

T_c values and lattice parameters for the samples of $UBe_{13-x}M_x$ (M= Ga,Cu,B).

		arc cast		5 days - 950°C	
M	x	a_o (Å)	T_c (K)	a_o (Å)	T_c (K)
-	0.0	10.254	0.81	10.253	0.83
Ga	0.12	10.262	0.60	10.253	0.83
Ga	0.35	10.277	0.39	10.264	0.64
Cu	0.05	10.257		10.257	<0.015
Cu	0.06	10.259		10.259	<0.015
Cu	0.11	10.261		10.261	
Cu	0.51	10.277		10.277	
B	0.10	10.246		10.246	<0.020
B	0.20	10.236		10.236	<0.020

As shown in Table I, the effects observed for the compositions containing Ga contrast sharply with those containing Cu or B. With increasing Ga content the lattice expands and the T_c drops gradually. Annealing appears to force the Ga out of the lattice as evidenced by the contraction of the lattice, the increase in T_c and the appearance of diffraction lines corresponding to UGa_3. The compositions containing Cu or B on the other hand show no T_c for measurements to 15 mK even for the lowest Cu and B content and no change in lattice parameter after the long term anneal.

The complete suppression of superconductivity by the substitution of such small amounts of Cu relative to the Be content suggested that a comparison of a low temperature specific heat measurement on such a sample with a similar measurement on pure UBe_{13} might be used to show what correlation, if any, exists between γ and T_c in UBe_{13}. Such measurements were therefore carried out.

The low temperature specific heat data for the $UBe_{12.94}Cu_{0.06}$ annealed sample and for the pure UBe_{13} are shown in Figure 1.

An examination of Figure 1 clearly shows that the two curves are identical indicating that the heavy fermion state exists in both samples at low temperatures and within the accuracy of the measurements the gammas are equal.

Thus, although we can postulate that their is no correlation between γ and T_c, we are unable to explain the difference in superconducting properties.

REFERENCES

(1) J.L. Smith, Z. Fisk, J.O. Willis, B. Batlogg and H.R. Ott, to appear in J. Appl. Phys.

(2) G.R. Stewart and A.L. Giorgi, Phys. Rev. B17 (1978) 3534.

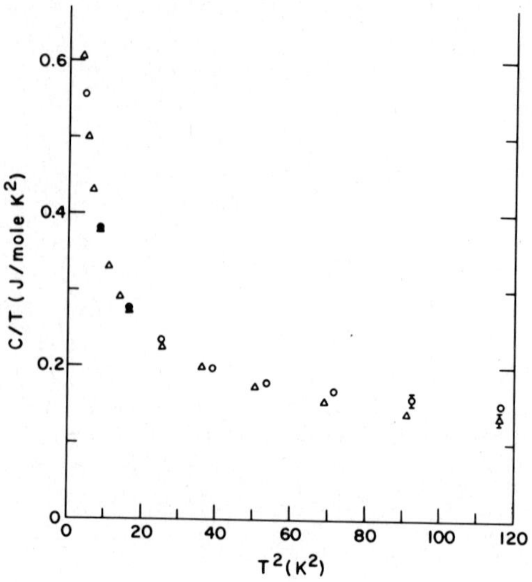

FIGURE 1
Low temperature specific heat of UBe_{13} (triangles) and $UBe_{12.94}Cu_{0.06}$ (circles).

LT-17 (Contributed Papers)
U. Eckern, A. Schmid, W. Weber, H. Wühl (eds)
© Elsevier Science Publishers B.V., 1984

COEXISTENCE OF SUPERCONDUCTIVITY AND INTERMEDIATE VALENCE IN $CeRu_3Si_2$

U. RAUCHSCHWALBE, U. AHLHEIM, U. GOTTWICK, W. LIEKE, and F. STEGLICH

Inst. f. Festkörperphysik, Technische Hochschule Darmstadt, D-6100 Darmstadt, F.R.G.

C. GODART[+], L.C. GUPTA[§], and R.D. PARKS

Department of Physics, Polytechnic Institute of New York, Brooklyn, New York 11201, U.S.A.

Thermopower and superconductivity measurements are reported on $CeRu_3Si_2$ and $Ce_{1-x}La_xRu_3Si_2$. For the compound, a mixed valence derived contribution to the thermopower is found. The transition temperatures $T_c(x)$ of the alloys interpolate smoothly between $T_c(CeRu_3Si_2)$ = 1.0K and $T_c(LaRu_3Si_2)$ = 7.2K.

The mixed-valence (MV) phenomenon in rare-earth systems, which arises from hybridization between localized 4f electrons and the conduction electrons, has been intensely investigated (1). In this paper, we address the relationship between MV and superconductivity in Ce-based compounds. These systems may be classified with respect to their spin-fluctuation temperature, which is a measure for the strength of the hybridization. While some of the strongly MV systems with $T_{sf} \simeq 1000K$, like $CeRu_2$ (2), show superconductivity, none of the prototypical MV systems with T_{sf} = 200 - 300K, like $CeSn_3$ and $CePd_3$ (1), is superconducting. On the other hand, the Kondo-lattice system $CeCu_2Si_2$ with $T_{sf} \simeq T_{Kondo} \simeq 10K$ exhibits "heavy-fermion" superconductivity below $\simeq 0.7K$ (3). Despite many quantitative differences between the superconducting states of $CeRu_2$ and $CeCu_2Si_2$, a few qualitative similarities can be stated: (i) Compared to the respective La homolog, the Ce compound shows higher T_c. (ii) Alloying with a sufficient ("critical") concentration of any kind of doping atoms results in a complete supression of superconductivity. (4,3)

Recently, we found (5) superconductivity below $T_c \simeq 1K$ in the compound $CeRu_3Si_2$. We also deduced from specific heat and magnetic susceptibility measurements that $CeRu_3Si_2$ is a MV compound with $T_{sf} \simeq 440K$, which is intermediate between $T_{sf} \simeq 270K$ for $CeSn_3$ and $\simeq 770K$ for $CeRu_2$ (5).

This conclusion is supported by the temperature dependence of the thermopower, shown in Fig. 1. The results of $LaRu_3Si_2$ reveal a complex diffusion thermopower which is negative below 180K and almost compensated by a positive phonon-drag contribution below 50K. A very similar phonon-drag term shows up in the data for $CeRu_3Si_2$. Here, a pronounced Ce-derived contribution to the diffusion thermopower is clearly visible. It is positive in the whole temperature range like for all MV Ce-based systems studied (6). Furtheron, it increases steadily, indicating that its expected (6) peak must occur well above T = 350K, in accordance with $T_{sf} \simeq 440K$ (5).

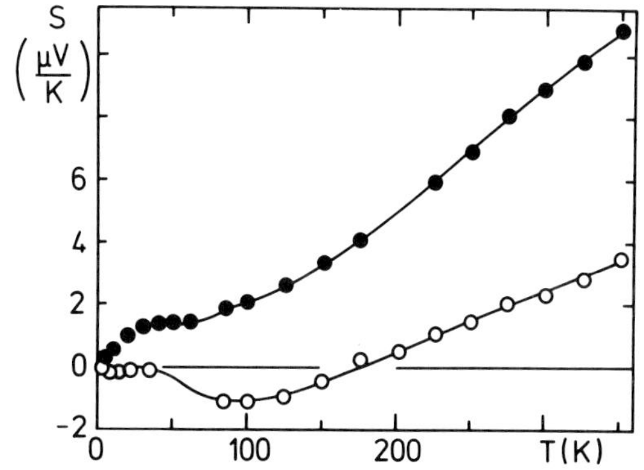

FIGURE 1
Thermopower as function of temperature for $CeRu_3Si_2$ (●) and $LaRu_3Si_2$ (o).

In Fig. 2, the concentration dependence of the transition temperature is shown for the quasi-binary system $Ce_{1-x}La_xRu_3Si_2$. In contrast to what one observes in $Ce_{1-x}La_xRu_2$ (2) and $Ce_{1-x}La_xCu_2Si_2$ (3), where T_c can be completely suppressed by a critical La-concentration $x_{cr} \simeq 40\%$ and 10% respectively, Fig. 2 shows superconductivity for all concentrations.

[+] Present address: E.R. 209-CNRS
1 Place A. Briand, F-92190 Meudon, France

[§] Present address: Tata Institute of Fundamental Research, Bombay 400 005, India

The transition temperatures of the alloys are smoothly interpolating between that of $CeRu_3Si_2$ and $T_c \approx 7.2K$, found for its La-homolog (7). Obviously, that rather high transition temperature of $LaRu_3Si_2$ becomes reduced when La is replaced by Ce, but one can afford a Ce concentration of 100at% without destroying superconductivity.

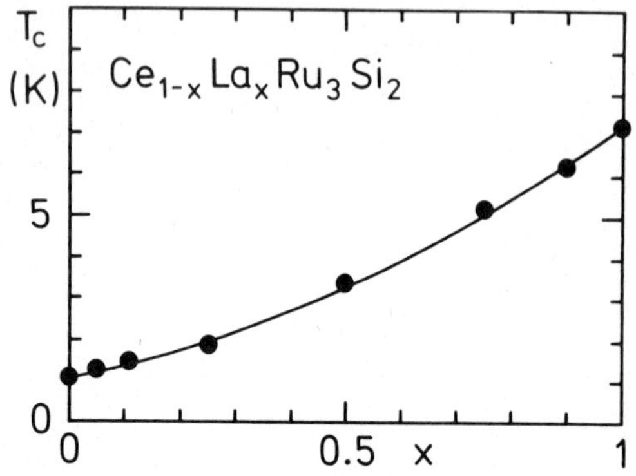

FIGURE 2
Transition temperature of $Ce_{1-x}La_xRu_3Si_2$ as function of La concentration x. T_c was determined inductively ($x \leq 0.5$) or resistively ($x > 0.5$).

In more quantitative terms, the initial T_c-depression (as Ce-concentration increases) is $8.6 \cdot 10^{-2}$K/at% Ce ≈ 51.5K/Ce. This is somewhat larger than 39.7K/Ce for (La,Ce)Ru_2 (2), but much smaller than 520K/Ce for (La,Ce)Sn_3 (8). Although this latter value is presumably enhanced by anisotropic exchange scattering (8), the dependence of T_c on Ce concentration in all of the three systems is in qualitative accord with a Ce-derived T_{sf} value which is much larger than T_c of the respective La compound; for, in this case the T_c vs. Ce concentration curve should have positive curvature, and its initial slope should increase with decreasing T_{sf} (9), as observed.

The results of this work allow some conclusions as to the superconductivity vs. mixed valence problem. First, in certain strongly MV systems superconductivity appears favored by the presence of the 4f electrons, as is clearly demonstrated by the re-appearance of superconductivity in the Ce-rich (Ce,La)Ru_2 systems (2). This is, as yet, not understood (10). Second, for all other Ce compounds coexistence of superconductivity and MV seems to require: (i) a favorable band structure (e.g. Ru-derived 4d electrons), causing superconductivity with relatively high T_c in the respective La (or Y) homolog, (ii) a sufficiently high spin-fluctuation temperature and (iii) a small effective concentration of Ce ions (16% in $CeRu_3Si_2$). Hence, we believe, there will be no superconductors found among the prototypical MV Ce-based compounds with T_{sf} = 200 - 300K. Further, the existence of superconductivity in the low-temperature phase of $CeCu_2Si_2$ with $T_{sf} \approx 10K$ cannot be understood on the basis of the MV concept. In fact, recent theoretical results (11,12) show that heavy-fermion superconductivity in this system results from the coherent action of the Kondo effect at the regular Ce sites and an electron-phonon interaction typical of rare-earth systems.

The work at Darmstadt was supported by the Sonderforschungsbereich 65 Frankfurt/Darmstadt and that at Polytechnic in part by the National Science Foundation under Grant No. DMR-8202726.

REFERENCES

(1) For a review, see: J.M. Lawrence, P.S. Riseborough and R.D. Parks, Rep. Progr. Phys. 44 (1981) 1.
(2) B. Hillenbrand and M. Wilhelm, Phys. Lett. 33A (1970) 61.
(3) H. Spille, U. Rauchschwalbe and F. Steglich, Helv. Phys. Acta 56 (1983) 165; and references cited therein.
(4) M. Wilhelm and B. Hillenbrand, Z. Naturforsch. 26a (1971) 141.
(5) U. Rauchschwalbe, W. Lieke, F. Steglich, C. Godart, L.C. Gupta and R.D. Parks, to be published.
(6) D. Jaccard and J. Sierro, in: Valence Instabilities, eds. P. Wachter and H. Boppard, (North-Holland, 1982) p. 409; S. Horn, W. Klämke and F. Steglich, ibid., p. 459.
(7) J.M. Vandenberg and H. Barz, Mat. Res. Bull. 15 (1980) 1493.
(8) W. Schmid and E. Umlauf, Commun. Physics 1 (1976) 67.
(9) see, e.g.: N.Y. Rivier and D.E. MacLaughlin, J. Phys. F.: Metal Physics 1 (1971) 248.
(10) J.W. Allen, S.J. Oh, I. Lindau, M.B. Maple, J.F. Suassuna and S.B. Hagström, Phys. Rev. B26 (1982) 445.
(11) H. Razafimandimby, P. Fulde and J. Keller, Z. Phys. B54 (1984) 111.
(12) N. Grewe, to be published.

LT-17 (Contributed Papers)
U. Eckern, A. Schmid, W. Weber, H. Wühl (eds)

COMPOSITION MODULATION MEASUREMENTS IN THIN 2D-SUPERCONDUCTING MULTILAYERS

Jean Claude VILLEGIER

LETI-IRDI 85 X 38041 Grenoble France

B. BLANCHARD

SEAPC-IRDI CENG 85 X 38041 Grenoble France.

O. LOBORDE

CRTBT-CNRS 166 X 38042 Grenoble France.

Composition modulation of Nb Al coherent multilayers is analysed using a selective etching technique removing individual layers. Resistive and superconductive properties are found to depend on the number of layers N while a strong critical field anisotropy $H_{//} / H_{\perp} > 2.1$ is observed.

1. Nb-Al MULTILAYERS DEPOSITION AND SELECTIVE ETCHING

The growth of coherent Nb-Al multilayers has been studied previously by Mc Whan et Al [1]. We intend to extend the study to a small number of layers ($1 < N \leqslant 10$). The films are deposited by DC magnetron sputtering [2] on various substrates : Si (100), SiO_2, Sapphire and resist coated Sapphire and have the same thickness ($d_{Nb}=d_{Al}=6 \pm 0.2$ nm). The speed of deposition is 1.5 nm/s and residual gas pressure is 10^{-7} torr, resistivity is low and thickness dependant : $\rho_{300K}(Nb) = 30\ \mu\Omega$.cm and $\rho_{300K}(Al) = 6.7\ \mu\Omega$.cm for 2.4 nm thick films. Selective etching is obtained for film thickness above 3 nm : Niobium is selectively etched on Al by RIE with SF_6 gas while Al films are disolved in an HCl solution. The end of attack is well controlled by sheet resistance measurements.

2. STRUCTURE AND COMPOSITION MODULATION

X-rays diffractions and XPS techniques have been used to investigate the structure of Nb-Al and Nb-Cu multilayers [1][3][4] , we also performed TEM diffraction sensitive to the cristalline orientations located in the plane of the film. RHEED and X rays camera patterns confirm the expected Nb (110) and Al (111) texture with random orientation in the plane of the films. The corresponding d spacing is d = 2.335 ∓ 0.005 A indicating that Nb (110) d spacing is expanded and Al (111) contracted. The perpendicular planes Nb (1$\bar{1}$0) and ($\bar{1}$10) are about 1% contracted and Nb (002) about 1.5% contracted. Such results confirm that Nb regions show an orthorhombic distorsion which indicate the presence of anisotropic coherency strains [1].

Grain sizes (20-50 nm) are of the order of multilayer thickness while Moire fringes are observed [5] . Diffraction lines at 1.49 Å and 3.02 Å (X-rays) and 3.07 Å (TEM) observed on all the substrates may be attribuated to an unknown $NbAl_x$ alloy. Composition modulation is observed by SIMS (figure 1) with a depth resolution of about 2 nm for 5.5 keV Xe ions bombarding the samples at 60° from the film normal.

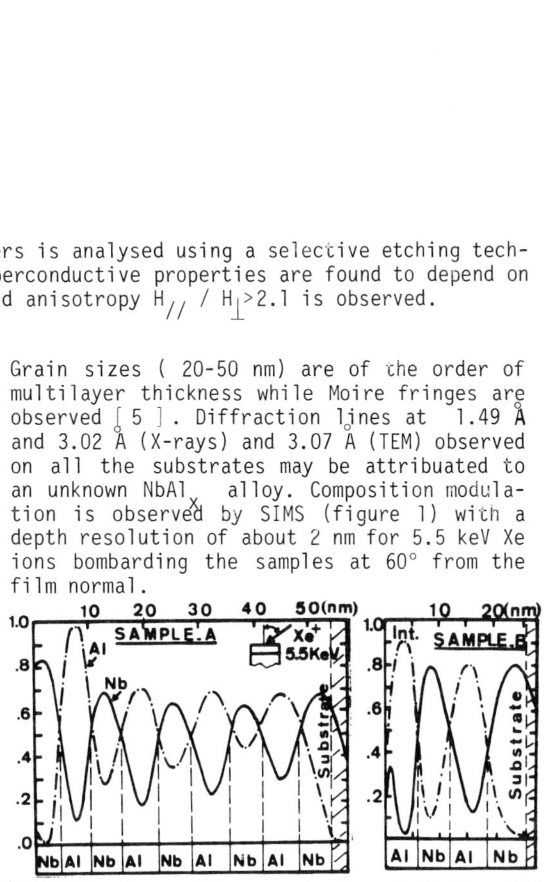

Figure I : Composition modulation of Nb-Al multilayers (Intensity versus depth during Xe sputtering)

Sample A has N = 9 films (T_c = 6.07 K) on a Sapphire substrate, sample B has N = 4 films after removing the 5 upper films on a Si + SiO_2 substrate. Oxide appears on about 2.3nm at the upper surfaces but oxygen level is very low (0.5 %)inside the multilayer with an increase in the first Nb film near the SiO_2 substrate (sample B). The modulation amplitude which represents approximately the composition [5] is high and conserved for the layers located near the substrate (sample B) while some ion mixing appears during Xe sputtering of sample A. The interdiffusion level can be directly observed after selective etching of the sample B surface where the Nb/Al signal amplitude gives approximately an alloy Nb Al_x ($x \simeq 2$) across thickness of 1.5 + .5nm. This is confirmed by XPS [5] and structure analysis [1].

3. RESISTIVE AND SUPERCONDUCTIVE PROPERTIES

Resistivity and critical temperature values, measured by a four-point method, depend upon the number of films N and thickness $d_T = N \times 6$(nm) of the multilayer (table 1).

FILMS NUMBER N D_T (NM)	9 54	8 48	7 42	6 36	5 30	4 24	3 18	2 12	1 6
ρ 300 K ($\mu\Omega$.CM)	27.0	27.8	30.7	32.2	34.2	36.7	52.2	54.0	70
ρ 10 K ($\mu\Omega$CM)	19.1	19.9	24.0	26.0		30.6			
$D\rho/DT_{300-100\ K}$ $\mu\Omega$CM.K^{-1}	2.57×10^{-2}					2.51×10^{-2}			
T_c (K)	6.08	5.75	4.40			2.67			

TABLE I

Results are found reproducible and T_c transition sharp (T_c 0.05K). Thermal contribution $\Delta\rho = \rho_{300\ K} - \rho_{10\ K} = 7.0 \pm 0.9$ $\mu\Omega$.cm is observed between the experimental Al and Nb values ($\Delta\rho_{Nb} = 14.6$ $\mu\Omega$.cm [2] and $_{Al} = 2.7$ $\mu\Omega$.cm) with a low N dependance for N > 3. Resistivity follows an experimental low : $\rho(T) = \rho_b\ (1 + 120/d_T) + 2.6\ 10^{-2}T$ for $d_T >$ 12 nm with ρ_b (bulk) = 6 $\mu\Omega$.cm indicating a low diffuse boundary scattering [5]. Critical temperature is approximately :

$T_c = 7.4 - 120/d_T$, d_T (nm). The ratio $\Delta T_c/\Delta\rho_{10K} = -0.17$ K. $(\mu\Omega.cm)^{-1}$

is independant of thickness (d_T) and corresponds to the T_c decrease in Nb films due to contaminations [2][6][7]. So mean free paths effects are found mainly responsible of resistive and superconductive dependence on films number in the observed range [5].

3.1 Anisotropic critical fields

The temperature dependence of critical fields $H_{//}$ and $H_\perp$ has been studied for N = 9 films with $T_c = 6.1$ K (figure 2) :
$H_\perp(T)$ is approximately linear on the whole T range :
$H_\perp(0)$ 22 kG and $T = T_c = 4$ kG.K^{-1}
So the coherence length can be estimated using the "dirty limit" relation [7].

$\xi^2(0) = (\phi_o/2\pi) \times (dH_\perp/d_T)_{T_c} \times T_c$, $\xi(0) = 12$ nm.
H// has a completely different behavior with a variation $H_{//} \sim (T - T_c)^{\frac{1}{2}}$ near T_c and an upturned curvature in the range 3.5 - 5 K. The ratio $H_{//}/H_\perp$ (T) increases continuously with the temperature from the value 2.1 at 1.9 K. Such a high value, indicating strong anisotropy, and the $H_{//}(T) \sim (T-T_c)^{\frac{1}{2}}$ near T_c has been found by Ruggiero et al [7] in the 2D behavior of Nb/Ge multilayers and interpreted by the Josephson coupling theory, valid near T_c.

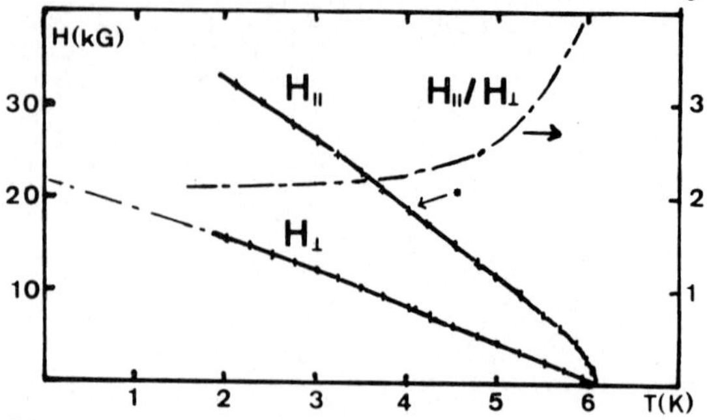

Figure II : Anisotropy of critical fields

Our films are in the 2D regime with respect of the film thickness $d = 6$ nm/ξ^2 in the whole T range while they behave like 3D at low temperature and 2D near T_c relatively to multilayer thickness d_T. More investigations are in course to analyse such thin coherent and anisotopic multilayers and their possible utilisation to make superconductive devices.

REFERENCES

(1) D.B. Mc Whan, M. Gurvitch, J.M. Rowell, L.R. Walker J. Appl. Phys. 54 (1983) 3886
(2) J.C. Villégier, J.C. Veber I.E.E.E.-Trans. Mag., MAG 19, (1983) 946
(3) I.K. Schuller Phys. Rev. Lett 44 (1980) 1597,
(4) M. Gurvitch, J. Kwo, Proc. ICMS-83 Adv. in Cryog. Eng. 30 (1984)
(5) J.C. Villégier et Al, to be published
(6) I. Banerjee, I.K. Schuller J Low Temp. Phys. 54 (1984) 501
(7) S.T. Ruggiero, T.W. Barbee, M.R. Beasley Phys. Rev. B 26 (1982) 4894

LT-17 (Contributed Papers)
U. Eckern, A. Schmid, W. Weber, H. Wühl (eds)

SUPERCONDUCTING Nb/Cu MULTILAYERS PREPARED IN A DUAL ELECTRON BEAM EVAPORATOR

Walter SEVENHANS, Jean-Pierre LOQUET*, Alain GILABERT+ and Yvan BRUYNSERAEDE

Laboratorium voor Vaste Stof-Fysika en Magnetisme, Katholieke Universiteit Leuven, B-3030 Leuven, Belgium
+ Physique du Solide, Université de Nice, Parc Valrose, 06034 Nice, France

Finely layered superconducting Nb/Cu composites have been fabricated by a sequential vapor deposition technique. The x-ray diffraction scans are characteristic for a metallic superlattice. Preliminary measurements of the superconducting parameters Tc, $Hc_{\parallel}$ and $Hc_{\perp}$ are reported.

1. INTRODUCTION

Artificially prepared layered metals, also called composition-modulated structures (CMS), are materials composed of thin normal or superconducting films separated by another metallic or semiconducting film. Recent work (1) has shown that the electrical, magnetic and structural properties of these multilayers are unique and determined by the variation of the individual layer thickness.

In this paper we describe an UHV evaporation system used for the preparation of these layered materials using electron-beam-gun sources and a rotating platform. The individual deposition rates could be accurately controlled by a quadrupole mass spectrometer. This technique was used to prepare Nb/Cu CMS, a material which has been studied by various groups using sputtering techniques for producing samples. The x-ray characterization of those Nb/Cu composites as well as preliminary measurements of the superconducting critical parameters Tc, $Hc_{\parallel}$ and $Hc_{\perp}$ are reported.

2. PREPARATION TECHNIQUES

Fig. 1 shows a schematic diagram of the stainless steel UHV system (base pressure $\simeq 10^{-10}$ Torr) in which the layered structures are prepared using two electron-beam-gun sources (6 and 10 kW). A vertical screen, placed between the sources, avoided cross contamination of the two material fluxes. The substrates are fixed to a rotating table and alternatively exposed to the evaporation sources. The composition wavelength λ and the composition symmetry is determined by the accurately controlled rotation speed of the table and evaporation rates of the two electron guns. This rate is measured by a programmable multichannel quadrupole mass spectrometer (Balzers QMG 511). The output signal per mass is fed into a home made rate controller, a feedback system which produces two control signals. The first signal controls the slow variations in the evaporation rate by changing the filament current of the electron gun; the second signal controls the fast rate variations (10 - 50 Hz) by changing the sweep amplitude of the electron beam. This arrangement produces an instantaneous evaporation rate which is constant within 3 - 5 %, depending on the rate. The average rate over a one second period is accurate within 0.5 %. An absolute value of the evaporation rate is obtained by calibrated quartz crystal thickness monitors. In order to decrease the pumpdown period a quick-access-port was installed through which the substrates can be fixed onto the turntable with a special manipulator. A temperature controlled substrate heater and cryopanels will be installed in the near future.

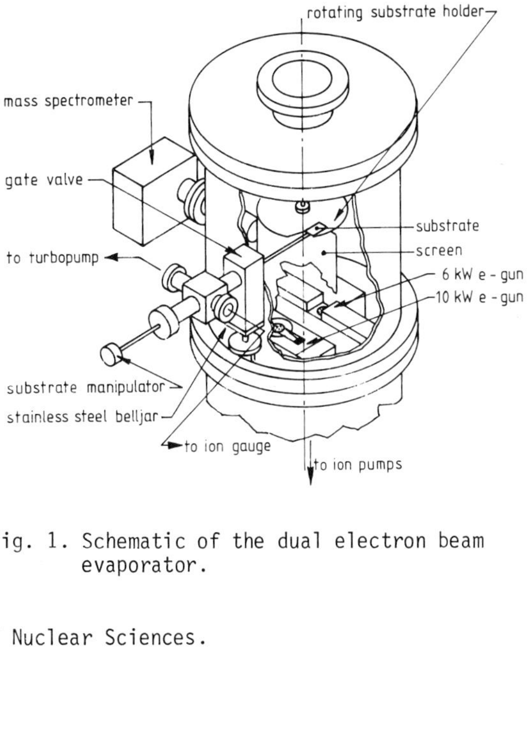

Fig. 1. Schematic of the dual electron beam evaporator.

---

* Research Fellow, Belgium Interuniversity Institute of Nuclear Sciences.

3. PROPERTIES OF Nb/Cu MULTILAYERS

The layered specimen, condensed on 0° sapphire substrates held at room temperature, was characterized by standard θ - 2θ x-ray diffraction measurements, normal to the layers. Fig. 2 shows a typical scan from a Nb/Cu sample (total thickness ≈0.5 um). The diffraction shows strong Nb (110) and Cu (111) Bragg peaks with higher order satellites, characteristic for metallic superlattices. The positions of the peaks are related to λ, the composition wavelength through the relation (2)

$$\lambda = \lambda_x(2(\sin\theta_{i-1} - \sin\theta_i))^{-1}$$

where λ_x is the wavelength of the CoK_α x-ray radiation and i refers to adjacent peaks.

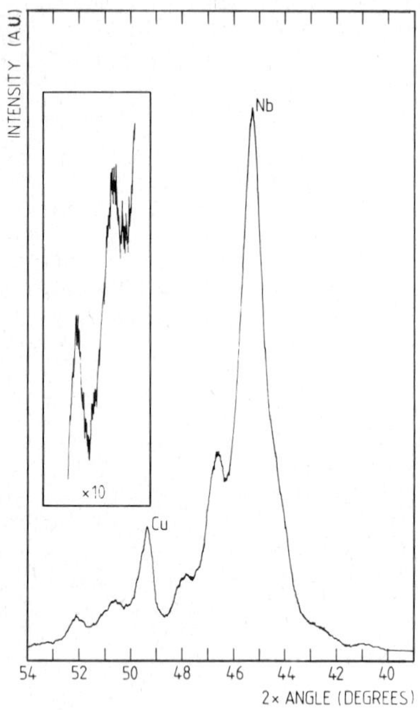

Fig. 2. The diffractometer scan of 58 layers Nb/Cu with a composition wavelength of 82 Å.

The Nb/Cu layer ($\rho300/\rho5$ = 12.7) has a low T_c= 4.65 K(<T_c(Nb)) due to the proximity effect. The temperature dependence of $H_{c\parallel}$ and $H_{c\perp}$ (Fig. 3) is typical for a layered system (3): i) the ratio $H_{c\parallel}/H_{c\perp}$ ≃ 2 (2D-coupled regime); ii) $H_{c\perp}$ has a small positive curvature near Tc; iii) $H_{c\parallel}$ is linear (the normal core of the vortices does not fit into a single Cu layer).

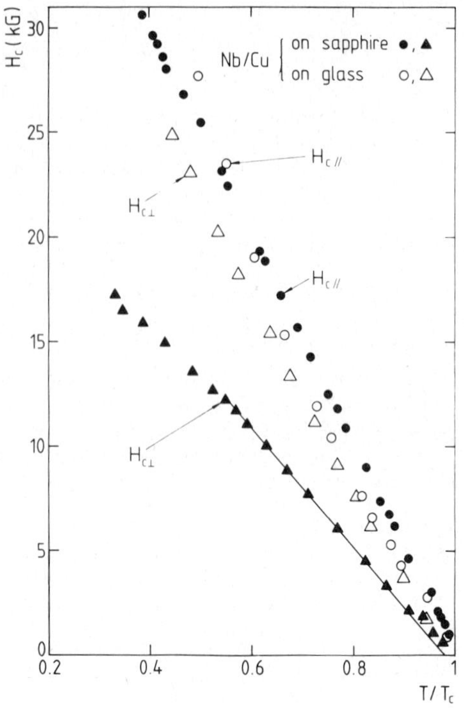

Fig. 3. Transition fields of a Nb/Cu sample on glass and sapphire substrates.

It should be remarked that in a Nb/Cu sample condensed simultaneously on a glass substrate poor layering was observed: i) absence of satellites in the x-ray scan; ii) T_c = T_c(Nb); iii) $H_{c\parallel}/H_{c\perp}$ ≃ 1 (isotropic regime). The difference in heat conductance between the glass and the sapphire substrate may influence the growth of the multilayer.

REFERENCES

(1) I.K. Schuller and C.M. Falco, in: Microstructure Science and Engineering, N.G. Einspruch, ed. (VLSI), vol. 4, pp. 183-205.

(2) I.K. Schuller, Phys. Rev. Lett. 44, (1980) 1597.

(3) I. Banerjee, Q.S. Yang, C. Falco and I.K. Schuller, Phys. Rev. B28, (1983) 5037.

LT-17 (Contributed Papers)
U. Eckern, A. Schmid, W. Weber, H. Wühl (eds)

A NEW NUCLEAR SPIN-LATTICE RELAXATION MECHANISM IN bcc ^{3}He

M. CHAPELLIER, M. BASSOU†, M. DEVORET, and J.M. DELRIEU

Universite d'Orsay, 91405 Orsay, and C.E.N.-Saclay, 91191 Gif/Yvette, France

N.S. SULLIVAN*

Department of Physics, University of Florida, Gainesville, FL 32611 USA

Measurements of the nuclear spin-lattice relaxation times in bcc helium three have shown the existence of a new relaxation process at low temperature with an approximately linear temperature dependence $T_1T \simeq$ const. Possible interpretations are presented for this new relaxation mechanism and unusual features of the vacancy induced relaxation seen at slightly higher temperatures.

1. INTRODUCTION

The nuclear spin-lattice relaxation times T_1 observed in solid ^{3}He have in most cases been successfully interpreted in terms of the different motions that occur in this quantum solid: tunneling of thermally activated vacancies, nuclear spin exchange, diffusion of ^{4}He impurities, coupling with phonons, etc. The overall variation of T_1 with temperature is related to the properties of the motions that determine the relaxation of the nuclear Zeeman energy (Z) to the thermal bath (B) via couplings to other energy reservoirs (exchange energy bath (E), vacancies (V), phonons (P) etc.) at different temperatures.

Typical results are shown in figure 1. In regime I the motion of thermally activated vacancies modulates the dipole-dipole interactions between the ^{3}He atoms and the temperature dependence of T_1 (1)-(4) can be used to deduce Φ, the energy of formation of the vacancies. This is the Z-VB relaxation regime where the bar designates the thermal bottleneck between the Z and V energy baths.

Regime II of figure 1 is attributed to the "mixing" of the Zeeman (Z) and exchange (E) baths via the dipole-dipole interactions. E is coupled tightly to the lattice via vacancies and phonons etc., and the bottleneck is the Z-E link which is temperature independent. In regime III there are insufficient vacancies (or phonons) to assure strong thermal contact to the bath and the weak link becomes the E-V (or E-P) coupling. We observe an exponential temperature dependence in this temperature range for a molar volume V_m = 23.25 cm^3 (fig. 1) and infer that the relaxation is due to thermally activated vacancies.

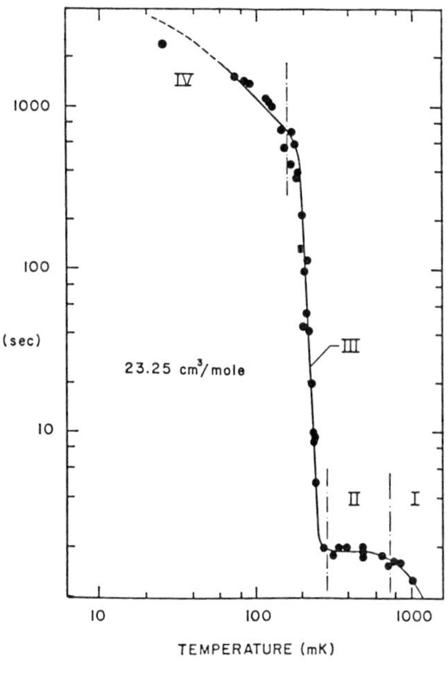

FIGURE 1
Typical results for temperature dependence of T_1.

For lower temperatures (regime IV) the exponential temperature dependence is not seen and for 23 < V_m < 24.6 cm^3 we have observed (5) an almost linear temperature dependence $T_1T = A(V_m)$ where A varies with density.

†Present Address: Faculte des Sciences de Fes, Maroc.
*Supported by N.S.F. - Low Temperature Physics - Grant DMR-8304322

2. EXPERIMENTAL CONSIDERATIONS

A capacitive pressure gauge monitored V_m and solid was formed from the liquid by a pressure pulse. Defects were removed by annealing for several hours. ^{4}He impurities were ~ 60 ppm.

The relaxation was measured using a 50 MHZ pulsed coherent crossed-coil spectrometer and was observed to be exponential except for small initial deviations at the lowest temperatures. This occurs when the Z and E baths come into equilibrium in a short time before they relax together slowly to B at longer times.

3. RESULTS AND DISCUSSION

Results for several molar volumes are collected in figure 2. Except for 24.8 cm^3/mole regimes II-IV can be distinguished. The drop in

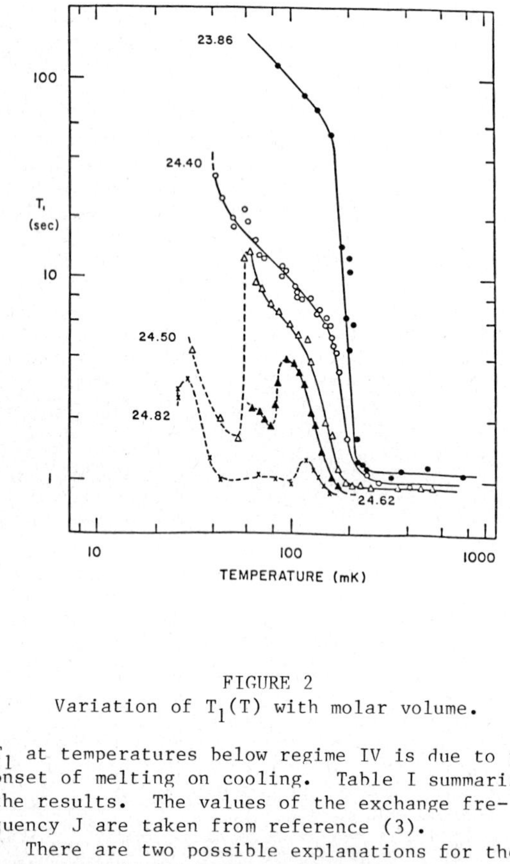

FIGURE 2
Variation of $T_1(T)$ with molar volume.

T_1 at temperatures below regime IV is due to the onset of melting on cooling. Table I summarizes the results. The values of the exchange frequency J are taken from reference (3).

There are two possible explanations for the new relaxation mechanism. The first is the one-phonon process for the coupling of the exchange bath to the lattice calculated by Griffiths (6) for organic radicals and Bernier (7) has suggested that it may apply to solid ^{3}He. The intrinsic rate (6) varies as J^4 and on accounting for the heat capacities of the Z and E baths, the observed T_1 should vary as J^{-6} for Larmor frequencies $\omega_L > J$. This is only approximately verified but there is considerable variation in the values of J cited in the literature and simultaneous measurements of J and T_1T for each sample are needed to verify this prediction.

TABLE I. Summary of results.

V_m (cm^3/mole)	$J/2\pi$ (MHz)	Φ (K)	(T_1T) (sK)
23.25	7.5	4.9±0.4	>120
23.86	10.0	3.9±0.4	10±2
24.40	13.0	2.4±0.3	0.98±0.15
24.44	13.5	2.2±0.2	0.91±0.10
24.50	13.6	1.4±0.2	0.65±0.10
24.58	14.0	1.4±0.3	0.58±0.15
24.62	14.4	0.8±0.2	0.77±0.25

The second mechanism suggested (8) to explain the T_1T behavior is related to the observation (2),(5) that both the vacancy formation energy Φ and the vacancy jump frequency ω_V vary smoothly toward zero as V_m increase. (Φ and log ω_V vary approximately as $(V_{mc}-V_m)^{1/2}$ as V_m approaches the melting curve minimum.) In this case one cannot exclude the existence of a small number of ground state vacancies at high molar V_m. The breaking of translational invariance would open a gap between a low energy band and the high energy band attributed to thermally excited lattice defects. The existence of the low energy band associated with highly delocalized ground state vacancies would lead to the observed T_1T behavior since the Raman scattering of low energy modes would determine the E-B coupling.

Further experiments are needed to distinguish between these possible mechanisms and studies of ultra pure samples and varying ^{4}He concentrations should also be carried out.

REFERENCES

(1) M.Bassou, thesis (Universite de Paris-Sud, Orsay, France, 1981).
(2) M. Chapellier, M. Bassou, M. Devoret, J.M. Delrieu and N.S. Sullivan, AIP Conf. Proceed. 103 (1983) 811
(3) R.A. Guyer, R.C. Richardson and L.I. Zane, Rev. Mod. Phys. 43 (1971) 532.
(4) N. Sullivan, G. Deville and A. Landesman, Phys. Rev. B11 (1975) 1858.
(5) M. Chapellier, M. Bassou, M. Devoret, J.M. Delrieu and N.S. Sullivan, submitted to Physical Review.
(6) R.B. Griffiths, Phys. Rev. 124 (1961) 1023.
(7) M. Bernier, to be published.
(8) N.S. Sullivan, J.M. Delrieu and M. Chapellier, Workshop on Ultra-Low Temp. Physics, LT-XVI, Los Angeles, CA 1981.

LT-17 (Contributed Papers)
U. Eckern, A. Schmid, W. Weber, H. Wühl (eds)

SPECIFIC HEAT MEASUREMENT OF bcc SOLID ^{3}He THROUGH THE NUCLEAR MAGNETIC ORDERING TEMPERATURE*

Anju SAWADA, Motoi KATO, Hideo YANO and Yoshika MASUDA

Department of Physics, Nagoya University, Chikusa-ku, Nagoya, Japan

The specific heat of bcc solid ^{3}He has been measured as a function of temperature in the temperature range including the nuclear magnetic ordering temperature for molar volume of 24.0 cm^3/mole. Specific heat measurement above the ordering temperature reveals that besides the known $1/T^2$ term there are a significant negative term $1/T^3$. The observed latent heat through the ordering temperature is discussed in connection with the first-order phase transition.

1. INTRODUCTION

In recent years, experimental studies on the magnetic properties of solid ^{3}He have been mainly performed in the ultralow temperature range near the nuclear magnetic ordering temperature, and provided much new informations. The ultra-low temperature studies have clarified the magnetic behavior anticipated from the experiments performed at much higher temperatures, where the high temperature series expansions of the spin-exchange Hamiltonian can be utilized. The understanding of consistency between information obtained in both temperature ranges is of great importance for a consistent theoretical picture.

It is expected that the specific heat measurements of solid ^{3}He will provide valuable knowledge on the exchange interaction. (1),(2) We have recently measured the specific heat and isochoric pressure of bcc solid ^{3}He in the high temperature range between 2 and 60 mK as a function of molar volume. (2) The results were analyzed using the current four-spin exchange model. So far the temperature range studied, however, has been limited to temperatures above the ordering temperature. We present here the first measurements of the specific heat of solid ^{3}He as a function of temperature in the temperature range including the nuclear magnetic ordering temperature for molar volume of 24.0 cm^3/mole.

The latent heat observed in the transition through the nuclear magnetic ordering temperature gives a fruitful information on the nature of the first-order phase transition. The specific heat measurements in the ordered state also gives a confirmation of prediction by the spin wave theory.

2. EXPERIMENT

The cryostat for specific heat measurement is shown in Fig. 1. This cryostat was originally made for the double nuclear demagnetization

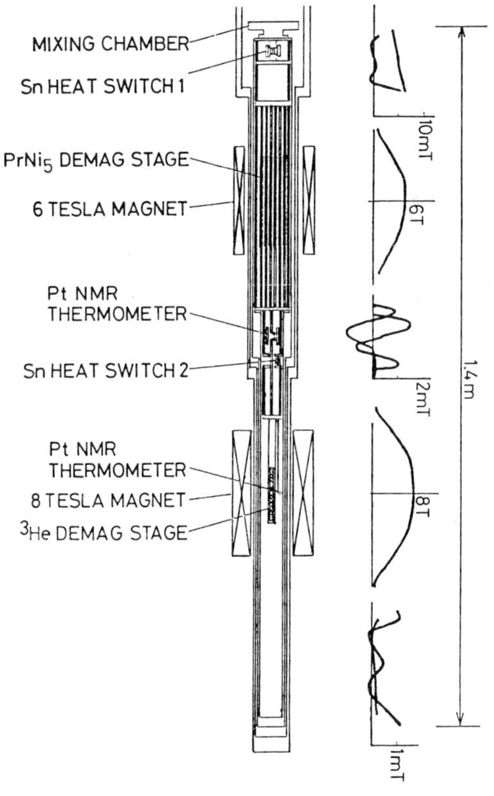

FIGURE 1
Cryostat for calorimetry and ^{3}He nuclear demagnetization.

cryostat, which is a cascade of two adiabatic demagnetization of a hyperfine enhanced nuclear stage of $PrNi_5$ and hcp solid ^{3}He. This is the first attempt to cool solid ^{3}He itself directly by nuclear cooling. The first cooling stage

* This work was supported in part by the Grant-in-Aid for Scientific Reseach of The Ministry of Education, Science and Culture, Japan.

consists of 34 arc-cast $PrNi_5$ rods weighing 1.1 kg (2.6 mole). The $PrNi_5$ rod has a cylindrical form of 6.4 mm in diameter and 100 mm in length, which was supplied by the Ames Laboratory. The ^3He cell for demagnetization was made from a silver rod of 10 mm in diameter which has a hole with 6 mm in diameter and 100 mm in length. The cell was packed with a sintered silver sponge composed of 700 A particles at a packing factor of 40%. The available volume was 1.701 cm^3 with a surface area of 40 m^2. $PrNi_5$ was precooled to 10 mK in a field of 5.4 T for about 2 days by the dilution refrigerator. After demagnetization of $PrNi_5$, solid ^3He was precooled at 0.7 mK in magnetic field of 5.0 T for about 4 days. Final temperature reached by demagnetization of hcp ^3He itself was only 0.5 mK, which was too high compared with the initial expectation. However, this is an important consequence for a possibility of the use of solid ^3He as a coolant in adiabatic demagnetization to study properties of solid ^3He. This experimental trial is still going on.

This cryostat was used for cooling solid ^3He with $PrNi_5$ in specific heat measurement. A pulsed NMR thermometer was used to measure the temperature of a ^3He cell. (3) Temperature was determined from the nuclear spin-lattice relaxation time T_1 of platinum, assuming the Korringa relation T_1T=29.6 sec·mK. The sample was performed at constant volume by using the blocked-capillary method. The ^3He used had a ^4He content less than 3 ppm. The specific heat measurements were performed by using the heat pulse method and by taking the warm up curve sometimes.

3. EXPERIMENTAL RESULTS

The experimental results of specific heat measurements are shown in Fig. 2. The result obtained in the range above the ordering temperature was fitted up to the third-order term of power of $\beta \equiv 1/T$ by

$$C=(R/4)(e_2\beta^2-e_3\beta^3+\ldots\ldots),$$

where e_2 and e_3 are constants independent of temperature. The values of e_2 and e_3 deduced from the present specific heat measurements are e_2= 7.4 mK^2 and e_3=12.2 mK^3, respectively. These values are in good agreement with our recent results obtained from both specific heat and isochoric measurements. (2) As a comparison, the specific heat curve calculated by making use of the values of e_2 and e_3 of reference 2 is depicted with the dot-dashed line in Fig. 2. As can be seen from the figure this calculated result reproduces quite well the present result. No depression in the specific heat curve was observed at temperatures close to the magnetic ordering temperature. This fact suggests that the possibility of existence of the significant positive higher terms, as expected from a gradual increase of the specific heat.

As clearly seen from Fig. 2, the specific

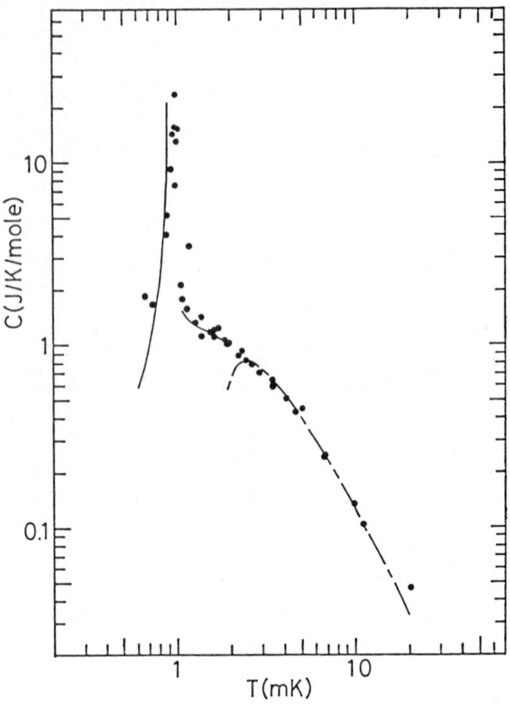

FIGURE 2
Specific heat of ^3He for 24.0 cm^3/mole in zero magnetic field. Solid lines show the results measured by the warm up curve.

heat curve diverges at about 0.95 mK, showing an appearance of the first-order transition. (4) The value of entropy jump derived from the latent heat accompanying the first-order transition was estimated to 0.4 Rln2. The total entropy integrated from 0 K to an infinite temperature was 0.93 Rln2. It is expected that the specific heat below T_N obeys to the T^3-dependence as predicted by the antiferromagnetic spin wave theory. (5) The temperature range in the present experiment is not wide enough to confirm this prediction precisely, at present.

We would like to thank S. Inoue for his technical assistance.

REFERENCES

(1) J. M. Dundon and J. M. Goodkind, Phys. Rev. Lett. 17 (1974) 1343.
(2) H. Fukuyama, Y. Miwa, A. Sawada and Y. Masuda, J. Phys. Soc. Jpn. 53 (1984) 916.
(3) A. Sawada. H. Yano and Y. Masuda, submitted to Rev. Sci. Instrum.
(4) T. Mamiya, A. Sawada, H. Fukuyama, Y. Hirao, K. Iwahashi and Y. Masuda, Phys. Rev. Lett. 47 (1981) 1304.
(5) K. Iwahashi and Y. Masuda,(in preparation).

LT-17 (Contributed Papers)
U. Eckern, A. Schmid, W. Weber, H. Wühl (eds)
© Elsevier Science Publishers B.V., 1984

NON-LINEAR SPIN DYNAMICS IN THE uudd PHASE OF SOLID ^{3}He

Makoto TSUBOTA and Toshihiko TSUNETO

Department of Physics, Kyoto University, Kyoto, Japan

The non-linear spin dynamics in the uudd phase of solid ^{3}He observed by the pulsed n.m.r. experiment is discussed. We point out the possible presence of the dissipative term similar to the Leggett-Takagi term in the case of superfluid ^{3}He. Non-linear motion induced by an r.f. field is studied by means of a numerical analysis.

1. INTRODUCTION

The spin dynamics of the uudd phase of solid ^{3}He is an interesting problem especially in the non-linear regime where the Zeeman energy is comparable to the dipole energy and deserves further studies. It was brought to our attention by the recent pulsed n.m.r. experiment by Kusumoto et al.[1] The observed free induction decay is evidently quite anomalous when the initial tipping angle exceeds certain critical value. One can think of various mechanisms for the observed rapid decay, such as instability of the uniform mode against generation of spin wave modes or spin current carrying away magnetic energy from the resonant magnetic domain to others with different $\hat{l}$-vectors (the normal to ferromagnetic uudd planes). These mechanisms which involve some kind of spatial non-uniformity will be discussed elesewhere. Here we will point out possibility of a dissipative mechanisms similar to the Leggett and Takagi term in the case of superfluid ^{3}He[2] and, assuming the motion to be uniform, try to disclose, by means of a numerical analysis, a few salient features of the non-linear spin dynamics appropriate to the pulsed n.m.r. experiment.

2. OCF EQUATIONS WITH A DISSIPATIVE TERM

We start with the Osheroff-Cross-Fisher (OCF) equations[3] for the total spin $\vec{S} = \vec{S}_1 + \vec{S}_2 + \vec{S}_3 + \vec{S}_4$ and the $\hat{d}$-vector which is the unit vector along $\vec{N} = \vec{S}_1 + \vec{S}_2 - \vec{S}_3 - \vec{S}_4$, where $\vec{S}_{1-4}$ are the statistical average of spin at the four sublattices:

$$d\vec{M}/d\tau = \vec{M}x(\hat{z} + \vec{h}) - \lambda(\hat{d}.\hat{l})(\hat{d} \times \hat{l}) \qquad (1)$$

$$d\,\hat{d}/d\tau = \hat{d} \times (\hat{z} + \vec{h} - \vec{M}) + \mu\hat{d} \times (\hat{d} \times \hat{l})(\hat{d}.\hat{l}) \quad (2)$$

Here the static magnetic field is taken along the $\hat{z}$-axis, $\vec{H}_0 = H_0\hat{z}$, and $\tau = \gamma H_0 t = \omega_L t$, $\vec{M} = \gamma\vec{S}/xH_0$, $\lambda = \Omega_0^2/\omega_L^2$ is the parameter for the dipole energy, Ω_0 being the zero-field resonance frequency, and $\vec{h} = \vec{H}_{r.f}/H_0$ is the r.f. field. The last term in (2), a dissipative term, is introduced purely phenomenologically and has the same form as the Leggett-Takagi term, a form allowed by the symmetry argument in the present case also. As seen from the observed temperature dependence of Ω_0, there are contributions of thermal magnons to the spin vectors $\vec{S}$ and $\vec{N}$. When out of equilibrium the part due to thermal magnons lags behind so that, for example, $\vec{S}$ is no longer orthogonal to $\vec{N}$. The deviations relax due to magnon-magnon collisions. The situation is, therefore, similar to the case of the superfluid. If the parameter μ is proportional to the magnon-magnon relaxation time, its temperature dependence should propably be T^3.

3. FORCED OSCILLATIONS AND TRANSITION TO IRREGULAR MOTION

In the pulsed n.m.r. experiment one usually applies an r.f. field resonant with the small oscillations. Since the motion becomes anharmonic with increasing tipping angle β, it is not clear how much tipping one induces by applying the pulse of a given duration. Therefore, we first study the effect of the r.f. field in the absence of the dissipative term. In this case $\vec{M}.\hat{d} = 0$, so that we use $|\vec{M}|$ and 3 Euler angles (α,β,γ) sepcifying an orientation of $\vec{M}$ and $\hat{d}$ as variables. The energy of the spin system $E = (1/2)M^2 - \vec{M}(\hat{z} + \vec{h}) + (\lambda/2)(\hat{d}.\hat{l})^2$ varies as

$$dE/dt = Mh\omega \cos[\omega t - \gamma(t)] \sin\beta(t) \quad ,$$

where we have taken $\vec{h} = h(\cos\omega t.\hat{x} - \sin\omega t.\hat{y})$. Due to the non-linearity the difference $\omega t - \gamma(t)$ develops with time so that the energy can start to decrease before the tipping angle β reaches π. With this remark in mind we show in Fig.1-a and 1-b the variation of β in the presence of the r.f. field with $h = 10^{-2}$ for the case $\theta = 0$ and $\theta = 1.19$ ($\cos\theta = \hat{z}.\hat{l}$), obtained by numerical integration. Here and in the following we choose the dipole constant λ equal to 0.62. These values for λ and h are appropriate to the experimental results of ref.1. In the first case β returns to 0 periodically but its maximum is only 0.8 rad.. In the second case the behavior is not periodic at all, perhaps chaotic. It should be interesting experimentally to measure this period for the periodic cases and also to look for the onset of chaotic behaviors. In Fig. 2-a and 2-b we give $\beta(t)$ for the r.f. pulse of duration $\tau = \omega_L t = 50$ and 70, respectively, for

$\theta = 1.19$. One can see that, when β becomes larger than roughly 0.6, the chaotic motion starts and persists after the turn-off of the r.f. field. Though this provides a nice example of chaotic motion, we are here not concerned with that aspect.[4] The average slope of $\alpha(t)$ for the periodic cases agrees with the theoretical value.[5]

4. FREE INDUCTION DECAY

Next we discuss the decay of induced spin motion due to the dissipation. When the dissipative term is included, the magnetic energy decreases as

$$dE/dt = -\lambda\mu[(\hat{d}.\hat{l})(\hat{d}\times\hat{l})]^2$$

after the r.f. field is turned off. We have computed E(t) for the three cases with the pulse durations 30, 50, and 70, two of which corresponds to Fig.2-a and 2-b. Note in Fig.3 that in the third case with the largest initial tipping, the energy decays most rapidly. Since the rate of change of E is proportional not to E itself but to the dipole energy, it is different for different orbits, and the crossing of the curves means that a different basin is reached by the orbit. Similar curves are obtained for the transverse component of the magnetization M sin β if we smooth its rapid oscillations. The fact that Fig.3 resembles Fig.1-a of ref.1 may be significant.

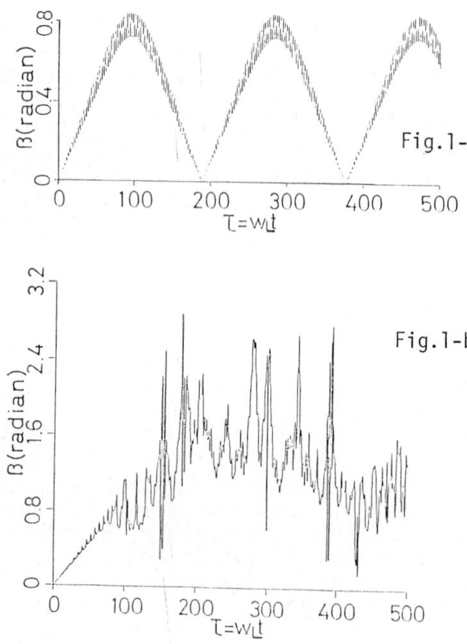

Fig.1 The tipping angle in the presence of the r.f. field for $\theta = 0$ (Fig.1-a) and $\theta = 1.19$ (Fig.1-b)

REFERENCES

1) T. Kusumoto, O. Ishikawa, T. Mizusaki and A. Hirai, Phys. Lett. 100A (1984) 201, and this Proceedings.
2) A.J. Leggett and S. Takagi, Phys. Rev. Lett. 34 (1975) 1424.
3) D.D. Osheroff, M.C. Cross and D.S. Fisher, Phys. Rev. Lett. 44 (1980) 792.
4) T. Katayama, Prog. Theor. Phys. 65 (1981) 1158. Y. Yamaguchi, Prog. Theor. Phys. 69 (1983) 1377.
5) H. Namaizawa, Prog. Theor. Phys. 67 (1982) 1989.

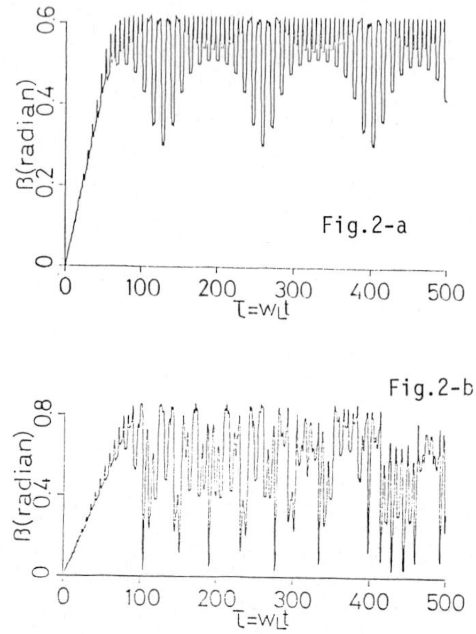

Fig.2 The tipping angle after the r.f. pulse of duration $\tau = 50$ (Fig.2-a) and 70 (Fig.2-b) for $\theta = 1.19$.

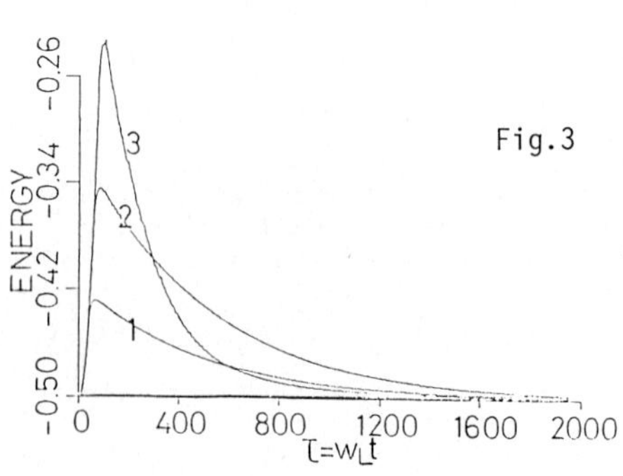

Fig.3 The free decay of the energy after the pulse of duration 30 (the curve 1), 50 (2) and 70 (3) for $\theta = 1.19$ and $\mu = 0.01$.

LT-17 (Contributed Papers)
U. Eckern, A. Schmid, W. Weber, H. Wühl (eds)

MULTIPLE EXCHANGE IN ^{3}He AND IN THE WIGNER SOLID

Michel ROGER

Orme des Merisiers, CEA-SACLAY, 91191 Gif-sur-Yvette, Cedex, France

We prove from the first principles that three particle exchange is preponderent in 3D hcp ^{3}He, and in the 2D triangular lattice, for ^{3}He and the electron Wigner Solid as well. In bcc ^{3}He we find dominant four and three atom exchanges with the same order of magnitude and similar volume dependences.

1. INTRODUCTION

A two parameter Hamiltonian, assuming that four and three particle exchange interactions dominate has provided a coherent explanation for all experimental data on nuclear magnetic order in bcc ^{3}He (1). The origin of large multiple exchange is purely geometric : the exchange of two atoms requires a large displacement of their neighbors in order to prevent hard core overlapping ; cyclic three and four atom exchange processes are favoured because they need a smaller disturbance of the surrounding.

The prediction of the exchange frequencies from the first principles is an huge task. Crude approximations have been proposed by Delrieu et al (1,2). However exact calculation can be performed if we restrict our ambition to the quasi-classical limit. We present here the determination of the first terms of a high density series expansion for exchange in solid ^{3}He. We also investigate the Wigner solid at low densities.

2. METHOD

We write the Schrödinger equation in reduced units $\vec{\rho} = \{\vec{\rho}_i\} = \{\vec{r}_i/a\}$, (a : nearest neighbor distance)

$$-\Delta\psi(\vec{\rho}) + g^{-2}V(\vec{\rho})\psi(\vec{\rho}) = E\psi(\rho) \; ; \; (E = E_0\, 2\, ma^2/\hbar^2)$$

g^2 represents the ratio between the kinetic and potential energies. For ^{3}He $V = \sum_{i<j} |\vec{\rho}_i - \vec{\rho}_j|^{-12}$ and $g^2 = \hbar^2(8m\varepsilon\sigma^2)^{-1}(a/\sigma)^{10}$; (we take the repulsive part of the Lennard Jones potential. With $\sigma = 2.6$Å and $\varepsilon = 10.2$K. For the Wigner solid $V = \sum_{i<j} |\vec{\rho}_i - \vec{\rho}_j|^{-1}$ and $g^2 = a_B/2a$ (a_B is the Bohr radius).

We use a WKB series expansion in powers of g. The lowest order represents the usual eikonal approximation. The leading term in the exchange frequency is exponential :

$$J_P = A_P \exp(-S_P/g) \text{ with } S_P = \int_{X_I}^{X_P} (V-E)^{1/2}\, dt \quad (1)$$

S_P is the action corresponding to the classical path of a point mass moving in the 3N dimensional configuration space under the action of a potential minus V. The extremities X_I and X_P of the path are the two permuted configurations $\{\vec{r}_i = \vec{R}_i\}$ and $\{\vec{r}_i = \vec{R}_{P(i)}\}$; $\vec{R}_i$ represents the lattice sites and P(i) the permutation P of the indices. The classical path minimizes S_P. It is symmetric with respect to the median hyperplane Σ equidistant from the configurations X_I and X_P. Let us call X_M the intersection of the path with Σ. We divide the half path into n+1 segments $X_I X_1, X_1 X_2, \ldots, X_n X_M$ and search for the configurations $\{X_i\}$ which minimize the action S_P. The minimum is obtained through successive approximations, with n = 0 to 4 using a steepest descend method in a 3nN dimensional space (see (3)). We take N ≃ 16 particles.

3. RESULTS

3.1 Two dimensional triangular lattice

In table 1 we compare the results : action S_P, half path length L (in unit a) for the two dimensional triangular lattice with two different potentials : i) $|\rho|^{-12}$ (^{3}He at high densities) ii) $|\rho|^{-1}$ (electrons at low densities). In both cases three particle exchange dominates. The hierarchy between various exchange processes depends essentially on the lattice geometry and not on the shape of the potential (provided it is repulsive). Fig.1a compares the variations of the three particle exchange frequency, in terms of the interatomic distance, with the experimental results (4) for ^{3}He. (The preexponential factor A_P is estimated within a factor

TABLE 1 : action and path length for 2 and 3D latt.

2D	Helium			Electrons		
	3	4	2	3	4	2
S_P	8.6	9.7	11.3	.87	.97	.97
L	1.07	1.13	1.15	.95	1.05	.85

(nb of exchanging particles)

3D	hcp ^{3}He				bcc ^{3}He			
	3	3'	2	4	4	3	2	4'
	J_T	J'_T	J'_2	K'_P	K_P	J_T	J_{NN}	K_F
S_P	9.3	9.9	10.9	11.1	8.5	9.1	10.2	10.6
L	1.00	1.00	.97	1.14	1.07	1.10	1.08	1.14

of a few unities by a crude approximation (see (3)).

3.2 Three dimensional hcp lattice

We apply the same method with $|\rho^{-12}|$ potential to compare all kinds of two three and four particle cycles involving first neighbors in solid hcp ^3He. Table 1 gives the results for the dominating processes. Triple exchange is still preponderent. The triangles in the basal plane of the hcp lattice (J_T) have a larger contribution than these out of the basal plane (J_T'). The next important processes are pair (J_2') and planar four atom exchange (K_p') out of the basal plane. They are however 10 times lower than J_T at the lowest densities of hcp ^3He. Triple exchange leads to an effective ferromagnetic first neighbor Heisenberg Hamiltonian (1) with anisotropic interactions $J_1=J_T+2J_T'$ for pairs in the basal plane and $J_1'=4J_T'$ for other pairs. The Curie Weiss temperature $\theta=3(J_T+6J_T')$ is positive. On Fig.1b we compare the variations of $J_{ef}=\theta/6$, as a function of a, to the values deduced from NMR (5). The preexponential factors are crudely estimated (3).

3.3 bcc lattice

Although the eikonal approximation is certainly unrealistic for the densities at which the bcc phase in ^3He is stable, it is interesting to predict the hierarchy of various exchange processes obtained through the first terms of our expansion in (a/σ). Taking the $|\rho^{-12}|$ potential we obtain (see table 1).

$$K_P > J_T > J_{NN} > K_F$$

planar 4 part. 3 particle 2 particle folded 4 part.

This is precisely the hierarchy assumed in the model hamiltonian which fits the experimental data. The results differ from these obtained by Avilov and Iordansky (6) in a similar WKB calculation : $J_{NN}>K_P>>J_T$. The reason is that they take a two small number of moving particles (N=4 for J_T). We obtain reliable values of Sp only for N>12 (we took N=16). Our calculation is the first which shows that among the two kinds of four atom cycles the planar type (K_P) dominates over the folded (K_F). The reason is purely geometric : in the critical configuration X_M (middle of the path) the number of atom pairs whose distances are appreciably reduced is 8 for Kp and 12 for K_F. As emphasised in §3.1, the hierarchy between various exchange processes is not very dependent of the potential shape ; hence we expect it is the same at lower densities where the contribution of the kinetic energy can be included, to some extend, in an effective potential (1-3).

The values of S_P for K_P, J_T, J_{NN} differ by less than 10%. At densities near melting where $S_P/g \simeq Ln(A_P/J_P) \simeq 11$ ($A_P \simeq \theta_D \simeq 20K$ and $K_P \simeq .4mK$) we expect only a factor of the order of $e^1 \simeq 3$ between these frequencies. For these three dominating processes the lengths of the exchange paths are almost equal (see Table 1). We believe this result is essentially geometric and subsists approximately at lower densities. For this reason we expect very similar volume dependences of K_P, J_T, J_{NN} in

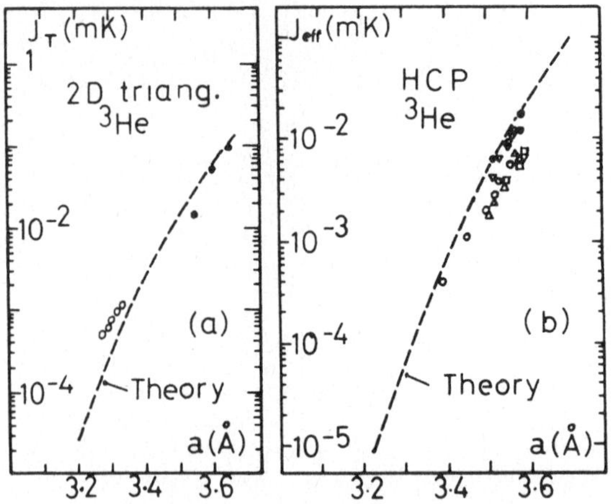

Fig.1 : Exchange vs interatomic distance

agreement with recent experimental results (7,8). Note that the (might be unrealistic ?) extrapolation to the densities near melting of the high density functional law (Rel.1) $J_P \sim \theta_D \exp(-S_P/g)$ with $g \sim a^5 \simeq V^{5/3}$ leads to

$$\partial \ln K_P / \partial \ln V \simeq \frac{5}{3} \ln(\theta_D/K_P) \simeq \frac{5}{3} \cdot 11 \simeq 18$$

($\simeq \partial \ln J_T / \partial \ln V$ within 10%) in perfect agreement with experiments (7,8).

CONCLUSION

From the first principles, we clearly justify the model Hamiltonian inferred from the experimental results on bcc ^3He (1) showing that :

a) several exchange processes (planar four particle, three particle and two particle exchanges) differing only by factors of a few unities play a fundamental role in the physics of bcc ^3He

b) these three exchange processes have very similar volume dependences

c) folded four particle exchange is negligible

We predict a ferromagnetic behaviour (positive Curie Weiss constant) in the two dimensional triangular lattice ; this result holds as well for ^3He adsorbed on a substrate and for the electron Wigner solid. We predict ferromagnetism with anisotropy in hcp solid ^3He.

ACKNOWLEDGMENTS. We thank JM Delrieu, N. Sullivan and J. Hetherington for fruitful discussions.

REFERENCES

(1) M. Roger, J.H. Hetherington and J.M. Delrieu, Rev. of Mod. Phys. 55 (1983) 1

(2) J.M. Delrieu and N.S. Sullivan, Phys. Rev. B23 (1981) 3197

(3) M. Roger, submitted to Phys. Rev.

(4) M. Richards, in Phase Transitions in Surface films, eds. J.G. Dash and J. Ruvald (Plenum Press, New York, 1980) p.187

(5) R.A. Guyer, R.C. Richardson and L.I. Zane, Rev. of Mod. Phys. 43 (1971) 532

(6) V.V. Avilov and S.V. Iordansky, J. of low temp. Phys. 48 (1982) 241

(7) T. Mamiya, A. Sawada, H. Fukuyama, Y. Hiro and Y. Masuda Phys. Rev. Lett. 47 (1981) 1304

(8) T. Hata, S. Yamasaki, M. Taneda, T. Kodama and T. Shigi, J. Magn. Magn. Mat. 31-34 (1983) 735

LT-17 (Contributed Papers)
U. Eckern, A. Schmid, W. Weber, H. Wühl (eds)

INITIAL SLOPE OF THE UPPER CRITICAL FIELD OF THE HEAVY FERMION SUPERCONDUCTOR UPt_3

J. O. WILLIS, Z. FISK, and J. L. SMITH

Materials Science and Technology Division, Los Alamos National Laboratory, Los Alamos, NM 87545 USA

J. W. CHEN, S. E. LAMBERT, and M. B. MAPLE

Department of Physics and Institute for Pure and Applied Physical Sciences, University of California, San Diego, La Jolla, CA 92093 USA

We have made low field measurements of the upper critical field as a function of temperature for all parallel and perpendicular configurations of magnetic field H, current I, and crystallographic orientation for UPt_3, the heavy fermion superconductor. The curves are quite linear for $H \parallel I$, with slopes of -4.0 T/K for I in the basal plane and -6.3 T/K for I along the c-axis. For $H \perp I$ the curves are concave upward near $H = 0$ and have a generally smaller slope than for the same field applied parallel to the current.

1. INTRODUCTION

The third heavy fermion superconductor UPt_3 (1) has increased interest in this new field because of its many differences from the other two known systems, $CeCu_2Si_2$ (2) and UBe_{13} (3). The presence of spin fluctuations (1,4) occurring for the same electrons responsible for superconductivity, and the qualitatively different resistance vs. temperature behavior for UPt_3 are two of the most important differences.

Recent measurements by Chen et al. (5) of the upper critical field H_{c2} as a function of temperature T for single crystal UPt_3 show substantial anisotropy with respect to the direction of the applied field H relative to the direction of the current I, which was along the hexagonal c-axis. In addition, the H_{c2} vs. T curve for $H \perp I$ is nonlinear near $H = 0$, being concave upward.

Because the details of the anisotropy are important to understanding the superconductivity, we have measured $H_{c2}(T)$ for all possible parallel and perpendicular configurations of I, H, and crystallographic orientation on the same sample of UPt_3.

2. EXPERIMENTAL

Flux-grown single crystals have a needle-like morphology, with the hexagonal c-axis parallel to the long direction. The crystal used for most of the measurements reported here was annealed for 1200°C for 40 hours and at 1100°C for an additional 20 hours. This crystal was actually a twin with a large basal plane cross section. It was necessary to select such a crystal in order to have sufficient resistance and size to attach electrical leads for the basal plane measurements. The voltage contacts, applied with silver epoxy, were in one twin, but the current contacts were made to both twins (except for I in the basal plane with $H \perp I$). Both twins had the same T_c, but they had differing superconducting transition widths ΔT_c. This was most apparent for the last mentioned configuration; however, for $H > 5$ mT the T_c's were approximately equal. We thus believe that the twinning has a very small effect on the determination of the critical field curves.

The measurements were made in dc fields up to 0.2 T using an ac current of 220 or 370 Hz. No self heating or temperature hysteresis was observed at the current level employed. A current of twice this magnitude resulted in a T_c depression of 5 mK. Data were taken at a cooling rate of 0.1 mK/s in constant applied fields. We define the superconducting transition temperature T_c as the temperature at which the sample resistance has fallen to half its normal state value.

The temperatures have been corrected for the magnetoresistance of the germanium thermometer employed; these corrections were always less than 3 mK. Corrections to the applied field caused by sample demagnetization were calculated to be less than 0.2%. The estimated temperature uncertainty is less than 2% and the magnetic field uncertainty is less than 5%.

This sample has a residual resistance ratio $(R(300K)/R(T_c^+))$ of 180. The zero field T_c is 0.486 K and the transition width ΔT_c is 8 mK, for the current along the c-axis, denoted as I_c. For the current in the ab-basal plane, denoted as I_{ab}, the T_c is the same, but the widths are a factor of 3 to 5 larger; this is reflected in the lower midpoint T_c values. The data for I_{ab} were taken using different current and voltage contact pairs for H ⊥ I and H ∥ I.

3. DISCUSSION

Figure 1 shows the four H_{c2} vs. T curves for the annealed single crystal. The fifth curve with H ⊥ I_c is for an unannealed crystal with a larger ΔT_c from another batch. The critical field curves are linear within the scatter of the data for H ∥ I_{ab} with a slope of -4.0 T/K and essentially linear for H ∥ I_c with a slope of -6.3 T/K; this is quite a large anisotropy with respect to the current direction. The value of the latter slope is identical to our previous result (5) on an unannealed crystal from another batch. The anisotropy with respect to field direction is also quite large. For example, at 0.2 T the T_c for H ⊥ I_c has been depressed 1.6 times as much as the T_c for H ∥ I_c.

The $H_{c2}(T)$ curves for H ⊥ I are nonlinear near T_c. For H ⊥ I_c, the departure from linearity appears to be sample dependent. The corresponding curve from our earlier work (5) exhibits an even greater departure from linearity than is seen in Fig. 1. A field-dependent ΔT_c may be part of the cause of this variation. The present annealed sample has an essentially field-independent ΔT_c. In contrast, ΔT_c changes from 20 to 50 mK between 0 and 0.2 T for the data shown for the unannealed sample in Fig. 1., and the ΔT_c for the sample of

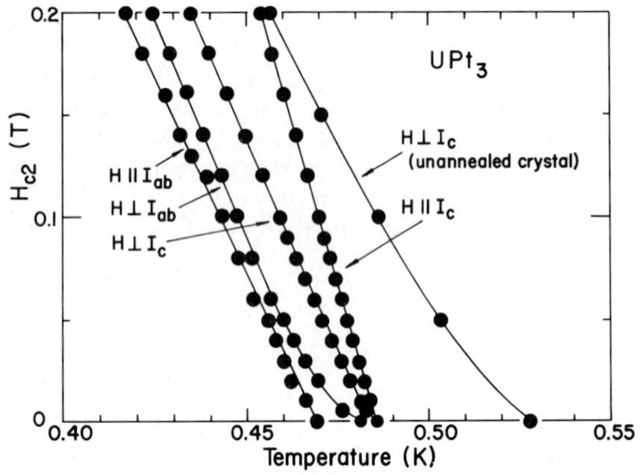

FIGURE 1
The upper critical field H_{c2} versus temperature for the four combinations of magnetic field H and current I shown, where I_c indicates current flow along the hexagonal c-axis and I_{ab} indicates current in the ab-basal plane, for the same single crystal of UPt_3. The fifth curve is for an unannealed sample from a different batch. The solid lines are only a guide to the eye.

our earlier work changes from 22 to 84 mK over the same field range. These enormous changes in ΔT_c tend to distort the shape of the H_{c2} vs. T curve near H = 0 where the depression of T_c is of the same order of magnitude as ΔT_c. Choosing T_c as the onset, the first decrease in resistance below the normal state value, of the transition helps to reduce this effect. Nevertheless, the curves remain nonlinear, but less so, and they appear to be approaching a common shape. Nonlinear H_{c2} vs. T curves for H ⊥ I near T_c may well be an intrinsic property of UPt_3.

4. CONCLUSIONS

The $H_{c2}(T)$ curves near T_c have been measured on a well-annealed single crystal of UPt_3. The curves for H ∥ I are linear within experimental error below 0.2 T. The slope dH_{c2}/dT at $T = T_c$ for H ∥ I_c is in good agreement with previous results. The curves for H ⊥ I are nonlinear and concave upward near H = 0. For I_c there are sample-dependent variations in the amount of the nonlinearity. This variation is reduced by using transition onsets rather than midpoints for those samples that show a large field-dependent ΔT_c. Even then, the curves remain concave upward near T_c.

ACKNOWLEDGEMENTS
This work was performed under the auspices of the U. S. Department of Energy, under Contract No. DE-AT03-76ER70227 at UCSD.

REFERENCES
(1) G. R. Stewart, Z. Fisk, J. O. Willis, and J. L. Smith, Phys. Rev. Lett. 52 (1984) 679.
(2) F. Steglich, J. Aarts, C. D. Bredl, W. Lieke, D. Meschede, W. Franz, and H. Schäfer, Phys. Rev. Lett. 43 (1979) 1892.
(3) H. R. Ott, H. Rudigier, Z. Fisk, and J. L. Smith, Phys. Rev. Lett. 50 (1983) 1595.
(4) P. H. Frings, J. J. M. Franse, F. R. de Boer, and A. Menovsky, J. Magn. Magn. Mater. 31-34 (1983) 240.
(5) J. W. Chen, S. E. Lambert, M. B. Maple, Z. Fisk, J. L. Smith, G. R. Stewart, and J. O. Willis, submitted to Phys. Rev. B.

LT-17 (Contributed Papers)
U. Eckern, A. Schmid, W. Weber, H. Wühl (eds)

COHERENCE AND SUPERCONDUCTIVITY IN KONDO LATTICE SYSTEMS

Norbert GREWE

Institut für Festkörperphysik der Technischen Hochschule Darmstadt, D-6100 Darmstadt, FRG

A theoretical development for the treatment of the impurity- and lattice Anderson models and its applications to the low temperature properties of Kondo systems ist described. It contributes to the understanding of coherence effects in the temperature dependent density of states and of the phenomenon of Heavy Fermion Superconductivity.

Recently, a renewed interest in the Kondo effect arose due to new experimental and theoretical progress. The most important experimental fact was the detection of the so called Heavy Fermion Superconductors, like $CeCu_2Si_2$ [1]. But also specific heat and thermopower data suggest an interesting low temperature behaviour of Kondo lattices [2]. Finally, it remained an open question why a system like $CeAl_2$ has a magnetically ordered ground state in contrast to $CeAl_3$ or $CeCu_2Si_2$. In the last years a powerful perturbational method has been developed [3] for the treatment of the Anderson model and its generalizations, which allows for a calculation of various thermodynamic and dynamical properties in the Kondo and Intermediate Valence regions [4-6]. In particular, it was possible to gain some insight into the behaviour of RE compounds which at low temperature should exhibit coherence effects [7]. The present short note intends to discuss some of these questions under a unified point of view.

It is useful to start from a look at the Anderson impurity in the Kondo regime, characterized by the following parameter values: local Coulomb repulsion $U \to \infty$, position of the local level ΔE far below the Fermi energy, i.e. $\Delta \equiv \pi V^2 N_F << -\Delta E$ (V = 4f-5d mixing strength, N_F = unperturbed density of conduction states at the Fermi level) and finite, and $\Delta E+U \to \infty$. Fig. 1 shows the local density of states which has been derived from the Greens function $F_\sigma(z)$ for one particle propagation on the local level. F_σ can be calculated approximately via a system of integral equations [8-10]. Obviously, the original resonance at $\omega = \Delta E$, which would be a Lorentzian with width Δ without the spin degree of freedom, is much flatter and broader now. A narrow resonance starts to grow slightly above the Fermi level at temperatures smaller than roughly $k_BT_K = N_F^{-1} \exp(-1/2\, N_F J)$ with $N_F J = \Delta/\pi|\Delta E|$. Its width Γ is of order k_BT_K and its height approximately $1/\pi\Delta$ at very low temperatures. Its maximum is slightly displaced above the Fermi level by an amount $\eta < k_BT_K$. Because of its strong temperature dependence this peak clearly is a many body phenomenon and has been ascribed to the Kondo effect [4-6].

A reasonable but simple approximation for F_σ near the Fermi level has been given in [11]:

$$F_\sigma(\omega+i\delta) \approx \zeta[\omega-\eta+i(\delta+\Gamma)]^{-1} \quad . \qquad (1)$$

In this expression, ζ is a function of temperature T which increases with decreasing T from 0 to Γ/Δ. Since in the nonmagnetic case one is considering here F_σ displays no symmetry breaking in spin space, the situation for each spin component is the same as in a simple resonant level model with energies rescaled by a factor ζ^{-1} and hybridization $\tilde{V}$ determined by $\tilde{\Delta} \equiv \pi N_F \tilde{V}^2 = \Gamma$. It is immediately obvious that going over to a regular lattice of ions in the frame of this approximation results as in the well known case of hybridized bands [10] in a (renormalized) conduction electron Green's function

$$G_{\underline{k}\sigma}(\omega+i\delta) = \left[\omega+i\delta - \varepsilon_{\underline{k}} - \frac{\zeta V^2}{\omega-\eta+i(\delta+\Gamma-\zeta\Delta)}\right]^{-1} \qquad (2)$$

for ω near the Fermi level. $G_{\underline{k}\sigma}$ describes the formation of a gap in the density of states as ζ goes from 0 to Γ/Δ, corresponding to $\Gamma-\zeta\Delta$ decreasing from Γ to 0. The band structure for each spin at T = 0 looks like that of a regular lattice of local levels with energy η hybridizing with a spinless band with strength $\sqrt{\zeta}\, V$. This picture in fact is supported by the more rigorous numerical calculations of Ref. [17], which are shown in Fig. 2. Part (a) contains the band density of states $\rho_\sigma^{(band)}(\omega) = -1/\pi \sum_{\underline{k}} \mathrm{Im}\, G_{\underline{k}\sigma}(\omega+i\delta)$, whereas the corresponding local density of states $\rho_\sigma^{(4f)}(\omega) = -1/\pi \sum_{\underline{k}} \mathrm{Im}\, F_{\underline{k}\sigma}(\omega+i\delta)$ is displayed in part (b). Here, $F_{\underline{k}}(z)$ is related to $G_{\underline{k}\sigma}(z)$ via exact equations of motion by $G = G^{(0)} + G^{(0)} F G^{(0)}$, $G^{(0)}_{\underline{k}\sigma} = [z-\varepsilon_{\underline{k}}]^{-1}$.

The dip in the Kondo resonance is a typical effect of coherence, which builds up in the lattice at low temperatures. In real systems, however, one cannot expect to find a real gap even at T = 0. This has various reasons: (a) a residual lifetime for the renormalized band electrons

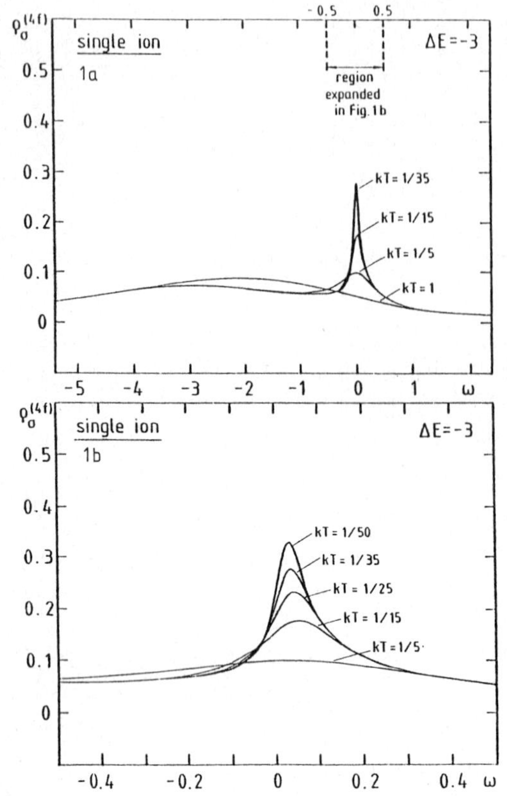

Fig. 1: Local one particle excitation spectrum of the Anderson impurity ($U \to \infty$) in the Kondo regime, $k_B T_K = 0.18$. The Fermi level lies at $\omega = 0$ and energies are given in units of the virtual level width $\Delta = \pi N_F V^2$.

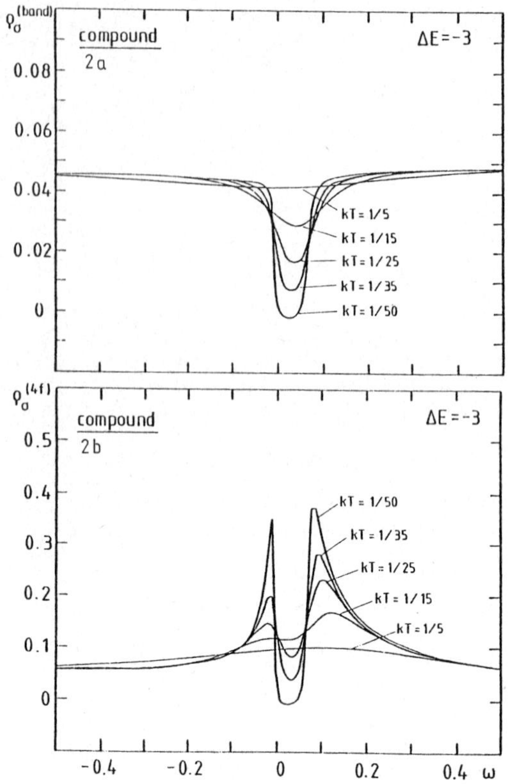

Fig. 2: Band (a) and local (b) one particle excitation spectrum of an Anderson lattice. Parameters are the same as in Fig. 1.

even near the Fermi level and at low temperatures, (b) a dispersion of the 4f-levels due to Umklapp scattering [12] or direct overlap, (c) asphericities of hybridization matrix elements contributing to V. Nevertheless, it is to be expected that a pseudogap remains in the density of states near T = 0. Such pronounced structures near the Fermi level should leave their trace in experimentally measurable quantities like specific heat and thermopower. This point is pursued in more detail in a different contribution to this conference [2].

The foregoing considerations were mainly based on certain types of perturbation processes which do not contain those which generate magnetic interactions between moments on different sites. The RKKY interaction, however, is hold responsible for the antiferromagnetic ground state of $CeAl_2$. So the discussion is only conclusive as far as this interaction energy is smaller than the binding energy gained from the coherent formation of Kondo singlets. This poses certain conditions on band structure, mixing strength V and on the energy ΔE of the local level. Since it seems a difficult task to include properly higher order interactions between sites into a tractable perturbational approach [13], no more detailed information is available at the moment.

Beside magnetism, superconductivity is a particularly exciting collective phenomenon occuring in Kondo lattices. There are special features which distinguish it from superconductivity in conventional materials. Most of all, the large specific heat coefficient indicates that the carriers involved are about two orders of magnitude heavier than free electrons [1]. Secondly, the quasiparticles near the Fermi level not only have to build up an appropriate condensate but at the same time have to quench the local moments on the ions, so that pairbreaking is largely suppressed. Recently, a model has been proposed, in which these two facts are intimately connected [12]. It assumes a BCS-like phonon mediated pairing of band electrons, which however heavily scatter at the ions and thereby at the same time acquire a large effective mass and produce local Kondo singlets. In the following, this kind of theory is discussed starting from the Anderson Hamiltonian considered above [11].

It is well known that changes of the 4f occupation number in Rare Earth ions strongly couple to the elastic degrees of freedom [14]. Such a coupling of the form c^+fb + h.c. has been used to derive an effective interaction between conduction electrons [11]:

$$U_{eff}(i\omega_m, i\omega_\ell; \underline{q}j, i\nu_p) = \frac{V^2}{2} \sum_{j=1}^{r} |g_{\underline{q}j}|^2 D^{(o)}_{\underline{q}j}(i\nu_p) \cdot$$

$$\cdot T_\uparrow(i\omega_m, i\omega_m - i\nu_p)\, T_\downarrow(i\omega_\ell, i\omega_\ell + i\nu_p) \quad . \quad (3)$$

For a Cooper pair with energy $i\omega_m + i\omega_\ell \to 0$ the product of $T_\uparrow$ with $T_\downarrow$ is the absolute square of the local scattering matrix T_σ for band electrons, which has a sharp maximum for $i\omega_m$ and $i\omega_m - i\nu_p$ both continued into the energy range of the Kondo peak. The phonon propagators $D^{(o)}_{qj}$ for $i\nu_p$ continued to a real frequency of this magnitude is approximately $-2/\hbar\omega_{qj}$, so that the following average static effective interaction is obtained for band electrons in the Kondo peak:

$$U_{eff} \approx -r\, V^2 \langle|g|^2\rangle \langle|T_\sigma|^2\rangle / \hbar\langle\omega\rangle$$

$$\approx -4\pi^2 r \langle|g|^2\rangle V^2 \langle\rho^{(4f)2}\rangle / \hbar\langle\omega\rangle \quad . \quad (4)$$

Due to the largely enhanced local density of states in this region ($\pi^2 V^2 \langle\rho^{(4f)}\rangle$ is roughly $1/\pi\, \Delta\, N_F \gg 1$) this interaction has an unusual strength compared to conventional situations.

There is however a compensating effect following from the repulsion and redistribution of band density of states in the region of the sharp Kondo resonance, which has mainly local character. One can recognize this also in Fig. 2(a). It leads to a reduced density of states for pair formation ($\underline{k}\uparrow, -\underline{k}\downarrow$) in the Kondo peak. The renormalized propagator for Cooper pairs,

$$H(\underline{q}, i\nu_\ell) = \frac{1}{\beta} \sum_m \frac{1}{N} \sum_{\underline{k}}{}' G_{\underline{k}\uparrow}(i\omega_m) G_{\underline{k}-\underline{q}\downarrow}(i\nu_\ell - i\omega_m) \quad , \quad (5)$$

therefore becomes much reduced, i.e. for $\underline{q} = 0$, $i\nu_\ell \to 0$:

$$H \approx 2\,\Gamma\, N_F \cdot N_F \ln \frac{\gamma}{2} \beta\Gamma \quad (\gamma = 1.14) \quad . \quad (6)$$

Compared to a conventional result there is a reduction in energy range $\hbar\omega_D \to k_B T_K$ and in amplitude $1 \to 2\,\Gamma\, N_F$. The results (4) and (6) together in the usual way give a Cooper instability at a critical temperature:

$$T_c = \frac{\gamma}{2} T_K \exp[-1/(g_o \cdot \chi)], \quad g_o = N_F \frac{2\, r \langle|g|^2\rangle}{\hbar\langle\omega\rangle} ,$$

$$\chi = \frac{4}{\pi} \frac{\Gamma}{\Delta} \quad . \quad (7)$$

Whereas χ roughly equals a few 10^{-1}, g_o is likely to be enhanced compared to values derived for conventional electron phonon coupling. Therefore $k_B T_c$ here can be related to $k_B T_K$ in the same way as usually to $\hbar\omega_D$.

A last point to mention is the possibility of pair breaking due to (virtual) excitations of local magnetic states. There exist experimental hints that this might play a role in $CeCu_2Si_2$ [2]. It seems desirable to include such effects in a general microscopic theory of Heavy Fermion Superconductivity in Kondo lattices.

REFERENCES

[1] F. Steglich, J. Aarts, C.D. Bredl, W. Lieke, D. Meschede, W. Franz and H. Schäfer, Phys. Rev. Lett. 43, 1892 (1979)

[2] C.D. Bredl, N. Grewe, F. Steglich, E. Umlauf, contribution to this conference

[3] H. Keiter and G. Morandi, to appear

[4] Y. Kuramoto, to appear

[5] N. Grewe, Z. Phys. B53, 271 (1983)

[6] P. Coleman, preprint

[7] N. Grewe, Sol. State Comm. 50, 19 (1984)

[8] H. Keiter and G. Czycholl, J.M.M.M. 31, 477 (1983)

[9] Y. Kuramoto, Z. Phys. B53, 37 (1983)

[10] N. Grewe, Z. Phys. B52, 193 (1983)

[11] N. Grewe, to appear

[12] H. Razafimandimby, P. Fulde and J. Keller, Z. Phys. B54, 111 (1984)

[13] N. Grewe and H. Keiter, Phys. Rev. B24, 4420 (1981)

[14] N. Grewe and P. Entel, Z. Phys. B33, 331 (1979)

LT-17 (Contributed Papers)
U. Eckern, A. Schmid, W. Weber, H. Wühl (eds)

THEORY OF HEAVY ELECTRON SUPERCONDUCTORS

T. M. Rice and K. Ueda

Theoretische Physik, ETH-Hönggerberg, 8093 Zürich, Switzerland

A microscopic model of an almost localized Fermi liquid is proposed for heavy electron superconductors. Several properties of the p-wave superconducting state that results are analyzed; spin-orbit coupling effects, critical magnetic fields, and Abrikosov vortex structures versus orbital singular structures in a magnetic field.

Analysis of the specific heat of the heavy electron metal UBe_{13} has led Ott and coworkers (1) to propose that it is a p-wave superconductor in the Anderson-Brinkman-Morel (ABM) state at temperatures $T<0.9K$. First we discuss a microscopic model which generalizes the almost localized Fermi liquid model of Brinkman and Rice (2). In the second part we discuss several properties of the p-wave superconducting states that result.

As a simple model for UBe_{13} we consider only the U 5f-bands after integrating out the Be states, and approximate them by a single band Hubbard model (interaction U: energy $\varepsilon(k)$). The many body wavefunctions in the Gutzwiller form are represented as a projection operator, determining the number of doubly occupied sites, acting on a Slater determinant of Bloch states. They are characterized by the fraction of doubly occupied sites d, average occupancy $1-\delta$ per site and a set of occupation numbers $\{n_{k\sigma}\}$ for the Bloch states. The internal energy is

$$E(d,\delta,\{n_k\}) = \sum_{k\sigma} q(d,\delta)\, n_{k\sigma}\varepsilon_k + Ud$$

In the almost localized limit $q\to 0$ as d, $\delta\to 0$. The limit $\delta\equiv 0$, $d\to 0$ has been successfully applied to ^{3}He (3,4) but we consider $d\equiv 0$ and $\delta\to 0$ more appropriate here and then the effective mass $m^*/m = q^{-1}\approx(2\delta)^{-1}>>1$. Note the ratio of the susceptibility enhancement χ/χ_0 to m^*/m is constant as $\delta\to 0$. We propose for the entropy a form

$$S = -\sum_{k\sigma} w_k\{n_{k\sigma}\ln n_{k\sigma}+(1-n_{k\sigma})\ln(1-n_{k\sigma})\}$$

The entropy weight function, w_k takes account of the fact that for $d<(1-\delta)^2/4$ the states with different $\{n_{k\sigma}\}$ are not orthogonal. We place the following restrictions on w_k;

a) $w_k\to 1$ as $k\to k_F$ to obtain Landau Fermi liquid theory as $T\to 0$

b)

$$\sum_k w_k = \frac{((1-\delta-2d)\ln(\frac{1-\delta}{2})-d\ +d\ln d+(d+\delta)\ln(d+\delta)}{(1+\delta)\ln\frac{1+\delta}{2} + (1-\delta)\ln\frac{1-\delta}{2}}$$

to obtain the correct entropy when $U>>T>>q\varepsilon_F$ and

c) $\sum_k w_k^{-1} = 2(1-2d-\delta)$ to obtain a Curie law for the susceptibility $[\chi(T) = \mu_0^2(1-2d-\delta)/T]$ when $U>>T>>q\varepsilon_F$ with a moment determined by the number of singly occupied sites. This extension to finite temperatures describes a smooth transition from the coherent Fermi liquid at T=0 to the state with incoherent spin fluctuations at $T>q\varepsilon_F$.

The possible forms of a p-wave superconductor have been analyzed in the context of ^{3}He for systems with spherical symmetry and no spin orbit coupling (3,5). U Be_{13} is in the limit of strong spin orbit coupling. The pairing takes place among the 4 degenerate states $\psi_{\alpha,\beta}(\pm k)$ related by time reversal and space inversion. If we assume that the basic interaction is not spin dependent then the pairing interaction has a form

$$\frac{1}{2}\sum V(k-k')I_{kk'}^{\gamma_1\gamma_4}\, I_{kk'}^{\gamma_2\gamma_3}\, a^+_{-k\gamma_1}a^+_{k\gamma_2}a_{k'\gamma_3}a_{-k'\gamma_4}$$

Because of the space inversion and time reversal symmetries

$$I_{kk'}^{\alpha\alpha} = I_{kk'}^{\beta\beta} = I_{kk'} = I_{k'k}$$

$$I_{kk'}^{\alpha\beta} = -\ I_{kk'}^{\beta\alpha} = J_{kk'} = -J_{k'k}\ .$$

In a case of an isotropic Fermi surface the coefficient $J_{kk'} = 0$ and we can expand $I_{kk'}$ in a form

$$I_{kk'} = I_0 + 3I_1\ k\cdot k' + \dots .$$

Therefore if the Fermi surface is not far from spherical we can write a pairing matrix for the ABM state of the form

$$\Delta_{\alpha\alpha'}(\hat{k}) = (\tfrac{3}{2})^{1/2}\ \Delta(\hat{\ell}_1.\hat{k}+i\hat{\ell}_2.\hat{k})\ \begin{pmatrix}1 & 0\\ 0 & 1\end{pmatrix};\hat{\ell}_1\perp\hat{\ell}_2$$

and the explicit corrections due to spin orbit coupling are small.

In the presence of an external magnetic field there are several possible states of an ABM superconductor. Ohmi et al (6) proposed a lattice of non-singular vortex states similar to Brinkman-Mermin-Ho structures as an alternative to the Abrikosov state. However, they considered a model without orbital anisotropy but in a cubic crystal there is an anisotropy energy $K_\ell(\ell_x^2\ell_y^2+\ell_y^2\ell_z^2+\ell_z^2\ell_x^2)$; $(\hat{\ell}=\hat{\ell}_1\times\hat{\ell}_2)$. Such a term has a drastic effect near H_{c1} giving an additional divergent contribution due to the need to misalign the $\hat{\ell}$ vector over large spatial regions. Therefore the Abrikosov state with singular vortices is favored. Further for $H \simeq H_{c2}$ the magnitude of the order parameter is small and the cost of making singular vortices is not large. We conclude that the state proposed by Ohmi et al (6) is unlikely to occur at any T, H.

We have examined the linearized Ginzburg-Landau equations that determine H_{c2} in the Abrikosov structure. When the $\hat{\ell}$-axis is oriented along the magnetic field the value of H_{c2} is largest. There is an enhancement of the critical field above the Abrikosov value $\sqrt{2}\ \kappa\ H_c$ by a factor $(1-\kappa\tau)^{-1}$. The parameter τ is defined by

$$\frac{T_{c\alpha,\beta}-T}{T_c-T} = 1 \pm \tau\ \frac{H}{\sqrt{2}H_c}$$

and determines the change of T_c for the $\alpha(\beta)$ pairing in the magnetic field. Although we expect $\tau \ll 1$ since $\kappa \gg 1$ this enhancement factor may be sizeable. If the Fermi surface is a sphere, there is no direct coupling term $\Delta^+_{\alpha\alpha}\Delta_{\beta\beta}$ so that the separate transitions at $T_{c\alpha}$ and $T_{c\beta}$ exist. The spin orbit coupling combined with the deviation from the spherical Fermi surface leads to a small coupling term so that the second transition may still be observable.

ACKNOWLEDGEMENTS

We are grateful to W.F. Brinkman and H.R.Ott for stimulating discussions.

REFERENCES

(1) H.R. Ott, H. Rudigier, T.M.Rice, K. Ueda, Z. Fisk and J.L. Smith (preprint)
(2) W.F. Brinkman and T.M. Rice, Phys. Rev. B2, 4302 (1970)
(3) P.W. Anderson and W.F. Brinkman in the Physics of Liquid and Solid Helium ed. by K.H. Bennemann and J.B. Ketterson (Wiley, N.Y.) II p. 177 (1978)
(4) D. Vollhardt, Rev. Mod. Phys. 5b, 99 (1984)
(5) A.J. Leggett, Rev. Mod. Phys. 47, 331 (1975)
(6) T. Ohmi, M. Nakahara and T. Tsuneto, Prog. Theor. Phys. 64, 1516 (1980)

LT-17 (Contributed Papers)
U. Eckern, A. Schmid, W. Weber, H. Wühl (eds)
© Elsevier Science Publishers B.V., 1984

REVERSIBLE CHANGE OF THE DENSITY OF TUNNELING STATES AFTER STRUCTURAL RELAXATION OF A GLASS

H. RÜSING and H.v.LÖHNEYSEN

2. Physikalisches Institut der Rheinisch-Westfälischen Technischen Hochschule, D-5100 Aachen and Sonderforschungsbereich 125 Aachen-Jülich-Köln, West Germany

The low temperature thermal conductivity of vitreous silica (Suprasil W) changes reversibly with heat treatment near the glass transition temperature. This arises from a reversible change of the density of tunneling states which can be interpreted in terms of the free volume theory of the glass transition.

It has been known for some time that structural changes in glasses can affect their low temperature properties which are dominated to a large extent by low energy excitations. These excitations can be described as two level systems (TLS) and are thought to arise from tunneling of atoms or groups of atoms between the two wells of a double well potential (1). Structural changes can be brought about e.g. by thermal treatment close to the glass transition temperature T_g (2,3) or by neutron irradiation (4,5). An attempt has been made to link the TLS to the glass transition in terms of the free volume model (6). In this model $n(E) \sim T_g^{-1}$ where n(E) is the TLS density of states which is constant at low energies ($E/k_B < 1$ K), $n(E) = n_0$. Such a behavior was indeed found for different glasses where T_g was varied by composition (7). In this paper we report on measurements on vitreous silica where the fictive temperature T_f was varied by thermal stabilization at a temperature T_a. T_f is the temperature where the glass is allowed to reach an equilibrium state before rapid quenching. For prolonged annealing, $T_a = T_f \approx T_g$. We chose thermal conductivity measurements to monitor changes in the TLS density. Our data are in qualitative agreement with the free volume model. An important result is the reversibility of the TLS density which shows that TLS can indeed be treated in the frame of a thermodynamic theory of the glass transition.

The vitreous silica sample (Suprasil W, < 1.5 ppm OH) was investigated in the as-received state (8), after annealing at T_a = 1330°C for 6 h in vacuum (< 10^{-4}mbar) and after an additional annealing at T_a = 900°C for 50 h. Each anneal was followed immediately by a quenching in water. The annealing times were sufficiently long, hence $T_a = T_f$. The thermal conductivity κ was measured with the standard steady heat flow technique with one heater and two thermometers. Care was taken to keep all dimensions, e.g. the distance between heater and thermometers, unchanged in order to minimize uncertainties in the geometry factor. In our sample with a rather large ratio of cross-section to length (1.13 cm^2 to 5.0 cm) this could otherwise pose a problem because the heat flow is not one-dimensional. The relative error in κ within the present series of measurements is thus confined to < 2 % while the agreement with other κ measurements on samples of different geometry is better than 10 % only.

Fig. 1 shows the thermal conductivity κ of the vitreous silica sample before and after the different heat treatments. κ shows the familiar behavior with $\kappa \sim T^m$, m < 2 below 1 K (in our case m = 1.8), and a temperature independent plateau between 5 and 10 K. After the 1330°C

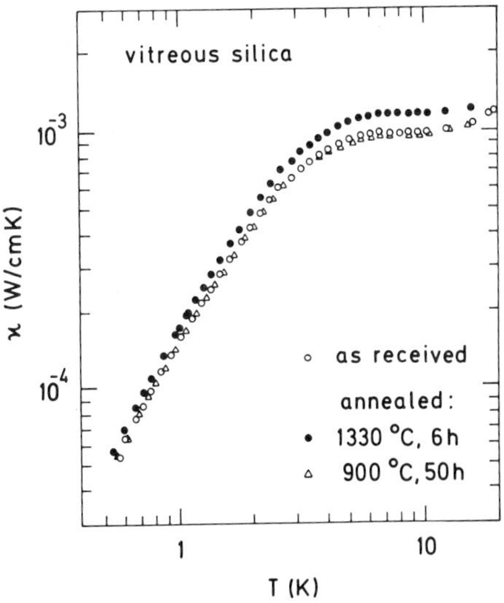

FIGURE 1
Thermal conductivity κ of vitreous silica (Suprasil W) as a function of temperature T in the as-received state and after different anneals.

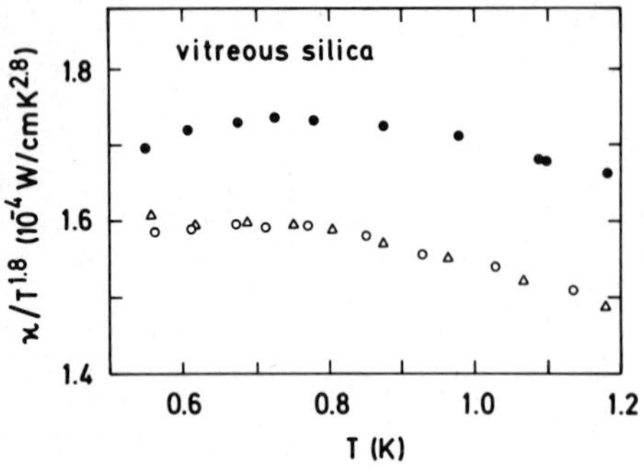

FIGURE 2
Thermal conductivity κ divided by $T^{1.8}$ as a function of temperature T. Symbols have the same meaning as in Fig. 1.

anneal, κ is enhanced by 10 % below 1 K and by 20 % in the plateau region. After the subsequent anneal at 900°C, κ recovers its original values except in the plateau region where it is even 3 % lower than initially. This shows that the thermally induced changes in κ are completely reversible. This is in contrast to metallic glasses where generally a large irreversible increase of κ is observed (3) followed by a smaller reversible change (9). In Fig. 2 the low temperature data are displayed on an enlarged plot. In order to expand the vertical scale, κ has been divided by $T^{1.8}$.

For a discussion we will only treat the region T < 1 K where the thermal conductivity is limited by phonon scattering from TLS, i.e. $\kappa \sim T^2/(\bar{n}_0 M_{av}^2)$ where M_{av} is a suitably averaged coupling constant of TLS to transverse and longitudinal phonons and $\bar{n}_0$ is the density of states of strongly coupled TLS (1). Specific heat measurements performed on the same sample (10) suggest that the total TLS density of states n_0 changes in a similar fashion as $\bar{n}_0 M_{av}^2$ after the 1330°C anneal. Hence we assume that M_{av} remains the same and the changes in $\bar{n}_0$, i.e. the relaxation time spectrum of the TLS remains unaltered. The reversibility observed in the TLS density of states strongly suggests that the TLS should be treated from a thermodynamic approach of the glass transition.

Unfortunately, the free volume model which links the glass transition and TLS, is not very specific as far as the prediction of annealing behavior is concerned. In this model, the TLS arise from tunneling of atoms into voids, i.e. an agglomeration of excess (or free) volume larger than the molecular volume. The equilibrium amount of the total free volume in an undercooled liquid is an increasing function of temperature, hence the total density N_T of TLS should increase with the annealing temperature T_a. A reversible decrease of κ, i.e. an increase of $\bar{n}_0 M_{av}^2$ with T_a in metallic glasses has been interpreted along these lines, with T_g remaining unchanged (9). On the other hand, the width of the TLS distribution n(E) should scale with the glass transition temperature, leading to $n(E) \sim T_g^{-1}$, if $N_T = \int n(E)dE$ is assumed constant (6). The present data on vitreous silica show a decrease of $\bar{n}_0 M_{av}^2$ with increasing T_a and hence are compatible with the second argument, as are the studies (5) on neutron-irradiated vitreous silica, from which $\bar{n}_0 M_{av}^2 \sim T_f$ was inferred, too. More experiments on different glasses as well as theoretical studies are needed to clarify this important point.

ACKNOWLEDGMENT
We thank Prof. W. Sander for valuable discussions.

REFERENCES
(1) Amorphous Solids: Low Temperature Properties, ed. W.A. Phillips (Springer, Berlin, 1981)
(2) R.L. Fagaly and J.C. Lasjaunias, J. Non-Cryst. Solids 43 (1981) 307 and refs. therein
(3) S. Grondey, H.v.Löhneysen, H.J. Schink, and K. Samwer, Z. Physik B 51 (1983) 287
(4) T.L. Smith, P.J. Anthony, and A.C. Anderson, Phys. Rev. B 17 (1978) 4997
(5) A.K. Raychaudhuri and R.O. Pohl, Solid State Commun. 44 (1982) 711
(6) M.H. Cohen and G.S. Grest, Phys. Rev. Lett. 45 (1980) 1271; Solid State Commun. 39 (1981) 143 and refs. therein
(7) A.K. Raychaudhuri and R.O. Pohl, Phys. Rev. B 25 (1982) 1310
(8) According to the manifacturer (Heraeus, Quarzschmelze, D-6450 Hanau) the samples were heat treated at 1080°C for 15 h. After slow cooling to 1000°C (with 0.4 K/min) the furnace was then switched off.
(9) E.J. Cotts, A.C. Anderson, and S.J. Poon, Phys. Rev. B 28 (1983) 6127
(10) H. Rüsing and H.v.Löhneysen, to be published

LT-17 (Contributed Papers)
U. Eckern, A. Schmid, W. Weber, H. Wühl (eds)

RESISTIVITY MINIMA IN $Fe_xNi_{80-x}B_{18}Si_2$ ALLOYS

R.KRSNIK, E.BABIĆ and H.H.LIEBERMANN*

Department of Physics, Faculty of Science, POB 162, Zagreb, Yugoslavia
*Allied Corp.,Metglas Products, Parsippany, New Jersy 07054, USA

Accurate measurements of the electrical resistivity (ρ) of amorphous $Fe_xNi_{80-x}B_{18}Si_2$ alloys ($0\leq x\leq 80$) have been analysed in terms of the electronic interaction with two level systems (TLS) and of interference effects. While the lnT dependence of ρ favours the first mentioned model the depth of the resistivity minima scales with $\rho^{5/2}$ lending strong support to electronic interference effects.

From the discovery of metallic glasses a small upturn in the resistivity at low temperatures has been a subject of considerable interest. This litle upturn with no practical importance clearly indicated the inadequacy of our knowledge of the electronic transport in amorphous alloys. Several models have been proposed to explain that phenomenon. None was able to explain all observations. However there were some common features in all results which indicated that at least a part of the effects was characteristic of the disorder. Two calculations satisfying this condition but based on different mechanisms have recently been published (1,2).Since these calculations predict different temperature dependences for the resistivity upturn ($\sqrt{T}$ and lnT respectively) it may seem easy to single out which one is appropriate. In what follows we show that the systematic analysis of $Fe_xNi_{80-x}B_{18}Si_2$ alloys indicated that this may not be the case.

The experimental details have been published elsewhere (3). In Fig.1 the resistivity below the minimum (T_m>10K) for all our alloys is shown vs lnT. The resistivity follows closely a-lnT law from the lowest temperature (1,7K) up to at least 5K. This form of plotting results is suitable for the comparison both with the magnetic (Kondo) and structural model (2) since they both predict a-lnT variation. The Kondo effect is observed in the metallic glasses where it could be expected also in crystalline systems i.e. either in the case of a small content of magnetic atoms in a nonmagnetic host (4) or in the complicated magnetic structures such as those near the magnetic percolation treshold (5). As our alloys are either paramagnetic (x=0) or good ferromagnets ($x\geq 20$) we believe that the Kondo effect does not contribute much to the observed lnT variation.

A detailed discussion of the resistivity due to interaction between the conduction electrons and the two level systems (TLS) has been given by Vladar and Zawadowski (2). They pointed out

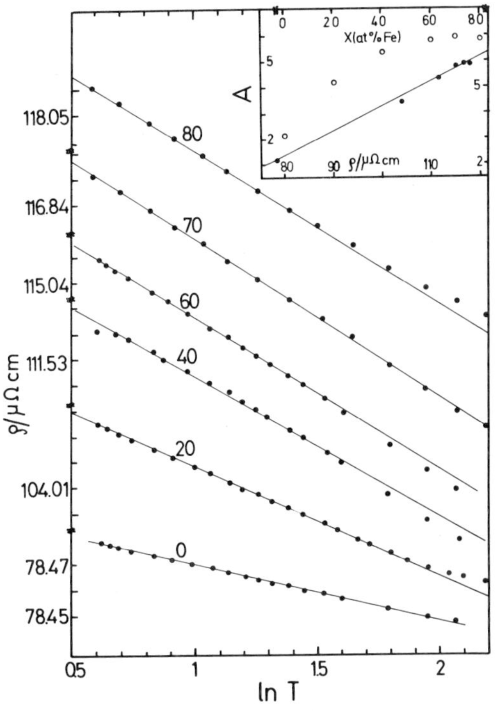

FIGURE 1
Resistivity of $Fe_xNi_{80-x}B_{18}Si_2$ alloys vs lnT. Numbers denote x. The inset: $A=\Delta\rho/\Delta(\ln T)$ in $10^{-2}\mu\Omega cm$ vs x and ρ_0.

that the ultrasonic experiments yield the coupling of conduction electrons to TLS which combined with the observed TLS densities ($\sim 10^{24}m^{-3}$) could just produce the observed resistivity upturns ($\Delta\rho\sim 10^{-9}\Omega m$). Furthemore they propose that the strong coupling may occur for tunneling atoms which have d-resonance near the Fermi level. Within the framework of their calculation (2) the size of the resistivity upturn and a slope of -lnT term are expected to depend both on the crossover temperature T_K (related to the coupling strengths) and to the density of TLS. Unfortunately none of these values are available for our alloys. However the absence of the saturation effects in resistivity at the lowest temperatures may indicate that T_K values do not vary much in our alloys. If that is a case a major contribution to the coefficients (A) of -lnT terms in resistivity may be due to variation in the density of TLS with x. Furthermore it seems reasonable to relate this density to resistivity since both are related to disorder.

In the inset to Fig.1 we plot $A=-\Delta\rho/\Delta(\ln T)$ both vs x and ρ_o (at T=1,7K). A linear variation of A with ρ_o seems to justify our conjectures. (For a more definite conclusion the resistivity data below 1,7K are neccessary). We note that A values extrapolate to zero for $\rho_o \sim 6\cdot10^{-7}\Omega$m what is reasonable since the glassy PdSi alloys with resistivities just above that value show extreemely weak resistance minima (6).

Altshuler and Aronov (1) have shown that the electronic interference effects in disordered alloys yield a-$\sqrt{T}$ contribution which may produce an upturn in the resistivity at low temperatures. Furthermore with reasonably selected parameters their calculation yields (1) the observed size of the resistivity upturns. Indeed a-$\sqrt{T}$ variation of resistivity has recently been observed in several Fe-Ni based amorphous alloys at the lowest temperatures (7,8). Because of the fact that in our temperature range -lnT and -$\sqrt{T}$ vary similarly we show in Fig. 2 ρ of our alloys vs $\sqrt{T}$. For the data at the lowest temperatures (T<4K) a fit to $\sqrt{T}$ is quite good. The positive deviation from the straight line at higher temperatures is probably due to positive contributions to resistivity. (Note that in Fig.1 the lnT plots of some alloys showed a negative deviation which may indicate a curvature similar to that (8) of $Ni_{80}P_{14}B_6$ alloy).

For a detailed comparison of our results with the predictions of the above model the electronic densities of states and the results of the magnetoresistance measurements are required. In the absence of these data we look however for a systematic trends which may indicate the applicabillity of that model. In the symplified version (1) the change in ρ in a fixed T interval is expected to be $\Delta\rho \propto \lambda\rho_o^{5/2}$ with λ the effective electronic interaction constant (f(x) in ref.8). In the inset to Fig.2 we plot $B=\Delta\rho/\Delta(\sqrt{T})$ and $C=B/\rho_o^{5/2}$ vs x. It is seen that C is nearly independent of x for x≳20 and is somewhat larger for x=0. This is indeed expected since alloys with x≳20 are good ferromagnets and x=0 is a paramagnet.

To summarize the above the low temperature resistivity of $Fe_xNi_{80-x}B_{18}Si_2$ alloys shows the

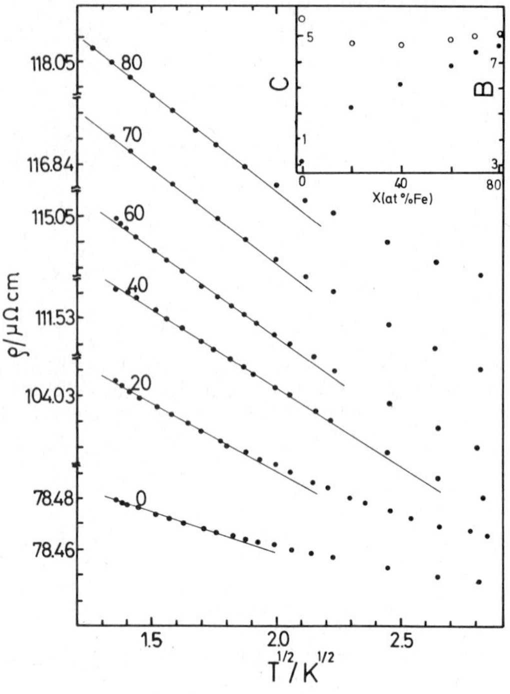

FIGURE 2
Resistivity of $Fe_xNi_{80-x}B_{18}Si_2$ vs $\sqrt{T}$. The inset: $B=\Delta\rho/\Delta(\sqrt{T})$ in 10^{-2} $\mu\Omega$cm/$\sqrt{K}$ and $C=B/\rho_o^{5/2}$ in $10^{-7}(\mu\Omega\text{cm})^{-3/2}/\sqrt{K}$ vs x.

upturn the size of which increases with x(and ρ_o). This upturn can be fitted well both to lnT and $\sqrt{T}$ law. The coefficient (A) of lnT law varies linearly with ρ_o possibly due to an increase in number of TLS. A extrapolates to zero for reasonable ρ_o. The coefficient of the $\sqrt{T}$ law scales well with $\rho_o^{5/2}$ indicating that the interference effects are dominant. Therefore a probable conclusion is that the interference effects are dominant but TLS also contribute to resistivity upturn (lnT law).

REFERENCES

(1) B.L.Altshuler and A.G.Aronov, Zh.Eksp.Teor. Fiz.77 (1979) 2028
(2) K.Vladar, A.Zawadowski, Phys.Rev.B28(1983)1956
(3) E.Babić, R.Krsnik and H.H.Liebermann, Recent Developments in Condensed Matter Physics, ed. J.T.Devreese et al (Plenum, New York, 1982) p.263
(4) R.Hasegawa, Phys.Letters 36A (1971) 175
(5) A.Hamzić, E.Babić and Ž.Marohnić, J.Physique 41 (1980) C8-694
(6) J.Kästner, H.J.Schink and E.F.Wassermann, Solid State Commun. 33 (1981) 527
(7) O.Rapp, S.M.Baghat and H.Gundmundsson, Solid State Commun. 42 (1982) 741
(8) E.Babić, A.Hamzić and M.Miljak, this volume

LT-17 (Contributed Papers)
U. Eckern, A. Schmid, W. Weber, H. Wühl (eds)

ANOMALOUS ELECTRICAL RESISTIVITY AND MAGNETORESISTIVITY OF TITANIUM-COPPER METALLIC GLASSES: EVIDENCE FOR WEAK ELECTRON LOCALIZATION*

J. WILLER† and R. R. HAKE

Department of Physics and Materials Research Institute, Indiana University, Bloomington, Indiana 47405, USA and Institute for Pure and Applied Physical Sciences, University of California, San Diego, La Jolla, California 92093, USA

Resistivity ρ and magnetoresistivity (mr) measurements on bulk amorphous alloys $Ti_{1-x}Cu_x$ ($x = 0.4, 0.5, 0.6$) in the range $1.1 \le T \le 35$ K and $0 \le H \le 13.6$ T show that above 20 K ρ is linear in T and nearly independent of H. Below about 4 K ρ varies approximately as $T^{1/2}$, suggesting a Coulomb interaction mechanism. On the other hand, in the same T range the magnetoresistive slope $\alpha_H \equiv \rho^{-1}(\partial\rho/\partial H)_T$ is <u>negative</u> for $x = 0.4$, suggesting weak electron localization. The observed shift of α_H to positive for $x = 0.5, 0.6$ may reflect dominance of Coulomb effects or the influence of spin-orbit interaction on weak localization.

1. INTRODUCTION

In a previous survey (1) of anomalous low-temperature (1-4 K) magnetoresistivity (mr) in various bulk, amorphous, metallic-glass Ti- and Zr-base alloys, it was found that $Ti_{0.5}Cu_{0.5}$ was unusual in displaying positive mr, i.e., a positive magnetoresistive slope $\alpha_H \equiv \rho^{-1}(\partial\rho/\partial H)_T$, in applied magnetic fields up to 13.5 T. In contrast, the mr of $Ti_{0.6}Cu_{0.4}$ was more typical of high-electrical-resistivity ($\rho \gtrsim 100\ \mu\Omega$cm) crystalline (2, 3) and amorphous (1, 4) alloys: positive mr at low H, changing to negative mr at high H. The low-H positive mr has been ascribed (1-4) to H-quenching of fluctuation or remnant superconductivity, while the high-H negative mr has been considered in terms of three dimensional (3D) weak localization (1, 3, 4) and spin fluctuations (1, 2). More recently Howson and Greig (5) observed positive mr varying as $H^{1/2}$ in $Ti_{0.5}Cu_{0.5}$ and

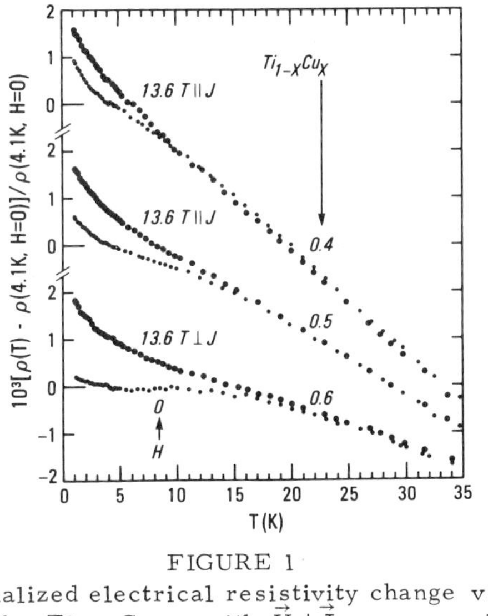

FIGURE 1
Normalized electrical resistivity change vs T. Data for $Ti_{0.6}Cu_{0.4}$ with $\vec{H} \perp \vec{J}$ are very close to those shown above for $\vec{H} \parallel \vec{J}$.

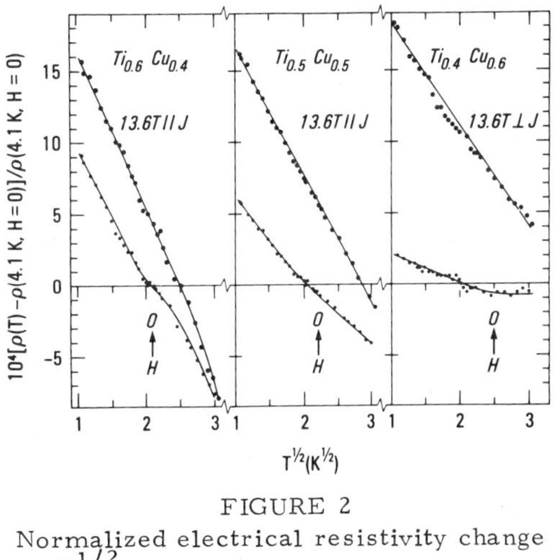

FIGURE 2
Normalized electrical resistivity change vs $T^{1/2}$.

*Supported by NSF DMR 80-24365, Indiana UROC 9-20-82-3, and NATO Postdoctoral Grant (J.W.) 315/402/3.

†Now at Siemens, Munich, Federal Republic of Germany.

$Ti_{0.35}Cu_{0.65}$ at $T < 1$ K and $H \leq 9$ T. They attributed the positive mr to electronic Coulomb interaction (6).

2. RESISTIVITY VERSUS TEMPERATURE

Figure 1 shows the normalized electrical resistivity change vs T for the same $Ti_{1-x}Cu_x$ specimens previously measured (1). Noteworthy features of the data are: (1) above 20 K $\rho(T)$ is nearly linear in T [implying inelastic scattering times $\tau_i \propto T^{-2}$ for the 3D weak-localization mechanism (7)] and not much affected by a 13.6 T applied field; (2) below 20 K an overall positive magnetoresistivity is displayed which becomes more prominent as T decreases and as Cu concentration x increases; (3) a peculiar leveling off of the $\rho(T)$ curves as T decreases from 20 K to about 6 K is especially pronounced at $H = 0$ and higher x; (4) the $\rho(T)$ curves (both for $H = 0$ and $H = 13.6$ T) steepen as T decreases below about 6 K. The above zero-H features are in agreement with the zero-H investigation of $Ti_{1-x}Cu_x$ (x = 0.5, 0.6) by Mizutani et al. (8)

Figure 2 shows the normalized electrical resistivity change vs $T^{1/2}$ over the range $T = 1-9$ K. Below about 4 K, $\Delta\rho/\rho \propto T^{1/2}$ for both $H = 0$ and $H = 13.6$ T. A $T^{1/2}$ variation in zero-H is expected from Coulomb interaction (6).

3. RESISTIVITY VERSUS APPLIED MAGNETIC FIELD

Figure 3 shows magnetoresistivity curves in the range 1.2-4.1 K and 0.75-13.6 T. Primary features of the data are: (1) the high-H magnetoresistive slope α_H becomes less positive as x decreases and is negative at $x = 0.4$, as previously shown (1); (2) the curves tend to become linear at high H rather than varying as $H^{1/2}$ as predicted by both Coulomb interaction (6) and spin-ignoring weak-localization (7) theories; (3) as indicated above, the $Ti_{0.6}Cu_{0.4}$ curves are similar to those observed (1-4) in other high-ρ alloys: positive mr at low H and negative mr at high H. Aside from Coulomb interaction, positive mr may also be due to the influence on weak localization of spin-orbit interaction (9,10) (increasing with average atomic number $\langle Z \rangle$ and thus x in $Ti_{1-x}Cu_x$). The $Ti_{0.6}Cu_{0.4}$ magnetoresistive curves are similar to those generated from equations of 3D weak-localization theory (9) which includes spin-orbit and Zeeman spin-splitting effects, and similar to those observed for "Au-dusted" Mg films (10). A more complete report on present and parallel investigations of higher superconducting T_c glasses (11) ZrCo, ZrNi, and ZrPd is in progress (12).

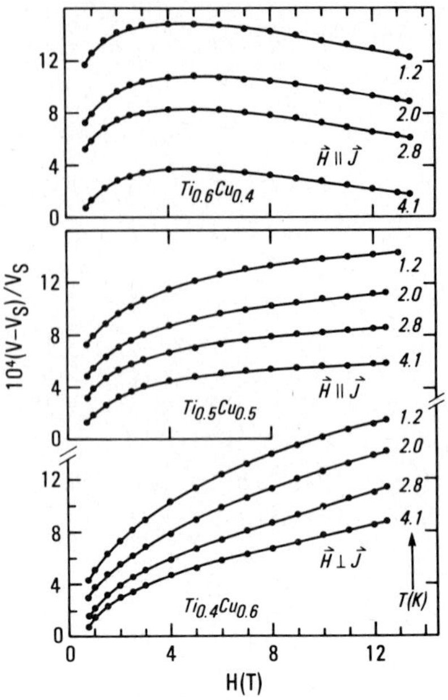

FIGURE 3

Magnetoresistivity curves. V is the resistive voltage and V_s is a constant nulling voltage so that $(V-V_s)/V_s \approx \Delta\rho(H)/\rho$. Data for $Ti_{0.6}Cu_{0.4}$ with $\vec{H} \perp \vec{J}$ are very close to those shown above for $\vec{H} \parallel \vec{J}$.

REFERENCES

(1) R. R. Hake, M. G. Karkut, and S. Aryainejad, Physica 107B (1981) 503.

(2) J. W. Lue, A. G. Montgomery, R. R. Hake, Phys. Rev. B11 (1975) 3393.

(3) S. Aryainejad and R. R. Hake, unpublished.

(4) R. R. Hake, S. Aryainejad, M. G. Karkut in Superconductivity in d- and f-Band Metals, eds. H. Suhl and M. B. Maple (Academic Press, New York, 1980).

(5) M. A. Howson and D. Greig, J. Phys. F13 (1983) L155.

(6) B. L. Al'tschuler and A. G. Aronov, Sov. Phys. JETP 50 (1979) 968. P. A. Lee and T. V. Ramakrishnan, Phys. Rev. B26 (1982) 4009.

(7) A. Kawabata in Anderson Localization, eds. Y. Nagaoka and H. Fukuyama (Springer-Verlag, Berlin, 1982), p. 122.

(8) U. Mizutani, N. Akutsu, T. Mizoguchi, J. Phys. F13 (1983) 2127.

(9) H. Fukuyama and K. Hoshino, J. Phys. Soc. Jpn. 50 (1981) 2131.

(10) G. Bergman, Phys. Rev. B28 (1983) 2914.

(11) J. Willer and R. R. Hake, Bull. Am. Phys. Soc. 29 (1984) 302.

(12) J. Willer, R. R. Hake, D. Sprinkle, unpublished.

LT-17 (Contributed Papers)
U. Eckern, A. Schmid, W. Weber, H. Wühl (eds)

LOGARITHMIC TEMPERATURE DEPENDENCE OF THE RESISTIVITY IN AMORPHOUS METALS

Yoshimasa ISAWA

Research Institute of Electrical Communication, Tohoku University, Katahira, Sendai 980, Japan

We show that a lnT dependence of the resistivity at low temperature for highly disordered amorphous metals naturally follows from the interplay between the electron-tunneling state interaction and the incipient Anderson localization. Our model also predicts that the lnT dependence is insensitive to the magnetic field.

1. INTRODUCTION

Cochrane et al.[1] have discovered a contribution to the low-temperature resistivity varying logarithmically with temperature, which is insensitive to the magnetic field, in a variety of amorphous metals. To account for the experimental findings, two theoretical models have been proposed: 1) Kondo type of s-d exchange scattering [2], 2) the scattering of the conduction electron from the tunneling states[1], often referred to as 2-level system(TLS). The TLS is a low energy excitation initially introduced by Anderson et al.[3] and Phillips[4] to explain the low temperature anomalies of insulating glasses. The situation, however, remains somewhat unclarified on the explanation of the lnT dependence.

We propose here an explanation for the lnT dependence, based on the interplay between the electron-tunneling state interaction and the incipient Anderson localization, which suggests that the lnT dependence of the resistivity should be a universal feature of highly disordered amorphous metals at low temperatures. The results obtained are consistent with the observed lnT dependences attributable to the consequence of the non-crystalline structure.

2. THE MODEL

The actual amorphous metals are highly disordered conductors. The mean free path of the conduction electron is very short (a few A), and $E_F\tau$ (E_F: the Fermi energy, τ: the elastic scattering time) is of the order of 1. For example, if the Fermi velocity is taken to be $v_F=10^8$(cm/sec) and the mean free path 3 A, we obtain $E_F\tau=1.35$ by use of free electron mass. Thus the amorphous metals will be in Weakly Localized Regime(WLR) in the terminology of the Anderson localization, which indicates that nearly free electron calculation is questionable.

Our total model hamiltonian is, with the addition of the one particle scattering potential together with the electron-TLS interaction, given by

$$H = H_0 + H_1 + H_2 , \tag{1}$$

$$H_0 = \sum_{k,\sigma} \left(\frac{k^2}{2m^*} - \mu\right) c^+_{k\sigma} c_{k\sigma} + \sum_i E_i S_{zi} , \tag{2}$$

$$H_1 = N^{-1} \sum_{k,k'} \sum_{i,\sigma} e^{i(k-k')R_i} (v_\perp S_{xi} + v_{/\!/} S_{zi}) c^+_{k\sigma} c_{k'\sigma} , \tag{3}$$

$$H_2 = u_0 \sum_{i,j} \delta(r_i - R_j^{(S)}) , \tag{4}$$

where H_0 represents the noninteracting conduction electron, and pseudo-spin 1/2 tunneling system with the energy separation E_i. The interaction hamiltonian H_1 between tunneling state and the conduction electron has both diagonal($v_{/\!/}$) and off-diagonal($v_\perp$) components. All k-dependences of $v_{/\!/}$ and $v_\perp$ are ignored. The N is the total number of atoms, R_i the position of the i-th tunneling state, and r_i and $R_j^{(S)}$ denote the position of the i-th electron and j-th scatterer, respectively. We assumed that the effects of disorder can be simulated by the short range potential(H_2). The (H_0+H_1) is exactly the same electron-TLS coupling hamiltonian[5],[6] as utilized for the analysis of singular many body effects which come from the divergent response of the conduction electron to localized TLS.

The effective interaction between conduction electrons mediated by TLS is given by, for the resonant term,

$$v_e(\omega) \equiv \frac{1}{N(0)} g_e(\omega) = -\left(\frac{v_\perp}{N}\right)^2 \left(\frac{N_i}{W}\right) \int_{-W}^{W} dX \frac{\mathrm{th}\left(\frac{\beta X}{2}\right)}{X - i\omega} , \tag{5}$$

where N(0) is the density of states of the conduction electron at the Fermi energy per spin, $\beta^{-1}=T$ the temperature, N_i the total number of TLS, and W the upper limit of the excitation energy of TLS. The process leading to eq.(5) is shown in Fig. 1. Note that $v_e(\omega)$ is attractive and short range. In eq.(5), we assumed that (i) the energy distribution is constant for $0 \lesssim E \lesssim W$ and zero otherwise, (ii) the relaxation time of TLS is long enough. The static value of $g_e(\omega)$ is given by

$$g_e(0) = -2N(0)\left(\frac{v_\perp}{N}\right)^2 \left(\frac{N_i}{W}\right) \ln(1.13W/T) , \tag{6}$$

where W>>T is assumed. The lnT dependence is essentially due to the broad distribution of the excitation energy in TLS. This is similar to the one observed in ultrasound velocity($\Delta v/v \propto \ln T$) of both insulating glasses[7] and amorphous metals [8], which has been explained by the resonant absorption process of phonons by TLS.

By taking into account the vertex correction [9],[10], the correction term of the lifetime proportional to $g_e(0)$($\propto \ln T$) is obtained from g_3- and g_4-processes(Fig.2), while the ω-dependence of $g_e(\omega)$ weakens the singularity in g_1- and g_2-processes(Fig.2).

For 3-dimensional system, the temperature dependent terms from g_1- and g_2-processes are proportional to $\sqrt{T}$ or $\sqrt{T}\ln T$, the magnitude of which is smaller than the lnT term in g_3- and g_4-processes by a factor of $(T\tau)^{\frac{1}{2}}$ (<<1). We can ignore g_1- and g_2-processes in the temperature range (T<20 K) of lnT dependence being observed[1],[11]. The correction to the conductivity,Δσ, is in the case of $2\pi T\tau_\varepsilon >> 1$,

$$\Delta\sigma/\sigma_0 = \frac{3\sqrt{3}}{4(E_F\tau)^2}[0.82 - 1.83(T\tau)^{\frac{1}{2}}]g_e(0) \quad , \quad (7)$$

where $\sigma_0 = ne^2\tau/m^*$ is the classical formula. The second term in the bracket is the $\sqrt{T}$-term[9],[12] well-known for 3-dimensional system in WLR, which can be ignored since $2\pi T\tau << 1$. Even in the case of $2\pi T\tau_\varepsilon << 1$, eq.(7) is valid except for small change in the coefficient of lnT as long as $\tau << \tau_\varepsilon$. Numerically we obtain

$$\Delta\sigma/\sigma_0 \cong 10^{-4}\ln T \quad , \quad (8)$$

and $g_e(0) \cong -1.48\times10^{-4}\ln(1.13W/T)$ provided that the following value of parameters is used: $k_F = 10^8(\mathrm{cm}^{-1})$, $E_F\tau = 1$, $N_i/WV = 10^{20}(\mathrm{eV}^{-1}\mathrm{cm}^{-3})$, $N/V = 10^{22}$ (cm^{-3}), $m^* = m_0$(free electron mass) and $v_\perp = 0.1$(eV). The higher order effect of $g_e(0)$ can be ignored since $g_e(0) << 1$ for $W \cong (10\sim100)$ K and $T \cong 1$ K. Considering the ambiguities in the value of parameters, the above estimate is consistent with the experimental results[1],[11], where the lnT dependences are attributed to the consequence of the non-crystalline structure.

For 2-dimensional system where $D(1/d)^2 >> 2\pi T$ and τ_ε^{-1} (D:diffusion constant, d:thickness of the film) are satisfied, we obtain

$$\Delta\sigma/\sigma_0 \propto -\ln(1/2\pi T\tau)\times\ln(1.13W/T) \quad . \quad (9)$$

The lnT term in eqs.(7),(9) is brought about by the whole part of correlation between wave-functions with different energy near the Fermi energy appearing as the precursor effect of the Anderson localization. There is only weak magnetic field dependences as long as $4D/\ell^2 << 1/\tau$ and $g\mu_B H << 1/\tau$ with the Larmor radius ℓ and the Bohr magneton μ_B. A numerical example of these conditions is $H << 1.2\times10^8$(Oe) and $H << 3.3\times10^6$(Oe) for $\tau = 1.73\times10^{-16}$(sec), $E_F\tau = 1$, g=2 and $m^* = m_0$. This implies that eq.(7) can explain the lnT term insensitive to the magnetic field since the field strength satisfies the above conditions in the actual experiments. It is in remarkable contrast with the singular behaviours found in transport phenomena of semiconductors in WLR[13],[14], which are associated with a part of reduction in correlation due to the temperature raise or the increase in the strength of the magnetic field.

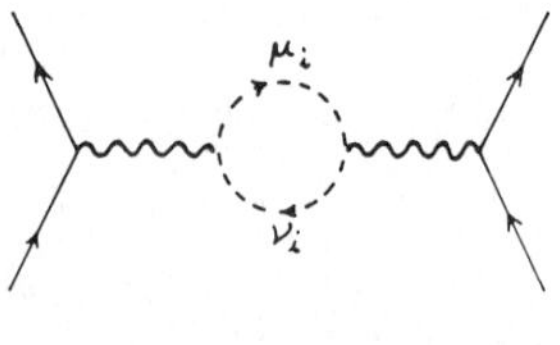

Fig.1 Effective interaction between conduction electrons mediated by the localized two level system. The broken lines and the solid ones denote the propagators for the two level system and the conduction electrons, respectively.

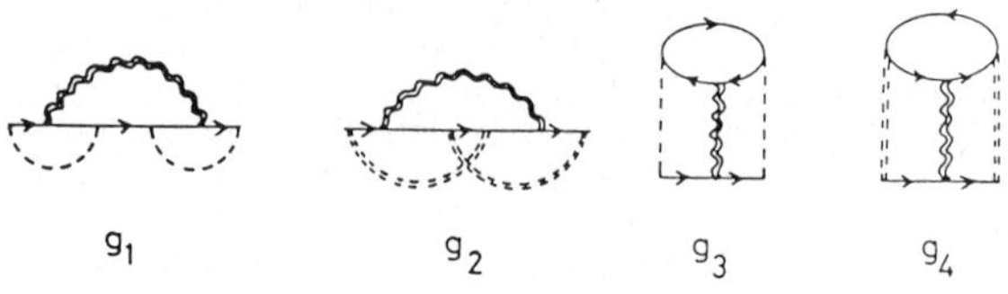

Fig.2 The four different processes: g_1,g_2,g_3 and g_4, for the self energy of the conduction electron in the presence of vertex corrections. The double wavy lines denote the effective interaction defined in Fig.1. The broken and double broken lines are the particle-hole and particle-particle diffusion propagators, respectively.

[1] R.W.Cochrane et al.:Phys.Rev.Lett,35(1975) 676.
[2] C.C.Tsuei:Amorphous Magnetism II,eds. Levy and Hasegawa(Plenum Press, New York,1977)181.
[3] P.W.Anderson et al.:Philos.Mag.25(1972)1.
[4] W.A.Phillips:J.Low Temp.Phys.7(1972)351.
[5] J.Kondo:Physica(Utrecht).84B(1976)40,207.
[6] K.Vladar and A.Zawadowski:Phys.Rev.B28(1982) 1564,1582,1596.
[7] L.Piché et al.:Phys.Rev.Lett.32(1974)1426.
[8] G.Bellessa et al.:J.Physique.Lett.38(1977) L-65.
[9] B.L.Altshuler and A.G.Aronov:Sov.Phys.JETP .50(1979)968.
[10] H.Fukuyama:J.Phys.Soc.Jpn.48(1980)2169.
[11] J.Kästner et al.:Solid State Commun.33(1980) 527.
[12] Y.Isawa et al.:J.Phys.Soc.Jpn.51(1982)3262.
[13] T.F.Rosenbaum et al.:Phys.Rev.Lett.45(1980) 1723.
[14] S.Morita et al.:Phys.Rev.B25(1982)5570.

LT-17 (Contributed Papers)
U. Eckern, A. Schmid, W. Weber, H. Wühl (eds)

RECENT RESULTS OF BIOMAGNETIC MEASUREMENTS WITH A DC-SQUID-SYSTEM

G. ALBRECHT[+], W. HABERKORN[++], G. KIRSCH[+++], H. NOWAK[+], and H.G. ZACH[+]

[+]PTI Jena, Academy of Sciences of the GDR, 6900 Jena, Helmholtzweg 4, GDR
[++]ZWG Berlin, Academy of Sciences of the GDR, 1199 Berlin, Rudower Chaussee 6, GDR
[+++]Friedrich-Schiller-University Jena, Sektion Physik, 6900 Jena, Max-Wien-Platz 1, GDR

A gradiometer-SQUID-system for biomagnetic measurements without any magnetic shielding is described. The sensitivity of 7×10^{-14} T Hz$^{-1/2}$ is sufficient for high resolution MCG. The response of a gradiometer loop to a magnetic multipole is investigated theoretically.

1. INTRODUCTION

In the last years some new fields of investigation have been developed in medical sciences where the application of biomagnetic measurements provide results not achievable by other methods.

Mostly fundamental research was carried out in a quiet environment or in large scale shieldings. By this method it is possible to detect very small signal amplitudes in real time measurements. We aimed - as Barbanera et al. (1) - at the development of a biomagnetic measurement system which is able to work in a normally noised environment without magnetic shielding. The required sensitivity of such an equipment is below 10^{-13} T Hz$^{-1/2}$. For brain studies or for the measurement of the heart's conduction system further noise rejection by electronic means is necessary.

These demands we fulfilled by using a dc-thin film SQUID with integrated input inductance (2) and a second order gradiometer.

2. INSTRUMENTATION

In connection with a quantitative description of the biomagnetic measurements it is necessary to investigate the effects of finite size of coils and finite base lengths of gradiometers used. The examination of the gradiometer response is also important for an analysis of the performance of various gradiometers.

In the following the biomagnetic field is represented by a multipole expansion (3) up to the octupole term. It is assumed that the multipole is located at the origin of the coordinate system. The flux through a single circular gradiometer loop perpendicular to the z-axis is investigated. In the case that the centre of the loop is placed on the z-axis we obtain the expression

$$\Phi = \frac{\mu_0 r^2}{2(z^2+r^2)^{3/2}} \left(A_{10} + \frac{3}{2}\frac{z}{z^2+r^2} A_{20} + \frac{1}{2}\frac{4z^2-r^2}{(z^2+r^2)^2} A_{30}\right).$$

Here, r is the radius of the loop. A_{10}, A_{20}, and A_{30} are dipole, quadrupole, and octupole coefficients, respectively (3). From this result follows that the effect of finite size of coils increases with increasing order of poles. For $z = r/2$ the octupolar contribution vanishes.

The complete analytical expression for the dipolar contribution is given by

$$\Phi = \frac{\mu_0 r}{(z^2+(r+a)^2)^{3/2}} \left(\frac{A_{10} r}{k'^2} E - \frac{2P}{ak^2} Q\right),$$

$$P = a^2 A_{10} - xzA_{11} - yzB_{11}, \quad Q = \frac{1+k'^2}{2k'^2} E - F,$$

$$k' = (1-k^2)^{1/2}, \quad k^2 = \frac{4\,ar}{z^2+(r+a)^2}, \quad a = (x^2+y^2)^{1/2}.$$

F and E are the complete elliptic integrals of the 1st and 2nd kind. x,y,z are the coordinates of the centre of the loop.

The response of a certain gradiometer can be obtained using the above results. The actual gradiometer used by us is a symmetrical second order gradiometer with a diameter of 26 mm. The base length was fixed at 50 mm. Based on our calculations (4) superconducting adjustable discs were used for compensating the imbalance which results from inaccuracy of manufacture. By using a modified pair of Helmholtz coils for balancing we achieved a balance of at least 10^{-5} for all three axes.

This gradiometer is connected with a flat dc-thin film SQUID with Nb-NbO_x-Pb tunnel junctions (5). The sensitivity of this SQUID gradiometer equipment is better than 7×10^{-14} T Hz$^{-1/2}$ and therefore sufficient for the investigation of biomagnetic phenomena.

The biomagnetic laboratory in Jena is located in a normal environment (fluctuation amplitude 10^{-7} T) without any shielding. All parts of the adjustable cryostat-holder and of the subject support are made of wood or plastic.

3. RESULTS OF BIOMAGNETIC MEASUREMENTS

Real time magnetocardiograms as one is shown in figure 1 were recorded with the equipment described above. In this case the position of the gradiometer was over the right ventricle.

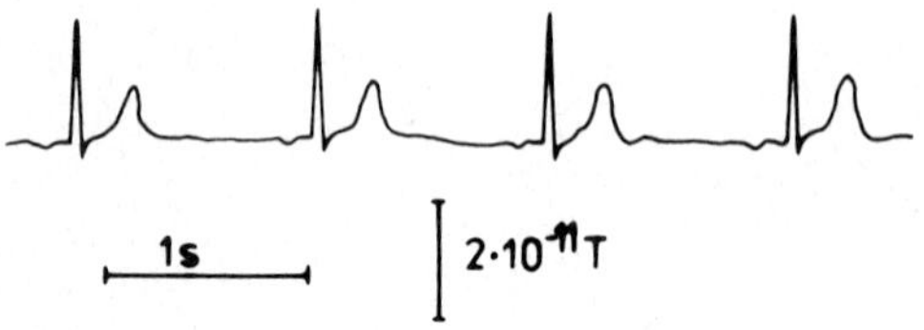

FIGURE 1

Measurements over the whole chest are displayed in a map. Starting from such a mapping different representations can be obtained. By so-called arrow maps vortex parts of the current density with field components perpendicular to the surface of the body become visible more directly. Such current distributions give no contribution to the ECG-recordings.

One advantage of biomagnetic measurements is the high local resolution. So it was possible to detect the very weak signals of the heart's conduction system by a special measurement technique. For such recordings it is necessary to

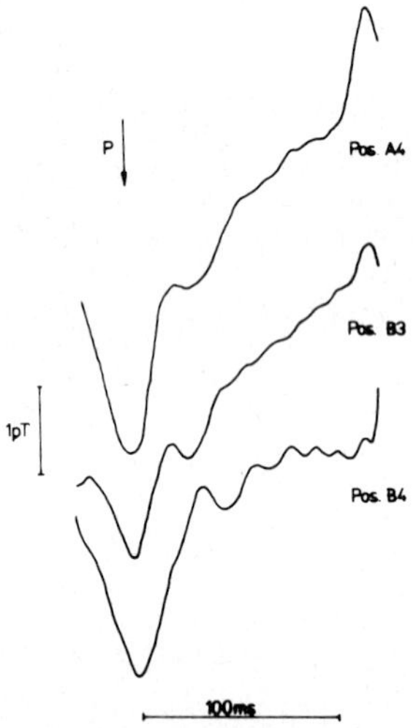

FIGURE 2

reduce noise by averaging technique. Because of the fact that QRS-triggering is more promising than P-wave triggering, we have averaged the recorded signals with an inverted time axis. Three curves shown in figure 2 are taken from different sites over the chest.

4. CONCLUSIONS

The interpretation of the above results needs a further co-operation with physicians. For a detailed quantitative analysis of the measurements more theoretical investigations are necessary. Finally, we remark that our equipment is also useful for the recording of evoked fields of the human brain.

ACKNOWLEDGEMENTS

We would like to thank our colleagues of the Department Detectorphysics of the University Jena for discussions and the manufacturing of SQUIDs.

REFERENCES

(1) S. Barbanera, P. Carelli, R. Fenici, R. Leoni, I. Modena, and G.L. Romani, IEEE Trans. Magn. MAG-17 (1981) 849.

(2) F. Dettmann, W. Richter, G. Albrecht, and W. Zahn, phys. stat. sol. (a) 51 (1979) K 185.

(3) F. Grynszpan, D.B. Geselowitz, Biophys. J. 13 (1973) 911.

(4) W. Haberkorn, G. Albrecht, H. Nowak, H.G. Zach, and G. Kirsch, Feingerätetechnik 31 (1982) 547.

(5) K.H. Berthel and F. Dettmann, Planar thin film dc SQUID/Fabrication and measuring system, this volume.

LT-17 (Contributed Papers)
U. Eckern, A. Schmid, W. Weber, H. Wühl (eds)

WELL COUPLED DC SQUID WITH EXTREMELY LOW 1/F NOISE

C. D. TESCHE, K. H. BROWN, A. C. CALLEGARI, M. M. CHEN, J. H. GREINER, H. C. JONES, M. B. KETCHEN, K. K. KIM, A. W. KLEINSASSER, H. A. NOTARYS, G. PROTO, R. H. WANG, AND T. YOGI

IBM T. J. WATSON RESEARCH CENTER, BOX 218, YORKTOWN HEIGHTS, NY 10598

A large number of extremely robust, thin film DC SQUIDs with excellent noise and coupling properties have been designed and fabricated. The lowest value of white noise observed corresponds to an energy factor of 350 $\hbar$. The low frequency (1/f) noise was reduced by careful operation of the device to less than 770 $\hbar$ at 0.1 Hz. This noise figure is an improvement of at least a factor of 60 over other devices with comperable coupling properties. The SQUID input coil inductance is 0.7 μH and the coupling efficiency α=0.9.

1. INTRODUCTION

Considerable effort has been expended in the development of planar high resolution DC SQUIDs. The potential applications of these devices are numerous, particularly as low noise magnetometers and of linear amplifiers operated at frequencies varying from a fraction of a Hertz to the MHz range. The devices reported on in this paper are of particular interest for three reasons. First, the noise and coupling properties of these SQUIDs are excellent. In particular, the low frequency noise is remarkably low. Second, the devices have been fabricated by an extremely well characterized and precisely controlled process. As a result, the devices are unusually robust and reliable. Finally, it was possible to produce a very large number of devices, opening the door to applications in which tens or hundreds of thin film devices will be used routinely. Examples are in the deployment of an array of superconducting inductive monopole detectors or in arrays of magnetometers for magnetoencephalography.

2. DEVICE STRUCTURE

The basic structure of the SQUID is similar to the planar coupling coil devices developed by M. B. Ketchen and J. Jaycox.(1) This structure provides tight coupling between the input coil and the SQUID. The SQUID loop is deformed into a large planar washer which is connected in series with a pair of shunted Josephson junctions. A 39 turn spiral coil is wound over the top surface of the washer and then dives down the central hole. An additional 39 turns are wound on the underneath side of the washer. The Josephson junctions are located in a square area below the washer. A single turn modulation coil is wound near the outside edge of the washer. In addition, a free floating rectangular groundplane is used directly over the slit in the washer in order to reduce the parasitic inductance introduced by the slit.

The devices were fabricated by a 12 level process developed for the fabrication of superconducting digital circuits.(2) The junctions are 2.5 micron wide niobium-lead alloy edge junctions. The SQUID loop is fabricated out of the same level as the junction electrode. The 39 turn input coil winding around underneath this loop is fabricated out of the niobium level usually used for a groundplane. The modulation coil, slit groundplane and the top 39 turns of the input coil are lead alloy. The washer is approximately 500 micron by 500 micron.

3. DEVICE CHARACTERISTICS

The isolated device characteristics were measured by coupling the SQUID through a transformer to a rf SQUID preamplifier.(3) The junction critical current is 25 μA, the shunt resistance 2.1 Ω, and the junction capacitance 0.2 pF. As a result, the junctions are strongly non-hysteretic, with a hysteresis parameter of β_c = 0.07. The loop inductance was determined from the modulation depth to be 96 pH. This corresponds to a screening parameter β = 2.4. The input coil inductance for a 78 turn coil is 0.7 μH, with a mutual inductance of 7.3 nH.

The SQUIDs were diced into blocks of four, then mounted on a ceramic header. The voltage leads were wire bonded with Al wire from the niobium pads on the chip to Cu header pins. The flux bias leads were wire bonded directly to the silicon wafer, adjacent to Pb pads on the chip. Silver paint was then used to complete the final connection. The dc SQUID was surrounded by a thick lead shield, then cooled to 4.2K in a low field environment ($B<1$ mG). The low frequency noise spectra were obtained with a HP 3582A spectrum analyzer. The contribution of the rf SQUID preamplifier circuit to the total observed noise was less than 20% in all cases.

The device characteristics were measured as a function of dc bias current and applied flux. Considerable structure was observed a low bias currents. In particular, regions of high noise were observed for bias currents below the critical current over a limited flux range. This behavior is to be expected for SQUIDs of this type, and is a result of the parasitic capacitance across the SQUID loop introduced by the floating slit groundplane.(4) The amount of structure seen is a function of the screening parameter. The excess noise regions were eliminated when the loop inductance was reduced to about 20 pH ($\beta = 0.5$) by shorting the input coil. The optimal operating bias for the SQUID with an open input coil was in the range of 60 to 90 μA, well above the region in which multivalued solutions exist.

The forward transfer function, dynamic resistance, and voltage noise spectral density were measured as a function of applied flux and bias current. A plot of the energy factor in units of h versus applied flux appears in the figure. The lowest value obtained was 315 $\hbar$ in

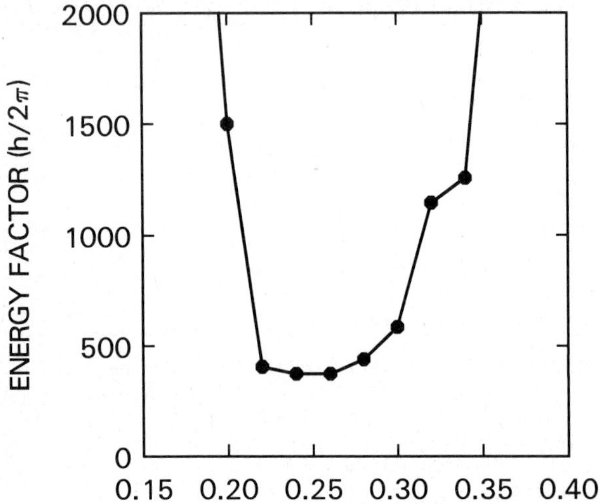

the white noise region. The 1/f noise was extremely low. The lowest values of low frequency noise was observed for a SQUID with energy factor 770 $\hbar$. No excess noise was observed down to 0.1 Hz. This corresponds to a 1/f noise contribution at 1 Hz of less than 77$\hbar$.

For comparison, the lowest value of 1/f noise reported previously is 500 $\hbar$ at 1 Hz.(5)

The source of the remarkable low frequency noise performance of these devices is not known at this time. However, since these devices were fabricated by the same processes developed for digital circuitry, the junctions and niobium electrode are of particularly high quality.(2,5) Thus it is likely that fluctuations associated with the individual junctions may have been strongly suppressed.

4. CONCLUSIONS

A highly robust and reliable DC SQUID has been designed and fabricated. The SQUID has excellent noise and coupling properties, and remarkably low 1/f noise, thus making it attractive for numerous practical applications.

The authors gratefully acknowledge the initiative, support and encouragement that P. Chaudhari, A. Malozemoff, B. Van Der Hoeven, C.C. Tsuei and C. C. Chi provided in this project.

REFERENCES

1. M. B. Ketchen and J. M. Jaycox, Appl. Phys. Lett. 35, 669 (1982).

2. K. H. Brown, T. Bucelot, M. M. Chen, J. H. Greiner, K. K. Kim, A. Kleinsasser, H. A. Notarys, G. Proto, B. J. Van Der Hoeven, R. H. Wang, D. J. Webb, and T. Yogi, unpublished results.

3. M. B. Ketchen and C. C. Tsuei, in SQUID '80, ed. H. D. Hahlbohm and H.Lubbig (Walter de Giruyter, Berlin 1980) pp227.

4. C. D. Tesche, J. Low Temp. Phys. 47,385 (1982).

5. J. Clarke, private communication.

6. T. Yogi and E. Callegari, to be published.

LT-17 (Contributed Papers)
U. Eckern, A. Schmid, W. Weber, H. Wühl (eds)

PULSED NMR USING A DC SQUID*

A.K.M. WENNBERG, L.J. FRIEDMAN, and H.M. BOZLER

Department of Physics, University of Southern California, Los Angeles, CA 90089-0484, USA

A DC SQUID has been used to directly observe the nuclear free induction decay signals from ^{3}He in very low magnetic fields. A design approach to minimize parasitic transients and noise is presented.

We have been pursuing several experiments that require the detection of NMR signals at very low frequencies, but with broad linewidths. These include the study of surface magnetism of ^{3}He at ultralow temperatures, where two dimensional magnetism may be observable with applied fields of less than 2 Oersted (1). In order to observe these signals, we need a detector with a wide bandwidth and low noise. Conventional pulsed NMR typically employs a resonant circuit in which precessing spins induce a voltage signal in a pick-up coil. Lowering the applied H_0 field results in a reduced S/N as both the equilibrium magnetization and the precession frequency of the spins decrease. In addition, the pulse recovery time for this type of arrangement follows a $\tau=2Q/\omega$ dependence, so that low frequency experiments are necessarily plagued by long recoveries. Any attempt to circumvent this problem by lowering the overall circuit Q further degrades the S/N. Both longitudinal (2) and transverse CW (4) NMR techniques involving SQUIDs (3) have been developed to circumvent some of these general problems.

Here we describe a technique for pulsed NMR, using a commercial dc SQUID (5). The technique differs from the previously used pulsed SQUID NMR techniques because in our case we study free induction decay signals, as in conventional pulsed NMR, rather than the average z-component of the magnetization.

In order to operate the SQUID with close to its intrinsic noise and to obtain rapid recovery times, we found that we had to overcome several difficulties. 1.) Due to the sensitivity of the SQUID, utmost care must be taken that noise is not brought into the cell via the leads from high temperatures. 2.)Johnson noise induced in the pickup curcuit would be a significant contribution to the total noise. 3.) Eddy currents set up in the normal metals by the rf-pulse induce transients on the output signal.

These problems cause several constraints on the design of the system. To solve the first problem one side of the rf magnet lead pair is blocked at 4.2 K with back to back diodes. Many types of signal diodes will turn on with 2-3 volts at 4.2 K. The leads below the diodes are completely shielded with superconductors.

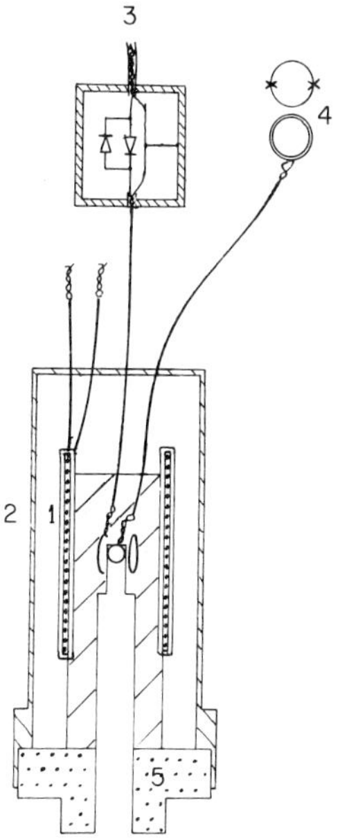

FIGURE 1
Schematic of the NMR cell. (1) Field trapping shield. A heater and solenoid are sandwiched between two superconducting layers.(2) Superconducting outer shield. (3) Pulsed rf input: The upper box contains blocking diodes, connected via superconductors to saddle coils around the sample. (4) Pick-up circuit for dc-SQUID. (5) Tower base for connection to nuclear demagnetization cryostat.

*Work supported by the National Science Foundation Grant Number DMR82-00661

Both the second and third problems originate from the presence of normal metals in the vicinity of the rf and pick-up coils. Any inductive coupling into normal metal leads to Johnson noise in the SQUID circuit and to L/R current transients following the rf pulse. Thus, in the present design we have sought to shield all normal metal from the cell. Figure 1 shows that the entire coil assembly is enclosed in a cylinder which is clad on the inside with superconductor. The tower base is also solder coated to conceal it from the SQUID pick-up coil. Those transients that remain may be controlled by fine tuning the pulse length. The object is to end the rf pulse at a phase which catches the normal metal currents at a zero crossing. In this manner we may change the sign of the transients or, preferably, arrange to minimize them. This combination of shielding and pulse length control has allowed us to produce recovery times approaching 30 µs, where the slew rate for the SQUID electronics (9.4×10^{-3} amps/sec when properly shielded) begins to limit performance. This slew rate deteriorates drastically if the SQUID is exposed to noise because of improper shielding methods. The noise measured under conditions producing the 30 µs recovery was approximately twice that specified (and observed) for the SQUID with shorted input leads (1.5×10^{-12} amps rms/$\sqrt{Hz}$ referred to input coil).

A typical free induction decay signal amplitude from ^{3}He at T=1 K and B_0=10 G (f≈33 kHz), is calculated to be 1.7×10^{-10} T. This will produce 8.4×10^{-9} amps in the SQUID input coil. Thus the expected signal to noise ratio is about 30. We have seen free induction decay signals in liquid ^{3}He using a preliminary design showing somewhat less than ideal pulse recovery, but with the expected signal to noise ratio.

Our present cell is currently being tested and will be used to study surface magnetism of ^{3}He in the limit of very low fields. We hope to first apply this technique to obtain both low field frequency and linewidth information on a 2-D magnetic system. We believe that we will be able to obtain satisfactory S/N between 5 kHz and 40 kHz. The lower frequency limit ultimatly depends on the noise of the system. The upper frequency limit is dependent on the slew rate in addition to the small signal bandwidth of our present SQUID electronics. Even for a small signal these limitations begin to degrade its response at 50 kHz. In addition, the slew rate limits our recovery following any rf pulse.

REFERENCES

(1) H.M. Bozler, D.M. Bates, A.L. Thomson, Phys. Rev. B 27 (Rapid Communications)(1983) 6992.
(2) R.A. Webb, Rev. Sci. Inst. 48 (1977) 1585.
(3) R.P. Gifford, R.A.Webb, and J.C. Wheatley, J. Low Temp. Phys. 6 (1972) 533.
(4) G.J.Enholm, J.P. Ekström, M.T. Loponen, and J.K. Soini, Cryogenics 19 (1979) 673.
(5) S.H.E. Corporation.

LT-17 (Contributed Papers)
U. Eckern, A. Schmid, W. Weber, H. Wühl (eds)

DC SQUID SMALL SIGNAL AMPLIFIERS FOR NMR

M. R. FREEMAN, M. L. ROUKES, R. S. GERMAIN, and R. C. RICHARDSON

Laboratory of Atomic and Solid State Physics, Cornell University, Ithaca NY 14853, USA

The application of SQUIDs to low temperature pulsed transverse NMR experiments has been re-evaluated and found to hold promise of higher sensitivity than can be achieved by other means. We describe one possible approach to using a SQUID for NMR, and make estimates of the sensitivity. In addition, we draw comparisons to the earlier, related work of Ehnholm et al.(1).

The energy sensitivity of state-of-the-art dc SQUIDs is now sufficiently good that they rank highly in the field of possible front-end amplifiers for nuclear magnetic resonance. Their real advantage appears in very low temperature applications, where matching to the Johnson noise from the pickup coil cannot be obtained with any other device. SQUIDs have been the heart of several novel methods of NMR detection during the past decade (1,3,4), but the most common type of NMR experiment (measurement of the tranverse spin precession at megahertz frequencies) has yet to benefit from this technology.

Figure 1 is a schematic of a simple dc SQUID-based tuned NMR spectrometer, intended for pulsed operation. The spin precession signal from the pickup coil L is inductively coupled into the SQUID through the pickup coil L_{in}. The crossed coils reduce the amount of r.f. excitation seen by the SQUID. Variations on this circuit may be required, depending upon the particular application. A more complete analysis, including parasitic inductance and capacitance of leads, distributed capacitance of inductors, and the interaction of the SQUID with the input circuit, can favor other configurations. In the following, we consider only the simplest case.

The emf induced in L by a precessing transverse magnetization is

$$V_{in} = (\mu_0\omega \;/\; 2a)\; N \;|\vec{M}|\; K \qquad (1)$$

where N is the number of turns in the coil, a the coil radius, M the effective macroscopic magnetic moment, and K a geometry dependent factor of order one. A limit on the sensitivity of the spectrometer is then obtained by comparison to the total noise voltage spectral density in the input circuit,

$$S_V(\omega) = 4k_B T_{sys} R \qquad (2)$$

where T_{sys} is the system noise temperature described below. The sensitivity can be conveniently expressed in terms of the minimum number of spins that can be detected, after a $\pi/2$ pulse, in a sample having a Curie susceptibility at temperature T (for $\hbar\omega \ll k_B T$):

$$N_{min} = \left(\frac{4k_B T}{\hbar\omega}\right)\left(\frac{4k_B T_{sys}\;\Delta f v}{K\gamma^2\hbar^2\omega\mu_0 Q}\right)^{\frac{1}{2}} \qquad (3)$$

Here Δf is the measurement bandwidth, γ the gyromagnetic ratio of the spins, and v an effective coil volume ($\sim 4\pi a^3/3$).

The system noise temperature is $T_{sys} = T_{coil} + T_{ns}$, where T_{ns} is the effective noise temperature of the dc SQUID. Important contributions to T_{ns} arise from both voltage and circulating current noise sources in the SQUID (5). T_{ns} is calculated in the limit $\omega\tau_{sq} \ll 1$, where $\tau_{sq} = L_{sq}/2R_s$ is the SQUID ring time constant (R_s is the junction shunt resistance). In this limit there is little screening by the SQUID loop, so we may neglect dynamic effects in L_{in}. We formulate the result in terms of the quantities $L_t = L + L_{in}$, $\omega_0 = 1/\sqrt{L_t C}$, $Q = \omega_0 L_t/R$, $\rho = L_{in}/L_t$, and the coupled energy sensitivity of the SQUID,

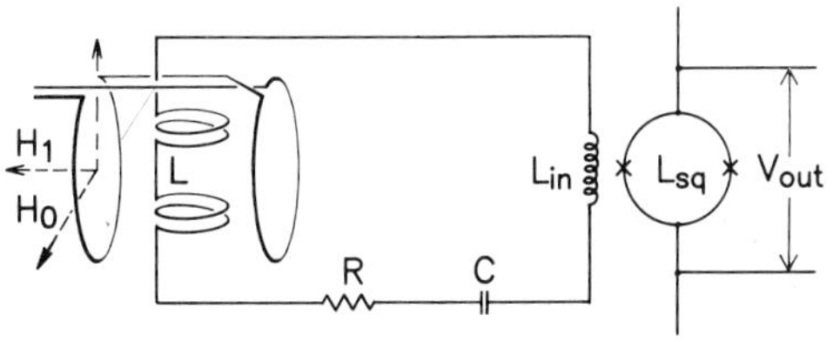

FIGURE 1
A simple series tuned tank circuit for dc SQUID detection of NMR. The resistor R represents the net dissipation of the pickup coil L and the capacitor C. The r.f. irradiation is applied through the coil orthogonal to L. The signal V_{out} must be fed into subsequent low noise stages, preferably cryogenically cooled FET amplifiers.

$$\varepsilon_c = \frac{8\,\gamma_V}{k_{sq}^2 \nu_\phi^2}\ \tau_{sq}\, k_B T_{sq}. \qquad (4)$$

k_{sq} is the coefficient of coupling of the input coil to the SQUID ring, ν_ϕ is the slope of the flux-voltage curve normalized to its idealized value $(4\tau_{sq})^{-1}$, $\gamma_V = 2V_n/(k_B T_{sq} R_S)$ is the reduced spectral density of the output noise voltage V_n (5), and T_{sq} is the temperature of the device. As $T_{sq} \rightarrow 0$, the ultimate energy sensitivity is limited by corrections of order h, which are neglected here.

The resulting SQUID noise temperature is

$$T_{ns} = \frac{\omega_0 \varepsilon_c}{2k_B Q\rho}\left[1 + Q^2\left(\frac{\omega}{\omega_0} - \frac{\omega_0}{\omega}\right)^2 + (Q\rho k_{sq}^2 \nu_\phi)\frac{\omega\gamma_J}{4\omega_0\gamma_V} - Q^2\rho k_{sq}^2 \nu_\phi\left(1 - \frac{\omega^2}{\omega_0^2}\right)\frac{\gamma_{VJ}}{\gamma_V}\right] \qquad (5)$$

where γ_J, γ_{VJ} are the reduced spectral densities of the circulating current and current-voltage correlation noise, respectively (5). T_{ns} is minimized on resonance for $(Q\rho)^{-1} = k_{sq}^2 \nu_\phi(\gamma_J/4\gamma_V)$ (~1 for a properly biased SQUID, but we note that in general the γ's are complicated functions of frequency and bias condition). The optimum signal for NMR is obtained by choosing L, with ω_0, L_{in}, R fixed, such that Q is as large as the required bandwidth will permit. The square bracketed factor in equation (5) varies from 2 to 3 over the resonant bandwidth ω_0/Q.

The dynamic range of a SQUID small signal amplifier is comparable to that of conventional low noise preamplifiers. Segments of the flux-voltage characteristic over which the SQUID is useful as a linear amplifier are ~$0.2\phi_0$ wide ($\phi_0 = 2 \times 10^{-15}$ Wb). With an intrinsic SQUID noise of $5 \times 10^{-7}\ \phi_0/\mathrm{Hz}^{1/2}$, one obtains a dynamic range of 113 dB in a 1 Hz bandwidth. The SQUID may therefore be used open loop, in order to avoid severe restrictions on the available frequency bandwidth and to ensure short pulse recovery times.

As an example, consider an experiment to be performed on surface layers of ^{3}He (Curie susceptibility) at $\omega = 2\pi \times 1.6$ MHz $= 10^7$ sec^{-1}. Let the SQUID have $L_{in} = 200$ nH and ε_c = 20 h at 1 K (ref. 6; for these SQUIDs $\tau_{sq} = 10^{-11}$ sec). Choosing $R = \omega L_{in} = 2\ \Omega$ gives $T_{ns} = 8$ mK. The resonance line is quite narrow, so we can use L = 5 μH, giving Q = 25 and $\Delta f \sim$ 50 kHz. Winding the pickup coil on a 1 cm diameter yields a minimum detectable number of spins $N_{min} \sim 10^{15}$ at 20 mK (on the order of 1 monolayer on 1 cm^2).

We now compare the present scheme to the earlier work of Ehnholm et al. (EELS, ref. 1). Briefly put, the comparison to be made is between an untuned rf SQUID magnetometer and a tuned dc SQUID magnetometer. A non-resonant input circuit was used by EELS for swept-frequency cw-NMR in low fields. They found that, for frequencies above 100 kHz, the signal-to-noise of the SQUID detection method was inferior to that obtained using an FET preamp. Their expression for the relative sensitivities of the two methods is, in our notation

$$\frac{N_{min}(\mathrm{SQUID})}{N_{min}\ (\mathrm{FET})} = \left(\frac{2Q_f \omega \varepsilon_c}{kT_{nf}}\right)^{1/2} \qquad (6)$$

where Q_f is the quality factor of the tuned FET input circuit, and T_{nf} is the amplifier noise temperature. They neglect the Johnson noise from the coils, a good approximation for temperatures < 4 K and $\varepsilon_c = 3 \times 10^5$ h, T_{nf} = 20 K as in EELS.

When the rf SQUID is replaced by a state-of-the-art dc SQUID, and the SQUID tank is resonated with the same Q as the FET circuit, the Johnson noise becomes an important consideration in many applications. We write

$$\frac{N_{min}(\mathrm{SQUID})}{N_{min}\ (\mathrm{FET})} = \left(\frac{T_{ns} + T_{coil}}{T_{nf} + T_{coil}}\right)^{1/2} \qquad (7)$$

where T_{nf} is typically in the Kelvin region, and T_{ns} on the milliKelvin scale. For optimized NMR in the μK regime, or for the possibility of quantum noise limited detection, the use of dc SQUIDs at mixing chamber temperatures appears promising.

ACKNOWLEDGEMENTS

This work was supported by NSERC, The SRC, the ONR, and NSF grants DMR-81-19847 and DMR-82-17727.

REFERENCES

(1) G. J. Ehnholm, J. P. Ekstrom, M. T. Loponen, and J. K. Soini, Cryogenics 19, 673 (1979).
(2) M. B. Ketchen, Proc. IEEE Mag-17, 387 (1981).
(3) E. P. Day, Phys. Rev. Lett. 29, 540 (1972)
(4) R. A. Webb, Rev. Sci. Inst. 48, 1585 (1978).
(5) J. Clarke, C. D. Tesche, and R. P. Giffard, JLTP 37, 405 (1979).
(6) M. B. Ketchen and J. M. Jaycox, App. Phys. Lett. 40, 736 (1982).

LT-17 (Contributed Papers)
U. Eckern, A. Schmid, W. Weber, H. Wühl (eds)

LOW TEMPERATURE SPIN LATTICE RELAXATION IN bcc ^{3}He

M.E.R. BERNIER

DPhG/PSRM,CEA,Orme des Merisiers, 91191 Gif-sur-Yvette Cedex, France

Two recent experiments have measured the spin lattice relaxation time in the bcc phase of solid ^{3}He for large molar volumes, 23.05 $cm^3 \leq V \leq$ 24.82 cm^3, at Larmor frequencies $\omega/2\pi$=50 MHz and 200 MHz. At low temperature (T < 200 mK) three relaxation mechanisms can be identified and they give a coherent picture of the relaxation.

The high temperature behaviour (T>300mK) of the spin lattice relaxation time T_1 has been widely investigated both in the bcc and hcp phases and the mechanisms responsible for the relaxation quantitatively described [1]. Some of the low temperature features have also been accounted for. The present paper deals with some puzzling results of two recent NMR experiments [2][3]made in the bcc phase at Larmor frequencies of 50MHz and 200MHz and molar volumes 23.05cm$^3 \leq V \leq$24.82cm^3. The measurements are shown to be consistent with important properties of the solid: (i) the possible existence of vacancy waves at temperatures close to 200mK for large molar volumes, (ii)the existence of dislocations also observed in sound attenuation experiments. (iii) the modulation of the exchange parameters describing the solid by long wavelengths phonons [6].

RELAXATION PROPERTIES

The solid ^{3}He can be described by a set of independent systems of excitations, each one characterized by its own temperature (Zeeman, exchange, phonons), coupled to one another by interactions. When ^{4}He atoms are present in the solid they store energy in elastic interactions, exchange positions with neighbouring ^{3}He atoms and are usually considered as a separate energy system with its own temperature. The best description of the atomic motion of ^{3}He atoms is given by a hamiltonian involving multiple spin exchange while the Heisenberg hamiltonian widely used at high temperature fails to account for the low temperature properties of the solid [4]. However, the characteristic frequencies of the various exchange interactions (2,3 or 4 spins) exhibit similar variations with the molar volume according to both experiment and theory. In the temperature range of the reported experiments, the exchange can then be satisfactorily characterized by a single parameter J taken equal to the Heisenberg parameter for the sake of simplicity.

A sketch of the variations $T_1(T)$ measured in [2] and [3] is given in fig.1 where four temperature regimes can be distinguished:regions I to III are observed in both experiments although region III is more clearly observed in [3] ; region IV is only observed in [2]. The relaxation in region I is known to be governed by the Zeeman exchange coupling and in the following we shall only deal with regions II,III and IV.

REGION II : In this region T_1varies rapidly with T, suggesting a relaxation of the exchange energy by a two phonons process on ^{4}He impurities or a modulation of the exchange by the motion of vacancies. The T_1 dependence on V observed in[2] is much faster than expected for a two phonons process, thus ruling out this mechanism. A modulation of the exchange interactions by vacancies would be consistent both with the strong temperature dependence of T_1 observed in [3] where the range of variation of T_1 is large in region II and with the strong molar volume dependence of T_1 in [2]. For such a process T_1 is given by:

$$(T_1)_V^{-1} = [k_e/(k_z+k_e+k_t+k_a)]\ (T_{e\ell})_V^{-1} \qquad (1)$$

where the heat capacity of an energy reservoir α at a temperature β^{-1} is defined by $C_\alpha=k_B\beta^2k_\alpha$. The indices z,e,t,a stand respectively for Zeeman, exchange of ^{3}He^{3}He atoms, exchange of ^{3}He ^{4}He atoms, elastic interactions between ^{4}He atoms. (1) If there are no vacancy waves in the solid

$$(T_{e\ell})_V^{-1} = (14/a^2)D_0\ \exp(-\phi/T) \qquad (2)$$

where a is the n.n. distance, D_0 the diffusion coefficient and the activation energy ϕ is equal to the formation energy of vacancies for a bcc lattice. (2) If the vacancy waves exist, $T_{e\ell}$ is now a function of the bandwidth Δ of the waves and if $T \ll \Delta$: [5][6]

$$(T_{e\ell})_V^{-1} = 1.66\ 10^{10}(T^{1.5}/\Delta^{0.5})\exp[-(\phi-\Delta/2)/T] \quad (3)$$

where T,ϕ and Δ are given in temperature units(K).

REGION III : In this region T_1 varies linearly with T within the experimental accuracy. This is observed in both experiments but appears more clearly in [3] for 23.25 cm$^3 \leq V \leq$24.82cm^3. In [2] this regime is difficult to separate from the low temperature regime IV where the characteristic time for the relaxation is temperature independent. However the relaxation in region III is always exponential while it becomes slightly non exponential in region IV as we explain later.

$V(cm^3)$	$J/2\pi$ (MH_Z)	$(T_1)_{pl}$ (sec)	$\phi(K)$ [5]	$\Delta(K)$	$\phi-\Delta/2$ (K)	$f_V(K)$ [8]	$\Delta(K)$	$f_V-\Delta/2$ (K)	Figure 2 label
23.05	6.54	15 000	7.05	3.2	5.45	6.12	1.1	5.57	I
23.52	9.56	1 800	6.65	3.9	4.7	5.26	0.8	4.86	II
23.91	13.06	650	6.35	4.5	4.1	4.6	0.7	4.25	III
24.59	18	50	6	5.7	3.15	3.32	0.2	3.22	IV

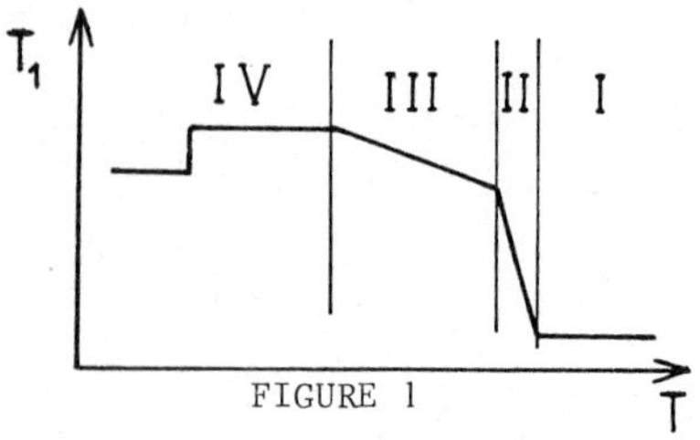

FIGURE 1

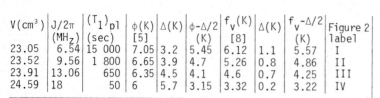

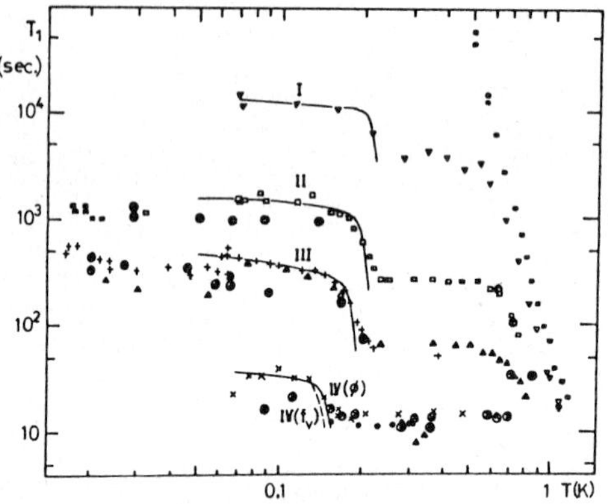

FIGURE 2

This behaviour is strongly suggestive of a one phonon process directly coupling the exchange energy to the lattice. The relaxation time for such a process would be [7] :

$$(T_1)_p^{-1} = [k_e/(k_z+k_e)]\ (T_{ep})^{-1} \qquad (4)$$

with $T_{ep}^{-1} \# \ 150\ (J/2\pi)^4\ T/\theta_D^5$ (5)

where θ_D and T are in temperature units (K), $J/2\pi$ in MH_Z and we assume a molar volume dependence of the exchange parameter $\partial LnJ/\partial LnV=18$. Whereas T_1 varies by a factor 500 when V varies from $23.25cm^3$ to $24.82cm^3$ at T=0.1K in region III of [3], the quantity $T_{ep}(J/2\pi)^4/\theta_D^5$ remains constant within ±35% in the same molar volume range. In order to fit the absolute value of T_1 the constant entering (5) should be 28 instead of 150, a small difference in view of the crudeness of the approximations.

REGION IV : In [2] the relaxation in region IV was satisfactorily explained by taking into account the diffusion of energy to the defects inside the solid, mainly dislocations which are strongly coupled to the lattice. According to the relative values of the characteristic diffusion time and the Zeeman exchange relaxation time weighted by the ratio of the Zeeman to the exchange heat capacities the relaxation could be nearly exponential or strongly non exponential. The characteristic time describing this mechanism $(T_1)_{p\ell}$ is independent of the temperature as observed experimentally. The dislocations have also been studied in sound attenuation and velocity experiments where the hysteresis loops observed when crossing the phase separation temperature are similar to the one described in [2]. The dislocation lengths deduced from sound experiments are in good agreement with the one deduced from NMR [9].

Assuming that the partial relaxation rates of the mechanisms governing the regions II, III and IV, are additive, $(T_1)^{-1}$ is then given by :

$$(T_1)^{-1} = (T_1)_{p\ell}^{-1} + (T_1)_p^{-1} + (T_1)_V^{-1} \qquad (6)$$

where we assume the existence of vacancy waves and use the experimental constant 28 instead of 150 in $(T_1)_p^{-1}$ in region III. A comparison of the experimental data with the predictions of (6) using the parameters of table 1 is given on fig2. In table 1, f_V stands for the values of the formation energy of vacancies obtained by Xrays methods. Let us point out that in region II the relaxation rate is governed by $\exp-(\phi-\Delta/2)$ and $(\phi-\Delta/2)$ is the quantity really measured. Using values of ϕ obtained in other experiments we deduce the estimates of the bandwidth of the vacancy waves given in table I.

[1] R.A.Guyer, R.C.Richardson, L.I.Zane, Rev.Mod. Phys. 43, 532 (1971)
[2] M.E.R.Bernier, G.Guerrier, Physica 121B, 202 (1983)
[3] M.Bassou, Thesis, Univ.of Paris (1981) M.Chapellier, M.Bassou, M.Devoret, J.M.Delrieu N.S.Sullivan, J.of Low Temp.Phys., to be published
[4] M.Roger, J.H.Hetherington, J.M.Delrieu, Rev. of Modern Phys. 55, 1 (1983)
[5] N.Sullivan, G.Deville and A.Landesman, Phys. Rev.B, 1858 (1975)
[6] M.E.R.Bernier, J.Low Temp.Phys. to be published
[7] R.B.Griffiths, Phys.Rev.124, 1023 (1961)
[8] S.M.Heald, D.R.Baer and R.O.Simmons, Phys. Rev.Lett., to be published
[9] I.Iwasa, N.Saito, H.Suzuki, J.de Phys. C5-37 (1981) J.R.Beamish, J.P.Franck, Phys.Rev.B, 28, 1419 (1983).

LT-17 (Contributed Papers)
U. Eckern, A. Schmid, W. Weber, H. Wühl (eds)

ANOMALOUS NMR FREE INDUCTION DECAY IN NUCLEAR ORDERED SOLID ^{3}He

Tadashi KUSUMOTO, Osamu ISHIKAWA, Takao MIZUSAKI and Akira HIRAI

Department of Physics, Kyoto University, Kyoto 606, Japan

The nuclear spin dynamics in nuclear ordered solid ^{3}He was studied by the pulsed NMR methods. The free induction decay behaviors were measured at three operating frequencies of 920, 1380 and 1840 KHz. The free induction decay signals were dependent on the tipping angle and were non-exponential. The tipping angle dependent frequency shift was observed at 1840 KHz.

1. INTRODUCTION

The nuclear spin ordering in bcc solid ^{3}He is observed at T = 1.0 mK on the melting curve. Osheroff, Cross and Fisher (OCF) (1) performed cw-NMR experiments on single crystals of bcc solid ^{3}He and observed the antiferromagnetic resonance spectrum. They proposed that the possible magnetic structure of the ordered nuclear spin system in low magnetic fields is of the sequence up-up-down-down (uudd) along the [100] direction, which is denoted by the unit vector $\hat{\boldsymbol{\ell}}$. They also proposed the following coupled equations of motion (OCF equations) for the magnetization $\mathbf{S}$ and the order parameter $\mathbf{d}$ corresponding to the sublattice magnetization.

$$\dot{\hat{\mathbf{d}}} = \hat{\mathbf{d}} \times (\gamma \mathbf{H} - \gamma^2 \chi_0^{-1} \mathbf{S}) \qquad [1]$$

$$\dot{\mathbf{S}} = \gamma \mathbf{S} \times \mathbf{H} - \lambda(\hat{\mathbf{d}}\cdot\hat{\boldsymbol{\ell}})(\hat{\mathbf{d}} \times \hat{\boldsymbol{\ell}}) \qquad [2]$$

where γ is the gyromagnetic ratio, χ_0 is the transverse susceptibility and $\hat{\mathbf{d}}$ is the unit vector of the direction of $\mathbf{d}$. Their cw-NMR spectrum is the solution of the above equations for the small oscillation about the equilibrium position with $\lambda = \chi_0 \Omega_0^2 / \gamma^2$, where Ω_0 is the antiferromagnetic resonance frequency at zero field.

The solid ^{3}He offers a unique opportunity to investigate the nuclear spin dynamics in the nuclear ordered spin system. In the previous note (2) we reported some data of the pulsed NMR experiments at one operating frequency of f_0 = 920 KHz. Here, we will report more detailed data as well as those at different operating frequencies.

2. EXPERIMENTAL METHODS

We observed the free induction decay signals in the single crystal of solid ^{3}He on the melting curve down to 0.6 mK. We always observed three kinds of domains in a single crystal and each domain is characterized by $\cos\theta$, where θ is the angle between the $\hat{\boldsymbol{\ell}}$-vector and the applied static magnetic field H_0. The tipping angle β_p is defined by $\gamma H_1 t_w$, where H_1 is the RF pulse field intensity and t_w is the pulse width. The β_p was changed by changing the H_1, with a fixed value of t_w of 24 µsec, and calibrated by the experiments on the normal Fermi liquid ^{3}He.

3. EXPERIMENTAL RESULTS AND DISCUSSIONS

Fig. 1 (A),(B) and (C) show the typical free induction decay signals at three operating frequencies, f_0 = 920, 1840 and 1380 KHz, respectively, for a sample with $\cos^2\theta$ = 0.044.

3.1. Some of the properties of the anomalous behaviors of the free induction decay signal at 920 KHz are summarized as follows:

(a) At small tipping angles, the free induction signals decayed nearly linearly in time with a small exponential tail at the end, and we call this decay as a linear decay. At large tipping angles, they decayed very rapidly, and we call this as a non-linear decay.

(b) By β, we denote the angle by which the magnetization is actually tipped. When we increased the β_p, the magnetization could be tipped following the relation of $\beta = \beta_p$, up to some critical value of β_p, β_c. If we increased the β_p beyond β_c, the magnetization did not follow as expected, and at the same time the non-linear decay started to appear. We measured the similar tipping angle dependence of the free induction decay for many different single crystals. We found the β_c decreased as $\cos^2\theta$ increased and the temperature was lowered. It seems likely that the β_c depends on the ratio of Ω_0^2/ω_L^2, where ω_L (= γH_0) is the larmor frequency.

(c) When the linear decay curves were extrapolated, curves with different β_p crossed nearly the same point on the time axis. This time is denoted by a full decay time T_F. The T_F did not depend on the temperature and decreased as $\cos^2\theta$ increased.

(d) We could not observe usual spin echoes.

The free induction decay time seemed to be determined by the usual magnetization recovery time, or the spin lattice relaxation time T_1.

(e) No tipping angle dependent frequency shift was observed within the accuracy of our experimental frequency determination, which was not so good due to shortness of the free induction decay time.

3.2 It is important to study the resonance frequency dependence of the free induction decay behaviors. At higher operating frequency, it is expected that the non-linear effect due to the second term in eq. [2] becomes less effective. Some results at f_0 = 1840 KHz are summarized as follows:

(a) The full decay time T_F for the small tipping angles at f_0 =1840 KHz was about a half of that at f_0 = 920 KHz.

(b) By increasing the β_p, the free induction decay became no longer shorter in the range of $\beta_p > 23^0$. The free induction signal could be observed even at the large tipping angles such as $\beta_p > 90^0$.

(c) When the operating frequency is increased, the resonance frequencies for three magnetic domains approach each other. The small ripples in the free induction decay signal in Fig. 1(B) come from the beat between the signals from the different domain, and at the tail the signal was mixed with the signal from the superfluid ^{3}He-B.

(d) The tipping angle dependent frequency shift of the resonance frequency was observed at 1840 KHz. The homodyned free induction decay signal for $\beta_p = 90^0$ is shown in the inset of Fig. 1(B). The frequency shift was about 18 KHz and independent of the tipping angle in the range between 75^0 and 120^0 for β_p. The frequency shift for $\beta_p < 75^0$ could not be determined accurately. The tipping angle dependence of the frequency shift in the ordered solid ^{3}He was calculated by Namaizawa (3). The result is given by the following formula, in the high field limit,

$$\omega-\omega_L = (\Omega_0^2/8\omega_L)\cdot\{(5\cos^2\theta-1)\cos\beta + \sin^2\theta\} \quad [3]$$

Quantitatively, if we assume that $\beta = \beta_p$, the estimated shift from eq. [3] is about twice as large as our observed shift. The disagreement may be attributed to the fact that $\beta \neq \beta_p$ at large tipping angles.

3.3 As shown in Fig. 1(C), the free induction decay at f_0 = 1380 KHz was very short, compared with the results at other operating frequencies. This may be because there is a resonance frequency of the lower spin wave mode just at a half of the operating frequency in this experimental condition, and in such case the three magnon process is effective.

3.4 Tsubota et al. (4) numerically solved the OCF equations with an RF pulse, in connection with our work. They showed that the motion of the nuclear spin becomes chaotic, when the tipping angle β_p exceeds a critical value. Their results seem to explain some of our experimental data satisfactorily.

REFERENCES

(1) D. D. Osheroff, M. C. Cross and D. S. Fisher, Phys. Rev. Lett. 44 (1980) 792.
(2) T. Kusumoto, O. Ishikawa, T. Mizusaki and A. Hirai, Phys. Lett. 100A (1984) 201.
(3) H. Namaizawa, Prog. Theor. Phys. 67 (1982) 1989.
(4) M. Tsubota and T. Tsuneto, Private communication and the present Proceedings.

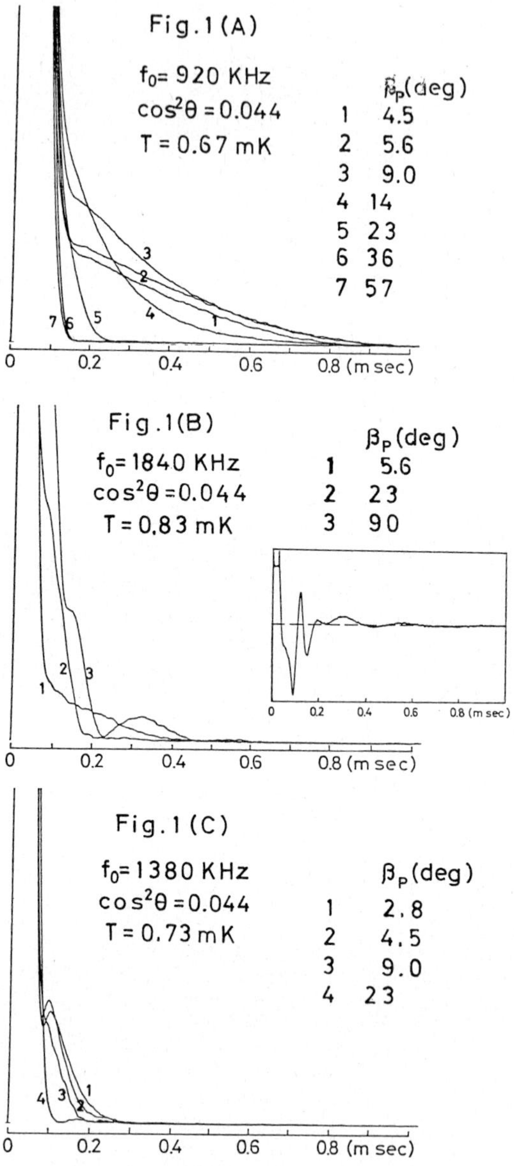

FIGURE 1

The tipping angle dependence of the free induction decay signals in nuclear ordered solid ^{3}He.

LT-17 (Contributed Papers)
U. Eckern, A. Schmid, W. Weber, H. Wühl (eds)
© Elsevier Science Publishers B.V., 1984

MAGNETIC SUSCEPTIBILITY ANOMALY IN LOW DENSITY bcc SOLID ^{3}He

W.P. KIRK, Z. OLEJNICZAK, P.S. KOBIELA, and A.A.V. GIBSON

Department of Physics, Texas A&M University, College Station, Texas 77843, USA*

The magnetic susceptibility of low-density solid ^{3}He was measured by a high-precision pulsed-NMR technique over a temperature range of 16 to 500 mK. Samples with molar volumes between 24.208 and 24.424 mL/mole displayed unexpected behavior below 50 mK. The effect increased with decreasing density and appeared to be associated with samples near melting.

1. INTRODUCTION

Quantum solids like solid ^{3}He provide useful systems to study many-body phenomena, whole atom-exchange processes, and the behavior of nuclear magnetism in matter. The magnetic behavior of this solid has been especially interesting, and a number of workers (1) have reported measurements of the magnetic susceptibility χ in the bcc phase at several intermediate densities (molar volumes between ≈20.0 to 24.0 mL/mole) and over a wide temperature range (T≈1mK to 1K).

Existing measurements of χ at low densities in the bcc phase are limited in accuracy. In this paper we report new measurements of χ for several low-density bcc phase samples over a temperature range of 16mK to 500mK. The magnetic susceptibility of these low density samples displayed unexpected behavior at temperatures below 50mK. This anomalous behavior increased with decreasing density and appeared to be associated with samples near the negative-slope portion of the melting curve.

2. EXPERIMENTAL TECHNIQUES AND RESULTS

All the measurements were done using the double sample-cell and Fourier-transform pulsed-NMR technique we developed earlier (1) to make high-precision magnetic susceptibility measurements on solid ^{3}He samples at intermediate densities (2). Thermal contact was achieved by solidifying the samples on 25 μm dia. copper wires tightly packed into the sample cell space. The helium volume-to-surface ratio was 6.3×10^{-4} cm.

The samples were solidified from ^{3}He gas with ≈27ppm ^{4}He impurity. The molar volumes of the samples were determined to within ±0.05% from the solidification pressures using capacitive strain gauges. The temperatures were established by the previously described method using the double-cell (1). As a cross check of the temperature scale, a secondary CMN thermometer with SQUID detection was also used.

In Fig. 1 we show the results plotted as $1/\chi T$ vs $1/T$. The 24.090 mL/mole sample displayed the expected linear Curie-Weiss behavior over the full temperature range. From the slope of the fitted straight line, the value of the Curie-Weiss temperature θ was found to be -1.56 ± 0.07 mK, in good agreement with our previous work (1). The sample was then reformed at a slightly higher molar volume of 24.208 mL/mole, which corresponded to a solidification pressure about 1 bar above the maximum (34.4 bar) in the melting curve. At about 50mK a distinct departure from the expected linear behavior in $1/\chi T$ was observed in this sample. From about 50mK to 25mK the susceptibility was

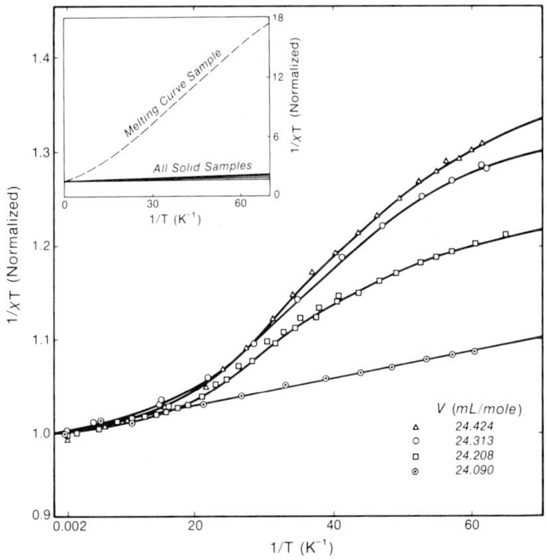

FIGURE 1
Magnetic susceptibility showing anomalous behavior in three low-density samples. Curves through data were fitted with a third order polynomial. Inset compares χ of melting curve sample with the four solid samples: note reduced resolution of ordinate.

* Research supported by NSF through grant #DMR 8205902.

observed to drop about 10% and then leveled off with an approximate 1/T dependence at lower temperatures.

Similar, more pronounced effects were observed in two other samples formed at yet lower densities. The 24.313 mL/mole sample corresponded to a solidification pressure about equal to the maximum in the melting curve, while the 24.424 mL/mole sample corresponded to a solidification pressure about 1 bar below the maximum. The anomalous behavior in the two higher molar volume samples occurred more gradually and began at temperatures somewhat higher than 50 mK.

At one stage of investigation, the 24.208 mL/mole sample was re-annealed (but not reformed) on the melting curve at high-T. The behavior of χ, as described above, reproduced down to 25 mK. Below 25 mK the data displayed an offset, ≈4% higher than shown by the 24.208 curve in Fig. 1. Study of offset behavior was not done for the other two samples.

Since these anomalous effects occurred in a temperature range close to melting conditions, we were careful to monitor the pressure strain gauge for evidence of sample melting. There was no evidence of melting. To further assure ourselves that melting was not the origin of the anomalous behavior, we formed a melting curve sample and measured its susceptibility by the same technique. The result of this measurement is shown by the dashed curve in the inset of Fig. 1. The striking difference between the melting curve sample and the all solid samples indicates that had the solid samples undergone melting, a rather different behavior would have been observed, and furthermore, the leveling off of the curves in Fig. 1 (or the appearance of an inflection point) should not have occurred.

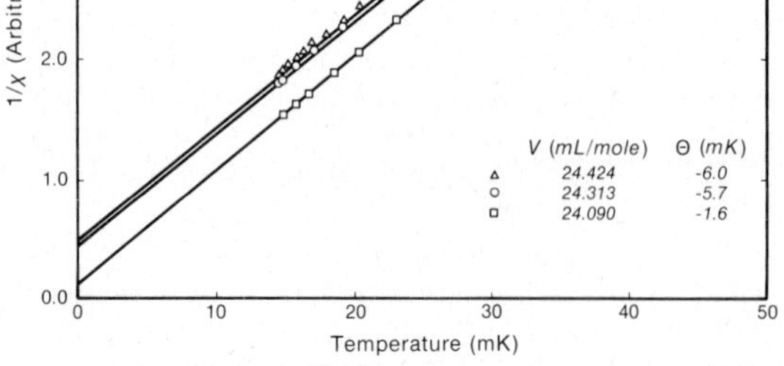

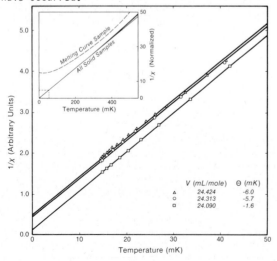

FIGURE 2
Magnetic susceptibility plotted as 1/χ vs T.

The possibility that an isotopic-phase-separation effect might have caused the anomalous behavior was considered. However, it can be shown through calculation that this cannot produce a large enough effect to explain the results since the ^4He impurity is so small.

The inset in Fig. 2 shows data for three of the solid samples and the melting curve sample plotted as 1/χ vs T. The data in the lower left corner of the inset has been replotted on a magnified scale in Fig. 2. The curves were obtained by fitting all the data over the entire temperature range (16mK to 500 mK). The intercept of these lines on the T-axis yielded the θ values noted in Fig. 2. The appearance of large negative θ values in those samples identified as being anomalous may offer a partial explanation as to why some of the early χ measurements on samples near 24.0 mL/mole produced similar large negative θ values.

3. DISCUSSION

In surveying other experiments for such an anomaly, we find that high-magnetic-field pressure measurements P(T,H) might indirectly yield observable effects. However, the dominant $(H/T)^2$ contribution to the pressure and the necessity for lower temperatures precludes the possibility of seeing the effect in the reported P(T,H) measurements (3). NMR spin-relaxation measurements, such as T_1 measurements, also might yield an effect, but the reported low-density T_1 measurements do not go low enough in temperature (4).

The next higher order term in a high-T series expansion (from a multiple exchange model) of χ cannot explain the anomalous behavior. A number of other mechanisms, such as vacancies, dislocations, polarons, order-disorder effects, etc. come to mind as possible candidates to explain the effect. However, the need for more experimental work is obvious . Clearly, as one lowers the density in a quantum solid, the kinetic energy and the zero-point motion of the particles about their equilibrium sites becomes more pronounced, but whether this leads to unusual and unexpected effects, other than what is known, has not been well explored.

REFERENCES

(1) W.P. Kirk, Z. Olejniczak, P. Kobiela, A.A.V. Gibson, and A. Czermak, Phys. Rev. Lett. 51, 2128 (1983). For earlier work, see references therein.
(2) Reference sample ≈21.0 mL/mole.
(3) C.T. Van Degrift, W.T. Bowers, P.B. Pipes, and D.F. McQueeney, Phys. Rev. Lett. 49, 149 (1982); W.P. Kirk and E.D. Adams, Phys. Rev. Lett. 27, 392 (1971).
(4) M. Chapellier, M. Bassou, M. Devoret, J.M. Delrieu, and N.S. Sullivan, J. Low Temp. Phys., to be published.

LT-17 (Contributed Papers)
U. Eckern, A. Schmid, W. Weber, H. Wühl (eds)

THERMODYNAMIC PROPERTIES OF bcc SOLID ^{3}He IN A LOW MAGNETIC FIELD*

Hiroshi FUKUYAMA**, Anju SAWADA, Yoshiyuki MIWA and Yoshika MASUDA

Department of Physics, Nagoya University, Chikusa-ku, Nagoya 464, Japan

The free energy of bcc solid ^{3}He has been experimentally determined from the specific heat and pressure measurements in the temperature range between 0.4 and 60 mK at several molar volumes in a low magnetic field (B=28 mT). The results show that the second order coefficient e_3 in the high temperature series expansion of the free energy is large and positive and that the ground state energy of the nuclear spin system is -(1.02±0.08) mK at 24.12 cm^3/mole.

1. INTRODUCTION

The thermodynamic properties of bcc solid ^{3}He below 100 mK are dominated by a freedom of the nuclear spins. Previously we reported the pressure change through the nuclear magnetic ordering temperature T_N at several molar volumes (1) and, subsequently, the specific heat and isochoric pressure measurements above 2 mK (2). In this study we have extended the range of molar volume in the pressure measurement using the same sample cell as ref. (1) and made a thermodynamic comparison between these measurements.

2. EXPERIMENTAL

In the cell I, employed for the pressure measurements in the temperature range from 0.4 to 16 mK, we made a gap between the diaphragm of the pressure gauge and the sintered silver (particle size∿700A) as small as possible, since a bulk sample gives long thermal relaxation time. Although this fabrication certainly allowed us to cool solid ^{3}He below 1 mK, it brought a necessity of volume correction. (1) A detailed description of the cell II, used for the specific heat and isochoric pressure measurements from 2 to 60 mK, has been reported elsewhere.(2) The important difference between them is that there remained a gap of 0.5 mm width between the diaphragm and the sintered silver in the cell II, in which the pressure data could be taken under the fixed volume condition. The temperature was measured by an NMR pulsed thermometer which was calibrated against the spin-lattice relaxation time of platinum.

3. RESULTS AND DISCUSSION

In a wide temperature range of $1/(60\ \text{mK})<1/T<1/(2T_N)$ the measured pressure P was well represented by $P=P_0+A/T-B/T^2$. Comparing the coefficients A and B measured in the cell I with those in the cell II, we can determine the correction

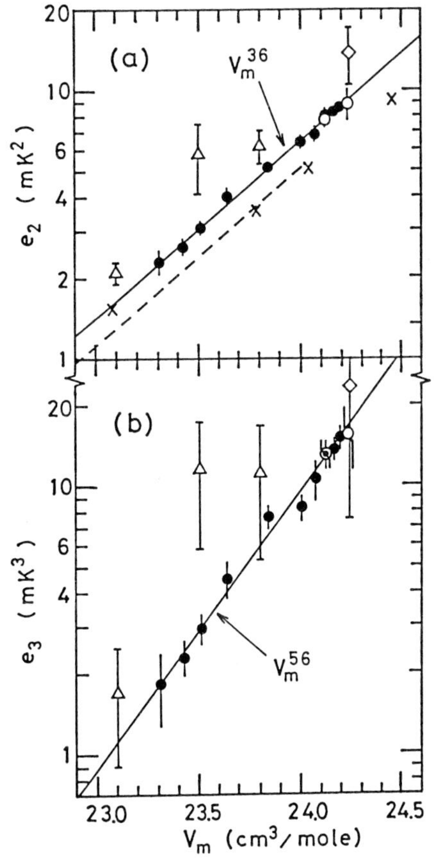

FIGURE 1
Comparison of e_2(a) and e_3(b) derived from pressure (●) and specific heat (○) measurements (2). The results in ref.(4) (dashed line), ref.(5)(x), ref.(8)(Δ) and ref.(9)(◇) are also included.

*This work was supported in part by the Grant-in-Aid for Special Project Research of The Ministry of Education, Science and Culture, Japan.
**Present address: The Institute for Solid State Physics, The University of Tokyo, Japan.

factor C_f in the cell I as C_f=2.34. When the temperature approaches to T_N, a positive deviation from the above expression becomes apparent indicating the existence of higher terms than $1/T^3$. Recently Sawada et al.(3) measured the specific heat as a function of temperature in the temperature range including T_N for molar volume of 24.0 cm^3/mole, and discussed the relation between both experiments.

The free energy F at high temperatures is expanded in powers of 1/T, without any hypotheses on the spin Hamiltonian, as

$$F=-T\ \ln 2-(e_2/8)/T+(e_3/24)/T^2-\cdots\cdot. \qquad (1)$$

e_2 and e_3 determined from the present experiments are summarized with those of other workers in Fig. 1. In this context we integrated A and B in volume assuming a simple power law $\propto(V_m)^\gamma$, in which V_m is molar volume. The consistency between our specific heat and pressure measurements is excellent. The solid lines in the figure represent $e_2=(7.7\pm0.4)(V_m/24.12)^{35.7\pm1.3}$ mK^2 and $e_3=(12.9\pm1.6)(V_m/24.12)^{56.3\pm3.7}$ mK^3. Our e_2 is 20 % larger than the previous high temperature results (4)(5) but has the same molar volume dependence as they have.

Below T_N the measured pressure change $p=P-P_0$ was roughly represented by the form of $p=p(T=0)-DT^4$, as expected from the antiferromagnetic spin wave theory. If we express the free energy F in the ordered phase as

$$F=F(T=0)-(\pi^2/90)(T^4/e_3'), \qquad (2)$$

the ground state energy F(T=0) and the temperature coefficient e_3' are expressed as $F(T=0)=-(1.02\pm0.08)(V_m/24.12)^{19.2\pm1.3}$ mK and $e_3'=(2.3\pm0.30)(V_m/24.12)^{51.1\pm3.4}$ mK^3. Our e_3' is in agreement with the result by Osheroff and Yu (6) at the melting density ($e_3'=3.3$ mK^3)within the experimental uncertainties. The volume dependence of e_3' indicates $e_3'\propto|J|^3$(J is an exchange energy), which is again consistent with the prediction by the spin wave theory. This is the first experimental determination of F(T=0).

The temperature variation of the free energy at 24.12 cm^3/mole deduced from Eqs. (1) and (2) by using the present values of e_2, e_3, F(T=0) and e_3' is shown with the solid line in Fig. 2. Recently Iwahashi et al. (7) calculated the free energy based on the spin wave approximation and proposed a parameter set of $j_t=-0.157$ mK and $K_p=K_F=-0.247$ mK as the most probable one satisfying the present value of F(T=0) as well as e_2 and e_3. Their result is also shown with a molecular field energy (-0.56 mK) for the same parameter set in Fig. 2. It would be noticed that an energy gain due to the zero-point motion of the spin is as large as the molecular field energy in this case.

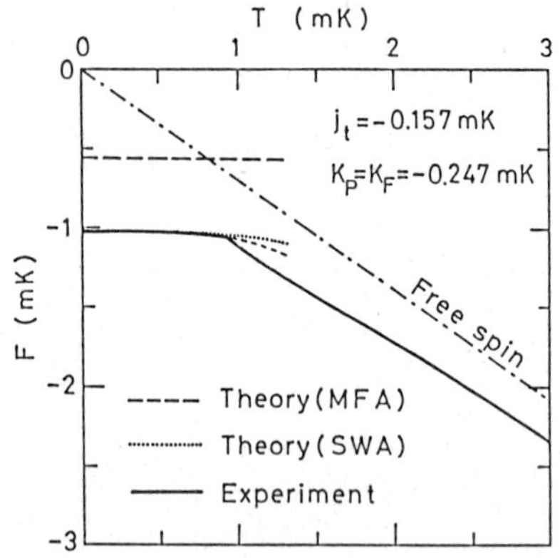

FIGURE 2
Free energy for 24.12 cm^3/mole; solid line shows experimental result; MFA, molecular field approximation; SWA, spin wave approximation (7).

ACKNOWLEDGEMENT

We wish to thank K. Iwahashi for useful suggestions in theoretical analyses and T. Mamiya for contributions in the early stage of experiments.

REFERENCES

(1) T. Mamiya, A. Sawada, H. Fukuyama, Y. Hirao, K. Iwahashi and Y. Masuda, Phys. Rev. Lett. 47 (1981) 1304.
(2) H. Fukuyama, Y. Miwa, A. Sawada and Y. Masuda, J. Phys. Soc. Japan 53 (1984) 916.
(3) A. Sawada, M. Kato, H. Yano and Y. Masuda, appear in this conference.
(4) M. F. Panczyk and E. D. Adams, Phys. Rev. 187 (1969) 321.
(5) D. S. Greywall, Phys. Rev. 15B (1977) 2604.
(6) D. D. Osheroff and C. Yu, Phys. Lett. 77A (1980) 458.
(7) K. Iwahashi, Y. Miwa and Y. Masuda, to be published.
(8) J. M. Dundon and J. M. Goodkind, Phys. Rev. Lett. 32 (1974) 1343.
(9) W. P. Halperin, F. B. Rasmussen, C. N. Archie and R. C. Richardson, J. Low Temp. Phys. 31 (1978) 617.

LT-17 (Contributed Papers)
U. Eckern, A. Schmid, W. Weber, H. Wühl (eds)

MEASUREMENT OF THE PRESSURE OF HCP SOLID ^{3}He AT LOW TEMPERATURES IN A MAGNETIC FIELD

T. MAMIYA*, D. G. WILDES, M. R. FREEMAN, and R. C. RICHARDSON

Laboratory of Atomic and Solid State Physics, Cornell University, Ithaca, New York 14853 U.S.A.

The sign of the exchange interaction and the possible absence of four-atom exchange in hcp solid ^{3}He can be determined by measurement of the pressure of solid ^{3}He at low temperatures in a magnetic field. An apparatus has been made to measure the temperature dependence of the pressure of hcp solid ^{3}He at constant volume.

The ordered phase of bcc solid ^{3}He is an antiferromagnetic state which is thought to occur as the result of a delicate balance between a ferromagnetic three particle exchange and a slightly larger antiferromagnetic four particle exchange. (1) When the crystal is cooled, the exchange interaction causes an increase in the pressure. When a large magnetic field is applied to bcc ^{3}He the exchange contribution to the pressure is decreased. (2)

In hcp ^{3}He the three particle exchange is expected to be dominant because of the triangular lattice structure. (3) The ordered phase is likely to be ferromagnetic and one might expect the pressure changes on cooling to be increased by the application of a magnetic field.

In the hcp phase the exchange interaction at 19.7 cm^3/mole is expected to be 100 times smaller than the exchange interaction on the melting curve. This implies a pressure change of only 30 μbar for crystals cooled from high temperatures to 0.7 mK. A similar pressure change, 30 μbar at 0.7 mK, can be achieved by applying a modest magnetic field, of order 1 kgauss.

In order to measure the pressure in magnetic fields up to 5 kgauss, we must use a material with a small nuclear heat capacity. Tungsten is a good candidate for this purpose because it has a high yield strength, a moderate elastic coefficient, and only 0.05% of the heat capacity per unit volume of copper. We have studied tungsten diaphragms made from a sheet and from two different rods. The diaphragms had an unexpectedly large hysteresis of more than 0.2% in the pressure range of 0 to 140 bar at 4K. A more serious problem was that all the diaphragms leaked after pressure cycling, possibly because of the brittleness of the material.

Our next choice was titanium, which has a nuclear heat capacity which is 1.2% that of copper. Our cell is made with a titanium diaphragm which shows a maximum hysteresis of less than 0.04% in the pressure range of 0 to 140 bar at 4K.

Figure 1 shows our cell designed for pressure measurements on hcp solid ^{3}He at constant volume. It is principally made from $Si_{.028}Ag_{.972}$ alloy. Pure silver is used parallel to the main piece of SiAg alloy in order to increase the thermal conductivity. The diaphragm is titanium, 1.4 mm. thick. The space above the diaphragm is sealed for use with a ^{4}He pressure-feedback system, as described by Wildes, et al. (4) The space between the capacitance electrodes is 25 μm. at zero bar. The capacitance changes from 29 to 40 pF when the ^{3}He pressure changes from 0 to 140 bar. The cell is assembled with eight titanium bolts. The pressure seals are made with 0.63 mm. indium O-rings.

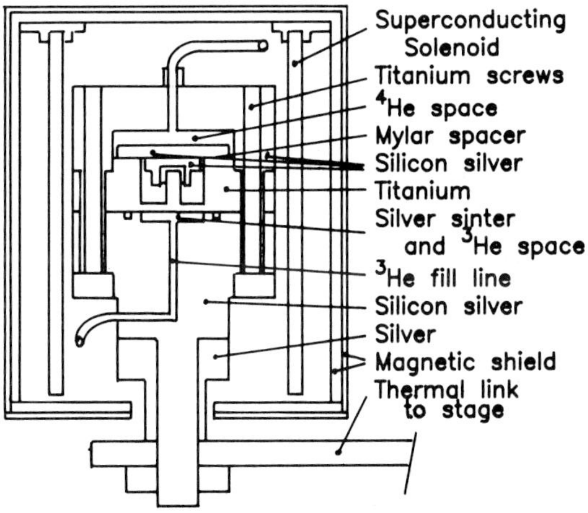

FIGURE 1
Diagram of experimental cell, superconducting solenoid, and magnetic shield cans.

The ^{3}He space is a disk of 10 mm. diameter and 2.5 mm. height. This space contains a sintered sponge made from a fine powder of 70 nm. silver. The gap between the sintered surface and the diaphragm is 13 μm. and is short enough for thermal equilibrium to be achieved in one hour at the lowest temperatures.

In order to confine the magnetic flux, a set of shield cans are installed outside the superconducting solenoid used for producing the magnetic field. The cans are made from a 0.8 mm. layer of mu-metal outside and a 1.6 mm. layer of iron inside. While calculations of the leakage field at the exterior surface of the shields suggested a value of 10^{-2} gauss outside for 5 kgauss at the center of the cell, the actual measurements at 4K showed a field of 5 gauss outside for 1.5 kgauss at the center. This leakage is presumably due to the sharply decreased permeability of the mu-metal at low temperatures.

ACKNOWLEDGEMENTS

One of us (TM) thanks the Yoshida Foundation for providing travel funding, and M. L. Roukes for assistance with preamplifier designs. This work was supported by National Science Foundation Grant DMR-81-19847 and by the Cornell Materials Science Center through the National Science Foundation Grant DMR-82-17227

REFERENCES

*Permanent Address: Department of Physics, Nagoya University, Chikusa-Ku 464, Japan

(1) M. Roger, J. H. Hetherington, and J. M. Delrieu, Rev. Mod. Phys. 55, 1 (1983).

(2) W. P. Kirk and E. D. Adams, Phys. Rev. Lett. 27, 392 (1971).

(3) J. M. Delrieu, M. Roger, and J. H. Hetherington, J. Low Temp. Phys. 40, 71 (1980).

(4) D. G. Wildes, M. R. Freeman, J. Saunders and R. C. Richardson, these proceedings.

LT-17 (Contributed Papers)
U. Eckern, A. Schmid, W. Weber, H. Wühl (eds)

PRESSURE MEASUREMENTS OF MAGNETICALLY ORDERED SOLID ^{3}He

K. UHLIG, E.D. ADAMS, G.E. HAAS,* R. ROSENBAUM,* Y. MORII,* and S.F. KRAL*

Department of Physics, University of Florida, Gainesville, Florida 32611 USA

The pressure of solid ^{3}He (v=24.15 cm^3/mole) has been studied as a function of magnetic field H and temperature T for 2 kOe < H < 6.425 kOe and 0.5 mK < T < 50 mK. We find a pressure discontinuity at the transition from the paramagnetic phase to the magnetically ordered high field phase, demonstrating that this is a first order phase transition. The discontinuity is ~ 0.5 that at the paramagnetic-to-antiferromagnetic phase boundary.

Three phases exist in solid ^{3}He at low temperatures, a paramagnetic phase (PP) for T>1.1 mK, an antiferromagnetic phase (AFP) for T<1.1 mK, and a high field phase (HFP) for H>4 kOe (Fig. 1).(1) The phase transitions from the PP to the AFP and from the HFP to the AFP are first order transitions (2,3); but there was still uncertainty about the order of the PP-HFP transition. We describe measurements of the pressure of solid ^{3}He (v=24.15 cm^3/mole) as a function of temperature at different magnetic

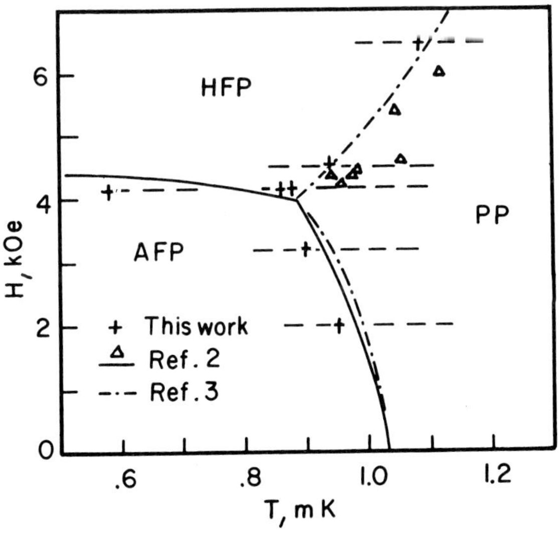

FIGURE 1
Magnetic phase diagram of solid ^{3}He at melting pressure (v=24.2 cm^3/mole), see Ref. 2. The phase transitions for our experiments (v=24.15 cm^3/mole) are shifted to slightly lower temperatures.

fields (see dashed lines in Fig. 1). From a pressure discontinuity at the PP-HFP transition we find that this is first order for the magnetic field range 3.98 < H < 6.425 kOe.

We used a nuclear demagnetization cryostat (0.6 mole of $PrNi_5$) to cool the ^{3}He sample to a final temperature of 0.5 mK. Temperatures were determined by means of pulsed NMR thermometry on ^{195}Pt wire calibrated against a ^{3}He-melting curve thermometer located in zero field.(4) The body of the cell, including the strain gauge, was made of coin silver. A layer of 700 Å silver powder was sintered on the face opposite the strain gauge. Because of the long thermal relaxation times in solid ^{3}He, only a small gap (~ 30 μm) was left for the ^{3}He sample between the silver sponge and the pressure transducer. The time constant for P(t) was ~ 3 hrs near 1 mK. Our warming rate caused by the 1 nW heat leak was ~ 1.2 μK/hr. Thus we calculate a temperature gradient within the ^{3}He of only 4 μK except during the phase transitions where it should be ~ 50 μK. A typical time spent in going through a transition was 100 hrs.

Figure 2 shows the pressure (uncorrected raw data) vs. 1/T for a field of 4.15 kOe. There are two pressure steps, the one at the lowest temperature occurs at the AFP→HFP transition, the other at the HFP→PP transition. These are spread out in temperature because of the time required to remove the latent heat. To show that the latter is actually a discontinuity in pressure, we kept the cell very near T_c for several days by doing small demagnetizations occasionally, as shown in the inset of Fig. 2. A small hysteresis is exhibited between the warming and cooling curves as a result of the temperature difference required to remove the latent heat. The data provide convincing evidence that the HFP→PP transition is first order.

*Present Addresses: GEH, Scientific Instruments, Inc., West Palm Beach, FL 33407; RR, Tel-Aviv University, Ramat-Aviv 69978, Israel; YM, Japanese Atomic Energy Research Inst., Tokai-mura, Japan; SFK, Empire State College, SUNY, Rochester, NY 14623.

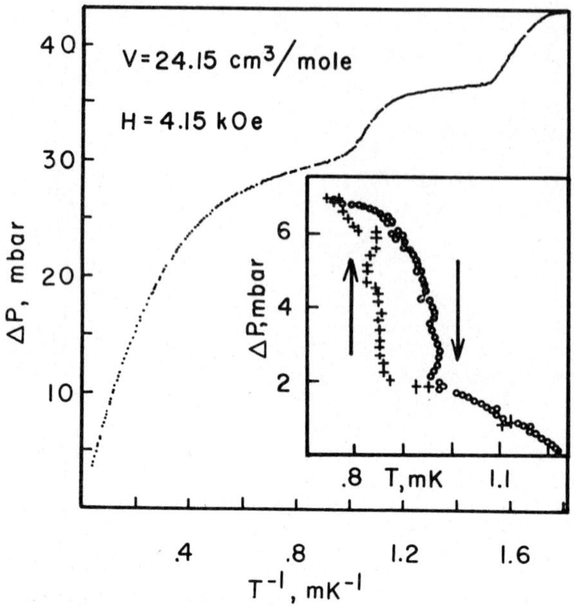

FIGURE 2
P versus T^{-1} diagram, for explanation see text. Inset: HFP→PP transition on warming and cooling.

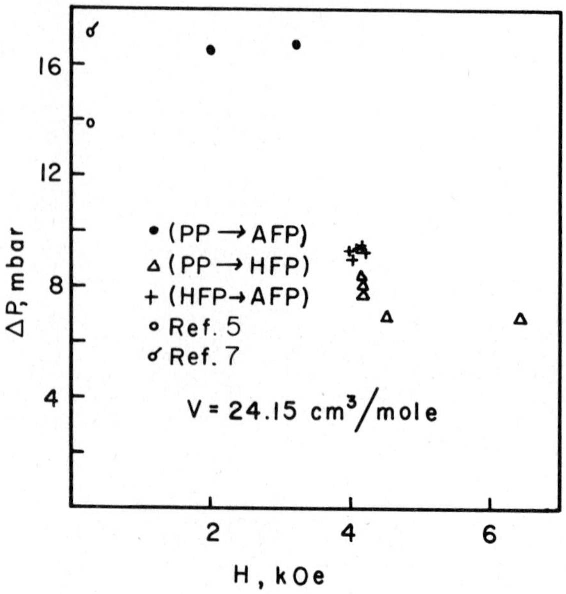

FIGURE 3
Pressure discontinuities versus magnetic field at three different phase transitions.

In Fig. 3 the pressure changes at various phase transitions are depicted versus the magnetic field. For $0 < H < 3.98$ kOe there is only one phase transition, AFP→PP (compare Fig. 1); for $3.98 < H \leq 4.15$ kOe we show ΔP at the AFP→HFP transition and for $3.98 < H < 6.425$ kOe at the transition HFP→PP. Theoretical predictions are that the first-order PP→HFP transition may end in a critical point at some high field.(1) We have made one measurement at 45.8 kOe, where the pressure change does not appear to be discontinuous.

The pressure changes which we have presented have been raw data, uncorrected for volume changes of the cell. The correction relating dP/dT to $(\partial P/\partial T)_V$ can be calculated, but with poor precision, from the mechanical properties of the cell. Instead, we have made the correction by using the high-temperature series expansion of P,

$$P = C_o + C_1 T^{-1} + C_2 T^{-2} + \ldots . \quad (1)$$

After fitting our data to this form, the correction factor was determined to be 1.56 by adjusting pressure changes so that C_1 agreed with other results (5,6). All our pressure changes in Fig. 2 should be multiplied by this correction factor 1.56.

The pressure changes ΔP at the phase transitions are related to the entropy discontinuity ΔS by a Clausius-Clapeyron equation

$$\Delta P = -\Delta S \, (dT_c/dv), \quad (2)$$

where dT_c/dv is the change in T_c with volume. For low fields, it has been found (7,8) that $vT_c^{-1}(dT_c/dv) \approx 17$. Assuming that this is valid near T = 1 mK at the fields involved here (and correcting ΔP for the volume change of the cell), we find ΔS(AFP→PP)=2.7 J/K•mole, ΔS(AFP→HFP)=1.5 J/K•mole, and ΔS(HFP→PP)= 1.3 J/K•mole.

We acknowledge useful conversations with Gary Ihas, Pradeep Kumar, and Neil Sullivan. Y.-H. Tang has assisted in taking some of the data. Support was provided by NSF, grant no. DMR-8312959. Partial support to one of us (KU) was provided by DFG, West Germany.

REFERENCES

(1) M. Roger, J.H. Hetherington and J.M. Delrieu, Rev. Mod. Phys. 55 (1983) 1.
(2) D.D. Osheroff, Physics 109 & 110 B+C (1982) 1461.
(3) R.B. Kummer, R.M. Mueller and E.D. Adams, J. Low Temp. Phys. 27 (1977) 319.
(4) W.P. Halperin, F.B. Rasmussen, C.N. Archie and R.C. Richardson, J. Low Temp. Phys. 31 (1978) 617.
(5) D. Wildes et al., private communication.
(6) M.F. Panczyk and E.D. Adams, Phys. Rev. 187 (1969) 321.
(7) T. Mamiya, A. Sawada, H. Fukuyama, Y. Hirao, K. Iwahashi and Y. Masuda, Phys. Rev. Lett. 47 (1981) 1304.
(8) T. Hata, S. Yamasaki, M. Taneda, T. Kodama and T. Shigi, Phys. Rev. Lett. 51 (1983) 1573.

LT-17 (Contributed Papers)
U. Eckern, A. Schmid, W. Weber, H. Wühl (eds)

CONSTANT-VOLUME PRESSURE MEASUREMENTS ON BCC SOLID ^{3}He BELOW 25 mK

D. G. WILDES, M. R. FREEMAN, J. SAUNDERS*, and R. C. RICHARDSON

Laboratory of Atomic and Solid State Physics, Cornell University, Ithaca, New York 14853 U.S.A.

We have measured the pressure of solid ^{3}He at molar volumes from 24.11 cm^3/mole to 23.18 cm^3/mole and for temperatures from 25 mK to 0.5 mK. The volume of our cell is regulated and constant. A simple rescaling of the pressure and temperature axes removes all of the volume dependence from our data.

We have measured the equation of state of solid ^{3}He in zero field at low temperatures. Our sample is a slab of bulk ^{3}He (< 75 ppm ^{4}He), 11 mm. dia. x 53 μm. thick, bounded by a sinter of 30 nm. copper powder (0.6 gm., 2 m^2/gm. and by a 1 mm. thick BeCu strain gauge diaphragm. The diaphragm position is sensed capacitively and regulated by a liquid ^{4}He pressure feedback system. As diagrammed in Fig. 1, changes in the capacitance are amplified by a regulator circuit and sent to a heater, to adjust the ^{4}He pressure and apply a restoring force to the back of the diaphragm. The diaphragm position depends principally on the differential pressure, so when the feedback loop is closed, changes in the ^{3}He pressure are matched equally by changes in the ^{4}He pressure, and the diaphragm does not move. The ^{4}He pressure is monitored by a separate strain gauge mounted nearby on the cryostat.

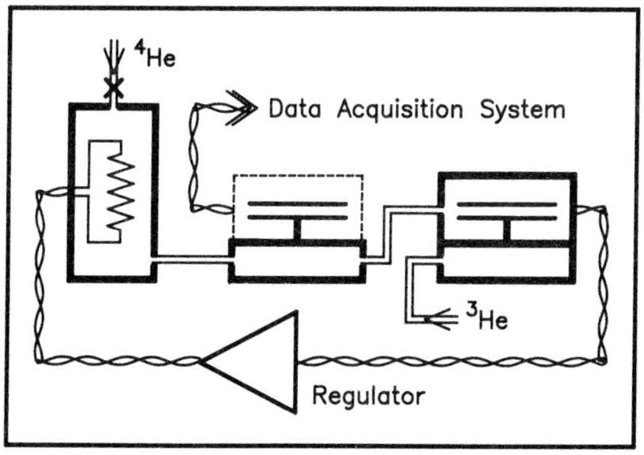

FIGURE 1
Schematic diagram of experimental cells. The ^{3}He and ^{4}He strain gauges are mounted on the demagnetization stage of the cryostat. The ^{4}He volume with heater and the fill line valve are immersed in the 4K bath.

Our temperature scale is based on the simultaneous comparison of several thermometers. Below 10 mK we use a SQUID magnetometer to measure the magnetic susceptibility of 95% La diluted CMN. Below 1.1 mK the LCMN is calibrated against the period of a torsional oscillator, which monitors the normal fraction density of superfluid ^{3}He at the saturated vapor pressure. The period vs. temperature behavior of this oscillator is well known from earlier measurements on this cryostat. (1) Above 1.1 mK the LCMN is calibrated against the ^{3}He melting curve, as measured by Halperin, et al. (2) Our melting curve scale is referenced to the fixed points at the superfluid A phase and solid ordering transitions, and is itself our most sensitive thermometer above 10 mK.

The raw data from the cryostat must be corrected to remove jumps due to intermittent shifts in the capacitance bridge electronics. The background temperature dependence of the reference capacitors must be subtracted. A fit is made to determine the "infinite temperature" pressure, P_0, from which the molar volume is calculated.

The minimum number of exchange parameters required to model the low temperature behavior of solid ^{3}He is not presently known. The model of Roger, Hetherington, and Delrieu (3) describes the interactions in bcc ^{3}He in terms of a mixture of three-spin exchange, J_t, planar four-spin exchange, K_p, and perhaps folded four-spin exchange, K_f. This model leads to a high temperature expansion for the pressure,

$$P-P_0=\frac{R}{8}\left[\frac{\partial\tilde{e}_2}{\partial V}\frac{1}{T}-\frac{1}{3}\frac{\partial\tilde{e}_3}{\partial V}\frac{1}{T^2}+\frac{1}{6}\frac{\partial\tilde{e}_4}{\partial V}\frac{1}{T^3}-\dots\right]$$

with $\tilde{e}_2$, $\tilde{e}_3$, and $\tilde{e}_4$ depending on different mixtures of the three- and four-spin exchanges.

If we assume a simple volume dependence,

$$J_t \propto V^{\gamma_{J_t}},\ K_p \propto V^{\gamma_{K_p}},\ \text{etc.}$$

*Permanent Address: Physics Department, University of Sussex, Brighton BN1 9QH, England

and, as suggested by Avilov and Iordansky, (4)

$$\gamma_{J_t} \simeq \gamma_{K_p} \simeq \gamma,$$

then

$$\tilde{e}_2 \propto V^{2\gamma},\ \tilde{e}_3 \propto V^{3\gamma},\ \tilde{e}_4 \propto V^{4\gamma}.$$

This suggests that the volume dependence of the pressure may be expressed as

$$P - P_0 \simeq \frac{R}{8}\left[\frac{\tilde{e}_2'}{T}\left(\frac{V}{24}\right)^{2\gamma-1} - \frac{\tilde{e}_3'}{3T^2}\left(\frac{V}{24}\right)^{3\gamma-1} + \frac{\tilde{e}_4'}{6T^3}\left(\frac{V}{24}\right)^{4\gamma-1}\right],$$

where $\tilde{e}_2'$ is the value of $\tilde{e}_2$ measured at V=24.0 cm^3/mole.

From fits to our measurements we estimate $\tilde{e}_2' \simeq 6.0$ (+/- 0.5) mbar · mk and $\tilde{e}_3' \simeq 9$ (+4/-3) mbar · mk^2. The uncertainties reflect principally the ambiguity of fitting to a truncated power series when the terms are of comparable magnitudes.

We have scaled our data according to:

$$P' = (P-P_0)/(V/24)^{\gamma-1},$$

$$T' = T/(V/24)^{\gamma},$$

with γ = 17 chosen by trial and error. Within the experimental uncertainties, all of our scaled data follow a single curve (Fig. 2).

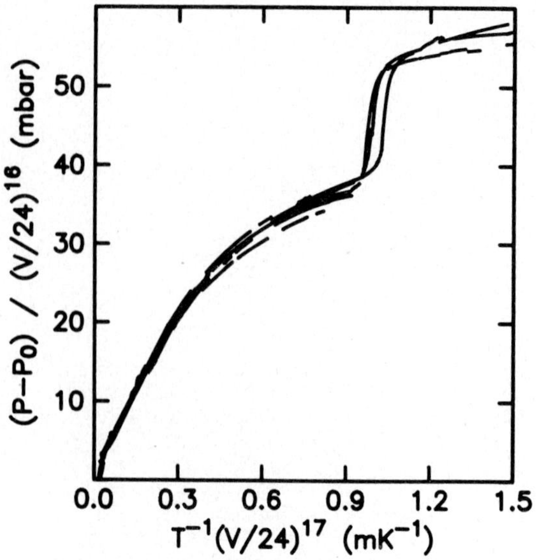

FIGURE 2
Equation of state of solid ^{3}He in zero magnetic field. Data for molar volumes 23.18 through 24.11 cm^3/mole, when scaled as indicated, follow a single curve.

ACKNOWLEDGEMENTS
We thank Dr. Takayoshi Mamiya for many useful discussions and suggestions during his sabbatic year at Cornell. This work was supported by NSF grant DMR-81-19847 and by the Cornell Materials Science Center through NSF grant DMR-82-17227.

REFERENCES
(1) J. Saunders, D.G. Wildes, J.M. Parpia, J.D. Reppy, and R.C. Richardson, Physica 108B, 791 (1981).
(2) W.P. Halperin, F.B. Rasmussen, C.N. Archie, and R.C. Richardson, J. Low Temp. Phys. 31, 617 (1978).
(3) M. Roger, J.H. Hetherington, and J.M. Delrieu, Rev. Mov. Phys. 55, 1 (1983).
(4) V.V. Avilov and S.V. Iordansky, J. Low Temp. Phys. 48, 241 (1982).

LT-17 (Contributed Papers)
U. Eckern, A. Schmid, W. Weber, H. Wühl (eds)
© Elsevier Science Publishers B.V., 1984

A QUANTUM SPIN-½ MODEL OF bcc ^{3}He

R. N. BHATT* and M. C. CROSS

AT&T Bell Laboratories, Murray Hill, NJ 07974, USA

We have performed an exact diagonalization of the exchange Hamiltonian of a finite, 16-spin model of bcc ^{3}He consisting of two interpenetrating cubes with periodic boundary conditions, for competing three-spin and planar four-spin exchange. Quantum effects are found to be much larger (upto 100% in ground state energy) than with pure Heisenberg exchange. However, the phase diagram has qualitative similarities with the mean field result. The comparison between theory and experiment is examined in light of the present results.

At temperatures below about 1 mK the nuclear spins of bcc ^{3}He are found to undergo a first order transition to a magnetically ordered phase which does not possess cubic symmetry (1,2). This observation led to the demolition of many proposed Hamiltonians for describing the nuclear spin degrees of freedom in the quantum solid, including the nearest-neighbour Heisenberg two-spin interaction. A considerable amount of analysis based on classical approximations of (intrinsically quantum) multiple-exchange Hamiltonians involving a small number of particles led Roger et al. (3) to propose the following Hamiltonian:

$$H = -J_t \sum_{ijk}(P_{ijk} + P_{ijk}^{-1} - \tfrac{1}{2}) + K_p \sum_{ijkl}(P_{ijkl} + P_{ijkl}^{-1} - \tfrac{1}{4}) \quad , \qquad (1)$$

where $P_{i_1 i_2 \ldots i_n}$ is the cyclic permutation operator for an n-spin ring composed of sites $i_1, i_2, \ldots, i_n$. The first term, the three-particle exchange is summed over all triangles in the lattice consisting of two nearest neighbours and one next-nearest neighbour. The four spin exchange is over all <u>planar</u> rings of four nearest neighbours. Fermi statistics implies $J_t, K_p > 0$, so that the triple exchange term is ferromagnetic and the four spin term is antiferromagnetic (4).

Roger et al. (3) showed that a classical analysis of Eq.(1) gave a sequence of ground states as a function of the ratio $K_p/J_t = R$. Starting from the ferromagnet at $R \gtrsim 0$, they obtained a ferrimagnet (a canted normal antiferromagnet, cnaf), and two antiferromagnetic phases, including the "uudd" phase deduced by Osheroff, Cross and Fisher (1) as the simplest structure consistent with the NMR results. The classical picture of the uudd structure consists of alternating pairs of (100) planes of spins pointing up and down, or its equivalent for the (010) or (001) planes. Further, a transition was predicted in finite magnetic field from the uudd phase to the cnaf with a finite moment when extrapolated to zero field, as seen in experiment. However, for values of J_t and K_p chosen to fit high temperature data (5), the transition field was found to be a factor of 3 to 4 larger than experiment. This (and other discrepancies) could conceivably be due to the classical approximation; indeed lowest order quantum corrections calculated by spin-wave analysis are found to be large ($\sim 70\%$).

As a first step towards investigating the nature of the ground state of the quantum problem, we have performed an exact numerical diagonalization of the Hamiltonian matrix of Eq. (1) for 16 spins on a bcc lattice, consisting of the unit cube of eight corner spins and that of the body centres, with periodic boundary conditions. This is the smallest size with cubic symmetry that allows for a uudd state and also does not have self-interaction of spins. By using the full lattice (translational and rotational) and spin symmetry of the system, the 65536 X 65536 matrix can be block diagonalized into blocks of size 53 X 53 or less, and rapid computation is possible. Prohibitively large computational time prevents such exact calculations for larger system sizes; consequently, we have little direct information on the magnitude of finite size effects. For the nearest neighbour Heisenberg Hamiltonian, however, the finite size ground state energy is much closer to the exact result than the classical approximation (6).

Fig 1 shows the energy of the ground state of the 16-spin system as a function of R in the absence of a magnetic field. We find a ferromagnetic ground state for $R < 1$, becoming unstable successively to a S=7 and then an S=6 state. At $R \simeq 1.45$, a singlet (S=0) state acquires the lowest eigenvalue, and the ground state remains a singlet for higher values of R. Also plotted in Fig 1 is the result for the classical (mean field)

* SERC Visiting Fellow, Department of Mathematics, Imperial College, London SW72BZ, UK.

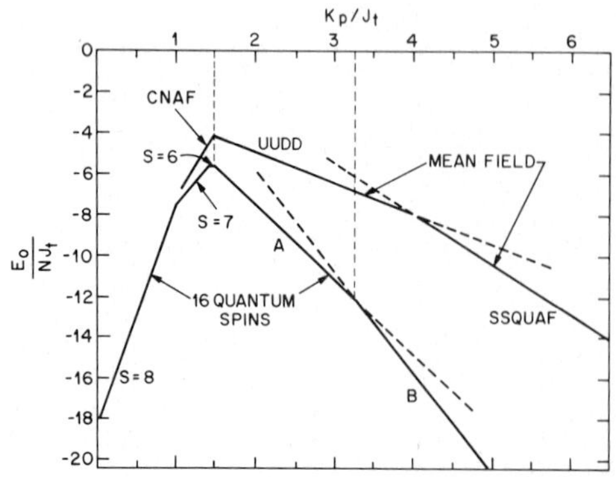

Fig 1: Ground state energy vs R for zero field.

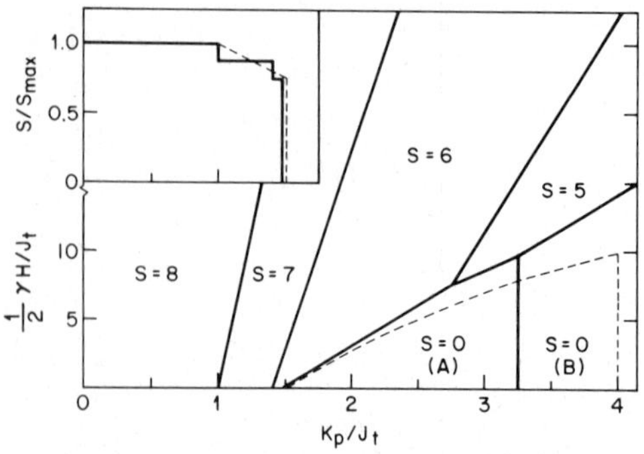

Fig 2: Ground state as function of R and field. Inset shows ground state spin vs R for H=0.

picture of the ground state. Results for $R < 1.5$ are quite similar, which is not unexpected since the classical picture of the ferromagnetic state is exact. The energy of the ferromagnetic state and the value of R for the transition out of it are given exactly for the infinite quantum system (if the transition is second order) both by the classical approximation and our 16-spin model.

As R increases beyond 3.3, our calculation suggests a transition to a different singlet ground state. (Actually, the levels do not cross as the two singlets have the same lattice symmetry; this small effect is not resolved in Fig 1). This is qualitatively consistent with the classical approximation where a transition to a second antiferromagnetic state, the simple square antiferromagnet (3) occurs, but at a higher R, R=4. The similarity can be substantiated by calculating the two spin correlation function in the ground state of our quantum system:

$$S(\bar{q}) = \langle \frac{4}{N^2} \sum_{ij} \bar{s}_i \cdot \bar{s}_j \, e^{i\bar{q}\cdot(\bar{r}_i - \bar{r}_j)} \rangle \qquad (2)$$

where $\bar{s}_i$ is the spin at site $\bar{r}_i$, and $\bar{q}$ is a wavevector consistent with our periodic boundary conditions. We find (6) the largest $S(\bar{q})$ corresponds to $\bar{q}=(2\pi/a)(100)$ for $R<3.3$ and $\bar{q}=(2\pi/a)(110)$ for $R>3.3$, with a rapid changeover, in accord with the classical picture of the transition. However, it should be remarked that the quantum energies are considerably below the classical one (by as much as a factor of 2) in the entire region $R>1.5$ which is of experimental relevance.

The finite field phase diagram for our system is shown in Fig 2. Of particular interest is the finite field transition from the uudd singlet (A) to the ferrimagnetic state. As can be seen, the transition field for our system is larger than the classical estimate (dashed), indicating that quantum fluctuations are not likely to rectify the discrepancy between experiment and theoretical value for the field obtained from fitting high T data with Eq. (1). This should be checked further by means of spin wave calculations.

We can use the eigenvalue spectrum to calculate any "thermodynamic" property of our 16-spin system. The most interesting result is that the specific heat shows two peaks for values of R when singlet A is the ground state (6). It is tempting to associate one with the latent heat at T_c, and the other with the peak seen above T_c in experiments, though we cannot rule out spurious features of the undeniably small size. However, a system with a strongly first order transition could possibly show some structure in the thermodynamic properties even on the small scale of our 16-spin sample.

In conclusion, we have found by an exact diagonalization of a finite 16-spin quantum system with three and four spin exchange, that quantum effects in the antiferromagnetic region are much larger than with two spin exchange. Despite this, the ground state phase diagram is qualitatively remarkably similar to the classical mean field one. Further, in bcc ^3He, the lack of change of transition field with quantum fluctuations may well imply a much stronger ferromagnetic tendency at low temperatures than is contained in the three spin-planar four spin Hamiltonian.

ACKNOWLEDGMENTS
Valuable discussions with D. S. Fisher and E. I. Blount, and computational help of W. Peterson are gratefully acknowledged.

REFERENCES
(1) D. D. Osheroff, M. C. Cross and D. S. Fisher, Phys. Rev. Lett. 44 (1980) 792; D. D. Osheroff, Physica 109&110B (1982) 1461.
(2) E. D. Adams, E. A. Schubert, G. E. Haas and D. M. Bakalyar, Phys. Rev. Lett. 44 (1980) 789.
(3) M. Roger, J. M. Delrieu and J. H. Hetherington, Phys. Rev. Lett. 45 (1980) 137; M. Roger, J. H. Hetherington and J. M. Delrieu, Rev. Mod. Phys. 55 (1983) 1.
(4) D. J. Thouless, Proc. Phys. Soc. 86 (1965) 893.
(5) C. T. Van Degrift, W. J. Bowers, P. B. Pipes and D. F. McQueeney, Phys. Rev. Lett. 49 (1982) 149; W. P. Kirk, Z. Olejniczak, P. Kobiela, A. A. V. Gibson and A. Gzermak, Phys. Rev. Lett. 51 (1983) 2128.
(6) M. C. Cross and R. N. Bhatt, to be published.

LT-17 (Contributed Papers)
U. Eckern, A. Schmid, W. Weber, H. Wühl (eds)

QUANTUM DESORPTION OF ^{3}He ATOMS FROM NEGATIVE IONS IN HeII[†]

G.G.Nancolas[*], R.M.Bowley[§] and P.V.E.McClintock[*].

Measurements have been made of the rate ν at which negative ions nucleate vortex rings in ultra-dilute ^{3}He/^{4}He solutions. It is deduced that, in strong electric fields, the average number of bound ^{3}He per ion is controlled by a new type of quantum desorption process; and that $\nu_0 \simeq 10^{-3}\ \nu_1 \simeq 10^{-6}\ \nu_2$, where ν_0, ν_1 and ν_2 are respectively the rates characteristic of ions with zero, one, or two bound ^{3}He atoms.

The rate ν at which negative ions nucleate quantised vortex rings in isotopically pure HeII (1) can be increased to a quite remarkable extent (2) by the addition of minute traces of ^{3}He, even when the ^{3}He/^{4}He ratio of the resultant solution, x_3, is no larger than the ca 2 x 10^{-7} of commercial well helium. Not only is the phenomenon of considerable intrinsic interest in its own right, but it would also appear to carry important implications for the theories (3,4) that are being developed to describe the microscopic mechanism through which vortices are created in the superfluid. It is therefore deserving of careful study. We present here a preliminary report of a detailed investigation of the ^{3}He effect in which we have measured ν for a fixed pressure of 23 bar, as a function of electric field, E, and temperature, T, for eight ultra-dilute solutions in the range $0 \leqslant x_3 \leqslant 1.7 \times 10^{-7}$. Full experimental details, a complete set of data and a discussion of its analysis will be presented elsewhere.

A small selection of our data, chosen to illustrate the general behaviour of $\nu(x_3)$, is shown in Figure 1. At the higher field, ν is a linear function of x_3 at both temperatures; whereas, for the lower field, $\nu(x_3)$ is highly non-linear, curving steeply upwards. This somewhat surprising behaviour is particularly noticeable at low temperatures. The data can be fitted (solid lines) by the power series

$$\nu(E,T) = \nu_0(E,T) + \nu'(E,T)\, x_3 + \nu''(E,T)\, x_3^2 \quad (1)$$

where, at least for the range of parameters currently employed, it is unnecessary to consider terms beyond x_3^2. The first term, ν_0, is simply the nucleation rate measured in pure ^{4}He, while the linear and quadratic terms represent additional contributions to the nucleation rate arising from the ^{3}He.

Despite encouraging developments in the theory (3,4), a detailed understanding of the vortex creation mechanism has yet to be achieved. We believe, however, that the mechanism responsible for the potent effect of ^{3}He on ν can be understood in terms of the trapping of ^{3}He atoms on the free surface provided by the bubble-like structure of the negative ion. This idea, originally proposed by Dahm (5) and supported by the mobility measurements of Kuchnir et al (6), has been developed by Shikin (7), who suggested that the ^{3}He atom becomes trapped in a spherical potential well with quantised energy levels. On this basis we have developed a model in which the average number of bound ^{3}He atoms per ion n_b, is determined by the competing processes of ^{3}He absorption and emission. At low ionic velocities (i.e. low E), these processes involve the corresponding emission (absorption) of a phonon, which carries off (provides) the binding energy; and n_b is consequently highly temperature-dependent. At higher ionic velocities it is also necessary to consider a novel velocity-dependent process whereby a ^{3}He atom is emitted together with a pair or rotons, above a critical velocity

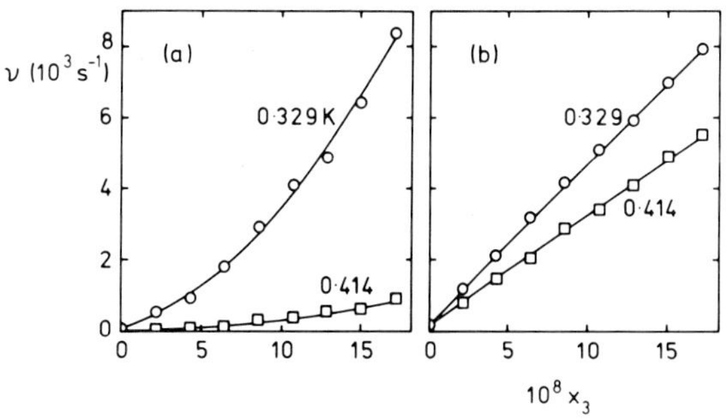

FIGURE 1
The measured nucleation rate ν as a function of x_3 for two temperatures and at electric fields of (a) 1.27 x 10^4 V m^{-1}, and (b) 9.5 x 10^4 V m^{-1}. The solid lines represent fits to equation (1).

† Work supported by the Science and Engineering Research Council, UK.

* Department of Physics, University of Lancaster, Lancaster, LA1 4YB, UK.

§ Department of Physics, University of Nottingham, Nottingham, NG7 2RD, UK.

determined by energy and momentum conservation. With the aid of this model we have been able to account for the form of $\nu(E,T,x_3)$ over the complete parameter range.

Returning to equation (1), we can see that ν' represents the nucleation rate (per unit concentration) for ions with one bound ^{3}He atom, in the limit $x_3 \to 0$. A detailed analysis (8) of $\nu'(E,T)$ implies that the effect of a single trapped ^{3}He atom is to decrease the critical velocity for vortex nucleation by ca 4 m s^{-1}, while simultaneously increasing the rate constant by a factor ca 10^3. The physical significance of ν'' is not immediately apparent, but it can be shown that its magnitude is determined principally by the proportion of ions that have *two* bound ^{3}He atoms. Values of $\nu''(E)$ for three temperatures are shown in Figure 2. On the basis of the model outlined above, the behaviour of $\nu''(E,T)$ can be understood in the following terms: at low fields n_b is determined by the phonon-related (thermal equilibrium) processes and consequently n_b, and hence also ν'', are independent of E but increase with decreasing temperature; as E increases ions are able to shed their bound ^{3}He atoms through the roton-related desorption process and hence ν'' decreases, becoming independent of T, and eventually falling to zero. The

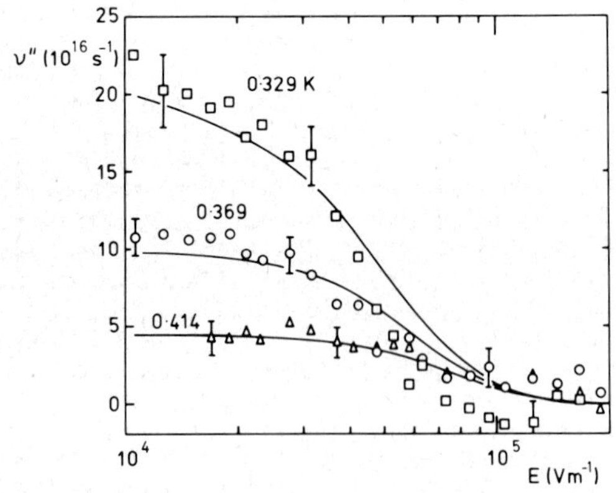

FIGURE 2
Values of the coefficient ν'' in equation (1), plotted as a function of electric field E for three temperatures. The solid lines represent fits to the theory discussed in the text.

solid lines of Figure 2 represent fits of the theory. It can be seen that agreement with the data is quite good except for relatively strong fields at the lowest temperature where, in reality, $\nu''(E)$ falls more rapidly than the theoretical curve. One of the adjustable parameters used in fitting the model to the data is the nucleation rate ν_2, characterising ions with two bound ^{3}He atoms. Although the scatter in the data is too large to permit accurate values of $\nu_2(E,T)$ to be extracted, we can estimate that ν_2 is of order 3×10^7 s^{-1}. Our fit to the data (Figure 2) entirely ignores the (likely) variation of ν_2 with E; which very probably accounts for the poor fit obtained at low temperatures.

It is of particular interest to note that ν_2 is larger by a factor of order 10^3 than ν_1, the nucleation rate appropriate to ions with single bound ^{3}He atoms, which is itself ca 10^3 larger than ν_0. This observation appears to rule out any suggestion that a bound ^{3}He atom acts merely as a localised site for the nucleation of vorticity since, if that were the case, we should have expected to find $\nu_2 = 2\nu_1$, rather than the ca 10^3 ν that we have in fact deduced.

It is our hope that these results will help to stimulate a fresh, sustained and ultimately successful theoretical attack on the hitherto intractable problem of how quantum vortices are created in HeII.

We acknowledge gratefully the many valuable discussions that we have enjoyed with R.J.Donnelly, C.M.Muirhead, P.C.E.Stamp and W.F.Vinen.

REFERENCES

(1) R.M.Bowley, P.V.E.McClintock, F.E.Moss, G.G.Nancolas and P.C.E.Stamp, Phil.Trans.R. Soc.Lond.A 307 (1982) 201.
(2) P.V.E.McClintock, F.E.Moss, G.G.Nancolas and P.C.E.Stamp, Physica 107B (1981) 573.
(3) R.M.Bowley, J.Phys.C : Solid St.Phys.17 (1984) 595.
(4) C.M.Muirhead, W.F.Vinen and R.J.Donnelly, Bull.Am.Phys.Soc.28 (1983) 676; and Phil. Trans.R.Soc.Lond., in print.
(5) A.J.Dahm, Phys.Rev.180 (1969) 259.
(6) M.Kuchnir, J.B.Ketterson and P.R.Roach, Phys.Rev.A6 (1972) 341.
(7) V.B.Shikin, Zh.Eksp.Teor.Fiz.64 (1973) 1414; translated in Soviet Phys.J.E.T.P.37 (1973) 718.
(8) R.M.Bowley, G.G.Nancolas and P.V.E.McClintock, Phys.Rev.Lett.52 (1984) 659.

LT-17 (Contributed Papers)
U. Eckern, A. Schmid, W. Weber, H. Wühl (eds)

ANOMALOUS BEHAVIOUR OF NEGATIVE IONS IN HeII*

T.ELLIS and P.V.E.McCLINTOCK

Department of Physics, University of Lancaster, Lancaster, LA1 4YB, UK.

When negative ions move through HeII under roton-emission-limited conditions in very weak electric fields: (a) their drift velocity falls below that predicted by the $E^{1/3}$ law followed accurately for $E > 400$ V m^{-1}; (b) their effective mass apparently rises with decreasing E. The physical origins of these unexpected phenomena are not understood.

1. INTRODUCTION

Isotopically pure HeII at very low temperatures provides a "mechanical vacuum" in which the drag on a moving object remains negligible, provided that a critical velocity is not exceeded. When the liquid is pressurised, it is found that the drift velocities $\bar{v}$ of negative ions can be raised above the Landau critical velocity v_L for roton creation, and it has been possible (1) to deduce precise numerical values of v_L from experimental measurements of $\bar{v}$ as a function of the applied electric field E. Studies of the inertial effects that occur when E is transiently reversed have permitted the effective mass m* of the negative ion to be determined (2). It has also been possible to follow the pressure dependence of these quantities from 25 bar down to 13 bar in the case of v_L (1), and down to 11 bar in the case of m* (3).

The measurements of v_L and m* were based on data for which $E \gtrsim 10^2$ V m^{-1}. The purpose of this paper is to report that, through improvements in our experimental technique (1), we have been able to extend our measurements to considerably lower values of E. In doing so, we have observed some quite unexpected field dependences of both $\bar{v}$ and m*, and it is these that we discuss below.

2. APPARATUS

The experimental cell has already been described (4,5). Provision is made for negative ions to be injected into a ca 1.5 ℓ sample of isotopically pure ^{4}He by applying a transient high voltage pulse to an array of field emission tips. The ions pass through three grids and then enter the region of uniform electric field across which their transit time is to be measured. Finally, they pass through a Frisch screening grid and are detected as they approach a collecting electrode. With the exception of the field emitters, all grids, field-homogenisers and other electrodes are gold-plated. The cell is cooled to ca 80 mK in a simple dilution refrigerator.

---

* Work supported by the Science and Engineering Research Council (UK)

3. DRIFT VELOCITIES IN VERY WEAK FIELDS

A number of improvements in our experimental technique (1) have enabled us to make precise measurements of $\bar{v}(E)$ down to $E \simeq 30$ V m^{-1}. A typical set of data, including this low field range, is shown in Figure 1 where $\bar{v}$ is plotted against $E^{1/3}$ so as to aid comparison with the equation

$$\bar{v} = v_L + AE^{1/3} \quad (1)$$

where A is a constant. It may be noted that equation (1), which is characteristic of roton pair emission (6), is followed accurately for $E \gtrsim 400$ V m^{-1}, enabling values of v_L to be deduced (1) by measurement of the intercept. For lower values of E, however, small but quite definite deviations from equation (1) become apparent. Their magnitude remains independent of the temperature of the cell provided that this is kept below ca 0.15 K, so that the effect is apparently not due to the scattering of thermal excitations. We have also established that the deviations are unaffected either by changes in

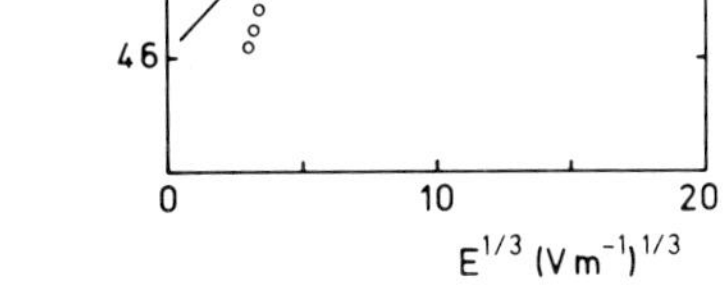

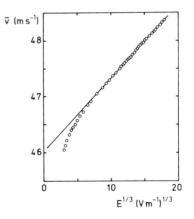

FIGURE 1
The drift velocity $\bar{v}$ of negative ions in HeII at 25 bar, 80 mK, plotted as a function of (electric field, E)$^{1/3}$.

the signal repetition rate or by the absolute magnitude of the signal. The form of the deviations, which can be fitted fairly well by adding to equation (1) an additional term in E^{-1}, suggests that they do not arise from the onset of single-roton emission (7).

4. THE EFFECTIVE MASS IN WEAK ELECTRIC FIELDS

From measurements of the delay τ in the ionic arrival time at the collector, caused by transient reversals of the electric field in the drift space, it is possible (2) to deduce the effective mass of the ions. Provided that the duration of the field reversal $\Delta t \leqslant \Delta t_c$, where $\Delta t_c = 2v_L m^*/eE$, the net delay is

$$\tau = (NeE/m^* \, v_L)\Delta t^2 \qquad (2)$$

where N is the number of field reversals that occur during the transit of the ions.

Extension of this technique to weak electric fields yields values of m* that apparently increase with decreasing E, as shown in Figure 2. These results should, however, be approached with considerable caution. This is because we have also found that τ falls significantly below the value of $2N\,\Delta t$ that is to be expected (2) when $\Delta t \geqslant \Delta t_c$, which would seem to imply that the dynamics of the ions are no longer identical with those of free particles in this regime. We have been unable to establish any significant correlation between the apparent m* and the signal magnitude, signal repetition rate or temperature of the cell.

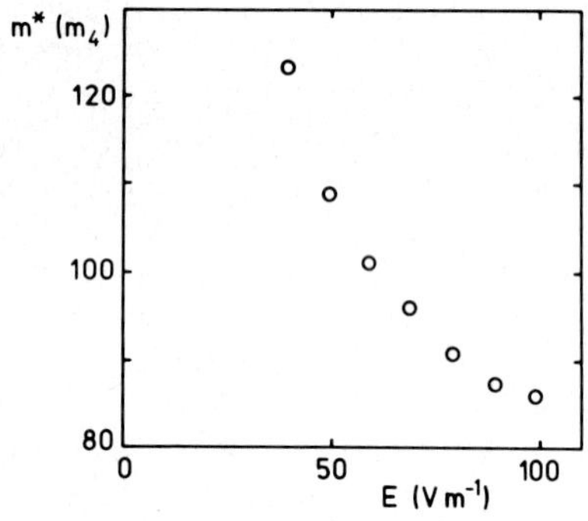

FIGURE 2
The apparent effective mass m* of negative ions in HeII at 25 bar, 80 mK, plotted as a function of the electric field E.

5. DISCUSSION

It seems quite probable *a priori* that the anomalies in $\bar{v}$ and m* may arise from some common cause, given that they both occur within the same range of E, but it is by no means obvious what that cause might be. Possible explanations include:

(i) Interactions between ions and the cloud of rotons that they themselves have created. These seem implausible, however, given that the anomalies are apparently independent of signal magnitude.

(ii) A form of inelastic scattering between the ions and a "background" of metastable remanent vortex lines (8) pinned within the drift space. This possibility is difficult to evaluate in a quantitative way because the ion/vortex interaction in the absence of damping does not appear to have been examined theoretically.

(iii) Changes in the self-energy of the ion (9) when it accelerates very slowly through v_L, conceivably leading to significant departures from free-particle dynamics in weak enough electric fields.

Further work, both experimental and theoretical, is needed to establish which of these explanations, if any, is the correct one.

ACKNOWLEDGEMENTS

It is a pleasure to acknowledge numerous stimulating discussions with R.M.Bowley and the continuing expert technical assistance of I.E.Miller.

REFERENCES

(1) T.Ellis and P.V.E.McClintock, Pressure dependence of the Landau critical velocity in HeII: a progress report, this volume.
(2) T.Ellis and P.V.E.McClintock, Phys.Rev.Lett. 48 (1982) 1834.
(3) T.Ellis, P.V.E.McClintock and R.M.Bowley, J.Phys.C: Solid St.Phys.16 (1983) L485.
(4) T.Ellis, C.I.Jewell and P.V.E.McClintock, Phys.Lett.78A (1980) 358.
(5) T.Ellis and P.V.E.McClintock, Physica 107B (1981) 569.
(6) D.R.Allum, P.V.E.McClintock, A.Phillips and R.M.Bowley, Phil.Trans.R.Soc.Lond. A 284 (1977) 179.
(7) F.W.Sheard and R.M.Bowley, Phys.Rev.B 17 (1978) 201.
(8) D.D.Awschalom and K.W.Schwarz, Phys.Rev.Lett. 52 (1984) 49.
(9) S.V.Iordanskii, Zh.eksp.teor.Fiz. 54 (1968) 1479; trans. in Soviet Phys. J.E.T.P. 27 (1968) 793.

LT-17 (Contributed Papers)
U. Eckern, A. Schmid, W. Weber, H. Wühl (eds)

FIELD-DEPENDENT ION TRAPPING ON REMANENT VORTEX LINES

D.D. Awschalom and K.W. Schwarz

IBM Thomas J. Watson Research Center, Yorktown Heights, NY 10598, USA

The ion current trapped by remanent vortex lines in superfluid ^{4}He has been measured as a function of injected current and electric field. The observed ion current initially increases linearly with the injected current and then saturates. The field dependence indicates that the observed signal is determined primarily by the strength of the field pulling the trapped ions along the vortex lines.

A recent experiment by the present authors {1} has confirmed the early conjectures of Feynman {2} and Vinen {3} that helium at rest contains metastably pinned quantized vortex lines. In this experiment (Fig. 1), two parallel plates spaced 1 cm apart were immersed in the superfluid with the aim of detecting any vortex lines pinned between them. Ions produced by the source S follow the electric field lines to the side electrode. Any ions trapped by the vortices, however, are led over to the collector C and detected by an electrometer. The lifetime edge for negative-ion trapping provides the necessary discrimination between the vortex-trapped ion current and any stray leakage currents that may be present. Comparison of the observed signal against ion-trapping measurements done in rotating helium {4,5} leads to the conclusion that there are on the order of 15 lines/cm^2 permanently pinned between the plates. This remanent vortex line density appears to be independent of the liquid's prior history, and is consistent with a rough theoretical estimate {1} of the maximum density which can exist in metastable equilibrium.

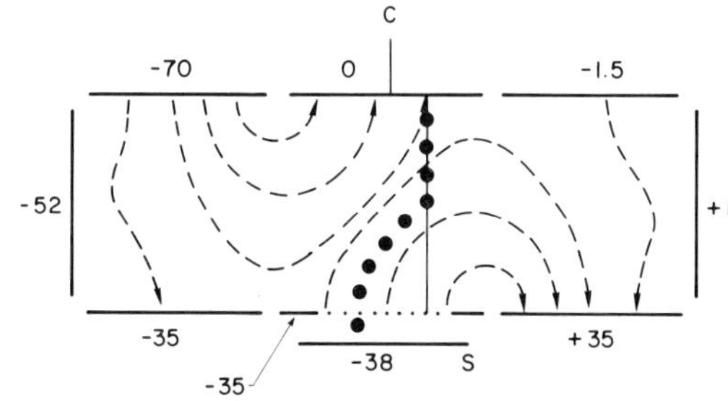

FIGURE 1. Cross-sectional view of the cell geometry showing normal biases. The distance between the top and bottom is 1 cm. The cell extends 6 cm into the plane of the figure. The collector sits in the middle of the top plate, and has an area of 2 cm^2. The negative field lines are shown as dashed lines. Dots show an ion trapped by a vortex pinned on the collector.

The analog of our experiment, in which the lines are saturated by a relatively high steady current and the trapped charges are bled off by a component of the electric field along the lines, has not been performed in rotating helium. Hence our estimate of the remanent line density should be regarded as very approximate. In order to obtain further insight, we have studied the trapped-ion signal I_t as a function of the injected current I_s and the nominal electric field E_n. The injected current can be varied by adjusting the voltage between S and the grid, while the strength of the electric field can be varied without changing its configuration by scaling the eight electrode potentials together. Data taken by varying these two quantities are shown in Fig. 2. These results were obtained after the superfluid had been at rest for several days. A similar study was done for the situation in which a high vortex line density is maintained between the plates by an ultrasonic transducer. Although the signal observed at C was of course much larger, it showed virtually the same dependence on I_s and the field. Thus the behavior in Fig. 2 appears to be characteristic of the ion-vortex trapping process in the present configuration.

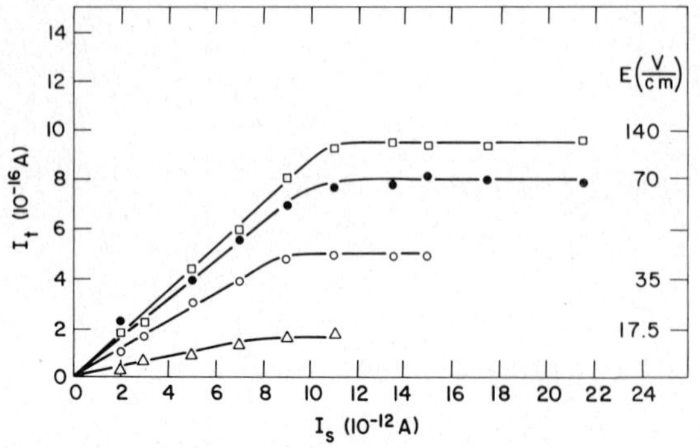

FIGURE 2. The electric field dependence of the vortex-trapped ion current I_t as a function of injected current I_s . The nominal electric field of 35 V/cm corresponds to the potentials shown in Fig. 1.

The observed signal is seen to increase with I_s, eventually saturating. This is qualitatively reasonable, although a quantitative treatment will have to deal with the fact that at the high current levels used in our experiment, space-charge limiting is expected to be important. Thus, as I_s increases one may visualize the trapping as taking place primarily in a growing region in front of the source grid where $E<<E_n$. As to the field dependence, one would normally expect an increase in E_n at fixed I_s to result in a lower charge density near the lines and hence a decrease of trapped current. This effect will be greatly reduced if the trapping takes place in a space-charge limited region. The fact that I_t is observed to increase strongly with E_n, however, leads us to conclude that I_t is primarily limited by the rate at which the field can clear the ions off the lines once they are trapped.

Quantitative interpretation of Fig. 2 is made difficult by the complicated field geometry, the high degree of space-charge limiting, and the lack of analogous experiments in rotating helium. To obtain a better estimate of the remanent vortex line density from Fig. 2 may require considerable analysis. It will therefore be important to repeat the measurements of Fig. 2 on a rotating table, thus obtaining a direct calibration our results against a known density of quantized vortices.

REFERENCES

{1} D.D. Awschalom and K.W. Schwarz, Phys. Rev. Lett. 52 (1984) 49.
{2} R.P. Feynman, Application of Quantum Mechanics to Liquid Helium , in: Progress in Low Temperature Physics, Vol. 1, ed. C.J. Gorter (North-Holland, Amsterdam, 1957) p. 17.
{3} W.F. Vinen, Critical Velocities in Liquid Helium II, in: Liquid Helium, Vol. 21, ed. G. Careri (Academic, New York, 1963) p. 336.
{4} D.J. Tanner, Phys. Rev. 152 (1966) 121.
{5} W.I. Glaberson, J. Low Temp. Phys. 1 (1969) 289.

LT-17 (Contributed Papers)
U. Eckern, A. Schmid, W. Weber, H. Wühl (eds)

STOCHASTIC NOISE OF AN ELECTRON CURRENT IN TURBULENT SUPERFLUID HELIUM-4

Charles W. SMITH

Department of Physics and Astronomy, University of Maine, Orono, ME 04469, U.S.A.*

The spectral density of the fluctuations of an electron current through a region of vortex-line turbulence in superfluid helium-4 shows a power law dependence upon frequency over several decades with slope - 4.0 ± 0.1. A simple conduction mechanism for charge trapped on vortex lines is proposed. The observed power law exponent follows from this conduction mechanism, recent microscopic theoretical ideas concerning specific statistical features of vortex-line turbulence and the general nature of the fluctuations in stochastically driven systems.

1. INTRODUCTION AND EXPERIMENTAL METHODS

We have measured the power spectral density of the fluctuations of an electron current through a region of vortex-line turbulence maintained by a constant thermal counterflow. Our ion probe employed a one Curie tritium source (a beta emitter) thermally bonded to a plane heater making up the closed end of a Lucite channel, 8.0 centimeters long and 3.0 centimeters in diameter. The ion drift velocity and the heat flux were directed toward the open end of the channel. The ion collector, an 85% open copper screen, was located half way down the channel. Appropriate guarding and field rings assured electric field uniformity in the channel. Time series records of the collector current were recorded at 14 bit resolution for intervals up to one hour. These records were Fourier transformed using a Hanning window and appropriate Nyquist sampling theorem constraints (1).

2. RESULTS AND DISCUSSION

The magnitude of the power spectral density of the current fluctuations was computed and plotted as a function of frequency. Figure 1 shows a log-log presentation of representative data taken at T = 1.57 K for a drift space field of 26.7 volts per centimeter, for three different heat flux values (corresponding to three different vortex-line density values) (2). The high frequency asymptotic behavior of the power spectral density is f^{α} where $\alpha = -4.0 \pm 0.1$ over several orders of magnitude. This power law was observed for the entire range of heat flux examined, from the onset of vortex-line production at 1.57 K for this channel at 1.2 milliwatts per square centimeter up to our temperature control limits at 20 milliwatts per square centimeter and over the entire range of drift space field, up to 100 volts per centimeter.

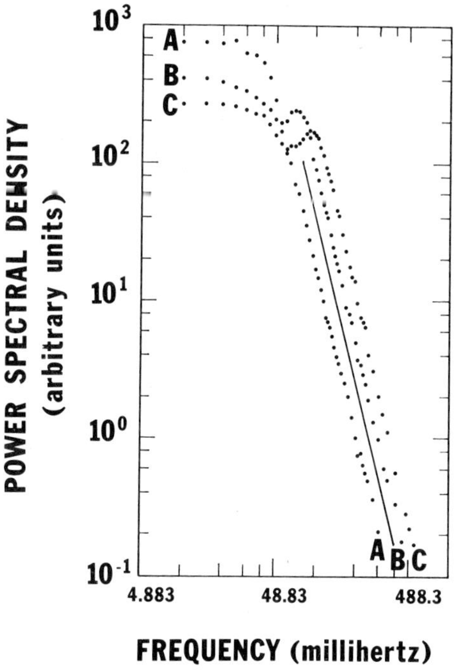

FIGURE 1
The power spectral density as a function of frequency at T = 1.57 K and drift space field 26.7 volts per centimeter is plotted parametrically with heat flux, A = 3.00, B = 6.75 and C = 12.0 milliwatts per square centimeter. A line with slope -4 is shown as reference.

*This work was supported in part by the National Science Foundation, DMR-8005358 in the Low Temperature Physics Program.

The system intrinsic noise was measured using positive ions at T = 1.57 K, which do not trap on vortex-line at that temperature and using electrons at T = 1.80 K, above the trapping life-time edge. The system noise was found to be several orders of magnitude below the signals shown in Figure 1 and not to obey a simple power law in frequency (3).

At fixed vortex-line density (fixed heat flux) we observe the power spectral density scales in a self-similar way with the mobility and the current-voltage curve of the ion probe. At fixed spectral density, the frequency of the fluctuations increases with heat flux. This is expected since the equilibrium vortex-line density increases with heat flux (2) and the average self-induced velocity of the vortex lines increases with vortex-line density (4).

We view the motion of a charge trapped on the moving tangle as a mass particle in a viscous medium undergoing a biased Brownian motion toward the collector. This process can be described by a nonlinear second order Langevin equation

$$\dot{\vec{V}} = -\gamma\vec{V} + \vec{F}(x) + \vec{f}(t) \qquad (1)$$

where $\dot{\vec{X}} = \vec{V}$ is the velocity of the charge on the vortex line, $\vec{f}(t)$ represents the influence of the stochastic dynamics on the tangle, $\vec{F}(x)$ represents the drift force and the gradients of the trapping potential and γ includes the nonconservative drag forces. It is only recently that this problem has been solved in general for the high frequency correlations (5). The appropriate Fokker-Planck operator leads to the asymptotic behavior of the power spectral density. For equation 1 we obtain an expression for the power spectral density, P(f) given by

$$P(f) = \frac{a}{f^4} + \frac{b}{f^6} + \ldots \qquad (2)$$

where a and b are independent of frequency. Equation 2 describes the high frequency power law behavior illustrated in Figure 1. From its derivation we expect that the power law exponent would be independent of drift space field, mobility and vortex-line density, as we in fact observe. Furthermore, the only restraint that this description places on the model for the vortex-line tangle is that the mechanisms which assure the randomness of the tangle must include stochastic processes. This is now thought to be the case. Independent measurements have shown (6) that the topological reconnections of vortex lines at line-line crossing events, which are essential to our microscopic understanding of the vortex-line tangle steady state and decay, are stochastic randomizing processes.

3. SUMMARY AND CONCLUSION

We observe a power law in frequency for the high frequency asymptotic behavior of the power spectral density of the noise of an electron current through a region of vortex-line turbulence maintained by thermal counterflow. The power law exponent was measured to be -4.0 ± 0.1 over several orders of magnitude in spectral density. This exponent is observed to be independent of drift space field, vortex-line density and mobility. A simple conduction mechanism for charge trapped on vortex line is proposed. This mechanism, together with our current understanding of the microscopic theoretical ideas about specific statistical features of the vortex-line tangle and the general nature of fluctuations in stochastically driven systems, leads to a power law exponent of -4.0. In addition, these results are compatible with two independent control experiments which were employed to measure the intrinsic system noise by eliminating the charge trapping process on the vortex lines.

ACKNOWLEDGEMENTS

I would like to thank M.J. Tejwani and J.A. Finkelstine for assistance in this study.

REFERENCES

(1) R.K. Otnes and L. Enochson, Digital Time Series Analysis (John Wiley and Sons, New York, 1972).
(2) W.F. Vinen, Proc. Roy. Soc. London, Ser. A 240, 114, 128 (1957), 242, 493 (1957) and 243, 400 (1958).
(3) C.W. Smith and M.J. Tejwani, in print.
(4) K.W. Schwarz, Phys. Rev. Lett. 38, 551 (1977) and Phys. Rev. B 18, 245 (1978).
(5) B. Caroli, C. Caroli and B. Roulet, Physica 112A, 517 (1982).
(6) F.P. Milliken, K.W. Schwarz and C.W. Smith, Phys. Rev. Lett. 48, 1204 (1982).

LT-17 (Contributed Papers)
U. Eckern, A. Schmid, W. Weber, H. Wühl (eds)

THE THEORY OF THE NUCLEATION OF VORTICITY BY NEGATIVE IONS IN VERY DILUTE SOLUTIONS OF ^{3}HE IN SUPERFLUID ^{4}HE

C.M. MUIRHEAD, W.F. VINEN AND R.J. DONNELLY*

Department of Physics, University of Birmingham, Birmingham B15 2TT, UK.

Recent theoretical work by the present authors on vortex nucleation by a moving negative ion in superfluid ^{4}He is extended to deal with nucleation in the presence of very small concentrations of ^{3}He. The results are in at least qualitative agreement with the recent experimental results of Bowley *et al*, which show that a single ^{3}He atom adsorbed onto the surface of the negative-ion bubble reduces the critical velocity for vortex nucleation by about 4 m s^{-1} and increases the nucleation rate by a factor of about 10^3.

1. INTRODUCTION

In a recent paper (2) we have developed a theory of the nucleation of vorticity by an ion moving in pure superfluid ^{4}He at a low temperature, the theory being to some extent an extension of earlier ideas (3,4). We showed by numerical computation of the appropriate energies and impulses that the most favourable vortex nucleus is probably in the form of a loop out of the side of the ion. Formation of this nucleus becomes energetically possible when the ionic velocity exceeds a certain critical value, U_c. However, nucleation tends to be inhibited by a potential barrier, the dimensions of which we were able to estimate. We suggested that in the absence of thermal activation (i.e. at the lowest temperatures) nucleation can occur by a direct quantum mechanical transition involving penetration of the barrier by quantum tunnelling, and we showed that this process might well occur at a rate consistent with experiment. At higher temperatures thermal activation over the barrier becomes possible by absorption of a single roton, and again this is consistent with experiment. Practically all the relevant experiments (5) relate to a situation in which relatively slow vortex nucleation is accompanied by rapid creation of rotons, and we discussed the extent to which these processes might interfere with each other. We concluded very tentatively that the two processes might be independent at ion velocities close to critical, but that at higher velocities the roton emission might inhibit the vortex nucleation. A fall-off in nucleation rate at high velocities is indeed observed (5), although it is possible that this can be explained also in terms of a shedding of the nucleating vortex from the ion at the very high electric fields required to maintain ionic velocities in this regime.

It has been known for many years that vortex nucleation by a moving negative ion is affected significantly by the presence of small amounts of dissolved ^{3}He, and it was suggested (6) that this effect might be associated with the adsorption of ^{3}He atoms on the surface of the negative-ion bubble. Recently it has been discovered that the nucleation process is sensitive even to minute concentrations of ^{3}He, and a detailed study (1) has yielded results that are consistent with the idea that the effect is indeed due in some way to the adsorption of ^{3}He atoms on the surface of the bubble, the adsorption of a single ^{3}He atom being sufficient to reduce the critical velocity by about 4 m s^{-1} and to enhance the nucleation rate by a factor of about 10^3.

2. THE THEORY OF THE EFFECT OF ^{3}HE

Our explanation of the effect of ^{3}He on the nucleation rate depends on the idea that a ^{3}He atom can be more strongly bound to the core of a vortex line than to the surface of the negative-ion bubble. The analysis in reference (1) leads to a value of the binding energy to the bubble of about 2.5K, while the theory of Ohmi & Usui (7) and a modified version of the theory of Senbetu (8) both yield a binding energy to the vortex of about 3K. When a negative ion with a trapped ^{3}He atom nucleates a vortex, the ^{3}He atom can be transferred to the vortex core, which leads to the release of an amount of

*Department of Physics, University of Oregon, Eugene, Oregon 97403, USA.

energy equal to the difference between the two binding energies. The availability of this extra energy could lead both to a reduction in the minimum velocity at which vortex nucleation becomes energetically possible (the critical velocity U_c) and, *via* a modification to the form of the potential barrier, to an increase in the nucleation rate at supercritical velocities.

We have attempted to modify our earlier theory and associated computations (2) to take account of these effects in a semi-quantitative way. In doing so we have allowed for the fact that the experimental results (1) relate to finite temperatures in the range 0.3 to 0.5K; this means that prior to nucleation the ^{3}He atom on surface of bubble is likely to be in one of the excited states described by Shikin (9), and this can make available a further supply of energy. We have therefore calculated the vortex nucleation rate for each initial state of excitation of the ^{3}He atom, and then taken an average with weighting factors proportional to the populations of the different initial states. We have assumed that the change in quantum mechanical transition probability governing the vortex nucleation rate occurs only because of a change in the dimensions of the potential barrier. This is unlikely to be accurately true; the requirement that nucleation be accompanied by the transfer of the ^{3}He atom from the ion to the vortex must presumably reduce the transition probability by an amount that we cannot estimate.

The results of our theoretical analysis are as follows. The adsorption of a single ^{3}He atom reduces the predicted critical velocity by about 4.5m s^{-1}. This result is in good agreement with experiment. At supercritical velocities the nucleation rate is increased by a factor of roughly 10^6. This is a factor of about 10^3 larger than is observed. The discrepancy may be due to our assumption that the rate is changed only because of the change in barrier dimensions; but it could also be due to quite small errors in the predicted changes in shape of the potential barrier (such errors would have little effect on the predicted critical velocity). Our calculations also lead to the prediction that the adsorption of a second ^{3}He atom should lead to a further large increase in the nucleation rate, by a factor of roughly 10^4. This is in agreement with experiment (McClintock, private communication).

A full account of this work will be published elsewhere.

ACKNOWLEDGEMENTS

We are very grateful to Dr. P.V.E.McClintock and his colleagues for showing us their experimental results and their analysis of them before publication.

REFERENCES

(1) R.M. Bowley, G.G. Nancolas and P.V.E. McClintock, Phys. Rev. Letters 52 (1984) 659.

(2) C.M. Muirhead, W.F. Vinen and R.J. Donnelly, Philos, Trans. Roy. Soc. London, A in print.

(3) R.J. Donnelly and P.H. Roberts, Philos. Trans. Roy. Soc. London, A271 (1971) 41.

(4) K.W. Schwarz and P.S. Jang, Phys. Rev. A8 (1973) 3199.

(5) R.M. Bowley, P.V.E. McClintock, F.E. Moss, G.G. Nancolas and P.C.E. Stamp, Philos. Trans. Roy. Soc. London, A307 (1982) 201.

(6) A.J. Dahm, Phys. Rev. 180 (1969) 259.

(7) T. Ohmi and T. Usui, Prog. Theor. Phys. 41 (1969) 1401.

(8) L. Senbetu, J. Low Temp. Phys. 32 (1978) 571.

(9) V.B. Shikin, Sov. Phys. JETP 37 (1973) 718.

LT-17 (Contributed Papers)
U. Eckern, A. Schmid, W. Weber, H. Wühl (eds)

STABILITY OF ELECTRONS ON SUPERFLUID ^{4}He FILMS

H. ETZ, W. GOMBERT, and P. LEIDERER

Fachbereich Physik, Johannes Gutenberg-Universität, 6500 Mainz, FRG*

The stability of 2-dimensional electron systems on saturated films of superfluid ^{4}He has been investigated. Electron densities up to 10^{11}cm^{-2} have been observed to be stable, nearly two orders of magnitude more than on bulk ^{4}He. The thickness of films charged to such high densities is found to be drastically reduced.

Electrons on liquid helium are an example of a particularly clean and nearly ideal 2-dimensional Coulomb system, which has revealed such interesting phenomena as the transition to a 2D electron solid (1). In the experiments performed to date these electrons behave like a classical Coulomb system, because the electron density accessible on bulk liquid helium is restricted to n 2×10^9cm^{-2}, a range where the Fermi energy is much smaller than the Coulomb energy.

The limit in the electron density on bulk helium is due to an electrohydrodynamic (ehd) instability which develops as a result of the coupling between excitations of the liquid surface and the electron system. A considerably higher density ought to be accessible when the electrons are supported by a thin helium film rather than bulk liquid (2,3). In the dispersion relation of the low frequency coupled plasmon-ripplon modes, given by (3)

$$\omega^2 = \{(\frac{3\alpha}{\rho d^4}+g)k+\frac{\sigma}{\rho}k^3 - \frac{4\pi e^2 n^2}{\rho}k^2 F(k,\varepsilon)\}\tanh(kd) \qquad (1)$$

the term originating from the van der Waals-interaction, α/d^4, is then orders of magnitude larger than the acceleration due to gravity, g, giving rise to an enhanced stability of the film. (Here ρ and σ are the density and surface tension of He, d is the film thickness, and the factor F is approximately equal to the dielectric constant ε of the solid substrate.)

The stability limit where according to eq. (1) ω drops to zero is plotted in Fig. 1. For a thickness of 300 Å - a typical value for a saturated film 1 cm above the bath level - it is expected that a density $n_c \sim 3\times10^{10}$cm^{-2} can be reached, one order of magnitude more than on bulk He.

In our experiments on the high-density behavior of the 2D electron - He film system we have used saturated films whose thickness d_0 before charging, as measured by ellipsometry, could be varied between 200 and 500 Å by changing the level of the bulk liquid. A small glow discharge served as an electron source. The charge density was obtained from the applied voltage U and a measurement of the electric field above the electron layer by means of a vibrating capacitor plate. In order to decide whether at a certain density the electrons were located above the film, or whether part of them had penetrated the liquid, U was reduced to zero or slightly negative bias. The electrons **on** the film, which are only weakly bound and mobile, then leave the film surface, whereas those charges which have punched the film are localized on the solid substrate and cannot be removed by small negative fields.

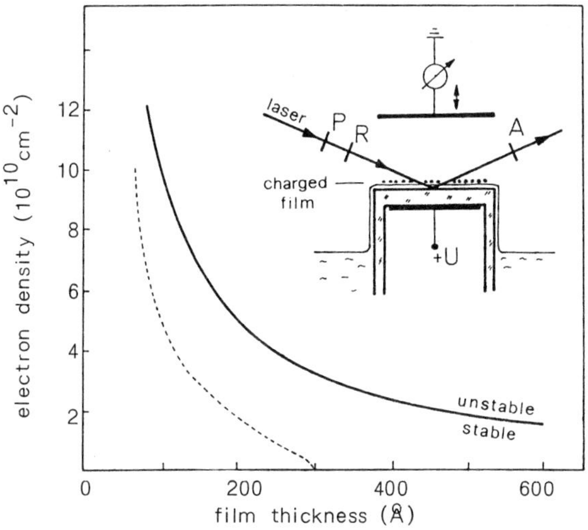

Fig. 1: Stability diagram of charged ^{4}He films. The solid line represents the locus of vanishing ripplon frequency (eq. 1) at the critical wave vector k_c, where the ehd instability sets in (after ref. (3)). The dotted line represents an example of a path for a saturated He film, given by eq. 2 as discussed below. The insert shows the principle of the experimental set-up. Ellipsometer components: P - polarizer, R - retardation plate, A - analyzer.

* Research supported by the Deutsche Forschungsgemeinschaft.

So far we have mainly studied insulating substrates, such as a polymer foil (ε=3.8) and glass (ε=6.5), which was polished mechanically or by flame treatment. It was found that electron densities up to a critical value $n_c \sim 10^{11} cm^{-2}$ are mobile* and can be completely removed again (6). For higher densities n the amount exceeding n_c is tightly bound to the substrate and can be neutralized only by a separate glow discharge cycle. The exact value of n_c varies slightly for different samples and is also affected by the quality of the sample surface as, e.g., minute traces of frozen air. The conclusion that the mobile part of the electrons is indeed stable **above** the He film is corborated by the observation that, once the film is boiled off, all electrons are localized and cannot be removed from the surface.

The high density of electrons attainable on saturated He films appears promising for further studies of 2D electron systems (7). Yet a comparison of our results with Fig. 1 is disturbing, at the first glance, because the experimental instability threshold is distinctly higher than anticipated for d $\gtrsim$200 Å on the basis of the ehd instability. What is more, a noticeable influence of the initial film thickness, which might by expected from Fig. 1, could not be observed. This apparent discrepancy is resolved when the curves plotted in Fig. 2 are considered: As a helium film is charged, its thickness is not constant, but decreases as a result of the electronic pressure according to

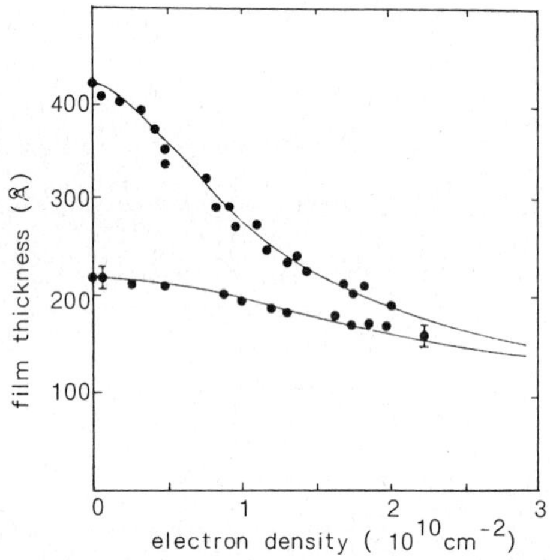

Fig. 2: Thickness d of charged saturated ^{4}He films as measured by ellipsometry. The temperature was 1.6 K, the initial thickness d_o of the uncharged films 220 and 420 Å, respectively. (The relatively low electron density reached in this experiment was not limited by an instability of the charges on the film, but by dielectric breakdown in the leads.)

$$d = (1/d_o^{\,3} + 2\pi n^2 e^2/\alpha)^{-1/3} \qquad (2)$$

(d_o is again the thickness of the uncharged film). The reduction in d in turn raises the instability threshold, calculated from eq. 1, the film thus stabilizes itself. As seen in Fig. 2 the data are in satisfactory agreement with the solid lines representing eq. 2. Since for high electron densities the curves converge, it is not surprising that the experimentally observed stability limit is not significantly influenced by the initial film thickness d_o.

The question remains which process finally leads to the loss of the electrons from the film surface and thus limits n_c. It is conceivable that, apart from the collective ehd instability, tunneling of individual electrons through the film into surface states of the substrate becomes important(2). Although for a homogeneous film on a perfectly flat substrate the tunneling rate is probably negligible even at the highest densities investigated here, any presence of surface roughness will locally decrease the film thickness and hence act as a "weak spot" in the film. In a addition, the film thickness will be reduced locally by the dimple forming underneath each electron. Measurements on carefully prepared samples will have to show the relevance of the respective mechanisms.

The authors thank Prof. A. Dahm and D. Cieslikowski for valuable discussions.

References

(1) For a recent review see, e.g., Yu.P. Monarkha and V.B. Shikin, Sov.J.Low Temp.Phys. 8 (1982), 279
(2) L.M. Sander, Phys.Rev. B11 (1975), 4350
(3) H. Ikezi and P.M. Platzman, Phys.Rev. 23(1981), 1145
(4) S.A. Jackson and P.M. Platzman, Phys.Rev. B24 (1981), 499; Phys.Rev. B25 (1982), 4886
(5) D. Cieslikowski, diploma thesis,Techn.Univ. Munich (1982)
(6) Electrons above He-films - at not quite as high densities - have already been investigated with different techniques by A.P. Volodin, M.S. Khaikin and V.S. Edelman,JETP Lett. 23 (1976), 478, and K. Kajita, J.Phys.Soc.Jpn. 52 (1983), 372
(7) F.M. Peeters and P.M. Platzman,Phys.Rev.Lett. 50 (1983), 2021

* The mobility is much smaller than on bulk He, however, probably due to the formation of polarons (4,5).

LT-17 (Contributed Papers)
U. Eckern, A. Schmid, W. Weber, H. Wühl (eds)

MULTIPLY CHARGED IONS IN LIQUID HELIUM AND IN HELIUM VAPOR

Scott T. HANNAHS and Gary A. WILLIAMS

Department of Physics, University of California, Los Angeles, California 90024, USA

Martti M. SALOMAA

Low Temperature Laboratory, Helsinki University of Technology, SF-02150 Espoo 15, Finland

We consider two multiply charged structures, multielectron bubbles in liquid He and multiply charged droplets in He vapor. The nonlinear coupling of radial and angular oscillation modes in multielectron bubbles is discussed. A model is introduced for the droplets; only saturated droplets are found stable with respect to angular oscillations.

Large multielectron bubbles enclosing $Z=10^6$ to 10^8 charges have been observed (1) in ^{4}He; they are generated by the capillary-wave instability of the charged He surface. We have shown that there exists a self-stabilizing region for low-amplitude oscillations of multielectron bubbles (2); here we consider this in more detail. We also discuss large multiply charged droplets in He vapor (3,4). We present a model for the droplets and examine their stability.

1. MULTIELECTRON BUBBLES

A first approximation to the structure of multielectron bubbles was proposed by Shikin (5). Minimizing the sum of the Coulomb energy $E_U=Z^2e^2/2\epsilon R$ and the surface tension energy $E_\sigma=4\pi R^2\sigma$, where R is the bubble radius, ϵ the dielectric constant, and σ the surface tension of ^{4}He, one obtains the classical "Coulomb" radius:

$$R_C = (Z^2e^2/16\pi\epsilon\sigma)^{1/3} \quad . \tag{1}$$

It is important to consider the stability of the multielectron bubbles with respect to small deformations. Shikin (5) pointed out that the equilibrium radius R_C coincides with the radius at which a charged liquid drop becomes unstable (6). However, a difference between liquid drops and the multielectron bubbles is the finite compressibility of the bubbles, which allows a purely radial oscillation; the nonlinear coupling between radial and angular oscillation modes yields a region of stability of the bubbles. Above a threshold oscillation amplitude a Rayleigh-Taylor instability was found (2).

A position on the multielectron bubble surface is represented by

$$r = R(t) + \sum_\ell a_\ell(t)P_\ell(\cos\theta) \quad , \tag{2}$$

where R(t) is the instantaneous average radius and the $a_\ell(t)$ are amplitudes of the angular oscillations having $\ell=2,3,\ldots$; these are assumed small: $a_\ell \ll R$, see Fig. 1. The radial mode obeys the Rayleigh equation

$$\frac{\ddot R}{R} + \frac{3}{2}\left(\frac{\dot R}{R}\right)^2 = \frac{1}{\rho}\left[\frac{Z^2e^2}{8\pi\epsilon R^6} - \frac{2\sigma}{R^3}\right] \quad , \tag{3}$$

while the angular modes satisfy

$$\ddot a_\ell + 3\frac{\dot R}{R}\dot a_\ell + a_\ell\left[\frac{\ell^2-1}{\rho}\left(\frac{(\ell+2)\sigma}{R^3} - \frac{1}{4\pi}\frac{Z^2e^2}{\epsilon R^6}\right) - (\ell-1)\frac{\ddot R}{R}\right]=0 \tag{4}$$

where ρ is the He density. For small oscillations about R_C, these reduce to equations for harmonic oscillators with eigenfrequencies $\omega_0^2=96\sigma^2\pi\epsilon/\rho Z^2e^2$ and $\omega_\ell^2=(\ell^2-1)(\ell-2)\omega_0^2/6$. The stability problem is the $\ell=2$ mode, which has $\omega_2=0$ for a bubble of radius R_C.

To investigate the stability, we solve Eqs. 3 and 4 for finite $R(t)-R_C$. The initial values are $\dot R(0)=\dot a_\ell(0)=0$. The variable parameter is the amplitude of the radial oscillations, $R(0)/R_C$. We have integrated Eqs. 3 and 4 numerically and Fourier-transformed the time records.

Fig. 1 represents the frequency shifts of the first three modes in a_2, a_3 and a_4 as functions of $R(0)/R_C$. Note first that for a finite $R(0)/R_C$, ω_2 shifts to a nonzero frequency, stabilizing the bubble oscillations. With further increase of the amplitude, a_2 is driven unstable as the mode frequency tends towards $\omega_0/2$. The instability (2) with frequency $\omega_0/2$ then sets in at $R(0)/R_C=1.063$.

However, we find that the surface oscillation a_3 becomes unstable <u>first</u> at the lower value $R(0)/R_C=1.054$, as in this case ω_3 moves down to zero with increasing $R(0)/R_C$. The frequency shift of a_4 (and higher modes) is slower and it becomes unstable at a much larger amplitude. While Eq. 4 may be approximated by a Mathieu equation, the stability criteria thus obtained are not quite accurate.

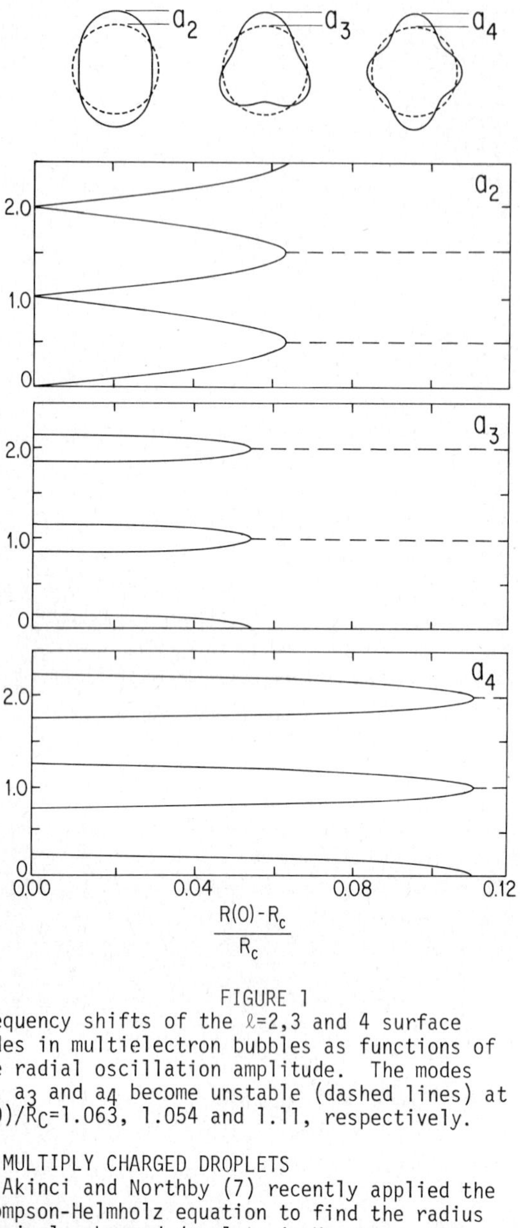

FIGURE 1
Frequency shifts of the ℓ=2,3 and 4 surface modes in multielectron bubbles as functions of the radial oscillation amplitude. The modes a_2, a_3 and a_4 become unstable (dashed lines) at $R(0)/R_C$=1.063, 1.054 and 1.11, respectively.

2. MULTIPLY CHARGED DROPLETS

Akinci and Northby (7) recently applied the Thompson-Helmholz equation to find the radius of singly charged droplets in He vapor. Here we introduce the following generalized Thompson equation for multiply charged positive ion droplets (3,4) in He vapor:

$$k_B T \ln \frac{p}{p_{sat}} = \frac{2\sigma}{nR} - \alpha \frac{Z^2 e^2}{2R'^4} - \frac{1}{8\pi} \frac{Z^2 e^2}{nR'^4} \quad . \qquad (5)$$

Here p_{sat} is the saturated vapor pressure, p the local pressure in the vapor, n the number density, and α the atomic polarizability. The charges lie below the liquid-vapor interface at the inner radius R' illustrated in Fig. 2.

Approximating R'=R yields at saturated vapor the "Coulomb" radius:

$$R_C = [(\frac{4\varepsilon-1}{\varepsilon+2}) \frac{Z^2 e^2}{16\pi\sigma}]^{1/3} \quad , \qquad (6)$$

which, except for the dielectric constant terms, coincides with Eq. 1. To estimate R-R' for large Z, we approximate the image potential with that of a plane surface and equate the image force to the force from the electrostatic potential of the spherical charge distribution;

$$R' = R[1+(\frac{\varepsilon-1}{2(\varepsilon+1)Z})^{1/2}]^{-1} \quad . \qquad (7)$$

For $Z=10^8$ with $R \approx R_C \approx 5.1\times10^6$Å, this yields the estimate $R-R' \approx 61$Å. For lower Z, it is important to take into account spherical geometry; the solution for R using Eq. 5 is shown in Fig. 2. Note the sensitive dependence of R for large droplets on pressure for p just less than p_{sat}. The analysis of the droplet stability results again in a regime of stability, however unsaturated droplets with $R<R_C$ disintegrate.

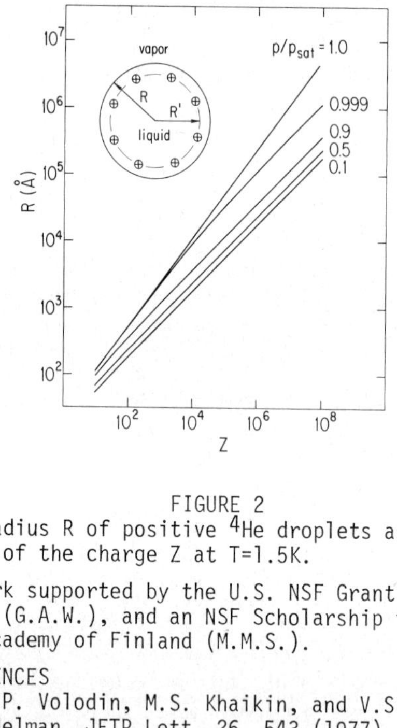

FIGURE 2
The radius R of positive ^{4}He droplets as functions of the charge Z at T=1.5K.

Work supported by the U.S. NSF Grant DMR 81-00218 (G.A.W.), and an NSF Scholarship through the Academy of Finland (M.M.S.).

REFERENCES

(1) A.P. Volodin, M.S. Khaikin, and V.S. Edelman, JETP Lett. 26, 543 (1977).
(2) M.M. Salomaa and G.A. Williams, Phys. Rev. Lett. 47, 1730 (1981).
(3) F.P.Boyle and A.J.Dahm, JLTP 23, 477 (1976).
(4) J.Gspann and H.Vollmar, JLTP 45, 343 (1981).
(5) V.B.Shikin, JETP Lett. 27, 39 (1978).
(6) Lord Rayleigh, Philos. Mag. 14, 184 (1882).
(7) G. Akinci and J.A. Northby, Phys. Rev. Lett. 42, 573 (1979).

LT-17 (Contributed Papers)
U. Eckern, A. Schmid, W. Weber, H. Wühl (eds)

VORTEX MOTION IN HELIUM FILMS

M. KIM and W.I. GLABERSON

Serin Physics Laboratory, Rutgers University, Piscataway, NJ 08854

The decay of persistent currents, associated with the motion of rotation-induced vortices in thin films, is observed. Decay associated with vortex motion out of films proceeds much more rapidly than that associated with vorticity entering films.

In a recent paper [1] we reported an experimental determination of vortex diffusivity in thin superfluid films near the Kosterlitz-Thouless transition temperature T_{KT}. We now discuss some results on vortex pinning, vortex creep and persistent currents in films at lower temperatures.

Our experimental technique involves the use of a circular third sound resonant cavity rotating about an axis perpendicular to the film, in which the resonance frequency and width are monitored as a function of film thickness, temperature and rotation speed. For relatively thin films and high temperatures the resonance Q is found to be independent of rotation history and strictly linear in the rotation speed and hence in the density of rotation-induced vortices. The vortex diffusivity can therefore be determined and is observed to be proportional to $(T/\sigma_s)^2$ where σ_s is the areal superfluid density, and is 0.4 $\hbar/m$ at T_{KT}. As the temperature is lowered below $\sim$0.3 T_{KT} the temperature dependence of the diffusivity changes, dropping much more rapidly and becoming unmeasureably small below $\sim$0.25 T_{KT}. It is in this lower temperature regime that hysteresis and persistent current effects become evident.

The particular third sound modes observed were well determined azimuthal modes. It can easily be shown that otherwise degenerate clockwise and counterclockwise travelling waves should have their degeneracy lifted in the presence of a persistent current. Waves travelling in the same sense as the persistent current, for example, should be doppler shifted upwards in frequency as observed by a detector in the substrate reference frame. For some reason, although as we shall discuss shortly we are convinced that substantial persistent currents were at times present in the cavity, we were never able to observe mode splitting. We note that Wang and Rudnick [2], in a somewhat different geometry, were also unable to observe rotation induced persistent currents using the technique of third sound mode splitting, although thermally induced persistent currents readily yield a doppler shift [3]. Third sound rides "on top" of the surface-averaged superfluid angular velocity and one can speculate that, because of the rough character of the surface, this velocity could be very close to that of the substrate even in the presence of a substantial persistent current. Related effects have been observed in fourth sound experiments [4], although not to the same extent, and would not necessarily be present in a thermally induced flow situation.

Because we were unable to observe mode splitting, our technique for measuring persistent currents involves an anomalous effect probably associated with capillary condensation between the quartz plates making up the third sound cavity. Below a critical temperature (well below the interesting vortex creep regime) we observe a resonance Q which does not depend on the state of motion of the cell, but does depend, quite reproducibly, on the difference between the angular velocity and that angular velocity the cell had when it was cooled through the vortex creep temperature regime. At low temperatures, the film has long term "memory" of its state of motion during cool down. It is therefore possible to calibrate the system so as to determine the effective persistent current in the cell.

The decay rate of the persistent current, presumably associated with vortex nucleation and/or creep [5,6,7], is determined as follows. The cell is cooled through the superfluid transition temperature to a target temperature in some state of rotation; the state of rotation is changed quickly and the cell is kept at the target temperature for some specified delay time; the cell is then cooled to a low temperature where the effective persistent current remaining is determined. Figure 1 is an isochronal map of the decay behavior, that is a plot of the effective persistent current remaining in the cell after a delay of 30 minutes as a function of the

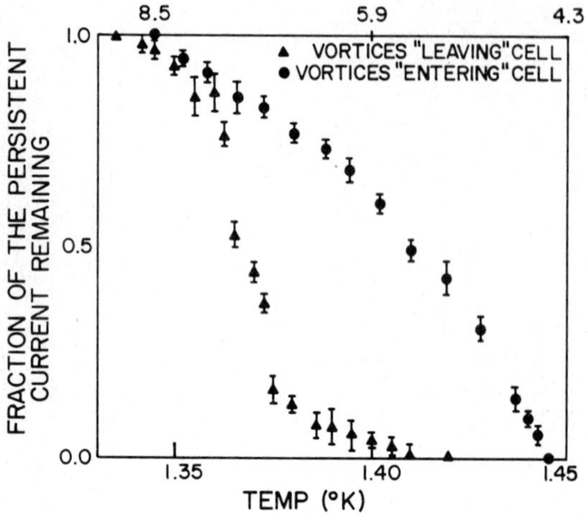

Figure 1.

target temperature. Because the sound cavity is sealed, the film thickness decreases as the temperature increases and the thickness in helium atomic layers is indicated at the top of the figure. The circles correspond to vortices entering the film (the cell is accelerated from rest to 8.4 rad/sec at the target temperature) and the triangles correspond to vortices leaving the film (the cell is cooled to the target temperature while rotating and then brought to rest). Figure 2 is an isothermal map of the decay behavior where the persistent current remaining in the cell is plotted as a

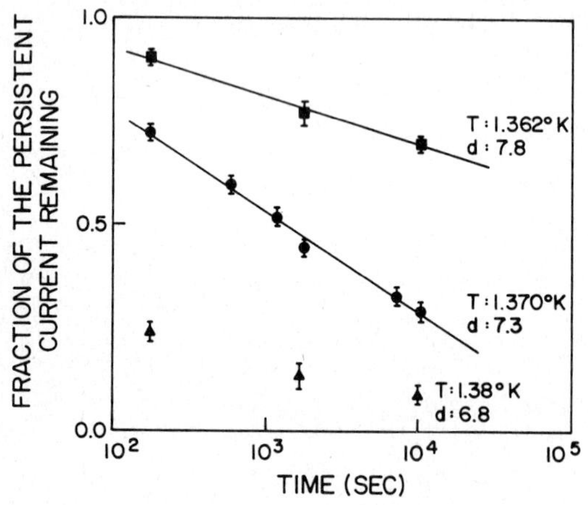

Figure 2.

function of the delay time for various target temperatures. At relatively low temperatures, the decay is logarithmic as is characteristic of vortex hopping from pinning site to pinning site [5]. For thinner films, particularly for decays associated with vortices entering the film, non-logarithmic time behavior is observed. The very clear difference between the vortex inflowing and outflowing decay behavior in Figure 1 suggests that in the latter the decay rate is limited by vortex hopping whereas in the former the decay rate is dominated by a much slower nucleation rate. We note that the decay behavior is in many respects, qualitatively similar to that observed by Ekholm and Hallock [3]. By consciously introducing vortices into the system by rotation, it may have been possible to disentangle vortex nucleation from vortex creep.

One serious problem with the suggestion that the vortex inflow decay is limited only by nucleation is the implication that the vortex distribution is then always reasonably uniform in this situation. It follows that at a given temperature the decay rate should only be a function of the persistent current. We find, however, that a persistent current prepared by accelerating from some nonzero rotation speed in which the cell was cooled down, decays much more rapidly than one prepared from a cell accelerated from rest that has been allowed to decay to the same effective persistent current. More work is clearly required for a satisfactory understanding of vortex motion in films.

REFERENCES

(1) M. Kim and W.I. Glaberson, Phys. Rev. Lett. 52 (1984) 53.
(2) T. Wang and I. Rudnick, in Low Temperature Physics LT-13 (Plenum Press, New York, 1972) p. 239.
(3) D.T. Ekholm and R.B. Hallock, Phys. Rev. B21 (1980) 3902.
(4) H. Kojima, W. Veith, E. Guyon, and I. Rudnick, J. Low Temp. Phys. 25 (1976) 195.
(5) D.A. Browne and S. Doniach, Phys. Rev. B 25 (1982) 136.
(6) J.L. McCauley, Phys. Rev. Lett. 45 (1980) 4677.
(7) L. Yu, Phys. Rev. B 23 (1981) 3569.

LT-17 (Contributed Papers)
U. Eckern, A. Schmid, W. Weber, H. Wühl (eds)

PRESSURE DEPENDENCE OF THE LANDAU CRITICAL VELOCITY IN HeII : A PROGRESS REPORT*

T.ELLIS and P.V.E.McCLINTOCK

Department of Physics, University of Lancaster, Lancaster LA1 4YB, UK.

Precise measurements of the Landau critical velocity v_L in HeII are reported for pressures P in the range $13 < P < 25$ bar. They agree within experimental error with predictions derived from accepted values of the roton parameters.

1. INTRODUCTION

The existence of the Landau critical velocity v_L is fundamental to an understanding (1) of HeII. A massive object moving at velocity v through pure HeII at very low temperatures, such that scattering processes can be ignored, can dissipate kinetic energy only through the creation of elementary excitations : a process which can occur if, and only if, $v \geqslant v_L$.

The magnitude of v_L is readily shown to be determined by the gradient of a line drawn from the origin to make a tangent with the dispersion curve near the roton minimum, so that

$$v_L = \left[\left(2\Delta\mu + \hbar^2 k_o^2\right)^{\frac{1}{2}} - \hbar k_o\right] \Big/ \mu \qquad (1)$$

where, to a good approximation, $v_L \simeq \Delta/\hbar k_o$. Here, Δ, μ and k_o are roton parameters representing, respectively, the energy, effective mass and wave vector of a roton at the minimum. An experimental measurement of v_L can therefore yield information about the roton region of the dispersion curve. Alternatively, a detailed comparison of the measured value of v_L with that predicted by equation (1), on the basis of accepted values of the roton parameters, may be regarded as a rather direct and fundamental test of Landau's excitation model of HeII.

The only technique so far devised (2) for measurement of v_L in the low temperature limit relies on the propagation of negative ions in pressurised (3) HeII. In the cases of all other probes, as well as that of negative ions at lower pressures, and that of the closely related process of superflow through a tube, the measured critical velocities are smaller than v_L: they correspond, not to roton creation, but to the formation (or expansion of pre-existing) quantum vortices. For sufficiently high pressures, however, and provided that the HeII has been isotopically purified, the measured drift velocities of negative ions are given (4) by

$$\bar{v} = v_L + AE^{1/3} \qquad (2)$$

where A is a constant and E is the applied electric field. The form of equation (2) indicates that rotons are apparently being created *in pairs* by the moving ion. Linear extrapolation (2) of a $\bar{v}(E^{1/3})$ plot to E = O can be used to determine v_L.

The known pressure dependence of the roton parameters (5) implies that v_L, too, will vary with the pressure P. Although vortex nucleation precludes the use of normal negative ions for measurement of v_L under the saturated vapour pressure, it is obviously desirable to extend the measurements of $v_L(P)$ to as low a pressure as possible. Our earlier measurements (2) were restricted to $19 < P < 25$ bar because the signal became too small to measure for $P < 19$ bar. The purpose of the present paper is to report that, through a number of improvements in technique, we have succeeded in pushing the lower bound of the measurements to P = 13 bar, thereby doubling the range of P over which v_L can be followed.

2. METHOD

The technique for determination of $\bar{v}$ has remained essentially as previously described (2,6). In order to be able to acquire reliable data in the pressure range below 19 bar, however, we have shaped the voltage pulses applied to the field emission ion sources in such a way as: (i) to minimise the production of charged vortex rings in the high field region near the emitters; and (ii) to minimise the extent to which charged vortex rings or lines "leak" from the latter region into the drift space, leading to a possible distortion of the local electric field and providing possible trapping centres for bare ions. In addition, we have used a digital cross-correlation technique to measure <u>changes</u> in drift velocity relative to a single, accurately determined, value of $\bar{v}$ at a relatively high electric field for each pressure. Full technical details of these procedures will be presented elsewhere.

---

* Work supported by the Science and Engineering Research Council (UK).

3. RESULTS

By use of the techniques outlined in §2, we have managed to acquire and analyse satisfactory signals down to about 13 bar, yielding the $v_L(P)$ measurements indicated by the points in Figure 1. The systematic uncertainty in v_L, arising mainly from the measurement of the cell length, is estimated at ca ± 0.5%; random errors are smaller than the sizes of the points as plotted. The full curve represents the theoretical pressure dependence of v_L, calculated from equation (1) on the basis of the Brooks and Donnelly tabulation (5) of roton parameters. Agreement between experiment and theory appears to be excellent; although the points do, in fact, fall systematically below the curve (not evident from figure).

A *caveat* must now be entered. As a result of our improved technique, we have been able to detect small departures from equation (2) in very weak electric fields. Their physical origin remains unknown. Explicit allowance can, of course, be made for the presence of what appears (7) to be an additional term, in E^{-1}, in equation 2; and the $\overline{v}(E)$ data can be re-fitted accordingly. Application of this procedure raises the fitted values of v_L only by ca 0.01 m s^{-1}, however: an increment that is negligible compared to other uncertainties in the measurements. We therefore believe that, notwithstanding the anomalous behaviour of $\overline{v}(E)$ at very low E, the experimental values of v_L given in Figure 1 are reliable.

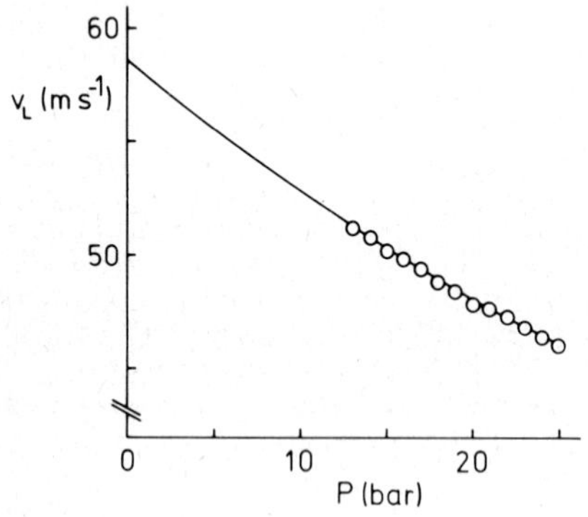

FIGURE 1
Values of the Landau critical velocity v_L for roton creation in HeII, deduced from $\overline{v}(E)$ data for E > 400 V m^{-1} on the basis of equation (2), and plotted as a function of pressure P. The full curve is a theoretical prediction derived by insertion into equation (1) of the roton parameters of reference (5).

4. CONCLUSION

Provided that allowance is made for uncertainties (8) in the roton parameters under pressure, our measured values of v_L agree within experimental error with those predicted by equation (1). It seems unlikely that it will ever be possible to push the measurements of v_L significantly below 13 bar through use of the normal negative ion. We are hoping (9), however, that it may in due course prove possible to obtain a single experimental value of v_L at the saturated vapour pressure by making use of the mysterious fast negative ion (10). Such a measurement is very much to be desired, not only to help to fill the obtrusive gap in the experimental $v_L(P)$ data of Figure 1, but also because the relevant roton parameters are known to much higher precision than those at elevated pressures, so that a more stringent test of equation (1) would then be possible.

ACKNOWLEDGEMENTS

We are much indebted to R.M.Bowley for many valuable discussions and to I.E.Miller for his continuing expert technical assistance.

REFERENCES

(1) L.D.Landau, J.Phys.Moscow 5 (1941) 71; 11 (1947) 91; translated in I.M.Khalatnikov, Introduction to the theory of superfluidity (Benjamin, New York, 1965) pp.185-206.
(2) T.Ellis, C.I.Jewell and P.V.E.McClintock, Phys.Lett.78A (1980) 358.
(3) G.W.Rayfield, Phys.Rev.Lett.16 (1966) 934.
(4) D.R.Allum, P.V.E.McClintock, A.Phillips and R.M.Bowley, Phil.Trans.R.Soc.Lond.A 284 (1977) 179.
(5) J.S.Brooks and R.J.Donnelly, J.Phys.Chem.Ref. Data 6 (1977) 51.
(6) T.Ellis and P.V.E.McClintock, Physica 107B (1981) 569.
(7) T.Ellis and P.V.E.McClintock, Anomalous behaviour of negative ions in HeII, this volume.
(8) R.J.Donnelly and P.H.Roberts, J.Phys.C: Solid St.Phys. 10 (1977) L683.
(9) V.L.Eden and P.V.E.McClintock, The "fast" negative ion in HeII, in Proc. 75th Anniversary Conf. on Liquid Helium, St.Andrews, 1983, to be published.
(10) C.S.M.Doake and P.W.F.Gribbon, Phys.Lett. 30A (1969) 251.

LT-17 (Contributed Papers)
U. Eckern, A. Schmid, W. Weber, H. Wühl (eds)

ADIABATIC FLOW OF HeII AND MOTION OF ^{3}He IN HeII

Mineo OKUYAMA, Toshimi SATOH, Takeo SATOH and Taiichiro OHTSUKA

Department of Physics, Faculty of Science, Tohoku University, Sendai 980, Japan

Taku SATO and Shinhachiro SAITO

The Institute for Iron, Steel and Other Metals, Tohoku University, Sendai 980, Japan

The flow phenomena of the superfluid HeII is investigated in a system consisting of a superleak, a chamber and a capillary in series. Of particular interest is the correlation between the thermal characteristic of the system and the flow state in the capillary, which is confirmed by the measurement of the pressure difference across the capillary. Further, the flow state in the capillary is probed with a nuclear magnetic resonance of ^{3}He atoms added to the system.

1. INTRODUCTION

A HeII flow system consisting of a superleak, a chamber and a capillary in series is a cooling device, in which the forced flow of the entropyless superfluid component drags away the entropy in the chamber through the capillary (1, 2, 3). Our previous investigation of the system suggests a correlation between the flow state in the capillary and the cooling characteristic of the device (4).

In order to confirm the correlation and gain insight into the flow state, the pressure difference across the capillary has been measured with a capacitance pressure gauge. A remarkable effect of ^{3}He added into the system was noticed in the previous study (4). Pulsed nuclear magnetic resonance has been applied to trace the motion of ^{3}He atoms in HeII.

2. RESULTS AND DISCUSSIONS

Figure 1 shows a typical behavior of the chamber temperature, T_A, as a function of the mean flow velocity, V, of HeII in the capillary. Measured pressure difference across the capillary, ΔP, is plotted in Fig.2(a) and 2(b) as a function of V. There are several important features to be noticed.

First, in Fig.2(a), the pressure difference appears when V exceeds a critical value about 3.2 cm/sec, which is assigned as the superfluid critical velocity V_{sc} in the capillary (i.d. 0.1mm). ΔP can be fitted well with a equation

$$\Delta P = A(V-V_{sc})$$

where the coefficient A is observed to be temperature independent. Assuming a Poiseuille flow of the normal component in the capillary with velocity $V-V_{sc}$, we obtain for the value of viscocity $\eta=14.7\times10^{-6}$ poise. This is just the order of the roton viscosity, which is known to be temperature independent. Therefore the flow state in the capillary may be considered to be a laminar flow state of rotons which move with the velocity $V_r=V-V_{sc}$. This ovservation means that,

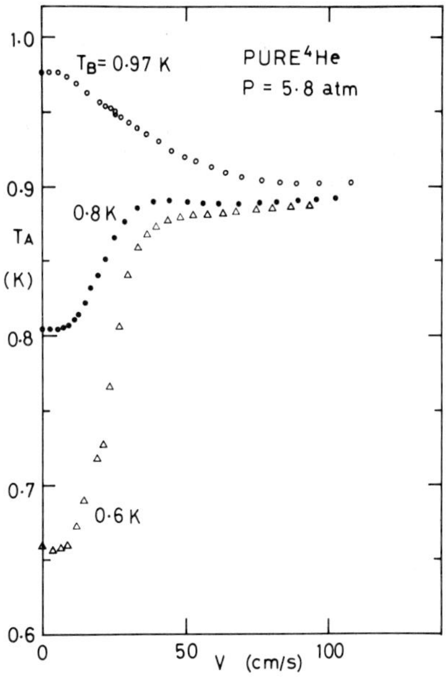

Fig.1 The chamber temperature T_A as a function of the net mass flow velocity V in the capillary at various bath temperature T_B.

in this velocity region, only rotons can be taken away from the chamber by the action of the flow of the superfluid component. The reason why V_r is equal to $V-V_{sc}$ is not clear.

Second, in the high velocity region shown in Fig.2(b), ΔP may be expressed as

$$\Delta P = B \cdot V^{1.75}$$

where the coefficient B turns out to be temperature dependent. Assuming the Blasius law for turbulent flow, we obtain for the viscosity $\eta=53\times10^{-6}$ poise, which nearly corresponds to the total viscosity of HeII at the temperature

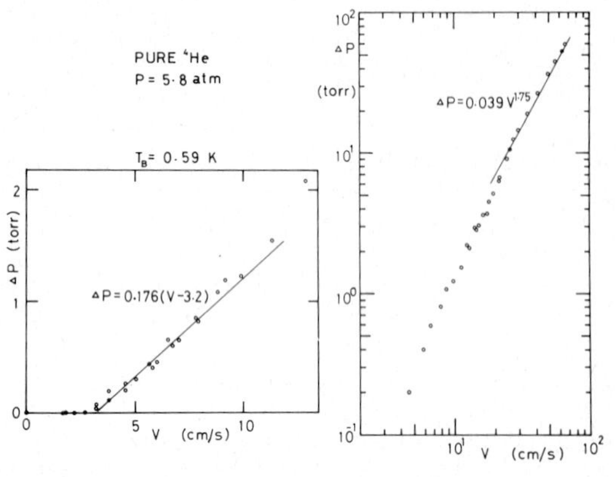

(a) (b)

Fig.2 Pressure difference across the capillary.
(a) Low flow velocity region
(b) High flow velocity region

concerned. Thus, the normal component in the capillary is considered to be in a fully developed turbulent state.

Based on the consideration given above, a comparison of Fig.1 with Fig.2(a) and 2(b) indicates that the system shows an appreciable thermal effect in the transitional region from the roton laminar flow state to the fully turbulent state. It does not show any noticeable thermal effect in the roton laminar flow state.

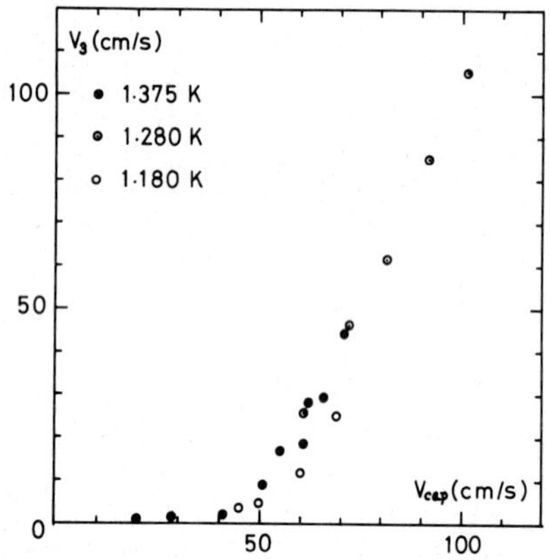

Fig.3 Velocity of ^{3}He atoms in the capillary plotted as a function of the mean flow velocity of ^{4}He at various bath temperature.

The NMR intensity of ^{3}He atoms in the chamber has been measured as a function of time at constant V. In this experiment, the chamber is thermally anchored tightly to the ^{3}He bath to eliminate the effect of the temperature dependence of the magnetic susceptibility of ^{3}He.

We have found that the intensity does not show a noticeable change until V reaches the transitional region mentioned above. The observed decrease in intensity is found to be fitted well with a single exponential function, from which the flow velocity of ^{3}He atoms, V_3, in the capillary can be deduced. Results are shown in Fig.3. It is noted that V_3 is much smaller than V in a wide range of V. This shows a clear contrast with the conjecture mentioned above that the normal component flows with the velocity nearly equal to V in the fully turbulent region. More precise measurements are underway to study the motion of ^{3}He atoms in the velocity region of $V<V_{sc}$ and/or the roton laminar flow state.

REFERENCES

(1) F. F. Simon, Physica 16 (1950) 753.
(2) J. F. Olijhoek, W. M. van Alphen, R. de Bruyn Ouboter and K. W. Taconis, Physica 35 (1967) 483.
(3) F. A. Staas and A. P. Severijns, Cryogenics 9 (1969) 422.
(4) Toshimi Satoh, H. Shinada and Takeo Satoh, Physica 114B (1982) 167.

LT-17 (Contributed Papers)
U. Eckern, A. Schmid, W. Weber, H. Wühl (eds)

MUTUAL FRICTION EVOLUTION NEAR THE He II - He I LAMBDA TRANSITION

A.G.F.DORSCHEIDT, T.H.K.FREDERKING*, H.VAN KEMPEN and P.WYDER

University of Nijmegen, The Netherlands ; *Univ. California, Los Angeles, California, USA

We have studied internal convection during counterflow in He II, in particular critical transport phenomena, up to the vicinity of the lambda point ($T = T_\lambda$) with a duct of 0.0137 cm diameter. This size requires an extension of the observation limit for temperature differences to the order of 1 to 10 micro-K . The results obtained support previous porous media data: The critical temperature gradient, suitably normalized, varies proportional to the superfluid density ratio, as T approaches T_λ .

1. INTRODUCTION

Internal convection processes in He II during low speed counterflow at zero net mass flow,are described quite well by the two-fluid model [1] [2] [3] . Axial heat transport rates in an insulated duct have been found to be remarkably large [4] with a ratio of the heat flux density (q) to the temperature gradient of $k_a = q/|\nabla T|$ exceeding the order of magnitude 100 W cm^{-1} K^{-1} . For low q and "high" T approaching T_λ , the two-fluid model predicts that $k_a \to \infty$, as the duct diameter (D) is raised to very large values. However, experiments impose *finite* observation limits on k_a and on the critical rate $q = q_c$ at the transition to vortex shedding regimes with the following behavior : 1. Generation of quantized Onsager - Feynman vortices ; 2. Interaction of the vortices with normal fluid (mutual friction [5]) ; 3. Vortex annihilation . As $T \to T_\lambda$, the superfluid critical velocity approaches zero , also $q_c \to 0$. One expects a very drastic reduction from a high k_a at ∇T_c to a low k_a once vortices have been created. As little is known about this, the present study has the purpose of resolving transition details by measuring q (∇T) in a duct of moderate diameter, possibly with an extension of k_a beyond present observation limits.

2. EXPERIMENTS

We have used a vertical stainless steel capillary, heated from the bottom (D= 0.0137 cm, test section length L = 1.9 cm ; $|\nabla T| = \Delta T/\Delta L$). Zero net mass flow has been imposed using vacuum insulation. Thus, the low speed convection rate has a k_a proportional to the thermal energy density (ρ S T) and inversely proportional to the shear viscosity (η_n) of normal fluid [2] :

$$k_a = (D^2/32)\, \rho^2 S^2 T / \eta_n = q/|\nabla T| \qquad (1)$$

(S entropy per unit mass of liquid, ρ liquid density) . Figure 1 presents experimental data: ΔT is shown versus the heater power $\dot{Q} = qD^2\pi/4$.

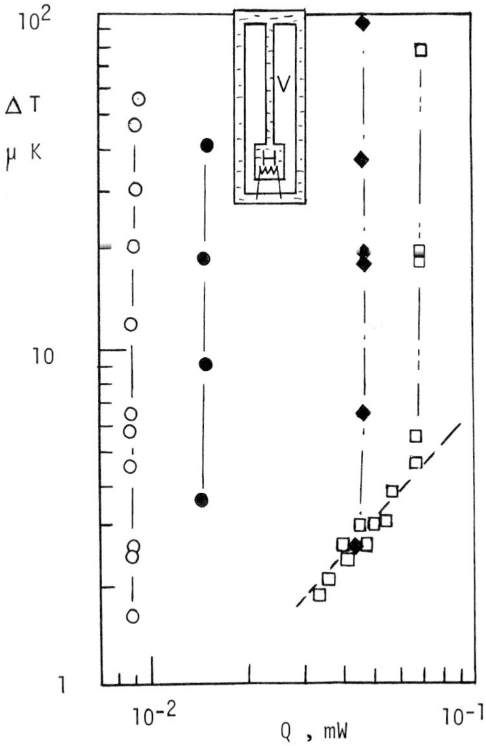

Figure 1. Temperature difference vs.heater power ; $t = T_\lambda - T$: o 3 mK; ● 6 mK; ◆ 30 mK ; □ 70 mK ; Lines are drawn to guide the eye ; Insert: Apparatus (schematically) ; H Heater; V Vacuum .

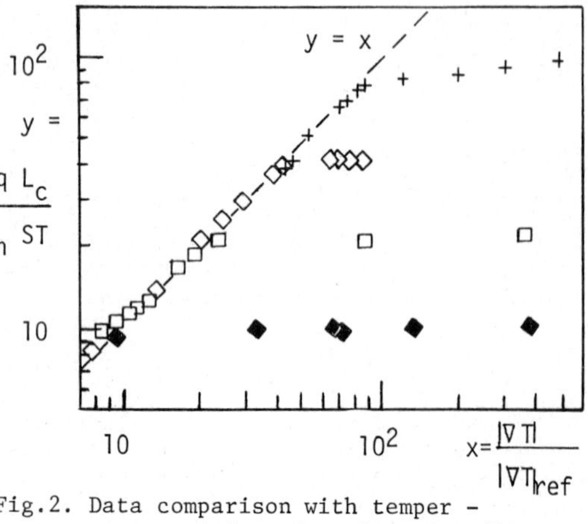

Fig.2. Data comparison with temperature - independent function (2); $t = T_\lambda - T$; ◆ 30 mK ; □ 70 mK ; ◇ 170 mK ; + 357 mK .

The insert (Fig.1) is a sketch of the apparatus. At short distances $t = T_\lambda - T$ from the λ-point, the linear regime is outside the observation limit. At t = 70 mK however, the data are in the subcritical regime. For the latter we introduce a normalized linear function. Its reference length $L_c = (K_p)^{1/2}$ is related to the permeability K_p of a previous porous media study [6] . For the present tube we have $K_p = D^2/32$, i.e. $L_c = 2.42 \times 10^{-3}$ cm. Multiplying Eq.(1) by $|\nabla T| L_c/(\eta_n ST)$ we obtain

$$q\,L_c/(\eta_n ST) = y = x \qquad (2)$$

The ratio x represents a normalized T-gradient

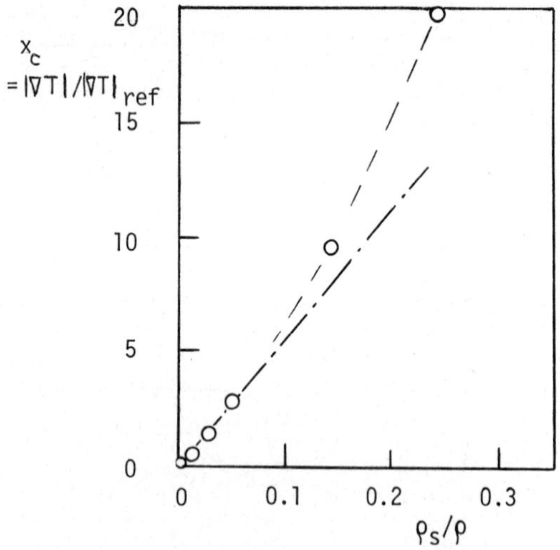

Fig.3. Critical temperature gradient (normalized) versus superfluid density ratio.

$x = |\nabla T|/|\nabla T|_{ref}$ with $|\nabla T|_{ref} = \eta_n^2/(L_c^3\,\rho^2\,S)$. There exists a critical value $x = x_c$:

$$x_c = |\nabla T_c|\;S\,\rho^2\,L_c^3/\eta_n^2 \qquad (3)$$

The largest k_a-value attained is of the order 10^5 W cm^{-1} K^{-1} . The non-linear function often is compared to the Gorter Mellink equation for mutual friction [5] with $m = d\log|\nabla T|/d\log q = 3$. The present m-values reach the order of 10, i.e. there is little variation in q in the data range covered for t < 70 mK.

3. DISCUSSION

We focus attention on the critical T-gradient. We note that function (2) is T-independent. Therefore, x_c is readily accessible. Figure 2 displays y(x) - data. For t < 70 mK extrapolation involves an x_c-error. Because of the large m-values, this error is small and within data uncertainty.

Critical values x_c have been determined previously for the porous medium of Ref.6 which shows $x_c \sim (\rho_s/\rho_n)$. Thus, as $T \rightarrow T_\lambda$ $(\rho_n \rightarrow \rho)$ we have an asymptote $(\rho_s/\rho) \sim t^{2/3}$ [7]. The present experimental results indeed show that x_c is proportional to (ρ_s/ρ) near T_λ (Fig.3) . Our duct represents a much better defined geometry. Thus, our data do support the previously found dependence of the normalized critical T-gradient on (ρ_s/ρ) .

ACKNOWLEDGEMENTS

The portion of the studies carried out at Univ. Calif. ,Los Angeles has been supported partly by NSF. S.W.K. Yuan provided input and assistance.

REFERENCES

[1] F. London, Superfluids, Vol.II,Wiley, New York, 1954.
[2] L.D. Landau and E.M. Lifshitz, Fluid Mechanics, Pergamon, London 1955.
[3] S.J. Putterman, Superfluid Hydrodynamics, North-Holland, Amsterdam 1974 .
[4] W.H. Keesom, B.F. Saris, L. Meyer, Physica 7 (1949) 817 .
[5] C.J. Gorter and H.J. Mellink, Physica 15 (1949) 285 .
[6] T.H.K.Frederking,H.Van Kempen, M.A.Weenen and P. Wyder, Physica 108 B (1981) 1129.
[7] J.R. Clow and R.D. Reppy, Phys. Rev.Letters 16 (1967) 887 .

LT-17 (Contributed Papers)
U. Eckern, A. Schmid, W. Weber, H. Wühl (eds)

OBSERVATION OF CRITICAL VELOCITY EFFECTS IN VIBRATING SUPERLEAK SECOND SOUND TRANSDUCERS

N. Giordano

Department of Physics, Purdue University, West Lafayette, IN 47907, USA

We have found that when a vibrating superleak transducer in ^{4}He is driven at high amplitudes, the magnitude of the generated second sound wave is not proportional to the driving force. In this nonlinear regime we also observe the generation of second sound at overtones of the driving frequency. Our results suggest that these nonlinearities occur when the velocity of the superfluid in the pores of the superleak becomes comparable to the critical velocity. These measurements appear to provide a unique probe of critical velocity effects.

1. INTRODUCTION

Since their invention [1] in 1969, vibrating superleak transducers (VSTs) have proven to be a very useful tool in the study of second sound in superfluids.[2,3] In recent experiments involving these transducers we have discovered unexpected nonlinearities in their behavior which appear to be due to critical velocity effects. These nonlinearities, which are most readily observed at temperatures near T_λ, do not appear to have been observed previously.

2. EXPERIMENTS

The experimental arrangement was the same as that described in detail in [3]. Two VSTs of standard design employing Nuclepore filters as superleaks were mounted at opposite ends of a cylindrical cavity. One VST was operated as a transmitter - it was biased at 90 VDC, and driven with an AC signal of amplitude $V_d < 20$ V(peak-to-peak), thus producing second sound at the drive frequency. The other VST was also biased at 90 VDC, and the resulting AC signal was detected with a lock-in amplifier or Fourier analyzed using a computer. At temperatures well away from T_λ the signal at the receiver was linearly proportional to the drive amplitude to within experimental error, which was typically a few percent. This is not surprising, since the VSTs themselves are essentially linear devices, [1] and such behavior has been observed by many previous workers.[2,4] However, near T_λ the signal at the receiver was no longer proportional to V_d, but saturated at high values of V_d. Moreover, this saturation occurred at lower values of V_d as T_λ was approached more closely.

Typical results for the receiver output as a function of drive frequency are shown in Fig. 1. Here we plot only the Fourier component of the receiver signal which is at the drive frequency The cavity was 2.0 cm long and the second sound velocity at this temperature is ~450 cm/s, so the resonance shown in Fig. 1 corresponds to the ~40th plane wave mode of the cavity. (We note

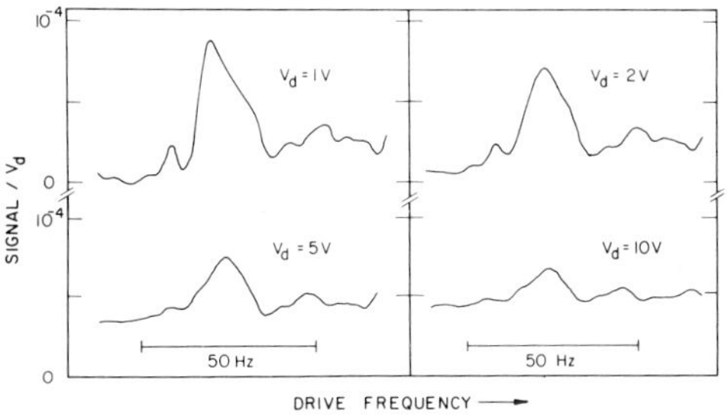

FIGURE 1
Receiver signal at the drive frequency, as measured with a lock-in detector, as a function of drive frequency for several different values of V_d. The drive frequency was near 4.5 kHz, the temperature was 5.5 mK below T_λ, and the superleak pores were 500 Å in diameter. Note that the signal has been normalized by V_d to facilitate comparisons of the results for different values of V_d, and that the receiver signal is given in V(rms).

that many non-plane wave modes were also excited, and some of these are also evident in Fig. 1; in fact the asymmetric line shape of this plane wave mode suggests that it partially overlaps one of the other modes.) For small drive signals the receiver output was proportional to the drive amplitude, but for large V_d the size of the resonance decreased relative to V_d, and the line shape also was distorted somewhat. These are clear indications of increased dissipation when the amplitude of second sound is large.

Figure 2 shows results obtained from Fourier analysis of the receiver signal. At low V_d the signal was dominated by a single component, which had the frequency of the driving force. However, as V_d was increased, components at integer multiples of the drive frequency became evident, and the magnitudes of these overtones grew rapidly with V_d. In addition, for a fixed V_d

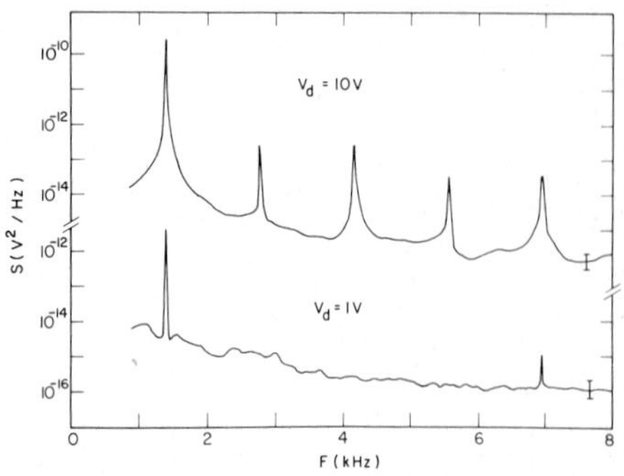

FIGURE 2
Receiver signal power, S, as a function of frequency for two different values of V_d as indicated. The conditions were the same as those given in the caption for Fig. 1, except that the drive frequency was 1.4 kHz.

the magnitudes of the overtones increased as $T \to T_\lambda$. The overtones were also observed with a lock-in arranged to detect at multiples of the drive frequency, and for large V_d were readily visible simply by examination of the raw signal with an oscilloscope.

3. DISCUSSION

The results shown in Figs. 1 and 2 strongly suggest that critical velocity effects become important at high drive amplitudes. Simple attenuation of second sound, etc., cannot explain our findings, since this would yield effects which scale linearly with V_d, and this is not observed. There are three locations in the system where the critical velocity induced dissipation could be occurring: in the superleak of the transmitting VST, in the superleak of the receiver VST, and in the cavity. If the critical velocity induced dissipation were confined to either the receiver VST or the cavity, then it should depend strongly on the overall level of second sound in the system, and hence on whether or not one is "on" a cavity resonance. However, this was not observed; instead, the sizes of the various overtones were ~ independent of whether or not the drive frequency corresponded to a cavity resonance. We therefore conclude that the dissipation is occurring in the transmitter VST. Intuitively this is not surprising since (1) one would expect the superfluid velocity, v_s, in the pores of the receiver superleak to be smaller than v_s in the transmitter, and (2) v_s in the transmitter superleak is larger than v_s in the cavity by a factor of α^{-1}, where α is the porosity of the superleak, which in our experiments was in the range 0.01 - 0.1.

Since the critical velocity vanishes at T_λ, it is to be expected that critical velocity effects will be important in VSTs when T approaches T_λ. However, such effects have not been observed by previous workers.[4] The reason for this is not entirely clear, but relatively minor differences between the VSTs used in the different experiments (such as in the tension in the superleak membrane, for example) could account for the difference which is observed.

Recent calculations of critical velocity effects in VSTs [5] yield results consistent with our experiments, in that the onset of dissipation and the associated overtones occur at values of the driving force which are in reasonable agreement with our findings. However, the calculations predict that the odd overtones (i.e., the overtones at odd multiples of the drive frequency) should be much larger than the even overtones. In contrast, Fig. 2 shows that this is definitely not the case. However, the theory as it now stands does not include the effect of the resonant cavity, and this may well be important. Another possibility, of course, is that the current theory of critical velocity effects in general [6] is in need of revision, but it would be premature to draw such a conclusion at this time. In any case, it seems clear that measurements of the type described here should prove to be a very useful tool in the study of critical velocities and related effects in ^{4}He.

ACKNOWLEDGEMENTS

I thank D. E. Beutler, S. Garrett, A. Tubis, and especially P. Muzikar for many useful discussions. This work was supported in part by the National Science Foundation through grant DMR 79-06716.

REFERENCES

[1] R. Williams, S.E.A. Beaver, J.C. Fraser, R.S. Kagiwada, and I. Rudnick, Phys. Lett. 29A (1969) 279; R.A. Sherlock and D.O. Edwards, Rev. Sci. Instrum. 41 (1970) 1603.
[2] See for example the references listed in [3].
[3] N. Giordano, J. Low Temp. Phys., 55 (1984) in press.
[4] D.S. Greywall and G. Ahlers, Phys. Rev. A7 (1973) 2145; G.G. Ihas and F. Pobell, Phys. Rev. A9 (1974) 1278.
[5] N. Giordano and P. Muzikar, this conference.
[6] J.S. Langer and M.E. Fisher, Phys. Rev. Lett. 19 (1967) 560.

LT-17 (Contributed Papers)
U. Eckern, A. Schmid, W. Weber, H. Wühl (eds)

THEORY OF CRITICAL VELOCITY EFFECTS IN VIBRATING SUPERLEAK SECOND SOUND TRANSDUCERS

N. GIORDANO and Paul MUZIKAR

Department of Physics, Purdue University, West Lafayette, IN 47907, USA

The theory of vibrating superleak second sound transducers is extended to include critical velocity effects. These become important when the speed of the superfluid in the pores of the superleak is large, and act to limit the amplitude of the generated second sound wave. They also lead to harmonic generation, i.e., the production of second sound waves with frequencies which are integer multiples of the driving frequency.

1. INTRODUCTION

There has recently been renewed theoretical interest in the properties of vibrating superleak second sound transducers (VSTs).(1-4) It has been shown that, among other things, these transducers are much more efficient at generating second sound than had previously been supposed. Some recent experiments involving the generation of second sound in ^{4}He using VSTs (5) have observed what appear to be critical velocity effects when the speed of the superfluid in the pores of the superleak becomes large. This has motivated us to extend the theory of VSTs to include critical velocity effects, and the results of some or our initial calculations are presented in this paper.

2. THEORY

When the superfluid velocity, v_s, is small, the chemical potential gradient within the superleak of a VST is negligible, (1,2) but when v_s is large, this gradient, which has been calculated by Langer and Fisher, (6) becomes significant. This leads to the following expression for the chemical potential difference across the superleak: (4,6)

$$\delta\mu_b-\delta\mu_c=\frac{1}{\rho L}\left[\rho_n c_2^2(x_m-x_s)-c_1^2(\rho_n x_m+\rho_s x_s)\right] - c_1 v+\frac{\rho_n c_2 w}{\rho_s}$$
$$= \mathrm{sign}(v_s-v_m)D\exp\left(-\frac{-B\alpha}{|v_s-v_m|}\right) . \qquad (1)$$

Here L is the spacing between the backplate and the superleak, x_m and x_s are the displacements of the normal and superfluid components (note that the normal component is assumed to be entrained in the superleak, so that the two have equal displacements), v and w are the center of mass and counterflow velocities just outside the superleak, the subscripts c and b refer to the chamber and backplate regions, and the other symbols have their usual meanings.(4) The last term in (1) describes the dissipation which occurs when the superfluid velocity is large. The constants B and D determine the size of the critical velocity and hence the size of the nonlinearities. The porosity of the superleak, α, enters here since in the pores of the superleak, the speed of the superfluid relative to the normal component is larger than that outside, v_s-v_m, by a factor of α^{-1}. We also need the equation of motion of the superleak,(3,4)

$$M_m^*\dot{v}_m+M_s^*\dot{v}_s=-Kx_m-A\rho c_1\left(\frac{c_1(\rho_n x_m+\rho_s x_s)}{\rho L}+v\right)+g(t). \qquad (2)$$

Here M_m^* and M_s^* are the effective masses of the superleak and superfluid in the superleak, K is the "spring constant" and A the area of the superleak, and g(t) is the driving force. Because of the awkward form of the nonlinear term in (1), we have solved these equations numerically using a standard fourth order Runge-Kutta method. There are a large number of parameters involved in (1) and (2), and it does not appear useful to attempt to cast these equations into a "dimensionless" form. Therefore, for the calculations described below, we have chosen parameter values appropriate for typical transducers (4) operating in ^{4}He.

Figure 1 shows results for the generated second sound wave, w, as a function of time for a driving force of the form $g(t)=A_o\sin(2\pi F_d t)$. The values of the constants B and D, which determine the size of the nonlinear term in (1), have been taken from the results of DC flow experiments.(7) For convenience, fig. 1 shows w/A_o so that results for different driving amplitudes can be compared directly. For small A_o, w is sinusoidal, and it is also proportional to A_o, as expected. However, as A_o is increased, critical velocity effects become important, and w exhibits a "clipped" waveform. This clipping results in a reduced value of w/A_o; that is, w no longer scales linearly with A_o, as it does for small drive amplitudes. In fig. 2 we show the Fourier transforms of the waveforms in fig. 1. At low drive amplitudes w consists of one component at the drive frequency, but for large A_o, when the clipping is evi-

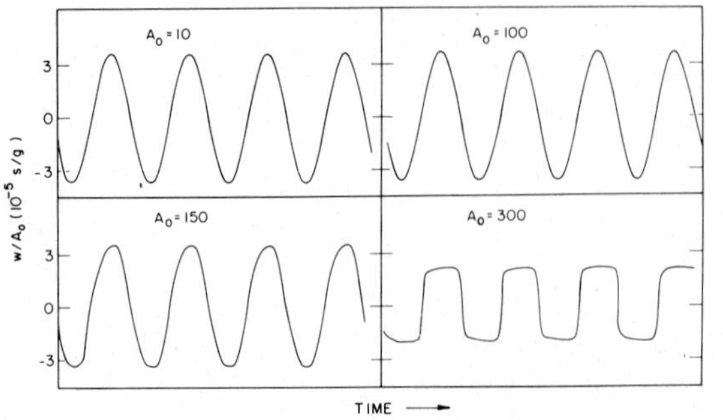

FIGURE 1
w/A_0 as a function of time for various values of A_0 (the units of A_0 are dynes). The following parameter values were used: F_d= 4kHz, c_2=450 cm/s, ρ_s/ρ=0.044, K=1×10^8 g/s^2, L=20μm, α=0.012, and the other parameters are appropriate for the transducers described in (4). These parameters correspond to a temperature 5 mK below T_λ, with a 500 Å Nucleopore filter as the superleak. The constants B and D were obtained from the parameters $\beta=1.3\times10^{-12}$cm^{3}g/s^3 and $f_0=5\times10^{27}$Hz/cm^3 given in (7).

dent, w also contains components at multiples of the drive frequency.

3. DISCUSSION

The qualitative features of our calculation agree fairly well with the available experiments.(5) Namely, the quantity w/A_0 does decrease at values of A_0 which are in order of magnitude agreement with our results, and the associated overtones of the driving frequency are also observed. However, a puzzling feature of the experiments (5) is that both even and odd overtones are observed; that is, w exhibits components with frequencies m times the driving frequency, where m is an integer, both even and odd. In contrast, the calculations, fig. 2, exhibit essentially only overtones with m odd. The reason for this discrepancy is not clear, but it should be noted that the experiments were performed in a resonant cavity. This may well be important, and we plan to extend our calculations to include the effects of a cavity. One goal of this work is to use the experimentally observed nonlinearities to actually test the theory of Langer and Fisher,(6) and, if appropriate, to provide an improved measure of the constants B and D.

Aside from comparing with experiment, our calculations are also of interest with regards to the possibility of observing chaotic behavior. Equations (1) and (2) describe a driven nonlinear dissipative system which is at least qualitatively similar to systems which are known to exhibit chaos.(8) So far we have not observed such behavior in our calculations, but it would clearly be very interesting if it were to occur.

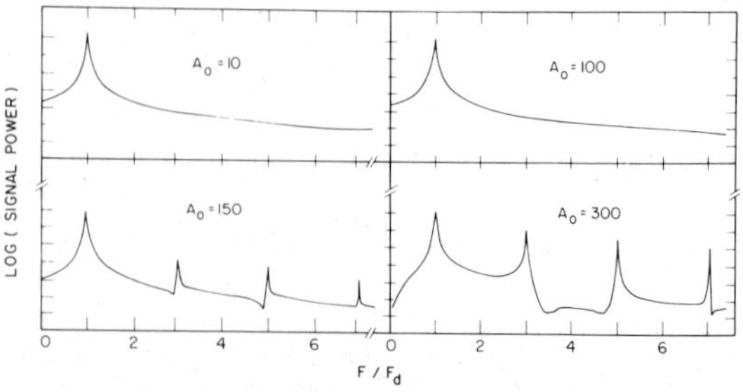

FIGURE 2
Fourier transforms of the waveforms shown in fig. 1. The tick marks on the vertical scales are separated by two decades in signal power.

ACKNOWLEDGEMENTS

We thank D.E. Beutler, S. Garrett, and A. Tubis for several useful conversations. This work was supported in part by the National Science Foundation through grants DMR 79-06716 and DMR 80-20249.

REFERENCES

(1) M. Liu and M.R. Stern, Phys. Rev. Lett. 48, 1842 (1982); ibid, 49, 1362 (1982).
(2) D.L. Johnson, Phys. Rev. Lett. 49, 1361 (1982).
(3) W.M. Saslow, Phys. Rev. B27, 588 (1983).
(4) N. Giordano, J. Low Temp. Phys., 55 (1984) in press.
(5) N. Giordano, this conference.
(6) J.S. Langer and M.E. Fisher, Phys. Rev. Lett. 19, 560 (1967).
(7) H.A. Notarys, Phys. Rev. Lett. 22, 1240 (1969).
(8) See for example, J.-P. Eckmann, Rev. Mod. Phys. 53, 643 (1981).

LT-17 (Contributed Papers)
U. Eckern, A. Schmid, W. Weber, H. Wühl (eds)

THERMOCIRCULATION EFFECT AND QUANTUM INTERFERENCE PHENOMENA IN SUPERFLUID HELIUM

V.L.Ginzburg and A.A.Sobyanin

P.N.Lebedev Physical Institute, Academy of Science USSR, 117924 GSP, Moscow,USSR

The circulation effect which must take place in a nonuniformly heated toroidal vessel with superfluid liquid (He II) under the conditions of non-zero complete superfluid velocity circulation is discussed. Some types of devices for observation of quantum interference phenomena in He II are proposed.

I. INTRODUCTION

When a temperature gradient is acting on a He II-filled toroidal vessel with two week-links (Fig. 1), there is to appear a superfluid motion involving the whole vessel. Such thermocirculation effect has been considered in refs 1-3 and was being studied in experiment (4).

In (1-3) a particular case was mainly treated where superfluid velocity circulation along a closed contour (of the type shown in Fig. 1) is zero. At the same time the case of non zero circulation seems to deserve special attention because some its specific features, such as the jumps in circulating flow velocity, hysteresis phenomena and residual superfluid flows, make it possible, in principle, to obtain a rich information about the nature of superfluid critical velocities in He II to say nothing of the possibility to measure directly the circulation quantum. Besides, the use of thermocirculation effect provides opportunities for observation of quantum interference phenomena in He II which are analogous to the ones in superconductors (5). These are the questions that we discuss here.

2. THERMOCIRCULATION EFFECT

If "week-links" of the vessel shown in Fig. 1 are circular capillaries

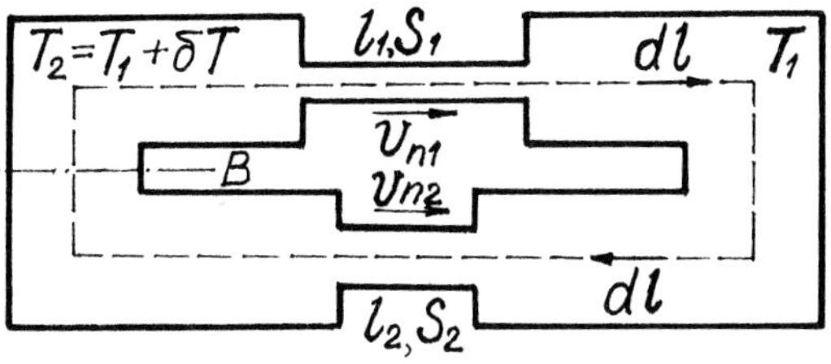

FIGURE 1

($i = 1,2$) with a radius r_i (of the cross-section $S_i = \pi r_i^2$) and length l_i, then in the presence of a temperature difference δT the normal part of the liquid flows through the capillaries with the velocities

$$v_{ni} = (\rho\sigma S_i / 8\pi\eta l_i)\,\delta T, \qquad (1)$$

where ρ helium density, σ entropy per unit mass and η normal fluid viscosity.

If the vessel ("ring") is open ("cut" along the AB line), then in each of the capillaries a mass flow is absent and the superfluid velocity is

$$v_{si}^{(0)} = -(\rho_{ni}/\rho_{si})\,v_{ni}. \qquad (2)$$

For a closed ring the quantisation condition

$$\oint \vec{v}_s\, d\vec{l} = (h/m)n, \qquad n = 0, \pm1, \pm2 \ldots (3)$$

is to be hold and in the vessel, along with counterflows of superfluid and normal components inside the links, a total circulating superfluid motion should to arise with the velocity (in the wide parts of the vessel)

$$v_{sc} = \frac{h}{m l_0}(n - A\delta T), \quad n = 0, \pm1, \pm2, \ldots \qquad (4)$$
$$A = \sigma\rho\rho_n S_2 / 16\pi^2\eta\rho_s .$$

When deriving the expression we assumed that the length l_0 of wide parts of the vessel is greate enough and $S_2 \gg S_1$.

If at $\delta T = 0$ there was no circulation, then with the appearence of a temperature difference the velosity v_{s0} is increasing according to eq. (4) with $n = 0$. For $A\delta T > 1/2$ the transition into the state with $n = 1$ is already admissible but a hysteresis should be observed generally. In fact the state with n = 0 would exist as a metastable one until at some part of the ring the velocity v_s reaches its critical value corresponding to the onset of vortex formation. From this it is clear already that the effect in question may be quite useful for studies of vortex nucleation kinetics. Besides, the fact that as a result of one or a series of the transitions, the circulation in the ring changes exactly by a integral number of quanta, provides a possibility, in principle, to measure single circulation quantum. The magnitude h/m of this quantum may be determined, for example, from measurements of residual superfluid flows which remain inside the ring after removal of temperature gradient if before this (in the course of heating) spontaneous transitions into the states with $n \neq 0$ took place.

The symplest way to notice experimentally the presence of circulating flow is from reaction of a freely suspended ring. In ref. 4 the oscillations of a ring suspended on an elastic filament were studied, under influence of a heat current created inside the ring. The results of ref. 4 are in agreement with the theory (1-3) but the sensitivity of these experiments was insufficient to reveal single circulation quanta.

3. SUPERFLUID QUANTUM INTERFEROMETERS

One of the most promising applications of thermocirculation effect consists in its use for observation of quantum interference phenomena in He II. To this end it is sufficient to place in the way of a thermally induced circulating flow a corresponding Josephson junction. The role of such a junction can be played, for example, by a diaphragm with a small orifice or a short capillary or a slit, in which the λ-transition temperature is shifted by means of electric field, etc. It should be emphasied that near the λ-point, where the order parameter coherence length increases rapidly, a corresponding orifice or a capillary may have quite macroscopic dimensions ($r \sim 10^{-4}$ cm at $T_\lambda - T \sim 10^{-6}$K).

The scheme of one of possible superfluid quantum interferometer which is analogous to the Mercereau et al's SQID (5) is shown in Fig. 2. Here two identical Josephson junctions D_1 and D_2 are placed into the lower wide part of the ring. Experiment should consist in measuring the maximal superfluid current I_{sm} passing through the interferometer as a function of the temperature difference δT between the upper and lower halves of the ring or as a function of the heat power Q delivered in the upper half. The difference δT plays here the same role as external magnetic flux for SQID's. A consideration similar to that used for SQID's (see ref.5) shows that I_{sm} is an oscillating function of δT or Q with the period

$$L_T = 2h\eta\rho_s / m\sigma\rho\rho_n(S_2 - S_1) \qquad (5)$$

or, if we deal with the dependence of I_{sm} on Q,

$$L_Q = h\sigma\rho\rho_{s2} T S_2 / 2m\rho_{n2}\ell_2, \quad S_2 \gg S_1 \qquad (6)$$

At $T \sim 1.5$K and $S_2/\ell_2 \sim 200$ cm the period $L_Q \sim 10^{-3}$W, so that for observation of oscillations of I_{sm} an appropriate choice of the geometry parameters of the ring and very small heat powers are required.

Another possible interferometer is a ring of the type shown in Fig.1 with one Josephson junction. The reaction of this ring is expected to be periodic in δT or Q with periods given by the same expressions as above.

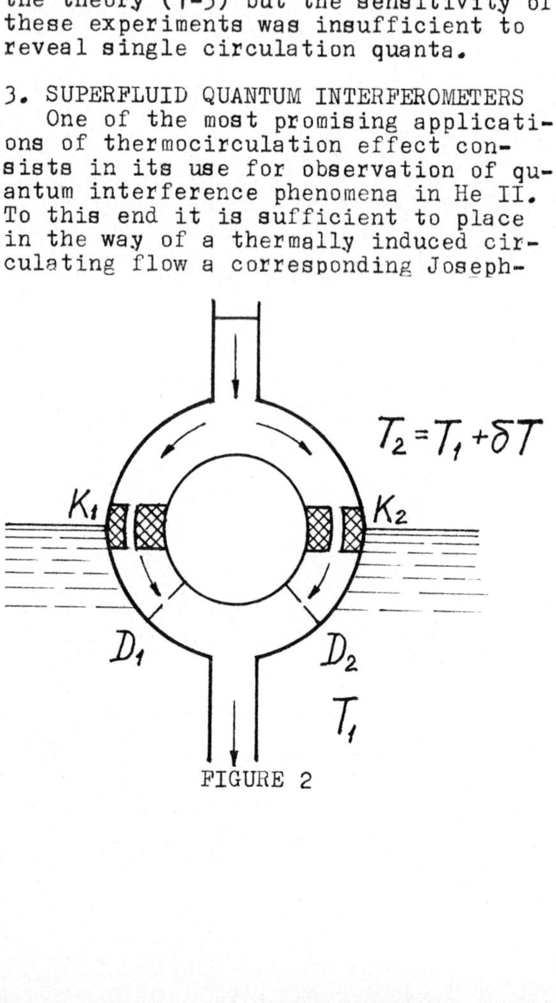

FIGURE 2

REFERENCES

(1) V.L.Ginzburg,G.F.Zharkov and A.A. Sobyanin,ZhETF Lett.,20(1974) 223.
(2) G.F.Zharkov and A.A.Sobyanin, ZhETF Lett., 20(1974) 163.
(3) V.L.Ginzburg,A.A.Sobyanin and G.F. Zharkov,Phys.Lett., 87A (1981)107.
(4) G.A.Gamzemlidze and M.I.Mirzoeva, ZhETF 79(1980)921; 84(1983) 1725.
(5) A.Barone and G.Paterno,Physics and Applications of the Josephson effect (Willey, N.Y., 1982).

LT-17 (Contributed Papers)
U. Eckern, A. Schmid, W. Weber, H. Wühl (eds)

THE ORDER-PARAMETER PHASE DIFFERENCE AT THE CRITICAL RATE OF FLOW OF SUPERFLUID ^{4}HE THROUGH A TINY ORIFICE

B.J. ANDERSON, B.P. BEECKEN, and W. ZIMMERMANN, JR.

Tate Laboratory of Physics, University of Minnesota, Minneapolis, Minnesota, U.S.A. 55455

In the course of a search for a Josephson effect in superfluid ^{4}He, we have made measurements by an ac method of the order-parameter phase difference existing between two reservoirs of superfluid ^{4}He connected by a tiny orifice at the critical rate of flow through the orifice. For orifices of somewhat irregular shape with mean diameters from 100 to 400 nm in nickel films $\lesssim$100 nm thick, we find for T < 1.0 K values of the critical phase difference $\Delta\phi_c$ from 7 to 38 times 2π.

1. INTRODUCTION

An unresolved question of considerable interest is whether Josephson effects can be observed in superfluid ^{4}He. Several years ago F.H. Wirth and one of us began a search for such effects, with the intent of using as weak links single pores whose lengths and diameters were of order 100 nm, two orders of magnitude smaller than had been used in many of the previous searches.[1] In this article we report direct measurements of the order-parameter phase difference at the critical rate of flow through several such pores. These measurements were made during a continued search for an ac Josephson effect and have an important bearing on the likelihood of observing such an effect.

Let $\Delta\phi$ be the difference in phase of the superfluid order parameter between two reservoirs connected by a pore. In the absence of vorticity, $\Delta\phi$ is well-defined and may be written

$$\Delta\phi = \frac{m}{\hbar}\int_1^2 \vec{v}_s \cdot d\vec{r}, \qquad (1)$$

where m is the mass of the ^{4}He atom, $\hbar$ is the reduced Planck constant, $\vec{v}_s$ is the superfluid velocity, and the integral runs from some quiescent point in reservoir 1 through the pore to a similar point in reservoir 2. Let $\Delta\phi_c$ be the value of $\Delta\phi$ at which phase slip and dissipation first appear at the pore, the quantity which we measure in the present experiments.

We believe that for an ac Josephson effect to be observable it is highly desirable for $\Delta\phi_c$ to be of order 2π. In this case, unlike the case in which $\Delta\phi_c$ is many times 2π, the number of final states with lower rates of flow to which phase slip can lead is limited to a very few, and the chance of synchronizing the phase slip with some external event is enhanced.

Shortening the length and diameter of the pore should tend to decrease $\Delta\phi_c$, as long as a compensating increase does not occur in v_{sc}, a representative critical value of v_s in the pore. If at low temperatures v_{sc} were determined by intrinsic microscopic considerations, in particular by the Landau criterion,[2] then the length and diameter of the pore would need to be reduced to values of the order of 1 nm, nearly that of the coherence length. On the other hand, if v_{sc} were given by an expression similar to that proposed by Feynman[3]

$$v_{sc} \sim (\hbar/md)\ln(d/a), \qquad (2)$$

where d is the diameter of the pore and a is the vortex core parameter, of order 0.1 nm, then $\Delta\phi_c$ would be of order 2π for any pore having length $\ell \lesssim d$ over a wide range of d.

The experimental situation is unclear. Critical velocities both less than and greater than those of Eq. (2) have been found. However, because these measured velocities were in general much smaller than the Landau critical velocity and tended to show less dependence on d than those of Eq. (2), it appeared that our criterion for $\Delta\phi_c$ might be satisfied by a very-small-diameter orifice in a film whose thickness was comparable to or less than the diameter, even though both of these lengths were considerably larger than the coherence length.

2. EXPERIMENT

The experimental cell consists of two chambers completely filled with superfluid ^{4}He connected by a weak-link orifice in parallel with a larger opening. This arrangement constitutes a double-ended superfluid Helmholtz resonator in which fluid oscillates between upper and lower chambers at a resonant frequency of the order of 1000 Hz. The resonance is excited by the use of a piezoelectric driver coupled to the flexible upper wall of the upper chamber. The resulting pressure fluctuations in the lower chamber are detected by sensing the flexure of the flexible lower wall of the lower chamber capacitively. The cell is coupled to a ^{3}He refrigerator and can be held at temperatures down to 0.65 K for several days. At 0.65 K the low-amplitude Q of the resonance is typically of the order of 10,000.

The onset of supercritical flow and dissipation in the weak-link orifice is detected by measuring the pressure amplitude as a function of volume drive amplitude at resonance. At low drive amplitudes, the characteristic is steep, reflecting a high Q. At a higher amplitude, there is seen an abrupt decrease in slope representing the onset of increased dissipation. At very much higher amplitudes, there is a further decrease in slope. We interpret the first decrease in slope to mark the onset of supercritical flow in the weak-link orifice and the second to mark the onset of supercritical flow in the larger opening. Confirmation of this interpretation comes from two runs without weak-link openings, in which only the higher-amplitude increase in dissipation was seen.

From such data we may calculate $\Delta\phi_c$ for the weak-link orifice. In the absence of vorticity, the superfluid equation of motion [2] may be integrated along the same contour used in Eq. (1) to yield

$$\frac{\hbar}{m}\frac{d\Delta\phi}{dt} + \frac{\Delta P}{\rho} = 0 , \qquad (3)$$

where ΔP is the pressure difference between chambers and ρ is the mass density. Here we have used Eq. (1) and assumed that temperature gradients and dissipative terms are unimportant. Since ΔP is undergoing sinusoidal oscillation, so is $\Delta\phi$. Thus $\Delta\phi_c$, which equals the critical amplitude of $\Delta\phi$, can be calculated from the critical amplitude of ΔP. The latter can be calculated from the critical amplitude of the pressure variation measured in the lower chamber, using a knowledge of the chamber volumes.

3. RESULTS

Including two orifices studied earlier by Wirth and Zimmermann,[1] a total of five weak-link orifices have been studied. A combination of photolithography, electron-beam lithography, and ion milling was used to make these orifices in nickel films $\lesssim 100$ nm thick which are free-standing over a diameter of ~ 20 µm.[4] Listed in TABLE 1, these orifices ranged from nearly circular to rather irregular in shape. The diameters are representative ones determined by electron microscopy. As shown in the table, $\Delta\phi_c/2\pi$ for these orifices ranged from 7 to 38, with no discernible relation between $\Delta\phi_c$ and size. Also shown in the table are rough estimates of v_{sc} inferred from $\Delta\phi_c$. All of the data were taken at T < 1.0 K, much at 0.65 K. The value of $\Delta\phi_c$ was found to be highly reproducible during a low-temperature run.

TABLE 1

Parameters for the five orifices studied.

	Diameters (nm)	$\Delta\phi_c/2\pi$	v_{sc} (m/s)
#1	90 x 100	18	9
#2	150 x 170	10	3.3
#3	170 x 200	38	13
#4	300 x 450	7	1.6
#5	290 x 470	27	6

Thus in conclusion, for several pores whose lengths and diameters are of the order of 100 nm we have found values of $\Delta\phi_c$ which are large in comparison to those favorable for the observation of a Josephson effect, despite the suggestions presented in Sec. 1. With the exception of orifice #4, for which $\Delta\phi_c/2\pi = 7$, none of the orifices has shown any suggestion of a Josephson effect.[1] We are continuing to study $\Delta\phi_c$ for other orifices in an effort to understand how it depends on orifice geometry and possibly other factors.

ACKNOWLEDGEMENTS

We wish to acknowledge our indebtedness in this work to Dr. F.H. Wirth. We wish also to acknowledge support by the National Science Foundation (Grant DMR 8112973) and by the Graduate School of the University of Minnesota. Our lithography was made possible with the assistance of Honeywell, Inc. and of the University's Electron Microscopy Center, Microelectronics Laboratory, and N.S.F. Regional Facility in Surface Analysis (Grant CHE 7916206).

REFERENCES

[1] F.H. Wirth and W. Zimmermann, Jr., Physica, 107B (1981) 579.

[2] I.M. Khalatnikov, Introduction to the Theory of Superfluidity (Benjamin, New York, 1965) Chaps. 1,8,9.

[3] R.P. Feynman, in Progress in Low Temperature Physics, Vol. I, edited by C.J. Gorter (North-Holland, Amsterdam, 1955) p. 46.

[4] B.P. Beecken, Fabrication of Small Orifices in Thin Films (M.S. thesis, University of Minnesota, 1984).

LT-17 (Contributed Papers)
U. Eckern, A. Schmid, W. Weber, H. Wühl (eds)

SUPERFLUID FLOW OF HELIUM II AT LOW VELOCITIES

G. MAREES and H. v. BEELEN

Kamerlingh Onnes Laboratorium der Rijksuniversiteit Leiden, Leiden, The Netherlands

A U-tube is used to study the properties of superfluid flow of helium II at low velocities. A high resolution is obtained by the use of a very long and narrow flow tube (glass, L=10.0 m, i.d.=217 μm). Both the chemical-potential and temperature difference along the tube are recorded during the decay back to zero from an imposed level difference and during the subsequent slightly damped U-tube oscillation. The results show that the flow resistance remains zero up to a critical superfluid velocity $v_{sc} \approx 1.6$ cm s^{-1} for all bath temperatures, confirming earlier results obtained with different measuring techniques by others.

1. INTRODUCTION

Recently Tough and coworkers have carried out high-resolution measurements on superfluid flow in glass tubes, extending down to very small superfluid velocities v_s (1),(2). The results revealed a very simple and systematic behaviour. It can be described by a mutual friction force which remains undetectably small up to a critical superfluid velocity v_{sc}, from where a rather steep transition sets in towards the cubic dependence on v_s valid for the higher velocities. For a circular glass tube (of length L=9.9 cm and inner diameter d=134 μm) it was found that $v_{sc} \approx 1.5$ cm s^{-1}, almost independent of the bath temperature in the measured range from 1.3 K to 1.9 K. It was further found that at the higher velocities, the proportionality factor is a universal function of temperature, independent of the geometry of the glass tubes.

The above results are in good agreement with the earlier findings of others. The cubic dependence at the higher velocities is always observed, also for metal tubes be it with a different proportionality factor (1). The resolution of the earlier experiments, however, was in general not good enough to resolve the behaviour at the smaller velocities and to establish the existence of a critical velocity; the clear evidence provided in ref. (2), could only be obtained with the standard method thanks to a highly improved measuring technique. Nevertheless, data down to very small flow velocities have been reported in ref.(3), where the required high resolution was obtained by the use of an extremely long flow tube (glass with L=106 m, d=170 μm). It was shown that in the range of superfluid velocities between 1.3-1.8 cm/s the flow resistance sharply drops to a very low value indeed.

Stimulated by the results of ref.(2) we recently decided to use the technique of ref.(3) to carry out an independent investigation on adiabatic superfluid flow in flow tubes of various dimensions. Some preliminary results, as obtained for a glass tube with L=10.0 m, d=217 μm are presented in this contribution.

2. THE EXPERIMENT

The flow device is essentially a U-tube, as is shown in the figure. Adiabatic flow conditions are achieved by means of the superleak and the vacuum can. The whole system is separated from the surrounding bath by the use of two filling lines, connected to the upper ends of the standpipes. The glass standpipes have a rectangular outer perimeter; two opposing sides are clad with a chromium layer on the outside, which enables the capacitive detection of the liquid levels. Thermal damping of the expected level oscillations is greatly reduced by the use of a large heat exchanger.

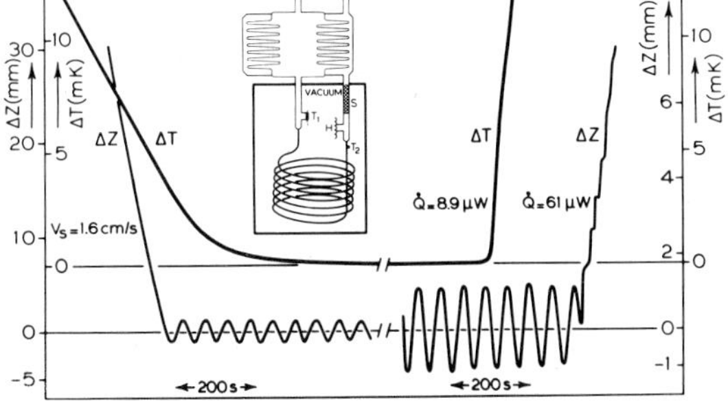

An initial level difference $\Delta z \equiv z_2 - z_1 > 0$ is imposed by switching on heater H. When H is switched off again, both Δz and $\Delta T \equiv T_2 - T_1 > 0$ decay back to zero. The typical run presented in the figure shows that after an almost straight decay a slightly damped U-tube oscillation in Δz results, which does not show up in ΔT. The temperature difference decreases directly back to zero, be it with a certain retardation time.
It is also shown in the figure that the oscillation in Δz is hardly affected by a small transport of normal fluid in the tube, induced by switching on heater H and registered by the immediate response of the temperature. Even when Δz starts to rise steadily, the oscillation is not immediately destroyed.

Up to now data have been collected at four bath temperatures (1.25 K, 1.40 K, 1.60 K and 1.80 K). Some runs have been taken in the opposite flow direction, i.e. they were started with $\Delta z < 0$. The results for both types of run are found to coincide.

3. PRELIMINARY ANALYSIS AND DISCUSSION

In the table various quantities derived from the calibrated recorder traces have been collected. The superfluid velocities v_{sd}, corresponding to the straight decays of Δz have been calculated according to

$$\rho_s v_s A_c = \tfrac{1}{2} \rho A_s \, \Delta\dot{z} \qquad (1)$$

where A_s (=5.1 mm^2) and A_c are the cross-sectional areas of the identical standpipes and of the tube. In eq.(1) the small counterflow of normal fluid, given by

$$\rho s T v_n A_c = -C\Delta\dot{T} \qquad (2)$$

(with C the heat capacity of the reservoir at T_2) has been neglected.

T_0 (K)	v_{sd} (cm s^{-1})	v_{nd} (cm s^{-1})	ΔT (mK)	τ_p (s)	τ (s)	α (10^{-4}s^{-1})
1.25	1.62±005	0.4	4.6	160	54.4	1.5
1.40	1.62	0.2	1.1	65	55.7	3.6
1.60	1.60	0.2	0.4	20	59.3	7.9
1.80	1.53	0.2	0.2	10	65.3	14.9

The values for v_{nd}, as calculated from eq.(2) are also given in the table, together with the corresponding temperature drops ΔT_{pd} that follow from Poiseuille's law. These values for ΔT_{pd} could very well agree with the values of ΔT that are observed at the moment when Δz crosses zero for the first time. Unfortunately the heat capacity C has not yet been determined with sufficient accuracy to decide from a comparison between the values of Δz, ΔT and ΔT_{pd} whether there is an additional pressure drop connected with the superfluid flow. It is mentioned that the observed subsequent retardation times for ΔT correspond well with the expected values τ_p for laminar counterflow.

Finally the table also gives the measured values for the period τ and the damping constant α of the U-tube oscillations. The values for τ agree within 1% with the expected values τ_0, calculated according to

$$\tau_0^2 = 4\pi^2 \frac{L}{2g} \frac{\rho A_s}{\rho_s A_c} \qquad (3)$$

while the values of α can be largely accounted for by the calculated value α_T of the thermal damping in the heat exchanger with thermal conductance K

$$\alpha_T = \frac{\rho_s s^2 T_0 A_c}{2KL} \qquad (4)$$

It is finally remarked that, only at the highest bath temperature, the oscillations did show up also in ΔT, be it with an amplitude $\lesssim 10\ \mu$K.

It can be concluded that the present results are in good agreement with that of refs.(1), (2) and (3). Particularly the agreement in the values found for v_{sc} with different tubes is remarkable. The present measurements will be continued and extended to flow tubes of other dimensions.

ACKNOWLEDGEMENTS

We would like to thank Mr. J. Dorrepaal for his assistance during the measurements.

REFERENCES

(1) R.A. Ashton, L.B. Opatowsky, and J.T. Tough, Phys. Rev. Letters 46(1981)658.
(2) M.L. Baehr, L.B. Opatowsky, and J.T. Tough, Phys. Rev. Letters 51(1983)2295.
(3) A. Hartoog and H. van Beelen, Physica 103B(1981)263.

LT-17 (Contributed Papers)
U. Eckern, A. Schmid, W. Weber, H. Wühl (eds)

THE EFFECT OF NORMAL FLUID FLOW ON THE TRANSITION TO SUPERFLUID TURBULENCE

J.T. TOUGH and Marie L. BAEHR

Department of Physics, The Ohio State University, Columbus, Ohio 43210, USA*

We have measured the critical velocity for the transition to superfluid turbulence in flows where both the normal fluid and superfluid velocities could be varied independently.

In a recent experiment (1) we demonstrated the surprising result that the transition from dissipationless superflow to superfluid turbulence occurs at a critical velocity V that is independent of temperature. For superfluid velocities $V_s \gtrsim 2V_c$ the dissipation associated with the turbulence is in quantitative agreement with a density of quantized vortex lines L_0 given by the Schwarz theory (2) of homogeneous turbulence. The modification of this theory (3) to account for the effect of flow tube boundaries does produce a critical velocity below which L_0=0, but does not give the temperature independent result observed.

In the Schwarz theory, vortex loops grow under the influence of the normal fluid but suffer topology-changing reconnections which keep the distribution random and lead to a steady state line density L_0. In the modified theory, vortices are also allowed to make surface reconnections at the tube boundaries. It is found that as the relative velocity is reduced the surface reconnection process dominates but yet is not sufficient to maintain the dynamical balance and the vortex tangle ceases to be self-sustaining. The actual value of the critical relative velocity is strongly dependent on the exact nature of the line-surface reconnection. Schwarz has shown that when the surface effect is inhibited the critical velocity is increased. He has suggested that this might be the situation when a normal fluid flow is present since the normal fluid velocity must vanish at the boundaries. We report here a systematic investigation of the critical velocity in flows where the average normal fluid velocity V_n, and the superfluid velocity V_s could be varied independently. Our results are in qualitative agreement with the suggestion of Schwarz.

The apparatus is similar to that used previously (1) and is shown very schematically in Figure 1. The flow tube is glass, 1.34×10^{-2} cm diameter and 10cm long. Superflow is produced by a fountain pump and an opposing normal flow is generated by the heater. Figure 2 shows the region of the (V_n,V_s) plane that is accessible in the experiments, and the angled line is the path in the plane that would be followed in a thermal counterflow experiment. The dissipation in the flow tube is determined from the temperature difference ΔT. Experiments are carried out by fixing V_n (on the thermal counterflow line) then increasing V_s (in the negative direction--the dashed vertical lines in Figure 2). For each V_n we observe a well-defined value

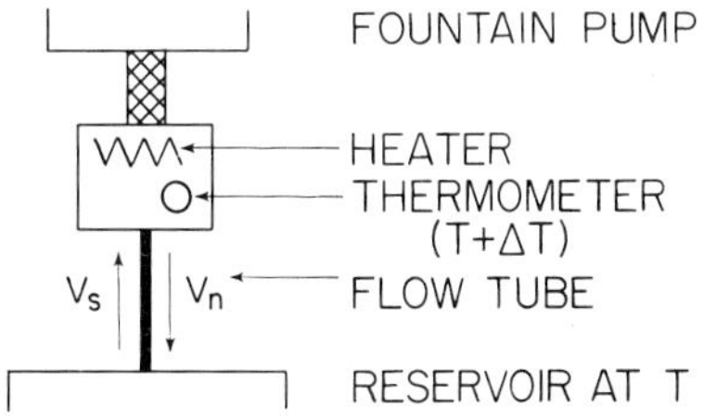

FIGURE 1
Schematic diagram of the apparatus.

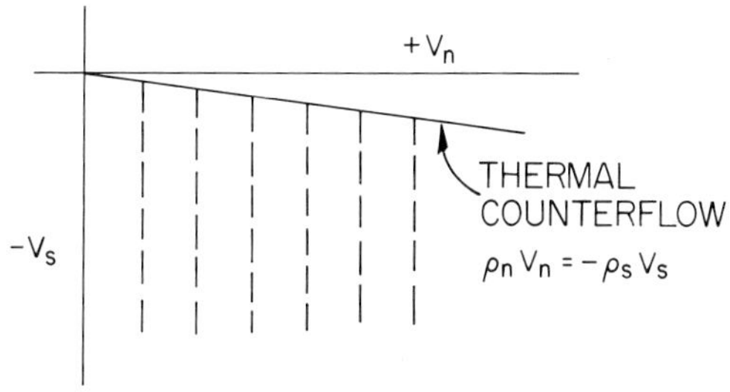

FIGURE 2
The region of the (V_n,V_s) plane accessible in these experiments is shown by the vertical dashed lines.

*Supported by National Science Foundation-Low Temperature Physics-Grant DMR8218052.

of V_s at which the vortex line density drops the zero and the dissipation is due entirely to the viscous normal fluid flow. The data are shown for 1.4K and 1.6K in Figure 3. The solid line is a simple linear fit to the data. The results indicate how the temperature independent critical velocity observed in superflow (V_n=0) continues into the (V_n,V_s) plane. If we boldly extrapolate this line above thermal counterflow into the region of positive V_s we obtain quantitative agreement with observations of Slegtenhorst and van Beelen (4) where the superfluid turbulence is observed to vanish abruptly (the "steep branch").

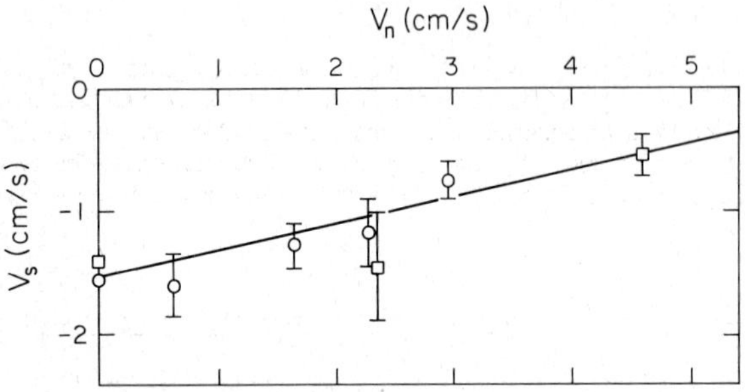

FIGURE 3
Points in the (V_n,V_s) plane where the vortex line density drops to zero. The solid line is given by Equation (1).

In the Schwarz model it is the relative velocity between the two fluids that supports the turbulent state. In the spirit of this model we can interpret our results in terms of a critical value of the relative velocity (V_n-V_s). The straight line in Figure 3 is then equivalent to a linear increase of V_c with V_n,

$$V_c = V_o + mV_n \tag{1}$$

where V_o=1.5cm/s and m=0.785. We note that Equation (1) gives a temperature independent description of the critical velocity in the (V_n,V_s) plane.

In conclusion, we have determined the dependence of the critical relative velocity on the average normal fluid velocity V_n. The results are in qualitative agreement with the suggestion of Schwarz that the normal fluid flow would increase V_c. Our data indicate that this is a linear, temperature independent effect.

REFERENCES

(1) Marie L. Baehr, L.B. Opatowsky and J.T. Tough, Phys. Rev. Lett. 51, (1983) 2295.
(2) K.W. Schwarz, Phys. Rev. Lett. 49, (1982) 283.
(3) K.W. Schwarz, Phys. Rev. Lett. 50, (1983) 364.
(4) R.P. Slegtenhorst and H. van Beelen, Physica 90B (1977) 245.

LT-17 (Contributed Papers)
U. Eckern, A. Schmid, W. Weber, H. Wühl (eds)

THE KONTOROVICH EFFECT IN UNSATURATED ^{4}He FILMS

M.G.M. BROCKEN, I. VAN ANDEL, D.P. STIKVOORT and H. VAN BEELEN

Kamerlingh Onnes Laboratorium der Rijksuniversiteit Leiden, Nieuwsteeg 18, 2311 SB LEIDEN, The Netherlands

Preliminary results of an investigation on the decrease in filmmass due to steady flow in unsaturated helium films are presented. The (quasi-)steady flow is obtained in our very low frequency Helmholtz oscillator. So far only a series of data at T_0 = 1.20 K down to coverages of ≈8.5 atomic layers has been taken. Fair qualitative agreement with Kontorovich's prediction applied to the case of unsaturated films is found. More data, extending to still smaller coverages and to other bath temperatures will shortly be collected, in order to investigate whether the observed deviations, found particularly at the larger coverages, are significant.

1. INTRODUCTION

A severe test for the applicability of a two-fluid description to the case of a superfluid ^{4}He film is the measurement of the so-called Kontorovich effect (1). Kontorovich predicted for saturated films, that bringing such a film into steady motion should result in a reduction of its mass. This rather subtle effect arises from the terms quadratic in the superfluid transport-velocity v_{so} that appear in the equation of motion and involves the boundary conditions for the moving film and the vapour. If for the case of unsaturated films the situation remains essentially unchanged, the reduction in the areal mass-density ρ_{f2} will be:

$$\Delta\rho_{f2}^{K} = (-\rho_{s2}/2\rho_{f2})v_{so}^2/(d\Omega/d\rho_{f2}) \qquad (1)$$

where ρ_{s2} is the areal superfluid density and Ω the Van der Waals potential per unit mass of helium in the film. The Kontorovich prediction has been verified experimentally for saturated films as well as for unsaturated films down to a mass coverage just below ten atomic layers (2). It is of interest to extend the investigations to still smaller coverages, in view of the expectation that the nature of the condensation process held responsible for the superfluid properties, will change drastically in the transition towards a two-dimensional behaviour.

2. MEASURING PRINCIPLE

We study the Kontorovich effect in our isothermal Helmholtz-oscillator device, consisting of two large, powder filled, filmreservoirs R_1 and R_2 connected by a long and narrow glass tube (see inset fig.2; L = 124 m, $2r_c$ = 0.41 mm) (3). A quasi-steady superfluid transport velocity is imposed by generating a Helmholtz oscillation in the device, i.e. a slowly varying periodic exchange of filmmass between the two reservoirs via the tube. According to prediction (1), the corresponding periodic variation of v_{so} will lead to an exchange of filmmass between the tube and the two reservoirs at twice the angular frequency ω_H of the Helmholtz oscillation:

$$\Delta M_f^K = \pi r_c L \frac{\rho_{s2}}{\rho_{f2}} \frac{v_{sH}^2 \sin^2\omega_H t}{(d\Omega/d\rho_{f2})} \qquad (2)$$

Expression (2) does not represent the full amount of mass that is expected to be exchanged between the helium inside the tube and in the reservoirs via the film. Quasi-steady variations in ρ_{f2} are accompanied by the corresponding variations in vapour density ρ_v, as has been verified so unambiguously for the case of isothermal third sound in unsaturated films (3). With the present tube this leads to a considerable contribution, such that the total exchange due to the Kontorovich effect is expected to be:

$$\Delta M_{tot}^K = \pi r_c L \frac{\rho_{s2}}{\rho_{f2}} \frac{v_{sH}^2 \sin^2\omega_H t}{(d\Omega/d\rho_{f2})} \{1 + \frac{r_c}{2} \frac{\rho_{vo}^2}{P_0}(\frac{d\Omega}{d\rho_{f2}})\} \qquad (3)$$

The slow variation in the amount of helium in each reservoir is registered capacitively as well as by the variation in the vapour pressure. A typical example of the recorder traces for ΔP_2 and ΔC_2 in reservoir R_2 during a Helmholtz oscillation, is shown in fig.1. The mass variations in the reservoirs, corresponding with those Helmholtz oscillations, are in opposite phase, so that the total amount of mass in the two reservoirs registers the Kontorovich effect. It is measured directly by adding the two independently obtained signals from both capacitors. Two separate capacitance bridges GR 1615A working at a different frequency, are employed. Besides, they are operated at different, adjustable, input voltages in order to correct for the difference in sensitivity of the two capacitors. A typical result is shown by the third recorder

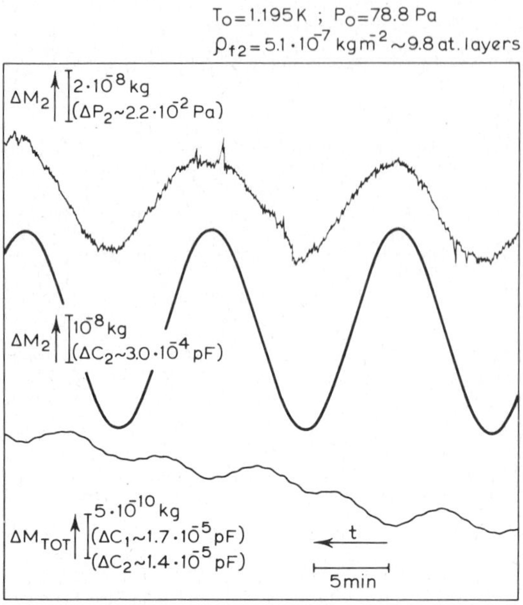

FIGURE 1

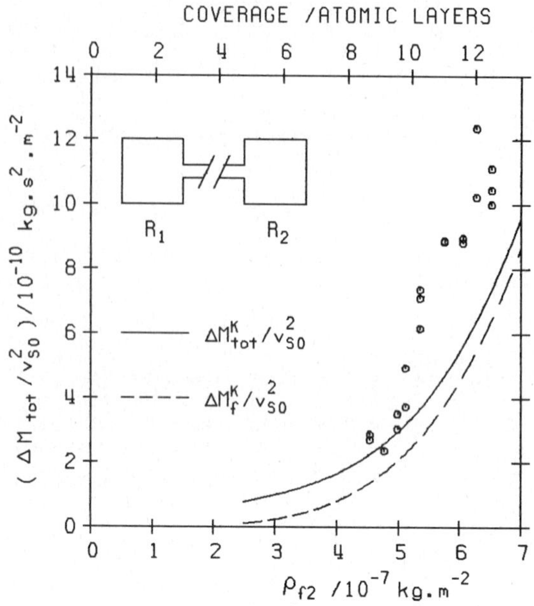

FIGURE 2

trace in fig.1. The small additional ripple that also shows up, corresponds to a highly suppressed standing wave of third sound in the tube. We have studied the frequency and damping of these waves in detail before, using almost the same measuring technique (4).

3. ANALYSIS AND DISCUSSION OF THE RESULTS

Fig.1 shows that ΔM_{tot} has indeed the character of ΔM^K_{tot} predicted by eq.(3). It is periodic at $2\omega_H$ and reaches its maximum simultaneously with the Helmholtz velocity*. A quantitative comparison is made in fig.2. This figure shows a plot of all available data for the reduced mass exchange $\Delta M_{tot}/v^2_{sH}$ versus masscoverage ρ_{f2} at bath temperature T_0 = 1.195 K. The amplitudes v_{sH} follow directly from the pressure amplitudes ΔP_{1H} and ΔP_{2H} as:

$$v_{sH} = (\Delta P_{1H} + \Delta P_{2H})/\omega_H L \rho_{vo} \quad (4)$$

Values for ρ_{vo} and ρ_{f2} are derived from the equilibrium pressure P_0 . For comparison the curves for $\Delta M^K_f/v^2_{so}$ and $\Delta M^K_{tot}/v^2_{so}$ calculated with relations (2) and (3) are also shown. Data for ρ_{f2} were used as obtained from the separate analysis of the Helmholtz oscillations and the third-sound phenomena in the tube, which proved to be in mutual agreement (3), (4).

Fig.2 shows that the observed effect is of the order of magnitude predicted by eq.(3), though the measured values are clearly too high, particularly at the larger coverages. It is remarked that a similar tendency was also observed in the analogous experiments with saturated films (5). Besides, also in ref.(2) a rather large effect was found, but this was shown to be in full agreement with the relatively weak Van der Waals force deduced from the measured speeds of third sound in the device. More data, taken at different bath temperatures and at variable Helmholtz amplitudes will be collected in order to investigate this effect in more detail.

Finally, as a tentative conclusion we can state that the Kontorovich prediction appears to maintain its value down to the lowest coverages of about 8 atomic layers investigated up till now. Extension of the measurements to still lower coverages is in preparation.

*The almost constant drift shown by ΔM_{tot} reflects the lowering of the helium bath, which results in a reduction of the vapour density in the filling lines.

REFERENCES

(1) V.M. Kontorovich, Zh. Eksp. Teor. Fiz. 30(1956)805 (Sov. Phys. JEPT 3 (1956)770).

(2) D.T. Ekholm and R.B. Hallock, J. Low Temp. Phys. 42(1981)339.

(3) G. Bannink, M.G.M. Brocken, I. van Andel and H. van Beelen, Physica 124B(1984)1.

(4) M.G.M. Brocken, I. van Andel and H. van Beelen, Proc. 75th Jubilee Conf. on ^{4}He, St. Andrews, Scotland, ed. J.G.M. Armitage (World Sci. Publ. Co. pte. ltd, Singapore, 1983).
M.G.M. Brocken et al. to be published.

(5) H.J. Verbeek, E. van Spronsen, H. Mars, H. van Beelen, R. de Bruyn Ouboter and K.W. Taconis, Physica 73(1974)621.

LT-17 (Contributed Papers)
U. Eckern, A. Schmid, W. Weber, H. Wühl (eds)

HIGH MAGNETIC FIELD NORMAL STATE PROPERTIES OF THE HEAVY FERMION SUPERCONDUCTOR UBe_{13}

G. R. STEWART, Z. FISK, and J. L. SMITH

Los Alamos National Laboratory, Los Alamos, N. M., USA 87545*

H. R. OTT

Laboratorium für Festkörperphysik, ETH, CH-8093 Zürich, Switzerland

F. M. MUELLER

Research Institute for Materials and Faculty of Science, 6525 ED Nijmegen, The Netherlands

We have measured the low temperature specific heat and resistivity of the heavy fermion superconductor UBe_{13} above T_c in fields to 11 T in order to better understand the unusual behavior of this interesting material. The increase of the specific heat in magnetic field that is observed appears to be reasonably consistent with a simple narrow band explanation for the rapid temperature variation of the low temperature specific heat in UBe_{13}, whereas a standard many body, paramagnon description would predict a *decrease* in the specific heat with increasing field. The magnetoresistance data for UBe_{13} are less well understood.

1. INTRODUCTION

Since the discovery by Steglich, et al. (1) of heavy fermion superconductivity in $CeCu_2Si_2$, with a T_c of 0.6 K, a specific heat γ (where $C \cong \gamma T + \beta T^3$) above T_c of 1100 mJ/mole-K^2, and an effective mass, m^*, of the conduction electrons of several hundred m_e, much interest has focussed on understanding why such highly correlated, high mass electrons should condense into the superconducting state. The recent discoveries of UBe_{13} (2) with properties similar to those of $CeCu_2Si_2$ and of UPt_3 (3), some of whose properties are like those of $CeCu_2Si_2$ and UBe_{13} (large $\gamma \sim 450$ mJ/mole-K^2, large m^*, and low, 0.54 K, T_c) but also with significant differences (in resistivity, critical field, and temperature dependence of C), has added a needed variety for intercomparison to this new field of study. Critical field measurements on these three materials indicate unusually high $-dH_{c2}/dT$ values at T_c for UBe_{13} {-257 kOe/K (2)} and $CeCu_2Si_2$ {-168 kOe/K (4)} and an unusual, increasing value of the $-dH_{c2}/dT$ value with increasing field for UPt_3 (5). In the present work, we report high field studies of the normal state specific heat and resistivity of one of these three materials (UBe_{13}) in order to further understand the unusual behavior of heavy fermion superconductors.

2. RESULTS

2.1. Specific heat

The low temperature specific heat of single crystal UBe_{13} from the same flux growth run as reported in (2) in 0 and 11 T is shown in Figure 1. If the large temperature dependent γ, which rises another factor of 2 at temperatures lower than shown in the figure, were due to many body correlation effects, i. e. paramagnons, one would qualitatively expect an applied field to suppress the low temperature increase in C/T ($\approx\gamma$ since the lattice contribution, βT^2, is small.) On the other hand, if the large increase at low temperatures in C/T is due to simply a narrow band (tens of K wide) then a straightforward model calculation using band splitting caused by the 11 T applied field would predict an increase in the specific heat above 3.5 K which is a maximum around 6 K and falling off at higher temperatures, with a decrease in the specific heat below 3.5 K. What is observed in Figure 1 is the maximum increase in field is seen around 3 K with the predicted falloff at higher

* Work at Los Alamos supported by the United States Department of Energy.

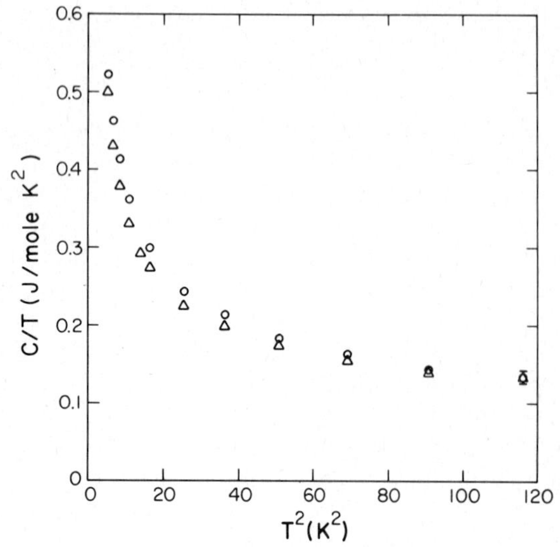

FIGURE 1
Zero field (Δ) and 11 T (o) specific heat data for a flux grown single crystal (2) of UBe_{13}. Note that the increase observed in C with 11 T goes to zero above 9 K and decreases below 3 K.

temperatures and a decreasing enhancement of C(H)/C(0) below 3 K, with, however, no observed net decrease of C(H) below C(0) in the measured temperature range. Clearly these data qualitatively agree with our simple narrow band model. However, a recent theoretical paper by Vollhardt (6) predicts an *increase* (of, admittedly, less than 1% in 18 T) in the specific heat of the paramagnon system ^3He with field. The intercomparability of ^3He and UBe_{13} is not obvious; however, a simple comparison of the product $\chi\gamma$, which appears in Vollhardt's treatment, for ^3He and UBe_{13} suggests a larger predicted increase for UBe_{13}. The validity of Vollhardt's model, and its application to UBe_{13}, if any, must await further work, both theoretical and experimental. The only other reported high field specific heat work on a heavy fermion material (7) was on $CeCu_2Si_2$, where the "Kondoesque" peak in the low temperature specific heat obscured the meaning of the results.

2.2. Resistivity

The resistivity versus temperature of single crystal UBe_{13} is shown in Figure 2 as a function of field. There is a large negative magnetoresistance which increases in magnitude at lower temperatures until at 1.229 K, $\Delta R/R$ is -42% in 11 T. These data, and other data omitted

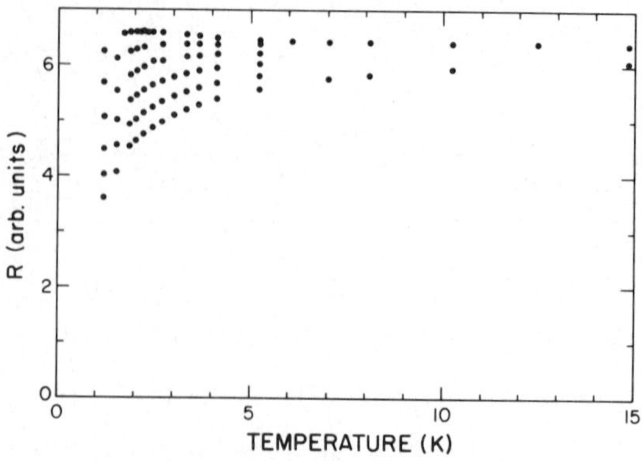

FIGURE 2
R versus T for a single crystal of UBe_{13} as a function of field. The upper set of dots is for H=0; each succeeding lower set of dots is for increasing field (H=3,5,7,9,11 T). Above 5 K, only the 0 and 11 T data are shown.

from the figure for the sake of clarity, may be plotted as $\Delta R/R$ versus H. Although it is not clear why, this family of $-\Delta R/R$ versus H curves can be mapped onto one another, with a maximum deviation in $\Delta R/R$ of 3%, by plotting $-\Delta R/R$ versus $H/(T+T^*)$, with $T^*=1$ K giving a better mapping than either $T^*=0$ or 2 K. Results not shown for $U_{0.974}Th_{0.026}Be_{13}$ display a similarly large negative magnetoresistance (-34% at 1.2 K in 11 T.) Data taken on $CeCu_2Si_2$ (7) also give a negative magnetoresistance at low temperatures, but of smaller magnitude (-4.5% in 11 T at 2 K)-at least in the direction measured. Thus, the large negative magnetoresistance in UBe_{13} seems quite anomalous. The relationship of these R(H) data to the low temperature specific heat results in 0 and 11 T remains unexplained at this time.

REFERENCES

(1) F. Steglich, et al., Phys. Rev. Lett. 43 (1979) 1892.
(2) H. R. Ott, et al., Phys. Rev. Lett. 50 (1983) 1595.
(3) G. R. Stewart, et al., Phys. Rev. Lett. 52 (1984) 679.
(4) U. Rauchschwalbe, et al., Phys. Rev. Lett. 49 (1982) 1448.
(5) J. W. Chen, et al., to be published and J. O. Willis, et al., this conference.
(6) D. Vollhardt,R.Mod.Phys.56(1984)99.
(7) G. R. Stewart, et al., Phys. Rev. B 28 (1983) 172.

LT-17 (Contributed Papers)
U. Eckern, A. Schmid, W. Weber, H. Wühl (eds)

PRESSURE-DEPENDENT RESISTIVE BEHAVIOR OF $YbBe_{13}$

J. D. THOMPSON, Z. FISK, and J. O. WILLIS

Los Alamos National Laboratory, Los Alamos, NM 87545, U.S.A.

We have studied the effect of hydrostatic pressures up to 16 kbar on the temperature-dependent resistance of $YbBe_{13}$. Anomalies in the low-temperature resistance are discussed in terms of an antiferromagnetically ordered ground state and resonant scattering off thermally populated crystalline electric field levels. A comparison is made to the resistive behavior of UBe_{13}. We find no evidence for superconductivity in $YbBe_{13}$ above 0.08 K.

1. INTRODUCTION

Ytterbium-based materials may exhibit either ordered or non-magnetic ground states, depending upon the 4f configuration. Mixing of the nearly degenerate $4f^n$ and $4f^{n-1}5d$ electronic configurations can lead to non-integral valency that is characterized by a temperature-independent magnetic susceptibility at low temperatures. Evidence for mixed valency has been found in many ytterbium-based compounds (1). In an early study, Bucher et al. (2) concluded from susceptibility and lattice parameter measurements that non-Curie-Weiss behavior in $YbBe_{13}$ at low temperatures was due to Yb intermediate valency. Subsequently, Heinrich et al. (3) found antiferromagnetic ordering at $T_N = 1.28$ K. Their magnetization and EPR data were analyzed in terms of a $J = 7/2$ multiplet (Yb^{3+}) split by cubic crystalline electric fields (CEF) into a Γ_7 ground state, with a doublet and quartet lying, respectively, 19 and 46 K above it.

Renewed interest in $NaZn_{13}$-crystal-structure compounds (to which $YbBe_{13}$ belongs) has been generated by the recent discovery of heavy-fermion superconductivity in UBe_{13} (4). Because of the controversy surrounding the low-temperature magnetic behavior of $YbBe_{13}$, its possible connection to heavy-fermion effects, and the need to identify characteristics that might lead to further discoveries of heavy-fermion behavior in this class of materials, we have studied the temperature-dependent resistance of $YbBe_{13}$ subjected to hydrostatic pressure.

2. EXPERIMENTAL PROCEDURE AND RESULTS

The temperature-dependent resistance of a flux-grown single crystal of $YbBe_{13}$ was measured using a standard four-probe ac technique. The single crystal was subjected to hydrostatic pressures up to 16 kbar generated in a self-clamping pressure cell (5). Resistance measurements at ambient pressure gave no indication for a superconducting transition above 80 mK; however, ac susceptibility below 2 K showed a maximum at 1.28 K, corresponding to T_N found by Heinrich et al. (3).

We show in Fig. 1 the resistance of $YbBe_{13}$ at four different clamp pressures. For reference, we estimate the resistivity at 100 K to be approximately 6 μΩ-cm. From T = 0.08 K, the resistance rises rapidly, reaching a plateau at 1.09 ± 0.005 K (at P = 0), followed by a well-defined maximum at 2.1 ± 0.1 K, then a minimum at 4.0 ± 0.1 K, above which is a large peak at 25 ± 0.5 K, and finally a broad minimum centered at 49 ± 1 K. Except for the plateau temperature T_p which increases monotonically from 1.09 to 1.17 K at 16 kbar, pressure has no detectable effect on the structure in R vs T. However, pressure does translate, approximately rigidly, the entire curve (1 to 300 K) to lower resistance.

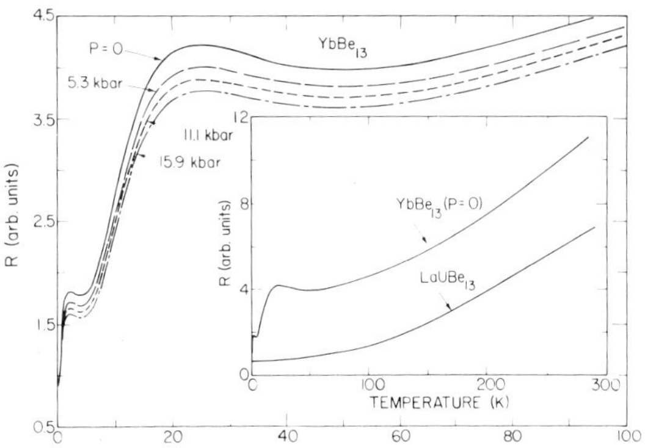

FIGURE 1
Resistance as a function of temperature for $YbBe_{13}$ at four different pressures (15.9 kbar corresponds to $\Delta V/V \approx -1.4\%$). The inset compares the resistance of $YbBe_{13}$ to that of non-magnetic $La_{.95}U_{.05}Be_{13}$.

3. DISCUSSION

The absence of a significant volume dependence in the features of R(T) argues strongly against mixed valency (since both the 4f level and the Fermi energy should be moderately pressure dependent) and favors the interpretation that the ground state is ordered antiferromagnetically. This viewpoint is supported further by the weak, positive pressure dependence of T_p, which, if associated with the suppression of spin-disorder scattering due to ordering, has the expected pressure dependence. The ineffectiveness of pressure to shift the resistance maxima also suggests that these originate from scattering off essentially stationary (with pressure) centers, e.g. CEF levels.

Thermal variations in the resistance of $YbBe_{13}$ are reminiscent of those in $Ce_{1-x}Th_xAl_3$ alloys (6). These alloys exhibit regions of negative $\partial R/\partial T$ that were interpreted in terms of a "Kondo sideband" theory (7) which models the scattering of conduction electrons by a CEF-split manifold of localized moments in terms of a Kondo exchange mechanism. Providing the CEF levels are split by more than k_BT, resonances in the scattering occur as the CEF levels become thermally populated. In this model, then, maxima in R vs T are expected at temperatures roughly equal to the energy separation of the CEF levels. Qualitatively, the resistance of $YbBe_{13}$ is consistent with this interpretation; although, quantitative discrepancies exist. For example, from the inset in Fig. 1, it is clear that the magnetic scattering contribution to the resitivity, while dominant at low temperatures, certainly does not exceed 6 $\mu\Omega$-cm. Such a small value is difficult to reconcile with the relatively large resistivity expected from resonant-type scattering such as observed in $(Ce,Th)Al_3$ alloys (6). In addition, positions of the resistive maxima do not correspond well to the calculated positions of the CEF levels. Nevertheless, significant deviations from Curie-Weiss behavior are apparent in $YbBe_{13}$ (3) only for $T \lesssim 35$ K, suggesting that the CEF splitting inferred by Heinrich et al. (3) may be overestimated.

It is interesting to compare the resistive behavior of $YbBe_{13}$ with that shown in Fig. 2 of isostructural UBe_{13}. The overall character of UBe_{13} differs markedly from $YbBe_{13}$ in that $\partial R/\partial T < 0$ for $T > 3$ K. However, two features, one at 2.35 K and one at $\sim$10 K, are similar to those in $YbBe_{13}$ at comparable temperatures, which might suggest the possibility of a common origin. We note, though, that preliminary inelastic-neutron-scattering experiments on UBe_{13} (8) show no evidence for CEF excitations in the required energy range, casting doubt on a "Kondo sideband" interpretation. Although arguments presented above favoring this interpretation for $YbBe_{13}$ require additional sub-

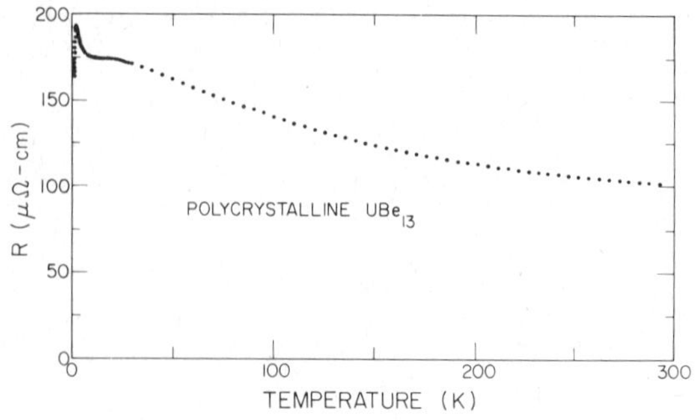

FIGURE 2
Resistivity versus temperature for UBe_{13} at ambient pressure.

stantiation, they are plausible and suggest that the resistive anomalies in UBe_{13} and $YbBe_{13}$ are probably not related. If they do have a common origin, an alternative explanation for these features must be found.

4. CONCLUSIONS

The suggestion that $YbBe_{13}$ is mixed valent appears inconsistent both with ac susceptibility measurements and with the lack of a noticeable volume dependence of the resistive features. Instead, an antiferromagnetically ordered ground state interpretation is favored, with no evidence for superconductivity above 0.08 K. Resistive anomalies in $YbBe_{13}$ are interpreted as arising from "Kondo sideband" scattering.

ACKNOWLEDGEMENTS

We thank J. L. Smith for the ac susceptibility measurement. Work was performed under the auspices of the U.S.DOE.

REFERENCES

(1) J. M. Lawrence, P. S. Riseborough, and R. D. Parks, Rep. Prog. Phys. 44 (1981) 1.
(2) E. Bucher et al., Phys. Rev. B 11 (1975) 440.
(3) G. Heinrich, J. P. Kappler, and A. Meyer, Phys. Lett. 74A (1979) 121.
(4) H. R. Ott et al., Phys. Rev. Lett. 50 (1983) 1595.
(5) J. D. Thompson, Rev. Sci. Instrum. 55 (1984) 231.
(6) K. H. J. Buschow et al., Phys. Rev. B 3 (1971) 1662.
(7) F. E. Maranzana, Phys. Rev. Lett. 25 (1970) 239.
(8) J. K. Kjems, H. R. Ott, and S. K. Sinha, unpublished results.

LT-17 (Contributed Papers)
U. Eckern, A. Schmid, W. Weber, H. Wühl (eds)
© Elsevier Science Publishers B.V., 1984

PRESSURE DEPENDENCE OF THE SUPERCONDUCTING TRANSITION TEMPERATURE OF $CeCu_2Si_2$ AND UBe_{13}

J. W. CHEN, S. E. LAMBERT and M. B. MAPLE*

Institute for Pure and Applied Physical Sciences, University of California, San Diego, La Jolla, CA 92093, USA

Z. FISK and J. L. SMITH†

Los Alamos National Laboratory, Los Alamos, NM 87545, USA

H. R. OTT‡

Laboratorium für Festkörperphysik, ETH-Hönggerberg, 8093 Zürich, Switzerland

The variation of the superconducting transition temperature T_c with hydrostatic pressure to 12 kbar was determined for $CeCu_2Si_2$ and UBe_{13}. Initially, T_c increases for $CeCu_2Si_2$, while for UBe_{13} a small depression ~ 0.01 K/kbar is observed.

Recently there has been considerable interest in the properties of "heavy fermion" superconductors, a small class of compounds which at present includes $CeCu_2Si_2$ (1), UBe_{13} (2) and UPt_3 (3). These materials are characterized by relatively low values of the superconducting transition temperature $T_c \sim 1$ K, large values of the electronic specific heat coefficient $\gamma \sim 1$ J/mole-K^2, correspondingly large discontinuities in the specific heat at T_c, and large values of the upper critical magnetic field and its initial slope (4-6). We have determined the effect of nearly hydrostatic pressure on T_c for $CeCu_2Si_2$ and UBe_{13}.

The polycrystalline $CeCu_2Si_2$ samples used in this study were prepared by arc melting stoichiometric amounts of the starting materials in an Ar arc furnace. Subsequent annealing at various temperatures for seven days of different pieces of a single ingot was carried out in silica tubes after wrapping the samples in Ta foil. Unannealed polycrystalline UBe_{13} was prepared in the same way. Preparation of single crystals of UBe_{13} has been described elsewhere (2). Bulk pieces of each sample were pressurized at room temperature in piston and cylinder clamps (7) using a 50:50 mixture of isoamyl alcohol and n-pentane as the pressure transmitting medium. The pressure was determined from the depression of the superconducting transition temperature of Sn (8). The ac susceptibility χ_{ac} was measured in a He^3-He^4 dilution refrigerator using a four wire ac bridge operating at 16 Hz. After subtracting $\chi_{ac}(T)$ for the same clamp containing no sample, T_c was defined as the 50% point of the transition and the transition width from the 10% and 90% points.

Displayed in Fig. 1 is the pressure P dependence of T_c determined for all of the samples investigated in this study. Considering first the data for $CeCu_2Si_2$, a strong increase of T_c is observed when P = 0 with increasing annealing temperature. This is consistent with previous work (9) which finds an enhancement of T_c upon annealing at somewhat higher temperatures (927-1127 °C). Increasing the pressure from P = 0 causes T_c to increase for all four samples, followed at higher pressure by a nearly constant T_c. The data for annealing temperatures of 800 °C and 900 °C are almost identical, perhaps due to some small variation in the composition of the arc melted ingot. The largest increase in T_c is observed for the unannealed specimen, indicating that the enhancement of T_c by pressure is directly

Research supported by: *US DOE DE-AT03-76ER-70227, †US DOE and ‡Schweizerische Nationalfonds zur Förderung der Wissenschaftlichen Forschung.

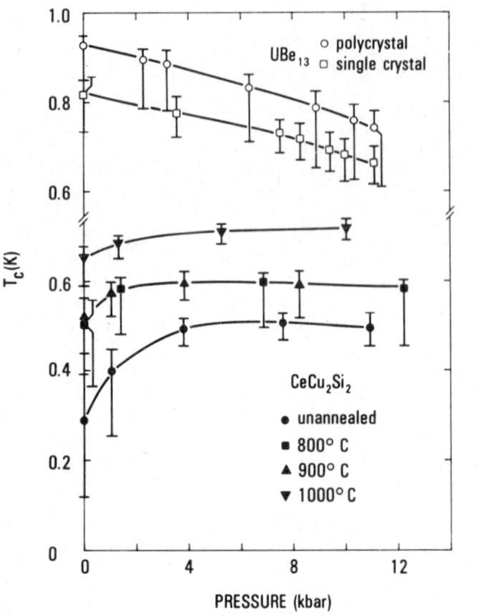

FIGURE 1
Superconducting transition temperature T_c vs pressure for $CeCu_2Si_2$ samples annealed at the three temperatures indicated for seven days. Also shown are data for polycrystalline and single crystal samples of UBe_{13} .

related to the degree of order in the crystal lattice. Previously reported $T_c(P)$ data for a single crystal specimen of $CeCu_2Si_2$ (10) are consistent with these results even though no superconductivity was observed until $P \geq 1$ kbar. The shape of $T_c(P)$ is similar to the data presented here with a very rapid initial increase of T_c followed by a nearly constant $T_c = 0.5$ K. This suggests that the single crystal was less ordered than all of the polycrystalline samples considered here. The complex metallurgy of $CeCu_2Si_2$ (4) makes a clear interpretation of these results difficult. However, an extrapolation of these data suggests that a well-annealed, ordered sample would exhibit a very small pressure dependence for $P \leq 12$ kbar. Finally, this work indicates that T_c is inversely proportional to the volume V of the unit cell for disordered samples at small pressures. For larger pressures, no clear trend is obvious. This is contrary to the assertion that $T_c \propto V$ based on studies of samples with varying composition (4).

Also displayed in Fig. 1 is $T_c(P)$ for polycrystalline and single crystal samples of UBe_{13}. The behavior of $T_c(P)$ is very similar for the two samples with T_c for the polycrystalline sample ~ 0.1 K higher than for the single crystal. The initial rate of depression of T_c is 0.016 K/kbar (polycrystal) and 0.012 K/kbar (single crystal), values comparable to those observed for many conventional superconductors.

We have presented measurements of the pressure dependence of T_c for $CeCu_2Si_2$ and UBe_{13}. For $CeCu_2Si_2$, an increase of T_c is observed initially, the size of which depends strongly on the annealing temperature. At higher pressure, T_c becomes nearly constant at a value which also increases with annealing. In contrast, a decrease of ~ 0.01 K/kbar is found for UBe_{13}. It can thus be concluded that despite many other anomalous properties, the influence of pressure on superconductivity in UBe_{13} is similar to that for conventional superconductors.

REFERENCES

(1) F. Steglich, J. Aarts, C. D. Bredl, W. Lieke, D. Meschede, W. Franz and H. Schafer, Phys. Rev. Lett. 43 (1979) 1892.
(2) H. R. Ott, H. Rudigier, Z. Fisk and J. L. Smith, Phys. Rev. Lett. 50 (1983) 1595.
(3) G. R. Stewart, Z. Fisk, J. O. Willis and J. L. Smith, Phys. Rev. Lett. 52 (1984) 697.
(4) W. Assmus, M. Herrmann, U. Rauchschwalbe, S. Riegel, W. Lieke, H. Spille, S. Horn, G. Weber, F. Steglich and G. Cordier, Phys. Rev. Lett. 52 (1984) 469.
(5) M. B. Maple, J. W. Chen, S. E. Lambert, Z. Fisk, J. L. Smith and H. R. Ott, to be published.
(6) J. W. Chen, S. E. Lambert, M. B. Maple, Z. Fisk, J. L. Smith, G. R. Stewart and J. O. Willis, to be published.
(7) D. Wohlleben and M. B. Maple, Rev. Sci. Instrum. 42 (1971) 1573.
(8) T. F. Smith, C. W. Chu and M. B. Maple, Cryogenics 9 (1969) 53.
(9) C. D. Bredl, H. Spille, U. Rauchschwalbe, W. Lieke, F. Steglich, G. Cordier, W. Assmus, M. Herrmann and J. Aarts, J. Magn. and Magn. Mat. 31-34 (1983) 373.
(10) F. G. Aliev, N. B. Brandt, V. V. Moshchalkov and S. M. Chudinov, Solid State Commun. 45 (1983) 215.

LT-17 (Contributed Papers)
U. Eckern, A. Schmid, W. Weber, H. Wühl (eds)

LOW TEMPERATURE PROPERTIES OF KONDO LATTICE SYSTEMS

C.D. BREDL, N. GREWE and F. STEGLICH

Institut für Festkörperphysik, Technische Hochschule Darmstadt, and SFB 65, D-6100 Darmstadt, FRG

E. UMLAUF

Zentralinstitut für Tieftemperaturforschung, Bayerische Akademie der Wissenschaften, D-8064 Garching

Experimental data for specific heat and thermopower of $CeAl_3$ and different samples of $CeCu_2Si_2$ for low temperatures and varying magnetic fields are discussed. Particular attention is paid to coherence effects and to the phenomenon of superconductivity via "Heavy Fermions". Attempts for a theoretical interpretation are based on a recent theory of Kondo lattices.

Kondo-lattice systems (KL) are nearly trivalent Ce intermetallic compounds which exhibit localized magnetic moments above their "Kondo temperature" $T_K \simeq$ a few K and intriguing low-temperature anomalies: Either they show magnetic ordering between reduced moments (1) as in $CeAl_2$ or "Heavy-Fermion" effects (2,3) as in $CeAl_3$ and $CeCu_2Si_2$. For example, in the latter systems a γT term in the specific heat was found below 1 K, whose coefficient $\gamma \geq 1$ J/mole-K^2 is larger than in any other metal. Moreover, in $CeCu_2Si_2$ this Fermi-liquid state becomes unstable below $T_c \sim 0.65$ K against a ("Heavy-Fermion") superconducting state (3).

In order to gain a physical understanding of these systems one needs information about the density of states (DOS) near the Fermi level ε_F. For this purpose we refer to recent calculations of the one particle excitation spectrum of the Anderson impurity and lattice models in the Kondo regime which are summarized in Fig. 1(a) and (b) (4-6). In the impurity case a narrow resonance builds up somewhat above the Fermi level below a temperature $k_BT_K = N_F^{-1} \exp(-1/2\, N_F\, J)$. In the lattice a pronounced fine structure of this peak is visible, which is due to coherence effects and can also be understood by Fermi-liquid theory (7). As a consequence of the many body nature of the Kondo effect these DOS are strongly temperature dependent, which has to be taken into account when applying formulas known from single free particle models. We nevertheless believe that qualitative conclusions may be drawn from an inspection of these pictures.

As a first application we refer to thermopower (TEP) measurements of Kondo compounds like $CeCu_2Si_2$ (8), $CeAl_3$ (9) and $CeAl_2$ (10) which show anomalous negative values at low temperature: see Fig. 1(c). A possible interpretation is that with decreasing T below T_K the density of states, which is piled up due to the Kondo effect in an asymmetric way with regard to ε_F, i.e. with its main part above ε_F (4), causes a pronounced particle-like behavior. An interesting question may be if at even lower temperatures the coherence effects could leave a visible trace in the TEP. This seems doubtful to us because of intrinsic temperature dependence of the DOS, Fig. 1 (b), and because of the fact that the gap seen in Fig. 1 (b) will in reality be smeared out to a certain extent by 4f dispersion, asphericities and lifetime effects.

A much simpler quantity to study is the specific heat. Here one can hope that well below T_K one might be able to detect a (somewhat smeared) fine structure corresponding to Fig. 1 (b) (11). In fact, a maximum in C(T)/T is detected at about 0.4 K in $CeAl_3$ and $CeCu_2Si_2$, see Fig. 2 (a) to (c) (12). If ε_F really is situated like shown in Fig. 1 (b), the effective DOS for T = 0 is relatively small. Its average over a width of k_BT around ε_F then increases with T when the peaks are approached. This would explain the low temperature increase in C(T)/T. Finally, when k_BT becomes so large that the peaks lie completely inside this range, or when the temperature induced decrease of the peaks dominates, the effective DOS and the corresponding specific heat again becomes smaller. As a test for these ideas measurements of impure systems are discussed below, which due to the lack of coherence should not show this specific heat maximum.

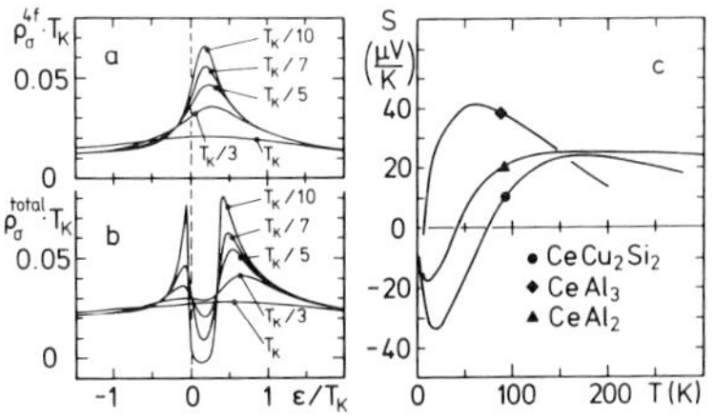

FIGURE 1

(a): Density of states for local one particle excitations on an Anderson impurity in the Kondo regime, taken from Ref. 5; (b): Total density of states for an Anderson lattice in the Kondo regime, taken from Ref. 5; (c): Thermopower results for three Kondo lattice systems, taken from Ref. 8 – 10.

These ideas are, in fact, supported by our results shown in Figs. 2 (d) and (e). They were obtained for two samples in which coherence was destroyed by substituting 20 a/o Ce by Y or La. At zero field, no peak in C(T)/T could be found down to our temperature limit of 60 mK. At finite field, a peak occurs for the Y-doped system and shifts upwards with field, as is known for dilute Kondo ions (12). On the contrary, the peak shifts towards lower temperatures in the lattice case.

An interesting variation in the C(T) dependence at differing magnetic fields between the stoichiometric $CeCu_2Si_2$ sample No. 11 (Ref. 13) and the $CeCu_{2.2}Si_2$ sample shows up in Fig. 2 (c) and (b). The latter ones are known (14,15) for their high T_c values ($\geq$ 0.6 K). In agreement with earlier results on stoichiometric high-T_c samples (13), one observes a pronounced specific-heat jump at B = 0 ($\Delta C/C_n(T_c) \simeq 1.0$) and a variation with field typical of a conventional superconductor. In low-T_c samples (T_c < 0.5 K), however, the specific-heat jump disappears (see Fig. 2 (c)) and, thus, the temperature dependence of the specific heat closely resembles that of the non-superconducting Kondo lattice $CeAl_3$. It has been shown (13) that this correlates with a moderate reduction of the Kondo temperature; if T_K is reduced by about 40%, T_c vanishes.

The observed temperature dependence of the low-field specific heats, $C_s(T)$, provides further evidence of a complex interplay between Kondo-lattice behavior and superconductivity. In particular, the non-exponential behavior seen in Fig. 2 (b) suggests a temperature-dependent energy gap as can result from finite DOS in the gap. The approximate equality of the two shaded areas shows that the experimental entropy is consistent with a BCS-like phase transition. We take this as an indication that a weak-coupling form of the theory already can give reasonable results (6,16). However, the competition between the formation of the superconducting gap and of the Kondo resonance on each Ce site could well result in a smearing of the gap. Such an effect would account for the anomalous low-T specific heat. In addition, Kondo-type pair breaking due to higher excitation energies also produces DOS in the gap, in particular upon decreasing T_K. It is highly desirable to investigate, if in such a way a gapless situation, with nearly the normal-state DOS (Fig. 2 (c)), can occur. The apparent difference in the low-field specific heats of Figs. 2 (b) and (c) seems to correlate with an interesting difference at overcritical fields, i.e. concerning the field dependence of the peak position, which we assign to DOS effects in the coherent Kondo lattice.

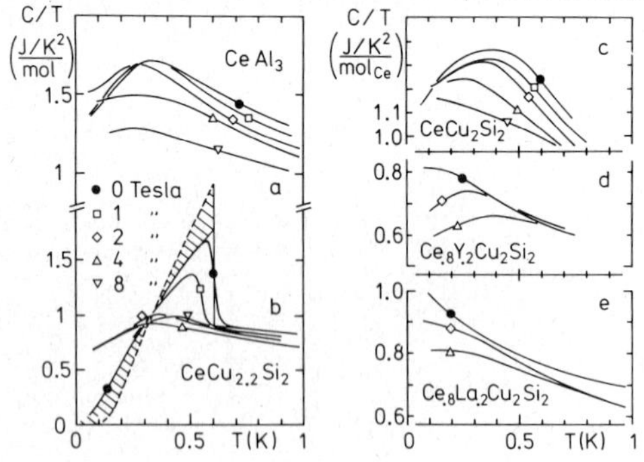

FIGURE 2

Specific heat results at differing magnetic fields. Dashed line in (b) calculated from BCS-theory. Sample shown in (c) is No. 11 of Ref. 13.

REFERENCES

(1) B. Barbara, J.X. Boucherle, J.L. Buevoz, M.F. Rossignol and J. Schweizer, Solid State Commun. 24 (1977) 481.

(2) K. Andres, J.E. Graebner and H.R. Ott, Phys. Rev. Lett. 27 (1975) 1779.

(3) W. Assmus, M. Herrmann, U. Rauchschwalbe, S. Riegel, W. Lieke, H. Spille, S. Horn, G. Weber, F. Steglich and G. Cordier, Phys. Rev. Lett. 52 (1984) 469; and references cited therein.

(4) N. Grewe, Z. Phys. B 53 (1983) 271.

(5) N. Grewe, Solid State Commun. 50 (1984) 19.

(6) N. Grewe, contribution to this conference.

(7) R.M. Martin, Phys. Rev. Lett. 48 (1982) 362.

(8) W. Franz, A. Griessel, F. Steglich and D. Wohlleben, Z. Phys. B 31 (1978) 7.

(9) P.B. Van Aken, H.J. Van Daal and K.H.J. Buschow, quoted in: A.K. Bhattacharjee and B. Coqblin, Phys. Rev. B 13 (1976) 3441.

(10) S. Horn, W. Klämke and F. Steglich, in: Valence Instabilities (Zürich), P. Wachter and H. Boppart (eds.) North-Holland (1982) 459.

(11) It would, however, be wrong to directly fit experimental data with the DOS for the bare electrons presented in Fig. 1. These data include all corrections between bare particles which show up as temperature dependence of their DOS. In a Fermi-liquid description they are à priori included.

(12) The same features have been found and qualitatively discussed in this spirit for other samples of these compounds, see: C.D. Bredl, S. Horn, F. Steglich, B. Lüthi and R.M. Martin, to be published. As described there, a nuclear contribution to the data has been subtracted, which arises in finite fields due to Zeeman splitting.

(13) C.D. Bredl, H. Spille, U. Rauchschwalbe, W. Lieke, F. Steglich, G. Cordier, W. Assmus, M. Herrmann and J. Aarts, J.M.M.M. 31-34 (1983) 373.

(14) M. Ishikawa, H.F. Braun and J.L. Jorda, Phys. Rev. B 27 (1983) 3092.

(15) H. Spille, U. Rauchschwalbe and F. Steglich, Helv. Phys. Acta 56 (1983) 165.

(16) H. Razafimandimby, P. Fulde and J. Keller, Z. Phys. B 54 (1984) 111.

LT-17 (Contributed Papers)
U. Eckern, A. Schmid, W. Weber, H. Wühl (eds)

LOW TEMPERATURE PROPERTIES OF $CeOs_2$

M. S. TORIKACHVILI and M. B. MAPLE

Institute for Pure and Applied Physical Sciences,
University of California, San Diego, La Jolla, CA 92093, USA

G. P. MEISNER

Materials Science and Technology Division,
Los Alamos National Laboratory, Los Alamos, NM 87545, USA

The binary compound $CeOs_2$ has been prepared in both cubic C15 and hexagonal C14 Laves phases. Measurements of ac magnetic susceptibility, electrical resistivity and specific heat have been performed on both phases. The C14 phase has been found to exhibit superconductivity with a superconducting transition temperature T_c of 1.1 K.

Several binary compounds with the general formula $REOs_2$ (RE = rare earth) have been reported to exist in either the C15- (cubic Laves phase-$MgCu_2$) or in the C14- (hexagonal Laves phase-$MgZn_2$) type structure (1). When prepared by arc melting in argon on a water cooled copper hearth, the light RE compounds $LaOs_2$ and $CeOs_2$ form in the C15 phase, while the heavier RE compounds form in the C14 phase. The $PrOs_2$ compound exhibits a mixture of both crystallographic phases (1). Superconductivity has been observed only in $LaOs_2$ and $LuOs_2$ with superconducting transition temperatures (T_c's) of 6.5 K and 3.5 K, respectively (1). The compounds with RE = Pr, Nd, Sm and Gd were found to be ferromagnetic.

Cannon et al. (2) later reported that pure C14 phases of $LaOs_2$, $CeOs_2$ and $PrOs_2$ could be prepared by appropriate annealing at high pressures. Lawson et al. (3) reported T_c's for the C14 and C15 phases of $LaOs_2$ of 5.9 K and 8.9 K, respectively. They also found that the contraction of the unit cell of $CeOs_2$ relative to the values expected from interpolation of the $LaOs_2$ and $PrOs_2$ compounds is comparable for the C14 and C15 phases.

Motivated by an interest in valence related phenomena in cerium compounds, we undertook ac magnetic susceptibility χ_{ac}, electrical resistivity ρ and specific heat C measurements on samples of $CeOs_2$ of both crystallographic structures.

Stoichiometric amounts of high purity Ce and Os were arc melted together in a Zr-gettered Ar atmosphere to obtain a pure C15 phase. We have found that annealing for a few days at temperatures in the range 600-700 °C, in subatmospheric pressures of Ar, promotes a complete transformation of the C15 phase into C14. Most of the experiments in this study were performed on a 6 gram arc melted sample. Pieces cut from this sample that were wrapped in Ta foil and annealed for 7 days at 700 °C in a sealed quartz tube filled with 150 mmHg of UHP Ar transformed to the C14 structure. Although X-ray powder diffraction analysis on samples of both phases did not indicate sizable amounts of impurities, small amounts of extra phases could be observed with a metallographic microscope.

The compound $CeOs_2$-C15 did not exhibit superconductivity down to 70 mK. However, $CeOs_2$-C14 underwent a superconducting transition at T_c = 1.1 K. This value of T_c was determined at the midpoint of the change in ac susceptibility between the normal and the superconducting states, for a powdered specimen, and correlates well with an abrupt drop to zero of the resistivity below 1.3 K, and with a jump observed in the specific heat between 1.02 K and 1.11 K.

The ρ vs T curves for both phases of $CeOs_2$ are displayed in Fig. 1. The shape of these curves is reminiscent of $CeSn_3$ (4) and other valence fluctuation compounds (5). The room temperature values of ρ for the C14 and C15 phases are, within 5%, 127 and 136 μΩ·cm, respectively. These relatively high values of ρ are presumably due to strong charge and/or spin fluctuations (5). For both phases the resistivity below ~ 40 K can be described by the expression $\rho_o + AT^n$ with $n \sim 2$ which is a

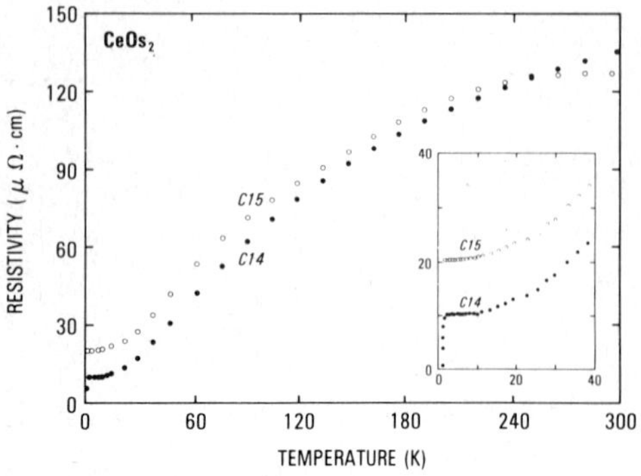

FIGURE 1
Resistivity vs temperature for the C14 and C15 crystallographic phases of $CeOs_2$.

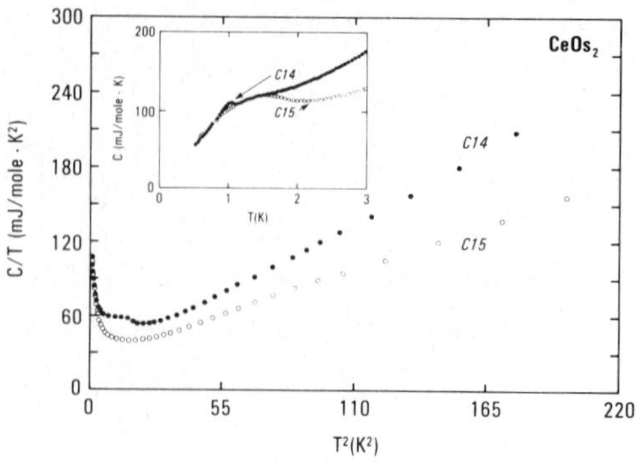

FIGURE 2
Specific heat C divided by temperature T vs T^2 for the C14 and C15 phases of $CeOs_2$. Shown in the inset are C vs T data up to 3 K.

typical behavior for a Fermi-liquid with a nonmagnetic ground state (5).

The behavior of C/T vs T^2 is displayed in Fig. 2. It is likely that the upturn of C/T at low temperatures, observed in both phases of $CeOs_2$, is due to magnetic impurities. Displayed in the inset to Fig. 2 are C vs T data between 0.5 K and 3.0 K, a range in which the background from the magnetic impurities is large. The jump in the specific heat exhibited by the C14 phase is approximately one third of the value expected for a typical BCS superconductor. If we disregard the upturn in the C/T vs T^2 data at low temperatures for both phases of $CeOs_2$, each set of data can be described by $C = \gamma T + \beta T^3$ between 6 K and 14 K, where γ and β are the electronic and lattice specific heat coefficients, respectively. The value of the Debye temperature θ_D can be obtained by $\theta_D = (12\pi^4 R/5\beta)^{1/3}$, where R is the universal gas constant. The values of γ for both phases are 22 ± 3 mJ/mole·K (C14) and 24 ± 3 mJ/mole. K (C15). The values of θ_D are 124 ± 2 K (C14) and 143 ± 2 K (C15). Whereas the electronic contribution to the specific heat is comparable for both phases of $CeOs_2$, the lattice contribution of the superconducting C14 phase is larger than the nonsuperconducting C15 phase.

ACKNOWLEDGMENTS
The research at UCSD and LANL was carried out under the auspices of the U.S. Department of Energy, under Contract No. DE-AT03-76ER-70227 at UCSD.

REFERENCES
(1) V. B. Compton and B. T. Matthias, Acta Cryst. 12 (1959) 651.
(2) J. F. Cannon, D. L. Robertson, H. T. Hall and A. C. Lawson, J. Less-Common Metals 31 (1973) 174.
(3) A. C. Lawson, J. F. Cannon, D. L. Robertson and H. T. Hall, J. Less-Common Metals 32 (1973) 173.
(4) B. Stalinski, Z. Kletowski and Z. Henkie, Phys. Stat. Solidi a 19 (1973) K165.
(5) J. M. Lawrence, P. S. Riseborough and R. D. Parks, Rep. Prog. Phys. 44 (1981) 1.

LT-17 (Contributed Papers)
U. Eckern, A. Schmid, W. Weber, H. Wühl (eds)

LATTICE VIBRATIONAL BEHAVIOR OF THE EXCHANGE ENHANCED SUPERCONDUCTOR U_6Fe

P. P. VAISHNAVA, C. A. STRELECKY, A. E. DWIGHT and C. W. KIMBALL[†]

Northern Illinois University, DeKalb, IL 60115, USA

F. Y. FRADIN[§]

Argonne National Laboratory, Argonne, IL 60439, USA

The temperature dependence of the Mossbauer effect has been studied at ^{57}Fe nuclei in the exchange-enhanced superconductor U_6Fe (T_C=3.76±0.17K). The nonharmonic variation in shift vs temperature is interpreted as due to changes in electronic structure below 100K. Phonon hardening below 100K is also inferred from the temperature dependence of the Debye-Waller factor.

Chandrasekhar and Hulm[1] and Hill and Mathias[2] found that the metallic uranium compounds U_6X (X=Mn, Fe, Co and Ni) are superconductors. These compounds i) possess a nearly temperature-independent paramagnetic susceptibility, ii) have a bcc tetragonal crystal structure which is unique to U, Np, and Pu compounds with magnetic 3d elements,[3] iii) exhibit no localized magnetic moment behavior, as does the Kondo lattice $CeCuSi_2$,[4] and iv) have high-field superconductivity and exchange-enhanced paramagnetism similar to that observed in materials exhibiting strongly interacting fermi-liquid behavior. All these properties are indicators of heavy-mass 5f-itinerant electrons playing an important role in determination of the physical properties of these materials. We have performed Mossbauer spectroscopic measurements at ^{57}Fe nuclei in U_6Fe to study the lattice vibrational properties and to determine the role of phonons in the superconducting behavior.

U_6Fe forms at a fixed stoichiometry (3.77wt%Fe) by peritectic reaction, liquid (∿5wt%Fe)+γ-U_{SS} (<0.5wt%Fe)+U_6Fe at 810°C. One 645mg ingot was prepared by triple arc melting (Ar+He atmosphere) high purity depleted uranium together with enough ^{57}Fe (90% enrichment) to yield an alloy of composition 0.3at% richer in uranium than U_6Fe to ensure absence of UFe_2. Portions of the ingots were wrapped in Ta foil, sealed in evacuated Vycor bulbs and annealed at 650°C and 700°C for 7 days.

The impurity phases which have been found to occur in this compound are UFe_2 and UO_2. X-ray diffraction patterns, using CoK_α radiation were made on -325 mesh powder, which was strained annealed at 650°C for 4 days. The films exhibited patterns that could be matched with the published pattern (JC PDS, card 15-145) for tetragonal U_6Fe. No UFe_2 was seen, but a small quantity of UO_2 was detected.

The superconducting temperature was measured in a resonant detector bridge at 900Hz. T_C was found to be 3.76K with ΔT_C of ±0.17K.

About 100mg of U_6Fe was mixed with 200mg of high purity alumina and sealed in a plastic material (1 inch diameter) to be used as an absorber for Mossbauer effect measurements. Transmission ^{57}Fe Mossbauer spectra using a 50mCi ^{57}Co in Rh matrix at 273.6K were recorded at various temperatures between 4.2 and 250K.

Spectra recorded between 4.2 and 250K were fitted with a symmetric quadrupole doublet. The area of Mossbauer absorption is related to the Debye-Waller factor in a harmonic solid by

$$-\ell nA = K^2 \langle r^2 \rangle \qquad (1)$$

where K is the wave vector of γ-radiation. The mean-square displacement $\langle r^2 \rangle$ is linear in temperature and proportional to θ_f^{-2} at high temperature. Therefore, the slope of $-\ell nA$ vs T is very sensitive to the Debye temperature. The temperature dependence of $-\ell nA$ vs T for U_6Fe is shown in Fig. 1. The solid lines are the least-squares fits to the Debye model. It is clear that phonon shifting occurs near 100K; the lattice hardens at low temperature.

The variation in shift with temperature is shown in Fig. 2. The centroid of the pattern, the Mossbauer shift, is a sum of two components, the thermal shift which is proportional to the integral over the lattice specific heat and the isomer shift which reflects the number and character of the valence electrons. The thermal shift is proportional to the mean-square

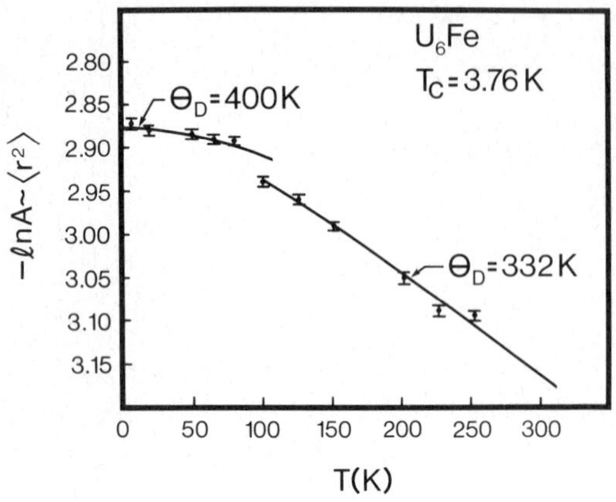

FIGURE 1
Recoilfree fraction (Debye-Waller factor) vs temperature. Solid lines are fits to Debye model.

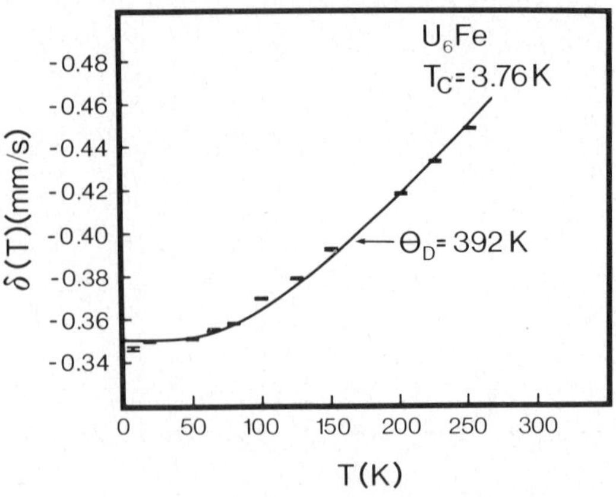

FIGURE 2
Temperature dependence of shift. Solid curve is based on Debye model.

velocity of the Mossbauer atom

$$\delta(T) \;\alpha\; \langle v^2 \rangle \qquad (2)$$

Using a Debye formulation for the Fe vibrational motion, $\delta(T)-\delta(0)$ is a measure of the change in thermal energy and determines a Debye temperature θ_δ. An average Debye fit over the entire range of temperature yields θ_δ=392K. It is evident that the variation in shift with temperature is not characteristic of harmonic behavior; there is a deviation from Debye behavior near 100K. When the data above and below 100K are analyzed separately with a Debye model, $\theta_\delta(T>100)\sim450$K and $\theta_\delta(T<100)\sim320$K are obtained. It appears that with decreasing temperature the lattice softens. The softening below 100K obtained from shift analysis is inconsistent with the results from the mean-square displacement and is probably due to a change in valence character of the Fe. The low temperature shift result indicates that there is a slight increase in the number of d-electrons.

As is usual, $\theta_f \;\alpha\; \langle r^2 \rangle$ is different from $\theta_\delta \;\alpha\; \langle v^2 \rangle$ due to the fact that these parameters are related to different moments of the Fe vibrational spectral density, $\langle r^2 \rangle$ weighting the low frequency phonons and $\langle v^2 \rangle$ the high frequency phonons.

Taking $\theta_f\sim400$K to reflect the average phonon behavior of Fe at low temperature, and using a superposition of Debye temperatures weighted by the number of Fe and U atoms, the Debye temperature for U is $\theta\sim80$K; that is, uranium is considerably softer than the θ=125K of the compound U_6Fe.[5]

REFERENCES

(1) B.S. Chandrasekhar and J.K. Hulm, J. Phys. Chem. Solids 7 (1958) 259.
(2) H.H. Hill and B.T. Mathias, Phys. Rev. 168 (1968) 464.
(3) D.J. Lam, J.B. Darby, Jr. and M.V. Nevitt, The actinides: Electronic structure and related properties, Vol. 2, ed. A.J. Freeman and J.B. Darby, Jr. (Academic, NY, 1974) Chapter 4.
(4) B.C. Sales and R. Viswanathan, J. Low Temp. Phys. 23 (1976) 449.
(5) L.E. DeLong, J.G. Huber, K.N. Yang and M.B. Maple, Phys. Rev. Letters 51 (1983) 312.

LT-17 (Contributed Papers)
U. Eckern, A. Schmid, W. Weber, H. Wühl (eds)

SUPERCONDUCTING PROPERTIES OF $EuMo_6S_8$ UNDER PRESSURE

M. Decroux (a), S.E. Lambert (a), M.B. Maple (a), R.P. Guertin (b) , R. Baillif (c) and Ø. Fischer (c)

(a) Department of Physics and Institute of Pure and Applied Physical Sciences, UCSD, La Jolla, CA, 92093, USA
(b) Department of Physics, Tufts University, Medford, MA 02155, USA
(c) Département de Physique de la Matière Condensée, Université de Genève,1211 Genève 4,Switzerland

Application of pressure higher than 13 kbar inhibits the structural transition in $EuMo_6S_8$ and makes it superconducting with a high critical temperature (T_c=12.2 K at 13.2 kbar). The bulk nature of this pressure induced superconductivity is demonstrated. The upper critical magnetic field measured at 14 kbar and 19 kbar can be described by a multiple pairbreaking theory in which a negative exchange interaction between the Eu^{2+} ions and the conduction electrons is included.

1. INTRODUCTION

In the series $Sn_{(1-x)}Eu_xMo_6S_8$ the superconducting critical temperature T_c is concentration independent until x=0.6 (1) and then suddenly drops to zero in a narrow concentration range. This step-like behavior of T_c is related to the appearance of a structural transition which was firstly discovered in $EuMo_6S_8$ (2). Under pressures larger than 7 kbar, $EuMo_6S_8$ was found to be superconducting at around 11 K (3,4). Recently, a detailed study of the pressure dependence of the structural transition (5) allowed to set up the rhombohedral-triclinic phase boundary curve. A minimum pressure of 13 kbar is needed to prevent the structural transition and above this pressure $EuMo_6S_8$ becomes superconducting (T_c=12.2 K at 13.2 kbar). But no flux expulsion at T_c was observed in this compound (6,7). Since this absence of the Meissner effect lefts some doubts as to the origin of this pressure induced superconducting state, we have measured the heat capacity under pressure on a melted $EuMo_6S_8$ sample.

2. HEAT CAPACITY UNDER PRESSURE

The heat capacity of $EuMo_6S_8$ at 15 kbar is presented on fig.1. A marked jump of the heat capacity is observed at the superconducting transition. Similar experiments carried out below 13 kbar did not show any anomaly related to a superconducting transition. From fig.1, we estimate that $\Delta C/\gamma T_c$ = 1 which implies, assuming the BCS value for this ratio, that 70% of the conducting system (heater, sample, thermometer and wires) are involved in the superconducting transition. This proves definitively the bulk nature of this superconducting state.

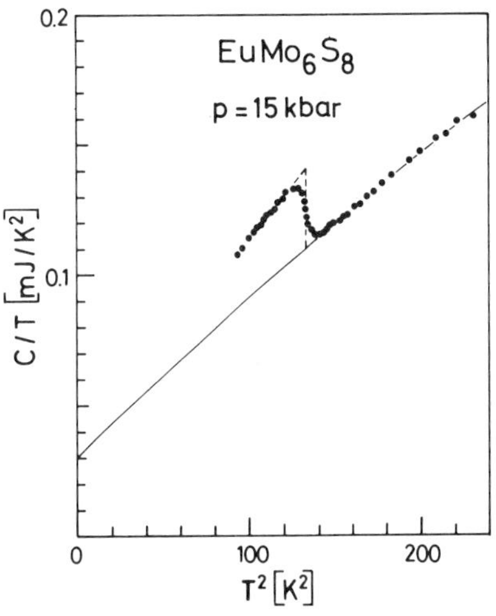

Fig.1 C/T vs T^2 for $EuMo_6S_8$ at 15 kbar

3. RESISTIVITY AND CRITICAL FIELD H_{c2}

The temperature dependence of the resistivity at 14 kbar is shown in fig. 2. The resistivity at room temperature is 750 μΩcm and decreases to 38 μΩcm just above T_c. If we assume that at high temperature the resistivity saturates at around 1 mΩcm, when the mean free path ℓ reaches the inter-atomic distance (6.55 Å), we can estimate that at T_c ℓ is larger than 180 Å. The temperature dependence of H_{c2} at 14 kbar and 19 kbar is displayed in fig.3 (H_{c2} is defined at 50 % of the resistive transition). The full lines give the theoretical predictions of a multiple pairbreaking theory where the spin-orbit scattering parameter λ_{so}, the Maki para-

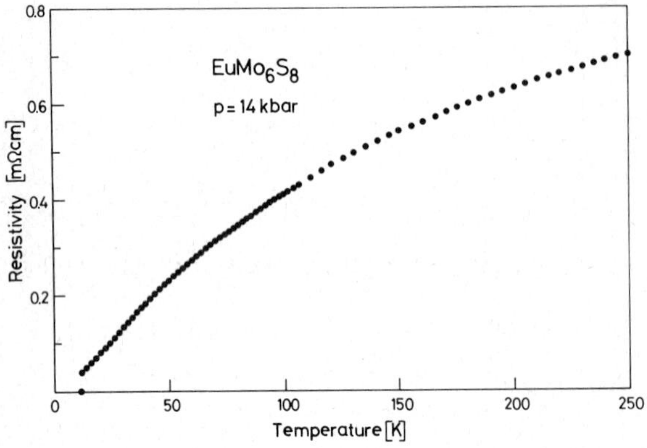

Fig.2 Resistivity vs temperature for $EuMo_6S_8$ at 14 kbar

meter α and the exchange field H_J were fixed by the best fit of the experimental results. But the anomalous behavior of $H_{c2}(T)$ makes the choice of this set of parameters almost unique within the experimental errors. These parameters, the mean free path ℓ at T_c and the coherence length ξ_0 (estimated in the clean limit) are reported in table I for the two pressures investigated. For both pressures ℓ is larger than ξ_0 indicating that the sample investigated is in the clean limit.

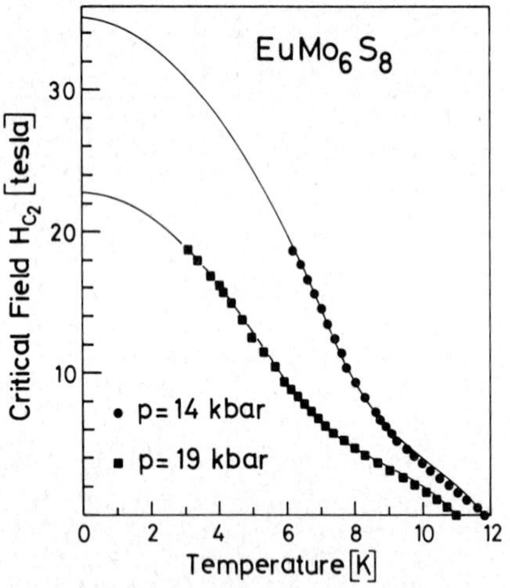

Fig.3 Critical field H_{c2} vs temperature at 14 kbar and 19 kbar

TABLE I

p [kbar]	T_c [K]	α	λ_{so}	H_J [tesla]	ξ_0 [Å]	ℓ [Å]
14	11.6	2.3	2.2	-35	35	180
19	11.0	1.6	1.1	-27	44	300

4. CONCLUSIONS

We have demonstrated in this work that the pressure induced superconductivity in $EuMo_6S_8$ is a bulk phenomenon, but surprisingly, no Meissner effect is observed. The idea of a conventional strong flux pinning mechanism associated with structure defects or impurities is in contradiction with the very large mean free path at low temperature and the assumption of an extreme type II superconducting material is not supported by the relatively modest orbital critical field found experimentally. Therefore the absence of sizeable flux expulsion at T_c remains to be explained.

The anomalous behavior of $H_{c2}(T)$ reveals the existence of a magnetic interaction between the localized moments of the Eu^{2+} ions and the conduction electrons

REFERENCES

(1) M. Decroux, H.W. Meul, C. Rossel, Ø. Fischer and R. Baillif, in Superconductivity in d-and f-Band Metals 1982, ed. by W. Buckel and W. Weber (KGK, Karlsruhe, 1982) p.167
(2) R. Baillif, A. Dunand, J. Muller and K. Yvon, Phys. Rev. Lett. 47 (1981) 672
(3) D.W. Harrison, K.C. Lim, J.D. Thompson, C.Y. Huang, P.D. Hambourger and H.L. Luo, Phys. Rev. Lett. 46 (1981) 280
(4) C.W. Chu, S.Z. Huang, C.H. Lin, R.L. Meng, M.K. Wu and P.H. Schmidt, Phys. Rev. Lett. 46 (1981) 276
(5) M. Decroux, M.S. Torikachvili, M.B. Maple, R. Baillif, Ø. Fischer and J. Muller, Phys. Rev. B28 (1983) 6270
(6) R.W. McCallum, W.A. Kalsbach, T.S. Radhakrishnan, F. Pobell, R.N. Shelton and P. Klavins, Solid State Commun. 42 (1982) 819
(7) M. Decroux, S.E. Lambert, M.S. Torikachvili L.D. Woolf, M.B. Maple, R.P. Guertin and R. Baillif, Phys. Rev. Lett. in print

LT-17 (Contributed Papers)
U. Eckern, A. Schmid, W. Weber, H. Wühl (eds)

MAGNETIZATION OF SPUTTERED $HoMo_6S_8$ FILMS

Toshizo FUJITA*, Sumio IKEGAWA+ and Taiichiro OHTSUKA

Department of Physics, Faculty of Science, Tohoku University, Sendai 980, Japan

The low-field magnetization of sputtered $HoMo_6S_8$ films was measured at low temperatures down to 25 mK. The temperature dependence of the observed magnetization depends substantially on the film thickness. The results are qualitatively consistent with the magnetic phase diagram theoretically predicted for films of ferromagnetic superconductors.

1. INTRODUCTION

Recently an interesting phase diagram has been theoretically investigated for films of reentrant ferromagnetic superconductors (1). As the film thickness is reduced, a sinusoidal magnetic order is predicted to coexist with superconductivity over a wider temperature interval above the reentrant transition than in bulk samples. Even a ferromagnetic superconducting state is predicted for a limited region of thickness comparable with the London penetration depth, λ_L .

In this paper, we report the low-field magnetization measurements on some film samples of $HoMo_6S_8$. In connection with the predicted phase diagram, discussion is focussed mainly on the characteristic temperature dependence of magnetization observed in the films with different thickness.

2. EXPERIMENTAL

Samples were prepared on sapphire substrates ($4 \times 16 \times 0.5$ mm^3) by an rf-sputtering technique similar to that employed by Banks et al (2). The details of the method, together with the characterization of the films obtained, will be described elsewhere. Therefore, only a brief outline is given here. The composite target used consists of 16 mm MoS_2 and Ho_2S_3 discs pressed from respective powders, arranged on a 100 mm disc of Mo metal. The composition of the obtained films was controlled by adjusting the arrangement of the discs. Since the as-sputtered films were found to be amorphous, the samples were sealed in a evacuated quartz ampoule together with some MoS_2 powder and were annealed at 1100°C for 3-4 hours. During the heat treatment, the MoS_2 powder added appeared to produce an appropriate vapour pressure of sulfur and to play an important role in suppressing sulfur deficiency frequently experienced in the annealed samples. The X-ray analysis indicated that the main component of the annealed films was the Chevrel phase $HoMo_6S_8$, although weak trace of Ho_2O_2S was detected as a second phase.

The upper superconducting transition temperature, which was resistively determined for the films with thickness of 0.5-4.0 µm, is appreciably higher than for bulk samples and ranges from 2.2 to 3.2 K.

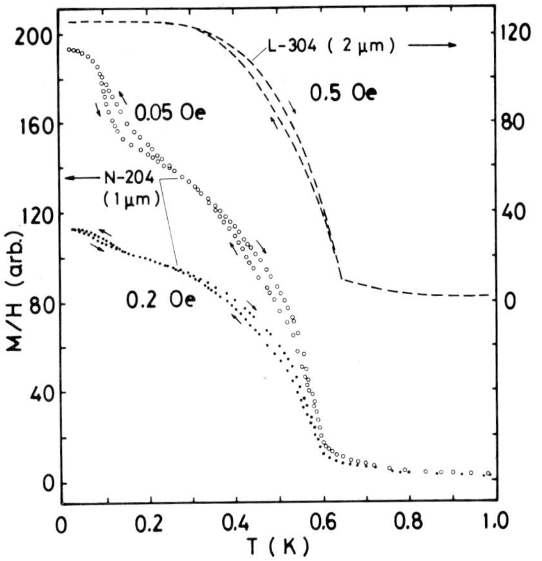

Fig.1 The normalized magnetization M/H of $HoMo_6S_8$ films observed in weak external fields as a function of temperature T. The left-hand scale is for the 1 µm film (N-204) and the right-hand scale for the 2 µm thick film (L-304). Although the scales are arbitrary, the relative values have been corrected for thickness and are valid for comparison.

* Present address: Department of Physics, Faculty of Science, Hiroshima University, Hiroshima 730, Japan.
+ Present address: Research & Development Center, Toshiba Corporation, Komukai-Toshiba-cho, Kawasaki-ku, Kawasaki 210, Japan.

An rf-SQUID magnetometer were used to measure the dc magnetization in low fields. A superconductive pickup coil of 10 turns was wound directly around the substrate to detect the magnetic signal parallel to the film plane. Another superconductive solenoid of 90 turns was wound to produce the dc external field. These were fixed on a copper holder and mounted in a dilution refrigerator cryostat. The sample and coil assembly was magnetically shielded with great care by a permalloy cylinder and a superconducting lead cover.

3. RESULTS AND DISCUSSION

Resistive measurements showed that superconductivity persists at least down to 50 mK in all our films with thickness of 0.5-4.0 μm. However, the dc magnetization was found to grow abruptly below 0.65 K. Of course, this cannot be interpreted simply as the coexistence of some magnetic order with superconductivity because superconducting paths are expected to be formed in inhomogeneous film samples from various origins. Here we base our main arguments on the magnetic data. Figure 1 and 2 show typical magnetization curves which depend on the film thickness.

The temperature dependence of magnetization for the 2 μm thick film (Fig.1) is very similar to the result reported for bulk samples (3). A definite increase in magnetization is recognized at T_m = 0.65 K. The measurements were performed very slowly for about 10 hours respectively on cooling and on warming between 1 K and 25 mK. The observed hysteresis suggests an intrinsic property such as a first order phase transition or very long relaxation time, as is the case for bulk samples.

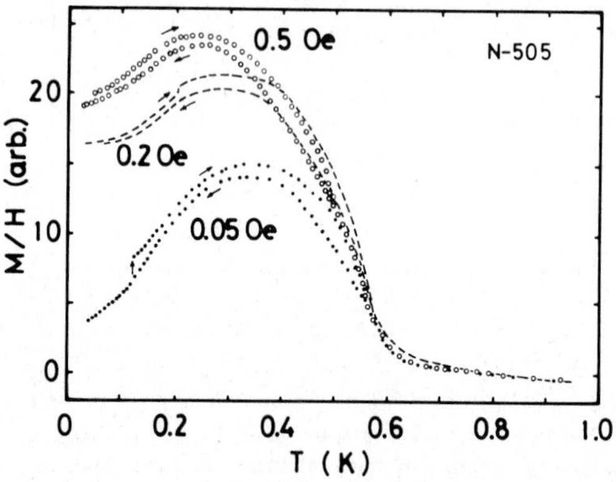

Fig.2 The normalized magnetization M/H for the 0.5 μm $HoMo_6S_8$ film measured in weak external fields of 0.05, 0.2 and 0.5 Oe as a function of temperature T.

In the film of 1 μm, the magnetization curve has a double structure suggesting two steps of phase transition. This characteristic behaviour can be seen more clearly when the external field is lowered. A puzzling feature is the reversed hysteresis observed around 0.15 K; the magnetization curve on warming shifts to lower temperature than the curve on cooling.

As shown in Fig.2, the magnetization curves for the film of 0.5 μm exhibit a relatively slow increase below 0.6 K and have a broad peak. The peak temperature, T_p, decreases with increasing magnetic field. Below T_p, a sudden jump was occasionally observed in magnetization only on warming, suggesting an effect of depinning.

According to the recent calculation by Takahashi and Tachiki (4), two types of temperature dependence are expected for magnetization of the films of ferromagnetic superconductors corresponding to the thickness-dependent magnetic phase diagram. In the case of films where the oscillatory magnetic order has a symmetric pattern across the film thickness, the ferro- or ferrimagnet-like magnetization increases monotonically with decreasing temperature. In the antisymmetric case, on the other hand, the magnetization curve exhibits a sharp peak in weak magnetic field.

Our experimental results are in a qualitative agreement with the theory if some speculation is added. An antisymmetric case is realized by our film whose thickness is 0.5 μm compared with λ_L (5) although the magnetization peak is broad due to the non-uniformity or imperfection of the sample. The 1 μm film is considered to be a symmetric case. The normalized magnetization, M/H, for this film is reasonably larger than that for the 0.5 μm film. The increase of magnetization around 0.15 K appears to suggest some transition; possibly the reentrant transition from the superconducting state to a normal ferromagnetic state. This transition is believed to shift to a higher temperature approaching 0.65 K in the 2 μm film. As a reasonable consequence, the thick films show a nearly bulk character. Further experiments are required to confirm the above interpretation using improved quality films.

REFERENCES

(1) A.Kotani, S.Takahashi, M.Tachiki, H.Matsumoto and H.Umezawa, Physica 108B (1981) 1043. S.Takahashi, A.Kotani, M.Tachiki, H. Matsumoto and H.Umezawa, J.Phys.Soc.Jpn.52 (1983) 989.
(2) C.K.Banks, L.Kammerdiner and H.L.Luo, J. Solid State Chem. 15 (1975) 271.
(3) J.W.Lynn, G.Shirane, W.Thomlinson, R.N.Shelton and D.E.Moncton, Phys.Rev.B24 (1981) 3817.
(4) S.Takahashi and M.Tachiki, Solid State Commun. 48 (1983) 9.
(5) J.A.Woollam and S.A.Alterovitz, Phys. Rev. B19 (1979) 749.

LT-17 (Contributed Papers)
U. Eckern, A. Schmid, W. Weber, H. Wühl (eds)

SEARCH FOR SUPERCONDUCTIVITY IN $BaMo_6S_8$

C. ROSSEL,* M. DECROUX and Ø. FISCHER

Département de Physique de la Matière Condensée, Université de Genève,
24, Quai Ernest-Ansermet, 1211 Genève, Switzerland

Due to the occurrence of a structural phase transformation at low temperature, $BaMo_6S_8$ is not superconducting. We tried to induce superconductivity in this compound by applying hydrostatic pressure on a pure melted sample, and by partially substituting Ba^{2+} with a monovalent or trivalent element in the series $Ba_{1-x}M_xMo_6S_8$ (M = K, Y, La).

Previous work has shown that the Chevrel phase compounds MMo_6S_8 (M = Eu, Ba, Sr, Ca) (1, 2) as well as $EuMo_6Se_8$ (3) are all non-superconducting at ambient pressure, due to a low temperature structural transformation from the rhombohedral to the triclinic phase at T_s, and the creation of a partial gap at the Fermi surface. This gap has also been associated to the possible competing effect of a charge density wave (CDW) transition. In pure melted $EuMo_6S_8$, the suppression of this structural transformation could be induced by means of pressure, leading to bulk superconductivity (4). The compound $BaMo_6S_8$ is expected to have a rather high density of states at the Fermi energy E_F, and thus to be a high T_c superconductor, if the rhombohedral phase could be retained at the lowest temperatures.

Two methods for suppressing the phase transition are the application of hydrostatic pressure and the partial substitution for Ba^{2+} by other ions of different valence. In the first case, the published data on $BaMo_6S_8$ under pressure were obtained on sintered powder samples and did not exhibit any bulk superconductivity (5, 6). Since the quality of the samples is important here, some $BaMo_6S_8$ was melted in a high pressure furnace, under 2 kbar of argon, at ~1800 °C. We applied various hydrostatic pressures, $0 \leq p \leq 16$ kbar, on one large grain extracted from the ingot, and measured its electrical resistance between 1.2 K and 300 K (Fig. 1). The large peak at $T_s = 180$ K is progressively reduced with increasing pressure, but its position shifts only slightly to lower T, with the initial rate $(dT_s/dp) = -0.22$ K/kbar. Thus, hydrostatic pressures of at least 30-40 kbar would be needed to see superconductivity in this compound. If we assume for $BaMo_6S_8$ an ideal T_c of 15 K, and the T_c depression rate $dT_c/dp = -0.2$ K/kbar, as usually observed in the isoelectronic compounds Pb-, Sn-, and $EuMo_6S_8$ (4), we might expect a T_c value of 7 to 9 K, above 30 kbar. Between 0 and 16 kbar, the resistance at 300 K of our sample has dropped by a factor of 4.3. In contrast to sintered $BaMo_6S_8$ (6), the resistivity decreases with applied pressure over the whole temperature range.

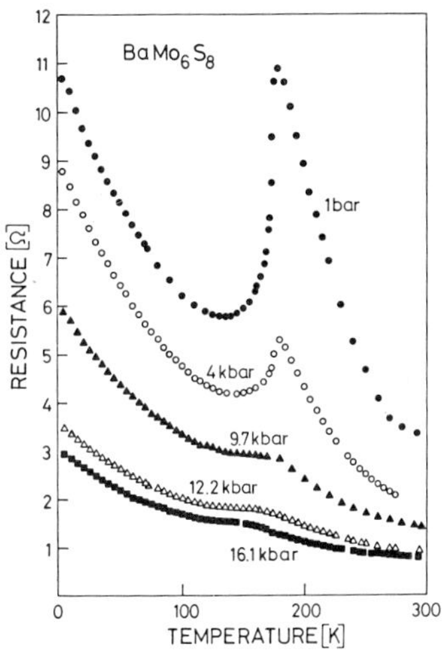

FIGURE 1
Electrical resistance vs temperature of melted $BaMo_6S_8$ at several pressures.

The second way of modifying the electronic structure at E_F was accomplished by substituting for Ba^{2+} either Y^{3+}, La^{3+} or K^+ in order to induce a charge transfer onto or from the

*Present address: Inst. for Pure & Appl. Sci., Univ. of Calif., San Diego, La Jolla, CA 92093, USA.

Mo_6 cluster, respectively. The samples of the series $Ba_{1-x}M_xMo_6S_8$ ($0 \leq x \leq 0.6$) were all hot-pressed at 1500°C under a uniaxial pressure of ~ 3 kbar. Presented in Fig. 2 are the $\rho(T)/\rho(300\ K)$ curves for $Ba_{1-x}Y_xMo_6S_8$. In spite of an overall similarity with Fig. 1, some striking features are found. For x = 0, the anomalous peak in $\rho(T)$ is located only at $T_s = 120$ K, and is much larger ($\rho(T_s)/\rho(300\ K) \cong 16$) than in the melted sample. By increasing x, it gradually disappears and its position shifts surprisingly towards higher temperatures. For x > 0.3 superconductivity appears above 1 K even though an anomalous bump remains at higher temperature. From ac susceptibility measurements, T_c rises from 1.63 K for x = 0.4 to 2.24 K for x = 0.6. To check our data, we made a low temperature X-ray diffraction analysis for the two samples x = 0 and x = 0.4. In the first case we could demonstrate that the structural transition occurs between 120 K and 60 K, by following the intensity and the splitting of the main lines (1), in good agreement with the drop of $\rho(T)$ below T_s. In the second case, no obvious change of the spectrum occurred down to 4.2 K. This indicates that the small bump in $\rho(T)$ around 190 K might be due to an inhomogeneous distribution of the Y atoms so that only a small amount of the transforming phase remains, to small quantity of Mo_2S_3, or to some collective phenomenon such as CDW transition.

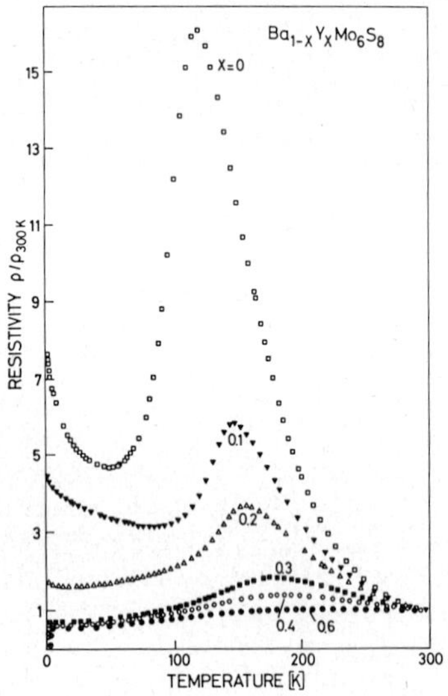

FIGURE 2
Normalized resistivity vs temperature in hot-pressed $Ba_{1-x}Y_xMo_6S_8$.

The high resistivity $\rho(300\ K) = 6.7\ m\Omega\cdot cm$ of the hot-pressed $BaMo_6S_8$, as well as the exponential increase of $\rho(T)$ measured between 600 K and 300 K, show that $BaMo_6S_8$ is semiconducting in the rhombohedral phase. But, even if the evidence for CDW is not clear in this phase, it cannot be ruled out, that the abrupt rise of $\rho(T)$ below 300 K might be due to a CDW transition above T_s, rather than to structural fluctuations, or to a normal semiconducting behavior. Very similar results were obtained with $M = La^{3+}$. By increasing x, the peak vanishes, moves to higher temperatures, and superconductivity shows up for x = 0.4. The situation is different for $M = K^+$. The peak is almost washed out at x = 0.2 without a superconducting transition, while at x = 0.3, $\rho(T)$ exhibits a smooth semiconducting-type curve. In summary, $BaMo_6S_8$ is a semiconductor at 300 K. It undergoes a structural transition between 180 K and 120 K, depending on the preparation. The effect of the transition is to reduce strongly the resistivity, what produces the anomalous peak in $\rho(T)$. If this peak is also the consequence of a CDW transition above T_s is not clear now. Hydrostatic pressure needs to be larger than about 30 kbar to generate superconductivity, nevertheless, substituting Ba^{2+} with x = 0.4 of Y^{3+} or La^{3+} leads to values of $T_c > 1$ K.

ACKNOWLEDGMENTS

The authors are grateful to Dr. R. Baillif and to Prof. M. B. Maple for help and comments on this work.

REFERENCES

(1) R. Baillif, A. Dunand, J. Muller and K. Yvon, Phys. Rev. Lett. 47 (1981) 672.
(2) B. Lachal, R. Baillif, A. Junod and J. Muller, Solid State Commun. 45 (1983) 849.
(3) C. Rossel, H. W. Meul, A. Junod, R. Baillif, Ø. Fischer and M. Decroux, Solid State Commun. 48 (1983) 431.
(4) M. Decroux, M. S. Torikachvili, M. B. Maple, R. Baillif, Ø. Fischer and J. Muller, Phys. Rev. B 28 (1983) 6270.
(5) W. A. Kalsbach, R. W. McCallum, V. Poppe, F. Pobell, R. N. Shelton and P. Klavins, in Proc. Int. Conf. on Superconductivity in d- and f-Band Metals, eds. W. Buckel and W. Weber (Karlsruhe, 1982), p. 185.
(6) P. H. Hor, M. K. Wu, T. H. Lin, X. Y. Shao, X. C. Jin and C. W. Chu, Solid State Commun. 44 (1982) 1605.

LT-17 (Contributed Papers)
U. Eckern, A. Schmid, W. Weber, H. Wühl (eds)

LOW TEMPERATURE SPECIFIC HEAT OF CHEVREL PHASE COMPOUND $AgMo_6S_8$

Masafumi FURUYAMA, Norio KOBAYASHI, Koshichi NOTO and Yoshio MUTO

The Research Institute for Iron, Steel and Other Metals, Tohoku University, Sendai 980, Japan

The specific heat of a single crystal of Chevrel phase compound $AgMo_6S_8$ has been measured from 1.5 to 9 K at 0 and 60 kOe. Various fundamental parameters are evaluated and this compound is confirmed to be a intermediately strong coupling superconductor. The lattice specific heat is explained by using the molecular crystal model. The existence of a soft mode is discussed.

Chevrel phase compounds are interesting superconductors with very large H_{c2} and high T_c. So far, many studies on properties of these compounds have been done [1]. However, the systematic explanation about electron-phonon interactions has not been obtained yet. We have measured the specific heat of a $AgMo_6S_8$ single crystal and evaluated various thermodynamic parameters, where we obtain some informations on the electron-phonon interactions. Anomalous behaviors of lattice specific heat are observed and analyzed.

The single crystal (~65mg wt) with the nominal concentration of $AgMo_6S_8$ was prepared from a single phase powder sample, which was annealed for three days at about 1300°C, in a high frequency induction furnace under the high argon pressure by the Bridgmann technique. The residual resistivity ρ_0 and residual resistance ratio $RRR=\rho(300K)/\rho_0$ are 25 $\mu\Omega\cdot$cm and 10.4, respectively. From the upper critical field measurement, the values of $H_{c2}(0)$ and $-dH_{c2}(T_c)/dT$ are estimated to be 44.5 kOe and 7.78 kOe/K, respectively. These values are almost similar to those given by the other groups. The specific heat was measured by means of thermal relaxation method in the temperature range 1.5 - 9 K. The magnetic field of 60 kOe which is larger than the value of $H_{c2}(0)$ was applied for the normal state measurements.

Fig.1 shows the specific heat in the form of C/T vs. T^2. From this result, the transition temperature T_c, its width ΔT_c and the specific heat jump at T_c, ΔC, are obtained as given in Table I. The specific heat of the normal metal at low temperature can generally be represented in the form, $C_n/T=\gamma+\beta T^2$, where γ and β are coefficients of the electron and lattice contributions, respectively. However, as can be seen, the normal state specific heat of our sample can not be expressed by the usual form. Such specific heat anomaly was often reported in other Chevrel phase compounds [2]. We assume that the value of γ is determined by an extrapolation to 0 K of the normal state data. The entropy of

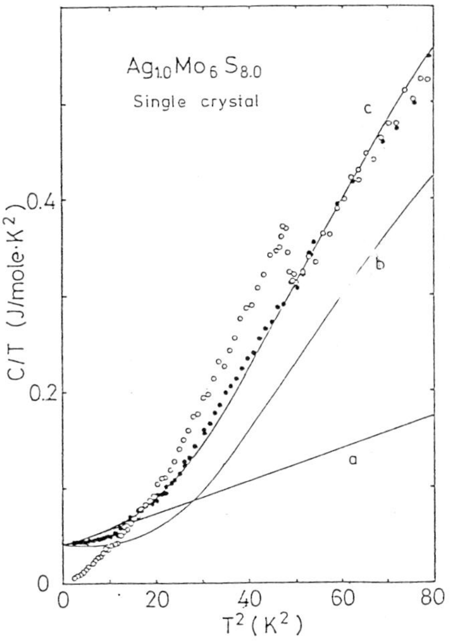

Fig.1 The measured specific heat and calculated specific heat. The open cirecles and the full ones indicate data in 0 and 60 kOe, respectively. The curves denoted by a, b and c represent the contributions of acoustic and optical modes, and total specific heat, respectively.

the normal state and that of the superconducting state are in good agreement at T_c within an accuracy of ±1%. The thermodynamic critical field $H_c(T)$ can be obtained from the difference of these entropies. The values of $H_c(0)$ and $-dH_c(T_c)/dT$ are also summarized in Table I.

Table I

T_c (K)	ΔT_c (K)	ΔC ($\frac{mJ}{mole\cdot K}$)	γ ($\frac{mJ}{mole\cdot K^2}$)	$H_c(0)$ (Oe)	$-\frac{dH_c(T_c)}{dT}$ (Oe/K)
6.96	0.13	560	41.0	880	250

Table II

	BCS	$AgMo_6S_8$	$Cu_{1.8}Mo_6S_8$	Mo_6Se_8	Nb
$\frac{\Delta C(T_c)}{\gamma T_c}$	1.43	1.95	2.63	2.28	1.87
$\frac{H_c^2(0)}{\gamma T_c^2}$	5.90	6.36	7.35	6.75	6.4
$-\frac{dh}{dt}\Big\vert_{t=1}$	1.74	1.97	2.11	2.05	1.87

The dimensionless thermodynamic quantities $\Delta C/\gamma T_c$, $H_c(0)^2/\gamma T_c^2$ and $-dh(1)/dt$ are listed in Table II, where $h(t)=H_c(T)/H_c(0)$ and $t=T/T_c$. These values for $AgMo_6S_8$ are larger than those of the BCS prediction. And also, the t^2-dependence of $D(t)=\{H_c(T)/H_c(0)-(1-t^2)\}$, which is the deviation of h(t) from the parabolic law, is different from that of the BCS prediction. In order to compare with other Chevrel phase compounds, the values of Mo_6Se_8 [3] and $Cu_{1.8}Mo_6S_8$ [4] are also included in Table II,together with ones for Nb. The deviation from the BCS prediction is comparable to Nb while it is smaller than for Mo_6Se_8 and $Cu_{1.8}Mo_6S_8$. These behaviors in the thermodynamic properties obviously indicate that the electron-phonon interaction is relatively strong, i.e. this compound is an intermediatly strong coupling superconductor. It seems that the strong coupling interaction is universal property in high T_c Chevrel phase compounds.

The normal state specific heat is decomposed into electronic and lattice parts, $C_n=\gamma T+C_l$. Low temperature lattice part is fitted by means of the molecular crystal model [5]. Namely, C_l can be expressed as follows;

$$C_l(T)=3C_a^D(T,\omega_a^D)+3C_o^E(T,\omega_o^E)+3C_t^E(T,\omega_t^E)+36C_i^E(T,\omega_i^E) \quad (1)$$

where the terms of the right side in (1) show the contributions from acoustic modes, optical modes associated with Ag atom, torsional modes of the Mo_6S_8 unit and internal modes, respectively. Then we assume that the acoustic modes are represented by the Debye model and the other modes by the Einstein one. The superscripts D and E show those models, and ω_a^D and so on represent the characteristic frequencies.

The results of fitting calculation are also shown in Fig.1 by solid curves, together with the contribution from each modes. As can be seen, the curve for total specific heat is in agreement with the experimental result. The deviation of the calculated specific heat from the experimental values is smaller than 5% throughout the whole temperature range. The characteristic frequencies ω_a^D and ω_o^E correspond to 9.1 meV (105K) and 4.1 meV (47.8K), respectively. These values are consistent with the experimental results of the inelastic neutron scattering on other Chevrel phase compounds, though the value of ω_a^D is slightly smaller than that estimated from the neutron experiment. Because we assume that the values for ω_t^E and ω_i^E are larger than those for ω_a^D and ω_o^E, the contributions from these modes to the total specific heat is negligible in the temperature range studied [6].

In the strong coupling superconductor, when the characteristic phonon with energy $\tilde{\omega}$ is dominant in the superconducting mechanism, the thermodynamic parameter $\Delta C/\gamma T_c$ depends on $\tilde{\omega}$ follwing Kresin and Parkhomenko's expression [7]. By use of their expression, the value of $\tilde{\omega}$ can be evaluated to be 6.9 meV (80K) from the experimental value of $\Delta C/\gamma T_c$. This value of $\tilde{\omega}$ is different from both of the characteristic phonon energies evaluated by the molecular crystal model. In our previous paper [3], we pointed out that the strong coupling character and the behavior of lattice specific heat in Mo_6Se_8 were explained by taking account of low-lying phonon modes attributed to the phonon softening observed by the inelastic neutron scattering experiment at 5 K and then the characteristic energy of these phonon modes was consistent with the value of $\tilde{\omega}$ estimated from Kresin and Parkhomenko's expression. Such soft phonons have been observed in several Chevrel phase compounds as well as Mo_6Se_8 [6]. Thus we also expect the existence of the softening modes with the broad phonon distribution in $AgMo_6S_8$. However, it is difficult for the present to separate the specific heat contribution from such modes because the contribution from the optical modes is too large.

In conclusion, the Chevrel phase compound $AgMo_6S_8$ is the intermediately strong coupling superconductor, in which we expect that the softening phonon modes with the energy of about 6.9 meV are mainly relating to the superconductivity.

REFERENCES

[1] Superconductivity in Ternary Compounds I ed. by Ø. Fischer and M.B. Maple (Springer-Verlag, 1982)

[2] N.E. Alekseevskii, G. Wolf, C. Hohlfeld and N.M. Dobrovolskii: J. Low Temp. Phys. 40 (1980) 479

[3] N. Kobayashi, S. Higuchi and Y. Muto: in Proc. 4th Conf. on Superconductivity in d-and f-Band Metals (Karlsruhe 1982) p.173

[4] S. Morohashi, K. Noto, N. Kobayashi and Y. Muto: Physica 108b (1981) 929

[5] S.D. Bader, G.S. Knapp and S.K. Sinha: Phys. Rev. Lett. 37 (1976) 344

[6] B.P. Schweiss, B.Render and R. Flükiger: in Ternary Superconductors ed. by G.K. Shenoy et al. (North Holland, New York 198) p.29

[7] V.Z. Kresin and V.P. Parkhomenko: Sov. Phys. Solid State 16 (1975) 2180

LT-17 (Contributed Papers)
U. Eckern, A. Schmid, W. Weber, H. Wühl (eds)

LATTICE VIBRATIONAL BEHAVIOR OF $SnMo_6S_8$ AND OXYGEN DEFECTED $SnMo_6S_{7.8}O_{0.2}$

C. W. KIMBALL and P. P. VAISHNAVA†

Northern Illinois University, DeKalb, IL 60115, USA

B. D. DUNLAP, F. Y. FRADIN, D. G. HINKS and G. K. SHENOY§

Argonne National Laboratory, Argonne, IL 60439, USA

The temperature dependence of the Mossbauer effect has been measured at ^{119}Sn sites in oxygen-free, $SnMo_6S_8$ (T_C=14.2K) and in oxygen substituted $SnMo_6S_{7.8}O_{0.2}$ (T_C=10.1K). The temperature dependence of the asymmetry of the quadrupole doublet for $SnMo_6S_8$ shows phonon anomalies near 50 and 120K. $\langle z^2 \rangle_T$ indicates contraction with increasing temperature between 4.3 and 50K. For the oxygen substituted compound a study of the variation of mean square displacement with temperature shows lattice softening near 120K.

The ternary molybdenum chalcogenides, $M_xMo_6X_8$ (M=Sn, Pb, Cu, etc., X=S, Se or Te) are of special interest because of their high superconducting critical temperature, very high critical magnetic field, and, in some compounds, the coexistence of superconductivity and magnetic ordering. One of the most extensively studied[1-5] Chevrel phases is $SnMo_6S_8$. The phase should be stoichiometric at 1-6-8; however, extensive variation in single phase composition, superconducting transition temperature, and structural parameters have been reported in the literature. Such variation in properties have been attributed to S or Sn vacancies or Mo interstitials, formed by different preparation techniques. Recently, Hinks et al.[4] have shown that oxygen can be incorporated in $SnMo_6S_8$. By means of neutron powder diffraction, x-ray diffraction and ac susceptibility measurements, a relationship between oxygen concentration, T_C and c/a was found that covers the entire range previously observed for these parameters.

In this paper we report a Mossbauer effect study of the lattice properties of ^{119}Sn in the stoichiometric Chevrel phase $SnMo_6S_8$ (high T_C=14.2K) and in an oxygen doped compound $SnMo_6S_{7.8}O_{0.2}$ (T_C=10.1K). It has been found that: (a) in the high T_C compound a quadrupole doublet with large asymmetry is observed. The temperature dependence of the mean-square displacements indicates anharmonic phonon behavior near 50 and 150K; (b) in the high T_C compound a temperature dependent shift is observed which indicates a large anharmonicity and mode softening; (c) the Mossbauer spectra for the oxygen doped sample can be fitted in terms of an asymmetric quadrupole doublet due to tin bonded only to sulphur (S-Sn-S) and a symmetric doublet due to Sn for which oxygen has substituted for sulphur (S-Sn-O). A study of the variation of $\langle r^2 \rangle$ with temperature shows lattice softening near 120K; (d) the variation in quadrupole coupling with temperature suggests a change in symmetry near 120K in oxygen doped compounds.

Mossbauer spectra of the two compounds were obtained as a function of temperature between 4.2K and 300K. The isomer shifts were measured with respect to $BaSnO_3$.

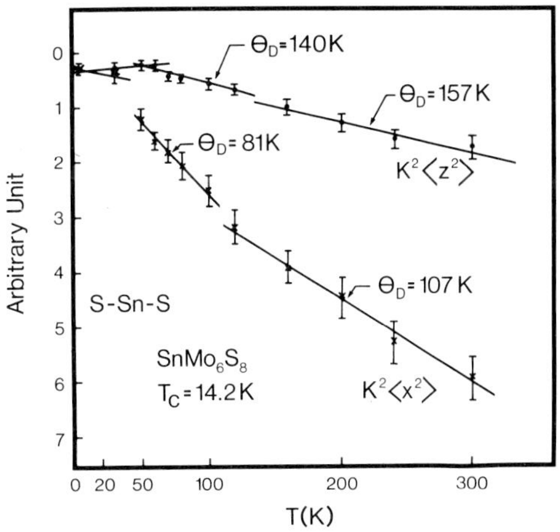

FIGURE 1
Temperature dependence of $\langle x^2 \rangle_T$ and $\langle z^2 \rangle_T$. Θ_D are from fits to Debye models.

† Based on work supported by the National Science Foundation (DMR 81-19475).
§ Based on work supported by the U.S. Department of Energy.

The Mossbauer spectra of $SnMo_6S_8$ exhibit an asymmetric quadrupole doublet. The ratio $R=I_\pi/I_\sigma$ of the intensities of the lines of the doublet yield the anisotropy of the Debye-Waller factor. The intensities I_π and I_σ are related to the mean-square displacements, in x and z directions respectively. Following the procedure described in Ref. 1, we obtain the temperature dependence of $\langle x^2\rangle_T$ and $\langle z^2\rangle_T$ shown in Fig. 1. It is evident that the slope of this curve changes at 50 and 120K. Both $\langle x^2\rangle_T$ and $\langle z^2\rangle_T$ indicate softening near 120K. $\langle z^2\rangle_T$ indicates contraction with increasing temperature between 4.2 and 50K. In order to characterize the thermal behavior the data were fitted to a Debye model. The Debye temperatures are shown in Fig. 1. We note that the changes in mean-square displacements near 120K are characteristic of a phase change. The observed behavior in the stoichiometric high T_c sample is consistent with the early observations of Bolz, Hauck and Pobell[2] on a lower T_c sample and with the inferences of Wagner and Freyhardt[4] from studies of the relationship of stoichiometry to lattice behavior.

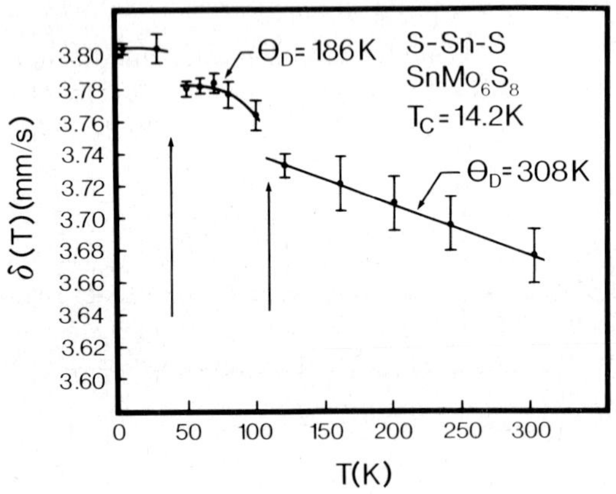

FIGURE 2
Temperature dependence of shift. Solid curves are based on Debye models.

The dependence of shift on temperature for the high T_c compound is shown in Fig. 2. From the plot it can be seen that the Debye temperature changes from θ_D=186K at 100K to 308K at 120K.

The local structure of $SnMo_6S_{7.8}O_{0.2}$ compound has been reported recently by Jorgensen and Hinks.[5] It is found that in the normal $SnMo_6S_8$ structure, Sn is located at the origin, halfway between the two S2 atoms on the threefold axis and is not strongly bonded to any of the sulphur neighbors. When the oxygen atom replaces one of the S2 atoms, the Sn atom moves off the origin and forms a covalent bond with oxygen. Jorgensen and Hinks[5] find that the Mo-Mo and Mo-S

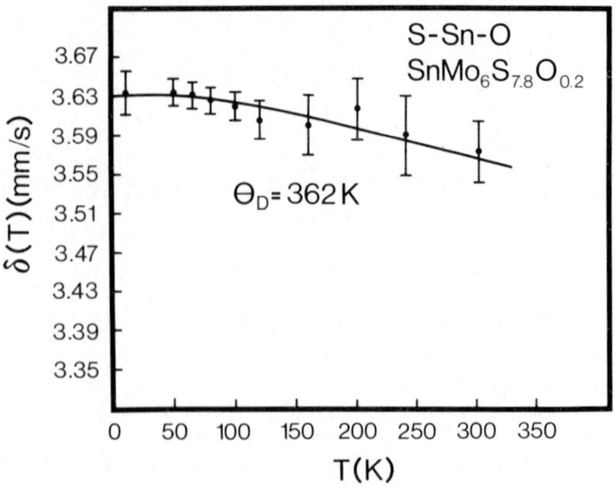

FIGURE 3
Shift vs temperature. Solid curve is based on Debye model.

bond lengths in Mo_6S_8 cluster are not dramatically changed by this oxygen defect. Accordingly the isomer shift should not change much in comparison to the normal $SnMo_6S_8$. Our isomer shift results (Fig. 3) confirm this inference. Unlike S-Sn-S in normal $SnMo_6S_8$ there are no temperature regions in which there are marked variations of the shift from the harmonic behavior.

REFERENCES

(1) C.W. Kimball, L. Weber, G. Van Landuyt, F.Y. Fradin, B.D. Dunlap and G.K. Shenoy, Phys. Rev. Letters 36 (1976) 412.
(2) J. Bolz, J. Hauck and F. Pobell, Z. Phys. B25 (1976) 351.
(3) D.G. Hinks, J.D. Jorgensen and Hung-Cheng Li, Solid State Comm. 49 (1984) 51.
(4) H.A. Wagner and H.C. Freyhardt, Proc. of the IV conference on superconductivity in d- and f-band metals, Karlsruhe, 1982, ed. W. Buckel and W. Weber (Kernforschungszentoum Karlsruhe, GmbH, Karlsruhe) 197.
(5) D.G. Hinks, J.D. Jorgensen and H.-C. Li, Phys. Rev. Letters 51 (1983) 1911.

LT-17 (Contributed Papers)
U. Eckern, A. Schmid, W. Weber, H. Wühl (eds)

THERMOELECTRIC POWER OF $PbMo_6S_8$ AND $EuMo_6S_8$

Geetha Balakrishnan*, R. Srinivasan@, V. Sankaranarayanan@, G. Rangarajan@, R. Janaki* and G.V. Subba Rao*

* Material Sciences Research Centre, Indian Institute of Technology, Madras 600 036, India
@ Low Temperature Laboratory, Department of Physics, Indian Institute of Technology, Madras 600 036 India

The thermoelectric power, S, of the Chevrel phase compounds $PbMo_6S_8$ and $EuMo_6S_8$ were measured on sintered pellets from liquid helium temperature to room temperature. In $PbMo_6S_8$ S is positive and is a sum of electron diffusion (S_e) and phonon drag (S_{ph}) parts. From S_e, using the rigid band model of Mott, the slope of the density of states curve is calculated. The result is compared with existing band structure calculation. In $EuMo_6S_8$ the thermoelectric power shows a sharp rise at around 100K which is the temperature at which a structural transformation takes place. The thermoelectric power reverses sign at 92K and 33K.

1. INTRODUCTION

Vasudeva Rao et al (1) reported recently the temperature variation of the thermoelectric power of selenium and tellurium substituted Chevrel phase compounds of the type $Cu_{1.8}Mo_6S_{8-y}(Se/Te)_y$. They analysed the data as the sum of an electron diffusion term S_e and a phonon drag term S_{ph}. From S_e, using the Mott-Wilson expression(2) they estimated the energy derivative of the logarithm of the density of states at the Fermi level E_F.

Band structure calculations of Nohl et al(3) provide information on the location of the Fermi level and shape of the number density of states curve in $PbMo_6S_8$. It is therefore of interest to study the thermoelectric power of this material.

$EuMo_6S_8$ does not become superconducting at ambient pressure. This has been attributed to an electronically driven structural phase transition to a triclinic structure of low density of states $N(E_F)$. A study of thermoelectric power in this material will also be of interest.

2. RESULTS

The preparation and characterisation of $PbMo_6S_8$ and $EuMo_6S_8$ have been reported elsewhere (4). The details of the experimental measurement have been described in (5). Measurements were made on sintered pellets.

Figure 1 shows the temperature variation of the thermoelectric power S of both the compounds.

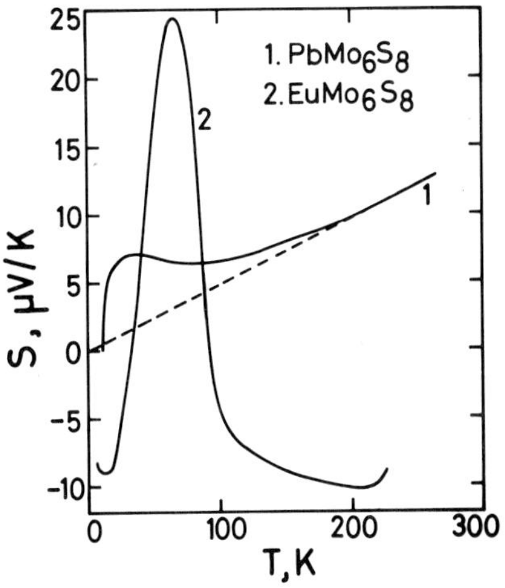

FIG. 1

3. DISCUSSION

a) $PbMo_6S_8$: The temperature variation of S is similar to that found by Vasudeva Rao et al (1) in the copper molybdenum chalcogenides. Towards the high temperature end of the curve, S varies linearly with temperature and extrapolates back to the origin as shown by the dotted line. This linear variation, $S_e=\alpha T$ is attributed to electron diffusion. $S-S_e=S_{ph}$ is the phonon drag part which shows a broad maximum around 35K. On the rigid band model of Mott (2).

$$\alpha = \frac{\pi^2 k^2}{3|e|}\left[\frac{d\ln N_d}{dE} - \frac{d\ln A}{dE}\right]_{E_F}$$

Here $N_d(E)$ is the number density of states in the d band and A the area of the Fermi surface. If we assume the second term to be negligible, then from the experimental value of α, $d\ln N_d/dE)_{E_F}$ comes out to be +2.05 per eV. From the work of Nohl et al(3), $N_d(E_F)$ is 0.94 electrons/spin, Mo atom eV. This would give $dN_d/dE)_{E_F}$ = +1.9 electrons/spin Mo atom $(eV)^2$. The positive sign of $dN_d/dE)_{E_F}$ will place E_F to the left of the peak in the density of states. The calculation of Nohl et al (3) places the Fermi level to the right of the peak in the density of states in the central valley of the $(t_{1u},e_g) \rightarrow$ e-band at which location dN_d/dE is very small. This discrepancy may be resolved if the area of the Fermi surface decreases with energy.

b) $EuMo_6S_8$: Lachal et al(6) measured the resistivity in arc melted and sintered pellets and the curves are reproduced in Fig.2.

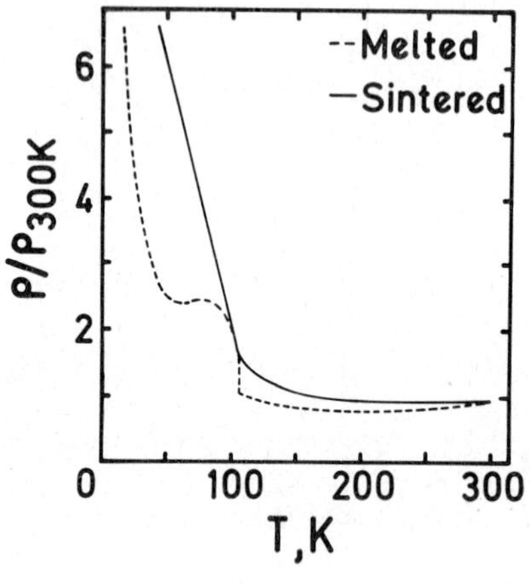

FIG.2

The arc melted sample shows an abrupt change in the slope of the resistivity curve at 109K. In the sintered pellet the transition is not so sharp and occurs at a slightly lower temperature. The measurement of specific heat gave evidence for a first order structural phase transition and X -ray diffraction indicated a transformation to a triclinic structure. The electronic specific heat in this structure is very low indicating a very low value of the density of states due to the removal of band degeneracy by the structural distortion.

The very rapid increase in thermopower from a negative value to a positive peak commences around 100K which is the same temperature at which the electrical resistivity shows a sharp rise . This behaviour in thermoelectric power arises from the reduction in the density of states due to the structural phase transformation. Since S must tend to zero as T goes to zero, one must expect a peak in S as the temperature is reduced. However the two reversals of sign of S still remain to be explained.

$EuMo_6S_8$ is known to become superconducting under pressure. One would then expect the peak in thermoelectric power to be reduced. Further experiments are required to verify this.

ACKNOWLEDGEMENTS

Thanks are due to the Department of Science and Technology, India, for the award of a research grant.

REFERENCES

(1) V. Vasudeva Rao, G. Rangarajan and R. Srinivasan, J. Phys. F. Metal Physics (In Print).

(2) R.P. Huebner, Thermoelectricity in Metals and Alloys in Solid State Physics, Vol.27 eds. H. Ehrenreich, F. Seitz and D. Turnbull. (Acad.Press, N.Y. London 1972) p.72.

(3) H. Nohl, W. Klose and O.K. Andersen, Superconductivity in Ternary Compounds in Topics of Current Physics, Vol.32 eds. Ø. Fischer and Maple (Springer Verlag 1982) p.165.

(4) A.M. Umarji, Ph.D Thesis (1980) Indian Institute of Technology, Madras, India.

(5) V. Vasudeva Rao, Ph.D Thesis (1983) Indian Institute of Technology, Madras, India.

(6) B. Lachal, R. Baillif, A. Junod and J. Muller. Solid State Commun. 45 (1983) 849.

LT-17 (Contributed Papers)
U. Eckern, A. Schmid, W. Weber, H. Wühl (eds)

MEASUREMENTS OF MAGNETIC SUSCEPTIBILITY AND SUPERCONDUCTING TRANSITION TEMPERATURE OF SOME PURE AND Mn-DOPED CHEVREL COMPOUNDS

M. E. REEVES, W. M. MILLER, D. M. GINSBERG, and F. C. BROWN

(Department of Physics and Materials Research Laboratory, University of Illinois at Urbana-Champaign, Urbana, Illinois 61801, USA)

We present the magnetic susceptibilities and the superconducting transition temperatures of some Chevrel compounds. We measured the magnetic susceptibilities of $SnMo_6S_8$, $PbMo_{6.35}S_8$, and $PbMo_{6.2}S_8$ samples. We measured the superconducting transition temperatures of $Mn_xSn_{1-x}Mo_6S_8$ samples with $0.0 \leq x \leq 0.13$. The temperature dependence of the magnetic susceptibility is explained as stemming from the transfer of electronic charge from the metal atoms to the Mo_6S_8 clusters. We also attribute an observed peak in the T_c versus x curve to charge transfer.

We have measured the magnetic susceptibilities and superconducting transition temperatures of some Chevrel samples. Sample preparation and characterization were carried out in the manner reported earlier (1).

The susceptibilities of the samples described in Ref. 1 were measured at various temperatures using an SHE SQUID susceptometer. The data showed a rise at low temperatures. We ascribed this rise to unknown magnetic impurities and fitted it with the Curie-Weiss law. (The size of the rise indicates about 1 to 5 atomic ppm. magnetic impurities.) This contribution was subtracted from the data, leaving the susceptibilities shown in Figure 1.

Yvon and Paoli (2) have pointed out the importance of electronic charge transfer from the metal atom (Sn or Pb) to the Mo_6S_8 cluster. We have observed a temperature dependence of the susceptibility which we believe to result from the effect of thermal expansion on the charge transfer. Jorgensen and Hinks measured the thermal expansion of tin and lead Chevrel compounds (3,4). As the sample is cooled from room temperature to cryogenic temperatures, the a-lattice constant decreases linearly by approximately 0.5%. Between room temperature and 100 K, the c-lattice constant decreases by 0.03% for the tin Chevrel compound and by 0.02% for the lead Chevrel compound. The c-lattice constant then increases by 0.03% for the tin and by 0.1% for the lead compound with further cooling. We note that as the lattice constant shrinks, the metal atom moves closer to the Mo_6S_8 clusters, transferring more charge.

The effect of this charge transfer depends on the shape of the electronic density of states (DOS) function. These materials are believed to have a peak in the DOS near the

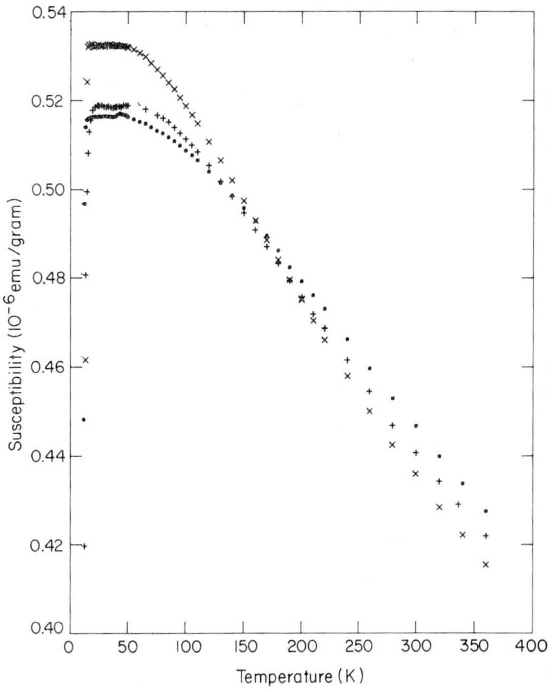

FIGURE 1
Susceptibility versus temperature for $SnMo_6S_8$ (circles), $PbMo_{6.35}S_8$ (plus signs), and $PbMo_{6.2}S_8$ (x's). Some of the data points have been removed for clarity. Also, a low-temperature Curie-Weiss contribution has been subtracted.

Fermi level (5). As charge is transferred to the cluster, the Fermi level rises, changing the DOS at the Fermi level. Another indication of a peak in the DOS is revealed by plotting superconducting transition temperature (T_c) versus valence electron concentration (VEC) (2). Yvon's calculations show that the VECs for tin and lead Chevrel compounds lie on the left side of the peak. This indicates that the Fermi level lies on the low-energy side of the peak in the DOS versus energy curve. Therefore, as more charge is transferred, the DOS and thus the Pauli paramagnetic susceptibility increases. This explains the temperature dependence of the susceptibility.

A quick examination of Figure 1 reveals several features that support our explanation of the susceptibility's temperature dependence. As the temperature is decreased to 100 K, the susceptibility increases linearly. This is consistent with both the a- and c-lattice constants shrinking linearly, allowing more charge to transfer from the metal atoms to the Mo_6S_8 clusters. Below 100 K, the c-lattice constant begins to grow as the temperature is lowered, lessening the charge transferred. The behavior of the susceptibility reflects this, increasing at a slower rate between 100 K and 40 K. The magnetic susceptibility is then constant down to the superconducting transition, near 15 K. This may indicate some transition near 40 K (6).

We have also measured the transition temperature of $Mn_xSn_{1-x}Mo_6S_8$. The manganese doping, x, was varied from 0.0 to 0.13. Optical microscopy, scanning electron microscopy, energy dispersive x-ray microanalysis, and x-ray absorption near edge structure (XANES) data all indicated that the manganese was incorporated into the Chevrel phase. The XANES data showed that the manganese was in a highly ionized valence state of 6+. This is consistent with the idea of charge transfer from the metal atom to the cluster (2).

T_c versus manganese concentration is shown in Figure 2. The data exhibit a peak in T_c at about x = 0.10. This peak can be explained in terms of the peak between VEC = 3.7 and VEC = 3.8 in the T_c versus VEC curve mentioned earlier. Using a valence of 6+ for the manganese, our peak in T_c versus x corresponds to a VEC of 3.73.

If the manganese atoms were magnetic, we would expect (7) $-dT_c/dx \gtrsim 1000$ K. The observed value is only about 23 K. The magnitude of dT_c/dx may be limited in two ways. The large separation between the manganese atoms and the molybdenum 4-d electrons, which are thought to be responsible for the superconductivity in Chevrel compounds, would limit the size of any exchange interaction (5). Also, manganese atoms in the 6+ valence state would lack all but one 3-d electron and would presumably be only slightly magnetic anyway. Furthermore, the increase in the VEC upon doping tends to increase T_c.

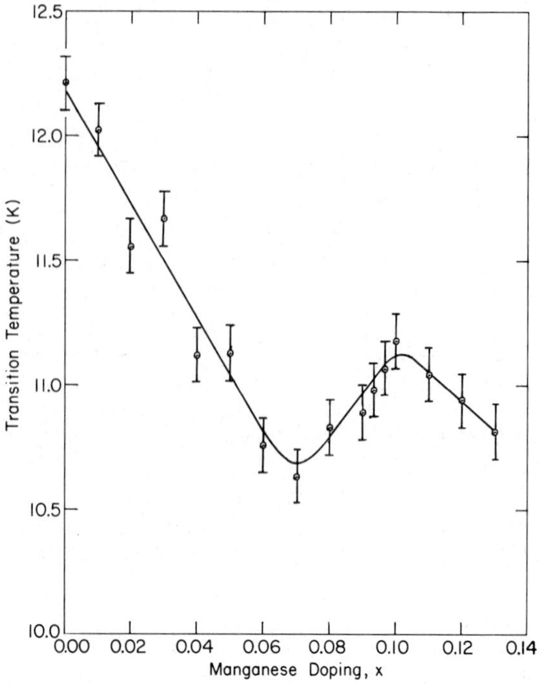

FIGURE 2
Transition temperature versus manganese content for $Mn_xSn_{1-x}Mo_6S_8$.

ACKNOWLEDGEMENTS

We thank Donald Holmgren and Apurba Roy for helpful advice on obtaining the magnetic susceptibility data. We also thank D. G. Hinks of Argonne National Laboratory for sharing his coefficient of thermal expansion data with us prior to publication. Finally, we acknowledge support by the U.S. Department of Energy, Division of Material Sciences, under Contract No. DE-AC02-76ER01198.

REFERENCES

(1) W. M. Miller and D. M. Ginsberg, Phys. Rev. B28 (1983) 3765.
(2) K. Yvon and A. Paoli, Sol. State. Commun. 24 (1977) 41.
(3) J. D. Jorgensen and D. G. Hinks, Bull. Am. Phys. Soc. 29 (1984) 230 AN-16.
(4) D. G. Hinks, private communication.
(5) O. Fischer, App. Phys. 16 (1978) 1.
(6) C. W. Kimball et al., Phys. Rev. Lett. 36 (1976) 412.
(7) D. M. Ginsberg, Phys. Rev. B10 (1974) 4044.

LT-17 (Contributed Papers)
U. Eckern, A. Schmid, W. Weber, H. Wühl (eds)

SUPERCONDUCTING PROPERTIES OF $Pb_{1-x}M_xMo_6S_8$ (M = Nb, Y AND La)

Shuzo TAKANO, Eisuke TOMITA and Shoichi MASE

Department of Physics, Kyushu University33, Fukuoka 812, Japan

The effects of the replacement of Pb atom in $PbMo_6S_8$ by Nb, Y and La ($Pb_{1-x}M_xMo_6S_8$) on T_c and $H_{c2}(T)$ were investigated. It was found that for M = Nb and Y, T_c increases to reach a maximum and decreases again, while for M = La, T_c continues to decrease with increasing x. It was also found that $-dH_{c2}(T)/dT$ at T = T_c decreases with x. The results may be related to the change of the width of the conduction band.

1. INTRODUCTION

Chevrel type compounds such as AMo_6S_8 (A = metal) have recently attracted one'sattention because of the high transition temperature, T_c and the high second critical field, $H_{c2}(T)$ (I). The clarification of their origins is not only physically interesting but also important from a point of view for constructing high field superconducting magnet. The previous experimental results have indicated that there are some regular rules between the position of A atom in the periodic table and magnitudes of T_c and $H_{c2}(T)$.

The object of the present paper is to investigate the correlation between T_c and $H_{c2}(T)$ and the kind of A atom. Among the Chevrel type compounds, $PbMo_6S_8$ has nearly the highest T_c. It was, then expected that the replacement of Pb atom by other atoms leads to appreciable effects on the density of states at the Fermi level, $N(E_F)$. In order to avoid the complexity due to the influence of 4f-electrons, as a substitute we chose Y ($4d^15s^2$) and La ($5d^16s^2$) with one less valence electrons than Pb ($6s^26p^2$) and also chose Nb ($4d^45s^1$) with one more valence electrons than Pb.

2. EXPERIMENTAL PROCEDURES

All samples were prepared by the powder sintering method at 1050 °C. The appropriate amounts of Pb, Mo, S, Nb, Y_2S_3 or La_2S_3 powder were weighed and mixed thoroughly, then lightly pressed for the subsequent heat treatment for 24 hours. The reacted material was again crushed and mixed. After repeating this process two or three times, the material was pressed up to 6 t/cm^2 by using an oil-press and a mould made of tungsten carbide with an inner diameter 7 mm and was finally heat-treated for 72 hours.

The resistivity was measured as functions of temperature near T_c and magnetic field up to 18 teslas supplied by a pulse magnet. The exponentially decaying time from the maximum field to zero field was about 80 msec so that the eddy current effect was not so serious.

The magnetic susceptibility χ was measured by an induction method with a Hartshorn bridge at 77 K. The sample had a form of a column with diameter 7 mm and height 20 mm. By using a Lock-in detection system, the susceptibility change as small as 10^{-7} emu/g could be detected.

In order to study the relation between the magnitude of T_c and the lattice structure, the $Pb_{1-x}Nb_xMo_6S_8$ samples were tested by the powder X-ray diffraction method.

3. EXPERIMENTAL RESULTS

Figure 1 shows the plots of T_c against x, where T_c is defined as temperature corresponding to the middle point of the resistivity change.

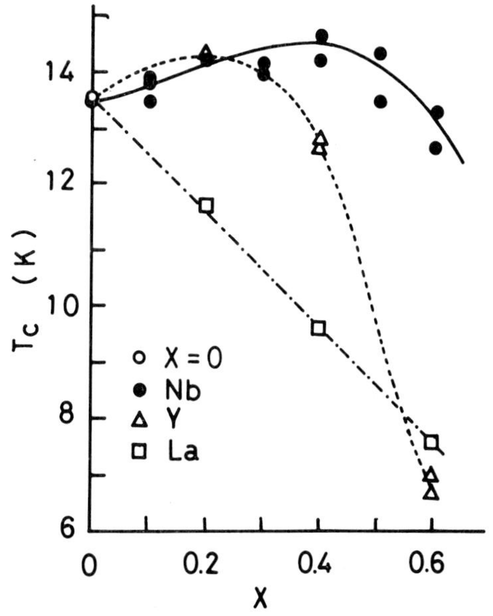

FIGURE 1
The x-dependence of the transition temperature T_c for $Pb_{1-x}M_xMo_6S_8$. o-$PbMo_6S_8$; ●-M=Nb; Δ-M=Y; □-M=La

In the case of $Pb_{1-x}Nb_xMo_6S_8$, T_c increases weakly with increasing x up to x = 0.4, and then turns to decrease with further increasing x. In the case of $Pb_{1-x}La_xMo_6S_8$, T_c decreases monotonously with increasing x, while for $Pb_{1-x}Y_xMo_6S_8$ the x-dependence of T_c is somewhat similar to that of $Pb_{1-x}Nb_xMo_6S_8$, but T_c attains the maximum at a smaller value of x. The width of the transition had the tendency to increase with increasing x for all three cases.

In Fig.2 the rhombohedral lattice parameter a_R, the rhombohedral angle α and the hexagonal unit cell volume V_H are plotted against x for $Pb_{1-x}Nb_xMo_6S_8$. It can be seen from Fig.1 and Fig.2 that there is a close correlation between T_c and V_H for $Pb_{1-x}Nb_xMo_6S_8$.

The slope of $H_{c2}(T)$ against T near T_c is shown in Fig.3 as a function of x. It can be seen that for all three cases $-dH_{c2}(T)/dT$ at T = T_c is a decreasing function of x. The value for x = 0 agrees well with the reported one (2). In the case of $Pb_{1-x}Y_xMo_6S_8$ for x = 0.6, $H_{c2}(T)$ was so small that a part of the resistivity already recovered even at 4.2 K.

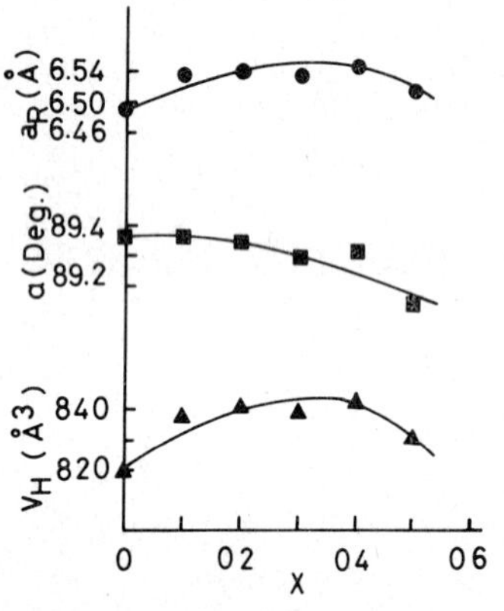

FIGURE 2
The x-dependence of the lattice parameters in $Pb_{1-x}Nb_xMo_6S_8$.

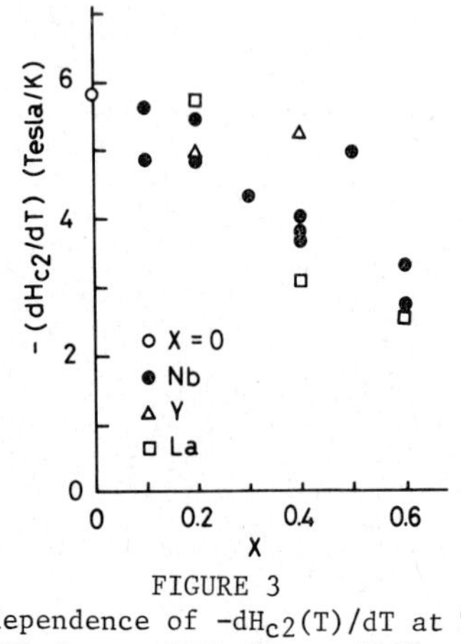

FIGURE 3
The x-dependence of $-dH_{c2}(T)/dT$ at T = T_c for $Pb_{1-x}M_xMo_6S_8$. o-$PbMo_6S_8$; ●- M=Nb; Δ-Y; □-M=La

The normal state resistivity ρ_n was almost an order of magnitude larger than that of a melted crystal $PbMo_6S_8$ (3), and ρ_n increased gradually with increasing x for all three cases. The measured ρ_n may depend on the degree of the packing of the small crystallites in the sample.

The magnetic susceptibility for $Pb_{1-x}Nb_xMo_6S_8$ at 77 K decreased with increasing x up to x = 0.6. It changed from 0.42×10^{-6} emu/g for x = 0 to 0.35×10^{-6} emu/g for x = 0.4.

4. DISCUSSIONS

We found the close correlation between T_c and V_H for $Pb_{1-x}Nb_xMo_6S_8$. The similar relation has already been found for $Pb_{1-x}Y_xMo_6S_8$ and $Pb_{1-x}La_xMo_6S_8$ (4). From these results we may consider that the change of $N(E_F)$ is responsible for the change of T_c in these cases. It seems that the valence of A atom itself does not play any dominant role, but the change of the width of the 4f-conduction band due to the change of the overlap of the wave functions localized in Mo_6S_8-cluster is important. Since E_F of $PbMo_6S_8$ is already located near the maximum of N(E) (5), the narrowing of the conduction band by replacing Pb atom by Nb and Y atoms makes $N(E_F)$ further large. On the other hand, since $-dH_{c2}(T)/dT$ at T = T_c decreased by about 35 % at x = 0.4 for $Pb_{1-x}Nb_xMo_6S_8$, the product $\ell\xi_0T_c$ must increase also by 35 % where ξ_0 is the coherence length given by $\xi_0 = 0.180(\hbar v_F/k_BT_c)$ and v_F the Fermi velocity. The mean free path $\ell = v_F\tau$ is estimated to be about 4 Å for $PbMo_6S_8$ from the resistivity data (3). As mentioned previously, ρ_n tended to increase with increasing x. If we assume that τ becomes shorter, the results presented in Fig.3lead to the conclusion that v_F increases rather largely with increasing x.

The decrease of χ seems to contradict with the above conclusion. This rather unexpected result may be due to the decrease in the g-value of the conduction electrons.

REFERENCES

(1) Φ. Fischer: Appl. Phys. 16 (1978) 1
(2) Φ. Fischer, H. Jones, G. Bongi, M. Sergent and R. Chevrel: J. Phys.C: Solid State Phys. 7 (1974) L450
(3) R. Flükiger, H. Devantay, J. L. Jorda, and J. Muller: IEEE Trans.Magn. 13 (1977) 818
(4) M. Sergent, R. Chevrel, C. Rossel and Φ. Fischer: J. Less comm. Met. 58 (1978) 179
(5) O. K. Andersen, W. Klose and H. Nohl: Phys. Rev. B17 (1978) 1209

LT-17 (Contributed Papers)
U. Eckern, A. Schmid, W. Weber, H. Wühl (eds)

DEPENDENCE OF HALL-EFFECT, SUPERCONDUCTIVITY AND ELECTRICAL RESISTIVITY ON ATOMIC ORDER IN CHEVREL PHASE $PbMo_6S_8$ FILMS

H. ADRIAN, G. HOLTER, L. SÖLDNER, and G. SAEMANN-ISCHENKO

Physikalisches Institut, Universität Erlangen-Nürnberg, Erwin-Rommel-Str. 1, D-8520 Erlangen, FRG

A strong dependence of the Hall-effect of high quality $PbMo_6S_8$ films on temperature and irradiation induced defects is reported. The data demonstrate the contribution of electron and hole-like states at E_F and the drastic influence of lattice defects on electronic properties which explains the strong T_c degradation in the framework of existing band structure calculations. The effects of isochronal annealing indicate the presence of two distinct annealing mechanisms.

1. INTRODUCTION AND EXPERIMENTAL

The main reason for the interesting properties of Chevrel-phase compounds including the occurrence of extremely high critical fields is the cluster character of the lattice structure which leads to high and narrow peaks in N(E) and a broad spectrum of phonon frequencies (1). From this point of view one expects a drastic influence of defects as shown by the strong degradation of T_c of $AgMo_6S_8$ and $PbMo_6S_8$ (2). In order to obtain more detailed information we measured the Hall-effect of $PbMo_6S_8$ at 15K as function of fluence irradiating with 20MeV S ions at low temperature (< 20K) and its temperature dependence before and after irradiation. Furthermore, first results of isochronal annealing up to 1000K are reported.

Single phase films (d ≈ 0.22μm, width 2mm) were prepared by magnetron sputtering as reported earlier (2). The midpoints of the resistive T_c's were up to 12.5K, $(dB_{c2}/dT)T_c$ was 4.2T/K, the resistance ratios ρ(290K)/ ρ(15K) were up to 2.1 with resistivities of about 200μΩcm at 300K.

2. RESULTS AND DISCUSSION

Before irradiation a strong temperature dependence of the Hall constant R_H, as shown in fig.1 was observed. R_H (15K) corresponds in a one-band-model to a hole concentration of $6.3 \times 10^{28} m^{-3}$(3). However, according to band structure calculations (4) electron-like states also contribute to the density of states at the Fermi-level $N(E_F)$. The origin of the temperature dependence could be the peaked structure of N(E) at E_F, from which one expects E_F to be slightly shifted to higher energy with temperature.

The strong degradation of T_c with fluence was reproduced as shown in fig.2. The corresponding behaviour of R_H(15K) as T_c decreases is displayed in fig.3. The general behavior clearly shows a minimum in R_H with decreasing T_c followed by a strong increase. It is evident that the lattice defects which depress T_c strongly influence the electronic properties. One has

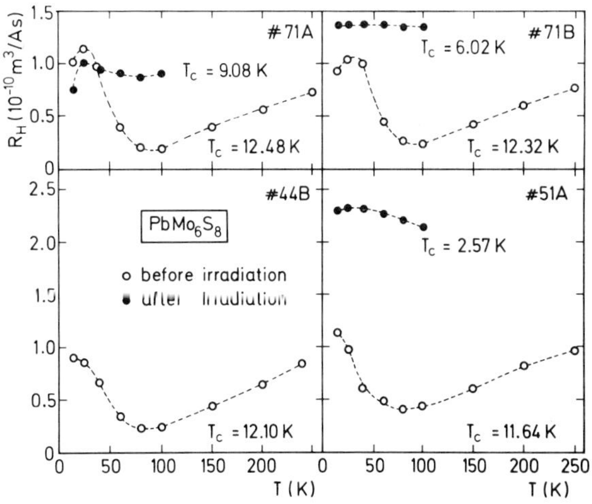

FIGURE 1
$R_H(T)$ before and after irradiation.

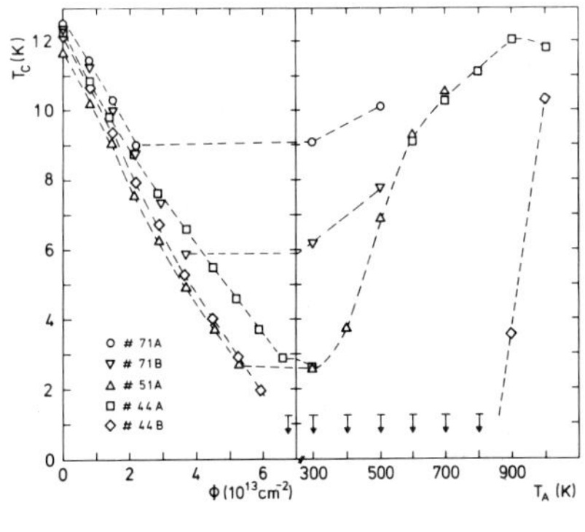

FIGURE 2
T_c vs. fluence Ø and annealing temperature T_A.

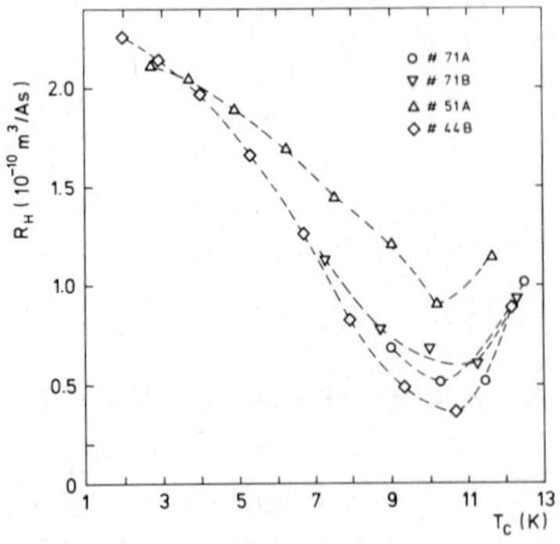

FIGURE 3
R_H(15K) as function of T_c.

to expect that irradiation leads to a smearing of N(E) due to strain fields around defects. This effect is demonstrated by the disappearance of the temperature dependence of R_H. Furthermore, sulfur vacancies and interstitials should shift E_F to higher energies (2). The strong increase of R_H at low T_c indicates a significant reduction of the carrier concentration which qualitatively explains the T_c degradation.

After irradiation the films were isochronally annealed. After low fluences (fig.1) recovery in T_c starts at annealing temperatures T_A as low as 400K. With increasing T_c the ρ(T)-curves (fig.4) gradually return to ρ(T) before irradiation. Contrary to this ρ of a film irradiated with high fluence ($Ø=9x10^{14}cm^{-2}$, $T_c<1.2K$, X-ray amorphous) dramatically increased with T_A up to 800K which indicates the opening of a gap at E_F. At T_A=900K the gap at E_F disappears as shown by the drastic change in the ρ(T)-behavior. Simultaneously very weak X-ray lines indicate the beginning of long range order and superconductivity below 3.56K

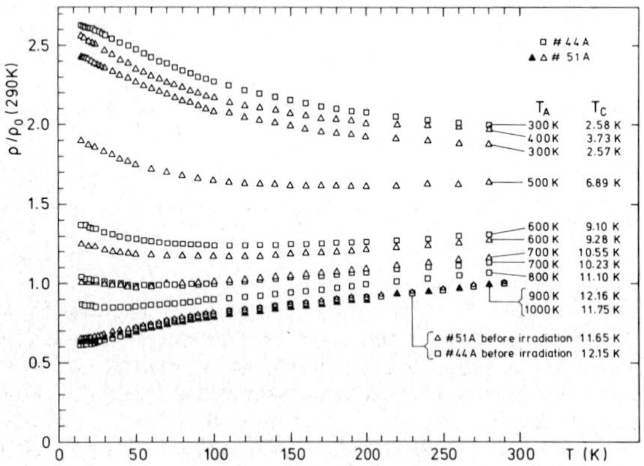

FIGURE 4
ρ(T) before irradiation and after annealing at T_A. The corresponding T_c's are indicated.

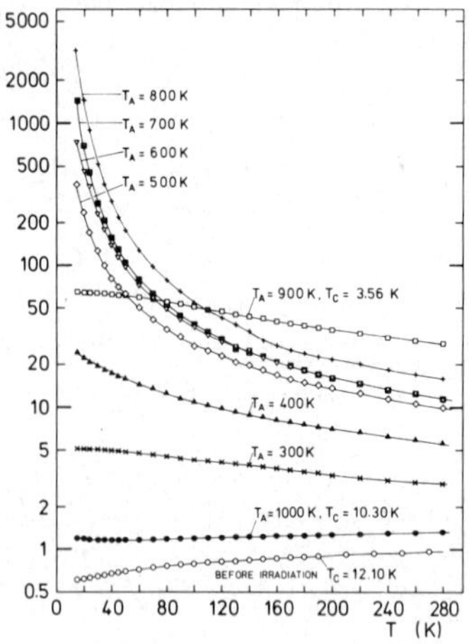

FIGURE 5
Log ρ vs. T of a film irradiated with high fluence after annealing at T_A.

reappears. Raising T_A another 100K increases T_c to 10.3K. The annealing data clearly suggest two different types of defects. After low fluences (about 0.02dpa) strain fields should be the main reason for the suppression of T_c. The recovery of T_c at 400K could then be explained by structural relaxation. After high fluences, however, the large concentration of sulphur defects should shift E_F into the gap between the Mo-d-states, which is expected to be partly filled by smearing of N(E). Structural relaxation reduces the defect-states in the gap, which explains the ρ(T) behavior.

3. CONCLUSIONS

R_H(T) clearly shows that electron and hole-like states contribute to N(E) in the vicinity of E_F. The data after irradiation show the strong influence of defects on electronic properties which explains the T_c degradation. Together with the annealing behavior of ρ and T_c the results support the defect model proposed earlier (2).

ACKNOWLEDGEMENTS

The authors thank their colleagues in Erlangen and acknowledge support by the BMFT.

REFERENCES

(1) for a review see: Superconductivity in Ternary Compounds I+II, eds. Ø. Fischer and M.B. Maple (Springer, Berlin, 1982)
(2) G. Hertel, H. Adrian, J. Bieger, C. Nölscher, L. Söldner, and G. Saemann-Ischenko, Phys. Rev. B27 (1983) 212
(3) J.A. Wollam and S.A. Alterovitz, Phys.Rev. B19 (1979) 749
(4) H. Nohl, W. Klose, and O.K. Andersen in ref.1, Vol. I, p. 165

LT-17 (Contributed Papers)
U. Eckern, A. Schmid, W. Weber, H. Wühl (eds)

NEW $PbMo_6S_8$ - BASED GRANULAR SUPERCONDUCTORS

E. CRUCEANU and L. MIU

Institute of Physics and Technology of Materials, Bucharest, Romania

New granular superconductors consisting of $PbMo_6S_8$ granules within a ductile matrix were obtained in form of wires by conventional cold drawing technique without subsequent annealing procedures. As matrix alloys either the constitutional element Pb or a Pb-30 w/o Bi composition were used. Some superconducting properties of these composite materials are reported and the mechanism of supercurrent transport is discussed.

1. INTRODUCTION

The high superconducting properties and especially the high critical field B_{c2} of $PbMo_6S_8$ gave an impetus to the search for techniques of producing this very brittle ternary compound in form of tapes or wires suitable for applications in the generation of magnetic fields. In this connection much effort was spent to develop methods to produce $PbMo_6S_8$ wires with high current carrying capacities. Among the first attempts one may mention the two step diffusion method used by Decroux et al. (1) to produce a $PbMo_6S_8$ surface layer on Mo wires and by Alekseevskii et al. (2) to produce $PbMo_6S_8$ on Mo ribbons. Luhman and Dew-Hughes (3) reported results on $PbMo_6S_8$ wires drawn in Ag tubes, while Seeber et al. (4) used Ta and Mo tubes filled with a mixture of MoS_2, PbS and Mo powder. With the aim to overcome the well known poor mechanical properties of $PbMo_6S_8$ we used for the brittle compound a ductile matrix which enables wire production by conventional metallurgical drawing techniques. As matrix alloys we used either the constitutional element Pb (5) or a Pb-Bi composition. In this way "x" in $Pb_xMo_6S_8$ may take values of 1 or higher with direct consequences on the critical temperature T_c of the samples (6). The aim of our work is to present the first results on the superconducting parameters of a composite material consisting of $PbMo_6S_8$ granules in a Pb or Pb-Bi matrix.

2. EXPERIMENTAL RESULTS

$PbMo_6S_8$ samples with different matrices and matrix/$PbMo_6S_8$ ratios were prepared from high purity Mo, S, Pb and Bi elements. After synthesis at 1100°C the composite material was put into a copper-cylinder with o. d. 15 mm which was reduced by conventional cold drawing technique to about 1 mm without subsequent annealing procedures. The following types of samples were studied:

I. $Pb/PbMo_6S_8$ = 3/1 (weight ratio)
II. $Pb\text{-}Bi/PbMo_6S_8$ = 3/1 - " -
III. $Pb\text{-}Bi/PbMo_6S_8$ = 2/1 - " -

T_c was measured by the four probe method; the critical current I_c was defined by a 1 μV/cm criterion and was measured always in a transverse ap-

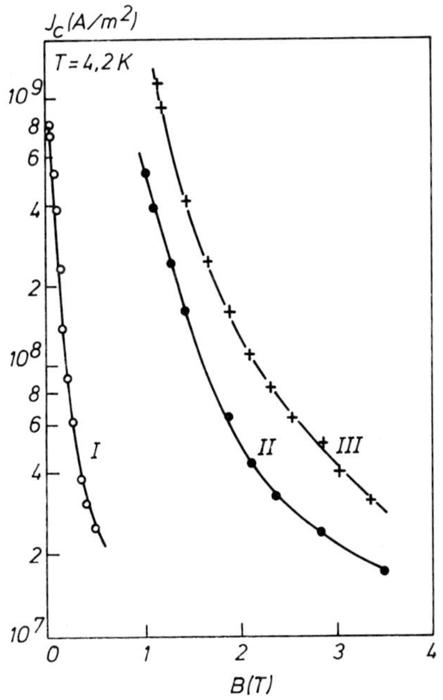

FIGURE 1
J_c for 3 types of wires vs. the external magnetic field H.

plied magnetic field H; the critical current density J_c was determined on the real superconducting cross-section. The variation of the critical current densities vs. the external magnetic field, measured at 4.2 K is shown in Fig. 1. In Fig. 2 we give the variation of J_c as a function of temperature T.

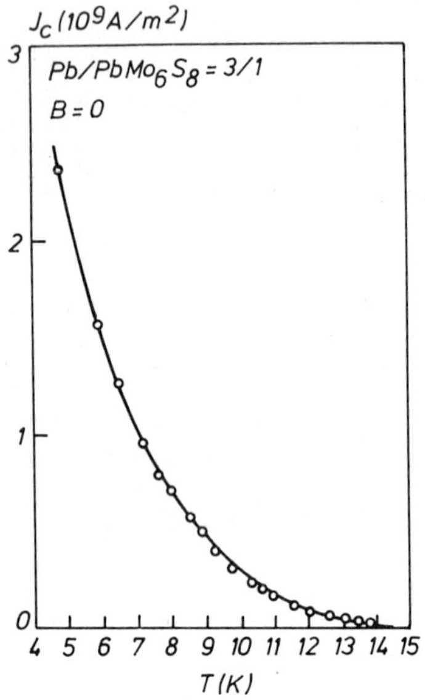

FIGURE 2
The critical current density at H = 0 as a function of temperature for the samples I.

3. DISCUSSION AND CONCLUSIONS

Samples I have T_c = 14.2 K (onset at 14.5 K), higher than the typical values situated around 12.5 K (6). This fact may indicate no defects at the Pb-site of $PbMo_6S_8$ crystal lattice. As a result of the alloying of the compound with Bi, T_c of the sample II and III was reduced to 13.7 K. Micrograph investigations together with T_c measurements demonstrated no reaction between the $PbMo_6S_8$ - based superconductors and the copper-jacket during the cold wire-drawing process.

As can be seen in Fig. 1, J_c of our wires are higher for the samples II and III obtained in a Pb-Bi matrix. Moreover, the samples III with a low matrix /$PbMo_6S_8$ ratio present higher J_c. At low values of the applied magnetic fields J_c are very high ($J_c > 10^9$ A/m^2) but a strong field dependence of J_c is observed.

The variation of J_c for the samples I as a function of T, for $T > T_{cM}$ (where T_{cM} is the critical temperature of the matrix) is well described by:

$$J_c \propto (1-T/T_c)^2$$

where T_c is the critical temperature of the composite.

Such a temperature dependence allows us to conclude that the transport of the supercurrent is due to the weak links between the $PbMo_6S_8$ grains through the matrix.

This conclusion is supported by the V-I characteristics at low electric field level ($E \leq 1\ \mu V/cm$), which are quite different from those observed in a classical type II superconductor. The field behaviour of J_c plotted in Fig. 1 is obviously related to the mechanism of the supercurrent in the $PbMo_6S_8$ wires produced by the present method.

Following to these preliminary results systematic investigations are needed in order to turn the reported technique into a useful method of producing high current carrying $PbMo_6S_8$ wires.

REFERENCES
(1) M. Decroux, O. Fischer and R. Chevrel, Cryogenics, 17 (1977) 291.
(2) N.E. Alekseevskii, M. Glinski, N. M. Dobrovolski and V.J. Tsebro, J.E.T.P. Lett. 23 (1976) 412.
(3) T. Luhman and D. Dew-Hughes, J. Appl. Phys. 49 (1978) 936.
(4) B. Seeber, C. Rossel and O. Fischer, $PbMo_6S_8$: A new generation of superconducting wires? in Ternary superconductors, eds. G.K. Shenoy, B.D. Dunlap, F.Y. Fradin (North-Holland, Amsterdam, 1981), p. 119 and IEEE Trans. Magn. 19 (1983) p. 402.
(5) E. Cruceanu, Proceed. National Conf. Phys. (Bucharest, 1982) p. 192
(6) M. Sergent, R. Chevrel, C. Rossel and O. Fischer, J. Less Comm. Metals 58 (1978) 179

LT-17 (Contributed Papers)
U. Eckern, A. Schmid, W. Weber, H. Wühl (eds)

SPECIFIC HEAT OF NON-SUPERCONDUCTING $PbMo_6S_8$

Alain JUNOD, Rémy BAILLIF, Bernd SEEBER, Øystein FISCHER and Jean MULLER

Département de physique de la matière condensée, Université de Genève, CH-1211 Genève 4, Switzerland

Grinding to 1 μm size reduces the superconducting temperature of $PbMo_6S_8$ from ~14 K to below 1.05 K. A size effect is excluded. No crystallographic changes are detected. Low temperature specific heat data are presented. Based on the low γ and the nearly unchanged phonon contribution, we conclude that the T_c degradation is, phenomenologically, an electronic effect.

1. INTRODUCTION

The Chevrel-phase compound $PbMo_6S_8$ has been the subject of numerous studies (see (1) and references quoted therein). This superconductor is characterized by a high transition temperature T_c ~ 12 - 15 K and an upper critical field in the vicinity of half a megagauss. These properties are somewhat sample-dependent suggesting a strong influence of metallurgical variables such as stoichiometry, homegeneity, strain etc.

During the progress of a program that aims at optimizing the critical current by reducing the particle size, it was realized that crushed $PbMo_6S_8$ eventually loses its superconducting properties (2). Neither the stoichiometry nor the room-temperature crystallographic structure differentiate the material thus obtained from the regular superconducting phase. Further, this degradation is reversible. A suitable heat treatment restores the usual critical temperature (3).

In a first step we shall compare the specific heats of the two modifications of $PbMo_6S_8$. The purpose is to point at the microscopic parameters that play a dominant role in the superconductivity of this compound. We shall then discuss the physical origin of the degradation.

2. METHOD AND RESULTS

The starting materials were molybdenum powder (99.95 % purity), sulfur lumps (99.999 %) and lead shots (99.9995 %). The mixture was reacted up to 900 °C in an evacuated silica tube. A subsequent treatment, 24 h at 1200 °C, was applied to obtain an homogeneous powder. The sample was finally gound in a ZrO_2 ball mill for 8 hours. Characteristics of the grain size distribution are : 98 % below 5.5 μm, 83 % below 2 μm, and 66 % below 1 μm. The powder was wrapped into a copper foil and the specific heat measured by standard semi-adiabatic techniques. Suitable corrections (<10 %) were allowed for internal time constant effects. The resulting accuracy is estimated to be 2 - 3 %. The data from 1.3 to 30 K are shown in Fig. 1, together with those of a superconducting sample (1).

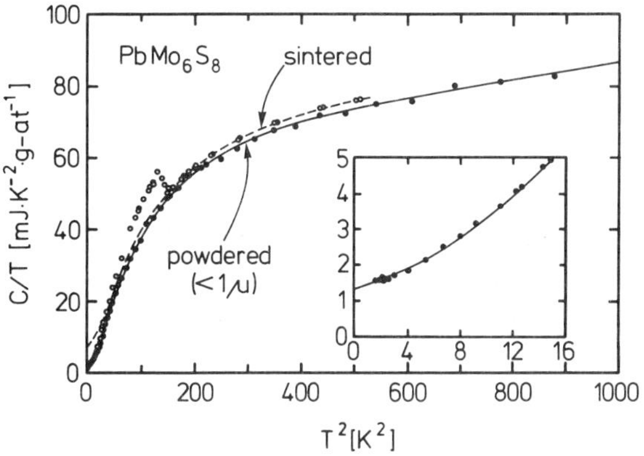

FIGURE 1
Specific heat of $PbMo_6S_8$ divided by the temperature vs. temperature squared in the domain 1.3 - 30 K. Full line : this work. Insert : same data below 4 K. Dashed line : superconducting sample, ref.(1).

3. DISCUSSION

There is no superconducting jump within the scatter of the data in the finely powdered sample. More sensitive a.c. susceptibility measurements down to 1.05 K have confirmed the absence of bulk superconductivity. These measurements showed, however, a faint continuous transition starting at 15 K and ending near 4.5 K. The amplitude of the full diamagnetic signal corresponds to 2.5 % of the total volume (calibration : Nb powder, 100 μm). For the semi-quantitative discussion to be given below, the sample may be considered as completely non-superconducting.

Fig. 1 further shows that γ is comparatively low in the non-superconducting modification, i.e. 1.30 ± 0.05 $mJK^{-2}g\text{-}at^{-1}$. The corresponding parameter is not easily extracted from the high temperature data of the superconducting phase, but admittedly lies in the range 5 - 9 $mJK^{-2}g\text{-}at^{-1}$ (1).

Fig. 1 finally shows that the phonon specific heat, which dominates over most of the temperature range, is very similar in both forms of $PbMo_6S_8$. A crude inversion of the specific heat was performed along the lines of ref.(4) to obtain more informations on the phonon spectrum $F(\omega)$. The following normalized spectrum

$$\frac{F(\omega)}{\omega^2} = \begin{cases} 0.2306 \times 10^{-6}\ K^{-3}, & 0 \le \omega \le 21.86\ K \\ 0.6950 \times 10^{-6}\ K^{-3}, & 21.86 \le \omega \le 71.55\ K \\ 0.0598 \times 10^{-6}\ K^{-3}, & 71.55 \le \omega \le 359.2\ K \end{cases}$$

together with $\gamma = 1.304$ $mJK^{-2}g\text{-}at^{-1}$, fits the data from 1.3 to 30 K within a r.m.s. error of 1.8 %.

The most prominent feature of this substitutional spectrum is a high density of $\omega^{-2}F(\omega)$ centered about $\hbar\omega/k \sim 40$ K. A similar but much narrower feature was found in superconducting $PbMo_6S_8$ (1). This peak, which is also found in the neutron scattering spectra of Schweiss et al (5), is thought to reflect the large vibrational amplitude of the heavy cation Pb, which is in turn related to a low Pb-(Mo_6S_8 cluster) interaction.

Average phonon energies derived from the spectrum above are given in Table I. They are almost identical to the corresponding figures of superconducting $PbMo_6S_8$.

Sample	Ref(4)	this work
Bulk T_c [K]	11.9	< 1.05
γ [mJK^{-2} g-at^{-1}]	5.0 ÷ 6.7	1.30
θ_o [K]	212	235
ω_E [meV]	4.1 - 4.5	3.4
$\{\int\omega^{-1}F(\omega)d\omega\}^{-1}$ [meV]	15.5	16.0
$\exp \dfrac{\int\omega^{-1}\ell n\omega F(\omega)d\omega}{\int\omega^{-1}F(\omega)d\omega}$ [meV]	12.3	12.2

Table I
Specific heat data for the two samples of $PbMo_6S_8$ discussed in text. ω_E is the geometrical center of the low-frequency phonon peak.

We conclude that the phenomenological cause of the suppression of T_c lies in the electron spectrum rather than the phonon spectrum. One could object that the bare electron density of states at the Fermi level may be the same for both modifications of $PbMo_6S_8$ and that the difference in γ is entirely due to a change of the renormalization parameter $1 + \lambda$. The reduction of γ by a factor of 5.5 ± 1.8 would however require a very high λ (7.2 ± 2.8) in the superconducting phase. This would not be consistent with the observed dimensionless thermodynamic ratios (e.g. $\Delta c/\gamma T_c$), to say the least.

This phenomenological answer does not solve however the fundamental problem of the physical origin of the T_c suppression. Why does a mechanical treatment have such a drastic effect on T_c and the band structure ? We now discuss possible answers. The dimensionality effect on the thermodynamics of small superconducting paricles has been investigated recently in Sn (6). Such a size effect can be rejected here, because the average grain size is two orders of magnitude above the coherence length. Further, the powder recovers its superconducting properties when it is heated (not sintered) at 800 ^oC for 24 hours.

The effect of possible defects on superconductivity is not well understood. We note that $PbMo_6S_8$, a hard material, is probably not prone to dislocations. The widths of the peaks in the X-ray diffraction pattern at room temperature are similar to those of superconducting $PbMo_6S_8$. The absolute intensities are lower though.

Keeping in mind the findings in well-crystallized Eu-, Ba-, Sr- and $CaMo_6S_8$ (7) the hypothesis of a structural phase transformation below room temperature should be considered. At the present time we have not succeeded in detecting any evidence for such a transformation.

REFERENCES

(1) B. Lachal, A. Junod and J. Muller, J. Low Temp. Phys. 55 (1984) 195.
(2) P. Müller and M. Rohr, Phys. Stat. Sol. (a) 43 (1977) K19.
(3) B. Seeber, C. Rossel, Ø. Fischer and W. Glaetzle, IEEE Trans. on Magnetics 19 (1983) 402.
(4) A. Junod, D. Bichsel and J. Muller, Helv. Phys. Acta 52 (1979) 580.
(5) B.P. Schweiss, B. Renker, E. Schneider and W. Reichardt in : Superconductivity in d- and f-Band Metals, Ed. D.H. Douglass (Plenum, New York, 1976), pp.189-207.
(6) N.A.H.K. Rao, J.C. Garland and D.B. Tanner, Phys. Rev. B 29 (1984) 1214.
(7) B. Lachal, R. Baillif, A. Junod and J. Muller, Solid State Commun. 45 (1983) 849.

LT-17 (Contributed Papers)
U. Eckern, A. Schmid, W. Weber, H. Wühl (eds)

EFFECT OF COPPER ADDITIONS ON THE COMPOSITION AND PROPERTIES OF THE $PbMo_6S_8$

E.M. SAVITSKY, M.I. BYCHKOVA, S.A. LACHENKOV

Baikov Institute of Metallurgy, USSR Academy of Sciences, Moscow

Interaction of copper with the $PbMo_6S_8$ compound at different temperatures and time of annealing has been investigated in the work. The analysis of the X-ray investigations and measurements of critical temperatures helps to draw a conclusion that Cu interacts actively with $PbMo_6S_8$.

1. INTRODUCTION

The present work was aimed at studying the effect of the small copper cation on the structure and superconducting properties of the $PbMo_6S_8$ compound.

2. EXPERIMENT

Specimens of the $PbMo_6S_8$ composition were prepared by direct synthesis. Corresponding quantities of copper were later on added to the $PbMo_6S_8$ powders prepared in this way to form $PbMo_6S_8 + Cu_x$, where $x=(0.2 \div 2.2)$ spaced at 0.2. The mixtures were heat treated at temperatures ranging from 1050°C to 200°C for 48-90 hours.

3. RESULTS OF EXPERIMENTS

The analysis of the diffraction patterns shows that the initial specimens consisted of $PbMo_6S_8$ with a slight quantity of the MoS_2 phase. This is in conformity to (1). The values of the hexagonal parameters a_h and c_h of the initial specimens were as follows: $a_h=(9.18 \pm 0.02)$Å, $c_h=(11.40 \pm 0.04)$Å. The T_c values were within the (12.5 - 13.0)K range. X-ray analysis showed that heat treatment of the $PbMo_6S_8+Cu_x$ mixtures at 1050°C and 800°C resulted in the growth of the a_h parameter Fig.1 It can be seen that interstitial solid solutions the $PbCu_yMo_6S_8$ (y-copper concentration in the interstitial phase) are formed at lower copper concentrations ($x=0.2 \div 0.4$). At higher copper concentrations $Cu_zMo_6S_8$ Pb, Cu_2S (z - copper concentration in the Chevrel phase) phases are formed which is connected with the process of the lead sub stitution in $PbMo_6S_8$. It has been found that as regards our $PbMo_6S_8+Cu_x$ mixtures, heat treated at 1050°C, 800°C and 600°C, these processes begin with x=0.8; 0.6 and 0.4, respectively. When X varies from 0 to 0.4, the T_{ci} (critical temperature at which transition into the superconducting state begins) value goes down Fig.1b. However, with $X > 0.4$, the T_{ci} value increases. This behaviour seems to be due to the formation of a small quantity of the $Cu_zMo_6S_8$ phase that cannot be detected by the X-ray analysis. For compositions ranging from X=1.2 to X=2.2, we observed three superconducting transitions from $T_{c.1}=(8.6-9.1)$K, $T_{c.2}=(6.7-6.9)$K and $T_{c.3}=(5.6-5.9)$K. Transition from $T_{c.1}=(8.6-9.1)$K may be due to the $Cu_{1.8}Mo_6S_8$ phase and transition from $T_{c.3}=(5.6-5.9)$K to the $Cu_{1.0}Mo_6S_8$ and

$Cu_{3.2}Mo_6S_8$ phases,(2),(3). Transition at (6.7-6.9)K is, obviously, due to the presence of free lead that was detected by X-ray analysis. For all these specimens, T_{ci} was 11.6 K. Let us point out that for X=2.2 the $PbCu_yMo_6S_8$ phase can not be detected by X-raying. X-ray analysis of $PbMo_6S_8$+ Cu_x mixtures heat treated at 400°C, 300°C and 200°C has shown that at 400°C copper is substituted or lead in $PbMo_6S_8$, Pb, Cu_2S phases are formed. At 300°C and 200°C the process of copper introduction into $PbMo_6S_8$ only is taking place. X-ray patterns of these specimens showed broadening and splittings of individual diffraction lines. Calculations have shown that this is due to the existence of $PbCu_yMo_6S_8$ phase with a broad copper concentration region.

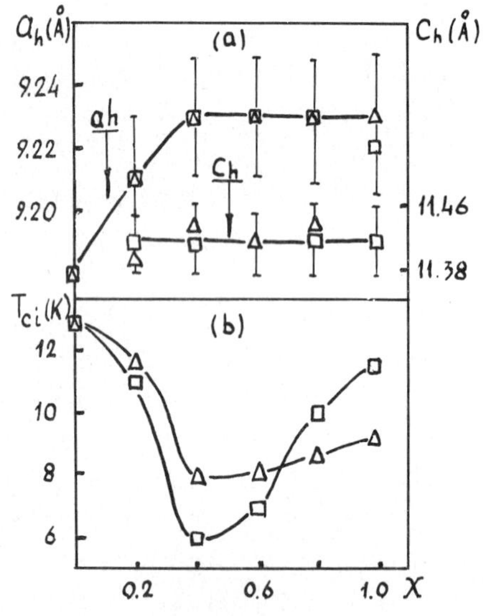

FIGURE 1 (a,b)

(a) T_{ci} for $PbMo_6S_8$ + Cu_x vs.X

(b) a_h, c_h for $PbMo_6S_8$ + Cu_x vs.X

(□ - 1050°C x 48 h, Δ-800°C x 48 h)

4.SUMMARY

1. Copper interacts actively with $PbMo_6S_8$. At low concentrations it forms an interstitial solid solution with the $PbMo_6S_8$ phase. At higher concentrations it is substituted for lead in $PbMo_6S_8$.

2. At 300°C and 200°C the process of copper introduction into $PbMo_6S_8$ only is taking place. It is significant that in this case the values of T_c remain high (~ 12 K).

3. Taking into account the fact that the a_h parameter grows together with copper concentration in the interstitial phase $PbCu_yMo_6S_8$ and that copper is a small cation the latter may be assumed to occupy second positions and as these positions are being taken the intercluster spacings dMo-Mo will be growing (4),(5).

4. The reasons for changes in T_c should, perhaps, be linked with changes in the intercluster spacing dMo-Mo and with copper occupying the second positions.

REFERENCES

(1) M.Sergent,R.Chevrel et al. J.Less Comm.Metals, 58(I978) p.180.

(2) R.Flukiger,A.Junod et al. Solid State Comm., 23(I977) p.699.

(3) M.Tovar,L.E.DeLong et al. Solid State Comm., 30(I979) p.551.

(4) E.S.Delk,M.J.Sienko. Solid State Comm., 31(I979) p.699.

(5) J.D.Jorgenser,D.C.Hinks. Ternary Supercond. Proc. Int. Conf. Lake, Geneva. Wise Sept. 24-26, I980. New York (I98I), p. 69-73.

LT-17 (Contributed Papers)
U. Eckern, A. Schmid, W. Weber, H. Wühl (eds)

SUPERCONDUCTIVITY OF CLUSTER COMPOUNDS

N.E. ALEKSEEVSKII

Institute for Physical Problems, the USSR Academy of Sciences, 117973 Moscow, The USSR

From investigations of the influence of 3-d metals, and particularly Fe on T_c of $Fe_xSnMo_6S_8$, it follows that $T_c(x)$ dependence can be described by the Abrikosov Gor'kov theory. The rise of x leads to an increase of dT_c/dP. The research on some beryllium compounds confirms the earlier data for WBe_{13} and UBe_{13}, and yields the new superconductor $ZrBe_{13}$ with T_c=1.3 K.

Cluster compounds are known to have an atom complex, where the distance between the atoms is shorter than the distance to the atoms of a neighbor complex.

Many superconducting systems can be regarded as cluster compounds. For example, Chevrel phases, Laves phases, tungsten and molybdenum bronzes and others. Perhaps the organic compounds which contain a big molecule-donor and rather a small acceptor could also be attributed to the cluster compounds.

As a rule, cluster compounds have a complicated phonon (1) spectrum. For example, in the case of Chevrel phases and some tungsten bronzes (2) the temperature dependence of the lattice specific heat does not follow Debye's law and can be presented as

$$C_P(T)=aC_E+bC_D,$$

where C_E and C_D are Einstein and Debye Terms of the lattice specific heat. Perhaps in some cases the existence of Einstein-type optical phonons can lead to the linear temperature dependence of electrical resistivity. Such a linear dependence was for example observed in Chevrel phases (1). The anomalous isotope effect of some Chevrel phase compounds may also be the result of a complicated phonon spectrum (3).

Recently, the investigation of molybdenum chalcogenides has led to some new results. When molybdenum sulfides were synthesized in a high pressure chamber (4), superconducting phases were obtained with the lattice different from the Chevrel phase. They have the monoclinic lattice of Mo_2S_3 type (space group $P2_m$) and rather high T_c. But H_{c2} is relatively small. After being annealed these samples transform into the common Chevrel phase and their critical field increases. The characteristics of these phases are presented in the Table.

The effect of 3d-metal impurities on the superconducting properties of some molybdenum chalcogenides has also been studied. The dependence of T_c on x in the system $Fe_xSnMo_6S_8$ was found (5,6) to follow the Abrikosov-Gor'kov (AG)

	T_c K	$H_{c2}(0)$ kG	$\frac{dH_{c2}}{dT}$ kG/K	$P_{syn.}$ kb	$T_{syn.}$ °C
$SnMo_6S_8$	14.4	65	6.5	30-70	1200-1400
$PbMo_6S_8$	15.5	70	7.5	30-70	1200-1400
$HgMo_6S_8$	6.3	40	9.1	70-80	1500-1700

formula. The derivative dT_c/dP grows rapidly when Fe is introduced. It increases by a factor 3 for x=0.04. In contrast to this result, Ni impurities only weakly affect T_c in $Ni_xCu_{1-x}Mo_6S_8$. In this case, $T_c(x)$ differs from theoretical AG dependence (7).

Chevrel phase compounds are known to have extremely high H_{c2} and dH_{c2}/dT. The high H_{c2} could be explained by assuming that such compounds have high electronic density of states N(0). This could be connected with the existence of flat parts on the Fermi surface. On the other hand very high H_{c2} could be connected with high electron effective mass. For example, in the case of $PbMo_6S_8$ $m^*=10m_o$ is found. Very recently data have been published about the superconducting properties of UBe_{13} (8). For this compound Ott et al. report T_c=0.85 K, dH_{c2}/dT=257 kG/K, γ=1.1 J/mol, and m^*=192 m_o. We have investigated some Be-compounds in 1962 and 1973 (9,10) and have found that WBe_{13} is a superconductor with T_c=4.1 K. Recently we have again investigated some Be compounds and have found, for example, that $ZrBe_{13}$ becomes superconducting at T_c=1.3 K, whereas $CeBe_{13}$ has T_c<0.1 K. For UBe_{13} we find T_c=0.55 K and for WBe_{13} T_c=4.1 K as earlier (10). From the investigations of $H_c(T)$ we get for UBe_{13} very high dH_{c2}/dT values (275 kG/K), in agreement with (8). But in the case of WBe_{13} and $ZrBe_{13}$, the dH_{c2}/dT values are very small (445 and 142 G/K, respectively). It must be noted that UBe_{13} and $CeBe_{13}$ have a cubic crystal lattice of the $NaZn_{13}$-type whereas WBe_{13} and $ZrBe_{13}$ have a tetragonal lattice.

The authors of (8) connect the peculiarity of the superconducting properties of UBe_{13} with the property of a Fermi liquid with extremely large effective mass. But it cannot be excluded that both the extremely high or low dH_{c2}/dT values are also connected with the specific properties of cluster compounds. In the cluster compounds it seems possible to change the bonding strength and also the charge transfer between the cluster and the other constituent of the compound and therefore to change gradually their properties. It is probable that the investigations of cluster compound will give many interesting new systems, some of which could be superconductors with unusual properties.

REFERENCES

(1) N.E. Alekseevskii, Cryogenics 20 (1980) 257.
(2) A.Y. Bevolo, H.R. Shanks, P.H. Sidles G.C. Danielson, Phys.Rev.B9 (1974) 3220.
(3) N.E. Alekseevskii, V.I. Nizhankovskii ZhETF Pis'ma 31 (1980) 63.
(4) N.E. Alekseevskii, V.V. Evdokimova, E.P. Khlybov, V.I. Novokshonov, ZhETF Pis'ma 35 (1982) 8.
(5) N.E. Alekseevskii, V.N. Narozhnyi ZhETF Pis'ma 35 (1982) 49.
(6) N.E. Alekseevskii, V.N. Narozhnyi, E.P. Khlybov, ZhETF 84 (1983) 1538.
(7) N.E. Alekseevskii, N.A. Tichonova, E.P. Khlybov, ZhETF Pis'ma 38 (1983) 272.
(8) M.R. Ott, M. Rudiger, Z. Fisk, J.L. Smith, Phys. Rev. Lett. 50 (1983) 1595.
(9) N.E. Alekseevskii, N.N. Michailov, ZhETF 43 (1962) 2110.
(10) N.E. Alekseevskii, V.M. Zakosarenko, DAN USSR 208 (1973) 303.

LT-17 (Contributed Papers)
U. Eckern, A. Schmid, W. Weber, H. Wühl (eds)
© Elsevier Science Publishers B.V., 1984

SUPERCONDUCTING, BAND AND EXCHANGE PARAMETERS OF MAGNETIC SUPERCONDUCTOR Y_9Co_7

A. Kołodziejczyk and C. Sułkowski+

Department of Solid State Physics, University of Mining and Metallurgy, AGH, Mickiewicza 30, 30-059 Cracow, Poland and +Institute for Low Temperature and Structure Research, Polish Academy of Science, 53-529 Wrocław, Poland

We have calculated some material parameters of the magnetic superconductor Y_9Co_7 from an analysis of our experimental data on the resistivity, upper critical field, magnetisation and specific heat basing on attainable theoretical relations.

1. INTRODUCTION

The magnetism of Y_4Co_3 intermetallic compound, later found to be rather Y_9Co_7 as a single phase, went through a revival when it was suggested (1) that the itinerant magnetic moment on some cobalt positions and more local moment on the other positions exist in contrary to the prediction that the compound should be Pauli paramagnet. More interest about Y_9Co_7 has arised since the discovery of superconductivity (2) with transition temperature T_S=2K, because later it was suggested as a new type of the coexistence of the weak itinerant ferromagnetism (T_C= 4.5K) and superconductivity (for a review see e.g. (3,4)). About thirty papers were published on the subject till now, but still one needs more details about microscopic parameters characterising both the magnetism and the superconductivity to clarify an origin of the coexistence.

2. EXPERIMENTALS

The specimen was arc-melted and chill-cast and then annealed at 850K for ten days and at 750K for three weeks. It was prepared from Rare Earth Product yttrium of purity 99,99% and Johnson-Mathey spec-pure cobalt. The X ray powder diffraction and metallographic observations did not show any second phase within 5% uncertainity.

We have measured the a.c. resistance by the standard four-point techique for a rod shaped specimen with dimensions 1x1.5x11 mm^3, from 1.55K up to room temperature and at 7.67Hz. For the same sample, temperature and magnetic field dependences of magnetisation were carried out by the standard extraction magnetometer, using superconducting coil, and within the temperature interval from 1.57K up to the Curie temperature.

3. RESULTS AND COMPARISON WITH THEORY

First of all, one can calculate some superconducting parameters on the basis of the Ginzburg-Landau-Abrikosov-Gorkov (GLAG) theory (5). For this purpose, we have made use of the resistivity data in Fig.1, and we have taken advantage with our previous specific heat measurements (6) and the upper critical field data (7). From these results we calculated the specific heat coefficient to be $\gamma = 2.525 \times 10^3$ $ergcm^{-3}K^{-2}$. From the slope of the upper critical field $(dH_{c2}/dT)_{T_S}$ = 0.23 TK^{-1} in the dirty limit of superconductivity (c.f. the Fig.2 of the paper (7)), we estimated the coefficient $n^{2/3}S/S_F$ which occurs in the GLAG formulae (c.f. the appendix of the paper (5)) to be 0.744×10^{15}, where n is the concentration of conduction electrons in cm^{-1} and S/S_F is the relative diameter of Fermi surface. The calculated superconducting parameters are listed in Table I.

Next we have calculated the density of states $N(E_F)$ from the γ value neglect-

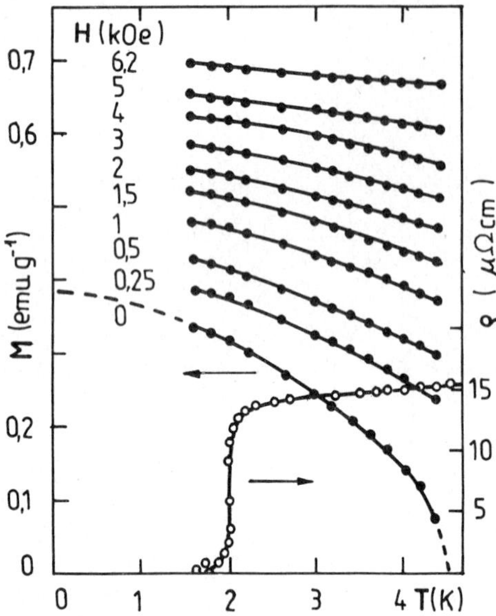

FIGURE 1

ing the electron-phonon enhancement factor of McMillan (8).

Next, we have calculated the band and exchange parameters in the framework of the single particle excitation model (9) making use of our magnetisation data in Fig.1 and (10) (see the table I).

It was concluded from the susceptibility and resistivity data above T_C (11) as well from magnetisation data below T_C (10,12) that there is a pretty large role of spin fluctuations, in the sense of the Moriya's theory (13), in the magnetic behaviour of Y_9Co_7.

TABLE I

SUPERCONDUCTING PARAMETERS
1. BCS coherence lenght ξ_o = 1180 Å
2. London penetration lenght λ_L = 890 Å
3. Gorkov parameter $\lambda = 0.88\xi_o/l = 9.15 \gg 1$, where the mean free path l = 115 Å
4. GLAG coherence lenght ξ_{GL} = 315 Å
5. GLAG penetration lenght λ_{GL} = 1750 Å
6. GLAG κ-parameters κ_{GL} = 5.6
7. Termodynamic critical field $H_c(0)$ = 423 Oe
8. Lower critical field $H_{c1}(0)$ = 92 Oe
BAND AND EXCHANGE PARAMETERS
1. $N(E_F)$ = 1.8 states eV^{-1}(atom-spin)$^{-1}$
2. $M(0,0)/N_A\mu_B = \zeta_o = 0.012 \ll 1$
3. Exchange splitting energy ΔE = 6.7 meV, with parameter α being α = 0.00167 (9)
4. Effective interaction parameter between itinerant electrons I = 0.56 eV
5. The criterion for existence of ferromagnetism $I\,N(E_F) = 1.00167 > 1$
6. Band parameters: $[dN(E_F)/dE]^2 = -2920\,eV^{-2}$; $d^2N(E_F)/dE^2 = -9650\,eV^{-2}$
7. Half width at the half maximum of the d- band Γ = 0.0145 eV
8. The degeneracy temperature T_F = 110K

It confirms the double, pure itinerant and more local, behaviour of magnetic moments in Y_9Co_7, coming from the three different types of Co positions in the unit cell (for the structure see e.g.(14)). The very weak itinerant component of the total magnetic moment dominates below T_C and most likely, that is why the very broad temperature interval of coexistence of superconductivity and magnetism exists.

REFERENCES

(1) E.Gratz, H.R.Kirchmayr, V.Sechovsky and E.P.Wohlfarth, J.M.M.M. 21 (1980) 197.
(2) A.Kołodziejczyk, B.V.B.Sarkissian and B.R.Coles, J.Phys.F 10 (1980) L333.
(3) B.V.B.Sarkissian, J.Appl.Phys. 53 (1982) 8070
(4) A.Kołodziejczyk, Proc.Int.Conf. on Cryogenic Fundamentals ICCF83, Cracow -Wrocław (1983) ed.G.Kozłowski, in print
(5) T.P.Orlando et al, Phys.Rev. B19(1979) 4545.
(6) A.Lewicki et al, J.M.M.M.36 (1983) 297.
(7) C.Sułkowski et al, J.Phys.F 13(1983) 2147.
(8) W.L.McMillan, Phys.Rev. 167(1968) 331.
(9) D.M.Edwards and E.P.Wohlfarth, Proc. Roy.Soc. A303 (1968) 127.
(10) A.Kołodziejczyk and C.Sułkowski, J.Phys.F, submitted for publication.
(11) A.Kołodziejczyk and J.Spałek, J.Phys. F, 14, in print.
(12) Y.Yamaguchi, Y.Nishihara and S.Ogawa, J.M.M.M. 31-34 (1983) 510.
(13) T.Moriya, J.M.M.M. 14 (1979) 1.
(14) R.Lemaire, J.Schweizer and J.Yakintos, Acta Cryst. B25 (1969) 710.

ACKNOWLEDGEMENTS. One of us (A.K.) thanks for financial help from Polish Acad.Scien.

LT-17 (Contributed Papers)
U. Eckern, A. Schmid, W. Weber, H. Wühl (eds)

RESISTIVITY MEASUREMENTS OF $Zr_{76}Fe_{24}$ AMORPHOUS ALLOYS

Olivier LABORDE*, Jean-Claude LASJAUNIAS and Gérard CHOUTEAU*

Centre de Recherches sur les Très Basses Températures,
* Also : Service National des Champs Intenses,
C.N.R.S., BP 166 X, 38042 Grenoble-Cédex, France.

We report resistivity measurements on $Zr_{76}Fe_{24}$ amorphous alloys (1.5 K<T<300 K). Taking new results from localization theory and experiments on related alloys into account, we interpret self consistently $\rho(T)$ for the whole range of T and for different thermal treatments.

We report resistivity measurements, between 1.5 K and room temperature, for two $Zr_{76}Fe_{24}$ amorphous samples. Both are subject to successive annealing and aging in order to induce structural relaxation. Details of the preparation and characterization of these sputtered zirconium-based alloys as also of the experimental techniques was described elsewhere (1,2). Magnetic properties of these alloys was previously reported (3). We use the same notation (samples 1 and 2) as in ref. (3). The two samples are obtained from the same batch, but they depict an appreciably different behaviour likely due to a small concentration shift for each sample from the nominal value.

The resistivity at room temperature is 190±30 μΩcm. For a same sample annealing induces a variation smaller than a few per cent. We indicate in table 1 the details of the thermal treatments for each sample and summarize the main results. Low temperature data, on which emphasis will be laid, are reported in fig. 1 and 2.

We can account for the thermal variation of $\rho(T)$ by considering four distinct contributions, each one dominates for a particular sample in a given range of T.

1) At high temperatures ($T \gtrsim 20$ K) for all samples, a Ziman term accounts for $\rho(T)$. The phonon diffusion leads to a decrease of the structure factor and of the resistivity when the temperature increases ; on annealing the amplitude of the thermal variation decreases as observed in Zr-Cu and Zr-Ni (2). Such a behaviour is found for sample 2, but is not so clear for sample 1.

2) Recently the resistivity of a lot of amorphous alloys was shown to behave as $-A\sqrt{T}$ at low temperature (4,5). This variation as well as the amplitude of A (A depends only of the absolute value of ρ) is in agreement with scaling theories of the metal-insulator transition. This $\sqrt{T}$ variation is apparent below 4 K for most of the samples (fig. 1) with the good value of A ; the result for related alloy $Zr_{40}Cu_{60}$ (6) with practically the same resistivity, is also shown.

3) The bending which occurs at the lowest T on curve 2-a is very similar to what is observed for $Zr_{50}Cu_{50}$ (6). It is ascribed to superconductive fluctuations which lead to detectable effect in amorphous superconductor, up to about 3 T_c (7). As long as contributions 2 and 3 are small R(T) can be written as

$$R(T) = R_\circ(1-8.2\times10^{-4}\sqrt{T}) \times\left[1 - \exp-(3-4.5\sqrt{\frac{T-T_c}{T_c}})\right] \qquad (1)$$

TABLE 1

	$100\times[R(4.2)/R(273)-1]$	$100\times[R(25)-R(273)-1]$	$\alpha(273)=-1/\rho\ d\rho/dT$	Curves
Sample 1				
As-sputtered	4.71±0.03	4.68±0.03	$1.4\times10^{-4}\ K^{-1}$	1-a
Aged 21 days	4.74	4.71	1.45	1-b
Annealed 1/4 hr 170°C	4.75	4.64	1.4	1-c
Annealed 1/4 hr 230°C	4.79	4.65	1.42	1-d
Aged 15 months	4.73	4.59	1.4±0.1	1-e
Sample 2				
As-sputtered	5.51±0.03	5.31±0.03	$1.5\times10^{-4}\ K^{-1}$	2-a
Annealed 1 hr 250°C	5.2		1.3±0.1	2-b

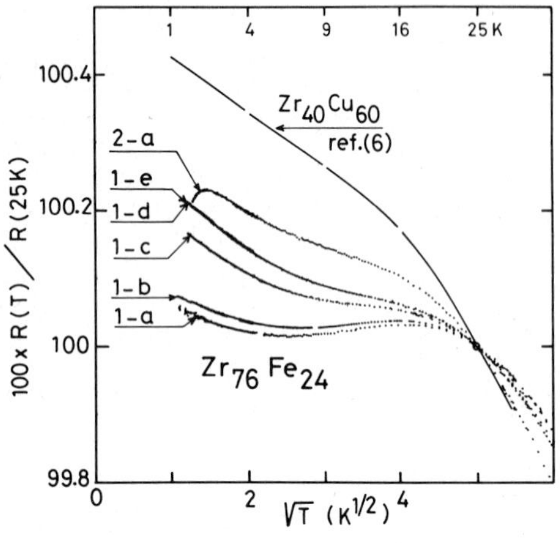

FIGURE 1
100×R(T)/R(25 K) versus $\sqrt{T}$ (see table 1).

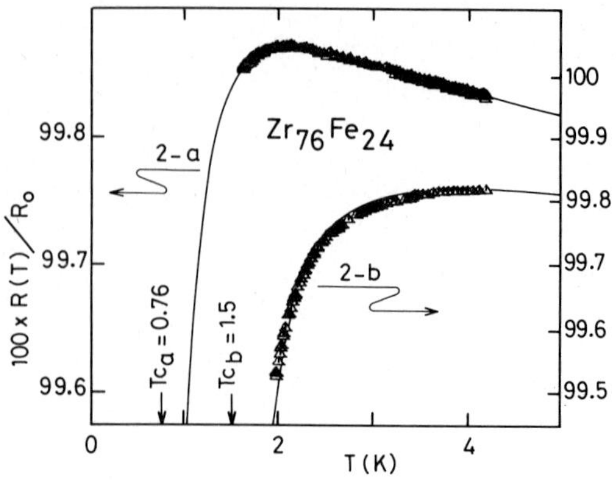

FIGURE 2
$100{\times}R(T)/R_o$ versus T. Mean-square fit of equation (1) and experimental data (see table 1).

where the value $A = 8.2{\times}10^{-4}$ is taken from $Zr_{40}Cu_{60}$ results, and the exponential term was shown to account accurately for superconductive fluctuations for $(T-T_c)/T_c \gtrsim 0.05$ in the case of two different amorphous systems (7).

As fig. 2 shows, expression (1) with only two free parameters R_o and T_c is in good agreement with experimental data of sample 2 at $T < 4.2$, with $T_c = 0.76$ K and 1.5 K respectively before and after the annealing. This last value is also in agreement with measurements at lower T (not used in the fit of equation (1)) which shows that the linear part of the resistive superconducting transition extrapolates at R = 0 for T = 1.57 K.

The superconductive origin of this contribution is reinforced by magnetoresistivity measurements. The decrease of ρ is suppressed by a field of a few kGauss.

4) The three previous contributions cannot explain the inflexion in curve 1-a which progressively disappears upon annealing. We have to introduce another term resulting from the occurence of superparamagnetic clusters when T is lowered. Magnetic measurements have shown that this type of order occurs in Zr-Fe (3) like in related $Zr_{40}Cu_{60-x}Fe_x$ amorphous alloys (8) and dislike for instance in Fe-Pd-Si which depicts Kondo-effect (9). This last magnetic behaviour can be ruled out for Zr-Fe because neither clear ℓnT law for $\rho(T)$ is observed nor negative magnetoresistivity. It is positive and very similar to that of non-magnetic Zr-Cu. At the moment, the theory is not so elaborated to explain qualitatively the experimental data. But in order to strengthen this interpretation, we note that $\rho(T)$ of $Zr_{40}Cu_{60-x}Fe_x$ for $x \gtrsim 8$ (8) behaves in a same way as Zr-Fe and further that magnetic hysteresis occur in both alloys. We also remark that the decrease on annealing of this negative contribution which is apparent at low T by the disappearance of the minimum can hide the decrease of the Ziman term which is seen in Zr-Ni, but not in sample 1.

We have self-consistently interpreted the resistivity of Zr-Fe amorphous alloys with concentration near the vanishing of the superconductivity by considering four contributions to $\rho(T)$. This provides a powerful tool to study the magnetic and the superconducting properties of this alloy. We have already shown that annealing can induce a large variation of the magnetic term and can increase the superconductive temperature contrary to non-magnetic alloys where a lowering of T_c is generally observed (2).

REFERENCES

(1) A. Ravex, J.C. Lasjaunias and O. Béthoux, J. Phys. F 14 (1984) 329.
(2) O. Laborde, A. Ravex, J.C. Lasjaunias and O. Béthoux, submitted to J. Low Temp. Phys.
(3) G. Chouteau, O. Béthoux and O. Laborde, J. Non-Cryst. Sol. 61-62 (1984) 1213.
(4) Ö. Rapp, S.M. Bhagat and H. Gudmundsson, Solid State Commun. 42 (1982) 741.
(5) R.W. Cochrane and J.O. Strom-Olsen, Phys. Rev. B 29 (1984) 1088.
(6) S.J. Poon, Solid State Commun. 45 (1983) 531.
(7) W.L. Johnson and C.C. Tsuei, Phys. Rev. B 13 (1976) 4827.
(8) F.R. Szofran, G.R. Gruzalski, J.W. Weymouth and D.J. Sellmyer, Phys. Rev. B 14 (1976) 2160.
(9) R. Hasegawa, J. Phys. Chem. Sol. 32 (1971) 2487.

LT-17 (Contributed Papers)
U. Eckern, A. Schmid, W. Weber, H. Wühl (eds)

TEMPERATURE DEPENDENCE OF THE VIBRATIONAL DENSITY OF STATES OF THE METALLIC GLASS $Zr_{75}Rh_{25}$

J.-B. SUCK

Kernforschungzentrum Karlsruhe, Institut für Nukleare Festkörperphysik, POB 3640, D-7500 Karlsruhe*

H. RUDIN and H.-J. GUENTHERODT

Institut für Physik der Universität Basel, Klingelberstr. 82, CH-4056 Basel, Switzerland

The vibrational density of states of the metallic glass $Zr_{25}Rh_{25}$ was determined at 6K and at 270K using neutron inelastic scattering techniques. The experiment was done at the HFR of the ILL in Grenoble. A softening of the vibrational spectrum on cooling is observed for the first time. The results are interpreted in terms of a new relaxation process in the glass after cooling or alternatively as a consequence of a strong electron phonon interaction in this superconducting metal-metal glass.

1. INTRODUCTION

Changes of the vibrational spectrum of a metallic glass have been observed in two cases up to now. On heat treatment of the glass below the crystallization temperature T_x a loss of intensity at the low energy side of the spectrum was found in the case of $Mg_{70}Zn_{30}$ (1). In the same glass on cooling to 6K a very slight hardening of the spectrum was observed which we attributed to the difference in density and in vibrational amplitude at 6K and at 270K (2). In both cases a hardening of the atomic dynamics of the metallic glass was observed.

Here we report first results from an investigation of the temperature dependence of the atomic dynamics of a metallic glass in which high electron phonon coupling is expected. From the metal-metal binary glassy alloys which can be produced by melt spinning techniques-a necessary condition for the large amount of sample needed for a neutron inelastic scattering experiment- $Zr_{75}Rh_{25}$ is one of the systems with the highest transition temperature T_c to the superconducting state found so far (3). In fact the T_c we measured with our sample was 4.5K.

2. EXPERIMENT

The melt spun ribbons were kept in a tight thin walled Al container under Ar gas. The experiment was done at the thermal neutron time-of-flight (t-o-f) spectrometer IN4 at the high flux reactor of the Institut Laue-Langevin (ILL) in Grenoble with an energy of the incident neutrons of 41.7 meV. T-o-f spectra were recorded from 56 detectors fixed in a distance of 4 m on a circle around the sample at 56 different scattering angles between 7 and 97 degrees. Neutron energy loss spectra were obtained for energy transfers between 4 and 32 meV and momentum transfers between 0.7 and 7 $Å^{-1}$. For energies lower than 4 meV the inelastic spectra could not be separated from the foot of the huge peak of the elastically scattered neutrons which had a FWHM of 1.6 meV. Besides the measurement of the sample the empty Al container (cylinder) and a Cd cylinder of the same dimension as the sample were measured at 6K and at 270K for a proper background subtraction. A Vanadium calibration run and an empty cryostat measurement (V-holder) were done at 270K.

3. DATA TREATMENT

The data treatment follows the procedure described in (4). All necessary corrections were applied to the data presented here except those for the finit resolution of the spectrometer and for multiple scattering contributions, which are expected to be very small due to the large absorption in the sample and its intercalation with Gd sheets every 6 mm.

For the self consistent correction for multiphonon contributions to the measured intensities the sum of 35 of the t-o-f spectra converted to the double differential scattering cross-section and weighted with the corresponding scattering angle was used. The resulting spectrum, from which the starting distribution for the iterative correction procedure is obtained, is shown in figure 1 after subtraction of 0.4 counts/chan. (6K) and 1.6 c/chan. (270K) additional background from these sums. More details of the correction procedure are given in (5).

4. RESULTS

The resulting generalized vibrational density of states are shown in figure 2.

---

* Present adress : Institut Laue Langevin, 156X, 38042 Grenoble Cédex, France

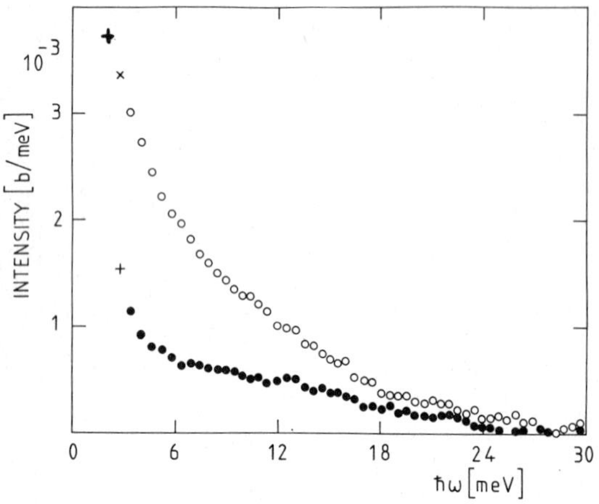

FIGURE 1
Weighted sum of the t-o-f spectra (given as $d^2\sigma/d\Omega dE$). Only the circles (● : 6K ; 0 : 270K) have been used to generate the frequency distributions.

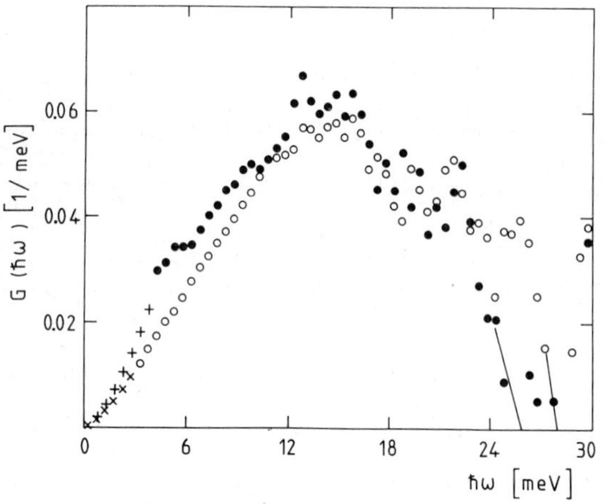

FIGURE 2
Generalized vibrational density of states for the metallic glass $Zr_{75}Rh_{25}$ measured at 6K(●) and at 270 K(0). The crosses represent the artificial spectra $\sim \omega^{3/2}$ inserted to bridge the low energy region where the inelastic spectra could not be separated from the foot of the elastic peak. The lines indicate the assumed end of the distributions.

In spite of the rather limited statistical accuracy of the data especially for energies above 12 meV due to the very high absorption in the sample (σ_a = 36.4 b, σ_{sc} = 6 b) the distributions clearly appear to be different. While the density of states measured at 270K extends out to approximately 28 meV the spectrum measured at 6K under identical experimental conditions is shifted to *lower* energies by about 1.5 meV, i.e. ≈ 5 % of the cut-off frequency. This is especially evident at the low energy side of the distribution.

From the second moment of the vibrational density of states the Debye temperature $\Theta(\infty)$ has been calculated. It decreases from 252K to 224K when the temperature of the glass is lowered from 270K to 6K.

5. DISCUSSION

To our knowledge a softening of the atomic dynamics of any glass on cooling has never been observed before. Depending on what one regards as the "original" spectrum possibly two mechanism could lead to the observed results.

A hardening of the spectrum after cooling to 6K and heating to 270K could be the equivalent to the loss of low energy modes observed after heat treatment of the metallic glass below T_x (1), i.e. a structural relaxation caused by the release of stresses on heating which were induced during the cooling procedure in the as quenched metal. Since we have done the 6K-measurement before the 270K-run we cannot exclude this possibility, though the glass had been kept at room temperature several weeks before the experiment.

If one regards the spectrum measured at 270K as being that of the as quenched sample near room temperature, then the observed softening could be ascribed to a strong electron phonon coupling near T_c in this metallic glass. Such a softening on cooling has been observed in some high T_c crystalline materials, however preferably in structures with appreciable anisotropic distribution of the metal atoms. For an isotropic and topologically disordered substance the observed softening is therefore rater unexpected.

ACKNOWLEDGEMENTS

One of us (J.-B. S.) would like to acknowledge the hospitality of the ILL where this work was done. It is a pleasure to thank H. Rietschel for a fruitful discussion and P. Reimann for the melt spinning of the sample.

REFERENCES

(1) J.-B. Suck, H. Rudin, H.-J. Güntherodt and H. Beck, J. Non-Cryst. Solids 61+62 (1984) 295
(2) J.-B. Suck, Proc. 4th Int. Conf. on Rapidly Quenched Metals (Sendai 1981) 407
(3) W.L. Johnson, in : Topics in Applied Physics Vol. 46 (Springer 1981) pp. 191-223
(4) J.-B. Suck, H. Rudin, H.-J. Güntherodt and H. Beck, J. Phys. C : Solid State Phys. 14 (1981) 2305
(5) J.-B. Suck and H. Rudin, in : Topics in Applied Phys. Vol. 53, (Springer 1983) pp.217

LT-17 (Contributed Papers)
U. Eckern, A. Schmid, W. Weber, H. Wühl (eds)

LOW FREQUENCY ELASTIC LOSS AND THERMOELASTIC EFFECT IN METALLIC GLASSES

H. Tietje, M. v.Schickfus*, E. Gmelin, H.J. Güntherodt**

Max-Planck-Institut für Festkörperforschung, Heisenbergstr. 1, D-7000 Stuttgart 80, Germany
* Institut für Angewandte Physik II, Universität Heidelberg
** Institut für Physik, Universität Basel

We have investigated the relaxational absorption and the Grüneisen parameter in the metallic glasses PdSiCu and PdZr at very low frequency through the elastocaloric effect. In both glasses a negative Grüneisen parameter is found at low temperatures. The absorption shows the temperature and frequency dependence predicted by the tunneling model.

The low temperature properties of amorphous solids are determined to a large extend by tunneling systems [1]. These tunneling systems interact with the phonons and, in the case of metallic glasses, with the conduction electrons [2]. The coupling of the tunneling systems to the electrons is much stronger than to the phonons, resulting in relaxation times τ of the tunneling systems which are 4 orders of magnitude shorter. The tunneling systems are characterized by the asymmetry Δ of the potential wells and by the tunnel splitting Δ_0 due to the overlap of the wavefunctions in the two wells. The distribution of the parameters Δ and Δ_0 leads to a broad distribution of the relaxation times τ of the tunneling systems [3]. This distribution diverges for $\tau \to \infty$ in a non-integrable manner, so that for a finite density of states some kind of a cut off has to be introduced for large values of τ or, equivalently, for small values of Δ_0.

We have extended our recent measurements of the temperature changes due to elastic absorption [4] at very low frequencies (5×10^{-3} to 30 Hz) to the metallic glasses $Pd_{0.775}Si_{0.165}Cu_{0.06}$ and $Pd_{0.3}Zr_{0.7}$, the latter being superconducting below 2.6 K, in order to test the distribution of relaxation times mentioned above. In this frequency range and at our measuring temperatures ($1K<T<6K$) the absorption is determined by the relaxation of the tunneling systems. If the tunneling model holds even for very long relaxation times τ, one finds for the absorbed intensity, the "internal friction":

$$Q^{-1} = (\pi \bar{P}\, \gamma^2)/(2\rho\, v^2) \qquad (1)$$

where $\bar{P}$ is the density of states of the tunneling systems, γ the deformation potential, v the velocity of sound and ρ the mass density. The predicted internal friction is thus independent of frequency and of temperature. By applying a mechanical stress σ of the form $\sigma = \sigma_0 \cos \omega t$, heat is produced in the sample:

$$\dot{q}_{abs} = (1/2Y)\cdot \omega\sigma_0^2\, Q^{-1}(1+\cos 2\,\omega t) \qquad (2)$$

Y being the Young's modulus.

A second contribution to the observed temperature change comes from the thermoelastic effect, determined by the Grüneisen parameter Γ [5]. Γ characterizes the non-harmonic behaviour of a material and is defined as $\Gamma = -\partial \ln E/\partial \ln V$, i.e. the change of energy associated with a change of volume. Under the influence of a periodic stress, the thermoelastically generated heat turns out to be:

$$\dot{q}_{th} = \Gamma\, c_V\, \chi/3\; T\, \omega\, \sigma_0 \sin \omega t \qquad (3)$$

where χ is the compressibility and c_V the specific heat of the sample.

In our experiments the internal friction and the Grüneisen parameter was measured through the elastocaloric effect: the periodic stress causes a temperature rise of the samples, either a ribbon of amorphous PdSiCu or PdZr. The samples were mounted in the vacuum chamber of a crystat and clamped at both ends to copper blocks at constant temperature. Uniaxial stress was applied from outside the crystat through a bellows-sealed stainless steel wire. The temperature variation of the sample was measured with an exposed-element germanium resistor using a Wheatstone bridge and recorded with a desktop computer.

As a typical experimental result, Fig. 1 shows the temperature variation of the PdSiCu sample at 4.35 K. Because of the small cross-section of the sample, the stress amplitude is relatively large but well below the yield stress of about 1000 MPa. From Fig. 1 it can be clearly seen that the temperature signal consists of both the absorptional part, described by eq. (2), and the thermoelastic contribution of eq. (3). From a fit to the temperature curves, we can deduce the Grüneisen parameter. Fig. 2 shows our results for the Γ of PdSiCu together with Γ values from thermal expansion [6,7]. Below 2.5 K our results lie between the results of the other authors, indicating a negative Grüneisen parameter of about $\Gamma = -2$. Obviously, there exists a great

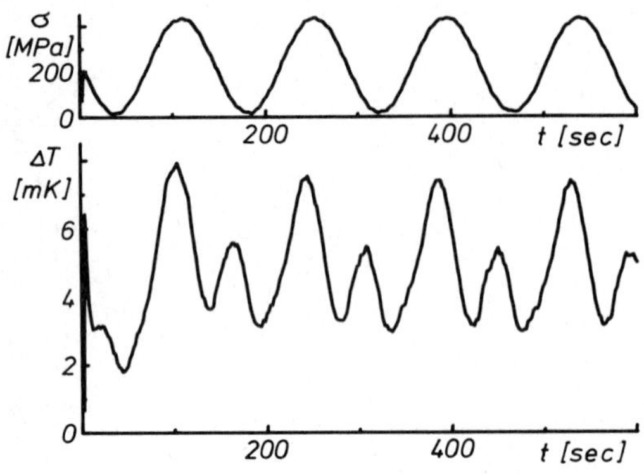

Figure 1
Temperature variation of a PdSiCu sample at 4.35 K. Stress is applied at t = 0 with a frequency of 7 mHz.

sample dependence of the Grüneisen parameter in PdSiCu. For PdZr, we find a Γ which is always negative. Our results fit to a dependence $\Gamma = -20/T$ in the temperature range 1.4 to 7 K.

The absolute value of $\dot{q}$, the heat generated by the absorption process, is determined by comparing the equilibrium temperature rise of the sample with that caused by an electric heater. In both PdSiCu and PdZr we find the linear frequency dependence predicted by eq. (2) at all temperatures. From the experimental data Q^{-1} can be determined. Fig. 3 show the temperature dependence of Q^{-1} for PdSiCu and PdZr. Only a very weak temperature dependence of the internal friction was detected. According to eq. (1), the coupling parameter $\bar{P}\gamma^2$ can be derived from the absorption data. We find $\bar{P}\gamma^2 = 7.6\cdot10^7$ erg/cm^3 for PdSiCu and $\bar{P}\gamma^2 = 2.9\cdot10^7$ erg/cm^3 for PdZr. Our value of $\bar{P}\gamma^2$ for PdSiCu is smaller than that found at 1kHz [8], but is in good agreement with ultrasonic data [9]. The value of $\bar{P}\gamma^2$ for PdZr is smaller by a factor of two than that found in ultrasonic measurements [10].

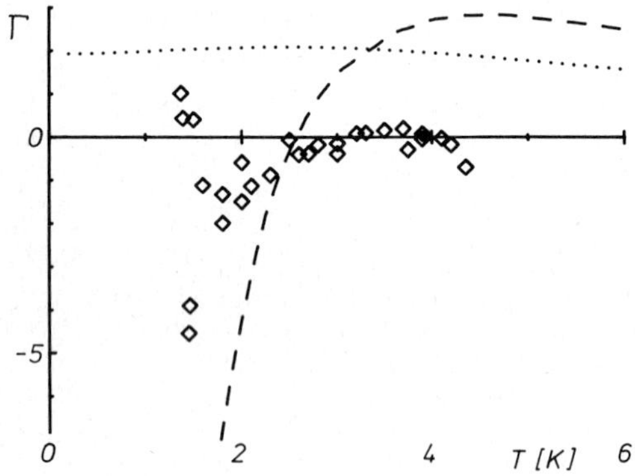

Figure 2
The Grüneisen parameter of PdSiCu as a function of temperature. ---: data of [6], •••: data of [7].

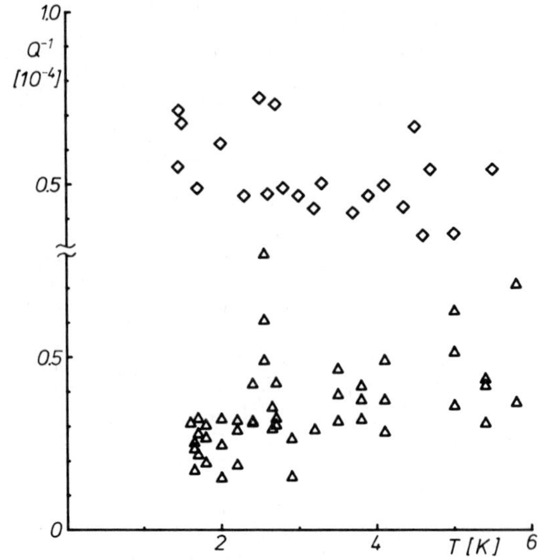

Figure 3
Temperature dependence of the internal friction Q^{-1} for PdSiCu (top) and PdZr (bottom).

In conclusion, our results confirm the predictions of the tunneling model in the metallic glasses at very low frequencies. Thus we have shown that the assumed distribution of relaxation times is still valid in the mHz-frequency range. Therefore a cutoff of the distribution must be well beyond the limits shown by our experiments.

REFERENCES

[1] review e.g.: Amorphous Solids, ed. W.A. Phillips (Springer Verlag, Berlin 1981)
[2] e.g.: J.L. Black in: Glassy Metals I, eds. H.J. Güntherodt, H. Beck (Springer Verlag, Berlin, 1981) pp. 167-190
[3] J. Jäckle, Z. Physik 257 (1972) 212
[4] H. Tietje, M. v.Schickfus, E. Gmelin, H.J. Güntherodt, Proc. 5th Int.Conf-Phonon Scattering in Cond. Matter (Stuttgart 1983)
[5] H. Tietje, M. v.Schickfus, E. Gmelin, J. Physique 43 (1982) C9-529
[6] W. Kaspers, R. Pott, D.M. Herlach, H. v.Löhneysen, Phys.Rev.Lett. 50 (1983) 433
[7] W.M. MacDonald, E.J. Cotts, A.C. Andersen, Phys.Rev.Lett. 51 (1983) 930
[8] A.K. Raychaudhuri, S.Hunklinger, J. Physique 43 (1982) C9-485
[9] G. Bellessa, O. Bethoux, Phys.Lett. 62A (1977) 125; the value given in this paper has to be increased by a factor of two
[10] G. Weiss, S. Hunklinger, H. v.Löhneysen, Physica 109&110B (1982) 1946

LT-17 (Contributed Papers)
U. Eckern, A. Schmid, W. Weber, H. Wühl (eds)

INTERFERENCE EFFECTS IN $Ni_{80}P_{14}B_6$ ALLOY

E.BABIĆ*,+ A.HAMZIĆ*+ and M.MILJAK+

* Physics Department,Faculty of Science,POB 162, 41001 Zagreb,Yugoslavia
\+ Institute of Physics of the University,POB 304, 41001 Zagreb,Yugoslavia

Low temperature resistivity and magnetoresistivity of $Ni_{80}P_{14}B_6$ alloy show the contributions due to the interference effects.The corresponding contribution to magnetic susceptibility is masked by the Curie-Weiss behaviour due to isolated spins, but could be traced after suitable correction.

Earlier we have shown (1) that the upturn in the resistivity of amorphous $Ni_{80}P_{14}B_6$ alloy at the lowest temperatures increases in the magnetic field.Such behaviour was in contrast to that of other $Fe_xNi_{80-x}P_{14}B_6$ alloys where the upturn in resistivity either decreased ($0<x\leq 20$) or remained unchanged ($x>20$) in the magnetic field.Furthermore in the magnetic fields lower than 3.5 T the magnetoresistance of $Ni_{80}P_{14}B_6$ followed roughly a H^{α} law with the noninteger exponent.These observations were at variance with the prediction of both structural and magnetic models for the resistance minimum in metallic glasses.

More recently it was pointed out (2) that the interference between the electron-electron and electron-impurity scattering could produce observable effects in the electronic and magnetic properties of the disordered systems at low temperatures.In particular it was found that the leading corrections to the conductivity and magnetic susceptibility should obey a $T^{0.5}$ law and that the magnetoresistance could be positive and very large compared to that due to Kohler's contribution.

In order to clarify this situation we performed a systematic investigation of the low temperature magnetoresistivity and magnetic susceptibility of $Ni_{80}P_{14}B_6$.In Fig.1 we plot the relative change in resistivity $\Delta\rho/\rho_m$ ($\Delta\rho = \rho(T) - \rho_m$ with ρ_m the resistivity at its minimum) of $Ni_{80}P_{14}B_6$alloy vs $T^{0.5}$ and lnT.It can be seen that at low temperatures $\Delta\rho/\rho_m$ can be fitted better with $T^{0.5}$ than with $\ln T$ dependence.While $\Delta\rho/\rho_m$ shows a S-shape curvature in lnT plot, there is a linear variation vs $T^{0.5}$ up to 7 K.At temperatures closer to the minimum (T_m = 13.1 K) the resistivity shows a positive deviation from a straight line which is a cosequence of the positive contributions to the resistivity.From a linear part of our plot we obtain the coefficient $A = 4.1\cdot 10^{-4}$ ($K^{-0.5}$).This value could be compared with that from the correction to the resistivity due to interference effects (3)

$$\Delta\rho/\rho=-T^{0.5}\ 2.5f(x)\rho e^2(2k_B)^{0.5}/6\pi^2\hbar^{1.5}D^{0.5} \qquad (1)$$

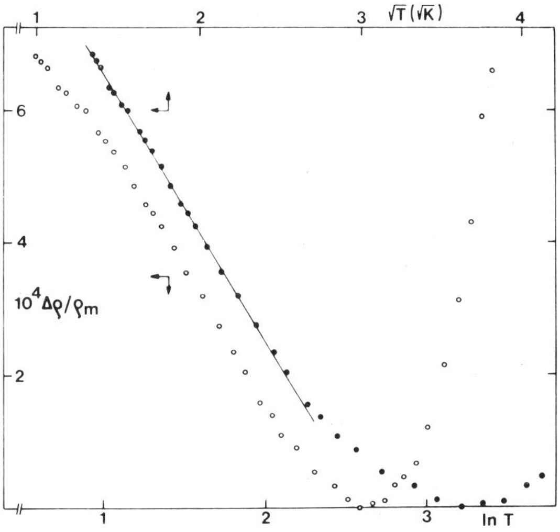

FIGURE 1
Relative change of resistivity vs $T^{0.5}$ and lnT

where D is the diffusion coefficient and $f(x)= 1-(3/2x)\ln(1+x)$ with $x=(2k_F/k_s)^2$.We calculated $D=1/e^2\rho N(E_F)$ from the experimental $\rho=98\cdot 10^{-8}$ Ωm and $N(E_F)$ value (4).Comparing the observed and calculated coefficients we find $f(x)=0.64$ ie. $k_F/k_s=1.6$, which we regard as very reasonable values.In principle k_F/k_s could be independently estimated, but we thought it was more straightforward to use the above x value in order to explain the Coulombic interaction contribution to the magnetoresistivity of our alloy.

In Fig.2 we present the magnetoresistivity data.The relative change in resistivity $\Delta\rho/\rho_0 = (\rho(H)-\rho_0)/\rho_0$ is plotted vs H^2 for three characteristic temperatures.The field dependence of $\Delta\rho/\rho_0$ can be separated in two regions:for H<4T there is pronounced deviation from a H^2 dependence,whereas for H>4T the data seem to follow a H^2 law.In order to investigate this deviation from a H^2 law in more details,we subtracted the apparent H^2 term.The resulting difference δ is plotted vs $H^{0.5}$ in the inset of Fig.2.

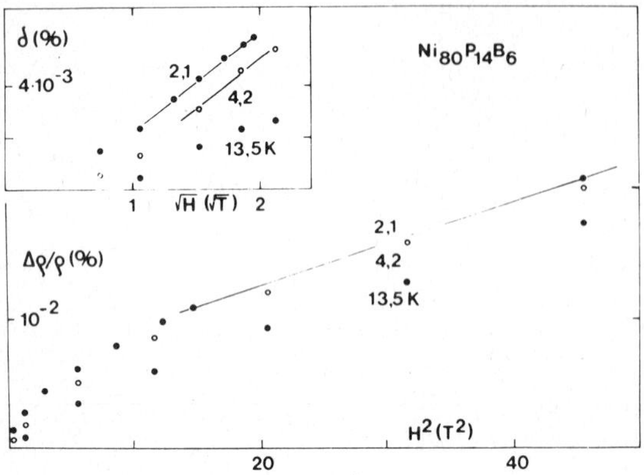

FIGURE 2
Normalised magnetoresistivity vs H^2. Inset: δ (see text) vs $H^{0.5}$

According to (5) the magnetoresistivity should be proportional to $H^{0.5}$ for $g\mu_B H \gg k_B T$. Consequently a $H^{0.5}$ dependence could appear at T=2K for H>1T and at higher fields for T>2K. Our results are consistent with this prediction. For $g\mu_B H \gg k_B T$ the correction to conductivity is

$$\Delta\sigma = H^{0.5} e^2 F(x) \mu_B^{0.5} / 4\pi^2 \hbar^{1.5} D^{0.5} \quad (2)$$

where $F(x)=(1/x)\ln(1+x)$. By using the same values for x and D as before, we calculate the coefficient of $H^{0.5}$ term to be $51(\Omega m)^{-1}$ which compares well with observed $40(\Omega m)^{-1}$ for T=2K. A similar conclusion is reached for T=4.2K. For 13K we are no longer in the $g\mu_B H \gg k_B T$ region.

The origin of H^2 contribution is not quite clear. The comparison with the nonmagnetic crystalline alloy indicates that the Kohler's term in our alloys should be 400 times smaller than the observed H^2 term. A possible explanation for this term is the development of magnetic moments in field similarly as in dilute PdNi (6). Hence our alloy could be regarded as containing isolated Ni moments in the isoelectronic host.

In Fig.3 we show the magnetic susceptibility (χ) results. At lower temperatures χ follows the Curie-Weiss (CW) law with Θ=1.6K (inset). At higher temperatures, χ increases monotonically with T. Subtracting the CW term, we obtain a weakly temperature dependent contribution χ_1 which extrapolates to $\chi_0=1.63\cdot10^{-6}$ emu/g. $\Delta\chi=\chi_1-\chi_0$ is shown at the bottom of Fig.3. Since $\Delta\chi$ is very small, its actual temperature dependence cannot be reliably established. If we assume $\Delta\chi/\chi_0=\beta T^{0.5}$ we obtain the coefficient $\beta\simeq5\cdot10^{-3}K^{-0.5}$. According to (2) the ratio of the coefficients of $T^{0.5}$ terms in χ and $\rho(T)$ should be $0.6\gamma^{-2}$ where γ is the Stoner enhancment factor. By using our χ_0 and $N(E_F)$ (4) we find γ=0.3. Thus the calculated ratio of the above coefficients is about 7, while the experimental one is 12.

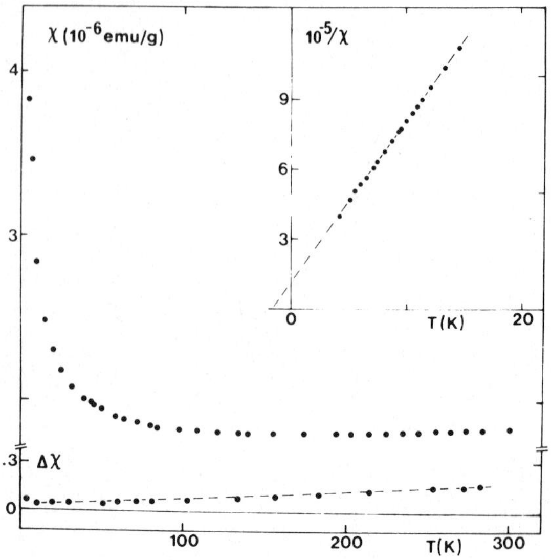

FIGURE 3
Magnetic susceptibility vs T. At the bottom $\Delta\chi$ (see text) vs T. Inset: χ^{-1} vs T

We also estimated the concentrations of the magnetic atoms in our alloy from the CW term. Assuming s=1/2 we obtain c=0,19 at%, and for s=1, c=0,07 at%. These values support our hypothesis of dilute magnetic atoms in the non-magnetic host.

To summarize we note that $\rho(T)$, $\rho(T,H)$ and χ measurements of $Ni_{80}P_{14}B_6$ alloy indicate the contributions due to interference effects at low temperatures. The analysis shows that the same set of parameters provides a semiquantitative description of all these contributions. In the case of $\rho(T,H)$ the agreement between the observed and calculated coefficient of $H^{0.5}$ term is very good. The isolated magnetic Ni atoms (seen via CW term in χ) could produce a H^2 term in $\rho(H)$.

REFERENCES

(1) A.Hamzić, E.Babić and Ž.Marohnić, J.Physique 41 (1980) C8-694
(2) B.L.Altshuler and A.G.Aronov, Zh.Eksp.Teor. Fiz. 77 (1979) 2028
(3) B.L.Altshuler, A.G.Aronov, D.E.Khmelnitskii and A.I.Larkin, Quantum Theory of Solids, ed.I.M.Lifshits (MIR, Moscow, 1982) p.181
(4) T.A.Donnely, T.Egami and D.G.Onn, Phys.Rev. B20 (1979) 1211
(5) P.A.Lee and T.V.Ramakrishnan, Phys.Rev. B26 (1982) 4009
(6) J.L.Genicon, F.Lapierre and J.Souletie, Phys. Rev. B10 (1974) 3976

LT-17 (Contributed Papers)
U. Eckern, A. Schmid, W. Weber, H. Wühl (eds)

LOW TEMPERATURE HALL EFFECT AND MAGNETORESISTANCE IN NiSiB METALLIC GLASSES

A. Schulte, G. Fritsch* and E. Lüscher

Physik-Department, TU München, D-8046 Garching, FRG

The temperature dependence of the Hall effect and the transverse magnetoresistance has been measured in the amorphous alloys $Ni_{76}Si_{12}B_{12}$, $Ni_{78}Si_8B_{14}$ as well as $Ni_{80}Si_{10}B_{10}$ near the resistance minima (T_{min} ≅25 K). In case of $Ni_{76}Si_{12}B_{12}$ and $Ni_{78}Si_8B_{14}$ an anomalous part, positive in sign, contributes to the Hall resistivity below ≅40 K. The alloy $Ni_{80}Si_{10}B_{10}$ is ferromagnetic below 110 K, showing a negative spontaneous Hall effect. The magnetoresistance detected at low temperatures is negative for the three alloys and increases with decreasing temperature. For the highest magnetic fields (6 Tesla) the electrical resistance continues to rise almost linearly with decreasing temperature below 4 K.

1. INTRODUCTION

At low temperatures metallic glasses -magnetic as well as non magnetic ones- frequently exhibit resistance minima. However, it is still not settled whether such a behaviour is of magnetic or of structural origin (1). We present Hall effect and magnetoresistance data for an amorphous system showing strong similarity in the temperature dependence of the resistivity, but qualitatively different behaviour for the anomalous part of the Hall resistivity. We conclude that in these cases the resistance minima are essentially magnetic in origin.

2. EXPERIMENTAL

The Hall effect (precision ≅1%) and the magnetoresistivity (precision $\cong 10^{-5}$) are determined with a standard dc-dc method. Voltages in the nanovolt range are measured with a Galvanometer amplifier which is connected to an integrating voltmeter. Side arms for the pressure contacts were cut into the samples made from melt-spun amorphous ribbons and measured as received. As the specimens are thin compared to their width the demagnetization factor is one to a very good approximation. Details of the experimental set up are given elsewhere (2).

3. RESULTS AND DISCUSSION

At room temperature the Hall resistivity ρ_H depends linearly on the magnetic field. The Hall coefficient R_H derived therefrom amounts to (-8.7±.2), (-7.7±.3) and (-10.1±.3)x10^{-11} $m^3A^{-1}s^{-1}$ for $Ni_{78}Si_8B_{14}$, $Ni_{76}Si_{12}B_{12}$ and $Ni_{80}Si_{10}B_{10}$, respectively. At low temperatures the magnetic field dependence of ρ_H turns out to be nonlinear for all three alloys due to an anomalous contribution, which is positive in sign for $Ni_{78}Si_8B_{14}$ and $Ni_{76}Si_{12}B_{12}$ but negative for $Ni_{80}Si_{10}B_{10}$.

Below about 110 K $Ni_{80}Si_{10}B_{10}$ shows ferromagnetism. The same is true for recrystallized samples of $Ni_{78}Si_8B_{14}$ and $Ni_{76}Si_{12}B_{12}$ exhibiting even higher Curie temperatures, as deduced from the occurence of hysteresis in the Hall resistivity (3).

The general behaviour of $Ni_{78}Si_8B_{14}$ and $Ni_{76}Si_{12}B_{12}$ is very similar. Their ordinary and their anomalous parts of ρ_H are of the same order of magnitude (Fig. 1). Hence, at high fields the normal part is dominating and the Hall resistivity changes from a positive to a negative slope. Starting from the relation (4) $\rho_H = R_0B + R_SM$, where R_0 and R_S denote the ordinary and the spontaneous Hall coefficient, M the magnetization and B the applied magnetic field, it has been shown (3) that $\rho_S = R_SM$ can be described by a Brillouin function, characteristic for superparamagnetism. The effective magnetic moment μ turns out to be ≅7 μ_B. The initial slope of the $\rho_H(B)$ curves obeys a Curie-Weiss law with T_Θ ≅0 K (Fig. 2). Therefore, we argue that clusters of atoms coupled ferromagnetically exist. However, we cannot exclude that their size is affected by the presence of trace iron impurities causing a positive value of ρ_S.

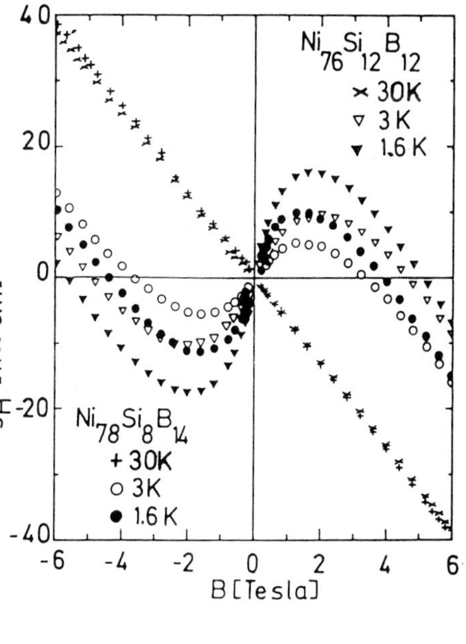

Fig. 1 The Hall resistivity $\rho_H(B)$

* Phys. Institut, HSBw München, D-8014 Neubiberg

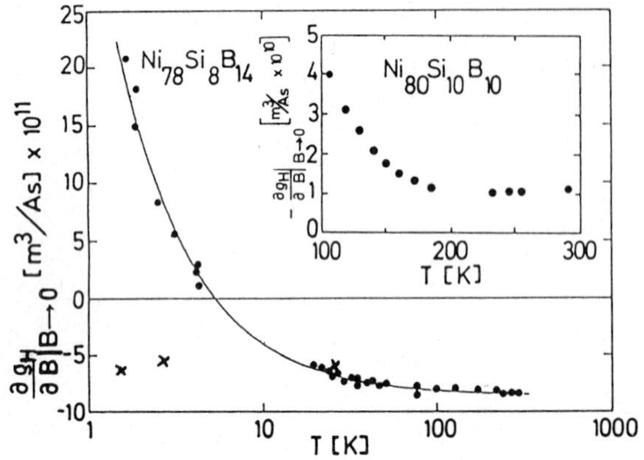

Fig. 2 The total Hall coefficient $\partial\rho_H/\partial B$ for low fields (•). The full line corresponds to a Curie-Weiss law. The normal part (x) is derived from the slope at high fields.

Near the Curie point of $Ni_{80}Si_{10}B_{10}$ ρ_S changes continuously from paramagnetic to ferromagnetic behaviour (see inset Fig. 2), probably due to the formation of superparamagnetic clusters larger than the mean free path of the electrons.

The magnetoresistivity is negative, independent from the sign of B. It is increasing with decreasing temperature. A tendency to saturation can be observed in higher fields indicated in Fig. 3. Within an accuracy of 10^{-5} no magnetoresistance could be detected above 30 K in $Ni_{78}Si_8B_{14}$ and $Ni_{76}Si_{12}B_{12}$. This finding emphasizes that the low temperature rise of the resistance is caused by an additional scattering mechanism. Its origin could be magnetic as indicated by the results of the Hall resistivity in all three alloys. Within the magnetic clusters all the spins are parallel, except for some defects caused by the amorphous structure due to the change from antiferromagnetic to ferromagnetic coupling

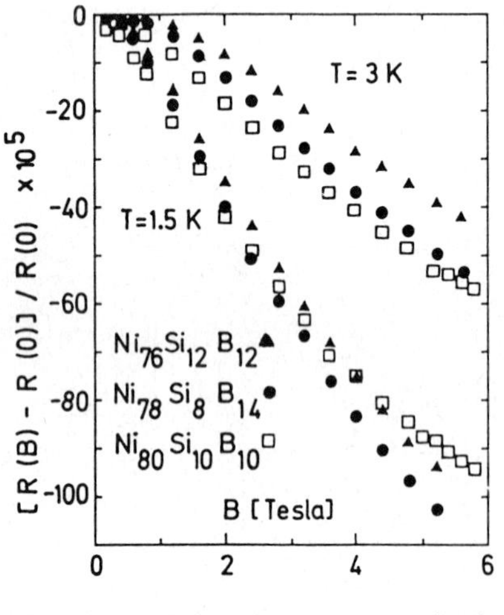

Fig. 3 The transverse magnetoresistivity

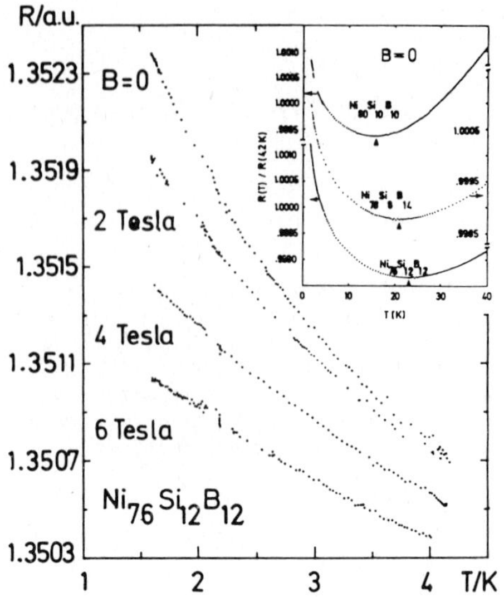

Fig. 4 The resistance R(T,B) ($\rho \cong 100\mu\Omega$cm)

with distance. Hence, we observe a defect resistance by elastic spin-spin scattering. In addition, a "Kondo" effect as discussed by Grest and Nagel (5) may be present. Both contributions decrease with rising temperature due to thermal smearing of the spin orientation. Applying magnetic fields does not alter the "Kondo" type contribution (5), but reduces spin disorder.

Below 4 K the resistivity increases almost linearly at 6 Tesla (see Fig. 4), a field which is not high enough to saturate the magnetoresistivity (see Fig. 3). Hence, additional contributions from weak localization or correlation effects of the electrons as investigated recently in 3-D amorphous alloys (6) are difficult to discuss here.

ACKNOWLEDGEMENTS

We would like to thank A. Eckert and A. Roithmayer for experimental assistance. We are also grateful to Dr. Hilzinger (Vacuumschmelze Hanau) for supplying us with the amorphous ribbons.

REFERENCES

(1) R. Harris and J. O. Strom-Olsen in: Glassy Metals II, eds. H. J. Güntherodt and H. Beck (Springer Verlag, Heidelberg, 1983)
(2) A. Schulte, A. Eckert, G. Fritsch, E. Lüscher, J. Phys. F (1984) in print
(3) A. Schulte, A. Eckert and G. Fritsch, Solid State Comm. (1984) in print
(4) C. M. Hurd, The Hall effect in metals and alloys (Plenum, New York, 1972)
(5) G. S. Grest and S. R. Nagel, Phys. Rev. B 19, (1979) 3571
(6) J. B. Bieri, A. Fert, G. Creuzet, J. C. Ousset, Solid State Comm. 49 (1984) 849

LT-17 (Contributed Papers)
U. Eckern, A. Schmid, W. Weber, H. Wühl (eds)

LOW TEMPERATURE SPECIFIC HEATS OF AMORPHOUS Cu-Ti ALLOYS

D.E. MOODY and T.K. NG

Department of Physics, University of Leeds, Leeds LS2 9JT, England, GB

Low temperature (1-4K) specific heat measurements are reported for a series of ten amorphous CuTi alloys, the titanium concentration varying from 27.5 to 67.5 atomic percent. The graphs of C/T vs T^2 all show slight upward curvature, and each set of results has been fitted to the equation $C = \gamma T + \beta T^3 + \delta T^5$. It is found that whereas γ increases uniformly with Ti concentration, the Debye temperature, corresponding to the coefficient β, is about 320 K for all the alloys measured.

1. INTRODUCTION

During the last few years there have been a number of investigations of the metallic glass systems CuZr and CuTi, both of which can be made relatively easily using melt spinning techniques. In these alloy systems the electronic density of states at the Fermi level is dominated by the d-states of Zr or Ti.

For CuZr alloys, numerical values of the density of states at the Fermi level can be obtained from the extensive measurements of low temperature specific heat which have already been published (1). For the CuTi series, only four such measurements have so far been reported, two by Mizutani, Akutsu and Mizoguchi (2) and two by ourselves (3). In this paper we present further results for the CuTi system, covering the concentration range from 27.5 to 67.5 atomic percent titanium.

2. MATERIALS

All the samples used in this investigation were prepared from JMC 'Specpure' materials using a single wheel melt spinning process (4), The specimens typically consisted of a pair of 30cm ribbons, each 2-3mm wide and up to 50μm thick, freshly made for the measurements.

3. MEASURING TECHNIQUE

The pulse calorimeter and cryostat have not been changed since our previous report (3), but the microprocessor system has been considerably developed. In addition, the accuracy with which the thermometer can be calibrated has been improved, as has the precision with which the thermal capacity of the sample holder is known. The provisional T-76 temperature scale (5) has been used in processing all the measurements.

A single 220 ohm Allen Bradley resistor has been used as both thermometer and heater. Repeated measurements have shown no apparent variation in specific heat which might be attributable to changing characteristics of the thermometer caused by cycling between 1.1 and 4.2 K (6), and although the apparatus does not allow us to test directly for thermometer variations during a single experimental run, these indirect tests indicate that any such variation is small enough to be ignored.

After cycling to room temperature and back, the thermometer characteristic typically changes by less than 1 mK at 1.1 K, and up to 5 mK at 4K. It was found that these variations could be described in terms of a single scaling factor for resistance, the calibration for one particular run having been fitted to a seven term polynomial of the form $(1/T) = \Sigma a_n \ell n^n R$. Although the use of a single scaling factor was arrived at independently from studies of the present series of thermometer calibrations, the principle is identical to the "reduced resistance" technique used in our laboratories some 25 years ago by Hoare and his co-workers (7).

4. TEMPERATURE VARIATION OF SPECIFIC HEAT

For all the samples of CuTi it was found that the conventional plots of C/T vs T^2 showed slight upward curvature, and all were therefore fitted to the equation $C = \gamma T + \beta T^3 + \delta T^5$ by minimising $\Sigma \left[(C - C_{calc})/C \right]^2$. Values of the coefficients γ, β and δ are given in table 1, together with the 95% confidence limits, rms deviation (rmsd), and Debye temperature Θ_D.

5. RELIABILITY OF THE RESULTS

As is shown clearly by the C/T vs T^2 llots in figure 1 of our earlier report, $Cu_{66}Ti_{34}$ was measured on the first "manual" version of the calorimeter, while $Cu_{43}Ti_{57}$ was measured with greater precision on the second run of the microprocessor controlled system. As expected, this later measurement is in excellent agreement with the γ values of table 1, whereas the first differs by 3% from the interpolated value from the present measurements.

Comparison with the work of Mizutani et al. is more difficult because these authors did not include a T^5 term in their analysis, and the differences between the two sets of values for γ and β are largely attributable to the use of different equations to describe the specific heat. We have made a direct comparison of our graphs of C/T vs T^2 with those given by these authors (2) in their figure 1, and find that there is excellent agreement for the $Cu_{50}Ti_{50}$ alloy, whereas for $Cu_{60}Ti_{40}$ our results show a systematic difference of about one percent. Since this only corresponds to a difference of about $1\frac{1}{2}$% in the compositions of the supposedly identical alloys, it seems to the writers that this is the likely source of the discrepancy.

The errors which we have quoted come directly from the statistical analysis, and their size in relation to the small rms deviations reflects the use of a three term equation. In fact, the largest error is probably associated with the uncertainties in compositions, and since the γ values lie on a smooth curve to within 0.1 mJ mol^{-1} K^{-2}, this figure may be taken as the true measure of the overall error.

It is more difficult to estimate the corresponding error in the Debye temperatures. The 95% confidence limits vary from ±6 to ±12 K, and since all the values quoted lie within a range of 20 degrees, we can only state that the Debye temperature is 320 ± 10 K over the range of concentration covered.

Since the value of the T^5 coefficient is, on average, only twice the 95% confidence limit, little can be said about its dependence on concentration. The average value of this coefficient, 0.6 μJ mol^{-1} K^{-6}, and the average Debye temperature of 320 K fit very well into the general correlation pattern noted by Mizutani and Massalski (8). We therefore feel justified in including the T^5 term to improve the precision with which γ can be determined.

TABLE 1

Conc % Ti	γ mJ	β μJ	δ μJ	rmsd (%)	Θ_D K
27.5	2.95±0.01	65±4	0.5±0.2	0.4	310
32.5	3.39±0.01	57±4	0.3±0.2	0.4	325
35	3.52±0.01	57±4	0.7±0.2	0.4	325
37.5	3.67±0.02	58±5	0.3±0.3	0.4	322
40	3.88±0.02	56±5	0.9±0.3	0.4	326
45	4.09±0.01	59±4	0.5±0.2	0.4	321
50	4.24±0.02	56±5	0.5±0.3	0.5	326
60	4.79±0.03	68±8	0.2±0.5	0.5	306
65	4.97±0.02	56±6	0.9±0.3	0.5	325
67.5	5.06±0.02	57±7	0.8±0.4	0.5	323

Concentrations are given as atomic percentage. The errors quoted for each molar coefficient are the 95% confidence limits.

6. DISCUSSION

The variation of γ with alloy concentration follows the approximately linear variation described by Mizutani et al. (2), but over the extended range from 30 to 70 atomic percent. Assuming that the contribution from two level states is similar to the average value of 0.05 mJ mol^{-1} K^{-2} found in CuZr alloys (1), almost the whole of the linear term in these CuTi measurements will be the usual electronic specific heat. The band structure determined by UPS studies (9) indicates that the variation in the electronic specific heat is associated with the fact the Fermi level lies in the midst of the Ti d-states.

To obtain numerical estimates of the density of states at the Fermi level in these alloys, it is necessary to estimate the values of the electron-phonon coupling constant λ. The usual method of estimating these is to make use of the McMillan equation (10) in conjunction with values of the superconducting transition temperature. These temperatures are not known in any detail for the amorphous CuTi alloy system, and although Gallagher (11) has shown that λ is about 0.6 for $Cu_{43}Ti_{57}$, we have not attempted to use this value, since the bare density of states can be easily calculated from the γ values when more direct estimates for λ become available.

ACKNOWLEDGEMENTS

We are indebted to the SERC for an award which covered the costs of the major components of the microprocessor based measuring system.

REFERENCES

(1) K. Samwer and H. von Löhneysen, Phys. Rev. B26 (1982) 107, and earlier references therein.
(2) U. Mizutani, N. Akutsu and T. Mizoguchi, J.Phys.F: Met.Phys. 13 (1983) 2127.
(3) D.E. Moody and T.K. Ng, Physics of Transition Metals (Inst.Phys.Conf.Ser. 55) (1981) 631.
(4) D. Pavuna, J. Mater. Sci. 16 (1981) 2419.
(5) M. Durieux, W.R.G. Kemp, C.A. Swenson and D.N. Astrov, Metrologia 15 (1979) 65.
(6) E.M. Forgan and S. Nedjat, Cryogenics 21 (1981) 681.
(7) F.E. Hoare, J.C. Matthews and B. Yates, Proc.Phys.Soc. B68 (1955) 388.
(8) U. Mizutani and T.B. Massalski, J.Phys.F: Met.Phys. 10 (1980) 1093.
(9) T. Mizoguchi, U. Gubler, P. Oelhafen, H.-J. Güntherodt, N. Akutsu and N. Watanabe, Proc.4th Int. Conf. on Rapidly Quenched Metals, Sendai, Vol. 2 (1981) 1307.
(10) W.L. McMillan, equation (30), Phys.Rev. 167 (1968) 331.
(11) B.L. Gallagher, J.Phys.F:Met.Phys.11 (1981) L207.

LT-17 (Contributed Papers)
U. Eckern, A. Schmid, W. Weber, H. Wühl (eds)
© Elsevier Science Publishers B.V., 1984

THERMOPOWER MEASUREMENTS AND THE ELECTRON-PHONON INTERACTION IN $La_{1-x}Al_x$ METALLIC GLASSES*

H. ARMBRÜSTER, R. DELGADO, and D.G. NAUGLE

Department of Physics, Texas A&M University, College Station, TX 77843, USA

C.L. TSAI

Institute for Chemical Analysis, Northeastern University, Boston, MA 02115, USA

W.L. JOHNSON and A.R. WILLIAMS†

W.W. Keck Laboratory, California Institute of Technology, Pasadena, CA 91125, USA

Measurements of the thermopower, S, for a series of $La_{1-x}Al_x$ metallic glasses ($.18 \leq x \leq .35$) over the temperature range 1.2-280K are presented. An increase in values of -S/T at low temperatures is the result of electron-phonon mass enhancement. Values of the electron-phonon coupling constant, λ, determined from these measurements are larger than those inferred from superconductivity. Their difference is at least four times larger than the estimated error.

Opsal, et al, (1) have pointed out that, although the effects of the electron-phonon mass renormalization cancel out in most electron transport coefficients, an enhancement of the low temperature thermopower, S, can be observed. In crystalline metals the "phonon drag" contribution hinders observation of this enhancement. For amorphous metals the "phonon drag" is absent since the scattering of the phonons by disorder is much greater than by electrons. Although the structure in the electronic density of states normally found in crystalline materials is washed out by disorder in an amorphous metal, Jäckle (2) has suggested that the fine structure near the Fermi energy that results from the electron-phonon interaction should persist since its energy scale is given by the Debye frequency which is only slightly affected by disorder. Gallagher (3) has interpreted increases in values of S/T at low temperatures from the corresponding values at temperatures, T, greater than the Debye temperature, θ_D, in several amorphous alloys to be the result of the electron-phonon interaction. He calculated the electron-phonon coupling constant, λ, from

$$1 + \lambda = \lim_{T \to 0} (S(T)/T)/(S(T)/T)_{T \gg \theta_D}.$$

The detailed temperature dependence of the normalized enhancement factor, $\lambda(T)/\lambda(0)$, for CuZr and CuTi alloy glasses is in agreement with calculations by Kaiser (4) of the electron-phonon enhancement based on phonon densities of states derived from neutron scattering.

We have measured S(T) with a differential technique (5) with a Pb reference material for a series of $La_{1-x}Al_x$ metallic glasses prepared by both the "melt spinning" and "splat-cooling" technique. Values of -S(T)/T are shown for some of the splat-cooled samples in Figure 1. The absolute thermopower of the Pb reference foil used in these measurements was measured against Nb_3Sn below its superconducting transition temperature and, unlike the Pb foil used for $La_{1-x}Ga_x$ measurements, its thermopower matches that reported by Roberts (6). For all the samples, the resistivities are greater than 100μcm and the temperature coefficients of resistivity are negative. The values of these quantities and T_c for samples prepared by the two different techniques are generally in good agreement. Differences can be attributed to small deviations from the nominal alloy concentration.

Values of λ determined from thermopower measurements of $La_{78}Ga_{22}$ (5) and $La_{1-x}Al_x$ are compared in Table I with λ' determined from superconductivity measurements for the same alloys with McMillan's formula (7),

$$\ln \frac{1.45\, T_c}{\theta_D} = -\frac{1.04(1+\lambda')}{\lambda' - \mu^*(1+0.62\lambda')}$$

The parameters used in this formula are also given. Values of θ_D for the $La_{1-x}Al_x$ al-

* Supported by NSF (DMR-8121439) and the Robert A. Welch Foundation at Texas A&M, by DOE (DE-AM03-765500767) at Cal. Tech. and by AFOSR (F49620-82-C0076) at Northeastern.
† Current address: Los Alamos National Laboratory, Los Alamos, NM 87545, USA.

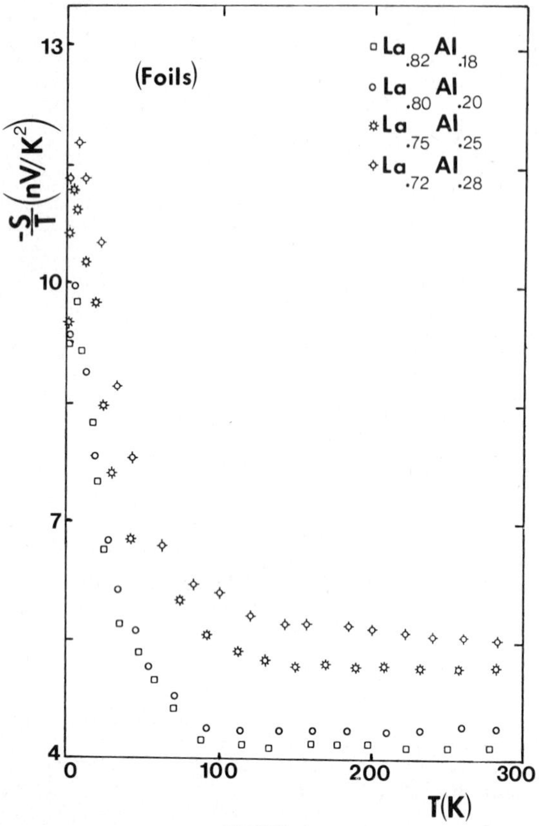

FIGURE 1
Values of -S(T)/T are shown as a function of temperature for several $La_{1-x}Al_x$ "splat-cooled" foils.

Table I: Comparison of λ and λ' and the McMillan parameters for La-based metallic glasses. $\mu^* = 0.13$. [a] Ref. (8), [b] estimated by scaling θ_D of $La_{1-x}Ga_x$ from Ref. (8), [m] is melt spun.

Alloy	T_c(K)	θ_D(K)	λ	λ'
$La_{73}Ga_{22}$	3.94[a]	109[a]	1.10	.85
$La_{82}Al_{18}$	4.210	118[b]	1.30	.85
$La_{80}Al_{20}$	3.904	119[b]	1.26	.82
$La_{75}Al_{25}$	3.436	120[b]	0.96	.77
$La_{75}Al_{25}$[m]	3.449	120[b]	0.96	.77
$La_{72}Al_{28}$	3.101	121[b]	1.05	.74
$La_{65}Al_{35}$[m]	1.822	124[b]	0.77	.62

loys were estimated by scaling measured values of θ_D for $La_{1-x}Ga_x$ (8) by the inverse square root of their mass densities. For $La_{1-x}Au_x$ alloys this scaling procedure produces at most a 5% error (8).

Note that in Figure 1 the low temperature enhancement of the thermopower becomes quite small as T approaches and exceeds θ_D. We conclude that this low temperature enhancement of the thermopower of La-based metallic glasses is primarily the result of the electron-phonon mass renormalization. Spin fluctuation enhancement of the thermopower in these alloys should not be important.

The values of the electron-phonon coupling constant, λ, determined from the thermopower measurements are higher than λ' estimated from the superconducting properties of these La-based metallic glasses. Uncertainty in determination of λ' results principally from the choice of μ^*. Reducing μ^* by 30% to 0.1 results in a reduction of λ' by 10%, which increases the discrepancy between λ' and λ. A 5% error in θ_D causes variations in λ' between 1 and 3%. Estimates of λ could vary by 0.1 due to extrapolation from T_c to T=0 and the scatter in the data as a result of the small thermal emf's measured at low temperatures. Therefore the difference between λ and λ' is at least four times larger than the estimated maximum error.

All samples exhibit a peak in the electron-phonon enhancement similar to CuZr and CuHf (9), consistent with Kaiser's theoretical results for a constant α^2 in the Eliashberg function $\alpha^2(\omega)F(\omega)$.

Different theoretical calculations of the enhancement factor, (1+aλ), vary from $0<a<2$. A quantitative analysis of the ratio $\lambda/\overline{\lambda'}$ may therefore provide information about model parameters used in the phonon enhancement theories (10). For a test of the theory, experimental values for the Debye temperature and electronic density of states are needed. They provide information about μ^*, the enhancement factor of the quasi particle mass, (1+λ), and the Thomas Fermi screening length. Heat capacity measurements of the samples are in progress. Estimates of λ from thermopower measurements in a metallic glass system, which does not exhibit superconductivity above 1.1K, $Ca_{1-x}Al_x$, will be discussed elsewhere.

REFERENCES

(1) J.L. Opsal, B.J. Thaler, and J. Bass, Phys. Rev. Letters 36 (1976) 1211.
(2) J. Jäckle, J. Phys. F10 (1980) L43.
(3) B.L. Gallagher, J. Phys. F11 (1981) L207.
(4) A.B. Kaiser, J. Phys. F12 (1982) L223.
(5) H. Armbrüster and D.G. Naugle, Sol. St. Comm. 39 (1981) 675.
(6) R.B. Roberts, Phil. Mag. 36 (1977) 91.
(7) W.L. McMillan, Phys. Rev. B167 (1968) 331.
(8) W.H. Shull, D.G. Naugle, S.J. Poon, and W.L. Johnson, Phys. Rev. B18 (1978) 3263.
(9) B.L. Gallagher, A.B. Kaiser, and D.Greig, J. Non. Cryst. Solids 61 & 62 (1984) 1231.
(10) Y.A. Ono and P.L. Taylor, Phys. Rev. B22 (1980) 1109.

LT-17 (Contributed Papers)
U. Eckern, A. Schmid, W. Weber, H. Wühl (eds)

THE PHONON-FRACTON DENSITY OF STATES. AN EXPLANATION OF THE PLATEAU IN THE THERMAL CONDUCTIVITY AND THE SPECIFIC HEAT BEHAVIOR OF AMORPHOUS MATERIALS

R. ORBACH and H. M. ROSENBERG*

Department of Physics, University of California, Los Angeles, CA 90024, U.S.A.

*Permanent address: The Clarendon Laboratory, Oxford OX1 3PU, U.K.

The phonon-fracton model which has been proposed for excitations in amorphous structures is shown to account for the main features of the specific heat and thermal conductivity behavior at and above the liquid helium region for samples of epoxy-resin of varying structural lengths.

1. INTRODUCTION

We explain in this paper how one can account for the behavior of the thermal conductivity and specific heat of glassy materials in the range 2-15 K by a consideration of the form of the phonon-fracton density of states curve.

The thermal conductivity of nearly all glassy materials exhibits a plateau in the liquid helium region. To account for this behavior we conjectured (1) that the vibrational excitations in the material change from being phonon (extended) to fracton (localized) states at a particular crossover length, or, equivalently, crossover frequency. Since this early analysis, experimental and theoretical advances have added substantial evidence in favor of this interpretation.

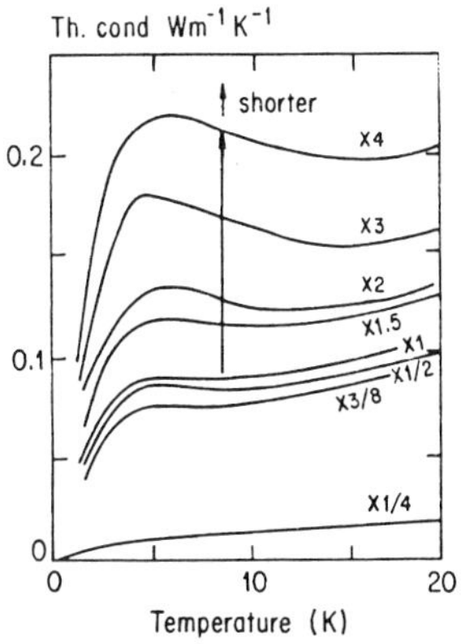

FIG. 1

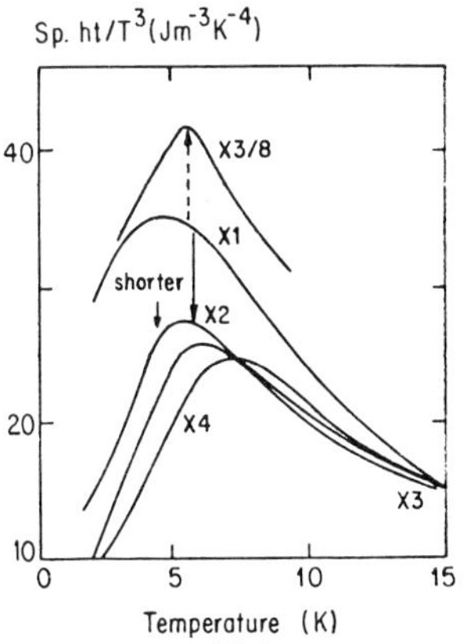

FIG. 2

2. EXPERIMENTAL AND THEORETICAL BACKGROUND

Experiments by Farrell, de Oliveira, and Rosenberg (2) have shown that the addition of ethylene diamine (EDA) hardener, in excess of the stoichiometric quantity, to MY 750 epoxy-resin *increases* the value of the thermal conductivity in the plateau region by a factor of up to three. This was accompanied by a *decrease* in the specific heat in the same temperature regime (2 to 15 K). (Curves taken from this work are reproduced in Figs. 1 and 2.) Derrida, Orbach, and Yu (3) have calculated the density of states, $N(\omega)$, for a percolating network within the effective medium approximation (EMA). They showed that $N(\omega)$ is proportional to ω^2 at low frequencies. These excitations are phonons, the states are extended (the relevant dimension is $d = 3$, the Euclidean dimension) and can transport heat. At a certain crossover frequency, corresponding to a characteristic length scale, $N(\omega)$ increases sharply (Fig. 3). Above the crossover frequency $N(\omega)$ levels off to a nearly constant value as calculated within the EMA. More generally $N(\omega) \propto \omega^{\bar{\bar{d}}-1}$ where $\bar{\bar{d}}$ is the fracton dimensionality. The excitations are fractons in the latter regime (Alexander and Orbach (3)). They are localized (the relevant dimension is $\bar{\bar{d}} \approx 4/3$, the fracton dimensionality, so that above crossover $N(\omega)$ increases with ω much more slowly than below) and so, whilst they contribute to the thermal energy, and hence to the specific heat, they cannot play a part in thermal conduction. The crossover from phonon to fracton behavior causes the rapid rise in $N(\omega)$ at crossover and occurs at a length scale which corresponds to some characteristic length of the structure. In the epoxy-resin this will be determined by the length of a molecular unit, i.e., the distance between cross links or some associated dimension.

3. DISCUSSION

It is believed that excess EDA in the epoxy resin decreases the size of the structural units. This would imply that the sharp increase in N(ω) will occur at higher frequencies as more EDA is added. This is simulated by the dashed curve in Fig. 3. If this be so, then, as we shall explain the behavior of the observed changes in thermal conductivity and the specific heat, upon addition of excess EDA, follow quite naturally without having to invoke any mechanism involving the low-lying two level states or additional scattering channels. (Indeed, no scattering is necessary. Spatial localization in the fracton region is a purely geometrical effect.) The thermal conductivity is determined only by the extended vibrational states, hence only by the number of phonons in the system, i.e., by the area under the ω^2 part of the N(ω) curve. This is smaller for the stoichiometrically hardened resin (shaded region of the curve) than for the specimen with excess hardener (shaded + cross-hatched area). Thus, those specimens with excess hardener will have a higher thermal conductivity, as is observed. However, the *total* area under the curve, up to a particular frequency ω_T (which will depend on the temperature of the measurement), decreases on going from the full curve to the dashed curve. This means that the specific heat will decrease with increasing excess EDA, also in accord with the experimental observations.

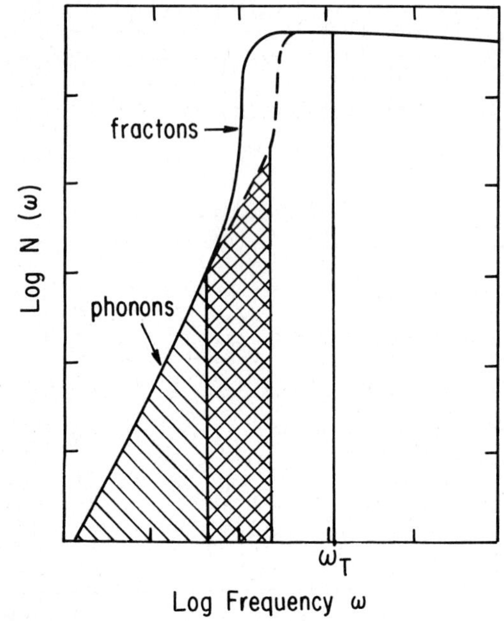

FIG. 3

4. FURTHER FEATURES

Two other qualitative features of the thermal conductivity and the specific heat measurements can also be accounted for by the phonon-fracton density of states model. These are as follows:

(a) The plateau region of the thermal conductivity begins at successively higher temperatures as more EDA is added, i.e., as the unit length of the polymer is reduced (Fig. 1). This occurs because a reduction in unit length would delay the onset of fractal behavior to shorter excitation wavelengths - higher frequencies - and so the temperature at which the plateau in the thermal conductivity begins will increase. Work on other epoxy polymers (5) shows the same tendency - polymers with longer units have a thermal conductivity plateau which starts at a lower temperature than those with shorter units.

(b) The specific heat curves, presented as plots of C/T^3 versus T (Fig. 2) also confirm the fracton model. These curves, which indicate deviations from the Debye T^3 law, show maxima at progressively higher temperatures as the unit molecular length decreases. This peak indicates the region where the density of states is rising most steeply. This again is in agreement with the N(ω) curves in which the steeper (crossover) region is shifted to successively higher frequencies as the phonon-fracton crossover frequency is increased.

5. CONCLUSIONS

The good qualitative agreement between the predictions of the phonon-fracton model and the measurements of the specific heat and thermal conductivity leads us to believe that the application of this model to amorphous systems is soundly based and that it should be generally applicable to all amorphous structures.

ACKNOWLEDGMENTS

This programme of research has been supported by the U. S. National Science Foundation under grant DMR 81-15542 and by the U. K. Science and Engineering Research Council.

REFERENCES

(1) S. Alexander, C. Laermans, R. Orbach, and H.M. Rosenberg, Phys. Rev. B 28 (1983) 4615.
(2) D. E. Farrell, J. E. de Oliveira, and H. M. Rosenberg, Proc. Fourth Intern. Conf. on Phonon Scattering, Stuttgart (1983), in print.
(3) S. Alexander and R. Orbach, J. Physique (Paris) Lettres 43 (1982) 625.
(4) B. Derrida, R. Orbach, and K.-W. Yu, Phys. Rev. B, in print.
(5) J. E. de Oliveira and H. M. Rosenberg, work in progress.

LT-17 (Contributed Papers)
U. Eckern, A. Schmid, W. Weber, H. Wühl (eds)

DYNAMICAL CONDUCTIVITY OF SIMPLE METALLIC-GLASSES

Mircea CRISAN and Ion KOSZTIN

Department of Physics, University of Cluj, 3400 Cluj, România

The Götze and Wölfe method for calculation of the frequency-dependent conductivity has been applied for a model of a metallic glass in which electrons are coupled to a local two-level system.

1. INTRODUCTION

Recently there has been considerable interest in the theory of metallic glasses. The most important problem for the microscopic theory of the metallic glasses is how the electrons interact with the tunneling levels.
One of the models which describes this interaction (and has been used by Fulde and Peschel to study crystalline-field effects in metals (1)) is defined by the Hamiltonian

$$H = \sum_k \varepsilon(k)\, C_k^+ C_k + E\sigma^z + (\vec{v}\vec{\sigma}) \sum_{k,k'} C_k^+ C_{k'}$$

with a local two-levels system described by a "pseudo"-spin (are the Pauli spin vector).

Black and Gyorffy (2) showed the existence of a divergence, quite different from that predicted by the Kondo model, in a simple model in which the conduction electrons experience a local time-dependent potential due to two - level tunneling states.
Keiter and Morandi (3) used the perturbation theory in order to calculate the electronic lifetime for a metallic-glass using the perturbation theory and obtained a logarithmic term and an exponentially small prefactor.

In this paper we will apply the memory function method (4) in order to calculate the time-life of the conduction electrons in a metallic-glass described by the Hamiltonian (1). The notations which will be used in this paper are the same with the notations from the reference (4).

2. DYNAMICAL CONDUCTIVITY

In order to obtain the dynamical conductivity we have to solve the equation of motion for the function

$$\phi(z) = \langle\langle A_i / A_i \rangle\rangle$$

where

$$J_i = \sum_{k,\sigma} V_{0i}(k)\, C_{k\sigma}^+ C_{k\sigma} \qquad V_{0i}(k) = \frac{\partial \varepsilon(k)}{\partial k_i}$$

The memory function M(z) defined as

$$M(z) = \left[\phi(z) - \phi(z=0)\right] / z x_0 \qquad x_0 = n/m$$

gives us the possibility to calculate the dynamic conductivity $\sigma(\omega)$ from the equation

$$\sigma(\omega \pm i0) = e^2 n/ma\left[i\omega \pm 1/\tau\right] / \left[\omega^2 \pm (1/\tau)^2\right]$$

where

$$a = \left(1 + \frac{\partial M'(\omega)}{\partial \omega}\right)^{-1} \qquad \omega \longrightarrow 0$$

$$\frac{1}{\tau} = a M''(0)$$

Following the method proposed in (4) we get for the imaginary part of the complex function $M(\omega)$ denoted by $M''(\omega)$ the expression

$$M''(\omega) = \frac{4\pi N^2(0)}{3n} \left\{ T\left[\langle(\vec{v}\vec{\sigma})^2\rangle \left(\frac{\varepsilon_0}{2T} \tanh\frac{\varepsilon_0}{2T} - \ln\cosh\frac{\varepsilon_0}{2T}\right) - \frac{v^2}{2}\tanh\frac{\varepsilon_0}{2T}\left(1 + \frac{1}{2}\tanh\frac{\varepsilon_0}{2T}\right)\right] + \omega\left[\langle(\vec{v}\vec{\sigma})^2\rangle \tanh\frac{\varepsilon_0}{2T} - \frac{v^2}{2} \tanh\frac{\varepsilon_0}{2T}\left(1 + \frac{1}{2}\tanh\frac{\varepsilon_0}{2T}\right)\right]\right\}$$

In the limit $\omega \longrightarrow 0$ and $a \simeq 1$ we get for $M''(0)$ the equation

$$M''(0) = \frac{8\pi N(0)}{3n}\left[\langle(\vec{v}\vec{\sigma})^2\rangle\left(\frac{\varepsilon_0}{2T}\tanh\frac{\varepsilon_0}{2T} - \ln\cosh\frac{\varepsilon_0}{2T}\right) + \frac{v^2}{2}\tanh\frac{\varepsilon_0}{2T}\right]$$

and the dynamic conductivity $\sigma(\omega \pm io)$ becomes

$$\sigma(\omega \pm io) = \frac{e^2 n}{m}\left[i\omega \pm \frac{8\pi N^2(0)}{3n}\left[\langle(\vec{v}\tau)^2\rangle\left(\frac{\varepsilon_0}{2T}\tanh\frac{\varepsilon_0}{2T} - \ln \cosh\frac{\varepsilon_0}{2T}\right) + \frac{v^2}{2}\tanh\frac{\varepsilon_0}{2T}\right]\right] \Big/ N(\omega^2, T)$$

From these results we can see that there are no divergencies in M"(o) in the second order approximation but using the corrections of superior order we expect to have logarithmic Kondo correction. However, as long as the type of the divergence is not clear we consider that this result is quite interesting because in the limit of low temperatures we still have a divergence given by the logarithmic term which is different from the Kondo divergence. On the other hand the result obtained for the relaxation time presents a similar behaviour (excepting the logarithmic term) with the result obtained by Black and Fulde (5) for the same problem but with the electrons in the superconducting state.

This method seems to be promising for the calculation of the dynamical properties of the metallic-glasses and especially for the amorphous superconductors (6).

3. CONCLUSIONS

We showed that the memory function method proposed by Götze and Wölfe can be applied for the calculation of dynamic conductivity of a metallic-glass. In the second order approximation we obtained a divergence different to the one, which appears at low temperatures.

REFERENCES

(1) P.Fulde and I.Peschel, Adv. Phys. 21 (1972) 1.

(2) J.L.Black and B.L.Gyorffy, J. Phys. (Paris) 39, C-6 (1978) 941.

(3) H.Keiter and G.Morandi, Phys. Rev. B22 (1980) 5004.

(4) W.Götze and P.Wölfe, Phys. Rev. B6 (1972) 1226.

(5) J.L.Black and P.Fulde, Phys. Rev. Lett. 43 (1979) 453.

(6) G.Weiss, S.Hunklinger and H.v. Löhneysen, Physica 110 B (1982) 1964.

LT-17 (Contributed Papers)
U. Eckern, A. Schmid, W. Weber, H. Wühl (eds)

ELASTIC LOW-ENERGY EXCITATIONS IN $LiTaO_3$

W.ARNOLD*, P.DOUSSINEAU**, A.LEVELUT**, and S.ZIOLKIEWICZ**

*Fraunhofer-Institute, Bldg.37, University, 6600 Saarbrücken, FRG
**Laboratoire d'Ultrasons, Université de Paris VI
Tour 13, 4 Place Jussieu, 75230 Paris-Cedex 05, France

We report measurements of the acoustic behavior of the crystalline material $LiTaO_3$ at low temperatures. We observe a logarithmic temperature dependence of the sound velocity similar to the one common in glasses. The effect is typically two order of magnitude smaller than in vitreous silica.

1.INTRODUCTION

Glasses, more generally speaking disordered systems, exhibit characteristic thermal, acoustic, and dielectric properties at low temperatures. The specific heat contains an almost linear term and the thermal conductivity varies quadratically with temperature. These effects are caused by elastic low-energy excitations whose energy spectrum is flat in the temperature range at least up to 10 K(1). These excitations scatter phonons resonantly which can be observed both in thermal conductivity and ultrasonic absorption and dispersion(2). The microscopic origin of the low-energy excitations is still an unresolved problem. The tunneling model, however, turned out to be a very powerful tool to describe many experimental observable quantities of these excitations (3).
In this model it is assumed that some part of the network can tunnel quantum-mechanically between two local equilibrium positions.
In the past few years increasing evidence has emerged that similar "glassy" properties can be observed in crystals. Such systems are polycristalline NbZr (4),superionic conductors(5,6), neutron irradiated quartz crystals(7), ferroelectric ceramics and crystals(8,9) and strongly doped alkali halides(10). In all these experiments the intriguing questions are: How much disorder is needed so that a cristalline material starts to exhibit glassy properties and what causes the disorder ?

2.EXPERIMENTAL RESULTS

Besides the measurement of the specific heat and thermal conductivity, acoustic dispersion measurements are a very sensitive tool to detect low-energy excitations(11). We have therefore undertaken such experiments in the ferroelectric materials $LiTaO_3$ and $LiNbO_3$.One set of $LiTaO_3$ crystals was manufactured by Crystal Technology in cylindrical form (10mm length, 5mm diameter) and they were single-domain. The other $LiTaO_3$ crystal examined was grown in our laboratory by the Cochralzki method and was multidomain. The $LiNbO_3$ crystal was also of cylindrical form (18mm length, 6mm diameter) as well as single-domain. The experiments were performed in a He-3 cryostat. The sound waves were generated by piezoelectric surface excitation in a non-resonant sample holder with the k-vector propagating parallel to the c-axis of the crystal. Frequencies employed were around 700 MHz.

Fig.1 shows a typical result in $LiTaO_3$ (sample from Cryst.Techn.). It displays the relative variation of sound velocity $(v(T)-v(T_o)/v(T_o)= \Delta v/v$ as a function of

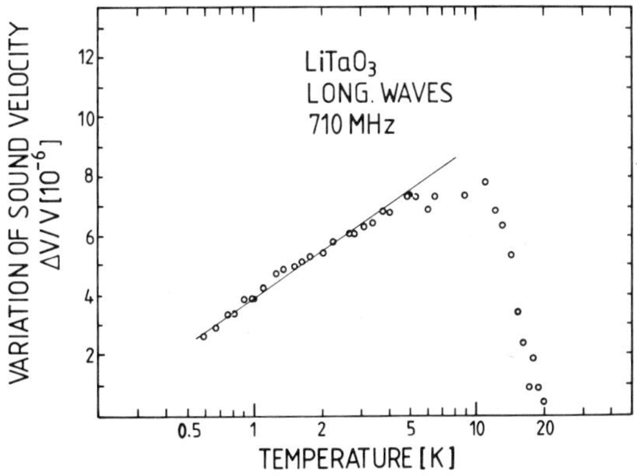

FIGURE 1
Relative variation of sound velocity in $LiTaO_3$ as a function of temperature. The logarithmic contribution is shown by the solid line.

temperature. Here, T_0 is a reference temperature. Between 0.5 K and 5 K $\Delta v/v$ increases logarithmically with temperature, passes through a maximum around 8 K, and then decreases rapidly. Similar results are obtained for the crystals grown in our laboratory. The slope of the logarithmic increase, however, is in this case considerably larger. The effect can be observed both for longitudinal and shear waves. Since $LiNbO_3$ is isomorphic with $LiTaO_3$, it is interesting to verify whether this behavior can also be observed in $LiNbO_3$. Within our experimental accuracy, we did not observe a logarithmic increase of $\Delta v/v$ with temperature in this material.

3. DISCUSSION

The experimental results can be described phenomenologically within the framework of the tunneling model. The logarithmic increase of $\Delta v/v$ is due to the resonant interaction of phonons with the tunneling states (TS)(2,11):

$$\Delta v/v = C_\alpha \cdot \mathrm{Ln}(T/T_0) \qquad C_\alpha = P\gamma_\alpha^2/\rho v_\alpha^2 \tag{1}$$

Here, γ_α is the TS-phonon coupling constant, ρ is the density, and α denotes the polarization. P is the density of states of the TS. In Table I the C_α-values as deduced from the measurements are listed. They are at least one order of magnitude smaller than the one found in insulating(1,2) and metallic glasses(12).

Table I

Material	C_L	C_T
$LiTaO_3$ (Cryst.Techn.)	$2.3\cdot10^{-6}$	$3.6\cdot10^{-6}$
$LiTaO_3$ (Paris VI)	$5\cdot10^{-6}$	--
$LiNbO_3$	$\leq 2\cdot10^{-7}$	--

At higher temperatures, the decrease of $\Delta v/v$ with temperature is caused by relaxational effects due to TS, but also an anharmonic contribution might play a role. In this context it is interesting to note that previous measurements have revealed an attenuation peak in $LiTaO_3$ in excess of the attenuation process caused by the well-known Akhiezer mechanism and multiphonon collision processes (13). These results have been interpreted as a classical Arrhenius-relaxation mechanism caused by some atomic motion due to the non-stochiometric crystal composition. The crystals used there were all grown with additional Ta in the melt. In attempting to correlate these results to our findings, we have replotted the data by Worley et al(13), and in fact the attenuation $\alpha(T)$ varies at the lowest temperatures as T^3 as is the case for the relaxational part of the ultrasonic attenuation caused by the TS(2). The magnitude of C_T necessary to fit $\alpha(T)$ is close to $3\cdot10^{-5}$, appreciably higher than our values but still much smaller than in vitreous silica. Furthermore, the distribution of relaxation times characteristic of the tunneling model had to be restricted (e.g. $0.1 < r < 1$). It is difficult to imagine, however, that the Ta atom with its large atomic mass should be the tunneling entity. We are presently extending our measurements to crystals grown under various conditions in the hope of identifying the origin of the TS in $LiTaO_3$.

4. SUMMARY

In conclusion, we should like to point out, that even crystals with well-defined crystallographic order can exhibit the acoustic anomalies of disordered systems at low temperatures.

REFERENCES

(1) "Amorphous Solids" Ed.W.A.Phillips Springer, N.Y.,(1981)

(2) S.Hunklinger and W.Arnold, Physical Acoustics 12,156(1976)

(3) P.W.Anderson,B.I.Halperin,C.M.Varma, Phil.Mag.25,1(1972);W.A.Phillips J.Low Temp.Phys.7,123(1972)

(4) L.F.Lou,Sol.St.Comm.19,335(1976)

(5) P.Doussineau,Ch.Frenois,R.G.Leisure, A.Levelut,J.Y.Prieur,J.Physique 41 1193(1980)

(6) T.Baumann,M.v.Schickfus,S.Hunklinger, J.Jäckle,Sol.St.Comm.35,587(1980)

(7) C.Laermans,Phys.Rev.Lett.42,250(1979)

(8) W.N.Lawless,Ferroelectr.43,223(1982)

(9) D.A.Ackermann,D.Moy,R.C.Potter, A.C.Anderson,Phys.Rev.B23,3886(1981)

(10) J.E.Benet,J.Pelous,R.Vacher J.Physique Lett.44,L433(1983)

(11) L.Piché,R.Maynard,S.Hunklinger, J.Jäckle,Phys.Rev.Lett.32,1426(1974)

(12) J.L.Black in: Metallic Glasses,Vol.I Eds. H.Beck and H.J.Güntherodt, Springer,N.Y.(1981)

(13) J.C.Worley,A.B.Smith,M.Kestigian J.Phys.Chem.Sol.31,1857(1970)

LT-17 (Contributed Papers)
U. Eckern, A. Schmid, W. Weber, H. Wühl (eds)

OBSERVATION OF TWO-LEVEL TUNNELING STATES IN AMORPHOUS SILICON BY SURFACE ACOUSTIC WAVES

H. TOKUMOTO, K. KAJIMURA, S. YAMASAKI, and K. TANAKA

Electrotechnical Laboratory, Umezono, Sakuramura, Niiharigun, Ibaraki 305, Japan

Temperature variation of velocity and attenuation of 330 MHz surface acoustic waves (SAW) have been measured in rf-sputtered amorphous silicon (a-Si) films. Evidence for the presence of the two-level states in a-Si is obtained conclusively by two experimental facts: (i) a logarithmic increase of sound velocity; (ii) a shoulder in the temperature dependence of attenuation.

1. INTRODUCTION

At very low temperatures amorphous solids exhibit anomalous behavior in their specific heat ($\sim T$; T is the temperature), the thermal conductivity ($\sim T^2$) and the acoustic wave propagation (a resonance absorption, $\Delta v/v \sim \ln(T)$; v is the sound velocity). These anomalies have been successfully explained by low energy excitations associated with the two-level tunneling states (TLS) which are assumed to exist commonly in amorphous solids (1). It is of particular interest to study whether TLS is also present in amorphous tetrahedral semiconductors a-Ge and a-Si (2). Recent measurements of the acoustic propagation (3) and the thermal conductivity (4) have revealed the existence of TLS with a very small density in a-Ge. In a-Si, however, no direct evidence for the existence of TLS has been reported so far. In this paper we present the first experimental evidence for TLS in a-Si by means of SAW.

2. EXPERIMENTAL

Experiments were carried out on polished 128° Y-cut $LiNbO_3$ SAW devices (35 × 7 mm^2) with interdigital transmitting transducer at the center and two interdigital receiving transducers at both ends spaced 15 mm apart from the center. Amorphous silicon films with a thickness of $\sim$5.7 μm were sputtered from a silicon target in an oxygen-free argon atmosphere of 10^{-2} Torr (background pressure: 10^{-6} Torr) onto a space (10 × 4 mm^2) between a pair of transducers. A space between another pair of transducers on the same substrate was left without a-Si films as a reference SAW path in order to eliminate the contribution of $LiNbO_3$ substrate. We prepared two types of samples: Sample I has never been exposed to air in order to avoid the oxygen contamination; Sample II was kept in air for more than 6 months. Sample I was kept in high-purity nitrogen gas atmosphere and was never exposed to air even during sample mounting onto a cryostat. Variation of SAW velocity and attenuation with temperature was measured below 30 K at an rf frequency of 330 MHz using standard ultrasonic and low temperature techniques.

3. RESULTS AND DISCUSSION

Figure 1 shows the temperature dependence of SAW velocity for two a-Si films I and II. Below 5 K, a clear logarithmic variation of the velocity with temperature was found for both samples. The logarithmic behavior is a specific feature of glasses due to the resonant interaction between acoustic waves and TLS. We obtained the slopes below 5 K as 3.17 × 10^{-6} and 4.65 × 10^{-6} for Sample I and II, respectively, from this plot.

In a film of thickness h, the logarithmic variation of SAW velocity is given by (5),

$$\Delta v/v = (n_0 M^2/\rho v^2) K[1-\exp(-2h/\lambda)] \ln(T/T_0),$$

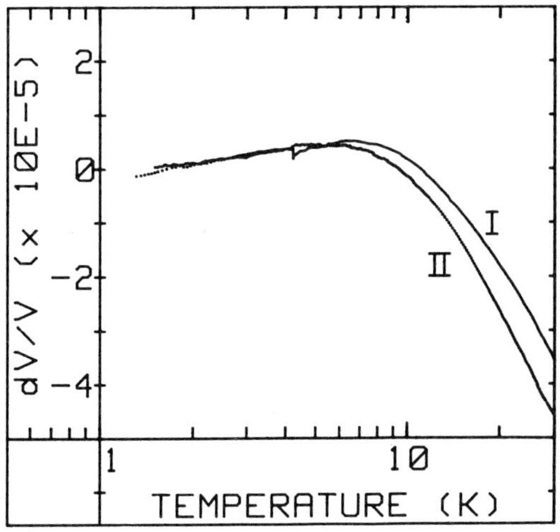

FIGURE 1
Variation of SAW velocity with temperature. Letters I and II attached to lines denote Sample I and II, respectively.

where n_0 is the density of states of TLS which is assumed to be constant for small energy splittings, M their coupling constant to SAW, ρ the mass density, λ SAW wavelength, and T_0 an arbitrary reference temperature. The constant K depends on the elastic properties of the substrate and film and it can be estimated as 0.03 for the present experimental conditions. Taking values of h = 5.7 μm, $v = 4 \times 10^5$ cm/sec, ρ = 2 g/cm^3, λ= 12 μm, the coupling parameter is deduced as $n_0M^2 = 3.51 \times 10^7$ and 5.15×10^7 erg/cm^3 for Sample I and II, respectively, at the energy of 1.1×10^{-18} erg (= 330 MHz), which are comparable with that (1.35×10^7 erg/cm^3) in a-Ge (3). These values are roughly an order of magnitude smaller than that ($\sim 2 \times 10^8$ erg/cm^3) found in a-SiOx and a-GeOx (6).

Oxygen atoms can incoorporate with a-Si and a-Ge and change their properties even in air (7). Moreover, oxygen impurities have been thought to be very effective in forming TLS in a-Si and a-Ge (6). Oxygen content of Sample I, however, is of the order of 10^{19} cm^{-3} which is three orders of magnitude smaller than that in a-SiOx and a-GeOx. If oxygen impurities form TLS, n_0M^2 should be as small as 10^5 erg/cm^3, which is not the present case. In addition Sample I and II give almost the same value of the coupling parameter in spite of the incoorporation of oxygen atoms in air as described above. These experimental facts indicate that the presence of TLS is of intrinsic nature in a-Si.

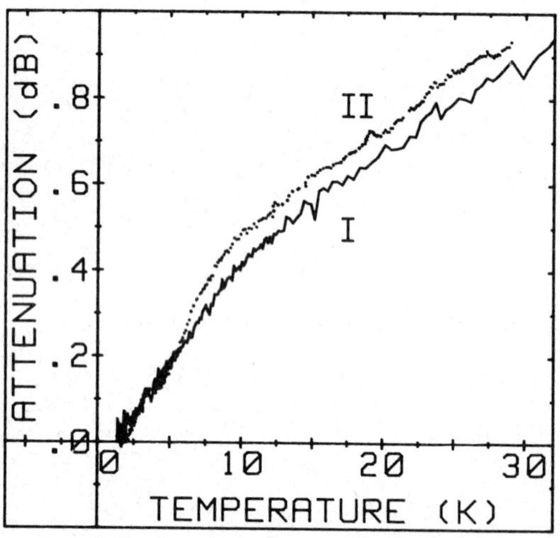

FIGURE 2
Variation of SAW attenuation with temperature. Letters I and II attached to lines denote Sample I and II, respectively.

In Fig. 2, we show the temperature dependence of SAW attenuation for Sample I and II. A shoulder was found near 10 K for both samples. The acoustic waves alter the level splitting of TLS and thus disturb the thermal equilibrium. Then TLS relax to the thermal equilibrium via phonon-assisted tunneling. The relaxation process results in the attenuation peak at $\sim$10 K which is superposed by a higher-temperature attenuation tail. This relaxation attenuation is also sensitive to the presence of oxygen impurities (6). Present samples I and II, however, exhibit the same behavior, indicating that the existence of TLS is intrinsic and is responsible for SAW attenuation in a-Si.

4. CONCLUSIONS

We have found a clear logarithmic variation of SAW velocity and a shoulder in SAW attenuation curve for both air-exposed and non-exposed a-Si films. These results lead to a conclusion that there exists TLS in the four-fold coordinated a-Si and a-Ge with a small density as well as the two-fold coordinated a-SiOx and a-GeOx. The failure of finding TLS in a-Si by Von Haumeder et al. (6) will be a consequence of lack of sensitivity since the film thickness of their sample is much less than SAW wavelength as Duquesne and Bellessa have claimed (3). Our samples with a thickness of $\sim$5.7 μm are thick enough to detect TLS.

REFERENCES

(1) See, for example, Amorphous Solids: Low Temperature Properties, ed. W.A. Phillips (Springer, Berlin, 1981).
(2) See, for example, H.v. Löhneysen, Proc. 10th Intl. Conf. on Amorphous and Liquid Semiconductors, ed. K. Tanaka and T. Shimizu (Tokyo, 1983), J. Non-Cryst. Solids 59&60 (1983) 1087.
(3) J.Y. Duquesne and G. Bellessa, J. Phys. C: Solid St. Phys. 16 (1983) L65.
(4) J.E. Graebner and L.C. Allen, Phys. Rev. Lett. 51 (1983) 1566.
(5) A. Tate, S. Tamura, and T. Sakuma, Solid State Commun. 30 (1979) 517.
(6) M. Von Haumeder, U. Strom, and S. Hunklinger, Phys. Rev. Lett. 44 (1980) 84; K.L. Bhatia and S. Hunklinger, Solid State Commun. 47 (1983) 489.
(7) K. Yokota, T. Kageyama, and S. Katayama, Jpn. J. Appl. Phys. 22 (1983) 370.

LT-17 (Contributed Papers)
U. Eckern, A. Schmid, W. Weber, H. Wühl (eds)
© Elsevier Science Publishers B.V., 1984

QUADRUPOLAR RESPONSE FOR TUNNELING EXCITATION OF OH^- ION IN NaCl

T.GOTO, E.KANDA, H.YAMADA*, S.SUTO*, S.TANAKA*, T.FUJITA* and T.FUJIMURA

Research Institute for Scientific Measurements and Department of Physics*, Tohoku University, Sendai 980, Japan

Elastic constants $(C_{11}-C_{12})/2$, C_{44}, C_{11} of NaCl doped OH^- ions measured with ultrasonic method reveal pronounced softenings below 50 K and show minimum near 0.8 K. These anomalous temperature dependences of elastic constants are accounted for the response of the quadrupolar moment of OH^- tunneling excitation in host NaCl lattice.

1 INTRODUCTION

The elastic behavior of the random system such as glass or amorphous solids in low temperature has been intensively investigated. The elastic softening and ultrasonic absorption below liquid helium temperature, and phonon echo in ultralow temperature are the characteristic phenomena which are not observed in crystalline solids.(1) In order to explain the low energy excitation in amorphous solids, the tunneling two level model has been introduced and succeeded to show elastic behaviors in low temperature.(2)

In the present report, we would like to show the low temperature elastic behaviors of NaCl crystals with the OH^- concentration below 650 ppm. The OH^- ions in the host NaCl lattice are substituted in the Cl^- sites and reveal the off-center instability with the six equivalent potential minimums along the (100) directions. Although the OH^- ions are randomly distributed in the host lattice, the system with low OH^- concentration could be treated by the one-ion problem without any interactions between OH^- ions. This one-ion tunneling excitation model may be regarded as a prototype model of the two level systems in amorphous solids.

In order to observe the excitation states of OH^- ion in NaCl, the various optical experiments have been performed so far. Millimeterwave experiments with the Ledatron spectrometer established the low energy states of OH^- ion as, ground state A_{1g} , exited T_{1u} with the energy of 1.85 cm^{-1} and excited E_g with 3.41 cm^{-1}.(3) The excitation energy levels and their symmerties govern the elastic properties of NaCl-OH^- crystals in low temperature as will be shown in following.

2 EXPERIMENTS AND RESULTS

The NaCl crystals were grown with the Kyropoulos method and the concentration of OH^- ions was determined by the ultraviolet absorption intensity. The ultrasonic experiments have been performed by a phase comparison method with the frequency of 10 MHz.

The temperature dependence of the elastic constants $(C_{11}-C_{12})/2$ and C_{44} of NaCl with 75, 230 and 650 ppm OH^- concentration are shown in Figs. 1 and 2 respectively. The $(C_{11}-C_{12})/2$ and C_{44} modes show the pronounced sofetning below 50 K. On the sample with 230 ppm OH^- concentration, the both elastic modes have minimum points near 0.8 K and slightly increase in the lower temperature below 0.8 K. The C_{11} mode also reveals the softening in low temperature, which causes mainly from the $(C_{11}-C_{12})/2$ mode and little contribution from the bulk modulus $(C_{11}+2C_{12})/3$. It should be noted that the amount of the softening is proportional to the OH^- concentration in the samples. This result suggests that the interaction between OH^- ions has no pronounced contribution on the low concentration samples below 650 ppm in the present experiments.

3 THEORY AND DISCUSSION

The tunneling state of OH^- ion has dipolar and quadrupolar moments, which are explained by the transition propabilites between the states. The quadrupolar moments, which are written with second rank tensors, belong to the same representations A_{1g} , E_g and T_{2g} as the strains. Therefore the interaction between the elastic strain and quadrupolar moment of the OH^- ionic state is possible. This strain - quadrupolar interaction is the microscopic origin of the elastic anomalies of NaCl-OH^- crystals and can be written as,

$$H_{int}=-g_\Gamma Q_\Gamma \varepsilon_\Gamma . \qquad (1)$$

Here Q_Γ is the quadrupolar operator belonging to the same symmetry of the strain and can be explained by the quantum coodinates x,y,z describing the tunneling state of OH^- ion. For example, the quadrupolar operators for $(C_{11}-C_{12})/2$ mode are written as, $Q_{Eg}{}^1=(2z^2-x^2-y^2)/\sqrt{3}$, $Q_{Eg}{}^2=x^2-y^2$. Note that the same interaction is introduced to explaine the elastic behavior of the 4f electronic system of the rare earth compounds.[4]

For the calculation of the transition propabilities $(Q_\Gamma)_{ij}=\langle\psi_i|Q_\Gamma|\psi_j\rangle$ of the quadrupolar operator, we need the eigen state ψ_i of the tunneling states. The calculation for the tunneling state with the off-center potential have been performed with a variation method by Yamada et al. and the resultant eigen function ψ_i, eigen value E_i and the numerical values the quadrupolar moment $(Q_\Gamma)_{ij}$ will be reported elsewhere. The free energy of the tunneling OH^- ions can be written as, $F_{ion}= -KTN\ln(\Sigma\exp(-\beta E_i))$, where N is the number of the OH^- ions in unit volume. The elastic energy for the strain ε_Γ is written as, $F_{lattice}=C_\Gamma^0\varepsilon_\Gamma^2/2$, where C_Γ^0 is the elastic constant without interaction with the OH^- ions. Including the strain-quadrupolar interaction of eq. (2) with the second order perturbation, the temperature dependence of the elastic constants $C_\Gamma(T)$ is obtained by the second derivative for the total free energy $F=F_{ion}+F_{lattice}$.

$$C_\Gamma(T)= \partial^2F/\partial\varepsilon_\Gamma^2=C_\Gamma^0-Ng_\Gamma^2\chi_\Gamma(T). \quad (2)$$

Here $\chi_\Gamma(T)$ is the strain susceptibility defined as,

$$-g_\Gamma^2\chi_\Gamma(T)=\langle\partial^2E_i/\partial\varepsilon_\Gamma^2\rangle -\beta\{\langle(\partial E_i/\partial\varepsilon_\Gamma)^2\rangle-\langle\partial E_i/\partial\varepsilon_\Gamma\rangle^2\}. \quad (3)$$

Here $\langle\ \rangle$ means the thermal average. Note that the first term is the van-Vleck contribution and the second one the Curie term, which are well known in the magnetic susceptibility.

As one can see from the above mentioned treatment, the elastic constants can be understood as the response for the quadrupolar moment of OH^- state. Because we have solved the motion of OH^- ion and obtained the matrix elements $(Q_\Gamma)_{ij}$, it is possible to calculate the temperature dependence of the strain susceptibility of eq. (3). With the coupling parameters $|g_{Eg}|=1.33\times10^4K/A^2\cdot$ion for $(C_{11}-C_{12})/2$ mode and $|g_{T2g}|=1.96\times10^4K/A^2\cdot$ion for C_{44}, we obtained the fitting curves as shown by the solid lines in Figs. 1 and 2. The back ground elastic constant C_Γ^0 is the result of the pure NaCl crystal. The quadrupolar moments have the order of values $\langle Q\rangle\simeq1A^2$, so that the coupling energy per ion is the order of $\langle Q\rangle|g_\Gamma|\simeq10^4K$/ion, which is extremely larger than that of 4f electronic states in rare earth compounds.

The theoretical fitting explains successfully the softening of elastic constants in low temperature. However, the minimum points are predicted to be at 1.1K, which do not agree with the expertimental minimum points near 0.8K. Moreover the enhancement of the elastic constants below 0.8K is smaller than that of the theoretical estimation. Although the origin of the disagreements in very low temperature below 1 K are still somewhat questionable, it might be accounted for the existance of the interaction between OH^- ion, which is excluded in the present treatment.

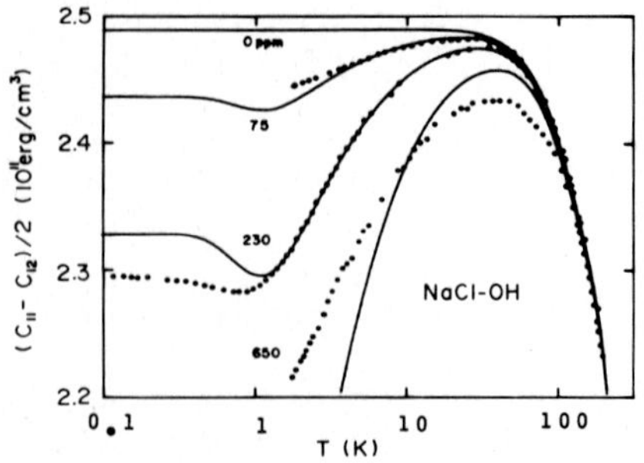

Fig.1 Temperature dependence of the elastic $(C_{11}-C_{12})/2$ mode of $NaCl-OH^-$ crystals.

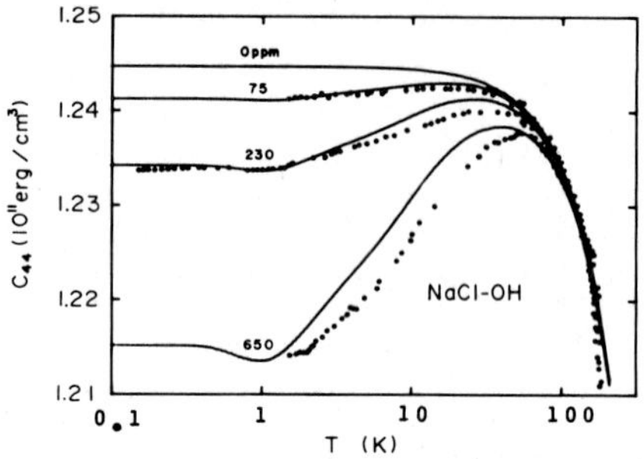

Fig.2 Temperature dependence of the elastic C_{44} mode of $NaCl-OH^-$ crystals.

In summary the elastic softenings of $NaCl-OH^-$ crystals in low temperature are explained by the quadrupolar response for the tunneling OH^- state. We believe that the present approach is also applicatable to the tunneling two level system of amorphous solids.

ACKNOLEDGEMENTS

We would like to thank Prof. M.Ikezawa and Dr. Y.Kayanuma for helpful discussion.

REFERENCES

(1) S.Hunklinger and M.v.Schickfus, Acoustic and dielectric properties of glasses at low temperature, in: Amorphous Solids, eds. W.A.Phillips (Springer, Berlin, 1981) pp. 81-103.
(2) P.W.Anderson, B.I.Halperin and C.M.Varma: Philos. Mag. 25 (1972) 1.
(3) S.Suto and M.Ikezawa : J.Phys. Soc. Japan, 53 (1984) 438.
(4) B.Lüthi : AIP Conf. 34 (1976) 7.

LT-17 (Contributed Papers)
U. Eckern, A. Schmid, W. Weber, H. Wühl (eds)

ULTRASONIC STUDY OF THE ELECTROLYTE-GLASS $LiCl \cdot 7H_2O$

G. KASPER, V. RÖHRING

Institut für Angewandte Physik II der Universität Heidelberg, D-6900 Heidelberg, FRG

Measurements of the ultrasonic absorption and the velocity of longitudinal waves of various frequencies have been carried out in $LiCl \cdot 7H_2O$ in the temperature range from 0.4K to 300K. The results are discussed on the basis of the tunneling model. From the temperature independent plateau of the absorption one can conclude that the density of states is constant up to at least 30K.

It is well established that the low temperature properties of amorphous materials and some highly disordered crystalline solids are governed by the existence of two-level systems (TLS) (1). These systems are assumed to originate from the tunneling motion of small groups of atoms between two nearly equivalent equilibrium configurations corresponding to the minima of asymmetric double-well potentials. The microscopic nature of these tunneling centers, however, is still unknown. The low temperature behaviour of the amorphous solids is to a large extent universal and independent of the chemical nature. Recently it has been proposed in several theoretical and experimental publications that these low temperature properties and the glass transition temperature T_g are related (2).

Concentrated ionic solutions of LiCl in H_2O exhibit a glassy state with a T_g around 140K (3). Since this transition temperature is much lower than that of ordinary glasses, $LiCl \cdot nH_2O$ solutions seem to be good candidates to test those ideas.

We have measured the temperature dependence of the ultrasonic attenuation and the velocity of longitudinal waves in ionic solutions at several frequencies.

Figure 1 shows the temperature dependence of the intensity absorption coefficient α for $LiCl \cdot 7H_2O$. The experimental points are a compilation of several independent runs and have an estimated accuracy of 0.05 cm^{-1}. It should be emphasized that no residual absorption has been subtracted. At the lowest temperatures α rises steeply and then bends to a plateau whose height varies linearly with frequency. Above 15K the absorption rises again leading to a huge viscoelastic peak near T_g, not discussed in this letter.

The low temperature behaviour is characteristic for an attenuation caused by a broad

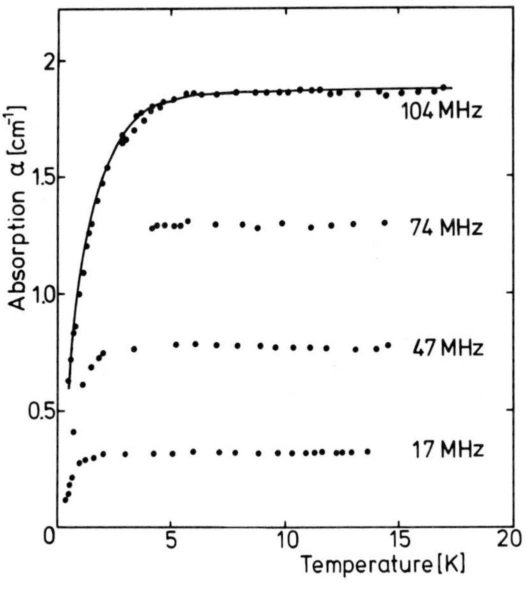

Fig.1: Temperature dependence of ultrasonic absorption in $LiCl \cdot 7H_2O$ glass. The solid line indicates the numerical fit based on the tunneling model.

spectrum of relaxation times (5). At relatively high temperatures in the plateau region we expect an absorption

$$\alpha = \frac{\pi}{2} C \frac{\omega}{v} \qquad (1)$$

with $C = \bar{P}\gamma^2/(\rho v^2)$.

Here $\bar{P}$ denotes the density of states, γ the coupling constant, $\rho = 1.2$ g cm^{-3} the mass density and $v = 4.0 \cdot 10^5$ cm/s the sound velocity.

Eq. 1 is valid in the limit $\omega\tau_m << 1$, where τ_m is the relaxation time of the fastest relaxing TLS of a given energy splitting. From the height of the plateau we obtain $C=(7.2\pm0.7)\cdot10^{-4}$.

An excellent numerical fit (see Fig. 1) of the absorption data below the plateau can be obtained by carrying out the integration of the well-known formula for the absorption of a distribution of tunneling systems (5).

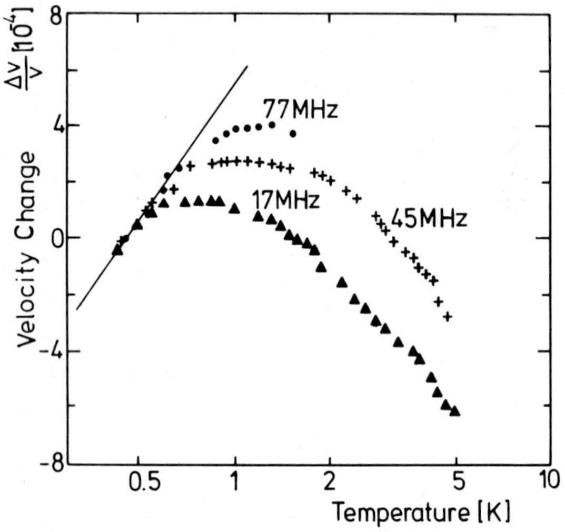

Fig.2: Relative variation of the longitudinal sound velocity in $LiCl\cdot7H_2O$ glass versus temperature, resolution $5\cdot10^{-6}$; straight line see text.

The absorption also leads to a variation of the sound velocity (Fig. 2). At very low temperatures glasses usually show a logarithmic variation of the sound velocity (6):

$$(v(T) - v(T_o))/v(T_o) = C \ln T/T_o \qquad (2)$$

where T_o is some arbitrary reference temperature. The straight line in Fig. 2 is the curve predicted by Eq. 2 using the value $C=7.2\cdot10^{-4}$ deduced from the plateau of the absorption.

Around 1K the velocity passes through a frequency dependent maximum, given by the condition $\omega\tau_m \simeq 1$. If the TLSs relax via one-phonon processes τ_m^{-1} scales with T^3. In this case the temperature T_m of the maximum should vary with frequency as $\omega^{1/3}$. An extrapolation to 20 GHz yields T_m = 7.5K which is in very good agreement with Brioullin scattering experiments carried out recently (7). Above T_m a logarithmic decrease of the sound velocity is expected which is probably masked by phonon processes of higher order for which the relaxation rates are proportional to T^n, where $n > 3$.

If $\bar{P}$ scales with T^{-1} an extrapolation of the data of fluoride glasses (8) to 140K will yield a density of states $\bar{P} = 8\cdot10^{32}\ erg^{-1}cm^{-3}$, a value much higher than reported so far for glasses. It should easily be observed in specific heat measurements.

In summary $LiCl\cdot7H_2O$ is the only glass where the ultrasonic absorption due to TLS is not masked by other absorption mechanisms up to a temperature of 20K. Since the dominant contribution to the absorption is due to TLS having an energy splitting E = 1.5 kT, we conclude from this observation, that the density of states is perfectly constant up to an energy E/k = 30K.

The authors gratefully acknowledge the stimulating discussions with S. Hunklinger.

(1) W.A. Phillips (ed.), Amorphous Solids, Topics in Current Physics (Springer Heidelberg, 1981)
(2) A.K. Raychaudhuri and R.O. Pohl, Phys.Rev. B25 (1982) 1310
M.H. Cohen and G.S. Grest, Sol. State Commun. 39 (1981) 143
(3) C.A. Angell, E.U. Sare, J. Donnella, and D.R. MacFarlane J.Phys.Chem. 85 (1981) 1461
(4) A. Elarby Aouizerat, et al. J. Physique 43 (1982) C9 - 205
(5) J. Jäckle, Z. Physik 257 (1972) 212
(6) S. Hunklinger and M. v. Schickfus in (1), pp. 81 - 105
(7) R. Vacher, M. Schmidt, private communication
(8) P. Doussineau, M. Matecki, and W. Schön, J. Physique 44 (1983) 101

LT-17 (Contributed Papers)
U. Eckern, A. Schmid, W. Weber, H. Wühl (eds)

OPTICAL ABSORPTION BY IMPURITIES IN AMORPHOUS HOSTS

K. KASSNER, P. REINEKER

University of Ulm, 7900 Ulm, Germany

The homogeneous optical linewidth of impurities in amourphous hosts is calculated using two different averaging procedures. It is shown that the usual method, which is not well-founded, yields acceptable results whereas the linewidths of a more plausible method do not agree with experiment. An improvement of the model is suggested.

1. INTRODUCTION AND MODEL

In recent years optical dephasing of guest molecules in glasses has been studied extensively (1). All theoretical descriptions of homogeneous optical linewidths of impurities in glasses, which are typically one to two orders of magnitude larger than in crystals, are based on characteristic degrees of freedom of glasses, the so-called two-level systems (TLSs).

The basic features of our model go back to Lyo and Orbach (2). The impurity is described by its two lowest energy levels E_α, E_β. We allow for an interaction between the impurity and a single TLS. Tunneling transitions between its two levels E_ℓ, E_u are mediated by a matrix element $W = \omega_0\, e^{-\lambda}$. It is assumed that the overlap parameter λ and the so-called asymmetry parameter $\Delta = E_u - E_\ell$ have a broad distribution over which physical properties of the system have to be averaged. The TLSs are coupled to the phonon bath. The Hamiltonian then reads:

$$H = H_1 + H_{12} + H_2 + H_{23} + H_3 \,, \tag{1}$$

$$H_1 = \sum_\rho E_\rho\, |\rho\rangle\langle\rho| \qquad \rho=\alpha,\beta \,, \tag{2}$$

$$H_2 = \sum_r E_r\, |r\rangle\langle r| + \frac{W}{2}\sum_{r\neq t} |r\rangle\langle t| \qquad r,t=u,l \,, \tag{3}$$

$$H_{12} = \sum_{r,\rho} V_{r\rho}\, |r\rangle|\rho\rangle\langle r|\langle\rho| \,, \tag{4}$$

$$H_3 = \sum_{q,s} \omega_{qs}\, b^+_{qs}\, b_{qs} \,, \tag{5}$$

$$H_{23} = \sum_{q,s,r} \frac{1}{\sqrt{N}}\, h^r_{qs}\, (b_{qs} + b^+_{-qs})\, |r\rangle\langle r| \,. \tag{6}$$

H_{12} represents the coupling between optical ion and TLS with coupling strength $V_{r\rho}$. The vibrations of the glass matrix are described by H_3. The b_{qs} are annihilation operators for acoustic phonons with wavenumber q in branch s and energy ω_{qs}. The interaction between the TLS and the phonon bath is given by H_{23}. The coupling matrix elements h^r_{qs} are proportional to the deformation potential.

Applying usual approximations (3), from the Liouville equation we derive equations of motion for the four relevant correlation functions of the dipole moment operator. We perform an analytically exact calculation of the eigenvalues of this system of equations, the real parts of which describe the linewidths and their imaginary parts the line positions.

2. AVERAGING PROCEDURES

Usually, the linewidth is averaged over the distribution of the TLSs (henceforth denoted as procedure no. 1), and the temperature dependence of this average is investigated. To our knowledge, no reasons are given in the literature (for a review see (3)) for this simple procedure. One might argue that some directly measurable quantity like the lineshape should be taken for the averaging (procedure no. 2). We give a tentative justification of procedure no. 1. Interacting with a TLS a state of the impurity decays with a certain relaxation rate. We hypothesize that the relaxation rates induced by the interaction with several TLSs sum up. This sum is the same as an average over a distribution whose normalization constant is given by the number of TLSs within an interaction volume. One must be aware, however, that reasoning this way means leaving the original model. For in the Hamiltonian the impurity is coupled to just a single TLS.

If one chooses to stick to the model, one has to average the lineshape (procedure no. 2) in a manner described below.

In principle, the averaging has to be done over all parameters connected with the TLSs. The averaged linewidth is then given by

$$\langle\Delta\omega\rangle = \int df\, dV_\alpha\, dV_\beta\, d\omega_o\, d\Delta\, d\lambda\; \tilde{P}(f,V_\alpha,V_\beta,\omega_o,\Delta,\lambda)\, \Delta\omega(f,V_\alpha,V_\beta,\omega_o,\Delta,\lambda) \,, \tag{7}$$

where f is a phonon coupling parameter determined by the h^r_{qs} in eq. (6).

For lack of knowledge about the distribution $\tilde{P}$ we suppose that it factorizes:

$$\tilde{P} = P_1(f,V_\alpha,V_\beta,\omega_o)\; P(\Delta,\lambda) \,. \tag{8}$$

It is generally accepted that the second distribution can be taken to be

$$P(\Delta,\lambda) = \begin{cases} P\neq 0 \text{ (constant) for } & \Delta_{min}<\Delta<\Delta_{max} \\ & \lambda_{min}<\lambda<\lambda_{max} \\ 0 \text{ elsewhere} & \end{cases} \tag{9}$$

The remaining parameters are replaced by effective values because P_1 is unknown.

Bearing in mind the hole burning experiment we average the lineshape in the following way

$$\langle g(\omega)\rangle \sim \int d\Delta\, d\lambda\; P(\Delta,\lambda)\; g(\omega,\Delta,\lambda)\; g(\omega_L,\Delta,\lambda). \tag{10}$$

Here $g(\omega,\Delta,\lambda)$ is the lineshape of a single impurity resulting from its coupling to a TLS described by the parameters Δ and λ. The contribution of this impurity to the lineshape of the hole is proportional to the spectral intensity of the single line at the observation frequency ω and - if the short burning time limit holds - to the absorption probability at the laser frequency which, in turn, is proportional to the height of the single line at this frequency.

3. NUMERICAL EVALUATION AND DISCUSSION

Fig. 1 shows the numerically averaged linewidth (according to procedure no. 1) of the most intensive of the four optical lines as a function of temperature for several values of the parameter λ_{min}. The representation is a log-log plot of reduced quantities which are related to the Debye energy. In this diagram a straight line with slope m corresponds to a function proportional to T^m.

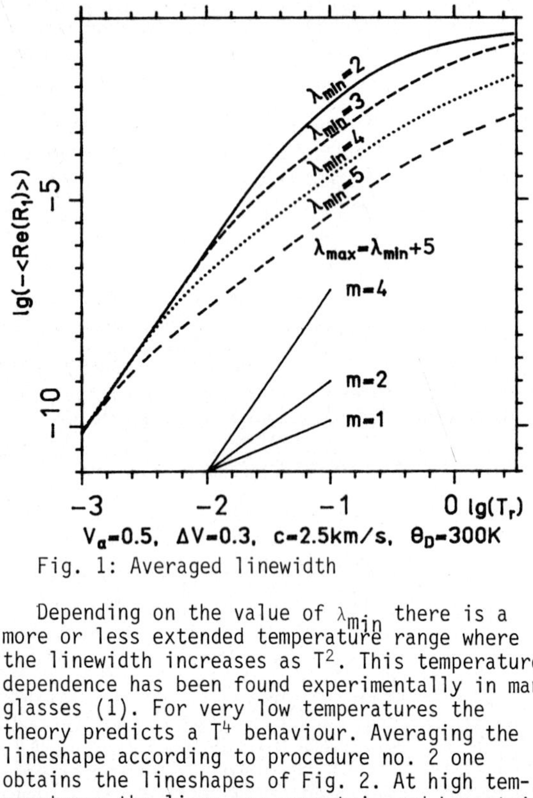

Fig. 1: Averaged linewidth

Depending on the value of λ_{min} there is a more or less extended temperature range where the linewidth increases as T^2. This temperature dependence has been found experimentally in many glasses (1). For very low temperatures the theory predicts a T^4 behaviour. Averaging the lineshape according to procedure no. 2 one obtains the lineshapes of Fig. 2. At high temperatures the lines are symmetric and Lorentzian, but they become more and asymmetric as temperature decreases. The linewidths are by far too small - their order of magnitude is 10^{-8} in units of the Debye frequency. The left and right halfwidths of these lines have been determined numerically. The smaller halfwidth shows a $T^{4.4}$ behaviour, whereas the larger one behaves like a power of T at high temperatures only. When temperature is decreased it goes through a minimum and then increases again.

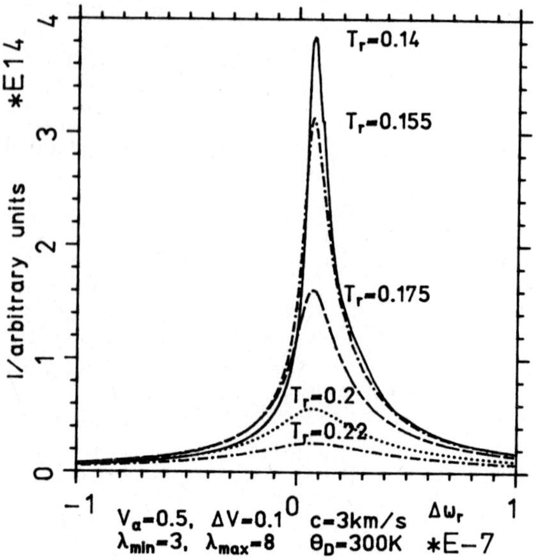

Fig. 2: Averaged lineshape for several temperatures (procedure no. 2)

Summarizing now, it has been shown that within a certain temperature range the usual averaging procedure (no. 1) yields a temperature dependence in agreement with experiments. Up to now, however, this procedure has not been justified consistently within the model. Furthermore, there exists no experimental evidence for the theoretically predicted T^4 behaviour at very low temperatures. On the contrary, if there is a change in the temperature dependence at all with decreasing temperature, it is towards a weaker temperature dependence (4,5).

The averaging of the lineshape, which is consistent within the model, does not explain the experimentally observed linewidths - at least as long as one adheres to parameter values of λ which are not too far from generally accepted ones. As an improvement of the model we suggest a generalized Hamiltonian, which contains the coupling of an impurity to several TLSs.

REFERENCES

(1) J. Friedrich and D. Haarer, Angew. Chem. 96 (1984) 96
(2) S. K. Lyo and R. Orbach, Phys. Rev. B22 (1980) 4223
(3) P. Reineker, H. Morawitz and K. Kassner, Phys. Rev. B (1984) in print
(4) J. Hegarty, M. M. Broer, B. Golding, J. R. Simpson and J. B. MacChesney, Phys. Rev. Lett. 51 (1983) 2033
(5) H. P. H. Thijssen, R. E. van den Berg, and S. Völker, Chem. Phys. Lett. 103 (1983) 23

LT-17 (Contributed Papers)
U. Eckern, A. Schmid, W. Weber, H. Wühl (eds)

OPTICAL DEPHASING BY TUNNELING SYSTEMS IN GLASS

B. GOLDING and M.M. BROER

AT&T Bell Laboratories, Murray Hill, N.J. 07974

A description of optical dephasing in inorganic glasses is presented which implicates low-energy atomic tunneling systems as the dominant contributor below 1K. Photon echoes have shown that the dephasing rate of Nd^{3+} in pure SiO_2 varies with temperature as $T^{1.3}$ between 0.1 and 1K. The behavior of the photon echoes and of the dephasing rate is quantitatively interpreted in terms of a spectral diffusion process which originates in the elastic dipolar coupling between active ions and tunneling centers.

Optical dephasing of paramagnetic ions and molecules in organic and inorganic amorphous solids at low temperatures ($\lesssim$100K) is characterized by unusual temperature dependences and enhanced relaxation rates compared to crystalline systems (1). For a large variety of impurity-host combinations the dephasing rate has been found to follow a T^m temperature dependence with m approximately 2. Several theories have attempted to explain this behavior by assuming interactions between the optical center, phonons, and atomic tunneling systems. However, the role of the intrinsic disorder in optical dephasing in these systems is not clearly understood.

We have studied (2) the dephasing of the $^4F_{3/2}(1) \rightarrow {}^4I_{9/2}(1)$ transition of Nd^{3+} doped SiO_2 glass between 0.05 and 1K. The dephasing rate T_2^{-1} is measured with two-pulse photon echoes in a novel geometry. The Nd^{3+} ions are imbedded in the SiO_2 core of a single mode optical fiber. This method has the intrinsic advantages of long interaction lengths, phase-matching and efficient heatsinking. The echo decays exponentially with a characteristic time constant related to T_2. The rate T_2^{-1} obeys a $T^{1.3}$ dependence between 0.1 and 1K as is shown in Fig.1. This figure also shows an example of higher temperature homogeneous linewidth data (3) of the same transition of Nd^{3+} doped in a borosilicate glass which shows the typical T^2 dependence above $\approx$20K. The echo results indicate that this T^2 dependence does not persist to low temperatures and constitute the first evidence for a different temperature dependence of the dephasing rate other than that at higher temperatures. This indicates the existence of a crossover in the temperature dependence above 1K.

The photon echo results show a striking similarity to the homogeneous linewidths of tunneling systems, as studied earlier with microwave acoustic methods(4) in silica based glasses. As can be seen in Fig. 2 , both the magnitude and the temperature dependence are comparable below 1K. A low tem-

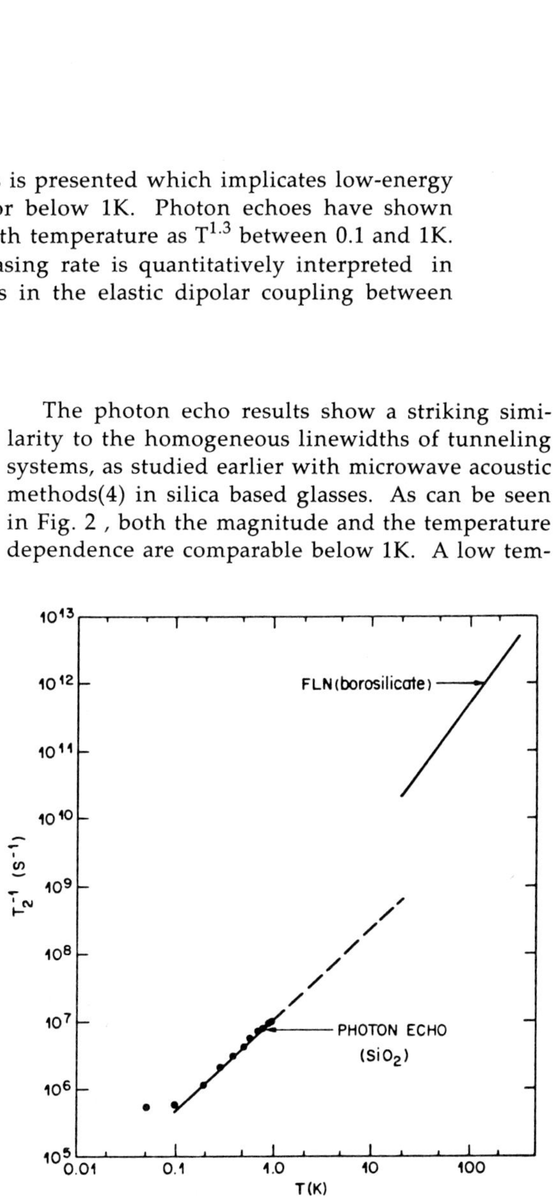

FIGURE 1

Optical dephasing rates for Nd^{3+} for over 3 decades in temperature. The high temperature results are derived from fluorescence line narrowing (FLN) spectroscopy(3) whereas the rates below 1K are obtained from two-pulse photon echoes. Note the high temperature T^2 behavior and the low temperature $T^{1.3}$ dependence.

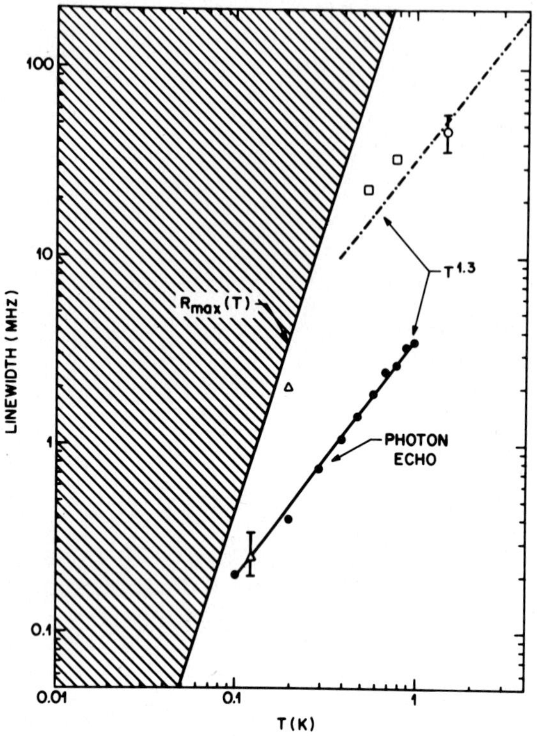

FIGURE 2

Linewidths at low temperature for (a) photon echoes resulting from 0.89μm excitation of Nd^{3+} in SiO_2 glass (solid circles) and (b) hole burning by microwave phonons of the atomic tunneling systems in silicate glasses (open symbols). Note the similar temperature dependence of both processes and the coincident widths at 0.1K. R_{max} is the maximum one phonon T_1^{-1} of tunneling systems and illustrates that measurements to the right of the boundary occur in a long-time spectral diffusion regime.

perature model (<10K) of dephasing has been developed(5) based on these similarities which invokes elastic coupling between the ion and the tunneling systems. This model assumes spectral diffusion among the tunneling systems to play a key role in the optical dephasing at these temperatures. A quantitative analysis of the echo decay for arbitrary multipolar interactions indicates that an elastic dipole-dipole coupling is consistent with the observed exponential decay and $T^{1.3}$ dependence in the long-time regime(5). Although this theory can explain the anomalous T^2 behavior at higher temperatures if higher order interactions are invoked, the contributions of tunneling systems to the linewidth in this regime are too weak to account for the observed linewidths. This is consistent with the suggested crossover and implicates other dephasing processes(6) as the dominant contributor to T^2 linewidths.

With these very low temperature photon echo results and the resemblance between the optical and acoustic dephasing rates we have provided the first quantitative evaluation of the role of tunneling systems in optical dephasing in an inorganic glass at low temperatures.

REFERENCES

1. P.M. Selzer, D.L. Huber, D.S. Hamilton, W.M. Yen, and M.J. Weber, Phys. Rev. Letters *36,* 813(1976)
2. J. Hegarty, M.M. Broer, B. Golding, J.R. Simpson, and J.B. MacChesney, Phys. Rev. Letters *51,* 2033(1983)
3. J.M. Pellegrino, W.M. Yen, and M.J. Weber, J. Appl. Phys. *51,* 6332(1980)
4. B. Golding and J.E. Graebner, Phys. Rev. Letters *37,* 852(1976), and in *Amorphous Solids,* edited by W.A. Phillips (Springer-Verlag, Berlin, 1981), Chap. 7
5. D.L. Huber, M.M. Broer, B. Golding, (to be published).
6. D.L. Huber, J. Non-Cryst. Solids, *51,* 241(1982).

LT-17 (Contributed Papers)
U. Eckern, A. Schmid, W. Weber, H. Wühl (eds)

MOLECULAR HYDROGEN IN a-Si:H

J. E. GRAEBNER, L. C. ALLEN, and B. GOLDING

AT&T Bell Laboratories, Murray Hill, New Jersey 07974, USA

Thermal measurements on a-Si:H for $0.05 < T < 5K$ show an enhanced specific heat and a slow release of heat, both of which are attributed to molecular hydrogen trapped in microvoids. These phenomena are evidence of a residual orientational ordering of ortho-H_2 ($J = 1$) and a spontaneous ortho-para conversion, respectively. The effects are largest for an annealing temperature of 500°C, which is high enough to break SiH bonds to produce H_2 but low enough to prevent escape of H_2 by diffusion.

1. INTRODUCTION

The low-temperature properties of the tetrahedrally-bonded amorphous semiconductors a-Ge and a-Si have been under intense scrutiny recently. A fundamental question is whether fourfold coordination permits the existence of the localized, two-level atomic tunneling systems which are found (1) at low temperatures in most glasses of lower coordination. Previous low-temperature investigations (2) of a-Ge and a-Si have not addressed the sample-dependent problems of microvoids and deviations from fourfold coordination (3). Thermal conductivity measurements (4) on carefully characterized samples, however, have established a correlation between the number of tunneling systems and the amount of microstructure present.

In the case of *hydrogenated* a-Si, the presence of a small amount of molecular H_2 has been inferred (5) from nmr measurements. Recently, a large and time-dependent specific heat C has been observed (6) below 5K which is qualitatively different from a typical tunneling-system contribution. This anomaly is accompanied (6,7) by a very slow release of heat $\dot{Q}$ from the sample with a time dependence indicative of a bimolecular process. Both $\dot{Q}$ and C decrease slowly over a period of many days after cooling to liquid helium temperatures. The heat is attributed to the spontaneous conversion of molecular H_2 from the ortho ($J = 1$) to the para ($J = 0$) state which proceeds (8) via the nuclear magnetic dipole interaction of two neighboring ortho molecules. The anomaly in C is attributed (6) to a tendency toward orientational ordering of ortho-H_2 due to an electric quadrupole interaction (8). The anomalous C and $\dot{Q}$ are found to depend sensitively on heat treatment, and in the present report we give a more detailed account of that dependence and the physical picture to which it leads.

2. RESULTS

The samples were prepared by plasma decomposition of silane with a substrate temperature of 230°C. The thermal measurements were performed with a thermal relaxation calorimeter. Further details are given elsewhere (6). The time dependence of $\dot{Q}$ was modeled by assuming a bimolecular ortho-para conversion described by

$$\dot{x} = -\alpha x^2, \tag{1}$$

where x is the fractional population of o-H_2 relative to the total H_2. By writing $\dot{Q} = -L\dot{x}N$, where N is the total number of *interacting* H_2 molecules per cm^3 and L is the energy ($k_B \times 170K$) of ortho-para conversion in solid hydrogen, one can fit the solution of Eq. 1 to $\dot{Q}(t)$ and determine both N and α. The results are plotted in Fig. 1. For comparison, we include (9,10) the rate of evolution $\dot{n}_H$ of hydrogen gas from similar samples during a slow heating, as well as the average density of spins n_s.

3. DISCUSSION

The evolution of hydrogen has been correlated with changes in the infrared absorption spectrum (9). The broad peak centered around 600°C is attributed to the breaking of SiH or SiH_2 bonds at sites located within the bulk of the sample, resulting in the formation of H_2. The dangling bonds left behind are detected as a rise in the spin density.

We interpret the peak in molecular hydrogen N, occuring at $T_A = 500°C$, as the result of a competition between the production of H_2 and its escape from the sample. At lower temperatures, less H_2 is produced, while at higher temperatures it escapes easily by diffusion. If the H_2 molecules were randomly located throughout the material, the average spacing between molecules would be ~20Å, which is far too large for the usual ortho-para conversion mechanism. We conclude that the H_2 is clustered. The fact that the ortho-para decay, as monitored by $\dot{Q}$, is bimolecular (6) for times up to 300 h or x as low as 0.1, indicates that the decaying H_2 is located in clusters containing initially a *minimum* of 10 [$= (x_{final})^{-1}$] H_2 molecules. The indirect measurement of Ref. 5 suggesting bimolecular decay to $x = 0.044$ implies a minimum cluster size of 25 H_2. These clusters could easily be located within the microvoids, which have typical dimensions on the order of

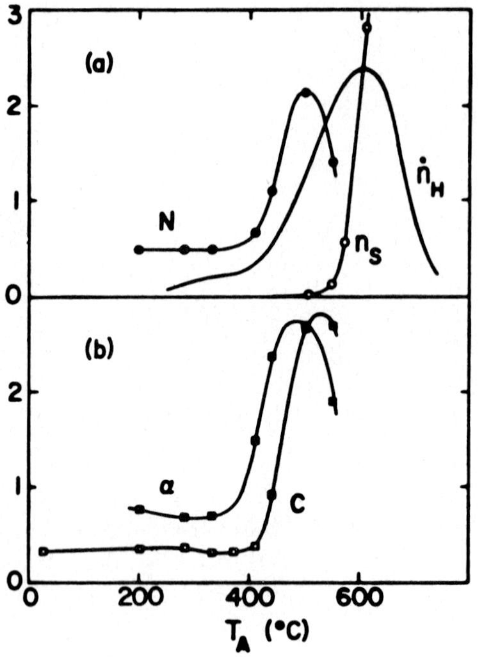

FIGURE 1
(a) Number N of interacting H_2 molecules per cm^3 in a-Si:H after a 15-minute anneal at temperature T_A. Also shown (from Refs. 9,10) is the density of spins n_s produced as hydrogen gas is evolved from the sample at a rate $\dot{n}_H$. (b) Bimolecular conversion rate α and heat capacity C (T = 1K) vs. T_A. Units are N ($10^{20}cm^{-3}$), n_s ($10^{17}cm^{-3}$), $\dot{n}_H$ (arbitrary units), α (% h^{-1}), and C (2×10^{-5}J/K per gram of a-Si:H).

tens of Angstroms. The nn distance in fcc solid H_2 at zero pressure is 3.8Å.

The ortho-para conversion rate constant α in solid hydrogen is sensitive to both the inter-molecular separation and the phonon spectrum. For zero-pressure solid hydrogen, $\alpha = 0.019\ h^{-1}$, increasing with increasing pressure. The fact that α in a-Si:H (Fig. 1b) at low T_A is one third the rate of solid hydrogen may be due to a somewhat (20%) larger nn separation because of interactions with the surrounding a-Si, and/or it may be due to a lower density of states for phonons in the cluster than in bulk hydrogen. The peak in α, which coincides with the peak in N, suggests a higher effective pressure of H_2. If the clustered H_2 for $T_A = 500°C$ is viewed as pressurized solid H_2 with a phonon spectrum equal to that in solid H_2, the conversion rate indicates (8) a density of 1.16 times the zero-pressure density, or an effective pressure of ~ 1 kbar.

The heat capacity at a representative temperature of 1 K (Fig. 1b) is clearly dominated by the presence of molecular hydrogen, as shown by the correlation between C and N. The temperature and time dependence of C, discussed in more detail elsewhere (6), also show a dominance by clusters of ortho-H_2 and suggest that the hydrogen undergoes a broadened phase transition to an orientationally ordered state (8).

ACKNOWLEDGEMENTS
We thank D. K. Biegelsen and M. Stutzmann for providing and characterizing the samples and for useful discussions.

REFERENCES

1. Amorphous Solids: Low-Temperature Properties, ed. W. A. Phillips (Springer-Verlag, Berlin, 1981).
2. See, for example, H. v. Löhneysen and F. Steglich, Phys. Rev. Lett. 39 (1977) 1420; M. von Haumeder, U. Strom, and S. Hunklinger, Phys. Rev. Lett. 44 (1980) 84; J. Y. Duquesne and G. Bellessa, J. Phys. C. : Solid State Phys. 16 (1983) L65.
3. J. C. Knights in: Amorphous and Liquid Semiconductors, eds. W. Paul and M. Kastner (North-Holland, Amsterdam, 1980) p. 159.
4. J. E. Graebner and L. C. Allen, Phys. Rev. Lett. 51 (1983) 1566, and Phys. Rev. B29 (May, 1984).
5. M. S. Conradi and R. E. Norberg, Phys. Rev. B24 (1981) 2285, and W. E. Carlos and P. C. Taylor, Phys. Rev. B25 (1982) 1435.
6. J. E. Graebner, B. Golding, L. C. Allen, D. K. Biegelsen, and M. Stutzmann, Phys. Rev. Lett. 52 (1984) 553, and in print.
7. Heat released from a-Si:H was also observed by H. v. Löhneysen, H. J. Schink, and W. Beyer, Phys. Rev. Lett. 52 (1984) 549, but they were unable to distinguish between a bimolecular and an exponential decay.
8. For a review, see I. F. Silvera, Rev. Mod. Phys. 52 (1980) 393.
9. D. K. Biegelsen, R. A. Street, C. C. Tsai, and J. C. Knights, in: Amorphous and Liquid Semiconductors, eds. W. Paul and M. Kastner (North-Holland, Amsterdam, 1980) p. 285.
10. M. Stutzmann and D. K. Biegelsen, Phys. Rev. B28 (1983) 6256.

LT-17 (Contributed Papers)
U. Eckern, A. Schmid, W. Weber, H. Wühl (eds)

SPECIFIC HEAT OF MOLECULAR SOLID HYDROGEN IN a-Si:H

H.v.LÖHNEYSEN, H.J. SCHINK

2. Physikalisches Institut der Rheinisch-Westfälischen Technischen Hochschule, D-5100 Aachen, and Sonderforschungsbereich 125 Aachen-Jülich-Köln, West Germany

W. BEYER

Institut für Grenzflächenforschung und Vakuumphysik der KFA Jülich, D-5170 Jülich, West Germany

The specific heat C of an amorphous Si:H film containing molecular hydrogen has been measured between 0.3 K and 2.5 K. While C of the film as deposited at 250°C (with 0.15 mol%H_2) resembles the behavior known from tunneling states in amorphous solids, annealing at 400°C leads to an increase of the H_2 concentration to 0.4 mol% and to an important enhancement of C. This large specific heat is attributed to orientational ordering of ortho-H_2 molecules in H_2 clusters.

The presence of molecular hydrogen in amorphous silicon prepared by glow-discharge of silane (1,2) has opened the possibility to study small solid hydrogen clusters in a restricted geometry provided by the a-Si:H. It has been found that the concentration of H_2 in a-Si:H can be varied easily over two orders of magnitude, i.e. between $2\cdot10^{18}cm^{-3}$ and $2\cdot10^{20}cm^{-3}$ (1) depending on deposition and annealing temperatures. This corresponds to 0.006 mol% and 0.4 mol% H_2 with respect to Si. In this paper we present specific heat results of a film prepared at 250°C (with 0.15 mol%H_2) and annealed at 400°C (with 0.4 mol%H_2). We find a large difference in the specific heats which we attribute to different average H_2 cluster sizes present in these films.

Fig. 1 shows the specific heat C of the as-prepared and annealed film. The film investigated here is sample 2 of Ref.(1). C for the as-prepared film can be represented by $C = aT+bT^3$ with $a = 2.0\cdot10^{-6}J/gK^2$ and $b = 1.5\cdot10^{-6}J/gK^4$ cf. solid line in Fig.1, although the large scatter in the data precludes a definite fit. The specific heat of the annealed film is largely enhanced. It shows a T^2 dependence below 0.8 K and levels off above that temperature. Even at the highest temperature (2.5 K), C of both films considerably exceeds the Debye contribution C_D.

Because of the heat release associated with the ortho-para conversion of H_2, the lowest temperature attained after 30 h was only 261 mK for the as-prepared film and 360 mK for the annealed film, even though the mixing chamber of our dilution refrigerator was at ∿30 mK (1). In fact, this (time-dependent) heat release established the presence of H_2 in a-Si:H (1,2).

The interpretation of the specific heat results is not straightforward because three possible contributions to C of a-Si:H must be considered: (i) by two-level tunneling states (TLS) akin to the amorphous state (4), (ii) by magnetic excitations arising from dangling bonds (5), (iii) by orientational ordering of o-H_2 molecules (6) in solid hydrogen clusters within the a-Si:H (2). We will treat these possible contributions in turn.

(i) From specific heat measurements of evaporated a-Ge without hydrogen it was concluded (5) that the density of TLS is small compared to that in vitreous silica. For an a-Ge film annealed at 350°C an upper limit of the corresponding linear specific heat coefficient, $a_{TLS} < 2\cdot10^{-7}J/gK^2$, was obtained. However, the structure of H-rich a-Si:H is quite different from that of a random fourfold coordinated network and may well support TLS (7). We note that the linear specific heat coefficient a observed in the as-prepared a-Si:H sample has the same magnitude as that of amorphous solids. Very recently, the specific heat of a-Si:H deposited at 120-180°C was reported (8) which shows a behavior quite similar to our as-prepared sample and which was attributed to TLS. Concerning the possible phonon scattering by TLS, contradictory results have been reported. Ultrasonic measurements performed on a-Si:H films showed no indication of TLS (9) while the thermal conductivity is compatible with the presence of TLS (8). In any case, while there might be a substantial contribution of TLS to C in the as-prepared sample, this contribution is negligible in the annealed sample with its much higher heat capacity (if anything, the TLS contribution to C generally decreases upon annealing).

(ii) A contribution to C of magnetic excitations arising from dangling bonds is possible

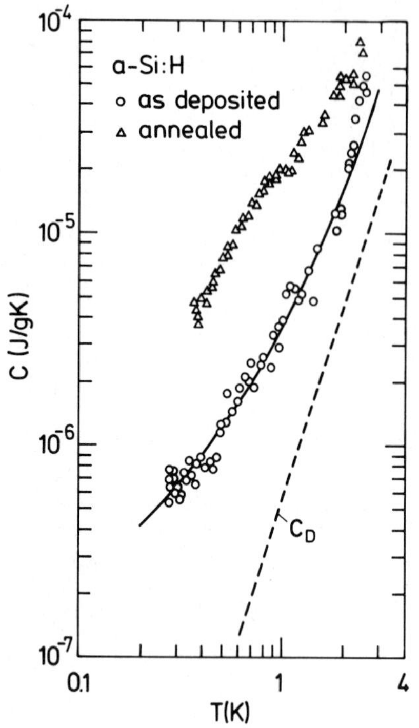

FIGURE 1
Specific heat of a-Si:H as deposited at 250°C and after annealing at 400°C for 10 min. C_D is the Debye contribution estimated from Brillouin scattering measurements with a similarly prepared a-Si:H sample (3).

in room-temperature deposited a-Si:H since there the spin density is rather high (7). Indeed, the specific heat of such a sample was found to decrease upon annealing in agreement with expectation (7). In the present sample which was deposited at elevated temperature, the spin density is more than an order of magnitude smaller, $N_S = 2.6\cdot10^{17} cm^{-3}$. The calculated entropy $S = (N_S/\rho)k_B \ln 2 = 1.2\cdot10^{-6}$J/gK is much smaller than the entropy change in our temperature range (0.25 to 3 K), i.e. $\Delta S \sim 6\cdot10^{-6}$J/gK if only the linear specific heat contribution of the as-prepared sample is taken into account. We mention that this argument has been used before (8) to rule out a spin contribution in a-Si:H prepared at elevated temperature. Of course, the argument applies a forteriori to the annealed sample.

(iii) In the light of the preceeding discussion, the specific heat of the annealed sample must be attributed largely to solid hydrogen clusters, i.e. the freezing of orientational degrees of freedom. Since C was measured 30 h after the initial cool down, the o-H_2 concentration is between $x_o = 0.49$ and 0.61 depending (1,2) on the conversion rate which was not measured for this particular sample. C of the annealed a-Si:H sample is about ten times smaller than expected for bulk solid hydrogen with roughly the same x_o (10). Incidentally, this is just the x_o range below which a quadrupolar glass phase without long range orientational orders occurs at very low temperatures (6). Apparently, the orientational specific heat is strongly reduced when decreasing the solid hydrogen "sample" dimensions. This conjecture is supported by comparing our two samples. A contribution by H_2 must be inferred for the as-prepared film, too, from the observation of a weak time dependence of C after prolonged waiting at low temperatures (2). Even if C of the as-prepared sample is entirely due to H_2 (a TLS contribution is also possible as discussed above), it is much smaller below 1 K (by a factor of 5) than the decrease of the H_2 concentration (with respect to the annealed sample). A smaller average cluster size in the as-prepared sample is suggested by the smaller conversion rate compared to bulk and also to a well annealed sample (1,2,8).

In conclusion, the specific heat of the as-prepared a-Si:H film indicates contributions by TLS and molecular hydrogen. The latter contribution is largely enhanced in the annealed film. This enhancement is attributed to the growth of H_2 clusters upon annealing.

REFERENCES
(1) H.v.Löhneysen, H.J. Schink, and W. Beyer, Phys. Rev. Lett. 52 (1984) 549
(2) J.E. Graebner, B. Golding, L.C. Allen, D.K. Biegelsen, and M. Stutzmann, Phys. Rev. Lett. 52 (1984) 553
(3) M. Grimsditch, W. Senn, G. Winterling, and M.H. Brodsky, Solid State Commun. 26 (1978) 229
(4) See e.g. W.A. Phillips (ed.), Amorphous Solids: Low Temperature Properties (Springer, Berlin, 1981)
(5) H.v.Löhneysen and H.J. Schink, Phys. Rev. Lett. 48 (1982) 1121
(6) See e.g. I.F. Silvera, Rev. Mod. Phys. 52 (1980) 393
(7) H.v.Löhneysen, J. Non-Cryst. Solids 59-60 (1983) 1087
(8) J.E. Graebner, B. Golding, L.C. Allen, J.C. Knights, and D.K. Biegelsen, Phys. Rev. B, in print
(9) K.L. Bhatia, M.v.Haumeder, and S. Hunklinger, J. Physique Colloq. 42 (1981) C4-365; Solid State Commun. 37 (1981) 943
(10) D.G. Haase and A.M. Saleh, Physica 107 B, (1981) 191

LT-17 (Contributed Papers)
U. Eckern, A. Schmid, W. Weber, H. Wühl (eds)

THE NBS TEMPERATURE SCALE IN THE RANGE 15 TO 200 mK

J. H. COLWELL, W. E. FOGLE, and R. J. SOULEN, Jr.

Center for Basic Standards, National Bureau of Standards, Washington, DC 20234

We have studied the reproducibility upon thermal cycling of several types of thermometers. A Josephson junction noise thermometer, a CMN thermometer, and an SRM 768 superconductive fixed-point device were very consistent, while a germanium and a carbon resistance thermometer showed significant irreproducibility.

1. INTRODUCTION

Several years ago a nuclear orientation thermometer and a noise thermometer were compared simultaneously in this laboratory leading to the establishment of the temperature scale NBS-CTS-1 that covered the range 10.5 to 520 mK (1). The inaccuracy of this scale was estimated to be less than 0.5% at the lower end and 0.2% at the upper end. This scale was maintained on a particular germanium resistance thermometer (GeRT), Ser. No. 1405, and the original data were smoothed by magnetic thermometry using a CMN single-crystal sphere. GeRT 1405 was subsequently used to assign temperatures to the superconducting transitions of the W, Be, Ir, $AuAl_2$, and $AuIn_2$ samples in the SRM 768 fixed-point devices produced by NBS (2).

It was soon found in calibrating SRM 768 devices that the reproducibility of the transition temperatures from run to run was even less than the stated accuracy of the NBS-CTS-1 scale. Either the superconducting transitions were not reproducible or the GeRT was unstable, the latter being suspected (2).

2. EXPERIMENTAL APPARATUS

Over the last two years a series of experiments has been conducted which clarifies this situation. The experiments incorporated a noise thermometer, an SRM 768 device, GeRT 1405, a carbon thermometer, and a CMN magnetic thermometer - all mounted on a copper platform suspended below a dilution refrigerator. In detail, the thermometric devices were:

2.1 R-SQUID Noise Thermometer. The noise thermometer is similar to the one used in deriving NBS-CTS-1 but contains several improvements (3). A faster data acquisition system helps to reduce the statistical imprecision. More importantly, a greater theoretical understanding of the device has led to improved technique and confidence. Temperatures can now be measured with an imprecision of 0.03% and an inaccuracy estimated to be 0.1%.

2.2. SRM 768, Ser. No. 7. This particular device is our laboratory reference standard and has been incorporated in all our calibration runs of other SRM 768 devices. The superconducting samples are enclosed in a set of compensated mutual inductance coils. Detection of the superconductive to normal phase change was effected by means of a conventional bridge. The imprecision of a measurement was typically a few μK.

2.3. CMN Thermometer. Powdered cerium magnesium nitrate was mixed with N-grease, gold-plated Cu wires and shaped into a sphere. A set of compensated Cu secondary coils mounted on a Cu primary coil surrounded the CMN sample. A conventional Hartshorn bridge was used to measure the ac susceptibility. The imprecision was approximately 10 μK at 10 mK and 50 μK at 200 mK.

2.4. GeRT Ser. No. 1405. The GeRT was enclosed in a gold-plated Cu pod with the current and voltage leads wrapped around the pod. Its resistance is ∿100 kΩ at 15 mK and ∿40 Ω at 4.2 K. The resistance was measured with an ac Mueller bridge driven at 100 Hz. Resolution is approximately 10 μK over the range 15 to 200 mK.

2.5. Carbon Resistance Thermometer. A thin (0.1 mm) wafer cut from a Speer 100-Ω resistor was sandwiched in epoxy and cigarette paper and clamped in a gold-plated Cu pod. The thermometer, C-G, has a resistance of ∿80 kΩ at 15 mK and ∿1 kΩ at 4.2 K. Resolution is approximately 10 μK at 15 mK and 30 μK at 200 mK.

In the series of runs described here (a numbered run identifies a complete cooling cycle from room temperature), both the CMN and noise thermometers did not participate in all the runs but there is considerable overlap.

3. DATA

In Table I we present the measurements of the transition temperatures (T_c's) of the five samples in the SRM 768 using the R-SQUID.

The numbers in parentheses in the body of the table are one-sigma values from counting statistics. At the bottom of each column the average temperature and its standard deviation expressed in mK and as a percentage are given. In spite of the fact that a different adjustment of the Josephson junction point contact was made

TABLE I
Noise Thermometer Measurements of the Superconductive T_c's in SRM 768, Ser. No. 7

Run#	W (mK)	Be (mK)	Ir (mK)	$AuAl_2$ (mK)	$AuIn_2$ (mK)
11	15.514(0.01)	22.642(.02)	---	159.79(.35)	203.96(.46)
12	15.574(.03)	22.653(.02)	---	159.76(.11)	---
13	15.538(.01)	22.678(.03)	99.251(.10)	159.90(.21)	204.05(.13)
14	15.599(.04)	---	---	---	---
15	---	---	---	---	---
16	15.538(.02)	22.666(.02)	99.239(.07)	159.64(.12)	204.03(.13)
17	---	---	---	---	---
18	---	---	---	---	---
19	15.534(.01)	22.593(.02)	---	---	---
20	---	22.654(.02)	99.203(.07)	160.00(.13)	204.00(.10)
21	---	22.724(.02)	---	---	---
22	15.584(.03)	22.768(.02)	99.249(.08)	159.84(.08)	204.02(.10)
	15.544(.029) (0.19%)	22.672(.053) (0.23%)	99.236(.022) (0.022%)	159.82(.12) (0.08%)	204.01(.03) (0.02%)

after each cooldown, the overall reproducibility of the three highest T_c's is better than 0.1%. For the two lowest T_c's, the reproducibility increases to 0.2%.

The comparison between the SRM 768 unit and the CMN salt is shown in Table II. The data for the eight runs are represented by the equation

$$M = \frac{7385.}{T+0.45} + M_0$$

where T is in mK and M is in μH. The coil constant, M_0, was found to vary slightly on each cooldown. The numbers in parentheses in Table II have the same meaning as those at the bottom of Table I and are similar. There appears to be no correlation, however, between the data in the two tables indicating that the scatter is random.

In the experiments reported here, the two resistance thermometers did show irreproducibilities in comparisons with the SRM 768 device. At the lowest temperatures, the resistance of 1405 approaches equilibrium exponentially after the initial cooldown with a relaxation time of 30 to 40 hours. At the T_c of W this drift is approximately 1 mK. At the three highest T_c's, 1405 is stable once cold, but exhibits variations from run to run of as much as 0.5 mK. We believe that these instabilities in 1405 introduced errors in the calibrations of SRM 768 devices. The results given in Tables I and II indicate that SRM 768, Ser. No. 7 may be used to correct all previous calibrations.

The carbon thermometer, C-G, was much more stable at low temperatures, having little if any time dependence. At higher temperatures it also showed a run-to-run instability comparable to that of 1405 and, in addition, exhibited small variations within runs where the resistance at a given temperature depended on the thermal history of the resistor.

TABLE II.
Magnetic Determination of T_c's in SRM 768, S/N 7

Run#	W (mK)	Be (mK)	Ir (mK)	$AuAl_2$ (mK)	$AuIn_2$ (mK)
12	15.60	22.66	99.17	159.85	204.23
13	15.54	22.60	99.21	159.78	204.12
14	15.57	22.66	---	159.82	203.95
15	15.57	22.60	99.26	159.78	203.89
16	15.51	22.66	99.26	159.75	203.95
17	15.63	22.66	---	---	---
18	15.60	22.69	99.19	159.92	204.00
19	15.63	22.73	99.21	159.85	---
	15.58 (.042) (0.27%)	22.66 (.043) (0.19%)	99.22 (.037) (0.037%)	159.82 (.058) (0.036%)	204.02 (.127) (0.06%)

4. CONCLUSION

A noise thermometer, a CMN thermometer and an SRM 768 device have been shown to be consistent to 0.2% at the two lowest T_c's and consistent to 0.1% at the three higher T_c's. A GeRT and a carbon thermometer were found to be significantly less stable. The practice of thermometry to the 0.1% level seems to be at hand for this temperature region if the proper thermometric devices are used.

5. REFERENCES

(1) R.J. Soulen, Jr. and H. Marshak, Cryogenics 20 (1980) 408.
(2) R.J. Soulen, Jr., Physica 109 & 110B (1982) 2020.
(3) R.J. Soulen, Jr., D. VanVechten, and H. Seppa, Rev. Sci. Instrum. 53 (1982) 1355.

LT-17 (Contributed Papers)
U. Eckern, A. Schmid, W. Weber, H. Wühl (eds)

A DIELECTRIC CONSTANT GAS THERMOMETER

K. GROHMANN and H. KOCH

Physikalisch-Technische Bundesanstalt, Institut Berlin, D-1000 Berlin 10, Germany

As has been shown by Gugan the conventional gas thermometry may be improved successfully by substituting its troublesome volume or density determination by dielectric constant determination of the measuring gas. This paper describes our new apparatus and reports on the first results of a test run which confirms the high efficiency of this new method.

1. INTRODUCTION

In 1980 Gugan and Michel (1) reported on their experimental results of Dielectric Constant Gas Thermometry (DCGT) which proved to be a powerful alternative to the conventional Constant Volume Gas Thermometry. The basic idea of DCGT is to substitute the molar volume V in the gas equation of state by the dielectric constant of the gas ε using the Clausius-Mosotti equation :

$$pV = RT \longleftrightarrow \frac{\varepsilon - 1}{\varepsilon + 2} = \frac{A_\varepsilon}{V}$$

The result is a relation between the easily measurable quantities p and ε,

$$P = \frac{RT}{A}\,\frac{\varepsilon - 1}{\varepsilon + 2}\qquad ,$$

which allowsto determine the temperature T, if the values of the molar polarizability A_ε of the gas and the gas constant R are known with sufficient accuracy. This method avoids the troublesome volume determination of the conventional gas thermometry which is complicated by the dead-spaces and by gas adsorption in the system. As has been pointed out in (1), the molecular interactions have to be included into a more detailed analysis by introducing the virial coefficients of the gas equation (B,C,..) and of the Clausius-Mosotti equation (b,c,..). In addition, the compressibility $\varkappa$ of the capacitor for the ε -determination has to be taken into account. With the abbreviations

$$\gamma = \varepsilon - 1 + \varkappa\,\varepsilon\,p \quad \text{and} \quad \mu = \gamma/(\gamma + 3)$$

this results in the series expansion for a DCGT isotherm

$$p = A_1\,\mu\,(1 + A_2\mu + A_3\mu^2 + \ldots).$$

By measuring the values p and μ at a constant temperature T, the constants

$$A_1 = \left(\frac{A_\varepsilon}{RT} + \frac{\varkappa}{3}\right)^{-1}$$

$$A_2 = (B-b)/A_\varepsilon \quad \text{and} \quad A_3 = C/A_\varepsilon^2$$

may be calculated byafit of the experimental results. If the constants A_ε/R and $\varkappa$ are known, the isothermal temperature may be determined as in a primary thermometer. If not, two measurements at known tmperatures T_1 and T_2 have to be performed to calibrate the DCGT system as a secondary thermometer.

As has been shown by Gugan in (2), the compressibility may be found from a surface fit of all measurements performed at different temperatures. Then we may use the DCGT as a primary thermometer using the calculation of Weinhold (3) for A_ε and the CODATA value (4) for R giving

$$A_\varepsilon/R = (6.221\,10 \pm 0.000\,19)\,10^{-8}\ \mathrm{K/Pa}$$

2. Apparatus

In general, our system has been constructed according to that described by Gugan (1). Therefore we confine here to that parts which have been changed:

2.1. Capacitor

In order to get a capacitor more stable against shock or vibrations we have constructed a system with a circumferential fixing of the electrodes instead of the mica washers used by Gugan. In addition, we have chosen copper beryllium instead of OFHC copper to get smaller temperature coefficients. Furthermore, the gap between the electrodes has been increased from 1.5 mm to 3 mm to reduce the possible influences of surface film effects.

2.2. Capacitance measurements

The capacitance bridge has been built up using inductive voltage dividers (IVD), Fig. 1: the main bridge divider A is a two-stage design (5) of dividing ratio $d_A = 0.5$. The capacitor C_X in the cryostat is compared with a capacitor C_R at room temperature of which the capacitance is kept constant to better than $\pm 1 \times 10^{-8}$ by immersing it into a thermostat controlled to $\pm 1 \times 10^{-4}$ K. The difference $C_X - C_R$ is measured by injecting a current across another room-temperature capacitor C_i using a commercial available IVD B (Sullivan Inc., F9200). The errors of both IVD's have been determined to be lower than $\pm 1 \times 10^{-8}$ at 1 kHz (6).

2.3. Pressure measurement

The pressure of the helim gas filled into the low temperature capacitor is measured out-

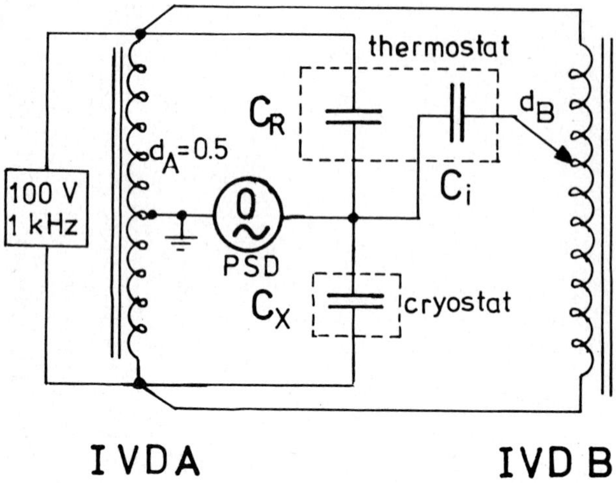

FIGURE 1
Capacitance bridge

side the cryostat using a pressure balance of RUSKA Instrument Corp., equipped with a special piston-cylinder assembly for the pressure range 0.006 - 0.3 MPa. The effective area of the piston has been determined at the PTB laboratory for Pressure Physics by intercomparison with the standard mercury barometer. The random error in this calibration has been ± 4 ppm. The pressure balance is connected to the capacitor by a tube of 9 mm inner diameter. Due to this large diameter a thermomolecular pressure correction is unnecessary. The temperature distribution along this tube is measured at 8 points and the aerostatic head correction is calculated for any measured pressure.

3. Preliminary test

In a preliminary test we have immersed our capacitor block directly into a liquid helium bath and determined the temperature by measuring the vapour pressure using the EPT-76 helium vapour pressure equations (7). The results of the fitting of 45 measuring points along a 4.2422 K isotherm are given in table 1. For comparison the results of Gugan from a surface fit procedure at a 4.2275 K isotherm are shown in the 2nd. column. In the 3th column the adopted best value for A_ε/R is given together with the virial coefficients extrapolated by calculation to our higher measuring temperature.

4. CONCLUSION

Our results are taken from a single third order fit without any constraints. The only assumption we have made is the value of the compressibility $\varkappa$ to be the same as in the Gugan experiment. It must be exspected, that the uncertainties will reduce if the results of other isotherms will be fitted together in a surface fit. Therefore the present results may be taken as a strong encouragement for proceeding with Dielectric Constant Gas Thermometry.

REFERENCES

(1) D. Gugan and G.W. Michel, Metrologia 16 (1980) 149
(2) D. Gugan in: Temperature, its measurement and control (Am. Inst. of Phys., New York 1982) 55
(3) P. Weinhold, J. Phys. Chem. 86 (1982) 1111
(4) E.R. Cohen and B.N. Taylor, J. Phys. Chem. Ref. Data 2 (1973) 663
(5) J.J. Hill and T.A. Deacon, IEEE Trans. Instr. Meas. IM-17 (1968) 269
(6) K. Grohmann, IEEE Trans. Instr. Meas. IM-25 (1976) 516 and IM-29 (1980) 496
(7) M. Durieux and R.L. Rusby, Metrologia 19 (1983) 67

TABLE 1
^4He polarizability and virial DCGT parameters

	this paper T=4.2422 K	Gugan (1) T=4.2275 K	Gugan,calc. T=4.2422 K	Theory (3) (4)
A_ε/R in K/Pa	6.221 75 ± .000 26	6.219 91 ± .000 07	6.221 21 ± .000 04	6.221 10 ± .000 19
B - b in $10^{-6}m^3/mol$	-78.92 ± .07	-79.85 ± .02	-79.50 ± .02	
C in $10^{-12}m^6/mol^2$	1072 ± 36	1265 ± 69	1260 ± 69	

LT-17 (Contributed Papers)
U. Eckern, A. Schmid, W. Weber, H. Wühl (eds)

NEW ^3He AND ^4He VAPOUR PRESSURE EQUATIONS

R.L. RUSBY

Division of Quantum Metrology, National Physical Laboratory, Teddington, Middlesex, UK

M. Durieux

Kamerlingh Onnes Laboratorium der Rijksuniversiteit, Leiden, The Netherlands

Vapour pressure equations for ^3He and ^4He are presented which are recommended by the International Committee of Weights and Measures (Sèvres, France) for deriving temperatures from vapour pressures. They supersede the "1958 ^4He Scale" and the "1962 ^3He Scale", from which they differ, between 0.2 K and 3 K, by roughly +0.2% in T. The definitive equations lnP(T) as well as approximate equations T(P) are given.

The International Committee of Weights and Measures has recommended the following equations for deriving temperatures T from measured vapour pressures P of ^3He or ^4He (1) (2)

^3He, 0.2 K to 3.3162 K

$$\ln(P/\mathrm{Pa}) = \sum_{i=-1}^{4} a_i (T_{76}/K)^i + a_5 \ln(T_{76}/K) \quad \text{(eq.1)}$$

^4He, 0.5 K to 2.1768 K

$$\ln(P/\mathrm{Pa}) = \sum_{i=-1}^{6} b_i (T_{76}/K)^i \quad \text{(eq.2)}$$

^4He, 2.1768 K to 5.1953 K

$$\ln(P/\mathrm{Pa}) = \sum_{i=-1}^{8} c_i t^i + c_9 (1-t)^{1.9} \quad \text{(eq.3)}$$

with t=T/5.1953 K. The coefficients are given in Table 1. In the formal recommendation temperatures are denoted T_{76} because, in fact, the vapour pressure equations relate P to temperatures on the "1976 Provisional 0.5 K to 30 K Temperature Scale" (EPT-76). The latter scale, which was approved by the International Committee of Weights and Measures in 1978, is presently the best-known approximation of the thermodynamic temperature in the 0.5 K to 30 K range (3). It is based on gas thermometry (2.6 K - 27 K), magnetic thermometry (0.5 K - 2.6 K) and thermodynamic calculations of ^4He and ^3He vapour pressures (0.5 K - 1.5 K) (4).

Pressures and temperatures at the λ-point and the normal boiling points and critical points are collected in Table 2.

We have also derived approximative equations, giving the temperature as a function of the vapour pressures (5). These equations, which were obtained by fitting to the same experimental data as used in the derivation of the definitive equations, are given here as normalised power series

$$T/K = \sum_{i=0}^{n} b_i x^i \quad \text{(eq.4)}$$

TABLE 1. Coefficients of the definitve ^3He and ^4He vapour pressure equations (eqs. 1, 2 and 3)

^3He: (eq.1) 0.2 K to 3.3162 K		^4He: (eq.2) 0.5 K to 2.1768 K	
a_{-1} =	-2.509 43	b_{-1} =	-7.418 16
a_0 =	9.708 76	b_0 =	5.421 28
a_1 =	-0.304 433	b_1 =	9.903 203
a_2 =	0.210 429	b_2 =	-9.617 095
a_3 =	-0.054 5145	b_3 =	6.804 602
a_4 =	0.005 6067	b_4 =	-3.015 4606
a_5 =	2.254 84	b_5 =	0.746 1357
		b_6 =	-0.079 1791

^4He: (eq.3) 2.1768 K to 5.1953 K	
c_{-1} =	-30.932 85
c_0 =	392.473 61
c_1 =	-2 328.045 87
c_2 =	8 111.303 47
c_3 =	-17 809.809 01
c_4 =	25 766.527 47
c_5 =	-24 601.4
c_6 =	14 944.651 42
c_7 =	-5 240.365 18
c_8 =	807.931 68
c_9 =	14.533 33

TABLE 2. Values of the ^{4}He λ-point and the normal boiling points and critical points of ^{3}He and ^{4}He on the old scales and the present one (see ref.(2))

	T_{58}	T_{76} (^{4}He)
^{4}He: T_λ	2.172 K	2.1768 K
P_λ	37.80 Torr	5041.8 Pa
		(37.817 Torr)
normal b.p.	4.215 K	4.2221 K
P_C	-	227.463 kPa
T_C	-	5.1953 K

	T_{62}	T_{76} (^{3}He)
^{3}He: normal b.p.	3.1905 K	3.1968 K
P_C	-	114.66 kPa
T_C	-	3.3162 K

where $x = (\ln(P/Pa) - A)/B$ with the constants A and B chosen so that the variable x lies within the range -1 to +1. In order to keep the equations simple, the ^{4}He range was not extended above 5.0 K and two equations, covering different temperature ranges, were derived for ^{3}He as well as for ^{4}He. Coefficients are given in Tables 3 and 4.

We should emphasize that the definitive equations remain the $\ln P(T)$ equations (eqs.

TABLE 3. Coefficients for the inverse vapour pressure equation for ^{4}He (eq.4)

2.1768 K to 5.0 K		1.0 K to 2.1768 K	
A = 10.3		A = 5.6	
B = 1.9		B = 2.9	
n = 7		n = 8	
i	b		b
0	3.146631	0	1.392408
1	1.357655	1	0.527153
2	0.413923	2	0.166756
3	0.091159	3	0.050988
4	0.016349	4	0.026514
5	0.001826	5	0.001975
6	-0.004325	6	-0.017976
7	-0.004973	7	0.005409
		8	0.013259

TABLE 4. Coefficients for the inverse vapour pressure equation for ^{3}He (eq.4)

0.5 K to 3.2 K		0.2 K to 2.0 K	
A = 7.3		A = 1.8	
B = 4.3		B = 8.2	
n = 9		n = 8	
i	b		b
0	1.053447	0	0.426055
1	0.980106	1	0.432581
2	0.676380	2	0.380500
3	0.372692	3	0.299547
4	0.151656	4	0.213673
5	-0.002263	5	0.149533
6	0.006596	6	0.099716
7	0.088966	7	0.044546
8	-0.004770	8	0.007914
9	-0.054943		

1,2,3). The inverse equations, which agree with the definitive equations within a tenth of a millikelvin (5) are, however, more practical since no iteration is required for deriving T from measured vapour pressures and are more easily used with small computers. The inverse equations can also be given as the sum of Chebyshev series, which is convenient for truncation in case deviations of more than 0.1 mK from the definitive equations are allowed (see ref.(5)).

A general revision of international practical temperature scales is planned at the end of the 1980s. A platinum resistance thermometer scale (i.e a revised IPTS-68) above 24.6 K the triple point of neon (or 13.81 K the triple point of hydrogen), a form of the EPT-76 between 4.2 K and 24.6 K (or 13.81 K) possibly to be realized with an interpolating gas thermometer, and the present ^{3}He and ^{4}He vapour pressure equations between 4.2 K and 0.5 K are envisaged to form the new international scale. The scale will probably not be defined below 0.5 K (1).

REFERENCES
(1) Comité Consultatif de Thermométrie, 14th Session (Bureau International des Poids et Mesures, Sèvres, France) 1982.
(2) M. Durieux and R.L. Rusby, Metrologia 19(1983)67.
(3) Metrologia 15(1979)65.
(4) M. Durieux, D.N. Astrov, W.R.G. Kemp, and C.A. Swenson, Metrologia 15(1979)57.
(5) R.L. Rusby and M. Durieux, submitted to Cryogenics.

LT-17 (Contributed Papers)
U. Eckern, A. Schmid, W. Weber, H. Wühl (eds)

HELIUM-4 SECOND VIRIAL IN LOW TEMPERATURE GAS THERMOMETRY: COMPARISON OF MEASURED AND CALCULATED VALUES

G. T. McCONVILLE

Monsanto Research Corporation-Mound*, Miamisburg, Ohio 45342 USA

Recent advances in low temperature gas thermometry down to 2.6K have led to the gas thermometer to be considered as a low temperature interpolation instrument for the International Practical Temperature Scale below 13.81K. Work on reducing the uncertainty in the low temperature second virial has appeared in Temperature, Vol. 5. The data of Berry, Plumb, Kemp, and Steur, et al, are compared here to values calculated using the HFDHE2 potential function of Aziz, et al. Below 10K the data fall significantly lower than the calculation. Modifications of the HFDHE2 potential show an increased well depth to 10.916K and the inclusion of a He_2 bound state energy of 1.9 mK describes Berry's data down to 2.6K. The agreement ranges from 0.3 cc/mole at 2.6K to 0.1 cc/mole at 20.28K.

1. INTRODUCTION

The gas thermometer offers a direct link between the practical and the thermodynamic temperature scales. For many years uncertainty in the values of the helium-4 second virial, B, limited the accuracy of the low temperature gas thermometer. This situation has been changed with the appearance of the work of Berry (1), Plumb (2), Kemp, et al (3) and Steur, et al (4) in the 5th volume of Temperature. In his Leiden PhD Thesis, Steur (5) compared a surface fit of second and third virials to the data and to the HFDHE2 calculation of Aziz, et al (6) and the HFIMD calculation of Feltgren, et al (7). The uncertainty in the gas thermometry caused by the uncertainty in B is less than 1 mK for an operating density of 1.33 kPa/K at T > 10K. Below 10K there is an increasing divergence between the data and the calculation. New calculations resolve this divergence.

2. POTENTIAL CALCULATIONS

The second virial was calculated using a Hartree-Fock Dispersion form (6) designated HFDHE2 with a simple one parameter damping function of the form of Ahlrichs, et al (8). The repulsive parameters A, α and the dispersion parameters C_n, n = 6,8,10 are taken from theoretical calculations. The damping parameter, D, was varied from the value of Ahlrichs, 1.28, to 1.241 in order to provide a ±0.05 cc/mole fit to Gammon's second virial (9) between 100 and 450K and a 0.3% fit to Haarman's thermoconductivity data (10) between 350 and 450K.

In Figure 1, the HFDHE2 calculation is used as a reference line. The vertical arrows indicate an uncertainty of ±1 mK in a gas thermom-

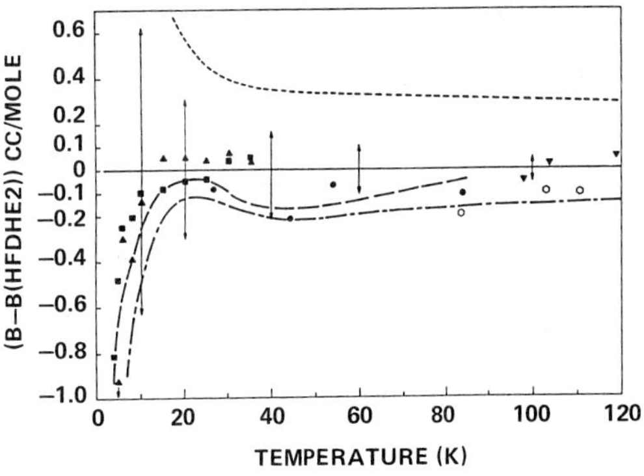

FIGURE 1
Comparison of measured and calculated helium-4 second virials: Data; ▼ Gammon, o Hall and Canfield, • Kemp, et al, ▲ Plumb, ■ Berry. Data fit; — – Kemp, ––– Steur. Calculation; —— HFDHE2, ---- Feltgren HFIMD.

eter at a density of 1.33 kPa/K (5). The calculation splits the difference between the data of Berry and Plumb between 15 and 30K. Below 10K the data are significantly below the calculation. Between 27 and 84K the values of Kemp, et al (3) average about 0.1 cc/mole below the calculation. Kemp's fit using his data and Berry's data and Steur's surface fit both are lower than the calculation by as much as 0.2 cc/mole down to 20K. Figure 1 also shows the

*Mound is operated by Monsanto Research Corporation for the U. S. Department of Energy under Contract No. DE-AC04-76-DP00053.

HFIMD calculation of Feltgren which is systematically higher than the data by about 0.4 cc/mole. The well depth, ε/k, for HFIMD is 10.74K and for HFDHE2 is 10.80K. A description of the data below 10K requires an ε/k greater than 10.80K.

The deviation of the data from the HFDHE2 calculation down to 2.5K is shown in Figure 2 on a larger scale. The data of Berry and Plumb taken from Plumb's table (2) are in good agreement down to 6K. Below 6K, Plumb's points are much lower. The error bars represent Berry's uncertainty at the temperatures of his measurements. The difference between the calculation and the data can be removed by modifying the potential function to produce a deeper potential minimum. The HFDHE2 function was modified in three ways but only one produced a decrease in B of less than 0.02 cc/mole at 298K. The dashed line in Figure 2 was obtained by replacing the HFDHE2 repulsive $A\exp(-\alpha_0 r)$ by $A\exp(-\alpha_1 r-\beta r^2)$ where $\alpha_0 = 4.50Å^{-1}$ and $\alpha_1 = 4.49Å^{-1}$ and $\beta = 0.01Å^{-2}$ and increasing D to 1.282. These changes produce a $\varepsilon/k = 10.916$ at $r_m = 2.9644Å$ and lead to a bound state energy of 1.9 mK. The addition of the bound state contribution to B is shown by the short dash curve in Figure 2. The bound state energy is larger than 0.84 mK calculated by Uang and Stwalley (11) for the HFDHE2 function.

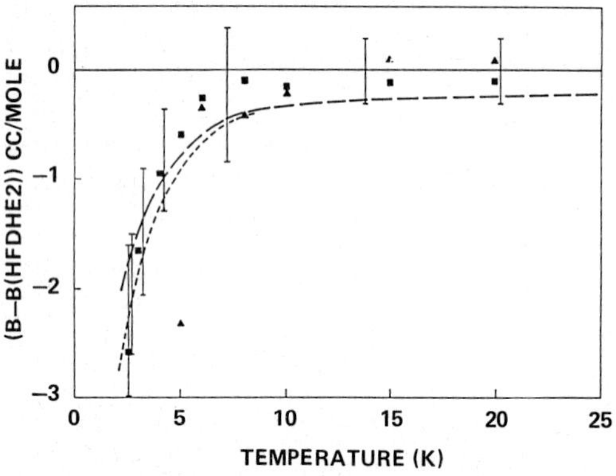

FIGURE 2
Deviation from HFDHE2 second virial: Data; ▲ Plumb, and ■ Berry, error bars represent Berry's uncertainty at his measured temperatures. Calculation: —— Modified HFDHE2, $\varepsilon/k = 10.916$ and $r_m = 2.9644$, --- addition of bound state contribution.

Table 1. Calculate second virial as a function of T.

Temperature(K)	B(cc/mole)
2.0	-194.56
3.0	-120.43
4.0	- 85.24
5.0	- 64.50
6.0	- 50.75
8.0	- 33.55
10.0	- 23.24
15.0	- 9.66
20.0	- 2.90
25.0	+ 1.13

3. CONCLUSION

By increasing well depth but leaving the repulsive part of the potential the same as the HFDHE2, agreement of the calculations with the data of Berry is found down to 2.6K. The new well depth produces a bound state energy of 1.9 mK. Addition of this contribution produces agreement between the calculation and the data at 2.6K of 0.3 cc/mole. Steur (5) has shown that use of the lower limit of Feltgren's HFIMD $\varepsilon/k = 10.94$K reproduces Berry's data to 0.3 cc/mole above 4K. The agreement between calculated and measured second virial, B, should allow a more accurate determination of third virial, C, from isotherm data for use in low temperature gas thermometry.

REFERENCES

(1) K. H. Berry in Temperature: Its Measurement and Control in Science and Industry, Vol. 5, ed. J. F. Schooley, (American Institute of Physics 1982) p. 21.
(2) H. H. Plumb, ibid, p. 77
(3) R. C. Kemp, L. M. Besley, W. R. G. Kemp, ibid, p. 33.
(4) P. P. M. Steur, J. E. van Dijk, J. P. Mars, H. ter Harmsel, M. Durieux, ibid, p. 25.
(5) P. P. M. Steur, PhD Thesis, University of Leiden (1983).
(6) R. A. Aziz, V. P. S. Nain, J. S. Carley, W. L. Taylor and G. T. McConville, J. Chem. Phys. 70, 4330 (1979).
(7) R. Feltgren, H. Kirst, K. A. Kohler, H. Pauly and F. Forello, J. Chem. Phys. 76, 2360 (1982).
(8) R. Ahlrichs, P. Penco, G. Scoles, Chem. Phys. 19, 119 (1976).
(9) B. E. Gammon, J. Chem. Phys. 64, 2556 (1976).
(10) J. W. Haarman, AIP Conf. Proc. 11, 193 (1973).
(11) Y. H. Uang, W. C. Stwalley, J. Chem. Phys. 76, 5069 (1982).

LT-17 (Contributed Papers)
U. Eckern, A. Schmid, W. Weber, H. Wühl (eds)

TEST OF SOME TEMPERATURE SCALES BELOW 1 K WITH ABSOLUTE THERMOMETRY

G. SCHUSTER, A. HOFFMANN, L. WOLBER, W. BUCK and J.-F. MARCH

Physikalisch-Technische Bundesanstalt, Institut Berlin, D-1000 Berlin 10, Germany

Three low temperature scales which are available through calibrated devices or numeric data have been tested with a noise and a nuclear orientation thermometer. The scales are NPL-TX1 recorded on rhodium-iron resistors, NBS-CTS1 recorded on superconductive fixed points, and the He-3 melting curve in the representation of Greywall et al. (1). Temperature deviations which are below 0.1 % at 1 K, and rise to 1 % near 10 mK stay, in general, near the expected limits, although they seem to be somewhat larger at several points in this range. In order to resolve these inconsistencies, the minimum of the He-3 melting-curve at about 0.3 K is proposed as a supplementary temperature reference point.

1. INTRODUCTION

In the course of investigations started to establish a low temperature scale at Physikalisch-Technische Bundesanstalt, the absolute thermometers developed for that purpose have been applied to a test of already existing scales. This provides additional information on the scales and may reveal inconsistencies in the thermometry involved which is a worth-while experiment because few independent checks on the scales have been done due to the tedious nature of the measurements.

The main part of the temperature range is covered by a noise thermometer. It operates with an rf resistive SQUID on the principles due to Kamper et al. (2). In order to arrive at measurement results with acceptable statistical certainty in shorter time intervals compared to the original design, the operating frequency and resolution have been increased such that an uncertainty of 0.1 % (1 σ) in temperature is obtained after approximately 1 h of data collection (3). The measurement integrity of the noise thermometer is checked at both ends of the temperature range: above 1 K in comparison with a qualitatively superior scale and below 50 mK, where it becomes more sensitive to non-thermal noise, in comparison with a nuclear orientation thermometer. This instrument is based on the temperature dependence of the γ-ray anisotropy of Co-60 in a cobalt single crystal. Its design follows the standard features found in literature (4).

2. MEASUREMENT PROCEDURES

During the measurements all participating temperature sensors were coupled to a copper block attached to the mixing chamber of a dilution refrigerator. To avoid alterations of the experimental arrangement, the refrigerator was operated as an evaporation cryostat (with He-4 circulation) at temperatures above 0.9 K. For each measurement, the block temperature was stabilized, and all data obtained during a time interval of at least 16 h were averaged.

At the upper end of the temperature range, comparison measurements were made with a rhodium-iron resistor calibrated at NPL between 0.5 K and 27.1 K. The measured resistance values were converted to temperatures using the data supplied with the calibration report. As the uncertainty of the NPL scale (5) is less than the uncertainty of the noise thermometer above 1 K, these measurements made only a check of the noise thermometer.

Measurements involving superconductive fixed points between 23 mK and 205 mK were made by stabilizing the cryostat at the midpoint of each superconducting-to-normal transition using the detector output as signal for the temperature controller. Due to the extremely fast response of the devices, this resulted in very stable temperatures. The superconductive fixed point device SRM 768-8(6), calibrated at NBS, was used with Cryoperm magnetic shields and operated with currents less or equal to those given in the calibration report.

The He-3 melting-curve thermometer operated in the usual manner with a capacitive pressure transducer. It was calibrated with a pressure balance and its indication corrected such that the pressure of the melting curve minimum coincided with the value of Greywall et al. (1). This method relaxed the accuracy requirements of the pressure calibration considerably (7). For the measurements each temperature was stabilized on the melting-curve using the pressure transducer output to feed the temperature controller. The pressure value itself was converted to temperature using the polynomial representation also published by Greywall et al. (1).

3. DISCUSSION

The measurement results are collected in Fig. 1 as relative temperature deviations. The scales recorded on rhodium-iron resistors above 0.5 K

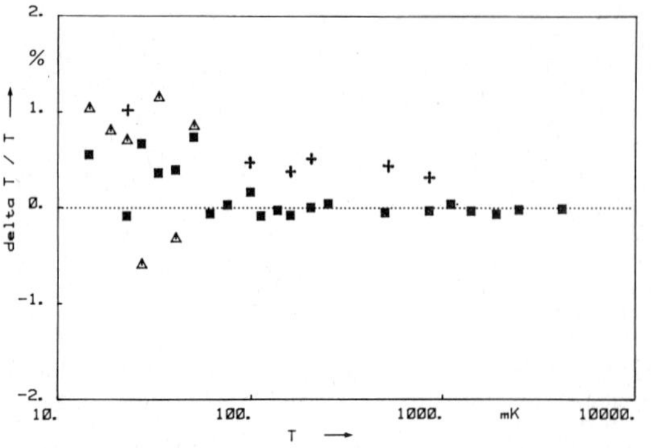

FIGURE 1
Relative differences between noise (squares) and nuclear orientation (triangles) temperatures and temperature scales. Crosses represent superconductive fixed point devices SRM 768 below, and SRM 767 above 0.3 K.

and the He-3 melting curve below 0.3 K have been taken as references because they are continuous in contrast to the discrete data supplied by the superconductive fixed points and the thermometers. It can be seen that - above 50 mK - the measured differences with respect to the continuous scales are less or equal to 0.1 %. This demonstrates that both scales present an excellent basis for realizing temperatures. The deviation below 50 mK, which increases to approximately 1 % at 10 mK, is not thought to be contradictory to the current accuracy status of the scale. An exception are the superconductive fixed points which have been measured lower than expected by approximately 0.4 %. These results are clearly outside the uncertainty limits assigned to the transition temperatures and indicate that there is some difference in realizing them at calibration and use. Since they are all affected by nearly the same relative shift, an individual analysis does not yield another clue to the problem.

4. CONCLUSION

In order to bridge the gap between the high temperature scale recorded on rhodium-iron resistors and the slightly contradictory results of the low-temperature He-3 melting-curve branch and the superconductive fixed points, the temperature of the melting curve minimum could be used as a further reference point. This point would be best defined by the zero crossing of the derivative of He-3 melting pressure with respect to temperature or, using the Clausius-Clapeyron equation, by the equality of the molar entropies of solid and liquid He-3. This reference point is clearly physically different from the superconductive fixed points and may, therefore, infer independent information. Its practical advantage is that no pressure calibration is required to realize it.

A test experiment made to demonstrate the feasibility had the results shown in Fig. 2. It

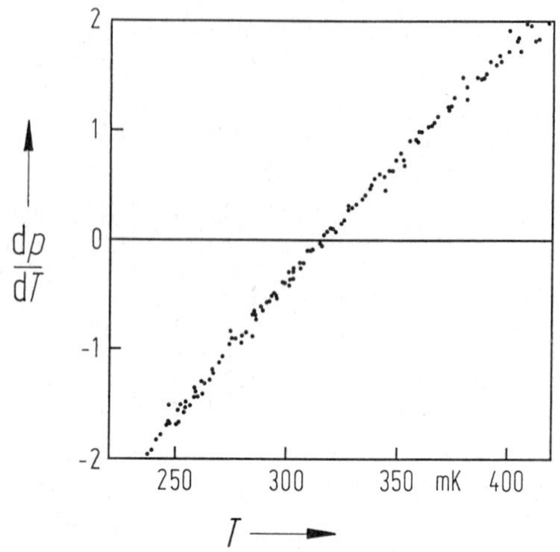

FIGURE 2
Slope of He-3 melting pressure (in arbitrary units) versus temperature near the melting curve minimum.

consisted of four sweeps through the pressure minimum both with increasing and decreasing temperatures. Pressure values were taken synchronously with data of the noise and a paramagnetic salt (CMN) thermometer. The CMN thermometer was calibrated with the noise values and the derivatives of pressure with respect to CMN temperatures were calculated. The zero crossing temperature was reproduced within +- 0.3 % for the individual sweeps which seems to be a sound basis for a determination of this point to better than 0.1 %.

REFERENCES

(1) D.S. Greywall and P.A. Busch, J. Low Temp. Phys. 46 (1982) 451.
(2) R.A. Kamper and J.E. Zimmerman, J. Appl. Phys. 42 (1971) 132.
(3) A. Hoffmann and B. Buchholz, UHF resistive SQUID noise thermometer at temperatures between 0.005 and 4.2 K. To be published.
(4) H. Marshak, J. Res. NBS 88 (1983) 175.
(5) R.L. Rusby and C.A. Swenson, Metrologia 16 (1980) 73.
(6) R.J. Soulen, Jr. and R.B. Dove, NBS Special Publication 260-62 (1979).
(7) G. Schuster and L. Wolber, Automated He-3 melting curve thermometer. To be published.

LT-17 (Contributed Papers)
U. Eckern, A. Schmid, W. Weber, H. Wühl (eds)

NEW RESISTANCE THERMOMETER WITH SMALL MAGNETIC FIELD DEPENDENCE FOR LOW TEMPERATURE MEASUREMENTS

Hideyuki DOI and Yoshimasa NARAHARA

Institute of Physics, University of Tsukuba, Ibaraki 305, Japan

Yasukage ODA and Hiroshi NAGANO

The Institute for Solid State Physics, University of Tokyo, Roppongi, Minato-ku, Tokyo 106, Japan

The posibility to use RuO_2 thick film chip resistors as thermometers for low temperature measurements (100 - 0.01K) in strong magnetic fields is discribed. The temperature dependences of resistance of the resistors above 100 ohm at room temperatures are negative, smooth and monotonic below 250 K. In 1 Kohm RuO_2 chip resistors (ALPS TBL type 1/8 W), which are suitable to measure the temperature range (10 - 0.01K), their temperature dependence of resistance is given by the relation $R = R_0 \exp[(T_0/T)^{\frac{1}{4}}]$ below 0.9 to 0.03 K. Their magnetoresistance $\Delta R(H)/R(0)$ is small and monotonic and at 8 Tesla is 0.4 % at 4.2 K and 2 % at 0.1 K. In 100 Kohm ALPS RuO_2 resistors suitable to the range (100 - 1K), their magnetoresistance $\Delta R(H)/R(0)$ at 8 Tesla is 0.7 % ($\Delta T(H)/T(0)$=1%) at 4.2 K, that is smaller than carbon glass thermometers'. No significant orientation dependence against the magnetic field directions was found between 4.2 K and 0.2 K. RuO_2 chip resistors have many other excellent properties; good reproducibility, small size, fast thermal response time down to the very low temperatures and low cost.

1. INTRODUCTION

Some kinds of resistance thermometers, Ge thermometers, carbon glass thermometers and carbon composition resistors, have been used as secondary thermometers for low temperature measurements. Ge thermometers are best in no magnetic fields in the temperature range below 30 to 0.3 K. Carbon glass thermometers have good reproducibility in heat cycles and considerable smaller magnetoresistance in strong magnetic fields and are able to be used in the wider temperature range up to 300 K, but not below 1 K. Carbon composition resistors have been most widely used as thermometers in the range below 4 to 0.01 K in spite of their less reproducibility, because of their advantages of high sensitivity, low heat capacity, fast thermal responce time and low cost. Magnetoresistance of carbon compsition resistors are rather large below 4.2K, except for the Speer carbon resistors' which are small above 0.5 K. However, the Speer resistors exhibit a complex behabior of the magneto-resistance with both temperature and field, and their magnetoresistances enlarge abruptly at the lower temperatures < 0.5 K).(1),(3)

RuO_2 thick film chip resistors, which are used in hybrid microcircuits, have small, smooth and monotonic magnetetic field dependences of resistance down to the lowest temperature of 0.1K where we have already made experiments, in addition to all of the advantages above mentioned about three kinds of the resistance thermometers.

2. RuO_2 THICK FILM CHIP RESISTORS

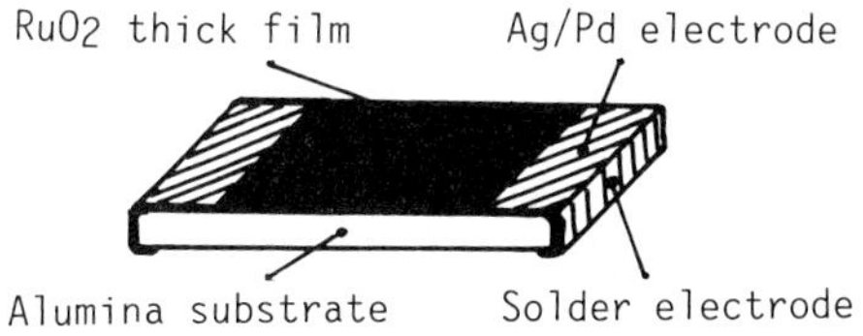

FIGURE 1
The structure of RuO_2 thick film chip resistors.

RuO_2 chip resisters are in the form of RuO_2 resistive thick film of (0.01-0.02) mm thickness with Ag/Pd thick film electrodes at both their ends. Ag/Pd electrodes at both sides are plated with solder and upper surfaces of the resistors are coated by glass. Dimensions of the resistor, length, width and thickness are 3.2, 1.6 and 0.5 mm, respectively.

3. EXPERIMENTAL METHODS AND RESULTS

Temperature measurements were made with a Ge thermometer, a CMN thermometer, Speer carbon resistors and NBS superconducting transition points (204.6, 162.7, 98.95, 23.06, 15.14 mK), which are mounted on the mixingchamber of a ^3He-^4He dilution refrigerator. The RuO_2 resistor was buried into a slit of 0.6 mm width cut in a copper block with Ge 7031 adhesive. Two such a block were prepared. One block was mounted on the mixingchamber near by the thermometers, where was always nearly zero magnetic fields. The other block was mounted on a copper plate thermaly connected to the mixing chamber and was located at the center of magnetic fields. Resistance measurements were made using a four terminal DC method and a four terminal AC potentiometric conductance bridge with low power disspation below 10^{-13} W.

The temperature dependence of resistance of 1 Kohm ALPS RuO_2 resistor has been measured in the temperature range from 300 down to 0.015 K, and it is negative and smooth below 250K down to very low temperatures. Its coefficient dR/dT is large enough to be used as a thermometer in the temperature range below 30 K. As shown in fig. 2, the temperature dependence of resistance below 0.9 to 0.03 K is well fitted by the formula

$$R = R_0\exp[(T_0/T)^{\frac{1}{4}}]$$

as same as the Speer resistor's.(2) This is a characteristic temperature dependence of the electrical conduction mechanism due to the variable range hopping in three-dimentional Anderson localized states and so the electronic states are considered to be strongly localized in this system.(3),(4)

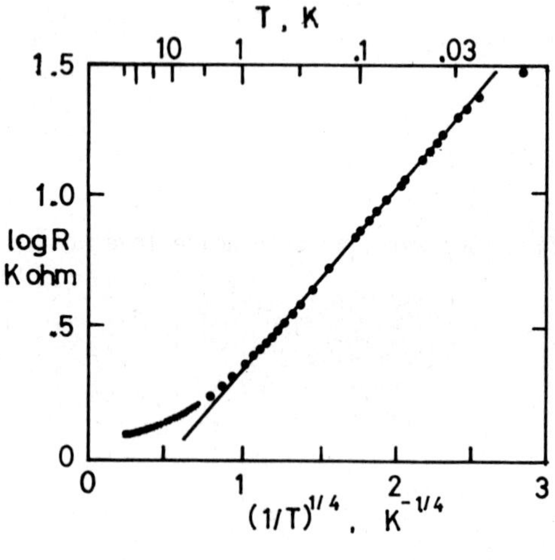

FIGURE 2
Resistance VS $T^{-\frac{1}{4}}$ or T for 1 Kohm ALPS RuO_2 chip resistors. The line represents the behabior $\ln R \propto T^{-\frac{1}{4}}$.

The magnetoresistance, $\Delta R(H)/R(0)$, of 1 Kohm ALPS RuO_2 resistor at several different temperatures below 4.2 K is shown in figure 3. It is a small, monotonic and positive magnetoresistance down to the lowest temperature where we have already made experiments. These data are combined with the temperature dependence in zero magnetic fields and the equivalent error $\Delta T(H)/T(0)$ in temperature is given. At 8 Tesla, the magnetic field induced error $\Delta T(H)/T(0)$ is 2.5 % at 4.2 K and 3 % at 0.1 K. It is comparable to the correspondence value of carbon glass thermometer, 2.5 % at 4.2 K. (5)

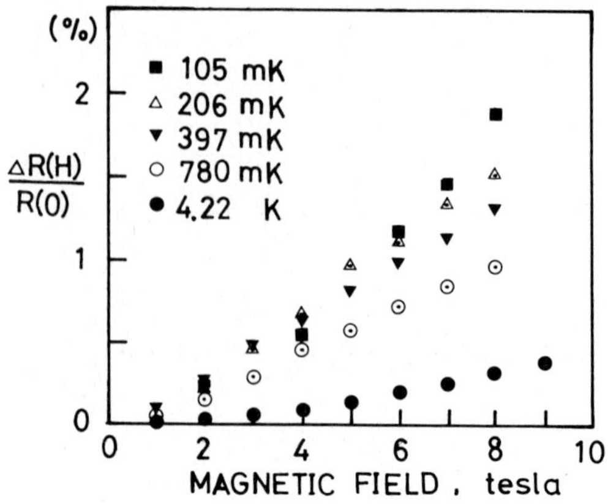

FIGURE 3
The magnetoresistance, expressed as $\Delta R(H)/R(0)$ in percents, is plotted as a function of the magnetic field H at various temperatures.

Eight commercial RuO_2 chip resistors have been subjected to fifteen thermal cycles for five months between 300 and 4.2 K and tested for stability at 4.2 K with using a Cryocal Ge thermometer as a reference one. Their deviations of resistance at 4.2 K were not beyond $\pm$ 0.05 %, which are comparable to our experimental errors.

4. CONCLUSION

By comparison with other thermometers at low temperatures in strong magnetic fields, the RuO_2 thick film chip resistor appears to offer an attractive alternative, especially below 0.5 K.

ACKNOWLEDGEMENTS

We are grateful to M. Matsuda and H. Izumida, ALPS Electric Co., Ltd. for kindly preparing the RuO_2 resistors and for their great interests in this work.

REFERENCS

(1) H.H. Sample and L.G. Rubin, Cryogencs 17 (1977) 597;H.H. Sample, L.J. Neuringer and L.G. Rubin, Rev. Sci. Instrum. 45 (1974) 64.
(2) A.C. Anderson, J.H. Anderson and M.P. Zaitlin, Rev. Sci. Instrum. 47 (1976) 407.
(3) Y. Koike, S. Morita, T. Fukase, N. Kobayashi, M. Okamura and N. Mikoshiba, J. Phys. Soc. Jpn. 52 (1983) 1111.
(4) N.F. Mott: Metal-Insulator Transition (Taylor and Francis, London,1974).
(5) J.M. Swartz, J.R. Gaines and L.G. Rubin, Rev. Sci. Instrum. 46 (1975) 1177.

LT-17 (Contributed Papers)
U. Eckern, A. Schmid, W. Weber, H. Wühl (eds)

A SUPERCONDUCTING PENETRATION DEPTH THERMOMETER

M.V. Moody, H.A. Chan, H.J. Paik and C. Stephens

Department of Physics and Astronomy, University of Maryland, College Park, Maryland 20742, USA

As is well known, the magnetic field penetration depth in a superconductor varies with temperature. This relationship is described by the empirical formula, $\lambda = \lambda_o (1 - t^4)^{-1/2}$, which holds quite well near T_c. When a superconducting coil containing a persistent current is placed near a superconducting plane, a change in the penetration depth will cause a change in the current. By sensing this current change with a SQUID amplifier, very small temperature changes may be detected. Using a crude geometry to test this principle, we have obtained a sensitivity of 1.4×10^{-7} K. A thin film design that would give a sensitivity of 1×10^{-9} K is proposed.

1. INTRODUCTION

One of the primary noise sources in superconducting gravity gradiometers is a result of the variation of the superconducting penetration depth, λ, with temperature [1]. The sensing scheme used in these gradiometers consists of a superconducting coil, containing a persistent current, located at a distance, d, from a superconducting proof mass. A change in the penetration depth is equivalent to a change in displacement of the proof mass with the inductance of the coil being proportional to $d + \lambda$ to first order. The gradiometer generally operates at a reduced temperature $t < 0.5$. However, the empirical formula describing λ near T_c ($t \gtrsim 0.8$) [2],

$$\lambda = \lambda_o (1 - t^4)^{-1/2}$$

indicates that this effect is much more significant as T approaches T_c. Thus, by choosing a geometry to minimize d, and operating at a temperature near T_c, a sensitive superconducting thermometer can be constructed. Such a thermometer might be useful in the study of critical phenomena. In the following sections, we describe an experiment which demonstrates this principle and propose a practical design for a sensitive thermometer.

2. EXPERIMENTAL DESIGN

In order to obtain a high sensitivity within a reasonable volume, a thin film geometry is necessary. Since a facility for producing such a geometry was not available, the experiment was designed mainly to test the principle of operation. The sensing element (see Fig. 1) consisted of single layer solenoid, 3.3 cm in length, wound tightly on a niobium core, 1.25 cm in diameter. The wire was also niobium and had a diameter of 0.076 mm and an insulation thickness of 0.010 mm. The calculated inductance was 2.5×10^{-6} H. In order to avoid a high current through the SQUID input coil the thermometer coil was coupled to the SQUID through a superconducting transformer. The SQUID used was an SHE Model 300. The noise specifications were: a white noise level of 1×10^{-11} A Hz$^{-1/2}$, a drift level of 2×10^{-11} A in 1 hour and a 1/f cut-off frequency of 0.01 Hz. A calibrated germanium thermometer was placed near the center of the solenoid in the niobium core. A manganin heater with a resistance of 110Ω was also wound around the niobium core and was magnetically shielded from the sensing coil. The experiment was placed inside a vacuum can with the sensing coil thermally isolated from the helium bath.

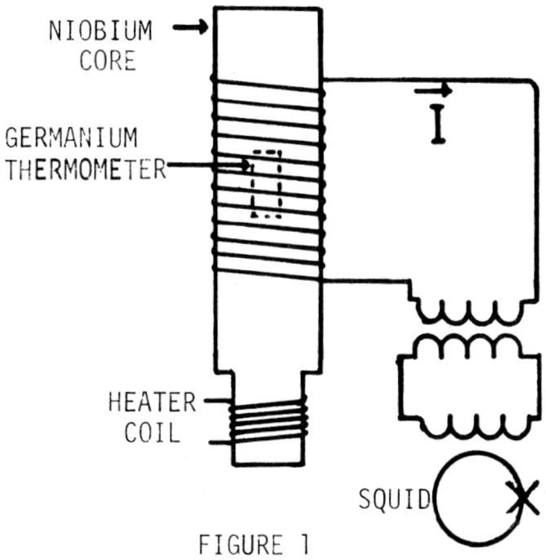

FIGURE 1

3. RESULTS

A least squares fit of the data was made to the equation

$$I = B(1 - T^4/C)^{-1/2} + D$$

The data was taken over the temperature range 8.818 to 9.204 K with a persistent current of 0.05 A gave the results, $B = 3 \times 10^{-5}$ and $C = (9.280)^4$. A Plot of this data versus $(1 - t^4)^{-1/2}$ is shown in Fig. 2. Using a simple model in which the magnetic field is uniformly distributed between the superconducting ground plane and the mid plane of the wire (the field is excluded from the wire) gives $B = 1 \times 10^{-4}$ A. However, due to an inefficient design of the transformer, the current transfer ratio was only 0.24 which results in a final B of 2.4×10^{-5} A. This discrepancy is most likely due to the oversimplification of the model. A similar fit to data in the 5 to 6 K range gives $B = 4.3 \times 10^{-5}$ A and $C = (9.50)^4$ which confirms that the equation is not as reliable for lower values of t.

Due to insufficient RF shielding, the SQUID noise level was about a factor of 10 greater than the specifications. This noise level was independent of the current stored in the sensing loop. Thus, the sensitivity in this experiment was 1.4×10^{-7} K $Hz^{-1/2}$ for t = 0.99 and $H = 0.25\, H_{c1}$, where H_{c1} is the magnitude of the first critical field at T = 0.

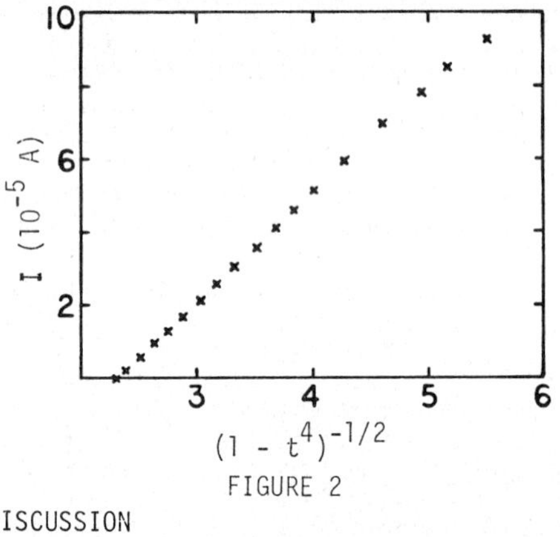

FIGURE 2

4. DISCUSSION

A more practical and more sensitive design for a superconducting penetration depth thermometer could be realized using thin films. For the purpose of calculation we have chosen a single layer coil with a pancake geometry. By choosing the parameters: area (A) = 10^{-4} m^2, $d = 10^{-6}$ m and turns density = 10^6 m^{-1}, an inductance of 2×10^{-6} H is obtained. This inductance would match the SHE SQUID input coil inductance. The maximum sensitivity (where the persistent current is limited by the critical field) for such a thermometer is given by,

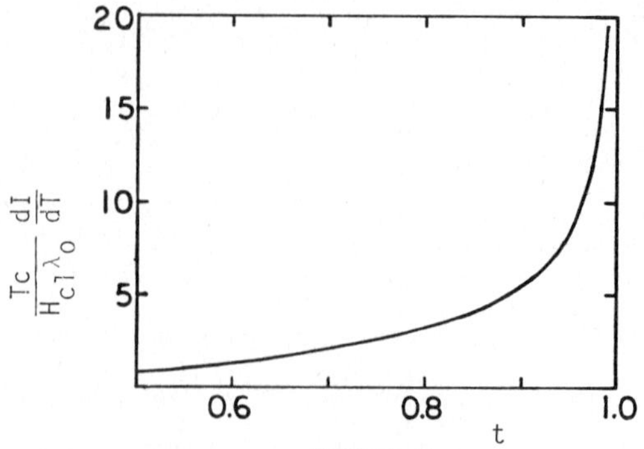

FIGURE 3

$$\frac{T_c}{H_{c1}\lambda_o}\frac{dI}{dT} = \left(\frac{\mu_o A}{d L_s}\right)^{1/2} t^3 (1 - t^2)(1 - t^4)^{-3/2}$$

where $L_s = 2 \times 10^{-6}$ H. A plot of $(T_c/H_{c1}\lambda_o)$ (dI/dT) versus t for the previously described geometry is shown in Fig. 3. The value of $H_{c1}\lambda_o/T_c$ is approximately 4×10^{-4} K A^{-1} for the superconducting elements. Thus, the maximum sensitivity is approximately 1×10^{-9} K $Hz^{-1/2}$ at t = 0.99 and drops to 1×10^{-8} K $Hz^{-1/2}$ at t = 0.5. This sensitivity could, of course, be improved by a more elaborate geometry such as multilayers or a higher turns density.

Unlike the paramagnetic salt thermometer developed by Lipa et al [3], the penetration depth thermometer has a nonlinear temperature dependence and has its highest sensitivity very near T_c. However, with the choice of the proper superconductor, this region could be made to be near the temperature of interest. Also, the thin film design might permit the deposition of a thermometer directly on an experimental sample, thus, reducing the resolution time.

The major contributions to power loss in this thermometer should be from eddy currents in the insulating material [4] and from hysteretic losses in the superconductor [5]. These effects should limit the Q of the circuit to approximately 10^6 to 10^8 resulting in a power loss of $< 10^{-14}$ W.

REFERENCES

(1) E.R. Mapoles, Ph.D. thesis, Stanford University, Stanford, California (1981).
(2) P.B. Miler, Phys. Rev. 113 (1959) 331.
(3) J.A. Lipa, et al, Physica 107B (1981) 331.
(4) B.W. Ricketts, J. Phys. E 9 (1976) 179.
(5) W.T. Beall, Jr. and R.W. Meyerhoff, J. Appl. Phys. 40 (1969) 2052.

LT-17 (Contributed Papers)
U. Eckern, A. Schmid, W. Weber, H. Wühl (eds)

SUPERCONDUCTIVE TEMPERATURE REFERENCE POINTS ABOVE 0.5 K

J. F. SCHOOLEY and J. H. COLWELL

Center for Basic Standards, National Bureau of Standards, Washington, DC 20234

Careful preparation and annealing of samples made from high-purity Pb, In, Al, Zn, and Cd have resulted in sample-to-sample T_c variations of less than 0.5 mK. Less-pure Nb samples, while more variable in T_c, still exhibit single-sample reproducibilities better than 0.2 mK. Temperature reference devices incorporating these six elements offer stable, high precision *in situ* calibration capability.

1. INTRODUCTION

Temperature reference devices based on superconductive transitions in samples of Pb, In, Al, Zn, and Cd have been available from the NBS since 1972 (1). Limitations in sample preparation techniques and in our reference thermometry, however, restricted the sample-to-sample T_c uniformity in these devices to ± 1 mK.

In the range 0.5 K to 10 K, rhodium-iron resistance thermometers (RIRT) used in conjunction with modern resistance bridges now offer computer-compatible thermometry with levels of imprecision no larger than $\pm$ 0.1 mK (2)(3). To be fully useful, fixed-point devices must offer similar temperature reproducibility levels.

Recently, we have re-addressed the question of the T_c reproducibility limit of well-annealed samples of high-purity bulk superconductors. Two quite independent aspects of temperature reproducibility are involved in this question:

1) The variation in T_c that is observed in a particular sample in the course of many cycles starting from room temperature. This property we choose to call the "sample T_c reproducibility";

2) The variation in T_c that is observed in the course of measuring many similar samples of a given element. This property we choose to call the "uniformity of T_c for that element."

The distinction between these two criteria is important for the following reason; whereas a certain level of T_c reproducibility can provide a stable calibration point for local laboratory work, a particular level of uniformity of an element's T_c can anchor a universal scale of temperature at that same level of uniformity.

We have prepared more than twenty devices, each containing one sample of Nb, Pb, In, Al, Zn, and Cd. We have measured the T_c's of each device using a RIRT calibrated on the EPT-76 scale (4). At least one, and often two, of five devices observed in a given measurement cycle were reference devices.

By re-measuring the same reference devices, we could determine the overall T_c reproducibility of their samples. By observing more than one reference device in repetitive measurements, we could hope to distinguish between variations in the measured T_c arising from drift in the thermometer calibration and variations arising from changes in individual samples. By measuring many devices, we expected to gain some knowledge about the uniformity of T_c for the elements involved, as well as to detect any dependence of T_c upon sample transition width.

2. FIXED-POINT DEVICES

Cylindrical samples ($\sim$2 mm diam. x 25 mm long) of Pb, In, Zn, and Cd were prepared by vacuum casting in borosilicate glass. Starting materials of these elements were supplied as 99.9999% pure ingots from the following sources: Pb (Cominco, Inc.); In (Indium Corp. of America); and Zn and Cd (NBS Office of Standard Reference Materials). Al was obtained as 2-mm-diam. wire of 99.9999% purity (Materials Research Corp.). All samples were annealed near their melting temperatures for periods of time ranging from 70-120 hours. The element Nb is available commercially only in moderate ($\sim$99.99%) purity. We have prepared samples by treating the 2-mm diam. wire supplied by the manufacturer (Materials Research Corp.)* in various ways, including: as-received, etches of several types, and vacuum annealing.

The samples were inserted into 3-mm deep holes in threaded Cu mounts and cemented in place with stopcock grease and electrically-conductive varnish. A pair of mutual inductance coils surrounded the samples in each device.

3. MEASUREMENTS

Five devices, a calibrated germanium thermometer, and the calibrated RIRT were screwed into a Cu block, with stopcock grease to aid thermal contact. The block was mounted *in vacuo* beneath a He-3 refrigerator. The cryostat was located in the center of three orthogonal, room temperature Helmholtz coils.

The device mutual inductances were monitored with a simple bridge circuit (1). Using an X-Y recorder, each superconductive transition was traced against the resistance of a calibrated germanium resistance thermometer so that the transition width W could be determined. W is defined as the temperature range spanned by the central 80% of a given transition.

After tracing each transition, the temperature of the copper block was adjusted so that the mutual inductance bridge registered the transition midpoint, which we arbitrarily chose to define as T_c. At least once during a measurement cycle, we adjusted the current in each Helmholtz coil pair so that a T_c was maximized, thus ensuring a minimum magnetic-field environment in the volume occupied by the samples.

While a particular sample was held at its transition midpoint, the RIRT resistance was measured with a computer-controlled resistance bridge (3). At least four, and as many as twelve, individual resistance measurements were averaged to obtain $R(T_c)$. The five samples of a given element were measured in sequence, often with the reference sample being measured both at the start and at the end of the sequence.

4. RESULTS

4.1 Sample T_c Reproducibility

The reference device used most frequently was code-named "Lima". Table I contains a summary of its T_c data. n is the number of measurement cycles for which the T_c of a given sample was observed; $\overline{T}_c$ is the mean of the $R(T_c)$ values converted to temperatures on EPT-76 Scale (4); and S.D. is one standard deviation of the set of n measurements.

TABLE I.

T_c Reproducibility of Reference Device "Lima"

Sample	n	$\overline{T}_c$, K (EPT-76)	S.D., mK
Nb	11	9.270 15	0.13
Pb	10	7.200 055	0.042
In	11	3.414 46	0.054
Al	12	1.181 04	0.033
Zn	10	0.850 22	0.049
Cd	12	0.520 01	0.083

Note that even for Nb, the sample T_c reproducibility compares favorably with the precision of present cryogenic thermometry.

4.2 Uniformity of Elemental T_c's

The uniformity of T_c for a given element was found to be independent of the transition width, so long as a certain width was not exceeded. These maximum allowable widths are 1.0 mK for Pb and Al, 1.1 mK for In, and 2.8 mK for Zn and Cd. No such criterion was found for Nb.

The range of T_c values found for the samples of the various elements that satisfied the width criterion are shown in Table II. In this table, N is the number of different samples of an element, and R is the range of T_c values found for that element.

TABLE II.

Uniformity of T_c for Six Elements

Element	N	R, mK
Nb	20	90.
Pb	19	0.14
In	17	0.52
Al	18	0.44
Zn	18	0.32
Cd	20	0.24

Examination of Table II shows that, whereas the superconductive transition temperatures of Nb samples appear to be too variable for present use in defining a universal temperature scale, a scale precision within ±0.3 mK is feasible in the range covered by the other elements studied (0.5 K to 7.2 K).

5. AVAILABILITY

In order to assist the low-temperature community in making use of the results described above, we are distributing the fixed-point devices prepared in this study through the NBS Standard Reference Materials program (5). More devices will be prepared if needed; all devices will be tested against the reference devices to ensure that they meet the range criterion of Table II and the width criterion mentioned in Section 4.2.

Various efforts are underway to provide improved values of T_c by thermodynamic measurements and to ascertain the T_c uniformity for other sources of the elements studied in this program.

REFERENCES

(1) J.F. Schooley, R.J. Soulen, Jr., and G.A. Evans, Jr., Cryogenics, April 1980, pp. 193-199.

(2) R.L. Rusby, <u>Temperature, Its Measurement and Control in Science and Industry 5</u>, J.F. Schooley, Ed., American Institute of Physics, New York, NY, 1982, pp. 829-833.

(3) R.D. Cutkosky, ibid., pp. 711-713.

(4) E.R. Pfeiffer and R.S. Kaeser, ibid., pp. 159-167.

(5) SRM 767a available from Office of Standard Reference Materials, NBS, Washington, DC 20234.

*Not to be construed as an endorsement.

LT-17 (Contributed Papers)
U. Eckern, A. Schmid, W. Weber, H. Wühl (eds)

REFERENCE RESISTANCE BASED ON THE QUANTUM HALL EFFECT

L. BLIEK, E. BRAUN, F. MELCHERT, P. WARNECKE (PTB, Braunschweig)
W. SCHLAPP, G. WEIMANN (FTZ, Darmstadt)
K. PLOOG, (MPI FKF, Stuttgart)
G. EBERT (TU München)
G. DORDA (Siemens FL, München)

Quantized Hall resistances in GaAs-GaAlAs heterostructures and Si-MOSFETs have been investigated. The influence of the sample temperature, the step number and the material was found to be less than 3 parts in 10^8. The experimental results justify the use of a quantized Hall resistance as a highly reproducible reference resistance.

In high magnetic fields and at low temperatures, the Hall resistance R_H of suitable two-dimensional conductors exhibits plateaus, for which the simple relation

$$R_{H,i} = \frac{1}{i} \cdot \frac{h}{e^2} \qquad (1)$$

appears to hold, with integer values for i. The plateaus are accompanied by a drop in the longitudinal sample resistance. Ideally, this resistance approaches zero.

Since its discovery (1), this so-called Quantum Hall Effect (QHE) has been studied extensively from a theoretical point of view. As yet, no reasons have emerged to expect any correction to equ. (1) but its validity has not been ultimatively proved either. Therefore, the suitability of the QHE for the maintenance and the realization of the unit of resistance as well as for the determination of fundamental constants, which is suggested by equ. (1), has to be proved experimentally.

In our experiments (2) silicon MOSFETs and GaAs-$Ga_{1-x}Al_xAs$ heterostructures have been used. It turned out that in Si only $R_{H,4}$ and in the heterostructures only $R_{H,2}$ could be measured with a sufficiently low uncertainty which is due to the different energy separation in the two Landau level systems. In all experi-

Experiment	Results	
Reproducibility of $R_{H,2}$ in GaAs	$\sigma_m = 0.8 \cdot 10^{-8}$	(N = 35)
Reproducibility of $R_{H,4}$ in Si	$\sigma_m = 3 \cdot 10^{-8}$	(N = 22)
Temperature dependence between 4.2 K and 1.2 K	$\sigma_m = 3 \cdot 10^{-8}$	(N = 14)
Ratio of $R_{H,2}$ in GaAs to $R_{H,4}$ in Si	$2(1 + 3 \cdot 10^{-8} \pm 3 \cdot 10^{-8})$	(N = 22 for Si) (N = 12 for GaAs)

Table 1 Summary of experimental results (σ_m standard deviation of the mean of N single measurements)

ments it had been carefully checked that the residual resistance of the samples did not exceed a few mΩ over an appreciable portion of the plateau.

As can be seen from table 1, in our experiments, there was no observable dependence of the quantized Hall resistances $R_{H,4}$ and $R_{H,2}$ on temperature, step number or material within a standard deviation of $3 \cdot 10^{-8}$. Since this is one order of magnitude lower than the uncertainty to which the ratio of the as-maintained unit at PTB and the SI-unit of resistance is known (3), we recommend the application of the QHE to establish a highly reproducible reference resistance.

By use of the relation

$$\alpha^{-1} = \frac{2}{\mu_0 c}\frac{h}{e^2} = \frac{4}{\mu_0 c} R_{H,2} \qquad (2)$$

for the fine-structure constant, we can express our results in terms of a value for α^{-1}:

$$\alpha^{-1} = 137.035\,992\,(1 \pm 2.7 \cdot 10^{-7}) \qquad (3)$$

As shown in Fig. 1 this value is in excellent agreement with values for α^{-1} determined by other experiments (4). This agreement strongly supports the validity of equ. (1).

REFERENCES

(1) K. v. Klitzing, G. Dorda, and M. Pepper, Phys. Rev. Lett. 45 (1980) 494
(2) L. Bliek et al., PTB-Mitt. 93 (1983) 21
(3) H. Bachmair et al., PTB-Jahresbericht 1982, 130
(4) L. Bliek et al., Phys. Bl. 39 (1983) 157

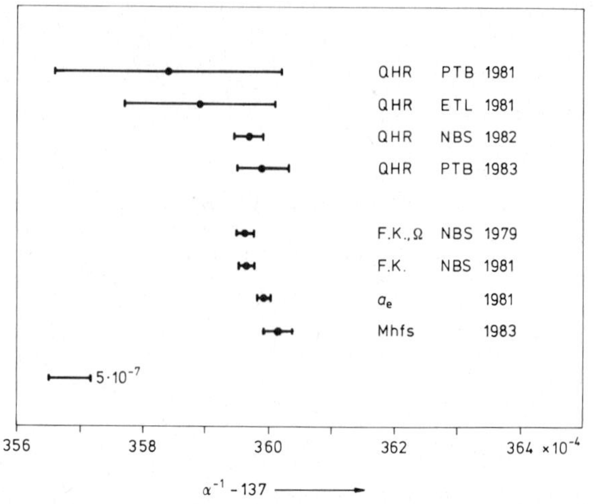

FIGURE 1
Values for the fine-structure constant α. QHR: measurements of quantized Hall resistances $R_{H,i}$ and the ratio of the as-maintained unit of resistance to the SI-ohm at three Standard Laboratories: ETL = Electrotechnical Laboratory, Japan; NBS = National Bureau of Standards, USA; and PTB, earlier data and present results. F.K., Ω: measurements of the gyromagnetic ratio of the proton in the low field $\gamma'_p(\text{low})_{NBS}$, the Josephson voltage-to-frequency ratio $(2e/h)_{NBS}$ and the ratio of the as-maintained ohm of the NBS to the SI ohm. F.K.: measurements of $(R_{H,i})_{NBS}$, $\gamma'_p(\text{low})_{NBS}$ and $(2e/h)_{NBS}$. a_e: based on the anomalous magnetic moment of the electron. Mhfs: based on the muonium hyperfine structure.

LT-17 (Contributed Papers)
U. Eckern, A. Schmid, W. Weber, H. Wühl (eds)

MILLIMETER WAVE INDUCED JOSEPHSON VOLTAGE STEPS ABOVE 1.0 VOLT WITH A 1474 JUNCTIONS ARRAY

Jürgen NIEMEYER, Johann H. HINKEN

Physikalisch-Technische Bundesanstalt, P.O. Box 3345, 3300 Braunschweig, Fed. Rep. of Germany

Richard L. KAUTZ

National Bureau of Standards, Boulder, Co 80303, USA

Series arrays of 100, 402, and 1474 Josephson tunnel junctions showed millimeter wave induced constant voltage steps up to >100 mV, 440 mV, and 1.2 V, respectively. The steps were stable and occurred near zero current bias. Their application in a standard with 1 Volt Josephson voltage seems possible.

1. INTRODUCTION

In Josephson voltage standards a resistive divider circuit is usually necessary to derive the output voltage of about 1 V from the precise microwave-induced Josephson voltage in the mV region. The calibration accuracy of the divider largely determines the accuracy of the output voltage. A higher Josephson voltage is desirable. The requirements regarding the divider then become less stringent or the divider could even be omitted. A series array of 20 Josephson junctions has therefore been used in (1). The individual current biasing of the junctions in (1) can be avoided by using tunnel junctions with zero or near zero current steps, (2) - (5). With this method and 70 GHz microwave irradiation on arrays with 54 Josephson junctions, voltages up to 34 mV have been obtained (6). In the following, experimental series arrays with voltages up to 1.2 V are described.

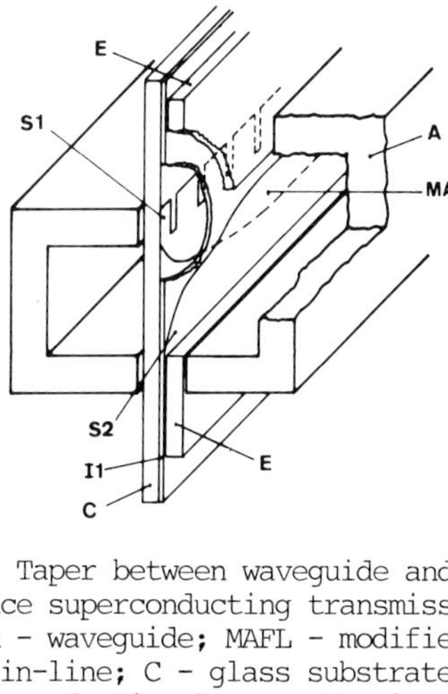

Fig. 1: Taper between waveguide and low-impedance superconducting transmission line. A - waveguide; MAFL - modified antipodal fin-line; C - glass substrate; S1, S2 superconducting layers; E - insulating glass plate, I1 - insulating layer

2. CIRCUIT

The millimeter wave integrated circuit with the junction array is coupled to an incident TE_{10} mode in rectangular waveguide by a planar taper as shown in fig. 1. The profile of the broadband taper is calculated according to (7), (8). The waveguide structur at the taper end is a modified antipodal fin-line, which in turn is transformed into a microstrip line of low impedance, see fig. 2. The circuit with the Josephson junctions was designed to have a mm-wave voltage dispersion across the junctions of less than 15 %. The junctions, about 25 μm x 50 μm in size, have critical currents of about 250 μA.

The films were structured by photolithography on 0.3 mm thick glass substrates. Layer S1 was niobium, covered by layer I1 (SiO, 0.25 μm thick). The layers forming the junction electrodes were PbInAu (S2) and PbAu (S3). Layer I2 (SiO, 0.25 μm thick), containing the windows for the junctions, was between S2 and S3. The junction oxide was made by low-pressure rf oxidation (9). The N1 layer for a matched termination consisted of an InAu alloy.

3. MEASUREMENTS

The specimens were measured at 4.2 K inside a liquid helium dewar. Measurement results with 100, 402 and 1474 junction arrays are shown in table 1.

The 100 junctions array showed stable zero current steps up to 80 mV and 100 mV at 70 GHz and 90 GHz, respectively. The step widths were ΔI = 50 μA and ΔI = 65 μA, respectively. With a small offset current common to all series-connected junctions, stable steps beyond 100 mV were observed.

The array with 402 junctions showed stable zero current steps up to 200 mV (ΔI = 50 μA) and 360 mV (ΔI = 80 μA) at 70 GHz and 90 GHz, respectively. With small offset currents the maximum stable steps rose to 360 mV and 440 mV (ΔI = 40 μA); respectively.

Fig.2: Layout of the circuit.

S1, S2, S3 - supercond. layers; N1 - normal conducting layer; insulating layer I1 between S1 and S2; insulating layer I2 between S2 and S3 except at the junction windows; MAFL - modified antipodal fin-line; ML - microstrip-line.

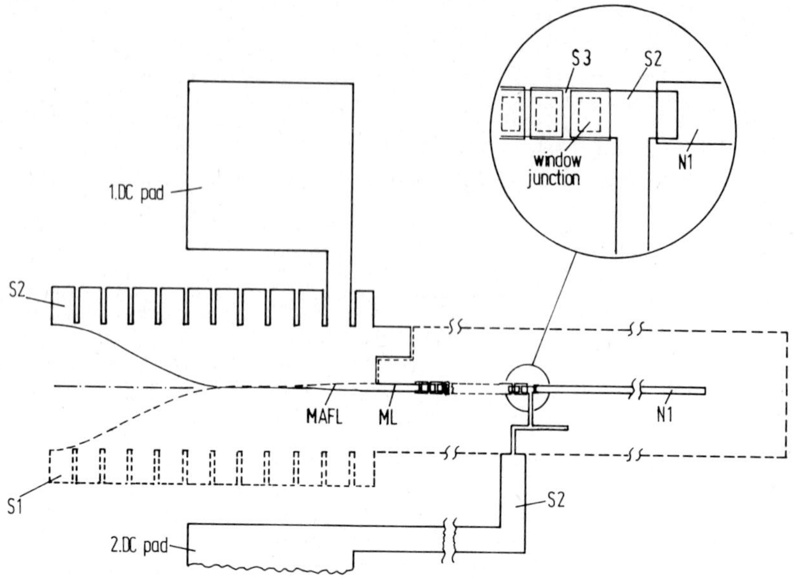

Table 1: Maximum millimeter-wave induced voltages measured across series arrays of Josephson tunnel junctions with zero and near zero current bias.

Number of junctions	Frequency	Maximum voltage for zero current	near zero current
100	70 GHz	80 mV	> 100 mV
	90 GHz	100 mV	>100 mV
402	70 GHz	200 mV	300 mV
	90 GHz	360 mV	440 mV
1474	90 GHz	1.0 V	1.2 V

In the 1474 junctions array, the junction chain was too long for the junctions to be placed in a straight line. The chain was therefore folded to form a meandertype structure with three parallel straight lines connected through two 180° bends. Although these bends may cause a stronger mm-wave voltage dispersion at the junctions than desired, the array at 90 GHz showed zero current steps up to 1.0 V and near zero current steps up to 1.2 V.

4. CONCLUSIONS

These experimental Josephson voltages higher than 1.0 V show that it should be possible to build up a Josephson voltage standard for 1 V without a resistive divider circuit. It must be checked, however, that step rounding does not introduce errors in the output voltage.

REFERENCES

(1) Koyanagi, M. et al., Proc. II. Int. Conf. Prec. Meas. Fund. Const., Gaithersburg/ U.S.A., June 1981.

(2) Levinsen, M.T. et al., Appl. Phys. Lett. 31 (1977) 776.

(3) Kautz, R.L., Costabile, G. IEEE Trans. Magn. MAG-17 (1981) 780.

(4) Cui Guang-ji et al., Joint Sino-Japanese Sem. on Jos. Eff. Phys. and Appl., Beijing/ China, Oct. 1983, Proc. p. 119.

(5) Costabile, G. et al., Intern. Cryogenic Materials Conf. Colorado Springs, 1983.

(6) Niemeyer, J. et al., to be published in IEEE Trans. Instrum. Meas.

(7) Hinken, J.H., Arch. Elektr. Übertr., 37 (1983) 375.

(8) Hinken, J.H. et al., to be published.

(9) Greiner, J.H. et al., IBM Journ. res. develop. 24 (1980) 195.

LT-17 (Contributed Papers)
U. Eckern, A. Schmid, W. Weber, H. Wühl (eds)

A SENSITIVE DIFFERENTIAL MANOMETER FOR MEASUREMENT OF SUPERFLOW IN ^{3}He

Gregory F. SPENCER and Gary G. IHAS*

Department of Physics, University of Florida, Gainesville, Florida, U.S.A.

A very sensitive differential manometer has been constructed for use in superfluid ^{3}He flow experiments. The manometer is a capacitive transducer with a flexible diaphragm which responds to pressure gradients. The device is non-metallic and has essentially zero internal open volume. Because its sensitivity depends primarily upon the tension in the diaphragm, the design and construction of the transducer allows for a wide variation of tension by changing a single construction parameter. In this manner, the sensitivity may be varied over several orders of magnitude as desired. At present, the most sensitive manometer has a resolution of $\delta P=5.5\times10^{-8}$ bar at room temperature at 1 bar ambient pressure, which increases at low temperature.

1. INTRODUCTION

One of the most dramatic properties of a superfluid is its ability to undergo superflow. In an open geometry, like a tube, superflow could appear without a pressure gradient across the length of the tube. In an experiment to study the flow of superfluid ^{3}He-B in a tube, a non-metallic capacitive transducer for use as a differential manometer has been constructed. The design is such that by variation of a single parameter in the construction the sensitivity range may be selected. Using the same apparatus, transducers have been built with sensitivities ranging over four orders of magnitude (10^{-4} bar to 10^{-8} bar). The most sensitive transducer at present can resolve $\delta P=5.5\times10^{-8}$ bar at room temperature. This sensitivity has been seen to increase during tests at 1 K, although it has not been calibrated. This particular transducer is not yet at the design limit for construction.

2. DESIGN AND CONSTRUCTION

The sensitivity of the manometer depends primarily upon the tension in the diaphragm. To be able to control this quantity, various fixtures have been built to control the strain in the diaphragm during construction. Consider a flat circular disk of overall radius R, with a step of height h cut at radius r<R. If a diaphragm of original radius R is laid across the disk and then stretched until it touches the lower step, the resulting strain in the diaphragm, assuming $h\ll(R-r)$, is $h^2[2R(R-r)]^{-1}$. The design of the manometer uses this simple relation to be able to calibrate the strain in the diaphragm during construction. By variation of the step height, the tension may be adjusted up or down.

The manometer is shown in Figure 1. The body consists of two circular disks made of Stycast 1266 epoxy (1) (diameter=2.54 cm, thickness=0.32 cm) which, after preparation, are epoxied together face to face. The front of each disk, corresponding ultimately to the interior walls of the manometer, has been machined to create a step on one disk, a bevel on the other, each at a 1.98 cm diameter. Before any polishing of the faces two thru-holes are drilled into each disk, one hole (0.05 cm) at the center to be used for a liquid connection, the other hole (0.07 cm) being placed just inside the bevel (or step) radius. This second hole is to make electrical connection to the electrode which is later placed on the face. To make this connection, a short channel (2-3 mm) is dug in the face using a file, going from the outer hole toward the center hole. The depth of this channel is approximately one-half the diameter of the copper wire to be used. The wire is inserted back-to-front and the leading end bent into an L shape. The wire is then laid to rest in the channel and epoxied into place. Upon drying, the front face is rough sanded until it is again flat, leaving the epoxied copper wire now exposed in the plane of the face. The disk is then polished in a lapping machine to achieve a mirror finish, the final step using 0.05 μm alumina. The resulting surface is specular and although not perfectly flat, its radius of curvature has been measured by optical interference to be at least 73 meters. Into the center hole a Cu-Ni capillary is epoxied for the liquid connection. Then, each disk is placed into a standard bell jar evaporator. Using a mask, a gold electrode is evaporated onto the front face, the electrode being circular (diameter=1.03 cm) with a finger extending to

*Work supported in part by U.S. NSF Grant DMR-8306579.

the inlaid copper wire to make contact. The electrode is estimated to be 3000 Å thick. This completes the initial preparation of each disk.

To epoxy a diaphragm to the stepped disk with the desired tension, a special clamping fixture is used. It consists of two identical aluminum rings with smooth faces whose ID's are made to snugly fit the OD of the step-disk. The two rings can be fastened together face-to-face using screws. A second fixture with a plunger can be fixed to one of the rings to allow gross placement of the step-disk with respect to the face of that ring. Since the two rings are used as a tensionless clamp to hold the diaphragm, the most important task is to place the stepped disk in the ring so that its outer circumference is flush with the ring face. In this manner, when the diaphragm is attached, the disk protrudes into the clamped diaphragm by exactly its step height to produce the desired strain. To assure the disk is flush with the ring face, a height indicator is used which can measure to ±2 μm. With the disk in place, a very small amount of epoxy is applied to the outer edge of the disk and allowed to set for 1/2 hour to become viscous. Next, an annular ring of mylar (OD=1.96 cm, ID=1.28 cm, thickness = 25 μm) is laid on top of the disk to be used as a spacer. The diaphragm material is an 8 μm Kapton film (2), which is laid on the ring and clamped into position in contact with the spacer and epoxy. Some migration of liquid epoxy does occur but if a small enough amount is used and sufficient time allowed before attaching the diaphragm, it does not migrate up onto the step or into the active region. After setting overnight, the disk with diaphragm attached is placed into the evaporator again using the same mask to produce a gold electrode on the diaphragm. Afterward a copper wire is attached using conducting paint to the gold finger which extends beyond the step. Next a second spacer is placed on the diaphragm and the beveled disk is brought into contact with it. The two halves are rigidly clamped together and epoxy is applied around the circumference into the bevel. In this manner, the spacings on both sides of the diaphragm are determined by the mylar spacers. This completes the construction of the manometer.

3. TESTING

Leak and sensitivity tests have been performed on several transducers at room temperature, 77 K and down to 1.6 K. The transducers are leak-tight to the outside world up to about 14 bar internal pressure. Because of the single thickness of Kapton film, leak checks across the diaphragm can only be performed at 4 K or lower, with the result that the diaphragms are in fact leak-tight. The diaphragms are susceptible to shock when performing a leak check across them and care must be taken to treat them gently. To check sensitivity, the manometer is connected via separate fill lines to a gas handling board at room temperature where a differential pressure gauge (10 Torr Baratron head) (3) is used. A differential pressure is created across the manometer, using the gauge to measure it, and the response of the manometer is measured capacitively using a standard bridge circuit. As the differential pressure is increased, the manometer saturates at about ΔP=10 Torr, probably due to contact of the diaphragm with the body. The highest sensitivity is of course around ΔP=0, the region for which the transducer was designed. The most sensitive manometer, employing a step height of 0.127 mm giving a strain of .028%, has a room temperature sensitivity of $\delta P=5.5\times10^{-8}$ bar at an ambient pressure of 1.15 bar. This sensitivity was apparently increased at 4 K but it was not possible to get an accurate calibration. When mounted on our research cryostat, the manometer will be calibrated against the viscosity of normal liquid ^{3}He.

Transducers using step heights of .025 mm and .050 mm are somewhat more difficult to make, particularly during the polishing phase. Also, a transducer with no step is being built. The only strain in the diaphragm will be due to the spacer. This transducer, if the epoxy migration can be controlled, will have 36 times less strain in the diaphragm and a correspondingly increased sensitivity.

REFERENCES

(1) Emerson and Cuming, Inc., Canton, Massachusetts 02021.

(2) DuPont Company, Polymer Products Department, Wilmington, Delaware 19898.

(3) MKS Instruments, 25 Adams Street, Burlington, Massachusetts 01803.

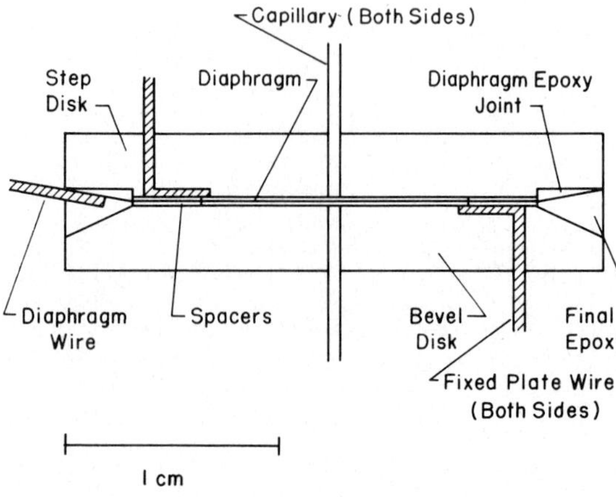

FIGURE 1
Cross-section of the manometer; the area around the spacers has been expanded for clarity.

LT-17 (Contributed Papers)
U. Eckern, A. Schmid, W. Weber, H. Wühl (eds)

$^3He/^4He$ ISOTOPIC RATIO MEASUREMENTS TO BELOW THE 10^{-12} LEVEL*

P.C.HENDRY and P.V.E.McCLINTOCK

Department of Physics, University of Lancaster, Lancaster LA1 4YB, UK.

A miniature helium isotopic concentration cryostat is described. By means of this device, the $^3He/^4He$ ratio of a given sample of helium gas can be increased by a measurable factor of up to ca 400, thereby enhancing by the same factor the ultimate sensitivity of $^3He/^4He$ ratio determination by mass spectrometry.

1. INTRODUCTION

Isotopically purified ^4He is increasingly being required in connection with a number of diverse applications including, for example: (a) investigations of quantised vortex ring nucleation in HeII (1); (b) experiments on quantum evaporation (2); (c) the storage of ultra-cold neutrons (UCN) in HeII (3); and (d) the long-term storage of ions in stable orbits in HeII at mK temperatures (4). Although a superfluid heat flush technique has been shown to be well capable of preparing ^4He of the required purity (5), there is no easy way to analyse samples of the helium being used in order to check for possible accidental contamination with ^3He at very low levels.

The difficulty arises because the admirable analytical service offered by the U.S.Bureau of Mines (6), based on mass spectrometry, does not quote $^3He/^4He$ ratios below 2×10^{-10}. This sensitivity, although impressive, is completely inadequate (by a factor of ca 100) for (c) and (d). It is also the case that the secondary heat flush used in the prototype batch-processing purifier (5) is inapplicable to the continuous flow machine (7) now in use so that, even at the time of production, it is impossible to ascertain whether or not the product is of the requisite purity.

We describe below a miniature isotopic concentration cryostat that we are developing specifically in order to overcome this problem, and we report and discuss some of our preliminary results.

2. METHOD

The aim is to take a fairly large sample (up to 40 ℓ at STP) of the gas to be analysed and then to concentrate all of the ^3He from it into a very much smaller volume (ca 60 cm^3). It is the latter sample that is then sent for analysis (6) by mass spectrometry.

The lower part of the insert for the concentrator cryostat is illustrated schematically in Figure 1. In operation, the liquid helium (a) in the glass dewar surrounding the insert is cooled to ca 1.1 K by direct pumping. The sample of helium to be analysed is then admitted via the 0.5 mm capillary (b), causing the temperature of the bath to rise to ca 1.4 K. As the sample condenses, it passes through the needle valve (c) and progressively (over a period of ca 40 minutes) occupies the vacuum-insulated pot (d). Slots (not illustrated) are cut in the needle so as to facilitate the passage of liquid or heat. During this process, a few mW of heat are supplied via the heater (e). When condensation is deemed to be complete, and the bath has returned to its base temperature, the needle valve (c) is closed.

Throughout the entire condensation procedure, a steady flow of heat passes down through the

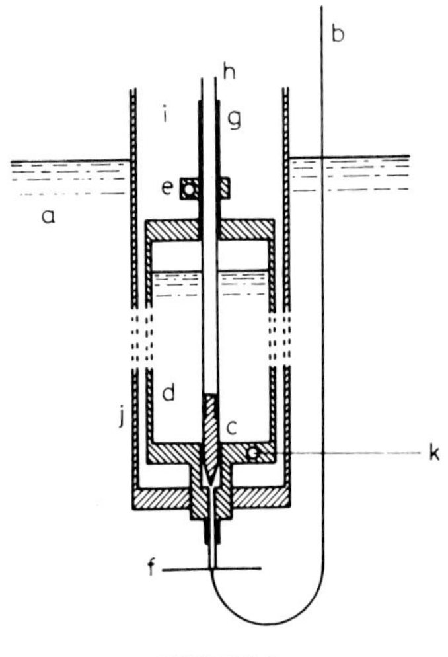

FIGURE 1
Lower end of the cryostat insert: a - HeII at ca 1.1 K; b - 0.5 mm OD capillary tube; c - needle valve assembly; d - helium pot; e - heater in copper block; f - copper disk; g - needle valve guide tube; h - needle valve operating rod; i - vacuum; j - vacuum can; k - thermometer.

*Work supported by the Science and Engineering Research Council (UK).

TABLE 1
Summary of the results of two experimental runs with the concentration cryostat

Starting pressure (bar)	Calculated concentration factor	$^3He/^4He$ ratio of product	Measured concentration factor
0.74	500 ± 18	$(3.2 \pm 0.2) \times 10^{-5}$	430 ± 30
0.33	247 ± 7	$(1.5 \pm 0.1) \times 10^{-5}$	197 ± 16

helium in the pot (d) and eventually, partly via the copper disk (f), into the main helium bath (a). In part, the heat comes down the needle valve guide tube (g) and operating rod (h) from room temperature, and in part from the heater (e). The corresponding flush of normal fluid component should prevent (5) any ^3He atoms that are present in the sample from entering (d). They may be expected to congregate, instead, at the coldest point in the system : that is, inside the copper tube fixed to the copper disk (f), illustrated in greater detail in Figure 2, and in the adjacent capillary tube (b). When the needle valve is finally closed, therefore, the sample is effectively divided into two sections such that (d) contains isotopically pure ^4He and all the ^3He is concentrated in the very much smaller volume below the valve.

The apparatus is then allowed to warm up and, in doing so, to expel the now divided sample to fill two containers of known volume at room temperature. From direct measurements of their respective final pressures, the concentration factor that should have been achieved is very easily calculated (8).

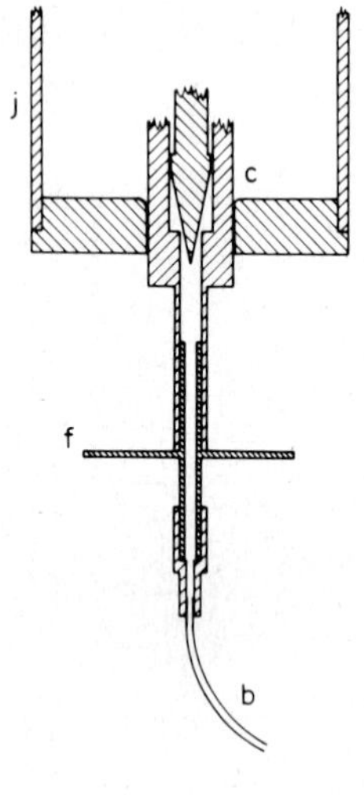

FIGURE 2
Expanded view of lower end of cryostat insert, showing its construction: the labelling of components is consistent with Figure 1.

3. RESULTS

In order to test the technique, we have operated the concentrator in conjunction with a helium sample of known (6) isotopic ratio (7.5 ± 0.1) x 10^{-8}. The results of two experimental runs, using different sample sizes, are summarised in Table 1. The starting pressure refers to the initial pressure of the sample in its 50 ℓ container; the calculated concentration factor is arrived at through measurement of the final pressures: the product $^3He/^4He$ ratio refers to analysis (6) of the portion of the sample evaporating from below the needle valve; and the measured concentration factor gives the increase in $^3He/^4He$ ratio over that of the initial sample. It will be noted that the measured and calculated concentration factors are in encouragingly good agreement, the difference in each case lying only just outside the respective error estimates.

4. CONCLUSION

The preliminary results reported above may be regarded as extremely promising. In effect the minimum measurable $^3He/^4He$ ratio has been reduced from 2 x 10^{-10} to 5 x 10^{-13}, which should certainly provide adequate sensitivity to support all the applications of isotopically pure ^4He that are currently in view.

ACKNOWLEDGEMENTS

It is a pleasure to acknowledge the invaluable technical assistance of G.Caley, I.Marsden and G.F.Turner.

REFERENCES

(1) R.M.Bowley, P.V.E.McClintock, F.E.Moss, G.G.Nancolas and P.C.E.Stamp, Phil.Trans.R. Soc.Lond.A 307 (1982) 201.
(2) M.J.Baird, F.R.Hope and A.F.G.Wyatt, Nature 304 (1983) 325.
(3) C.Jewell, B.Heckel, P.Ageron, R.Golub, W.Mampe and P.V.E.McClintock, Physica 107 B (1981) 587.
(4) W.F.Vinen, private communication.
(5) P.V.E.McClintock, Cryogenics 18 (1978) 201.
(6) All our helium samples have been analysed by Mr.D.E.Emerson of the U.S. Bureau of Mines, Amarillo, Texas.
(7) T.H.Ngan, J.C.H.Small and P.V.E.McClintock, Physica 107 B (1981) 597.
(8) Full details of these calculations, together with a description of the gas handling system, will be presented elsewhere.

LT-17 (Contributed Papers)
U. Eckern, A. Schmid, W. Weber, H. Wühl (eds)
© Elsevier Science Publishers B.V., 1984

HIGH PRESSURE ANOMALY OF SUPERCONDUCTIVITY OF TIN AND PRESSURE SCALES

Dirk-Roger SCHMITT and Wolfgang GEY

Institut für Technische Physik der Technischen Universität Braunschweig, Mendelssohnstraße 2, 33oo Braunschweig, Federal Republic of Germany

A new high pressure apparatus (10 GPa) has been used for a detailed investigation of the pressure shift $\Delta T_c(p)$ of the superconducting transition of Pb, Sn, and amorphous ZrPd. Mutual comparison uncovers a small kink in $\Delta T_c(p)$ for Sn at about 3 GPa. Comments on existing pressure scales are made.

Recently, the quality of low temperature-high pressure-data has been significantly improved by the use of a diamond cell with a liquid pressure medium (at room temperature) and a SQUID system for the detection of bulk superconductivity via flux exclusion (1). Although, of course, the liquid freezes upon cooling the pressure stays extremely homogenous. This is manifested, e.g., by very sharp lead transitions which correspond to a pressure gradient across the sample of less than 0,08 GPa up to pressures of 10 GPa. This upper pressure bound may further be extended without quality loss by warming the pressure device during pressurizing, thus keeping the liquid's viscosity low. Recording the superconducting transition in a dc-type fashion by monitoring the exclusion of a small auxiliary magnetic field with a superconducting pick up coil and a SQUID can provide a reproducibility of $\Delta T_c(p)$ of better than 2 mK.

The accuracy achieved here would seem to be of little use at present, as there are the well known pressure calibration problems for pressures in excess of 1 GPa. It is, however, possible to draw important conclusions by directly comparing the pressure shifts of different superconductors which are mounted in the same pressure cell, i.e. are subject to the same pressure. Two lines may be followed:

- i: Search for possible irregularities in $T_c(p)$. Here exact knowledge of the pressure is less important; however, independent checks on at least three materials are required.
- ii: Information may be obtained on the reliability of existing pressure calibrations for $p > 1$ GPa, which invariably rest on assumptions.

An example for the first approach is shown in Fig. 1. In the upper part comparison is made between Pb and Sn. Clearly a kink appears, which, when the conventional Pb-calibration is used (2) is located near 3 GPa. It is not clear however, whether the kink belongs to Pb or Sn, or both. For a decision amorphous materials would seem to carry little anomalous $T_c(p)$ structure, thus reducing the number of needed independent checks to unity.

The middle and the lower parts of Fig 1 show direct comparisons of a-$Zr_{76}Pd_{24}$ with Sn and Pb, respectively. Although a-ZrPd is anomalous in a sense as it shows a positive, progressive $T_c(p)$-variation, the kink is clearly seen for the ZrPd-Sn-comparison, while for ZrPd-Pb no deviation from a smooth trace is detectable. Thus the observed irregularity is most likely due to tin.

We do not comment on the origin of this anomaly which amounts to a change in the $T_c(p)$-slope of approximately 30 % at 3 GPa. We rather note that tin has been used as a key material for an extrapolation of the pressure scale beyond 1 GPa: It was postulated that T_c of tin would vary linearly with the volume change $\Delta V/V$ (3). Unless this volume change exhibits an analogous kink, which appears unlikely, this concept must be abandoned.

Eiling and Schilling have used their $T_c(p)$ calibration curve, which is based on the resistance of lead, to present their tin measurements up to 4,5 GPa (4). Inspection of these tin data, after transformation into $T_c(\Delta V/V)$ (5), shows essentially a linear variation with $\Delta V/V$. One thus concludes, in the above sense that this calibration is to be reexamined.

REFERENCES

(1) J. Ottow, Dissertation, Technische Universität Braunschweig, (1983).
(2) A. Eichler and J. Wittig, Z. angew. Phys. 25 (1968) 319.
(3) T. F. Smith and C. W. Chu, Phys. Rev. 159 (1967) 353.
M. J. Clark and T. F. Smith, J. Low Temp. Phys. 32 (1978) 495.
(4) A. Eiling and J. S. Schilling, J. Phys. F.: Met. Phys. 11 (1981) 623.
(5) T. F. Smith et al., Cryogenics 9 (1969) 53.

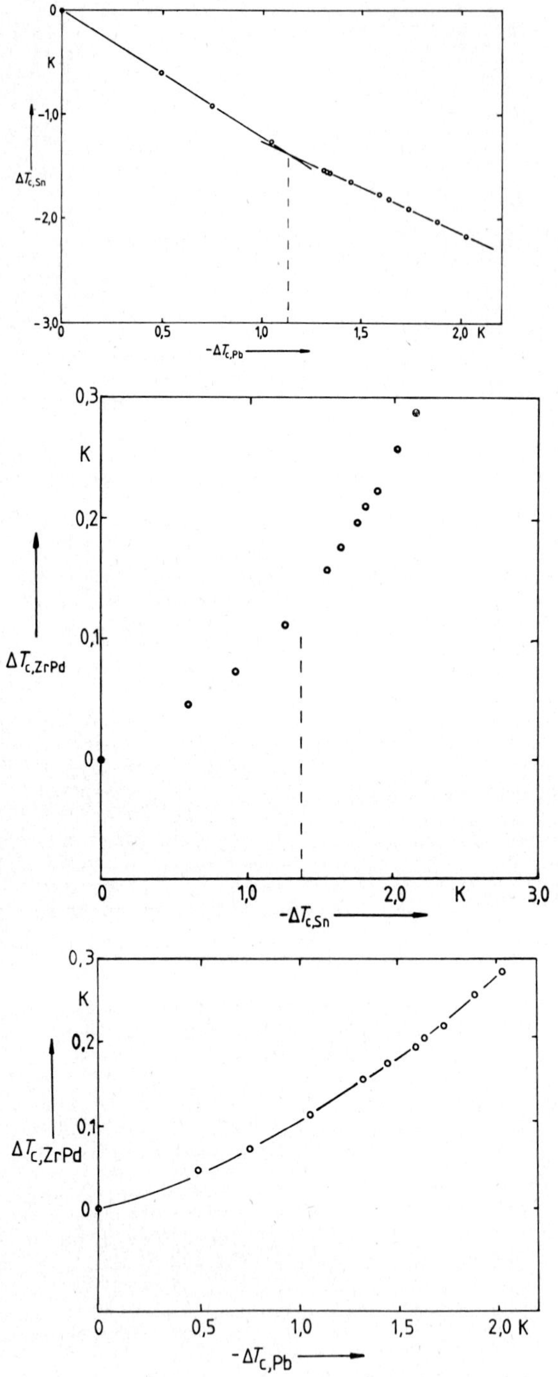

Fig. 1.: Direct comparison of the pressure shifts ΔT_c of the transition temperature of the three superconductors Pb, Sn, and $Zr_{76}Pd_{24}$ which were mounted in the same pressure cell.

LT-17 (Contributed Papers)
U. Eckern, A. Schmid, W. Weber, H. Wühl (eds)
© Elsevier Science Publishers B.V., 1984

A SUPERCONDUCTING RECTANGULAR CAVITY RESONATOR FOR THE MEASUREMENT OF HIGH DC VOLTAGES

Wolfgang LUCAS, Klaus SCHON, Johann Heyen HINKEN

Physikalisch-Technische Bundesanstalt, Braunschweig, Fed. Rep. of Germany

A superconducting rectangular cavity resonator made of niobium is used as a speed filter for electrons which are accelerated by voltages between 40 kV and 100 kV. As only electrons with well-defined velocities can pass the resonator by a small aperture this effect is utilized for the measurement of high dc voltages.

1. INTRODUCTION

In practice, high dc voltages are usually measured by means of resistive voltage dividers which should be calibrated by a standard measuring device. In general the standard device consists of a precision divider, and relative uncertainties of a few parts in 10^5 for the measurement of a 100 kV voltage are achievable and generally satisfactory. However, due to the same operating principle for both the standard divider and the one to be tested, unrecognized systematic errors may occur.

Realizing this fundamental disadvantage, many metrological institutes have made attempts to develop alternative methods for the precise measurement of high dc voltages. In this paper, the use of a superconducting rectangular cavity resonator is described, referring the measurement of high dc voltages to that of frequencies, i.e. of the base unit "second" which, in principle, can be measured with high accuracy.

2. MEASURING PRINCIPLE AND EXPERIMENTAL SET-UP

The method is based on the calculable interaction of accelerated electrons with the microwave field inside a rectangular cavity resonator /1, 2, 3/. The experimental set-up is shown in Fig. 1. Monoenergetic electrons are emitted and accelerated by the electron gun 1 according to the applied voltage U between 40 kV and 100 kV. The electrons enter the rectangular cavity resonator 2 in its mid-axis by a small aperture and are deflected by the microwave field of a H_{10p} mode. Only electrons with well-defined velocities will leave the resonator on its electron-optical axis by a second aperture, whereas electrons with other velocities are deflected in such a manner that they hit the resonator walls. The rectangular cavity resonator therefore acts as an electron speed filter (ELFI). The electrons leaving the resonator on its mid-axis are detected by the Faraday cup 3 which produces a reading of the ammeter.

The accelerating voltages for electrons which pass the resonator are calculable and read /1,2, 3/:

$$U_n = \frac{c_o^2}{e/m_o}\left[\left(1 - k_n \frac{f_{o1}^2}{f_{o2}^2 - f_{o1}^2}\right)^{-\frac{1}{2}} - 1\right] \quad (1)$$

where

$$k_n = \frac{p_2^2 - p_1^2}{(2n + p_1)^2} \quad \text{for } n = 1, 2, 3 \ldots \quad (2)$$

Thus a set of voltage calibration values can be obtained by the measurement of two resonance frequencies of the H_{10p} mode inside the resonator, f_{o1} and f_{o2}, and knowledge of the elementary charge per electron rest mass, e/m_o. p_1 and p_2 designate the number of half waves corresponding to f_{o1} and f_{o2}, and c_o is the speed of light which was defined in 1983.

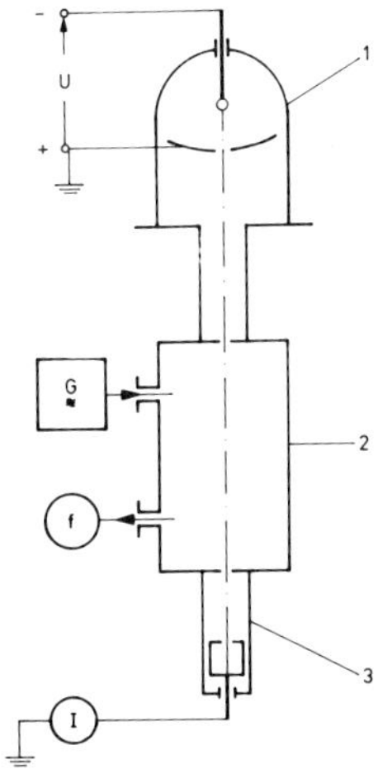

FIGURE 1: Experimental test set-up

In the preliminary experimental investigations a rectangular resonator made of copper was used as ELFI at room and LN_2 temperatures. According to (1), five non-equidistant calibration values could be determined in the range between 40 kV and 100 kV. The relative uncertainty was estimated to be $4 \cdot 10^{-4}$ /1, 2, 3/.

In order to reduce the total uncertainty to a few parts in 10^5 the investigations were continued using a superconducting cavity resonator with a considerably higher unloaded Q factor.

3. IMPROVEMENTS ACHIEVED WITH THE SUPERCONDUCTING RECTANGULAR CAVITY RESONATOR

The rectangular cavity resonator is made of niobium (T_c = 9.2 K) and has the same nominal dimensions (60 mm x 30 mm x 400 mm) as the copper one previously used /1, 2, 3/. The diameters of the microwave coupling holes and probes are 1 mm and 0.4 mm, respectively. The penetration depths of both coupling probes are adjustable from outside the cryostat in order to obtain the optimum position for achieving a high loaded Q factor.

The resonator is operated in a feed-back loop together with a travelling wave tube amplifier, a phase shifter, a variable attenuator and a filter for the resonant frequency. Due to the high Q factor of the superconducting resonator the instability of the resonant frequency was ascertained to be less than $1 \cdot 10^{-8}$.

By the logarithmic decrement method Q factors between 10^7 and 10^8 were measured. This is an improvement of three orders of magnitude compared to that of the copper resonator at LN_2 temperature. According to theory, the transversal electromagnetic field components E_y and H_x by which the electrons are deflected increase with the square root of the Q factor, thus increasing the selectivity of ELFI by the same amount. For the copper and the superconducting niobium resonator, Table 1 gives the unloaded factor Q_o, the electrical field strength E_y and the maximum deflection y_e of those electrons at the end face of the resonator which are parallel to its mid-axis.

Resonator	T in K	Q_o	$\hat{E}_y$ in Vm^{-1}	$\hat{y}_e$ in µm
Cu	300	$2.1 \cdot 10^4$	$6.9 \cdot 10^4$	13
Cu	77	$5.9 \cdot 10^4$	$1.2 \cdot 10^5$	22
Nb	4.2	$4 \cdot 10^7$	$2.9 \cdot 10^6$	540

Table 1: Characteristics of the resonators

Apart from the higher Q factor the superconducting resonator shows further advantages. Magnetic fields from outside the resonator, e.g., the magnetic earth field and low-frequency fields of the 50-Hz power supply, cannot penetrate the resonator and cannot influence the electron beam inside. This facilitates the adjustment of the beam, though magnetic interference is still possible inside the normal conducting part of the tube between the electron gun and the resonator (Figure 1). Moreover, the distortion of the microwave field due to the coupling devices is considerably reduced because the probes do not penetrate the superconducting resonator as deeply as they do the resonator with normal conduction.

If the coupling probes are regarded as small antennas of length l, the relative field distortion $\Delta E/E$ at the intersections of the resonator axis and the antenna axis can be estimated from:

$$\frac{\Delta E}{E} \sim \frac{1}{\sqrt{Q_o}} \tag{3}$$

For the superconducting resonator with $Q_o = 10^8$ and l = 1 mm, this relative field distortion is less than 10^{-5} and can therefore be neglected.

4. TEST RESULTS

Figure 2 shows the relative current difference versus voltage where U_s is the reading of the regulator for the accelerating voltage U, I_o and I_D are the detector currents without and with resonator excitation. From the resonance frequencies measured

$$f_{o1} = 4.50664 \text{ GHz and } f_{o2} = 4.82322 \text{ GHz}$$

the voltages U_n for n = 9, 10, 11 were calculated by (1) and marked on the voltage scale.

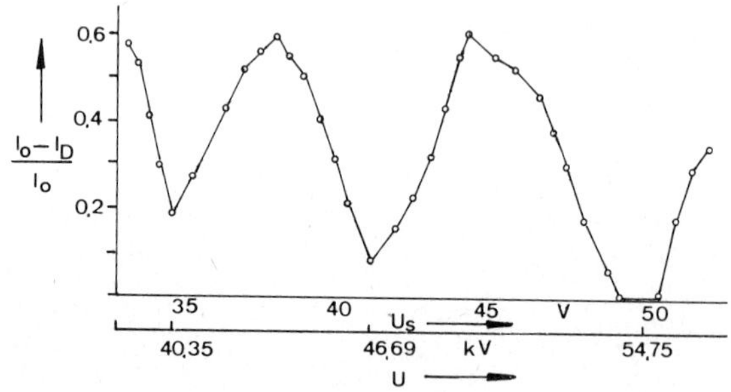

FIGURE 2
Relative difference of the detector current versus accelerating voltage at LHe temperature. Aperture diameter d = 100 µm

5. CONCLUSIONS

The experimental results using a superconducting resonator agree well with theory and demonstrate again the suitability of the method for establishing calibration points of the voltage scale between 40 kV and 100 kV. As expected, the selectivity of the superconducting resonator is higher than that of the copper one at 77 K with the same aperture. Improvements in adjusting the electron beam will enable the use of smaller apertures to increase the selectivity.

A thorough investigation of the total measuring uncertainty has not yet been completed. With the achievement of a few parts in 10^5, this method will be suitable for the dissemination of high dc voltages.

REFERENCES

/1/ B.Schulz,Dissertat.TU Braunschweig, 1981
/2/ D.Peier,B.Schulz,Metrologia 19(1983) 9-13
/3/ D.Kind,W.Lucas,D.Peier,B.Schulz
IEEE Transactions Vol.IM-32 (1983), 8-11

LT-17 (Contributed Papers)
U. Eckern, A. Schmid, W. Weber, H. Wühl (eds)
© Elsevier Science Publishers B.V., 1984

THE SURFACE ANISOTROPY OF SOLID ^{4}He IN THE REGION OF A ROUGHENING TRANSITION

Y. CARMI*, L.S. BALFOUR and S.G. LIPSON

Physics Department, Technion - Israel Institute of Technology, Haifa, Israel

Experiments are described in which flat single crystals of hcp solid ^{4}He are grown on a horizontal surface. The profile of the crystals is investigated as a function of temperature. From the results the anisotropy of the surface tension can be deduced. The results show the development of a singularity at the roughening temperature, 0.8K, on the $(10\bar{1}0)$ face. In general, the method will lead to a complete three-dimensional γ-plot for ^{4}He when the work is completed.

The recent discoveries of several roughening transitions on the surface of solid ^{4}He crystals [1] have led to a spate of theoretical models [2] which claim to describe the equilibrium surface structure in detail. They give different predictions about the surface properties, in particular the surface tension, in the region of the roughening transitions. So far, none of these predictions has been verified experimentally. In this paper we shall describe measurements of the anisotropy of the surface tension of hcp ^{4}He crystals as a function of temperature in the region of one of the three roughening temperatures.

The conventional description [3] of the anisotropy of surface tension is in terms of the γ-plot, which is a polar graph of the surface tension γ as a function of the orientation of the surface. The Wulff construction then allows the shape of the ideal equilibrium crystal to be calculated in the absence of gravity. In the presence of gravity (which is an important consideration in helium crystals larger than 1 mm) there is no known way of calculating the crystal shape in three dimensions, although two-dimensional constructions exist [4]. The reverse process, the deduction of the surface-tension anisotropy, is likewise possible in the absence of gravity, and a method applicable in the presence of gravity in two dimensions has been suggested although its application is problematic [5].

In our experiments we grow flat helium crystals on a horizontal plane surface and observe their shapes from above after equilibrium has been reached. In the plane of observation gravity has no effect. If the direction of observation referred to the crystal axes is denoted by the vector $\hat{n}$, then the profile of the observed crystal represents the zone of orientations normal to $\hat{n}$. The profile of the crystal is therefore related to the section of the γ-plot normal to $\hat{n}$ by the Wulff construction, and since $\hat{n}$ is vertical, gravity does not distort the shape [6].

At high enough temperatures the surface-tension γ is an isotropic constant. As the temperature falls, the γ-plot is expected to acquire the symmetry of the crystal, and at the roughening temperature for a certain crystallographic orientation a singularity (a dimple or cusp) develops in the γ-plot in that orientation. It is the critical behaviour of the singularity which is of fundamental interest.

In the experiment, crystals are grown from standard quality ^{4}He in an optical cell cooled by a ^{3}He refrigerator. The base of the cell is a horizontal mirror, and the crystals are observed optically from above using He-Ne laser light and are photographed on 35-mm film. (A symmetrical lens system giving unit magnification was chosen since it cannot distort the image). The crystals are grown by cooling helium at an initial pressure greater than the melting pressure at low temperatures (25 atmos.). Once there is solid inside the cell, it can be manipulated by decreasing and increasing the pressure until a single crystal is obtained, which has a suitable orientation and is out of contact with the cell perimeter. Such a crystal equilibrates to the form of a thin horizontal plate of thickness equal to the capillary length $a = (\gamma/\Delta\rho\, g)^{\frac{1}{2}}$ (about 1.4 mm). The horizontal projection of the crystal is then investigated as a function of temperature. Since crystals grown this way have random orientations, it will eventually be possible to determine the whole three-dimensional γ-plot given enough patience and time.

At high temperature the crystal profile is a circle. As we cool down, two qualitatively different situations can occur. Firstly, if $\hat{n}$ is in a symmetry zone, the zone normal to $\hat{n}$ will pass through the singularity which should develop at the roughening temperature for that orientation. The crystal profile will therefore develop facets at that temperature. The crystal shape can be observed around the roughening temperature and the development of the singularity investigated. Secondly, if $\hat{n}$ is not normal to a

---
*In partial fulfillment of the requirements for the D.Sc. degree of Technion.

high symmetry direction, the orthogonal zone will pass through no singularities, and therefore the crystal profile will not facet, although the circular shape may become slightly distorted. This situation occurs in an overwhelming proportion of experiments, for obvious statistical reasons.

We can illustrate the results with examples of both situations, although the full investigation is still in progress. Fig.1 shows a crystal for which $\hat{n}$ is $(11\bar{2}0)$ which is normal to both (0001) and $10\bar{1}0$), which have roughening transitions previously reported at 1.1K and 0.85K. The temperatures shown for this crystal are above and below the lower transition; one facet is present in all examples, whereas the other one disappears at the lower roughening temperature. Fig.2 shows a different crystal having the same orientation observed above both transitions. Fig.3 shows a crystal for which $\hat{n}$ is close to (0001) at 0.6K, which is below both transitions; but the profile is unfacetted because $\hat{n}$ is not exactly in a symmetry zone. Fig.4 shows the γ-plots for the crystal shown in Fig.1 at three temperatures [7]. Further details will be published when a larger number of experiments has been analysed.

ACKNOWLEDGEMENTS

This work is partially supported by a grant from the H. Gutwirth Foundation.

REFERENCES

[1] J.E. Avron, L.S. Balfour, C.G. Kuper, J. Landau, S.G. Lipson & L.S. Schulman, Phys. Rev. Lett. 45, 814 (1980).
S. Balibar and B. Castaing, J. Phys. Lett. 41, L329 (1980).
P.E. Wolf, S. Balibar and F. Gallet, Phys. Rev. Lett. 51, 1366 (1983).
[2] C. Jayaprakash, W.F. Saam and S. Teitel, Phys. Rev. Lett. 50, 2017 (1983).
N. Cabrera and N. Garcia, Phys. Rev. B 25, 6075 (1982).
[3] C. Herring, in "Structure and Properties of Surfaces" ed. R. Gower & G.S. Smith (U. of Chicago Press, Chicago, 1955).
[4] J. Avron, S. Taylor and R.K.P. Zia, J. Stat. Phys. 34, 0 (1984).
[5] E. Arbel and J.W. Cahn, Surface Science 66, 14 (1977).
[6] We have not yet devised a rigorous proof of this point, but the following remarks can be made: (a) The crystal surface is always convex and so it has a uniquely defined perimeter. (b) All points on the perimeter have vertical tangents, or (c) do not have tangents at all. Case (c) will occur when there is a discontinuity in the orientation of the surface at the edge of the facets, which occurs for some crystals such as Pb (see J. Heyrand and J-J Metois, J. Cryst. Growth 50, 571 (1980)).
[7] We should point out that in Fig.4 the radius of the curved parts of the γ-plots is temperature independent, and that this value is 0.16 erg.cm^{-2} (see Landau et al., Phys. Rev. Lett. 45, 31 (1980)).

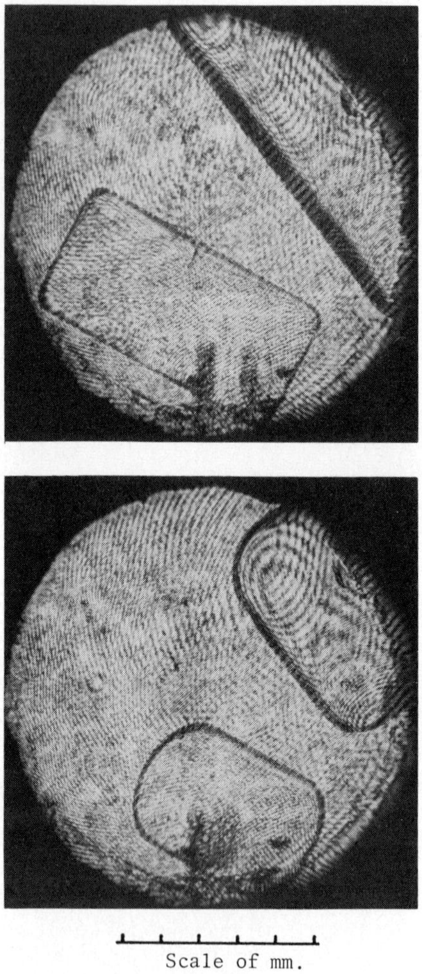

FIGURE 1

Crystal profiles viewed from the $(11\bar{2}0)$ direction: (a) at 0.58K and (b) at 0.75K. The lower crystal (all four facets visible) is oriented exactly in $(11\bar{2}0)$; the upper crystal (in contact with the cell periphery) has the $(11\bar{2}0)$ axis tilted by about 10° to the NE.

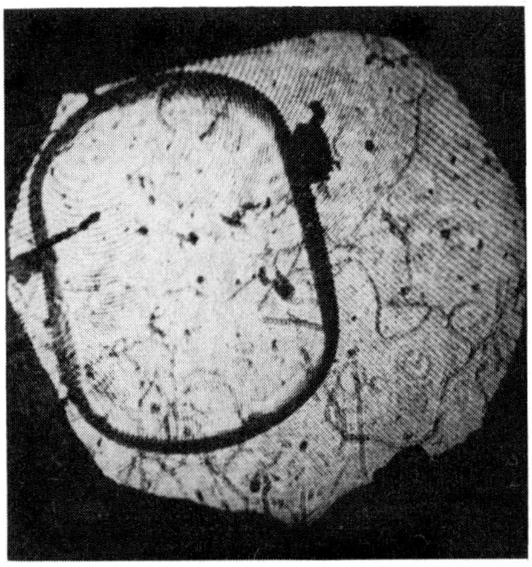

FIGURE 2

A second crystal viewed from (11$\bar{2}$0) at 1.25K in a different cell from Fig. 1. Various dirt particles spoil the photograph.

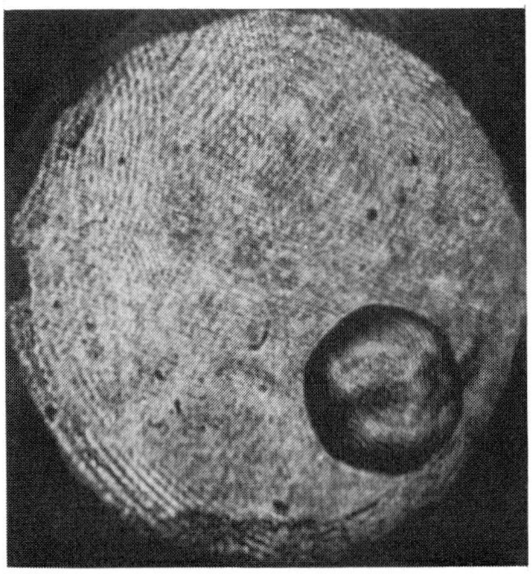

FIGURE 3

A crystal viewed from a direction close to (0001) showing incomplete faceting at 0.55K.

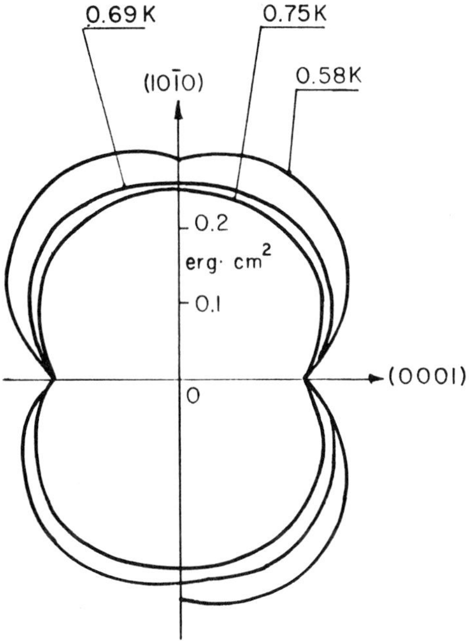

FIGURE 4

γ-plots for the crystal in Fig. 1 at three temperatures, showing development of the (10$\bar{1}$0) cusp at the roughening temperature.

LT-17 (Contributed Papers)
U. Eckern, A. Schmid, W. Weber, H. Wühl (eds)

MOBILITY OF THE LIQUID SOLID ^{3}HE INTERFACE

L. PUECH, G. BONFAIT, D. THOULOUZE and B. CASTAING.

C.R.T.B.T.-C.N.R.S., B.P. 166X, 38042 GRENOBLE Cedex, France.

We present two experiments which show a high mobility for the ^{3}He liquid-solid interface, in disagreement with predictions. The first one shows that the pressure remains close to its equilibrium value during rapid melting. Correlatively the Kapitza resistance of this interface measured between 45 mK and 75 mK is found to be higher than predicted by the acoustic mismatch theory, suggesting an influence of the mobility on the phonon transmission, as in the ^{4}He case.

1. INTRODUCTION

The recent interest in the growth kinetics of Helium crystals has been triggered by the paper of Andreev and Parshin (A.P.)(1). In their opinion, the rough liquid solid interface of ^{4}He has an infinite mobility at 0 K. At finite temperature, this mobility is limited by the bulk excitations. This general picture has been confirmed by the experiments although the details of the interaction between the rotons and the interface is always controversial (2).

Following the same ideas, A.P. predicted a low mobility is to be expected for the rough liquid solid interface of ^{3}He, due to the large momentum p_F of the quasi particles :

$$\Delta\mu = \mu_S - \mu_L \simeq p_F v$$

where v is the velocity of the interface and μ the chemical potential per atom.

We present here two different experiments which both show that this mobility is in fact much higher. The first one is a direct measurement of the pressure drop during rapid melting : it provides only a minimum value which is 200 times the A.P. estimation. The second one is a measurement of the Kapitza resistance of the ^{3}He liquid-solid interface, which has been theoretically (3) and experimentally (4)(5) shown, in the case of ^{4}He to be affected by the high mobility of the interface.

2. EXPERIMENTAL CELL

The cell (Fig. 1) is the same as that used in the previous ^{4}He experiment (5). The crystal is grown between the electrodes of a cylindrical capacitor, which allows to monitor the interface position. Four Matsushita 68 Ω resistors of the same batch act as thermometers. The pressure is measured with a capacitive gauge placed in a box connected to the top of the cell which temperature can also be regulated.

The ^{3}He used contained 30 ppm of ^{4}He. Once the crystal occupies the bottom of the cell, it is possible to regulate its temperature with the bottom heater and R_4 owing to the high thermal conductivity of solid ^{3}He.

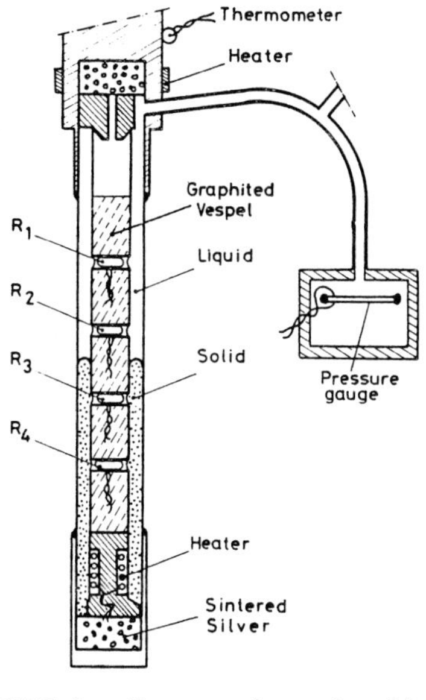

FIGURE 1 : The experimental cell.

3. THE RAPID MELTING

We start with a crystal 5 cm high (i.e. up to R_1), at a temperature as close as possible to the melting curve minimum. Fast melting is obtained by opening a valve at room temperature. The simultaneous records of level, pressure and temperature are presented in Fig. 2.

It is clear that, within the accuracy, no change occurs neither in the temperature, nor in the pressure (6) during the melting of the crystal. After melting, the temperature rise corresponds to the adiabatic decompression of the liquid.

With the A.P. estimation, such an interface velocity ($\sim$0.7 cm/s) would have needed a pressure drop of about 50 m bar. It is clearly less than 0.2 m bar.

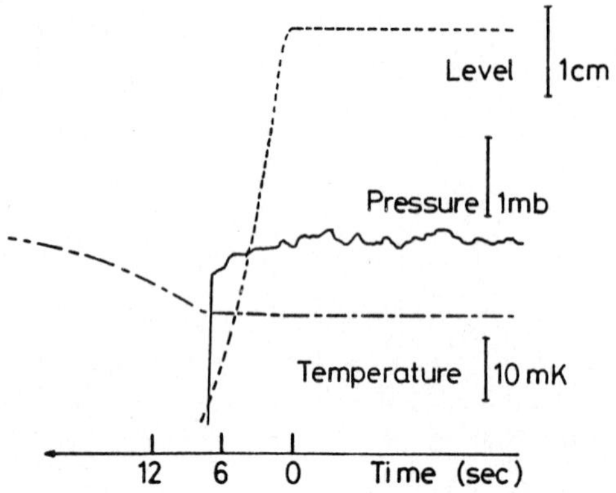

FIGURE 2 : Evolution of the pressure and the temperature during rapid melting.

4. THE KAPITZA RESISTANCE

Such a mobility is sufficient to affect the phonon transmission and therefore the Kapitza resistance that we obtained in two different ways :

- We directly measured the temperature gradients and extrapolated them to the interface. For the position of the latter the capacitance was of no aid due to the dilatation of the cell with the increasing pressure. We used the fact that the mass of ^3He was constant (the capillary is blocked). So the position of the interface being known at a particular pressure, it could be calculated at any other pressure. The passage of the interface on each thermometer gave the calibration and the verification of the method.
- During the slow cooling down the crystal was melting slowly ($\sim$ 1 mm/hour). With a constant heating at the bottom of the cell, a thermometer crosses rather rapidly from the solid to the liquid temperature when the interface passes on it. This "jump" gave us another estimation of the Kapitza resistance. In this latter case (Fig. 3) it was necessary to take into account the flux of entropy across the interface, due to the slow melting.

In both cases, measurements were impossible above 90 mK when the thermal resistance of the interface is smaller than that of 1 mm of liquid.

5. DISCUSSION AND CONCLUSION

The values obtained (Fig. 3) are to be compared to the result of the acoustic mismatch theory. In the similar case of ^4He, it gives

$$R_K\ T^3 \sim \frac{0.1\ \text{cm}^2\,\text{K}^4}{\text{W}}$$

(dashed line). Despite the large error bars, it is clear that the phonon transmission is altered by the mobility of the interface. However $R_K T^3$ has not the temperature behavior it shows in the

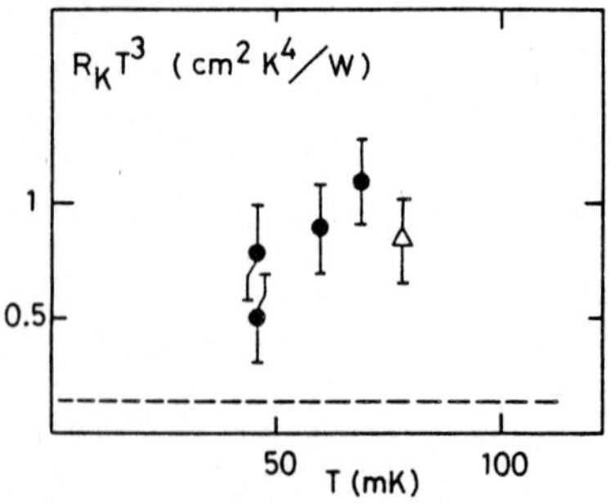

FIGURE 3 : The Kapitza resistance R_K :
• Extrapolating the temperature gradients
Δ Jump of the resistance.

^4He case ($R_K T^3 \propto T^{-2}$). If R_K is always controlled by the acoustic transmission, and thus by the mobility of the interface, this mobility would lead to a lowering $\delta p \simeq 5$ Pa ($= 5 \times 10^{-2}$ mb), for a typical interface velocity of 1 cm/s. This estimation is in agreement with the negative result of the first experiment. Both experiments thus coherently show that the ^3He liquid solid interface mobility is much higher than the Andreev and Parshin estimation.

REFERENCES

(1) A.F. Andreev and A. Ya. Parshin, Zh. Eksp. Teor. Fiz. 75 151 (1978) (Sov. Phys. J.E.T.P. 48 763 (1978)).
(2) B. Castaing, J. de Phys. Lettres 45 233 (1984) and references there in.
(3) V.I. Marchenko and A. Ya. Parshin, Pis'ma Zh. Eksp. Teor. Fiz. 31 767 (1980) (J.E.T.P. Lett. 31 724 (1980)).
(4) T.E. Huber and H.J. Maris, J. Low Temp. Phys. 48 463 (1982).
(5) L. Puech, B. Hebral, D. Thoulouze and B. Castaing, J. de Phys. Lettres 43 809 (1982).
(6) The small change in pressure at the limit of the noise agrees with the increase of liquid height between the interface and the pressure gauge.

LT-17 (Contributed Papers)
U. Eckern, A. Schmid, W. Weber, H. Wühl (eds)

CHARGE-IMBALANCE IN SUPERCONDUCTOR-INSULATOR-NORMAL METAL TUNNEL JUNCTIONS*

Yeouchung YEN and Thomas R. LEMBERGER**

Physics Department, Ohio State University, Columbus, OH 43210 USA

The low-voltage dc resistance of a Superconductor-Insulator-Normal metal (SIN) tunnel junction contains contributions from the insulating layer and from the charge-imbalance generated in the S electrode by the measuring current. Measurements on very low resistance junctions are in excellent quantitative agreement with a modified theory that includes the postulate that the resistance of the insulating layer is independent of temperature. In this paper, we report measurements on junctions with resistances that are up to two hundred times larger than those reported earlier. The data agree with the modified theory over the entire resistance range studied.

1. INTRODUCTION

A recent theory (1) predicts that the charge imbalance that is generated in the superconducting electrode of an SIN tunnel junction can relax through a novel process involving tunneling into the normal electrode, in addition to bulk scattering processes. The tunneling process is inelastic and is characterized by a rate:

$$1/\tau_{tun} = 1/2N(0)e^2\Omega R_i(T_c) \tag{1}$$

where 2N(0) is the density of electron states per unit volume, e is the electron charge, Ω is the area A of the junction times the thickness d of the S electrode, and $R_i(T_c)$ is the resistance of the insulating layer at temperature $T=T_c$.

An analytic result of the theory, valid for $T \approx T_c$, shows how the tunneling process affects the low-voltage dc resistance of the junction, $R_j(T)$:

$$R_j(T)=R_i(T_c)+\left[\tau_{in}/2N(0)e^2\Omega\right](4k_BT_c/\pi\Delta) \tag{2}$$

$$1/\tau_{in} = 1/\tau_E + 1/\tau_{tun}, \tag{3}$$

where $1/\tau_E$ is the electron-phonon scattering rate and Δ is the order parameter. From Eq. (2), $R_j(T)$ diverges as $(T_c-T)^{-1/2}$ with a magnitude that is determined by $1/\tau_{in}$. Thus, $R_j(T)$ should reflect the existence of the tunneling process if $1/\tau_{tun}$ is significant compared with $1/\tau_E$, i.e., $\tau_E/\tau_{tun} \gtrsim 1$.

Measurements have been reported (2) on junctions in which $R_i(T_c)$ was so small that $30 \lesssim \tau_E/\tau_{tun} \lesssim 200$. For $T \gtrsim 0.97T_c$, the data were in quantitative agreement with Eqs. (1)-(3). For lower temperatures, $0.85 \lesssim T/T_c \lesssim 0.96$, $R_j(T)$ was smaller than expected from a numerical extension of the theory. However, the theory fitted the data within about 1% when the theory was modified with the postulate that the resistance of the insulator, $R_i(T)$, was independent of T. The modification was made by subtracting the BCS result for $R_i(T)$ from the calculated value of $R_j(T)$, and then adding $R_i(T_c)$ to $R_j(T)$.

In this paper, we report measurements of the low-voltage dc resistance of junctions for which $1.0 \lesssim \tau_E/\tau_{tun}$ 200. The purpose of the measurements is to test the range of resistances over which the modified theory is valid.

2. DATA

The sample configuration is shown in Fig. 1. Each sample was made by evaporating films onto a room temperature glass substrate. The order of evaporation and the film thicknesses were: Al(200Å), SiO(~500Å), CuAlFe(~3000Å), Pb(~1000Å), SiO(1500Å), Pb(4000Å). The Cu alloy

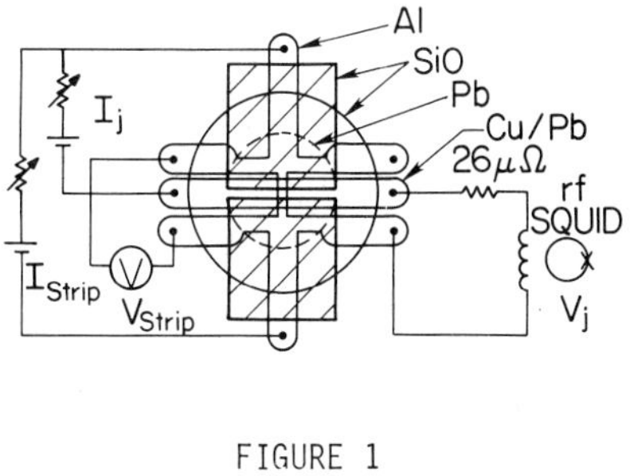

FIGURE 1
Sample configuration.

---

*Research supported by the Low Temperature Physics Program of the National Science Foundation under Grant No. DMR-8300254.
**Alfred P. Sloan Foundation Fellow

was ~98% Cu, 1%Al and 1%Fe. The Fe was necessary to prevent a supercurrent from flowing between the Al film and the Pb film on top of the Cu alloy film. The area of the junction was ~330μmx330μm. Film thicknesses were determined by a thickness monitor with a vibrating quartz sensor. The important physical parameters for our samples are listed in Table I. Values of $1/\tau_E$ were estimated from the approximate relation (3), $1/\tau_E = (T_c/1.2K)^3/12ns$. The mean-free-path was determined from (4) $\rho\ell=4x10^{-16}\Omega m^2$.

Table I. Sample parameters.

Sample	ℓ(Å)	d(Å)	T_c(K)	$R_i(T_c)$	$1/\tau_E(10^8 s^{-1})$
1	105	200	1.405	770 μΩ	1.34
2	129	200	1.404	375 μΩ	1.33
3	63	163	1.432	63.2μΩ	1.42
4	103	200	1.321	3.52μΩ	1.11

The junction resistance R_j was measured with a SQUID used in a feedback mode. For voltages less then 50nV the current-voltage characteristics were linear. (The critical current of the Al film was measured, but will not be discussed here.) Because R_j diverges as $(T_c-T)^{-\frac{1}{2}}$ near T_c, it is convenient to present the results in terms of the conductance, $G_j=1/R_j$, squared since G_j^2 extrapolates linearly to zero at T_c.

Figure 2 shows measured values of $g_j^2 \equiv [R_j(T_c)/R_j(T)]^2$, where $R_j(T_c)$ and T_c were chosen for each sample for a best fit to the modified theory, (solid curves). The temperature dependence of R_j is quite sensitive to the value of τ_E/τ_{tun} for $0.01 \lesssim \tau_E/\tau_{tun} \lesssim 10$, so that the values of $R_j(T_c)$ and τ_E/τ_{tun} determined for each sample in this way were accurate to ±20%. To check the theory quantitatively, the ratio τ_E/τ_{tun} was estimated from Eq. (1) and from the sample parameters in Table I. The 'best fit' values and the estimated values are given in Fig. 2, and they are in excellent agreement.

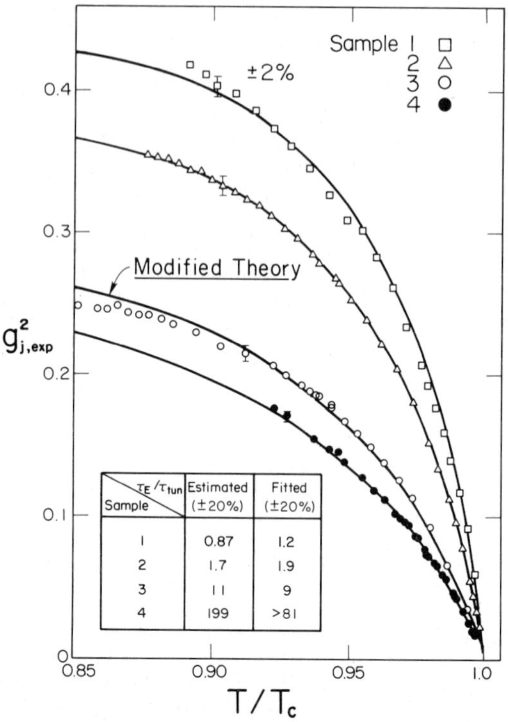

FIGURE 2
Plot of square of the normalized conductances vs. the reduced temperatures.

3. CONCLUSION

The modified theory accurately describes the low-voltage dc resistance of SIN tunnel junctions for $T/T_c \gtrsim 0.85$, and for $1.0 \lesssim \tau_E/\tau_{tun} \lesssim 200$, thereby confirming the basic validity of the theory. However, it is unknown why the modified theory fits the data so much better than the unmodified theory, when both the insulator resistance and the charge-imbalance resistance follow directly from the same tunneling Hamiltonian formulation. The agreement is probably not coincidental. The modified curves differ from the unmodified curves by up to 25% at $T/T_c=0.85$. It seems unlikely that such a large correction could be made with such high accuracy by such a simple modification for all values of T/T_c and τ_E/τ_{tun} that were studied.

REFERENCES

(1) T.R. Lemberger, to be published in Phys. Rev. Lett. 52, (1984).
(2) T.R. Lemberger and Y. Yen, submitted to Phys. Rev. B.
(3) C.C. Chi and J. Clarke, Phys. Rev. B 19, 4495 (1979).
(4) J. Romijn, T.M. Klapwijk, M.J. Renne, and J.E. Mooij, Phys. Rev. B 26, 3648 (1982).

LT-17 (Contributed Papers)
U. Eckern, A. Schmid, W. Weber, H. Wühl (eds)

TWO-DIMENSIONAL IMAGING OF HOTSPOTS IN SUPERCONDUCTING TUNNEL JUNCTIONS BY LOW TEMPERATURE SCANNING ELECTRON MICROSCOPY*

R. GROSS, M. KOYANAGI[+§], H. SEIFERT, and R. P. Huebener

Physikalisches Institut II, Universität Tübingen, D - 7400 Tübingen, FRG

Hotspots can be generated in superconducting tunnel junctions through dissipation of Joule energy. They can be imaged by scanning the junction surface with the electron beam of an electron scanning microscope equipped with a low temperature stage. It is shown that caused by inhomogeneities of the tunneling barrier the local current density can become very high leading to hotspots at these parts of the junction at a relatively low total tunneling current.

INTRODUCTION

Scanning electron microscopy performed at cryogenic conditions provides an interesting tool for investigating spatial and temporal structures in superconducting samples (1). By this method one can obtain a two-dimensional "voltage image" of superconducting tunnel junctions showing the spatial distribution of the quasiparticle tunneling current density and of the energy gap (2,3). In the same way low temperature scanning electron microscopy (LTSEM) can be applied for two-dimensional imaging of hotspots generated in a thin-film superconductor by Joule hating (4). In both cases the signal used for imaging is the change of the sample voltage caused by the electron beam irradiation. The spatial resolution of this imaging technique is limited by the thermal healing length. Due to the thermal skin effect, the resolution can be improved remarkably by high frequency modulation of the beam intensity and synchronous signal detection (5,6). Here we report the imaging of hotspots in superconducting tunnel junctions which are generated at points of locally higher tunneling current density caused by inhomogeneities of the tunneling barrier.

EXPERIMENTAL PROCEDURES

The samples investigated are PbIn/Pb cross-line or window tunnel junctions deposited on sapphire substrates of 1 mm thickness and 20 mm diameter. Patterning of the geometrical configuration is performed by the usual lift-off method employing AZ 1350 J. The oxide barrier is formed by rf plasma oxidation or thermal oxidation. The base- and counter-electrodes are typically 2000 Å and 4000 Å thick, respectively. The junction areas were rectangles of about 40 x 60 μm^2. The specific resistance ranged from 10^{-4} to 10^{-6} Ωcm^2.

The samples are mounted on the liquid-He stage (7) of the SEM so that the junction surface can be exposed directly to the primary electron beam and the back of the substrate is in direct contact with the liquid He. The beam parameters were: voltage = 26 keV, current ≈ 10 - 100 pA, beam diamter ≈ 0,1 µm. The electron beam is modulated between 1 kHz and 10 MHz by applying a periodic rectangular pulse to the beam blanking unit with a 50 % duty factor. For signal detection we use a lock-in technique.

RESULTS

The pictures shown in Fig. 1a-c are obtained with a PbIn/Pb cross-line junction. The tunnel area was 25 x 55 μm^2, the oxide barrier was formed by rf plasma oxidation. The experiments were performed at a He-bath temperature of 4.2K. The photographs were obtained from many linear scans in the horizontal direction across the tunneling area using the y-modulation presentation. The signal in y-direction is proportional to the change of the junction voltage generated by the irradiation with the electron beam.

In Fig. 1a the sample is current-biased in the thermal regime. In this case the voltage signal is proportional to the local quasiparticle tunneling current density. A detailed discussion of the signal response of a superconducting tunnel junction and its dependence upon the junction bias current has been given in (3). The picture shows that the junction has an extremely high current density at the middle of the lower edge. When the junction is current biased in the gap regime (Fig. 1b) the image indicates that the gap is reduced at this point. This gap reduction is mainly due to Joule heating at the

---

* Supported by a grant of the Stiftung Volkswagenwerk
+ Permanent address: Electrotechnical Laboratory, 1-1-4 Umezono, Sakura-mura, Niikarigun, Ibaraki, 305 JAPAN
§ Supported by the Alexander von Humboldt Stiftung
•

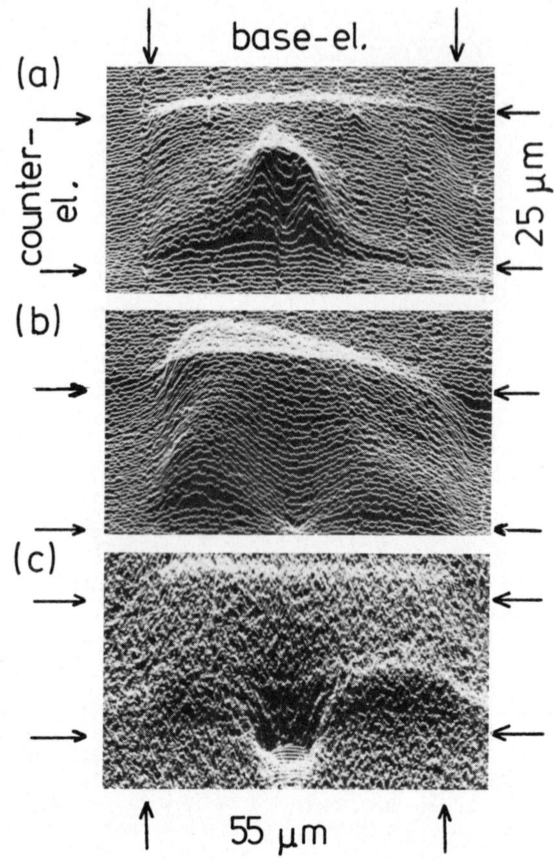

Fig. 1: Two-dimensional voltage images of a PbIn/Pb cross-line junction obtained at a bias current in the thermal regime (a), in the gap regime (b) and above the gap regime (c). The voltage signal is superimposed in vertical direction on the horizontal scans. The boundary lines of the junction electrodes are indicated.

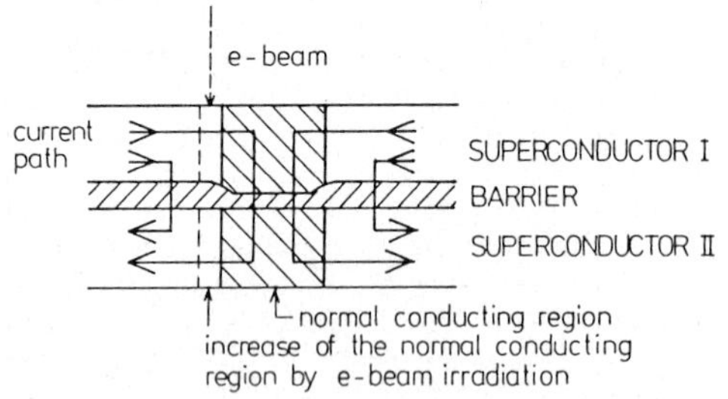

Fig. 2: Tunnel junction with hotspot and e-beam irradiation. The irradiation increases the current path through the normal region.

location of high tunneling current density. The gap reduction due to injection of excess quasiparticles can be estimated to be less than 2 % (3). If the bias current is increased further above the gap regime the voltage signal decreases rapidly (3). But now the Joule heating at the point of high current density is strong enough to generate a hotspot. Here the change of the junction voltage due to the beam irradiation is much higher and has the opposite sign.

Fig. 2 gives a qualitative explanation of this behavior. At the point of reduced barrier thickness the electrodes are normal conducting due to heat dissipation. When the electron beam hits the border of the normal region the heat supplied by the beam causes a shift of the boundary into the superconducting region. This increases the length of the current path through the normal region. When the junction is current biased as in our case the voltage increases. This case is shown in Fig. 1c.

In order to make the junction boundary visible, we have used a large voltage sensitivity of the lock-in amplifier. Therefore, the amplifier is overloaded in the region of the hotspot and due to its recovery time the signal right to the hotspot is slightly enhanced. The power dissipation at the bias point used in Fig. 1c is 40 µW corresponding to an average power density of 2,9 W/cm^2. With a heat transfer coefficient between junction and substrate of about 8 $Watt/cm^2K$ we estimate an average temperature rise of less than 0,5 K, and the generation of a hotspot is not to be expected. However, taking into account the strongly increased current density at this part of the junction (see Fig. 1a), the generation of the hotspot at this point is not surprising.

CONCLUSION

Hotspots in superconducting tunnel junctions can be imaged by LTSEM. These hotspots are generated (due to the locally enhanced tunneling current density) at an average heat dissipation much smaller than expected from the heat-transfer coefficient. This effect must be taken into account in experiments dealing with high quasiparticle injection.

REFERENCES

(1) J.R. Clem and R.P. Huebener, J. Appl. Phys. 51 (1980) 2764
(2) P.W. Epperlein, H. Seifert, and R.P. Huebener, Phys. Lett. 92A (1982) 146
(3) H. Seifert, R.P. Huebener, and P.W. Epperlein, Phys. Lett. 95A (1983) 326
(4) R.Eichele, L.Freytag, H.Seifert, R.P.Huebener, and J.R.Clem, J.Low Temp.Phys.52 (1983) 449
(5) H. Seifert, R.P. Huebener, and P.W. Epperlein, Phys. Lett. 97A (1983) 421
(6) H.Pavlicek, L.Freytag, R.P.Huebener, and H.Seifert, IEEE Trans.Magn.Vol.MAG-19, 1296 (1983)
(7) H. Seifert, Cryogenics 22 (1982) 657

LT-17 (Contributed Papers)
U. Eckern, A. Schmid, W. Weber, H. Wühl (eds)

NONEQUILIBRIUM PHENOMENA IN NORMAL METAL/SUPERCONDUCTOR POINT CONTACTS

F. Hohl[+], G. Voss

II. Physikalisches Institut, Universität zu Köln, D-5000 Köln 41, West Germany

We have studied the dV/dI-characteristics of Mo/Ta point contacts without and with microwave irradiation in the temperature interval $1.5 \leq T \leq 4.2$ K. The experimental results are discussed in terms of the conversion of normal current to supercurrent at a normal metal/superconductor interface and in terms of photon-assisted tunneling.

1. INTRODUCTION

In this paper we present the results of an experimental study of the properties of Mo/Ta {normal metal (N)/superconductor (S)} point contacts. We report especially on the response of those contacts to 140 GHz-microwave radiation.

Similar N/S systems have been studied experimentally several times in the past {Cu/Sn (2,4); Cu/Nb (1,3,4,6); Cu/Ta (1,5)}. Until recently however there was no adequate theoretical description of the observed phenomena.

We demonstrate that the recent theories of Artemenko, Volkov and Zaitsev (7) and Blonder, Tinkham and Klapwijk (8) provide a first good semiquantitative description of the *static* giaeverlike (3) dV/dI-characteristics (in the absence of electromagnetic radiation). *Dynamic* dV/dI-characteristics (in the presence of electromagnetic radiation) are then analyzed in terms of the well known Photon-Assisted Tunneling (PAT)-theory (11).

2. EXPERIMENTAL RESULTS AND DISCUSSIONS

Within the temperature range $1.5 \text{ K} < T < 4.2$ K Mo is normal conducting and Ta superconducting. By alloying of Mo into Ta we have produced superconducting samples, whose transition temperatures were lower than the T_c of pure Ta {T_c(Ta) = 4.48 K}.

Fig. 1 shows the *static* dV/dI-characteristic, of a $Mo/Ta_{0.99}Mo_{0.01}$ point contact at T = 1.6 K (solid curve). The transition temperature of the $Ta_{0.99}Mo_{0.01}$-sample determined experimentally from the temperature dependence of the electrical resistivity was $T_c(Ta_{0.99}Mo_{0.01}) = 4.0$ K. With this value we obtain a gap energy $\Delta(1.6 \text{ K}) = 0.60$ meV. With the model of Blonder, Tinkham and Klapwijk (8) the dashed curve in Fig. 1 is calculated for the parameters Z = 0.58, $R_{N/N} = R_\infty$ and $\Delta = 0.60$ meV. For low DC voltage ($|V_o| < \Delta/e$) the measured and the calculated curves agree quite well, but there are striking discrepancies in the voltage regime $|V_o| > \Delta/e$.

As pointed out in (8) in clean metallic N/S microconstrictions the conversion of normal current to supercurrent at quasiparticle excitation energies $E = e|V_o| < \Delta$ is dominated by the scattering processes of *Andreev reflection* (9). Consequently, the energy dependence of Andreev reflection can be directly measured in an N/S point contact experiment.

On the other hand, in the regime $|V_o| > \Delta/e$

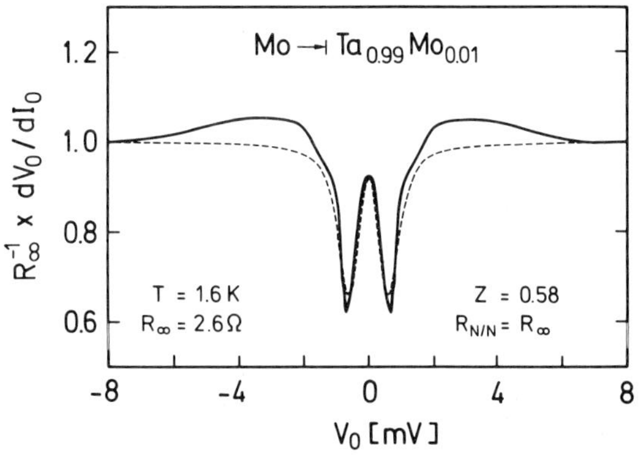

Fig. 1: Normalized differential resistance as a function of DC voltage for a $Mo/Ta_{0.99}Mo_{0.01}$ point contact in the absence of microwaves at T = 1.6 K (solid curve). $R_\infty = 2.6\ \Omega$ is the differential resistance of the contact at high DC voltage.

The dashed curve represents the theory of Blonder, Tinkham and Klapwijk (2) for Z = 0.58, $R_{N/N} = R_\infty$ and $\Delta(1.6 \text{ K}) = 0.985\ \Delta(0 \text{ K}) = 1.76\ k_B T_c\ (Ta_{0.99}Mo_{0.01}) = 0.60$ meV.

[+]present address: INTERATOM, Bergisch-Gladbach West Germany

the current flow across the N/S interface generates a *charge imbalance* (9,10) in the superconducting contact region. This nonequilibrium state can decay only via *inelastic quasiparticle/phonon scattering* processes characterized by a time constant τ_{Q^*} which is typically of the order of 10^{-10}s. We believe that the observed excess differential resistance indicates the presence of such relaxation processes.

Fig. 2 shows the effects of microwave radiation on the dV/dI-characteristic of a $Mo/Ta_{0.99}Mo_{0.01}$ point contact at 1.6 K. With our microwave frequency f_1 = 140 GHz the single photon energy $\hbar\omega_1 = 2\pi\hbar f_1$ = 0.58 meV is comparable to the energy scale Δ of the DC nonlinearity, i.e. we are approaching the quantum regime of direct video detection (11). Note that the differential resistance at zero bias shows an oscillating behaviour and that with increasing microwave power more and more nonlinearities appear.

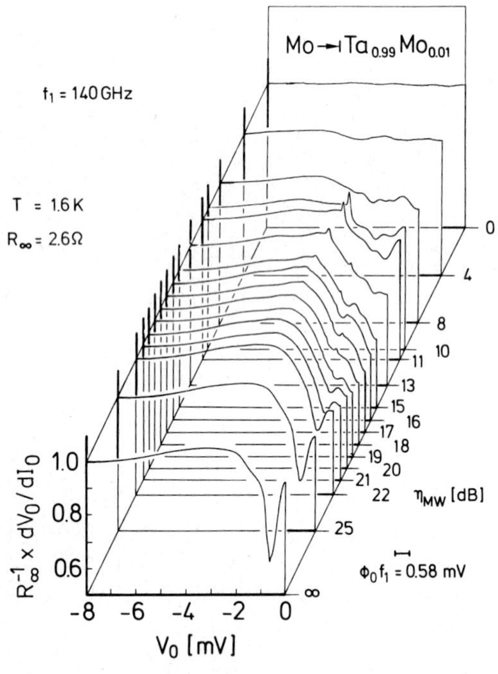

Fig. 2: dV/dI-characteristics of a $Mo/Ta_{0.99}Mo_{0.01}$ point contact in the absence ($\eta_{MW} = \infty$ dB, same curve as shown in Fig. 1) and in the presence of microwaves (from left to right increasing microwave power) at T = 1.6 K.

It turns out that while at low voltage experiment and PAT-theory are in satisfactory agreement, in the regime $|V_o| > \Delta/e$ the microwave-induced effects are only partly reproduced by the theoretical curves. In particular, the sharp structures which we observe in the experiment at sufficiently high microwave power (η_{MW} < 13 dB) cannot be explained within the framework of the PAT-theory.

Since in our experiments the energy of a single photon is lower than that necessary to break up a Cooper-pair, and, since on the other hand, the relation $f_1\tau_{Q^*} > 1$ is fulfilled, such effects can possibly be traced to microwave-induced gap enhancement (12).

REFERENCES

(1) H. Happ, U. Kaiser-Dieckhoff
Phys. Lett. 29A, 161 (1969)
(2) I. Taguchi, H. Yoshioka
J. Phys. Soc. Japan 27, 1074 (1969)
(3) U. Kaiser-Dieckhoff
"Voltage Current Characteristics of Superconducting to Normal Metal Point Contacts in an RF-Field"
in "Superconducting Quantum Interference Devices"
eds.: H.D. Hahlbohm, H. Lübbig
(Walter de Gruyter; Berlin, New York; 1977)
(4) R.W. van der Heijden, J.H.M. Stoelinga
H.M. Swartjes and P. Wyder
Solid State Commun. 39, 133 (1981)
(5) V.N. Gubankov, N.M. Margolin
Pis'ma Zh. Eksp. Teor. Fiz. 29, 733 (1979)
{JETP Letters 29, 673 (1979)}
(6) G.E. Blonder, M. Tinkham
Phys. Rev. B27, 112 (1983)
(7) S.N. Artemenko, A.F. Volkov and A.V. Zaitsev
Solid State Commun. 30, 771 (1979)

A.V. Zaitsev
Zh. Eksp. Teor. Fiz. 78, 221 (1980)
{Sov. Phys. JETP 51, 111 (1980)}
(8) G.E. Blonder, M. Tinkham and T.M. Klapwijk
Phys. Rev. B25, 4515 (1982)
(9) see, for example, A.B. Pippard
"Normal-Superconducting Boundaries"
in (I), pp. 341-352
(10) see, for example, J. Clarke
"Charge Imbalance"
in (I), pp. 353-422
(11) see, for example, J.R. Tucker
IEEE J. Quantum Electron. QE-15, 1234 (1979)
(12) E.D. Dahlberg, R.L. Orbach and I. Schuller
J. Low Temp. Phys. 36, 367 (1979)

(I) *"Nonequilibrium Superconductivity, Phonons and Kapitza Boundaries"*
edited by K.E. Gray
NATO advanced study institutes series
Series B, Physics; volume 65
(Plenum Press; New York; 1981)

LT-17 (Contributed Papers)
U. Eckern, A. Schmid, W. Weber, H. Wühl (eds)

SUBHARMONIC ENERGY GAP STRUCTURE AND EXCESS CURRENT IN NIOBIUM POINT CONTACTS

J. BINDSLEV HANSEN *, M. TINKHAM and M. OCTAVIO †

Physics Dept. and Division of Applied Sciences, Harvard University, Cambridge, Mass. 02138, USA

We report experiments on Nb-Nb point contacts to test recent theories for the excess current and the subharmonic energy gap structure in the IV-curves of superconducting weak links. We find good agreement between theory and experiment using a single fitting parameter to account for both phenomena.

Excess current (EC) at high voltage ($eV > 2\Delta$) and subharmonic energy gap structure (SGS) around $eV = 2\Delta/n$ are observed in the IV-characteristics of superconducting metallic weak links and nonideal,"leaky" tunnel junctions. These features of the IV-curve which have been studied in the last twenty years (see (1) for refs.) have now been explained in terms of Andreev reflection (AR) (1-5). Modelling the weak link as a superconductor-normal metal-superconductor (SNS) microconstriction the EC has been shown to stem from the extra quasiparticle conductance through both NS interfaces due to AR of quasiparticles in the gap region and the concomitant addition of electron pairs to the superconductor. Assuming that there is no interference between scattering processes at the two interfaces the total EC, I_{exc}, is simply the sum of the contributions from the two NS boundaries. For the symmetrical, purely metallic SNS contact the result is (2,3)($\Delta \ll kT \ll eV$): $I_{exc} = 8\Delta/3eR_N$ where R_N is the normal state resistance of the contact. If there is a tunnel barrier at the NS interface, electrons undergo normal reflection as well as AR and transmission. A dimensionless barrier strength, Z, is introduced in the model. In the normal state the reflection coefficient is simply $Z^2/(1+Z^2)$ and $R_N = R_0(1+Z^2)$ where R_0 is the geometrical (Sharvin) resistance. In the SNS contact the SGS can be attributed to multiple AR that allow a quasiparticle to gain neV ($=2\Delta$) in energy by traversing the normal region n times as is alternates between having electron-like and hole-like character (1). Recently, Octavio et al. (5) used a self-consistent Boltzmann equation approach to calculate the SGS for a SINIS (I = insulator) contact with some elastic scattering at the interfaces. This is the model we shall be using in the following for comparison with our measurements of the EC and the SGS in Nb-Nb point contacts. In Fig.1 we have plotted the computed values of I_{exc} as a function of Z both for the SINIS model and for the simple case of two independent NS interfaces. We shall only think of Z as a phenomenological parameter, Z_{eff}, that "measures" the elastic scattering in the contact and other deviations from the clean 1-D SNS contact.

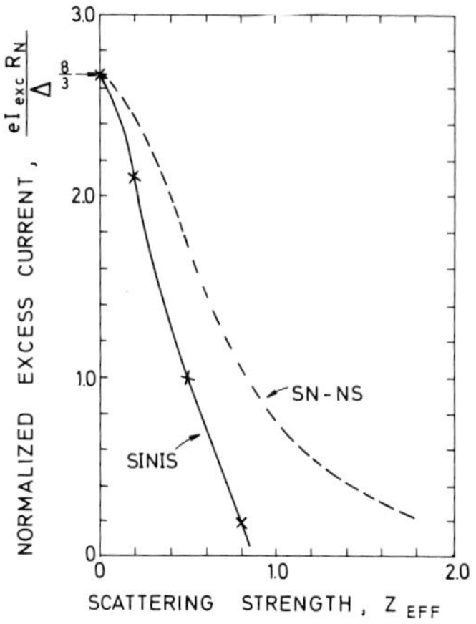

FIGURE 1
Normalized I_{exc} vs. Z for the SINIS model (5) and for two independent NS interfaces (3). For $Z \to \infty$ I_{exc} (SINIS) $\to -4\Delta/3eR_N$ (6).

The Nb-Nb point contact was chosen for this study due to its 3-D geometry and the wide temperature range for Nb accessible in a ^{4}He cryostat. The 3-D geometry is essential in order to minimize the potential complications due to non-equilibrium phenomena in the banks:quasiparticle charge imbalance and heating (7). The electrochemically sharpened wire (99.5% Nb with $\ell \approx 150$ Å) and the polished flat (99.99% Nb) were prepared as described in ref.(4). The contact was mounted on a temperature-stabilized Cu block inside a vacuum can immersed in liquid helium. Bellows on top of the can transmitted the mechanical pressure from a differential screw to the contact while only introducing a negligibly small heat

* Present address: Physics Laboratory I, H.C.Ørsted Institute, Univ. of Copenhagen, Copenhagen, Denmark.
† Permanent address: Fundacion Instituto de Ingeniera, Apartado 1827, Caracas, Venezuela.

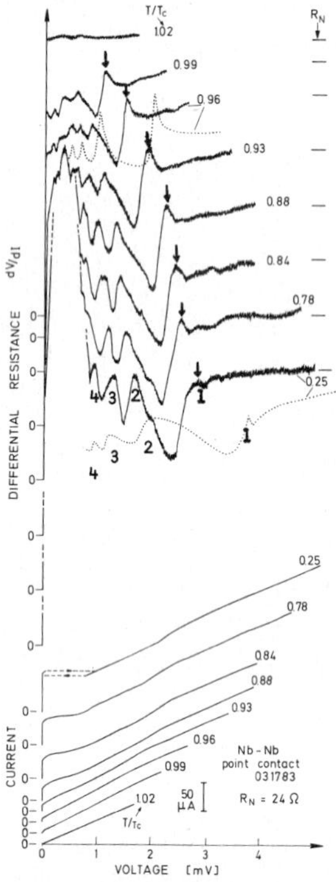

FIGURE 2
Experimental IV- and dV/dI-V-curves for a 24Ω Nb-Nb point contact. The dotted dV/dI-V curves are computed on the basis of the SINIS model (5).

leak. In this way the temperature of the point contact could be varied between 1.3 K and 9.2 K ($= T_c$ (Nb)) without affecting the mechanical setting of the contact. During a single run 4-10 useful contacts could be formed by adjusting the pressure on the point. Over 10 runs 65 different contacts with $0.1\,\Omega < R_N < 2k\Omega$ were studied. With increasing pressure the point contacts typically progressed through three qualitatively different regions (4): I) $R_N \geq 100\,\Omega$ (tunnel junction, no EC, only structure at $eV = 2\Delta$); II) $1\,\Omega < R_N \leq 100\,\Omega$ (mostly metallic, possibly in the Sharvin limit, EC and SGS observed); and III) $R_N < 1\Omega$ (metallic, contact dimension much greater than ℓ, the mean free path, EC and SGS observed but behavior dominated by heating). For region II contacts (40 in all) the EC and SGS data can nearly all be accounted for by the SINIS model. For the normalized I_{exc} we found in this regime: $0.4 < eI_{exc}R_N/\Delta < 1.4$. The experimental curves for a 24 Ω contact reproduced in Fig.2 are representative of the contacts in region II. Fig.3 shows the measured I_{exc} vs. the measured, heating depressed value of $2\Delta/e$ given by the voltage at the 2Δ peak in dV/dI. From this graph we find $eI_{exc}R_N/\Delta = 0.67 \pm 0.04$ and Fig.1 (SINIS) yields $Z_{eff} = 0.58 \pm 0.02$. Using this value of Z_{eff} we have computed the dV/dI-V curves for $T/T_c = 0.96$ and 0.25 (dotted in Fig.2). Apart from the

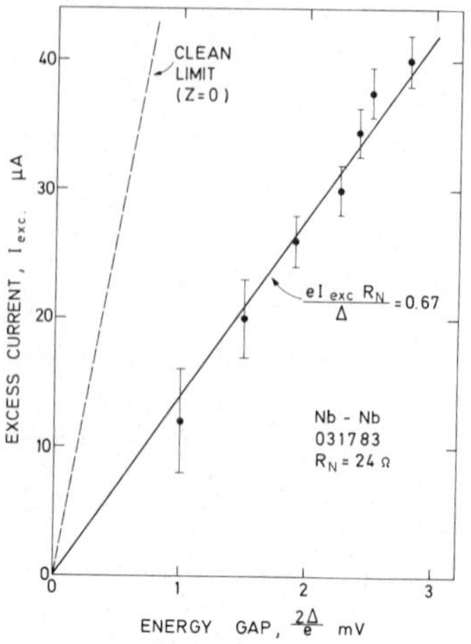

FIGURE 3
Measured I_{exc} vs. measured value of $2\Delta/e$ (from $2\Delta/e$ peak in the SGS, see arrows in Fig.2).

Josephson effect (at low V) and the heating induced distortion (at high V) the experimental curves agree well with the computed ones with respect to the detailed, temperature-dependent shape of the SGS peaks. The effect of heating on the measured curves is clearly seen as a non-linear compression along the voltage axis and a corresponding enhancement of the peak height.

We conclude that using a single, phenomenological fitting parameter, Z_{eff}, the SINIS model (5) can account for both of the Andreev reflection induced phenomena – the excess current and the subharmonic energy gap structure – in small, metallic Nb-Nb point contacts. The model of the S(N)S' microconstriction could, however, still be improved by including distributed elastic scattering, heating and the Josephson effect into the theoretical treatment.

ACKNOWLEDGEMENTS
We thank Greg Blonder for stimulating discussions. This work was supported in part by NSF, ONR and JSEP. One of us (JBH) thanks the Carlsberg Foundation and NATO for grants.

REFERENCES
(1) T.M.Klapwijk, G.E.Blonder and M.Tinkham, Physica 109-110B (1982) 1657.
(2) S.N.Artemenko, A.F.Volkov and A.V.Zaitsev, JETP Lett. 28 (1978) 589, Sov. Phys. JETP 49 (1979) 924 and A.V.Zaitsev, Sov. Phys. JETP 51 (1980) 111.
(3) G.E. Blonder, M.Tinkham and T.M.Klapwijk, Phys. Rev. 25B (1982) 4515.
(4) G.E.Blonder and M.Tinkham, Phys. Rev. 27B (1983) 112.
(5) M.Octavio, M.Tinkham, G.E.Blonder and T.M. Klapwijk, Phys. Rev. 27B (1983) 6739.
(6) G.E.Blonder, private communication.
(7) M.Peshkin and R.Buhrman, Phys.Rev.28B(1983)161.

LT-17 (Contributed Papers)
U. Eckern, A. Schmid, W. Weber, H. Wühl (eds)

MAGNETIC SUSCEPTIBILITY OF <u>Cu</u>Mn: CURIE-WEISS PARAMETERS AND THE EFFECT OF ANNEALING*

R. A. Fisher, E. W. Hornung, N. E. Phillips, and J. Van Curen

Department of Chemistry, University of California, Berkeley, California 94720, and Materials and Molecular Research Division, Lawrence Berkeley Laboratory, Berkeley, California 94720

The dc field-cooled susceptibilities of <u>Cu</u>Mn samples with concentrations between 0.082 and 1.280 at.% have been measured within the range $2 \leq T \leq 350$K and $1 \leq H \leq 50000$ Oe. The features associated with T_g depend strongly on sample preparation. Analysis of the high temperature data gives the Weiss constants, Δ, which also depend on sample preparation, and the effective magneton numbers, p, which do not. Accurate values of g have been derived.

The specific heat associated with spin-glass ordering in <u>Cu</u>Mn has been investigated as a function of both concentration and heat treatment (1). In connection with the calorimetric studies, the field-cooled, dc susceptility, χ, was measured on the same samples, and those results are reported here.

Samples of a total mass of the order of 100g or more were prepared by melting degassed 6-9's Cu and 5-9's Mn under vacuum in sealed high-purity quartz tubes. The melt was quenched by plunging the tube into water. Samples investigated in the resulting condition are designated "unannealed". The "annealed" samples were prepared from the unannealed samples, usually after the specific heat had been measured, by heating in an argon atmosphere at 950-975°C for 2 days. Samples were taken from the top and bottom of the ingots, dissolved, and analyzed by atomic absorption analysis. The Mn concentrations, obtained by comparison with standard solutions prepared by dissolving the pure Mn and Cu, were typically 5-10% lower than calculated from the weights of the starting materials. The analytical procedure was shown to give concentrations accurate to within 2%. Samples for χ measurements were 5mm x 5mm right circular cylinders cut from the larger ingots used for specific heat measurements.

The susceptibility measurements were made in an SHE magnetometer that had been checked with an NBS Pt susceptibility standard and, between 0.5 and 200 Oe, with a Sn sphere. Values of χ are believed to be accurate to within 1%. After each change of field the sample temperature was raised to at least $3T_g$ and the field then held constant until the measurements in that field were completed.

Typical low-temperature data for an annealed sample are shown in Fig. 1 as χ vs T and in Fig. 2 as $(\partial\chi/\partial T)_H$. These curves show the

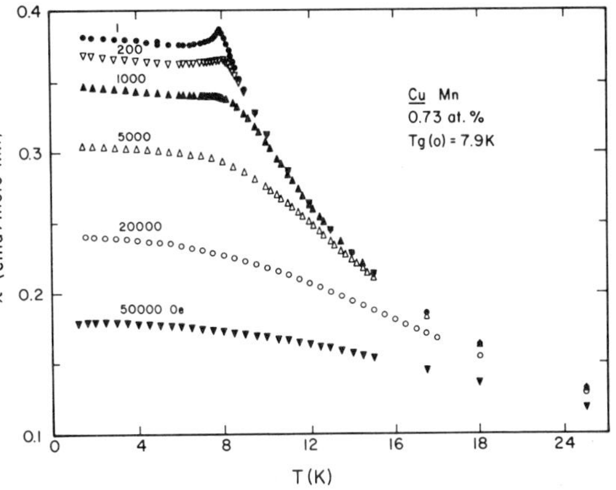

FIGURE 1. Typical low-temperature χ data for an annealed sample.

characteristic features associated with spin-glass ordering. As shown in Fig.3, these sharp features are not displayed by the unannealed sample--there is still a conspicuous change in slope of χ vs T near 5K, but the relatively sharp peak in χ that occurred near 8K in the annealed sample is washed out. No doubt the difference between the annealed and unannealed samples is associated with a difference in atomic short range order, but the available data on atomic order and its correlation with heat treatment does not provide a basis for a detailed interpretation.

For $T \geq 4T_g$ (T_g is the temperature of the cusp in the low-field χ of the annealed samples) and for H/T not too high, χ can be represented by a Curie-Weiss relation,

$$\chi - \chi_i = C/(T+\Delta), \qquad (1)$$

* Work supported by Office of Energy Research, Office of Basic Energy Sciences, Materials Sciences Division of the U. S. Department of Energy under Contract No. W-7405-ENG-48.

where χ_i is the value of χ for pure Cu. For T>50K and H/T<1000 Oe/K, the data fit Eq. (1) with rms deviations ≤ 1%, as illustrated in Fig. 4 for c = 0.73 at.%. To within the experimental error the values of C are the same for annealed and unannealed samples, but there is a significant difference in the values of Δ.

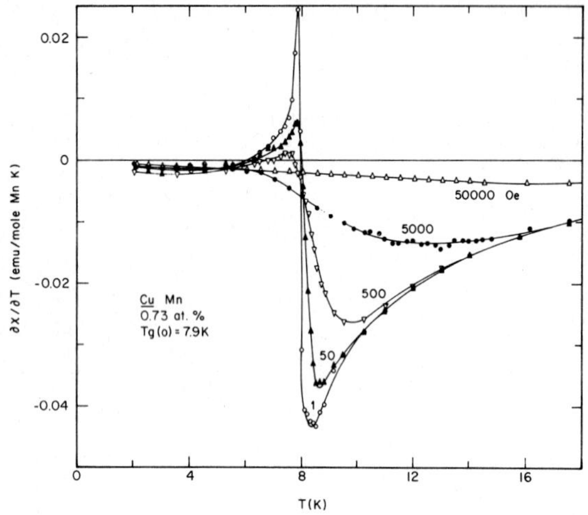

FIGURE 2. $(\partial\chi/\partial T)_H$ for an annealed sample.

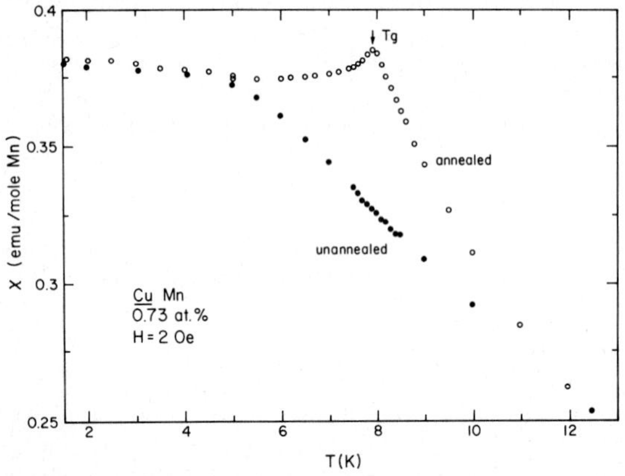

FIGURE 3. Effect of annealing on χ near $T = T_g$.

Values of parameters determined by the Curie-Weiss fits and of T_g are tabulated in Table I and plotted as a function of c in the inset to Fig. 4. The effective magneton numbers, p, derived from

$$C = g^2S(S+1)\mu_B^2/3R \equiv p^2\mu_B^2/3R, \qquad (2)$$

are higher than earlier reported values. The 1% discrepancy with a recent measurement (2) is within experimental error but the discrepancy ranges up to 10% for others (3). The larger discrepancies may reflect errors in c in the other measurements. If either g or S is known, Eq. (2) can be used to calculate the other from p. From the entropy associated with spin ordering (4) in CuMn, S=2.5, and this value has been used to derive the values of g in Table I, which are in good agreement with values derived from NMR studies (5).

TABLE I. Values of Parameters Derived from χ.

c(at.%)	p	g	Δ(K)	T_g(K)
Annealed samples				
0.082	5.08	1.72	1.7	--
0.247	5.00	1.69	1.0	--
0.265	4.92	1.66	1.2	3.64
0.279	--	--	-	3.89
0.490	5.12	1.73	0.2	5.54
0.730	5.04	1.70	-1.3	7.9
1.280	5.20	1.76	-9.4	11.9
Unannealed samples				
0.247	5.01	1.69	0.6	--
0.730	5.03	1.70	-3.1	--

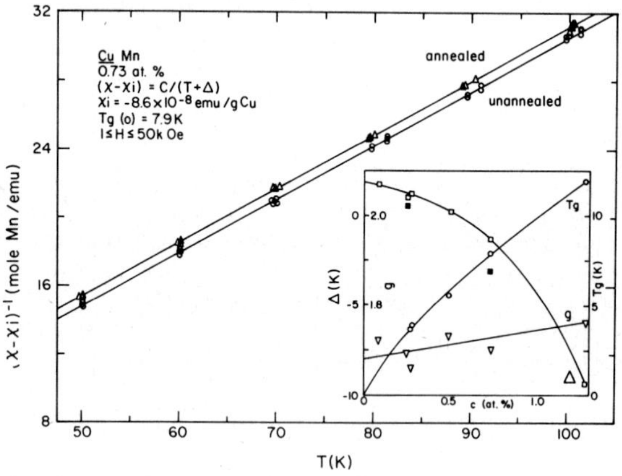

FIGURE 4. Typical Curie-Weiss fit and c dependence of the derived parameters. The solid symbols in the inset represent unannealed samples.

REFERENCES

(1) G. E. Brodale, R. A. Fisher, W. E. Fogle, N. E. Phillips, and J. Van Curen, this volume.
(2) A. F. J. Morgownik and J. A. Mydosh, Solid State Commun. 47 (1983) 321.
(3) C. M. Hurd, J. Phys. Chem. Solids 28 (1967) 1345; E. C. Hirschkoff, O. G. Symko, and J. C. Wheatley, J. Low Temp. Phys. 5 (1971) 155; J. M. Franz and D. J. Sellmyer, Phys. Rev. B8 (1973) 2083; F. W. Smith, Phys. Rev. B14 (1976) 241; S. Nagata, P. H. Keesom, and H. N. Harrison, Phys. Rev. B19 (1979) 1633.
(4) W. E. Fogle et. al., to be published.
(5) H. Alloul, F. Hippert, and H. Ishii, J. Phys. F: Metal Phys. 9 (1979) 725.

LT-17 (Contributed Papers)
U. Eckern, A. Schmid, W. Weber, H. Wühl (eds)

SCALED MAGNETIZATION IN NICKEL VS. SCALED "EQUIVALENT MAGNETIZATION" IN CuMn SPIN-GLASS

Jean-Jacques PREJEAN, Jean SOULETIE

CRTBT, CNRS, BP 166 X, 38042 Grenoble Cédex, France

We compare the scaled magnetization of nickel in the paramagnetic regime over the ferromagnetic transition with the equivalent quantity (the scaled susceptibility) in CuMn spin-glass. The fields $T-T_c$ and H (resp. $(T-T_c)^2$ and H^2) are normalized to T rather than to T_c which extends considerably the regime where good scaling is obtained.

1. THE SCALED MAGNETIZATION IN NICKEL

The static scaling hypothesis states that the Gibbs potential is a generalized homogeneous function of the fields t and η :

$$t = (T-T_c)/T_c, \quad \eta = \mu H/kT_c \qquad (1)$$

Differentiating with respect to η we obtain the scaled magnetization

$$M = t^{\beta} \; f\left(\frac{\eta}{t^{\gamma+\beta}}\right) \qquad (2)$$

and the susceptibility diverges like

$$\chi = C(T-T_c)^{-\gamma} \qquad (3)$$

This expression imposed itself to the experimentalists faced to the failure of mean field theory in a χ^{-1} vs. $(T-T_c)$ plot near T_c. The subsequent success of scaling ideas had the additional effect to fix this definition of the fields t and η and of the exponent γ and to accredit the belief that the range where scaling ideas are practical (critical regime) is narrow. We find it more natural to represent the anomalous behaviour of the magnetization at T_c in a diagram of M vs. H/T (fig. 1) which conserves the high T limit : the Langevin function La(H/T). Considering this plot without the prejudice induced by the mean field result, we are struck by the fact that it is the Curie constant which blows up at T_c. Rather than η and t, this suggests to use the non linear fields

$$t' = (T-T_c)/T \quad , \quad \eta' = \mu H/kT \qquad (4)$$

which do not affect the results near T_c but insure the finiteness of the Curie constant in the high T limit ($t' \to 1$) where

$$\chi T = (T/(T-T_c))^{\gamma} \to \quad \chi = (T-T_c^{MF})^{-1} \qquad (5)$$

with the non-trivial prediction $T_c^{MF} = \gamma T_c$. Evidence in CuMn (2), Ni(3) and in model systems (4) shows that with the new variables the critical regime is not subject to restrictions as strong as previously suspected. η' and t', besides, are the natural variables of the theory : among many arguments we reproduce one which we owe to P. Peretto. Starting with a Landau hamiltonian of the form

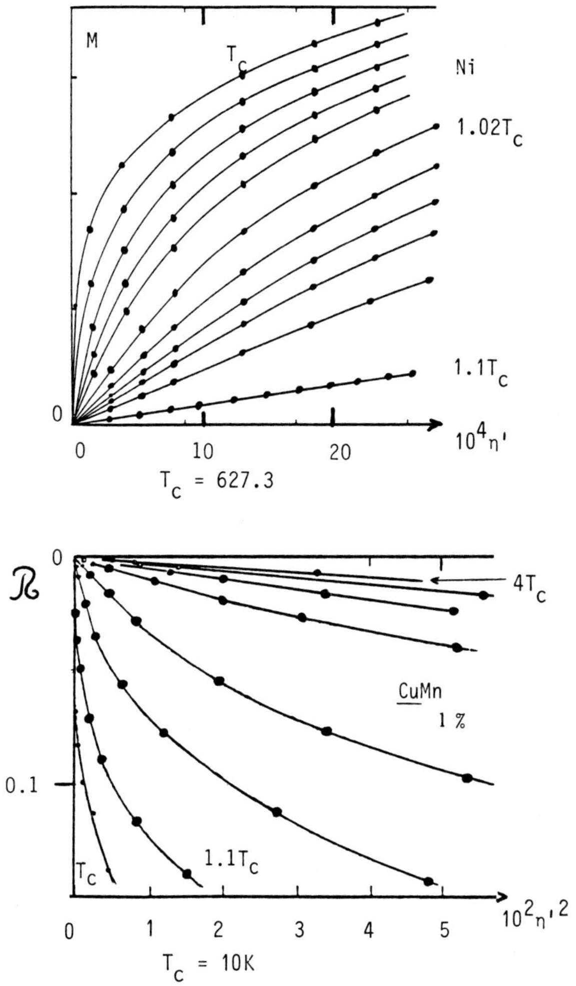

Fig. 1 - Plot of the magnetization of nickel vs. H/T and of the "equivalent magnetization" of CuMn $= MT/L_1H - 1$ vs $(H/T)^2$ showing the divergence of the first order (resp. third order) Curie constants at T_c.

$$-T^{-1}\int dx[J\psi^2(x)+\Delta\psi^2(x)]+\int dx(a\psi^2(x)+b\psi^4(x)$$

which we Fourier transform into

$$\int dk(r-r_c+k^2)\psi^2(k)+u\int dk_1..\int dk_4\delta(\Sigma k)\psi_{k_1}\cdots\psi_{k_4}$$

with $r = J/T$ and $r_c = J/T_c$ which defines T_c, if we apply renormalisation group theory it is r and not T which is renormalised. The scale variable normalised by the critical variable is $(r-r_c)/rc = (T-T_c)/T$. Similarly, the magnetic field introduces a term $T^{-1}\int\psi(k)H(k)\,dk$ and the scale variable is $H/T - H_c/T$ i.e. H/T.

Fig. 2 shows our ultimate check of the scaling law (eq. 2) using all the data of Weiss and Forrer(1) in nickel ($T_c < T < 1.4\ T_c$) corrected for a small T independent contribution $\chi_0 H$ (3). Previous checks (5) with the same data were restricted to a range $T < 1.02\ T_c$ where the authors optimised the exponents $\gamma = 1.31$, $\beta = .4$ which we conserved.

2. THE SCALED "EQUIVALENT MAGNETIZATION" in CuMn

The scaling approach to the spin-glass transition has to account for the fact that the order parameter $q = \langle S^2 \rangle$ is the square of a moment. The associated field must be H^2 and not H. Differentiating G with respect to H^2, we obtain an "equivalent magnetization" which is a susceptibility $\mathcal{R} = dG/dH^2 = dG/dH \times dH/dH^2 \sim M/H$. Eq.2 then yields the Suzuki-Chalupa result (6) which, with our choice of variables ($\eta \to \eta'$, $t \to t'$), becomes

$$\mathcal{R} = \frac{M}{L_1 H/T} - 1 = \left(\frac{T-T_c}{T}\right)^{\beta} f\left[\left(\frac{\mu H}{kT}\right)^2\left(\frac{T}{T-T_c}\right)^{\gamma+\beta}\right] \quad (6)$$

A plot of $\mathcal{R}$ vs. $(H/T)^2$ (fig. 1) stresses pathological features, analog to those observed with M(H.T) in nickel. Expanding M in terms of odd powers of H/T :

$$M = a_1L_1(H/T)+a_3L_3(H/T)^3+a_5L_5(H/T)^5+\ldots \quad (L_1 = 1/3) \quad (7)$$

we obtain from the divergence of the non-linear Curie constants a_3, a_5 the equivalent information which we would obtain from a_1 and a_3 in a ferromagnet. a_3 and a_5... in eq. 7 are normalised by the coefficients L_3, L_5... of La(H/T) and tend to one in the paramagnetic limit. From a study of $a_3(t) = t^{-\gamma}$, $a_5(t) = t^{-(2\gamma+\beta)}$ we derived $\gamma = 3.2$, $\beta = .95$, hence the scaled "equivalent magnetization" shown fig.2 for $T_c<T<4T_c$(2).

3. DISCUSSION

In order to maintain a parallel requested by dimensional analysis, we are tempted to introduce with the new field $\eta^* \sim H^2$ a field t^*

$$t^* = ((T-T_c)/T)^2 \qquad \eta^* = (\mu H/kT)^2 \quad . \qquad (8)$$

Then eq.6 transforms back into eq.2 with new exponents $\gamma^* = \gamma/2$ and $\beta^* = \beta/2$ of a more conventional size. In particular they are comparable to those determined in nickel.

The two scaled curves look very similar in the Log-Log representation of fig. 2 in terms of

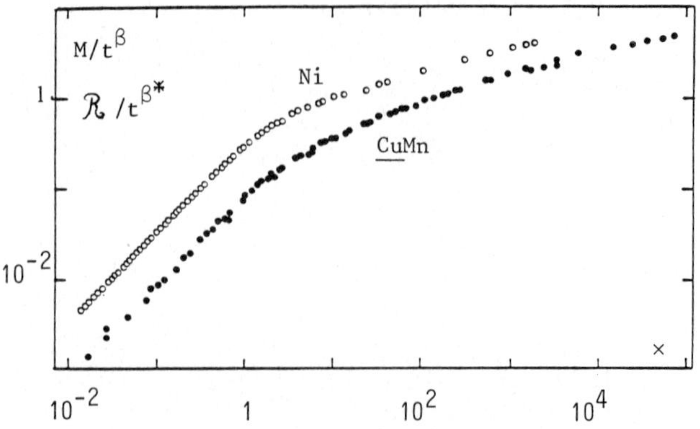

Fig. 2 - The scaled magnetization M/t^{β} of nickel and the scaled "equivalent magnetization" of CuMn $\mathcal{R}/t^{*\beta^*}$ vs.x. In nickel $x = \eta'/t'^{\gamma+\beta}$ and in CuMn (heavy dots) $x = \eta^*/t^{*\gamma^*+\beta^*}$.

the dimensionless units η', t' and η^*, t^*. Notice, however, that in the paramagnetic limit, M tends to $La(\eta')$, an odd function of η', while $\mathcal{R}$ tends to $La(H/T)/L_1H/T - 1$ with odd and even terms in η^*.

Two major points, we believe, emerge from this discussion :i) the use of the natural variables η' and t', considerably extends the so-called critical range where scaling ideas are practical ; ii) the evidence which we provide for the existence of a spin-glass transition in CuMn, is about as convincing as the existing magnetic evidence that nickel is a ferromagnet (see fig. 2).

REFERENCES

(1) P. Weiss and R. Forrer, Ann. Phys. (Paris), 5 (1926), 153.
(2) R. Omari, J.J. Préjean and J. Souletie, J. de Physique, 44 (1983), 1069.
R. Omari, J.J. Préjean and J. Souletie, Heidelberg Colloquium on spin glasses. Lecture Notes in Phys. 192 eds. J.L. van Hemmen and I. Morgenstern (Springer 1983), p. 70-78.
(3) J. Souletie and J.L. Tholence, Sol. Stat. Comm. 48 (1983), 407.
(4) M. Fahnle and J. Souletie, submitted to J. Phys. C.
(5) A. Arrot and J.E. Noakes, Phys. Rev .Lett. 19 (1967), 786.
(6) M. Susuki, Progr. Theor. Phys. 58 (1977), 1151.
J. Chalupa, Sol. Stat. Comm. 22 (1977), 315.

AKNOWLEDGMENTS

The authors with to thank P. Peretto, R. Omari, J.L. Tholence, J. Odin and J.P. Renard.

LT-17 (Contributed Papers)
U. Eckern, A. Schmid, W. Weber, H. Wühl (eds)

MODEL FOR NONEQUILIBRIUM SUPERCONDUCTING DOUBLE JUNCTION STRUCTURES

W.J. Gallagher

IBM T.J. Watson Research Center, Yorktown Heights, NY 10598

We develop a simple model for nonequilibrium double tunnel junction superconducting sandwich structures. The model consists of coupled Rothwarf-Taylor equations and effective temperature (T^*) characterizations of quasiparticle distributions. Calculated quiteron static and dynamic characteristics are in fair agreement with experimental results.

1. INTRODUCTION

Several attempts have been made to combine two superconducting tunnel junctions in a manner such that nonequilibrium effects would result in transistor-like device behavior (1-3). These efforts resulted in demonstrations of current gain and claims of power gain, but none led to a demonstration of any circuit functions. Since there is no device model for these structures, it is not even possible to reliably extrapolate experimental results and estimate circuit performance. Moreover, without theoretical guidance the task of developing these double junction devices into useful ones is a formidable materials and fabrication effort of unknown benefit. In this paper we describe a tractable model that contains the essential physics of these double junction sandwich devices. We illustrate the model by calculating static and dynamic characteristics for the quiteron devices described in reference (2).

2. FORMULATION

To go beyond a purely empirical description (4) of the electrical properties of the double junction devices, a model must contain microscopic information on the densities of quasiparticle states and the occupancy of the states. In the Gray-effect transistor (1) and the quiteron (2) at least two of the films are driven out of equilibrium. Most microscopic models for nonequilibrium superconductors consider only one film out of equilibrium(5-7). Recently, Frank et al. (8) in a microscopic model of the Gray-effect transistor allowed more than one film to go out of equilibrium, but not to the point of significant gap suppression. Extending these microscopic models to the case where two or three films can be driven significantly out of equilibrium is difficult and would result in a complicated description.

Here we rely on results from these microscopic theories to formulate a simpler description. Our description has three components: quasiparticle and phonon coupling, gap suppression, and quasiparticle distribution functions. We use coupled Rothwarf-Taylor equations (9) to describe the couplings in and between the total numbers of quasiparticles N_i^{qp} and phonons N_i^{ph} in each film i. Chang and Scalapino (5) have studied the microscopic justification for the Rothwarf-Taylor equations. They showed that the coefficients that Rothwarf-Taylor assumed to be constant depend on integrals of phonon and quasiparticle distribution functions. The coefficients are thus weakly dependent on details of the distributions. A number of microscopic models (10) indicate that the gap suppression is rather well described by

$$\Delta_i/\Delta_{0i} = 1-2.5n_i^{qp} \qquad (1)$$

where Δ_i is the superconducting gap of film i with zero temperature gap Δ_{0i} and with normalized quasiparticle density $n_i^{qp} = N_i^{qp}/4N_i(E_F)\Delta_{0i}A_i d_i$. Film i has thickness d_i, area A_i, and density of states at the Fermi level $N_i(E_F)$. Chang et al. (11) have shown that the quasiparticle distribution is reasonably described by a Fermi distribution at an elevated temperature T^* when there is strong phonon trapping. This should be the case here. We determine the effective temperature by

$$T_i^*/T_{Ci} = \Delta_i/[\Delta_{0i}\tanh^{-1}(\Delta_i/\Delta_{0i})] \qquad (2)$$

where T_{Ci} is superconductor i 's transition temperature.

The coupled Rothwarf-Taylor equations are

$$\dot{N}_i^{qp} = \frac{I_i}{2e} - R_i N_i^{qp^2} + 2\tau_{B_i}^{-1} N_i^{ph} \; ; \; i = 1,2,3 \qquad (3)$$

$$\dot{N}_i^{ph} = \frac{1}{2} R_i N_i^{qp^2} - \tau_{B_i}^{-1} N_i^{ph} - \frac{s_i \eta_{i2}}{4d_i} N_i^{ph} + \frac{s_2 \eta_{2i}}{4d_2} N_2^{ph} - \frac{s_i \eta_{ai}}{4d_i}[N_i^{ph} - N_i^{ph}(T_{ai})] \; ; \; i = 1,3 \qquad (4)$$

$$\dot{N}_2^{ph} = \frac{1}{2} R_2 N_2^{qp^2} - \tau_{B_2}^{-1} N_2^{ph} - \frac{s_2(\eta_{21}+\eta_{23})}{4d_2} N_2^{ph} + \frac{s_1 \eta_{12}}{4d_1} N_1^{ph} + \frac{s_3 \eta_{32}}{4d_3} N_3^{ph}. \qquad (5)$$

Here for each film i, I_i is the net injection rate for quasiparticles, RN_i^{qp} is the quasiparticle recombination rate, τ_{B_i} is the phonon pair breaking time, and s_i is the speed of sound. The Kapitza transmission coefficient for phonons incident from film i into film j is η_{ij}, with $\eta_{ai} = \eta_{01}(\eta_{43})$ for $i = 1(3)$. $N_1^{ph}(T_{a1})$ and $N_3^{ph}(T_{a3})$ are the equilibrium concentrations of phonons in films 1 and 3 when they are at the substrate and bath temperatures, respectively.

I_1, I_2, and I_3 are determined by conventional tunneling integrals generalized to apply for superconductors at different effective temperatures. Note that the quasiparticle currents are not equal to the electrical currents. Some electrical current involves quasiparticle transfer between films and some is associated with quasiparticle creation in both films.

3. STEADY STATE CHARACTERISTICS

Our calculated current-voltage characteristics for the Nb-Nb-PbAuIn quiteron structures fabricated in reference (2) are given in Figure 1. We have adjusted film areas beyond the

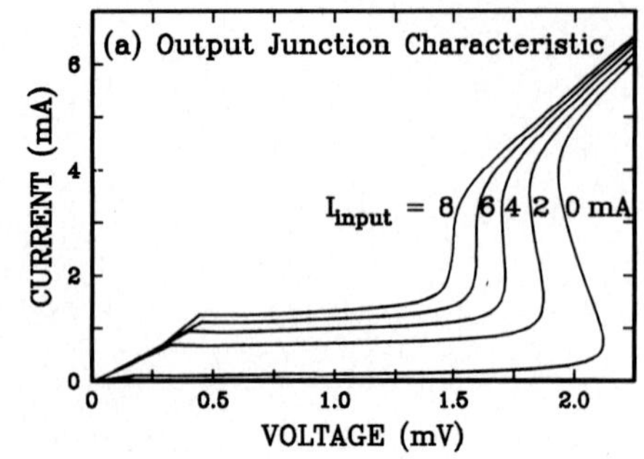

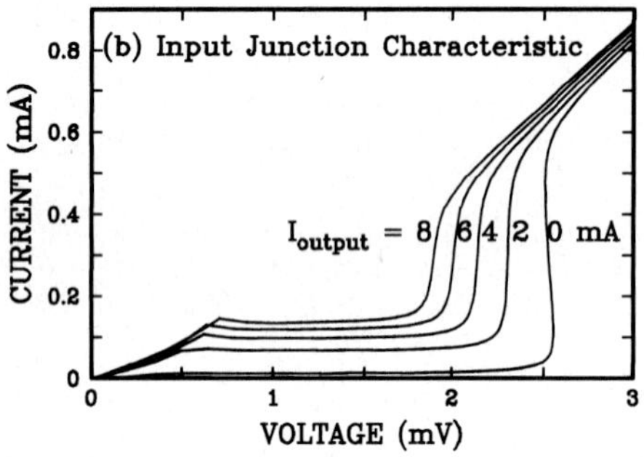

FIGURE 1
(a) Quiteron output (thin Nb-PbAuIn) junction characteristics at several input injection current levels. (b) Input (Nb-thin Nb) characteristics at several output current levels. Input and output junction resistances were 3 and 0.3 ohms, effective film areas were 400 and 100 μ^2 for the outer films and center film, respectively. Kapitza transmission coefficients (12) going from substrate to Nb, to thin Nb, to PbAuIn, and then to He were taken as 0.8, 1, 0.9, and 0.2, respectively. Going from He to PbAuIn, to Nb, to Nb, to substrate they were 0.2, 0.3, 1, and 0.4. These and other materials parameters were taken from references (12-14). For the thin Nb we used the experimentally observed T_C of 7.5 K and Δ_0 of 1.16 meV.

actual junction areas to account for the diffusion of quasiparticles into an area greater than the junction area. The adjustment was made until we achieved approximate agreement with the experimental gap reduction due to self-injection only. No other parameters were adjusted.

In Figure 1(a) we show show the calculated thin Nb-PbAuIn "output" junction characteristic at several levels of injection from the Nb-Nb "input" junction. As the injection from the Nb-Nb junction is increased, there is increased subgap current, increased gap suppression, and a broadening of the width of the current step at the sum gap. All of these features are in qualitative agreement with the experiment. Looking more closely, however, we see that the calculated gap suppression increases slower than linearly with increasing current while the experimental characteristics in reference (2) increase approximately linearly with injection current. The weaker than observed injection effects can be attributed to the neglect in the Rothwarf-Taylor picture of the energy dependence of the quasiparticle distribution.

In Figure 1(b) we show the effect that varying the Nb-PbAuIn "output" junction current has on the characteristics of the Nb-Nb "input" junction. The effect of the "output" current on the "input" characteristic is of the same magnitude as the effect of the "input" on the "output." That this should be the case is obvious from the the structure of equations (3-5), and, indeed, from the symmetric structure of the device and the absence of unidirectionality in the current flow mechanisms. The consequences of this symmetry for the utility of the quiteron as a transistor-like device are severe, as Frank has noted (4).

4. DYNAMIC CHARACTERISTICS

We can also calculate dynamic responses using our model. We find a number of time constants. For instance, if we current bias the output junction in the sum-gap region and instantaneously change the current in the input junction by δi_2, then the change in output junction voltage is given by

$$d\frac{(eV_{23})}{dt} = -2.5\Delta_{20}\frac{dn_2^{qp}}{dt} = -\Delta_{20}/\tau_G^* \qquad (6)$$

where the gap response time is $\tau_G^* = 3.2N_2(E_F)At_2e/\delta I_2$. For our case $\tau_G^* = 4.9$ ns for $\delta I_2 = 5$ ma. In the subgap region the large dynamic resistance enhances the speed of the (voltage) response by a factor $0.8R_{SG}/R_{NN}$. Experimentally this factor is between 5 and 20. There are other time constants in our model, though they are slower than the above. The slowest response time is that of film 3, tens of nanoseconds, and is unfortunately substantially coupled to the electrical properties of the device. Responses on all three of these times scales were observed experimentally (15).

CONCLUSIONS

We have developed a tractable description of double junction sandwich structures that allows us to estimate device properties in terms of known superconducting and phonon interface properties. The model gives fair agreement with experimental static and dynamic quiteron properties. Application of our model to the Gray-effect transistor is also possible.

REFERENCES

(1) K.E. Gray, Appl. Phys. Lett. 32 (1978) 392.
(2) S.M. Faris, S.I. Raider, W.J. Gallagher, and R.E. Drake, IEEE Trans. Mag. 19 (1983) 1293.
(3) B.D. Hunt and R.A. Buhrman, IEEE Trans. Mag. 19 (1983) 1155.
(4) D.J. Frank, to be published.
(5) J.J. Chang and D.J. Scalapino, Phys. Rev. B15 (1977) 2651.
(6) J.R. Kirtley, D.S. Kent, D.N. Langenberg, S.B. Kaplan, J.J. Chang, and C.C. Yang, Phys. Rev. B22 (1980) 1218.
(7) J.L. Paterson, J. Low Temp. Phys. 33 (1978) 285.
(8) D.J. Frank, A. Davidson, and T. Klapwijk, Bull. Am. Phys. Soc. 29 (1984) 521.
(9) A. Rothwarf and B.N. Taylor, Phys. Rev. Lett. 19 (1967) 27.
(10) W.H. Parker, Phys. Rev. B12 (1975) 3667.
(11) J.J. Chang, W.Y. Lai, and D.J. Scalapino, Phys. Rev. B20 (1979) 2739.
(12) S.B. Kaplan, J. of Low Temp. Phys. 37 (1979) 343.
(13) S.B. Kaplan, C.C. Chi, D.N. Langenberg, J.J. Chang, S. Jafarey, and D.J. Scalapino, Phys. Rev. B14 (1976) 4854.
(14) C.C. Chi, M.M.T. Loy, and D.C. Cronemeyer, Phys. Rev. B23 (1981) 124.
(15) W.J. Gallagher, unpublished data.

LT-17 (Contributed Papers)
U. Eckern, A. Schmid, W. Weber, H. Wühl (eds)
© Elsevier Science Publishers B.V., 1984

DYNAMIC PROPERTIES OF A TRIANGULAR LOOP CONTAINING THREE DAYEM BRIDGES

S. HOSOGI* and T. AOMINE**

* International Institute for Advanced Study of Social Information Science, Fujitsu Limited, Numazu 410-03, Japan
** Research Institute of Fundamental Information Science, Kyushu University, Fukuoka 812, Japan

Properties of a superconducting triangular loop with three Dayem bridges have been studied experimentally and calculated by computer simulations. Threshold currents at which voltage appears from two to three bridges are explained qualitatively by the extention of the flux quantum states with quantum number equal to ±2. The existence of one or two flux quanta is confirmed in the voltage state with two bridges by calculations.

1. INTRODUCTION

Current-voltage (I-V) characteristics in a loop containing three Dayem bridges being located at the centre of each side of a triangle were investigated in previous works (1, 2). Either of the two currents, which are passed into the loop from independent current sources, is increased under a constant value of the other. According to the I-V characteristics, there are three dc voltage states; complete zero voltage state (CZVS) in which all the bridges have no voltage, partial voltage state (PVS) in which two bridges have voltage and the other remains to have no voltage, complete voltage state (CVS) in which all the bridges have voltage.

The experimental values of the threshold current between CZVS and PVS are found to be in fairly good agreement with the calculated ones (1, 2). On the other hand, results of experiment and calculation for the current corresponding to the threshold between PVS and CVS have not been given yet. In this report we present them, their comparison and results of computer simulation for dynamic properties of the loop.

2. EXPERIMENTAL

Methods of sample fablication and measurement of the I-V characteristics are the same as used in previous works (1, 2). Each bridge made of Sn is 1 μm long, 1 μm wide and 0.2 μm thick. Superconducting films connecting the bridges have a width of 10 μm and thickness of 0.2 μm.

3. SIMULATION MODEL

An equivalent circuit of the triangular loop is shown in Fig. 1, where L_i's (i=1, 2, 3) are inductances of triangular sides and Ic and Ig are applied currents.

The equations for the phase difference φ_i of

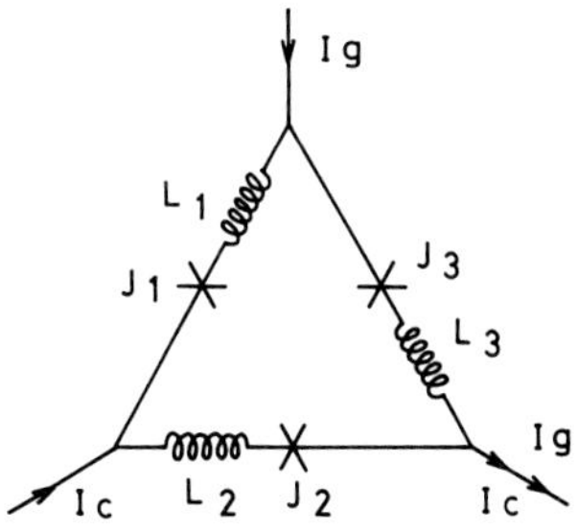

FIGURE 1
Equivalent circuit.

each bridge can be obtained by the use of Resistively Shunted Junction (RSJ) model (a capacitance is taken to be zero) as follows:

$$\dot{\varphi}_1 + (I_{01}R_1/\Phi_0)\sin\varphi_1 + (R_1/L)(\varphi_1+\varphi_2-\varphi_3) = R_1(L_2 Ic - L_3 Ig)/\Phi_0 L$$

$$\dot{\varphi}_2 + (I_{02}R_2/\Phi_0)\sin\varphi_2 + (R_2/L)(\varphi_1+\varphi_2-\varphi_3) = R_2\{(L_1+L_3)Ic + L_3 Ig\}/\Phi_0 L$$

$$\dot{\varphi}_3 + (I_{03}R_3/\Phi_0)\sin\varphi_3 + (R_3/L)(-\varphi_1-\varphi_2+\varphi_3) = R_3\{L_2 Ic + (L_1+L_2)Ig\}/\Phi_0 L\ ,$$

where I_{0i} and R_i are the critical current and resistance, respectively, $L=L_1+L_2+L_3$ and Φ_0 is a flux quantum divided by 2π.

4. RESULTS AND DISCUSSION

Simulations have been performed for the parameters with both identical (case 1) and nonidentical (case 2) values. When Ic=0, the calculated I-V characteristics show that dc voltage arises at all the bridges simultaneously for case 1 and that PVS arises for case 2, which is the case for the experimental circumstances. The current range of PVS is found to be several times larger than the difference of the critical currents of the bridges. When Ic≠0, PVS exists outside flux quantum states. In order to obtain the thresholds between PVS and CVS in case 2, the I-V characteristics have been calculated. The circuit parameters are chosen as follows; $I_{01}=I_{03}=0.23$ (mA), $I_{02}=0.21$ (mA), $R_1=R_3=0.14$ (Ω), $R_2=0.155$ (Ω), $L_1=L_2=L_3=15.6$ (pH).

The thresholds between PVS and CVS obtained from the experiments (solid circles) and the simulations (dotted lines) are shown in Fig. 2, together with the experimental thresholds between CZVS and PVS (open circles). In the figure, the flux quantum states (solid lines) calculated by using a linear approximation (3) are written for ease to the interpretation of the results. The thresholds obtained from the experiment and calculation are in fairly good agreement with the extrapolated lines of flux quantum states with quantum number n=±2. This fact is explained as follows: In CZVS, thresholds of the flux quantum states are influenced by the bifurcated supercurrents, which are determined from the inductance, but in PVS both supercurrent and quasi-particle current are associated with the bifurcated current. Therefore, the threshold of PVS generally depends on both resistances and inductances. In the present case, however, the ratios between corresponding resistance and inductance are nearly constant for all bridges, and hence there is the relation between the thresholds of PVS and of flux quantum state with the maximum quantum number of external flux quantum states.

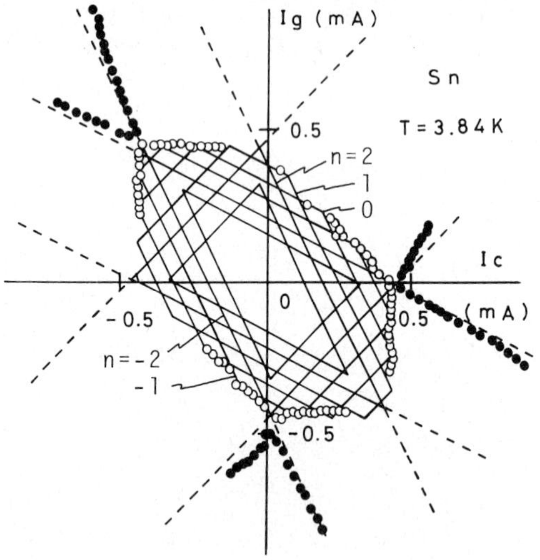

FIGURE 2
The thresholds between PVS and CVS (● : exp. and --- : cal.). The calculated flux quantum states (——) and experimental thresholds between CZVS and PVS (○) are also given.

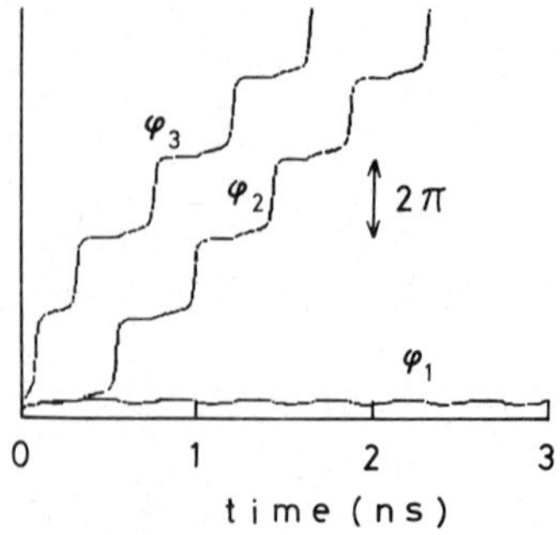

FIGURE 3
The phase differences vs. time curves in PVS.

To see the dynamic behaviour of flux quanta in PVS, the time dependence of the phase difference of each bridge is computed and the results are shown in Fig. 3. The figure shows that the phases, φ_2 and φ_3, of the bridges having dc voltage increase with two characteristic times. On the other hand φ_1 remains finite within $\pi/2$. As a flux quantum enters into the loop at the bridges that the phases change a value by 2π, it can be seen that one or two flux quanta exist alternatively in time even when the circuit is in PVS.

5. SUMMARY

We have studied the dynamic properties of the triangular loop containing three Dayem bridges by experiments and computer simulations. The thresholds of PVS agree with the extrapolated lines of the flux quantum states with n=±2. One or two flux quanta exist even when the circuit is in PVS.

ACKNOWLEDGEMENTS

We would like to thank Dr. T. Kitagawa for suggesting the present triangular loop. We are also grateful to Mr. M. Hidaka for experimental assistance.

REFERENCES

(1) T. Aomine, M. Hidaka, K. Matsuo and T. Kitagawa, Proceedings on Symposium on Superconducting Quantum Electronics (Tokyo, 1983) 92.
(2) T. Aomine, M. Hidaka, K. Matsuo and T. Kitagawa, Bull. Informatics and Cybernetics 21 (1984) 85.
(3) E.E. Schulz-DuBois and P. Wolf, Appl. Phys. 16 (1978) 317.

LT-17 (Contributed Papers)
U. Eckern, A. Schmid, W. Weber, H. Wühl (eds)

SELF-BIASING LOGIC/MEMORY CELL WITH WIDE MARGIN

Yoichi OKABE, Atsuki INOUE and Takuo SUGANO

University of Tokyo, 7-3-1, Hongo, Bunkyo-ku, Tokyo, Japan

A new type of Josephson logic/memory cell is presented. The fundamental cell consists of two superconducting loops coupled by a Josephson junction controlled by a magnetic field. The cell, initialized by injecting a single flux quantum, is self-biasing and thus needs no bias current. In principle, this cell can be operated by a small input signal. A wider margin in terms of signal is expected for this device. The proper operation of the cell was confirmed by computer simulation and the results are presented.

1. Introduction

Various types of Josephson logic circuits have been proposed recently but most have the drawback of requiring bias current (1) which is difficult to regulate due to the poor controllability of the Josephson critical current. In this report, we propose a new type of Josephson logic/memory cell which is initialized by a single flux quantum, needs no bias current and operates with a relatively small input signal.

2. Fundamental Cell Configuration

The circuits configuration of the fundamental cell is shown in Fig. 1. I_s is the input signal, I_b is the clock signal. The basic cell consists of two superconducting loops coupled by Josephson junctions and is controlled by using a magnetic field. Upon initialization with a single flux quantum the flux will be trapped in the circuit forever. Linearizing the I-Φ relation between two junctions, we can represent the cell operating modes on the I_s-I_b plane as shown in Fig. 2. L_3 is chosen sufficiently small so that the Gibbs free energy will be stable inside and outside the region of $|\phi_1|\leq\pi/2$, $|\phi_2|\leq\pi/2$(2). Thus, in this way the switching from mode Ⅲ to mode I and from mode Ⅲ to mode Ⅱ will be continuous. Mode I and mode Ⅱ overlap each other, however, near the origin so that both modes will be stable in that region.

3. Cell Operation

From the above, mode I or mode Ⅱ is stable when $I_s=I_b=0$: each mode can be thought of as corresponding to a logic "1" or "0". Given the input signal I_s and the clock signal I_b, let I_s fall to zero before I_b is removed in order to establish a cell mode (see Fig. 2). The cell will switch into either mode I or mode Ⅱ depending on the sign of the input signal. In principle, then, if the junctions J_1 and J_2 are assumed to be of low-capacitance the cell can be operated with a small signal.

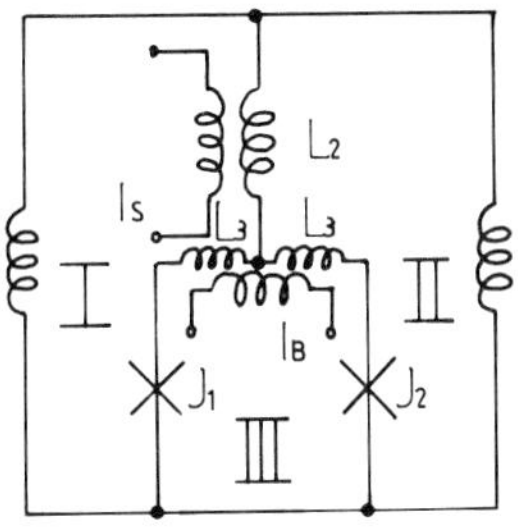

Fig.1 Equivalent circuit of the presented logic/memory cell. Here Is is the input signal and Ib is the clock signal.

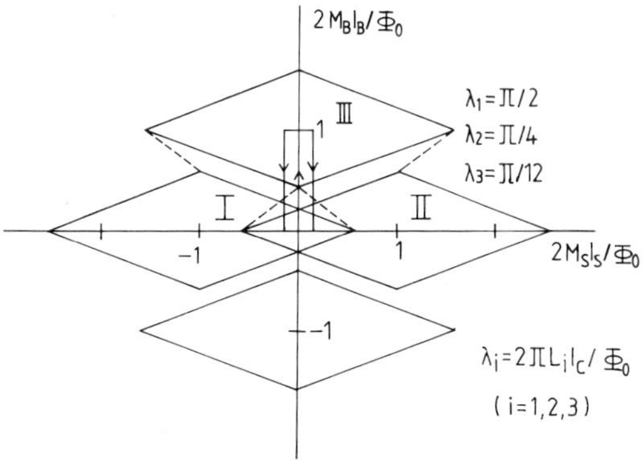

Fig.2 Cell operating modes on Is-Ib plane. Modes I,Ⅱ,Ⅲ correspond to a single flux in loops I,Ⅱ,Ⅲ of Fig.1.

4. Dynamic Simulation

The cell dynamics were simulated on a computer by using the RSJ model with the capacitance C=0, corresponding to an ideal bridge. Simulation results are shown in Fig. 3. Here, the critical current of the Josephson junctions I_c=0.1 [mA], the normal resistance R_n=26 [Ohms] and the inductances L_i are the same as in Fig. 2. The cell is thus seen to drop down to a particular state depending on the sign of the input signal.

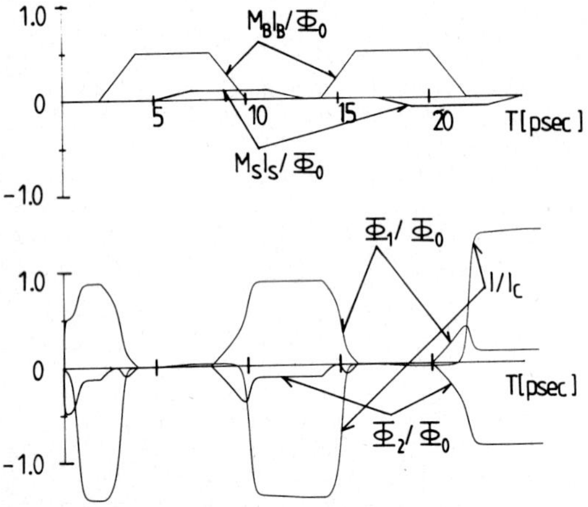

Fig.3 Simulation results of the operation of the cell.

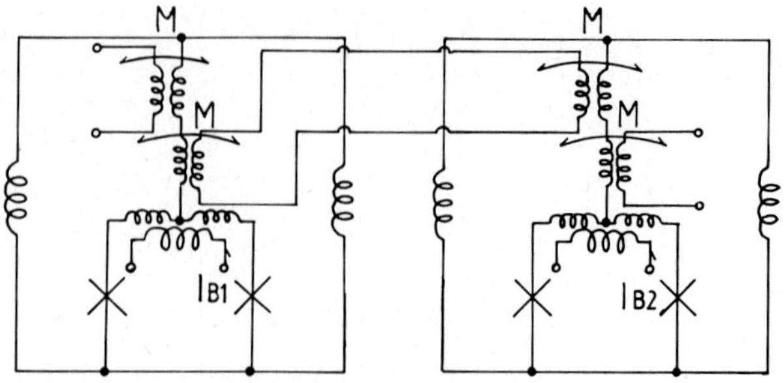

Fig.4 The method of connection between two cells. Here M is the mutual inductance.

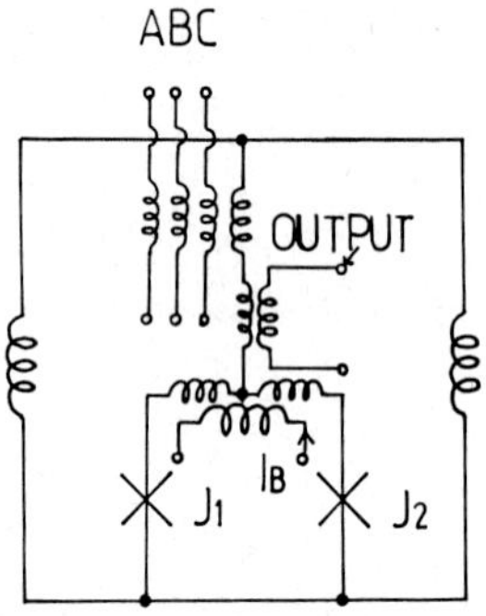

Fig.5 The configuration of an AND/OR logic. Here A, B, are the input signal and C is the cell function select signal.

5. Interconnecting and Driving the Cells

Adjacent cells are interconnected by using an inductive loop to couple the branch with L_2 in it, as shown in Fig. 4. Simulation results confirm that the current cannot flow through that branch while there is a clock signal I_b. By using the three phase clock, once the state of a cell has settled, the preceeding cells will have no effect on it. Thus by choosing a small mutual inductance M, one can ensure uni-directional signal propagation.

6. Logic Gates

Logic "1" and "0" can be thought of as corresponding to the direction of the current in the branch which contains L_2. The basic operation of the cell is the majority logic, then we need cell function signal C besides two input signal A, B to construct AND/OR logic(Fig. 5). When C equals logic "1", the cell operates as an OR logic. On the contrary, when C equals "0", the operation is a logical product. On the other hand, an INVERTER logic can easily be constructed by changing the direction of the input signal.

7. Conclusion

A new type of Josephson logic cell has been proposed. The cell, initiallized with a single flux quantum, is self-biasing and hence needs no bias current. Operation of the cell was simulated on a computer. The cell, in principle, can be operated with small signal. The uni-directional propagation of the signal is obtained by using a 3-phase clock. This cell can be used to construct memory cells as well as logic cells.

Acknowledgement

We would like to thank all my colleagues who had a useful discussion with us.

References

(1) H.Tamura, Y.Okabe and T.sugano IEEE Trans. Electron. Devices vol. ED-27 2035 1980

(2) E.O.Schulz-DuBois and P.Wolf Appl. Phys. 16 317 1978

LT-17 (Contributed Papers)
U. Eckern, A. Schmid, W. Weber, H. Wühl (eds)

THRESHOLD CHARACTERISTICS OF MUTUALLY COUPLED SQUIDs

C. C. CHI, L. KRUSIN-ELBAUM, C. C. TSUEI, C. D. TESCHE, K. H. BROWN, A. C. CALLEGARI, M. M. CHEN, J. H. GREINER, H. C. JONES, K. K. KIM, A. W. KLEINSASSER, H. A. NOTARYS, G. PROTO, R. H. WANG, and T. YOGI

IBM T. J. Watson Research Center, Yorktown Heights, NY 10598

The threshold characteristics of two nested SQUIDs has been investigated. Due to the flux quantization requirement of the larger SQUID loop, the threshold current of the smaller SQUID loop exhibits a large abrupt jump as the externally applied flux increases to a level beyond $\phi_o/2$ in the larger SQUID loop, when the larger SQUID is not biased. The threshold curve of the smaller SQUID changes dramatically when the larger SQUID is biased near its maximum critical current. The potential digital application of the coupled SQUID system is discussed.

The phenomena of two mutually coupled nonhysteretic dc SQUIDs have previously been studied. (1) The most striking observations were the strong voltage locking and the synchronization of their responses to the applied magnetic flux modulation. The potential energy of such a system has been explored theoretically (2) to explain the observed abrupt jumps in the voltage vs. applied flux curve when one or both SQUIDs are biased just above their respective critical currents. These interesting phenomena strongly suggest that the coupled SQUID system might have a very interesting threshold behavior and potential digital applications.

To experimentally study the threshold characteristics of coupled SQUID systems, a device schematically shown in Fig. 1, was fabricated by IBM Josephson Technology group. The Josephson junctions in this device were niobium edge junctions with PbInAu counterelectrodes. The circuit consisted of two nested SQUID loops with two Josephson junctions in each loop. The outer and inner loops are refered to as SQUID 1 and SQUID 2 respectively in this paper. The critical currents of each junction in SQUID 1 and 2 were designed to be about 0.05 mA and 0.15 mA respectively. For reasons explained below, the junctions in SQUID 1 were nonhysteretically shunted with 2Ω AuTi thin film resistors, while the junctions in SQUID 2 were not shunted. The nesting geometry, shown in Fig. 1, ensures that the two SQUIDs are mutually coupled and SQUID 2 has about one half of the self-inductance of SQUID 1. The mearsured total critical currents were 0.33 mA and 0.10 mA for SQUID 1 and 2 respectively. The self inductances estimated from the modulation depth of the critical current of each SQUID were about 19 pH and 10 pH respectively.

The threshold curve of the SQUID 2 is defined to be the maximum critical current of the SQUID 2 as a function of the applied magnetic flux. Experimentally, the bias current I_2 was ramped up past the critical current and then back down to zero at about 10 kHz rate, while the applied flux was modulated by an ac current I_m at a slower repetition rate about 50 Hz. The voltage of SQUID 2 was monitored with a threshold detector which produces a fast pulse of about a few microsecond duration each time when the SQUID 2 was switched from the superconducting state to the resistive state. The fast pulses from the threshold detector were used to turn on the electron beam of an oscilloscope. When the electron beam was turn

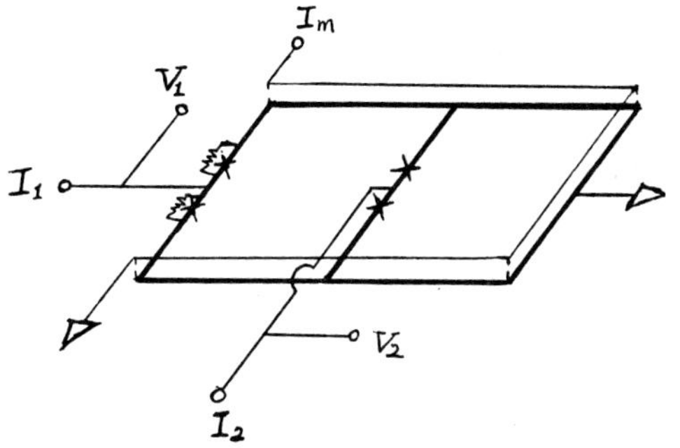

Figure 1
Schematic diagram of two nested SQUIDs with modulation loop overlayed on top of the larger SQUID loop (SQUID 1).

on, the currents I_2 *and* I_m were measured on the vertical and horizontal axis of the oscilloscope. The examples of the threshold curves displayed on the scope screen are shown in Fig. 2. Also shown in Fig. 2 is an I-V curve of SQUID 1. The corresponding bias points of SQUID 1 for the threshold curves shown are indicated on the I-V curve. Due to the ambient magnetic fields and the magnetic flux trapped in the superconducting ground plane near this device, the exact point corresponding to $I_m = 0$ in the threshold curve does not have any physical significance. The approximate periodicity of the threshold curve shown in Fig. 2 corresponds to a superconducting flux quantum, ϕ_0, in the SQUID 2. The origin of the absence of the exact periodicity is due to the mutual coupling between the two SQUIDs and the fact that the ratio of the two SQUID inductances is not exact 2 to 1. The most noticeable features of these threshold curves are the abrupt jumps on the first two traces (a) and (b) in Fig. 2. Qualitatively, this is due to the flux quantization in the SQUID 1. When the applied flux increases from zero, the induced supercurrent in the SQUID 1 generates a flux opposite to the applied flux in order to maintain the zero flux state. Thus, the SQUID only sees a very small amount of net flux. When the applied flux exceeds $\phi_0/2$ in the SQUID 1, it becomes energetically more favorable to switch the circulation current direction to admit one flux quantum. The exact switching position would depend on the product of the critical current (i_0) and the loop inductance (L). For $i_0 L < \phi_0$, the switching point is reasonably close to $\phi_0/2$ according to our potential-energy calculation. When the circulation current in SQUID 1 changes sign, SQUID 2 is suddenly exposed to a larger amount of flux than that externally applied. Thus, its threshold drops abruptly at that point. This screening-antiscreening argument is valid only when SQUID 1 is not biased or biased not too close to its maximum critical current. On the other hand, when SQUID 1 is biased near or even above its maximum critical current, a small applied dc flux induces an ac circulation current in SQUID 1. Through the inductive coupling, this ac circulation current is responsible for the rapid drop, although not an abrupt one, in the threshold current in SQUID 2, shown as trace (c) of Fig. 2. The amplitude of the induced ac circulation current decreases, as the average voltage of the SQUID 1 increases with the increasing bias current above its maximum critical current. That explains why the initial slope in the threshold curve, shown as trace (d) of Fig. 2, is less steep as the bias current of SQUID 1 is increased. When the bias current of SQUID 1 is extremely high, say more than 10 times higher than the maximum critical current, then SQUID 2 becomes decoupled from SQUID 1 and has the usual smooth threshold curve expected for an isolated 2-junction SQUID as shown in Fig. 3. Note that in this case, the periodicity in the threshold curve of SQUID 2 becomes exact.

Numerical calculations of the threshold curves of such a coupled SQUID system have been done. The results are in good agreement with the experimental data and will be published elsewhere.

For potential digital application of this device, it is obvious that the output port should be on SQUID 2. Thus, it is desirable to leave the junctions in SQUID 2 unshunted so that a maximum amount of current can be delivered to a load resistance which is about equal to the normal-state resistance of the junctions. Both modulation current and the bias current of SQUID 1 can be used as input currents to switch the device. This capability makes the coupled SQUIDs more suitable for a logic "and" gate than the single SQUID. The advantages of shunting the junctions in SQUID 1 with small resistance are (i) less vulnerable to transient noise due to the nonhysteretic nature of the shunted junctions; (ii) no isolation problem usually associated with injection devices (3) because the shunting resistance keeps the voltage drop across SQUID 1 small; (iii) damping of the unwanted LC resonances in SQUID 2 through the mutual coupling to the heavily damped SQUID 1.

The 3 to 1 critical current ratio of the two SQUIDs makes this device capable of having a large current and power gain. If more gain or more input or control ports are needed, one can extend the 2-SQUID system to a n-SQUID system, in which three n-SQUID loops are nested together with progressively larger critical currents for the smaller SQUID loop. Such a "domino" coupled SQUID system can be quite compact for high packing density and capable of having very large current and power gain.

The authors would like to acknowledge helpful discussions with M. B. Ketchen, E. P. Harris, S. B. Kaplan and W. H. Chang.

REFERENCES

(1) C. C. Chi, L. Krusin-Elbaum and C. C. Tsuei, Physica **108B**, 1085 (1981).

(2) L. Krusin-Elbaum, C. C. Chi and C. D. Tesche, Bull of APS **27**, 266 (1982) and unpublished results.

(3) T. R. Gheewala, IBM J. Res. Develop. **24**, 130 (1980).

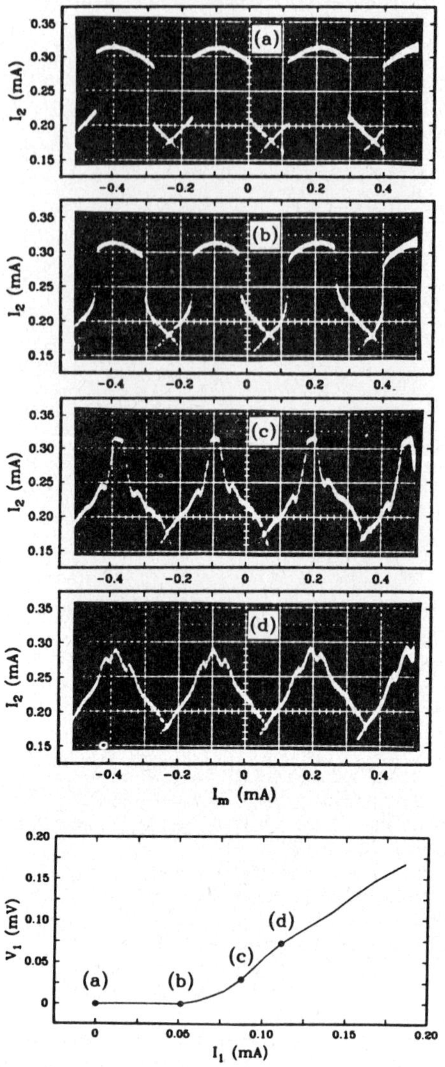

Figure 2
Threshold curves of SQUID 2 , (a) to (d), with corresponding bias points denoted on the I-V curve of SQUID 1. The dc I-V curve of SQUID 1 was subtantially smeared due to the ac modulation current. Without the ac modulation, the maximum critical current was very close to the bias point (c).

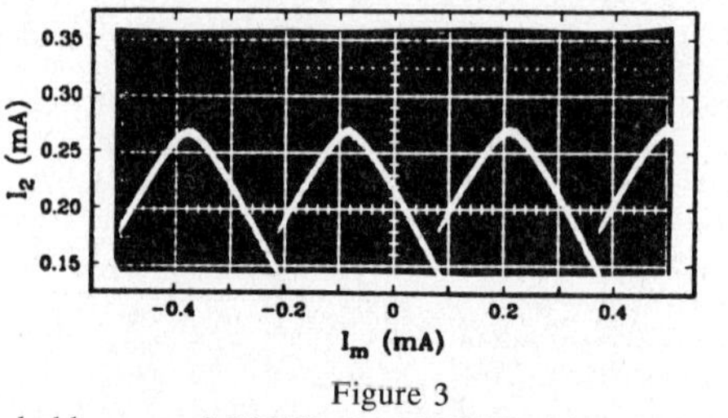

Figure 3
Threshold curve of SQUID 2 as SQUID 1 becomes decoupled because of the extremely high bias current.

LT-17 (Contributed Papers)
U. Eckern, A. Schmid, W. Weber, H. Wühl (eds)
© Elsevier Science Publishers B.V., 1984

ROLE OF ELECTRON DEPOLARIZATION IN GAS PHASE THREE-BODY RECOMBINATION OF SPIN-POLARIZED HYDROGEN

D. A. Bell, G. P. Kochanski, L. Pollack, H. F. Hess,[7] D. Kleppner, and T. J. Greytak

Physics Department and Center for Materials Science and Engineering
Massachusetts Institute of Technology, Cambridge, MA. 02139 U.S.A.

Measurements of three-body recombination in the gas reveal two separate mechanisms: thresholdless dipole three-body events and recombination due to a finite density of atoms depolarized by dipole two-body events.

Three-body recombination due to the electronic dipole interaction has been predicted(1) and observed(2,3) in spin-polarized atomic hydrogen. In a recent experiment(4) we used the temperature dependence of the recombination rate to obtain values for the gas and surface three-body rate constants, L_g and L_s. These measured values were close to the predictions, but we found the magnetic field dependence of the three-body rate to be weaker than predicted and opposite in sign. The field dependence was measured at a temperature where the surface contribution dominated the three-body rate. We report here a measurement of the magnetic field dependence of the three-body rate when it is determined predominantly by the gas rate, L_g. Kagan et al.(1) have discussed the role of electronic dipole interactions in depolarizing (i.e. relaxing) electronic spins during pair-wise collisions. Sprik et al.(3) observed the contribution of this process to the two-body decay. We report here first observations of this process in the three-body decay.

The experimental technique is described in references (2) and (4). A gas of hydrogen in the b state (a - d are the hyperfine states in order of increasing energy) is compressed and the decay of the gas density n(t) at constant volume is monitored. The derivative of the density is fit to

$$-\dot{n} = 2Gn^2 + Ln^3 , \qquad (1)$$

where G and L are the rate constants for two-body relaxation and three-body recombination. G and L were measured at 0.60K and 0.35K. At both temperatures G is determined by processes in the gas. L is dominated by gas phase processes at 0.60K and by the adsorbed surface phase at 0.35K.

G consists of two terms: $G = G^{ba} + G^{bc}$. G^{ba} represents nuclear relaxation from b to a and varies with magnetic field as $(1+16.7/B)^2$ where B is the field in tesla. G^{bc} represents electronic relaxation from b to c, a thermally activated process proportional to $\exp(-2\mu B/kT)$. Once in the a or c state, an atom recombines so rapidly that the overall recombination rate is limited by the relaxation rate.

Figure 1 shows the measured values of G. We have fit the 0.60K data to

$$G = G(\infty)(1+16.7/B)^2 + \alpha e^{-2\mu B/kT} . \qquad (2)$$

The resulting fit (solid line) corresponds to $G(\infty)=5.6(2)\times10^{-22}cm^3s^{-1}$, $\alpha=10(1)\times10^{-16}cm^3s^{-1}$. $G(\infty)$ is in good agreement with previous measurements and with the value 5.5×10^{-22} calculated theoretically(5). The value of α is in good agreement with the theoretical prediction (1) and with the value $8(4)\times10^{-16}$ measured by Sprik et al.(3).

The contribution of G^{bc} to G is much smaller at 0.35K. We have fit the 0.35K data by using the value of α found above and determining the value of $G(\infty)$ appropriate to this temperature. We find $G(\infty)=5.0(4)\times10^{-22}cm^3s^{-1}$. The theoretical value(5) at this temperature is 4.4×10^{-22}. The resulting fit is a solid line in the figure; the dashed line corresponds to this value of $G(\infty)$, but $\alpha=0$.

L can be divided into two parts:

$$L = L_0 + K^{(3)}(n_c/n) \qquad (3)$$

L_0 arises from the thresholdless dipole three-particle recombination(1). $K^{(3)}$ is the three-

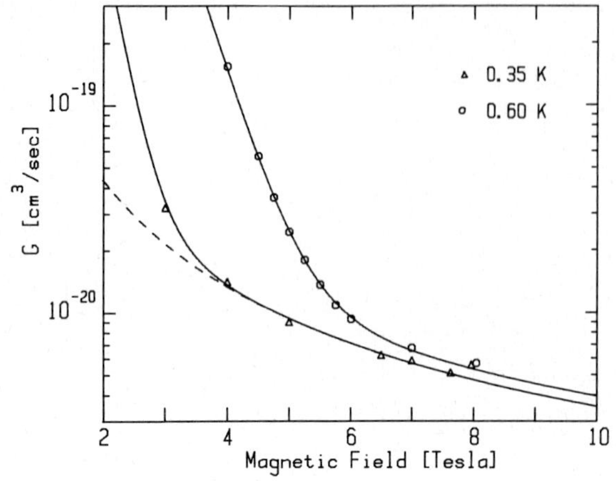

Fig. 1. The measured two-body relaxation rate constant as a function of magnetic field. The values at high field are determined by nuclear relaxation. The sharp increase at low field results from electronic relaxation.

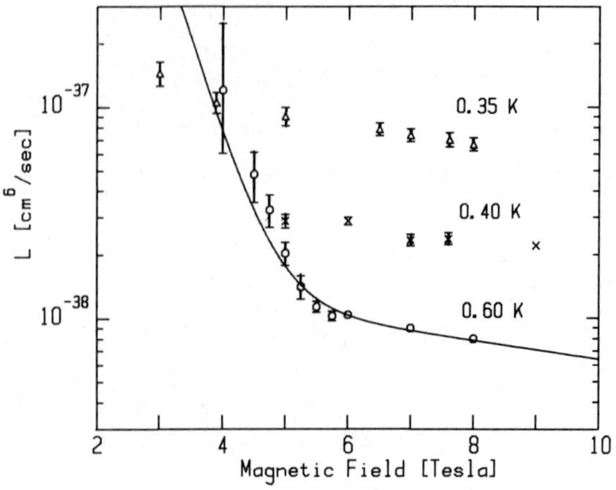

Fig. 2. The measured three-body recombination rate constant as a function of magnetic field. At high field or low temperatures the dipole recombination of three b state atoms dominates. The sharp increase at 0.6K at low fields indicates recombination with a c state atom.

body rate constant for recombination in the gas due to the finite density of c state atoms, $n_c=(G^{bc}/K^{(2)})n$. $K^{(2)}$ is the effective gas two-body rate constant due to recombination of the c atoms on the walls.

Figure 2 shows the measured values of L. At 0.35K the second term in Eq.3 is negligible and the data show L_0 to be a weakly decreasing function of the magnetic field from 3 to 8T. This is consistent with the behavior we reported previously(4) at 0.40K; those data are reproduced for comparison. The results at 0.60K have been fit to

$$L = L_0 + \gamma e^{-2\mu B/kT} \qquad (4)$$

where $\gamma = K^{(3)}(\alpha/K^{(2)})$. We have done a fit assuming L_0 and γ to be independent of field; however, we find that a better fit (solid line) can be achieved if L_0 is assumed to have the field dependence exhibited by the 0.35K data. We find $L_0(B=8T)=7.9(4)\times10^{-39}cm^6s^{-1}$, $\gamma=5(2)\times10^{-34}cm^6s^{-1}$.

We have used the results of Morrow(6) for the wall recombination of unpolarized hydrogen to calculate that in our geometry $K^{(2)} = 1.3(2)\times10^{-14}cm^3s^{-1}$. This, together with our measured values of α and γ, gives $K^{(3)} = 7(3)\times10^{-33}cm^6s^{-1}$. Morrow(6) obtained $K^{(3)} = 2.0(3)\times10^{-33}$ for H+H+He → H_2+He in an unpolarized gas. Our $K^{(3)}$ should be larger than this value, due to the increased probability of singlet pairing in the c state recombinations.

This work was supported by the NSF through grants DMR-8304888 and DMR-8119295.

REFERENCES

(1) Yu. Kagan, I. A. Vartanyantz, and G. V. Shlyapnikov, Zh. Eksp. Teor. Fiz., 81, 1113 (1981) (Sov. Phys. JETP, 54, 590 (1981))
(2) H. F. Hess, D. A. Bell, G. P. Kochanski, R. W. Cline, D. Kleppner, and T. J. Greytak, Phys. Rev. Lett. 51, 483 (1983)
(3) R. Sprik, J. T. M. Walraven, and I. F. Silvera, Phys. Rev. Lett. 51, 479 (1983) and 51, 942 (1983)
(4) H. F. Hess, D. A. Bell, G. P. Kochanski, D. Kleppner, and T. J. Greytak, Phys. Rev. Lett. (in print)
(5) R. M. C. Ahn, J. P. H. W. v. d. Eijnde, and B. J. Verhaar, Phys. Rev. B27, 5424 (1983)
(6) M. Morrow, dissertation (unpublished), U. British Columbia (1983)
(7) AT&T Bell Laboratories Scholar

LT-17 (Contributed Papers)
U. Eckern, A. Schmid, W. Weber, H. Wühl (eds)

OBSERVATION OF SPIN WAVES IN SPIN POLARIZED HYDROGEN

B. R. JOHNSON*, N. BIGELOW, J. S. DENKER, L. P. LÉVY§, J. H. FREED, and D. M. LEE

Laboratory of Atomic and Solid State Physics and Baker Laboratory of Chemistry, Cornell University, Ithaca, NY 14853, USA

We report the results of continuing experiments on the spin transport properties of a dilute quantum gas, spin polarized hydrogen H↓. Spin wave resonances are prominent features of the pulsed Fourier transform NMR spectrum. For small tipping angles, the dependence of the spectrum on polarization and temperature are found to be in good qualitative agreement with theory. Preliminary results are presented on the large tipping angle spectrum, which exhibits a number of features not observed in the small angle spectrum.

For our experimental conditions, $H_0 = 7.7$ Tesla (in the z direction) and $T < 0.8$ K, only the two lowest hyperfine states |b> = |↓⇓> and |a> = |↓⇑> - η|↑⇓> of H↓ are populated. Here ↓ denotes the electronic spin and ⇓ the nuclear spin. The term "spin polarized hydrogen" refers to the electron spin, and means that the upper two hyperfine levels are thermally inaccessible. Throughout the rest of this paper the term "spin" will refer to the nuclear spin, or more precisely to the pseudospin in the |a> - |b> two level system. The admixture η promotes recombination of |a> state atoms into molecular hydrogen, creating a large nuclear spin polarization (1) which in turn produces a large molecular field. The cryostat and spectrometer are identical to the one described in earlier work (2, 3).

Exchange effects have been observed to play a significant role in the spin transport properties of spin polarized hydrogen, H↓ (3). The observed effects have been interpreted in terms of collective nuclear spin oscillations in this rarefied quantum gas. For small NMR tipping angles, the equation of motion for the transverse spin density $s_+(\vec{r}, t)$ in the rotating frame is (4, 5)

$$i \frac{\partial s_+}{\partial t} = \delta(\gamma H_0)\, s_+ + i D_0 \frac{1 - i\mu P_z}{1+\mu^2 P^2} \nabla^2 s_+ \qquad (1)$$

where we are using the notation of reference 6. This equation, subject to appropriate boundary conditions, predicts the appearance of discrete resonances in the NMR spectrum at frequencies corresponding to damped standing spin wave modes. For large tipping angles the equation of motion becomes highly non-linear, as discussed by Lhuillier and Laloë (4) and by Lévy and Ruckenstein (5).

Figure 1 shows typical small-angle (≈10°) spectra -- broad resonances with several narrow lines superimposed. Note that as T decreases, causing |P| and presumably (7) |μ| to increase, the lines become narrower, closer together and more prominent. This is in good qualitative agreement with equation 1 which predicts that the spin wave lifetime should scale as |μP|. In the presence of a linear field gradient, the separation between lines should scale as $|\mu P|^{-1/3}$. Here we neglect the weak variation of D_0 with temperature (3, 7). The lack of variation of the overall width of the spectrum with T is consistent with confinement of the spin waves by $\delta(\gamma H_0)$, as predicted (5).

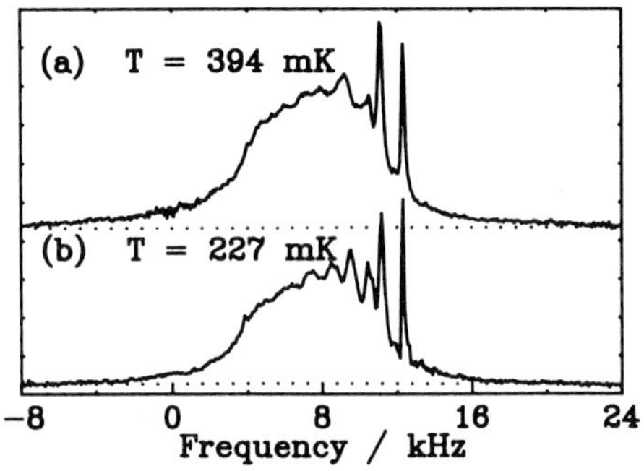

Figure 1. Temperature dependence of small-angle spectum; $n \approx 4\times10^{16} cm^{-3}$

Drawing an analogy between equation 1 and the Schrödinger equation, confinement of the spin waves means that the applied "potential" due to $\delta(\gamma H_0)$ is large in comparison to the spin wave "kinetic energy" corresponding to the term containing the Laplacian. The sign of μP_z determines whether the spin wave modes will be confined in the region of most positive or most negative $\delta(\gamma H_0)$. The narrowest lines (which have the smallest "kinetic energy" and the

*Present address: Department of Physics, University of Massachusetts, Amherst, MA 01003
§Present address: AT&T Bell Laboratories, Murray Hill, NJ 07974

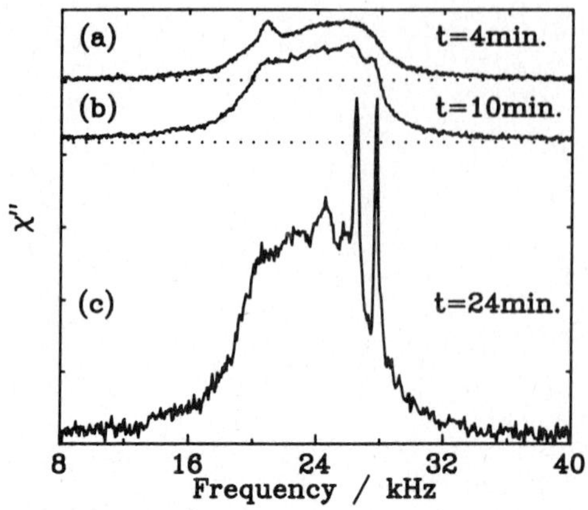

Figure 2. Polarization dependence of small-angle spectrum at T≈400mK. Loading began at t=0 and stopped at t=15min. Polarizations were in the ratio +.1 : -.13 : -.6 (a:b:c).

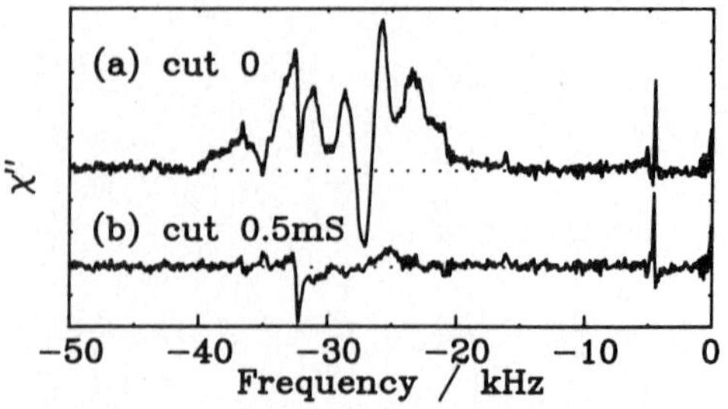

Figure 3. Large tipping angle (≈85°) spectrum. In (b), the first 0.5mS of the FID was deleted before Fourier transforming. T ≈ 395mK, n ≈ $10^{16}cm^{-3}$.

longest lifetime) should appear on the high frequency side of the spectrum when μP_z is positive whereas they should appear on the low frequency side when μP_z is negative.

When H↓ atoms are loaded into the cell, P_z is initially positive due to the thermal Boltzmann distribution between the |a> and |b> states, although it soon inverts due to preferential recombination of |a> state atoms. The change of sign of P_z during this time can be verified by observing the sign of the initial voltage of the FID. Spectra obtained under the condition of positive P_z have the lines flipped to the low frequency side of the spectrum; the edge frequencies of the spectrum are not shifted. This is illustrated in figure 2. This result further confirms the applicability of equation 1 and the confinement of the spin waves by $\delta(\gamma H_0)$. The increase in |P| is accompanied by an increased integrated spectral intensity, reflecting the overall increase of magnetization with time.

For large tipping angles the spectrum is markedly different from the small angle spectrum. As seen in figure 3a, the spectrum consists of jagged peaks and deep troughs spread over a frequency range characteristic of the expected inhomogeneous linewidth. In contrast to the small angle spectra, these spectra show no indication that the spin waves are confined near a wall. Although most of the lines are broader than typical lines in the small-angle spectra, we sometimes observe some extraordinarily sharp lines, such as the one at the left of the resonance and the very remarkable line about 20 kHz beyond the right side of the resonance in figure 3a. Sharp lines of this sort are commonly seen in the large-angle spectra, but we are not completely certain what conditions are required to produce them. The 200 Hz linewidth of the peak is instrumentally limited. If the first 0.5 msec of the FID is deleted (to eliminate short-lived phenomena) the peak survives (see figure 3b) with almost the same intensity as in the original spectrum. It is clear that the linearized theory (equation 1) does not apply to these spectra.

In conclusion, exchange effects in a dilute H↓ gas cause damped spin waves to play an important role in spin transport properties. The small NMR tipping angle behavior continues to be in good agreement with the theory. There are remarkable features in the large-angle spectra which may also be explained by molecular field effects, but a better understanding of the full equations (4, 5, 6) of motion of the spin density will be required.

We thank L. S. Goldner, K. Feldman and K. Earle for valuable assistance and A. E. Ruckenstein for illuminating discussions. This work was supported by NSF grant #DMR-8305284. LPL is a General Motors Fellow.

REFERENCES

(1) B. W. Statt and A. J. Berlinsky, Phys. Rev. Lett. 45, 2105 (1980). For a review of H↓, see I. F. Silvera, Physica 109 & 110B, 1499 (1982); T. J. Greytak and D. Kleppner, Proc. of Les Houches, July 1982, to be published.
(2) B. Yurke, J. S. Denker, B. R. Johnson, N. Bigelow, L. Lévy, J. H. Freed and D. M. Lee, Phys. Rev. Lett. 50, 1137 (1983). B. Yurke, Ph.D. thesis, Cornell University, 1983 (unpublished).
(3) B. R. Johnson, J. S. Denker, N. Bigelow, L. P. Lévy, J. H. Freed and D. M. Lee, Phys. Rev. Lett. (in press).
(4) C. Lhuillier and F. Laloë, J. de Phys. 43, 197 (1982) and 225 (1982).
(5) L. P. Lévy and A. E. Ruckenstein, Phys. Rev. Lett. (in press).
(6) J. S. Denker and N. Bigelow, Spin Transport in H↓, this volume.
(7) C. Lhuillier, J. de Phys. 44, 1 (1983).

LT-17 (Contributed Papers)
U. Eckern, A. Schmid, W. Weber, H. Wühl (eds)

COMPRESSION AND EXPLOSION OF SPIN-POLARIZED HYDROGEN

T. TOMMILA, S. JAAKKOLA, M. KRUSIUS, K. SALONEN, E. TJUKANOV

Wihuri Physical Laboratory and Department of Physical Sciences, University of Turku, SF-20500 Turku, Finland

A liquid ^{4}He-driven bellows system for the compression of spin-polarized atomic hydrogen (H↓) is described. Compression sweeps at different rates show the ultimate density of the H↓ sample to reach 10^{18} cm^{-3} before an explosive recombination takes place in H↓ bubbles $\gtrsim$ 1 mm^3 at a sample cell temperature of 370 mK and in a 7.5 T polarizing field.

1. INTRDUCTION

Spin-polarized atomic hydrogen was recently pressurized to densities $n \sim 10^{18}$ cm^{-3} (1). Two recently predicted new decay processes (2) were detected and their rate constants measured in these experiments: 1) The 3-body dipolar recombination in any hyperfine state of H at a rate proportional to n^3, and 2) the relaxation from the electron and nuclear polarized state *b* to the electron-spin-flipped state *c*. At the end of a compression thermally highly excited H_2 molecules heat up the dense atomic sample. This triggers the appearance of *c*-state atoms, which leads to a chain reaction-like recombination due to the exponential temperature dependence of the *b*→*c* process. We have undertaken a study of the critical parameters of the explosion and present here some preliminary observations.

2. COMPRESSION DEVICE AND SAMPLE CELL

Fig. 1 shows a hydraulic bellows-operated device with which liquid helium can be raised to the sample cell in order to compress the H↓ sample into a disc-shaped 0.95 mm gap between two Kapton foils in the top part of the cell. The upper foil with gold-plated surfaces acts as the pressure transducer of a capacitive manometer used to monitor the H↓ pressure during the accumulation and decay of the sample. The hydrostatic pressure acting on the compressed sample is measured with a level sensing coaxial capacitor from the height of the liquid column above the disc-shaped space. The *upper* surface of the lower foil is also gold-plated and in addition acts as a substrate for a calibrated carbon-film bolometer which monitors the recombination heating. By measuring the capacitance between the two foils the volume of the compressed H↓ bubble is determined.

A 30 cm^3 charge of liquid helium is condensed into the outer bronze bellows of the compression device. A 2.5 bar pressure applied to the inner bellows is sufficient to fill the cell. This pressure is supplied from a ^{4}He gas cylinder through a line thermally anchored to the different cooling stages of the cryostat. The rate of the helium displacement is controlled by regulating the gas stream into the cryostat with a precision needle valve. The compression device has proved to work reproducibly, with a linear displacement and a negligible heat load to the sample chamber at temperatures above 0.2 K.

3. RESULTS AND DISCUSSION

Fig. 2 shows the time variation of the bolometer signal (B), the helium level (L) and the volume of the sample (V) in a typical compression which terminates in a thermal explosion of a 7 mm^3 H↓ bubble at an ultimate critical density $n_c \sim 10^{18}$ cm^{-3}, corresponding to a 4 mm ^{4}He hydrostatic head above the top of the sample space. The explosion manifests itself in an abrupt release of heat, expansion of the sample volume, a drop of the helium level in the level indicator and also as an instantaneous pressure increase in the foil-manometer. The transient in the level indicator occurs as the newly warmed liquid in the cell attracts superfluid by the thermomechanical effect. In this case a sample of n_H = $1.1 \cdot 10^{16}$ cm^{-3} was initially accumulated and then pressurized at a rate of 13 mm^3/s for 180 s. The cell temperature is stabilized to (360 ± 1) mK,

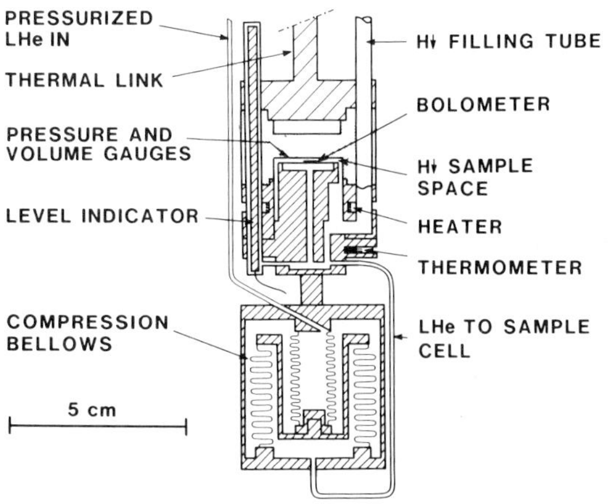

Fig. 1. H↓ sample cell and He compression bellows.

but directly prior to the explosion the bolometer, located in the immediate vicinity of the bubble, records a temperature rise to 370 mK. The subsequent recombination spike, traced on a very non-linear scale, shows the bolometer to reach a temperature of about 700 mK. The descending portion of curve L is caused by depressurization of the cell to clear it for a new compression.

The results from a number of compression experiments have been compiled in Fig. 3 where the H↓ pressure $P_c = n_c k_B T$ before the explosive recombination event has been plotted as a function of bubble volume. Different compression rates with displacements from 0.4 to 18 mm^3/s have been used in collecting this data at a cell temperature of 360 mK. Nevertheless P_c appears to settle to about 6 Pa for bubbles > 1 mm^3. For large bubbles the measurements are more straightforward. The bubble pressure can be determined either from the pressure manometer trace or from the hydrostatic pressure record with the He level meter while the total number of atoms is read from the thermal response of the bolometer. On the other hand recombination heating, thermal equilibrium in the dense and poorly conducting H↓ gas, and thermal contact to the liquid He heat sink become more problematic in large bubbles. They are bounded by the two foils in the flat cylindrical region and shrink only in their radial dimension. The thermal contact area to the liquid He bath is limited to the curved surfaces on the cicumference and to the liquid below the 25 μm thick bottom foil. Thus thermal equilibrium is largely controlled by the vertical temperature gradient determined by the conduction and contact resistances from the gas through the foil to the liquid.

Smaller bubbles lose contact with the bottom foil and approach spherical shape when the surface tension contribution to the bubble pressure starts to build up. This process leads to the increase in P_c in Fig. 3 below 1 mm^3. Small bubbles have been prepared by opening the entry to the entry to the H↓ fill tube after a large large recombination blast and allowing the remaining atoms in the fill tube enter the cell. This means that the small bubbles have been purified of a-state atoms by selective recombination over some 15 - 30 min before their compression is started. Therefore their lower initial recombination heating rate, in addition to their much reduced relative thermal contact and conduction resistances, enhance their stability during compression. The P_c values in Fig. 3 represent lower bounds since the resolution of the volume measurement has so far been limited to an upper bound estimate for bubbles < 0.1 mm^3. Finally it should be noted that these experiments demonstrate that in well-controlled conditions the explosive recombination evolves in a reproducible, deterministic fashion.

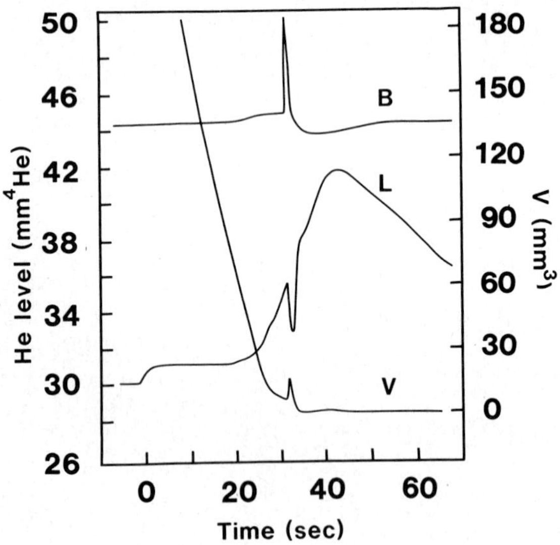

Fig. 2. Recorder traces from a compression at T(ambient) = 360 mK and 7.5 tesla: V, volume of H↓ sample; L, liquid He level; B, bolometer signal. Time t=0 is set at the point when liquid helium enters the flat cylindrical 300 mm^3 space on the top of the cell.

REFERENCES

(1) R. Sprik, J.T.M. Walraven, I.F. Silvera, Phys. Rev. Lett. 51 (1983) 479 ; H.F. Hess, D.A. Bell, G.P. Kochanski, R.W. Cline, D. Kleppner, T.J. Greytak, ibid. 51 (1983) 483.
(2) Yu. Kagan and G.V. Shlyapnikov, Phys. Lett. 88A (1982) 356.

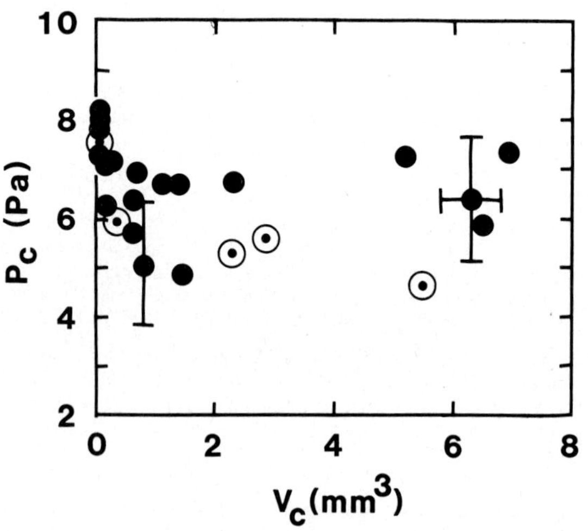

Fig. 3. Pressure of H↓ sample as a function of its volume at the point when the explosion sets off. The magnetic field is 4.5 T (⊙) or 7.5 T (●).

LT-17 (Contributed Papers)
U. Eckern, A. Schmid, W. Weber, H. Wühl (eds)

LOW TEMPERATURE RECOMBINATION AND RELAXATION MECHANISMS OF ATOMIC HYDROGEN AND ATOMIC DEUTERIUM

R. Mayer and G. Seidel

Physics Department, Brown University, Providence, RI 02912 USA

The kinetics of atomic hydrogen and atomic deuterium at low fields and temperature in the range of .5 to .7K has been investigated in an ESR spectrometer. The electronic relaxation of atomic hydrogen is attributed to the interaction of atomic hydrogen on the surface of superfluid helium with magnetic impurities on the microwave cavity walls. Although a flux of deuterium was measured at low temperatures, no atomic deuterium could be accumulated. These results imply a recombination constant at least of 5×10^{-11} cm^3/sec. Assuming recombination to occur on the surface and the adsorbtion energy of atomic deuterium on helium of 2.5 K, a zero field recombination cross-length is found to be greater than 200 Å, consistent with previous results.

1. INTRODUCTION

The low temperature recombination and relaxation mechanisms of atomic hydrogen at zero and high magnetic fields have been extensively studied and described by a number of groups (1,2). The kinetics of atomic H is understood at a qualitative level although detailed agreement between theory and experiment is occasionally absent. Naively, one would expect atomic deuterium to behave like atomic hydrogen when differences in mass, statistics, hyperfine interactions, etc. are taken into account.

Unlike atomic hydrogen, only one published experimental report (3) exists discussing the low temperature recombination mechanisms of atomic deuterium. Silvera and Walraven accumulated atomic deuterium in a field of 8.0 T, detected the atoms by bolometric techniques, and observed the density decay as a function of temperature. They attributed the decay to the recombination of two atoms on a superfluid helium surface and inferred an adsorbtion energy of 2.5 K for a deuterium atom on a superfluid surface. A long asymptotic tail observed in the density decay was attributed to the presence of atomic hydrogen and not to double polarized deuterium. The measured recombination rates can be used to compute a zero field cross-length as has been done in the case of atomic hydrogen. For D the cross-length is calculated to be 200 Å, where as for H the comparable number is .5Å.

2. RELAXATION OF ATOMIC HYDROGEN

The apparatus for producing H atoms and transporting them to a cell below 1 K having He coated walls has been described previously (4). The relaxation time of the ESR transition at 9.4GHz was observed by first saturating the transition and then measuring the recovery of the signal as a function of time. The recoveries were essentially exponential with time and only weakly dependent on density and temperature in the range .5 to .7 K. Furthermore, the rates varied from run to run, the values for T_1 falling within the range

$$.18\ \text{sec}^{-1} < 1/T_1 < .46\ \text{sec}^{-1}$$

In addition to the two ESR transitions associated with the atomic hydrogen, a broad (300 G), intense absorption centered at a resonance corresponding to g=2.11 was observed. Because of its intensity and lack of temperature dependence at low temperatures, it is presumed to be associated with ferromagnetic impurities on the cavity walls.

The recoveries of the absorption signal is attributed to the ferromagnetic impurities on the cavity wall. The H atoms adsorbed on the helium film are free to move on the plane of the surface and thereby experience a time varying field due to the localized impurities. Chapman and Bloom (5) have developed a theory for the relaxation of a magnetic dipole moving parallel to a plane of paramagnets. When that theory is applied to this problem, the measured density of ferromagnetic impurities appears reasonable to explain the observed relaxation.

At high magnetic fields the simultaneous polarization of electrons and nuclei, double polarization, (6) occurs due to different recombination rates from the various hyperfine states. If relaxation and recombination were dominated by the electronic spin-exchange mechanism, then double polarization should occur (7) under the conditions of these measurements where the electronic polarization is only 2.

The fact that the intensity of the two ESR transitions decayed at the same rate once filling of the cavity ceased indicates spin exchange is not the dominant relaxation mechanism in this experiment.

3. ATOMIC DEUTERIUM

Although densities of $2x10^{14}$ cm^{-3} of H could easily be accumulated between .5 and .7 K, no ESR signal from atomic D was observable, indicating densities of less than $1x10^{11}$ cm^{-3}. The flux entering the low temperature cell was measured calorimetrically and found to be 10^{14} sec^{-1} for atomic deuterium. From the minimum detectable density and the flux, a minimum recombination coefficient K = $5x10^{-11}$ cm^3/sec is estimated where $\frac{dn}{dt} = Kn^2$ and n is the gas density. The standard formula for the wall recombination is

$$K = \lambda v \left(\frac{A}{V}\right)^2 \Lambda e^{2E/kT}$$

where the symbols have their usual meaning (1). If the adsorption energy, E, of D on a He film is taken to be 2.5 K, one finds the zero field cross-length to be greater than 200 Å. This value, while improbably large, is consistent with the results of Silvera and Walraven (3), if one assumes that the recombination rate scales with field as determined by the admixture of spin states by the hyperfine interaction (1).

Although no concentration of D was observable, H atoms were found to be present at densities the order of 10^{13} cm^{-3} resulting from the trace amounts of hydrogen in the 99.5% pure D_2 source material.

In a separate experiment, the ESR of H and D were studied at 4 K and above. The flux of D at 4 K was found to be somewhat lower than that of H under comparable conditions. The more strongly varying dependence of the deuterium flux on temperature is taken to be on indication of a higher binding energy of D to the surface than H. The difference in flux of H and D cannot account for the absence of an ESR signal of D in the cavity below 1 K.

The origin of the anomalously large recombination rate for D atoms is not understood. A number of considerations are discussed briefly below but none appear to offer an explanation of the problem.

Papoular (8) has suggested that two D atoms can form a dimer. For the existence of dimers and their recombination to explain these low field results an enoromously large cross-length for dimer formation on the surface would be required.

Is it possible that D atoms tunnel through the superfluid film to the underlying wall? Although the energy of a D atom is much less than that of an H atom in liquid He (10 K compared to 40 K) (9), the tunneling is inconsequential.

To place limits on the density of D atoms from ESR observations requires an estimate of the spin temperature. The wall relaxation mechanism studied with H is sufficient to thermalize the D spins.

Selection rules fail to explain the difference in recombination rates between H and D. The formation of HD, for which there are no symmetry restrictions, must occur much slower than does D_2 given the measurable concentration of H atoms in the presence of a larger flux of D atoms.

Stwalley (10) has suggested that resonant enhancement of the recombination of D atoms can occur as a consequence of the existence of excited states close to the continuum. Since D_2 and HD have similar states near the continuum, resonant enhancement could only be important for D_2 and not for HD if subtleties of the interactions are important in determining recombination rates.

Should the recombination be occurring in the gas with a ^4He atom taking the role of the third body, the equivalent two-body cross sections would have to be large indeed. While the cross-section for D-He scattering is known to be substantially larger than for H-He (11), one does not see that it is sufficient to explain the enormous difference between the recombination rates of D and H.

A resolution of this problem may have to await detailed calculations of the interactions between D atoms at low velocities.

This work was supported in part by NSF grant DMR 83-16559.

REFERENCES

(1) I. F. Silvera, Physica 109 & 110B (1982) 1499.
(2) W. Hardy, Physica 109 & 110B (1982) 1964.
(3) I. F. Silvera and J. T. M. Walraven, Phys. Rev. Lett. 45 (1980) 1268.
(4) R. Mayer, A. Ridner, and G. Seidel, Physica 108B (1981) 937.
(5) R. Chapman and M. Bloom, Can. J. Phys. 54 (1976) 861.
(6) R. W. Cline, T. J. Greytak, and D. Kleppner, Phys. Rev. Lett. 47 (1981) 1195.
(7) R. Mayer, PhD thesis, Brown University unpublished (1984).
(8) M. Papoular, J. Low Temp. Phys. 50 (1983) 253.
(9) R. A. Guyer and M. D. Miller, Phys. Lett. 42 (1979) 1754.
(10) W. Stwalley, Phys. Rev. Lett. 37 (1976) 1628, (and references).
(11) J. P. Toennies, W. Welz, and G. Wolf, Chem. Phys. Lett. 44 (1976) 5.

LT-17 (Contributed Papers)
U. Eckern, A. Schmid, W. Weber, H. Wühl (eds)

PEAK EFFECT IN TWO DIMENSIONAL COLLECTIVE FLUX PINNING

P.H. KES and R. WÖRDENWEBER

Kamerlingh Onnes Laboratory, University of Leiden, P.O.Box 9506, 2300 RA Leiden, The Netherlands

C.C. TSUEI

IBM Thomas J. Watson Research Center, P.O.Box 218, Yorktown Heights, New York 10598, U.S.A.

It is demonstrated that the onset of the peak effect in amorphous superconducting films with two dimensional collective pinning coincides with a structural transition from an elastically to a plastically distorted flux line lattice. The distortions are produced by fluctuations of the collective pinning force on individual flux lines.

Recently, the predicted (1) two dimensional collective pinning (2D CP) has been experimentally established in thin amorphous Nb_3Ge films in a perpendicular magnetic field (2). A typical example is shown in Fig.1 where the volume pinning force F_p is plotted vs $b=B/B_{c2}$ for the three thinnest samples at $T/T_c \equiv t \approx 0.50$. Three regimes are distinguished: (i) for $0.1<b<b_{ST}$ (3) the 2D CP theory (thick drawn line) fits the data very well. We call this the *elastic* regime, because theory assumes elastic distortions of the flux line lattice (FLL) due to the collective action of randomly distributed, weak pins. (ii) For $b_{ST}<b<b_p$ the *peak* effect shows up; the data clearly lie above the theoretical prediction. (iii) For $b>b_p$ the *amorphous* limit of the 2D CP theory fits the data very well (2). The amount of order in the FLL follows from the correlation length R_c, also shown in Fig.1 relative to the FLL parameter a_0. For all thicknesses d of the samples (60 nm<d<2.9 μm) and at all temperatures of the measurements (0.4<t<0.9) it was observed that the peak starts at values of $R_c/a_0 \approx 15$. Several reasons for the origin of the peak has been discussed in Ref.2. It was suggested that the increase of disorder of the FLL, as illustrated by the decrease of R_c/a_0 for $b \to 1$, softens the shear modulus c_{66} of the FLL. In a plot of c_{66} vs R_c/a_0 (Fig.5 of Ref.2) the softening showed up as a fairly sharp drop of c_{66} at $R_c/a_0 \approx 15$. It remained an open question why the transition occurred so suddenly. Brandt (4) concluded from his computer simulations that the onset of the peak might by related to a structural transition from elastic to plastic deformations. The enhancement of the disorder and the related reduction of c_{66} beyond b_{ST} (Ref.2) would then come about by the generation of topological defects, mainly flux line dislocations. This picture agrees well with the time dependent effects and field history effects we observe around b_{ST}. Several authors (5,6) recently proposed the same mechanism for the peak effect in three dimensional flux pinning.

For a thin film in a perpendicular field the criterion for an elastic instability in the FLL due to a single very strong pin has been given among others by Brandt (4). At fields $B \gtrsim 0.3\ B_{c2}$:

$$u(o) \approx a_0/2\pi \qquad (1)$$

where u(o) is the distortion of the FLL by the action of a line force f_1 in the origin. In the

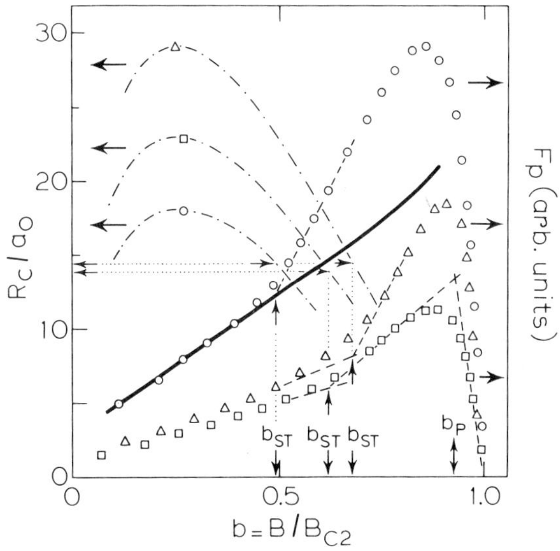

FIGURE 1
F_p and R_c/a_0 vs B/B_{c2} for three Nb_3Ge samples: d=60nm, t=0.46 (circles); d=93nm, t=0.50 (triangles); d=170nm, t=0.55 (squares). Thick drawn line: 2D CP theory fitted to data below b=0.4; b_{ST} and b_p are defined by the linear extrapolations (dashed lines). It is clearly demonstrated that $R_c/a_0(b_{ST}) \approx 15$, independent of thickness.

elastic continuum approximation (7) u(o) is given by

$$u(o) = \frac{5}{16\pi}\frac{f_1}{c_{66}} \ln\left(\frac{w}{a_o}\right) \tag{2}$$

where the appropriate cut offs are chosen at w, the width of the sample (4 mm), and a_o.

For a very dense system of collectively acting, randomly distributed, *weak* pins no criterion for a structural transition has been formulated yet. In this case the net forces (per unit length) due to many pins per individual flux line form a distribution, supposedly Gaussian, with a variance given by

$$\sigma_f = [(n_v\Phi_o/B)\langle f^2\rangle/d]^{\frac{1}{2}} \tag{3}$$

Here n_v is the density of pins (per unit volume) and $\langle f^2\rangle$ is the actual elementary interaction squared and averaged over a lattice cell. We propose here to identify σ_f with f_1 since, if σ_f fulfils the criterion for single line pins given by Eqs.(1) and (2), there is a good chance to find somewhere in the FLL a fluctuation big enough to cause a local elastic instability. The latter may generate topological defects upon a uniform movement of the FLL through the pins. Combining the expression derived in Ref.2

$$\frac{R_c}{a_o} \approx \left[\frac{2\pi}{\ln(w/R_c)}\ \frac{c_{66}^2 d}{n_v\langle f^2\rangle}\right]^{\frac{1}{2}} \tag{4}$$

with Eqs.(1)-(3), we find

$$\left(\frac{R_c}{a_o}\right)_{ST} \approx 1.46\ \frac{\ln(w/a_o)}{[\ln(w/R_c)]^{\frac{1}{2}}} \tag{5}$$

at the structural transition. Inserting a typical experimental value of 40 nm for a_o, we obtain $(R_c/a_o)_{ST} \approx 5.4$. This result corresponds qualitatively very well with our experimental observation of a practically constant value of R_c/a_o at the onset of the peak. Quantitatively it is also close. The computer simulations (4) at $b \approx 1$ yield a lower threshold for elastic instabilities by a factor of 2, which would give $(R_c/a_o)_{ST} \approx 11$. The statistical character of our argument may well account for a factor of 2 as well.

In order to compare the temperature dependence in more detail we computed $1-b_{ST}$ from Eqs.(1) - (3) using Eqs.(1) and (12b) of Ref.2 for c_{66} and $n_v\langle f^2\rangle$, respectively. This quantity, defined as $(1-b_{ST})_{theor}$, turns out to be smaller than $(1-b_{ST})_{expt}$ by a factor of ≈ 3. To summarize: $(1-b_{ST})_{expt}/(1-b_{ST})_{theor} = 3.0\pm0.1$, 3.3 ± 0.1, 3.3 ± 0.3, and 3.5 ± 0.3 for t=0.5, 0.7, 0.8, and 0.9, respectively. From this result one may perhaps conclude that at higher temperatures the inelastic instabilities occur at slightly lower fields compared to the values predicted by the present model. This may indicate that thermal activation contributes slightly to the creation of elastic instabilities.

The thickness dependence can be checked by

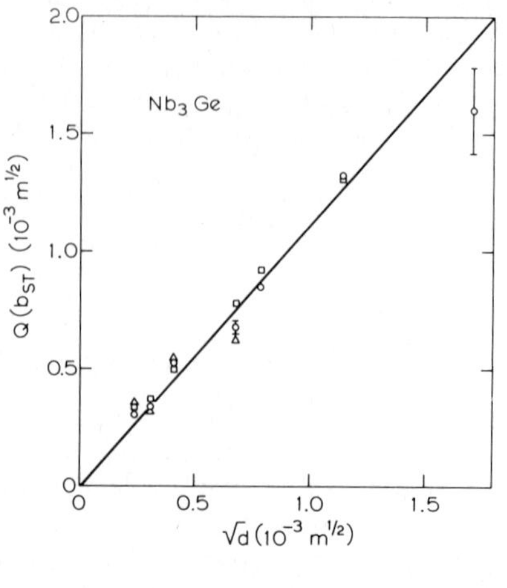

FIGURE 2
The quantity $Q(B_{ST})$ (defined in text) vs $d^{\frac{1}{2}}$ for $t \approx 0.5$ (squares), t=0.7 (circles), and t=0.8 (triangles). The error bars denote a shift of b_{ST} of ±0.01.

plotting the quantity $\ln(w/a_o)^2 \cdot [n_v\langle f^2\rangle]^{\frac{1}{2}}/c_{66}$, abbreviated by Q, determined at the experimental b_{ST}'s versus $d^{\frac{1}{2}}$. In such a plot (Fig.2) the data should demonstrate a linear dependence which indeed is found.

Finally, the relative distortion of the FLL necessary to produce an elastic instability can be estimated from the experimental b_{ST}'s to be: $u/a_o \approx 0.047$ (see Eqs.(B19) and (B20) of Ref.2).

REFERENCES

(1) A.I. Larkin and Yu.N. Ovchinnikov, J. Low Temp. Phys. 34 (1979) 409.
(2) P.H. Kes and C.C. Tsuei, Phys. Rev. Lett. 47 (1981) 1930; Phys. Rev. B28 (1983) 5126.
(3) $b_{ST}=b_{RN}$ in reference 2.
(4) E.H. Brandt, Phys. Rev. Letter 50 (1983) 1599; J. Low Temp. Phys. 53 (1983) 41 and 71.
(5) Yu.N. Ovchinnikov, Zh. Ekop. Teor. Fiz. 82 (1982) 2020 and 84 (1983) 237 [Sov. Phys. JETP 55 (1982) 1162 and 57 (1983) 136].
(6) J.E. Evetts, extended abstract for Topical Workshop on Flux Pinning in High-Field Superconductors, Reinhausen (near Göttingen, Germany) March, 1982 (unpublished).
(7) I.S. Sokolnikoff, Mathematical Theory of Elasticity (McGraw Hill, New York, 1956), p. 289, 337.

In summary, fluctuations of the collective pinning force on individual flux lines produce distortions large enough to create elastic instabilities in the FLL. The criterion for this structural transition explains why the onset of the peak effect occurs at the same value of the relative correlation length R_c/a_o *irrespective thickness or temperature.*

LT-17 (Contributed Papers)
U. Eckern, A. Schmid, W. Weber, H. Wühl (eds)

COLLECTIVE PINNING SIMULATED ON THE COMPUTER

Ernst Helmut BRANDT

Max-Planck-Institut für Metallforschung, Institut für Physik, Stuttgart, F.R. Germany

Flux-line pinning in Type-II superconductors is simulated on a computer. For weak, randomly positioned point pins the two-dimensional collective pinning summation of Larkin and Ovchinnikov is confirmed. For stronger pins a simular formula holds in spite of plastic deformation of the flux-line lattice. For strong pins a modified direct summation is approached.

The summation of pinning forces acting on the flux lines (FLs) in type-II superconductors is complicated by the fact that in order to get a finite average pinning force one should have a correlation between the positions of pins ($\vec{r}_i^p$) and vortices ($\vec{r}_i$). This correlation is established by elastic or plastic deformation of the flux-line lattice (FLL) caused by the pins. Summation theories should thus account for the *elastic and plastic* properties of the FLL, the size, strength f_p, density n_p, and arrangement of the pins and for the dimensionality of the problem: 2D for weak pins which do not bend the FLs appreciably in a thin specimen, 3D for strong pins in thick specimens. The problem of a reliable prediction of the maximum loss-free current density j_c in a given material is, therefore, far from being solved (1, 2). Only in one case there appears to be good agreement between a well founded theory [the 2D version of (3)] and experiment [weak pinning in thin films of amorphous Nb_3Ge and Nb_3Si (4)].

In order to check the range of applicability of the summation (3), which was derived for elastically deformed FLL, and to get practice for later 3D simulations, we simulate on a computer 2D pinning (5). In the simplest case we relax $N_v \leq 1254$ FLs interacting with eachother and with $N_p \leq 4N_v$ randomly positioned point pins by model potentials of variable amplitude and range:

$$\exp(-|\vec{r}_i-\vec{r}_j|^2/R_v^2), \quad A_p\exp(-|\vec{r}_i-\vec{r}_j^{\,p}|^2/R_p^2).$$

We start from an ideal triangular or amorphous FLL and use periodic boundary conditions with periodicity lengths $L_x \approx L_y$. The *dynamic* pinning force $F_{dyn}(\dot{X})$ is calculated by introducing viscose forces $-\eta\dot{\vec{r}}_i$ and shifting the mean x-coordinate X of the FLs with constant velocity $\dot{X}$. The *static* pinning force F_{stat} is obtained as follows [for previous methods cf. (6)]:

From their equilibrium state (X=0) all FLs are shifted along x by dX = 0.001a (a=mean FL spacing), then a homogeneous force is applied on all FLs which just compensates the mean pinning force F, and finally the FLL is relaxed again. This relaxation leaves X practically unchanged since the total force on the FLL is zero. By repeating this procedure we get a *force-displacement curve* F(X). Typically F(X) exhibits an "elastic" increase and a "plastic" saturation where F fluctuates randomly about its average value $\bar{F}=\langle F(X)\rangle=j_c\phi_o$ (ϕ_o=flux quantum) (Fig. 1).

We have investigated in detail the function $\bar{F}(A_p)$ and its dependence on the shear modulus of the FLL c_{66} and on the relative strength $|A_p|$, range R_p/a, and density $N_p/N_v \approx n_p a^2$ of the pins both for attractive and repulsive pins ($A_p<0,>0$). For the discussion we define some auxiliary quantities: The *maximum force* a pin can exert on a FL is $f_{max}=\sqrt{2/e}\,A_p/R_p$ (Gaussian potential). The *actual force* on a given pin is $\vec{f}_i$; its quadratic average over the volume and over X is $f_p^2 \equiv \langle\langle|\vec{f}_i|^2\rangle\rangle$. One has $\bar{F}=n_p\hat{x}\langle\langle\vec{f}_i\rangle\rangle$ ($\hat{x}$ unit vector along x). Characteristic pin amplitudes are the "threshold" $A_{thr}=8\pi c_{66}R_p^2/0.89\,\ln N_v$ ($R_p \lesssim 0.3a$) [for $-A_p<A_{thr}$ very dilute pins should yield $\bar{F}=0$ in finite systems (1)] and the critical value $A_c=8.7R_p c_{66}/\sqrt{N_p}$ ($R_p \lesssim 0.25a$) [for $|A_p|<A_c$ the pin-caused FL displacement remains $<R_p$ in our finite periodicity volume L_xL_y and thus

$\bar{F}=0$ according to (3)].

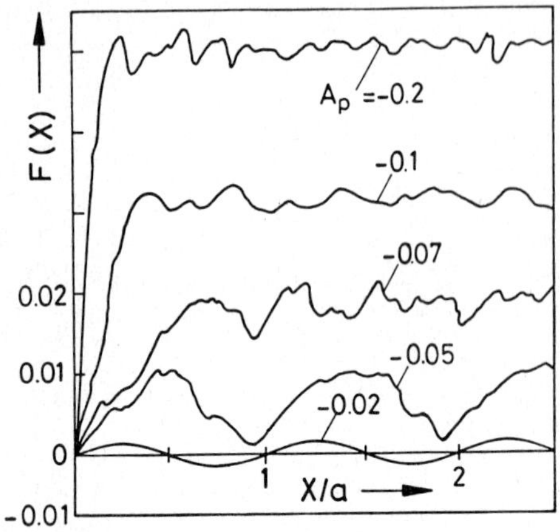

FIGURE 1
Force-displacement curves $\bar{F}(X)$ for various strengths of attractive pins. $N_V=780$, $N_p=210$, $R_V=0.6a$, $R_p=0.25a$.

The condition $A_c/A_{thr} \approx \ell n N_V/\sqrt{N_p} \ll 1$ required for a reliable check of collective pinning (3) in an only elastically deformed FLL, and also the requirement of complete relaxation for each value of X, means very large numerical effort. The cases of medium and strong pinning in a plastically deformed FLL, for which analytical summation formulae do not exist, are, however, computed easily.

We give some of our results for dense pins ($N_p \gtrsim N_V$), Fig.2. For $|A_p| \ll A_c$ we get $\bar{F}=0$ provided the relaxation is complete. For $|A_p| > A_c$ there is a region where the 2D summation (3) applies, namely for ideal FLL

$$\bar{F} = 0.11\sqrt{\ell n(N_V/33)}\; n_p A_p^2 / R_p c_{66} \, . \qquad (1)$$

If we start with amorphous FLLs Eq.(1) still applies but with prefactors larger by factors 1.3 to 3. For strong pinning $\bar{F}(A_p)$ approaches the modified limit of direct summation, $\hat{F}_d=(N_p/N_V)f_p$ rather than the maximum conceivable value $F_d=(N_p/N_V)f_{max} \ll \hat{F}_d$.

Fig.2 shows that for dense, weak pins one has $f_p \sim A_p$. For dilute or strong pins relaxation of the FLL leads to a gradual saturation of f_p at large values of $|A_p|$ such that simple laws $\bar{F} \sim A_p^2$ or $\bar{F} \sim A_p$ do not hold.

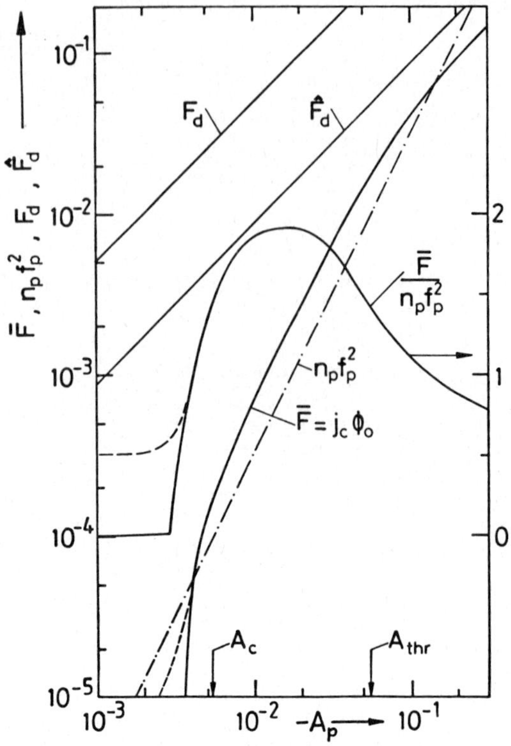

FIGURE 2
The linear and square averages $\bar{F}=j_c\phi_o$ and $n_p f_p^2$ of the pinning forces, and the direct summations F_d and $\hat{F}_d$ (see text), plotted versus the pin amplitude A_p for attractive pins. The figure is half schematically, extrapolated to the large system $N_V=N_p=10^4$. Also plotted is the ratio $\bar{F}/n_p f_p^2$ (linear right scale). The dotted line is for incomplete relaxation and corresponds to a not static but slowly drifting FLL.

REFERENCES

(1) A.M. Campbell and J.E. Evetts, Adv. Phys. 21 (1972) 199
(2) E.J. Kramer, J. Appl. Phys. 49 (1978) 742
(3) A.J. Larkin and Yu.N. Ovchinnikov, J. Low Temp. Phys. 34 (1979) 409
(4) P.H. Kes and C.C. Tsuei, Phys. Rev. Lett. 47 (1981) 1930, Phys. Rev. B 28, 5126
(5) E.H. Brandt, Phys. Rev. Lett. 50 (1983) 1599, J. Low Temp. Phys. 53 (1983) 41, 71
(6) J.R. Appleyard, J.E. Evetts, and A.M. Campbell, Sol. St. Comm. 14 (1974) 567

LT-17 (Contributed Papers)
U. Eckern, A. Schmid, W. Weber, H. Wühl (eds)

THE ELEMENTARY PINNING POTENTIAL OF VORTICES TO SMALL OBJECTS IN TYPE II SUPERCONDUCTORS

E.V. THUNEBERG

Research Institute for Theoretical Physics, Helsinki University, 00170 Helsinki 17, Finland

J. KURKIJÄRVI

Department of Technical Physics, Helsinki Technical University, 02150 Espoo 15, Finland

D. RAINER

Physikalisches Institut, Universität Bayreuth, 8580 Bayreuth, West Germany

Flux pinning is studied by the quasiclassical method. We give results on a quasiparticle scattering mechanism of elementary pinning which outweighs the traditional excluded volume effect by the factor coherence length divided by the spatial extent of the pinning center.

Flux pinning has been a glaring exception to the virtually quantitative success of the BCS theory in problems of superconductivity. There are two reasons for this. First, it is hard to bridge the gap between a critical current measurement and the elementary pinning force of a defect on a flux line. Second, and more fundamentally, it has been difficult to calculate a microscopic effect such as pinning via quasiparticle scattering within the framework of the traditional BCS-Gorkov theory. The calculation has become feasible with the advent of the quasiclassical method (1,2,3). Quasiparticle scattering leads to strong pinning of small impurities, up to one hundred times stronger than the avoided loss of condensation energy within the volume of the pinning center. Pinning by quasiparticle scattering is not contained in the conventional Ginzburg-Landau (GL) theory. If one calculates the pinning applying the boundary condition of a vanishing order parameter gradient at the defect, one finds a pinning energy proportional to the volume of the defect. This prediction is far too small. One may not use the usual Ginzburg-Landau boundary condition at small objects. There is a shortcut derivation of the quasiparticle scattering effect, however, which is valid in the GL region and which requires no new microscopic analysis. The idea is to take advantage of an impurity averaged result to obtain the energy of a single localized defect.

One looks at Gorkov's derivation of the GL theory in the case of uniformly distributed impurities (4). The GL free energy functional is given by

$$\Omega = \int d^3R\{a|\Psi|^2+\frac{1}{2}b|\Psi|^4+\frac{\chi(\alpha)}{4m}|(i\hbar\nabla+\frac{2e}{c}\bar{A})\Psi|^2 + \frac{1}{8\pi}(B-B_e)^2\}$$

The coefficients a and b are independent of the impurity parameter $\alpha = 0.882\ \xi_0/\ell_{tr}$, because the Gorkov impurity function

$$\chi(\alpha) = \frac{8}{7\zeta(3)} \sum_{n=0}^{\infty} \frac{1}{(2n+1)^2(2n+1+\alpha)}$$

appears explicitly in the gradient term. The addition of one localized impurity (transport cross section σ_{tr}) changes the transport mean free path ($\ell = (n\sigma)^{-1}$) and thereby changes the coefficient of the gradient term in the energy functional. For small impurities, the change in the order parameter can be neglected, and the energy change emerges as

$$\delta\Omega = -\frac{\hbar^2}{4m}|(\nabla-i\frac{2e}{c\hbar}\bar{A})\Psi|^2\ 0.882\xi_0\sigma_{tr} \times\frac{8}{7\zeta(3)} \sum_{n=0}^{\infty} \frac{1}{(2n+1)^2(2n+1+\alpha)^2}$$

According to this extremely simple result, the energy change from a single defect is proportional to the square of the order parameter gradient at the defect. The gradient term and the bulk terms are the same order of magnitude in a vortex. The pinning energy proportional to the square of the gradient is therefore on the order of the condensation energy in the volume $\xi_0\sigma$. This result holds for all impurity contents of the metal (provided $\ell \gg k_F^{-1}$). Pinning grows monotonically less strong the dirtier the metal. At large values of α the last factor of the above expression varies like α^{-2}, but the steep variation is partly cancelled by the gradients of Ψ which grow larger. Near B_{c2}, the net dependence is according to α^{-1} at constant B/B_{c2}.

At lower temperatures, a full microscopic calculation (5) consists of finding only the divergent (which is difficult to avoid numerically) solution of the first order linear differential transport-like equation of Eilenberger around a vortex. The order parameter there has been calculated by Pesch and Kramer (6). To first order in σ/ξ_0, the divergent solutions suffice to find the modification to the physical (quasi-classical) Green's function arising from the presence of the pinning impurity. The impurity is described by a model scatterer such as a hard sphere. Its (quasiclassical) t-matrix is calculated for a number of phases reflecting the size of the impurity. As the method works in any order parameter configuration, the free energy (potential) associated with the presence of a defect can be calculated as a function of the position of the impurity. We find that the pinning potential scales, as a function of temperature, roughly like $(1-T/T_c)^2$. Its absolute value at the vortex is around 11 in units $(\xi_0\sigma_{tr}N(0))(T_c-T)^2$ and it decays within roughly 1.5 lengths $\xi_0/(1-T/T_c)$.

The physical effect that leads to the strong pinning is relaxing the requirement of phase matching on the part of the "quasiparticle" scattered by the pinning center. The requirement is next to impossible to meet near or at a vortex and normally reduces the amplitude of the order parameter.

As we can calculate the pinning by a single impurity at all order parameter configurations, we can do it at arbitrary fields in particular. At fields close to H_{c2}, Thuneberg has exploited the smallness of the order parameter to work out, at all temperatures, the dependence of pinning on several parameters, among them the results already quoted on the variation as a function of the background dirtiness of the metal(7). The pinning potential depends linearly on the magnetic field. It has pronounced maxima in space in addition to the minima at the vortex cores. Its dependence on the pure metal GL-parameter is weak.

The agreement between the present theory and a couple of recent pinning experiments (8,9) is good.

REFERENCES

(1) G. Eilenberger, Z. Phys. 214, 195(1968).
(2) A.I. Larkin and Yu.N. Ovchinnikov, Zh. Eksp. Teor. Fiz., 55, 2262(1968).
(3) J.W. Serene and D. Rainer, Phys. Rep. 101, 221(1983).
(4) L.P. Gorkov, J. Exptl. Teoret. Phys. 37, 1407(1959).
(5) E.V. Thuneberg, J. Kurkijärvi and D. Rainer, to be published in Phys. Rev. B.
(6) W. Pesch and L. Kramer, J. Low Temp. Phys. 15, 367(1974).
(7) E.V. Thuneberg, to be published.
(8) H.R. Kerchner, D.K. Christen, C.E. Klabunde, S.T. Sekula and R.R. Coltman, Jr., Phys. Rev. B27, 5467(1983).
(9) G.P. van der Meij and P.H. Kes, to be published.

LT-17 (Contributed Papers)
U. Eckern, A. Schmid, W. Weber, H. Wühl (eds)

FLUX-LINE PINNING BY THE GRAIN BOUNDARY IN NIOBIUM BICRYSTALS

H. R. KERCHNER,* D. K. CHRISTEN,* A. DAS GUPTA,* and S. T. SEKULA,* Oak Ridge National Laboratory, Oak Ridge, Tennessee and B. C. CAI+ and Y. T. CHOU,+ Lehigh University, Bethlehem, Pennsylvania

The interaction between the single grain boundary in bicrystals of niobium and the lattice of magnetic flux lines in the superconducting mixed state was investigated by measuring the critical current. Flux lines are found to be pinned at the grain boundary when the two are aligned to within about 1°. The data imply that both the intrinsic anisotropy of niobium and the electron scattering by the boundary contribute to the pinning.

In recent years interest in flux-line pinning by grain boundaries has been rekindled for a variety of reasons. First, the grain boundary/cell wall structure is a major source of pinning in both Nb_3Sn and NbTi conductors, which presently dominate the superconducting-magnet market. Second, the theoretical work of Zerweck (1) and later of Yetter, et al.(2), gave new hope that the observed pinning in polycrystals could be understood quantitatively. Third, the production of bicrystals of niobium by groups in the United States (3) and the Soviet Union (4) has made it possible to perform experiments on individual grain boundaries. We report here new measurements of the critical current on the grain boundary in niobium bicrystals.

We have investigated the critical current of the grain boundary in two bicrystals grown by techniques described elsewhere (3). Both samples were degassed at high temperatures in ultra-high vacuum in order to reduce or eliminate pinning due to precipitates of interstitial compounds. They were subsequently oxidized at 400°C in order to reduce the surface pinning (5). Current-voltage relations were measured at 4.2 K in a conventional iron-core electromagnet. The samples were about 3-mm in diameter and had a 10-mm gauge length (30-mm total length). The dc voltage was measured by using a MOSFET reversing switch and lockin amplifier as described elsewhere (5). Although other procedures were sometimes used, typically a constant current larger than the critical value was applied at each field of interest, and the dc flux-flow voltage was measured at 0.1° intervals as the applied magnetic field was rotated. Elsewhere we also measured the inductive and resistive transitions of each sample in zero field.

The flux-flow voltages observed always dipped sharply when the angle Γ between the grain boundary and the applied field was less than 1°-2°. Outside this range the voltage was nearly independent of Γ and almost proportional to the current (indicating a small bulk critical current). The voltage dips (critical current peaks) showed fine structure that depended on the applied field and the magnitude and direction of the applied current. These observations indicate that the grain boundaries are not perfectly flat but instead consist of flat segments misoriented by small angles less than 1°. It was possible to match curved flux lines roughly to the segmented boundary as evidenced by the fact that a current level could be found (different for each sample and weakly field dependent) that produced a single, symmetric, and almost structure-free, voltage dip. We deduced the critical current of the boundary $I_b(\Gamma)$ by the following procedure: First, the flux-flow resistance was measured at an angle away from the voltage dip but close enough to $\Gamma=0$ to eliminate errors due to anisotropy in the resistance. Second, the voltage at large Γ was extrapolated through the voltage dip. Finally at each angle Γ, the difference between the measured voltage and the extrapolated smooth curve was divided by the flux-flow resistance.

The measured pinning force per unit area of grain boundary, $F_b(\Gamma) = BI_b(\Gamma)/2a$, is summarized for two samples in Figs. 1 and 2. The magnetic induction B was deduced from the measured upper critical field of these samples and a previously published study of the magnetization of niobium (6). The sample radius is a. Both the peak pinning force $F_b(\Gamma_p)$ and the angle-integrated force $\int F_b(\Gamma)d\Gamma$ are shown. Although both quantities depend on the magnitude and direction of the applied current, this dependence of the angle-integrated force is only very weak except for the direction dependence of sample II (Fig. 2). Evidently, the overall curvature of the grain boundary causes the total pinning force in these samples to depend on current. It produces

*Research sponsored by Union Carbide Corporation under contract W-7405-eng-26 with the U. S. Department of Energy. +Research sponsored by the National Science Foundation under Grant DMR79-15799.

a small rectification effect because both the peak force and the angle-integrated force peak when the flux-line curvature matches the grain boundary curvature. However, this picture cannot fully account for the rectification observed in sample II, where the angle-integrated force changes by more than a factor of two with current direction at low and intermediate fields. The major portion of this rectification must arise from the intrinsic anisotropy of the material. Since the field is parallel to the same crystal axis in both grains, no such intrinsic effect can be present in sample I.

Both samples show evidence of superconductivity on the grain boundary at applied fields well above the upper critical field H_{c2}. Since we could not observe superconductivity (we measured nonzero dc resistance) at temperatures above the bulk critical temperature (determined from the inductive transition in a very low field), we conclude that electron scattering by the grain boundary must be responsible for the high-field superconductivity on the boundary. Moreover, the smooth field dependence of $F_b(\Gamma)$ at H_{c2} suggests that the same mechanism accounts for the pinning below H_{c2} in sample I. The pinning force shows a change in slope at H_{c2} in sample II, probably indicating the presence of two pinning mechanisms in the mixed state--one of them (the anisotropy contribution) going to zero at H_{c2}. By comparing the two samples and by examining the field dependences of $F_b(\Gamma)$ we can thus roughly separate the anisotropy contribution and the electron-scattering contribution to the grain-boundary pinning.

We have investigated flux-line pinning by the grain boundary in two different niobium bicrystals. Two mechanisms for the pinning can be identified qualitatively: one is due to electron scattering by the grain boundary, and the other arises from anisotropy of the material. No extant theory can account quantitatively for our observations. The electron-scattering theory of Refs. (1)and (2) give only a small effect for samples as clean as ours. (Impurity parameters are .03 and .15 for samples I and II, respectively.) The recent theory of Thuneberg et al.(6) may account for our observations although the calculations have not yet been performed for a grain boundary. Theories of the anisotropy contribution (7, 8) do not predict a rectification as large as we see.

REFERENCES

(1) G. Zerweck, J. Low Temp. Phys. 42 (1981) 1.
(2) W. E. Yetter, D. A. Thomas, and E. J. Kramer, Philos. Mag. B 46 (1982) 523.
(3) B. C. Cai, A. Das Gupta, and Y. T. Chou, J. Less Common Metals 86 (1982) 145.
(4) L. Ya. Vinnikov, E. A. Zasavitskii, and S. I. Moskvin, Zh. Eksp. Teor. Fiz. 83 (1982) 2225 [Sov. Phys. JETP 56 (1982) 1288].
(5) H. R. Kerchner, D. K. Christen, and S. T. Sekula, Phys. Rev. B21 (1980) 86.
(6) E. V. Thuneberg, J. Kurkijärvi, and D. Rainer, Phys. Rev. Lett. 48 (1982) 1853.
(7) D. Dew-Hughes and M. J. Witcomb, Philos. Mag. 26 (1972) 73.
(8) G. S. Mkrtchyan and V. V. Schmidt, Zh. Eksp. Teor. Fiz. 68 (1975) 186 [Sov. Phys. JETP 41 (1975) 90] .

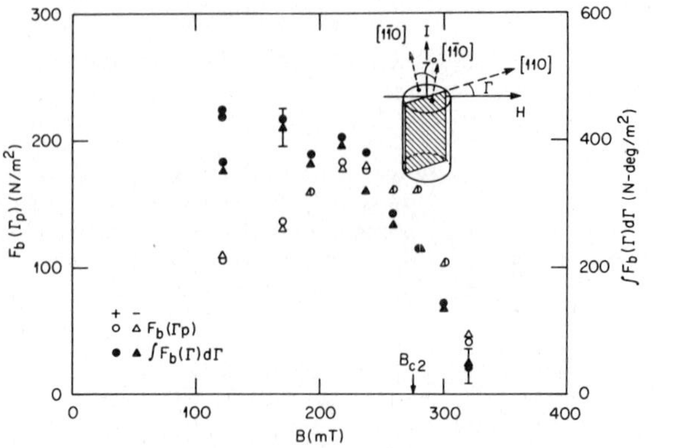

FIG. 1. The peak value and the angle-integrated pinning force per unit-area of grain boundary vs the magnetic induction B for sample I at 10A total current. The induction B_{c2} at the upper critical field is indicated on the abscissa. The inset shows the orientations of the two grains, the boundary, the applied magnetic field and the positive current direction (corresponding to a positive pinning force).

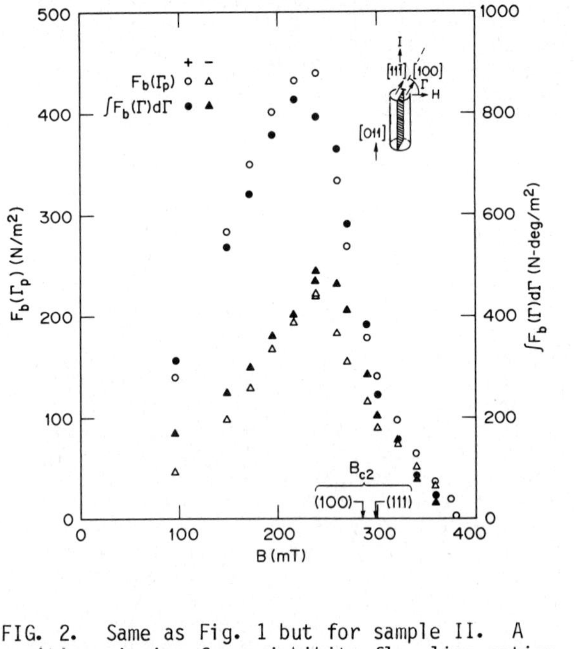

FIG. 2. Same as Fig. 1 but for sample II. A positive pinning force inhibits flux-line motion from the [100] grain into the [11$\bar{1}$] grain.

LT-17 (Contributed Papers)
U. Eckern, A. Schmid, W. Weber, H. Wühl (eds)

INTERACTION OF ADSORBED ATOMS WITH PHONON PULSES

H. C. Basso, W. Dietsche, and H. Kinder

Physik Department E 10, TU München 8046 Garching, W. Germany

We prepared clean Si surfaces at low temperatures by laser annealing. Submonolayers of several species of atoms were deposited onto these surfaces. We found that in some cases the reflection of phonon pulses was strongly affected with only 0.2 monolayers. This effect coincided with the onset of the anomalously large Kapitza transmission.

It became clear in recent years that the phonon reflection and transmission processes at interfaces are only in rare cases governed by the accoustic properties. Particularly conspicious is the Kapitza transmission (1) between solids and liquid Helium which is up to a 100 times larger than theory predicts unless perfect surfaces are used (2). Neither the underlying phonon processes nor the nature of the irregularities which cause them are presently understood. The goal of this research was, therefore, to study the influence of small amounts of deposits on the phonon processes at the surface. Our investigative tool is the observation of the phonon reflection from a test surface.

The experimental set-up is shown in Fig. 1.a. A Si crystal, 15 mm in diameter, 4 mm in thickness, and (100) orientation is sealed to a vacuum chamber. The chamber could be filled with He if desired. The inner surface of the crystal served as the test surface and could be modified by high-power ruby-laser pulses which were irradiated through the window on the opposite side. The whole set-up was immersed in liquid Helium.

Phonons were generated at the outer side of the Si crystal by heating a Sn film with HeNe-laser pulses. The phonons were reflected from the test surface and detected by an Al tunnel junction detector. The size of the measured

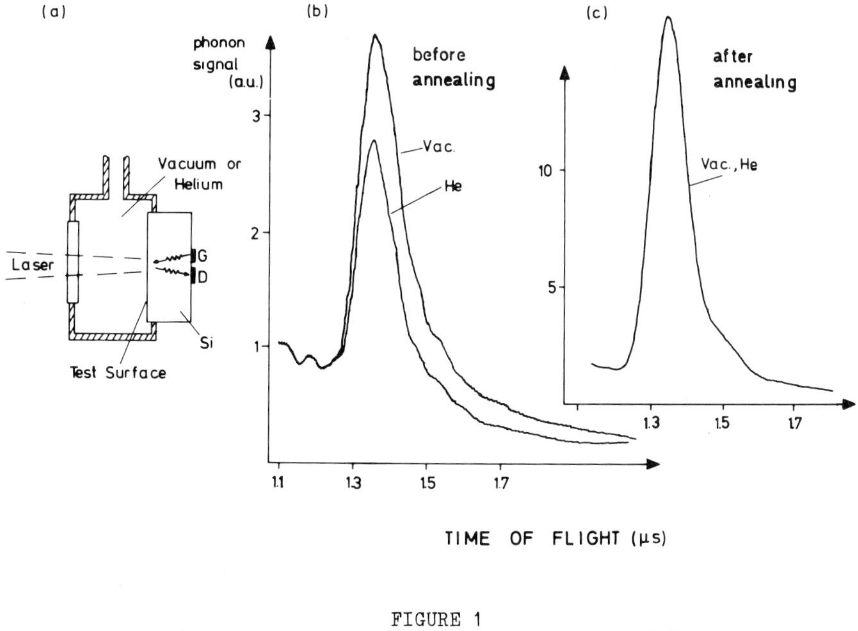

FIGURE 1
(a) Experimental set up. (b) Phonon echoes before laser annealing. (c) The echoes after annealing

phonon signal depended on the losses at the test surface.

The ruby laser provided pulses of 40 ns duration. We focused the laser so that in an area of 0.12 mm^2 the annealing-threshold energy of 1 J/cm^2 was exceeded. Above this threshold a thin (<5000 Å) surface layer is known to melt and after cooling to recrystallize perfectly without surface impurities (3). Optical scanning was used to anneal larger areas.

Experimental results are shown in Fig. 1.b and 1.c. The intensities of the reflected phonons are plotted vs. their respective times of flight. Due to phonon focusing transverse phonons were exclusively detected. In Fig. 1.b the trace "vac" was obtained before the annealing and with the chamber being evacuated. The poor quality of the "as received" polished surface became evident after filling the chamber with He (trace marked "He"). A large decrease of the signal due to the anomalous Kapitza transmission was observed.

The traces in Fig. 1. c. were taken after an 8 mm^2 area in the center of the test surface had been laser-annealed. This time no anomalous Kapitza transmission is observed signaling a perfect surface. Equally striking is the significant increase in signal size, probably indicating the transition from diffusive to specular reflection.

The availability of the high power laser provided an elegant technique to evaporate submonolayers of different metals onto the annealed surface (see insert of Fig. 2). Before the experiment metals films of 500 A thickness were evaporated onto the inner side of the window. The central area was kept clear, thus the films did not interfere with the annealing process. After the annealing of the Si surface, the laser could be directed onto one of the films. One laser pulse evaporated the whole metal in the illuminated area. From geometric considerations we concluded that 0.2 monolayers were deposited in the central region of the test surface with one pulse. So far, we were able to deposit and remove by annealing three different metal species in one experimental run.

Results for Au are shown in Fig. 2. The top trace was obtained after the annealing. Strikingly, the deposition of just 0.2 monolayers was already sufficient to reduce the signal significantly. Adding more Au decreased the pulse even further and a tail developed. The lowermost trace was taken with He in the chamber at this Au coverage. The large anomalous Kapitza transmission had returned!

Similar results were obtained if Al was deposited instead of Au. With Sn, however, the situation was different. No change from the "after annealed" trace was observed for Sn deposits of up to one monlayer. At the same time almost no anomalous Kapitza transmission was induced by the Sn.

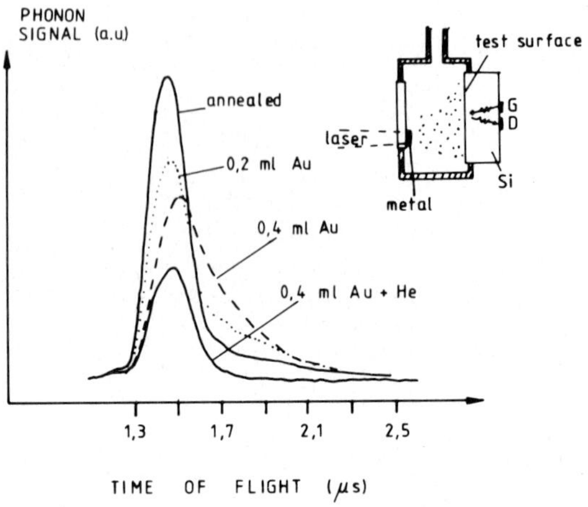

FIGURE 2

Effect of depositing submonolayers of Au onto a laser-annealed surface. Insert: set-up to evaporate the Au.

In conclusion, we have demonstrated that laser annealing became an important tool for studying the interaction of phonons with submonolayer deposits. Large effects were observed with just 0.2 monolayers. This corresponds to a density of 10^{14} atoms/cm^2. It is interesting to note that the two-level model by H. Kinder (4) requires almost the same value ($2.10^{14} cm^{-2}$) to explain an anomalous transmission of similar size.

REFERENCES

(1) For a review see: A. F. G. Wyatt: in Nonequilibrium Superconductivity, Phonons, and Kapitza boundaries, ed. K. E. Gray (Plenum, 1981) pp 31-72.
(2) J. Weber, W. Sandmann, W. Dietsche, and H. Kinder, Phys. Rev. Lett. 40, 1469 (1978).
(3) For reviews see: J.M. Poate and J. W. Mayer (Editors), Laser Annealing of Semiconductors (Academic, New York, 1982).
(4) H. Kinder, Physica 107 B, 549 (1981).

LT-17 (Contributed Papers)
U. Eckern, A. Schmid, W. Weber, H. Wühl (eds)

MAGNETIC HEAT CONDUCTANCE BETWEEN A CMN SINGLE CRYSTAL AND LIQUID ^{3}HE

D. MAREK, A.C. MOTA, J.C. WEBER and F.J. van HOUT

Laboratorium für Festkörperphysik, ETH-Z, CH-8093 Zürich, Switzerland

We report measurements of the thermal boundary resistance between a single crystal of cerium magnesium nitrate (CMN) and liquid ^{3}He. The contribution of the magnetic coupling to the heat conductance is much stronger than previously observed for powdered CMN-liquid ^{3}He. Below 8 mK the magnetic resistance can be described by $R_m \propto T^5$ and $R_m \propto H^{-2}$ for low fields.

The discovery in 1966 by Abel, Anderson, Black and Wheatley [1] that the thermal boundary resistance between cerium magnesium nitrate (CMN) and liquid ^{3}He follows $R \propto T$ below 20 mK, has stimulated numerous investigations into this and other systems. Unfortunately, after the success of the magnetic coupling theory in explaining this resistance [2], little progress has been made. One of the reasons for this situation is connected with some conflicting experimental results after Abel et al. [1], that failed to show the magnetic coupling. This fact has been traced [3] to high levels of ^{4}He at the interfaces under study.

For convenience, most of the Kapitza resistance experiments have been done using powders immersed in liquid helium. However, size and surface effects are then rather difficult to control. Very little work has been reported on Kapitza resistance using bulk solids in contact with liquid ^{3}He at millikelvin temperatures. Although such experiments are desirable, one is confronted with very long time constants and with even more stringent requirements on the liquid ^{3}He purity. It is known [4] that in order to observe a magnetic coupling, the ^{4}He surface coverage must not exceed a fraction of a monolayer.

We report measurements of the thermal Kapitza resistance between a single crystal of cerium magnesium nitrate and pure liquid ^{3}He. The sample consisted of a freshly grown, clean crystal with a mass $m = 6.92 \times 10^{-3}$ g, and a surface area of 1.26×10^{-5} m^2. As a means to keep such a small surface free from ^{4}He, we have used an artifice that has proved to be very successful in previous experiments [5]. Liquid ^{3}He with an impurity level of 5 ppm of ^{4}He was allowed to enter into the measuring cell through a sintered silver body with an estimated surface area of 40 m^2. In order to efficiently clean the surfaces inside the cell, a small valve was installed in the ^{3}He filling line above the cell. Removal of residual gases was carried out then through a provisional low impedance pumping line attached to this valve and not through the long and thin tubing of the ^{3}He filling line. During this operation the CMN specimen was kept wellbelow the ice point to avoid dehydration.

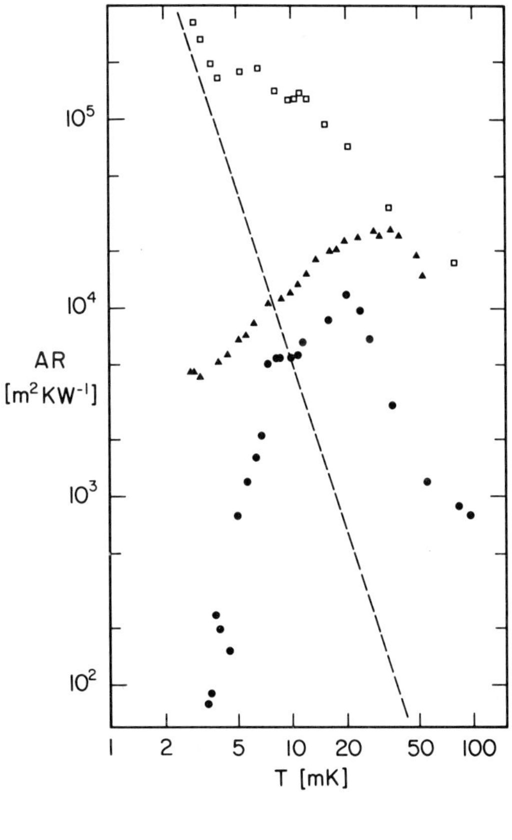

FIG 1. Thermal resistance vs temperature. CMN single crystal-liquid ^{3}He (•), CMN powder-liquid ^{3}He (▲) and CMN powder-liquid ^{3}He-^{4}He (□) Dashed line: $AR_K T^3 = 52 \times 10^{-4}$ m^2 K^4 W^{-1}.

A standard adiabatic demagnetization stage containing 7.5 g of CMN powder was included in the upper part of the cell. Samples were measured in the lower part. Thermal resistances were obtained from relaxations after a small magnetic pulse was applied. ac and dc susceptibilities were detected with SQUIDs. For comparison with the single crystal we made measurements of R in a powdered CMN specimen in contact with liquid ^3He and with a solution of 6% ^3He in ^4He. The powdered CMN consisted of 22.1×10^{-3} g of freshly grounded crystals with diameters $40 \leqslant \emptyset \leqslant 45$ μm packed into a right circular cylinder with equal diameter and height (d = h = 3 mm). Within the experimental errors, we observed pure exponential decays, as expected from the selected particle sizes.

The single crystal was 0.75 μm thick in the c-direction and between 2 to 3 mm wide in the other directions. External fields were applied in a direction perpendicular to the c-axis.

Figure 1 shows the measured thermal resistance AR as function of temperature for the three cases. Here A is the estimated area and $R = \tau/C$, with τ the measured time constant and C the specific heat of CMN. In all cases $C_{CMN} \ll C_{He}$. Magnetic temperatures measured with the single crystal were corrected to account for the different demagnetization factor. From a calibration against a CMN powder thermometer (h = d) done inside the mixing chamber in the temperature range 8-30 mK we concluded that

$$T^*_{powder} = T^*_{crystal} + \Delta$$

with $\Delta = 0.8 \pm 0.2$ mK. For $T < 6$ mK we have used the relationship between T^*_{powder} and Johnson noise temperature [6].

The dashed line in Fig. 1 is the normal Kapitza boundary resistance $AR_K T^3 = 52 \times 10^{-4}\ m^2K^4W^{-1}$ as obtained from the low temperature values of AR for a CMN single crystal immersed in a solution of 6% ^3He in ^4He [3]. For the CMN powder in a similar solution (□) we observe, below 4 mK, $R \propto T^{-3}$. Above 4 mK the data is dominated by the "bottlenecked" spin-lattice resistance in the salt, R_b. This resistance is in series with the normal boundary resistance R_K [3].

For the CMN powder-pure liquid ^3He (▲) we obtain a result similar to Abel et al. [1], $R_m \propto T$ below 20 mK. At higher temperatures the magnetic Kapitza resistance R_m becomes bigger than $R_b + R_K$.

Finally, for the single crystal-pure liquid ^3He (•) at the lowest temperatures we observe surprisingly low values of AR. Below T = 8 mK the data can be described as approaching $R_m \propto T^5$. This strong temperature dependence has never been observed before and cannot be obtained with the present theories for magnetic coupling.

We have made some preliminary measurements of the magnetic field dependence of R_m. For the single crystal-liquid ^3He sample we observe below 10 mK clearly $R_m \propto H^{-2}$ in fields up to 100 Oe followed by a minimum, and then a rapid increase to values well above the zero field value for R_m. Here we take $H = (H_a^2 + H_i^2)^{1/2}$ with H_a the applied external field, and H_i the internal field in CMN.

Similar measurements for the CMN powder-liquid ^3He sample show, below 10 mK, an extremely weak dependence of R_m on H up to 100 Oe. This very weak field dependence of R_m as well as the weaker temperature dependence is probably connected with the fact that CMN is highly anisotropic ($g_\perp = 1.84$ and $g_\parallel \simeq 0$). Furthermore, in the process of crushing and tightly packing the sieved powder, we are left with a seriously damaged surface which is very different from the surface of a single crystal as grown in the liquid solution.

In conclusion, we have measured the thermal resistance between a small CMN single crystal and very pure liquid ^3He. Below T = 8 mK we obtain $R_m \propto T^5$ at H = 0, and $R_m \propto H^{-2}$ for fields below 100 Oe. More work is planned in order to understand this extremely efficient magnetic contribution to the Kapitza conductance.

ACKNOWLEDGEMENTS

This work was partly supported by the Schweizerischer Nationalfonds zur Förderung der wissenschaftlichen Forschung.

REFERENCES

[1] W.R. Abel, A.C. Anderson, W.C. Black and J.C. Wheatley, Phys. Rev. Lett. 16 (1966) 273

[2] A.J. Leggett and M. Vuorio, J. Low Temp. Phys. 3 (1970) 359

[3] M. Jutzler and A.C. Mota, J. Low Temp. Phys. 55 (1984) 439

[4] W.C. Black, A.C. Mota, J.C. Wheatley, J.H. Bishop and P.M. Brewster, J. Low Temp. Phys. 4 (1971) 391

[5] M. Jutzler and A.C. Mota, Physica 107B (1981) 553

[6] R.A. Webb, R.P. Giffard and J.C. Wheatley, J. Low Temp. Phys. 13 (1973) 383

LT-17 (Contributed Papers)
U. Eckern, A. Schmid, W. Weber, H. Wühl (eds)

LONG AND NARROW OVERLAP JOSEPHSON JUNCTIONS WITH A REALISTIC BIAS CURRENT DISTRIBUTION

M.R. SAMUELSEN* and S.A. VASENKO

Department of Physics, Moscow State University, Moscow 117234, U.S.S.R.

The maximum supercurrent through a long and narrow overlap Josephson tunnel junction as a function of an applied magnetic field is calculated numerically for a bias current distribution as the one in a long thin superconducting strip. Also the first zero field step in the dc IV-characteristic is calculated.

1. INTRODUCTION

Zero field steps in a long Josephson tunnel junction are ascribed to a resonant motion of fluxons back and forth along the Josephson transmission line[1]. In the overlap junction - where the bias current is fed into the junction perpendicular to the long direction - most theoretical treatments have assumed a homogeneous current distribution along the junction. Most overlap junctions consists of an interruption by an oxide layer of a long thin superconducting strip. According to the shortening principle[2] the real current distribution in an overlap junction is then as the current distribution in the long and thin superconducting strip[3]. This current distribution has singularities at the ends of the junction making the situation very similar to a mixture of homogeneous overlap and inline current feed[4]. This non-homogeneous current distribution naturally affects the fluxon motion and the stability of the static state (zero voltage state) also when an external magnetic field is applied.

2. THE MODEL

We consider a long and narrow rectangular junction of length L and width W ($L \gg \lambda_J \gg W$). The fluxon motion is assumed to be governed by the perturbed sine-Gordon equation[1,4]

$$\phi_{xx} - \phi_{tt} = \sin\phi + \alpha\phi_t + \eta(x) \qquad (1)$$

where $\phi(x,t)$ is the quantum phase difference between the two superconductors of the Josephson junction. The spatial variable x is measured in units of the Josephson penetration depth $\lambda_J = (\hbar/2eJ\mu_o d)^{\frac{1}{2}}$ and the time t in units of the reciprocal maximum Josephson plasma frequency ω_o^{-1}, $\omega_o = (2eJ/\hbar C)^{\frac{1}{2}}$. Here J is the maximum Josephson current density, d the magnetic thickness of the oxide layer, and C is the capacitance per unit area. In the loss term $\alpha = G(\hbar/2eJC)^{\frac{1}{2}}$ where G is the shunt conductance per unit area.

The force term in Eq. 1 $\eta(x)$ is given by[3]

$$\eta(x) = \frac{\eta_o \ell}{\pi} \frac{1}{\sqrt{x(\ell-x)}} \qquad (2)$$

where ℓ is the normalized length of the junction (L/λ_J) and η_o is the normalized current density through the junction $\eta_o = I_{dc}/JWL$, I_{dc} being the current.

An external magnetic field H_{ext} applied in the plane of the junction and perpendicular to the long direction imposes the following boundary conditions[1]

$$\phi_x(0,t) = \phi_x(\ell,t) = \kappa \qquad (3)$$

where κ is the normalized magnetic field $\kappa = H_{ext}/\lambda_J J$.

3. NUMERICAL RESULTS

In order to find solutions to Eqs. 1 and 3 we have used an implicit finite difference method with $\Delta x = 0.1$ and $\Delta t = 0.05$ (which turned out to be sufficient). For the force $\eta(x)$ the expression Eq. 2 has been used except at the ends of the junction where values are chosen to make the numerical integral of the force equal to $\eta_o\ell$.

The result for the maximum supercurrent versus the magnetic field is shown in Fig. 1 as points for (b) ℓ=10 and (c) ℓ=20. For comparison we also show the analytical result for a homogeneous current distribution[5], the dotted curve (a). The full curves are perturbational theoretical results for mixed homogeneous overlap-inline current fed[4] with overlap parameter (b) y = 0.72 (c) and y = 0.76 (best fit). If the overlap parameter is determined alone from the maximum (κ=0) the results are (b) y=0.71 and (c) y=0.74. It is seen that the numerically determined points agree very well with the theoretical curves provided a proper overlap parameter y is chosen. Only static states without fluxons

* Permanent address:
Physics Laboratory I, The Technical University of Denmark, DK-2800 Lyngby, Denmark.

in the junction has been determined.

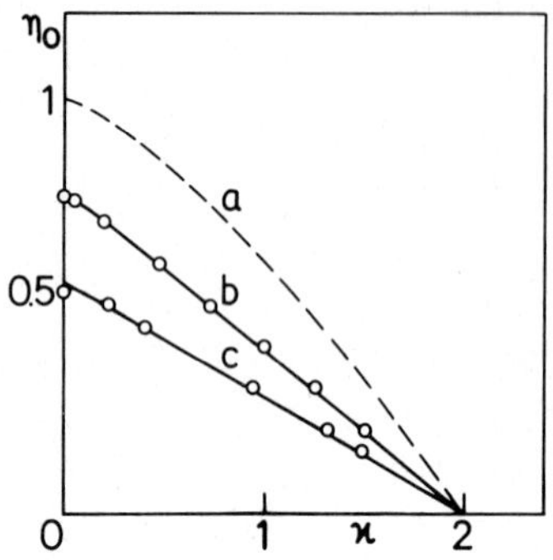

FIGURE 1
The maximum supercurrent η_o versus the applied magnetic field κ. (a) the dotted curve is the analytic result for a homogeneous current distribution[5]. For (b) ℓ=10 and (c) ℓ=20 the points are the numerical results and the solid curves are the perturbational theoretical curves for mixed inline-overlap[4]: (b) y=0.72 and (c) y = 0.76.

The first zero field step: current versus average velocity is shown in Fig. 2 for ℓ=10 and α=0.1. From the average velocity u_{av} we calculate the voltage $V_{dc} = (\hbar\omega_o/2e)(2\pi u_{av}/\ell)$[1,4]. The maximum bias point on the first zero field step was η_o=0.715. This should be compared to the maximum supercurrent without magnetic field η_o = 0.76 Fig. 1 (b). The two full curves in Fig. 2 are perturbational theoretical for the homogeneous overlap (a) and inline (b)[1,4]. It is seen that the numerical result for large η_o is close to (a) and as the current η_o decreases the numerical result approaches (b). The perturbational curve for mixed homogeneous overlap and inline with y > 0.64 (min. of $\eta(x)$) is very close to the curve (a)[4]. The discrepancy seems to be due to some reflection losses for small η_o not included in the perturbation treatment in Ref. 4.

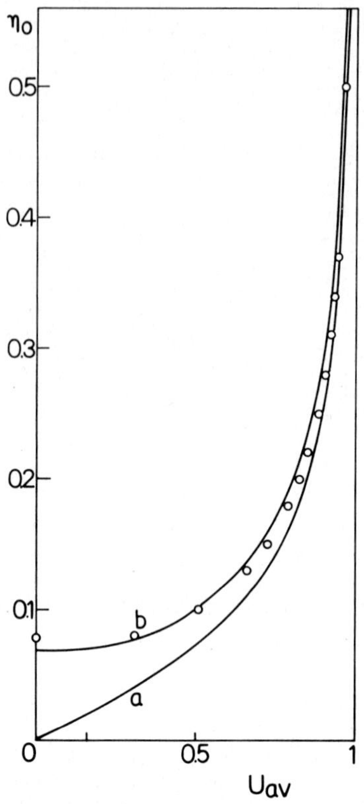

FIGURE 2
The first zero field step. The applied current η_o versus the average velocity of the fluxon u_{av} (voltage). The points are the numerical result. For η_o=0.715 (not shown) the zero field step ceases to exist (switches). The solid curves are theoretical based on perturbation theory for (a) overlap and (b) inline geometry.

4. CONCLUSION

The maximum supercurrent through a long Josephson tunnel junction as a function of an applied magnetic field and the first zero field step in the dc IV-curve has been calculated numerically with a (realistic) current distribution as that in a long thin superconducting strip. The results has been compared to a perturbation calculation where the current distribution was approximated as a constant plus two delta-functions at the ends and a good agreement is found especially for the maximum supercurrent versus magnetic field.

REFERENCES

(1) O.A. Levring, N.F. Pedersen and M.R. Samuelsen, J.Appl.Phys. 54, 987(1983).
(2) A. Barone, F. Esposito, K.K. Likharev, V.K. Semenov, B.N. Todorov and R. Vaglio. J.Appl.Phys. 53, 5802(1982).
(3) V.L. Newhouse, J.W. Bremer and H.H. Edwards, Proc. IRE 48, 1395(1960).
(4) O.H. Olsen and M.R. Samuelsen, J.Appl.Phys. 54, 6522(1983).
(5) M. Yu. Kupriyanov, K.K. Likharev and V.K. Semenov, Sov. J. Low Temp. Phys. 2, 610 (1976) [Fiz. Nizk. Temp. 2, 1252(1976)].

LT-17 (Contributed Papers)
U. Eckern, A. Schmid, W. Weber, H. Wühl (eds)

ESTIMATION OF THE CRITICAL-CURRENT DENSITY DISTRIBUTION IN JOSEPHSON TUNNEL JUNCTIONS

K. YOSHIDA, N. UCHIDA, K. ENPUKU, T. TANAKA and F. IRIE

Department of Electronics, Kyushu University 36, Fukuoka 812, Japan

A method is given to estimate the spatial distribution of the critical current density in Josephson tunnel junctions when the variation of the current density is small. The present method is valid for the estimation of the even part of the critical current.

1. INTRODUCTION

Spatial distribution of the critical-current density in Josephson tunnel junction, $j_c(x)$, gives information on the oxide-barrier of junctions[1-5], and is believed to degrade the reproducibility of the junction critical current. For the estimation of $j_c(x)$, an indirect method based on the magnetic field dependence of the maximum Josephson current, i.e., the threshold curve, has so far been developed[1-3]. Recently the direct observation of $j_c(x)$, such as electron-[4] or laser-beam [5] scanning are proposed.

Although the threshold curve is a simple and very sensitive indicator of $j_c(x)$, it is known [2] that the determination of $j_c(x)$ is not unique in general. In this paper we show theoretically that the estimation of the even part of $j_c(x)$ is possible in the case when the variation of $j_c(x)$ is small and apply this method to experimental results on the aging effect of the critical current density distribution.

2. PRINCIPLE

We consider a rectangular junction with length L and width W, and express the critical current density $j_c(x)$ as

$$j_c(x) = j_{c,o} + \Delta j_c(x), \tag{1}$$

where $j_{c,o}$ is a constant and we assume $|\Delta j_c/j_{c,o}| << 1$, i.e., the variation of $j_c(x)$ is small. When the magnetic field H is applied in the direction of junction width, the maximum Josephson current I_c is given as a function of H as [1]

$$I_c^2(\beta) = [W\int_{-L/2}^{L/2} j_c(x)\cos(\beta x)\,dx]^2 + [W\int_{-L/2}^{L/2} j_c(x)\sin(\beta x)\,dx]^2 \tag{2}$$

where $\beta = 2\pi\mu_0 Hd/\Phi_0$, μ_0 is a permeability, d is a magnetic thickness, and Φ_0 is the flux quantum.

Substituting Eq.(1) into Eq.(2), and neglecting the terms proportional to $|\Delta j_c|^2$ in a sense of perturbation, we obtain

$$I_c^2(\beta) = I_{c,o}^2(\beta) + 2WI_{c,o}(\beta)[\int_{-L/2}^{L/2}\Delta j_{c,e}(x)\cos(\beta x)\,dx] \tag{3}$$

with

$$I_{c,o}(\beta) = j_{c,o}LW\sin(\beta L/2)/(\beta L/2). \tag{4}$$

where $\Delta j_{c,e}(x) = [\Delta j_c(x) + \Delta j_c(-x)]/2$ is the even part of $\Delta j_c(x)$, and $|I_{c,o}(\beta)|$ corresponds to the ideal threshold curve in the case of $\Delta j_c(x) = 0$.

Assuming $j_c(x) = 0$ outside the junction, i.e., $|x| > L/2$, we can obtain the even part of $\Delta j_c(x)$ from the Fourier-transform of Eq.(3) as

$$\Delta j_{c,e}(x) = \frac{1}{2\pi W}\int_0^\infty \frac{I_c^2(\beta) - I_{c,o}^2(\beta)}{I_{c,o}(\beta)}\cos(\beta x)\,d\beta, \tag{5}$$

Equation (5) means that the even part of $j_c(x)$ can be estimated from the experimental values of $I_c(\beta)$ when $|\Delta j_c/j_{c,o}| \ll 1$. Especially when the variation of $j_c(x)$ is even symmetrical with respect to x, Eq.(5) gives a good approximation of $j_c(x)$.

3. COMPUTER SIMULATION

In the calculation of Eq.(5), the spatial resolution of $j_c(x)$ depends on the numer of sampling points of $I_c(\beta)$. It is shown, from the numerical simulation, that the spatial resolution of 10% can be obtained if the values of $I_c(\beta)$ are taken up to 10 periods with 8 sampling points in each period, i.e., up to $\beta L/2\pi = 10$ with increment $\Delta(\beta L/2\pi) = 1/8$. It is also shown that Eq.(5) gives a good estimation of $\Delta j_{c,e}$, within an error of about ten percents, if the variation of $|\Delta j_c/j_{c,o}|$ is less than 25%.

Figure 1 shows an example of the numerical simulation of the estimation of $j_c(x)$ with Eq.(5). The solid line in Fig.1(a) is the assumed distribution of $j_c(x)$ in the case of symmetrical variation. The resultant $I_c(\beta)$ is calculated from Eq.(2), and is shown in Fig.1(b) by circles. The broken line in Fig.1(a) shows the spatial distribution of $j_c(x)$ which is estimated from these value of $I_c(\beta)$ with Eq.(5). As shown in Fig.1(a), the estimated distribution of $j_c(x)$ agrees well with the assumed one, which shows the validity of Eq.(5). The $I_c(\beta)$ curve calculated from the estiamted $j_c(x)$, which is shown in Fig.1(b) by a broken line, is also in good agreement with the original one.

In the case when the variation of $j_c(x)$ is not symmetrical, only the even part of $j_c(x)$ can be estimated with Eq.(5). In this case, the agreement between the $I_c(\beta)$ calculated from the estimated $j_c(x)$ and the original one is not so good as in the case of symmetrical variation of $j_c(x)$. Especially, the discrepancy becomes large in the region where $I_c(\beta)$ becomes lacal minima. Therefore, we can see the degree of symmetry of $j_c(x)$ by comparing the $I_c(\beta)$ calculated from the estimated $j_c(x)$ with the original one.

3. EXPERIMENT

As an example of the estimation of $j_c(x)$ with Eq.(5), we study the change of $j_c(x)$ due to the so-called aging effect. In Fig. 2(a), we show the experimental $I_c(\beta)$. The triangles in Fig. 2(a) show $I_c(\beta)$ which are measured 4 days after the fabrication, while circles show those after 10 days. As shown in Fig.2(a), $I_c(\beta)$ reaches the ideal threshold curve with the aging.

The spatial distribution of $j_c(x)$ is estimated with Eq.(5), and is shown in Fig.2(b). The solid and broken lines are obtained from $I_c(\beta)$ which are indicated by circles and triangles in Fig.2(a), respectively. As shown in Fig. 2(b), $j_c(x)$ is initially small near junction edges, and then becomes uniform with the aging. The initial reduction of $j_c(x)$ near junction edges may show the non-uniformity of the oxidation process in the fabrication. It must be mentioned that $j_c(x)$ is nearly symmetrical in the present sample, since it can be shown that $I_c(\beta)$, which are calculated from the estimated $j_c(x)$ shown in Fig.2(b), are in good agreement with the experimental ones shown in Fig.2(a).

5. CONCLUSION

We show theoretically and experimentally that the even part of $j_c(x)$ can be estimated from the threshold curve when the variation of $j_c(x)$ is small. The present method can be a usefull tool for oxide-barrier diagnostics.

REFERENCES

[1] R.C.Dynes et al., Phys. Rev. B 3(1971) 3015.
[2] H.H.Zappe, Phys. Rev. B 11(1975) 2535.
[3] A.Barone et al., Phys. Stat. Sol.(a) 41 (1977) 393.
[4] H.Seifert et al., Phys. Lett. 95A(1983) 326.
[5] M.Schenermann et al., Phys. Rev. Lett. 50 (1983) 74.

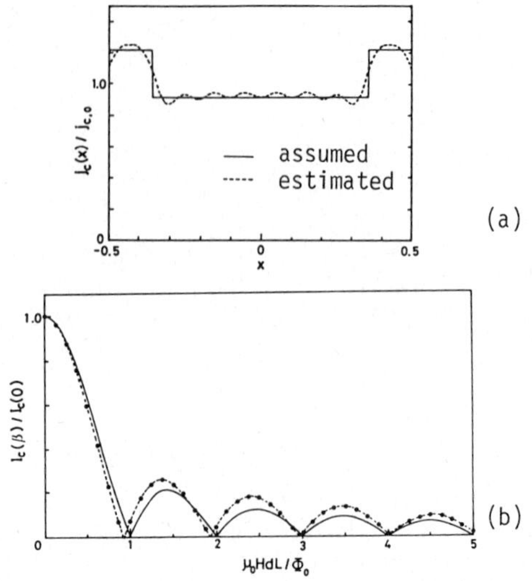

Fig. 1 Numerical simulation of the estimation of $j_c(x)$ with Eq.(5). (a) Spatial distribution of $j_c(x)$. (b) Threshold curve. Circles and the broken line show $I_c(\beta)$ calculated from the assumed $j_c(x)$ and the estimated one, respectively. The solid line represents $I_{c,o}(\beta)$.

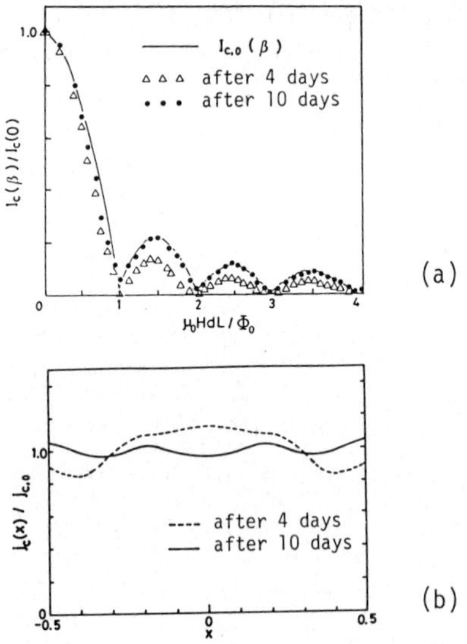

Fig. 2 Change of $j_c(x)$ due to the aging effect. (a) Experimental $I_c(\beta)$. (b) $j_c(x)$ estimated with Eq.(5).

LT-17 (Contributed Papers)
U. Eckern, A. Schmid, W. Weber, H. Wühl (eds)

THE EFFECT OF A SCANNING LASER BEAM ON A DYNAMIC VORTEX MODE OF A JOSEPHSON JUNCTION*

S.C. MEEPAGALA, W.D. SHEN, P.K. KUO, and J.T. CHEN

Department of Physics, Wayne State University, Detroit, Michigan 48202, U.S.A.

We have found that a focused laser beam irradiated on a Josephson tunnel junction can enhance the magnitude of a vortex step in its current-voltage characteristic. Qualitatively, the spatial variation of the enhancement, with an opposite sign, seems to be correlated to the distribution of the zero-voltage current.

1. INTRODUCTION

One of the major tasks in studying the dynamic properties (1) of a Josephson junction is to determine the distribution of the bias current in the following equation (2,3)

$$\frac{\partial^2\phi}{\partial x^2} + \Gamma\frac{\partial\phi}{\partial t} - \frac{1}{\bar{c}^2}\frac{\partial^2\phi}{\partial t^2} = \frac{1}{\lambda_J^2}\sin\phi + \gamma$$

where ϕ is the phase of the Josephson junction, Γ the damping factor related to the quasiparticle and the surface currents, $\bar{c}$ the speed of electromagnetic waves in the junction, λ_J the Josephson penetration depth, and γ is the normalized bias current generally assumed to be constant. Recently, a laser scanning technique has been successfully used to probe the distribution of the zero-voltage dc Josephson currents (4,5). In order to understand the nature of the bias current at non-zero voltage we have used the same technique to study the response of the zero-field steps (6-8) in the presence of a scanning laser beam. Some preliminary results are to be reported here.

2. SAMPLE AND EXPERIMENT

The sample was Pb-Pb oxide-Pb tunnel junction of the overlap geometry, prepared by conventional thermal evaporation. The dimension of the junction area is 0.35 mm x 0.12 mm. The maximum Josephson current density, j_o, at 4.2 K is approximately 150 A/cm^2 which is equivalent to $L/2\pi\lambda_J \simeq 1.5$. This means that one or two vortices (fluxons) can be formed in the junction in the absence of an external magnetic field. Indeed, we have observed two vortex branches in the I-V characteristic; one near 80 μV and the other near 160 μV. Using $\bar{c} = 2\times10^7$ m/s, we obtained the plasma voltage $V_p = \Phi_o\bar{c}/2\pi\lambda_J = 177$ μV, which is in good agreement with the observed cut-off voltage.

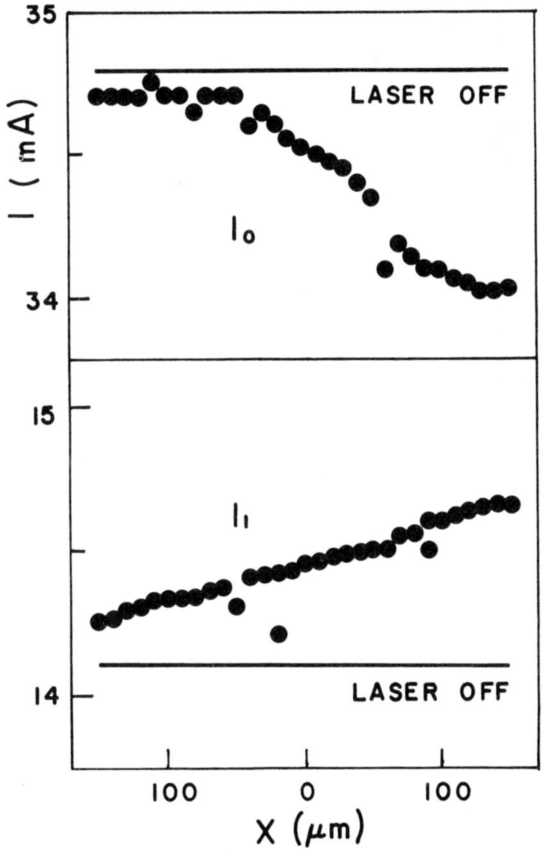

FIGURE 1
I_o and I_1 versus the laser beam position.

* This work is supported by the National Science Foundation through Grant No. DMR 80-24052.

The laser experiment was carried out at 1.7 K. As the laser beam (of diameter about 50 μm) was scanned across the junction, the maximum values of both the zero-voltage current, $I_o(x)$, and the first vortex branch, $I_1(x)$, were recorded as a function of beam position x.

3. RESULTS

Figure 1 shows the $I_o(x)$ and the $I_1(x)$ in the presence of a focused laser beam.[1] The horizontal lines represent the values of I_o and I_1 without the laser. The asymmetrical current distribution shown in $I_o(x)$ indicates that there is a small amount (equivalent to one-half vortex) of residual field of unknown origin. However, this asymmetry was not due to the nonuniformity of the oxide barrier since

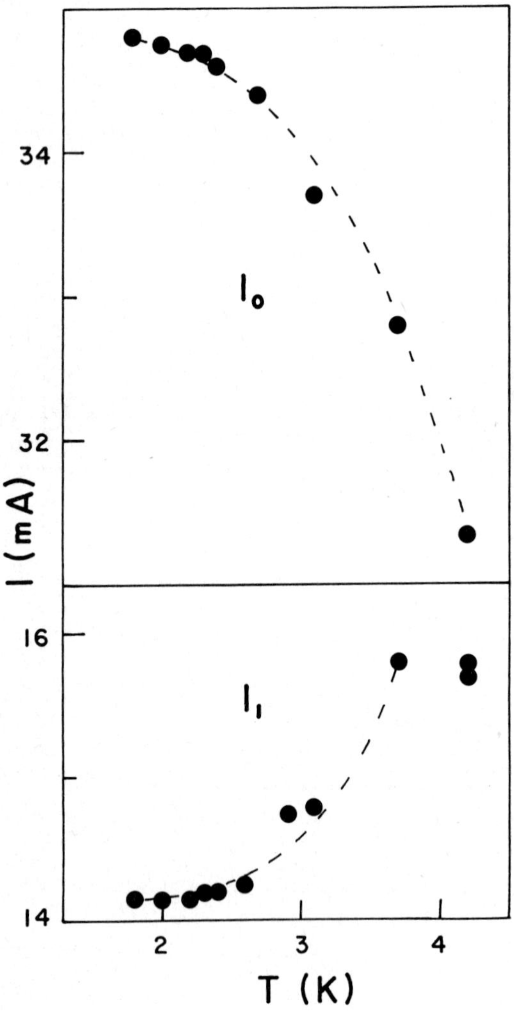

FIGURE 2
I_o and I_1 as a function of temperature.

the gap-voltage variation in the presence of a scanning laser did not show such an asymmetry. Several remarks can be made about the effect of a scanning laser on I_1. First, the voltage position of the first vortex branch was unaffected by the laser. Second, the change of I_1 is opposite to that of I_o in sign. Third, the spatial dependence of I_1^o is similar to that of I_o, i.e. the larger effects occur to the same side.

4. DISCUSSION

It was suggested in Ref. (4) that the change in $I_o = \int j_o \sin\phi dA$ is due to a local change of current, $\Delta I_o = \Delta j_o \sin\phi \cdot \delta A$, where δA is the area heated by the laser. Since the heating reduces j_o, i.e. $\Delta j_o < 0$, a negative ΔI_o means that the local current represented by $\sin\phi$ term is positive. Likewise, a positive ΔI_o means that the $\sin\phi$ term is negative. The sign of the observed $\Delta I_1(x)$ suggests that we cannot simply interpret I_1 as $j_o \int \sin\phi_1 dx$ where $\phi_1(x)$ is the solution for the one vortex mode. According to the phenomenological model (7), the dc current of the vortex branches should be related to the loss mechanisms. As shown in Fig. 2, we found experimentally that I_1 decreases as the temperature is lowered while I_o has an opposite temperature dependence. This is consistant with the picture that I_1 is determined by the loss factor rather than simply proportional to $j_o \sin\phi$. The nonuniform variation of ΔI_1 shown in Fig. 1 indicates that the distribution of I_1 may be sensitive to magnetic field. This suggests that the γ term, which represents a bias current should not be a constant. The verification of this point requires a sample of perfect symmetry and a detailed study of the magnetic-field dependence.

REFERENCES

(1) B.D. Josephson, Adv. Phys. 14 (1965) 419.
(2) G. Costabile, R.D. Parmentier, B.Savo, D.W. McLaughlin, and A.C. Scott, Appl. Phys. Lett. 32(9) (1978) 587.
(3) H. Kawamoto, Prog. Theor. Phys. 70 (1983) 1171.
(4) James R. Lhota, M. Scheuermann, P.K. Kuo, and J.T. Chen, Appl. Phys. Lett. 44(2) (1984) 255.
(5) M. Scheuermann, James R. Lhota, P.K. Kuo, and J.T. Chen, Phys. Rev. Lett. 50 (1983) 74.
(6) J.T. Chen, R.F. Finnegan, and D.N. Langenberg, Physica 55 (1971) 413.
(7) T.A. Fulton and R.C. Dynes, Solid State Commun. 12 (1973) 57.
(8) T.V. Rajeevakumar, John X. Przybysz, J.T. Chen, and D.N. Langenberg, Phys. Rev. B21 (1980) 5432.

LT-17 (Contributed Papers)
U. Eckern, A. Schmid, W. Weber, H. Wühl (eds)

ZERO-FIELD STEPS AND FISKE MODES IN LONG JOSEPHSON JUNCTIONS

M.CIRILLO, G.COSTABILE, S.PACE, B.SAVO

Dipartimento di Fisica, Università di Salerno
I-84100 Salerno, Italy

The magnetic field response of a long Josephson junction when d.c. current-biased on zero-field steps (ZFS) reveals significant differences from the measured magnetic field dependence of the Fiske step (FS) amplitude. In addition we find that the different behaviors are well suitable for qualitative modelling respectively in terms of fluxon dynamic propagation (ZFS) and cavity mode excitations (FS).

1. INTRODUCTION

A long, one-dimensional, overlap Josephson tunnel junction shows in the current-voltage (I-V) characteristic two fundamental families of d.c. current singularities: Fiske steps (FS) and Zero-field steps (ZFS). Fiske steps appear when an external magnetic field is applied along the direction perpendicular to the extended dimension of the junction, at voltages $V_n = n\,\Phi_o\,(\overline{c}/2L)$ ($n = 1,2,3,\ldots$, $\Phi_o = (2\times10^{-7})$ Gcm^2 is the flux quantum, $\overline{c}$ is the electromagnetic propagation velocity in the junction and L is the junction length). Zero-field steps, on the other hand, show the maximum current amplitude in zero external magnetic field and exhibit voltages $V_n = 2n\,\Phi_o\,(\overline{c}/2L)$. Thus, even order FS appear at the same voltage as ZFS. In the light of results coming from microwave emission experiments Dueholm et al.(1) suggested that ZFS and FS could be manifestations of the same underlying phenomenon. This possibility has been consistently studied from the numerical point of view over the past couple of years (1,2,3). The data that we present here bring a deeper insight in this argument: it is found, in fact, that, especially for very long junctions, the magnetic field behavior of FS differs significantly from that exhibited by stable ZFS, which can be easily distinguished from the former.

2. EXPERIMENTAL BACKGROUND

The procedure for the fabrication of our samples is the same as that described in ref.(4). The experiments were carried out on Nb-NbOx-Pb Josephson tunnel junctions of overlap geometry. A long solenoid provided the external magnetic field; all the experiments were performed at 4.2 K. Two kinds of d.c. measurements were used during the experiments. In the first method, we started from a stable working point on the step and increased either the bias current or the field until the junction switched to a different voltage. This technique must be employed in particular for fully measuring the magnetic field behavior of the ZFS (see next section). The second method consisted of sweeping at low frequency the I-V characteristic having fixed the magnetic field amplitude. The latter was recorded together with the current amplitude of the d.c. singularities. This technique is well suited to plot the FS amplitudes vs. the external magnetic field. Several overlap junctions have been tested all over the range $1<\ell<10$ (where $\ell = L/\lambda_j$ and λ_j is the Josephson penetration depth (5,6)).We report here the results obtained from two samples, 141H14 and 141H24, having $\ell = 5.7$ and $\ell = 8.8$ respectively; they show clearly the typical features we have observed in junctions having $\ell>3$ (the behavior of junctions having $\ell<3$ has been considered in detail in recent papers (see e.g. ref. (5)).

3. RESULTS AND DISCUSSION

Fig. 1 shows the magnetic field behavior of the zero-voltage current and of the ZFS observed on the junction 141H14. The ZFS were at integer multiples of 30 μV. An interesting feature is evident in fig. 1. In the I-B plane the range of existence of the 1st ZFS (as well as the 4th and the 5th), delimited by the dotted curve joining triangles, is confined within a small region. To this, we must add that the transitions from the ZFS at a constant B are to the quasi-particle branch of the I-V characteristic whenever the

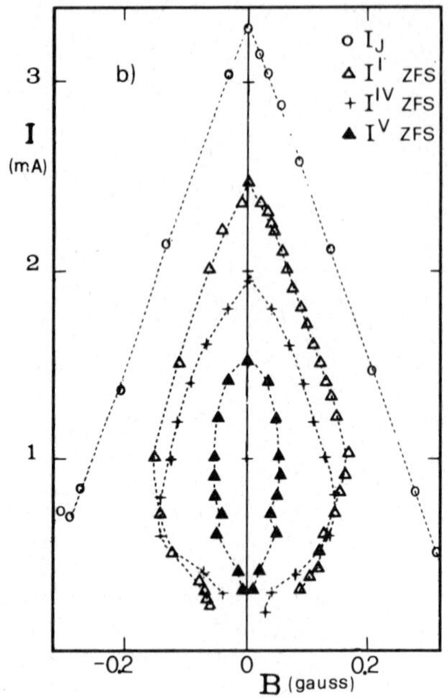

FIGURE 1
Measurements on the sample 141H14. This junction had ℓ=5.7, with L=970 μm and λj=170 μm.

bias is increased, while they are to the zero voltage state whenever the bias is decreased. The different switching mechanism indicates that in the first case the switch is caused by the excessive velocity increment at the positive input end of the junction; in the second case, the insufficient energy of the fluxon during the process of reflection at the negative input end is responsible for the decay into zero-voltage state oscillations (breathers and plasma waves). When the value of 0.2 gauss is exceeded (consider still fig. 1) no stable singularities are observable until we increase B by several tenths of gauss, passing the threshold, after which stable steps at voltages $V_n = n \times 15\mu V$ will appear(5). As the voltages of these steps are spaced like Fiske resonances in small junctions, it seems obvious to call them Fiske steps. Also we observe in fig. 2, relative to the junction 141H24, that the modulation of the heights of these FS with increasing magnetic field is completely analogous to that measured in small junctions (5,6). The limited range of existence of the ZFS for relatively small B value and the agreement between the descriptions of FS in higher magnetic field in both short and long junctions seem to support the hypothesis that the FS observed in high B field are not described by the simple model based

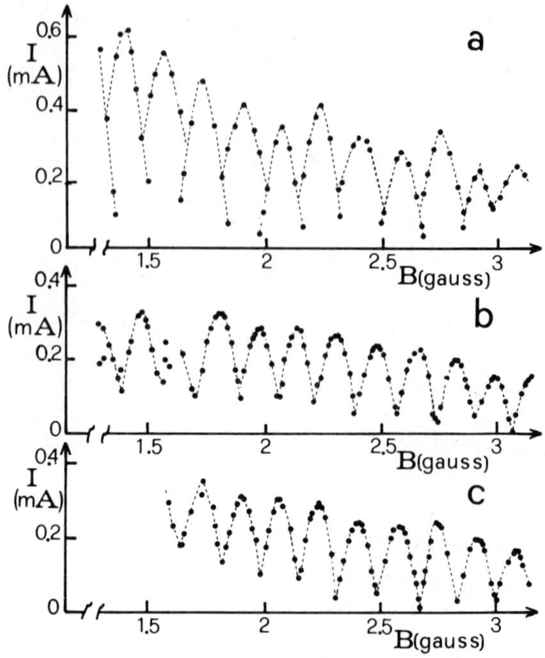

FIGURE 2
Josephson current (a) and 1st and 2nd FS (resp. b and c) magnetic field behavior for the sample 141H24.

on asymmetric fluxon propagation (1). In conclusion the effect of the external magnetic field on a long junction can be analyzed dividing the I-B plane into three regions: the first, in which fluxon propagation (and ZFS) takes place; the second, characterized by instability (7); the third, characterized by high-B value and high step stability were small junction-like behavior is found.

REFERENCES

(1) B. Dueholm, E. Joergensen, O.A. Levring, J. Mygind, N.F. Pedersen, M.R. Samuelsen, O.H. Olsen and M. Cirillo, Physica 108 B (1981) 1303.

(2) S.N. Erné, A. Ferrigno and R.D. Parmentier, Phys. Rev. B27 (1983) 5440.

(3) M. Radparvar and J.E. Nordman, IEEE Trans. Mag. MAG-19 (1983) 1017.

(4) V. Lacquaniti, G. Marullo and R. Vaglio, IEEE Trans. Mag. MAG-17 (1981) 812.

(5) M. Cirillo, A.M. Cucolo, S. Pace, B. Savo, J. Low Temp. Phys.,54 (1984) 489.

(6) A. Barone and G. Paternò, Physics and applications of the Josephson effect (J. Wiley, New York, 1982).

(7) M. Cirillo et al. : "Possible observation of chaotic intermittency....", preprint 1984

LT-17 (Contributed Papers)
U. Eckern, A. Schmid, W. Weber, H. Wühl (eds)

THEORY OF DYNAMIC POLARIZATION OF LIQUID ^{3}He

D. L. Stein, S. A. Langer,† and K. DeConde

Joseph Henry Laboratories of Physics, Princeton University, Princeton, N. J. 08544

A method is proposed for obtaining large non-equilibrium spin polarizations of liquid ^{3}He. Calculations based on spin transfer between a substrate and the liquid ^{3}He via the solid ^{3}He layer at the interface show a faster liquid polarization rate when insulators are used as a substrate rather than metals.

For the past several years, the subject of spin polarized liquid ^{3}He (^{3}He↑) has received much attention,[1] but experimentally it has not been possible to maintain appreciably polarized ^{3}He for times longer than a few minutes.[2] The straightforward application of a static magnetic field H leads to small polarizations of the order of $\mu_N H/k_B T_F$ where μ_N is the nuclear magnetic moment, k_B is Boltzmann's constant, and T_F is the Fermi temperature ($\sim$1K) of the liquid ^{3}He Pauli paramagnet. For a field of 1T, this leads to a polarization of only 10^{-3}. In this paper we describe a proposal for a technique that might produce non-equilibrium, steady-state polarizations two to three orders of magnitude larger.[3,4]

At the interface between liquid ^{3}He and any substrate, a solid layer (approximately 3-4Å thick) of ^{3}He atoms plates the surface due to the van der Waals attraction between the ^{3}He atoms and the substrate. Recent experiments indicate that the spins in this solid layer couple strongly to localized spins in insulators[5] and to conduction electron spins in metals.[6] Since the coupling between the solid layer and the bulk liquid is certainly larger[6] than that to the substrate, the solid layer may serve to couple spins in the substrate with those in the liquid. We propose to use this indirect coupling to pump polarization from the substrate to the liquid via a process analogous to the Overhauser effect.[7] Overhauser showed that in a metal with a hyperfine interaction between conduction electron and nuclear spins, one can enhance the nuclear spin polarization by a factor of $\mu_B/\mu_N \sim 10^3$ by saturating the conduction electron spin resonance, provided that other nuclear spin couplings are small. A similar effect can produce large nuclear polarizations in insulators with localized spins. We have found theoretically[4] that under certain conditions a new application of this effect can be used to transfer polarization from a substrate to liquid ^{3}He, leading to an enhanced ^{3}He polarization.

For such a procedure to succeed, the rate of polarization transfer to the liquid must exceed the fast spin relaxation rate to the walls.[8,9] The rate for polarization transfer may be calculated using second-order time-dependent perturbation theory. The probability per unit time for a transition from an initial state $|i\rangle$ to a final state $|f\rangle$ is given by

$$dW = \frac{2\pi}{\hbar} |V_{fi}|^2 \delta(E_i - E_f)\, \rho_f$$

where ρ_f is the density of available final states and

$$V_{fi} = \sum_m \frac{\langle f|\, H\, |m\rangle \langle m|\, H\, |i\rangle}{E_m - E_i} \tag{1}$$

includes a sum over intermediate states $|m\rangle$. The Hamiltonian H that appears in (1) differs for metals and insulators. For metals, it describes the interaction of a solid layer atom with the conduction electron gas and with the liquid ^{3}He and may be written

$$H^{(metal)} = \sum_{\substack{\vec{k}\sigma \\ \vec{k}'\sigma'}} \left(\frac{g_o}{N_e}\right) \vec{I}\cdot\vec{S}^{(e)}_{\sigma\sigma'}\, a^*_{\vec{k}\sigma}\, a_{\vec{k}'\sigma'} + \sum_{\substack{\vec{k}\sigma \\ \vec{k}'\sigma'}} \left(\frac{J}{N_{He}}\right) \vec{I}\cdot\vec{S}^{(\ell)}_{\sigma\sigma'}\, c^*_{k\sigma}\, c_{k'\sigma'} \tag{2}$$

where $\vec{S}^{(e)}$, $\vec{S}^{(\ell)}$, and $\vec{I}$ are Pauli spin operators for the conduction electrons, liquid ^{3}He, and solid ^{3}He, respectively; $a^*_{\vec{k}\sigma}$ [$a_{\vec{k}\sigma}$] and $c^*_{\vec{k}\sigma}$ [$c_{\vec{k}\sigma}$] are creation [annihilation] operators for electrons and liquid ^{3}He atoms, and N_e and N_{He} are the numbers of electrons and liquid ^{3}He atoms. The coupling energies are $g_o \sim 2 \times 10^{-6}$ eV and $J \sim 10^{-5}$ eV.[4] The Hamiltonian for the coupling to insulators is obtained by replacing the Pauli operator $\vec{S}^{(e)}_{\sigma\sigma'}\, a^*_{\vec{k}\sigma}\, a_{k'\sigma'}$ for the conduction electrons with a Pauli operator for localized spins, and by replacing the coupling g_o with a new coupling $V_o \sim 10^{-7}$ representing the magnetic dipole interaction between an electron and a ^{3}He nuclear moment separated by about 1Å.

The polarization rates for the liquid with either metallic or insulating substrates were calculated in Ref. 4, and are displayed in Fig. 1. The rate for the metallic substrates is

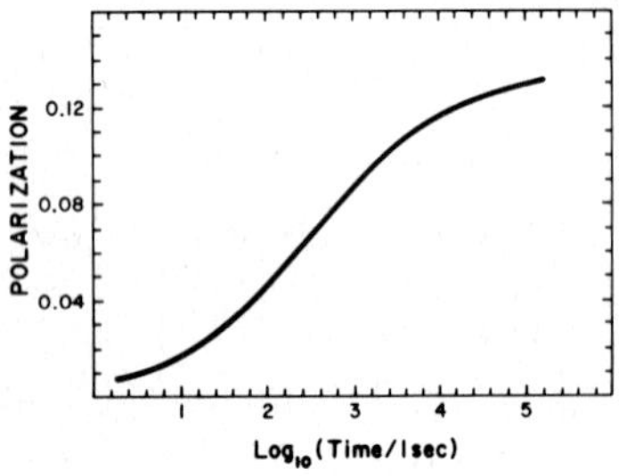

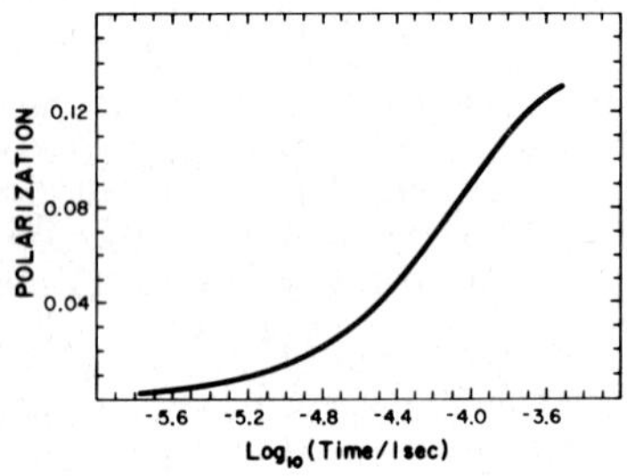

Fig.1: a) ^3He polarization as a function of time for metallic substrates at 1 mK and $\mu_B H/k_B=.1°K$. No wall relaxation terms have been included, and the surface-to-volume ratio $N_A/N_{He} = 1$ (see text). The Overhauser effect predicts a maximum polarization of 13.7 %. b) ^3He polarization as a function of time for insulating substrates at 10 mK; all other parameters as in (a).

roughly seven orders of magnitude slower than that for insulating substrates, primarily due to the fact that only a fraction T/T_F of the conduction electrons are available for spin flipping processes. However, the curves in Fig. 1 do not include the depolarizing effect of other spin relaxation processes at the container walls. Although these processes are poorly understood [8,9], this effect can be estimated by using a rough empirical formula for the wall relaxation rate $T_{wall}^{-1} \sim \frac{\eta v_F}{d}$, where d is the volume-to-surface ratio of the ^3He cell, v_F the liquid ^3He Fermi velocity, and η the probability that a ^3He spin flips in one collision with the walls. Reported values for η with various substrates include $\eta \sim 10^{-6}$ for Pt and $\eta \sim 10^{-8}$ for Al_2O_3 and grafoil.[8] These values imply depolarizing rates of $10^5 \times \frac{N_A}{N_{He}}$ sec^{-1} for Pt and $10^3 \times \frac{N_A}{N_{He}}$ for grafoil,

where N_A is the number of ^3He atoms localized on the substrate surface. Comparison with Fig. 1 indicates that although the polarization rate for metallic substrates may be too small, enhanced polarizations might be attainable with insulating substrates.

Even with insulating substrates, the enhancement will only be seen for a limited range of temperatures. Below roughly 65 mK, the effect will be suppressed because the magnetic interaction energy between adjacent electron moments (separated by $\sim 3Å$) will be greater than the thermal energy, leading to ordering. Similarly, the effect will be suppressed if the static applied field H is so large that the Zeeman energy exceeds the thermal energy, again leading to ordering. Furthermore, for fields $H \gtrsim J/\mu_B \sim$ several hundred Gauss, intermediate virtual processes involving exchange between the liquid and solid ^3He will be suppressed due to large energy denominators, slowing the rate of polarization transfer to the liquid. This restriction means that the nuclear polarization may be greatly enhanced over the equilibrium value, but it cannot approach one.

This work was supported in part by NSF Grants DMR 8020263 and DMR 8301218.

†Present address: Laboratory of Atomic and Solid State Physics, Cornell University, Ithaca, N.Y. 14893.

REFERENCES

1. B. Castaing and P. Nozières, J. Phys. 40, 257 (1979)
2. G. Frossati, J. Phys. 41, C7-95 (1980)
3. M. Chapellier, J. Phys. Lett. 43, L609 (1982)
4. S. Langer, K. DeConde, and D.L. Stein, submitted to J. Low Temp. Phys.
5. L.J. Friedman, P.J. Millet, and R.C. Richardson, Phys. Rev. Lett. 47, 1078 (1981).
6. T. Perry, K. DeConde, J.A. Sauls, and D.L. Stein, Phys. Rev. Lett. 48, 1831 (1982).
7. A. W. Overhauser, Phys. Rev. 92, 411 (1953).
8. H. Godfrin, G. Frossati, B. Hebral, and D. Thoulouze, J. Phys. 41, C7-275 (1980)
9. J.F. Kelly and R.C. Richardson, Low Temperature Physics LT-13, 167 (1972)

LT-17 (Contributed Papers)
U. Eckern, A. Schmid, W. Weber, H. Wühl (eds)

OBSERVATION OF SPIN WAVES IN GASEOUS ^{3}He↑

P.J. NACHER, G. TASTEVIN, M. LEDUC, S.B. CRAMPTON (*) and F. LALOË

Laboratoire de Spectroscopie Hertzienne de l'E.N.S. 24, Rue Lhomond 75231 Paris Cedex 05 France

The polarization of the nuclear spins introduces significant changes in the transport properties of a ^{3}He gas at low temperature. In particular, in a polarized gas of ^{3}He (^{3}He↑), "identical spin rotation effects" take place during collisions and result in an oscillatory evolution of the spins (spin waves) which changes the properties of spin diffusion in a ^{3}He↑ gas in the temperature range $2 < T < 6$K. The ^{3}He nuclear spins are polarized by optical pumping with a colour-center laser. Nuclear polarizations of the order of 50% are obtained in a relatively dense gas ($n \sim 10^{18}$cm^{-3}) at $T \lesssim 4$K. Spin waves are observed through NMR frequency shifts in magnetic field gradients. Measurements of the coefficient μ, which characterizes the quality factor of the spin oscillations, are in good agreement with numerical calculations.

1. INTRODUCTION

Spin polarized helium (^{3}He↑) and hydrogen (H↑) exhibit interesting quantum properties at low temperatures. For dilute polarized gases, the major changes introduced by the nuclear spin polarization concern the transport properties of the gas (heat conduction, viscosity, spin diffusion). These phenomena are a pure consequence of particule indistinguishability and can be calculated ab initio from interatomic potentials (1). In particular, the so called " identical spin rotation effect" during collisions between two identical particles modifies the spin diffusion and results in the existence of (damped) spin waves at low temperatures. If M is the nuclear polarization ($-1 < M < 1$), one defines a coefficient μ such as μM is the "quality factor" of the spin oscillations. According to the predictions of (1) for ^{3}He↑ , $|\mu| \lesssim 1$ for $2 < T < 6$K. Consequently the observation of spin waves in ^{3}He↑ requires relatively large polarizations, which can be obtained by optical pumping. This technique was first developped for ^{3}He 20 years ago (2). Using a laser and cryogenic coatings, we recently extended it to higher M and lower temperatures and we report here the observation of spin waves.

2. POLARIZING ^{3}He BY LASER OPTICAL PUMPING.

Nuclear orientation of ^{3}He results of a two step process : the optical pumping of the $2\ ^3S_1$ metastable level of helium at the wavelength $\lambda = 1.08\mu$ ($2^3S_1 - 2^3P$ transition), followed by metastability exchange collisions which

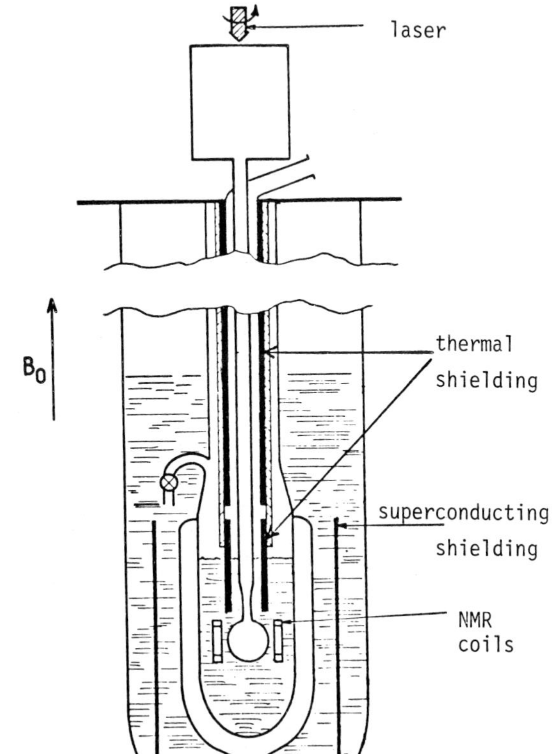

Figure 1.

strongly couple the nuclear polarizations of the 2^3S_1 and ground states. As a pumping source, we use a tunable colour center laser ($F_2{}^{+*}$ in NaF) delivering several hundred milliwatts at the appropriate wavelength. At room temperature such a laser provides M up to 0.7, for densities close to 10^{16} atoms/cm^3 (3). The method can be extended to low temperatures by the use of two cells at different temperatures, coupled by a narrow pipe where spin diffusion occurs (see fig 1.). The upper cell at 300K is pumped by the laser. The lower cell is immersed in a liquid helium bath. Hydrogen is added to ^{3}He in order to provide a coating of solid frozen hydrogen in the lower part of the cell interior, preventing wall nuclear relaxation. The whole cell is put in an homogeneous magnetic field B_0 of 4G. The polarization M of the lower cell, monitored by NMR techniques, can reach 0.5 at 4.2K for densities of the order of 2×10^{18} atoms/cm^3 ; no substantial loss of M is observed when the temperature is decreased to 1.2K.

3. OBSERVATION OF SPIN WAVES

The waves are excited in the lower cell by applying a gradient of magnetic field δB_0 .

A small transverse magnetization M_+ is produced by a NMR pulse ($M_+ << M$). If the spin polarization were uniform over the sample, no "identical spin rotation effect" would occur. The role of δB_0 is to generate a gradient of spin magnetization and to couple M_+ to an oscillating spin wave, thus causing a frequency shift $\Delta\omega$ for M_+. One obtains (1) :

$$\Delta\omega = \frac{\mu M}{T_2}$$

where T_2 is the transverse relaxation time. The order of magnitude of $\Delta\omega$ is so small ($\frac{\Delta\omega}{2\pi} \sim 1$ to 10 mHz) that a great stability of the field B_0 is required. Therefore, a vertical superconducting cylinder is used to reduce the field drifts by a factor of $\sim 10^4$ (see fig 1). Much care is taken to eliminate systematic errors in the measurements, primarily those due to radiation damping, (even below maser threshold, the reaction of the pick-up coils can easily produce frequency shifts larger than $\Delta\omega$).

The values of μ obtained are shown in fig 2, as compared to C. Lhuillier's calculations (solid curve) (1).

4. CONCLUSION

We have shown that identical spin rotation effects exist and can be measured in a spin polarized ^{3}He gas (4). Recent results of other groups show that spin waves can also be observed in spin polarized H↑ (5), as well as in solutions of ^{3}He in superfluid ^{4}He (6) (7). We are presently extending the temperature range and improving the precision of our μ measurements in gaseous ^{3}He, as well as preparing new experiments to investigate other transport properties of ^{3}He ↑, such as heat conduction.

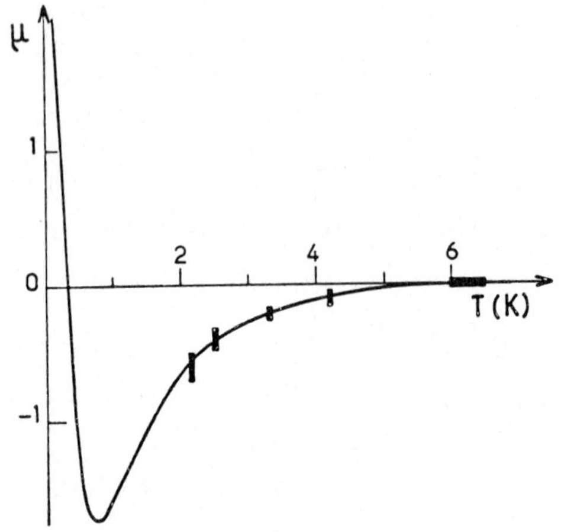

Figure 2.

(1) C. Lhuillier et F. Laloë , J. Physique 40 (1979) 239.
C. Lhuillier J. Physique 44, (1983), 1
V. Lefèvre-Seguin, P.J. Nacher et F. Laloë J. de Physique 43 (1982), 737.

(2) F.D. Colegrove, L.D. Schearer and G.K. Walters, Phys. Rev. 132, 2561 (1963)

(3) P.J. Nacher, M. Leduc, G. Trenec et F. Laloë J. de Physique Lettres 43, L525 (1982).

(4) P.J. Nacher, G. Tastevin, M. Leduc S. Crampton and F. Laloë to be published in Journal de Physique Lettres.

(5) B.R. Johnson, J.S. Denker, N. Bigelow, L.P. Lévy, J.H. Freed and D.M. Lee article submitted to Phys. Rev. Letters.

(6) J.R. Overs-Bradley , H. Chocolacs, R.M. Mueller, Ch. Buchal, M. Kubota and F. Pobell, Phys. Rev. Lett. 51, (1983) 2120.

(7) W.J. Gully and W.J. Mullin, article submitted to Phys. Rev. Letter.

LT-17 (Contributed Papers)
U. Eckern, A. Schmid, W. Weber, H. Wühl (eds)

A MEASUREMENT OF THE MELTING CURVE LOWERING FROM STRONGLY POLARIZED LIQUID ^3He

G. BONFAIT, L. PUECH, A.S. GREENBERG, G. ESKA*, B. CASTAING and D. THOULOUZE

Centre de Recherches sur les Très Basses Températures, CNRS, BP 166 X, 38042 Grenoble-Cédex, France.
* Permanent address : Walther Meissner Institute für Tieftemperaturforschung, Garching. W. Germany.

By rapid melting of a 55 % polarized solid ^3He, we obtain a strongly polarized liquid solid mixture. We have measured a lowering ΔP of the liquid solid coexistence curve as large as 1 bar at 100 mK due to this polarization. The dependence of ΔP on the global magnetization m is explained by a dendritic melting which ensures the magnetization homogeneity in each phase. ΔP has the right m^2 dependence for small m, but the slope $d\Delta P/dm^2$ is three times the expected value. This could be explained by a rotation of the magnetization on melting.

1. INTRODUCTION

It is now well established (1,2) that one can obtain highly polarized liquid ^3He or a liquid solid mixture by rapid decompression of a polarized solid. The instantaneous equilibrium is then described (1) by an effective magnetic field H_{eff}. A consequence of the increase of liquid energy when polarized is a lowering of the melting curve. We report here the first measurement of this depressed melting pressure. We have observed a lowering ΔP as large as 1 bar at 100 mK, which corresponds to H_{eff} of the order of 90 Tesla.

2. EXPERIMENTAL ARRANGEMENT

The Pomeranchuk cell (PC) is similar to the type used in previous experiments (3) and is located in the mixing chamber of a dilution refrigerator which cools to 3 mK. The ^3He volume is about 8 cm^3. It contains a small horizontal N.M.R. coil tuned to 215 MHz (6.6 T). The magnetization m is measured by sweeping the N.M.R. line every 1.6 s and integrating the C.W. signal with a computer. The observed magnetization m is thus a combination of the liquid and the solid's :

$$m = (1-x_s)m_L + x_s m_s = x_{eff} m_s$$

where x_s is the fraction of atoms in the solid phase, $m_i = M_i/M_{oi}$, M is the molar magnetization, and M_o the saturated one.

A capacitive pressure gauge is placed in the upper part of the PC, together with a heater. The principal thermometer is a Matsushita 68 Ω carbon resistor located in the lower part of the PC.

3. DECOMPRESSION

The cell is first filled with 100 % solid close to the minimum of the melting curve, at a pressure slightly above melting (30.5 bar), the ^4He pressure being 0. The precooling typically takes one day with the ^3He capillary blocked. At the lowest temperature (6 mK), the ^4He pressure is increased at a rate ∿ 1 bar/hour. We end with 100 % solid at 6 mK : a magnetization of 55 % in 6.6 T.

The decompression is obtained by opening the ^4He valve at room temperature. The subsequent evolution of the ^3He pressure versus the inverse of the resistance is sketched in fig. 1 : After a large drop (∿ 3 s) too fast for the various time constants of the measuring system, the pressure reaches a regime which can be analyzed:

AB corresponds to the decrease of the magnetization m. Its relaxation time is T_1∿12 s to be compared to T_1∿300 s when the whole solid is melted. At B the magnetization has reached its equilibrium value (∿ 100 s). All the decompressions gave the same final temperature (105±3 mK).

BC : The slow mechanical relaxation of the resistor during the cooling back of the cell (30 min.) is also seen in zero field decompressions. Whether or not we substract it from $\Delta P(t)$

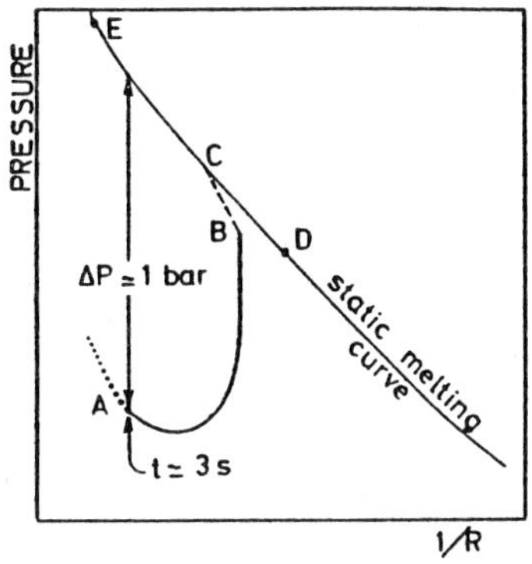

FIGURE 1
The pressure versus the inverse of the resistance during a high field decompression.

does not change significantly the results analyzed here.

CDE : After each decompression, the cell is first heated (CD) ; then the static melting curve pressure is calibrated versus the resistor R during the cooling (∿ 1 hour). The part DE reproduces to within ±10 mbar.

As it is shown, the pressure P of the polarized liquid solid mixture always lies well below the corresponding static melting pressure P_T at the same temperature. In fact the parameter of interest here is $\Delta P = P_T - P$.

The interdependence between ΔP and the magnetization is clearly shown by the reproducibility of the ΔP versus m^2 curve plotted in fig. 2. We used this plot since in a complete equilibrium situation ΔP would be proportional to m^2 in the linear regime. When a heat pulse is applied with the heater, the equivalency of P_T deduced from R, and P are restablished in less than 2 s. The value $\Delta P \cong 1$ bar is obtained 3 s after the decompression and thus is significant.

4. DISCUSSION

The fact that ΔP is related to the global measured magnetization is first puzzling, since ΔP should depend on the magnetization of both phases at the interface. The explanation is most likely the following : Before the decompression, the polarized solid (55 %) is in equilibrium with the nearly unpolarized liquid (3 %). In the early stage of the decompression, a nearly equal number of up and down spins starts to the liquid. This produces over-polarization of the solid at the interface (1) and gives rise to H_{eff} which in turn lowers the equilibrium pressure. But due to slow spin diffusion, the bulk solid remains moderately polarized, and thus stays well below its melting pressure. This breaks the solid into parts of size d ∿ 1 μ. Indeed, for such a size, the diffusion becomes very efficient in homogeneizing the solid magnetization, before the surface tension could limit the process. In these conditions the whole liquid and solid phases are in equilibrium under an effective magnetic field H_{eff} which slowly relaxes to the external field.

In fig. 2, the experimental slope is $d\Delta P/dm^2 = 6\pm1$ bar (as $x_{eff} \cong 0.95$, $m_s^2 = 1.1\ m^2$). This is three times larger than expected from the thermodynamical calculation, using the known values of the liquid and solid susceptibilities : $d\Delta P/dm_s^2 = 1.9$ bar at T = 105 mK.

This discrepancy can hardly be reduced by simple experimental arguments. It could be explained by a rotation of the magnetization in the early stage of the decompression, due either to demagnetizing fields (appearing during the dendritic melting) or to a precession around a strong local exchange field.

5. CONCLUSION

To fully exploit the results of this experiment one has to determine the exact relation between the observed m and the solid magnetization at the interface. It is the first time however, that a lowering of the melting curve as large as 1 bar has been obtained due to the polarization.

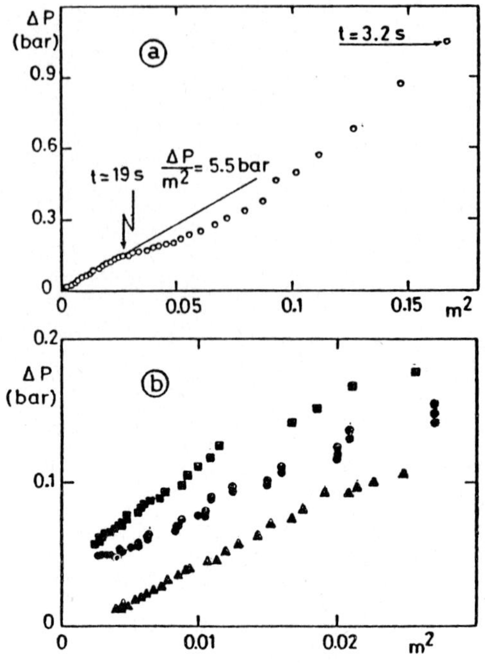

FIGURE 2
a) The whole relaxation of the magnetization as P versus m^2.
b) The end of this relaxation for three different decompressions, showing the reproducibility of the slopes. The results are shifted for clarity.

REFERENCES

(1) B. Castaing and P. Nozières, J. Physique 40 (1979) 257.

(2) G. Schumacher, D. Thoulouze, B. Castaing, Y. Chabre, P. Segransan and J. Joffrin, J. Physique-Lettres 40 (1979) L143.
M. Chapellier, G. Frossati, F.B. Rassmussen, Phys. Rev. Lett. 42 (1979) 904.

(3) H. Godfrin, G. Frossati, A.S. Greenberg, B. Hébral and D. Thoulouze, Phys. Rev. Lett. 44 (1980) 1695.

LT-17 (Contributed Papers)
U. Eckern, A. Schmid, W. Weber, H. Wühl (eds)

EXPERIMENTS ON A PARTIALLY POLARIZED DILUTE MIXTURE

W.J. GULLY, W.J. MULLIN, M. MCGURRIN, G. SCHMIEDESHOFF

Laboratory for Low Temperature Physics, Hasbrouck Laboratory, University of Massachusetts, Amherst, MA 01003*

Several Experiments on a partially polarized 370 ppm ^3He-^4He mixture have been made. Spin rotation effects were found in a spin diffusion experiment. Spin waves were sought but were not detected, and an attempt to measure the viscosity of the mixture will be described.

Measurements have been made on a dilute mixture at high fields (8.9 Tesla) and low temperatures (10 mk). Under these conditions polarizations in excess of 30% were attained.

The solute ^3He atoms in a mixture constitute a weakly interacting gas of quasiparticles. The transport properties of a dilute gas are determined by binary collisions. When the deBroglie wavelength of the quasiparticles exceeds their scattering length, particle exchange effects during collisions become important (1). Since these effects depend upon the spin orientations of the particles the transport properties become polarization dependent. This "two particle" picture holds until the degeneracy temperature is approached at which time it must be replaced by the molecular field description (2) of Fermi liquid theory.

The first experiment to be discussed will be a measurement of the spin diffusion of the mixture (3). A standard two pulse spin echo measurement was made at 290 Mhz. Several unusual effects, caused by an exchange induced rotation of colliding spins, were found in the spin echo behavior. The deflection of the spins by the rotation effect led to a reduced diffusion rate when the system was highly polarized. This can be seen in the figure, which shows several spin echo amplitudes recorded after various delay times. The diffusion constant is proportional to the slope of a line drawn through the data. Ordinarily constant, the slope of the line in this case depends on the height of the echo, which is a measure of the local polarization. Initially the polarization is high and the effective diffusion rate is low. After the signal has substantially decayed, the diffusion rate has increased to a value comparable to that expected for an unpolarized mixture (4).

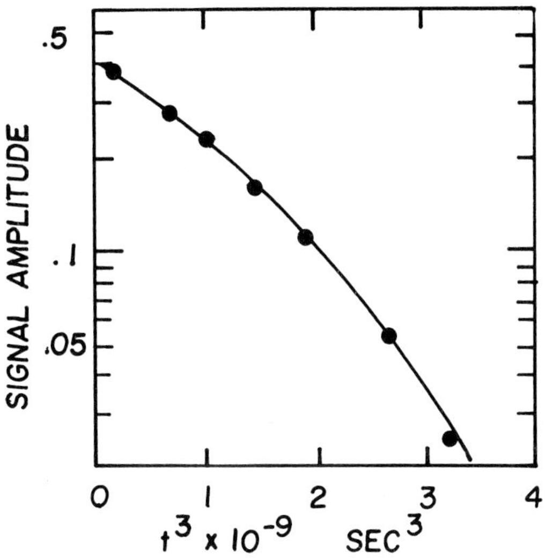

FIGURE 1
290 Mhz spin echo data taken at 32 mk with an applied field gradient of 0.26 Tesla/meter. t is the time after the 90° pulse the echo occurred.

Another unusual property observed was the rotation of the direction of the magnetization of the spin echo about the direction of the magnetic field for initial tip angles less than 90°. This is a direct manifestation of the rotation of spins during collisions. The rotation was detected by mixing the echo signal with a signal from an oscillator set to the Larmor frequency. The phase shifts grew larger the longer the system was allowed to diffuse, and were typically ~ $\pi/2$ after the echo had decreased to ~ 1/e of its initial height.

*Work supported by the National Science Foundation, Low Temperature Physics DMR-8120600,

These effects were found to be in reasonable agreement with theoretical predictions from their inception at high temperature (60 mk) down to about twice the Fermi temperature. Below 25 mk, the polarization dependence of the diffusion constant peculiarly decreased. Finally, at the Fermi temperature there was a noticeable discrepancy between the parameters inferred from the shape of the decay curve and those from the phase shift of the echo, qualitatively at odds with the theoretical framework. The latter data indicates that μM_0, the parameter introduced in reference (1) as a relative importance of the spin rotation effects in the diffusion, should be larger than the decay rate suggests.

The experiments at the lowest temperatures were difficult because of long relaxation times and the presence of a heat leak. Improved experiments are underway to determine if the deviations were experimental artifacts or were the intrinsic behavior of the half polarized, half degenerate mixture.

Experiments were carried out between 15 mk and 30 mk in search of spin waves. The free induction decays following small tipping pulses were analyzed for collective modes. In contrast to the result in spin polarized hydrogen (5), none were found. We feel that the absence can be explained by the smaller diffusion constant, a result of the lower thermal velocities at millikelvin temperatures, and the large residual magnetic field gradient present on our cell. Both these factors limit the size of a region participating in the spin wave mode. To be phase locked the rate of phase diffusion over a region of length L ($D/\mu_0 ML^2$) must be comparable to the differential precession rate (γGL). In our case the resulting modes have a physical extent an order of magnitude less than those in the H↓ experiment, resulting in an unobservably small filling factor and frequency shift, even though the relative importance of the spin rotation effects (i.e., μM_0) was comparable.

An attempt was made to measure the viscosity of the mixture in a field with a NbTi vibrating wire viscometer. However in the presence of the field the motion of our viscometer, a loop fixed at one end, became quite complicated. A large increase in frequency occurred which we attribute to the torque applied to pinned fluxlines as they rotated with the wire. This was accompanied by a degradation of the Q and loss of sensitivity of the wire.

REFERENCES

(1) C. Lhuillier and F. Laloë, J. Phys. (Paris) 43, 197, 225 (1982).
(2) A.J. Leggett, J. Phys. C12, 448 (1970); also L. Levy and A. Ruckenstein, preprint; K. Miyake and W.J. Mullin, to be published.
(3) W.J. Gully and W.J. Mullin, to be published.
(4) C. Ebner, Phys. Rev. 156, 155 (1967).
(5) B.R. Johnson, J.S. Denker, N. Bigelow, L.P. Levy, J.H. Freed, and D.M. Lee, preprint.

LT-17 (Contributed Papers)
U. Eckern, A. Schmid, W. Weber, H. Wühl (eds)

ELECTRON-PHONON COUPLING IN Nb_3Sn and V_3Si: A TIGHT-BINDING APPROACH

Leonard M. KAHN

Department of Physics, University of Rhode Island, Kingston, RI 02881

John RUVALDS

Department of Physics, University of Virginia, Charlottesville, VA 22901

The electron-phonon coupling for V_3Si and Nb_3Sn is calculated using a method based on the localized description of superconductivity proposed by Appel and Kohn. The calculation includes π-band contributions and the effects of second nearest neighbors. The coupling is weak for both materials, and the small values of $\lambda \sim 0.2$ are attributed to the narrow bandwidths. The sensitivity of the results to changes in the parameters is examined.

1. INTRODUCTION

The role of the electron-phonon (e-p) coupling in A-15 compounds is controversial, partly because the anomalies in the transport properties tend to mask the contributions associated with e-p scattering (1). Furthermore, the large variation in the e-p coupling required to account for the diverse transition temperatures in A-15 superconductors has not been explained.

Our calculations are based on the tight-binding formalism of Appel and Kohn (2) which should be ideally suited to the narrow energy bandwidths of the A-15 compounds. We utilize the computational scheme of Birnboim and Gutfreund (3) in treating the most recent APW band structure calculations for Nb_3Sn and V_3Si (4). These calculations have been fitted by a non-orthogonal tight-binding scheme (5) yielding parameters which enter our calculations of the e-p coupling.

II. TIGHT-BINDING FORMALISM

The primary goal of our study is the calculation of the e-p coupling which enters in the BCS theory of superconductivity in the form of the parameter

$$\lambda = \frac{N(o)I^2}{M\langle\omega^2\rangle} \qquad (1)$$

where N(0) is the electronic density of states at the Fermi energy, M is the ion mass, and $\langle\omega^2\rangle$ is the averaged phonon frequency. The e-p interaction is defined as an average over the Fermi surface given by

$$I^2 = \int\int d\vec{k}d\vec{k}'\,|g(\vec{k},\vec{k}')|^2 \qquad (2)$$

where

$$g(\vec{k},\vec{k}') = \langle\vec{k}|\sum_i V(\vec{r}-\vec{R}_i-\vec{u}_i)|k'\rangle \qquad (3)$$

is the matrix element of the potential associated with the deformation u_i within the modified tight-binding approximation (MTBA). The potential and the wave functions are determined by the instantaneous position of the ions and this feature is incorporated in the MTBA.

As described in Ref. (3) the essential input to the calculation of the e-p coupling is the gradient of the two-center integrals, J. To calculate these terms we use the scheme of Ashkenazi and Weger (6) to parametrize the two-center integrals analytically and fit these parameters to the values arrived at in band structure calculations. The derivatives of J are then computed analytically.

It is tempting to associate large λ values, i.e., $\lambda > 1$, with the high density of states, N(0), expected in the narrow band transition metal compounds. However, a careful investigation of the coupling in the tight-binding approximation demonstrates that the matrix element, I^2 is greatly reduced by the narrow bands and the net result is an expression for λ which is proportional to the bandwidth. This conclusion, first emphasized by Frolich and Mitra (7), suggests that the e-p coupling may be surprisingly small in the narrow band A-15 compounds. Entin-Wohlman (8,9) used a tight-binding scheme, including three-center integrals, to determine $N(0)I^2$ for V_3Si, and found that $N(0)I^2$ can be very small even for substantial values of N(0). In her calculation, Entin-Wohlman used a band structure calculation of Weger and Goldberg (10) and ignored next nearest neighbor interactions. The energy bands of Weger and Goldberg are quite different from the results of recent APW calculations and the advent of these more recent band structure results provide a primary motivation for the present study.

III. PRESENT CALCULATION

Our computations differ from previous tight-binding results for A-15 compounds in several respects. The primary distinction is our use of band structures for Nb_3Sn and V_3Si obtained from

TABLE I–CONTRIBUTIONS TO ELECTRON-PHONON COUPLING

	$I^2_{\sigma\sigma}$	$I^2_{\pi\pi}$	$I^2_{\delta_1\delta_1}$	$I^2_{\sigma\sigma 2}$	$I^2_{\pi\pi 2}$	$I^2_{\delta_1\delta_1 2}$	$N(0)I^2$	λ
V_3Si	0.001	0.011	0.002	$8.53*10^{-5}$	$1.63*10^{-3}$	$3.83*10^{-3}$	1.07	~0.14
Nb_3Sn	0.001	0.016	0.002	$1.67*10^{-4}$	$1.11*10^{-3}$	$2.51*10^{-3}$	1.03	~0.14

refined APW calculations (4) which have been previously analyzed in terms of relevant tight-binding parameters (3). Also we develop a higher order parametrization of the transfer integrals to assure a more precise fit to the bands, and we include contributions from π-symmetry orbitals.

Including the contribution from next nearest neighbors, the contribution from the two center integrals to the quantity $N(0)I^2$ is given by

$$N(0)I^2 = \frac{1}{6}N(0)\{[F_\sigma^2 I_{\sigma\sigma}^2 + F_\pi^2 I_{\pi\pi}^2 + F_{\delta_1}^2 I_{\delta_1\delta_1}^2] + [F_\sigma^2 I_{\sigma\sigma 2}^2 + F_{\delta_1}^2 I_{\delta_1\delta_1 2}^2 + F_{\delta_1} F_\sigma I_{\sigma\delta_1}]\} \quad (4)$$

Following Ref. (11) $\omega(V_3Si)$ = 285K and $\omega(Nb_3Sn)$ = 210K. The fractional densities of states at the Fermi energy, F_i, for the σ, π and δ, orbitals are given in Ref. (5) as 0.38, 0.22, and 0.16 respectively for both compounds. The values of the parameters in Eq. 4 are given in Table 1. The I^2's are given in units of $(ev/\mathring{A})^2$ and $N(0)I^2$ is in units of $(ev/\mathring{A}^2)$.

IV. DISCUSSION

Several points regarding our results deserve special attention. Comparison of our results with those of Ref. (9) indicate the importance of the band structure which is parametrized. The most significant additional contribution comes through the inclusion of the π-orbitals. This effect is in keeping with the results of Ref. (9). The importance of the additional expansion terms can be seen by noting that the root-mean square deviation of the present parametrization from the "exact" result is 4% while that for Ref. (9) is 12%. Furthermore the proper ordering of the magnitudes of the transfer integrals for the π and σ bands could not be obtained unless the additional terms were included in the present calculation.

Since this work rests substantially on the parametrization of the transfer integrals, it is important to ascertain the sensitivity of the results to changes in the fitting parameters. Changing the parameters by 20% while demanding a good fit (RMS deviation of 5.5%) caused only a 5% decrease in the value of $N(0)I^2$. Other variations in parameters yield comparable variations in results. This gives us confidence that the results at which we arrive are characteristic of the band structure rather than the choice of parameters.

Also, we have found that altering a parameter which describes the spatial extent of the atomic wavefunction, which appears in the two-center integrals, significantly affects the value of $N(0)I^2$, when all other parameters are held constant. Increasing the extent of the wavefunctions tends to broaden the energy band and it is found that the resulting value of $N(0)I^2$ increases. This is in keeping with the tight-binding arguments which suggest that narrow band materials have decreased e-p coupling.

Finally, it may be enlightening to extend the tight-binding method to calculate the e-p coupling to V_3Ge and similar A-15 compounds which bear many similarities to Nb_3Ge for example, but nevertheless have an anomalously low superconducting transition temperature.

ACKNOWLEDGEMENTS

We are indebted to S. Schnatterly for stimulating our interest in the A-15's, and we thank B. Klein, D. Papaconstantopoulos for many helpful comments. Discussions with H. Gutfreund and M. Weger have been particularly enlightening and we appreciate many useful communications from O. Entin-Wohlman.

REFERENCES

(1) I. Tutto, L.M. Kahn, and J. Ruvalds, Phys. Rev. B20 (1979) 952.
(2) J. Appel and W. Kohn, Phys. Rev. B4 (1971) 2162; B5 (1972) 1823.
(3) A. Birnboim and H. Gutfreund, Phys. Rev. B9 (1974) 139; B12 (1975) 2682.
(4) B.M. Klein, L.L. Boyer, D.A. Papaconstantopoulos, and L.F. Mattheiss, Phys. Rev. B18 (1978) 6411.
(5) L.F. Mattheiss and W. Weber, Phys. Rev. B25 (1982) 2248.
(6) J. Ashkenazi and M. Weger, J. Phys. Chem. Solids 33 (1972) 631.
(7) T.K. Mitra, J. Phys. C2 (1969) 52.
(8) O. Entin-Wohlman, Physica 104B (1981) 383; Z. Phys. B42 (1981) 119.
(9) O. Entin-Wohlman and M. Weger, Synthetic Metals 5 (1983) 217; J. Phys. F in print.
(10) M. Weger and I.B. Goldberg in Solid State Physics, Vol. 28 eds. H. Ehenreich, F. Seitz, and D. Turnbull (Academic Press, N.Y. 1973)
(11) B.P. Schweiss, B. Renker, E. Schneider, and W. Reichart in Superconductivity in d- and f-Band Metals, ed. D.H. Douglass (Plenum, N.Y. 1976) p. 189.

LT-17 (Contributed Papers)
U. Eckern, A. Schmid, W. Weber, H. Wühl (eds)

DEGRADATION OF T_c IN DISORDERED A-15 COMPOUNDS*

H. GUTFREUND, M. WEGER and O. ENTIN-WOHLMAN**

The Racah Institute of Physics, Hebrew University, Jerusalem, Israel

Recently it was proposed by Anderson, Muttalib and Ramakrishnan that the decrease of T_c with disorder in A-15 compounds can be attributed to an enhancement of the effective Coulomb repulsion, resulting from localization. We propose that the enhanced Coulomb repulsion can also be related to the quasi-one-dimensional nature of the A-15 compounds.

It has been known for some time that the transition temperature T_c of high T_c A-15 compounds like V_3Si, Nb_3Ge and similar materials, is unusually sensitive to disorder and that the depression of T_c can be correlated in a "universal" manner with the resisitivity, regardless of what caused the disorder (1). In a recent publication Anderson, Muttalib and Ramakrishnan (2) (to be referred as AMR) have attributed this depression to an increase in the Coulomb pseudopotential μ^* in a strongly disordered system. Their argument concerns the scale size dependence of the diffusion associated with the onset of Anderson localization. The charge density fluctuations diffuse more slowly over a scale size which increases with disorder and therefore interact more strongly, leading to an increase in μ^*. The starting point of AMR is the expression for the Coulomb kernel in the strong-disordered regime

$$K^c(\omega) = \mu + 2V_x \int dt\ P(t) \cos \omega t, \qquad (1)$$

where V_x is the local screened Coulomb potential, μ is the interaction parameter, $\mu = N(\varepsilon_F)V_x$ where $N(\varepsilon_F)$ is the density of states. In (1), $P(t)$ is the diffusional autocorrelation function between charge fluctuations, that is, it is the probability that an electron has not diffused away in a time t. It changes from the classical $t^{-3/2}$ behavior to a slower t^{-1} decrease as one crosses to the localization region where the diffusion is nonclassical. The reduced diffusion over a length scale which increases with increasing disorder results in a retarded and enhanced Coulomb repulsion. However, AMR find that the critical resistivity at which localization occurs in a free electron gas is much higher (by a factor of 25) than the resistivities required to interpret the experimental T_c vs. ρ dependence.

In the present note we wish to point out a different possible cause for a reduced decay of the charge density fluctuations. We propose that the linear chain structure of the transition metal atoms in A-15 compounds (3) implies a one-dimensional character of diffusion along the chains. This leads to an increase in μ^* which is correlated with the right order of magnitude of the resistivity. We start from eq. (1) with an autocorrelation function P(t) appropriate for coupled linear chains. For a pure 1-d system P(t) is given by $(4\pi Dt)^{-1/2}$, where D is the diffusion constant along the chains. In real systems the 1-d divergence of Fourrier transform is suppressed by the escape of electrons from one chain to another in a characteristic time $\tau_\perp$ (4). We take the limit $(\omega\tau_\perp)^2 \ll 1$ (in V_3Si $\omega_D\tau_\perp < 0.3$ (3,5)) and obtain from eq. (1)

$$K^c = \mu + \frac{V_s}{\hbar v_F S}\sqrt{\frac{\tau_\perp}{\tau_\parallel}}, \qquad (2)$$

Here we have used the free electron diffusion constant, $D = v_F^2\tau_\parallel$, where $\tau_\parallel$ is the back scattering time along the chains, and S is the cross section of the unit cell perpendicular to the chains.

Finally, we point out that $\tau_\perp$, the tunneling time between chains, is generally related to $\tau_\parallel$. In a coupled chain system (6)

$$\frac{\hbar}{\tau_\perp} = \frac{t_\perp^2\tau_\parallel}{\hbar}, \qquad (3)$$

where $t_\perp$ is the transfer integral between chains. This expression for the transition probability between chains is the analog of the "golden rule". It shows that a decrease in $\tau_\parallel$ enhances the lifetime on a chain–intrachain scattering interferes with interchain tunneling. In A-15 compounds the

---

* This research was supported by a grant from the National Council for Research and Development, Israel and the K.f.K., Karlsruhe, Germany

** Permanent address: Department of Physics, Tel-Aviv University, Israel

situation is more complicated. Here the transfer integrals $t_\perp$ between individual atoms on two perpendicular chains are rather large, but the one-dimensional character follows in part from a cancellation effect due to destructive interference. For example, between the overlap of the lobes of an atomic wavefunction on one chain with the lobes of wavefunctions on two atoms on a perpendicular chain (3) or between π-π and π-σ scattering (7). This results in a small effective $t_\perp$. We expect eq. (5) to be valid as long as $(k_F\ell)>2$.

Using this expression in eq. (2) we find

$$\mu' = \mu + \Delta\mu = \mu + \alpha \frac{\hbar}{t_\perp \tau_\parallel}, \qquad (4)$$

where $\alpha = V_x/hv_FS$. Assuming that $\rho = \tau_\parallel^{-1}$, we can correlate the effect of disorder on T_c, through the effect of ρ on $\Delta\mu$ in eq. (4). We take for V_3Si: $\lambda = 1.2$, $\omega_D = 330K$, $\mu = 0.3$ and $\ln(\varepsilon_F/\omega_D) \cong 2$, and find that when T_c changes from 14K to 5K, $\Delta\mu$ has to increase by about a factor of 10. In this range the resistivity changes by a factor of 5. The details of the numerical estimates are described in ref. (8). It is also found there that the order of magnitude of $\Delta\mu$ is in accord with reasonable values of $\tau_\parallel$ and $t_\perp$.

The linear chain structure of the A-15 compounds and its implications for the energy band structure and the shape of the Fermi surface are described extensively in refs. (3,5,9). Experimentally the anisotropic nature manifests itself in a number of ways: a) A very anisotropic depression of T_c due to uniaxial stress (10); b) Anisotropic tunneling into a single crystal of Nb_3Sn (11); c) Positron annihilation experiments (9,12); d) One-dimensional fluctuations above T_c in Nb_3Al (13).

We agree with AMR that it is unrealistic to explain the strong depression of T_c with ρ by the smearing of the peaks in the density of states with increasing disorder. Our conclusion is that the strong depression of T_c can be related to the effect of disorder on μ^*, and that the increase in μ^* implied by the quasi-one-dimensional nature of the A-15 compounds is in agreement with the T_c vs. ρ data. We should point out, however, that a recent preprint (14) reports a measurement indicating that μ^* is hardly effected by disorder.

REFERENCES

1. L.R. Testardi, J.M. Poate and H.J.L. Levinstein, Phys. Rev. B15, 2570 (1977); A.K. Ghosh, H. Weissman, M. Gurevitch, H. Lutz, O.F. Kammerer, C.L. Shead, A. Goland and M. Strongin, J. Nucl. Mater. 72, 70 (1978); A.R. Sweedler, D.E. Cox and S. Moehlecke, J. Nucl. Mater. 72, 50 (1978).

2. P.W. Anderson, K.A. Muttalib and T.V. Ramakrishnan, Phys. Rev. B28, 117 (1983).

3. M. Weger and I.B. Goldsberg, Solid State Physics 28, 1 (1973) (Turnbull, Seitz and Ehrenreich, Eds., Academic Press, N.Y.); M. Weger, J. Less Common Metals 62, 39 (1978).

4. G. Soda, D. Jerome, M. Weger, J.M. Fabre and L. Giral, Solid State Comm. 18, 1417 (1976).

5. A.T. Van Kessel, H.W. Myron and F.M. Mueller, Phys. Rev. Lett. 41, 181 (1978); F.Y. Fradin and D. Zamir, Phys. Rev. B7, 4861 (1978); O. Entin-Wohlman and M. Weger, Synthetic Metals 5, 217 (1983).

6. M. Weger, J. Physique 39, C6-1456 (1978).

7. L.F. Mattheiss and W. Weber, Phys. Rev. B25, 2248 (1982).

8. H. Gutfreund, M. Weger and O. Entin-Wohlman, in print.

9. T. Jarlborg, A.A. Manual and M. Peter, Phys. Rev. B27, 4210 (1983).

10. M. Weger, B.G. Silbernagel and E.S. Greiner, Phys. Rev. Lett. 13, 521 (1964).

11. V. Hoffstein and R.W. Cohen, Phys. Lett. A29, 603 (1969); M. Weger, Solid State Comm. 9, 107 (1971).

12. S. Samuelov, M. Weger, I. Nowik, I.B. Goldberg and J. Ashenazi, J. Phys. F11, 1281 (1981); S. Berko and M. Weger, Phys. Rev. Lett. 24, 55 (1970).

13. T. Maniv, Solid State Comm. 26, 115 (1978).

14. J. Geerk, H. Rietschel and V. Schneider, preprint.

LT-17 (Contributed Papers)
U. Eckern, A. Schmid, W. Weber, H. Wühl (eds)

TUNNEL SPECTROSCOPY ON SUPERCONDUCTING Nb_3Sn WITH ARTIFICIAL TUNNEL BARRIERS

U. SCHNEIDER, J. GEERK and H. RIETSCHEL

Kernforschungszentrum Karlsruhe, Institut f. Nukleare Festkörperphysik, P.O.B.3640,Karlsruhe,FRG

Tunnel junctions on $Nb_{1-x}Sn_x$ (x=0.18 ... 0.25) were prepared using artificial tunnel barriers which showed no proximity effects. Second derivative measurements revealed a fine structure in $\alpha^2F(\omega)$ at an energy of 5-7 meV. Whereas the tunnel spectra of Sn deficient A15 Nb-Sn could be analyzed by the standard McMillan-Rowell procedure, the data of stoichiometric samples exhibited characteristic deformations which are referred to the influence of an electronic density of states which is strongly structured in the range of phonon energies near the Fermi level.

1. INTRODUCTION

High-T_c superconductivity in A15 compounds is conventionally associated with a sharp peak in the electronic density of states (EDOS). If a sharply structured EDOS exists near ε_F, the standard Eliashberg equations have to be modified. Regarding the evaluation of tunneling data, it has recently been pointed out (1,2) that these modifications may strongly effect the extracted Eliashberg function $\alpha^2F(\omega)$.

Several tunneling experiments on Nb_3Sn have already been published (3-5). But in all quoted results, in the reduced density of states (RDOS) deformations could be observed which were caused by the occurrence of proximity effects in the junctions. For a proper analysis of such data, Arnold and Wolf (6) modified the McMillan-Rowell inversion procedure to allow for proximity effects.

Stimulated by the progress in tunneling into Nb and Nb_3Ge (7,8), we prepared Nb_3Sn junctions with aluminum based artificial barriers which were virtually free of proximity effects.

2. EXPERIMENTAL

The Nb_3Sn films were prepared by magnetron sputtering on hot (1000°C) sapphire substrates. Thin Al or AlZr overlayers were sputtered in situ on top of the freshly deposited A15 film at a substrate temperature of 200°C. Oxidation was performed at RT and in ambient air for 10 min and gave rise to reasonable junction resistances of 100 Ω- 1kΩ . In most cases Pb was chosen as a counterelectrode. First and second derivative measurements were taken on all junctions and numerically combined to yield a smooth RDOS . Due to the restricted space the reader is referred to a forthcoming publication of the authors on a detailed description of sample preparation and data evaluation.

3. RESULTS AND DISCUSSION

Compared to overlayers of pure Al we found that AlZr mixtures as overlayer on top of the Nb_3Sn film resulted in a strong improvement of the junctions concerning proximity effects. Junctions prepared with this technique had highly reproducible characteristics such as tunneling resistance and extremely low subgap conductivities.

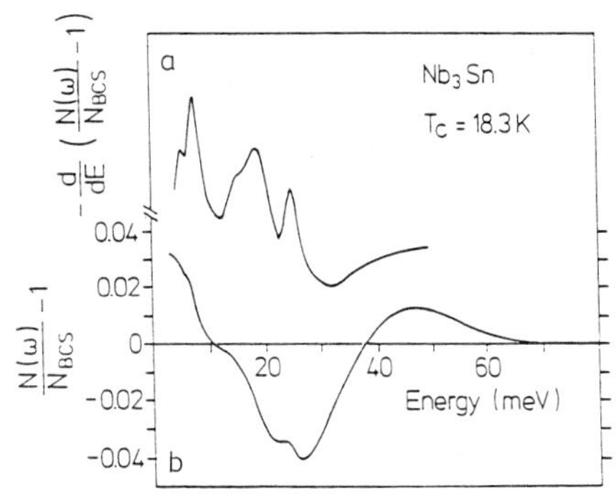

FIGURE 1
(a) Second and (b) first derivative data

Fig. 1a shows the derivative of the reduced density of states (DRDS) as a function of energy. The DRDS shows peaks where the Eliashberg function $\alpha^2F(\omega)$ shows peaks or shoulders. Besides the peaks at 8, 20 and 26 meV which are already known as peaks in $\alpha^2F(\omega)$ we observe a peak at about 5 meV. This peak shifts to about 7 meV for Sn-deficient A15 NbSn with a T_c of 5 K. Phonon dispersion calculations of Weber (9) predict an anomaly of the lowest LA branch of the λ_1 phonons which should give rise to a structure in $\alpha^2F(\omega)$ at about 6 meV.

Whereas the DRDS traces of a junction are nearly independent of the preparation technique, the shape of the RDOS is highly sensitive to proximity effects. The strength of the proximity effect and hence the progress of our AlZr overlayers on Nb_3Sn can be exhibited by a quite simple argument: Because of particle number conservation a RDOS(N_{red}) obtained from a proximity free junction must follow the relation

$$\int_{\Delta_o}^{\omega_c} N_{red} \cdot N_{BCS} \, d\omega = 0.$$

ω_c is a cut-off frequency far above the phonon energies of Nb_3Sn (80 meV), Δ_o is the energy gap. We found that the quantity P defined as

$$P = \int_{\Delta_o}^{\omega_c} N_{red} \cdot N_{BCS} d\omega / \int_{\Delta_o}^{\omega_c} |N_{red}| \cdot N_{BCS} \, d\omega$$

is convenient to compare the proximity qualities of junctions with different strengths of the RDOS. Proximity junctions always yield P<0, for example P=-0.5 for Nb_3Sn junctions as published so far in literature. Due to the AlZr overlay technique we were able to produce junctions with $P=0\pm0.04$. In the following, only data for such junctions will be discussed.

We found that the RDOS for Sn-deficient samples could be analyzed by the standard McMillan-Rowell (MMR) inversion as has been known since now only for simple nontransition metals like Pb or Sn or Nb junctions with Al overlayers.

For stoichiometric samples, however, no agreement between experimental and MMR generated spectra could be achieved. The reason for that is that the experimental curve shows a pronounced overswing in the energy range between 25 and 70 meV (Fig. 1b). Calculations of the RDOS along the lines of Ref. (1) demonstrate that the influence of a sharply peaked EDOS near the Fermi level results in a very similar overswing of N_{red} at energies above 25 meV.

Since presently no algorithm is available which allows the inversion of Eliashberg equations as used in Refs. (1,2), we roughly analyzed our data by calculating the RDOS using the computer program of Ref. (1) for a parametrized $\alpha^2F(\omega)$.

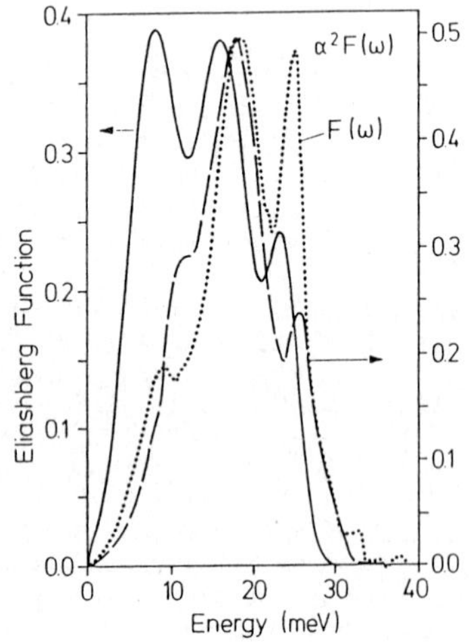

FIGURE 2
Eliashberg functions of a 6 K sample (--) and an 18 K sample (——) and F(ω) (...) .

Fig. 2 compares the Eliashberg function of a $Nb_{.81}Sn_{.19}$(T_c=6 K) sample with the result of a stoichiometric sample. Included in the figure is the phonon density of states of stoichiometric Nb_3Sn (10). We observe very strong electron-phonon coupling near the low energy peak at 8 meV. The coupling strength is somewhat reduced for the Sn-deficient material.

Concerning a proximity analysis following Arnold and Wolf, we found that the overswing of the RDOS as calculated in Ref. (1) can also be fitted by the second proximity parameter of Arnold and Wolf describing Andreev oscillations. Thus, as two kinds of deformations are present in the RDOS of a stoichiometric Nb_3Sn proximity junction, a simple proximity analysis of such a junction will yield incorrect results. For example, the shape of the Eliashberg function is strongly dependent on the strength of the proximity effect. Such a behaviour has already been observed by Geerk et al. on Nb_3Ge junctions (8).

REFERENCES

(1) G. Kieselmann and H. Rietschel, J. Low Temp. Phys. 46 (1982) 27.
(2) J.P. Carbotte, in: Superconductivity in d- and f-Band Metals, eds. W. Buckel and W. Weber (Kernforschungszentrum Karlsruhe 1982) 487.
(3) L.Y.L. Shen, Phys. Rev. Lett. 29 (1972)1082.
(4) E.L. Wolf, J. Zasadzinski, G.B. Arnold, D.F. Moore, J.M. Rowell and M.R. Beasley, Phys. Rev. B 22 (1980) 1214.
(5) D.A. Rudman and M.R. Beasley, Bull. Am. Phys. Soc. 26 (1981) 211.
(6) E.L. Wolf, J. Zasadzinski, J.W. Osmun, and G.B. Arnold, Solid State Commun. 31 (1979) 321 and references therein.
(7) J. Geerk, M. Gurvitch, D.B. McWhan, J.M. Rowell, Physica C 109 (1982) 1775.
(8) J. Geerk, J.M. Rowell, P.H. Schmidt, F. Wüchner, W. Schauer, in: Superconductivity in d- and f-Band Metals, eds. W. Buckel and W. Weber (Kernforschungszentrum Karlsruhe 1982) 23.
(9) W. Weber, in: Superconductivity in d- and f-Band Metals, eds. W. Buckel and W. Weber (Kernforschungszentrum Karlsruhe 1982) 15.
(10) B.P. Schweiß, B. Renker, E. Schneider, W. Reichardt, in: Superconductivity in d- and f-Band Metals, ed. D.H. Douglass (Plenum Press, New York 1976) 189.

LT-17 (Contributed Papers)
U. Eckern, A. Schmid, W. Weber, H. Wühl (eds)

SUPERCONDUCTIVITY IN HYDROGENATED AND DISORDERED A15 Nb_3Ge

C. NÖLSCHER*, H. ADRIAN*, R. MÜLLER*, W. SCHAUER+, F. WÜCHNER+, and G. SAEMANN-ISCHENKO*

*Physikalisches Institut, Universität Erlangen-Nürnberg, Erwin-Rommel-Str.1, D-8520 Erlangen, FRG
+Kernforschungszentrum Karlsruhe, ITP, P.O. Box 3640, D-7500 Karlsruhe, FRG

Superconductive T_c's of Nb_3Ge layers were decreased by hydrogenation or by ion-irradiation. The correlations of T_c vs. resistivity ρ and T_c vs. $dH_{c2}/dT|_{T=T_c}$ are distinctly different for hydrogenated and irradiated samples, but the derived correlations of T_c vs. the coefficient of the electronic specific heat γ are very similar. An analysis showed that at T_c=5K the density of states at the Fermi level $N(E_F)$ has decreased in both cases to about 50%, independent from a reasonably assumed change in the lattice stiffness or the matrix element of electron-phonon coupling $\langle J^2\rangle$. Contrary to Tsuei the data suggest that $\langle J^2\rangle$ is less important than $N(E_F)$ for high T_c in Nb_3Ge.

In Nb_3Ge the high T_c has its origin in a high electron-phonon coupling parameter λ as has been derived from tunneling measurements (1). According to McMillan λ is a product of $N(E_F)$, $\langle J^2\rangle$ and the inverse square of a mean phonon frequency $\langle w^2\rangle^{-1}$ (2). In order to get information about the importance of each parameter we varied T_c by two different means: hydrogenation and irradiation.

The Nb_3Ge layers have been prepared by coevaporation on polished hot sapphire substrates (3) to thicknesses of 300-430nm with resistance ratios of 2.0-2.7, inductively measured midpoint T_c's of 19-21K and Nb/Ge ratios of 2.7±0.3. The samples contained about 10±5at% Nb_5Ge_3. 17 layers were irradiated with 20 MeV sulphur ions. 10 layers were hydrogenated in a H_2 glow discharge at 1kV, 0.1 Torr and were subsequently annealed for at least 10 hours at 100°C in high vacuum. This resulted in rather homogeneous samples as determined by ERDA (4) and X-ray diffraction. Annealing at 500°C fully recovered T_c and ρ.

Fig.1 and 2 show the correlations of T_c vs. ρ(24K) and $B'_{c2}:=dB_{c2}/dT|_{T\approx T_c}$. The irradiation data agree with the "universal" T_c-ρ behavior but the hydrogen data similar to Nb_3Sn (5) appreciably deviate. Despite the large differences in T_c vs. ρ and T_c vs. B'_{c2}, the T_c vs. γ correlations are very similar. γ (in mJ $mole^{-1}$ K^{-2}) was deduced from B'_{c2} (in T/K), ρ (in $\mu\Omega$cm) and T_c (in K) using the coherence length formula (6):

$$\eta\gamma = 884\ B'_{c2}/(\rho + 8.78\cdot10^{-3}\gamma T_c(1+\lambda))$$

For the strong coupling correction $\eta(T_c/\langle w\rangle)$ we used a function of Ref. (7) yielding η=1.34 at T_c=21.5K. To take into account the clean limit correction we assumed $\rho\cdot l$=const. The mean free path in the saturation regime (ρ=140$\mu\Omega$cm) was adjusted to l=0.43nm in order to get a γ of 32.4mJ $mole^{-1}$ K^{-2} at T_c=21.5K which is an

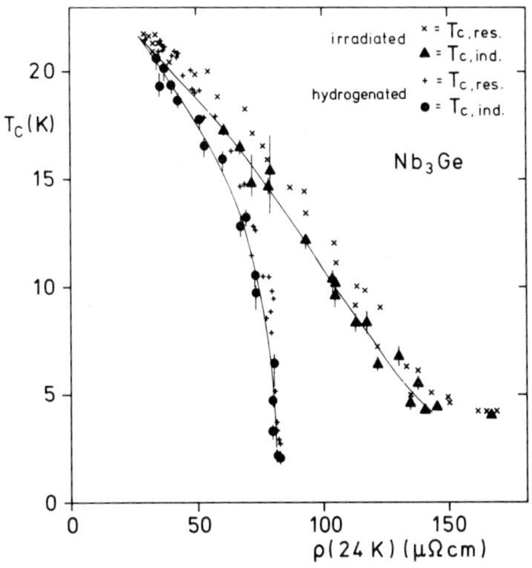

FIGURE 1
Inductive transition (10%,50%,90%) vs. ρ(24K).

average from specific heat measurements (2 and references therein). λ was estimated from the T_c formula of Allen and Dynes. For undamaged Nb_3Ge the parameters were adjusted to the results of tunneling measurements (1) to yield a T_c of 21.5K(λ=1.9, μ^*=0.07, w_{log}=125.5K, $\langle w^2\rangle$=170K). To analyse the data for the damaged samples we assumed μ^*=const. and $w^2_{log}/\langle w^2\rangle$= const. Together with $\gamma \sim N(E_F)(1+\lambda)$ one has 2 equations with the experimental parameters T_c, B'_{c2}, and ρ. If assumption is made about one of the three free parameters $N(E_F)$, $\langle w^2\rangle$ or $\langle J^2\rangle$ the two others can be calculated as function of T_c. From fig.3 we see that the assumption $N(E_F)$= const. is reasonable only for $T_c \gtrsim$17K, below that $\langle w^2\rangle$ and $\langle J^2\rangle$ would diverge. The assumption $\langle w^2\rangle$=const. leads to a decrease of $\langle J^2\rangle$

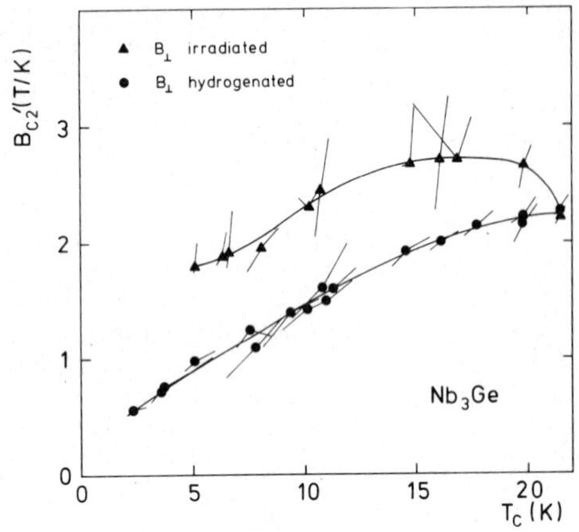

FIGURE2

B'_{c2} of the 10%, 50% and 90% points of the resistively measured superconductive transitions.

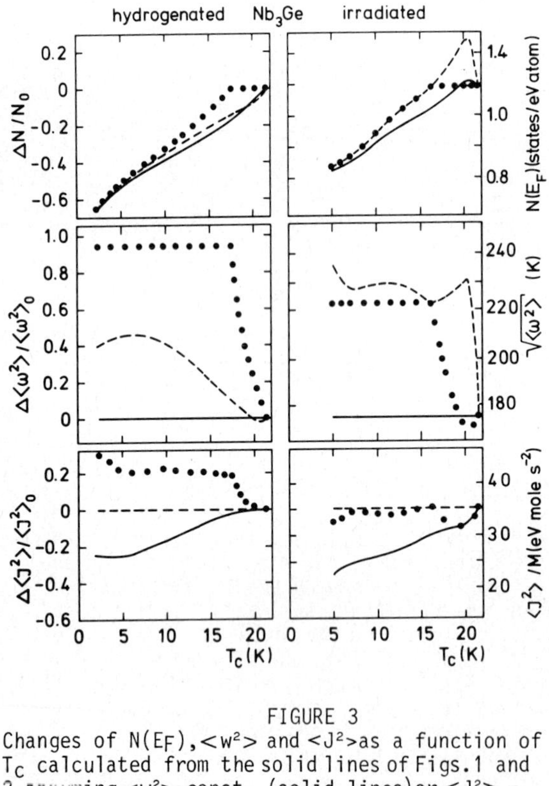

FIGURE 3

Changes of $N(E_F)$, $<w^2>$ and $<J^2>$ as a function of T_c calculated from the solid lines of Figs.1 and 2 assuming $<w^2>$-const. (solid lines) or $<J^2>$ = const. (dashed lines). For the dots we assumed $N(E_F)$=const. down to 16 resp. 17.5K, then $<w^2>$ was held constant.

which is about the same as deduced from the theory of Appel (8) for irradiated Nb_3Ge. An even stronger decrease of $<J^2>$, as demanded by Tsuei (9), inevitably leads to a phonon softening. From tunneling measurements instead we would expect an increase of w_{log}^2 of about 65% (1). This is in agreement with our measurements assuming either $<J^2>$=const. or $N(E_F)$=const. The discrepancy to Tsuei may come from the fact that our samples are cristalline and not amorphous as those of him. The 50% decrease of $N(E_F)$ at 5K is independent from all 3 assumptions made (fig.3) and is, even absolutely, in agreement with broadening of calculated band structures (10).

ACKNOWLEDGEMENTS

We like to acknowledge the help of our colleagues during the irradiation experiments and financial support by the BMFT.

REFERENCES

(1) K.E. Kihlstrom, D. Mael, T.H. Geballe, Phys.Rev. B29 (1984) 150
(2) P.B. Allen, R.C. Dynes, Phys.Rev. B12 (1975) 905
(3) B. Krevet, W. Schauer, F. Wüchner, K. Schulze, Appl. Phys. Lett. 36 (1980) 704
(4) C. Nölscher, K. Brenner, R. Knauf, W. Schmidt, Nucl.Instr.Meth. 218 (1983) 116
(5) C. Nölscher, P. Müller, H. Adrian, M. Lehmann, G. Saemann-Ischenko, Z.Physik B41 (1980) 291
(6) H. Wiesmann, M. Gurvitch, A.K. Ghosh, H. Lutz, O.F. Kammerer, M. Strongin, Phys.Rev. B17 (1978) 122
(7) D. Rainer, G. Bergmann, J. Low Temp.Phys.14 (1974) 501
(8) J. Appel, Phys.Rev. B13 (1976) 3203
(9) C.C. Tsuei, A-15 compounds and their amorphous counterparts, in: Superconductivity in d- and f-Band Metals, eds. H. Suhl, M.B. Maple (Academic Press, New York, 1980) pp. 233-246
(10) L.F. Mattheiss, L.R. Testardi, Phys.Rev.B20, (1979) 2196

LT-17 (Contributed Papers)
U. Eckern, A. Schmid, W. Weber, H. Wühl (eds)

ELECTRON LOCALIZATION AND INTERACTION EFFECTS IN ALUMINUM FILMS AT TEMPERATURES JUST ABOVE THE SUPERCONDUCTING TRANSITION*

James M. GORDON, C. J. LOBB, and M. TINKHAM

Department of Physics, Harvard University, Cambridge, MA 02138 USA

Magnetoconductance measurements on thin Al films at temperatures down to $1.02T_c$ are analyzed in terms of the localization and electron interactions theories. The divergent electron-electron attraction strength, β, and a new inelastic scattering rate due to the presence of superconducting fluctuations, $1/\tau_n$, are found to be in good agreement with theoretical predictions.

1. INTRODUCTION

In the past few years, magnetoconductance (MC) measurements have been used to answer fundamental questions concerning the origin of the inelastic scattering rate in thin films (1,2,3). More recently, the connection between $1/\tau_i$ and the superconducting pair-breaking strength, δ, has been used to explore the effects of disorder on the superconducting transition (3,4).

In this paper we present results of an extension of earlier MC measurements on Al films (2) to temperatures very close to T_c. We have found evidence for a new inelastic scattering rate, due to the presence of Cooper-pairs, which appears to diverge as T approaches T_c from above.

2. THEORY

The low-field magnetoconductance in our films is dominated by contributions from localization effects, as well as those due to Maki-Thompson (MT) and Aslamazov-Larkin (AL) superconducting fluctuations. The localization contribution (5) is given by

$$\frac{\Delta G}{G_{00}}\Big|_{loc} \equiv \frac{d[\sigma'(B,T)-\sigma'(0,T)]}{e^2/2\pi^2\hbar}$$

$$= [\tfrac{3}{2}\Psi(\tfrac{1}{2}+\tfrac{B_2}{B})-\tfrac{1}{2}\Psi(\tfrac{1}{2}+\tfrac{B_i}{B})-\tfrac{3}{2}\ln\tfrac{B_2}{B} + \tfrac{1}{2}\ln\tfrac{B_i}{B}] \qquad (1)$$

where d is the film thickness and $B_2=B_i+4B_{so}/3$. The characteristic fields are defined by the relation $B_x = \phi_0/4\pi D\tau_x$, where x may denote elastic (o), inelastic (i) or spin-orbit (so) scattering. Ψ is the digamma function. The magnetoconductance due to the MT fluctuation diagram has been considered by Larkin (6). It is

$$\frac{\Delta G}{G_{00}}\Big|_{MT} = -\beta(\tfrac{T}{T_c})[\ln\tfrac{B}{B_i}+\Psi(\tfrac{1}{2}+\tfrac{B_i}{B})] \qquad (2)$$

where $\beta(T/T_c)$ represents the electron-electron attraction strength and diverges as $1/\ln(T/T_c)$ as T approaches T_c. Finally, the AL term, which is significant only at extremely small values of the reduced temperature $\varepsilon=(T-T_c)/T_c$, is given by (3,7)

$$\frac{\Delta G}{G_{00}}\Big|_{AL} = -\frac{\pi^2}{16\varepsilon^3}(\frac{B}{\tilde{H}_{c2}(0)})^2 \qquad (3)$$

where we have introduced the characteristic field $\tilde{H}_{c2}(0)=\phi_0/2\pi\,\xi^2(0)$ which depends upon the coherence length $\xi^2(0)=\pi\hbar D/8k_BT_c$.

3. EXPERIMENTAL RESULTS AND DISCUSSION

Samples studied were 70-200 Å thick strips (45mmX.088mm) made by evaporating 99.999% Al onto room temperature substrates, patterned for lift-off lithography. Perpendicular field MC measurements were made with a low frequency, three-terminal bridge, employing a PAR 124 lock-in amplifier, which offered a resistance resolution of better than 1 part in 10^6.

At each temperature we fit the MC curves to theory by varying $\beta(t)$, $B_i(T)$ and B_{so}. Figure 1 shows experimentally determined values of $\beta(t)$ for three films at temperatures $T<1.75T_c$. The good fit of our experimental points to Larkin's theoretical curve at these temperatures, where the MT term [Eq. (2)] is generally much larger than either the AL or localization terms, is a convincing argument for the applicability of Eq. (2) to MC data at temperatures near T_c.

Using diffusion constants extracted from critical field data and the characteristic fields $B_i(T)$, from MC measurements, we have determined the temperature dependence of the inelastic rate. This is shown in Fig. 2 for two of our samples. For temperatures sufficiently far above T_c ($T\gtrsim 2T_c$), $1/\tau_i$ is well

*Research supported in part by NSF grants DMR-79-04155 and DMR-80-20247

described by a sum of terms representing ee and ep scattering. A best fit of our $T>2T_c$ data to such a model, described in detail elsewhere (2,3), is shown as a solid line in Fig. 2. It is clear that as T approaches T_c from above a new inelastic scattering mechanism adds a rapidly growing term, which we refer to as $1/\tau_n$, to the ee and ep rates.

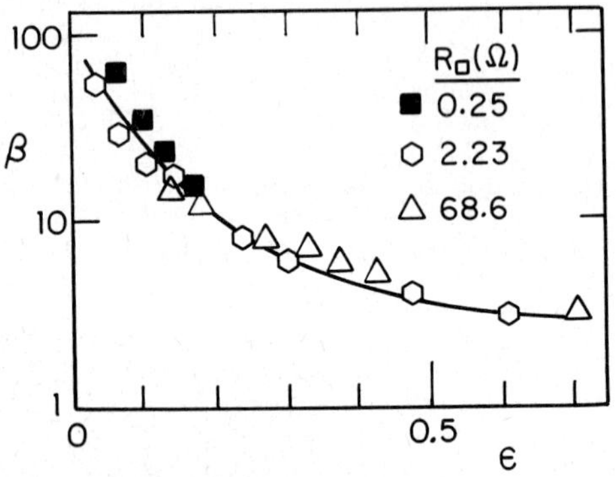

FIG. 1: Electron-electron attraction strength parameter β versus reduced temperature $\varepsilon = (T-T_c)/T_c$.

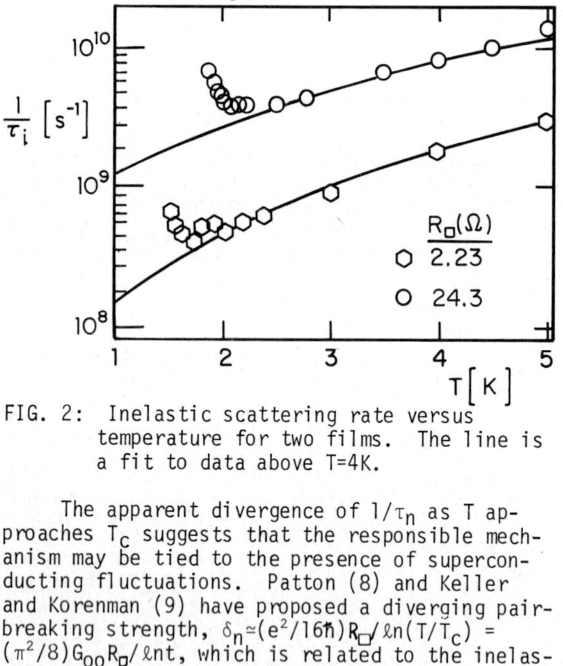

FIG. 2: Inelastic scattering rate versus temperature for two films. The line is a fit to data above T=4K.

The apparent divergence of $1/\tau_n$ as T approaches T_c suggests that the responsible mechanism may be tied to the presence of superconducting fluctuations. Patton (8) and Keller and Korenman (9) have proposed a diverging pair-breaking strength, $\delta_n \simeq (e^2/16\hbar)R_\square/\ell n(T/T_c) = (\pi^2/8)G_{00}R_\square/\ell nt$, which is related to the inelastic scattering rate through the relation $\delta_n = \pi\hbar/8k_BT\tau_n$ (3,10). This provides us with an expression for the electron inelastic scattering rate, due to the presence of Cooper-pairs,

$$1/\tau_n = A(TR_\square/\ell nt) \qquad (4)$$

with $A \simeq (\pi k_B/\hbar)G_{00} = 5.1\times10^6$ $(K\text{-}s\text{-}\Omega)^{-1}$.

In Fig. 3 we have plotted the product $(1/T\tau_n)(\ell nt/R_\square)$ versus ε, with $1/\tau_n$ defined as the deviation of experimentally determined values of $1/\tau_i$ (points in Fig. 2) from the high temperature fit (solid line in Fig. 2). Equation (4) predicts that all of the points in Fig. 3 should fall on a universal value A, approximately equal to $(\pi k_B/\hbar)G_{00}$, on the vertical axis. Indeed, for $\varepsilon<.2$ data from all of the samples fall at about $A=2.5\times10^6$. Further above T_c, $1/\tau_n$ falls off more rapidly than in Equation (4), as might be expected, due to the instability of superconducting fluctuations far above T_c(8).

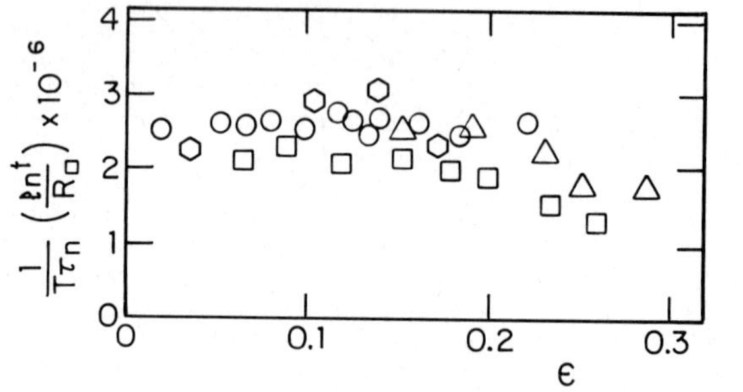

FIG. 3: Test of predicted temperature dependence of the divergent inelastic rate, $1/\tau_n$, for four films with $R_\square$=2.23Ω, 24.3Ω, 38.3Ω, 68.6Ω.

REFERENCES

(1) For example, G. Bergmann, Z. Phys. B48, 5 (1982) and references within.
(2) J. M. Gordon, C. J. Lobb and M. Tinkham, Phys. Rev. B28, 4046 (1983) and references therein.
(3) J. M. Gordon, C. J. Lobb and M. Tinkham, to appear in Phys. Rev. B.
(4) A. F. Hebard and M. A. Paalanen, preprint.
(5) S. Maekawa and H. Fukuyama, J. Phys. Soc. Japan 50, 2516 (1981).
(6) A. I. Larkin, Pis'ma Zh. Eksp. Teor. Fiz. 31, 329 (1980) [JETP Lett. 31, 219 (1980)].
(7) E. Abrahams, R. E. Prange and M. J. Stephen, Physica 55, 230 (1971).
(8) B. R. Patton, Phys. Rev. Letters 27, 1273 (1971).
(9) J. Keller and V. Korenman, Phys. Rev. B5, 4367 (1972).
(10) H. Ebisawa, S. Maekawa and H. Fukuyama, Solid State Comm. 45, 75 (1983).

LT-17 (Contributed Papers)
U. Eckern, A. Schmid, W. Weber, H. Wühl (eds)

LOCALIZATION AND SUPERCONDUCTIVITY IN NARROW METALLIC WIRES

P. SANTHANAM, S. WIND and D. E. PROBER

Section of Applied Physics, Yale University, New Haven, CT 06520, USA

The theory of Magnetoresistance (MR) due to localization for quasi-one dimensional systems has been extended to include spin-orbit scattering and Maki-Thompson (MT) superconducting fluctuations. Experimental measurements on Aluminum wires support the theory quite well.

1. INTRODUCTION

For narrow wires of width W and thickness d such that W, d << ℓ_i ($=\sqrt{D\tau_i}$, where τ_i is the inelastic scattering time and D is the diffusion constant) Altshuler and Aronov[1] calculated the localization contribution to the MR for small magnetic fields in the absence of spin-orbit scattering. Our experimental study in Aluminum wires[2] necessitated a theory for quasi-one dimensional systems with spin-orbit scattering and superconducting fluctuations properly included. We report here the extension of the existing theory to include these effects and the experimental confirmation of the new predictions. We ignore magnetic scattering in our discussions[3] and assume elastic scattering time, $\tau \ll \tau_i$, τ_{so}, τ_{so} being the spin-orbit scattering time.

2. CALCULATION OF CONDUCTIVITY

2.1 Results of Altshuler and Aronov

The quantum correction to the dc conductivity due to localization is[1]

$$\Delta\sigma(\bar{r},\bar{r}') = -\frac{2e^2}{\pi} D\, C(\bar{r},\bar{r}') \qquad (1)$$

with the Cooperon amplitude $C(\bar{r},\bar{r}')$ given by

$$\hbar\{D[-i\bar{\nabla}-\frac{2e}{\hbar c}\bar{A}]^2 + \frac{1}{\tau_i}\}C(\bar{r},\bar{r}') = \delta(\bar{r},\bar{r}') \qquad (2)$$

and the boundary condition

$$[-i\bar{\nabla} - \frac{2e}{\hbar c}\bar{A}]_n C(\bar{r},\bar{r}') = 0$$

for the component normal to the sample. Altshuler and Aronov found by using perturbation theory for a wire with its length along x axis and width W along y axis in a field $\bar{H} = H\hat{z}$ that

$$C(y,y') = \sum_{q_x} C_{q_x}\phi_{q_x}(y)\phi_{q_x}(y')$$

where for small H

$$C_{q_x} = \frac{1}{\hbar[Dq_x^{\,2} + D(\frac{2eH}{\hbar c})^2\frac{W^2}{12} + \frac{1}{\tau_i}]} \qquad (3)$$

and $\phi_{q_x}(y)$ are the eigenfunctions of the operator in Eq. 2. Eq. 1 allows the evaluation of conductivity. One obtains for the fractional change of resistance due to localization at a given temperature and magnetic field,

$$\frac{\Delta R}{R}^{Loc}(T,H) = \frac{R_\square}{(\pi\hbar/e^2)}\frac{\sqrt{\hbar c/4e}}{W\sqrt{H_i}}\left[1 + \frac{H^2}{48H_iH_W}\right]^{-1/2}$$

$$= f_1(H,H_i) \qquad (4)$$

in terms of sheet resistance $R_\square$ and characteristic fields $H_i = \hbar c/4eD\tau_i$ and $H_W = \hbar c/4eW^2$

2.2 Validity of the small field approximation

In a calculation of nucleation of superconductivity for a slab in a parallel field, Saint James and deGennes[4] solved an equation identical to that for which $\phi_{q_x}(y)$ is a solution. They found that the correction, ε, to their ground state eigenvalue goes as the square of the magnetic field for values of $\varepsilon \lesssim 0.8$ (in their dimensionless units) or equivalently in our case the low field result for

$$(\frac{H}{H_W})^2 \frac{1}{192} \lesssim 0.8 \text{ or } H \lesssim 12.5\, H_W$$

This implies for a sample of width 0.4 μm a limit of H ∿ 80G for using Eq. 4.

3. EFFECT OF SPIN-ORBIT SCATTERING

3.1 Narrow widths W,d << ℓ_i, ℓ_{so} ($=\sqrt{D\tau_{so}}$)

When spin-orbit scattering is not negligible the correlator has two terms,[5] corresponding to triplet and singlet wave functions respectively:

$$C_{q_x} = \frac{3}{2}\frac{1}{\hbar[Dq_x^2 + D(\frac{2eH}{\hbar c})^2\frac{W^2}{12} + \frac{1}{\tau_2}]} - \frac{1}{2}\frac{1}{\hbar[Dq_x^2 + D(\frac{2eH}{\hbar c})^2\frac{W^2}{12} + \frac{1}{\tau_i}]} \qquad (5)$$

with $\tau_2^{-1} = (4/3)\tau_{so}^{-1} + \tau_i^{-1}$. Evaluation of the change in fractional resistance yields

$$\frac{\Delta R^{Loc}}{R}(T,H) = \frac{3}{2} f_1(H,H_2) - \frac{1}{2} f_1(H,H_i) \quad (6)$$

where $H_2 = \hbar c/4eD\tau_2$ and $f_1(H,H_i)$ is the 1-D form.

3.2 Intermediate widths $\ell_{so} << W << \ell_i$, and $d << \ell_i$, ℓ_{so}

The correlator sum for the triplet term will be two-dimensional as in the case of wide films,[3] since $\sqrt{D\tau_2} << W$, while the singlet term will be one dimensional, so that

$$\frac{\Delta R^{Loc}}{R}(T,H) = \frac{3}{2} f_2(H,H_2) - \frac{1}{2} f_1(H,H_i) \quad (7)$$

where $f_2(H,H_2) = \frac{-R_\square}{2\pi^2\hbar/e^2}\psi(\frac{1}{2}+\frac{H_2}{H})$ is the 2D form.[3]

4. MAKI-THOMPSON SUPERCONDUCTING FLUCTUATIONS

We observe that the parameter $\beta(T/T_c)$ introduced by Larkin[6] is independent of localization dimension[5] and the MT contribution is not affected by spin-orbit scattering. We therefore find for samples with $W << \sqrt{D\tau_i}$, for small H, when $k_B(T-T_c) > \hbar/\tau_i$ that

$$\frac{\Delta R^{MT}}{R}(T,H) = -\beta(\frac{T}{T_c}) f_1(H,H_i) \quad (8)$$

5. EXPERIMENTAL RESULTS

MR is defined as $\frac{\delta R}{R} = \frac{\Delta R}{R}(T,H) - \frac{\Delta R}{R}(T, H=0)$. From the discussion above it follows that

$$\frac{\delta R}{R} = \frac{\delta R^{Loc}}{R} + \frac{\delta R^{MT}}{R}$$

if we ignore other contributions to MR.[2,3] Hence,

$$\frac{\delta R}{R} = \frac{3}{2}[f_1(H,H_2)-f_1(H=0,H_2)] - (\beta+\frac{1}{2})[f_1(H,H_i)-f_1(H=0,H_i)] \quad \text{for } W<<\ell_i,\ell_{so} \quad (9a)$$

and

$$\frac{\delta R}{R} = \frac{3}{2}[f_2(H,H_2)-f_2(H\to 0,H_2)] - (\beta+\frac{1}{2})[f_1(H,H_i)-f_1(H=0,H_i)] \quad \text{for } \ell_{so}<<W<<\ell_i \quad (9b)$$

We illustrate the predictions of the theory for intermediate widths, Eq. (9b), for our sample S562, an aluminum wire of width 0.4 μm and thickness 150Å. The parameter $\beta(T/T_c)$ for this analysis is taken from Ref. 6. The MR data for S562 at 15K and 20K can be fitted to 2D theory[3] satisfactorily with a choice of H_{so} = 18G (independent of temperature) and reasonable temperature dependent values for H_i for $\ell_i << W$. ℓ_{so} from these fits is 0.3 μm. In contrast, for T ≲ 12K, the experimental data cannot be fit to the 2D theory for *any* set of parameters H_i and H_{so}. We show in Fig.1 the experimental data at 4.5K. We expect the inelastic scattering rate to be the same as in 2D films since the sample width is larger than the characteristic lengths for the inelastic scattering mechanisms under consideration, dirty-limit electron-electron scattering and electron-phonon scattering. ℓ_i(4.5K)∼1μm from the 2D results.[3] Hence the data at 4.5K should be a good test of Eq. (9b) above. (ℓ_{so} is assumed to be temperature independent, as found in Ref. 3) In Fig. 1 we show the fit to the expression 9b with H_{so} = 18G and H_i=2.7G. We see that the fit is quite good and infer a value of ℓ_i=0.8μm. For comparison we also show the corresponding 2D prediction as a dashed line with the same choice of parameters for H_i and H_{so}. We have also verified the theoretical prediction of Eq.9a in narrower samples.[2]

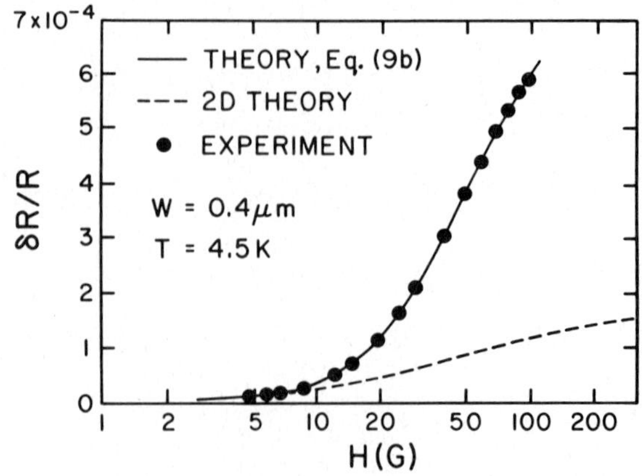

Fig. 1. Normalized Magnetoresistance for sample S562. Fitting parameters are H_i=2.7G and H_{so}=18G

6. CONCLUSION

We have extended the theory for magnetoresistance of quasi 1-D systems to include spin-orbit scattering and Maki-Thompson fluctuations. The agreement with experiment is very good.

ACKNOWLEDGEMENTS

We thank J. W. Serene for useful discussions. This work was supported by NSF Grant DMR 8207443.

REFERENCES

(1) B.L. Altshuler and A.G. Aronov, JETP Lett., 33 (1981) 499.
(2) P. Santhanam, S. Wind and D.E. Prober (to be published).
(3) P. Santhanam and D.E. Prober, Phys. Rev. B, 29 (1984) 3733.
(4) D. Saint James and P.G. deGennes, Phys. Lett. 7 (1963) 306.
(5) B.L. Altshuler, A.G. Aronov, A.I. Larkin and D.E. Khmelnitskii, Sov. Phys. JETP, 54 (1981) 411.
(6) A.I. Larkin, JETP Lett. 31 (1980) 219.

LT-17 (Contributed Papers)
U. Eckern, A. Schmid, W. Weber, H. Wühl (eds)
© Elsevier Science Publishers B.V., 1984

INELASTIC LIFE-TIME OF CONDUCTION ELECTRONS IN 3D COPPER BASED ALLOYS

Wolfgang ESCHNER and Wolfgang GEY

Institut für Technische Physik der Technischen Universität Braunschweig, Mendelssohnstraße 2, 3300 Braunschweig, FRG

Peter WARNECKE

Physikalisch-Technische Bundesanstalt, 3300 Braunschweig, FRG

Magneto-conductance measurements on massive wires of Cu-Ge solid solutions in the temperature range 2,1 K < T < 76 K are reported. The data are analyzed with the theory of weak localization which has been extended to three dimensions. Good agreement is achieved. The structure of the magneto-conductance curves is similar to that known for thin films, however weaker by a factor 100. The measurements yield the inelastic life-time τ_i and the spin-orbit coupling for copper alloys; τ_i obeys a T^{-p} law with p = 3,05. This indicates that the electron-phonon-interaction dominates τ_i in bulk material.

1. INTRODUCTION

At low temperatures the conduction electrons of a metal posess a long inelastic life-time. If disorder is introduced, the resultant quantum interferences known as "weak localization", yield anomalies in the temperature dependence of the resistance. Also there is a pronounced structure in the magneto-conductance in the presence of spin-orbit interaction.

For two dimensions these properties have been studied extensively (1,2). In thin films the inelastic life time τ_i is reported to obey a T^{-p} law with p of order 2. It is not clear which interaction processes are responsible for τ_i.

For three dimensions, weak localization effects are expected to be much weaker and in fact have not been studied quantitatively as yet. Also, similar anomalics of the resistance caused by electron-electron interaction (3, 4) may become important. Using an extremely sensitive resistance bridge (resolution 3 x 10^{-8}) it was possible to analyze the extremely small magneto-conductance anomalies with an accuracy sufficient for a $\tau_i(T)$ analysis up to temperatures as high as 76 K.

2. THEORY

It is not difficult to extend the calculation for two dimensions (5) to three dimensions. For a given temperature we obtain for the magneto-conductance due to localization (all symbols having their usual meaning):

$$\Delta\sigma(B) = e^2/2\pi^2\hbar\sqrt{eB/\hbar}\,\{3/2\, f_3(B/B_1) - 1/2 f_3(B/B_2)\}$$

$f_3(x)$ see ref.(4) (1)

$$B_n = \hbar/4eD\tau_n$$

$$1/\tau_n = 1/\tau_i + 4/\tau_{so} \quad ; \quad 1/\tau_2 = 1/\tau_i \; .$$

3. EXPERIMENT

Solid solutions of Ge, Si, and Ga in copper were produced. These alloys scatter conduction electrons rather strongly; their physical properties are very well known. Samples were drawn into wires of 0.3 mm diameter, i.e. five orders of magnitude larger than the electron's mean free path. The magneto-conductance was measured for alloys with different degrees of disorder at fixed temperatures (2.1 K to 76 K) up to magnetic inductions of 8 T. It was proved that $\Delta\sigma(B)$ was completely isotropic.

4. RESULTS

Fits to equation (1) up to 1 T; i.e. far beyond the magneto-conductance minimum (~0,1 T) were made. Good agreement was obtained. Fig. 1 shows the temperature dependence for $Cu_{89,3}Ge_{10,7}$ of the inelastic life-time and of the spin-orbit time. While $\tau_{so} \approx 6\times10^{-12}$ s stays constant as expected, τ_i varies between $4,5\times10^{-9}$ s and $1,5\times10^{-13}$ s and obeys a T^{-p} law with p = 3,05 for the relatively wide temperature range 2.1 K < T < 76 K. This is in contrast with p = 1,65 for nobel metal films (2) and, for the first time, emphasizes the action of phonon processes in limiting τ_i in a wide temperature range for 3D metals.

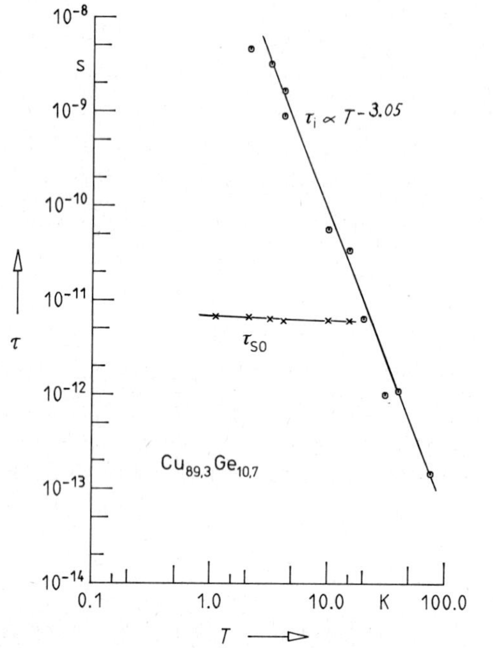

Fig. 1.: The variation of the inelastic electron life-time τ_i and the spin-orbit scattering time τ_{so} with temperature.

As for the dependence of τ_i on disorder, we find for all alloys the puzzling result that τ_i increases with disorder or resistivity ρ. τ_{so} however, is inversely proportional to ρ, as expected.

REFERENCES

(1) H. Fukuyama, Surface Science 113 (1982) 489.

(2) G. Bergmann, Z. Phys. B - condensed Matter 48 (1982) 5.

(3) B. L. Altshuler and A. G. Aronov, Sov.Phys. JETP 50 (1979) 968.

(4) B. L. Altshuler et al., Sov. Phys. JETP 54 (1981) 411.

(5) S. Hikami et al., Progr. Theor. Phys. 63 (1980) 707.

LT-17 (Contributed Papers)
U. Eckern, A. Schmid, W. Weber, H. Wühl (eds)

KAPITZA CONDUCTANCE BETWEEN SOLID ^{4}He AND ^{3}He-^{4}He SOLUTIONS†

R. M. BOWLEY*, M. J. GRAF, and H. J. MARIS

Department of Physics, Brown University, Providence, RI 02912, USA

We show that measurements of the Kapitza conductance between solid ^{4}He and solutions of liquid ^{3}He-^{4}He can provide information about the mobility of the interface. Preliminary experiments indicate that for a ^{3}He concentration of 5 x 10^{-4} the interface retains the high mobility observed previously for the pure ^{4}He system.

1. INTRODUCTION

Andreev and Parshin (1) proposed that the interface between liquid and solid ^{4}He at low temperatures should have several unique properties. The most dramatic of these is the high mobility of the interface. They proposed that at T=0 melting or freezing can occur very rapidly, and without dissipation. At finite T the growth rate of the solid is only limited by the damping of the interface caused by interactions with phonons and rotons. Andreev and Parshin's ideas have been confirmed by many experiments, including studies of melting-freezing waves (2), measurements of sound transmission across the interface (3), and investigations of the Kapitza conductance (4-6) and related phenomena (7).

For solid helium in contact with a dilute solution of ^{3}He in liquid ^{4}He, the situation is not so clear. There is some evidence (8,9) that even a ^{3}He concentration as low as 10^{-4} may significantly alter the nature of the interface, by promoting the formation of facets. If the interface is facetted, there will be a drastic reduction in its mobility. In this paper we consider two interrelated problems. Firstly, does ^{3}He really reduce the mobility of the interface? Secondly, are there ways in which the ^{3}He can contribute to heat flow across the interface?

2. THEORY

Suppose that the interface becomes immobile (I) when ^{3}He is added. The transmission coefficient t_{SL}^{I} of phonons from the solid to the liquid is then given by standard acoustic-mismatch theory. Thus, it is of the order of 1, since liquid and solid helium are acoustically similar. There will be a small transmission of energy to the ^{3}He quasiparticles, of order of magnitude

$$t_{S3}^{I} \sim x_3 \, (kT/m_3^*)^{1/2}/c_S \qquad (1)$$

where $(kT/m_3^*)^{1/2}$ is the characteristic velocity of the ^{3}He, x_3 is the ^{3}He concentration, and c_S is the solid sound velocity.

Consider now what happens if the surface is mobile (M). The transmission coefficient for phonons going from the solid to the liquid is reduced considerably (4,5), to a value of order

$$t_{SL}^{M} \sim \left(\frac{\alpha kT}{h\rho_L c_L^3}\right)^2 \left(\frac{\rho_L}{\rho_S-\rho_L}\right)^4 \qquad (2)$$

where α = surface energy, $kT/\hbar$=average phonon frequency, ρ_L and ρ_S are the densities of liquid and solid respectively, and c_L is the liquid sound velocity. However, because of the high mobility of the interface, the amplitude of oscillation of the surface is larger than the amplitude of the incident phonon by a factor of $\rho_S/(\rho_S-\rho_L)$. This enhancement increases the coupling to the ^{3}He so that

$$t_{S3}^{M} \sim t_{S3}^{I} \left(\frac{\rho_S}{\rho_S-\rho_L}\right)^2 \qquad (3)$$

Thus, there is an enhancement of the order of 100 compared to (1).

3. EXPERIMENT

We have measured these transmission coefficients with an apparatus similar to that used by Huber and Maris (4) for pure ^{4}He. A cell contains liquid and solid helium, and pulses of heat can be applied to the solid. The induced temperature changes in the liquid and solid (for an example, see Fig. 1) can be analysed to find t_{SL} and t_{S3}. In addition, information can be obtained about the rate of energy transfer between the liquid phonons and the ^{3}He. We

*Permanent address: Department of Physics, University of Nottingham, NG7 2RD, UK.

†Work supported in part by the National Science Foundation through grant no. DMR-8304224.

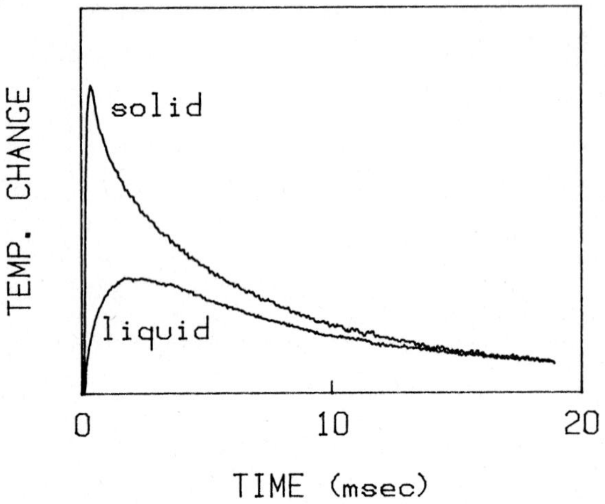

Fig. 1. Temperature changes induced in the liquid and solid by means of a heat pulse applied to the solid. The initial temperature is 0.17K.

have made preliminary measurements for $x_3 = 5 \times 10^{-4}$ at temperatures between 0.1 and 0.27K. Our results give

$$t_{SL} = (6.4 \pm 0.4)\ T^2 \tag{4}$$

This is close to the results obtained previously (4,6,7) for the pure ^4He system, and thus indicates a mobile interface. For t_{S3} we find a magnitude which is close to t^M_{S3} given by Eq. (3), i.e. approximately 2 orders of magnitude larger than what is expected for an immobile interface.

Our results therefore strongly indicate that the interface is still mobile when ^3He is present. We are currently extending these experiments to investigate a range of ^3He concentrations. A detailed account of these results, together with a quantitative version of the theory of the phonon transmission will be published elsewhere.

REFERENCES

(1) A. F. Andreev and A. Y. Parshin, Sov. Phys.-JETP 48, 763 (1978).
(2) K. O. Keshishev, A. Y. Parshin, and A. V. Babkin, JETP Lett. 30, 56 (1979).
(3) B. Castaing, S. Balibar, and C. Laroche, J. Phys. 41, 897 (1980).
(4) T. E. Huber, and H. J. Maris, Phys. Rev. Lett. 47, 1907 (1981), J. Low Temp. Phys. 48, 463 (1982).
(5) H. J. Maris, and T. E. Huber, J. Low Temp. Phys. 48, 99 (1982).
(6) L. Puech, B. Hebral, D. Thoulouze, and B. Castaing, J. Phys. Lett. 43, L809 (1982).
(7) P. E. Wolf, D. O. Edwards, and S. Balibar, J. Low Temp. Phys. 51, 489 (1983).
(8) J. Landau, S. G. Lipson, L. M. Määttänen, L. S. Balfour, and D. O. Edwards, Phys. Rev. Lett. 45, 31 (1980).
(9) J. E. Avron, L. S. Balfour, C. G. Kuper, J. Landau, S. G. Lipson, and L. S. Schulman, Phys. Rev. Lett. 45, 814 (1980).

LT-17 (Contributed Papers)
U. Eckern, A. Schmid, W. Weber, H. Wühl (eds)

UNIVERSAL JUMP OF THE CURVATURE OF ^{4}He CRYSTALS AT THE ROUGHENING TRANSITION

F. GALLET, P.E. WOLF and S. BALIBAR

Groupe de Physique des Solides de l'Ecole Normale Supérieure, 24 rue Lhomond, 75231 Paris Cedex 05, France

We measured the curvature of hcp ^{4}He cristals in the c-direction at the first roughening transition (T_R = 1.2 K) and showed that it remains finite at this temperature. The obtained value is in agreement with the predictions of the renormalization calculations, i.e. $2/\pi$ in reduced units.

Several experiments (1-4) have established the existence of three different roughening transitions, where facets appear on the equilibrium shape of hcp ^{4}He crystal. At the same time, this transition has been the object of contradictory theoretical studies. On the one hand, Chui and Weeks (5) pointed out the duality between the roughening transition and the general 2-D Kosterlitz-Thouless model of transition. By renormalization techniques, one can establish a simple relation between the roughening temperature T_R and the surface tension $\tilde{\alpha}$ at the same temperature, for a given surface orientation (6) :

$$\frac{1}{\tilde{\alpha}} = \frac{2}{\pi} \frac{d^2}{kT_R}$$

In this formula, d is the lattice period in the considered direction. It is equivalent to say that the curvature of the interface in this direction is expected to jump from a finite value above T_R (rough surface) to zero below (facetted surface). On the other hand, Andreev (7) showed that, in a Landau type theory, the curvature of the interface vanishes continuously through the roughening transition. In order to decide between these predictions, we set up an experiment for measuring the radius of curvature in the c-direction of a hcp ^{4}He crystal, near the corresponding roughening transition ($T_R \simeq 1.2$ K). Like in ref.(4), we suspended in our experimental cell a box with glass walls and a 0.83 mm hole drilled in its bottom plate. When a crystal is grown up to a height H above the bottom of the box, one forms a meniscus on the hole, at the equilibrium (fig.1). For a given pressure difference across the interface (determined by H), the local curvature of the meniscus is inversely proportional to the surface tension. The c-axis of the crystal is vertical and coincides with the hole axis, so that the c-facet appears in the profile of the meniscus. After taking a photograph of the magnified profile (×20), we digitize it over an angular range of about ± 15° around the c-axis, and

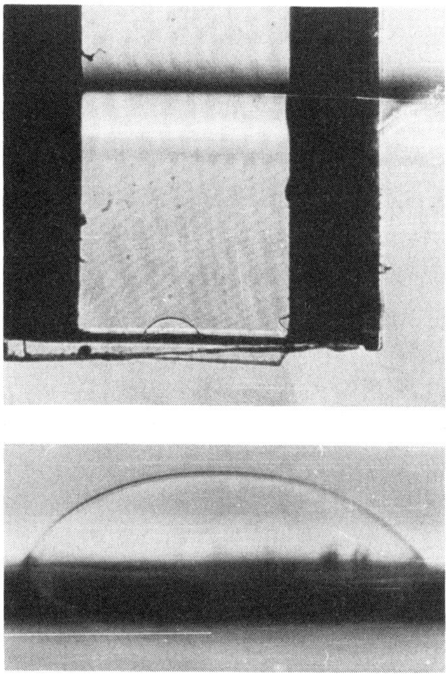

FIGURE 1
Two photographs of our experimental cell. On the upper one, the dark horizontal line crossing the top of the transparent box is the solid-liquid interface outside. A bent meniscus at the equilibrium stands on the bottom hole. The lower one is an enlarged view of a meniscus at 1.245 K (rough surfaces). The c-axis is vertical.

$^+$ Laboratoire associé au Centre National de la Recherche Scientifique numerically fit it with an expression of type $z(x) = ax^2 + bx^4$. From the value of the coefficient a, we deduce the surface tension in the c-direction. The accuracy ($\sim$ 10%) is limited by our optical resolution on the profile ($\pm$ 3 μm).

The dimensionless quantity $kT_R/d^2\tilde{\alpha}$ is plotted versus temperature on fig.2. In the range 1.2-1.35 K, it remains roughly constant and equals the expected value $2/\pi$, within our precision. Between 1.1 and 1.2 K, the curvature seems to vanish continuously, but several arguments indicate that this decay is unphysical, and only due to our method of analysis. Namely, the roughening temperature for the c-facet has been measured by analyzing the growth kinetics (8) : it is 1.20 ± 0.01 K. More precisely, the mobility in the c-direction varies by two orders of magnitude between 1.1. and 1.2 K. As shown on fig.3, the crystal growth involves a threshold below 1.20 K. This threshold is characteristic of the presence of a facet and vanishes at 1.20 K. Above 1.20 K, the mobility gets close to that of adjacent rough surfaces. We are convinced of the existence of a facet (of zero curvature) on our profile between 1.13 and 1.20 K, but its size is too small to be visible : in this domain of temperature, in fact, we measure a curvature averaged on flat and round parts.

Our measurements of the curvature for the c-direction indicate that it remains finite at the roughening transition. This behaviour cannot be explained by a Landau-type theory. On the contrary, the relation between the roughening temperature and the surface tension deduced from renormalization calculations seems well verified by the experiment. We are presently improving the precision on the determination of the profile, in order to check the predictions of the renormalization theory concerning the critical behaviour of the surface tension close to T_R.

REFERENCES

(1) S. Balibar and B. Castaing, J. Phys. Lett. 41 (1980) L 329.
(2) J.E. Avron et al, Phys. Rev. Lett. 45 (1980) 814
(3) K.O. Keshishev, A.Y. Parshin and A.B. Babkin, Zh. Eksp. Teor. Fiz. 80 (1980) 716 (Sov. Phys. JETP 53 (1981) 362).
(4) P.E. Wolf, S. Balibar and F. Gallet, Phys. Rev. Lett. 51, 15 (1983) 1366.
(5) S.T. Chui and J.D. Weeks, Phys. Rev. B 14, 11 (1976) 4978.
(6) D.S. Fisher and J.D. Weeks, Phys. Rev. Lett. 50, 14 (1983) 1077 - C. Jayaprakash, W.F. Saam and S. Teitel, Phys. Rev. Lett. 50, 25 (1983) 2017 - P. Nozières, Cours au Collège de France, Paris (1984), unpublished.
(7) A.F. Andreev, Zh. Eksp. Teor. Phys. 80 (1981) 2042 (Sov. Phys. JETP 53, 5 (1981), 1063)
(8) P.E. Wolf, F. Gallet and S. Balibar, "Growth kinetics of facets...", this volume.

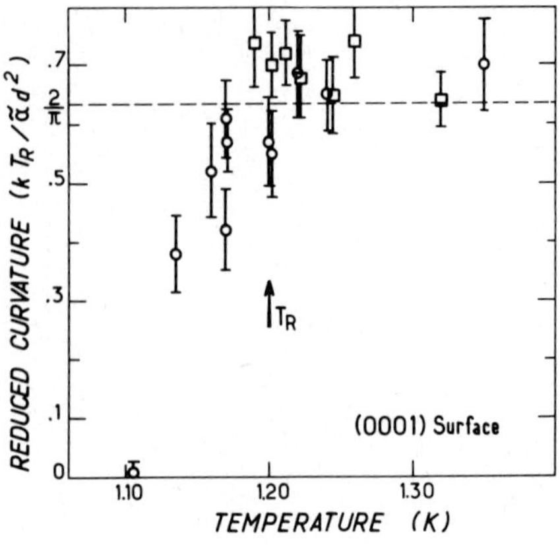

FIGURE 2
Reduced curvature of the meniscus in the c-direction versus T. The broken line represents the theoretical value $2/\pi$ at the roughening transition, calculated by Jayaprakash, Saam and Teitel. d is 2.99 Å.

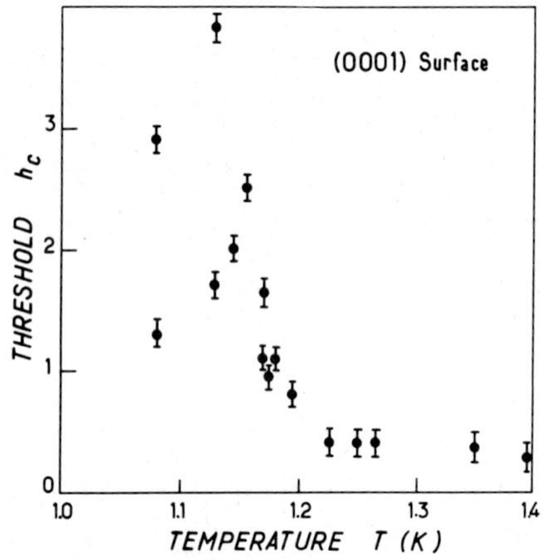

FIGURE 3
Minimum pressure difference necessary to observe the growth in the c-direction (it is expressed in terms of the interface height - in millimeters - outside of the box). The residual height h_0 above 1.20 K is the capillary depression of the meniscus in the box.

LT-17 (Contributed Papers)
U. Eckern, A. Schmid, W. Weber, H. Wühl (eds)

GROWTH KINETICS OF hcp ^{4}He FACETS

P.E. WOLF, F. GALLET and S. BALIBAR

Groupe de Physique des Solides de l'Ecole Normale Supérieure, 24 rue Lhomond, 75231 Paris Cedex 05, France*

We report here the first measurements of the growth kinetics of facets of hcp ^{4}He in contact with superfluid. We show that the facet mobility increases near its roughening transition to approach that of rough orientations. It is also enhanced at low temperature, implying that it results from the motion of steps, which become more mobile as the temperature decreases.

1. INTRODUCTION

The interface between solid hcp and superfluid ^{4}He presents three roughening transitions at 1.2, 0.9 and 0.35K for respectively the [0001] (c), [1$\bar{1}$00] (a) and [1$\bar{1}$01] orientations. Other orientations stay rough to the lowest reached temperature (1). Because of its purety and the superfluidity of the liquid, the growth kinetics of solid ^{4}He are driven by the interface and not by macroscopic transport in the bulk. For a rough interface, the growth velocity v is proportional to the difference $\Delta\mu$ between the solid and liquid chemical potentials and the mobility $k = v/\Delta\mu$ increases as the temperature decreases (2,3,4). The growth of smooth interfaces, i.e. facets, was only known bo te quite slower. We report here the first detailed study of the growth of (c) and (a) facets.

2. EXPERIMENT

Experiments have been performed in the range 0.1K-1.5K. We use cryostats with windows and nucleate crystals with horizontal facet by Keshishev's method (2). A 4mm×4mm transparent box is suspended in the cell. Its top is opened and a hole is drilled in its bottom. As long as the level outside stays below a critical height, a meniscus can remain on the hole (6). Above, it becomes unstable and grows, the outside level being fixed. In a first stage, the meniscus becomes widely facetted and fills the box width. Then the interface in the box relaxes to a stable height. In both stages, the vertical velocity v is controlled by the facet, the difference h of levels between the facet and the outside interface providing the necessary $\Delta\mu$. Per unit mass : $\Delta\mu = [(\rho_\ell - \rho_S)/\rho_S]\, g(h - h_0)$ where ρ_ℓ and ρ_S are the liquid and solid densities and h_0 the capillary depression in the box. By measuring h versus time, we get v(h) or v($\Delta\mu$); values of $(h - h_0)$ range from 0.1 to 10 mm (i.e. $10^{-7} < \Delta\mu/kT < 10^{-5}$ at 1K).

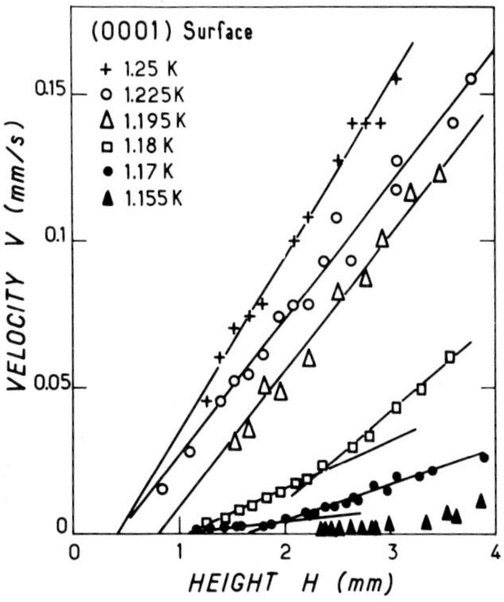

Figure 1 : Growth caracteristic of a c-surface. Full lines are guides to the eyes.

3. RESULTS AND DISCUSSIONS

Figure 1 shows v(h) for a given c-facet at several temperatures near its roughening transition. The precision on v is about 15%. Growth is only observed beyond a critical value h_c. Above 1.2K, h_c is the capillary depression. Below 1.2K, despite some dispersion, it increases continuously as the temperature decreases. We used this property in ref.(6) to determine the roughening transition temperature value $T_R = 1.20 \pm 0.01$ K. Moreover, beyond h_c, v is a reproducible and linear function of $\Delta\mu$ above 1.2K. On the contrary, below 1.2K, v is first linear in $(h - h_c)$ but then bends upwards, the slope above the threshold varying somewhat from a crystal to another. Figure 2 shows the strong increase of the mobility $k = dv/d(\Delta\mu)$near the roughening transition (k is precisely measured in the linear regime by a semilog plot of $(h - h_c)$ versus time) as well as the weak anisotropy of the kinetics above T_R. Finally, below 1K, h_c seems to increase strongly, while k remains of the order 10^{-5}s/cm down to 0.5K.

---

* Laboratoire associé au Centre National de la Recherche Scientifique.

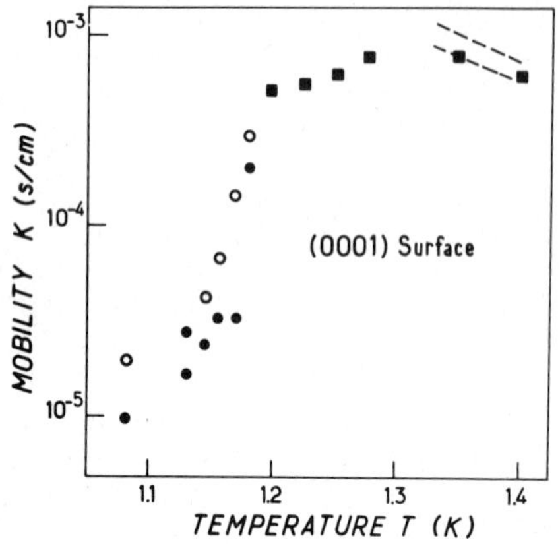

Figure 2 : Mobility of the c-surface versus temperature. Full and open circles respectively refer to the mobility above the threshold and at higher $\Delta\mu$, below T_R; squares to the mobility above T_R; the broken lines to rough surfaces (4).

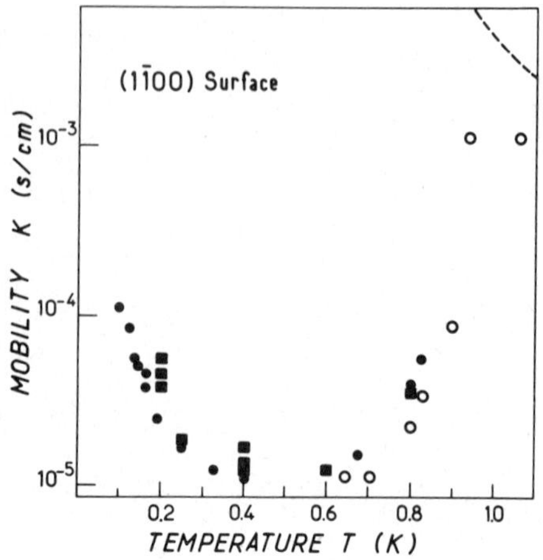

Figure 3 : Mobility of a a-surface versus temperature. The broken line is the interpolated rough surface mobility (5).

The study of the a-facet appeared more difficult. In some cases, v varies linearly with $\Delta\mu$ above the threshold, which is smaller than for the c-facet and increases as temperature decreases. At a given temperature, the mobility may vary by a factor of two (Fig.3, full squares). However, in most cases, the facet stops far below h_c, because the interface contains other facets which anchor on the side walls. To avoid this problem, other data were obtained by measuring, for a given external height, the facet velocity in the first stage of growth, before the meniscus touches the walls. We plot the mobility k versus temperature on fig.3 (full and open circles refer to different experiments). We get values in good agreement with the previous ones. Moreover, we see an unexpected increase of k at low temperature. The increase near the a roughening-transition is also observed.

We thus show that the facets mobility is quite lower (six orders of magnitude at 0.3K) than that of rough surfaces. By analogy with other crystals, we interpret our results by the motion of steps issued from screw dislocations emerging on the facet. In this case, the threshold increases with the step energy E_s, and the dislocation density n. This explains the dispersion of our results (n may vary from one crystal to another). The threshold dependence on temperature and orientation is understood from the value of E_s, which is smaller for a than for c facets and increases with decreasing temperature. As temperature decreases, the mobility increase shows that, like rough surfaces, steps become more mobile. However, it takes place in a temperature range where the rough surfaces mobility saturates (2), and is rather like T^{-2} than T^{-4}.

This may indicate that phonons limit the mobility of rough surfaces and steps in different ways. The strong increase of the mobility near roughening transitions is most probably due to the vanishing of the step energy, which enhances both the dislocations activity and two dimensional nucleation. Estimations of step energies, together with the temperature range over which k increases, favour the first mechanism.

4. CONCLUSION

The facets mobility increases both at low temperature and near their roughening transition. Our observations qualitatively agree with a growth mechanism due to steps motion. Further measurements at higher $\Delta\mu$ should allow a detailed understanding of the growth processes.

REFERENCES

(1) P.E. Wolf, S. Balibar and F. Gallet, Phys. Rev. Lett. 51, 15 (1983) 1366 and references therein.
(2) K.O. Keshishev, A. Parshin and A.B. Babkhin, Zh. Eksp. Teor. F80 (1981) 716. (Sov. Phys. JETP, 53 (1981) 362).
(3) B. Castaing, S. Balibar and C. Laroche, J. Phys. (Paris) 41 (1980) 897.
(4) J. Bodensohn, P. Leiderer and D. Savignac, in : Proceedings of the 4th International Conference on Phonon Scattering in Condensed Matter (1983).
(5) B. Castaing, J. Phys. Lett. (Paris), 45 (1984) L233.
(6) F. Gallet, P.E. Wolf and S. Balibar, Universal jump of the curvature of ^{4}He crystals..., this volume.

LT-17 (Contributed Papers)
U. Eckern, A. Schmid, W. Weber, H. Wühl (eds)

LAYER BY LAYER GROWTH OF SOLID ^{4}HE ON GRAFOIL

V. Gridin, Y. Eckstein, and E. Polturak

Department of Physics, The Technion, Haifa, Israel

We performed direct measurements of the growth of solid ^{4}He on Grafoil from the pressurized liquid phase, using a combination of differential adsorption and acoustics. We found that a layer by layer growth does take place, confirming earlier suggestions. In contrast with previous work, our results show that solid growth starts at very low pressures and that the Franchetti relation doesn't describe the solid thickness at all temperatures.

In the course of an investigation of the propagation of 4th sound in Grafoil, Maynard et al.[1] have discovered a series of minima of the resonant frequency of their sound cell as the pressure was swept at constant temperature. This novel effect was interpreted by them in terms of a coupling between the sound (pressure) wave to a wavelike growth and melting of surface solid. The series of minima were taken to represent the layer by layer growth of solid which is predicted by modern theories of multilayer growth.[2] This interpretation relied on the periodicity of the frequency minima when plotted vs. nominal surface solid thickness, θ, calculated from the so-called Franchetti[3] relation

$$\theta = \frac{\alpha}{(P_M - P)^{1/3}} \qquad (1)$$

here, P is the pressure, P_M is the bulk melting pressure and α is the Van der Waals constant in layer × bar$^{1/3}$ units. This relation was never experimentally checked, and when one takes into account the spread in the values of α found in the literature, the interpretation of the results becomes less than certain. To this end, we decided to perform a direct adsorption experiment combined with acoustics in the same cell.

In trying to measure the amount of solid adsorbed from the liquid phase, one has to separate each quantity of ^{4}He added to the cell into two parts: one that solidifies on the surface and one that compresses the liquid, thereby raising the ambient pressure. This is not made any easier by the closeness of the molar volumes of the liquid and solid ^{4}He. To get around this problem, we decided to use two cells in a differential arrangement. Both cells had nearly the same open volume, but whereas the reference cell had no surface area to speak of, the other cell contained ∿ 20 gm of Grafoil discs with a surface area in excess of 400 m^2. This cell was also equipped with two capacitive sound transducers for simultaneous acoustic measurements.To eliminate spurious effects due to pressure drifts, identical lines connected the cells to the gas handling system. After bringing the cells to a pressure near solidification, metered amounts of ^{4}He were removed from each cell in turn, such as to produce the same pressure change in both cells. By taking the difference between the amounts removed from each cell (corrected for the small volume difference between the two cells) substracted out the liquid phase contribution. The residuals represent changes in surface solid coverage, and summing these yielded the adsorption isotherm. We point out that this method relies only on thermodynamics.

In the figure, we show the adsorption isotherm obtained at about 1K along with the frequency of one of the cavity resonances. Experimental precision is better than the size of a data point. There is indeed an excellent correlation between changes of slope of the adsorption isotherm and changes of the resonant frequency. Therefore, the basic idea of Maynard et al.[1] regarding a correlation between 4th sound and growth of surface solid is correct. There are some important differences between their data and ours. We found that the adsorption isotherm and the acoustic data indicate that solid adsorption takes place at all pressures, whereas there is no indication of solid growth in the acoustic data of Ref. 1 below ∿ 15 bars. Second, they interpret the narrow dips in their data as taking place at mid layer, whereas we found that the narrow (in pressure) features occur at full layers, and they can be either minima or maxima of the frequency. This is allowed by the model of Adler et al.[4], but not by the model in Ref. 1. We repeated our experiment using a different sample of Grafoil having a different porosity. We found an excellent reproducibility of the adsorption isotherm, but the acoustic results looked considerably different. It therefore appears that the coupling of 4th sound to the surface is not sufficiently understood and

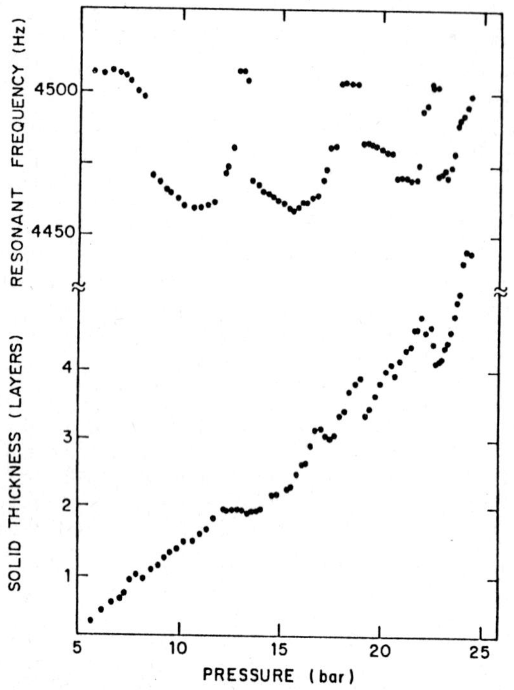

FIGURE 1
Adsorption isotherm at 1K (lower trace) and the acoustic resonant frequency (upper trace) vs. pressure.

the direct method of study is preferable. The layer by layer solid growth is of considerable interest to the testing of the modern theories of multilayer adsorption. A detailed interpretation of the isotherm itself will be presented elsewhere.

Finally, we remark that the Franchetti relation can be fitted to our data only if we assume that 6 layers of solid are adsorbed on the surface in the liquid filled cell already at P = 0. This result, one layer higher than obtained in thick film studies[5], indicates a considerably higher Van der Waals constant, α, than used in Ref. 1. Furthermore, we found that the same constant does not apply at a different temperature, and therefore Eq. 1 cannot be trusted to convert pressure to solid thickness.

We acknowledge many fruitful discussions with Charles Kuper, Hanan Shechter, Stephen Lipson and Joan Adler.

REFERENCES

(1) J.D. Maynard, G.J. Gelatis and R.A. Roth, Physica 107B, 243 (1981), S. Ramesh and J.D. Maynard, Phys. Rev. Lett. 49, 47 (1982).
(2) M.J. de Olivera and R.B. Griffiths, Surf. Sci. 71, 687 (1978).
(3) S. Franchetti, Nuovo Cimento 4, 1540 (1956).
(4) J. Adler, C.G. Kuper and L.S. Shulman, Jour. Low Temp. Phys. 52, 73 (1983).
(5) M. Bienfait, J.G. Dash and J. Stoltenberg, Phys. Rev. B21, 2765 (1980), S.F. Polanco, J.H. Quateman and M. Bretz, J. Phys. (Paris) 39, C6-344 (1978).

LT-17 (Contributed Papers)
U. Eckern, A. Schmid, W. Weber, H. Wühl (eds)

DENDRITIC GROWTH OF CRYSTALS OF DILUTE ^3He-^4He MIXTURES

L.S. BALFOUR* and S.G. LIPSON

Physics Department, Technion - Israel Institute of Technology, Haifa, Israel

We have observed the growth of crystals of dilute ^3He-^4He mixtures from supercooled superfluid. The planar interface between the solid and liquid becomes unstable under certain conditions and exhibits dendritic or cellular growth. The relationship between the growth rate and the dendrite dimensions is compared with the predictions of stability analysis.

Recently there has been renewed interest in the problem of pattern selection at the solid-liquid interface during solidification. (1-3) Various models have been proposed to account for the planar, cellular or dendritic morphologies developed by the interface. This paper presents experimental observations on the dendritic mode developed by the solid-super-fluid interface of dilute He^3/He^4 mixtures. Dilute solutions of He^3 in He^4 provide one with a convenient system in which the morphology of the solid-liquid interface during solidification can be observed optically and whose properties can be modified by changing the ^3He concentration. At temperatures above 0.9K the planar morphology is observed to be the stable one for small undercoolings. For larger undercoolings the planar interface is unstable and growth of the solid takes place in the form of dendrites. The dendritic morphology was observed in the temperature range between 0.9K and 1.6K for three different concentrations: 0.25%, 0.9% and 2%, where the solid mixture has a b.c.c. structure and the liquid is a superfluid along the freezing curve as indicated by the phase diagram (Fig.1).

EXPERIMENTAL SET UP

The sample dendrites are grown in an optical cell made of a flat copper cylindrical body one end of which is sealed by a mirror disc and the other by transparent optically flat glass. The diameter of the cell is 10 mm and the separation of the windows is 3.2 mm. The cell is mounted at the end of a long tube such that a parallel beam of light sent down the tube illuminates the sample, and the light reflected from the mirror returns up the same tube and is imaged into a 35 mm camera. Dendritic growth experiments have been carried out in two similar optical cells attached to two different cryostats, one an ordinary pumped He^4 bath cryostat and the other a He^3 cryostat capable of reaching temperatures around 0.35K.

GROWTH OF DENDRITES

A necessary condition for the growth of dendrites is that the superfluid should be supercooled initially. In either apparatus the initial supercooling of the liquid is achieved by overpressuring the liquid above the equilibrium freezing-curve pressure of a particular mixture at a certain temperature. In the He^4 cryostat the optical cell is immersed in the pumped liquid and the cell filling line contained a resistance wire inside it right up to the entrance of the cell. A voltage pulse applied to this resistor achieved the necessary supercooling to initiate dendritic growth. In the He^3 apparatus the optical cell is thermally connected to a pumped He^3 pot and the necessary supercooling is achieved by applying a pressure pulse to the gaseous mixture at the room temperature end of the filling line. The exact overpressure achieved in the cell in each case is determined interferometrically.

RESULTS

At low supercoolings a ring of solid is observed to grow out from the walls towards the centre of the cell. This was always observed in the He^4 apparatus, but only occasionally in the He^3 cryostat. At larger supercoolings the planar interface (manifested as a ring of solid) is unstable and dendrites grow out of it at a much faster rate ahead of the general solidification front. Quite a large number of dendrites begin to grow initially from different points along the planar interface, but only few continue to grow for larger periods. This could either be due to anisotropic growth rates of the differently oriented dendrites, or due to local variations in the supercooling of the adjacent liquid. The whole process takes a veryshort time, typically a second or two and hence an automatic camera at 4 frames/sec was used to record the growth. Typical dendrite tip velocities are 3 mm/sec.

---

*In partial fulfillment of the requirements for the D.Sc. degree of the Technion (I.I.T.) (L.S.B).

The dendrites grown consist of a primary stalk which thickens as the growth proceeds. Also a frame by frame observation of the directed growth shows that usually there is a slight sway of the advancing tip about the general growth direction, which could possibly be due to convective effects in the liquid. The formation of secondary branches from the primary stalk is observed to be a function of He^3 concentration in the mixture. Mixture of 0.9% He^3/He^4 concentration exhibits a branchless mode of dendritic growth (Fig.2) whereas those observed at the other two concentrations exhibit side branching (Figs.3,4).

The dendrites grown show remarkable alignment and the primary stalks are seen to be arranged in parallel rows. The dendrite stems are spaced periodically, and the period is a function of concentration, temperature and supercooling. This aspect has yet to be investigated. It is also observed that, at the end of the dendritic mode of growth, when the liquid is no longer supercooled, melting of the dendrites takes place, the side branches being the first to melt followed by the primary stalk and finally the planar interface.

It appears from the theoretical work (1-3) that the radius of curvature of the dendrite tip, the growth velocity and the supercooling can be related. These parameters can be measured in the present experiments and the results compared with those obtained for other physical systems (4,5).

Various stability analyses suggest that $\alpha d_o/VR^2$ is equal to a constant which depends somewhat on the model chosen. Here, V is the dendrite tip growth velocity and R its radius of curvature, α is the thermal diffusion constant and d_o the capillary length (6). This expression has been confirmed for some dendrites, but a complete study is still in progress. It should be emphasized that the ^3He-^4He system, unlike other systems used for studying growth instabilities, allows α and d_o to be changed within a wide range.

REFERENCES
(1) J.S. Langer, Rev. Mod. Phys. 52 (1980) 1.
(2) J.S. Langer and H. Müller-Krumbhaar, Acta Metall. 26 (1978) 1691, 1999, 1697.
(3) R.C. Brower, D.A. Kessler, J. Koplik & H. Levine, Phys. Rev. Lett. 51 (1983) 1111.
(4) M.E. Glicksman, R.J. Schaefer & J.D. Ayers, Metal. Trans. A 7 (1976) 1747.
(5) S.C. Huang & M.E. Glicksman, Acta Metall. 29 (1981) 701.
(6) The capillary length $d_o = \gamma T_m C_p/L^2$ where γ is the surface tension (assumed provisionally, and probably incorrectly, to be equal to that for pure ^4He), T_m is the freezing temperature, C_p the specific heat of the fluid and L the latent heat of freezing (also assumed to be that of pure ^4He). L in particular is very temperature dependent and α is concentration dependent.
(7) C. le Pair, K.W. Taconis, R. de Bruyn Ouboter, P. Das and E. de Jong, Physica 31 (1965), 764.

ACKNOWLEDGEMENTS
This work is partially supported by a grant from the H. Gutwirth Foundation.

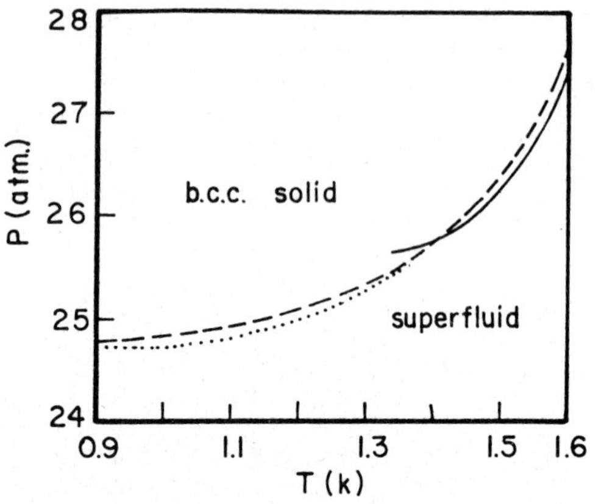

FIGURE 1

Phase diagram of ^3He-^4He mixtures at three different concentrations - 0.25%, -- 0.9%, (our experiment) ··· 2.77% (7).

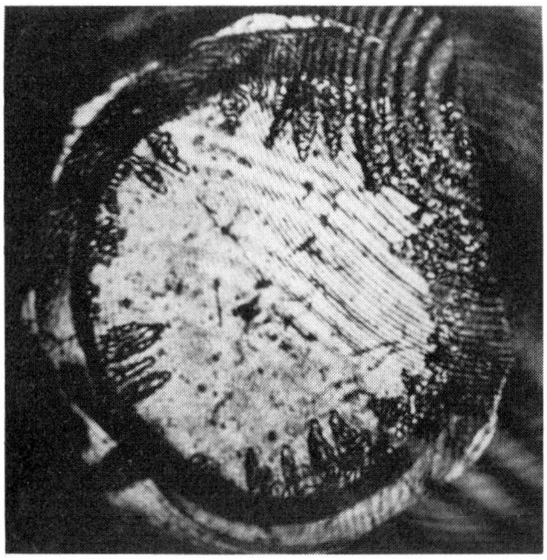

FIGURE 2a

Branchless dendrites of 0.9% mixture grow out of planar surface at 1.34K.

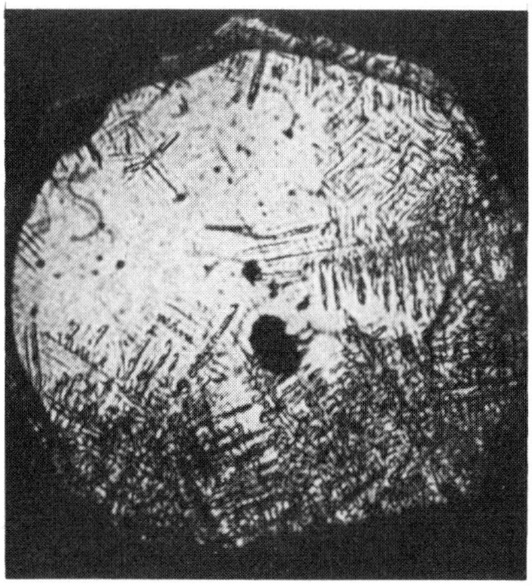

FIGURE 3

Dendrites with pronounced side branches of 0.25% mixture grown at T = 1.44K. The dark band parallel to the circumference shows planar solid.

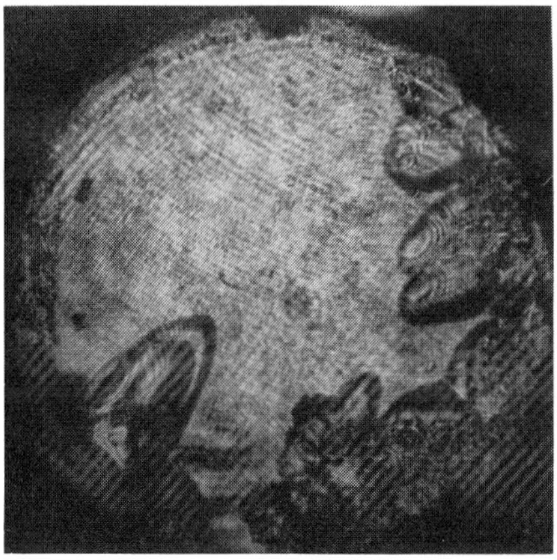

FIGURE 2b

Branchless dendrites of 0.9% mixture grown at T = 0.88K.

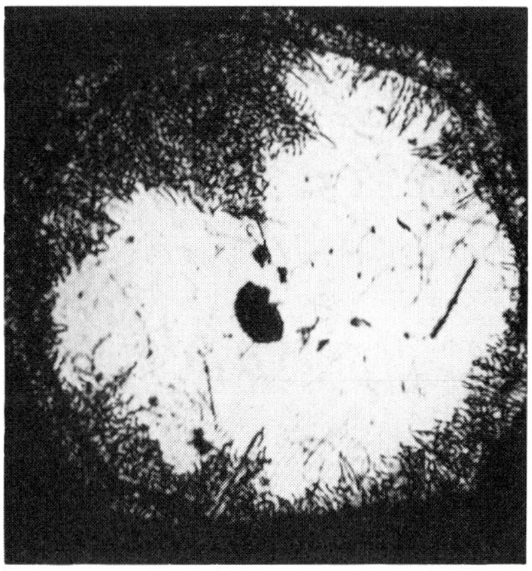

FIGURE 4

Dendrites with side branches of 2% mixture grown at T = 1.43K. The dark band parallel to the circumference shows planar solid.

LT-17 (Contributed Papers)
U. Eckern, A. Schmid, W. Weber, H. Wühl (eds)

ULTRASONIC STUDY ON MELTING AND FREEZING OF HELIUM-4

Yosio HIKI, Yasuhiro KANAYA, and Fujio TSURUOKA

Tokyo Institute of Technology, Oh-okayama, Meguro-ku, Tokyo 152, Japan

Melting of ^{4}He single crystals and freezing of liquid ^{4}He have been studied by measuring the attenuation of MHz ultrasound in the material. A remarkable supercooling is always observed in the freezing liquid. A large increase of attenuation occurs in the crystal near a temperature lower than the melting point. The latter behavior could be related with crystal dislocations.

1. INTRODUCTION

Studies of ultrasonic attenuation in ^{4}He crystals have firstly been started by the present authors (1). A large amount of experimental results have been accumulated, and they were able to be interpreted in a systematic manner (2, 3). It was found that the main origin of the acoustic loss in MHz frequency range in the crystals at temperatures above 1 K was the vibration of crystal dislocations. We intend to develop the study to investigate the phenomena of melting and freezing of ^{4}He, since there are several theoretical suggestions that the solid-liquid transition could be related with dislocations in meterials (4).

2. EXPERIMENTAL METHOD

Experiments were made on the melting of hcp ^{4}He single crystals and the freezing of liquid ^{4}He specimen by measuring the ultrasonic attenuation with a pulse reflection method. The specimen crystal used was prepared in a sample cell by cooling normal liquid helium under a constant pressure of 32.5 atm. A 5-MHz X-cut transducer and a reflecting plate were assembled inside the cell, and the sound attenuation was measured with longitudinal sound at frequencies of 5, 15, 25, 35, and 45 MHz. The absolute value of sound attenuation in the specimen was determined on an oscilloscope by adjusting the calibrated exponential curve to coincide with the envelope of the ultrasonic pulse echoes. The electronic apparatus adopted in the measurements consisted of an ultrasonic generator and receiver (Matec model 6000 + 750 or 760) and a synchronizer and exponential generator (model 1204A). Another apparatus used was an automatic attenuation recorder (model 2470A), by which relative values of attenuation derived from the height of a pulse echo were recorded on a chart. The recording of attenuation was made with increasing or decreasing the specimen temperature slowly. The temperature was controlled by pumping the liquid helium in the cryostat, and was measured by mercury and oil monometers and also monitored by a carbon thermometer.

3. EXPERIMENTAL RESULTS

The variation of ultrasonic attenuation α with temperature T in the melting and freezing ^{4}He can be seen in Figure 1. The overall features are summarized as follows. (a) As the temperature is increased, the attenuation in solid helium increases gradually and then very steeply. The ultrasonic pulse echoes disappear and the attenuation becomes immeasurable at a temperature around 1.78 K (= T_a). (b) Pulse echoes representative of liquid helium appear at a temperature around 1.89 K (= T_b). The attenuation decreases rapidly and becomes nearly constant with the increasing temperature. (c) When the liquid helium is cooled, the liquid pulse echoes continue to exist at temperatures below T_b. This may be due to the supercooling of the liquid, which is always observed in the present experiment. The attenuation increases gradually, and the pulse echoes disappear at a temperature around 1.80 K (= T_c). (d) The pulse echoes of solid appear at a temperature around 1.72 K (= T_d), and the attenuation decreases as the temperature is lowered.

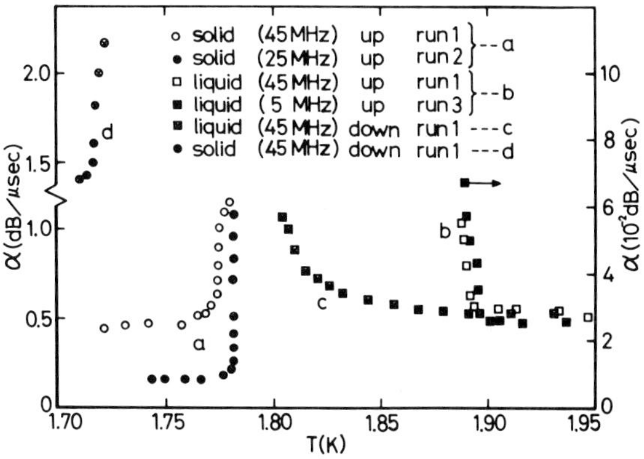

FIGURE 1
Temperature dependence of ultrasonic attenuation α in melting and freezing ^{4}He.

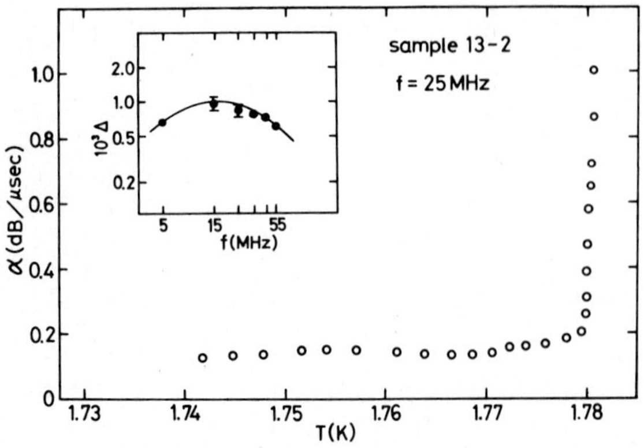

FIGURE 2
Temperature dependence of ultrasonic attenuation α in the premelting region, and dependence of decrement Δ on sound frequency f.

After referring to the method of preparing helium crystals (2) and the PVT relations in ^{4}He (5), the temperature T_b is considered to be the melting point of the solid specimens. It is interesting to see that the ultrasonic attenuation markedly increases near a temperature T_a which is lower than the melting point T_b. This seems to be a kind of premelting phenomena, and has mainly been studied in the present investigation. Single crystals of good quality were used in the experiments for that purpose. An example of detailed attenuation-vs-temperature data in the premelting region is shown in Figure 2. The inserted figure shows the dependence of decrement Δ (= $\alpha c/f$, where α is the attenuation in Np/cm, c is the sound velocity in cm/sec, and f is the frequency in Hz) on the sound frequency. The data were taken at a temperature around 1.70 K before the premelting experiment was started.

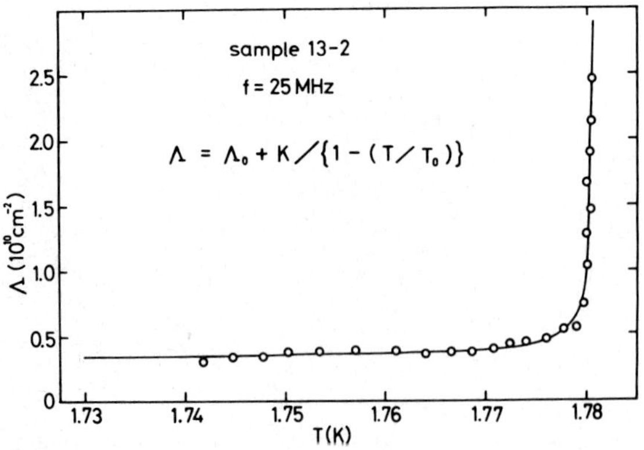

FIGURE 3
Parameter fit for temperature dependence of dislocation density Λ in the premelting region.

4. ANALYSIS

The origin of the sound attenuation or mecanical vibrational loss in ^{4}He crystals in MHz frequency range was shown to be the overdamped resonance of dislocations (2). The decrement then can be represented as

$$\Delta = \Delta_0\omega\tau/(1 + \omega^2\tau^2),$$

where ω is the angular frequency of sound; Δ_0 and τ are parameters containing such quantities as the dislocation density Λ, the pinning length L, the damping constant B, and also known quantities related with the elasticity and the orientation of the specimen crystal. The curve drawn in the Δ-vs-f data in Figure 2 is the fitted theoretical one obtained by adjusting the two parameters. The good fitting indicates that the sound attenuation in the present specimen crystal is really due to the dislocation damping.

In the present case of ^{4}He crystals, the pinning length L and the damping constant B can be calculated as functions of temperature (2). Then the dislocation density Λ can be obtained. The determined dependence of the dislocation density on the temperature is shown in Figure 3. It was found that the temperature dependence could well be represented as

$$\Lambda = \Lambda_0 + K/\{1 - (T/T_0)\}.$$

The curve in Figure 3 was obtained by adjusting three parameters Λ_0, K, and T_0 to fit the above formula to the data. The fitted values were: $\Lambda_0 = 3.4 \times 10^9$ cm^{-2}, $K = 4.0 \times 10^6$ cm^{-2}, and $T_0 = 1.781$ K. Experiments made on several other specimen crystals also revealed the same temperature dependence with nearly the same values of K and T_0.

We are considering that dislocations are spontaneously generated by a thermal fluctuation in the crystal below and near a characteristic temperature (T_0). A quantitative discussion will appear in the near future.

ACKNOWLEDGEMENT
The authors express their thanks to M. Yamanaka and J. Nishijima for their help in the experiments.

REFERENCES
(1) Y. Hiki and F. Tsuruoka, in Proceedings of the Fourteenth International Conference on Low Temperature Physics, Vol. 1, eds. M. Krusius and M. Vuorio (North-Holland, Amsterdam, 1975) p. 479.
(2) F. Tsuruoka and Y. Hiki, Phys. Rev. B20 (1979) 2702.
(3) Y. Hiki and F. Tsuruoka, Phys. Rev. B27 (1983) 696.
(4) F. R. N. Nabarro, Theory of Crystal Dislocations (Clarendon, Oxford, 1967).
(5) E. R. Grilly and R. L. Mills, Ann. Phys. (N.Y.) 18 (1962) 250.

LT-17 (Contributed Papers)
U. Eckern, A. Schmid, W. Weber, H. Wühl (eds)

FREEZING, MELTING, AND SUPERFLUIDITY OF ^{4}HE IN VYCOR

E.D. ADAMS, K. UHLIG, Yi-Hua TANG, and G.E. HAAS

Department of Physics, University of Florida, Gainesville, Florida 32611, USA

The solidification of ^{4}He in Vycor glass and in sintered Ag powder has been studied between 0.8 and 2.5 K. At 0.8 K, the melting pressure is elevated by 12 bar in Vycor and by 0.3 bar in the Ag sinter. In each material a range of melting pressures, corresponding to different pore sizes, is observed. The λ-transition in the liquid in Vycor is determined up to solidification pressure.

Several studies have been made of the solidification and superfluidity of ^{4}He in confined geometries.(1-3) Experiments on helium in Vycor glass have indicated that it remained a liquid at pressures of ~ 15 bar above the bulk melting pressure.(2,3) However, an accurate phase diagram has not been available because of uncertainties in the pressure of the helium in the Vycor.

We have measured pressure vs. temperature along isochores and the cooling rate of helium in Vycor to obtain a phase diagram up to P= 60 bar and T=2.5 K. There is a range of pressures where both solid and liquid exist, which, at T=0.8 K, extends from 36.3 to 38.0 bar. Similar behavior is seen in the Ag sinter, but at only ~ 0.3 bar above bulk melting. The λ-transition of helium in Vycor has been followed to solidification pressure.

The cell is depicted in Fig. 1 (inset). A short time constant for measurement of the pressure of ^{4}He in Vycor above bulk melting was made possible by grinding the Vycor into a powder with grains ≤ 74 μm, thereby reducing the distance for pressure communication. The powder loosely filled the cell to a filling factor of 0.5, allowing individual grains to move with the bulk helium surrounding them and to transmit pressure to the capacitive transducer.(4)

A layer of 700 Å sintered Ag powder provided thermal contact to the helium. The cell was mounted on the still of a dilution refrigerator with a weak thermal link, to allow slow cooling with a time constant ~ 500 times that for equilibrium within the cell. Temperatures were measured with a carbon resistor calibrated against the ^{4}He melting pressure.(5) The cooling rate $|dT/dt|$ is related to the heat current along the thermal link $\dot{Q}$ and the effective heat capacity by $\dot{Q}(T) = C_{eff}\,|dT/dt|$. The qualitative behavior of (dT/dt) indicates C_{eff} and the state of the helium in the cell.

After the cell was filled at a temperature above the bulk melting point,(5) the blocked fill line kept the quantity of helium constant. Then as the cell slowly cooled along an isochore, P(t) and T(t) were recorded. Also, P(T) was measured for both warming and cooling, while regulating the temperature.

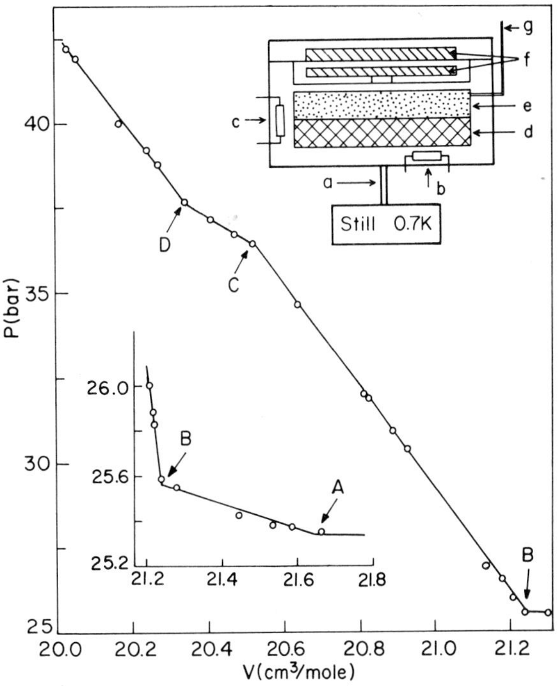

FIGURE 1
Inset, the cell: a- thermal link, b- heater, c- thermometer, d- Ag sinter, e- Vycor powder, f- capacitance pressure gauge, g- fill line. Main figure: P vs. v at T=0.8 K.

The state of the helium at T=0.8 K is seen from the isotherm P(v) (constructed from the isochores) in Fig. 1. At large v, P is constant at 25.3 bar, the bulk melting pressure. As v is decreased to point A (inset), solidification begins in the Ag sinter, with the freezing pressure increasing by 0.27 bar when the sinter is just filled with solid at point B. Solidification does not occur in the Vycor until the pressure is 36.3 bar, point C, with all solid above 38.0 bar. The slope of the isotherm between points A and B and between C and D indicates the elevation of melting pressure with decreasing pore size.

From the behavior of P(t) and T(t) along isochores, (not shown) the λ-transition or the freezing point of the Vycor helium was detected. Isochores with v > 20.5 cm^3/mole (point C, Fig. 1) showed a minimum in P(t), followed by inflection points in both P(t) and T(t), indicating the λ-point. The λ-temperature determined by this procedure is probably above that for the onset of superflow.(6) At smaller v, freezing occurred in the Vycor helium, marked by a rapid drop in P(t) and a decrease in the cooling rate. The reproducibility of the freezing point depended upon complete melting of the Vycor helium, indicating that the liquid was supercooled. The λ-line and freezing curve are shown in Fig. 2.

Because of the pore-size dependence of the melting pressure, liquid and solid coexist over a surface bounded by the freezing curve (all solid) and the melting curve (all liquid). The melting point T_m was indicated in P(T) data taken on warming by an increase in time constants for pressure equilibrium, a large C_{eff}, and a discontinuity in dP/dT. The melting points are shown in Fig. 2, where the lower end

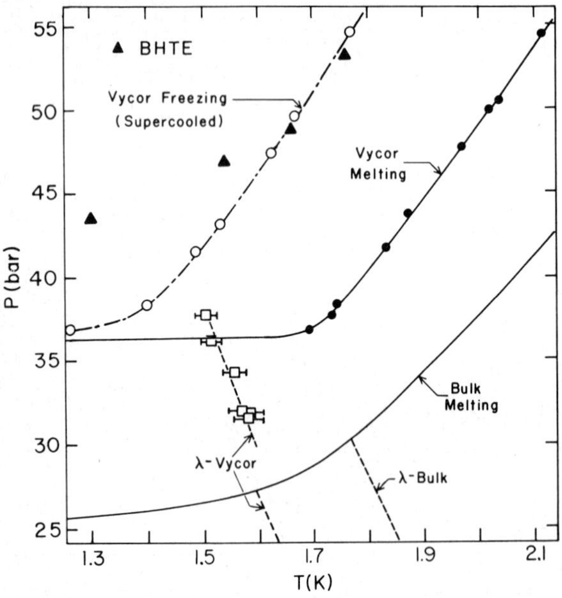

FIGURE 2
P-T phase diagram of helium in Vycor and of bulk helium. Triangles, Ref. 3; all other points, this work.

of the line has been drawn to coincide with point C of Fig. 1.

The equilibrium freezing line (all solid) $T_f(P)$ must intersect the T=0.8 K isotherm of Fig. 1 at point D. However location of $T_f(P)$ was not possible because of the supercooling. As shown in Fig. 1, the freezing and melting pressures at 0.8 K differ by only 1.7 bar. Thus the equilibrium freezing line is probably near the melting line.

The phase diagram for helium in Vycor (and bulk helium) is shown in Fig. 2. The freezing points of BHTE, which should be the same as our $T_{f,sc}$, are shown for comparison. We also determined the molar volume of the Vycor helium and found that relative to bulk helium, the liquid has a higher density, is less compressible, and changes volume less on freezing (e.g. Δv = 0.55 ± .1 cm^3/mole at P=37 bar vs. 1.44 for the bulk).

The model which has been applied to freezing at elevated pressures in confined geometries employs the lack of wetting of the surface by solid helium and the interfacial tension between the liquid and solid $\alpha_{\ell s}$.(1,3,7) Thus, to nucleate a seed crystal of radius r limited by the pore size, requires an excess pressure

$$\Delta P = 2\alpha_{\ell s} v_s/\Delta v_{\ell s} r.$$

Using our observed Δv at P=37 bar and T=1.3 K (instead of Δv_{bulk}), we obtain $\alpha_{\ell s}$ = 0.05 erg/cm^2, a factor of three less than the observed $\alpha_{\ell s}$ = 0.15. Further tests of this model could be made if $\alpha_{\ell s}$ were available at higher temperatures where greater elevation of the freezing pressure occurs.

We acknowledge useful discussions with Bob Guyer, Gary Ihas, and Pradeep Kumar. This work was supported by the National Science Foundation, grant number DMR-8312959. One of us (KU) acknowledges partial support from DFG, West Germany.

REFERENCES

(1) E.N. Smith, D.F. Brewer, Cao Liezhao and J.D. Reppy, Physica 107B (1981) 585.
(2) D.F. Brewer, Cao Liezhao and C. Girit, Physica 107B (1981) 583.
(3) J.R. Beamish, A. Hikata, L. Tell and C. Elbaum, Phys. Rev. Lett. 50 (1983) 425.
(4) G.C. Straty and E.D. Adams, Rev. Sci. Instrum. 40 (1969) 1393.
(5) E.R. Grilly, J. Low Temp. Phys. 11 (1973) 33.
(6) D.F. Brewer, J. Low Temp. Phys. 3 (1970) 205.
(7) S. Balibar, D.O. Edwards and C. Laroche, Phys. Rev. Lett. 42 (1979) 782; J. Landau, S.G. Lipson, L.M. Määttänen, Edwards, Phys. Rev. Lett. 45 (1980) 31.

LT-17 (Contributed Papers)
U. Eckern, A. Schmid, W. Weber, H. Wühl (eds)

SOUND PROPAGATION IN A SUPERLEAK AND CRYSTAL GROWTH OF ^{4}He

Makio UWAHA

Institut Laue-Langevin, BP 156X, 38042 Grenoble Cedex, France

Solid ^{4}He layers are formed on the surface of a superleak owing to the strong van der Waals attraction. Effects of crystal growth on the fourth sound propagation are studied in the framework of hydrodynamics. It is found that with fast crystal growth the sound velocity decreases drastically when the solid thickness is increased.

We consider a superfluid in a very narrow channel between two parallel plates which are covered with layers of solid ^{4}He formed by their van der Waals potentials $U(z)$ (Fig. 1). When a sound wave propagates in this system, the increase of pressure causes crystal growth of the solid. If the crystal growth is fast enough, the equilibrium between the solid and the liquid is maintained and the sound velocity will be modified. The extent of crystal growth is controlled by the strength of the substrate potential and therefore the change of the sound velocity depends on the thickness of the solid layer.

The propagation of long wavelength sound ($k\zeta \ll 1$) is determined by a continuity equation, an equation for entropy conservation (we neglect dissipative processes), an equation for superfluid velocity v, and a growth rate equation at the interface :

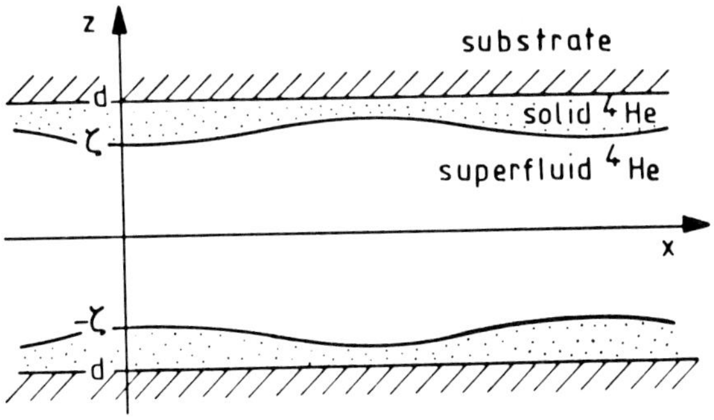

FIGURE 1
The superleak consists of two parallel plates, which are covered with solid ^{4}He formed by their van der Waals potentials. The upper half and the lower half are symmetrical. The solid thickness $d-\zeta$ is much smaller than the width $2d$.

$$\frac{\partial}{\partial t}(n_2\zeta) - n_1\frac{\partial\zeta}{\partial t} + \frac{\partial}{\partial x}(n_s\zeta v) = 0, \qquad (1)$$

$$\frac{\partial}{\partial t}(n_2 s_2\zeta) - n_1 s_1\frac{\partial\zeta}{\partial t} = 0, \qquad (2)$$

$$\frac{\partial v}{\partial t} + \frac{1}{m}\frac{\partial}{\partial x}\left(\mu_2 + \frac{1}{2}m v^2\right) = 0, \qquad (3)$$

$$\frac{1}{K}\frac{\partial\zeta}{\partial t} + \left(\mu_2 + \frac{1}{2}m v^2 - \mu_1\right)\Big|_\zeta = 0, \qquad (4)$$

where 1 and 2 represent solid and liquid respectively, n is the number density, μ the chemical potential per atom, and K the crystal growth coefficient defined as in Ref. [1]. The normal fluid component of the ^{4}He liquid is completely locked to the solid wall since ζ is much shorter that the mean free paths of thermal excitations. We have assumed $\partial n_1/\partial P=0$ and $\partial s_1/\partial T=0$ for simplicity. The interfacial tension and the interfacial inertia can be neglected in the long wavelength region.

The dispersion relation of sound waves obtained from linearized equations of (1)-(4) has the simple structure

$$(\omega^2-k^2c_t^2) + g(\omega)(\omega^2-k^2c_f^2) = 0, \qquad (5)$$

where c_f is the usual fourth sound velocity. The coupling factor $g(\omega)$ is defined by

$$g(\omega) = \frac{1}{R}\left[\frac{n_1-n_2}{n_2}\frac{\zeta}{mc_o^2}\left|\frac{\partial U}{\partial z}\right| - i\omega\frac{\zeta}{mKc_o^2}\frac{n_1}{n_2}\right], \qquad (6)$$

$$R = \left(\frac{n_1-n_2}{n_2}\right)^2 + \left(\frac{n_1}{n_2}\right)^2 (s_1-s_2)^2\frac{\partial n_2}{\partial P}\left(\frac{\partial s_2}{\partial T}\right)^{-1}, \qquad (7)$$

where c_0 is the first sound velocity. The other sound velocity in (5) is given by

$$c_t^2 = \frac{n_s}{n_2}\frac{1}{m}\left(\frac{n_1 s_1 - n_2 s_2}{n_2}\right)^2 \left(\frac{\partial s_2}{\partial T}\right)^{-1} \frac{1}{R} \ . \qquad (8)$$

If the solid grows very slowly ($K \to 0$) or the substrate potential is very strong ($|\partial U/\partial z| \to \infty$), the second term in (5) dominates and we have the usual fourth sound propagation. In the opposite case ($K \to \infty$ and $|\partial U/\partial z| \to 0$) the sound velocity goes to zero at $T = 0$: the change of pressure is completely absorbed by immediate growth of the crystal. At finite temperatures, however, the sound mode is recovered : the change of entropy density on crystallization or melting produces a change of chemical potential, which accelerates the superfluid. In Fig. 2 we show the sound velocity as a function of temperature and the potential strength (equivalent to the thickness of solid layer) for the case of fast growth ($K \to \infty$). The sound velocity decreases drastically when the solid thickness is increased. The transition between the slow growth and the fast growth regime is expected when the crystal growth coefficient is such that the second term of (6) is of order unity.

Maynard et al. [2] performed fourth sound resonance measurements with a grafoil ring and found an oscillation of the resonance frequency, which is attributed to layer-by-layer crystal growth. Systematic decreases of resonance frequency with the increasing solid thickness observed in the experiment agrees qualitatively with our results with large K. The measured values of the growth coefficient K in other experiments [3] all correspond to large K and are consistent with the above result. (If there are facets the situation may be different : $K = 0$ below some critical chemical potential difference [4]. In that case imperfections in the system play important roles). No resonance was observed when the solid was very thick ($\gtrsim$ 20 Å). As we expect drastic changes of the sound velocity and the mechanism of sound propagation in this region, further experimental studies are very desirable.

Detailed argument and calculation including effects of steady liquid flow will be given elsewhere [5].

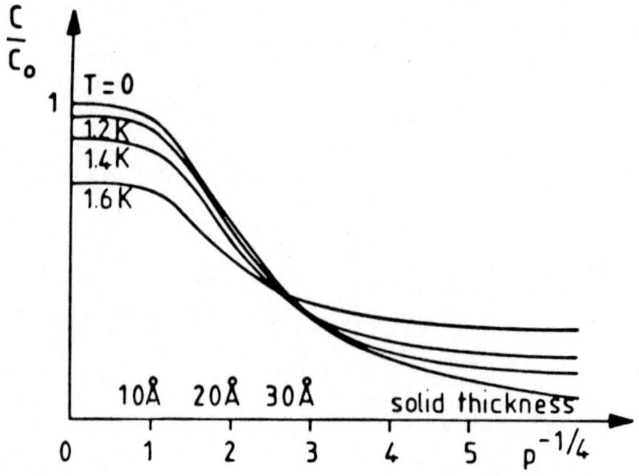

FIGURE 2
Sound velocity c for $K \to \infty$. It becomes smaller as the potential, $p = (\zeta/mc_0^2)|\partial U/\partial z|$, becomes weaker (the solid layer is thicker). At finite temperatures c remain finite even when $p \to 0$.

REFERENCES

[1] A.F. Andreev and A. Ya. Parshin, Zh. Eksp. Teor. Fiz. 75 (1978) 1511 (Sov. Phys. JETP 48 (1978) 763).

[2] J.D. Maynard, G.J. Jelatis and J.A. Roth, Physica 107B (1981) 243.
S. Ramesh and J.D. Maynard, Phys. Rev. Lett. 49 (1982) 47.

[3] B. Castaing, S. Balibar and C. Laroche, J. Physique 41 (1980) 897.
K.O. Keshishev, A. Ya Parshin and A.V. Babkin, Zh. Eksp. Teor. Fiz. 80 (1981) 716 (Sov. Phys. JETP 53 (1981) 362).
J. Bodensohn, P. Leiderer and D. Savignac, Proceedings of the 4th International Conference on Phonon Scattering in Condensed Matter.

[4] P.E. Wolf and S. Balibar, private communication.

[5] M. Uwaha, submitted to J. Physique.

LT-17 (Contributed Papers)
U. Eckern, A. Schmid, W. Weber, H. Wühl (eds)

SOUND TRANSFORMATION AT THE SOLID ^{4}He-He II INTERFACE

N.E. DYUMIN, V.N. GRIGOR'EV and S.V. SVATKO

Institute for Low Temperature Physics and Engineering, UkrSSR Academy of Sciences, Kharkov, USSR

Experiments have revealed transformation of the first and second sounds at the interface between solid ^{4}He and He II. The temperature dependence of the coefficient of transformation of the first sound into second at the interface has been measured. The growth coefficient K was found to vary roughly as exp (Δ/T), where $\Delta \approx 7.2$ K.

The surface properties of helium crystals are interesting to study primarily because the solid ^{4}He-He II interface is of quantum nature (1). Owing to delocalization of growth steps on the crystal surface, the interface is highly mobile, causing a variety of new effects proper to quantum crystals only (2). Reference (3) considered the problem of the interaction of a first sound pressure wave with the solid ^{4}He-He II interface and revealed that because of the high mobility of the interface, in addition to a transmitted and reflected waves, there results a second sound in He II. The study of this effect provides a new means to investigate the features of quantum crystal growth. We intended to find mutual transformation of the first and second sounds at the interface of solid ^{4}He and He II and obtain information on the kinetic growth coefficient.

The experimental cell represented a thick-walled beryllium bronze cylinder of 10 mm inner diameter and 30 mm high. On the top and bottom of the cell, similar capacitive gauges were mounted to generate and detect the first and second sounds. The main part of the gauges was a low-porous two-layer diaphragm acting as the mobile element. The cell bottom was connected through a copper heat lead with a ^{3}He refrigerator. Some part of the high pressure filling capillary was in thermal contact with the ^{3}He refrigerator. All the arrangement was surrounded by a vacuum jacket. The crystal growth took place at a constant ^{4}He mass.

The bottom gauge, inside the crystal grown, excited a pressure wave train (first sound) in the solid ^{4}He. The top gauge, in the liquid, registered two sound modes. One of these (fast) had the velocity which was superposition of the first sound in the solid and the liquid, and the other (slow) superposition of the first sound in the solid and the second sound in the liquid. Figure 1 represents a typical oscillogram of signals of the slow mode (hcp phase, T $\approx$ 1.2 K). The presence of several echo signals suggests that the solid ^{4}He-He II interface is flat. In some instances the functions of the emitter and the receiver were interchanged: the top gauge was used as the emitter (with the second sound mostly emitted in the liquid) and the bottom one as the receiver. This did

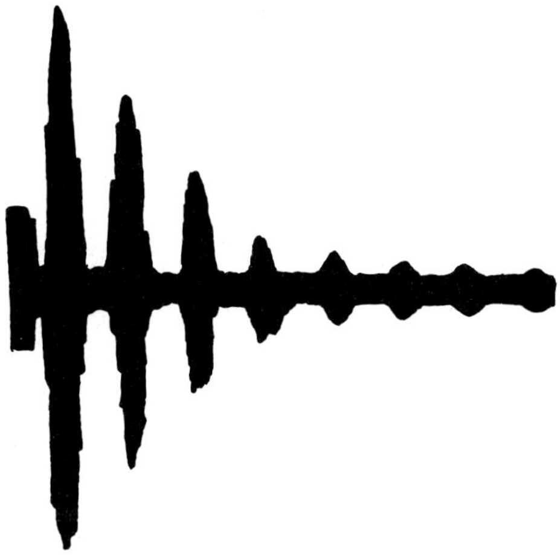

FIGURE 1

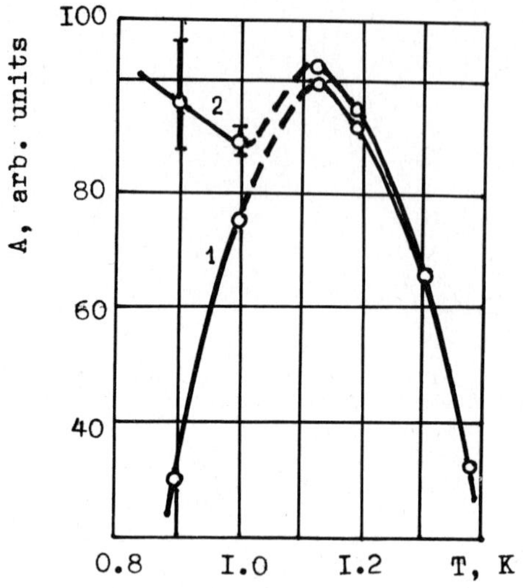

FIGURE 2

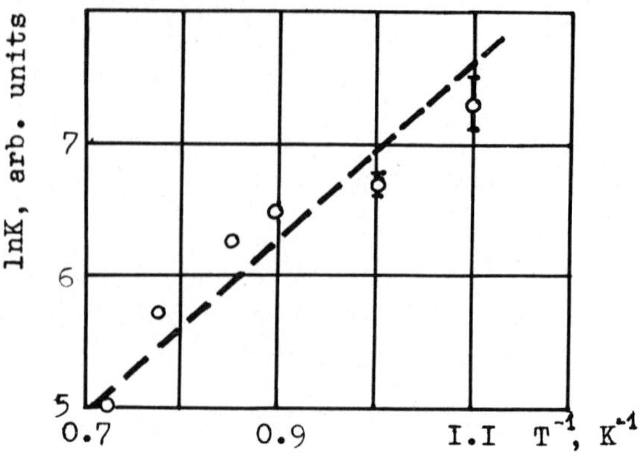

FIGURE 3

not much affect the whole picture.

The experiments were run both on bcc and on hcp solid ^{4}He at temperatures from 1.7 to 0.9 K. The mutual transformation of the first and second sounds was observed on the bcc phase (5); the transformed second sound signal amplitude increased with lowering temperature. Note that this amplitude was in the hcp phase several times larger than in the bcc phase, which enabled measurements of the transformation coefficient as a function of temperature between 1.4 and 0.9 K. A typical temperature dependence of the second sound signal amplitude is shown in Fig. 2 (curve 1). At low temperature, the correction for second sound absorption becomes significant. Curve 2 allows for this (the error values indicated are due to inaccurate data on second sound absorption).

From the experimental data, with using the formula for the coefficient of transformation of the first sound into the second at the interface (3), information on the kinetic growth coefficient was obtained:

$$K = \frac{T'}{P'} \frac{\rho_2 C_v U_2}{L} \frac{\rho_1 - \rho_2}{\rho_1 \rho_2} \left[1 + \frac{KL^2}{\rho_2 C_v U_2 T} \right]^{-1}$$

where T' is the temperature variation at the interface, P' the pressure variation, ρ_1 and ρ_2 the solid and liquid phase densities, C_v the liquid He specific heat at a constant volume, L the melting heat, U_2 the second sound velocity in He II, and T temperature. Estimates using absolute K values (2) extrapolated to high temperatures indicate that the second term between the brackets in the equation is negligible as compared to unity. It was also assumed by the calculation of K that P' is throughout the temperature range constant for a constant emitter power and the received signal amplitude is proportional to T'. The so found growth coefficient is $K \sim \exp(\Delta/T)$, where $\Delta \approx 7.2$ K, i.e. the same as the roton gap in He II at a pressure of 25 atm. (Fig. 3 represents the 7.2 K slope by a dashed line).

Note a curious anomaly of the transformed signal amplitude at 1.1-1.2 K revealed by experiments. As is known, in this temperature range the roughening transition on the ^{4}He crystal surface was found, and we hope that study of the sound transformation near the singular points will result in more data on the above phenomena.

REFERENCES

(1) A.F. Andreev and A.Ya. Parshin, ZhETF 75 (1978) 1511.

(2) K.O. Keshishev, A.Ya. Parshin and A.V. Babkin, ZhETF 80 (1981) 716.

(3) B. Castaing and P. Nozieres, J. de Physique 41 (1980) 701.

(4) V.N. Grigor'ev, N.E. Dyumin and S.V. Svatko, Fiz. Nizk. Temp. 9 (1983) 649.

LT-17 (Contributed Papers)
U. Eckern, A. Schmid, W. Weber, H. Wühl (eds)

THE MORPHOLOGY OF THIN FILMS OF SOLID ^{4}He DEPOSITED ON SAPPHIRE AND THE TRANSFORMATION hcp$\rightleftarrows$fcc ^{4}He

J.P. FRANCK, J.T. GLEESON, K.E. KORNELSEN, and K.A. McGREER*

Department of Physics, University of Alberta, Edmonton, Alberta Canada T6G 2J1

Thin layers of solid helium were grown on sapphire single crystal substrates at elevated pressures (roughly 50 to 900 MPa). The morphology of the films depends on the crystal structure. Below 113 MPa (in the hcp phase) a system of parallel slip bands is observed, whereas above 113 MPa (in the fcc phase) a polygonal structure similar to a helium froth is found. Melting of this froth shows grain boundary melting by fluid ^{4}He. Grain boundaries observed in hcp ^{4}He are not wetted by the fluid. Near the triple point at 113 MPa and 15 K one can deposit both phases side by side. In such structures the transition fcc→hcp ^{4}He can be observed, it proceeds by the parallel motion of low energy boundaries. Similar visual transition observations were also made on carefully grown crystals at higher pressures (157 MPa).

The experiments were conducted in the optical high pressure cell described by Franck and Daniels (1). The windows were sapphire single crystals of diameter 4.76 mm, with an optically accessible diameter of 1.5 mm. The sapphire hexogonal c-axis was within 15' of arc normal to the deposition plane. The experiments used high purity helium, impurities (including ^{3}He) being below 1 ppm.

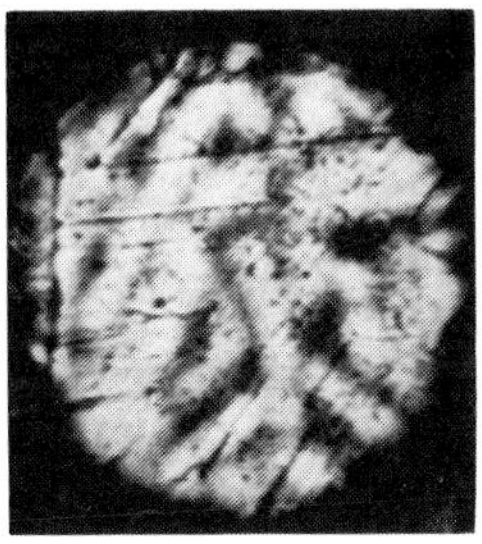

FIGURE 1

Banded morphology of an hcp ^{4}He film. The diameter of the field of view in this and the following figures is 1.5 mm. 85 MPa

The thin helium films on the sapphire windows were grown by slowly lowering the temperature below the melting line at constant pressure. It was found that a crystallization front sweeps across the windows, in its wake the thin layer thickens somewhat and develops grain boundaries. The crystallization front could be stopped, and moved both forward and backwards by judicious cooling or heating. The observed grain boundary system shows parallel banding in the hcp phase along slip or twin boundaries. In Fig. 1 we show a structure of this kind. Melting of these structures, as shown in Fig. 2, does not result in wetting of these boundaries. In the fcc phase the grain boundary structure resembles a helium froth, as shown in Fig. 3. The structure

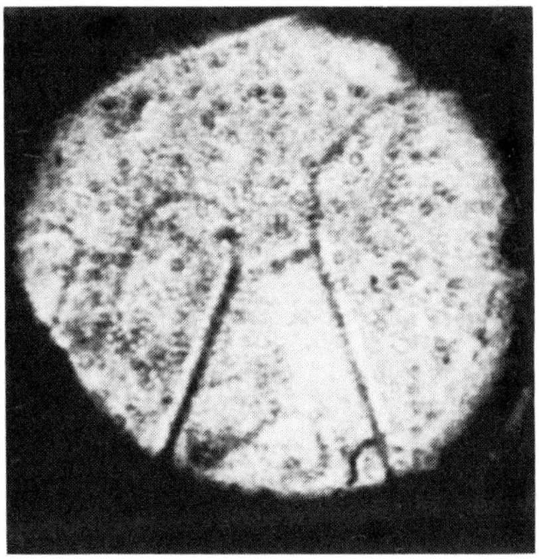

FIGURE 2

Melting of an hcp ^{4}He structure. The fluid covers the left upper part, low energy grain boundaries as seen in the lower half. 110 MPa

is determined by mechanical equilibrium conditions at the grain boundary vertices. The size distribution of grains is well described by the theoretical distribution derived by Lifshitz and Slezov (2) and Hillert (3):

$$(1) \qquad P(u) = \frac{8u}{(2-u)^4} \cdot \exp \frac{2u}{2-u}$$

where u=d/<d> is the reduced effective grain diameter. The more recently proposed exponential distributions (Louat (4)), or log-normal

*Supported by grants from the Natural Sciences and Engineering Research Council of Canada.

distributions (Srolowitz et.al. (5), Weaire and Kermode (6)), do not fit our data. This may be due to the two-dimensional character of our films.

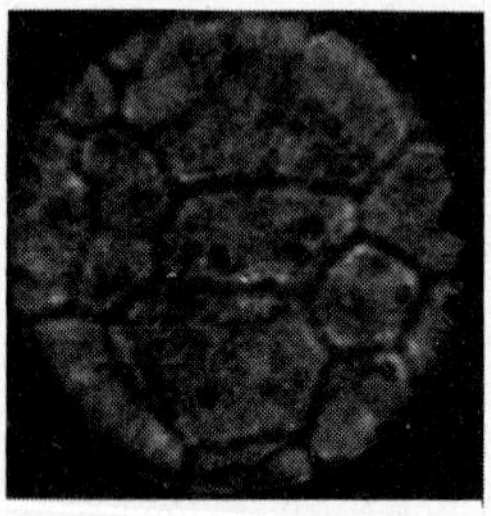

FIGURE 3

fcc ^4He grain boundaries. 193 MPa

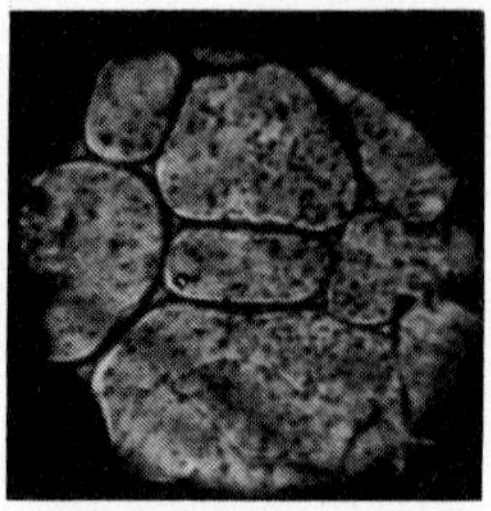

FIGURE 4

Appearance of fluid at the vertices and grain boundaries in fcc ^4He. 193 MPa

When the polygonal fcc He films are melted, one observes grain boundary melting and wetting of the grain boundaries. This is shown in Fig. 4. In this case one can observe the dihedral angle between fcc and fluid ^4He. From this one can find the ratio of surface energies between two grains, and between a grain and fluid, using the mechanical equilibrium condition (see Fig. 5):

(2) $$\gamma_{gg}=2\gamma_{gf}\cdot\cos\frac{\theta}{2}\ .$$

One finds that $\gamma_{gg}\approx 2\gamma_{gf}$, i.e. one is at the limit of complete wetting.

When the solid helium films are grown near the triple point at 15 K and 113 MPa, one finds hcp and fcc material side by side. Thermal holding leads to the elimination of the hcp material and its transformation to fcc ^4He. This process takes place by the parallel movement of low energy boundaries.

The contrast necessary to observe these grain boundaries was probably produced through small angle refraction at surface grooves along the grain boundaries. In order to have this contrast, we found that the observations had to be made in close proximity to the melting line. When the temperature was lowered by more than 50 mK below the melting line, one or several fronts were seen to cross the sapphire surface reducing the contrast to almost complete invisibility. Heating back to within 50 mK of

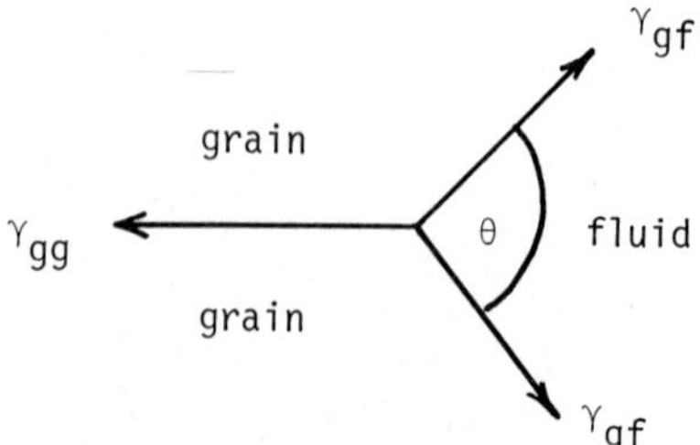

FIGURE 5

Mechanical equilibrium of a fluid-grain boundary vertex.

the melting line made the contrast reappear. It is possible that this effect is connected with a transition between wetting and non-wetting of the sapphire surface by the solid.

When crystals are carefully grown at higher pressures (157 MPa), completely filling the pressure cell, one can also observe the hcp$\rightleftarrows$fcc transition. Contrast between the phases in this case is obtained by an elliptometric method. In good crystals we have observed the transition proceeding by parallel motion of interfaces, similar to the behaviour observed near the triple point. The transition is readily observed in either direction. In some cases we were able to observe the thermo-elastic transition.

The observed characteristics of the transition are strikingly similar to those found in metals and alloys which undergo a martensitic transition between the hexagonal and cubic close-packed phases, as eg. Co and CoNi (7). This is further evidence for the martensitic character of the hcp-fcc transition in helium.

REFERENCES

(1) J.P. Franck and W.B. Daniels, Phys. Rev. B24 (1981) 2456.
(2) I.M. Lifshitz and V.V. Slezov, Zh. Eksp. Teor. Fiz. 35 (1958) 479, Sov. Phys. JETP 8 (1959), 331.
(3) M. Hillert, Acta Met. 13 (1965) 227.
(4) N.P. Louat, Acta Met. 22 (1974) 721.
(5) D.J. Srolovitz, M.P. Anderson, G.S. Grest and P.S. Sahui, Scripta Metall. 17 (1983) 241.
(6) D. Weaire and J.P. Kermode, Phil. Mag. B48, (1983), 245.
(7) Lee, e.g., E. deLamotte and C. Altstetter, in: "The Mechanism of phase transformations in crystalline solids", Proc. of an Intern. Symposium by the Inst. of Metals, Manchester (1968), publ. by The Inst. of Metals, London 1969.

LT-17 (Contributed Papers)
U. Eckern, A. Schmid, W. Weber, H. Wühl (eds)

PHASE CHANGE OF HELIUM ADSORBED ON Y ZEOLITES

N. WADA, H. KATO, H. SHIRATAKI, S. TAKAYANAGI†, T. ITO*, and T. WATANABE

Department of Physics, Faculty of Science, Hokkaido University, Sapporo 060, Japan
†Physics Division, Sapporo Branch, Hokkaido University of Education, Sapporo 064, Japan
*Research Institute for Catalysis, Hokkaido University, Sapporo 060, Japan

The heat capacity is measured for ^{3}He and ^{4}He adsorbed on the narrow void channel of Y zeolite crystals (Na-Y and H-Y). The isotope effect and the dependence on Na^+ and H^+ cations are observed. The critical phase change of ^{4}He and ^{3}He in Na-Y is discussed by considering the zero point energy and the van der Waals potential due to the electric field of cations.

1. INTRODUCTION

The behavior of helium system in restricted geometries is one of great interests in recent years. There have been many studies for helium adsorbed on graphite(1), porous Vycor glass(2) and some substrates(3). Recently we reported on a ^{4}He system adsorbed on Na-Y zeolite crystal having a regular narrow void channel(4). It was suggested that unsaturated ^{4}He in Na-Y zeolite behaves as a semiquantum liquid and that it makes a phase transition at lower temperature. Here, we report the heat capacities of ^{3}He and ^{4}He adsorbed on Na-Y and H-Y zeolites that have Na^+ and H^+ cations on the wall of the channel, respectively.

2. SAMPLES

Na-Y zeolite, $Na_{56}[(AlO_2)_{56}(SiO_2)_{136}]\cdot 250H_2O$ (5), has a cubic unit cell with a dimensions as large as 25 Å. This zeolite has a large void volume about 50 vol.%. In each unit cell, there are 8 void cages (about 13 Å in diameter) linked through apertures (about 8 Å in diameter) forming a diamond-like structure. The cations Na^+ cancel the minus charges of the aluminosilicate framework. H-Y zeolite is Y zeolite exchanged Na^+ with H^+.

3. EXPERIMENTAL RESULTS

The heat capacities, C, of ^{3}He as well as ^{4}He adsorbed on Na-Y are measured below the quantity adsorbed $n \sim 4$ atoms/cage. The experimental results of C/n versus temperature, T, of n=1.73 for both isotopes are shown in Fig. 1. The result for ^{4}He has been reported before(4). In both cases, below the critical temperature T_c, C/n steeply decrease with decreasing T, where T_c (=2.55 K) for ^{3}He is lower than T_c (=2.75 K) for ^{4}He. The n dependences of T_c for ^{3}He and ^{4}He are shown in Fig. 2. In both cases, T_c decreases with increasing n. As shown in Fig. 1, above T_c, C/n for ^{3}He is smaller than that for ^{4}He. Both are almost proportional to T, i.e. $C/n=\alpha T$, where $\alpha=2.0$ J/K^2/mol for ^{4}He and $\alpha=1.8$ J/K^2/mol for ^{3}He. On the contrary, in the case of H-Y zeolite, no critical change is observed in the measuring temperature range. In Fig. 3, C/n of ^{4}He in H-Y is compaired with that in Na-Y. The result for ^{3}He shows a similar T dependence, but C/n of ^{3}He is several percent smaller than that of ^{4}He.

4. DISCUSSION

At first, we consider the zero point energy of helium isotopes in Y zeolites. The zero point energy of ^{3}He is larger than that of ^{4}He. The zero point energy of ^{4}He (or ^{3}He) is inversely proportional to the volume therein. The void space of Y zeolite is composed of the large cages (about 13 Å in diameter) and the apertures (8 Å) that connect the cages to each other. Then, at low temperature, ^{4}He (^{3}He) atoms are bounded into the large cage rather than into the apertures so as to lower the zero point energy. The binding potential for ^{4}He is estimated to be about $2\sim3$ K(6). Next, we consider the van der Waals potential due to

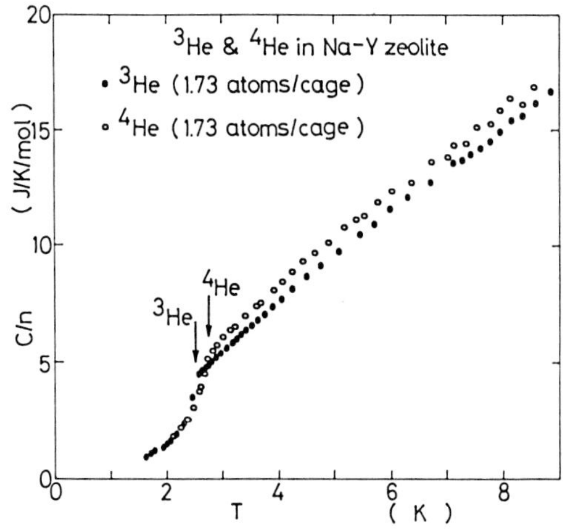

Fig. 1: Heat capacities per mole of ^{3}He and ^{4}He adsorbed on Na-Y zeolite.

the electric field of cations. This van der Waals potential tends to localize a helium atom at a position,i.e. the effective space for a helium atom is reduced, and this makes the zero point energy higher. In other words, there is a competition between the van der Waals potential that localizes helium atom and the zero point energy.

The T dependence of C/n is quite different between that for Na-Y and for H-Y (Fig. 3). In the case of Na-Y, T_c of ^{3}He is lower than that of ^{4}He at each n (Fig. 2). This difference may indicate that the increase of the zero point energy due to the isotope effect reduces the dominant van der Waals potential. In the range below T_c the helium atoms seem to be localized by the van der Waals potential. And also, the n dependence of T_c may be in favor of this supposition. Above T_c, the adsorbed helium seems to move along the channel. The linear T dependence of C/n may show a semiquantum liquid(7). In the case of H-Y and Na-Y in the temperature range above T_c, C/n of ^{3}He is smaller than that of ^{4}He at each T. The isotope effect shows a larger binding potential for ^{3}He than that for ^{4}He. It may be supposed that the binding potential due to the zero point energy is dominant in these cases.

5. CONCLUSION

The heat capacities of the helium systems in Y zeolites show the isotope effect and the dependence on the cations (Na^+ and H^+). The heat capacities of ^{4}He and ^{3}He in Na-Y zeolite at the quantity adsorbed below about 4 atoms/cage show the phase change at the critical temperature T_c. Whereas, the phase change is not observed for ^{4}He and ^{3}He in H-Y zeolite. The experimental results suggests that the heat capacities of helium systems in Y zeolites depend on the zero point energy and on the van der Waals potential due to the electric field of cations.

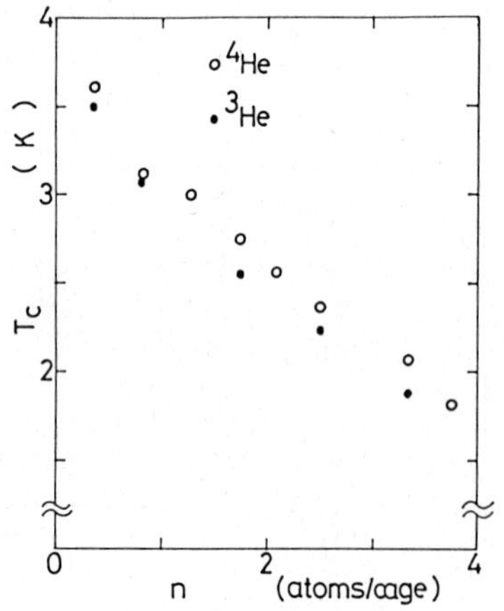

Fig. 2: Critical temperatures T_c for ^{4}He and ^{3}He in Na-Y zeolite as a function of quantity adsorbed n.

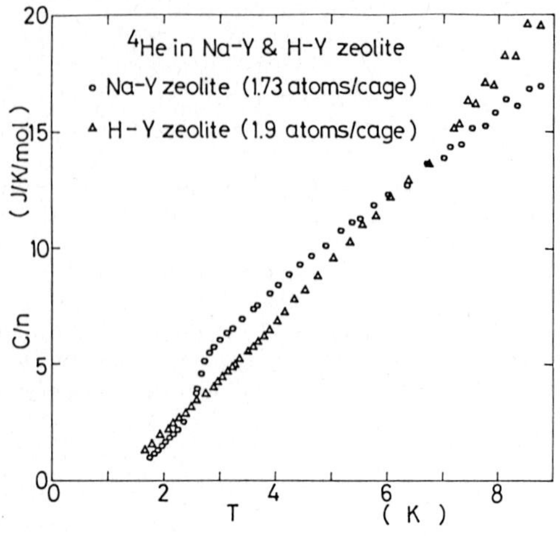

Fig. 3: Heat capacities per mole of ^{4}He in Na-Y and H-Y zeolites.

ACKNOWLEDGEMENT

The authors acknowledge Dr. K. Miyake of Nagoya University for stimulating discussions.

REFERENCES

(1) J.G. Dash, Films on Solid Surfaces (Academic Press, New York, 1975).
(2) R.H. Tait and J.D. Reppy, Phys. Rev. B20 (1979) 997.
(3) N. Wada et al., Heat capacity of ^{4}He adsorbed on the one-dimensional void channel of K-L zeolite, this volume.
(4) N. Wada, T. Ito and T. Watanabe, J. Phys. Soc. Jpn. 53 (1984) 913.
(5) D.W. Breck, Zeolite Molecular Sieves (John Wily and Sons, New York-London, 1974).
(6) K. Miyake, private communication.
(7) A.F. Andreev, JETP Lett. 28 (1978) 556.

LT-17 (Contributed Papers)
U. Eckern, A. Schmid, W. Weber, H. Wühl (eds)

HEAT CAPACITY OF ^{4}He ADSORBED ON THE ONE-DIMENSIONAL VOID CHANNEL OF K-L ZEOLITE

N. WADA, H. KATO, S. SATO, S. TAKAYANAGI†, T. ITO*, and T. WATANABE

Department of Physics, Faculty of Science, Hokkaido University, Sapporo 060, Japan
†Physics Division, Sapporo Branch, Hokkaido University of Education, Sapporo 064, Japan
*Research Institute for Catalysis, Hokkaido University, Sapporo 060, Japan

We have measured the heat capacity of ^{4}He adsorbed on the narrow one-dimensional channel of K-L zeolite crystal as a function of quantity adsorbed and temperature. The linear temperature dependence of the heat capacity is observed for the quantity adsorbed about 8 atoms/unit cell between 1.5 K and 9 K. It suggests that ^{4}He adsorbed on the channel behaves as a semiquantum liquid in this temperature range.

1. INTRODUCTION

In recent several years, there has been a considerable interest in ^{4}He adsorbed on the plane surface of pyrolitic graphite(1) and on the pore of Vycor glass(2). Especially, the recent study of ^{4}He on Vycor(3) is extremely attractive one for a superfluidity. On the other hand, we have investigated ^{4}He and ^{3}He on zeolites having different crystal structures (4,5). In this paper, we have studied ^{4}He adsorbed on a one-dimensional void channel of K-L zeolite by means of the heat capacity measurement.

2. SAMPLE

The crystal structure(6) of the synthetic K-L zeolite, $K_xNa_y[(AlO_2)_9(SiO_2)_{27}]\cdot 22H_2O$ ($x \approx 9$ and $y \approx 0$), is hexagonal with unit-cell dimensions a=18.4 Å and c=7.5 Å. The aluminosilicate framework composes the channel along the c-axis. The water is removed by baking the zeolite in a vacuum at about 400°C. The cations K^+ remain in the channel, and cancel the minus charges of the aluminosilicate framework. The one-dimensional void channels are not connected with each other through any hole. The free diameter of the void channel is periodically modulated in every 7.5 Å along the c-axis. The largest section is about 13 Å in diameter, and the smallest one has minimum and maximum diameters of 7.1 Å and 7.8 Å. In other words, void cages are linked one-dimensionally.

3. RESULTS AND DISCUSSIONS

The ^{4}He adsorption on the dehydrated K-L zeolite is saturated at the quantity of n=17 atoms/unit cell at the temperature 4.2 K. The heat capacity of ^{4}He per mole at T=3.0 K and 5.0 K is shown in Fig. 1 as a function of quantity adsorbed n. Each C/n-n curve has a minimum at the quantity adsorbed n_0=4 atoms/unit cell. Below n_0, C/n increases linearly with decreasing n. These facts indicate that the ^{4}He atoms interact with each other. Figure 2(a) shows C/n vs T at quantities adsorbed below n_0. A steep increase of C/n with increasing T is observed for $n<n_0$ around 4 K, e.g. between 4 K and 5 K for n=0.67. Figure 2(b) and (c) shows the heat capacity above n_0. At $n \approx 8$, the heat capacity is proportional to the temperature, i.e. $C/n=\alpha T$ (=1.0 J/K^2/mol). When n is increased or decreased from about 8, the C/n-T curve deviates down from the linear temperature dependence.

The property of ^{4}He in K-L zeolite seems to be different below and above n_0. To explain the behavior of ^{4}He adsorbed less than n_0, we consider as follows. Figure 1 indicates that the heat capacity C/n for $n=n_0(=4)$ is almost zero below about 3 K. It may be supposed that a definite region (definite sites in each cage) is filled with four ^{4}He atoms. There is no thermal excitation in each definite region below about 3 K. When n is reduced, a vacancy or a 'hole' is formed in the definite region. The motion of the 'hole' is excited by a thermal energy. The heat capacity C/n increases in proportion to the number of the

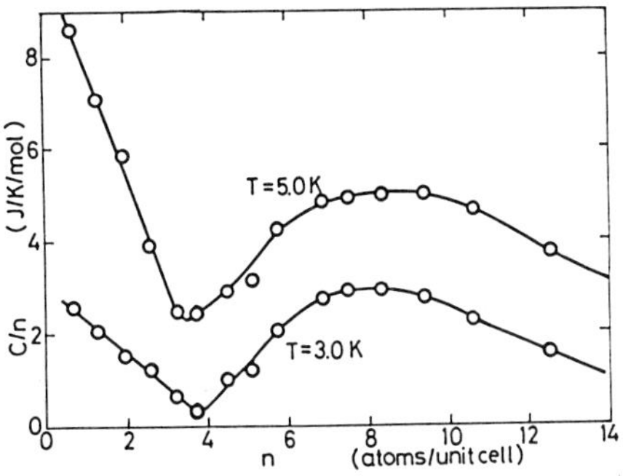

Fig.1: Heat capacity of ^{4}He per mole adsorbed in K-L zeolite at T=3.0 K and 5.0 K as a function of quantity adsorbed, n.

'hole', which is supposed by the dependence of C/n on n ($n<n_0$) at T=3.0 K in Fig. 1. As shown in Fig. 2(a), C/n steeply increases around 4 K. It may be due to the begining of a collective motion of the 'hole' from a definite region to the next one.

When n is increased above n_0 until about 8, C/n increases at each temperature. The excess helium atoms adsorbed more than n_0 move freely through the channel, and attack to the atoms trapped in the definite regions. Then, near n=8, it seems that the all ^4He atoms move freely along the channel in the measuring temperature range. The heat capacity has the simple relation, $C/n=\alpha T$ ($\alpha=1.0$ J/K^2/mol). It is well known that the bulk ^4He remains liquid state at much lower temperatures than the Debye

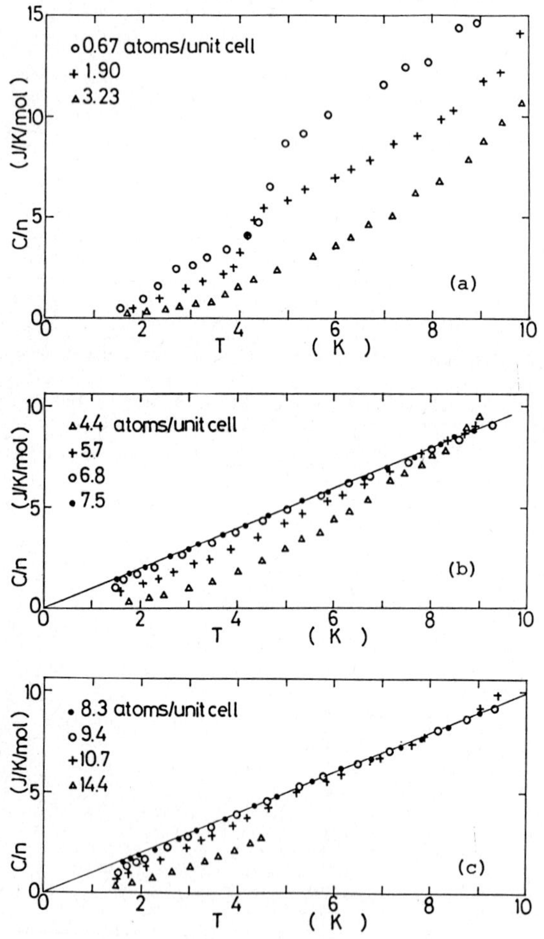

Fig. 2(a)-(c): Heat capacity of ^4He per mole as a function of temperature for several quantities adsorbed.

temperature Θ because of a large zero-point oscillation. In the lower temperature region ($T<T_c$), the liquid sets into a superfluid or a solid. In the temperature range $T_c<T<\Theta$, heat capacity has the following simple relation(7,8).

$$C/n = \frac{\pi^2}{6} \nu RT ,$$

where R is the gas constant and ν a certain constant. The magnitude ν is of the order of z/U, where z is a number of vacant equilibrium positions neighboring with one atom and U is the interaction energy between the neighboring atoms. The liquid in this temperature range is named a semiquantum liquid. The value $\alpha(=\pi^2\nu R/6)$ for bulk ^4He liquid is 2.3 J/K^2/mol in the temperature range 1.5 K<T<9 K at the pressure 25 atms. The magnitude of $\alpha=1.0$ J/K^2/mol for ^4He in K-L zeolite is about a half of that for bulk ^4He. The linear dependence of C/n on T for ^4He in K-L zeolite seems to indicate a semiquantum liquid in the one-dimensional channel. At higher quantity n than 8, the free space for ^4He atom is reduced, consequently C/n diminishes.

4. CONCLUSIONS

(a) In the case for $n<n_0$ (=4 atoms/unit cell), it is supposed that helium atoms are localized below about 3 K in a definite region of each cage.
(b) For $n>n_0$, the helium atoms seems to move freely in the channel in the measuring temperature range (1.5<T<9 K).
(c) Especially at $n\sim 8$, the adsorbed helium seems to be a semiquantum liquid because of the linear temperature dependence of C/n.
The study of this helium system at further low temperatures is now under way.

REFERENCES

(1) J. Yuyama and T. Watanabe, J. Low Temp. Phys. 48 (1982) 331.
(2) R.H. Tait and J.D. Reppy, Phys. Rev. B20 (1979) 997.
(3) B.C. Crooker, B. Hebral, E.N. Smith, Y. Takano, and J.D. Reppy, Phys. Rev. Lett. 51 (1983) 666.
(4) N. Wada, T. Ito and T. Watanabe, J. Phys. Soc. Jpn. 53 (1984) 913.
(5) N. Wada et al., Phase change of helium adsorbed on Y zeolites, this volume.
(6) R.M. Barrer and H. Villiger, Z. Krystallogr. 128 (1969) 352.
(7) A.F. Andreev, JETP Lett. 28 (1978) 556.
(8) A.F. Andreev and Yu. A. Kosevich, Sov. Phys. JETP 50 (1979) 1218.

LT-17 (Contributed Papers)
U. Eckern, A. Schmid, W. Weber, H. Wühl (eds)

THE SHIFTED SOLIDIFICATION CURVE OF HELIUM IN SMALL PORES

A.L. THOMSON, D.F. BREWER and S.R. HAYNES

School of Mathematical and Physical Sciences, The University of Sussex, Brighton, BN1 9QH, U.K.

Measurements are presented of the variation of pressure in a sample chamber containing a vycor porous glass filled with a fixed amount of helium. The normal solidification curve is observed for helium surrounding the glass but a quite separate region of solidification is observed for the helium inside our ~65 Å pores. This is shifted by several tenths of a degree to lower temperatures and raised by about 10 bars to higher pressures compared with the normal curve. This phenomenon, which confirms our previous observations has been seen for both isotopes of helium.

1. INTRODUCTION

The presence of a solid surface can influence liquid in contact with it even to the extent of altering the temperature at which the liquid solidifies. This may be most readily observed in porous materials where supercooling of several liquids has been observed (1). Previous work on ^{4}He (2,3,4) has been able to infer that supercooling occurs in small pores but no pressure measurements were made which might allow the shifted melting curve to be determined. We present here the results of pressure measurements on both isotopes of helium when they are contained within the pores of vycor glass.

2. EXPERIMENTAL METHOD

The method used was to measure the pressure of our sample helium as a function of temperature at constant density, using a Straty-Adams (5) type capacitive strain gauge mounted onto our sample chamber. The chamber itself contained a stack of four discs of porous vycor glass each approximately 1mm thick and 20mm in diameter. The pores of our vycor glass have a diameter of about 65Å and the dead space around the vycor was measured to contain about 14 per cent of our total helium sample. The pressure data were supplemented by heat capacity measurements also taken at constant density but these are not discussed in detail here.

3. RESULTS

Displayed in figure 1 are data obtained from five separate experimental runs in which the ^{4}He sample was maintained at constant density once solidification of the bulk helium present had occurred. In each case the closed symbols represent data taken while cooling the sample, and the open symbols data taken while warming the sample. For all of the runs, the sample pressure follows the bulk solidification curve while helium surrounding the vycor solidifies.

Let us consider first the data taken at the

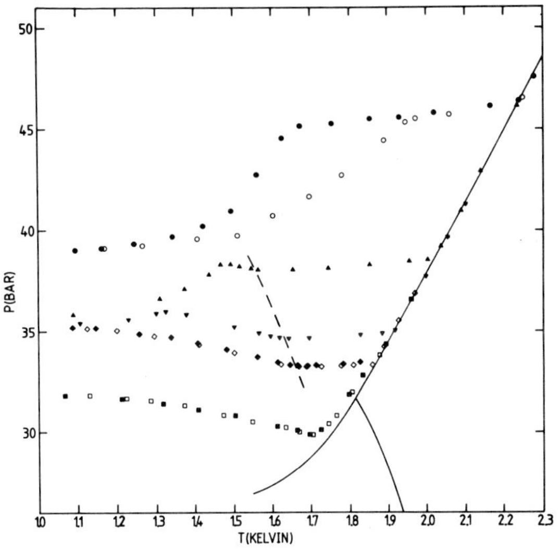

FIGURE 1
Pressure inside the sample chamber versus temperature with ^{4}He filling the vycor pores. Solid symbols represent experimental points taken while cooling, open symbols while warming. The solid lines show the solidification curve for bulk helium as well as the locus of density maxima in the bulk liquid. The dashed line indicates where our pressure data have minimum values indicating of density maxima for the liquid inside the pores.

highest pressures which involved an initial pressure of 57 bars. As the sample chamber is cooled down from 4.2K there is observed at first a gradual reduction in pressure due to the positive expansion coefficient of the liquid both inside and outside the vycor glass. Once the helium surrounding the vycor begins to solidify the pressure drops sharply following the normal bulk solidification curve. When this process is complete the pressure varies much more slowly and it is mainly this region which is displayed on the graph. We see a pressure plateau in the temperature region immediately below the solidification curve but at lower temperatures a fairly rapid drop in pressure is observed. This we take as evidence for the solidification of the helium inside the vycor glass and this appears to cease at the lowest temperatures of our experiments where the pressure is observed again to be constant. When the temperature was moved in the opposite direction it is interesting to see that the same curve is not traced out. The observed pressure rise is seen to be qualitatively similar but more gradual. This hysteresis may well be due to a difference in the curvature of the surfaces on which solidification takes place while cooling compared with the surfaces involved when melting takes place on warming.

The data at the lowest pressures in figure 1 correspond to a starting pressure at 4.2K of 31 bars. Here the behaviour of the pressure is quite different. It is seen to increase monotonically towards the lowest temperatures indicating that the sample has a negative expansion coefficient. This is precisely what is expected for liquid ^{4}He where a density maximum is observed close to its superfluid transition. This behaviour of the pressure suggests that the helium inside the vycor glass does not solidify at the pressures measured in this experimental run even although they are above the normal solidification curve. Furthermore, there is no hysteresis observed upon warming and this can be taken again as an indication of the absence of a first order change of phase. The data in figure 1 which occurs at pressures intermediate between those already discussed display behaviour consistent with both. They show further evidence of the increase of pressure due to the presence of a superfluid and the drop in pressure produced when solidification occurs.

So far our discussion has involved only the heavy isotope, ^{4}He. We have also performed experiments on ^{3}He and find that it displays behaviour which is qualitatively quite similar. At the highest pressures the same kind of variation of pressure is observed consistent with solidification inside the vycor pores accompanied by a similar hysteresis. If one accepts the idea that supercooling is associated with the angle of contact of the liquid-solid meniscus with a solid surface then it would appear that a similar situation exists for both ^{3}He and ^{4}He. This would indicate that solid ^{3}He has the same lack of ability to 'wet' a glass surface as does the heavier helium isotope(6).The behaviour of ^{3}He at pressures just above its solidification curve is also similar to ^{4}He. The pressure rises as the temperature is lowered indicating the existence of a negative expansion coefficient and furthermore no hysteresis is observed as the sample is warmed. Bulk liquid ^{3}He also possesses a negative expansion coefficient at the lowest temperatures.

4. CONCLUSION

In conclusion it can be stated that for both isotopes of helium the liquid phase exists inside small pores over a considerably larger region of the P-T diagram than it does in the bulk phases.

ACKNOWLEDGMENTS

In making the experimental measurements we wish to acknowledge the enthusiastic assistance of Clifford Wiltshire and of Anders Mattsson.

REFERENCES

(1) A.A. Antoniou, J. Phys. Chem. 68, 2754 (1964). G. Litvan and R. McIntosh, Can. J. Chem. 41, 3095 (1963). K.A. Jackson and B. Chalmers, J. Appl. Phys. 29,1178 (1958).
(2) A.L. Thomson, D.F. Brewer, T. Naji, S. Haynes, and J.D. Reppy, Physica 107B, 581 (1981).
(3) D.F. Brewer, Cao Liezhao, C. Girit and J.D. Reppy, Physica 107B, 583 (1981).
(4) J.R. Beamish, A. Hikata, L. Tell and C. Elbaum, Phys. Rev. Lett. 50, 425 (1983).
(5) G.C. Straty and E.D. Adams, Rev. Sci. Inst. 40, 1393 (1969).
(6) S. Balibar, D.O. Edwards and C. Laroche, Phys. Rev. Lett. 42, 782 (1979).

LT-17 (Contributed Papers)
U. Eckern, A. Schmid, W. Weber, H. Wühl (eds)

PHONON SCATTERING ON DISLOCATIONS AND RECOVERY PROCESSES IN PLASTICALLY DEFORMED ^{4}He CRYSTALS

A.A.Levchenko, L.P.Mezhov-Deglin

Institute of Solid State Physics USSR Academy of Sciences
Chernogolovka, Moscow district USSR

The activation energy recovery processes in deformed ^{4}He crystals, containing up to 0,5% ^{3}He impurity atoms, coincides with the activation energies of the diffusion of ^{3}He impurity atoms and positive charges in specimens with the same density. The activation energy falled more than three times when concentration of ^{3}He in ^{4}He increased up to 5%.

Bending of crystals grown from pure ^{4}He at pressures 26, 31, 50 and 84 atm at low temperatures $T_B < T_0 \simeq 0,4$ K leads to a considerable decrease of the coefficient of thermal conductivity. The measurements (1) have shown that the additional resistance W_{add} which is due primarily to scattering of phonons by newly induced dislocations was proportional to $W_{add} \sim T^{-3}$. The cross section of the phonon-dislocation scattering calculated from W_{add} at T_0 and the density of dislocations (estimated by the bending radius) exceeded up to 500 times the known theoretical values for the case of scattering on static dislocations. This indicates that the main mechanism is the scattering of phonons by vibrating dislocations (flutter-effect).

The investigations of the recovery phenomena in weakly deformed (~1%) crystals (2) have shown that the dependence W_{add} at $T=T_0$ from the annealing time t at fixed temperature $T_{ann} > T_0$ may be described by an expression of the form

$$W_{add}(t)/W_{add}(0) = 1 - A\ln(1+\frac{t}{\tau}) \quad (1)$$

where W_{add} (0) is the additional resistance before annealing (t=0), A is a constant for the sample, and τ is a characteristic relaxation time dependent from T nearly exponentially. Control measurements on the same sample have demonstrated that a change of a degree of deformation changes significantly the value of W_{add} but practically does not affect A and τ in the expression (1). From known τ (T) it is possible to estimate the magnitude of the activation energy of the recovery processes in solid helium. The values obtained coincide with the activation energy of point defects (^{3}He atoms, positive charges) in the samples of the same molar volume. It may be supposed so that the diffusion of dislocations and point defects in the thermally activated regions is controlled by the same mechanism, most likely by their interactions with vacancies (equilibrium concentration of vacancies in pure helium decreases exponentially with temperature lowering).

The addition from 0,005 to 0,5% ^{3}He atoms into the ^{4}He gas leads to systematical lowering of the thermal conductivity of the samples at low temperatures. But it affected insignificantly the dependence of W_{add} (T) and W_{add} (t) (see fig.1) and the values of the

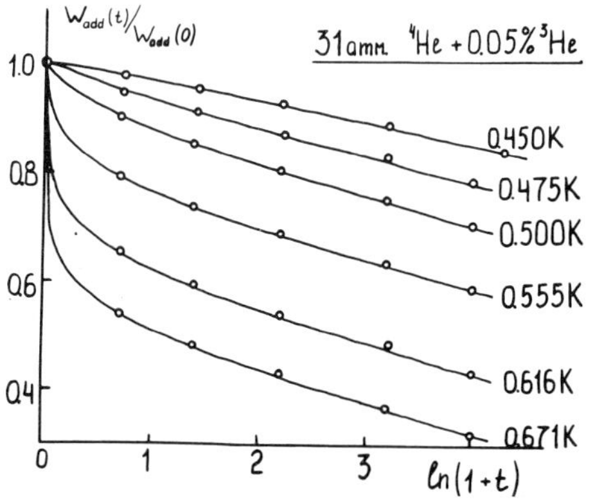

Fig.1
The relative resistance as a function of holding time at different T_3 (time in min. P=31atm ^{4}He+0,05%^{3}He). Solid curves are the approximations of eq.(1).

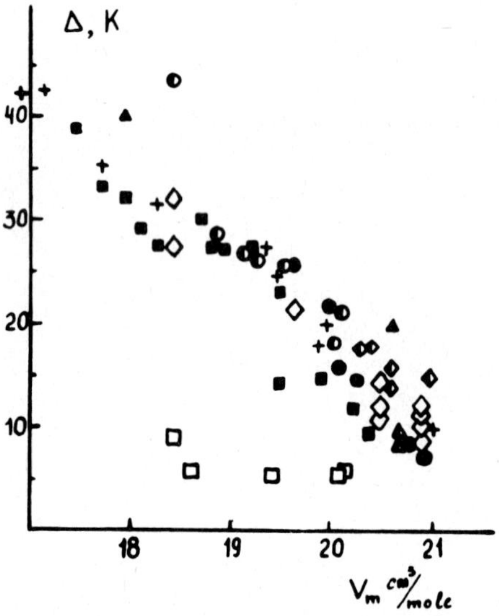

Fig.2
Effective activation energy of recovery processes. ◇ -pure He, □ - $^4He+5\%\,^3He$ and of diffusion of positive charges (●, ▲, ■, +), He impurity atoms (◐, △, ◈) as a function of molar volume.

activation energy of the recovery processes. When the concentration of ^3He in ^4He increased up to 5% it turned out that the temperature dependencies of W_{add} (T) and τ (T) have changed significantly. At $T < 1$ K $W_{add} \sim T^{-2}$, and calculated from τ (T) values of the activation energy falled more than three times at 84 atm (fig.2) though the magnitude of τ near T_o changed slightly. Those results are unexpected enough and may be connected with changes either in vacancy system at $T < 1$ K or in dislocations motion mechanism in crystals with impurities.

REFERENCES

(1) A.A. Levchenko, L.P. Mezhov-Deglin, Zh.Eksp.Teor.Fiz.,(1982) 82, 278.
(2) A.A. Levchenko, L.P. Mezhov-Deglin, Pis'ma Zh.Eksp.Teor.Fiz.,(1983) 37, 173.

LT-17 (Contributed Papers)
U. Eckern, A. Schmid, W. Weber, H. Wühl (eds)

SPECIFIC FEATURES IN THE MOTION OF POSITIVE CHARGES IN SOLID ^{4}HE (1,2)

A.I.Golov, V.B.Efimov, L.P.Mezhov-Deglin

Institute of Solid State Physics USSR Academy of Sciences
Chernogolovka, Moscow district USSR

The direct measurements of positive charge velocities in ^{4}He crystals grown at 31 atm have revealed the existence of knees on $V_+(T)$ curves in weak electrical fields $E \leqslant 2.10^4$ V/cm at temperatures near 0.8 K. In high fields at low temperatures the charge velocity $V_+(E)$ after a monotonic growth has dropped several times near $E \sim 5.10^4$ V/cm and then weakly depended on E. All of this might be attributed to changes on mechanisms of the charge diffusion.

We reported earlier (3) on the observation of knees on the curves describing the temperature dependences of currents through a diode with a radioactive source in solid He. in weak electric fields with a mean strength $E \lesssim 2.10^4$V/cm a nearly exponential decrease of the current with decreasing temperature was followed by an increase. The current reached a maximum and then began to decrease rapidly again. the results of direct measurements of the velocities of positive charges using the time-of-flight technique have shown that these knees are due to a velocity change.

The time-of-flight was determined by the maxima positions on the J(t) curves, describing the dependence of the current through a diode on the time at a step switch-on of the voltage. the source-collector gap was d=0.3 mm in our experiments, that is the mean field wherein the charges were moving was $E \approx U/d$ = 33 U V/cm. The sign of the charges withdrawn from the plasma near the source was determined by the polarity of the applied voltage. Fig.1 shows the results of the measurements of the velocity of positive charges V_+ as a function of temperature T in weak fields $E \leqslant 2.10^4$V/cm in one of the samples grown at a pressure of 31 atm. The numerals along the curves denote the values of the applied voltage U.At the same pressure the temperature dependences $V_+(T)$ in a wide temperature range and a relative height of the maxima on the $V_+(T)$ curves in the region of the knees can vary noticeably from sample to sample. The temperatures at which the knees were observed, however, coincident, and the shapes of the $V_+(T)$ curves were, in general, similar.

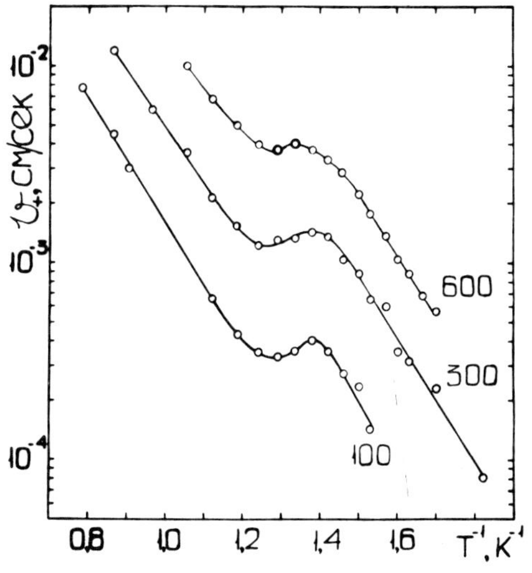

Fig.1

In the fields $E \leqslant 10^4$V/cm the knee temperature is field independent, at $E > 2.10^4$V/cm, as can be seen by the figure, the knee was shifted towards higher temperatures. The knees on the $V_+(T)$ curves in weak fields might be attributed to a change of the mechanisms of charge mobilities with decreasing temperature. We could not measure the velocities of negative charges in a wide temperature range in the same sample, since the velocities of negative charges and the corresponding current values decreased with decrea-

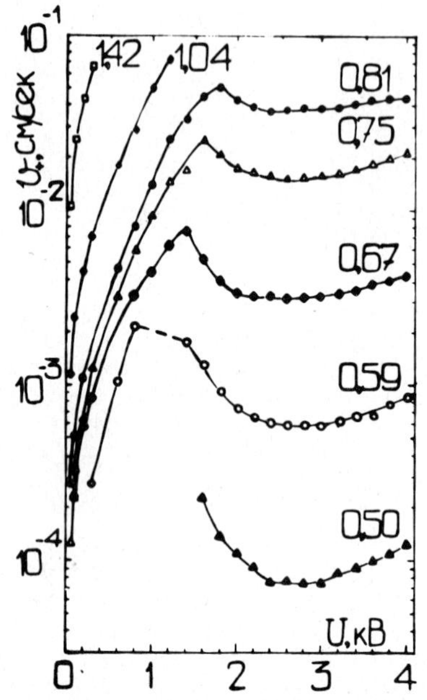

Fig.2

sing temperature much faster than those of the positive ones. But the $J_-(T)$ curves in the constant field $E \lesssim 2 \cdot 10^4$ V/cm did exhibit the knees. The investigations of the charge velocity dependences from applied fields $V_+(E)$ have shown that at low temperatures the velocity of positive charges after a monotonic growth dropped by several times at fields near $5 \cdot 10^4$ V/cm and next it was weakly dependent on E (fig.2). In the high fields ($\sim 10^5$ V/cm) the value $V_+(T)$ at constant E decreased nearly exponentionally with lowering the temperature. As contrasted from positive charges the velocity of negative charges in the same fields increased monotonically with the field up to $1,3 \cdot 10^5$ V/cm.

Concurrent with the known observations of the charge velocity decrease with an increase of the accelerating field in the superfluid liquid, the drop of $V_+(E)$ in solid helium in high fields may be attributed to the generation of dislocation loops which then capture the charges and move together with these charges. Nosanow and Titus (4) were the first to note the possibility that dislocations can be generated during the motion of the charges. The experiments dealing with the investigations of recovery in ^{4}He crystals (5) have shown that the mobility of freshly induced dislocations as high, which makes this mechanism plausible.

REFERENCES

(1) A.I.Golov, V.B.Efimov, L.P.Mezhov-Deglin, JETP Lett, 38 (1983) 58.
(2) L.P.Mezhov-Deglin, V.B.Efimov, A.I.Golov, Fiz.Nizk.Temp., 10 (1984) 99.
(3) V.B.Efimov, L.P. Mezhov-Deglin, Fiz.Nizk.Temp., 4 (1978) 397.
(4) J.H. Nosanov, W.J.Titus, J.Low Temp.Phys., 1 (1969) 73.
(5) A.A.Levchenko, L.P.Mezhov-Deglin, JETP Lett., 37 (1983) 173.

LT-17 (Contributed Papers)
U. Eckern, A. Schmid, W. Weber, H. Wühl (eds)

Diffusion of ^3He Atoms in Phase-Separated Solid Helium

Izumi IWASA and Hideji SUZUKI

Department of Physics, University of Tokyo, Bunkyo-ku, Tokyo 113, Japan

Phase separation of a solid solution of ^3He in hcp ^4He was studied by measuring the pressure. The relaxation time of the pressure in warming runs was found to be almost independent of temperature and its magnitude was about 10 min. The results suggest that the diffusion constant is independent of both temperature and ^3He concentration along the phase-separation line, which is inconsistent with the impurity wave model.

1. INTRODUCTION

NMR studies on diffusion of ^3He impurity in hcp ^4He crystals (1), (2) show that the diffusion constant is independent of temperature and inversely proportional to the concentration of ^3He at temperatures below 0.8 K. This behaviour is regarded as evidence for the impurity wave (impuriton). Another method for investigating impurity diffusion is the pressure measurement at temperatures of phase separation (3). A relaxation time can be determined by changing the temperature of the phase-separated solid helium typically by 10 mK and observing the following relaxation of pressure at a constant temperature. The relaxation time is expressed by

$$\tau = D^{-1}n^{-2/3} , \qquad (1)$$

appart from a numerical factor of order unity, where D is the impurity diffusion constant in the matrix and n the number density of the precipitates formed by the phase separation. For a ^3He-rich solution with 1 % ^4He, τ has been found to be temperature independent (4). If the impurity behaves wave-like, τ may strongly depend on temperature. In this paper we report new results of pressure measurements on hcp ^4He crystals containing about 1 % ^3He. Our purpose is to understand the dynamical process of the phase separation by comparing the results with those of ^3He-rich solutions and to extract the diffusion constant in order to see whether the impurity wave picture of the NMR studies is adequate.

2. EXPERIMENTAL METHOD

The pressure of solid helium was measured with a Straty-Adams type capacitive strain gauge. The sample space was 3 mm thick and 16 mm in diameter. The carbon thermometer was calibrated against the melting temperature of pure ^3He between 75 and 800 mK using the strain gauge. The concentration of ^3He in the mixture gas was analyzed to be x = 0.84 %. The crystals were grown at nearly constant pressure with the filling capillary open, and then annealed by cycling the temperature between the melting point and about 1 K a few times.

3. RESULTS

The ^3He-rich precipitates are either hcp, bcc or liquid depending on the pressure, temperature and other conditions. The phase transformation between these states causes a pressure change in addition to the pressure change due to phase separation and complicates the analysis. The cooling run in Fig. 1 (run 1) shows that the pressure increases at temperatures between 200 and 100 mK due to phase separation and that below 100 mK it further increases due to melting of the ^3He precipitates. The pressure and temperature of melting are in good agreement with the melting line of pure ^3He. On warming (run 2), the phase transformation and the mixing cannot be separated clearly.

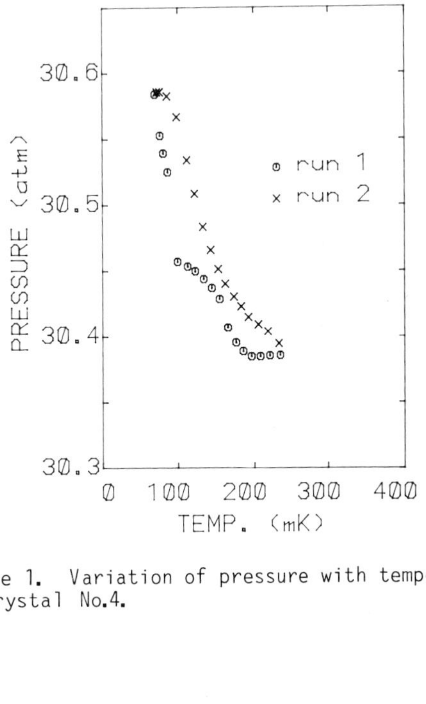

Figure 1. Variation of pressure with temperature in crystal No.4.

The pressure change due to phase separation in another crystal at a higher pressure is shown in Fig. 2. The value of the pressure increase is 0.06 atm in run 1 (rapid cooling) as well as in run 3 (slow cooling). The relaxation time is shown in Fig. 3. Because of the long relaxation times in the cooling runs the pressure continues to increase in warming runs (runs 2 and 4) at temperatures below 100 mK.

4. DISCUSSION

The pressure increase ΔP at constant volume is

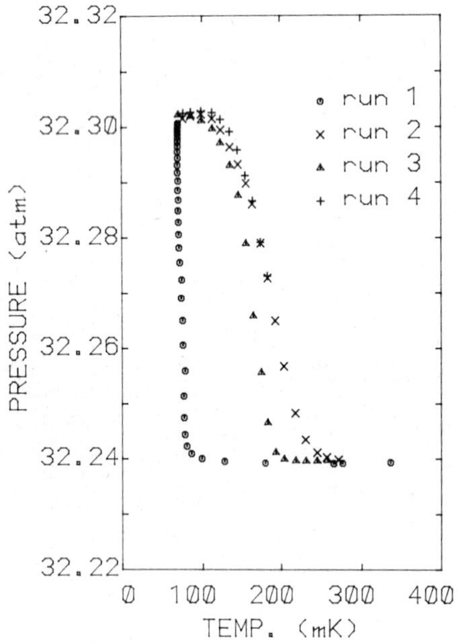

Figure 2. Variation of pressure with temperature in crystal No.5.

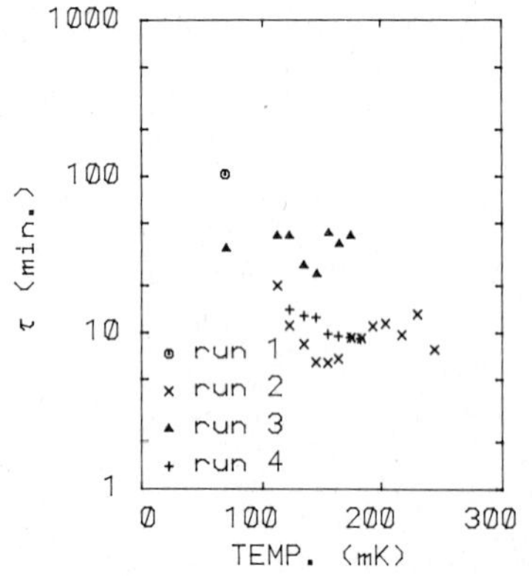

Figure 3. Relaxation times obtained for crystal No.5 (the same crystal as in Fig.2).

related to the volume change ΔV at constant pressure as follows:

$$\Delta P = B(\Delta V/V_M) \,, \tag{2}$$

where B and V_M are the bulk modulus and molar volume of the matrix (i.e. hcp ^{4}He). The expected values of ΔP are 0.05, 0.07 and 0.23 atm for hcp, bcc and liquid ^{3}He as the precipitates, respectively, using plausible values of ΔV with x = 0.84 %, B = 290 atm and V_M = 20.5 cm^3/mole. The observed values are 0.2 atm in Fig. 1 (with melting) and 0.06 atm in Fig. 2 (without melting). We consider that most precipitates formed in the crystal in Fig. 2 are bcc because ΔP is not sensitive to the cooling method and τ is longer in the cooling runs than in the warming runs.

The phase separation temperature T_{PS} is determined to be 230 mK from run 2 in Fig. 2. From the regular solution theory (5), on the other hand, we expect T_{PS} = 156 mK at x = 0.84 %. Such a discrepancy has been observed by Panczyk et al. (6) in ^{4}He-rich solutions.

According to the NMR experiments on dilute ^{3}He in hcp ^{4}He, the diffusion constant of ^{3}He is independent of temperature below 0.8 K and inversely proportional to x. The lowest temperature of the measurements of D, however, is about 0.4 K and the diffusion constant below T_{PS} is not known experimentally. Theoretically, if the diffusion constant of ^{3}He in the matrix of hcp ^{4}He depends solely on x, we expect a temperature dependence of D through x because x varies with T along the phase separation line. In this case the relaxation time should strongly depend on temperature as

$$\tau \propto D^{-1} \propto x(T) \,. \tag{3}$$

Namely, τ should increase with T. The observed relaxation time, on the other hand, is almost independent of temperature in the warming runs as shown in Fig. 3, which suggests the temperature independence of D below T_{PS}. Such a result is in conflict with what is expected from the impurity wave model. The present results are rather similar to the case of the ^{3}He-rich solutions where τ is independent of temperature and its magnitude is about 10 min.

REFERENCES

(1) M.G. Richards, J. Pope, P.S. Tofts and J.H. Smith, J. Low Temp. Phys. 24 (1976) 1.
(2) A.R. Allen, M.G. Richards and J. Schratter, J. Low Temp. Phys. 47 (1982) 289.
(3) M. Uwaha, J. Phys. Soc. Jpn. 48 (1980) 1921.
(4) I. Iwasa and H. Suzuki, to be published in Proc. 4th Int. Conf. Phonon Scattering in Condensed Matter, Stuttgart, 1983.
(5) D.O. Edwards, A.S. McWilliams and J.G. Daunt, Phys. Rev. Lett. 9 (1962) 195.
(6) M.F. Panczyk, R.A. Scribner, J.R. Gonano and E.D. Adams, Phys. Rev. Lett. 21 (1968) 594.

LT-17 (Contributed Papers)
U. Eckern, A. Schmid, W. Weber, H. Wühl (eds)

THE QUANTUM DIFFUSION OF ^3He IN THE HCP PHASE OF ^3He - ^4He SOLID SOLUTIONS AT LOW TEMPERATURES

V.A.MIKHEEV, V.A.MAIDANOV and N.P.MIKHIN

Institute for Low Temperature Physics and Engineering, UkrSSR Academy of Sciences, Kharkov, USSR

The NMR experimental data on the diffusion coefficient D of ^3He impurity atoms in a wide range of ^3He concentrations, densities and temperatures are presented. A sharp decrease in D at low temperatures suggests a strong localization of impurity atoms, in full agreement with theory.

One of the most interesting features of the quantum diffusion of atoms(1) in crystals is that by Kagan's predictions an essentially complete localization of diffusing atoms in perfect lattice is possible at low temperatures. According to Kagan's theory (2), in quantum crystals the diffusing particles of one type in the matrix of atoms of other type are elastically interacting defects; this produces a random shift, $\delta\mathcal{E}$, of energy levels at neighbouring sites and, if the shift value is higher than the energy gap of impurity quasi-particles (impuritons), then for $T \to 0K$ a spatial localization of particles takes place. As the temperature rises, the localization gradually vanishes owing to the level fluctuations caused by the impuriton-phonon interaction; for $T \gg \delta\mathcal{E}$, when the two-phonon processes are effective, the diffusion coefficient D increases with temperature as $D \sim T^9$. The prediction has been recently confirmed by the experimental data on ^3He diffusion in ^3He - ^4He solid solutions (3). It should be noted that the ^4He lattice phonons have different effects on the diffusion processes (Fig. 1). For low ^3He concentrations (curves 1,2), when the band-type diffusion is realized, the phonons act as an additional resistance and $D \sim T^{-9}$

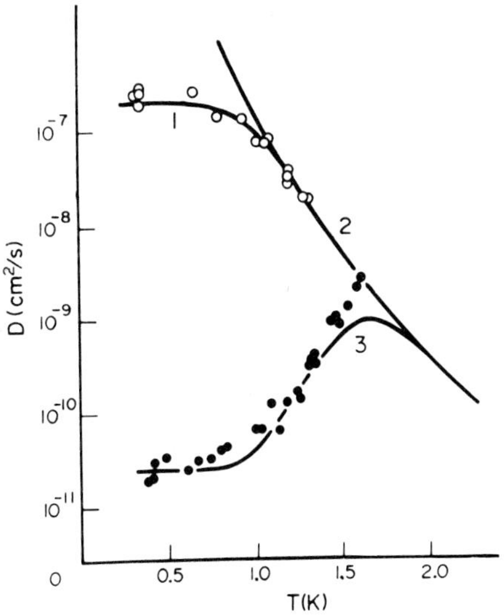

FIGURE I

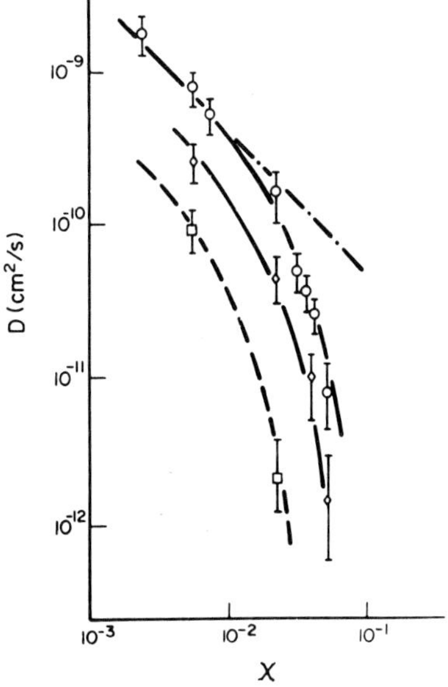

FIGURE 2

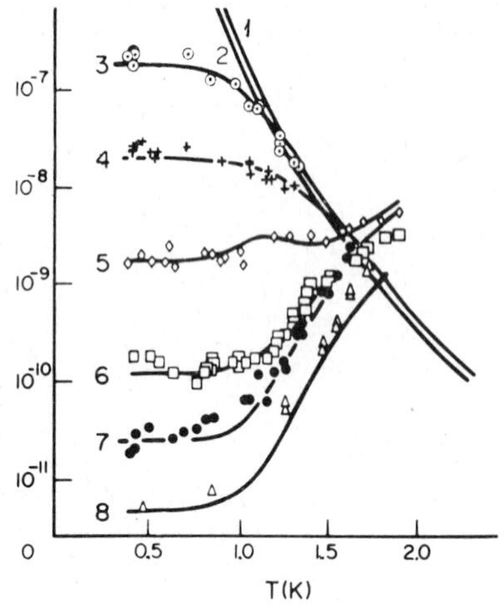

FIGURE 3

while at high ^3He concentrations in the strong localization region (curve 3) the band is broken down and the phonons induce delocalization producing a channel for ^3He motion. At high temperatures under real conditions, the diffusion increases further because of an additive contribution by vacancies.

The low temperature behaviour of D in the strong localization region is of great interest for the ^3He concentration close to a critical one, x_c. In this case the critical dependence of D should be of the form (4,5):

$$D = \frac{\text{const}}{x}\left(\frac{x_c - x}{x_c}\right)^t \qquad (I)$$

The experimental data on the concentration dependence of ^3He diffusion coefficient for three values of molar volume (20.7, 20.5 and 19.9 cm^3/mole) are shown in Fig.2. It is seen that at higher ^3He concentrations there occurs a pronounced tendency for localization and the D value decreases then by two orders of magnitude. As the density increases, the diffusion coefficient decreases and the band motion is disturbed at lower concentrations. The curves shown in Fig.2 are described by Eq.(I) with t=I.7+0.3. It should be noted that the critical index, t, is similar to that of electrical conductivity under the metal-dielectric transition (6).

The localization theory (4,5) is confirmed by the data shown in Fig. 3 where the temperature dependences measured in the whole studied range of x, V and T are correlated with the generalized theoretical dependence (4,5):

$$D = \frac{za^2}{3\hbar} J^2 \left\{ \frac{Q(x)}{\Omega_x + \Omega_p(T)} + \frac{p(T)\,[1 - Q(x)]}{(\delta\mathcal{E})^2 + \Omega_p^2(T)} \right\} \qquad (2)$$

Here z is the coordination number, a is the interatomic distance, $Q = (x_c - x)^t / x_c^t$; $\delta\mathcal{E} = \alpha U_0 x^{4/3}$; $\Omega_x = 2.3 \cdot A x J^{1/3} \cdot U_0^{2/3}$; $\Omega_p(T) = 10^6 \cdot B (T/\Theta)^9 \cdot \Theta$; Θ is the Debye temperature; numerical constants A and B were obtained by correlating theoretical (7) and experimental (8) data on the band diffusion. For calculation we used the values of elastic interaction constant of impurities, U_0, and tunnelling amplitude, J, and their dependences on V obtained elsewhere (7,9,10), and α = 0.35. Curves I and 2 present calculated data on D (the interaction between ^3He atoms being absent) for V = 2I and 20.7 cm^3/mole, respectively. Curves 3 through 8 present the results calculated by Eq.(2) for x = 0.006% and 0.5% (V = 2I cm^3/mole); 0.25%, 2.I7%, 4.0% (V = 20.7 cm^3/mole); 4.98% (V = 20.5 cm^3/mole), respectively, with taking into account an additive contribution of the vacancy diffusion $2 \cdot 10^{-5} \exp(-W/T)$. The W value was obtained at T > I.4 K with x=4.98%. The correlation between the measured data both from the present work and Ref.(8) for x < 2%, and the calculated results indicates a good agreement between theory and experiment.

REFERENCES

(I) A.F.Andreev and I.M.Lifshitz, Sov. Phys. JETF 29 (I969) II07.
(2) Yu.Kagan. In: Defects in Insulating Crystals (Springer Verlag, Berlin-Heidelberg, I98I) p.I7.
(3) V.A.Mikheev, V.A.Maidanov and N.P.Mikhin, Solid St.Comm. 48 (I983) 36I.
(4) Yu.Kagan, L.A.Maksimov, Phys.Lett. 95A (I983) 242.
(5) Yu.Kagan and L.A.Maksimov, Zh.Eksp. Teor.Phys. 84 (I983) 792.
(6) B.I.Shklovskii and A.L.Efros, Electronic Properties of Doped Semiconductors (Nauka, Moscow,I979).
(7) V.A.Slusarev, M.A.Strezhemechnyi and I.A.Burakhovich, Fiz. Nizk. Temp. 3 (I977) I229.
(8) B.N.Esel'son, et al. Sov. Phys. JETF 47 (I978) I200.
(9) M.G.Richards et al. J. Low Temp. Phys. 24 (I976) I.
(I0) A.R.Allen, M.G.Richards and J.Schratter, J. Low Temp. Phys. 47 (I982) 289.

LT-17 (Contributed Papers)
U. Eckern, A. Schmid, W. Weber, H. Wühl (eds)

DISPERSION OF THE SPIN-POLARIZED ATOMIC HYDROGEN FLOATING ON THE SUPERFLUID HELIUM

Keshav N. SHRIVASTAVA

School of Physics, University of Hyderabad, Hyderabad 500 134, India

The interaction of the spin-polarized hydrogen atoms in the bulk as well as on the surface of the superfluid helium with the third-sound ripplons is investigated. The dispersion relation is found to show a kink at a particular wave vector at which new quasiparticles appear.

We wish to report that the spin-polarized atomic hydrogen floating on the superfluid helium interacts with the third-sound ripplons giving rise to a novel boson wave. The dispersion of this boson wave has a kink at a certain wave vector at which new quasiparticles propagate. This phenomena is distinguished from the Bose-Einstein condensation.

The potential energy shows a minimum for H↓-H↑ distance of 0.74 Å, H↓-H↓ pair distance of about 4 Å and He-He pair distance of ~2.8Å. A dense chain of H↓ consists of 0.25 x 10^8 atoms/cm and hence a sheet of H↓ atoms consisting of about 0.62 x 10^{15} atoms/cm^2 forms a dense fluid. For the superfluid helium, the dispersion relation is given by $E_p = \Delta+(p-p_o)^2/2\mu$ with $\Delta \simeq 8.65$ K $\mu = 0.17\ m_4$ and $p_o = 1.9\times10^8 cm^{-1}$ so that the distance between ^{4}He atoms in the superfluid state is of the order of 1.6 Å. Therefore for the H↓ fluid in a cooperative state the inter-atomic distance may be of the order of ~2.3 Å. Hence the H↓ matrix floats on the superfluid helium surface. The H↓ atoms dip in the helium to create a third-sound ripplon and the H↓ coming out from the bulk helium destroys a ripplon as suggested by Yaple and Guyer[1]. The Bose-Einstein condensation and hence the superfluidity in the spin-polarized atomic hydrogen, H↓ fluid may occur at high magnetic fields and ultra-low temperatures.

We describe a phenomena which is distinct from the Bose condensation. In this effect the composite bosons created as a result of H↓-third sound interaction show a divergence in the single-particle energy at a given wave vector.

The hamiltonian of the system of H↓ floating on the superfluid helium is given by

$$\mathcal{H} = \sum_p \hbar\omega_p\, a_p^\dagger a_p + \sum_q \varepsilon_q\, b_q^\dagger b_q + \sum_k E_k\, c_k^\dagger c_k + \sum_{kpq} A(k,p,q)(a_p + a_{-p}^\dagger) b_q^\dagger c_k \delta_{k,q-p} + \text{H.c.} \quad (1)$$

The operators $a_p^\dagger(a_p)$, $b_q^\dagger(b_q)$ and $c_k^\dagger(c_k)$ are the creation and annihilation operators for the third-sound ripplons, the H↓ atoms on the surface and the H↓ atoms in the bulk superfluid helium, respectively. The H.c. indicates the hermitian conjugate of the previous term. The matrix element of the interaction A(k,p,q) is given by Yaple and Guyer. The single-particle energy for H↓ on the surface and those in the bulk superfluid is given by

$$\varepsilon_q = \frac{\hbar^2 q^2}{2m} - \varepsilon_B,\quad E_k = E_o + \frac{\hbar^2 k^2}{2m} \quad (2)$$

where ε_B is the binding energy of H↓ on the superfluid helium surface and E_o is the energy required to dissolve H↓ in liquid helium. The dispersion of the third-sound ripplons is given by

$$\omega_p^2 = \left(\frac{c_3^2}{d} + \sigma\frac{a^3 p^2}{m_4}\right) p \tanh(pd) \quad (3)$$

Here c_3 is the velocity of the third-sound ripplons, d the depth of the helium film, σ the conductivity, m_4 the mass of the helium per atom and the interactomic separation in the superfluid is of the order of a ≃3.6Å. The number densities are

$$n_p = \langle a_p^\dagger a_p \rangle,\quad m_q = \langle b_q^\dagger b_q \rangle,\quad N_k = \langle c_k^\dagger c_k \rangle \quad (4)$$

for the ripplons, the H↓ on the surface and the H↓ in the bulk superfluid helium, respectively. The Fourier transform of the unperturbed Green function for surface H↓ atoms is,

$$G_o = \langle\langle b_q | b_q^\dagger \rangle\rangle = (E-\varepsilon_q)^{-1} \quad (5)$$

and the self energy is calculated to be,

$$\Sigma = \sum_p A^2(k,k+p)\left(\frac{1+n_p+N_{k+p}}{E-E_{k+p}-\hbar\omega_p} + \frac{n_p - N_{k+p}}{E-E_{k+p}+\hbar\omega_p}\right) \quad (6)$$

The Dyson equation leads to the Fourier transform of the Green function as

$$G = G_o[1-G_o\Sigma]^{-1} \tag{7}$$

The zero-temperature dispersion of the H↓ on the surface is given by

$$1 - G_o\Sigma = 0 \tag{8}$$

Using (2) and $\omega_p = c_3 p$ for ripplons in the zero-conductivity approximation[8],

$$A_{k,k+p} = B(p)\, D_k \tag{9a}$$

with

$$B_p = \left(\frac{\hbar\omega_p d}{2An_4 m_4 c_3^2}\right)^{1/2} \tag{9b}$$

so that the dispersion (8) is found to be

$$E = \varepsilon_k + \Sigma_p \frac{B^2(p) D_k^2}{\varepsilon_k - E_{k+p} - \hbar\omega_p} \tag{10}$$

Introducing a two-dimensional integration for the ripplons we find that

$$E = \frac{\hbar^2 k^2}{2m} - \varepsilon_B - \frac{D_k^2 d\phi_k}{2\pi c_3 n_4 m_4} \tag{11a}$$

where the double integral is given by

$$\phi_k = \int_o^{p_r}\int_1^{-1} p^2 dp[\hbar^2 p^2/2m) + pc_3 + \varepsilon_B^o + (pk\hbar^2/m)\cos\theta]^{-1} d(\cos\theta) \tag{11b}$$

with p_r as the ripplon cutoff wave vector and $\varepsilon_B^o = E_o + \varepsilon_B$. Upon evaluating the angular integral in the above we find that

$$\phi_k(+) = \frac{m}{\hbar^2 k}[\phi_k(-) - \phi_k(+)] \tag{12a}$$

$$\phi_k(\pm) = \int pdp\, \ln[\frac{\hbar^2 p^2}{2m} + p(c_3 \pm \frac{\hbar^2 k}{m}) + \varepsilon_B^o] \tag{12b}$$

The analytical solution of the above integral is found to be

$$\begin{aligned}\phi_k(\pm) = {} & \frac{b_\pm}{2c} - \frac{1}{2}p_r^2 + \left(\frac{b_\pm^2}{4c^2} - \frac{\varepsilon_B^o}{2c}\right)\ln \varepsilon_B^o \\ & + \left(\frac{1}{2}p_r^2 - \frac{b_\pm^2}{4c^2} + \frac{\varepsilon_B^o}{2c}\right)\ln(\varepsilon_B^o + p_r b_\pm + cp_r^2) \\ & + \left[\frac{b_\pm}{4c^2}(b_\pm^2 - 2\varepsilon_B^o c) - \frac{\varepsilon_B^o b_\pm}{2c}\right](b_\pm^2 - 4\varepsilon_B^o c)^{-\frac{1}{2}} \\ & \ln\{[2cp_r + b_\pm - (b_\pm^2 - 4ac)^{\frac{1}{2}}][2cp_2 + b_\pm \\ & + (b_\pm^2 - 4ac)^{\frac{1}{2}}]^{-1}[b_\pm + (b_\pm^2 - 4\varepsilon_B^o c)^{\frac{1}{2}}] \\ & [b_\pm - (b_\pm^2 - 4\varepsilon_B^o c)^{\frac{1}{2}}]^{-1}\}\end{aligned} \tag{13}$$

where $c = \hbar^2/2m$ and $b_\pm = c_3 \pm \hbar^2 k/m$. From (11) we find that E varies first as k^2 and then has divergence at

$$k_o = \pm[m\varepsilon_B^o/(\hbar^2 p_r) + mc_3/\hbar + \frac{1}{2}p_r] \tag{14}$$

There are two other divergences in (13), one when the effective binding energy of the surface H↓ atoms is zero, $\varepsilon_B^o = 0$, and the other when the size of the H↓ is zero, $1/m = 0$. These divergencies are of no interest. However at (14) the group velocity of the H↓ on the surface is infinite. Therefore we predict a kink in the dispersion relation as depicted in Fig.1. This means that new bosons propagate in the system.

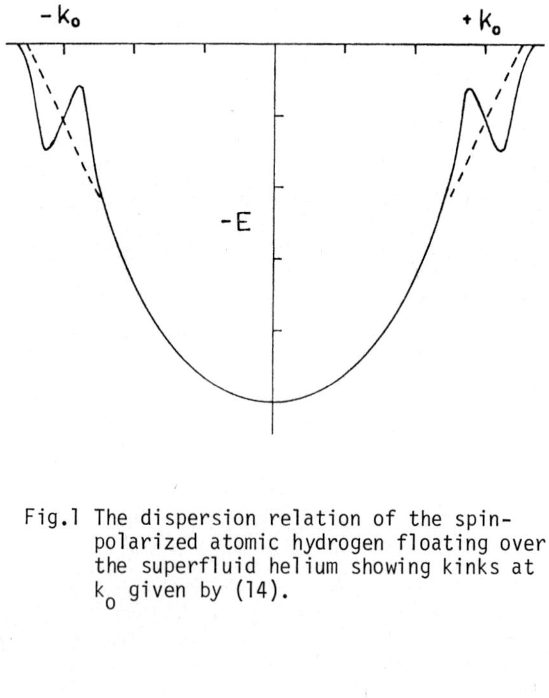

Fig.1 The dispersion relation of the spin-polarized atomic hydrogen floating over the superfluid helium showing kinks at k_o given by (14).

REFERENCES

(1) J.A. Yaple and R.A. Guyer, Phys. Rev. B27, 1629 (1983), D.S. Zimmerman and A.J. Berlinsky, Canad J. Phys. 61, 508 (1983).

LT-17 (Contributed Papers)
U. Eckern, A. Schmid, W. Weber, H. Wühl (eds)

THIRD SOUND INDUCED PROTON-SPIN-ATOMIC INTERACTION IN SPIN-POLARIZED HYDROGEN ON SUPERFLUID HELIUM

Keshav N. Shrivastava

School of Physics, University of Hyderabad, P.O. Central University, Hyderabad 500 134, India

We predict a new interaction through which the proton spin flips as the spin-polarized hydrogen atom dips in the superfluid helium. This interaction leads to the proton spin-lattice relaxation time varying as the square root of temperature, $\tau \sim T^{1/2}$, which is in accord with the experimental measurements.

We wish to report a new interaction through which the proton spin flips by the dip of the spin polarized hydrogen atom in the superfluid helium. The predicted proton spin-lattice relaxation time varies as the square root of the temperature in accord with the recent experimental observations.

We consider the interaction of the proton spin with the third sound ripplon waves for the spin-polarized hydrogen atoms which are floating on the superfluid helium surface as,

$$\mathcal{H}_1 = \sum_p [\lambda_p I_+ (a_p + a^\dagger_{-p}) + \lambda^*_p I_- (a^\dagger_p + a_{-p})] \qquad (1)$$

where $I_\pm$ are the proton spin operators and $a^\dagger_p$, a_p are the third-sound ripplon creation and annihilation operators, respectively. For the spin-polarized hydrogen atoms, H↓, floating on the superfluid helium the atomic dip is described by Yaple and Guyer by the interaction.

$$\mathcal{H}_2 = \sum_{kpq} A_{kpq} (a_p + a^\dagger_{-p}) b^\dagger_q c_k \delta_{q,k+p} + \text{H.c.} \qquad (2)$$

where $b^\dagger_q$, b_q and $c^\dagger_k$, c_k are the boson creation and annihilation operators for the H↓ on the surface and the H↓ in the bulk superfluid helium respectively. The H.c. indicates the hermitian conjugate of the previous terms.

$$\mathcal{H}_0 = \hbar\omega_H I_z + \sum_p \hbar\omega_p a^\dagger_p a_p + \sum_p \hbar\omega_{-p} a^\dagger_{-p} a_{-p} + \sum_q \varepsilon_q b^\dagger_q b_q + \sum_k E_k c^\dagger_k c_k \qquad (3)$$

where the first term is the nuclear Zeeman interaction with the proton resonance frequency $\omega_H = \gamma H$ with $\gamma = g_N \mu_N/\hbar$. Here g_N is the nuclear gyromagnetic ratio, μ_N is the nuclear magneton and H is the external magnetic field. The single particle energies for the surface and the bulk H↓ atoms are

$$\varepsilon_q = \frac{\hbar^2 q^2}{2m} - \varepsilon_B, \quad E_k = \frac{\hbar^2 k^2}{2m} + E_0 \qquad (4)$$

respectively with ε_B as the surface binding energy. The energy E_0 is required to dissolve one H↓ in liquid helium. We define $\varepsilon^0_B = \varepsilon_B + E_0$. The hamiltonian thus defined from (1),(2) and (3) becomes,

$$\mathcal{H} = \mathcal{H}_0 + \mathcal{H}_1 + \mathcal{H}_2 \qquad (5)$$

We apply a canonical transformation[3] to (5) so that the proton spin operators are introduced in the swiming process of H↓ in the helium liquid. The transformed hamiltonian appears as

$$\mathcal{H} = e^{-S}\mathcal{H}e^{S} = \mathcal{H}_0 + \mathcal{H}_1 + \mathcal{H}_2 + [\mathcal{H}_0, S] + [\mathcal{H}_1 + \mathcal{H}_2, S] \qquad (6)$$

which is calculated to be

$$\mathcal{H} = \mathcal{H}_0 + 4\sum_p \lambda^2_p \hbar\omega_p [(\hbar\omega_p)^2 - (\hbar\omega_H)^2]^{-1} (I_+ I_- + I_- I_+) + H_{sn} \qquad (7)$$

where the H↓-swiming nuclear interaction is

$$H_{sn} = \sum_{kp} [\phi_1(k,p) I_- b^\dagger_{k+p} c_k + \phi_2(k,p) I_- b_{k+p} c^\dagger_k + \phi^*_1(k,p) I_+ c^\dagger_k b_{k+p} + \phi^*_2(k,p) I_+ c_k b^\dagger_{k+p}] \qquad (8)$$

where

$$\phi_1(k,p) = F_1(k,p) + G_1(k,p) \qquad (9a)$$

$$\phi_2(k,p) = F_2(k,p) + G_2(k,p) \qquad (9b)$$

$$F_1(k,p) = 2\lambda^*_p A_{k,p,k+p} \hbar\omega_p [(\hbar\omega_p)^2 - (E_k - \varepsilon_{k+p})^2]^{-1} \qquad (9c)$$

$$G_1(k,p) = 2\lambda^*_p A_{k,p,k+p} \hbar\omega_p [(\hbar\omega_p)^2 - (\hbar\omega_H)^2]^{-1} \qquad (9d)$$

$$F_2(k,p) = 2\lambda^*_p A_{k,p,k-p} \hbar\omega_p [(\hbar\omega_p)^2 - (E_k - \varepsilon_{k+p})^2]^{-1} \qquad (9e)$$

$$G_2(k,p) = 2\lambda^*_{-p}A^*_{k,p,k+p}\hbar\omega_p[(\hbar\omega_p)^2 -(\hbar\omega_H)^2]^{-1} \quad (9f)$$

The expression (8) describes a new interaction. The first term shows that the nuclear spin flips down when an atom of H↓ is destroyed in the bulk helium and created on the surface with $\phi_1(k,p)$ as the corresponding matrix element.

The second term in (8) similarly describes the process of diping of H↓ accompanied by the proton spin flip. The H↓ on the surface disappears and appears in the bulk with its nuclear spin changed with a certain probability determined by $\phi_2(k,p)$. The probability of the proton spin flip from the state +½ to the state -½ along with the destruction of H↓ in the bulk and appearance of the same on the surface from the first term in (8) is calculated to be,

$$P_-^{(1)} = \frac{2\pi}{\hbar}\sum_{kp}\phi_1^2(k,p)(m_{k+p}+1)N_k\,\delta(\varepsilon_{k+p}-E_k-\hbar\omega_H) \quad (10)$$

where the number density of the surface bosons is $m_k = b_k^\dagger b_k$ and that in the bulk is $N_k = c_k^\dagger c_k$. From (4), the argument of the δ-function in (10) is found to possess zeros for two values of the ripplon wave vector

$$p_{1,2} = k\,\xi_\pm \quad (11a)$$

$$\xi_\pm = [-1\pm\{1+\frac{2m}{\hbar^2k^2}(\varepsilon_B^o+\hbar\omega_H)\}^{1/2}] \quad (11b)$$

The summation over the ripplon wave vector p in (10) is replaced by a two-dimensional integral and that over k, the wave vector of the bulk H↓ atoms by a three-dimensional integral so that we obtain,

$$P_-^{(1)} = \frac{16mVA}{\hbar^3\pi^2}\sum_\pm\int\lambda^2_{p_{1,2}}A^2_{k,p_{1,2}}(\hbar\omega_{p_{1,2}})^2 (m_{k+p_{1,2}}+1)\xi_\pm N_k\{[c_3^2\xi_\pm-(\hbar\omega_H/k)^2]^2 [1+(\frac{2m}{\hbar^2k^2})(\varepsilon_B^o+\hbar\omega_H)]^{1/2}k^2\}^{-1}dk \quad (12)$$

where the summation over two terms, one with p_1 and positive sign and the other with p_2 and negative sign is indicated. Upon making the change of variables, $x = \hbar^2k^2/(2mk_BT)$ integral in (12) reduces to,

$$P_-^{(1)} = \frac{\hbar}{2(2mk_BT)^{1/2}}\sum_\pm\int\frac{f_\pm(x)dx}{(e^x-1)x^{3/2}} \quad (13a)$$

where

$$f_\pm(x) = [16mVA\lambda^2_{p_{1,2}}A^2_{x,p_{1,2}}(\hbar\omega_{p_{1,2}})^2 (m_{x,p_{1,2}}+1)\xi_\pm]\{\pi^2\hbar^3[c_3^2\xi_\pm-2mk_BTx\omega_H^2] [1+(k_BTx)^{-1}(\varepsilon_B^o+\hbar\omega_H)]^{1/2}\}^{-1} \quad (13b)$$

The probability (13a) varies as the inverse square root of temperature at low temperatures. The second term of (8) also gives the probability, $P_-^{(2)}$, of the downward transition from the state $|+\frac{1}{2}\rangle$ to $|-\frac{1}{2}\rangle$, which can be obtained from (13) by changing the sign of ω_H, using ϕ_2 instead of ϕ_1, writing $e^x/(e^x-1)$ instead of $(e^x-1)^{-1}$ and $m_{x,p_{3,4}}$ in place of $(m_{x,p_{1,2}}+1)$ for which $p_{3,4}$ are obtained from $p_{1,2}$ of the result (11) by replacing $\hbar\omega_H$ by $-\hbar\omega_H$. The total probability of the downward transition of the nuclear spin thus $P_- = P_-^{(1)}+P_-^{(2)}$. Similarly, the probability of the upward transition is $P_+ = P_+^{(1)}+P_+^{(2)}$ where $P_+^{(1)}$ can be written from $P_-^{(1)}$ by replacing $(e^x-1)^{-1}$ by $e^x(e^x-1)^{-1}$ in (13a) and $m_{x,p_{1,2}}+1$ by $m_{x,p_{1,2}}$ (13b) and $P_+^{(2)}$ can be obtained from $P_-^{(2)}$ by similar replacements. The proton spin-lattice relaxation time is then defined as, $\tau = (P_-+P_+)^{-1} = G^{-1}$ which is predicted to vary as the square root of temperature, in agreement with that measured by Yurke et al[4]. We estimate that

$$G = 0.56\times10^{-20}T^{-1/2}cm^3/s$$

for T < 0.22 K as predicted.

REFERENCES

(1) J.A. Yaple and R.A. Guyer, Phys. Rev. B27, 1629 (1983).
(2) D.S. Zimmerman and A.J. Berlinsky, Can. J. Phys. 61, 508 (1983).
(3) A. Suguna and K.N. Shrivastava, Phys. Rev. B22, 2343 (1980).
(4) B. Yurke, J.S. Denker, B.R. Johnson, N. Bigelow, L.P. Levy, D.M. Lee and J.H. Freed, Phys. Rev. Letters 50, 1137 (1983).

LT-17 (Contributed Papers)
U. Eckern, A. Schmid, W. Weber, H. Wühl (eds)

HYDROGEN ATOM ADSORPTION ON SOLID NEON

K.E. Anderson, S.B. Crampton, K.M. Jones, G. Nunes, Jr., and S.P. Souza

Department of Physics and Astronomy, Williams College, Williamstown, MA 01267, U.S.A.*

A compact liquid helium temperature atomic hydrogen maser has been developed and used to study the ground state hyperfine resonance of hydrogen atoms stored using solid neon coatings over the temperature range 6 K to 11 K. Preliminary results are consistent with two-dimensional gas adsorption of hydrogen atoms on neon surfaces and a value of the binding energy in agreement with theory.

1. INTRODUCTION

Techniques for producing and storing atomic hydrogen using liquid helium coated surfaces (1-3) are not easily extended to temperatures over 1 K because the density of helium vapor impedes hydrogen atom mobility. Hydrogen atoms (H) can be stored in containers coated with solid molecular hydrogen (H_2) at temperatures up to 6 K, but strong H binding to the H_2 surface makes it difficult to study anything other than the binding itself (4). Solid neon surfaces bind H about as strongly as solid H_2, but the lower neon vapor pressure allows H storage at temperatures up to 11 K, where the effects of surface binding are much diminished. We have developed an apparatus for studying the ground state hyperfine resonance of spin polarized H stored in containers coated with solid neon. We plan to study the gas phase collision cross sections (5) and to investigate applications to frequency metrology (6), in addition to studying the interactions of H with the surface. We report here preliminary results for frequency shifts and broadening due to surface adsorption and the energy with which H is bound to neon while adsorbed.

2. COMPACT LOW TEMPERATURE ATOMIC HYDROGEN MASER

The apparatus resembles our earlier apparatus (4) except that H emerging from the dissociator through the 2 mm ID orifice pass into a 5 mm by 1.5 cm long "accomodator" and are then focused by a hexapole state-selecting magnet into the storage bottle, instead of simply bouncing down a H_2 coated pyrex tube from the source to the bottle. The accomodator is cooled to ≈5K by a copper heat conduction path to the liquid helium bath outside the vacuum envelope. Semi-circular baffles ensure that any impurities are trapped and that the H are thermalized to the accomodator temperature. A nest of three 0.13 mm thick magnetic shields around the cavity reduce the rms magnetic field gradient in the bottle due to the magnet fringing field to ≈1 mOe. Bottle, cavity, and shields are enclosed in an outer vacuum can containing helium exchange gas, so as to allow control of the storage surface temperature over the range 4.2 K to ≈15 K. Short rf pulses set the atoms radiating, and the decaying signals are detected and stored by a microprocessor-based data acquisition system.

3. BEAM PERFORMANCE

Detected intensity of $F=1, m_F=0$ ground state H focused by the magnet into the .7 cm diameter storage bottle entrance tube is plotted against accomodator temperature in Figure 1.

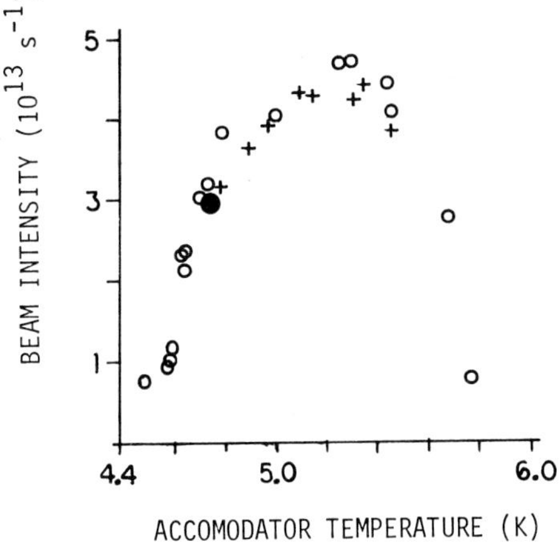

Figure 1

Beam intensity is calculated directly from amplitudes of the free induction decays and level population recovery rates. Accomodator temperature varies because of H recombination on its surface, plus any heat leaking from the dissociator or supplied by a heater. The circles represent data for a variety of dissociator pressures and no supply of extra heat. The

* Supported by ONR Contract N00014-80-C-0240 and NSF Grant PHY-8120579 A01.

crosses represent data for dissociator pressure held at the value that gave the large dot with no extra heating, but with the accomodator temperature varied by supplying extra heat. We believe that the flux increases with temperature at low temperatures because of decreased recombination during shorter adsorption times (7). The flux falls off at high temperatures because the increasing H_2 restricts the H flow, causing more collisions with the accomodator surface. The rough coincidence of circles and crosses indicates that the H density in the accomodator is near saturation at these densities. A larger diameter bottle entrance would accept several times as much of the focused flux. An accomodator design that allowed fewer collisions would likely increase the focused flux.

4. NEON SURFACES

Solid neon surfaces are prepared by freezing about 0.1 mole of neon on the surface at 4.2 K and then annealing the surface by heating above 12 K for a few minutes. The reproducibility of the effective surface area from one preparation to another can be monitored by the frequency shifts due to surface adsorption. The surfaces produced so far are less reproducible by a factor ten than the solid H_2 surfaces produced in the earlier apparatus. Frequency shifts observed for a particular surface are plotted as the log of frequency shift times$\sqrt{\text{Temperature}}$ against inverse temperature in Figure 2. The degree to

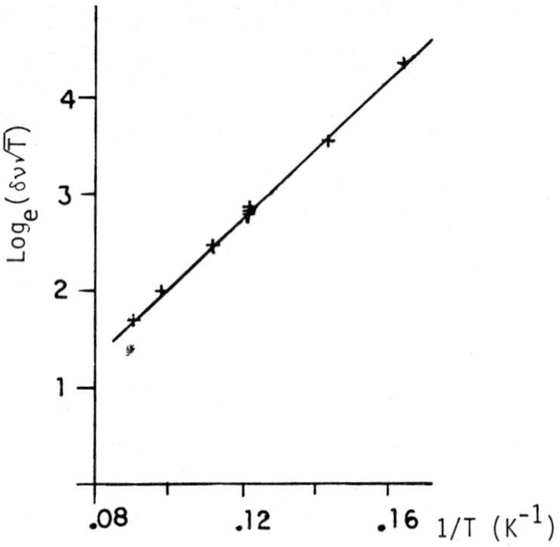

Figure 2

which these data fit a straight line tests the two-dimensional gas model for binding to the surface and also the stability of the effective surface area as more and more of the surface evaporates. The points at $1/T = .122\ K^{-1}$ were taken before and after points at higher and lower temperatures. They indicate good stability over temperature and time. A binding energy of 34(2) K is consistent with these data and with other data for effective surface areas as much as 40% larger or smaller than that indicated by these data. Within the present uncertainty this binding energy agrees with the previously measured binding energy for H on solid H_2 and with the theoretical calculations of C. Lhuillier and colleagues (8).

Within present uncertainties the radiative decay rates and level population recovery rates are equal and vary from about 70 s^{-1} at 7 K to 15 s^{-1} at 11 K, where they are dominated by effusion from the storage bottle. At the lower end of the temperature range they are much higher than would be expected by extrapolating the rates previously observed for H on H_2 to higher temperatures.

5. CONCLUSIONS

We find a binding energy for H adsorbed on solid neon about equal to the binding energy of H on solid H_2. The frequency shifts due to H collisions with solid neon surfaces at 10 K are a few times larger than those found for H colliding with liquid helium surfaces at 0.5 K (3). Lack of reproducibility of the frequency shifts and decay rates from one surface preparation to another and the large size of the current decay rates compared to those found earlier for fairly smooth solid H_2 surfaces suggest that some further improvement in the overall size of the frequency shifts and decay rates is likely.

REFERENCES

(1) I.F. Silvera and J.T.M. Walraven, Phys. Rev. Lett. 44, 164 (1980).
(2) R.W. Cline, T.J. Greytak and D. Kleppner, Phys. Rev. Lett. 47, 1195 (1981).
(3) M. Morrow, R. Jochemsen, A.J. Berlinsky and W.N. Hardy, Phys. Rev. Lett. 46, 195 (1981) and Phys. Rev. Lett. 47, 455 (1981).
(4) S.B. Crampton, J.J. Krupczak and S.P. Souza, Phys. Rev. B25, 4383 (1982).
(5) S.B. Crampton, W.D. Phillips and D. Kleppner, Bull. Am. Phys. Soc. 23, 86 (1978).
(6) A.J. Berlinsky and B. Shizgal, Can. J. Phys. 58, 881 (1980).
(7) S.B. Crampton, J. Phys. (Paris) Colloq. 41, C7-249 (1980).
(8) C. Lhuillier in Rapport de l'Activitie Pour l'Annee 1982-83, Laboratoire de Spectrosco[illegible] Hertzienne de l'E.N.S., 24, rue Lh[illegible]nd, 75231 Paris Cedex 05, France (u[illegible]blished), p. 34.

LT-17 (Contributed Papers)
U. Eckern, A. Schmid, W. Weber, H. Wühl (eds)

ORIENTATION DEPENDENCE OF SURFACE THREE-BODY RECOMBINATION IN SPIN-POLARIZED HYDROGEN (8)

D. A. Bell, G. P. Kochanski, D. Kleppner, and T. J. Greytak

Physics Department and Center for Materials Science and Engineering
Massachusetts Institute of Technology, Cambridge, MA. 02139 U.S.A.

The three-body dipole recombination rate on ^{4}He surfaces oriented parallel to the applied magnetic field is found to be equal to that previously measured on surfaces perpendicular to the field.

In a recent experiment(1) we measured the surface three-body rate constant L_s for the thresholdless dipole recombination process(2) in spin-polarized hydrogen. We suggested that the large value of L_s could account for effects previously attributed to an anomalously large surface two-body nuclear relaxation rate(3,4,5). We demonstrate this contention here by analyzing a set of data originally taken to study the nuclear relaxation process. We show that the decay rates are consistent in magnitude and temperature dependence with three-body recombination. These data were taken in a cell whose walls were parallel to the magnetic field, whereas our recent measurements used walls which were perpendicular to the field. A comparison of the results from the two different experiments allows us to examine the orientation dependence of L_s.

The experimental apparatus and procedure are described in reference (3). Hydrogen is trapped in a vertical copper tube whose walls are covered with a saturated film of superfluid ^{4}He. It is confined from below by a pool of liquid ^{4}He, and from above by the magnetic field gradient of the surrounding superconducting magnet (B=11T) used to polarize the electron spins. After the gas has become doubly polarized with most of the atoms in the b state (a - d are the hyperfine states in order of increasing energy), the time rate of change of the density n(t) is measured. In these experiments $\dot{n}$ was determined at essentially a single density rather than over the wide range of densities accessible to compression experiments.

The density should decay according to

$$-\dot{n} = 2Gn^2 + Ln^3 . \qquad (1)$$

G is the rate constant for nuclear relaxation from b to a. The contribution to G from collisions in the gas has been calculated by Ahn et al.(6) and is in excellent agreement with experimental values which we have measured(1,7). Their results can be approximated in our range of temperatures and at a field of 11T as $G=3.2\times10^{-21}(1+.84(T-.50))$ cm^{3}s^{-1}. Their calculations show that the contribution to G from collisions on the surface should be negligible. We have used Eq. 1, our measured values of $\dot{n}$, and the above expression for G to extract L as a function of temperature from 0.14 to 0.32K. The result is shown in Fig. 1.

L is related to the gas and surface rate constants by

$$L = L_g + (A/V)L_s\Lambda(T)^3 e^{3E_b/kT} \qquad (2)$$

where A/V is the area to volume ratio, $\Lambda(T)$ is the thermal de Broglie wavelength, and E_b is the binding energy of H on liquid ^{4}He. A/V varies from 4.27cm^{-1} at 0.32K to 4.75 at 0.14K, due to the softness of the magnetic confinement. Using E_b = 0.99K(1) and $L_g = 5.24\times10^{-39}$cm^{6}s^{-1} (the values of Ref. (7) were extrapolated from 8 to 11T using the measured weak field dependence), we find that a fit of Eq. 2 to the data gives $(L_g/L_s)^{1/2} = 7\times10^{-8}$cm. This is in reasonable agreememt with the value $(L_g/L_s)^{1/2} = 6\times10^{-8}$cm found earlier(1) however, some corrections must be made before an accurate comparison is possible.

When a three body dipole-dipole recombination occurs between atoms in the lower two hyperfine states, there is a 93% chance that the third atom will be flipped into the c state(2). This atom will recombine in about 1 second in the gas, or a few nanoseconds on the surface, thereby removing two more atoms from the sample. However, the field gradients which serve to confine the a and b states work to expel the c

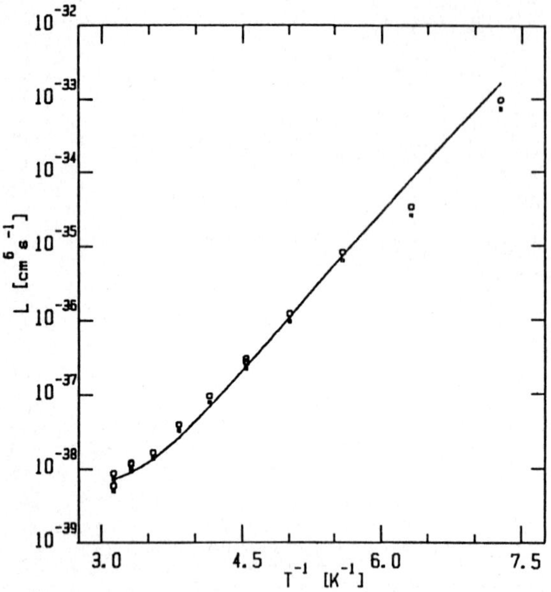

Fig. 1. Temperature dependence of the dipole recombination rate. Dots are without correction for particle escape, circles are corrected data. The line is a one parameter fit to the corrected data.

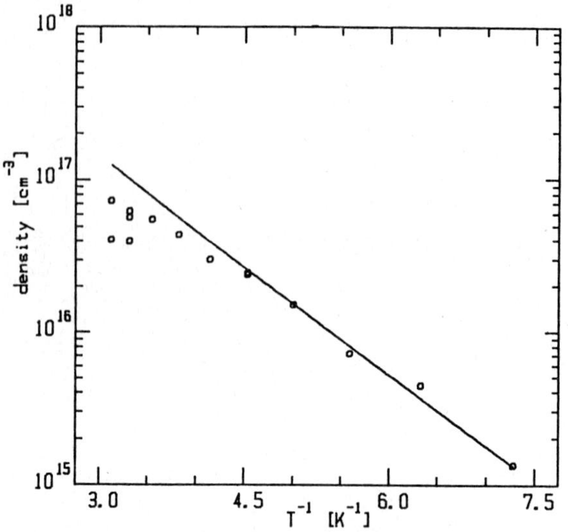

Fig. 2. Volume density n at which $\dot{n}$ was measured versus inverse temperature. The line indicates the volume density that would result if the surface density were constant at $8.4 \times 10^{11} cm^{-2}$.

atom, and unless it recombines or relaxes to the b state on the way out, it will be lost, resulting in a net removal of three rather than four atoms per event. We have modeled the diffusion of c atoms out of the cell, and calculated the probabilities of escape and relaxation. P(escape) is found to vary from 0.5 at our highest density to 0.998 at low density. P(relaxation) was smaller than 0.08. The mean number of atoms removed per event varied from 2.94 to 3.30. The result for a completely closed volume in a compression cell is 3.86. The data of Fig. 1 were adjusted to take this effect into account and were fit again, varying only L_s. We find that $(L_g/L_s)^{1/2} = 6 \times 10^{-8}$cm, identical to the value obtained in ref. (1).

We conclude that the surface three-body dipole recombination can not be strongly anisotropic. Comparison of these results with those of Ref. (1) indicate little, if any, difference in the recombination rates for surfaces parallel and perpendicular to the applied magnetic field.

Decay rate data similar to this was origionally interpreted(3) as arising from a surface contribution to G. As explained in Ref.(1), this misinterpretation can be understood if the decay rates $\dot{n}$ were measured under conditions of constant surface density. Figure 2 shows that this was indeed the case for this set of data.

We would like to acknowledge the major role of Dr. Richard Cline in the experiments on which this discussion is based. This work was supported by the NSF through grant DMR-8304888.

REFERENCES

(1) H. F. Hess, D. A. Bell, G. P. Kochanski, D. Kleppner, and T. J. Greytak, Phys. Rev. Lett. (in print)

(2) Yu. Kagan, I. A. Vartanyantz, and G. V. Shlyapnikov, Zh. Eksp. Teor. Fiz., 81, 1113 (1981) (Sov. Phys. JETP, 54, 590 (1981))

(3) R. W. Cline, T. J. Greytak, and D. Kleppner, Phys. Rev. Lett. 47, 1195 (1981)

(4) R. Sprik, J. T. M. Walraven, G. H. van Yperen, and I. F. Silvera, Phys. Rev. Lett. 49, 153 (1982)

(5) B. Yurke, J. S. Denker, B. R. Johnson, N. Bigelow, L. P. Levy, D. M. Lee, and J. H. Freed, Phys. Rev. Lett. 50, 1137 (1983)

(6) R. M. C. Ahn, J. P. H. W. v. d. Eijnde, and B. J. Verhaar, Phys. Rev. B27, 5424 (1983)

(7) D. A. Bell, G. P. Kochanski, L. Pollack, H. F. Hess, D. Kleppner, and T. J. Greytak, this conference

(8) AT&T Bell Laboratories Scholar

LT-17 (Contributed Papers)
U. Eckern, A. Schmid, W. Weber, H. Wühl (eds)

THERMAL ACCOMMODATION OF ATOMIC HYDROGEN ON SATURATED ^{4}HE FILM

K. SALONEN, S. JAAKKOLA, M. KARHUNEN, E. TJUKANOV, T. TOMMILA

Department of Physical Sciences and Wihuri Physical Laboratory, University of Turku, SF-20500 Turku, Finland

Bolometric measurements on the energy accommodation of H↓ gas on a saturated ^{4}He film are presented. For surface temperatures around 0.4 K the accommodation coefficient is observed to increase from about 0.2 to above 0.3 with decreasing average energy of the incident H↓ atoms.

1. INTRODUCTION

Thermal accommodation of gases at surfaces is characterized by the energy accommodation coefficient $\alpha = (T_r-T_i)/(T_s-T_i)$, where T_i and T_r refer to the average kinetic energies ($2k_BT$) of incident and reflected gas particles and T_s is the temperature of the surface. Experimental results are presented here for the temperature dependence of α of spin-polarized atomic hydrogen (H↓) gas on a saturated ^{4}He film. This system offers an example of atomic collisions with a surface in the extreme quantum mechanical limit and of gas particles possessing a well-defined surface state as adatoms which do not dissolve into the liquid.

2. EXPERIMENTAL METHOD

The measurements were carried out with a carbon film bolometer suspended in a sample cell by thin superconducting wires and heated above the ambient (sample cell) temperature T_0 by a bias current i. Fig.1 shows the measured bolometer surface temperature T_s during a typical H↓ experiment at T_0 = 350 mK when a relatively high bias current has been used. At t = 0 accumulation of H↓ into the sample cell is started and the bolometer begins to cool due to improving thermal contact with the environment. At t = 2 min the density, as measured with a capacitive thin-foil pressure gauge, has risen to $n_H = 3.8\cdot 10^{15}$ cm^{-3} and the H↓ inlet is closed off. Then, after a decay of 11.5 min, the remainder of the sample is destroyed with a short heat pulse. This bolometric technique can be considered as a simple means to continuously monitor H↓ densities with a sensitivity better than 10^{13} cm^{-3}.

Assuming $T_i = T_0$ and the incident and reflected particle fluxes to be equal one may write for the net heat flux transferred from the surface by H↓ gas

$$\Delta P_H = (\alpha/2)n_H A\langle v(T_0)\rangle k_B(T_s-T_0),$$

where A is the area of the heated surface and $\langle v(T_0)\rangle$ the average velocity of the atoms. In order to observe ΔP_H we measured the U-i curves of the bolometer during H↓ experiments, i.e. as a function of n_H, at different cell temperatures stabilized to within $\Delta T_0/T_0 \leq 0.1$ %. The surface temperature T_s is not equal to the average temperature of the carbon. However, T_s corresponding to each (U,i) pair could be calibrated

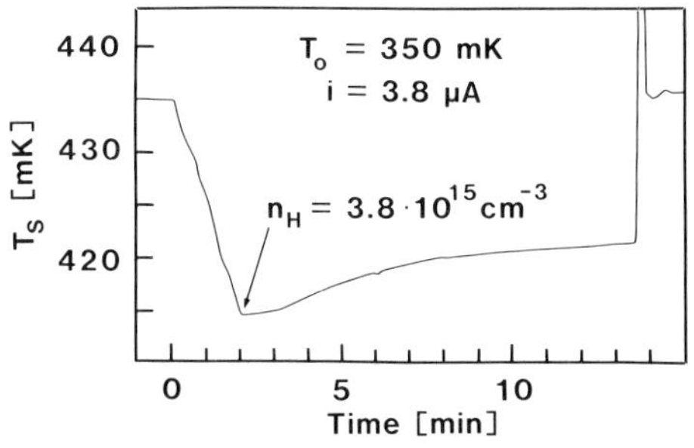

Fig.1. Bolometer surface temperature during a typical H↓ experiment.

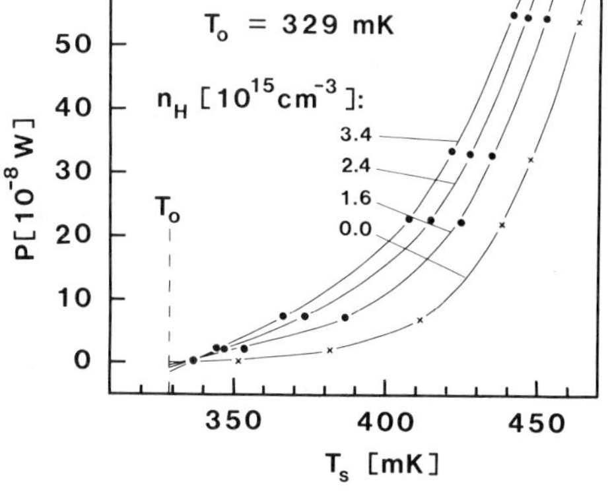

Fig.2. Measuring power P as a function of T_s for various hydrogen densities.

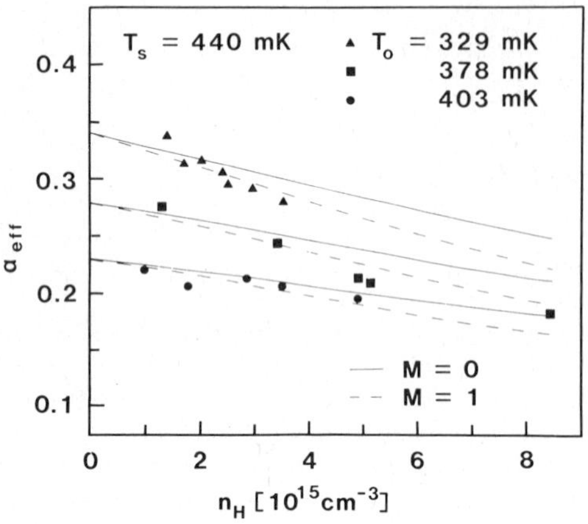

Fig.3. Density variation of the effective accommodation coefficient for T_S = 440 mK at three ambient temperatures. For theoretical curves, see text.

against the cell thermometer when the cell was full of superfluid helium. Graphs like those in Fig.2 could be constructed in order to determine the changes of power, ΔP, necessary to maintain T_S at various constant values. As shown by Fig.2, especially for higher H↓ densities ΔP does not extrapolate to zero at $T_S = T_0$, but to a small negative power. Obviously this power represents the surface recombination heat flux which has to be added to ΔP in order to extract the final ΔP_H.

3. RESULTS

Because of collisions in the gas the assumption $T_i = T_0$ becomes decreasingly justified with increasing n_H. Therefore the procedure described above yields a density-dependent effective accommodation coefficient α_{eff} from which the true α can be obtained in the limit $n_H \to 0$. Fig.3 shows the experimental α_{eff} points measured for T_S = 440 mK at three ambient temperatures as a function of n_H. The curves represent the density dependence of α_{eff} expected on the basis of an approximate kinetic model including the so-called thermal jump at surfaces (1). The jump depends on the true α which has been used here as a fitting parameter to the data. The mean free path in the unbounded space has been taken from Lhuillier´s calculations (2) of the thermal conductivity of H↓ which depends on the nuclear polarization M. Due to their poor S/N-ratio the experimental points do not however reproduce unambigiously the effect of the time evolution of M during a measurement.

4. CONCLUSIONS

In addition to the statistical scatter of the experimental data a total systematic error of at most 30 % may arise from, e.g., inaccuracies in the determination of A and the calibration of the pressure gauge. However, we may state that in the neighbourhood of 0.4 K the surface will thermalize slow atoms more effectively than fast ones. Such a behaviour is not predicted by quantum mechanical calculations based on the distorted wave Born-approximation, but could be due to a polarization response of the surface to the adatom as suggested by Knowles and Suhl (3). Our values of α agree with the one observed earlier at 0.4 K (4) but are clearly higher than the experimental (5) and theoretical (6) results for the sticking probability of H on liquid helium. Therefore one could conclude that, contrary to ^{4}He atoms impinging on a cold liquid ^{4}He surface (7), the probability of diffuse inelastic scattering is not small for H↓.

REFERENCES

(1) See e.g. E.H. Kennard, Kinetic Theory of Gases (McGraw-Hill, New York, 1938) p. 312.
(2) C. Lhuillier, J. Phys. (Paris) 44 (1983) 1.
(3) T.R. Knowles and H. Suhl, Phys. Rev. Lett. 39 (1977) 1417.
(4) K.T. Salonen, Isaac F. Silvera, J.T.M. Walraven and G.H. van Yperen, Phys. Rev. B25 (1982) 6002.
(5) R. Jochemsen, M. Morrow, A.J. Berlinsky and W.N. Hardy, Phys. Rev. Lett. 47 (1981) 852.
(6) D.S. Zimmerman and A.J. Berlinsky, Can J. Phys. 61 (1983) 508.
(7) V.U. Nayak, D.O. Edwards and N.H. Masuhara, Phys. Rev. Lett. 50 (1983) 990.

LT-17 (Contributed Papers)
U. Eckern, A. Schmid, W. Weber, H. Wühl (eds)

CRITICAL WINDOW IN THE TRANSMISSION OF HYDROGEN ATOMS TO LOW TEMPERATURES

T. TOMMILA, S. JAAKKOLA, M. KRUSIUS, K. SALONEN, E. TJUKANOV

Department of Physical Sciences and Wihuri Physical Laboratory, University of Turku, SF-20500 Turku, Finland

The transparency of the hydrogen atom flow channel from a room temperature dissociator to the spin-polarized hydrogen experiment becomes critical in the temperature interval 0.8 - 30 K. The construction and the operating characteristics of this section are discussed.

The most frequently employed technique for preparing spin-polarized hydrogen (H↓) relies on a room temperature hydrogen dissociator and a beam line feeding the atoms to the low temperature H↓ accumulation system. In this approach large recombination losses often make experiments impossible. H atom transport presents no problems at high temperatures where a teflon tube has proved to reduce surface recombination to an acceptable level. Similarly at low temperatures a superfluid ^4He film coating efficiently suppresses recombination. However, these two limiting regions are separated by a temperature gap from 0.8 to about 30 K where losses become critical.

We present one solution to bridging the critical region with moderately low cooling power requirements. This construction has proved to operate reliably once it was recognized that the operating parameters have to be adjusted within a narrow window.

The H flow system is composed of one integral vacuum space which runs centrally through a $^3He/^4He$ dilution cryostat. The critical region, shown in Fig. 1, is cooled with a recirculating ^3He refrigerator with a useful cooling power of $(28T^2/K^2 - 7)$mW in its operating range of 0.5 - 0.8 K. The H beam line consists of a vacuum cover tube which extends from room temperature into the vacuum jacket. The tube ends in a copper vessel which is cooled to 0.60 K with its interior walls lined with a saturated ^4He film. From this ^4He film reservoir (FR) the flow channel continues with a smaller diameter of 5 mm to H↓ sample cell. The flow impedance of this tube has to be high enough to curb the cryopumping action at the low temperature end and to secure a saturated ^4He vapor in the FR.

The H atoms are fed from a room temperature dissociator in a double walled vacuum insulated beam tube (BT), into which an 8 mm ID teflon hose has been inserted. Its lower end fits tightly inside a nylon tube which thermally discon-

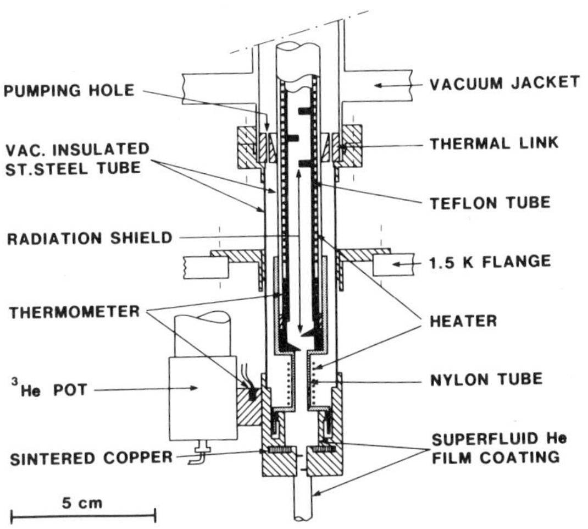

Fig. 1. Construction of the H flow channel.

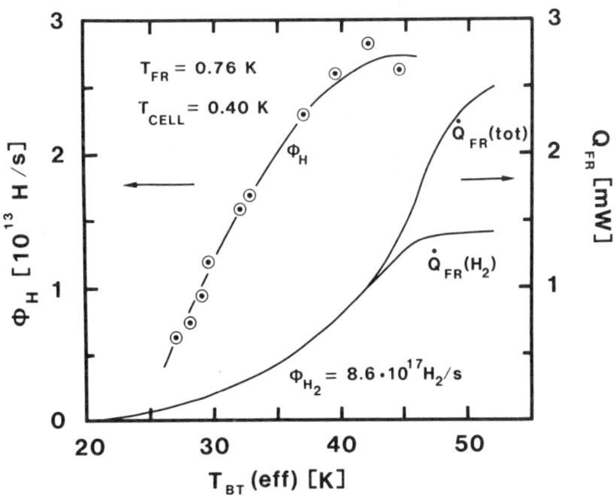

Fig. 2. The effect of teflon beam tube temperature T_{BT}(eff) on beam transmission: $\dot{Q}_{FR}$, heatload to the film reservoir as measured with a constant H_2 flux from the dissociator, and Φ_H, H atom flux to the H↓ cell.

nects the warm BT from the cold FR. The narrow section of the nylon tube (18 mm long, 5 mm IDx 0.3 mm wall) has to support a 30 K temperature difference. Thus the axial temperature distribution in the nylon tube and the lowest radiation shield in the teflon tube determine the recombination losses and also the blocking of the flow channel with H_2 snow. In order to control these processes heater wires have been wound around the lower half of the teflon hose and the neck of the nylon tube. The space between the BT and its cover tube is pumped during accumulation conditions. This removes a heat leak of 1 mW to the FR caused by a thermal bridge via the ^4He vapor. In the proper accumulation conditions the refluxing He in the central beam space produces a heat load of up to 1 mW to the FR.

The adjustment of the axial temperature profile of the flow channel occurs in two steps. First we establish the required heating of the lower half of the teflon tube by determining the penetration of H_2 to the FR. The FR heat load $\dot{Q}_{FR}$ is used as a measure of the BT transparency. The measurement is performed as a function of the effective temperature $T_{BT}(eff)$ of the lower half of the BT as recorded with a Cu wire resistance thermometer wound around the teflon hose. The result is shown in Fig. 2 where we have plotted $\dot{Q}_{FR}(tot)$, generated by the H_2 flux and the electrical heating, and $\dot{Q}_{FR}(H_2)$, caused by the H_2 flux only. A flux of $9 \cdot 10^{17}$ H_2/s from the dissociator was used. Note that the slope of $\dot{Q}_{FR}(H_2)$ is abruptly reduced at $T_{BT}(eff) \simeq 47$ K where the H_2 flux fully penetrates to the FR. In Fig. 3 this conclusion has been checked by monitoring $\dot{Q}_{FR}$ at $T_{BT}(eff) = 47$ k as a function of the H_2 flux from the dissociator. The result is indeed linear with a slope yielding 94 K/H_2, to be compared with the sublimation energy of 90 K/H_2.

Next we adjust the temperature distributions of the lowest teflon radiation baffle and of the nylon tube while keeping $T_{BT}(eff) = 47$ K. This is accomplished by determining the H flux Φ_H to the sample cell with a foil manometer. The flux Φ_H does not depend on the cell temperature in the range 0.3 - 0.5 K in a polarization field of 7.5 T within the required resolution. Fig. 4 shows Φ_H as a function of T_{FR} at constant H atom flux from the dissociator while the nylon tube is heated by varying amounts. It is found that optimum transmission through the critical section is achieved when T_{FR} is raised to 0.72 K. If T_{FR} is increased to above 0.85 K the transmission declines because of an increasing deterioration of the film coating of the FR.

A cross check of the two adjustment steps is reproduced in Fig. 2 where Φ_H to the H↓ cell is plotted as a function of $T_{BT}(eff)$ while the nylon tube heating is regulated such that $T_{FR} = 0.76$ K. As is to be expected the low temperature edge of the transmission window is now shifted lower in temperature if compared to a H_2 flux as illustrated by $\dot{Q}_{FR}(H_2)$ in Fig. 2.

The standard procedure for collecting a H↓ sample consists of the following opeeations. First $T_{BT}(eff)$ is adjusted to 47 K which raises T_{FR} from 0.60 K to 0.64 K. Next the H beam is switched on which further boosts T_{FR} to 0.68 K. Finally the nylon tube heating is switched on which adjusts T_{FR} to 0.73 K and secures transmission to the H↓ cell. Right after a cool down from room temperature the BT may require an exposure of order 15 min to the H flux until the effects from poisonous catalytic spots on the walls have been removed, a H_2 coverage has been established and a measurable flux is observed in the H↓ cell.

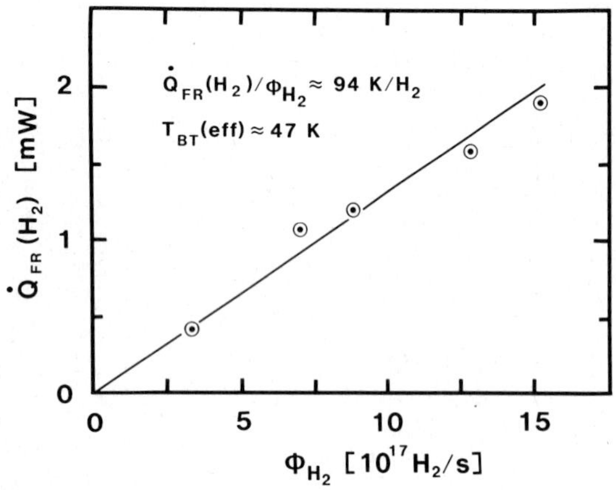

Fig. 3. Heat load to the film reservoir as a function of the H_2 flux from the dissociator.

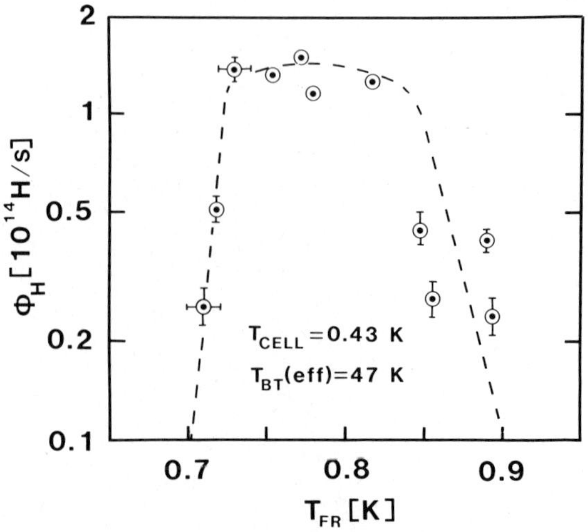

Fig. 4. H atom flux to the H↓ cell as a function of the film reservoir temperature.

LT-17 (Contributed Papers)
U. Eckern, A. Schmid, W. Weber, H. Wühl (eds)

SPIN TRANSPORT IN H↓, A DILUTE SPIN-POLARIZED QUANTUM GAS

J. S. DENKER and N. BIGELOW

Laboratory of Atomic and Solid State Physics, Cornell University, Ithaca, New York 14853 USA

This is a pedestrian and pedagogical introduction to the theory of spin transport in spin polarized gases. It discusses the "identical spin rotation" effect and the resulting macroscopic equation of motion. The emphasis is on spin polarized hydrogen, although most of the analysis applies equally well to polarized ^3He gas and ^3He diluted in ^4He.

Spin transport in H↓ is very remarkable. There are spectacular collective effects, including spin waves (1).

One expects cooperative behavior in degenerate systems such as liquid ^3He or the electrons in metals. By the same token, one might have expected H↓ to be relatively boring since it is an ideal gas -- the atoms are very far apart, interacting very rarely. Furthermore, even when two H↓ atoms do collide, the interaction Hamiltonian is essentially independent of spin. One might have expected the evolution of the spins in H↓ to be independent and overdamped -- the NMR signal would have been dominated by T_2* and the transverse relaxation (as measured by spin echos) would have been dominated by simple diffusion.

In fact (2), a collision between two polarized atoms causes the two spin vectors ($\underset{\sim}{s}_1$ and $\underset{\sim}{s}_2$) to precess about their resultant ($\underset{\sim}{s}_1 + \underset{\sim}{s}_2$), as in figure 1. They always precess in the same direction (e.g. clockwise) so the effects of many collisions are cumulative. In this way the collisions, rare as they are, can transmit information about the spin state of the gas in distant parts of the cell. This couples all the spins in the equation of motion.

Figure 2 shows two ways in which a |b> atom can scatter against another |b> atom. Scattering of |a> by |b>, etc., is similar; the scattering amplitudes for all such diagrams are equal. At low temperatures where the thermal de Broglie wavelength of the atoms is large compared to their hard-sphere diameter, the two diagrams for |b>-|b> scattering (and not for |a>-|b> scattering) will interfere. Parallel spins will be scattered more than antiparallel

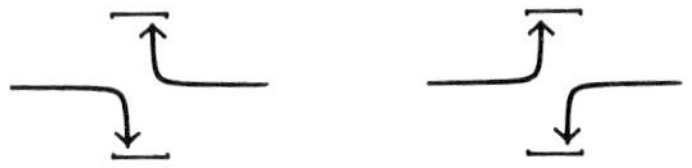

Figure 2.

spins. (Note that H↓ is a boson; for fermions, the interference would be destructive and the scattering would vanish for parallel spins). At high temperatures (short de Broglie wavelengths) the two processes in figure 2 can be distinguished by their impact parameters. There is no interference, and the scattering rates are independent of spin.

Now consider a box containing a large number of cold H↓ atoms (the scatterers) all in the |b> state, and a single incident atom |Φ> (the scatteree) in some superposition of |a> and |b>. The |b> component of |Φ> will suffer a different scattering phase shift than the !a> component. That means that the spin of the scattered atom will undergo a rotation through some angle θ, since passing through the box is the same as multiplying |Φ> by $\exp(i\theta\sigma_z/2)$. Here σ_z is a Pauli matrix, the generator of rotations about the z axis in spin space: $\sigma_z|a\rangle \equiv +|a\rangle$ and $\sigma_z|b\rangle \equiv -|b\rangle$. This rotation can be ascribed (4) to a "molecular field" in the box. It is not a magnetic field per se, although it has the same vector character as a magnetic field. Its value at each point is proportional to the spin density at that point.

Let us now discuss a fluid with number density $n = n(\vec{r}, t)$, spin density $\underset{\sim}{s} = \underset{\sim}{s}(\vec{r}, t)$, and spin current $\tilde{J} = \tilde{J}(\vec{r}, t)$. The local density matrix is then $\rho = n1 + \underset{\sim}{s}\cdot\underset{\sim}{\sigma}$, and the polariza-

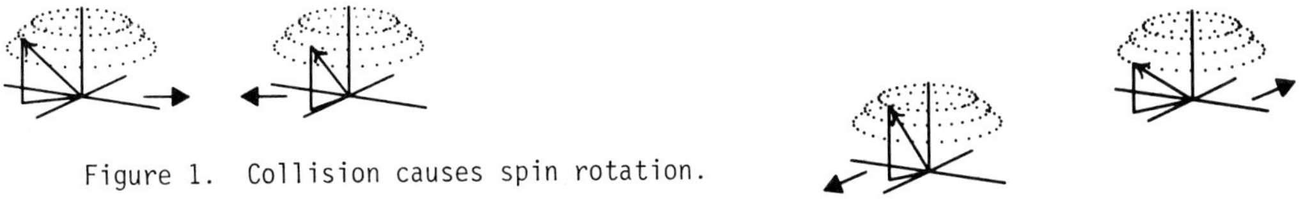

Figure 1. Collision causes spin rotation.

tion is $\underset{\sim}{P} = \underset{\sim}{s}/n$. The equations of motion are:

$$\frac{\partial s_\nu}{\partial t} + \varepsilon_{\nu\eta\lambda}(\gamma H_0+\alpha s)_\eta\, s_\lambda + \nabla_i\, J_{i\nu} = 0 \qquad (1a)$$

$$\frac{\partial J_{i\nu}}{\partial t} + \varepsilon_{\nu\eta\lambda}(\gamma H_0+\alpha s)_\eta J_{i\lambda} + \frac{\nabla_i\, \delta s_\nu}{\beta m} + \frac{J_{i\nu}}{\tau} = 0 \qquad (1b)$$

where $\beta = 1/k_BT$ and m is the particle mass; τ and α are discussed below. We have written out the cross products in terms of the completely antisymmetric tensor $\varepsilon_{\mu\eta\lambda}$ and have used Greek indices for spin-space vectors, and Latin indices for real-space vectors. Most of the Larmor precession can be removed from both equations by transforming into the rotating frame, but if the applied field $\vec{H}_0$ is non-uniform there will remain a term $\delta\{\gamma\vec{H}_0(\vec{r})\}$. The only effect of the molecular field in equation 1a is the term $\alpha\, \underset{\sim}{s}\times\underset{\sim}{s}$, which obviously vanishes, so 1a becomes just the familiar expression for local spin conservation. In equation 1b, the precession of the spin current in the molecular field does not vanish. The last two terms of 1b express the fact that the current is driven by the spin density gradient, but decays with some relaxation time τ.

These two terms can be understood by considering a fluid in equilibrium in a potential $\Phi(x)$. The current is zero because the drive due to the density profile $\exp(-\beta\Phi)$, balances the forces $-d\Phi/dx$. We write $\delta\underset{\sim}{s}$ in the driving term to indicate the deviation of $\underset{\sim}{s}$ from its equilibrium value.

One of the nice things about H↓ is that τ and the parameter α (which determines the strength of the molecular field) can be calculated (2,3,4) from first principles, using the known atomic scattering cross sections. This is in contrast with the Fermi liquid theory of ^{3}He, where the molecular fields are described by a whole series of phenomenological parameters that are difficult to calculate.

The next step depends on a slightly delicate distinction of time scales, similar to the Born-Oppenheimer approximation. It turns out that the time dependence of $\underset{\sim}{J}$ in eq. 1b is quite rapid compared to the time dependence of $\underset{\sim}{s}$. For times on the order of τ, we can neglect $\delta(\gamma H_0)$ and take $\underset{\sim}{s}$ to be quasi-constant, while $\underset{\sim}{J}$ relaxes on a time scale τ to a quasi-stationary value which satisfies the equation:

$$-D_0\nabla_i\, \delta s_\nu = J_{i\nu} + \varepsilon_{\nu\eta\lambda}\, \mu\, P_\eta\, J_{i\lambda} \qquad (2)$$

where we have identified the ordinary diffusion constant $D_0 = \tau/\beta m$. We define the parameter $\mu = n\tau\alpha$, which is dimensionless and essentially independent of density. It measures how rapidly the spin current precesses in the molecular field, relative to the momentum relaxation rate. It turns out that μ is quite large ($\approx$10) for H↓. This means roughly that collisions are ten times more effective at changing the atom's internal spin coordinates than changing its momentum. When μ becomes large compared to 1, it totally changes the character of eq. 2. The quasi-stationary current becomes nearly perpendicular to the driving spin density gradient, and smaller in magnitude by a factor of $\vec{\mu}$.

It is possible to solve equation 2 for $\underset{\sim}{J}$ in terms of $\underset{\sim}{s}$:

$$J_{i\nu} = \frac{-D_0}{1+\mu^2P^2}(\nabla_i\, \delta s_\nu - \mu\varepsilon_{\nu\eta\lambda}P_\eta\nabla_i\, \delta s_\lambda + \mu^2P_\nu P_\eta\nabla_i\, \delta s_\eta) \qquad (3)$$

This is equivalent to equation 25 of reference (5). It applies inside a bulk sample; if the boundaries are important then one can add additional terms to the RHS of equation 3 to enforce the boundary conditions. Substituting equation 3 into 1a gives a closed equation of motion for $\underset{\sim}{s}$. The resulting equation has remarkable consequences, many of which are not completely understood.

Let us consider the simple Ansatz:

$$|s_z(\vec{r},t)| = \text{const} \gg |s_+(\vec{r},0)| \qquad (4a)$$

where $s_+(\vec{r}, t) \equiv s_x(\vec{r}, t) + is_y(\vec{r}, t) = s_+(\vec{r}, 0)\exp(i\omega t - \Gamma t)$

This Ansatz permits many simplifications in the equation of motion for $\underset{\sim}{s}$, which becomes:

$$i\frac{\partial s_+}{\partial t} = \delta(\gamma H_0)s_+ + iD_0\frac{1-i\mu P_z}{1+\mu^2P^2}\nabla^2 s_+ \qquad (4b)$$

This is the spin wave equation. It is used, for instance, to describe a NMR experiment where the tipping angle is small. Comparing the real and imaginary parts of the coefficient in front of the Laplacian in equation 4b, one can see that the Q of the spin wave, $Q = \omega/\Gamma$, will be on the order of $|\mu P|$. This is in contrast to the spin waves in ferromagnets where the Q can be made arbitrarily large, but is analogous to spin waves in a ferromagnet subjected to an applied magnetic field. In any case $|\mu P|$ is large enough in H↓ that the spin waves are quite sharp and long-lived.

Practically all of the ideas in this paper have previously appeared in the literature. Please see the references cited in (1, 4, 6, and 7). It is a pleasure to acknowledge useful discussions with A. E. Ruckenstein, L. P. Lévy, D. M. Lee and J. H. Freed. This work was supported by NSF Grant No. DMR-8305284.

REFERENCES

(1) B. R. Johnson, J. S. Denker, N. Bigelow, J. H. Freed and D. M. Lee, Phys. Rev. Lett. (in press).

(2) C. Lhuillier and F. Laloë, J. Physique 43, 197 (1982); 43, 225 (1982).

(3) C. Lhuillier, J. Physique 44, 1 (1983).

(4) L. P. Lévy and A. E. Ruckenstein, Phys. Rev. Lett. (in press).

(5) A. J. Leggett, J. Phys. C3, 448 (1970).

(6) B. R. Johnson, N. Bigelow, J. S. Denker, L. P. Lévy, D. M. Lee and J. H. Freed, Observation of Spin Waves in Spin Polarized Hydrogen, this volume.

(7) J. S. Denker, N. Bigelow, D. Thompson, J. H. Freed and D. M. Lee, Spin Echoes in Spin Polarized Hydrogen, this volume.

LT-17 (Contributed Papers)
U. Eckern, A. Schmid, W. Weber, H. Wühl (eds)

SPIN ECHOES IN SPIN POLARIZED HYDROGEN

J. S. DENKER, N. BIGELOW, D. THOMPSON, J. H. FREED and D. M. LEE

Laboratory of Atomic and Solid State Physics, and Baker Laboratory of Chemistry, Cornell University, Ithaca, New York 14853 USA

We have conducted preliminary spin echo experiments in spin polarized hydrogen. The heights of the echoes follow a remarkable pattern, and we often see a regular series of additional peaks in the FID envelope which resemble multiple echoes. We conjecture that molecular fields, with due regard for boundary conditions, could account for most of the observed effects.

A typical ideal spin echo experiment would consist of a $\pi/2$ pulse applied at time $t = 0$ and a π pulse applied at $t = \tau$, forming an echo at $t = 2\tau$. Another π pulse applied at $t = 2\tau + \tau'$ forms an echo at $t = 2\tau + 2\tau'$.

Figure 1a shows some typical H↓ spin echo signals. There are normal echoes at $t = 2\tau$ and $t = 4\tau$, followed by several prominent peaks in the FID envelope at multiples of τ thereafter. This is the basic pattern of the data, which can be seen over a fair range of magnetic field gradient (≈0 to 2 G/cm) and spacing (τ < 300 μsec). This pattern is also observed in $\pi/2$-π pulse experiments. Sometimes the recurrent peaks appear to ride atop a nonzero but gently varying background which may be due to spin waves excited by imperfections in the π pulses.

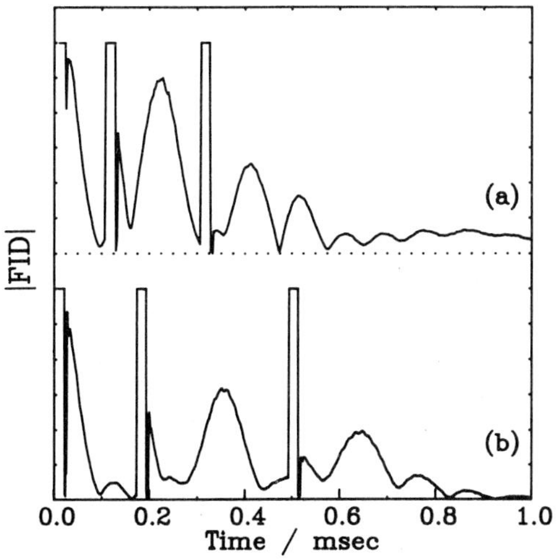

Figure 1. Typical H↓ spin echoes resulting from a $\pi/2$ - π - π pulse sequence. T ≈ 245 mK, $n \approx 4.3\times10^{16}$cm^{-3}, G ≈ 23 kHz/cm.

A variation from the simple pattern can be seen in figure 1b: although the first echo forms at the expected time, the second echo forms slightly earlier than expected, and then the multiple echoes form at progressively shorter intervals, all slightly shorter than τ and τ'. Sometimes the reverse is true: the second echo forms late and the intervals between echoes are progressively longer.

Often there is a prominent FID immediately following the first π pulse, as in figure 1b. This is despite the fact that we adjusted the π pulses to minimize the FID following an isolated π pulse. We do not observe any echo-like structures following an isolated π pulse. Note that FID following the second π pulse is much smaller, although we are reasonably certain that the two pulses were identical.

The magnitude of H_1 is the same for all pulses; the tipping angle was determined by the pulse duration. Nominally, all tipping pulses were applied about the same axis, although there was a slight drift of the tipping axis in the x--y plane. Magnet drift may have detracted from the accuracy of a few of the tipping pulses.

We do not have a clear picture of what causes the recurrent peaks. They may not be genuine multiple echoes, but we emphasize that the close correlation between peak position and τ is observed in a large fraction of FIDs, over the entire range of the data. It appears that these peaks are unrelated to the multiple echoes seen in superfluid ^{3}He (1). The theory (2) which explains the multiple echoes in solid ^{3}He (3) predicts no multiple echos for a $\pi/2$-π sequence, and it is unlikely that imperfections in our π pulses could give rise to the large effects we see. Perhaps in the absence of boundary effects (see below) we wouldn't see the additional peaks.

It is instructive to study the height of the first echo as a function of τ. We start with the hypothesis that the decrease of echo height is due to spin diffusion in a gradient $G = \vec{\nabla}(\gamma H_0)$. The formula of Torrey (4) is $s_+(2\tau)/s_+(0) = \exp\{-(2/3)\tau^3 G^2 D_{eff}\}$, where we are using the notation of reference (5). D_{eff} is some effective diffusion constant (see below). We have analyzed our data by plotting $\ln[s_+(2\tau)/s_+(0)]$ versus $\tau^3 G^2 D_{eff}$. If we neglect molecular field effects and set D_{eff} equal to the diffusion constant D_0 determined by other means (6), the data points do not fall on a straight line. Furthermore, they consistently fall at least a factor of 10 below the prediction of Torrey, i.e. the observed decay of transverse magnetization is anomalously fast.

We can try to account for molecular field effects by using a reduced diffusion constant $D_{eff} = D_0/(1+\mu^2 P^2)$. This collapses all the data from a variety of G and μP values onto a fairly straight line. This supports the idea that the echoes decay because of dephased spins diffusing in a gradient -- as opposed to some other process, e.g. diffusion limited surface relaxation. It also indicates that the molecular field has important effects on the diffusion. The fact that it is the slope (not the intercept) that is anomalous indicates that imperfections in the tipping pulses can not account for the decay.

On the other hand, the reduced diffusion constant makes the theoretical decay rate even **slower** than the observed rate. If one adjusts the value of D_0 to get agreement with the observed ratios of (echo/FID), then the theory would predict a much **faster** decay rate of the ratios of (second echo/first echo) than we observe. The full theory of Leggett and Rice (7) predicts an entirely different form of the echo decay anyway, which we can approximate by $1-[s_+(2\tau)/s_+(0)]^2 = \exp[-(2/3)\tau^3 G^2 D_{eff}]$. Our data does not fit this form.

There are three factors which complicate the analysis of this situation: the large molecular fields, the nonlinear effects of the large tipping angle, and the effects of the boundaries. Removing any one of these factors produces a relatively tractable system: Robertson (8) discussed spin echoes in the presence of boundaries, Lévy and Ruckenstein (9) discussed the effect of molecular fields and boundaries (for small tipping angles), and Leggett and Rice (7) and Leggett (10) discussed spin echoes in the presence of molecular fields. Leggett's equations do not have any boundary terms, which corresponds to the case of very distant or perfectly absorbing walls. Our studies (6) of spin waves indicate that reflecting boundary conditions are appropriate for our cell, and the large values of μ and D_0 for H↓ cause boundary effects to be important farther from the boundary than in ^{3}He.

We conjecture that the observed echo behavior might be explained by including appropriate boundary terms in eq. 3 in ref. (5).

For example, consider spins that have been tipped into the x--y plane, and subjected to a gradient $G_{xz} = \partial(\gamma H_z)/\partial x$, that is, H_0 is parallel to z and its z component varies with x. The dephased spins will have a gradient $\vec{\nabla}\underset{\sim}{s}$ in the x--y plane and perpendicular to $\underset{\sim}{s}$. Equation 3 of reference (5) predicts a spin current that carries z magnetization in the x direction. Except at the walls, this current is uniform and divergence free, contributing nothing to the equation of motion of $\underset{\sim}{s}(\vec{r}, t)$. For reflecting boundary conditions, though, there will be an accumulation of s_z at the +x boundary and an accumulation of $-s_z$ at the -x boundary. Applying a π pulse rotates the triad $(\underset{\sim}{s}, \vec{\nabla}\underset{\sim}{s}, \underset{\sim}{J})$ by π, reversing the direction of the current, but also reversing the sign of the accumulated s_z. Therefore the accumulation continues to grow while the echo is forming.

This process causes the magnetization to rotate out of the x--y plane. This reduces the size of the received NMR signal, but does not necessarily destroy the magnetization. Estimates of the size of this effect indicate that it could be significant in our experimental situation. It produces a very inhomogeneous distribution of spin density, producing a very inhomogeneous molecular field that can affect the equation of motion of the spins in very complicated ways.

It is clear that much more work needs to be done before the spin dynamics of H↓ are fully understood. We are continuing our investigations.

We thank K. Earle, L. Goldner, K. Feldmann, and M. Kantor for valuable assistance with the experiments, and L. P. Lévy and A. E. Ruckenstein for valuable discussions. This work was supported by NSF grant #DMR-8305284

REFERENCES

(1) G. Eska, H.-G. Willers, B. Amend and W. Wiedemann, Physica 108B, 1155 (1981).
(2) M. Bernier and J. M. Delrieu, XIXth Congres Ampére 123 (1976).
(3) O. Avenel, P. Berglund, M. Bernier, J. M. Delrieu, E. Varoquaux, C. Vibet, XIXth Congres Ampére 127 (1976).
(4) H. C. Torrey, Phys. Rev. 104, 563 (1956).
(5) J. S. Denker and N. Bigelow, Spin Transport in H↓, a Dilute Spin Polarized Quantum Gas, this volume.
(6) B. R. Johnson, J. S. Denker, N. Bigelow, L. P. Lévy, J. H. Freed and D. M. Lee, Phys. Rev. Lett. (in press).
(7) A. J. Leggett and M. J. Rice, Phys. Rev. Lett. 20, 586; 21, 506 (1968).
(8) B. Robertson, Phys. Rev. 151, 273 (1966).
(9) L. P. Lévy and A. E. Ruckenstein, Phys. Rev. Lett. (in press).
(10) A. J. Leggett, J. Phys. C3, 448 (1970).

LT-17 (Contributed Papers)
U. Eckern, A. Schmid, W. Weber, H. Wühl (eds)

STATIC PROPERTIES OF SPIN-POLARISED ^3He - ^4He MIXTURES

John R OWERS-BRADLEY, Roger M BOWLEY and Peter C MAIN

Physics Department, University of Nottingham, University Park, Nottingham, NG7 2RD, UK

We have calculated the polarisation and osmotic pressure of very dilute mixtures of ^3He in ^4He as a function of temperature and magnetic field including the effect of interactions.

1. INTRODUCTION

In a dilute mixture of ^3He in ^4He it is accepted that the ^3He quasiparticles can be treated as a gas of weakly interacting fermions but there is some doubt about the exact form of the dispersion relation and also the interaction itself and how it should be parameterised.

If very dilute solutions were studied (say < 5000 ppm ^3He) then it would be possible to treat the interaction as being constant. Strictly this also requires that the temperature is sufficiently low (< 100 mk) so that the thermal momentum of the quasiparticles is small. This also ensures that any p^4 term in the dispersion relation can be ignored.

The other result of reducing the concentration is to lower the fermi energy. By applying a large magnetic field it is possible to produce relative energy shifts of the different spin states larger than the fermi energy. In a magnetic field non-interacting quasiparticles have an energy given by

$$\varepsilon^0(\underline{p},\sigma) = \frac{p^2}{2m} - \frac{\hbar\gamma B}{2}\sigma \qquad (1)$$

where $\sigma = \pm 1$, γ is the gyromagnetic ratio and B the magnetic field.

Here we have developed a theory that is valid over the entire temperature range and which can also be used for higher concentrations (up to X = 0.01) though obviously practical limitations in the magnetic field mean it would not be possible to achieve very high polarisations in this case.

2. CALCULATIONS

Even though the ^3He concentration may be small the interaction between the quasi-particles still has to be taken into account. This was not done by Mullin and Miyake (1) in their calculation of the polarisation.

The Fermi statistics determine that for s wave scattering the up spins interact only with the down spins. We include this interaction, with amplitude V_0, in the single particle energy in the form

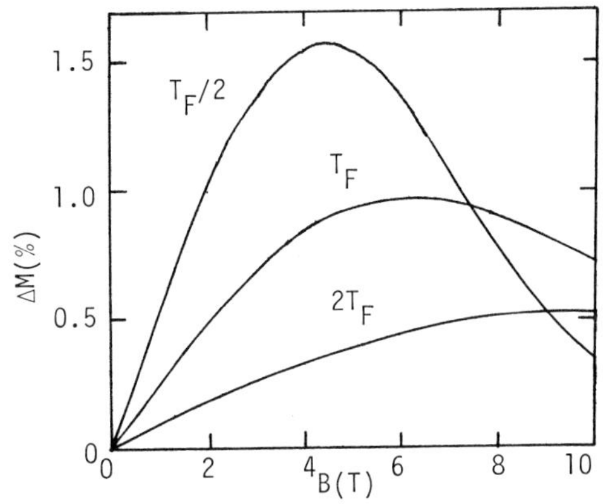

FIGURE 1
The change in the polarisation due to the presence of interactions plotted as a function of field at three temperatures. T_F = 5.56 mk.

$$\varepsilon(\underline{p},\sigma) = \varepsilon^0(\underline{p},\sigma) + V_0 \sum_{\sigma'} n_{\sigma'}(1-\delta_{\sigma,\sigma'}) \qquad (2)$$

where $\sigma' = \pm 1$. The number of particles with spin σ, n_σ is readily found in terms of the total number, n_3 and the fermi temperature.

$$n_\sigma = \sum_p n_{\underline{p},\sigma} = \frac{3}{4} n_3 \left(\frac{T}{T_F}\right)^{3/2} F_{\frac{1}{2}}(\eta_\sigma) \qquad (3)$$

where

$$F_{\frac{1}{2}}(\eta) = \int_0^\infty \frac{dz\, z^{\frac{1}{2}}}{e^{z-\eta}+1} \qquad (4)$$

and

$$\eta_\sigma = \frac{1}{k_B T}\left(\mu_\sigma + \frac{\hbar\gamma B\sigma}{2} - V_0 \sum_{\sigma'} n_{\sigma'}(1-\delta_{\sigma\sigma'})\right) \qquad (5)$$

Here μ_σ is the chemical potential for the spin system σ. In thermal equilibrium $\mu_1 = \mu_{-1} \equiv \mu$ provided we are interested in times long compared with the dipole relaxation time.

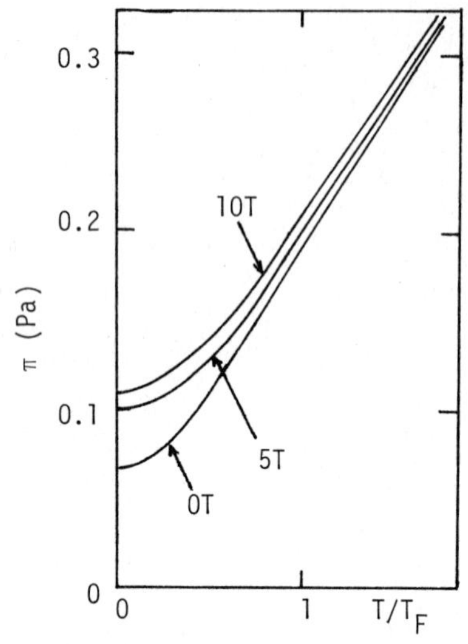

FIGURE 2
The osmotic pressure plotted versus temperature for three magnetic fields. T_F = 5.56 mk.

Denoting $\sigma = +1$ as + and $\sigma = -1$ as - the chemical potential is found by solving

$$\frac{4}{3}\left(\frac{T_F}{T}\right)^{3/2} = F_{1/2}(\eta_+) + F_{1/2}(\eta_-) \quad (6)$$

The polarisation M is now defined as

$$M = \frac{n_+ - n_-}{n_+ + n_-} = \frac{n_+ - n_-}{n_3} \quad . \quad (7)$$

The equations (5) and (6) have to be solved self-consistently for M in terms of T, B, X and V_o. This has been done numerically using the tables of the fermi integrals $F_{\frac{1}{2}}(\eta)$ compiled by McDougall and Stoner (2). Without the interaction term our polarisations are identical to those of Mullin and Miyake (1). Figure 1 shows the effect of the interaction term on the polarisation of an $X = 10^{-4}$ (T_F = 5.56 mk) solution as a function of magnetic field for $T = \frac{T_F}{2}$, T_F and $2T_F$. V_o is taken as $-0.12\ V_c$ where $V_c = 1.73 \times 10^{-50}\ \mathrm{J\ m^3}$ (3). The factor 0.12 is determined from a fit to the second-sound data of Greywall and Paalanen (4) and will be discussed elsewhere (5). The effect of the interaction is to reduce the polarisation at a given magnetic field.

Next we examine the osmotic pressure. For polarised systems this is given by (5)

$$\pi = \frac{2}{3} k_B T n_3 \frac{F_{3/2}(\eta_+) + F_{3/2}(\eta_-)}{F_{1/2}(\eta_+) + F_{1/2}(\eta_-)} + V_o n_+ n_- \quad (8)$$

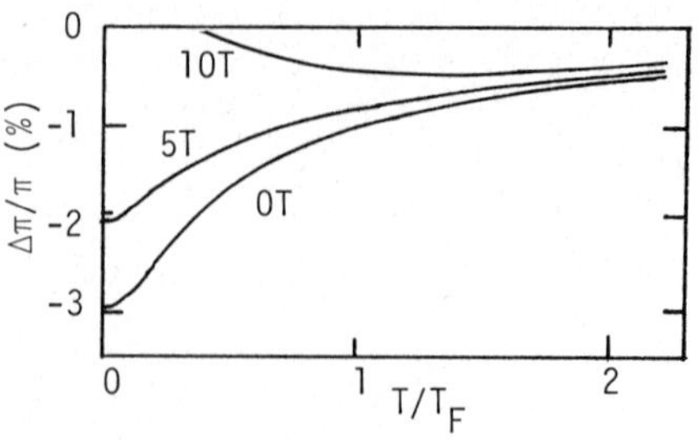

FIGURE 3
The relative change in the osmotic pressure due to interactions plotted against temperature for three magnetic fields. T_F = 5.56 mk

where $F_{3/2}$ is defined by McDougall and Stoner (2).

In figure 2 we have plotted the osmotic pressure, including the interaction term for an $X = 10^{-4}$ solution as a function of temperature for magnetic fields 0T, 5T and 10T. Figure 3 shows the relative effects of the interaction under the same conditions. The term in Eqn (8) which includes V_o explicitly tends to reduce π (V_o is negative), but the interaction also reduces the polarisation thereby tending to increase π through the first term. Thus, in zero field the osmotic pressure is reduced by $\sim$3% at $T \simeq 0$, whereas at 10T the two effects more or less cancel out.

3. DISCUSSION

The changes due to polarisation should certainly be resolvable in the osmotic pressure (4,5) and some results have already been reported (6). However the interaction has a much smaller effect and has the added disadvantage that it cannot be switched on and off. To detect these effects would appear to be a very difficult experimental problem. We are more hopeful that second sound experiments which have a high resolution can be used to deduce the value of the interaction and we are investigating this property within the same framework as for the properties treated above.

REFERENCES

(1) W.J. Mullin and K. Miyake, J. Low Temp. Phys. 53 (1983) 313
(2) J. McDougall and C. Stoner, Trans. Roy. Soc. London A237 (1938) 67.
(3) C. Ebner and D.O. Edwards, Physics Reports 2 (1970) 77.
(4) D.S. Greywall and M.A. Paalanen, Phys. Rev. Lett. 46 (1981) 1292 and Proc. LT-16 Physica 109 + 110 (B+C) 1575 (1981).
(5) J.R. Owers-Bradley, R.M. Bowley and P.C. Main to be published.
(6) W.J. Gully and G.M. Schmiedeshoff, Bull. Am. Phys. Soc. 28 (1983) 675.

LT-17 (Contributed Papers)
U. Eckern, A. Schmid, W. Weber, H. Wühl (eds)

ON TEMPERATURE VARIATIONS DURING ^{3}He POLARIZATION EXPERIMENTS IN POMERANCHUK CELLS

Q. GENG and F.B. RASMUSSEN

Physics Lab. I, H.C.Ørsted Institute, Universitetsparken 5, DK-2100, Copenhagen,Denmark*

Simple model calculations have been performed in relation to temperature changes in decompression experiments with Pomeranchuk cells, aiming at the production of spin polarized liquid ^{3}He. Comparison with reported experiments indicates that thermal contact with the surroundings is too strong for our models to apply.

1. INTRODUCTION

More than five years ago, Castaing and Nozières (1) suggested spin polarizing liquid ^{3}He by a thermodynamic method and studying properties of the strongly polarized liquid (and solid) by experiments performed rapidly in comparison with the relaxation.

The feasibility of the idea was demonstrated experimentally shortly after (2,3), but since then, to our knowledge, only two experiments (4,5) to probe other properties than relaxation times have been reported. As an aid in the planning and interpretation of this type of thermodynamic experiment, we have performed some simple model calculations of temporal and spatial temperature variations. In this paper we present representative examples. Regarding the "violent" nature of the experiments we consider only some clean, extreme cases or use very crude approximations.

2. FINAL TEMPERATURE

2.1. Spin relaxation

In the upper part of Fig.1 we show the Fermi seas of free, independent nuclear up and down spins for a case with initial polarization $m_0 = 0.5$. The externally applied field is considered to be small in relation to the energy difference shown, and so the liquid magnetization will be negligibly small after relaxation. We assume that during relaxation the surplus energy is converted to heat. By numerical integration of the product of the Fermi distribution function with the density of states shown we have fitted the chemical potential to the requirement that the total energy stays constant. In this way we have obtained the temperature T_r at the end of a relaxation starting from the sharp (zero temperature) distribution shown. T_r is proportional to the Fermi temperature T_F, but does not depend on any other system parameter except the parabolic shape of the density of states. This means that any <u>changes</u> in the interaction energies, in particular in the spin dependent part, have been neglected. For the

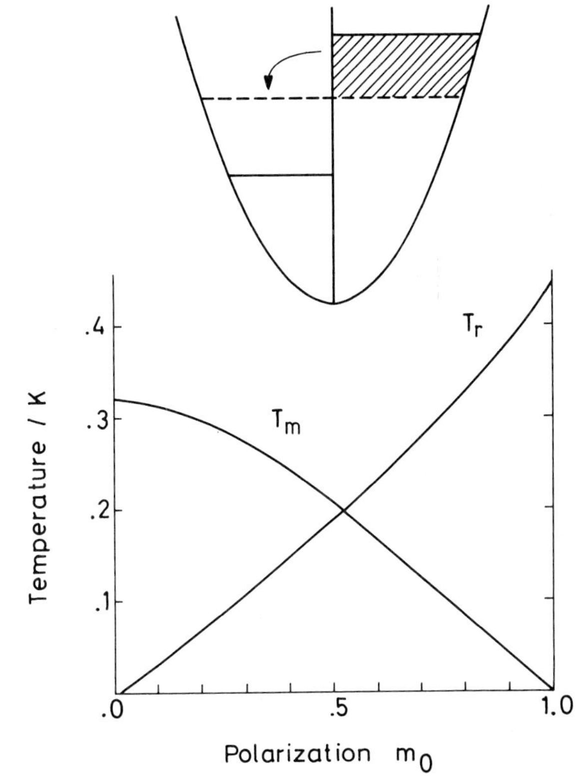

FIGURE 1
Final temperatures. T_r: after complete relaxation of polarization m_0. T_m: after isentropic melting of solid ^{3}He with polarization m_0.

*Work supported by the Danish Research Council for Natural Science.

plot of T_r in Fig.1 we have taken $T_F = 1$ K.

2.2 Latent heat of melting

If the polarized liquid is produced by melting of polarized solid, the latent heat will cause an initial temperature rise. For comparison with the effect of relaxation we show (Fig. 1) the temperature T_m expected right after an isentropic melting, considering solid ^{3}He to be an ideal paramagnet and using experimental data (6) for the liquid entropy.

In a real experiment the two effects will obviously combine and a temperature rise of 300 mK or more is to be expected. The time development may depend on m_0, however,

3. TEMPERATURE VARIATION WITH TIME

3.1. Homogeneous, isolated sample

Considering again the model presented in 2.1 we let the magnetization decay exponentially with a characteristic time t_1. We assume energy to be released at the same rate as if the Fermi sea levels stayed "flat" as they approach each other:

$$dE/dt = G\, T_F^{5/2}((1+m)^{2/3}-(1-m)^{2/3})m/t_1, \quad (1)$$

where the constant G contains the effective mass. Letting this energy heat the liquid we obtain the full curves in Fig.2, using the specific heat of normal liquid ^{3}He (6).

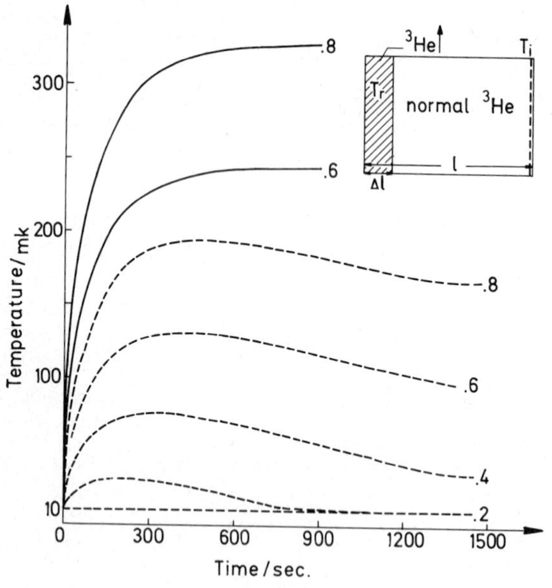

FIGURE 2

Temperature rise versus time. Full curves: isolated sample. Dotted curves: sample as inserted. t_1=600 sec., x=0.13. m_0 is given for each curve.

3.2 Inhomogeneous, unisolated sample

The insert in Fig.2 shows a much more complicated case: the initial polarization m_0 is restricted to a fraction x of the sample volume, located at one end of the cell. The other end is kept at a fixed temperature T_i by means of an infinitely good heat exchanger connected to a thermal reservoir. The power in Eq.(1) will now heat the polarized and the normal liquid, and part of it will be conducted away. The changed behaviour is shown by dotted lines in Fig.2. The thermal conductivity used was K = (120/T) microwatt/m, and for mathematical simplicity we considered only a linear temperature variation through the cell.

4. DISCUSSION

We have tried to extend the above considerations in order to understand better some of the observations in (4) and (5). We find clear discrepancies: the final temperatures were lower than predicted, and the temperature rise from relaxation was not observed, so far. Probably, thermal contact with the mixing chamber has to be interrupted before these effects become observable.

REFERENCES

(1) B. Castaing and P. Nozières, J. Phys.(Paris) 40 (1979) 257.
(2) G. Schumacher, D. Toulouze, B. Castaing, Y. Chabre, P. Segransan and J. Joffrin, J. Phys. Lett. (Paris) 40 (1979) L143.
(3) M. Chapellier, G. Frossati and F.B. Rasmussen, Phys. Rev. Lett. 42 (1979) 904.
(4) M. Chapellier, M. Olsen and F.B. Rasmussen, Physica 107B (1981) 31.
(5) G. Bonfait, L. Puech, A.S. Greenberg, G. Eska, B. Castaing and D. Thoulouze, submitted to Phys. Rev. Lett.
(6) D.S. Greywall and P.A. Busch, Phys. Rev. Lett. 49 (1982) 146.

LT-17 (Contributed Papers)
U. Eckern, A. Schmid, W. Weber, H. Wühl (eds)

MEASUREMENTS OF SPIN WAVES IN NORMAL LIQUID ^{3}He

N. MASUHARA,* D. CANDELA,* R. COMBESCOT,† D.O. EDWARDS,* R.F. HOYT,*
H.N. SCHOLZ,* and D.S. SHERRILL*

*Department of Physics, The Ohio State University, Columbus, OH 43210, USA
†Groupe de Physique des Solides de l'Ecole Normale Supérieure, 75231 Paris Cédex 05, France

Earlier observations of standing spin-wave resonances in cw NMR experiments on normal liquid ^{3}He are extended to a rectangular coil geometry, resulting in a particularly clear mode structure. The experimental data are compared with numerical solutions of the Leggett theory.

A normal Fermi liquid in a magnetic field can support propagating spin waves, provided the quasiparticle relaxation time τ_D (which varies with temperature as T^{-2}) is much longer than $\omega_o{}^{-1}$, where ω_o is the Larmor frequency.[1-4] The closely related Leggett-Rice effect[3,4] (a variation with temperature and spin-tipping angle of the effective spin diffusion coefficient) was verified by Corruccini et al.[5] for both pure ^{3}He and a 6.4% ^{3}He-^{4}He solution. More recently, resonances due to standing spin-waves have been observed in normal ^{3}He liquid[6] and in ^{3}He-^{4}He mixtures.[7] For ^{3}He these resonances appeared as a series of sharp peaks in the NMR absorption line, for temperatures below about 5mK and pressures below 10bar. In this paper we present an extension of these earlier experiments to a rectangular geometry (which simplifies the mode structure), along with a theoretical analysis.

For cw NMR experiments in a small rf field $\vec{H}_1(\vec{r})$, Leggett's theory[4] leads to the following equation for the resonant part of the transverse magnetization $\psi = M^+ - \chi_o H_1^+$, where $M^+ = M_x + iM_y$ (χ_o is the static susceptibility, and a time dependence $\exp(i\omega t)$ has been assumed):

$$iD\nabla^2\psi(\vec{r}) + \Omega_o(\vec{r})\psi(\vec{r}) = \omega[\psi(\vec{r}) + \chi_o H_1^+(\vec{r})]$$

Here D is a complex diffusion coefficient given by

$$D = \frac{(v_F{}^2/3)(1+F_o^a)\tau_D}{1 + i\omega_o\lambda\tau_D} = \frac{D_o}{1 + i\omega_o\lambda\tau_D}$$

where v_F is the Fermi velocity, $\lambda = (1+F_o^a)^{-1} - (1+F_1^a/3)^{-1}$, and F_o^a and F_1^a are Fermi-liquid parameters. D_o is the zero-field spin diffusion coefficient. The static field distribution determines the local Larmor frequency $\Omega_o(\vec{r})$ and its average ω_o. We have solved this equation by expanding ψ in a Fourier series and then solving for the eigenmodes and frequencies of the resulting matrix equation, trying both zero-spin-current boundary conditions (nor-

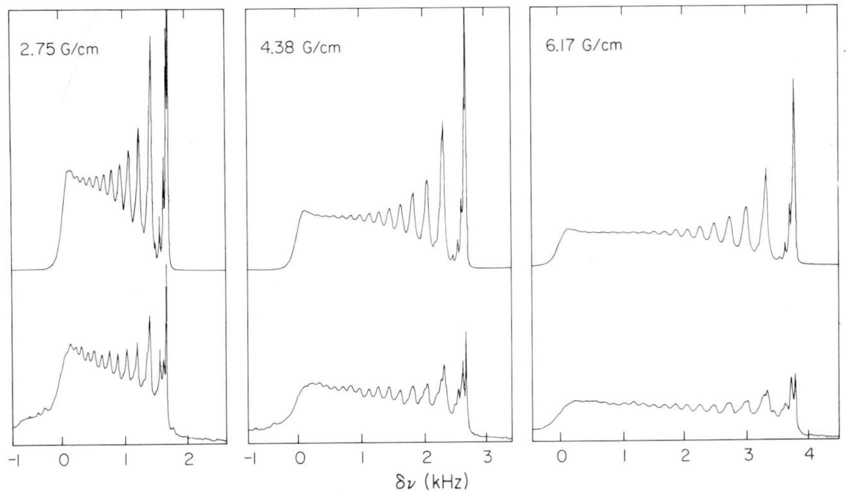

Fig. 1. Theoretical (top) and experimental (bottom) NMR absorption line shapes at T = 1.01mK, for three different values of the field gradient applied in the z direction. (The superfluid transition temperature is 1.00mK on this temperature scale.)

mal component of $\nabla\psi = 0$) and zero-spin boundary conditions ($\psi = 0$).

Figure 1 compares theoretical and experimental line shapes at a Larmor frequency of 2MHz, at zero pressure. The NMR coil was wound on an open-ended quartz former 2.8mm long and with a 2×2mm square hole, the axis of which was perpendicular to the static field (z-axis). A 6.1mm i.d. epoxy tower contained the coil and ^{3}He. Figure 2 compares theoretical and experimental spin wave peak positions, for a number of different values of the field gradient applied in the z direction. Only the main series of peaks (corresponding to quantum numbers for the z direction) has been plotted in this graph.

For the theoretical line shapes we assumed the sample to extend over 2×2×6.1mm^3 and the local Larmor frequency to be $\Omega_o = A + Bz + C(z^2 - x^2/2 - y^2/2)$. The parameters A, B and C for both main and gradient coils were highly constrained by the high temperature line shapes, leaving only $\tau_D T^2$ and λ freely adjustable. The best fit to the data over a range of applied gradients was obtained when zero-spin-current boundary conditions were assumed. This fit, shown in Figures 1 and 2, resulted in $\tau_D T^2 = (3.0\pm0.6)\times10^{-7}$sec mK^2 and $\lambda = 2.41\pm0.12$, using the values[8] $v_F = 5.99\times10^3$cm/sec, $F_o^a = -0.700$. When scaled to the same values for v_F and F_o^a, the results of Corruccini et al. for the Leggett-Rice effect[5] are $\tau_D T^2 = (3.26\pm0.33)\times10^{-7}$sec mK^2 and $\lambda = 2.20\pm0.11$. Our results must be considered preliminary, since an investigation of possible systematic errors has not been completed. Aside from information on the physical parameters of liquid ^{3}He, the overall agreement between predicted and observed line shapes provides a convincing confirmation of the theory of spin currents in a normal Fermi liquid when $\omega_o\tau_D >> 1$.

ACKNOWLEDGEMENT
This work was supported by the U.S. National Science Foundation under grant No. DMR 7901073.

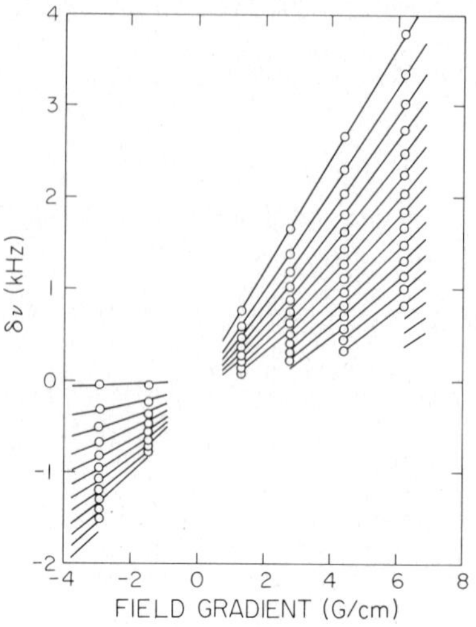

Fig. 2. Locations of main spin wave peaks as a function of applied field gradient, at T=1.01mK. Lines are theory, circles are experiment.

REFERENCES
(1) V. P. Silin, Zh. Eksperim. i Teor. Fiz. 33, 1227 (1957) (Soviet Phys. JETP 6, 945 (1958)).
(2) P. M. Platzman and P. A. Wolff, Phys. Rev. Lett. 18, 280 (1967).
(3) A. J. Leggett and M. J. Rice, Phys. Rev. Lett. 20, 586; 21, 506 (1968).
(4) A. J. Leggett, J. Phys. C 3, 448 (1970).
(5) L. R. Corruccini, D. D. Osheroff, D. M. Lee, and R. C. Richardson, J. Low Temperature Phys. 8, 229 (1972).
(6) H. N. Scholz, Ph.D. thesis, Ohio State University, 1981 (unpublished).
(7) J. R. Owers-Bradley, H. Chocholacs, R. M. Mueller, Ch. Buchal, M. Kubota, and F. Pobell, Phys. Rev. Lett. 51, 2120 (1983).
(8) D. S. Greywall, Phys. Rev. B 27, 2747 (1983).

LT-17 (Contributed Papers)
U. Eckern, A. Schmid, W. Weber, H. Wühl (eds)

SPIN-LATTICE RELAXATION TIMES OF THERMODYNAMICALLY POLARIZED LIQUID ^{3}He

R.C.M.DOW and G.R.PICKETT

Department of Physics, University of Lancaster, Lancaster LA1 4YB, United Kingdom.

We report a series of spin-lattice relaxation time measurements for liquid ^{3}He at ∿200 mK. The results are derived from the exponential decay of the magnetization after rapid decompression of solid polarized by a 6 T field at 20 mK. The measured τ_1 is found to be dependent on pressure and very weakly temperature dependent.

1. INTRODUCTION

We present a series of measurements of the spin-lattice relaxation time in polarized liquid ^{3}He. The liquid is polarized thermodynamically by depressurization from the polarized solid: Noziere's method (1). The solid is polarized by brute force cooling in a dilution refrigerator in a field of 6 T, c.f. Schumacher et al. (2).

2. THE EXPERIMENTAL SYSTEM

The experimental arrangement is shown in Fig.1. An epoxy cell of volume ∿20 cm^3 is thermally linked to the mixing chamber of a 4.5 mK dilution refrigerator via a sintered silver pad and a high purity silver link. The experimental volume contains a Speer carbon resistor for monitoring the temperature and a vibrating wire viscometer for observing the solid to liquid transition. Outside the cell there is a two-turn NMR transmitter coil which is inductively coupled via a miniature coaxial 50Ω transmission line to a sweepable signal generator. The NMR signal from the ^{3}He is received by an open circuit partial-turn antenna coupled via a second miniature 50Ω line. The signal is amplified at room temperature by a high gain Trontek r.f. amplifier. The NMR is operated in continuous mode with the frequency being repeatedly swept through the ^{3}He absorption line. In the 6T field used the central frequency is ∿200 MHz. The decompression-filling line to the cell is connected to a switchable variable volume so that the pressure after decompression may be continuously adjusted from 0 to 27 bar.

3. THE EXPERIMENT

We cool solid ^{3}He under 38 bars' pressure to a temperature of around 20 mK, achieving a solid polarization of around 25% in a 6T field. Rapid depressurization via the fill line leaves the liquid with an initial residual magnetization of between 20% and 70% of the immediately preceding solid value. This instantaneous loss of magnetization on decompression varied enormously in a completely unpredictable manner and could not be correlated with any experimental parameter (and was independent of the rate of depressurization). The relaxation

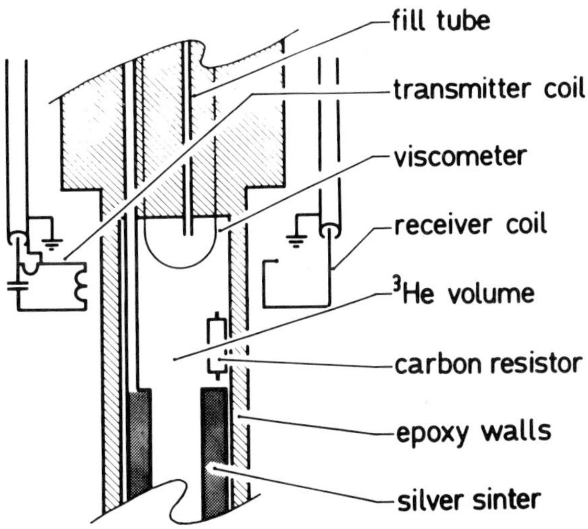

FIGURE 1
The experimental cell showing the NMR coils, viscometer, fill-line and carbon thermometer.

time, τ_1, for the polarized state is inferred from the exponential decay of the liquid polarization as monitored by the cw NMR resonance amplitude. The relaxation time is typically of the order of 300 seconds. After decompression the temperature is followed by the carbon resistor, and the pressure can be monitored via the external pressure gauge and indirectly by the vibrating wire viscometer.

4. THE RESULTS

After decompression the final pressure stabilizes in about 10 seconds. The viscometer resonance appears within 2 seconds showing that the full solid-liquid transition has been completed in this time. The NMR then monitors the exponential decay characteristic of relaxing polarized liquid ^3He (2),(3), with τ_1 varying between 100 and 400 seconds. During this time the carbon resistor shows a very rapid rise of the temperature of the ^3He to the minimum in the melting curve. The temperature then falls back to about 200 mK over 150 s as the cell and mixing chamber equalize temperatures. The cell subsequently cools to 150 mK over the next ten minutes as the refrigerator begins to cool again. The measurements made at low final pressure display a relaxation time which falls as the temperature decreases, in contrast to the results at high final pressure which appear to be temperature independent. This behaviour is characteristic of bulk relaxation combined with a contribution to the relaxation from the boundaries, as seen by Beal and Hatton (4) for measurements on the unpolarized liquid. (The surface area of our cell is very large since it contains two sintered silver pads.) This effect (which is seen as a non-exponential decay) is,

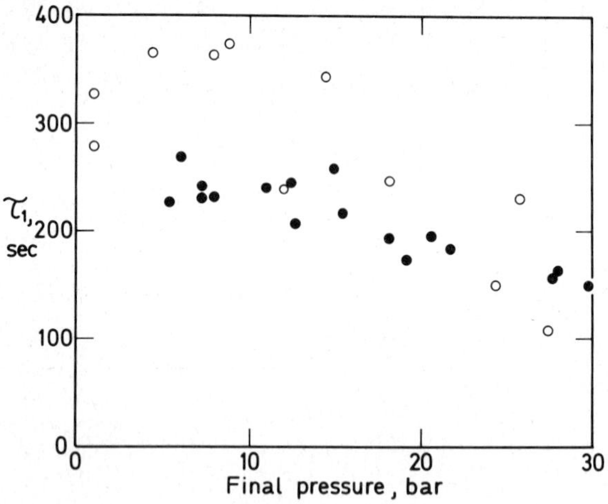

FIGURE 2
The measured spin-lattice relaxation values as a function of final pressure. Filled circles represent measurements where the instantaneous magnetization loss on decompression was less than 60% (see text).

however, not partcularly marked being to some extent lost in the noise of the tail of decay of the magnetization. We can thus characterize τ_1 for times relatively short after the depressurization where temperature effects do not seem to be significant and display the result simply as a function of pressure. The results, plotted in this way are shown in Fig.2, where filled points represent measurements for which the instantaneous magnetization loss on decompression is less than 60% of the immediately preceding solid value. The points for losses greater than 60% lie systematically higher for the lower final pressures. The results show qualitative agreement with the results taken in a similar way by Chappellier et al. (3) and also with many other studies of relaxation in the unpolarized liquid.

5. CONCLUSION

Our τ_1 measurements are consistent with the temperature and pressure dependencies expected for a mixed relaxation environment. The bulk relaxation rate is proportional to temperature and is also enhanced at high pressure where the quasiparticle separation is reduced. Relaxation at the walls is presumed to be temperature independent but the probability of a particle reaching the wall is not, giving an effective boundary relaxation rate inversely proportional to temperature. At the higher pressures these two contributions become comparable and thus the temperature dependence tends to be smoothed out. The boundary relaxation is not of primary importance since our measured τ_1 values are not significantly different from those measured in a similar way where special care has been taken to minimise boundary effects. We note that τ_1 values we measure at low final pressure and associated with a large ($\sim$75%) initial loss of polarization are consistently larger and compare well with the values expected for pure bulk relaxation. Finally we note that we were unable to detect any polarization-dependence of the melting curve, which is perhaps not surprising given the small polarizations used.

REFERENCES

(1) B.Castaing and P.Nozières, J. de Phys. 40, (1979) 257.
(2) G.Schumacher, D.Thoulouze, B.Castaing, Y.Chabre, P.Segransan and J.Joffrin, J. de Phys. 40, (1979) L-143.
(3) M.Chapellier, G.Frossati and F.B. Rasmussen, Phys. Rev. Lett. 42, (1979) 904.
(4) B.T.Beal and J.Hatton, Phys. Rev. 139, (1965) A1751.

LT-17 (Contributed Papers)
U. Eckern, A. Schmid, W. Weber, H. Wühl (eds)

SPIN RELAXATION OF GASEOUS POLARIZED ^3He ON A SURFACE COATED BY A THIN ^4He FILM

Marc HIMBERT, Jacques DUPONT-ROC

Laboratoire de Spectroscopie Hertzienne de l'Ecole Normale Supérieure
24, rue Lhomond, 75231 Paris Cedex 05 - France

The spin relaxation of a low density polarized ^3He gas occurs essentially by interaction of the atoms with the wall of the cell containing the sample. We measured the relaxation time T_1, in the temperature range between 0.5 K and 1 K, in a situation where a solid hydrogen coating, covered with a ^4He film, is condensed on the pyrex wall. We obtained reproducible results for the variations of T_1 with temperature, with the amount of ^4He and with the thickness of the hydrogen coating. They give interesting, although still incomplete, information on ^3He-^4He films.

1. INTRODUCTION

In order to observe experimentally the interesting properties of polarized ^3He at low temperature (1) one has to use containers with walls producing only a weak relaxation for the nuclear spins. This requires the use of non-magnetic materials, and a minimization of the binding energy of the ^3He atoms on the wall. ^4He films meet the last condition. A preliminary coating ("undercoating") made of solid H_2 can be used to increase the distance between the ^3He atoms and the pyrex wall.

Even in those optimal conditions, the spin relaxation of a low density ^3He gas (10^{17} cm^{-3}) is still determined by the wall. Hence the relaxation time T_1 is directly related to the dynamics of ^3He atoms on or inside the ^4He film. Although the physics of ^3He-^4He films has been actively studied, during the last years (2) (3) (4), their structure is not fully understood. The method described here may provide additional informations.

We report results of T_1 measurements in the temperature range 0.5 K to 1 K. Similar experiments at higher ^3He density and higher temperature are in progress at Sussex University (5).

2. EXPERIMENTAL TECHNIQUE

A known mixture of ^3He, ^4He and H_2 gas is introduced at room temperature in a spherical pyrex bulb (3 cm in diameter), which has previously undergone carefull cleaning, baking and pumping. The cell is sealed, then placed inside a ^3He refrigerator and cooled. As the temperature decreases, the H_2 and ^4He coatings are successively condensed on the wall. The spins are polarized by optical pumping at 0.7 K, in a moderate magnetic field (14 Gauss). The cell contains typically $N = 7.10^{17}$ ^3He atoms. The decay of this initial polarization (3%) is monitored, at a fixed temperature, by N.M.R. using an adiabatic fast passage technique. In this way we measure the spin relaxation time T_1 of the whole sample. Measured values range from 20 to 1600s.

The relaxation time of the atoms which are adsorbed on the wall is certainly much shorter, but in a situation of rapid exchange between the adsorbed phase and the gas phase, each atom stays only a short time on the wall. The relaxation time T_1 is related to the relaxation time T_a of the N_a adsorbed atoms by

$$\frac{1}{T_1} = \frac{1}{T_a} \times \frac{N_a}{N}$$

Thus, variations of T_1 may be due either to a change of N_a or to a variation of T_a.

We have undertaken a systematic investigation of T_1 as a function of T, varying the amount of ^3He and H_2 in the cell. Since each sample is contained in a similar but different cell, reproducibility of the T_1 values is not necessarily straightforward. In our experiment, we believe that the measurements of T_1, which are made with an accuracy of typically $\pm$ 10%, are reproducible within limits of the same order of magnitude.

3. INFLUENCE OF TEMPERATURE AND ^4He

Each curve on figure 1 shows the variation of T_1 versus 1/T, for a fixed hydrogen undercoating and a given amount of ^3He and ^4He.

When the temperature is lowered from 1 K to 0.5 K, T_1 first increases rapidly, then reaches a kind of plateau, only weakly decreasing. Both regions are well described by straight lines. The condensation temperature of ^4He in bulk liquid would correspond to $T^{-1} = 1.35$ K^{-1}. In fact, a monolayer is likely to be already condensed at higher temperature, of the order of 1 K. The first part of the T_1 variation can certainly be interpreted as a gradual exclusion of the ^3He atoms from the vicinity of the wall, but we have no quantitative model to explain the straight line and its slope.

The plateau is unexpected: since T_1 scales

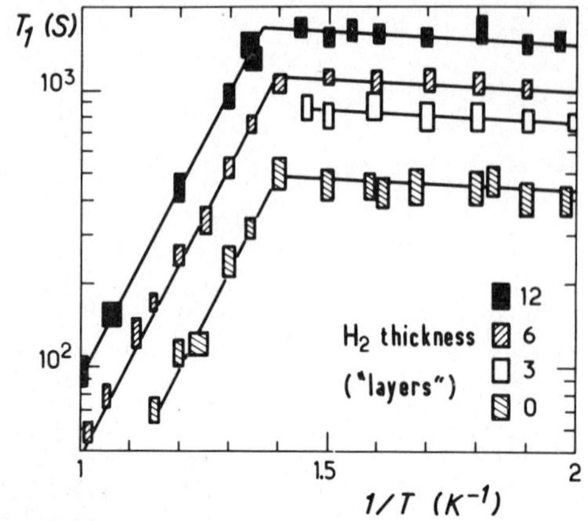

Figure 1:
Longitudinal relaxation time T_1 (logarithmic scale) versus the inverse temperature 1/T in the range 1 K to 0.5 K for different sealed cells filled with the same amount of ^3He and ^4He (7.10^{16} at cm^{-3}), and increasing amounts of H_2.

as $1/N_a$, it would seem reasonable that the adsorption energy of the ^3He atoms on the ^4He film ($\sim$ 5 K) or inside the film ($\sim$ 2.8 K) should govern the variation of T_1 in this region. The slope of the plateau is much smaller (less than 0.5 K). Furthermore, this plateau seems to be independent of the amount of ^4He in the cell. We have varied the ^4He gas filling pressure from 0.3 torr to 4 torr. The same plateau is reached, although at various temperatures. Assuming a uniform film and an area of the inner surface of the cell equal to the geometrical one, the thickness of the ^4He film has been varied from 6 to 70 "layers" (1 layer = 3.6 Å). However, these values may be wrong by a large factor if capillary condensation occurs somewhere in the cell.

4. CHANGING THE THICKNESS OF THE "UNDERCOATING"

The four curves on figure 1 correspond to increasing quantity of H_2 in the cell, and presumably to increasing thickness of the solid hydrogen at low temperature, ranging from 0 to 12 "geometrical layers" (1 layer = 3.4 Å). At a given temperature (0.53 K) the variation of T_1 as a function of the H_2 coating thickness is shown on figure 2.

From these data, we conclude that:

(i) The magnetic field responsible for the spin relaxation has its origin in or on the pyrex wall, since increasing the distance to the pyrex reduce the relaxation.

(ii) Extrapolation of the straight line of fig. 2 to $T_1=0$ may indicate that ^3He atoms are excluded from the part of the ^4He film closest to the wall, on a thickness of the order of 4 or 5 layers. This is consistent with conclusions obtained from third sound measurements (2).

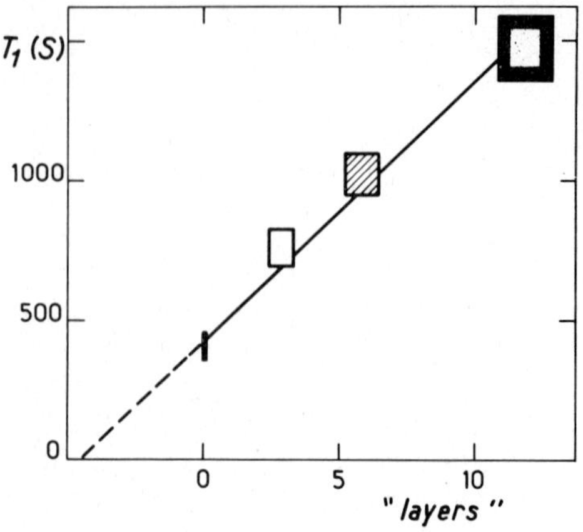

Figure 2:
Relaxation time at 0.53 K versus the thickness of the H_2 "undercoating" for the cells corresponding to Fig.1.

The linear variation of T_1 with H_2 thickness may seem surprising. Actually, simple relaxation models may provide such a simple law. For instance, the relaxation time of an ensemble of spins (gyromagnetic ratio γ_n) undergoing a diffusive motion (diffusion coefficient D) in a plane at a distance d from a half space of randomly distributed magnetic moments $\vec{\mu}$ (density ρ_μ) is given by (6):

$$\frac{1}{T_a} \propto \vec{\mu}^2 \gamma_n^2 \rho_\mu \int d^2q \; q \; e^{-2dq} \frac{1}{Dq^2} \propto \frac{1}{d}$$

However, the relevance of such a model for describing the experimental situation is not completely established.

5. CONCLUSION

Our first experimental results on the spin relaxation of ^3He on a wall coated with a ^4He film indicate that interesting informations may be gained on the relaxation mechanism, and on the physics of ^3He - ^4He films.

REFERENCES

(1) C. Lhuillier, F. Laloë; J. Physique 43, (1982) 197 and 225.
(2) F.M.Ellis, R.B. Hallock, Phys. Rev. B 29 (1984) 497.
(3) J.P. Laheurte et al. J. Physique Lett. 42 (1981) L 197.
(4) B. Bhattacharrya, F.M. Gasparini , Phys. Rev. Lett. 49 (1982) 919.
(5) M.G. Richards, C.P. Lusher, M.F. Secca private communication.
(6) V. Lefèvre, Thesis, Paris (1984)
(7) R.C. Albers, J.W. Wilkins, J. Low Temp. Phys. 34 (1979) 105

LT-17 (Contributed Papers)
U. Eckern, A. Schmid, W. Weber, H. Wühl (eds)

INTERNAL ENERGIES OF THE ^{3}HE PHASES*

C.N. ARCHIE and K.S. BEDELL

Department of Physics, State University of New York at Stony Brook, Stony Brook, N.Y. 11794, U.S.A.

Accurate predictions for the phase diagram of polarized ^{3}He depend not only upon promising liquid calculations but on unknown large molar volume solid behavior. Conversely, transient measurements other than melting pressure suppression may be better able to obtain internal energy information on the highly polarized liquid.

A revealing way for comparing internal energies of different phases is by plotting them versus molar volume as is done with experimentally derived values and others for the liquid and solid phases of ^{3}He in the figure. The Maxwell double tangent construction technique can then be used to determine the melting equilibrium conditions. This follows from the equality of the phases' Gibbs free energies (not their internal energies) at melting. The actual internal energy difference between the two phases can readily be inferred from this graph even if the zero of energy is not known for either phase since the double tangent line must have a slope equal to the zero temperature melting pressure of P_m=34.394 bar.(1) This, of course, is simply a graphical way of expressing the thermodynamic equality

$$U_s - U_\ell = P_m(V_\ell - V_s) \qquad (1)$$

where U_s and U_ℓ are the molar internal energies at zero temperature of the solid and liquid respectively and V_s and V_ℓ are the corresponding molar volumes. This molar volume difference, 1.314 cm^3, and the melting pressure P_m have been fairly well known for over a decade.(2) It is therefore surprising that numerous authors compare their theoretical work with values of U_s-U_ℓ substantially different from the result of Eqn. 1, 0.544 K in temperature units.(3) Partly at fault is an often quoted but inaccurate early experimental estimate for the solid internal energy at melting. (4)

To determine the absolute internal energies for the liquid and solid phases at melting at

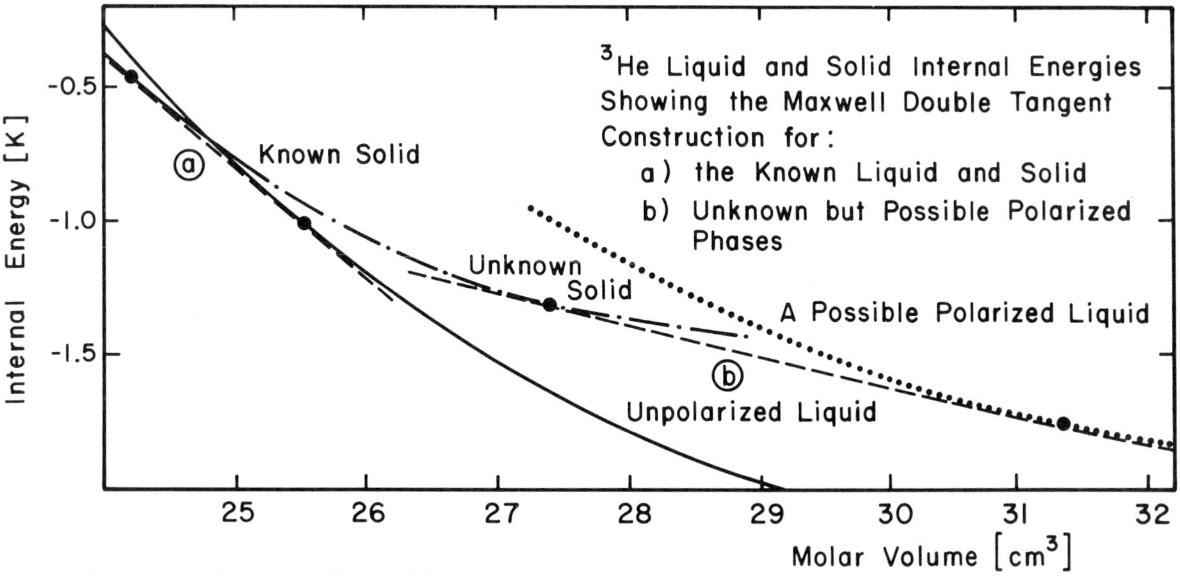

FIGURE 1. Uses and Abuses of Maxwell Double Tangent Construction.

*The material is based upon work supported by the National Science Foundation under Grant No. DMR-8218993 and by USDOE under Contract No. DE-AC02-76ER13001.

absolute zero, we begin in the gas phase at 1 K. Let $U_g(P,T)$ denote the gas internal energy at pressure P and temperature T. Clearly $U_g(0,1\ K) = 3/2\ RT$ where R is the gas constant. The gas internal energy at saturated vapor pressure P_{sv} is 1.4584 K, determined by using the gas virial coefficients.(5) The internal energy of the liquid $U_\ell(P_{sv},1\ K) = -2.1102$ K, determined via the gas-liquid version of Eqn. 1 and SVP data. Using recent low pressure liquid specific heat measurements (6), the liquid internal energy essentially at absolute zero and zero pressure $U_\ell(0,0)$ is found to be -2.4860 K. Finally with zero temperature inferred liquid molar volumes (7), the liquid internal energy at melting, $U_\ell(P_m,0)$ becomes -1.0115 K.

To obtain the internal energies' molar volume dependencies, we have parameterized the liquid and solid molar volume data. Because of the expected increase in the liquid internal energy upon polarization, the molar volume dependence for volumes larger than the experimentally determined unpolarized value of the solid internal energy is particularly important. For this purpose, we have used confirmed solid molar volumes along the melting curve to several hundred millikelvin in the solid data base.(1,2) The temperature dependence of this can be shown to be negligible. The resulting parameterizations along with their strict ranges of validity follow:

$$U_p(V)/R = \sum_{i=0}^{3} A_{pi} V^i \qquad (2)$$

$A_{\ell 0}=66.9776$ K	$A_{s0}=107.1876$ K
$A_{\ell 1}=-6.02045$ K/cm^3	$A_{s1}=-11.35610$ K/cm^3
$A_{\ell 2}=0.174892$ K/cm^6	$A_{s2}=0.4042160$ K/cm^6
$A_{\ell 3}=-1.69942\times10^{-3}$ K/cm^9	$A_{s3}=-4.90779\times10^{-3}$ K/cm^9

$$25.3\ \mathrm{cm}^3 < V_\ell < 28\ \mathrm{cm}^3$$
$$V_s < 25\ \mathrm{cm}^3$$

The actual polarization dependence of the liquid internal energy should be very revealing with regard to the microscopic nature of this strongly interacting Fermi liquid. The extreme possibilities have often been characterized as nearly ferromagnetic or nearly solid behavior. (8,9) A recent calculation for the polarized liquid (10) is shown in the figure by a dotted line. Since the exchange interaction in the solid is only on the order of a few millikelvin we can safely ignore the difference between polarized and unpolarized solid. We note that the relevant energy scale for discussing the phase transition is the liquid-solid internal energy difference at melting, 0.544 K, rather than any absolute internal energy. In light of this, the calculation shows a major change in the liquid internal energy which could result in a large change in the melting pressure, possibly suppressing the liquid altogether. To resolve this issue, the solid internal energy needs to be known well beyond the regime of validity of the experimentally derived result, Eqn. 2. Naively extending Eqn. 2 for the solid produces complete liquid suppression. On the other hand, the equally plausible extension shown in the figure predicts a melting pressure around 10 bars. Clearly without better knowledge of the solid for molar volumes much larger than 25 cm^3, quantitative conclusions about melting based solely on liquid calculation must be viewed with skepticism.

Experimental determinations of the polarization dependence of the liquid internal energy (more accurately, the free energy F=U-TS) seem possible by various techniques. Recently, rapid partial melting of 55% polarized solid produced results that have been interpreted as major drops in the melting pressure.(11) Providing this is a correct interpretation, melting pressure changes of a few bars or less can be combined with the above solid internal energy data to provide the change in the liquid free energy for the given polarization. We wish to discuss another promising technique for determining free energy changes, especially in light of the inevitable polarization relaxation and concurrent self-heating which occurs after rapid melting of polarized solid. The method calls for the simultaneous monitoring of the magnetization decay, dM/dt, and self-heating, dT/dt, in a thermally isolated cell, in which case

$$\frac{\partial^2 F}{\partial M \partial T} = \frac{c_M}{T}\,\frac{dT/dt}{dM/dt} \qquad (3)$$

where c_M is the molar specific heat at constant magnetization. Consequently, if as part of a program of measuring the magnetization dependent specific heat of polarization-decaying liquid, the self-heating rate is also mapped out, then the polarization dependences of the free energy and the internal energy can be inferred.

REFERENCES

(1) W.P. Halperin, F.B. Rasmussen, C.N. Archie, and R.C. Richardson, J. Low Temp. Phys. 31 (1978) 617.
(2) E.R. Grilly, J.Low Temp.Phys. 4(1971) 615.
(3) H.R. Glyde and V.V. Goldman, J. Low Temp. Phys. 25(1976)601; and references therein.
(4) R.C. Pandorf and D.O. Edwards, Phys. Rev. 169 (1968) 169.
(5) T.R. Roberts, R.H. Sherman, & S.G. Sydoriak, J. Res. Nat'l. Bur. Std. 68A (1964) 567.
(6) D.S. Greywall and P.A. Busch, J. Low Temp. Phys. 46 (1982) 451.
(7) J.C. Wheatley, Rev.Mod.Phys. 47 (1975) 467.
(8) K. Levin and O.T. Valls, Phys. Reports 98 (1983) 1-56.
(9) Dieter Volhardt, Rev.Mod.Phys. 56(1984)99.
(10) E. Manousakis, S.Fantoni, V.R.Pandharipande, & Q.N. Usmani, Phys.Rev. B 28(1983) 3770.
(11) G. Bonfait, L. Puech, A.S. Greenberg, G. Eska, B. Castaing, and D. Thoulouze, in print.

LT-17 (Contributed Papers)
U. Eckern, A. Schmid, W. Weber, H. Wühl (eds)

PINNING ON GRAIN BOUNDARIES BY TRANSFORMATION OF THE FLUX LINE LATTICE

Silvester TAKÁCS

Electrotechnical Institute of the Electro-Physical Research Centre,
Slovak Academy of Sciences, 842 39 Bratislava, Czechoslovakia

The free energy difference between the hexagonal and the quadratic flux line lattice (FLL) near surfaces is calculated and its influence on pinning on grain boundaries is studied. Hence, one may also obtain the interaction range of flux lines (FLs) with the surface. It is approximately the penetration depth λ. We give an estimation of the maximum volume pinning force in superconductors with grain boundary pinning due to this mechanism (especially in Nb_3Sn). Other peculiarities of pinning caused by this transition mechanism are sketched.

1. INTRODUCTION

The Labusch criterion for pinning of FLs by defects (interaction forces K above the threshold K_o) can hardly be fulfilled with known interaction mechanisms. Other mechanisms were considered (e.g. with defect combinations) to suppress the failure of pinning theories in explaining the experiments (1–5).

In practically important A–15 superconductors, the pinning on grain boundaries is supposed to be dominant, but the actual pinning mechanism is still an open question. Pinning studies on grain boundaries are therefore of large practical interest. The most effective mechanism seems to be due to anisotropy of the electron mean free path and GL parameter κ (6). The transition of the FLL from hexagonal to quadratic structure (7,8) can also contribute to pinning. As many FLs participate in it, Labusch's criterion could be fulfilled (2).

2. CALCULATION OF THE FLUX LINE LATTICE NEAR SURFACES AND BOUNDARIES

The general calculations of the FLL are complicated, as the positions of the FLs are unknown (they should minimize the Gibbs' free energy). Due to the surface barrier (9) one would expect a position dependent FL spacing near the surface (10). However, this seems to be not the case for ideal surface (11).

We suppose some amorphous region of dimension a_o near the boundary, whose interaction energy with both lattices is nearly the same. We do not calculate the free energies by images of FLs like in (11). Instead of this, we use the results for the self-energy of FLs near surfaces (9) and for the interaction energy (10) of two FLs $U=U_{ij}+U_{ji}$ in thin films, which can be reformulated for the half-plane. In relative units (9) we have

$$U=\frac{8\pi}{\kappa^2}\int_{-\infty}^{\infty}\frac{dk}{p}\, e^{-pv_2}\,\mathrm{sh}(pv_1)\cos(ku)\ ,$$

where $v_1=v_2$ are the x-coordinates of the FLs from the surface, u is their distance in y-direction (field $H_o \| z$). The sum of the interaction energies U_q for the quadratic lattice can be summed up as coming from vortices in the same row perpendicular to the surface (g_1), in the same row parallel to the surface (g_2) and from other rows (not containing g_1) parallel to the surface (g_3):

$$U=\frac{8\pi}{\kappa^2}\int_{-\infty}^{\infty}\frac{dk}{p}\,(g_1+g_2+g_3)\ ,$$

$$g_1=e^{-A_s}\sum_{i=1}^{s-1}\mathrm{sh}(A_i)+\mathrm{sh}(A_s)\sum_{i=s+1}^{N}e^{-A_i}\ ,$$

$$g_2=\left(1-e^{-2A_s}\right)\sum_{m=1}^{N}\cos(kma)\ ,\quad A_i=p(a_o+ia)\ ,$$

$$g_3=2g_1\sum_{m=1}^{N}\cos(kma)\ ,\quad p=(k^2+1)^{1/2}\ ,$$

where i and s mean the numbers of the rows. The finite sums appearing above can all be summed up, as e.g.

$$\sum_{i=1}^{n}e^{-p(a_o+ia)}=\frac{1-e^{-nap}}{2\,\mathrm{sh}(ap/2)}\,e^{-p(a_o+\frac{a}{2})}$$

The summing up is more complicated for the hexagonal lattice, but it can be done favorably by dividing into two

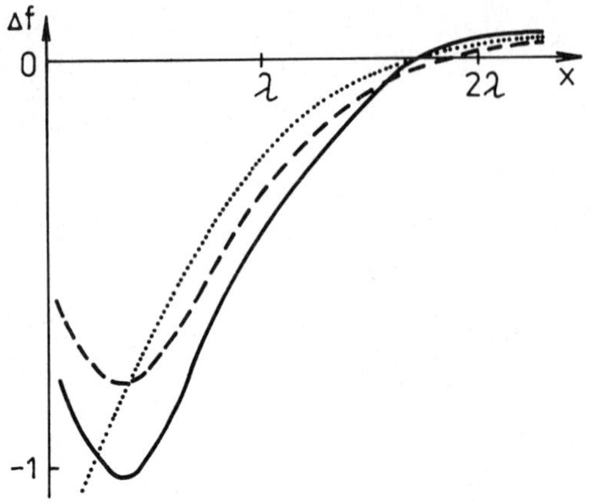

FIGURE 1
The difference of the free energy Δf of the quadratic and hexagonal FLL vs. distance x from the amorphous region of dimension $a_o = \lambda$ (dotted curve) and 0.2λ (other curves) for fields $b = B/B_{c2} = 0.8$ (dashed curve) and 0.6 (other curves).

quadratic sublattices plus the interaction between these two sublattices.

The difference between free energies of both lattices is given in Figure 1 in dependence on the distance from the amorphous region. The interaction of all flux lines with mutual distance $\lesssim 8\lambda$ was considered. The results are nearly independent on the GL parameter for $\kappa > 20$. For $a_o \gtrsim 0.5\lambda$, there is no minimum in Δf.

In spite of the repulsive surface-FL interaction, the surface is attracting the lattice. Although the energies of the lattices are sensitive to changes of the lattice spacing, Δf changes only little. Therefore, the lattice parameter need not be determined precisely.

3. THE PINNING FORCE

The results can be applied to A–15 superconductors with grain sizes much larger than λ. As the grain boundary density per area is $n_p = 1/d^2$, one can define an effective pinning density $n_{eff} = 1/d^2 t\xi$ with $t \approx 1$. The energy gain ΔE of the volume $d\lambda\xi$ near the boundary and the corresponding elementary interaction force K (as Δf changes approximately on the distance of λ) are

$$\Delta E = E_o\, d\lambda\xi \ , \quad K = E_o\, d\xi \ , \quad E_o \doteq \mu_o H_c^2/2 \ .$$

The grain boundary pinning is very strong (see below), we use therefore the direct summation formula. Then, we obtain for the volume pinning force

$$F_p = K\, n_{eff} = E_o/d \ .$$

Due to the changing direction of the Lorentz force with respect to the grain boundary, approximately half of the surface is active for pinning. For Nb_3Sn is then $F_p \approx 7\times10^{10}\,N/m^3$, which is nearly the value for the best commercial Nb_3Sn. It corresponds to $j_c \approx 1.5\times10^{10}\,A/m^2$ at 5T. For smaller grains we would obtain F_p nearly constant. The increase of j_c with further decreasing of d cannot be explained by this mechanism. However, defects near boundaries can act more effectively with this mechanism.

The transversal R_c and longitudinal L_c correlation length (4,12) is

$$R_c/a = 8\pi a c_{44}^{1/2} c_{66}^{1/2}/n_{eff}K^2, \quad L_c = R_c(c_{44}\, c_{66})^{1/2}.$$

For Nb_3Sn we have $R_c/a = \alpha/80 \ll 1$, $L_c/a = \kappa\alpha/80 \approx 1$ with $a = \alpha\xi$. The FLs are therefore pinned very strongly. This results in strong bending of FLs, not allowed by the elastic theory. One has to have in mind, however, that our transition means already plastic deformation.

Due to the strong bending of FLs, another force is acting on them. This can contribute to the change of j_c (13) of the form $j_c \sim 1/(B+B_o)$, which is known to appear especially in Nb_3Sn. Quantitatively, however, one can hardly obtain values of $B_o > 1T$ due to this mechanism.

REFERENCES

(1) E.J. Kramer, J.Nucl.Mat. 72(1978)5.
(2) R. Labusch, ibid., p. 28.
(3) E.H. Brandt, J. Low Temp. Phys. 26(1977)709.
(4) A.I. Larkin, Yu.N. Ovchinnikov, J. Low Temp. Phys. 34(1979)409.
(5) S. Takács, phys. stat. sol. (a) 74(1982)437.
(6) W.E. Yetter, D.A. Thomas, E.J. Kramer, Phil. Mag. B 46(1982)523.
(7) A.I. Larkin, Yu.N. Ovchinnikov, Zh. eksp. teor. fiz. 80(1981)2334.
(8) Yu.N. Ovchinnikov, Zh. eksp. teor. fiz. 82(1982)2020.
(9) S. Takács, Z. Phys. 199(1967)495.
(10) S. Takács, Czech. J. Phys. B 33 (1983)1248.
(11) E.H. Brandt, J. Low Temp. Phys. 42(1981)557.
(12) E.H. Brandt, J. Low Temp. Phys. 53(1983)412.
(13) H.C. Freyhardt, Habilitation Thesis, Universität Göttingen, 1976.

LT-17 (Contributed Papers)
U. Eckern, A. Schmid, W. Weber, H. Wühl (eds)

ON THE PINNING BY GRAIN BOUNDARIES IN SUPERCONDUCTING Nb-BASED FILMS

V.M.PAN, V.G.PROKHOROV, G.G.KAMINSKY, K.G.TRETIATCHENKO

Institute Of Metal Physics, Kiev 252142, USSR

Dependence of elementary pinning force f_{pl} versus impurity parameter α has been measured for Nb films doped with oxigen, vanadium and germanium. It was shown to have non-monotonic nature with a peak at α_p=4. Experimental data are in qualitive accordance with the ESFP model modified by the account of a finite grain size and transparativity of boundaries.

I. INTRODUCTION

It is known that grain bondaries are dominant pins in practical second type superconductors. The strongest interaction between flux line lattice and such defects must occur in polycrystal films tending to have column structure. So they can be used as the model to study flux pinning by grain boundaries. The object of our study was to ascertain the influence of impurities in polycrystal Nb films on the pinning volume force.

2. EXPERIMENT

Samples under study were fabricated by electron beam evaporation, the rate of deposition has been 10Å/sec. 850°C sapphire substrates were used. The pressure in a vacuum chamber was 10^{-4}Pa during evaporation. From the individual source another component (V or Ge) was deposited, its rate of evaporation been adjusted to obtain given concentration of impurity.

X-Ray and electronographical analysis have shown that crystal lattice parameter in films corresponded to the value usual for pure Nb: a=3.301-3.292Å, mean grain size was ~ 0.1μ.

Electric measurements were carried out in liquid helium by the conventional four-probe d.c. method (1). The direction of a current was always orthogonal to the applied magnetic field. A value of the current resulting in 2μV/cm voltage across a specimen were chosen as a critical. The upper critical magnetic field H_{c2} was defined by extrapolation of the linear section of experimental $J_c^{1/2}B^{1/4}$(H) curve.

Concentration of impurities in Nb films varied by the following means: air oxidation (2h at 200°C and natural aging during a year (2)), Ge and V doping during evaporation of the film.

All films had T_c=9.2K. Residual specific resistance ρ_0 was measured at T=T_c. The mean free path L was defined by the use of relation $L\rho_0=3.1\ 10^{-12}$ $\Omega\,cm^{-2}$ (3).

Dependence of the maximum elementary pinning force f_{pl} per unit length of a flux line on the impurity parameter α =0.882ξ_0/L, where ξ_0 is a coherence length, is shown at Fig.1.

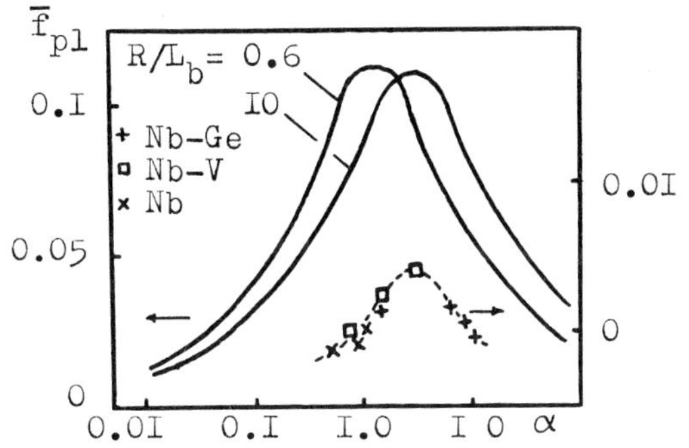

FIGURE 1

3. DISCUSSION

Experimental dependence $f_{pl}(\alpha)$ has non-monotonic nature with a peak at α_p=4 and can not be explained neither by the mechanism of interaction between deformation field of grain boundaries and FLL stress field, nor by the mechanism connected with anisotropy of the upper critical field H_{c2} (4), but qualitively conforms with the calculation of ESFP model.

However, Zerweck (5) predicted the peak position α_p=1.4. Yetter et al. (6, 7) using another empirical formula for ξ(L) and taking into account second order FLL energy terms have obtained another value α_p=I0. The single grain boundary was considered by these authors. Consideration of a finite transparativity of boundaries was not completed. It is attractive to modify ESFP model taking into account a finite grain size and a probability of scattering β.

Similar (6) we suppose diffuse isotropic scattering of electrons at the boundaries. The probability of the electron surviving without scattering up to a distance r,

$$P(r,\theta)=\begin{cases} \exp(-r/L_b) & , \quad r<s \\ \delta(r-s)L_b\beta\exp(-s/L_b), & \quad r=s \\ (I-\beta)\exp(-r/L_b) & , \quad r>s \end{cases} \qquad (I)$$

where L_b is mean free path in a boundary-free medium, s is distance to the boundary depending on the direction. If a grain is imaged as a sphere with a radius R (Fig.2),

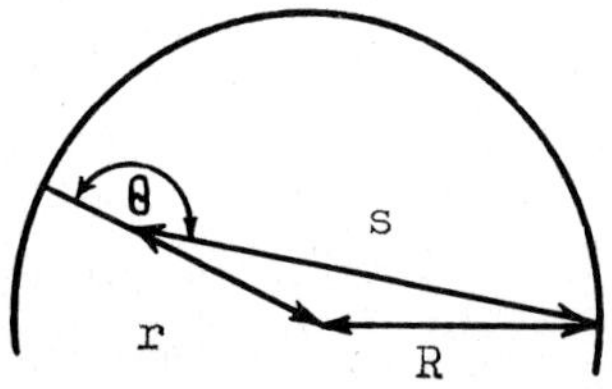

FIGURE 2

$$s=-r\cos\theta+\sqrt{R^2-r^2\sin^2\theta} \qquad (2)$$

Averaging for all directions local mean free path is,

$$L=L_b-\frac{Lb}{2}\exp(-R/L_b)\Big[\frac{Lb}{r}sh\frac{r}{L_b}+ch\frac{r}{L_b}+ +\frac{R}{r}sh\frac{r}{L_b}\Big]+\frac{R^2-r^2}{2rL_b}\Big[Ei\,(-\frac{R-r}{L_b})- -Ei(-\frac{R+r}{L_b})\Big] \qquad (3)$$

We have regarded a single flux line since results are qualitively independent on the account of the lattice (7). At this approach,

$$f_{pl}=(\xi_b/\xi_o)(d\xi/dr) \qquad (4)$$

To calculate $\xi(L)$ at α~I empirical function from (6) has been used, but with another coefficients.

Computed dependences $f_{pl}(\alpha)$ for different grain sizes are shown at Fig.I with solid lines. The peak position proved to be dependent on the value R/L_b. At $R/L_b \gg I$ α_p=3.7 corresponds to the experimental value. It should be noticed that we have measured the mean volume value $\langle L \rangle$ but not L_b involved in theory.

In spite of qualitive accordance of theoretical and experimental curves values f_{pl} differ for an order. Such difference seems to be caused by using of direct summation of pinning forces of each flux line to calculate f_{pl} from experimental data. Perhaps, grain boundaries in our films had probability of scattering noticeably smaller than unit. Dependence f_{pl} versus β is shown at Fig.3.

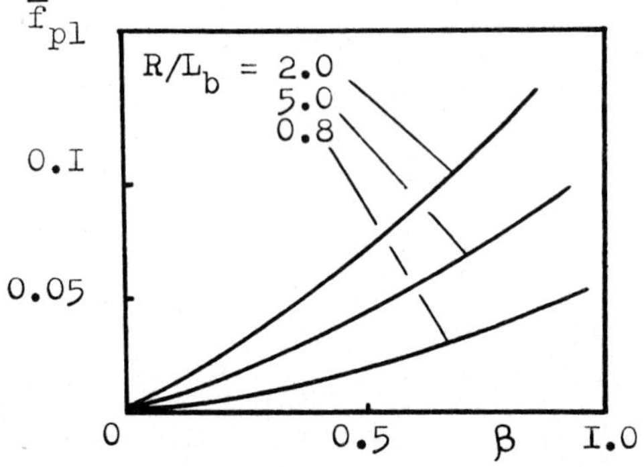

FIGURE 3

REFERENCES
(I) V.M.Pan et al. Fiz.Metody Issl.Met. (Naukova Dumka, Kiev, I98I) pp.36-4I
(2) V.M.Pan et al. Fiz.Nizk.Temp.(USSR) 6 (I980) 968
(3) N.E.Alekseevsky et al Fiz.Met. and Metallov. (USSR) 34 (I974) 63
(4) E.J.Kramer, Advances In Cryogenic Engineering Materials, ed. by R.P.Reed and A.F.Clark (Plenum Press, New York - London, I98I) v.28
(5) G.Zerweck, J.Low Temp.Phys. 42 (I98I) I
(6) W.E.Yetter et al. Phil.Mag. 46B (I982) 523
(7) W.E.Yetter and E.J.Kramer, J.Mat. Sci. I7 (I982) 2792

LT-17 (Contributed Papers)
U. Eckern, A. Schmid, W. Weber, H. Wühl (eds)

THRESHOLD CRITERION AND SINGLE-PARTICLE PINNING ON He VOIDS IN Nb-Ti ALLOYS

Petko VASSILEV

Faculty of Physics, University of Sofia, 1126 Sofia, Bld. A Ivanov 5, Bulgaria

A single-particle pinning is observed in $Nb_{0.83}Ti_{0.17}$ with He voids having a sufficiently high $d_i/\xi(t)$ ratio for to be above the threshold level for direct summation of the elementary forces of interaction f_{pi}.

1. INTRODUCTION

By construction master curves $Q=F_p/N$, Kramer (1) have pointed out that neither the direct $/Q=f_p/$ nor the statistical $/Q=C\cdot f_p^2/$ summation of the elementary interaction forces f_p can be used for adequate description of the global pinning force $F_p=J_c\cdot B$ in real type superconductors even if we ignore that the direct summation is applicable to pins far below the Labusch threshold f_t (2). The threshold may be lowered by introducing the spatial dispersion of the tilt modulus $\tilde{C}_{44}$ (3), supercenters or perturbations of the periodic well. Kramer (1) proposed to consider the effect of FL dislocations, resulting to an effective FLL complience with a spectrum of local threshold forces f_{ti}. Plastic deformation and peak-effect arise when the deformation u_i caused by one pinning center becomes of the order of FLL constant a_o. Then, in the framework of the Larkin-Ovchinnikov theory (4), the direct summation can be applied to pins with $f_{pi}>\tilde{f}_t=a_o^2(\tilde{C}_{44}\cdot C_{66})^{1/2}$. The density of such strong pins

$$N^+ = \sum_i n_i^+$$

vary with $h=H/H_{c2}$ and $t=T/T_c$ at least for voids, precipitates and loops, always having some natural size dispersion $n_i=n(d_i)$, while f_p as well as $f_p/\tilde{f}_t$ is strongly dependent on d.

In this work, a confirmation of the main features of the L.O. theory is presented for He voids with $0.8<d_i/\xi(t)<5$, introduced in a high κ-material as $Nb_{0.83}Ti_{0.17}$,

2. PINNING ON NOBLE GAS VOIDS

Because of the absence of a proximity effect the absolute values of f_{pi} of noble gas voids are higher than that of precipitates or loops with equivalent size (5); for small cubic voids

$$f_{pi}=\frac{2\pi\mu_o}{3a_o}H_c^2(1-h)\ d_i^3\ ,$$

where $H_c=H_{c2}/\kappa\sqrt{2}$ is the thermodynamic critical field. The deformation $u_i\approx 0.18(d_i/\xi(t))^3\cdot a_o\cdot h^{1/2}\cdot(1-h)^{-1/2}$ for $h\to 1$ differs from that for $h\to 0$, which is $u_i\approx 0.15(d_i/\xi(t))^3\cdot a_o\cdot(1-h)^{1/2}$, similarly to $f_{pi}/\tilde{f}_t$. The threshold then will be satisfied for voids with $d_i>d_c$, where $d_c(h,t)=g(h)\cdot\xi(t)$; the universal curve g(h) is shown in

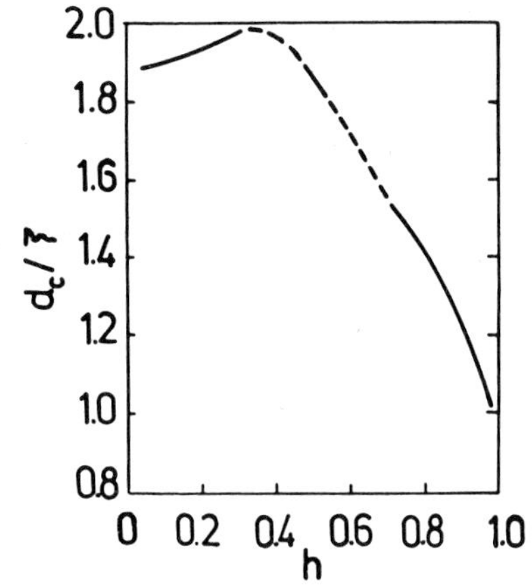

FIGURE 1
The values of $g(h)=d_c(h,t)/\xi(t)$ versus h.

Fig. 1/dashed line obtained by extrapolation of C_{66}/. Since the dimensionless parameter $d_i/\xi(t)$ plays a very important role, ths size dispersion $n(d_i)$ cannot be ignored. It is useful to take $d_i=i\delta$, where δ is the step of histogram. If all voids with $d_i>d_c(h,t)$ act with a maximum pinning force f_{pi} then

$$F_{cal}=\sum_{i>i_c(h,t)} n_i f_{pi} \sim \Delta V_c^+(h,t)\cdot H_{c2}^{5/2}h^{1/2}(1-h).$$

Here $\Delta V_c^+(h,t)$ is the volume of all voids above the threshold, taken in unit volume of the material; it can be easily determined from the experimental curve

$$\Delta V_c(d_k)=\sum_{i>k} n_i\ d_i^3 \ \text{ if } k=d_k/\delta.$$

In order to have smooth curves $\Delta V_c^+(h,t)$ it is recommanded to have δ as small as possible. We assume that voids below the threshold take part in a weak collective pinning so that their role can be ignored.

3. COMPARISON WITH THE EXPERIMENT

in low κ materials d/ξ is small and threshold is very difficult to be attained. In earlier works on $Nb_{0.97}Ti_{0.03}$ (6) and $Nb_{0.08}Ti_{0.02}$ (7)

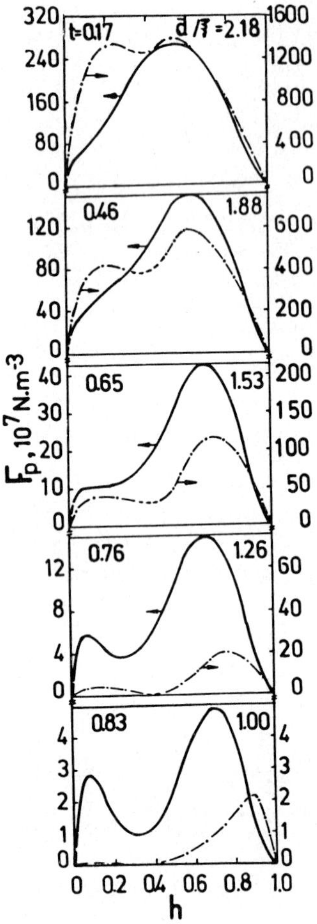

FIGURE 2
The global pinning force plotted versus h for a sample of $Nb_{0.83}Ti_{0.17}$: experimental (solid lines) and calculated (dashed) curves for different values of $\bar{d}/\xi(t)$.

we had difficulties with the formation of precipitates during annealing after the He^+-implantation, resulting to large decrease of H_{c2} and κ. In the last serial experiments (8,9) on $Nb_{0.83}Ti_{0.17}/\kappa\sim 15, \xi\ (0)\sim 75$ Å, C_{He}=0.05-1.00 at.%, T_{anneal}=1100-1260°C/these difficulties did not exist; the changes of H_{c2} and κ were below 5-6%. The transition temperature T_c remained practically unchanged, precipitates not observed, while the critical current density and F_p drastically increased due to a dominant pinning on He voids having a cubic shape and quite regularly distributed. The voids density

$$N = \sum_i n_i$$

varied from $1\cdot 10^{15}$ to $15\cdot 10^{15}$ cm^{-3} at mean size $90\leq\bar{d}<260$ Å and size dispersion from 50 to 400 Å.

Because of the small thickness /10 mkm/ and high residual resistance it was possible to obtain correct data from t=0.17 to t=0.97 for $0<h\leq 1$. Typical results are presented in Fig. 2 for a sample with C_{He}=0.15 at.%, annealed at 1100°C, for which $N=2\cdot 10^{15}$ cm^{-3}, $\bar{d}$=170 Å, $\xi(0)\sim 76$ Å and $\Delta V_c(d=0)=9.8\cdot 10^{-3}$ cm^3 /taken for 1 cm^3 of the material/ at a size dispersion $50\leq d_i\leq 350$ Å. As seen, at smaller $\bar{d}/\xi(t)$ a valley with the minimum at about $h\sim 0.3$ appears; it corresponds with the maximum of the effective modulus. The expected second peak-effect (10,4) is at $h\sim 0.05\lessapprox H_c/H_{c2}$. Be increasing $\bar{d}/\xi(t)$ the valley disappears and the well known peak-effect /near H_{c2}/ shifts to lower h. As seen from the quantitative and qualitative comparison of F_p and F_{cal}, the correlation of their shape with the parameter $d_i/\xi(t)$ shows that the density of strong pins above the threshold is not constant. This was observed for a large diversity of voids systems. It is noticable that for $t\to 1$ and smaller h only very large voids with a little statistical weight are able to direct summation. However, in the case of low t and h, the accuracy is sufficiently high; it seems that the disagreement is connected with the impossibility of each strong pin to act with a maximum f_p. The mean pin efficiency, formally introduced as $\langle w\rangle=F_p/F_{cal}$ decreases by decreasing h below 0.4, as predicted by L.O. theory.

4. CONCLUSIONS

Single-particle pinning on strong He voids in high κ NbTi alloys is observed. A direct summation of f_{pi} together with a consideration of the real size dispersion and mean pin efficiency allows to obtain agreement between experimental and calculated values of F_p.

REFERENCES

(1) E.J. Kramer,J.Appl.Phys. 49 (1978) 742.
(2) R. Labusch,Crystal Lattice Defects 1 (1969)1.
(3) E.H. Brandt,J.Low Temp.Phys. 26 (1977) 709.
(4) A. Larkin, Yu. Ovchinnikov,J.Low Temp. Phys. 34 (1979) 409.
(5) E.J. Kramer, H.C. Freyhardt, J.Appl.Phys. 51 (1980) 4930.
(6) P. Vassilev et al.Fiz.Met.i Metalov. 51 (1981) 309.
(7) P. Vassilev et al.Fiz.Met.i Metalov. 55 (1983) 1092.
(8) P. Vassilev et al.Fiz.Met.i Metalov. to be published.
(9) P. Vassilev, I.N. Goncharov, Preprint JINR, to be published.
(10) I.N. Goncharov, Preprint JINR P8-10498, Dubna, 1977. in: Materially XX th Vsesoyuznovo soveshtniya" Fizika niskich temperatur, Chernogolovka, 1978, part 3, p. 123.

LT-17 (Contributed Papers)
U. Eckern, A. Schmid, W. Weber, H. Wühl (eds)

COLLECTIVE FLUX LINE PINNING AND RELATED HISTORY-EFFECTS

H. KÜPFER, R. MEIER-HIRMER and M. BRANDENBUSCH

Universität Karlsruhe, Institut für Experimentelle Kernphysik and Kernforschungszentrum Karlsruhe, Institut für Technische Physik, D-7500 Karlsruhe, Federal Republic of Germany

H. SCHEURER

European Communities Joint Research Centre, Petten, Netherlands

The volume pinning force F_p is investigated as a function of the density of weak interacting defects. With increasing defect concentration and increasing field the quadratic dependence of F_p vanishes abruptly and an almost linear relation dominates. In the transition area F_p shows large History-effects. The collective weak pinning interaction is the stable mechanism in that area. Whereas the unsteady flux line lattice configuration which results in high critical currents requires a stabilization obtained at increasing fields and/or increasing defect concentration.

1. COLLECTIVE PINNING

The loss-free currents in the mixed state of type II superconductors are caused by crystal defects. The elementary interaction between flux lines and defects prevent the flux line lattice (FLL) from dissipative motion by compensating the Lorentz force up to the critical volume pinning force F_p. The summation of the elementary pinning forces to F_p is an unsolved problem. This initiates us to investigate the relation between F_p and the concentration of weak interacting defects in the high-κ A15 superconductor V_3Si.

Stoichiometric and non-stoichiometric specimens were irradiated with fast neutrons at irradiation conditions which result in a proportionality between radiation induced defect density and fast neutron fluence ϕt. We measured F_p in dependence of the magnetic field B, at fluences between 10^{19} m^{-2} and 10^{22} m^{-2}, and at temperatures at which the upper critical field B_{c2} has a constant value independent of the fluence. From a comparison between specimens of different stoichiometry we eliminate the influence of the increase of κ at higher fluences. The exponent n (ϕt, B/B_{c2}) of the relation $F_p \sim (\phi t)^n$ was obtained comparing F_p of neighbouring fluences. This quantity shown in figure (1) is about 2 at low fluences and below reduced fields marked with arrows. The second power points to a collective interaction in this area as proposed by Larkin and Ovchinnikov (1). At higher fields and/or higher fluences the elastic behaviour of the FLL may become nonlocal which results in a rapid decrease of the correlation volume i.e. in a steep rise of F_p leading to a maximum of n. Beyond this maximum the influence of the defects on F_p is reduced, n becomes about 1 and decreases with raised field and/or fluence. This dependence of F_p on the defect density points then to a single particle pinning in spite of the weak interacting defects.

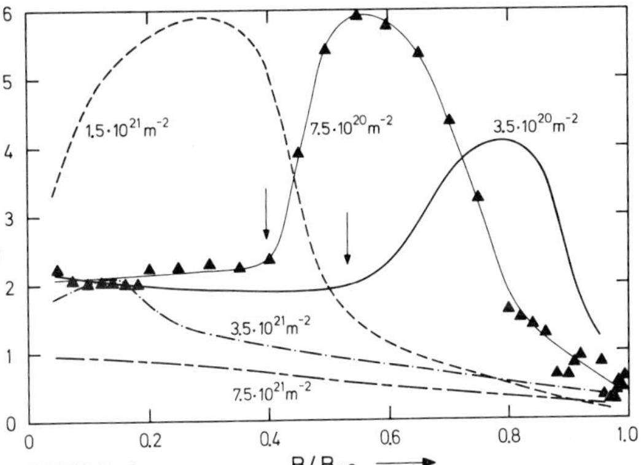

FIGURE 1
Exponent n of the relation $F_p \sim (\phi t)^n$ in dependence on the reduced field B/B_{c2} at different fluences. The obtained values are shown only for the fluence of $7.5 \cdot 10^{20}$ m^{-2}.

2. FLUX LINE LATTICE HISTORY-EFFECTS

The different power of the concentration dependence of F_p and the difference of F_p itself by about two order of magnitude indicates the existence of two distinct interactions which are competitive in certain field and defect density areas. In order to get more information we investigate the stability of the two mechanisms especially in the transition region of them where n becomes a maximum.

It is well known, that a different degree of order in the FLL may result in different volume pinning forces. This History-effect means that F_p is not a unique function of field and temperature. A different distortion of the FLL can be achieved if different paths are taken to adjust field and temperature in the B-T plane. A perfect accomodation of the FLL on the defect structure is obtained if the temperature is adjusted coming from the normal state at a constant field. Another possibility of achieving different degrees of order offers the ac-amplitude h_o which is a variable parameter in our inductive measurement method. With increasing amplitude the disorder in the structure of the FLL may increase also because the movement of flux is raised. A decreasing order causes an increase of F_p via a lowered shear modulus of the FLL. This was deduced from measurements by Küpfer and Gey (2) as well as from computer simulation by Brandt (3). Path dependence and amplitude dependence are coupled with each other.

We observe this common increase of F_p with increasing ac-amplitude in the field region close to B_{c2} and at fields sufficiently below the transition region. But the striking point is the observation of the opposite behaviour at fields where n changes violently. As an example the amplitude dependence of F_p is given in figure (2) in the insertions for 3 fields. At 0.95 reduced field $F_p(h_o)$ shows the common amplitude effect. The pinning force density is a unique function of the amplitude as indicated by the arrows. This is not the case at 0.35 reduced field at which the high F_p state becomes unstable. Measurements of F_p starting from a small amplitude up to a certain maximum value followed by a subsequent decrease of the amplitude always result in a decrease of F_p. With decreasing B/B_{c2} there is a smooth change from the "reversible" amplitude dependence of the History-effect across the "reversible" but opposite effect at 0.88 reduced field to the "irreversible" opposite effect at lower fields. This behaviour opposite to the History-effect is observed if the sample was heated to a temperature above T_c and then slowly cooled in field to the desired temperature without an ac-field. Measuring F_p after taking this path results in values much larger than that ones obtained after the field had been raised or lowered from its value as demonstrated in figure (2). This last possibility leads to low but stable and reproducible F_p values used in the calculation of figure (1). The opposite amplitude dependence of F_p may be coupled with an opposite path dependence of F_p if one attributes less disorder to that FLL created if the sample is cooled in field. In the fluence and field area far away from the transition a decrease in the order of the FLL results in an increase of F_p in contrary to the transition region at which a decrease of order initiates the transition to the collective mechanism.

From these investigations we can answer the question whether pinning is enhanced or degraded by defects in the FLL. A decrease in the order of the FLL increase F_p due to a softening of the

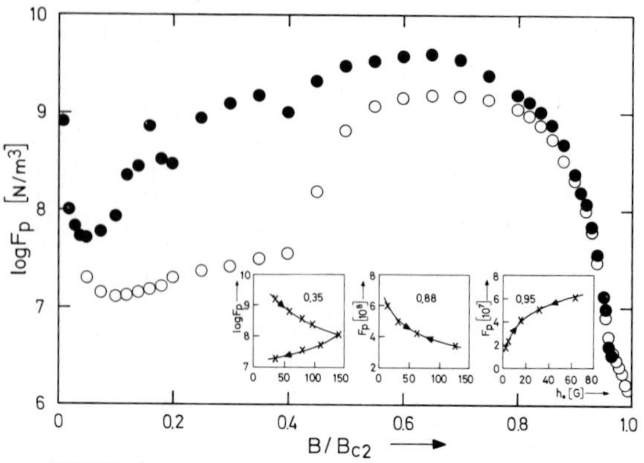

FIGURE 2
F_p in dependence of reduced field measured if the sample is cooled at a constant field (●) and if the field is swept at a constant temperature (○). The insertions show the different amplitude dependences of F_p at 3 reduced fields. F_p is given in N/m^3. At this fluence of 10^{21} m^{-2} the unstable transition behaviour dominates at all fields except close to B_{c2}.

FLL moduli. But if two competitive mechanisms are present, FLL defects cause the transition from the less stable high F_p state to the energetic favourable low one.

(1) A.I. Larkin and Yu. N. Ovchinnikov, J. Low Temp. Phys. 34 (1979) 409.
(2) H. Küpfer and W. Gey, Philos. Mag. 36 (1977) 859.
(3) E.H. Brandt, J. Low Temp. Phys. 53 (1983) 41.

LT-17 (Contributed Papers)
U. Eckern, A. Schmid, W. Weber, H. Wühl (eds)

THE DIFFUSION OF POSITIVE MUONS INVESTIGATED BY MUON SPIN ROTATION IN TYPE-II SUPERCONDUCTORS

E. H. Brandt and A. Seeger

Max-Planck-Institut für Metallforschung, Institut für Physik,
Postfach 800665, D-7000 Stuttgart 80, Germany

The strong effects of the diffusivity D of positive muons on the spin precession of positive muons in the mixed state of Type-II superconductors allow us to deduce D experimentally. Increasing D leads first to the suppression of the van-Hove singularity associated with the magnetic-field maximum in the flux-line centres, then to an enhanced damping of the main precession frequency, and finally to a shift and motional narrowing of the spin-precession frequency distribution.

The knowledge of the diffusivity D of positive muons (μ^+) as a function of temperature T is of great interest for the theory of quantum diffusion in crystals and for the study of crystal imperfections by means of μ^+ trapping (1). Information on D may be deduced by means of μ^+ spin-rotation (μ^+SR) experiments from the relaxation of the transverse spin polarization $P_2(t)$, caused by the dephasing of an ensemble of μ^+ spins rotating in a spatially varying magnetic field $\vec{B}(\vec{r})$. Under favourable circumstances the Gurevich technique (2), in which the fluctuations of $\vec{B}(\vec{r})$ originate from the nuclear magnetic moments, is suitable for $D \lesssim 10^{-13}\ m^2s^{-1}$. In Type-II superconductors larger μ^+ diffusivities may be studied by taking advantage of the spatial variation of $B(\vec{r})$ in the flux-line lattice (FLL) in a manner analogous to the determination of diffusivities by NMR in pulsed magnetic fields. The magnetically inhomogeneous superconducting states provide us with magnetic-field gradients and curvatures that otherwise are unattainable by means of static external fields. The success of the proposed method, however, depends critically on our capability to calculate $B(\vec{r})$ in the FLL reliably (4).

In terms of the muon lifetime $\tau_\mu = 2.2\cdot10^{-6}$ s and the spacing d of a triangular FLL with average field $\bar{B} = \langle B(\vec{r})\rangle = (2/\sqrt{3})\phi_o/d^2 < B_{c2}(T) \approx B_{c2}(0)\cdot(1-T^2/T_c^2)$ ($\phi_o = 2.07\cdot10^{-15}\ Tm^2$ = fluxoid quantum; B_{c2} = upper critical field, T_c = critical temperature of the superconductor), the order of magnitude of the accessible diffusivities may be characterized by $\bar{D} = (d/2\pi)^2\tau_\mu^{-1}$. In pure Nb with $B_{c2}(0) = 0.43$ T, $T_c = 9.25$ K we have for $\bar{B} = B_{c2}$

$$\bar{D} = 6.5\cdot10^{-11}(1-T^2/T_c^2)^{-1}\ m^2s^{-1}\ .$$

In μ^+SR experiments spin-polarized μ^+, obtained from the decay of positive π-mesons, are implanted with polarization transverse to $\vec{B}(\vec{r})$. The time-differential decay rates $\dot{N}_j(t)$ (j=1...4) of the positrons emitted by the decaying muons are recorded in positron telescopes arranged transversely to $\vec{B}$. The complex spin polarization $P^+(t) = P_x+P_y$ is extracted from

$$\dot{N}_j(t) = \dot{N}_j(0)\exp(-t/\tau_\mu)\,\mathrm{Re}\{1+a_jP^+(t)\}+R_j$$

[$\dot{N}_j(0), a_j, R_j$ denote constants, Re{...} the real part] and compared with the spatial average $P^+(t)$ of the solution $P^+(\vec{r},t)$ of the Bloch-Torrey equation (3)

$$[\frac{\partial}{\partial t} + i\gamma B(\vec{r}) - D\nabla^2]P^+(\vec{r},t) = 0 \qquad (1)$$

($\gamma = 8.516\cdot10^8$ rad $T^{-1}s^{-1}$ = gyromagnetic ratio of the muons) with initial condition $P^+(\vec{r},0) \equiv$ const. Different approximations are appropriate depending on whether the diffusivity to be deduced lies in the regimes $D \gg \hat{D}$, or $D \ll \hat{D}$, where $\hat{D} = \gamma|M|(d/2\pi)^2$ estimates the diffusivity causing maximum damping of $P^+(t)$ in a specimen with magnetization M.

a) In the motional-narrowing regime $D \gg \hat{D}$ we may use

$$P^+(t) = P^+(0)\exp(-i\gamma\bar{B}-\gamma^2D^{-1}\sum_K{}' B_{\vec{K}}^2|\vec{K}|^{-2}).$$

Here $B_{\vec{K}}$ denote the Fourier coefficients of $B(\vec{r})$ associated with the reciprocal lattice vectors $\vec{K} \neq 0$ of the FLL.

b) For $D \approx \hat{D}$ the comparison is best based on the normalized Fourier transform

$$\tilde{P}(\omega) = 2\mathrm{Re}\{\int_0^\infty \exp(-i\omega t)[P^+(t)/P^+(0)]dt\}$$

as obtained by numerical integration of Eq.(1) (cf. Fig.1).

c) Extrapolation of the experimental data on Nb (5) indicates that in the temperature range accessible to the present technique $\hat{D} \gg D > 10^{-13}\ m^2s^{-1}$. In the limiting case D=0 the Fourier transform $\tilde{P}(\omega)$ exhibits three van-Hove singularities corresponding to B_{min}, B_s, and B_{max} [minimum, saddle point, and maximum of the magnetic-field distribution, of which the first two are indistinguishable in μ^+SR experiments (Fig.1)]. Diffusivities $D \ll \hat{D}$ may be determined from the jump of $\tilde{P}(\omega)$ at $\omega = -\gamma B_{max}$. From the analytical solution of Eq.(1) [comp.(6)] we get for $\omega \approx -\gamma B_{max}$

$$\tilde{P}(\omega) = \frac{4\pi^2\bar{B}}{\phi_0\gamma\alpha_{max}} h\left(\frac{\omega+\gamma B_{max}}{\gamma\alpha_{max}D}\right); \qquad h(\Omega) \equiv \frac{1}{4} + \frac{\sinh(\pi\Omega/2)/2}{\cosh(\pi\Omega/2)+\cos(\pi\Omega/2)} + \sum_{n=0}^{\infty} \frac{(2n+1)^2\Omega^2/\pi}{(2n+1)^4+\Omega^4/4} .$$

Here α_{max} is the curvature of $B(\vec{r})$ at the flux-line centres. Note that from the amplitude and width of this jump (see inset of Fig.1) and the knowledge of the macroscopic quantity $\bar{B}$, α_{max} and D may be determined separately.

A detailed analysis (7) of the effects of various experimental inadequacies (lack of constancy of applied field and/or temperature, statistical fluctuation of the counting rates $\dot{N}_j(t)$) indicates that for the determination of small D the sensivity of this method is best at low temperatures and small magnetization (i.e. for applied fields close to B_{c2}). With large enough total count numbers ($\approx 10^8$) it should be possible to determine, in pure Nb, μ^+ diffusivities down to about $2\cdot 10^{-13}\ m^2s^{-1}$, i.e., to almost reach the Gurevich regime (2). The upper limit of the method, given by case a) is $D \approx 10^{-5}\ m^2s^{-1}$. Provided it occurs in the accessible temperature range, the onset of coherent μ^+ diffusion may thus be studied in detail. From the application of the present method to high-purity Nb (planned by the STUTTGART–HEIDELBERG μ^+SR collaboration) a deeper understanding of quantum diffusion may be expected.

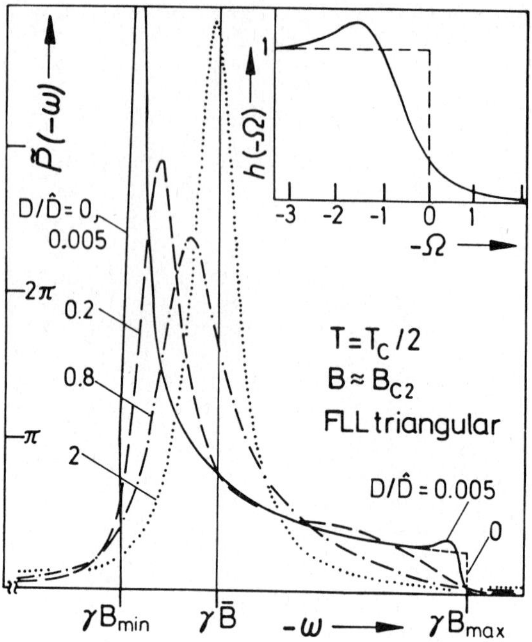

FIGURE 1
The Fourier-transformed polarization in the mixed state of pure superconductors as computed from Eq.(1) with $B(\vec{r})$ taken from (4). The case D = 0 represents the magnetic field distribution; the van-Hove infinity occurs at the saddle-point field B_s. The function $h(-\Omega)$ enlarges the region $\omega \approx -\gamma B_{max}$.

REFERENCES
(1) A. Seeger, Hyperf. Int. 17/18/19 (1984) 75, 982
(2) I.I. Gurevich, E.A. Meleshko, I.A. Muratova, P.A. Nikol'skii, V.S. Roganov, V.I. Selivanov, and B.V. Sokolov, Phys.Lett. 40A (1972) 143
(3) H.C. Torrey, Phys.Rev. 104 (1956) 563
(4) E.H. Brandt, phys.stat.sol. (b) 64 (1974) 467, 77 (1976) 105
(5) T. Aurenz, K.-P. Arnold, K.-P. Döring, M. Gladisch, N. Haas, D. Herlach, W. Jacobs, M. Krause, M. Krauth, H. Orth, H.-E. Schaefer, K. Schulze, and A. Seeger, Hyperf. Int. 17/18/19 (1984) 191
(6) A. Seeger, Phys.Lett. 77A (1979) 259
(7) E. H. Brandt and A. Seeger, to be published

ACKNOWLEDGEMENT
The support of the STUTTGART–HEIDELberg μ^+SR collaboration by the Bundesministerium für Forschung und Technologie (BMFT), Bonn, and stimulating discussions with Drs. D. Herlach and M. Gladisch are gratefully acknowledged.

LT-17 (Contributed Papers)
U. Eckern, A. Schmid, W. Weber, H. Wühl (eds)

DYNAMICS OF COMMENSURATE AND INCOMMENSURATE PHASES OF A 2D LATTICE OF SUPERCONDUCTING VORTICES

F. PATTHEY, G.-A. RACINE, C. LEEMANN, H. BECK and P. MARTINOLI

Institut de Physique, Université de Neuchâtel, CH - 2000 Neuchâtel, Switzerland

A study of the complex rf vortex response of superconducting films with periodically modulated thickness is presented. Structures emerging at well defined vortex densities in both the real and imaginary parts of the response are shown to arise from a commensurate-incommensurate phase transition of the vortex lattice triggered by solitons.

Modulated structures whose period is incommensurable with that of the underlying lattice have been studied in a variety of condensed-matter systems. In particular, the very existence of commensurate (C) and incommensurate (I) phases has been demonstrated for a 2D lattice of vortices in thin superconducting films whose thickness is periodically modulated in one direction (1). Here we study the anisotropic dynamic response of C and I vortex phases to a small oscillating driving force $\vec{f}$.

At low temperatures the phase diagram of 2D crystals exposed to a periodic 1D force field is determined only by soliton excitations which trigger the CI-transition (1,2). For an incompressible lattice the I-phase is characterized by a 1D soliton superlattice, of period P, varying under 45^0 with respect to $\vec{q}_0$, the wave vector of the thickness modulation. For an infinite 2D vortex crystal at T = 0 the CI-phase transition, where $P \to \infty$, occurs when the deviations from a matching configuration $\vec{q}_0 = \vec{g}_{mn}$ ($\vec{g}_{mn}$ is a reciprocal vortex lattice vector) are such that the mismatch $\delta = 1 - (g_{mn}/q_0)$ reaches $\delta_c = (2/\pi)(\Delta/\mu)^{1/2}$, Δ being the strength of the cosine potential and μ the shear modulus of the vortex lattice.

To study the dynamic response of the pinned vortex medium, we rely on a modified version of the technique devised by Fiory and Hebard (3). With the vortex crystal in a C-phase, this method does not excite transverse modes of an infinite lattice and coupling to shear modes only arises from the finite size of the sample and/or from residual random pinning. In an I-phase the 1D periodic sequence of solitons breaks the translational symmetry of the vortex lattice, thereby allowing intrinsic coupling of the driving field to transverse modes of the soliton superlattice. Thus, structures due to the CI-transition at $\delta = \delta_c$ are expected in the complex vortex impedance of modulated layers as the magnetic field B is swept across B_{mn}, the value defining perfect matching (1).

The experiments were performed on a modulated ($2\pi/q_0$ = 0.73 μm) Al-film (T_c = 2.05 K, d = 500 Å, R_n = 25 Ω). As shown by the insert of Fig. 2, excitation of the vortex lattice was provided by two coils, D_1 and D_2, placed on top of the film and driven in opposition by a 5 μA rf-current I_{rf}. The voltage V due solely to the film response was phase sensitively detected by a receive coil R placed between D_1 and D_2 and lying in a plane perpendicular to the film. Because of the 3 Hz field modulation used to discriminate against vortex motion independent pick-up, the detected signals are proportional to $\partial V/\partial B$. To study the angular dependence of the response, the R-coil was oriented either parallel or orthogonal to D_1 and D_2 and the whole arrangement rigidly rotated with respect to $\vec{q}_0$. We denote by (α,β) a coil configuration such that $\vec{f}$ acts in the the α-direction while the R-coil detects the response arising from the projection of vortex motion along the β-direction.

Data taken at 3 MHz and at different temperatures are shown in Fig. 1 for the (y,y)- configuration. Pronounced structures show up in both components of $\partial V/\partial B$ around B_{10}. These structures gradually disappear at higher temperatures. In Fig. 2 signals corresponding to

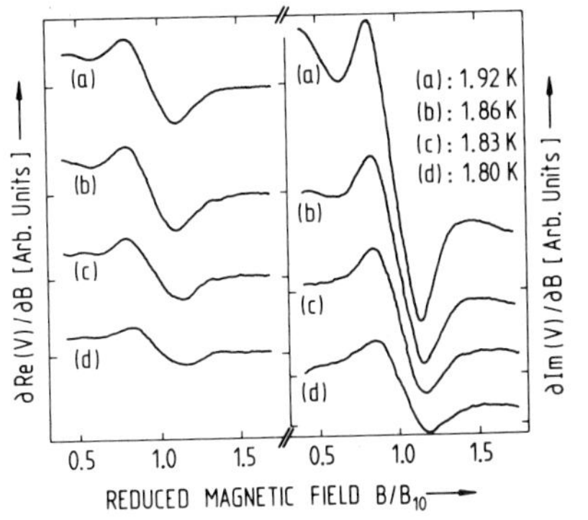

FIG. 1 : Field derivatives of the vortex response at 3 MHz. Coil configuration (y,y).

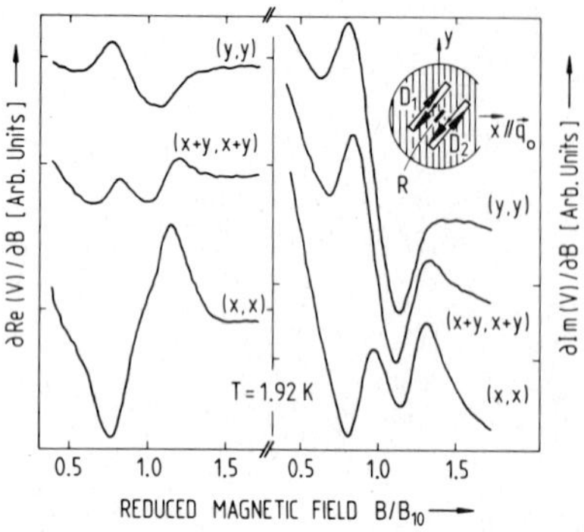

FIG. 2 : Field derivatives of the complex vortex response at 3 MHz for three different coil configurations.

three different coil configurations are shown for T = 1.92 K. A striking feature emerging from these data is the appearance, as the coils are progressively rotated from the (y,y) towards the (x,x)-configuration, of a large new signal which washes out almost completely the structure resulting from the CI-phase transition in $\partial Re(V)/\partial B$ and generates additional structure in $\partial Im(V)/\partial B$. The fact that it does not show up at lower temperatures indicates that it cannot be associated with the type of CI-phase transition described above.

Our calculation of the vortex lattice response has been done in the continuum approximation (1,2). The equation of motion for the displacement field $\vec{w} = \vec{w}_{st} + \vec{u}$ is linearized in the deviation $\vec{u}$ from the static deformation $\vec{w}_{st}$. The finite size L of the sample has been accounted for by allowing for boundary solitons (BS) in a C-phase (4). The I-phase and the C-phase with BS are matched where $P(\delta) = L$. In practice, the appropriate L may be considerably shorter than the actual sample length, approximately of the order of the mean distance between lattice imperfections which interrupt the coherence of the soliton lattice.

The equation of motion of the dissipative (viscosity η) vortex continuum (Lamé coefficients λ,μ) driven by $\vec{f}$ in the periodic potential $U(\phi) = U(w_x + \delta x)$ has the form

$$\eta(\partial\vec{u}/\partial t) = O\vec{u} + \vec{f} \quad , \qquad (1)$$

where $O = (\lambda + \mu)\nabla\nabla\cdot + \mu\nabla^2 - \hat{q}_0 U''(\phi_{st})\hat{q}_0\cdot$.Its solution $\vec{u}(\vec{r},t)$ can be expressed in terms of the eigenfunctions of the operator O.

Using London's and Maxwell's equations, the voltage V due exclusively to vortex motion is shown to be given by an integral over Fourier components of the deviation, $\delta\vec{\Phi}$, of the fluxoid field $\vec{\Phi}$ (3,5) from its static value :

$$\delta\vec{\Phi}(\vec{q},t) = \phi_0(\hat{q} \times \hat{z})[\hat{q}\cdot\vec{u}(\vec{q},t)] \quad . \qquad (2)$$

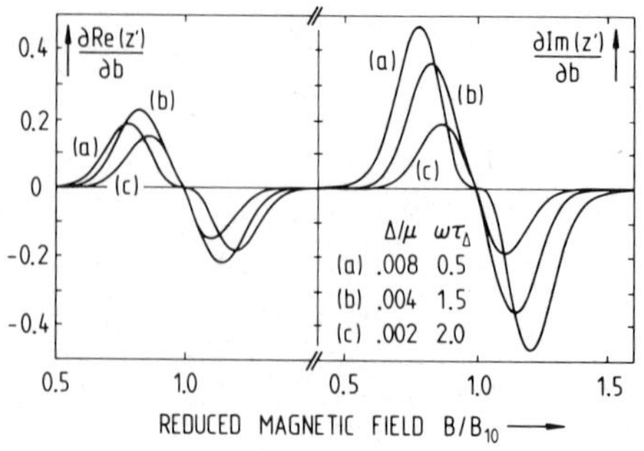

FIG. 3 : Theoretical field derivatives of $z' = Z'/R_f$. $L = 30(2\pi/q_0)$, $b = B/B_{10}$.

Introducing the solution of Eq. (1) into Eq. (2), V can be approximately expressed by $V = GZ'I_{rf}$, where G describes geometry and orientation of D_1, D_2 and R and the impedance $Z'(B,T,\omega)$ contains the vortex dynamics.

For practical calculations we have chosen a piece-wise parabolic potential U. Thus, in the I-phase $U''(\phi_{st})$ yields a Kronig-Penney (KP) potential with distance P between the "spikes". For the C-phase with BS we use periodic boundary conditions, a procedure resulting again in a KP-potential with P replaced by L. The eigenfunctions of O are characterized by a Bloch wave vector $\vec{K}$. If the geometry dependent quantities vary slowly, only small $\vec{q}$ are needed in Eq. (2) and the relevant $\vec{K}$ are the reciprocal lattice vectors $\vec{Q}$ corresponding to the periodic potential in O. For an incompressible lattice ($\lambda \to \infty$) the $\vec{Q}$ = O-mode doesn't contribute to Z'.

Theoretical field derivatives of Z' are shown in Fig. 3. As shown in Ref. 1, rising temperatures can be simulated by decreasing values of Δ/μ and of the relaxation rate $\tau^{-1} = q_0^2\Delta/\eta$. Comparison with the data of Fig. 1 shows good qualitative agreement for the B-dependence as well as for the relative magnitude of the real and imaginary parts of the signal. When T is high enough (1), the structures in Z' around B_{mn} disappear, as observed experimentally. Details of the temperature and angular dependences, however, need further analysis.

REFERENCES

(1) P. Martinoli, M. Nsabimana, G.-A. Racine, H. Beck and J.R. Clem, Helv. Phys. Acta 55 (1982) 655.
(2) V.L. Pokrovsky and A.L. Talapov, Phys. Rev. Lett. 42 (1979) 65.
(3) A.T. Fiory and A.F. Hebard, AIP Conf. Proc. 58 (1980) 293.
(4) S.E. Burkov and V.L. Pokrovsky, J. Low Temp. Phys. 44 (1981) 423.
(5) J. Pearl, Vortex Theory of Superconductive Memories, Thesis (Polytechnic Institute of Brooklyn, New York, 1965).

LT-17 (Contributed Papers)
U. Eckern, A. Schmid, W. Weber, H. Wühl (eds)

TIME-RESOLVED MEASUREMENTS OF THE FLUX-FLOW VOLTAGE IN CURRENT-BIASED THIN-FILM SUPERCONDUCTORS

B. MÜHLEMEIER, J. PARISI, R.P. HUEBENER, and W. BUCK*

Physikalisches Institut II, Universität Tübingen, D - 7400 Tübingen, FRG
* Physikalisch-Technische Bundesanstalt Braunschweig und Berlin, Institut Berlin,D-1000 Berlin,FRG

The temporal structure of the instantaneous voltage generated during both the transition from the superconducting to the current-induced dissipative flux-flow state and the reverse transition is measured. By employing a high-resolution signal-averaging technique, we were able to detect individual multiquantum flux tubes, thereby yielding a voltage resolution of better than 10 nV at a recording bandwidth of 25 MHz (corresponding to a time resolution of 10 ns). The experimentally obtained voltage profiles agree reasonably with the behavior expected from theoretical model calculations.

INTRODUCTION

Thin-film type-I superconductors exhibit an intrinsic step structure in the current-voltage characteristic. From magnetooptical experiments (1) it is known that each voltage step is related to the onset of flux flow perpendicular to the current direction. A constricted geometry, which localizes the first flux channel generated by an increasing current, allows the investigation of a single flux channel. Using this geometry, it could be shown by stroboscopic observation of the flux-flow process that each flux channel consists of a train of multiquantum flux tubes moving from their nucleation site at one edge of the film to the opposite edge. If the nucleation takes place on both sides of the film (symmetric case), flux tubes of opposite orientation annihilate each other near the middle of the path, otherwise (asymmetric case) each flux tube nucleated at one edge of the film crosses the entire film and leaves it on the opposite edge.

Since thin-film superconductors exhibit an inhomogeneous current distribution and an energy barrier (2) against flux entry, one expects an inhomogeneous velocity distribution of the moving flux tubes. This strongly influences the temporal structure of the voltage generated by the flux motion.

Here, we report on time-resolved measurements of the flux-flow voltage during the onset of the current-induced resistive state including the complete cycles of switching this state on and off. Thereby, we could resolve voltage pulses generated by individual flux tubes containing from about 10 to 100 flux quanta. In some cases the temporal voltage structure has been resolved for flux tubes containing as little as 5 flux quanta.

THE EXPECTED VOLTAGE PROFILE

The time-dependent voltage generated by a moving flux tube was calculated by Clem (3) and Buck et al. (4). It is predicted to consist of a sharp peak at the nucleation time, a very small and nearly constant voltage during flux movement, and another even higher peak during the annihilation of the flux tube.

This calculation has been performed for the case of an ideal flux-free superconductor crossed by a single flux tube at a current above the critical current. Model calculations (5) taking into account the repulsive interaction between simultaneously existing flux tubes and the temporal profile of the transport current used in our experiments, showed that the nucleation voltage pulse may nearly vanish. This results mainly from the fact that the nucleation takes place at a current nearly equal to the critical current. So we expect the voltage profile of a single flux tube to start with a small step, holding that level for a relatively long time, and ending with a sharp peak. The total voltage signal of many flux tubes should be the superposition of the individual profiles.

EXPERIMENTAL TECHNIQUE

The experiments have been performed on an In sample with constricted geometry (thickness 2µm, width 45µm, length 3µm) at temperatures from 2.04 to 3.34 K. The sample, surrounded by two superconducting magnetic shields, was in direct contact with the liquid He bath in a glass cryostat mounted in a rf screening box.

Because of the small voltage amplitude and the broad bandwidth of the expected signal, we had to apply a signal-averaging technique for the time-resolved measurements. The signal-averaging system (6) consists of a transient recorder BIOMATION 8100 coupled to a real-time computer PDP 11/04 via a CAMAC I/O-register. Beyond the pure averaging process, the software includes the possibility to perform differential measurements enabling the reduction of coherent interferences. The maximum bandwidth of the system is 25 MHz corresponding to a time resolution of 10 ns, and the voltage resolution attained in our experiments was about 5 nV at a measurement time of few minutes.

The current applied to the sample was composed of a dc current and a triangular current pulse. This current profile allows an effective use of the pick-up suppression facility of the measurement system. Further, it enables a simple control of the origin of the measured voltage: a shift of the dc current must shift the total voltage profile in time without large variations of the waveform.

RESULTS

In Fig. 1a we present a typical measured profile of the time-dependent voltage generated during the current-induced transition from the superconducting to the dissipative flux-flow state. Such an experimentally resolved voltage structure clearly indicates the voltage peaks expected from the annihilation process, whereas the small voltage step associated with the nucleation process is still buried below the noise level. In some experiments, however, such a voltage step can be noticed (see Fig. 1b). Here the signal to noise ratio is further improved by increasing the number of recordings averaged. Due to the necessary temporal stability restricting the total measurement time, the recording period is accordingly reduced. The important parameters are summarized in the figure legends.

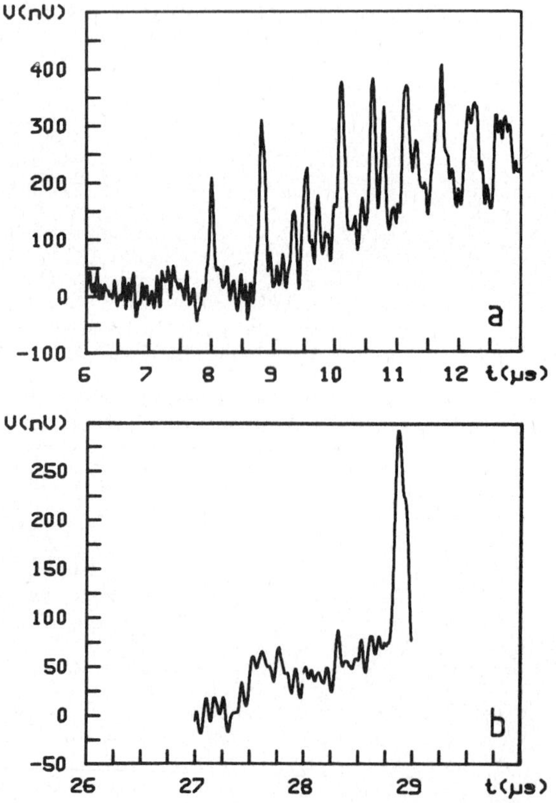

Fig. 1: Temporal voltage structures at the onset of the flux-flow state (a) flux-tube size about 70 flux quanta (T=3.24K); averaged cycles 12000 (b) flux-tube size about 50 flux quanta (T=3.00K); averaged cycles 160000.

Beside the onset of the flux-flow voltage, the process of switching back to the superconducting state is of interest. In Fig. 2 we concentrate on both transitions by generating only a small number of individual flux tubes during one cycle. Here the maximum transport current is taken only slightly above the critical current resulting in relatively short supercritical current pulses. For comparison, Fig. 2 (lower curve) shows the corresponding zero voltage line measured with same resolution at permanently subcritical currents.

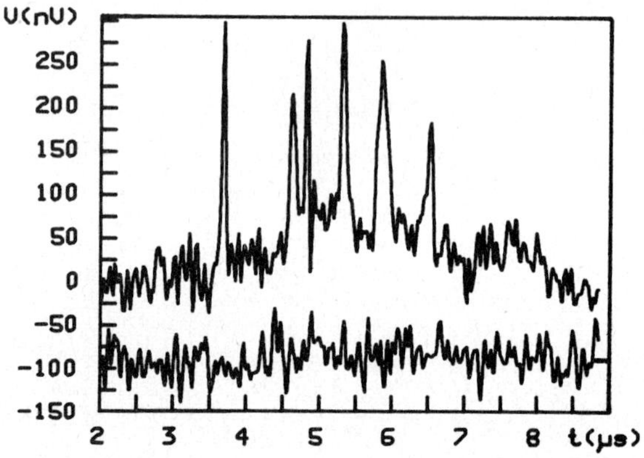

Fig. 2: Temporal voltage structure of a complete flux-flow cycle (top) in comparison with the corresponding zero voltage line (displaced at bottom); flux-tube size about 30 flux quanta (T=3.24K); averaged cycles 16000.

Finally, we note that all experimentally measured voltage profiles can be well reproduced by theoretical computer simulations (5).

ACKNOWLEDGEMENT

This work was supported by a grant of the Deutsche Forschungsgemeinschaft.

REFERENCES

(1) D.E. Chimenti, H.L. Watson, and R.P.Huebener, J. Low Temp. Phys. 23 (1976) 303
(2) J.R. Clem, R.P. Huebener, and D.E. Gallus, J. Low Temp. Phys. 12 (1973) 449
(3) J.R. Clem, J. Low Temp.Phys. 42 (1981) 363
(4) W.Buck, J.Parisi, and B. Mühlemeier, J. Low Temp. Phys. 55 (1984) 51
(5) W. Buck, J. Parisi, and B. Mühlemeier, Computer Simulation of Flux-Tube Motion in Thin-Film Superconductors, this volume
(6) B. Mühlemeier, J. Parisi, and W. Buck, J. Phys. E: Sci. Instrum. 17 (1984) 78

LT-17 (Contributed Papers)
U. Eckern, A. Schmid, W. Weber, H. Wühl (eds)

NONLINEARITY IN THE FLUX-FLOW BEHAVIOR AT HIGH VELOCITIES

S. GAUSS, W. KLEIN, R. P. HUEBENER, and J. PARISI

Physikalisches Institut II, Universität Tübingen, D - 7400 Tübingen, FRG

The nonlinear flux-flow behavior predicted by Larkin and Ovchinnikov has been observed in thin films of Al, Sn, and In. The following values of the energy relaxation time τ_E were found from our results: Al: $(2-3)\cdot 10^{-9}$ s; Sn: $6\cdot 10^{-10}$ s; In: $1.0\cdot 10^{-10}$ s.

As pointed out by Larkin and Ovchinnikov (1) (LO), an interesting regime of nonlinear flux-flow behavior in the mixed state (set up by an applied magnetic field) appears at high flux-flow velocities. In their theory Joule heating or pair breaking due to the applied current do not play any role. The nonlinearity arises from the finite energy relaxation time τ_E of the electrons causing appreciable changes of the quasiparticle distribution during the vortex motion. This effect is strongest near the critical temperature T_c, where the order parameter is most sensitive to small changes in the distribution function. Due to the vortex motion, the number of normal excitations decreases within the vortex core and increases on the outside. Consequently, the vortex diameter and the viscous damping coefficient decreases with increasing velocity v. This effect results in an upward curvature of the voltage-current characteristic, described by the equation (1)

$$\frac{V}{1+(\frac{V}{V^*})^2} + c\cdot V\cdot(1-t)^{1/2} = (I-I_c)\cdot R_f, \qquad (1)$$

where V, t, I, I_c, and R_f denote voltage, reduced temperature T/T_c, current, critical current, and flux-flow resistance, respectively. c is a constant of order unity. At the critical voltage $V^* = v^* \cdot B \cdot L$ the V(I) curve bends backward (v^* = critical flux-flow velocity, B = applied magnetic field, L = gauge length between voltage probes). For current-biased operation, the sample is then expected to switch into another state (like the normal state). For the critical velocity v* LO found (1)

$$v^* = \frac{D^{1/2}[14\zeta(3)]^{1/4}(1-t)^{1/4}}{(\pi\tau_E)^{1/2}} \qquad (2)$$

with $D = v_F \cdot l/3$. Here v_F is the Fermi velocity, l the electron mean free path, and $\zeta(3)$ the Riemann zeta-function of 3. We see that the critical velocity v* is directly related to the inelastic relaxation time τ_E.

The predictions of the LO theory have been verified experimentally by Musienko et al. (2). These authors investigated thin films of Al, Sn, and the alloy Sn + 6 % In using current-biased operation. The thickness of their specimens ranged between 300 and 500 Å. Near T_c Musienko et al. (2) confirmed the temperature dependence of v* predicted by Eq. (2). Most importantly, from their data they obtained values of τ_E in reasonable agreement with the expected values.

We have measured the nonlinear flux-flow behavior and the critical velocity v* in thin films of Al, Sn, and In. The films were evaporated on sapphire substrates of 20 mm diameter and 1 mm thickness. The important sample parameters are listed in Table 1. For minimizing effects due to Joule heating, the film thickness was taken as small as possible compatible with metallic conductivity. During evaporation, all samples were oxygen doped, in order to obtain a high electric resistance (3). During the experiments the samples were in direct contact with the liquid-He bath. We have used both current- and voltage-biased operation. During the latter, a small shunt resistor was connected to the samples. Our experiments were extended to a lower temperature range than the data reported by Musienko et al. (2). Further details are given elsewhere (4).

Table 1: sample parameters

Sample	Al	Sn1	Sn2	In
length (mm)	5.48	4.43	1.67	1.61
width (μm)	1120	105	92	98
thickness (Å)	315	300	300	815
R(300K)/R(4.2K)	1.2	6.14	6.8	16.3
resistivity (4.2K) (μΩcm)	19.3	3.12	3.55	0.78
T_c (K)	1.80	3.82	3.97	3.44

A typical series of nonlinear V(I) curves for different temperatures is shown in Fig. 1 (sample Sn1; voltage bias; B = 3.95 G). At the critical voltage V* all curves show a sharp bend

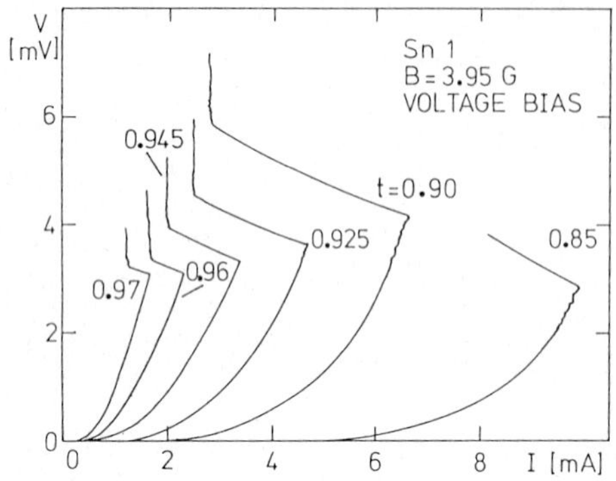

Fig. 1: V(I) curves for sample Sn1 at different temperatures

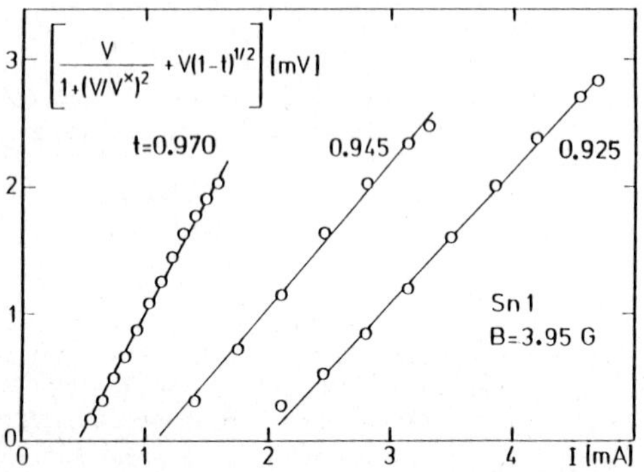

Fig. 2: Plot of the lhs of Eq. (1) versus current at different temperatures

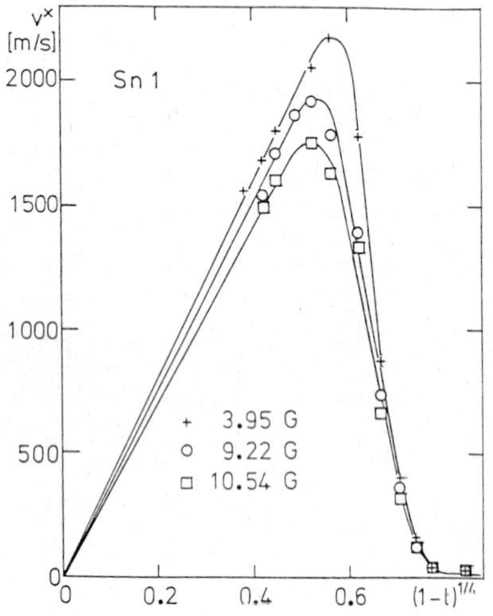

Fig. 3: Critical velocity v^* versus $(1-t)^{1/4}$ for sample Sn1 and different magnetic fields

and turn backward. Figure 2 contains a plot of the expression on the lhs of Eq. (1) up to the critical voltage V^* versus current for three temperatures (sample Sn1; B = 3.95 G). Here V^* was taken from curves such as shown in Fig. 1, and we have used c = 1. We see that the prediction of Eq. (1) is reasonably satisfied. A typical plot of the critical velocity $v^* = V^*/BL$ versus $(1-t)^{1/4}$ is shown in Fig. 3 for sample Sn1 and different magnetic fields. We see that the proportionality predicted by Eq. (2) seems to be satisfied up to $(1-t)^{1/4} = 0.5 - 0.6$ similar to the results reported by Musienko et al. (2). However, at reduced temperatures below 0.87 - 0.94 the OL theory appears to be invalid. For the other samples we obtained results similar to those shown in Fig. 1 - 3.

As the most important result the inelastic relaxation time τ_E can be found for the different samples from plots such as shown in Fig. 3 in combination with Eq. (2) (restricting ourselves to the regime $t \gtrsim 0.9$). The following values were obtained: Al sample: $\tau_E = (2-3)\cdot 10^{-9}$s; sample Sn1: $\tau_E = 6\cdot 10^{-10}$s; In sample: $\tau_E = 1.0\cdot 10^{-10}$ s. Our value for sample Sn1 is in excellent agreement with the value reported by Musienko et al. (2) for Sn, whereas τ_E of our Al sample is slightly smaller than the value reported previously (2) for Al. Our value for In agrees well with the result of Hsiang and Clarke (5). From Fig. 3 we note that the critical velocity v^* shows some dependence upon the applied magnetic field (corresponding to an increase of τ_E with increasing B). Since the LO theory is valid only in the low-field limit, our values for τ_E given above refer to the low-field range of our measurements. It is possible that the finite vortex distance at finite magnetic fields results in a reduction of the critical velocity below the value given by the LO theory. This may explain the increase of τ_E with increasing B, which we found when calculating τ_E from Eq. (2).

ACKNOWLEDGEMENTS

One of the authors (R.P.H.) is pleased to acknowledge stimulating discussions with I.M. Dmitrenko, L.E. Musienko, and V.G. Volotskaya.

REFERENCES

(1) A.I. Larkin and Yu.N. Ovchinnikov, Zh. Eksp. Teor. Fiz. 68 (1975) 1915 [Sov. Phys. JETP 41 (1976) 960].
(2) L.E. Musienko, I.M. Dmitrenko, and V.G. Volotskaya, Pis'ma Zh. Eksp. Teor. Fiz. 31 (1980) 603 [JETP Lett. 31 (1980) 567].
(3) R. Eichele, W. Kern, and R.P. Huebener, Appl. Phys. 25 (1981) 95.
(4) S. Gauss, Thesis, University of Tübingen, 1983 (unpublished)
(5) T.Y. Hsiang and J. Clarke, Phys. Rev. B21 (1980) 945

LT-17 (Contributed Papers)
U. Eckern, A. Schmid, W. Weber, H. Wühl (eds)

ASYMMETRIC FLUX-FLOW BEHAVIOR IN SUPERCONDUCTING MULTILAYERED COMPOSITES

A.M. KADIN, R.W. BURKHARDT, J.T. CHEN*, J.E. KEEM, S.R. OVSHINSKY

Energy Conversion Devices, Inc., 1675 West Maple Road, Troy, Michigan 48084, U.S.A.

Multilayer composites have been fabricated, with 20-400 alternating layers of refractory metal (Nb, Mo, W) and semiconductor (Si, Ge) of thickness 10-50 Å/layer. At low temperatures, for magnetic fields parallel to the layers, some samples exhibit flux-flow-like characteristics with a marked asymmetry for the critical depinning currents of opposite polarities. This suggests that for flux flow perpendicular to the layers, one direction is preferred over the other.

Recently, advanced synthetic capabilities have developed to the point where it has become feasible to artificially produce compositionally modulated superstructures of a wide variety of materials over a wide range of modulation periods (1,2). We have been fabricating periodically modulated structures of refractory materials, typically alternating between a transition metal and a semiconducting element, made by sputtering alternately one material and then the other in a periodic fashion onto a smooth substrate. Crystalline epitaxy has not been required; in fact, some of the best-layered samples have consisted of amorphous material in the layers, and the "superlattice" is actually the only true lattice evident from x-ray diffraction. These structures have been fabricated with coherent repeat spacings d covering the range from 10 to 100 Å, for up to hundreds of layer pairs. Similar structures have also been used in our laboratory for the development of x-ray optical devices, as quasi-Bragg reflectors for a wide range of d-spacings (3).

The superconducting samples discussed here consist of multilayers of Mo-Si, W-Si, and Nb-Si. The sample geometry consisted of photolithographically defined "dumbbells",with a narrow section 3.8 mm x 0.3 mm, and overall thickness from 300 Å to 1μm. Most of the samples were fabricated by rf magnetron sputtering from two targets onto an unheated substrate (most commonly an oxidized Si wafer) that was transported past them. Alternatively, several samples were made using an ion-beam deposition system with a rotating target assembly. The layer spacing was determined by low-angle x-ray diffraction, and agreed with the total measured thickness divided by the number of rotations during the deposition. Auger depth profiling was also routinely carried out, and qualitatively confirmed the compositional modulation. For the best samples, the root-mean-square surface roughness of the layers, as estimated from the sharpness of the x-ray diffraction via a simple model, was less than 3 Å.

We will concentrate here on the behavior in large magnetic fields of one particular sample, #851-5c, which consisted of 50 bilayers of alternating 7 Å Nb and 8 Å Si. This sample had a normal-state resistance of 310 ohms (it was amorphous, with a temperature coefficient of resistivity close to zero), with a critical temperature of 3.23 K and a transition width (5% to 95% of the resistance) of 30 mK. The parallel critical field extrapolated to zero temperature was 81 kOe.

At low temperatures, for parallel fields that were large but well below the critical field, I-V curves such as those in Fig. 1. were observed. These exhibited a fairly sharp critical current followed by a linear resistance region. These are precisely the sort of characteristics that are usually associated with flux flow. Since the magnetic field and the current are mutually perpendicular, this corresponds to vortices lying parallel to the layers and moving perpendicular to them.

The remarkable aspect about these curves is that the critical depinning current I_c (defined by extrapolating from the linear region) can be dramatically different for positive and negative currents, with magnitudes I_c^+ and I_c^-, differing by a factor of two or even more. Furthermore, this asymmetry reverses completely with the polarity of the magnetic field. This suggests that it may be easier to initiate vortex motion in one direction than in the other, i.e., that the pinning force is asymmetric. For the sample here, the "easy" direction (lower critical current) corresponds to vortex motion from the top of the film towards the substrate. This asymmetric behavior was not

*Permanent Address: Department of Physics, Wayne State University, Detroit, Michigan, U.S.A.

present for other relative orientations of the current and field. The detailed dependence on the fabrication parameters is being investigated.

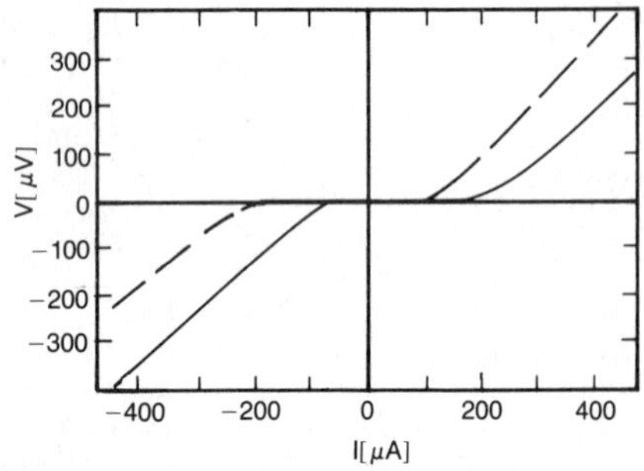

Fig. 1: I-V characteristics for a Nb-Si multilayered structure at T=1.99 K for a parallel field of magnitude H=6.7 kOe. Positive and negative field orientations are indicated by dashed and solid lines respectively.

The ratio $f = I_c^-/I_c^+$ (defined for the field in the positive direction) is plotted as a function of field in Fig. 2. It never goes below unity (for any T), but it does exhibit an interesting oscillation down to one and then back up as the field is increased. In contrast, the average critical current $\overline{I}_c = (I_c^- + I_c^+)/2$ exhibits no striking features, but rather a gradual decrease with increasing field.

One may speculate as to the physical basis for this asymmetry. It is conceivable that the Si layers may act as pinning centers, and an asymmetry might arise from the difference between alternate interfaces (i.e., Nb deposited on Si may be different from Si on Nb). This is conceptually similar to an asymmetric critical current that was earlier observed (4) using an array of scratches on a superconducting foil, where the scratch profiles themselves were asymmetric. On the other hand, the very small spacing (smaller than the superconducting coherence length $\xi \simeq$ 70 Å) would seem to argue against this. We cannot rule out at present that an explanation involving the relative ease of flux entry at the upper and lower faces of the film. A nonuniform field or current, although unlikely, might also cause an asymmetry. Investigations are continuing.

Finally, such an asymmetry will cause rectification under application of an ac current, and we have observed such rectification up to frequencies of at least 5 MHz. We note, however, that we have also observed a small induced dc voltage at high frequencies very near the resistive transition, where no dc asymmetry exists. This latter phenomenon, we believe, is different from the asymmetric flux-flow problem, and is more likely to be associated with the inverse ac Josephson effect in the effective inhomogeneous superconductor that may exist close to $T_c(H)$. Such effects have also been seen in non-layered films (5).

In conclusion, we have observed asymmetric flux flow in a superconducting multilayered composite. We speculate that this may be related to an asymmetric materials profile across the layers, but alternative explanations cannot yet be ruled out. This effect produces a rectified dc voltage under application of an ac current, but is distinct from the Josephson-like rf-induced dc voltage that may appear at the resistive transition, in the absence of a dc asymmetry, in microscopically inhomogeneous or granular samples.

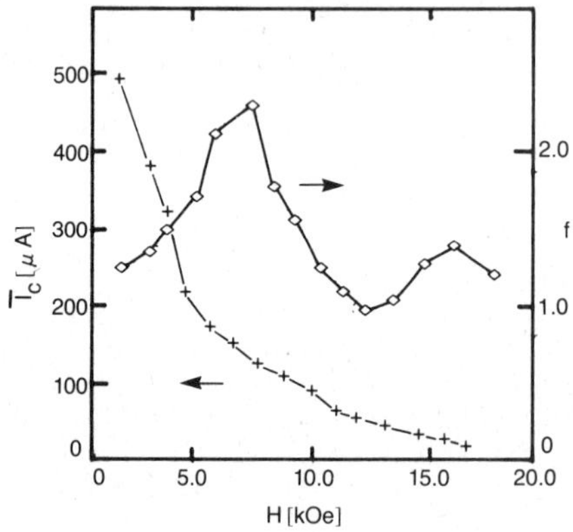

Fig. 2: Field dependence of critical current average $\overline{I}_c$ (+) and ratio f (open squares) for the sample of Fig. 1 at T=1.99 K. The lines are a guide to the eye.

REFERENCES

(1) S.T. Ruggiero, T.W. Barbee, and M.R. Beasley, Phys. Rev. B26 (1982) 4894.

(2) I. Banerjee, Q.S. Yang, C.M. Falco, and I.K. Schuller, Phys. Rev. B28 (1983) 5037.

(3) Comercially available as Ovonyx™ elements from Energy Conversion Devices.

(4) D.D. Morrison and R.M. Rose, Phys. Rev. Lett. 25 (1970) 356.

(5) H. Sadate-Akhavi, J.T. Chen, A.M. Kadin, J.E. Keem, and S.R. Ovshinsky, Bull. Am. Phys. Soc. 29 (1983) 363, and to be published in Solid State Commun.

LT-17 (Contributed Papers)
U. Eckern, A. Schmid, W. Weber, H. Wühl (eds)

COMPUTER SIMULATION OF FLUX-TUBE MOTION IN THIN-FILM SUPERCONDUCTORS

W. BUCK, J. PARISI* and B. MÜHLEMEIER*

Physikalisch-Technische Bundesanstalt, Institut Berlin, D-1000 Berlin, FRG
* Physikalisches Institut II, Universität Tübingen, D-7400 Tübingen, FRG

The motion of flux tubes across a superconducting thin-film strip during the dissipative current-induced flux-flow state is simulated on a digital computer. Only if the complete Gibbs free-energy barrier model is taken into account and the interaction between simultaneously existing flux tubes is included the calculated voltage as a function of time agrees reasonably well with experimental results obtained from experiments with high voltage and time resolution.

1. INTRODUCTION

In type-I superconducting thin-film strips a resistive state appears for transport currents above the critical current which is often called current-induced resistive state (CIRS) (1). This state is visibly shown to be a dynamic state (2), where flux tubes containing up to more than one hundred flux quanta move perpendicular to the current direction from their nucleation site on one sample edge towards their annihilation on the opposite sample edge in the so-called asymmetric case (3).

2. MODEL DESCRIPTION

The thin-film sample strip is approximated by an elliptical cylinder of infinite axial length as shown in fig. 1 of Ref. (3). A flux tube is represented by a small normal cylinder as long as the local sample thickness and with its axis pointing perpendicular to the sample width. The equation of motion is derived from the balance of the forces acting simultaneously on a flux tube. The Lorentz force F_T (explicit formulas see Ref. (4)) resulting from the actual transport current acts as driving term. For a single flux tube only the viscous drag force F_d has to be added to F_T and the sum is set to equal zero. From the resulting differential equation the time evolution of the flux tube coordinate as well as the instantaneous flux-tube velocity is evaluated by iterative integration with a standard Runge-Kutta routine. Following Ref. (5) the flux-flow voltage is calculated from the velocity and a geometry dependent factor for the case with the leads of the measuring circuit kept far from the sample surface. Applying a smoothly increasing transport current (fig. 1a) the voltage generated by one single flux tube (fig. 1b) clearly shows two sharp voltage peaks caused during nucleation and annihilation and a flat region in between. The central part and the annihilation peak in fig. 1b of Ref. (6) seems to be reproduced quite well. The calculated nucleation peak, however, shows a quite different shape, since nucleation is simply included by the condition for the transport current to be larger than an arbitrarily given critical value.

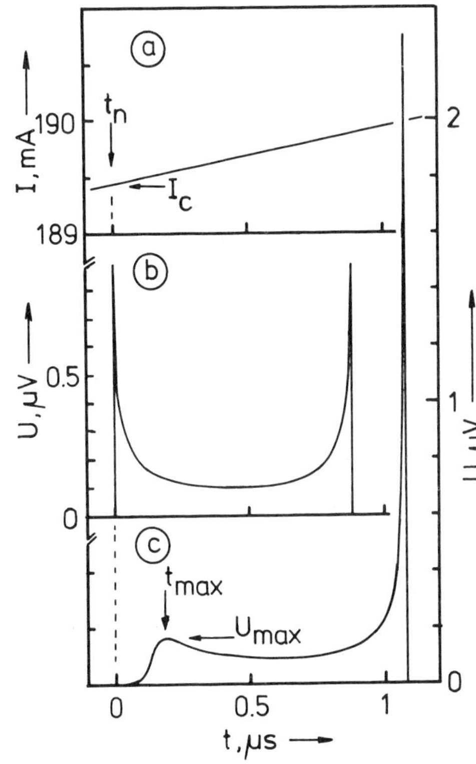

FIGURE 1
Temporal voltage profile generated by a single flux tube containing 89 flux quanta (sample width: 40 µm, sample thickness: 2 µm, T = 2.03 K); (a) transport current, (b) voltage structure without barrier; (c) same as (b) but including barrier.

2. GIBBS FREE-ENERGY BARRIER

Therefore, the Gibbs free-energy barrier must be included which arises from two additional energy contributions, namely the loss of condensation energy combined with the magnetic field energy both inside the normal volume of the flux tube and from the distortion of the magnetic field outside by the self-field of the flux tube. The derived barrier forces F_{in} and F_{out} (4) both tend to keep the magnetic flux outside the superconducting film. Consequently, flux entry is impeded during nucleation, whereas flux exit is considerably accelerated. Therefore, the observed voltage starts with a horizontal slope at the nucleation time t_n (fig. 1c), since the driving force is almost completely compensated by the barrier forces. The value of transport current, at which nucleation starts, is then defined as critical current I_c, which now comes out from theory and is no longer an arbitrary variable.

Immediately after nucleation the flux tube is strongly accelereated by the current density concentrated near the sample edge. This causes together with the geometrical factor mentioned above a small voltage peak at t_{max}. Afterwards the voltage smoothly follows flux-tube velocity. At the opposite sample edge both the flux-tube velocity and the geometry factor increase drastically towards infinity forming the sharp annihilation peak. Its height depends on the cutoff coordinate chosen according to the normal edge structure which is assumed to be of the size of nearly one flux tube radius (3).

4. COMPARISON WITH EXPERIMENT

After smoothing the fluctuations in fig. 1b of Ref. (6) a sharp voltage step at 27.3 μs is revealed followed by a small maximum at 27.7 μs both indicating the nucleation of a flux tube. After a flat region the large annihilation peak appears at 28.9 μs. This voltage trace is in excellent qualitative agreement with the calculated result of fig. 1c and it seems to be the first time that such a single event has been detected by direct electrical methods.

Usually more than one flux tube are generated (6). If the preceding one is still existing during the nucleation of the following the repulsive interaction force F_{int} (4) between them has to be included into the equation of motion. This means an acceleration to the preceding and a retardation to the succeeding flux tubes. Then, the single diffential equation has to be replaced by a set of equations coupled via the interaction term. The voltage pattern now arises from the sum of the contributions of all simultaneously existing flux tubes (fig. 2). Again, the qualitative agreement with the experimental curves is obvious.

5. CONCLUSIONS

Even with parameters roughly adapted to the experimental data the achieved agreement between simulation and experiment becomes surprisingly well, although deviations from sample geometry, material inhomogeneities, and other pinning effects are neglected at all.

Finally, we note that the simulation can simply be applied to type II superconductors by reducing the flux-tube content to one flux quantum.

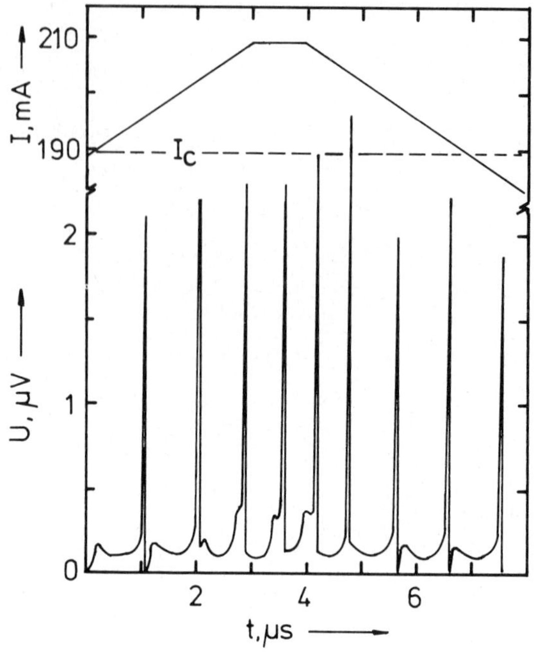

FIGURE 2
Temporal evaluation of current bias (upper trace) and flux-flow voltage (lower trace) with several flux tubes. At left and right part only one flux tube is existing inside the sample. At central part two flux tubes exist simultaneously.

REFERENCES
(1) R.P. Hübener, Magnetic Flux Structures in Superconductors (Springer, Berlin, 1979), p. 215.
(2) D.E. Chimenti and J.R. Clem, Phil. Mag. B 38 (1978) 635.
(3) W. Buck, K.-P. Selig and J. Parisi, J. Low Temp. Phys. 45 (1981) 21, and Refs. cited there.
(4) W. Buck, J. Parisi and B. Mühlemeier, J. Low Temp. Phys. 55 (1984) 51.
(5) J.R. Clem, J. Low Temp. Phys. 42 (1981) 363.
(6) B. Mühlemeier, J. Parisi, R.P. Hübener and W. Buck, Time-Resolved measurements of the flux-flow voltage in current-biased thin-film superconductors, this volume.

LT-17 (Contributed Papers)
U. Eckern, A. Schmid, W. Weber, H. Wühl (eds)

THEORY OF THE DOUBLE CRITICAL STATE IN TYPE-II SUPERCONDUCTORS

John R. CLEM and Antonio PEREZ-GONZALEZ

Ames Laboratory-USDOE* and Department of Physics, Iowa State University, Ames, Iowa 50011, USA

A double critical-state theory is described by which hysteretic states of a type-II superconductor in an applied magnetic field that varies in both magnitude and direction can be calculated. In this theory, if the magnitude of the electrical current density component perpendicular to the magnetic induction $\vec{B}$ exceeds the corresponding critical value $J_{c\perp}$, depinning occurs, and an electric field component $E_\perp$ perpendicular to $\vec{B}$ appears; if the magnitude of the current density component parallel to $\vec{B}$ exceeds the corresponding critical value $J_{c\parallel}$, flux-line cutting occurs, and an electric field component $E_\parallel$ parallel to $\vec{B}$ appears. Solutions of the double critical-state model are presented for a hysteretic type-II superconducting slab subjected to a parallel applied magnetic field that is constant in magnitude but varying in direction.

1. INTRODUCTION

The hysteretic behavior of a type-II superconductor subjected to a parallel applied magnetic field that is fixed in direction but varying in magnitude usually is well described by the critical-state theory reviewed in Ref. (1). When the applied field varies in both magnitude and direction, however, there are not enough equations within this theory to determine the magnetic induction $\vec{B}$ uniquely. In a series of important experiments on the magnetic states of a type-II superconducting disk that was rotated relative to a parallel applied magnetic field, LeBlanc and coworkers (2-4) demonstrated some of the inadequacies of the usual critical-state theory and proposed an additional empirical critical-state equation, which permitted the calculation of hysteretic states that were in reasonably good agreement with those found experimentally. Clem (5) has interpreted this new equation in terms of a critical current at the threshold of flux-line cutting (intersection and cross-joining of adjacent nonparallel vortices). In the present paper, we extend the treatment of Ref. (5) and formulate the resulting double critical-state model for slab geometry in Sec. 2. We then apply the theory to one simple case in Sec. 3. Finally, we discuss extensions to more complicated cases in Sec. 4.

2. FORMULATION OF THE THEORY

Consider a high-κ, irreversible type-II superconducting infinite slab with surfaces at x=0 and X. Application of a time-varying, parallel magnetic field and transport current yields a time-varying, parallel magnetic induction at the two surfaces, $\vec{B}_a(0,t)$ and $\vec{B}_a(X,t)$. This in turn generates inside the superconductor the following macroscopic (averaged over a few intervortex spacings) quantities: magnetic induction $\vec{B}(x,t)$, current density $\vec{J}(x,t)$, and electric field $\vec{E}(x,t)$, all of which depend only upon the spatial coordinate x and the time t. For simplicity we assume that (a) $\vec{B}=\mu_0\vec{H}$ over the most important field range, (b) the length scales for spatial variation of $\vec{B}$, $\vec{J}$, and $\vec{E}$ are much larger than the weak-field penetration depth λ, and (c) the boundary conditions at x=0 and X are $\vec{B}(0,t)=\vec{B}_a(0,t)$ and $\vec{B}(X,t)=\vec{B}_a(X,t)$ (i.e., there are no surface barriers against flux entry or exit). In this geometry, it is convenient to focus on the magnitude B and direction $\hat{\alpha}$ of $\vec{B}$ by writing $\vec{B}=B\hat{\alpha}$, where $B=|\vec{B}|$ and

$$\hat{\alpha} = \hat{y}\sin\alpha + \hat{z}\cos\alpha. \tag{1}$$

We here and henceforth suppress the dependence upon the space and time coordinates (x,t).

In general, the current density $\vec{J}$ has components $J_\parallel$ and $J_\perp$ parallel and perpendicular to $\vec{B}$: $\vec{J}=J_\parallel\hat{\alpha}+J_\perp\hat{\beta}$, where $\hat{\beta}=\hat{\alpha}\times\hat{x}$. Ampere's law, $\vec{J}=\vec{\nabla}\times\vec{H}$, yields (with neglect of the displacement current)

$$J_\parallel = \mu_0^{-1} B\,\partial\alpha/\partial x, \tag{2}$$

$$J_\perp = -\mu_0^{-1}\,\partial B/\partial x. \tag{3}$$

Similarly, the electric field $\vec{E}$ has components $E_\parallel$ and $E_\perp$ parallel and perpendicular to $\vec{B}$: $\vec{E}=E_\parallel\hat{\alpha}+E_\perp\hat{\beta}$. Faraday's law, $\vec{\nabla}\times\vec{E}=-\partial\vec{B}/\partial t$, yields

$$\partial E_\parallel/\partial x = B\,\partial\alpha/\partial t + E_\perp\,\partial\alpha/\partial x, \tag{4}$$

$$\partial E_\perp/\partial x = -\partial B/\partial t - E_\parallel\,\partial\alpha/\partial x. \tag{5}$$

*Operated for the USDOE by Iowa State University under contract No. W-7405-Eng-82 and supported by the Director for Energy Research, Office of Basic Energy Sciences.

For time variation of $\vec{B}_a(0,t)$ and $\vec{B}_a(X,t)$ sufficiently slow that eddy currents are negligible, local vortex configurations generating $\vec{B}$ are assumed to be governed by the following double critical-state model: metastable stationary distributions of B, in which $E_\perp=0$, are always such that $J_\perp$ obeys

$$|J_\perp| \leq J_{c\perp}(B,T), \tag{6}$$

where $J_{c\perp}(B,T)$ is the function describing the transverse critical current density at the threshold for vortex depinning from a distribution of bulk pinning centers. Equation (6) is equivalent to the statement that the magnitude of the Lorentz force $F_{Lx}=J_\perp B$ never exceeds the maximum volume pinning force $F_p(B,T)\equiv J_{c\perp}(B,T)B$. Similarly, metastable stationary distributions of α, in which $E_\parallel=0$, are always such that $J_\parallel$ obeys

$$|J_\parallel| \leq J_{c\parallel}(B,T), \tag{7}$$

where $J_{c\parallel}(B,T)$ is the function describing the longitudinal critical current density at the threshold for the onset of flux-line cutting in the vortex array (5). Using Eq. (2), we can reexpress Eq. (7) as

$$|\partial\alpha/\partial x| \leq k_{c\parallel}(B,T), \tag{8}$$

where

$$k_{c\parallel}(B,T) \equiv \mu_0 J_{c\parallel}(B,T)/B \quad . \tag{9}$$

Flux redistribution occurs when $J_\perp>J_{c\perp}$, in which case $E_\perp>0$ and $J_\perp E_\perp>0$; when $J_\perp<-J_{c\perp}$, we have $E_\perp<0$ and again $J_\perp E_\perp>0$. Flux redistribution also occurs when $J_\parallel>J_{c\parallel}$, in which case $E_\parallel>0$ and $J_\parallel E_\parallel>0$; when $J_\parallel<-J_{c\parallel}$, we have $E_\parallel<0$ and again $J_\parallel E_\parallel>0$. Regardless of their spatial dependences, both $E_\perp$ and $E_\parallel$ are continuous functions of x, as can be shown from Eqs. (4) and (5). We denote a zone in which (a) only flux transport occurs ($E_\perp\neq0$ but $E_\parallel=0$) as a T zone, (b) only flux cutting occurs ($E_\parallel\neq0$ but $E_\perp=0$) as a C zone, (c) both flux cutting and transport occur ($E_\parallel\neq0$ and $E_\perp\neq0$) as a CT zone, and (d) neither flux cutting nor transport occurs ($E_\parallel=0$ and $E_\perp=0$) as an O zone.

Starting with some metastable, stationary magnetic state, changes in $\vec{B}_a(0,t)$ and $\vec{B}_a(X,t)$ initially cause changes in B and α only close to the surface, as in the usual critical-state model. As the flux redistribution region penetrates more deeply into the specimen, Eqs. (2)-(7) must be solved simultaneously to obtain the resulting profiles of B and α.

3. MODEL CALCULATIONS

As an example of the use of the above theory, we assume for simplicity that $J_{c\perp}$ and $k_{c\parallel}$ are constants independent of B. We consider an initial state such that $\vec{B}(x,0) = \vec{B}_a(0,0) = \vec{B}_a(X,0) = B_0\hat{z}$. We seek $\vec{B}(x,t)$ near the surface x=0 when $\vec{B}_a(0,t) = B_0\hat{\alpha}_s(t)$, where

$$\hat{\alpha}_s = \hat{y}\sin\alpha_s + \hat{z}\cos\alpha_s \tag{10}$$

and $\alpha_s = \omega t$. We discuss only the initial behavior as α_s first increases from zero. The solutions of Eqs. (2)-(7) then are as follows:

(a) Both flux-line cutting and flux transport occur in the CT zone $0 \leq x < x_v$. Here, $B=B_0(1-k_{c\perp}x)$, $k_{c\perp}=\mu_0 J_{c\perp}/B_0$, $\alpha=\omega t-k_{c\parallel}x$, $\alpha_v=\omega t-k_{c\parallel}x_v$, $J_\perp=J_{c\perp}$, $J_\parallel=-J_{c\parallel}(B,T)$,

$$E_\perp = \frac{\omega B_0}{k_{c\parallel}}\left[(1-k_{c\perp}x) - \cos\alpha - \left(\frac{k_{c\perp}}{k_{c\parallel}}\right)\sin(\alpha-\alpha_v)\right], \tag{11}$$

$$E_\parallel = \frac{-\omega B_0}{k_{c\parallel}}\left\{\sin\alpha + \left(\frac{k_{c\perp}}{k_{c\parallel}}\right)\left[1-\cos(\alpha-\alpha_v)\right]\right\}. \tag{12}$$

The boundary at x_v is determined from $1-k_{c\perp}x_v=\cos\alpha_v$.

(b) Only flux-line cutting (no flux transport) occurs in the C zone $x_v \leq x < x_c$. Here, $B=B_0\cos\alpha$, $\alpha=\omega t-k_{c\parallel}x$, $J_\perp=(k_{c\parallel}B_0/\mu_0)\sin\alpha$, $J_\parallel=-J_{c\parallel}(B,T)$, $E_\perp=0$, and

$$E_\parallel=-(\omega B_0/k_{c\parallel})\sin\alpha \quad . \tag{13}$$

(c) Neither flux-line cutting nor flux transport occurs in the O zone $x_c \leq x$. Here, $B=B_0$, and $\alpha=J_\perp=J_\parallel=E_\perp=E_\parallel=0$.

4. EXTENSIONS TO MORE COMPLICATED SITUATIONS

As in the case of the usual critical-state model, analytic results are obtainable only with simple models for the B dependence of $J_{c\perp}$ and $J_{c\parallel}$. We have developed a computer program which numerically solves Eqs. (2)-(7) for arbitrary B dependence of $J_{c\perp}$ and $J_{c\parallel}$, as well as for arbitrary trajectories of $\vec{B}_a(0,t)$ and $\vec{B}_a(X,t)$. Sample results will be published later.

ACKNOWLEDGEMENTS

We are grateful to Dr. M. A. R. LeBlanc for stimulating discussions and correspondence.

REFERENCES

(1) A. M. Campbell and J. E. Evetts, Adv. Phys. 21 (1972) 199.
(2) R. Boyer and M. A. R. LeBlanc, Sol. State Commun. 24 (1977) 261.
(3) R. Boyer, G. Fillion, and M. A. R. LeBlanc J. Appl. Phys. 51 (1980) 1692.
(4) J. R. Cave and M. A. R. LeBlanc, J. Appl. Phys. 53 (1982) 1631.
(5) J. R. Clem, Phys. Rev. B26 (1982) 2463.

LT-17 (Contributed Papers)
U. Eckern, A. Schmid, W. Weber, H. Wühl (eds)

ON THE FLUX PENETRATION INTO ANTIFERROMAGNETIC SUPERCONDUCTOR WITH INDUCED FERROMAGNETISM

Tomasz KRZYSZTOŃ

Institute for Low Temperature and Structure Research, Polish Academy of Sciences,
ul. Próchnika 95, 53-529 Wrocław, Poland

An induced ferromagnetism in the antiferromagnetic superconductor has been explained as the consequence of magnetic structure of the vortex lines. The structure appears in the external magnetic field sufficiently high to flip-over the spins from antiferromagnetic configuration in the normal core. Two-stage flux penetration process has been postulated depending on the strength of the external field. In the first stage flux penetrates in the form of the entirely antiferromagnetic vortex lines, next a new surface energy barrier appears when the spins in the core flip-over.

In the recently formulated model [1] we have explained the induced ferromagnetism in certain kind of the antiferromagnetic superconductors as the consequence of magnetic structure of the vortices. The structure begins to develop when the external magnetic field is sufficiently high to flip-over the spins from the ground antiferromagnetic configuration in the normal core.

The spatial distribution of the magnetic field decreasing from the center of the vortex induces then various magnetic phases. If the spins in the core are in the paramagnetic (P) configuration the structure is most complex and consists of three coaxial domains in which paramagnetic, spin-flop (SF) and antiferromagnetic (A) (A) phases coexist with superconductivity.

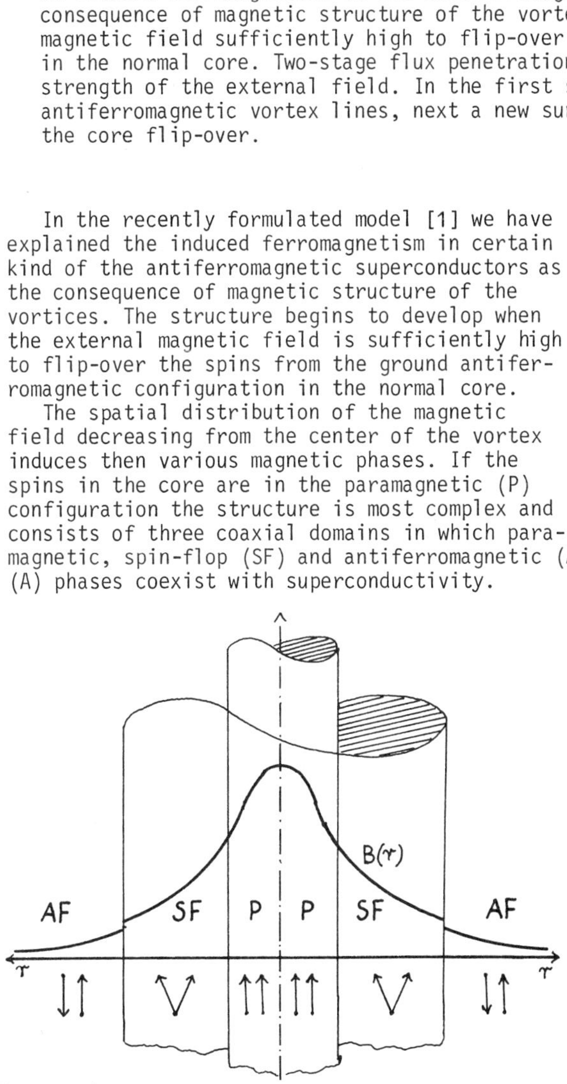

FIGURE 1
The magnetic structure of the isolated vortex line and the magnetic induction distribution in the appropriate magnetic phase.

There are, at least, two different types of vortex lines in the model depending on the magnetic field intensity in the core, higher or lower than H_T (flip-over field). Thus, two-stage flux penetration process has been postulated. First, when the magnetic field intensity in the core $H < H_T$ the magnetic flux penetrates in the form of entirely antiferromagnetic vortices. Next, when $H = H_T$ a new surface energy barrier develops strongly influencing flux penetration. Above the second entry field H_{en} the flux again starts to penetrate but in the form of the vortices with magnetic structure. Without loss of generality we have considered simplified model, we believe most probable in $DyMo_6S_8$, in which induced ferromagnetism is produced by the (SF) phase alone (simulated by the phase with constant magnetization (P) phase). In this case the vortex core is the axis of the cylindrical (P) domain of radius r_o and the remaining part of the vortex is in the (A) phase. Our starting point is the Ginzburg-Landau theory modified by the presence of antiferromagnetic order with the single ion anisotropy [2]. In this framework we have obtained [3] the formulae for distribution of the magnetic induction of the single vortex which have the following form

$$b_A = \mu_o H_T K_o(r_o\lambda_P^{-1}) K_o(r_o\lambda_P^{-1}) \quad , \tag{1}$$

$$\begin{aligned} b_P = {} & [1-I_o(r_o\lambda_P^{-1})]^{-1}\{[r_o\lambda_P^{-1}B_T I_1(r_o\lambda_P^{-1}) - \\ & - \varphi_o(2\pi\lambda_P^2)^{-1}\, I_o(r_o\lambda_P^{-1}]\, K_o(r_o\lambda_P^{-1}) + \\ & + [\varphi_o(2\pi\lambda_P^2)^{-1} K_o(r_o\lambda_P^{-1}) + \\ & + B_T r_o\lambda_P^{-1} K_1(r_o\lambda_P^{-1}) - B_T]\, I_o(r_o\lambda_P^{-1})\} \quad , \end{aligned} \tag{2}$$

where λ denotes the penetration depth in the appropriate phase, $B_T = \mu_o H_T + 2M_o$ (M_o - magnetization of the sublattice), φ_o - the flux quantum, K_o, K_1, I_o, I_1 - the modified Bessel functions.

In order to obtain critical entry field H_{en} we have to consider the Gibbs free energy G of the vortex line assumed to be crossing (y - x) plane at $x = x_v$ near the planar surface (y - z) of the superconductor. The external magnetic field H_o and the vortex line are parallel to the z-direction. The slope $\partial G/\partial x$ at $x_v = r_o$ becomes zero for $H_o = H_{en}$. the boundary conditions at the surface require the normal component of the persistent current to be zero and magnetic field field H to be continuous. We fulfil these conditions by adding the image line (with the same magnetic structure) of the opposite sign at $x = -x_v$. In this way we may find the magnetic field distribution of the vortex near the surface of the superconductor

$$B_P = b_P(x - x_v) - b_A(x + x_v) + \mu_o H_o \exp(-x_v/\lambda_A), \quad (3)$$

$$B_A = b_A(x - x_v) - b_A(x + x_v) + \mu_o H_o \exp(-x_v/\lambda_A). \quad (4)$$

The last term describes penetration of the external magnetic field in the absence of the vortex. The Gibbs free energy in the London approximation has the following form

$$2\mu_o G = \lambda_P^2 \int dS_1 [(B_P - 2M_o - 2\mu_o H_o) \times \mathrm{rot}(B_P - 2M_o)] + \lambda_A^2 \int d(S_2 + S_3)\ [(B_A - 2\mu_o H_o) \times \mathrm{rot}\ B_A] \quad , (5)$$

where S_1, S_2, S_3 denote the surface of the (P) and (A) phases of the vortex and the planar surface of the superconductor respectively. Performing calculations in the spirit of de Gennes calculations [4] and neglecting terms of the order of r_o/λ we may find that

$$H_{en} = H_T \lambda_A [2 r_o \ln(\lambda_A r_o^{-1})]^{-1} \{(\varphi_o - \pi \xi_P^2 B_T)(4\pi\lambda_A^2 \mu_o H_T)^{-1} \ln(\lambda_A r_o^{-1}) + 1\} \quad (6)$$

where ξ denotes coherence length.

We can use the above formula to describe the process of flux penetration into $DyMo_6S_8$ [5] which is the magnetic superconductor developing induced ferromagnetism in the fields higher than 200 (Oe). Simple estimation shows that for this compound an additional surface energy barrier should occur in the range between 200 and approximately 260 (Oe) = H_{en} .

References

[1] T. Krzysztoń, J. Magn. Magn. Mat. 15-18 (1980) 1572.
[2] T. Krzysztoń, G. Kozłowski and P. Tekiel, Acta Phys. Polon. A56 (1979) 49.
[3] G. Kozłowski, T. Krzysztoń and P. Tekiel, phys. stat. sol.(b) 102 (1980) K23.
[4] P.G. de Gennes, Superconductivity in metals and alloys. W. Benjamin, New York, 1966.
[5] W. Thomlinson, G. Shirane, D.E. Moncton, M. Ishikawa and Ø. Fischer, J. Appl. Phys. 50 (1979) 1981; Phys. Rev. B23 (1981)4455. M. Ishikawa and J. Muller, Solid State Commun. 27 (1978) 761.

LT-17 (Contributed Papers)
U. Eckern, A. Schmid, W. Weber, H. Wühl (eds)

MAGNETIC FLUX DIFFUSION INTO LONG TYPE-I SUPERCONDUCTING CYLINDERS: AN EXPERIMENTAL STUDY

K. ASWATHY, G. RANGARAJAN, R. SRINIVASAN

Low Temperature Laboratory, Department of Physics, Indian Institute of Technology, Madras, India

B.K. MUKHERJEE

Department of Physics, Royal Military College of Canada, Kingston, Ontario, Canada, K7L 2W3

We have measured the total time required for the penetration of magnetic flux into a cylindrical specimen of pure indium, a Type-I superconductor, when an overcritical magnetic field is suddenly applied. Whereas previously reported measurements covered small excess fields, our measurements extend over magnetic fields up to 15 times the critical field and are in reasonably good agreement with the generalized theory recently proposed by Gauthier and Rochon.

When an axial magnetic field H_a greater than the critical field H_c is suddenly applied to a long cylindrical superconductor (of negligible demagnetization coefficient), it takes a finite time τ for flux to penetrate the cylinder and for the flux density to reach the applied value throughout the specimen. Pippard(1) and, independently, Lifshitz(2) presented the first theoretical analysis of this method of destroying superconductivity in cylinders. They assumed that: (i) the rate of inward collapse of the superconducting-normal boundary would be determined by electromagnetic damping, (ii) the magnetic field on the boundary would always have the critical value H_c, and (iii) the process would end when superconductivity was completely destroyed and the field has reached the critical value on the axis of the wire. They found that the total time τ required would be $\tau = \mu_0\sigma a^2/4p$ (in S.I. units) where a is the radius of the specimen, σ is its conductivity in the normal state and p is the reduced field: $p = (H_a - H_c)/H_c$. Faber(3) showed that for higher values of p a better approximation would be:

$$\tau = \frac{\mu_0\sigma a^2}{4} \cdot \frac{(1+p)}{p} \qquad (1)$$

Faber(3) measured τ in tin cylinders for values of p up to about 0.2 and Ittner(4) determined τ in hollow tantalum cylinders for values of p up to 5. They found that except for very small p, experimental values of τ differed considerably from the values given by expression (1) and they attempted to explain the discrepancies on the basis of thermal effects. Several attempts have since been made to find a more complete theoretical solution (5,6). Recently Gauthier and Rochon(7) have suggested that for moderate to large values of p it is important to consider a second stage of flux penetration which begins when the field on the axis of the cylinder reaches H_c (i.e. at the point where the Pippard-Faber calculation stops) and continues until the field throughout the cylinder equals the applied field H_a. They have given a complete theory of the transition and have calculated τ as the sum of τ_1 and τ_2 the times required for the two stages of the penetration process. Gauthier and Rochon have also calculated the rate of flux entry into the specimen as a function time, $\phi(t)$.

We are carrying out an experimental investigation to measure τ and $\phi(t)$ in a range of pure and impure indium and tin specimens for values of p up to 15. In this preliminary report we present our measurements for a 99.9999% pure indium specimen.

Our specimens consist of vacuum cast cylindrical rods of 30 mm length and 3 mm diameter. A search coil consisting of 250 turns of SWG46 copper wire is wound carefully around the middle of the specimen along a 6 mm length. The specimen is placed along the axis of a solenoid of 140 mm length and 7 mm inside diameter which produces the magnetic field H_a with a rise time of 10 milliseconds. The entire assembly is mounted in a liquid helium bath whose temperature can be controlled by the usual method of varying the vapour pressure by means of a rotary pump and a needle valve. The temperature of the bath is measured by a germanium resistance thermometer mounted near the specimen. When a current is suddenly switched on in the solenoid to produce a field greater than H_c, flux penetrates the superconductor and a voltage is induced in the search coil. A Siemens electronic fluxmeter integrates the voltage to give the flux ϕ which is recorded as a function of time in a storage oscilloscope. τ is then the time when the saturation value of flux is reached. Due to the presence of noise and drift, it is difficult to precisely determine the instant when the flux is saturated. Therefore the measurement is repeated a number of times with up to 150 data points being taken for a given specimen at any temperature; an average $\tau(p)$ curve may then be drawn as shown in Figure 1 for our pure indium specimen at 2.8 K.

The conductivity of the specimen has been measured by the eddy current decay method of Bean et al.(8) and found to be

$$\sigma = (1.5 \pm 0.15) \times 10^{11} \text{ S/m} .$$

This value of σ has been used to define the reduced transition time $\tau_r = \tau/(\mu_0 \sigma a^2)$ for the purpose of comparison with theory in Figure 2. In this figure the broken line is a curve of expression (1) whereas the full line represents the time required for the penetration of 98% of the magnetic flux according to the Gauthier-Rochon theory - this curve is comparable but not identical to curve e in Figure 1 of reference (7) which corresponds to 95% of flux penetration. We note that the curve based on the Gauthier-Rochon model lies very close to our experimental values even for large p and it is clear that electromagnetic damping represents by far the most important factor in the flux diffusion process. The small remaining discrepancy may be due to thermal effects as discussed by Faber(3) and Ittner(4). We are extending our measurements to specimens of lower conductivities in which thermal effects are likely to be more important. We shall also be reporting later on the agreement between the experimental and theoretical curves of $\phi(t)$.

We are grateful to Dr. P. Rochon for providing us with the theoretical curve for 98% flux penetration and we wish to acknowledge the useful discussions with Dr. N. Gauthier. BKM wishes to thank Professor R. Srinivasan and Dr. G. Rangarajan for their kind hospitality during a sabbatical year spent at the Low Temperature Laboratory, I.I.T., Madras during which this work was initiated.

REFERENCES

(1) A.B. Pippard, Phil. Mag 41 (1950) 243.
(2) E.M. Lifshitz, J. Exp. Theor. Phys. USSR 9 (1950) 834.
(3) T.E. Faber, Proc. Roy. Soc. A219 (1953) 75.
(4) W.B. Ittner, Phys. Rev. 111 (1958) 1483.
(5) J.B. Keller, Phys. Rev. 111 (1958) 1497.
(6) J.C. Swihart, J. App. Phys. 34 (1963) 851.
(7) N. Gauthier and P. Rochon, "Generalized Theory of Magnetic Field Diffusion in Superconducting Cylinders", this volume.
(8) C.P. Bean, R.W. DeBlois and L.B. Nesbitt, J. App. Phys. 30 (1959) 1976.

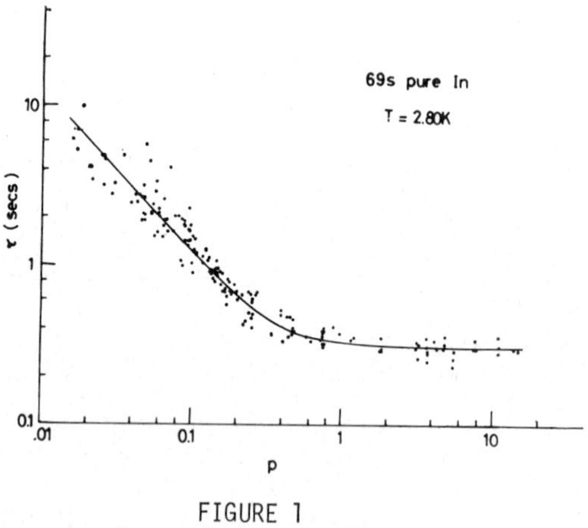

FIGURE 1
Measured values of the transition time τ plotted as a function of the excess field parameter p for our pure indium specimen at T = 2.80 K.

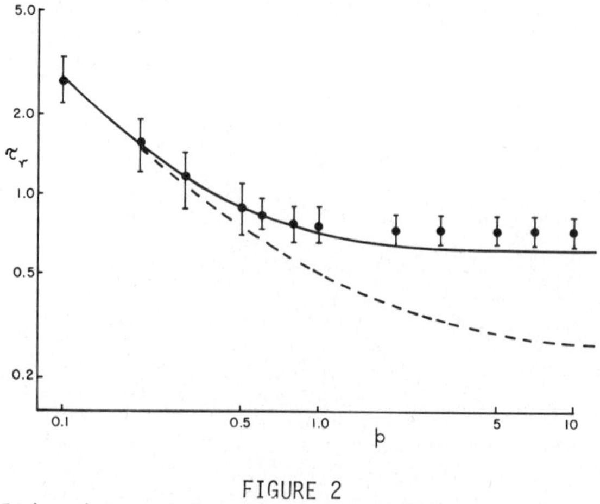

FIGURE 2
Reduced transition time τ_r as a function of p: experimental points for our indium specimen at 2.8 K compared with curves of expression (1) (broken line) and the Gauthier-Rochon $\tau_r(0.98)$ (full line).

LT-17 (Contributed Papers)
U. Eckern, A. Schmid, W. Weber, H. Wühl (eds)

NONHOMOGENEOUS MIXED STATE IN TYPE II SUPERCONDUCTORS

Przemysław TEKIEL

Institute for Low Temperature and Structure Research, Polish Academy of Sciences,
ul Próchnika 95, 53-529 Wrocław, Poland

The characteristic length $\tilde{\delta}$ of the mixed state in type II superconductors is related to several properties of the mixed state. The relation between the Campbell pinning length and $\tilde{\delta}$ is given. The critical state equation for simple pinning system is obtained and the surface impedance of the mixed state is derived.

When a macroscopic current is flowing through the mixed state of type II superconductors then the magnetic induction vector is a function of coordinates. In the longitudinal geometry (current direction parallel to magnetic field direction) the transport current penetrates into homogeneous mixed state of an ideal superconductor with a characteristic depth $\tilde{\delta} = \lambda[H/(H - B)]^{1/2}$ (see (1)), where H is the thermodynamic magnetic field and B is the magnetic induction. λ is the London penetration depth. $\bar{B}$ changes its direction only within the layer of thickness of $\tilde{\delta}$ near the surface and it is constant far away from the surface i.e. the mixed state is homogeneous there. The length $\tilde{\delta}$ is much larger than λ for considered range of magnetic field H (or **B**) values ($H_{c1} \ll H \ll H_{c2}$). H_{c1} and H_{c2} are the lower and upper critical fields, respectively. When the macroscopic current direction is perpendicular to $\bar{B}$ direction then the mixed state cannot be in an equilibrium state without an external barrier. In an ideal superconductor the surface barrier exists. Due to this barrier the sufficiently small current can flow in layers of thickness of $\tilde{\delta}$ near the surfaces and the system can be in a state of metastable equilibrium (see (2)). The length $\tilde{\delta}$ charactierizes the space dimension of a small perturbation of mixed state homogeneity. In the following we show that the existence of this length influences various properties of the mixed state.

Let's consider a long thick cylindrical sample of a type II superconductor with Ginzburg--Landau parameter κ much larger than unity. The sample is placed in the magnetic field H_e ($H_{c1} \ll H \ll H_{c2}$) directed parallel to the sample axis. The radius of the cylinder is assumed to be so large that all averaged quantities depend only on x (we introduce the frame of references with the z-axis parallel to the sample axis and with the x-axis perpendicular to the surface). We assume that inside the sample pinning centres are present. The centres are the planar voids parallel to the sample surface. The distance between neighbouring void's edges are sufficiently small so that we can replace this system by a system of slabs separated by very narrow vacuum gaps (see (3)). The slab thickness - 2ℓ we assume to be much larger than λ to apply the macroscopic description. Because of surface barriers the values of magnetic field on both sides of i- slab are different. This situation is equivalent to that in (2) i.e. the slab placed in external field H_o parallel to the surface and with current I (per unit length in z-direction) flowing in y- direction. Then we can calculate the free energy density in i-subsystem (slab). Within the framework of the London theory we have

$$F = F_o + \frac{\lambda^2}{8\pi}\,\bar{B}\cdot\mathrm{rot}\,\mathrm{rot}\,\bar{B} \quad , \qquad (1)$$

where F_o is the value of F for homogeneous mixed state (see (4)), but with $B = B(x)$. The expression (1) has been obtained under assumption that there exists the characteristic length $\tilde{\delta} \gg \lambda$.

The total thermodynamic potential $\tilde{F}_i$, of our subsystem has the form

$$\tilde{F}_i = \int dv\left[\frac{\lambda^2}{8\pi}(\mathrm{rot}\,\bar{B})^2 - \bar{B}\bar{M}\right] - \int \bar{P}\cdot d\bar{s} \quad , \qquad (2)$$

where $|4\pi\bar{M}| = |\bar{B} - \bar{H}| \approx \frac{H_{c1}}{2\ln\kappa}\ln\frac{H_{c2}}{B}$ and $\bar{P} = \frac{\lambda^2}{8\pi}\bar{B}\times\mathrm{rot}\,\bar{B}$.

Integration in the last term of (2) is performed over the slab surface. The condition of total magnetization conservation can be written as follows

$$\int (M - M_o)\,dv = \mathrm{const} \quad , \qquad (3)$$

where the index - o denotes the corresponding value for the case I = 0. Minimizing (2) with condition (3) we obtain

$$\tilde{\delta}^2 \cdot \frac{d^2}{dx^2} \Delta B - \Delta B = 0 \quad , \tag{4}$$

where $\Delta B = B - B_0$. The solution of this equation has the form

$$\Delta B = \Delta B_m \frac{sh(x/\tilde{\delta})}{sh(\ell/\tilde{\delta})} \quad , \tag{5}$$

where $|\Delta B_m| = 2\pi I/c$. In order to determine I_{max} - the maximum value of I we take into consideration the surface barrier (see (5)) which extends over distance of order of λ. The order of magnitude of the barrier against the flux exit is equal to $H_{c1}|M_0|\lambda/\ln\kappa$ (in terms of density of $\tilde{F}$). Comparing this value with P and taking into account (5) we obtain

$$\frac{\lambda \cdot B_0 \cdot \Delta B_{max}}{8\pi \cdot \tilde{\delta} \cdot th(\ell/\tilde{\delta})} = \frac{H_{c1}|M_0|}{\ln\kappa} \quad . \tag{6}$$

Defining $J = I_{max}/2\ell$ we obtain from (6) the critical state equation. For the case $2\ell >> \tilde{\delta}$ we have

$$J_c = \frac{cH_{c1}(4\pi|M|)^{1/2}}{2\pi\ell \cdot B^{1/2}\ln\kappa} \quad . \tag{7}$$

For the case $\lambda << 2\ell << \tilde{\delta}$ we have

$$J_c = \frac{2c \cdot H_{c1} \cdot |M|}{\lambda \cdot B \ln\kappa} [1 - 4\pi|M|\ell^2/3B \cdot \lambda^2] \quad . \tag{8}$$

It is equivalent to $J_c \sim [B + B_1]^{-1}$, where

$$B_1 = \left[H_{c1}\ell^2 \cdot \ln\left(\frac{H_{c2}}{B}\right)\right]/6 \cdot \lambda^2 \cdot \ln\kappa \quad . \tag{9}$$

When a sufficiently small ac field is superimposed on the field H_e then the flux motion is reversible (see (6)). From our point of view it means that the state of the system at each moment is determined by minimum of $\Delta\tilde{F}$ - the deviation of the potential $\tilde{F}$ from its value for the system without ac field. This potential is the sum of $\tilde{F}_i$ - the potential of i- subsystem, which is limited by neighbouring nets of pinning centres. In order to calculate $\tilde{F}_i$ in first approximation we can use Eqs. (2) and (5), where ΔB denotes the deviation of B from the mean value B_i in this system. Denoting $b(i) = B_i - B_0$, where B_0 is the mean value of B in i- sybsystem without ac field and assuming 2ℓ to be much less than the characteristic length of the problem we can replace the sum of $\tilde{F}_i$ by an integral. Then the deviation $\Delta\tilde{F}$ becomes the function of $b(x)$. Minimizing $\Delta\tilde{F}$ and using the condition of total magnetization conservation we obtain

$$\lambda'^2 \cdot \frac{d^2 b}{dx^2} - b = 0 \quad , \tag{10}$$

where

$$2 \cdot \lambda'^2 = \ell \cdot [\ell \cdot sh^{-2}(\ell/\tilde{\delta}) + \tilde{\delta} \cdot cth(\ell/\tilde{\delta})] \quad . \tag{11}$$

The solution of Eq. (10) has the form

$$b = b(0) \cdot \exp(-x/\lambda') \quad . \tag{12}$$

Our assumption: $2\ell << \lambda'$ is equivalent to $\lambda << 2\ell << \tilde{\delta}$. Using the last condition and Eq. (11) we obtain $\lambda' = \tilde{\delta}$. So, the pinning length λ' under certain conditions is equal to $\tilde{\delta}$. However, the length $\tilde{\delta}$ can reveal itself also in the responce of the mixed state of an ideal superconductor to ac field of a small amplitude. For sufficiently large frequency of ac field the surface impedance of the mixed state is determined by the skin depth connected with the effective conductivity $\tilde{\sigma} = \sigma \cdot H_{c2}/B$, where σ is the normal state conductivity. Let's assume that the frequency fulfils the following relationship

$$\tilde{\delta} << \lambda_{sk} << D \quad , \tag{13}$$

where $\lambda_{sk} \sim [\sigma\omega]^{-1/2}$ is the skin depth and D is the sample thickness. ω is the frequency. Then using the Maxwell equations and the following relationship between macroscopic supercurrent and the ac field: $\tilde{b} + 4\pi c^{-1} \text{ rot } \tilde{j}_s \tilde{\delta}^2 = 0$, for the surface impedance Z we obtain

$$(4\pi)^{-1} c Z = -i c^{-1} \tilde{\delta}\omega + 2\pi c^{-3} \tilde{\delta}^3 \omega^2 \tilde{\sigma} \quad . \tag{14}$$

REFERENCES

(1) P. Tekiel, Phys. Lett. 93A (1983) 499.
(2) P. Tekiel, Phys. Stat. Sol. 121 (1984) K91.
(3) J.R. Clem, Proc. LT13, eds. K.D. Timmerhaus, W.J.O'Sullivan and E.F. Hammel (Plenum Press, New York, 1974) p. 102.
(4) P.de Gennes, Superconductivity of metals and alloys, (Benjamin, New York, 1966).
(5) F.F. Ternovskii and L.N. Shekhata, Sov. Phys. - JETP, 35 (1972) 1202.
(6) R. Meier-Hirmer, H. Kupfer and A.M. Campbell, Physica 107B (1981) 437.

LT-17 (Contributed Papers)
U. Eckern, A. Schmid, W. Weber, H. Wühl (eds)

CHAOTIC BEHAVIOUR OF THE PINNING-VORTEX INTERACTION IN SUPERCONDUCTORS

I. KIRSCHNER and K. MARTINÁS

Department for Low Temperature Physics, Roland Eötvös University, Budapest, Hungary

It is shown that the effective pinning in one dimensional case can be handled in a deterministic way, but the solution of the actual non-linear differential equations leads to a chaotic behaviour of the pinning-vortex interaction. This method provides the concept of the correlation length too, which is in agreement with Larkin-Ovchinnikov theory extrapolated to one dimension.

The effective pinning in type II superconductors is basically a 3-dimensional effect, nevertheless to investigate the fundamental features of it the 1-dimensional modelling also provides some useful informations (1), (2), (3), (4), (5), (6).

Some 1-d calculations, similarly to the 3-d ones, yield the threshold criterion. In an earlier paper (7) we have shown, that there is a possibility for synchronization of the vortex lattice to the pinning system and in case of very dilute pinnings the result is the direct summation. Here we will show, that if the collective pinning is handled as a deterministic system, the position of the nth vortex can not be determined by the position of the first, if n is large enough.

On the other hand this leads to the concept of correlation length, since the position of the nth vortex is independent from the position of the first one although it is affected by the highest pinning force due to the kth pinning centre is the nearest to it.

To prove this statements, at first we show, that the solution is really chaotic and so the information entropy production is higher than zero, after it we determine the correlation length.

It means that the chaotic features of the pinning-vortex interaction follows from the solution of the non-linear differential equation system which describes the position of the vortices.

If the pinning problem is considered as the mapping of unit interval onto itself, then in a 1-d model the equilibrium position of the ith vortex is determined by the equation

$$C\,\frac{r_{i+1}+r_{i-1}-2r_i}{\frac{1}{2}(r_{i+1}-r_{i-1})^2} = f_p(r_i,R_k)\,, \qquad /1/$$

where we have used the first neighbourhood approximation. In this formula C is the elastic constant, r_i is the position of the ith vortex, R_k is the position of the kth pinning centre and f_p is the pinning force acting on the ith flux line

$$f_p(r_i,R_k) = \begin{cases} 0, & \text{if } |r_i-R_k| > a_i/2 \\ f(x_i^k)\,, & \text{if } |r_i-R_k| < a_i/2, \end{cases} \qquad /2/$$

where $a_i = r_{i+1} - r_i$ is the lattice constant.

In this way we introduced the mapping variable x_i^k, which can be changed between 0 and 1:

$$x_i^k = \frac{r_i - R_k + a_i/2}{a_i}\,. \qquad /3/$$

By the help of this the equation /1/ can be written in the form of

$$\delta\left(\frac{1}{a_i^k}\right) = \frac{1}{C}\,f_p\,(x_i^k)\,, \qquad /4/$$

which shows that a change in the lattice constant occurs due to the existence of the pinning forces.

On the base of this considerations the position of the ith vortex is

$$r_i = r_{i-1} + a_{i-1} - \delta\left(\frac{1}{a_{i-1}}\right) a_{i-1}^2 \qquad /5/$$

and so the mapping parameter will be

$$x_i^k = x_j^{k-1} + (i-j)\,a_{j-1} - (i-j)\,\delta\left(\frac{1}{a_j}\right) a_j^2 - (R_k - R_{k-1})\,, \qquad /6/$$

where j is the index of the vortex bound by (k-1)th pinning centre and $(R_k - R_{k-1})$ is the distance between (k-1)th and kth pinning centres, which is a random quantity but fixed to the working sample.

Equation /6/ is a map of the unit interval onto itself, which can be characterized by the information parameter λ measuring the average change of the information over the entire interval 0-1:

$$\lambda = \int_0^1 P(x)\,\Delta H(x)\,dx, \qquad /7/$$

where $\Delta H = \log Q_f/Q_i$ is the logarithm of the ratio of states distiguishable before and after certain time interval and P (x) is the probability density (8).

If $\lambda > 0$, then after some iteration steps N, there is a possibility for the complete loss of the initinal information H_{ini}, where

$$N = \frac{H_{ini}}{\lambda}. \qquad /8/$$

The calculation shows, that in a linear approximation

$$\lambda \approx \frac{1}{C^2}\left(\frac{R}{a_i}\right)^2\left(\frac{df_p}{dx}\right)^2 > 0, \qquad /9/$$

which provides the full loss of information after N steps ($\sim 1/\lambda$).

The physical meaning of this result is that the position of the jth vortex being bound by kth pinning centre is independent from the 1st.

In the case of pinning problem it means that after N pinning centres the vortices can be handled as independent ones, beacuse the position of the first vortex does not restrict the position of the vortex being at the Nth pinning centre and so it results in a natural correlation length

$$L = N\,R, \qquad /10/$$

where R is the average distance between the pinning centres.

The direct suming up of the information loss for several steps yields the following expression for the positional uncertainty of the ith vortex:

$$\Delta x_i = \Delta x_o + \sum_{j<i} (i-j)\frac{1}{C}\frac{df_p}{dx}\Delta x_j. \qquad /11/$$

Calculating the average response due to the displacement of the first vortex

$$(\Delta x_i)^2 = (\Delta x_o)^2 + \sum_{j<i}\frac{(i-j)^2}{C^2}\left(\frac{\overline{df_p}}{dx}\right)^2(\Delta x_j)^2 \qquad /12/$$

and taking into account that $\frac{df_p}{dx} \neq 0$ only in the vicinity of the pinning centres, this results in

$$(\Delta x_i)^2 \gtrsim (\Delta x_o)^2 + \frac{1}{6C^2}N_i(N_i+1)(2N_i+1)\frac{df_p}{dx}R^2(\Delta x_o)^2, \qquad /13/$$

where $N_i = \frac{ia}{R}$ the number of pinning centres between the first and the ith vortices.

If $\Delta x_i > 1$ (in the a_i measuring units), then the position of the ith vortex is independent from that of the first one, and in this way the system is chaotic.

This can be interpreted so, that the vortices feel each other inside a distance of order

$$L = ia, \qquad /14/$$

which is just the correlation length, where a is the average lattice constant.

Although the exact value of the correlation length can be obtained as a function of Δx_o, nevertheless it really depends on the physical parameter of the material in the form of

$$L \approx \left(\frac{f_p}{C}\right)^{-2/3} R^{1/3}. \qquad /15/$$

This result agrees with the Larkin-Ovchinnikov formula (9) corresponding to one dimensional situation.

(1) A.M.Campbell, Phil.Mag.37B /1978/ 149,
(2) K.Martinás,Phys.Lett. 82A /1981/ 369,
(3) I.Kirschner and K.Martinás,Journ.Low Temp.Phys. 47 /1982/ 105,
(4) S.Takács, Acta Phys.Hung.53 /1982/337,
(5) E.H.Brandt, Journ.Low Temp.Phys., 53 /1983/ 41,
(6) T.Matsushita and K.Yamafuji, Journ. Phys. Soc. Japan, 48 /1980/ 1885,
(7) I.Kirschner and K.Martinás, Acta Phys. Hung. 53 /1982/ 347,
(8) V.I.Oseledec, Trans.Mosc.Math.Soc. 19 /1968/ 197,
(9) A.I.Larkin and Yu.N.Ovchinnikov,Journ. Low Temp. Phys. 34 /1979/ 409.

LT-17 (Contributed Papers)
U. Eckern, A. Schmid, W. Weber, H. Wühl (eds)

THE VORTEX LATTICE MELTING PHENOMENON IN DUAL SUPERCONDUCTING FILMS

L.I. GLAZMAN* and N.Ya. FOGEL' **

*Kharkov State University, Kharkov, USSR;
**Institute for Low Temperature Physics and Engineering, UkrSSR Academy of Sciences, Kharkov, USSR

The variations in current-voltage characteristics of a superconducting transformer caused by a two-dimensional melting of the vortex lattice are studied. The melting produces a considerable suppressing of the vortex drag effect and a vanishing of the square-root singularity in I-V characteristics which is typical of a dc transformer under usual conditions.

The vortex lattice of a thin superconducting film can undergo a two - dimensional Kosterlitz-Thouless-type melting transition (1,2). When the melting occurs in a superconducting transformer (a sandwich of two superconducting films separated by the dielectric interlayer that eliminates any tunnelling of Cooper's pairs), the I-V characteristic of such a structure changes radically. These changes can be of great aid both in detecting the effect of two-dimensional melting and in studying the critical behaviour in the vicinity of melting point T_M.

We consider the effect of melting in one of the films and in both the films. When describing the effect of vortex drag due to the magnetic interaction between vortices in different films, the vortex motion in the melted lattice is treated in terms of the diffusion approach. In the vortex liquid, thermal fluctuations are essential. Under these conditions a single vortex can surmount the periodic potential associated with the unmelted vortex lattice in the primary film due to the thermal fluctuations. The vortex slippage results in suppressing the drag effect and in varying the dependence of secondary voltage V_2 on current I_1, passing through the primary film, for any value of I_1. After obtaining the average velocity of a vortex interacting with the moving unmelted primary lattice, the following relation for the secondary voltage V_2 can be derived:

$$V_2 = \frac{1}{2}\left(\frac{u_0}{2T}\right)^2 \left[\left(\frac{\sqrt{3}\eta_2 a_0^3 c V_1}{4\pi T \Phi_0 L_1}\right)^2 + 1\right]^{-1} V_1 \qquad (1)$$

Here $V = LH\Phi_0 I_1/\eta_1 c^2 w$; $\eta_{1,2}$ are the viscosity coefficients for vortices in each of the films; $a = (\Phi_0/H)^{1/2}(4/3)^{1/4}$; c is the velocity of light; L and w are the length and width of the films, respectively; H is the magnetic field; Φ_0 is the magnetic flux quantum; u_0 is the barrier separating two equivalent positions of the same vortex in the primary periodic potential. It is easy to verify that the condition $u_0 \ll T$ is valid.

The I-V characteristics of a superconducting transformer are sketched in the figure. The square-root singularity is typical of the unmelted vortex lattices in both films (3) (curve 1). The singularity is smeared after the melting transition (curve 2). Curve 3 corresponds to the completely melted secondary lattice.

In the vicinity of T_M, the lattice breaks down into separate blocks which retain the crystalline order. The block

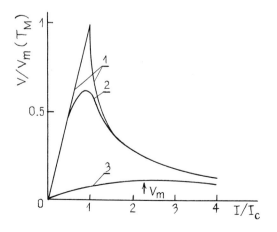

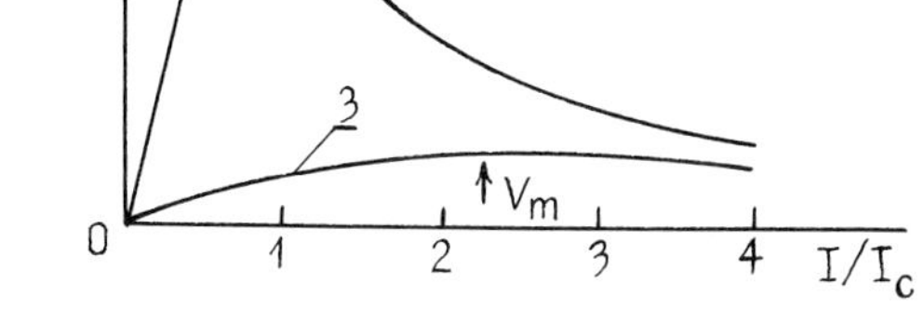

size is given by the correlation length (4):

$$\xi(T) = a_0 \exp\left\{ b \left[\frac{(T_c - T_M)}{(T - T_M)} \frac{T_M}{T_c} \right]^{\nu} \right\}$$

where $\nu = 0.37$, $b \sim 1$, T_c is the superconducting transition temperature. For small ξ, i.e. for $1 \ll N \ll T/u_0$ ($N \sim (\xi/a)^2$ is the number of vortices in a block), the estimation of V_2 results in Eq.(1) with the substitutions $u_0 \rightarrow Nu_0$ and $\eta_2 \rightarrow N\eta_2$. The square-root singularity of I-V characteristics vanishes even in the critical temperature range $T \gtrsim T_M$. For $T > T_M$, the maximum $V_m(T)$ value of voltage V_2 is diminished in comparison with the $V_m(T_M)$ value:

$$V_m(T_M) - V_m(T) \sim (T_M/u_0)^{1/3} [\xi(T)/a_0]^{1/3} V_m(T_M).$$

For the case of vortex lattice melting in both films, the solution can be obtained in the pair correlation approximation for vortex positions. The dependence $V_2(V_1)$ is in qualitative agreement with the law given by Eq.(1). In the linear portion of the curve $V_2(V_1)$, the ratio V_2/V_1 for both melted lattices appears to be smaller than that for a solid primary lattice. Nevertheless, in both the cases $V_2 \propto H$. The dependences $V_2(V_1)$ and $V_2(H)$ obtained in this communication are in good agreement with the experimental data on aluminium superconducting transformer (5); this permits the experimental data to be described with taking the effect of vortex lattice melting into account.

REFERENCES

(1) D.S. Fisher, Phys. Rev. B22 (1980) 1190.
(2) B.A. Huberman and S. Doniach, Phys. Rev. Lett. 43 (1979) 950.
(3) J.R. Clem, Phys. Rev. B19 (1974) 898.
(4) D.R. Nelson and B.I. Halperin, Phys. Rev. B19 (1979) 2457.
(5) V.Yu. Tarenkov, A.I.D'yachenko, V.V. Stupakov, Fiz. Tverd. Tela, 24 (1982) 2569.

LT-17 (Contributed Papers)
U. Eckern, A. Schmid, W. Weber, H. Wühl (eds)

SOFT MODES IN Nb_3Ga

L. PINTSCHOVIUS, W. REICHARDT, E. AKER[+], C. POLITIS and H.G. SMITH*

Kernforschungszentrum Karlsruhe, Inst. f. Nukl. Festkörperphysik, P.O.B. 3640, Karlsruhe, FRG
*Oak Ridge National Laboratory, Oak Ridge, Tennessee 37830, USA

Selected phonon branches of non-stoichiometric Nb_3Ga have been investigated at 295 K and 90 K to see if there is a tendency towards a structural instability like in Nb_3Sn and V_3Si. Indeed such a tendency has been found, but not towards a cubic-to-tetragonal phase transition. In contrast to Nb_3Sn and V_3Si, the long-wave length acoustic phonons show very little variation with temperature, whereas a pronounced softening has been observed for those modes which correspond to a "buckling"-type motion of the Nb chains.

It has been known for many years that the high T_c A15 compounds Nb_3Sn and V_3Si undergo a cubic-to-tetragonal phase transition at 45 K and 21 K, respectively. The question arises if all A15 compounds with high T_c are structurally unstable at low temperatures. There is no evidence from structural investigations in favor of this hypothesis, but precursor effects of a structural phase transition, i.e. a phonon softening, have indeed been observed in temperature dependent measurements of the phonon density of states for all high T_c A15 compounds so far investigated (1,2). Little has been known in this respect for Nb_3Ga, which has one of the highest T_c's of the A15 family.

We investigated the temperature dependence of selected phonon branches in non-stoichiometric Nb_3Ga (T_c=12 K) by inelastic neutron scattering. The sample consisted of 5 co-aligned very small single crystals with a total volume of 0.05 cm³. The measurements were performed at the reactors FR2 (Karlsruhe), HFIR (Oak Ridge) and ORPHEE (Saclay).

Measurements of the phonon density of states on a polycrystalline sample of the same composition as the single-crystals revealed that there occurs a phonon softening at low temperatures like in other A15 superconductors (see Fig. 1). The effect is less pronounced than in Nb_3Sn (1), which is not astonishing in view of the moderate T_c of our sample. The softening extends down to the lowest frequencies accessible by this measurement, which suggests that the acoustic branches are involved, as in the case of Nb_3Sn and V_3Si (3-5). However, the measurements of the slopes of the acoustic branches on the single crystals did not show a frequency decrease upon cooling to 90 K (see Figs. 2-4). In particular, the TA 110 branch with polarization 110 does not soften at low temperatures, whereas in Nb_3Sn and V_3Si the slope of this branch goes to zero when approaching the phase transition. On the other hand, the TA 111

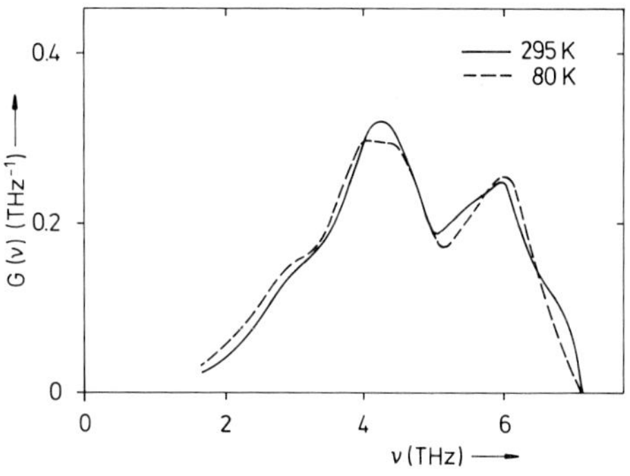

FIGURE 1
Generalized phonon density of states of $Nb_{3.1}Ga_{0.9}$ at 295 K and 80 K.

branch does show a softening, but in contrast to Nb_3Sn the softening is strongest at the zone boundary.

The zone boundary mode is the common end point of the acoustic and an optic branch starting from the $\Gamma_{15'}$ mode. It has been observed that the whole branch decreases in frequency on cooling by about the same amount throughout the Brillouin zone. An analysis of the Eigenvectors reveals that the pattern of vibration is very similar from Γ to R for this branch: The Nb atoms vibrate perpendicular to the chains, whereas the Ga atoms are completely or nearly at rest.

When compared to the low T_c compound Nb_3Sb (T_c=0.2 K) the frequencies of the $\Gamma_{15'}$- and R_{1-2}-modes in $Nb_{3.1}Ga_{0.9}$ are already very low at room temperature, i.e. 3.3 THz and 3.6 THz as against 4.8 THz (6). At low temperatures the differences become even more pronounced, for in Nb_3Sb the temperature dependence is very small. This supports the idea that the low frequency of these modes in Nb_3Ga is mainly due to electron-phonon coupling effects. In the

+Present address: ILL Grenoble, France

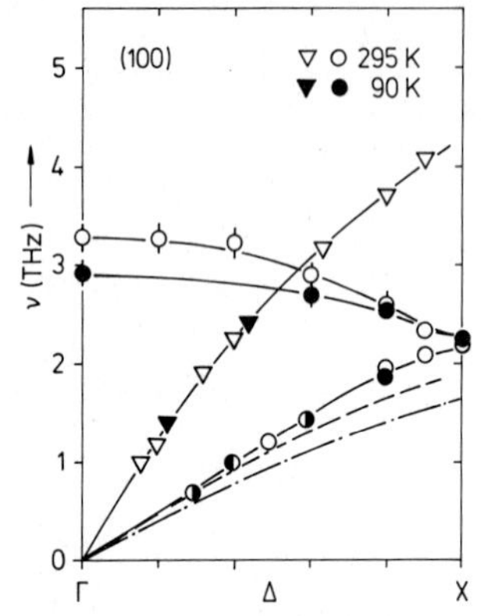

FIGURE 1
Phonon dispersion relations of low frequency modes in $Nb_{3.1}Ga_{0.9}$ at 295 K and 90 K. The dashed and chain-lined curves refer to the TA[100] branch in Nb_3Sn at 295 K and 46 K (after ref. 4).

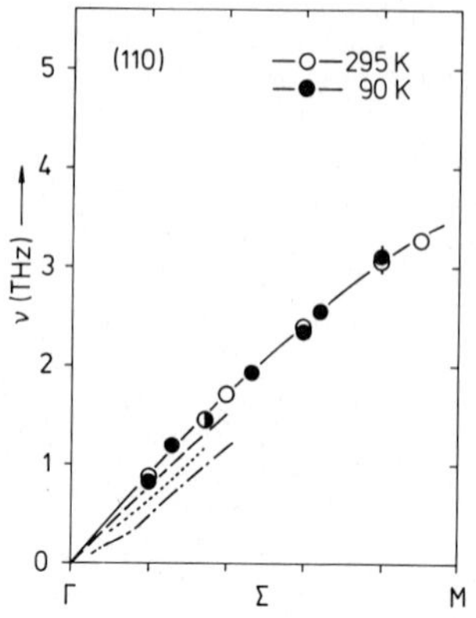

FIGURE 2
Phonon dispersion relation of TA[110] phonons with polarization $[1\bar{1}0]$. The dashed, dotted and chain-lined curves refer to data taken on Nb_3Sn at 295 K, 120 K and 62 K, respectively (after ref. 4).

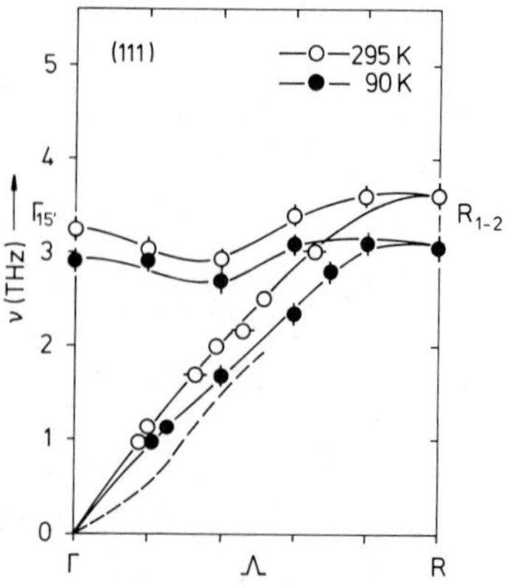

FIGURE 3
Phonon dispersion relations of the two lowest Λ_3 branches at 295 K and 90 K. The dashed lines refer to data measured at 65 K on Nb_3Sn (after ref. 5).

stoichiometric compound which has a much higher T_c (T_c = 21 K) and hence a much stronger electron-phonon coupling, it may well be that the $\Gamma_{15'}$- or the R_{1-2}-frequency goes to zero on cooling, inducing a transition to a phase with "buckled" Nb chains.

Recently Weber (7,8) has predicted the phonons of Nb_3Sn from a microscopic theory. Although the electronic structure of Nb_3Ga and Nb_3Sn is somewhat different, we think that Weber's main result applies also for Nb_3Ga: He reported that the electron-phonon coupling is strong for those modes, where the Nb atoms vibrate in longitudinal or transverse direction against their nearest neighbours within the chains, whereas the electron-phonon coupling is weak for modes which involve a vibration of the Nb sublattice against the Sn sublattice. Our findings are in excellent agreement with Weber's results: Whereas the frequency of the "strongly coupled" modes $\Gamma_{15'}$ and R_{1-2} decreases by 10% between 295 K and 90 K, normal hardening of about 3% has been observed for the "weakly coupled" modes Γ_{15} and R_4.

REFERENCES

(1) B.P. Schweiss, B. Renker, E. Schneider and W. Reichardt, in: Superconductivity in d- and f-Band Metals, ed. D.H. Douglass (New York 1976) p. 189.
(2) P. Müller, N. Nücker, W. Reichardt, A. Müller, in: Superconductivity in d- and f-Band Metals, eds. W. Buckel and W. Weber (Kernforschungszentrum Karlsruhe 1982)p.19.
(3) G. Shirane, J.D. Axe, and R.J. Birgerau, Sol. State Comm. 9, 397 (1971).
(4) J.D. Axe and G. Shirane, Phys. Rev. B 8, (1973) 1965.
(5) G. Shirane and J.D. Axe, Phys. Rev. B 18, (1978) 3742.
(6) L. Pintschovius, H.G. Smith, N. Wakabayashi, W. Reichardt, W. Weber,G.W. Webb and Z. Fisk, Phys. Rev. B 28 (1983) 5866.
(7) W. Weber, in: Superconductivity in d- and f-Band Metals, eds. W. Buckel and W. Weber (Kernforschungszentrum Karlsruhe 1982) p.15.
(8) W. Weber, in: Electronic Structure of Complex Systems, Nato-A.S.I. Series, eds. P. Phariseau and W. Timmerman (Plenum Press, New York) 1984.

LT-17 (Contributed Papers)
U. Eckern, A. Schmid, W. Weber, H. Wühl (eds)
© Elsevier Science Publishers B.V., 1984

ELECTRON PHONON INTERACTION IN V_3Ge

W. REICHARDT[*], H.G. SMITH and Y.K. CHANG

Solid State Division, Oak Ridge National Laboratory, Oak Ridge, Tennessee 37830, USA

[*]Kernforschungszentrum Karlsruhe, Institut für Nukleare Festkörperphysik, Postfach 3640,
D-7500 Karlsruhe, Federal Republic of Germany

We observed anomalous structures in some phonon branches of V_3Ge which are caused by electron phonon interaction. These features suggest a lattice instability with respect to zigzag elongations of the V atoms within the chains.

The influence of electron phonon interaction on the lattice vibrations in Nb_3Sn has recently been studied by Weber (1). He predicted various anomalous features in the phonon dispersion curves.

V_3Ge is an ideal counterpart to Nb_3Sn due to several reasons:
(i) The ratios between the atomic masses of the constituents are very similar. Thus differences in the interatomic forces can be observed directly in the measured dispersion curves without the necessity to use lattice dynamical models.
(ii) The electronic configurations are identical and the electronic densities of states are very similar.
(iii) Whereas in Nb_3Sn the Fermi energy E_F lies very close to a sharp peak in the density of states, the Fermi energy of V_3Ge is shifted by about 5 mRy towards higher energies into a region of lower density of states (2).

Unfortunately V_3Ge is a rather unfavorable candidate for a coherent neutron scattering experiment as V is an almost completely incoherent scatterer. Nevertheless we were able to measure all acoustic and the lowest optic branches in the main symmetry directions. The measurements were performed on the triple axis spectrometer HB2 at the Oak Ridge HFIR. The sample ($T_c \sim 6K$) of about 0.5 cm³ was prepared by zone melting.

Results obtained at 300 K for the $\zeta 00$ and $\zeta\zeta\zeta$ directions are presented in Fig. 1. Some phonon branches exhibit structures which are absent in the low T_c compound Nb_3Sb (3) (T_c = .2 K): A deviation from the sine-like behavior in the optic Δ_5 branch and dips in the Λ_1 and Λ_2 branches. It is tempting to identify the structure in the Λ_1 branch with the dip predicted for Nb_3Sn by Weber's calculation. A recent search for this feature in Nb_3Sn was not successful, because in the experiment the intensity died out halfway to the zone boundary (4). This experimental difficulty did not exist for V_3Ge, because the nearly negligible coherent cross-section of V leads to more favorable interference conditions

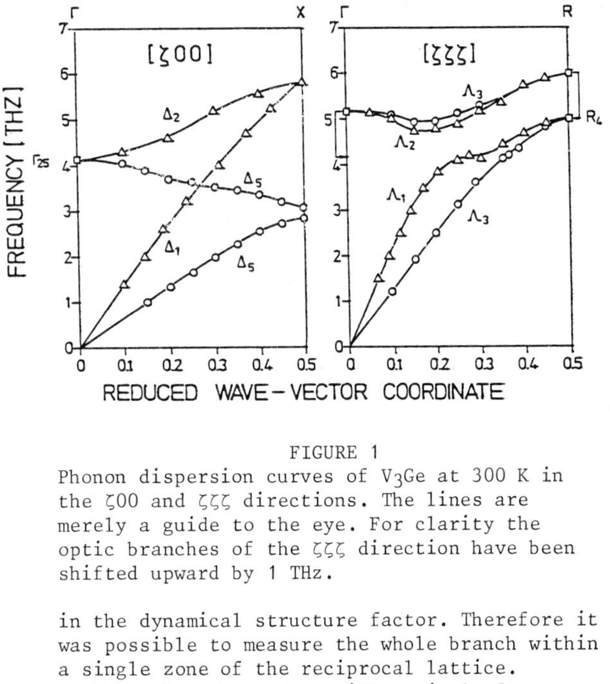

FIGURE 1
Phonon dispersion curves of V_3Ge at 300 K in the $\zeta 00$ and $\zeta\zeta\zeta$ directions. The lines are merely a guide to the eye. For clarity the optic branches of the $\zeta\zeta\zeta$ direction have been shifted upward by 1 THz.

in the dynamical structure factor. Therefore it was possible to measure the whole branch within a single zone of the reciprocal lattice.

We analyzed our data with a simple force-constant model. It turned out that the dip in the Λ_2 branch could only be described satisfactorily by including additional interactions mediated by massless shells at the V positions. The influence of a single negative tangential force-constant between shells of nearest V neighbors on some phonon branches is demonstrated in Fig. 2. Besides reproducing fairly well the dip in the Λ_2 branch it reduces substantially the phonon frequencies of the Λ_1 branch in the region where the dip has been observed. Furthermore the optic Δ_1 branch is changed to a more linear behavior in accordance with the experiment.

Although our model is certainly too simplified for a quantitative description of all de-

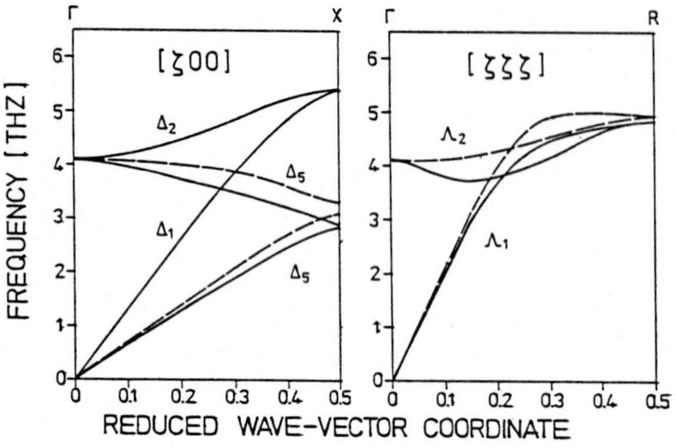

FIGURE 2
Some phonon branches of V_3Ge calculated with the model described in the text (full lines). The dashed curves were obtained without the negative transverse shell-shell force constant.

tails, the analysis suggests, that the observed structures in the dispersion curves of V_3Ge may be traced back to one single origin, an instability of the lattice with respect to zigzag elongations of the V atoms in the chains.

Our r-space analysis is in accordance with Weber's results, which predict in Nb_3Sn two types of renormalization effects due to electron phonon interaction. The anomalous temperature dependence of the Γ_{12} frequency and of the shear elastic constant c' is associated with the existance of a sharp peak in the electronic density of states very close to E_F.

This appears to be a unique feature of Nb_3Sn and possibly V_3Si. The other mechansim, which is of more general nature among the A15 compounds, arises from a large intrachain coupling between $d(\sigma) \equiv (3z^2-r)$ and $d(\pi) \equiv (xz),(yz)$ electrons, due to zigzag displacements of V atoms perpendicular to the chain directions z. It strongly renormalizes the frequencies of the chain buckling modes.

A comparison with recent unpublished results of Nb_3Sn (5) shows that at 300 K the dispersion curves of the two compounds are very similar apart from a scaling factor due to the different atomic masses. This refers also to the anomalous structures in the dispersion curves. From this we must conclude that phonon renormalization by σ-π intrachain coupling does not depend crucially on shifts of E_F relative to the sharp peak in the electronic density of states.

REFERENCES

(1) W. Weber, in "Superconductivity in d- and f-Band Metals 1982" (Eds. W. Buckel and W. Weber), Kernforschungszentrum Karlsruhe, p. 15.
(2) B.M. Klein et al., Phys. Rev. B 18, 6411 (1978).
(3) L. Pintschovius et al., Phys. Rev. B 28, 5866 (1983).
(4) J.D. Axe and G. Shirane, Phys. Rev. B 28, 4829 (1983).
(5) L. Pintschovius et al., to be published.

LT-17 (Contributed Papers)
U. Eckern, A. Schmid, W. Weber, H. Wühl (eds)

ELASTIC CONSTANTS OF A15 - Nb_3Ge

P. MÜLLER (a), U. BUCHENAU (b), N. NÜCKER, B. RENKER (c), A. MÜLLER (d)

(a) Physik-Department E21, TU München, D-8046 Garching, FRG
(b) Institut für Festkörperforschung, KFA Jülich, D-5170 Jülich, FRG
(c) KfK Karlsruhe, Institut für Nukleare Festkörperphysik, D-7500 Karlsruhe, FRG
(d) Research Laboratories, Siemens AG, D-8520 Erlangen, FRG

Phonon frequency distributions and elastic moduli of polycrystalline, high T_c Nb_3Ge were measured by inelastic neutron scattering as a function of temperature. The observed softening of $c' = 1/2(c_{11} - c_{12})$ is less pronounced than in other high T_c A15 compounds.

1. INTRODUCTION

Nb_3Ge is still the superconductor with the highest critical temperature known to date. The strong electron-phonon coupling effects (1) seem to be connected with the extreme instability of this compound at the stoichiometric composition. In an earlier experiment (2) we observed a pronounced softening of the low energy part of the phonon density of states upon cooling. Therefore, more detailed information, especially on the shear mode instability, known from other high T_c A15 materials, is desirable. However, because Nb_3Ge is obtainable only in polycrystalline form, the well established ultrasonic method is not applicable in this situation. With the now available advanced neutron scattering methods we were able to measure the elastic moduli $c' = 1/2\ (c_{11} - c_{12})$ and c_{44} as a function of temperature.

2. METHOD

The (coherent) neutron scattering law $S(Q,\omega)$ (Q : momentum transfer, $\hbar\omega$:energy transfer) can provide information on phonon properties of polycrystals in the following two limiting cases:

- Averaging over a sufficient Q-range at high Q-values leads to the incoherent approximation, which yields phonon frequency distributions (2, 3).
- The inelastic intensity at sufficiently low energy transfers near Debye-Scherrer rings allows the determination of elastic constants (4).

In this case the one-phonon intensity $I_1(Q,\omega)$ normalized to the intensity $I_0(G,0)$ of the neighboring elastic reflection G is given by (4):

$$\frac{I_1(Q,\omega)}{I_0(G,0)} = \frac{kT}{16\pi^3} \int dq\ \sin\theta\ d\theta d\varphi \sum_{n=1}^{3} \frac{(\vec{G}\vec{e}_{qn})^2}{c_n^{\,2}\rho} \frac{R_1}{R_0}$$

($\vec{e}_{qn}$: Eigenvector of the phonon mode n with wave vector q and velocity c_n; ρ : Density; R_1/R_0 : Weighing factor for the two different instrumental resolution functions).

In a plot of this ratio at constant ω vs. Q, intensity maxima at the different G's are observed, since the main contribution results from the transversal phonons. According to $\omega = c_n q$, these structures are broadened at higher $\hbar\omega$ (c.f. FIGURE 1). Softening of certain modes shows up as intensity increase and change of the width of these phonon Debye-Scherrer rings.

3. EXPERIMENTS

The Nb_3Ge sample (80g, $T_c \approx 21K$) was prepared by chemical vapor deposition onto Hastelloy substrates (5). After removing of the substrate, the sample was characterized by X-ray and neutron diffraction. Details of sample characteristics are given elsewhere (2, 5). Down to 20K, no martensitic transformation with a change of lattice spacing d greater than $\Delta d/d = 7\cdot 10^{-4}$ was observed (2). Inelastic neutron scattering experiments were carried out at the time of flight spectrometer IN4; Institut Laue-Langevin, Grenoble, with 9.05meV and 39.4 meV neutrons at 300K, 100K and 20K. Special care had to be taken to suppress any intensity contributions from spurious reflections of the monochromators and from scattering from the cryostat shieldings.

4. RESULTS

The measured phonon spectra were similar to those reported from an earlier experiment (2). Figure 1 shows the structures in the inelastic intensity arround the (200), (210) and (211) reflections at different energy transfers. The full line indicates a least square fit to the data points using the formula given above. At lower temperatures no pronounced softening was observed. Therefore, because of the small thermal occupation factor, the low temperature

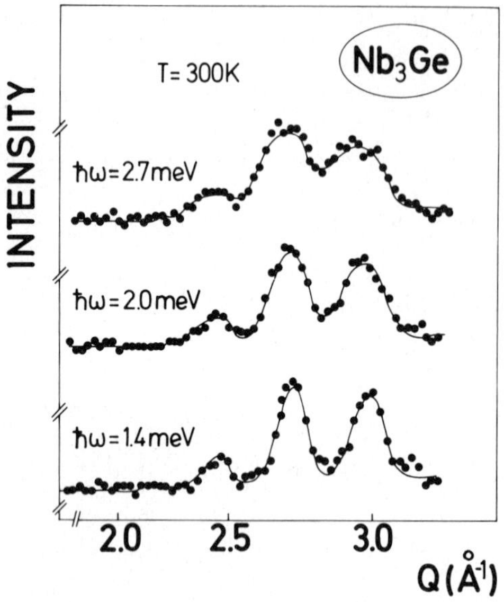

FIGURE 1
Inelastic intensity (arbitrary units) at different constant energy transfers vs. momentum transfer near the (200), (210) and (211) elastic reflections. Elastic constants were determined from a least square fit (full line, see text).

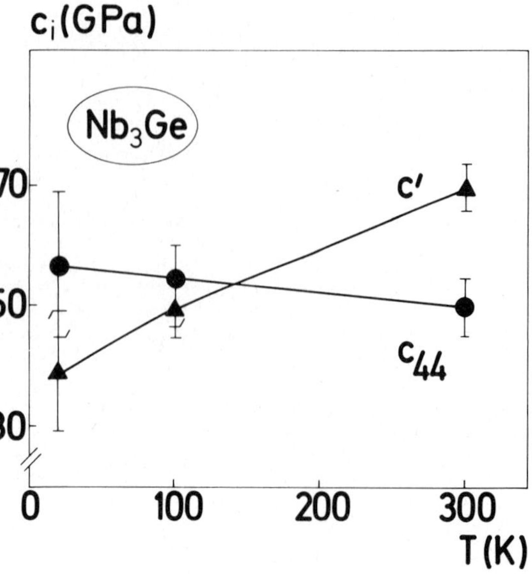

FIGURE 2
Elastic moduli c' (triangles) and c_{44} (points) vs. temperature. The error bars indicate the variation of the elastic constants obtained from fits at different energy transfers.

measurements became very difficult. Figure 2 shows c' and c_{44} as a function of temperature. The error bars indicate the variation of the elastic constants obtained from diffent energy transfers. Compared to Nb_3Sn, the softening of c' is relatively weak. Also, c_{44} is approximately constant. Because of the low contribution of longitudinal phonons to the inelastic intensity at small $\hbar\omega$, c_{11} could not be determined with sufficient accuracy.

5. SUMMARY AND CONCLUSIONS

In an inelastic neutron scattering experiment we measured phonon frequency distributions and the elastic moduli c' and c_{44} as a function of temperature. Obviously the softening of the phonon spectrum at lower energies is not accompanied by a pronounced softening of the shear elastic constants upon cooling. Neutron irradiation experiments on V_3Si show clearly, that small amounts of damage, where T_c remains unchanged, suppress the martensitic transformation but do not affect the shear elastic mode annomaly (6). Therefore, the relatively weak softening of c' in Nb_3Ge cannot be attributed to sample imperfections. As far as elastic annomalies are concerned, Nb_3Ge doesn't fit into the scheme of other high T_c A15's like Nb_3Sn and V_3Si.

ACKNOWLEDGEMENT

Financial support from the Bundesministerium für Forschung und Technologie is gratefully acknowledged.

REFERENCES

(1) K.E. Kihlstrom, T.H. Geballe; Phys. Rev. B 24 (1981), 4101.
J. Geerk, Private communication.
(2) P. Müller, W.Reichardt, N. Nücker, A. Müller Superconductivity in d- and f-Band Metals, ed. W. Buckel, W. Weber (KfK Karlsruhe, 1982) p. 19.
(3) B.P. Schweiß, B. Renker, E. Schneider, W. Reichardt; Superconductivity in d- and f-Band Metals, ed. D.H. Douglass (plenum press, New York, 1976) p.189.
(4) U. Buchenau; Solid State Commun. 32 (1979) 1329.
(5) A. Müller; Z. Metallkde 71 (1980) 507.
(6) A. Guha, M.P. Sarachik, F.W. Smith, L.R. Testardi; Phys. Rev. B 18 (1978) 9.

LT-17 (Contributed Papers)
U. Eckern, A. Schmid, W. Weber, H. Wühl (eds)

MEASUREMENT OF THE SUPERCONDUCTING ENERGY GAP IN SINGLE CRYSTAL V_3Si

Seizo MORITA, Syozo IMAI, Seiji YAMASHITA, Nobuo MIKOSHIBA, Naoki TOYOTA*, Tetsuo FUKASE* and Takashi NAKANOMYO*

Research Institute of Electrical Communication, Tohoku University, Sendai 980, Japan
* The Research Institute for Iron, Steel and Other Metals, Tohoku University, Sendai 980, Japan

We measured the energy gap of single crystal V_3Si using a bridge-type point-contact Josephson junction. From dc I-V curves, we obtained that the energy gap of Nb-V_3Si bridge-type junction was $\Delta(Nb)+\Delta(V_3Si)=3.93$meV for [100] and 4.53meV for [111] crystal orientation of V_3Si, respectively. Assuming $2\Delta(Nb)=2.9$meV at 4.2 K and $T_c(V_3Si)=17$ K, we had $\Delta(V_3Si)=2.48$meV and $2\Delta(V_3Si)/kT_c(V_3Si)=3.4$ for [100], and $\Delta(V_3Si)=3.08$meV and $2\Delta(V_3Si)/kT_c(V_3Si)=4.2$ for [111]. The values of the energy gap $\Delta(V_3Si)$ and $2\Delta(V_3Si)/kT_c(V_3Si)$ for [111] crystal orientation of V_3Si are quite large compared to the previous result obtained by Moore et al. The energy gap anisotropy, $\Delta[111]>\Delta[100]$, is qualitatively opposite to the previous result obtained for Nb_3Sn.

1. INTRODUCTION

Measurement of the energy gap Δ in an A-15 superconductor will be of great interest, because the ratio $2\Delta/kT_c$ gives a measure of how strong-coupling the superconductor is, and, above all, the anisotropy of the energy gap on the Fermi surface gives an important clue to the microscopic understanding of its high transition temperature from its energy band structure.

Large energy gap anisotropy in Nb_3Sn was observed in ref.[1], using point-contact junctions. However, the result should be taken with caution, since $2\Delta/kT_c$ in any crystal orientation is much smaller than even the weak-coupling (BCS) limit.

In this paper, we measured the energy gap of a single crystal V_3Si, by forming a point-contact junction on (100) or (111) planes of V_3Si crystal.

2. EXPERIMENTAL RESULTS AND DISCUSSIONS

2.1. Surface Treatment of V_3Si

The samples used in this experiment have the resistance ratio 15∿30 [2], and were cut in a slice with a spark cutter from V_3Si crystal rod. To remove damaged region near the surface, we polished mechanically the surface of the samples with #4000 carborundum, 3μm, 1μm and 0.3μm alumina in regular order. After ultrasonic cleaning, the sample was chemically etched with an immersion etching method. Several acids were tried to etch the sample. Then, we formed a point-contact junction by pressing the tip of a Nb needle on the V_3Si surface.

2.2. Measurement of the Energy Gap

In Fig.1, we show dc I-V curves of point-contact junctions with (100) and (111) planes of V_3Si samples, respectively. The dc I-V curves show that the junctions are typical bridge-type point-contacts with small but clear gap structure. (Unfortunately, we could not fabricate tunnel-type point-contacts with V_3Si [3], which are most suitable to the energy gap measurement, because we could not form the oxide with high quality). We define the gap voltage $V_G=[\Delta(Nb)+\Delta(V_3Si)]/e$ by $V_G=(V_g+V_g')/2$, where V_g and V_g'

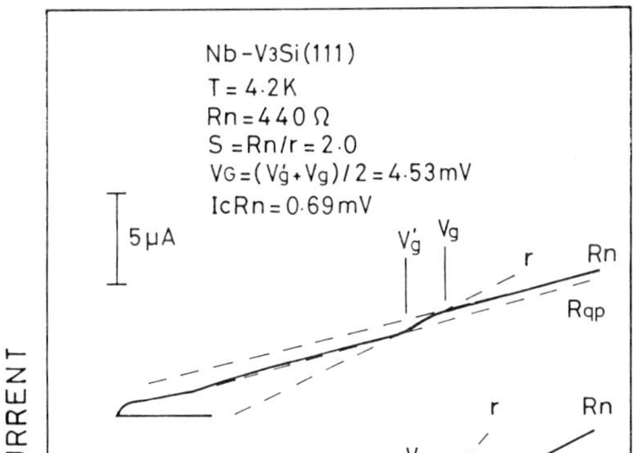

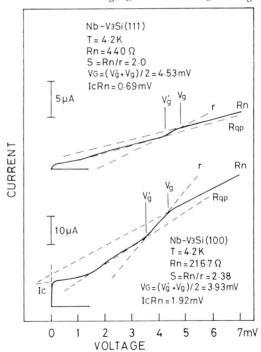

Fig.1 dc I-V curves of Nb-V_3Si bridge-type Josephson junctions, which showed the largest gap values of $V_G=[\Delta(Nb)+\Delta(V_3Si)]/e$ for [100] and for [111] crystal orientation of V_3Si.

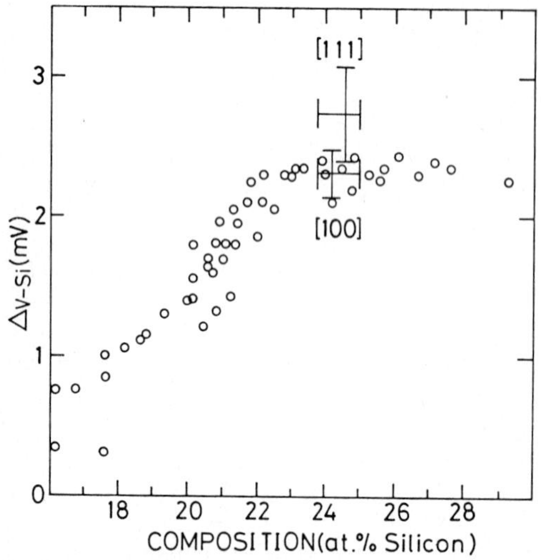

Fig.2 Superconducting energy gap (o) plotted vs composition of a series of V-Si depositions obtained by Moore et al.[3] is compared with $\Delta(V_3Si)$ in single crystal V_3Si (+) determined by the present experiment.

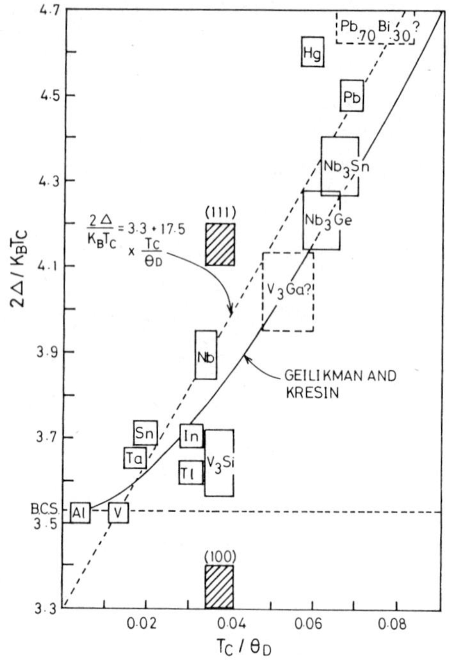

Fig.3 $2\Delta/kT_c$ vs T_c/Θ_D [Θ_D: the Debye temperature] obtained by Moore et al.[3]. Hatched region shows the present result for [100] and [111] crystal orientation of V_3Si.

are determined as shown in the figure 1. We observed a large scatter in the magnitude of V_G, each time a point-contact is formed by detaching or pressing the Nb needle, from or on the same V_3Si sample. From a series of experiments using Nb-Sn point-contact junctions [4], we found that the energy gap in the bridge-type junction is smaller than the intrinsic (bulk) value, mainly owing to heating around the constriction. Nevertheless, we found [4] that, when many point-contacts are formed, some of the highest values give the energy gap nearly equal to the intrinsic value. Therefore, we assume in this experiment that the highest energy gap observed using V_3Si sample is nearly equal to the intrinsic value at that surface (or crystal orientation) of the V_3Si sample.

The dc I-V curves in Fig.1 give the highest V_G for each orientation of V_3Si sample. The sample with (111) plane was etched with dilute HNO_3, and that with (100) plane was obtained with only mechanical polishing. We found that the energy gap (at the surface) is very sensitive to the kinds of the acid used in chemical etching, details of which will be reported elsewhere. From Fig.1, we obtained $\Delta(Nb)+\Delta(V_3Si)$ =3.93meV for [100] and 4.53meV for [111], respectively. Assuming $2\Delta(Nb)$=2.9meV at 4.2 K and $T_c(V_3Si)$=17 K, we have $\Delta(V_3Si)$=2.48meV and $2\Delta(V_3Si)/kT_c(V_3Si)$=3.4 for [100], and $\Delta(V_3Si)$= 3.08meV and $2\Delta(V_3Si)/kT_c(V_3Si)$=4.2 for [111].

2.3. Comparison with the Previous Results

The energy gap of deposited V-Si films was measured by Moore et al.[3] varying the V to Si composition ratio. In Fig.2, we plot our result (+) for comparison with their result (o). Atomic composition of our (bulk) sample is estimated from the lattice constant determined by X-ray diffractmeter [5]. The vertical lines show the magnitude of the scatter in V_G corresponding to the highest 4 data points for [111] and 6 data points for [100]. Fig.2 shows that the energy gap for [100] agrees with the previous result [3], while that for [111] is considerably larger than the previous result [3].

Moore et al.[3] summarized the correlation of $2\Delta/kT_c$ with T_c/Θ_D, as is shown in Fig.3. Rectangles with hatch are from our present result. This figure shows that the energy gap measured in ref.[3], corresponding to the average on the crystal orientation, lies between the two values for [111] and [100] orientations.

Finally, we note that the energy gap anisotropy, $\Delta[111]>\Delta[100]$, is qualitatively opposite to the previous result obtained for Nb_3Sn [1].

[1] V.Hoffstein and R.W.Cohen, Phys.Lett. 29A (1969) 603.
[2] Y.Muto et al., J.Low Temp.Phys. 34 (1979) 617.
[3] D.F.Moore et al., Phys.Rev.B 20 (1979) 2721.
[4] S.Morita et al., Jpn.J.Appl.Phys. 21 (1982) 71.
[5] H.A.C.M.Bruning, Philips Research Reports 22 (1967) 349.

LT-17 (Contributed Papers)
U. Eckern, A. Schmid, W. Weber, H. Wühl (eds)

ELECTRON TUNNELING INTO Nb_3Al USING Al_2O_3 TUNNELING BARRIERS

J. GEERK and W. BANGERT

Kernforschungszentrum Karlsruhe, Institut f. Nukleare Festkörperphysik, P.O.B. 3640,Karlsruhe,FRG

Tunnel junctions on Nb_3Al with artificial Al_2O_3 tunneling barriers have been prepared by magnetron sputtering. The maximum energy gap of the Nb_3Al films was 2.55 meV. Detailed second derivative studies show that the phonons in the energy region between 11 and 14 meV are shifting to higher energies with decreasing Al content. The best junctions of this study showed virtually no proximity effect but the tunnel data of high T_c junctions (15 K) could not be satisfactorily described by the conventional Eliashberg equations.

1. INTRODUCTION

The first tunneling experiments on Nb_3Al were reported by Kwo and Geballe (1), who used SiO_2 as artificial tunnel barrier. Their junctions show excellent quality with respect to the material properties of the A15 film but are suffering from proximity effects probably due to a layer of depressed T_c between the SiO_2 barrier and the Nb_3Al material. Experience has shown that for tunneling spectroscopy the Al_2O_3 tunnel barrier is the most favourable known so far. Stimulated by the progress we had with this type of barrier on Nb_3Ge (2) and Nb_3Sn (3) we applied our preparation technique to Nb_3Al which is of special interest in cause of its nearly localized Al phonon mode (4) and in order to compare the results with our recent Nb_3Sn data (3).

2. EXPERIMENTAL

Thin film tunnel junctions of the type Nb_3Al-Al_2O_3-Pb have been prepared by magnetron sputtering as described in Ref. (3). Additional to I-V and dV/dI measurements also d^2I/dV^2 traces were taken on all the junctions in the normal and superconducting state of the Nb_3Al film. The integral functions of the second derivatives were fitted to the first derivatives to generate reduced density of states (RDOS) data containing all measurable fine structure. X-ray and RBS-measurements were performed to confirm the exclusive presence of the A15 phase and to determine the Nb-Al ratio.

3. RESULTS AND DISCUSSION

The I-V characteristics of our tunnel junctions measured with the Pb counterelectrode in the superconducting state showed typical leakage currents of 0.1% or better at voltages below the energy gap Δ_o of Pb. With increasing Al content of the A15 NbAl an excess current probably due to tunneling into a normal conducting phase appeared at voltages above the Pb energy gap. This excess current reached 13% for A15 NbAl with Δ_o=2.55 meV the maximum gap we achieved in

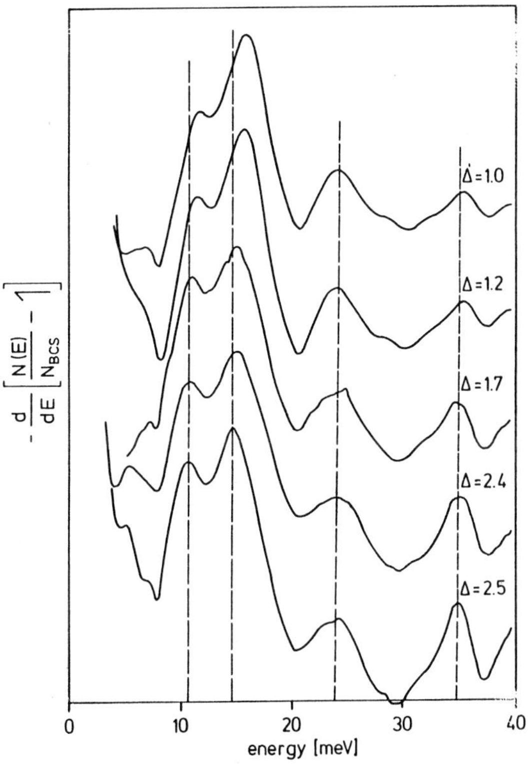

FIGURE 1
The derivative of the reduced density of states versus energy of several junctions on A15 NbAl with different Al content. Δ_o=2.5 meV 25±2 at%, Δ_o=2.4 meV 24 at%, Δ_o=1.7 meV 21 at%, Δ_o=1.2 meV 18 at%, Δ_o=1.0 meV 16 at%.

our study. Fig. 1 shows the derivative of the reduced density of states (DRDS) versus voltage measured from the gap edge for 4 tunnel

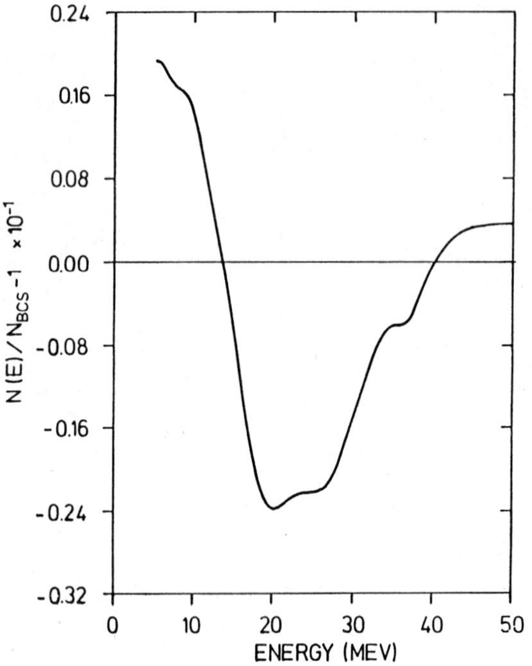

FIGURE 2
The reduced density of states of a $Nb_3Al-Al_2O_3$-Pb tunnel junction, Δ_o=2.45 meV.

junctions with different Al content of the A15 NbAl. Generally the DRDS curve shows peaks where there are shoulders or peaks in the Eliashberg function. We see that with increasing gap there is considerable softening of the peaks near 11 and 14 meV. The Al localized mode as determined from neutron scattering (4) appears near 36 meV. An interesting feature are the small structures at low energies near 5 meV. Also the first derivative data of high T_c junctions of Kwo and Geballe (1) show a step-like behaviour near this energy. Fig. 1 shows that this low energy mode, which was also found in Nb_3Sn (3) is almost not dependent on the energy gap Δ_o.

Fig. 2 shows the RDOS of our best tunnel junction prepared on high T_c A15 NbAl. The quantity P as defined in Ref. (3), which is zero for non-proximity and strongly negative for typical proximity junctions turns out to be -0.06 ± 0.04 for this junction. Below 4 meV, where the tunnel data are not measurable in cause of normalizating difficulties the data were set constant to the value of the first measured data point. This small value of P indicates that there is virtually no proximity effect present in the tunnel junction. Attempts to invert the data of Fig. 2 by the conventional McMillan-Rowell (MMR) program failed in cause of a too steep increase of the data between 25 and 40 meV and a strong "overswing" above 40 meV. The proximity inversion procedure of Arnold and Wolf yields d/l = 0 (d is the thickness of the proximity layer and l is an effective sampling depth) which is consistent with $P \cong 0$. The data of Al deficient A15 NbAl (Δ_o=1 meV) are very near to be described by the MMR program. All these findings are very similar to our Nb_3Sn results (3) so our conclusion is also here, that in the vicinity (on the scale of typical phonon frequencies) of the Fermi edge of high T_c A15 NbAl there are strong structures in the electronic density of states giving rise to an "overswing" in the RDOS as outlined in Ref. (3). A structured electronic density of states of Nb_3Al near the Fermi edge is in accordance with recent specific heat and magnetic susceptibility data from Junod et al. (5).

REFERENCES

(1) J. Kwo and T.H. Geballe, Phys. Rev. B 23 3230 (1981).
(2) J. Geerk, J.M. Rowell, P.H. Schmidt, F. Wüchner, W. Schauer, Superconductivity in d- and f-Band Metals (ed. W. Buckel and W. Weber, Karlsruhe 1982).
(3) U. Schneider, J. Geerk and H. Rietschel, this volume.
(4) B.P. Schweiss, B. Renker, E. Schneider, and W. Reichardt, in: Superconductivity in d- and f-Band Metals, ed. by D.H. Douglass (Plenum, New York, 1976) 189.
(5) A. Junod, J.-L. Jorda, M. Pellizone, and J. Muller, Phys. Rev. B 29 (1984) 1189.

LT-17 (Contributed Papers)
U. Eckern, A. Schmid, W. Weber, H. Wühl (eds)

ELECTRONIC STRUCTURE OF HIGH-T_c A15 SUPERCONDUCTORS FROM ELECTRON ENERGY-LOSS SPECTROSCOPY

Th. MÜLLER-HEINZERLING, W. WEBER, J. FINK and J. PFLÜGER*

Kernforschungszentrum Karlsruhe, Institut für Nukleare Festkörperphysik, Postfach 3640,
D-7500 Karlsruhe, Federal Republic of Germany

Several A15 superconductors with a high T_c have been studied by high resolution electron energy-loss spectroscopy. From the valence spectra the transition probability is obtained by a Kramers-Kronig analysis and compared to the theoretical joint density of states. In the fine structure of the loss spectra near core edges, we find the theoretically predicted structure of the unoccupied density of states.

1. INTRODUCTION

Because of their unusual properties, especially their high transition temperature T_c for superconductivity, the A15 compounds have attracted much attention both theoretically and experimentally (1). The key to most of the anomalies is their complex electronic structure near the Fermi level.

Calculations of the bandstructure and density of states (DOS) are very complicated due to the complexity of the A15 structure; experimental work suffered from preparation problems due to the high sensitivity of the (partly unstable) structure and from lack of sufficient resolution to verify the sharp structures predicted by theory.

We have studied several high-T_c superconductors by high resolution electron energy-loss spectroscopy (ELS) with a 170 KV transmission ELS-spectrometer. Information on their electronic properties can be gained by ELS in two ways: From a Kramers-Kronig analysis of the valence spectra (0 to 40 eV) the dielectric properties and a transition probability related (2) to the joint density of states (JDOS) have been obtained. By measurements at higher energy losses near characteristic edges from inner shell excitations one can study the fine structure above the edges, which should reflect the density of unoccupied states.

Calculations of DOS and JDOS curves for the A15 compounds have been based on tight binding fits (3,4) to the self-consistent APW calculations of Klein et al. (5).

2. SAMPLE PREPARATION

We have investigated samples of Nb_3Al, Nb_3Ge and Nb_3Sn. The 700 Å thick films were prepared by cosputtering resp. coevaporation onto Molybdenum substrates and characterized by T_c-measurements, X-ray diffraction and Rutherford backscattering. Subsequently the substrates were dissolved in $FeCl_3$ solution and the films placed on standard copper electron microscope grids.

3. RESULTS AND DISCUSSION

3.1 Valence Spectra

The main features of the JDOS of all Nb-based A15's as predicted by theory *assuming constant transition matrix elements and ignoring selection rules* are two peaks located near 5 and 12 eV (see Fig. 1 for Nb_3Sn as an example). In reality the matrix elements are *not* constant but are expected to vary smoothly with energy so that structure in the experiment derived transition probability should correlate with structure in the JDOS.

Comparing the transition probability with the JDOS (Fig. 1), one finds good agreement for the first peak, but strong suppression of the second peak in the experimental results. The probable reason for this discrepancy is that

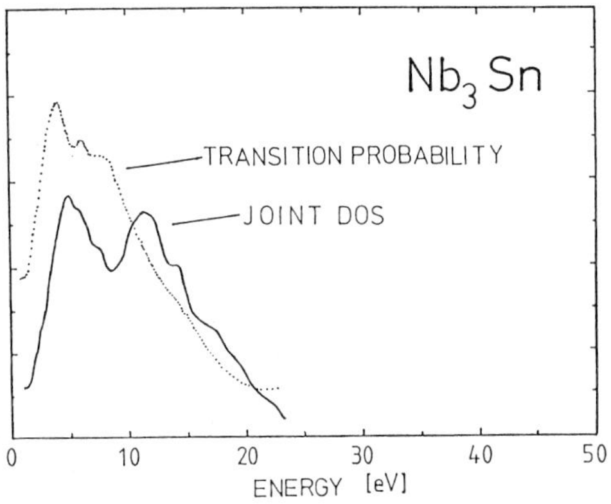

FIGURE 1
Joint density of states for Nb_3Sn after (4) (——) and transition probability from a Kramers-Kronig analysis of ELS valence spectra (···).

*Present address: Inst. f. Festkörperforschung, KFA Jülich, Postfach 1913, D-5170 Jülich, FRG

the second peak results mainly from dipole forbidden s→d transitions, so that the matrix elements are much smaller than those for the first peak (resulting from p→d and intersite d→d transitions). Results for the other compounds are essentially similar.

3.2 Core Level Spectra

The fine structure of the loss function above the Nb 3d edges can be expected to reflect the unoccupied DOS of p-symmetry. Comparing the experimental results for Nb_3Sn with appropriately broadened theoretical curves (3) (shifted and added to itself to represent the spin-orbit split double edge), one finds good agreement for the peak positions in the applicability range of the calculations (up to ∿ 10 eV above the Fermi energy) (Fig. 2). Peak heights are again changed by matrix element effects.

For Nb_3Ge the agreement is similar.

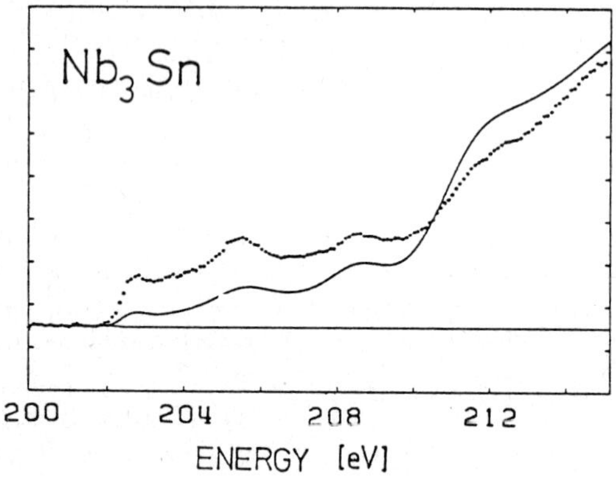

FIGURE 2
Broadened density of unoccupied states of p-symmetry in Nb_3Sn after (3) shifted by 3.06 eV and added to itself with an intensity ratio of 2:3 (——) and ELS spectrum above Nb 3d edge with 0.6 eV resolution after background subtraction (···).

Of special importance for the properties of the A15 compounds is the sharp peak in the DOS at the Fermi energy. The unoccupied high energy tail of this peak can be seen as the first peak in our spectra. We have measured this region with an improved resolution of 200 meV. For Nb_3Sn we find agreement for shape, width and relative height with theoretical predictions (Fig. 3) (absolute heights are adjusted).

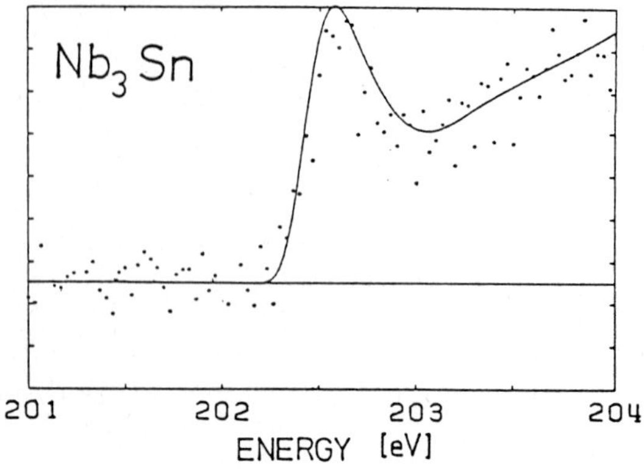

FIGURE 3
Same as figure 2, but detail with a better resolution (200 meV).

4. CONCLUSIONS

Theoretical predictions for the joint density of states and the unoccupied density of states in Nb_3Sn have been verified. In particular, the unoccupied part of the prominent peak in the density of states at the Fermi energy, which is the reason for most of the unusual properties of this material, has been confirmed with an experimental resolution of 200 meV. Results for Nb_3Ge show similar results, for Nb_3Al and others (V_3Si, Nb_3Ga) work is in progress.

ACKNOWLEDGEMENTS

We thank U. Schneider for providing Nb_3Sn-samples and G. Linker for help with the X-ray diffraction and Rutherford backscattering measurements.

REFERENCES

(1) For a review see e.g. L.R. Testardi in "Physical Acoustics" (Ed. W.P. Mason and R.N. Thurston), Academic Press, New York (1973), Vol. X, 193.
L.R. Testardi, Rev. Mod. Phys. 47 (1975) 637.
(2) M.G. Bell, W.Y. Liang, Adv. Phys. 25 (1976) 53.
(3) L.F. Mattheiss, W. Weber, Phys. Rev. B 25 (1982) 2248.
(4) W. Weber, unpublished results.
(5) B.M. Klein, L.L. Boyer, D.A. Papaconstantopoulos, Phys. Rev. B 18 (1978) 6411.

LT-17 (Contributed Papers)
U. Eckern, A. Schmid, W. Weber, H. Wühl (eds)

THE INFLUENCE OF NEUTRON IRRADIATION ON THE PARAMAGNETIC SUSCEPTIBILITY AND THE UPPER CRITICAL FIELD OF V_3GE SINGLE CRYSTALS

T. SOLLEDER, H. KRONMÜLLER, U. ESSMANN

Max-Planck-Institut für Metallforschung, Institut für Physik
P.O.B. 80 06 85, 7000 Stuttgart 80, F.R. Germany

We have measured the paramagnetic susceptibility χ and the upper critical field B_{c2} of several V_3Ge single crystals as a function of neutron irradiation up to fluences of 20 x 10^{18} n/cm^2 (E > 1 MeV). The susceptibility of unirradiated samples shows a weak temperature dependence in the range between the transition temperature T_c and 350 K with a flat maximum at T = 65 ± 6 K. This effect is reduced by increasing neutron doses. A preliminary analysis of the linear behaviour of the upper critical field up to 2 Tesla yields a considerable increase of ($-dB_{c2}/dT$) at T_c with increasing fluences. The temperature and dose dependence of the susceptibility can be understood within the framework of models based on a high density of electronic states (DOS).

1. INTRODUCTION

Superconducting A15 compounds with high transition temperatures are very sensitive to atomic disorder induced by irradiation. The influence of disorder on T_c and the other superconducting parameters can be explained in terms of a one dimensional density of states model (1-4). This model is based on a high and narrow peak in the DOS near the Fermi energy E_f, which is lowered and smeared out by irradiation with fast particles, e.g. neutrons. Measurements of the magnetic susceptibility which is directly related to the DOS, $N(E_f)$ at E_f are very suitable to investigate the influence of atomic disorder on the DOS. Furthermore the abrupt change in the susceptibility occuring at the transition from the normalconducting into the superconducting state can, although being sometimes ambiguous, be used to determine B_{c2}.

Investigations by Guha et al. (5) in the high T_c compound V_3Si yielded the expected reduction of χ and an increase in ($-dB_{c2}/dT$) at T_c with increasing neutron doses. The purpose of this work was to examine if a similar behaviour can be found in the low T_c compound V_3Ge ($T_{c,unirradiated}$ = 6.0 K), and if the DOS models can be extended to this class of A15 materials.

2. EXPERIMENTAL PROCEDURE AND RESULTS

V_3Ge crystals were prepared by two different zone melting procedures, by RF induction and electron bombardement, after sintering and inductive melting of stoichiometric mixtures of the two metals (6). Due to the large difference between the vapor pressures of Vanadium and Germanium it was difficult to achieve the stoichiometric composition and accordingly the microstructure of the different samples varied considerably. The A15 structure and composition have been examined by X-ray analysis.

The irradiation of the crystals by fast neutrons with fluences between 0.3 and 20 x 10^{18} n/cm^2 (E > 1 MeV) was performed at the FRM Munich at a temperature of about 80°C.

The temperature dependence of $\chi(T)$ was measured in a commercial squid magnetometer in the normalconducting range between the transition temperature and 350 K at a magnetic induction of 1 Tesla. The absolute accuracy of the susceptibility measurements is given by 2%. A characteristic feature of the temperature dependence of $\chi(T)$ of the unirradiated samples is a flat maximum at T = 65 ± 6 K and a continuous decrease above this temperature. The maximum and the decrease of χ with increasing T becomes less significant and vanishes finally at highest doses. This behaviour is shown in Fig. 1 where the mass susceptibility χ_g of five selected samples with different neutron doses $\Phi \cdot t$ is plotted versus temperature T.

Irradiation with doses of 20 x 10^{18} n/cm^2 lowers the susceptibility values at 100 K by about 20% and yields a considerable reduction of the temperature dependence of χ.

The upper critical field B_{c2} was measured by observing the change from the paramagnetic behaviour of the normalconducting state to the superconducting diamagnetism at T_c as a function of the applied field. This change is, probably due to pinning effects, sometimes ambiguous. As a preliminary result we obtain a linear relation for B_{c2} vs. ($T - T_c$) up to 2 Tesla, and an appreciable increase of ($-dB_{c2}/dT$) at T_c with increasing doses (see Fig. 2).

3. DISCUSSION

The measured temperature and dose dependence of the paramagnetic susceptibility in the low T_c compound V_3Ge is in qualitative agreement with results obtained for the high T_c compound

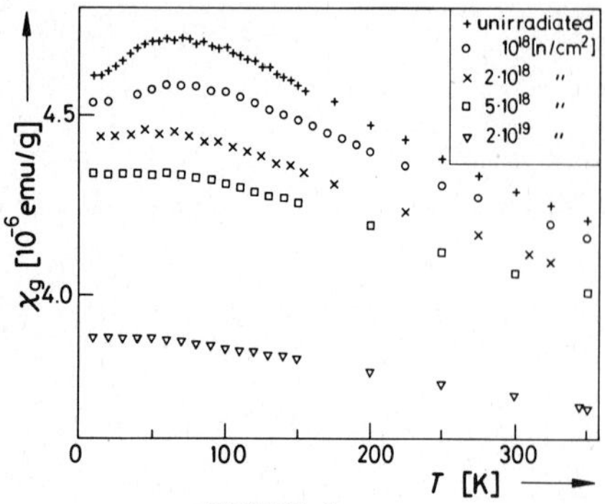

FIGURE 1
Temperature and dose dependence of the susceptibility χ_g of five V_3Ge single crystals.

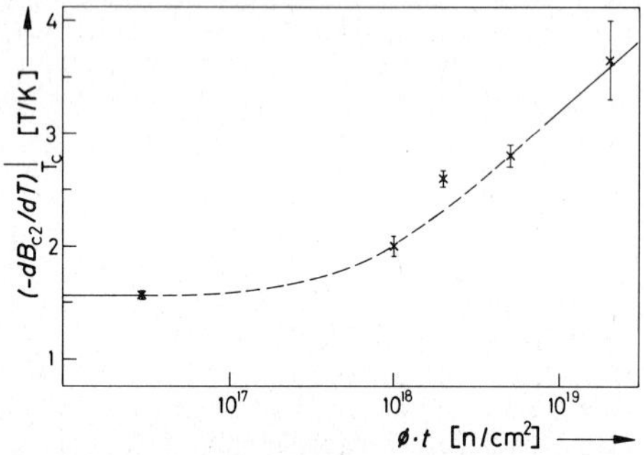

FIGURE 2
$(-dB_{c2}/dT)$ at T_c of V_3Ge single crystals versus the fast neutron fluence ϕt.

V_3Si (5).

In order to show, that the decrease of χ can be related to the accompanying T_c reduction (6) via the density of states we can make the crude assumption that the susceptibility corresponds to the Pauli susceptibility $\chi_g \propto N(E_f)$. For T_c we use McMillan's formula (7) in the limit $\mu^* < \lambda < 1$

$$T_c \cong 0.7\, \Theta_D \exp\{-(1+\lambda)/\lambda\}\ .$$

Θ_D denotes the Debye temperature and μ^* an electron repulsion potential. The electron phonon coupling constant λ is given by $\lambda = A\cdot N(E_f)$.

The constant A is equivalent to the BCS interaction parameter V. Accordingly we obtain the following relation between T_c and χ_g:

$$\ell n\, T_c = \ell n\, \Theta_D - 1.36 - 12\, \mu_B^2/(a^3\cdot A\cdot\chi_g)\ ,$$

where μ_B denotes Bohr's magneton and a the lattice parameter. For a test of this relation our experimental results for T_c are plotted in Fig. 3 vs. $1/\chi_g$ as measured at 100 K. The experimental points are in fair agreement with the linear relation and yield as results Θ_D = 346 K and A = 0.069 eV.

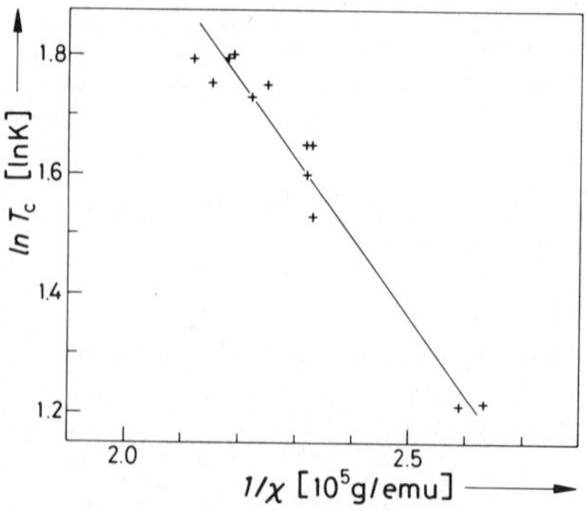

FIGURE 3
Logarithm of the critical temperature T_c after varying neutron doses plotted versus the reciprocal susceptibility χ_g as measured at 100 K.

The fact, that the derived relation is fulfilled by the experiments verifies the assumption that the reduction of the critical temperature as well as of the susceptibility is caused by a density of states effect and not by a phonon effect.

For a correct understanding of the results shown in Fig. 1 we should use for the susceptibility the expression $\chi = \chi_{orb} + \chi_{spin}(c,T)$, with χ_{orb} as a temperature and defect independent contribution of orbital paramagnetism, and $\chi_{spin}(c,T)$ given by (8)

$$\chi_{spin}(c,T) = \frac{2\mu_B^2 \int N(E,c)\cdot(-df/dE)\, dE}{1 - J\int N(E,c)\cdot(-df/dE)\, dE}\ .$$

Here f(E,c) = Fermi distribution, J = electron repulsion term, c = defects' concentration. Using for the DOS the one dimensional linear chain model proposed by Labbé and Friedel (9,10) we obtain a good qualitative understanding of the temperature dependence (including the maximum) and of the dose dependence of χ (1,11,12). With the modified model of the DOS as proposed by Fähnle (4) it should be possible to obtain a quantitative fit of our results.

REFERENCES

(1) R.E. Somekh, J. Phys. F5 (1975) 713.
(2) M. Fähnle and H. Kronmüller, Comm. on Phys. 1 (1976) 91.
(3) M. Fähnle and H. Kronmüller, J. Nuclear Materials 72 (1978) 249.
(4) M. Fähnle, J. Low Temp. Phys. 46 (1982) 3.
(5) A. Guha et al. Phys. Rev. B18 (1978) 9.
(6) R. Conrad, Thesis Uni. Stuttgart (1980).
(7) W.L. McMillan, Phys. Rev. 167 (1968) 331.
(8) A.M. Clogston, Phys. Rev. 136 (1964) A8.
(9) J. Labbé and J. Friedel, J. Phys. Radium 27 (1966) 153.
(10) J. Labbé and E.C. van Reuth, Phys. Rev. Lett. 24 (1970) 1232.
(11) J. Labbé, Phys. Rev. 158 (1967) 647 & 655.
(12) A. Junod, J. Phys. F8 (1978) 1891.

LT-17 (Contributed Papers)
U. Eckern, A. Schmid, W. Weber, H. Wühl (eds)

NONEQUILIBRIUM STATES IN A15 TYPE COMPOUNDS AFTER LOW TEMPERATURE IRRADIATION

René FLÜKIGER

Kernforschungszentrum Karlsruhe, Institut für Technische Physik, Postfach 3640,
7500 Karlsruhe, Federal Republic of Germany

A mechanism describing the homogeneous decrease of the long range order parameter in A15 type compounds after low temperature irradiation with high energy particles is proposed. It is based on the occupation of nonequilibrium or "virtual" sites created by the instability of single 6c vacancies, recently found by Welch et al. by pair potential calculations. The partial occupation of these virtual sites renders A↔B exchanges over several lattice distances by focused replacement collision sequences possible. The occurrence of the latter definitively attributes the observed decrease of T_c in typical A15 compounds after irradiation to a decrease of the order parameter.

1. INTRODUCTION

The decrease of T_c in A15 type compounds after low temperature irradiation with high energy particles is generally attributed to the lowering of the long range order parameter (1). However, no mechanism leading to a homogeneous decrease of the degree of ordering in the whole irradiated sample has been proposed so far. The question arises whwther it is possible that a homogeneous decrease of the order parameter, S, implying site exchanges over several lattice spacings can occur during irradiations at temperatures $T<150^{o}C$, where no thermal diffusion takes place.

A new mechanism, based on the occupation of nonequilibrium or "virtual" sites, will be proposed in the present paper.

2. THE "VIRTUAL" SITE IN THE A15 STRUCTURE

The strong covalent bonding between neighbouring A atoms on the 6c sites of an A15 compound A_3B leads to highly nonspherical atomic shapes. If a 6c vacancy is created, it is highly improbable that both A atoms adjacent to this vacancy will remain in their original "overlapped" configurations. Relaxation effects are expected to occur.

Welch et al. (2) have recently shown by pair potential calculations that a single vacancy on a 6c site is unstable. These authors (2) found that the state of lower energy corresponds to a cofiguration where one of the two A atoms neighbouring the single 6c vacancy is shifted towards a new site which is equidistant from the next two A atoms on the same chain (see Fig. 1). Each "split" vacancy (or negative crowdion) corresponds to the occupation of a nonequilibrium state, which will be called "virtual" site in the following.

3. A↔B EXCHANGES ALONG FOCUSING DIRECTIONS

A↔B exchanges over several lattice spacings, a necessary condition for the homogeneous decrease of the degree of ordering after low temperature irradiation, are only expected to occur along certain focusing directions of the lattice. For the ordered A15 lattice prior to irradiation, it has been shown by Pande (3) that focused replacement collision sequences are unlikely, even in the <102> direction.

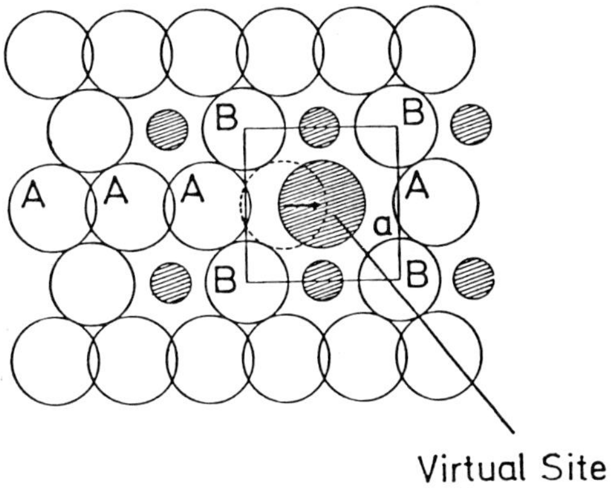

FIGURE 1
Occupation of a "virtual" site by an A atom as a consequence of a radiation induced 6c lattice vacancy in an A15 compound.

The occupation of a small number (<0.3%) of virtual sites, however, fundamentally changes the situation. As shown in Fig. 2, the new atomic sequence in the <102> direction around a virtual site occupied by an A atom would be oABAoABAAABAo instead of oABAoABAo, the sequence in the unirradiated state. The virtual site can be occupied either by A or by B atoms, now constituting a "bridge" between ABA sequences and allowing the A↔B exchanges over several lattice spacings. As noted by Schneider (4), the energy transmitted at primary events is sufficient to move all the atoms of the crystal, while for a homogeneous

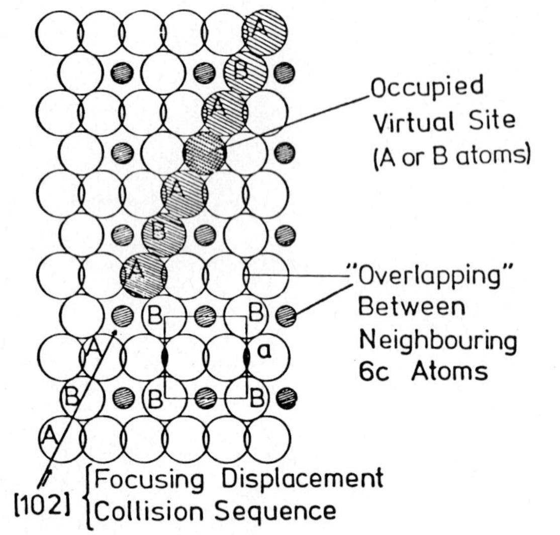

FIGURE 2
The occupation of the virtual site in the A15 structure by an A or a B atom leads to the sequences oABAoABAAABAoABA or oABAoABABABAoABAo, respectively, in the <102> focusing directions.

distribution of the latter, focused replacement collision sequences are essential.

4. LATTICE EXPANSION IN IRRADIATED A15 COMPOUNDS

A comparison of the A15 lattice parameter changes after low temperature irradiations shows that the increase, $\Delta a/a$, is considerably smaller for transition B elements, e.g. Os, Ir or Pt than for nontransition B elements. In particular, $\Delta a/a$ for the systems Nb_3Ge (5) and Mo_3Ge (6) is very similar after corresponding doses, $\sim$1.2%. This shows that there is no simple correlation between $\Delta a/a$ and the ratio r_A/r_B between atomic radii. In a more detailed paper, the author (7) has shown that the correlation between $\Delta a/a$ and r_A/r_B previously proposed in Ref. 1 is only apparent. In reality, it is based on a coincidence, the considered systems Nb_3Ge, Nb_3Al and Nb_3Pt all having ratios $r_A/r_B > 1$. It is easily seen that this correlation does not hold for cases like V_3Ga, where $r_A/r_B < 1$. It follows that the spherical approximation is not valid, and that $\Delta a/a$ is rather due to the particular electronic configuration of the B element, which influences the repulsive forces between the complex around an occupied virtual site.

The two next neighbours of an atom on a virtual site are B atoms, their distance to the latter being a/2 in the direction perpendicular to the page plane in Fig. 1. Thus, the occupation of this nonequilibrium site not only leads to very short interatomic distances (and even overlapping) between A and B atoms, but also between B atoms. In the case of the occupation of the virtual site by a B element, this would lead to BBB sequences.

The occurrence of BBB sequences, possible after low temperature irradiation only, could furnish the key for understanding the causes of the lattice expansion observed in A15 compounds. Indeed, nontransition elements ordinarily do not crystallize in close packed structures, in contrast to transition elements. It could thus be concluded that the larger lattice expansion of irradiated A15 compounds containing a nontransition B element is a consequence of stronger repulsive forces in the BBB sequences. A similar reasonment could be made for static displacements, for which less data are known.

CONCLUSIONS

The proposed site exchange mechanism for A15 compounds after low temperature irradiation leads to the following conclusions:

a) a small number (< 0.3%) of virtual sites, a consequence of unstable 6c vacancies produced by primary collision events, renders A↔B site exchanges by focusing replacement collision sequences in the <102> direction possible. This is a necessary condition for the homogeneous decrease of the long range atomic order parameter over the whole crystal,

b) the occurrence of focused replacement collision sequences definitively attributes the causes for the observed initial decrease of T_c in typical A15 compounds after low temperature irradiation to a decrease of the order parameter,

c) a correlation between atomic disordering as produced by both, quenching from high temperatures or low temperature irradiation can be established. The common point between these two processes is the occupation of virtual sites, a necessary condition for the occurrence of diffusion processes in the A15 structure, as well at high as at low temperatures. The main difference, however, resides in the fact that the virtual sites are still occupied after low temperature irradiation, in contrast to the quenched case , thus leading to the observed lattice expansion and static displacements.

ACKNOWLEDGMENTS

The author would like to thank R. Schneider and G. Linker for communicating the irradiation data on Nb_3Ir prior to publication and O. Meyer, E.L. Haase and W. Schauer for critical discussions.

REFERENCES

(1) A.R. Sweedler, D.E. Cox and S. Moehlecke, J. Nucl. Mater. 72 (1978) 50
(2) D.O. Welch, G.J. Dienes, O.W. Lazareth and R.D. Hatcher, IEEE Trans.Magn. MAG-19 (1983) 889
(3) C.S. Pande, phys. stat. sol. a) 52 (1979) 687
(4) R. Schneider, PhD Thesis, 1984, KfK Karlsruhe
(5) J. Pflüger and O. Meyer, Sol. State Comm. 32 (1979) 1143
(6) M. Lehmann, H. Adrian, J. Bieter, G. Saemann E.L. Haase, Sol. State Comm. 39 (1981) 145
(7) R. Flükiger, KfK Bericht Nr. 3622, 1983, Kernforschungszentrum Karlsruhe, to be published

LT-17 (Contributed Papers)
U. Eckern, A. Schmid, W. Weber, H. Wühl (eds)

THE INFLUENCE OF RADIATION INDUCED DISORDERING ON THE SUPERCONDUCTING TRANSITION TEMPERATURE OF Nb_3Ir FILMS

R. SCHNEIDER (a), G. LINKER (a), O. MEYER (a), M. KRAATZ (a) and F. WÜCHNER (b)

Kernforschungszentrum Karlsruhe (a) Institut für Nukleare Festkörperphysik, (b) Institut für Technische Physik, P.O.B. 3640, D-7500 Karlsruhe, Federal Republic of Germany

Evaporated Nb_3Ir films were irradiated with protons and He ions to study the influence of the induced defect structures on the superconducting transition temperature T_c. A T_c reduction was observed for energies deposited into nuclear collisions below 1 eV/atom. Beyond this threshold T_c increased and reached values above the initial transition temperature depending on the irradiation conditions. The T_c depression is connected to a decrease of the long range order parameter. The T_c enhancement is accompanied by the formation of a defect structure consisting of static displacements of the lattice atoms (p-irradiation) and in addition by a partial amorphization (He-irradiation). The maximum T_c value of 5.7 K was measured in totally amorphized films.

1. INTRODUCTION

It is well established that particle irradiation of high T_c A15-type superconductors leads to a strong degradation of the transition temperature. In low T_c A15-type superconductors like Mo_3Ge or Mo_3Si the opposite effect namely a T_c enhancement under irradiation has been reported (1,2). A similar observation was made in irradiated Nb_3Ir bulk samples where the defect structure with respect to static displacements of the lattice atoms and amorphization was determined by channelling effect measurements (3). In this contribution we report results from an extension of the latter study to thin film experiments with an emphasis on the determination of the Bragg-Williams long range order parameter S(4) which could not be assessed by the channelling measurements. T_c changes in A15-type superconductors are often discussed in connection with variations of the order parameter, i.e., with formation of antisite defects (5). It was therefore of interest to include this parameter into the discussion of defect structures accompanying the T_c changes in irradiated low T_c A15-type superconductors.

2. EXPERIMENTAL

Single phase A15 Nb_3Ir films with typical thicknesses of 200 nm have been prepared by simultaneous electron beam evaporation of niobium and iridium from two sources onto sapphire substrates. T_c of the stoichiometric films was 1.8 K. The films were irradiated with 300 keV protons and He ions at RT such that the particles came to rest in the substrate and that a homogeneous energy deposition into nuclear collisions was achieved throughout a film. Structural analysis of the films before and after irradiation was performed employing a Guinier thin film diffractometer.

3. RESULTS

The transition temperature of p-irradiated films determined resistively and inductively is shown in Fig. 1 as a function of the energy Q deposited by the bombarding particles into nuclear collisions. Q (1 eV/atom corresponds to

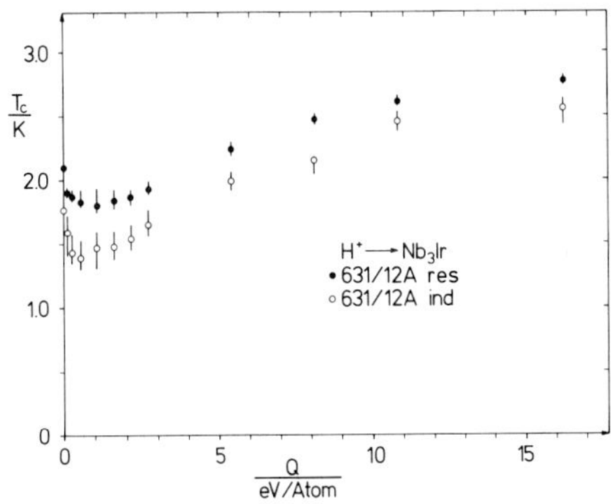

FIGURE 1
Resistively and inductively determined transition temperature T_c as a function of the energy Q deposited into nuclear collisions for proton irradiated Nb_3Ir films. Note the decrease of T_c for $Q \lesssim 1$ eV/atom.

approximately 0.02 displacements per atom) may be considered as an integral measure of disorder. For small Q values ($Q \lesssim 1$ eV/atom) T_c decreases by about 20%. This observation, though less pronounced, was also made for the He irradiations. It is a new effect not detected in the previous investigations which agrees with the "normal" behaviour of high T_c A15-type superconductors. This T_c degradation coincides with

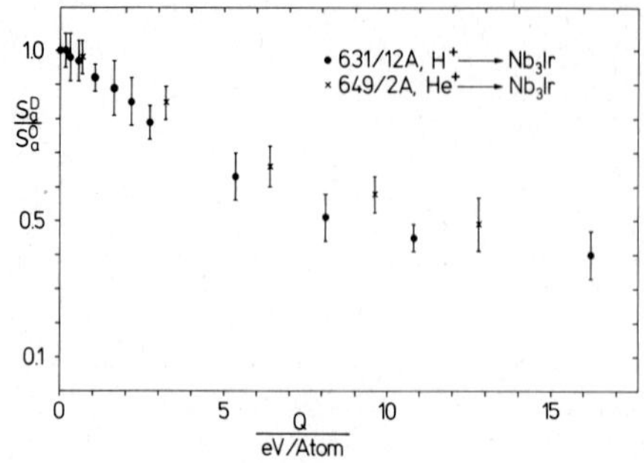

FIGURE 2
Ratio of the order parameters before (S_a^o) and after (S_a^D) irradiation as a function of Q for p and He ion irradiated Nb_3Ir films.

a decrease of the order parameter S by about 10% thus supporting a general conclusion that the production of antisite defects in A15 structured materials always should lead to a T_c depression (5).

The ratio of the order parameters before (S_a^o) and after (S_a^D) irradiation (determined from the intensity change of selected reflection groups in the X-ray diffraction spectra) is plotted in Fig. 2 as a function of Q for proton and He irradiated samples (the absolute value of S_a^o has been estimated to be 0.96 ± 0.04). The order parameter decreases continuously with progressive irradiation, however, a total disordering could not be achieved at RT. A saturation ($S_a^D \cong 0.3$) was observed at Q values of about 65 eV/atom achieved by He ion irradiation only since p-irradiation with fluences corresponding to such high Q values would mechanically destroy the films. The saturation is thought to be due to a temperature dependent equilibrium between radiation induced disordering and radiation assisted thermal ordering.

Though S decreased continuously, T_c revealed an opposite behaviour for Q values above 1 eV/atom and reached saturation values of 2.8

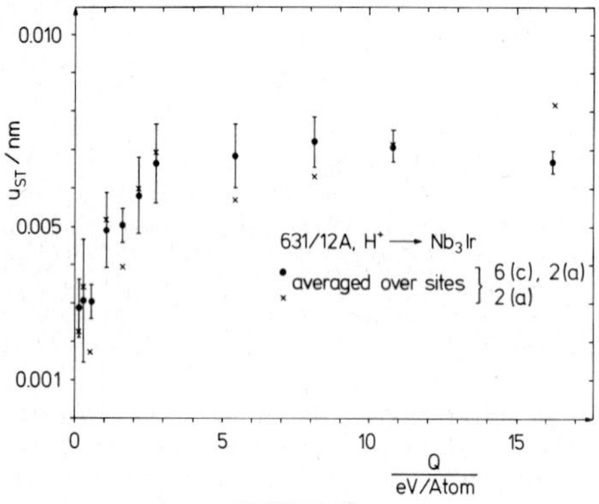

FIGURE 3
Static displacement values u_{st} as a function of Q determined in p-irradiated Nb_3Ir films.

and 3.7 K for p and He irradiation, respectively. The T_c increase coincides with growing displacement amplitude values of a defect structure consisting of small displacements of the lattice atoms also detected in other irradiated A15 compound thin films (6-8). The quantitative values of these displacements u_{st} have been determined by an analysis of the X-ray intensity data in modified Wilson plots (9) and are shown as a function of Q in Fig. 3. We note that u_{st} is constant in the Q range where T_c decreases and that the region of the steep T_c rise coincides with the steepest increase of the displacement values. The higher T_c saturation value for the He irradiated films probably is due to a third defect component namely amorphized parts of material which could be detected in the X-ray spectra. It was not detected for p-irradiated films (the detection sensitivity was about 4%). A total amorphization of the Nb_3Ir films however was only possible with the He irradiations of targets kept at liquid nitrogen temperature. The T_c of the amorphous films was 5.7 K in agreement with the previous findings in bulk Nb_3Ir (3).

4. SUMMARY

In summary, we have found a T_c depression in Nb_3Ir films irradiated at low ion fluences which is accompanied by a decrease of the long range order parameter. The T_c enhancements at higher irradiation fluences are correlated with a defect structure consisting of small displacements of the lattice atoms and in addition with a partial or total (depending on the irradiation conditions) amorphization of the film in accordance with the findings in irradiated Nb_3Ir bulk samples.

REFERENCES

(1) M. Gurvitch, A.K. Ghosh, B.L. Gyorffy, H. Lutz, O.F. Kammerer, J.S. Rosner and M. Strongin, Phys.Rev.Lett.41(1978) 1616
(2) M. Lehmann and G. Saemann-Ischenko, Phys. Lett. 87A (1982) 369
(3) O. Meyer, R. Kaufmann and R. Flükiger, Superconductivity in d- and f-Band Metals, eds. W. Buckel and W. Weber (Kernforschungszentrum Karlsruhe, 1982) 111
(4) W.L. Bragg and E.J. Williams, Proc. Royal Soc. A145 (1934)699
(5) R. Flükiger, KfK Reports 3622 and 3623, Kernforschungszentrum Karlsruhe 1984
(6) J. Pflüger and O. Meyer, Solid State Comm. 32 (1979) 1143
(7) O. Meyer and G. Linker, J. Low Temp. Phys. 38 (1980) 747
(8) U. Schneider, G. Linker and O. Meyer, J. Low Temp. Phys. 47 (1982) 439
(9) G. Linker, Nucl. Instr. Meth. 182/183 (1981) 501

LT-17 (Contributed Papers)
U. Eckern, A. Schmid, W. Weber, H. Wühl (eds)

CRITICAL CURRENT DENSITY AND UPPER CRITICAL FIELD IN SPUTTERED Nb_3Ge FILMS

Mitsumasa SUZUKI, Takeshi ANAYAMA, Koshichi NOTO* and Kazuo WATANABE*

Department of Electrical Engineering, Faculty of Engineering, Tohoku University, Sendai, Japan
*The Research Institute for Iron, Steel and Other Metals, Tohoku University, Sendai, Japan

Superconducting critical current densities, J_c, of sputter-deposited Nb_3Ge films have been measured in magnetic fields(up to 15 T) either parallel or normal to the film surface. J_c's at 15 T are in excess of 1×10^5 A/cm² for most of films with T_c above 20 K, and higher in parallel field. The highest J_c is $\sim4.8\times10^5$ A/cm² for a film of $\sim$2500 Å in thickness. The H_{c2}(4.2K) of films is examined from the flux pinning theory of Kramer on the basis of the results of $J_c(H)$, and the relation of H_{c2} to T_c is presented.

1. INTRODUCTION

A15 Nb_3Ge is one of the promising materials for use in the hydrogen temperature range and for production of magnets generating magnetic fields higher than 20 T [1,2]. Recent work [3, 4,5] has been concentrated on developing H_{c2} and J_c from the practical viewpoint of high field applications. However, properties of Nb-Ge films prepared by various thin-film deposition techniques are sensitively influenced by their deposition parameters, because the high-T_c phase is thermodynamically unstable. This difficulty involved in preparing the high-T_c phase, which causes compositional inhomogeneity and precipitation of the second phase in films, prevents exact evaluation of inherent H_{c2} and J_c. In this report, we present the results of J_c measured in magnetic fields up to 15 T at 4.2 K for sputter-deposited Nb_3Ge films. The difference of J_c among films deposited from targets with different compositions is investigated, and the thickness dependence of J_c is examined. The H_{c2} of films is estimated by applying the Kramer theory [6] to the present $J_c(H)$ results.

2. EXPERIMENTAL

Nb_3Ge films were prepared on sapphire substrates from compound targets by dc sputtering [7]. The Nb/Ge atomic ratios of targets are 3.0, 2.8 and 2.6(designated as target No.1, 2 and 3 hereafter). The T_c and J_c of all the samples were measured by a four-probe resistance method. The applied field was oriented either parallel($H_{/\!/}$) or normal($H_\perp$) to the film surface in a plane normal to the current direction. The crystal structures of some samples were examined by X-ray diffraction analysis.

3. RESULTS AND DISCUSSION

Our effort has focused on preparing films with T_c above 20 K. As shown in Table 1, almost all the samples, except for the three samples deposited from target No.1, indicate $T_{c,end}$ above 20 K. According to X-ray diffraction analysis, films deposited from target No.1 show almost single A15 phase. For target No.2 traces of the hexagonal Nb_5Ge_3 phase seem to be present. The hexagonal Nb_5Ge_3 phase are clearly identified as the minor phase in films deposited from target No.3. The A15 lattice parameter of films investigated in this study falls in the range $\sim$5.14 to $\sim$5.15 Å without regard to the target composition.

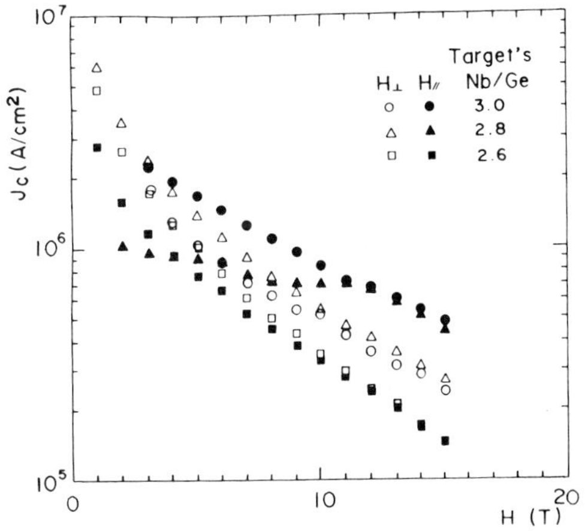

Fig.1 J_c versus external magnetic field.

Fig.1 shows dependences of J_c on $H_{/\!/}$ and $H_\perp$ for the three highest J_c samples obtained from each of the three targets. For samples 543 and 42 obtained from target No.1 and 2, different field dependences of J_c are seen with regard to the applied field direction. In case of $H_\perp$, the J_c of the two samples decreases more rapidly with increasing field intensity in comparison to that obtained for $H_{/\!/}$. In the high field range, J_c's obtained for $H_{/\!/}$ are higher than those for $H_\perp$. This tendency was always seen for any other samples deposited from target No.1

and 2. On the other hand, sample 343 deposited from target No.3 shows generally lower J_c's and besides little difference between the J_c's measured in each of the two field directions.

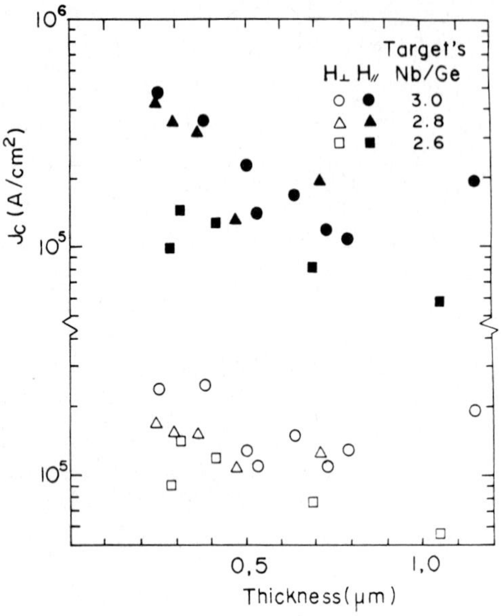

Fig.2 J_c(15T) versus the film thickness

Table 1. Properties of samples

Sample	Thickness (Å)	T_c(K) start	T_c(K) end	H_{c2}(T)	Target's Nb/Ge
543	2500	21.3	20.9	32.0	3.0
603	3800	21.4	21.1	32.0	3.0
623	5000	20.8	20.3	27.0	3.0
573	5300	20.9	20.3	26.5	3.0
613	6100	20.5	18.9	24.5	3.0
733	6400	20.5	19.9	27.0	3.0
583	7300	20.9	20.4	27.5	3.0
593	7900	20.7	19.3	27.5	3.0
743	11500	20.8	20.2	30.0	3.0
42	2400	22.1	21.1	31.0	2.8
22	2900	22.6	21.4	31.0	2.8
32	3600	22.2	21.1	31.5	2.8
93	4700	21.9	20.6	25.0	2.8
113	7100	22.1	21.0	28.5	2.8
373	2800	20.9	20.0	22.5	2.6
343	3100	21.1	20.1	26.5	2.6
292	4100	21.6	20.6	25.5	2.6
312	6900	21.6	20.6	26.0	2.6
323	10500	21.3	20.6	26.0	2.6

J_c's measured at 15 T are given in Fig.2. Most samples, except for the three samples obtained from target No.3, show J_c's higher than 1×10^5 A/cm^2. Higher J_c's are obtained for samples of 2000-3000 Å in thickness. As the film thickness is increased, J_c tends to decrease, especially more clearly for $H_{//}$. According to SEM observation of the film surface, a 2400 Å thick sample with a higher J_c seems to be composed of columnar grains 1000 Å in columnar diameter [8].

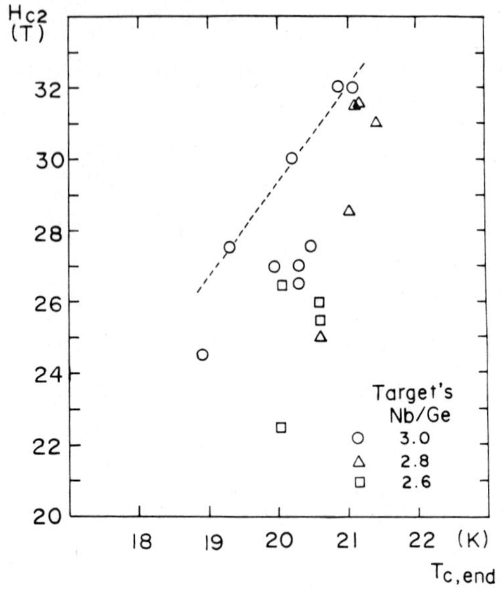

Fig.3 H_{c2}(4.2K) as a function of $T_{c,end}$

The H_{c2}(4.2K) values estimated by applying the Kramer theory to the results of $J_c(H)$ for $H_\perp$ are presented in Table 1. Fig.3 shows the relation between the H_{c2} and $T_{c,end}$. Increase in H_{c2} with increasing $T_{c,end}$ from ~18 to ~21 K is seen. In particular, samples prepared from target No.1 seem to have relatively higher H_{c2}'s. For sample 42 obtained from target No.2 $H_{c2}(0)$ was calculated from the relation $H_{c2}(0) = -0.693T_c(dH_{c2}/dT)_{T_c}$ [9], where directly measured value of dH_{c2}/dT (~2.3 T/K) was used. This $H_{c2}(0)$ of 33.6 T is somewhat higher than H_{c2}(4.2K)=31.5 T obtained from Kramer plots.

References

[1] J.R.Gavaler, Appl.Phys.Lett.23 (1973) 480.
[2] S.Foner, E.J.McNiff,Jr., J.R.Gavaler and M.A.Janocko, Phys.Lett.47A (1974) 485.
[3] A.I.Braginski, G.W.Roland and A.T.Santhanam, IEEE Trans.Magn.MAG-15 (1979) 505.
[4] R.T.Kampwirth, IEEE Trans.Magn.MAG-15 (1979) 502.
[5] M.P.Maley, L.R.Newkirk, J.D.Thompson and F. A.Valencia, IEEE Trans.Magn.MAG-17 (1981) 533.
[6] E.J.Kramer, J.Electron.Mater.4 (1975) 839.
[7] M.Suzuki, N.Suzuki and T.Anayama, Jpn.J. Appl.Phys.21 (1982) 840.
[8] M.Suzuki, H.Ouchi and T.Anayama, Jpn.J.Appl. Phys.22 (1983) L307.
[9] N.R.Werthamer, E.Helfand and P.C.Hohenberg, Phys.Rev.147 (1966) 295.

LT-17 (Contributed Papers)
U. Eckern, A. Schmid, W. Weber, H. Wühl (eds)

STUDY OF TEXTURES IN HIGH-T_c Nb_3Ge FILMS BY ATOMIC RESOLUTION ELECTRON MICROSCOPY

Y. KITANO,[+] H.-U. NISSEN (ETH Zürich, Labor. f. Festkörperphysik, CH-8093 Zürich)
W. SCHAUER, D. YIN*(Kernforschungszentrum Karlsruhe, Inst. f. Techn. Physik, D-7500 Karlsruhe)

Thin films consisting of high-T_c Nb_3Ge and Nb_5Ge_3 with both ($D8_8$)- and ($D8_m$)-structure were prepared by coevaporation and investigated by high resolution 200 kV transmission electron microscopy. Orientation relations between grains are reported.

1. INTRODUCTION

The critical superconducting properties of Nb_3Ge thin films are strongly related to the microstructure (cell parameter, defect structure) and composition (Nb:Ge-ratio, impurity content, additional phases). High resolution transmission electron microscopy (HREM) is very suitable to analyze non-periodic features of the structure which cannot be studied by diffraction methods. In this paper, HREM has been applied to investigate phase composition and growth morphology of NbGe thin films prepared by coevaporation.

2. EXPERIMENTAL

Nb_3Ge films were prepared by evaporating Nb and Ge using two electron guns. Film composition (Nb:Ge-ratio) and thickness were determined by Rutherford backscattering. X-ray diffraction using the Seemann-Bohlin geometry gave information on additional phases and the A15 cell parameter. Samples selected for transmission electron microscopy (TEM) are part of a film with $T_c \cong 21$ K (inductive onset), cell parameter of 5.142 Å and approximately stoichiometric composition (∿ 26 at % Ge). - Films having ∿ 2500 Å thickness were deposited at 800°C onto a molybdenum foil substrate (0.2 mm thickness). Preparation for TEM consists of four steps (1): a. covering the Nb_3Ge film side by a protective laquer spray, b. chemically removing the Mo foil by a 30 w% $FeCl_3$ water solution, c. thinning the remaining film down to ≦ 1000 Å in $FeCl_3$ + HF, d. removing the laquer by acetone.

Specimens were investigated using a JEOL JEM 200 CX transmission electron microscope operated at 200 kV. The microscope is equipped with a high resolution pole piece for the objective lens having a spherical aberration constant of 1.2 mm. Specimens were precisely oriented parallel to a reciprocal plane using a ± 10° top entry tilting goniometer. Images were recorded at magnifications of 250,000 or 360,000 after inserting a 40 μm objective aperture at the center of the diffraction pattern, corresponding to a limiting value of 0.68 $Å^{-1}$.

+ Y.K. now: Dep. Mat. Sc., Hiroshima Univ., Hiroshima, Japan; * D.Y. now: Dep. of Phys., Beijing Univ., Beijing, China.

3. RESULTS AND DISCUSSION

Stoichiometric A15 Nb_3Ge prepared by non-equilibrium preparation techniques such as coevaporation or sputtering forms a metastable phase beyond the A15 phase boundary of the equilibrium phase diagramm (2). The competing phases at stoichiometry (25 at % Ge) are the hexagonal Nb_5Ge_3 ($D8_8$) and, as a minor constituent, the tetragonal Nb_5Ge_3 ($D8_m$) phase. Though procedures are known to stabilize the A15 phase and to suppress the adjacent Nb_5Ge_3 phases (3), the latter are usually present in small amounts unless a Nb-rich off-stoichiometric composition is chosen. The grain fabric in Fig. 1 shows both the A15 and the hexagonal Nb_5Ge_3 phase. Selected area diffraction patterns (SAD) as well as optical diffractograms made from the structure images show that the small rounded grains in extinction position belong to the hexagonal Nb_5Ge_3 phase. As demonstrated earlier, these grains have a large influence on flux pinning (4,5). A structure image of Nb_3Ge in the [001]-projection is shown in Fig. 2. Black dots surrounded by

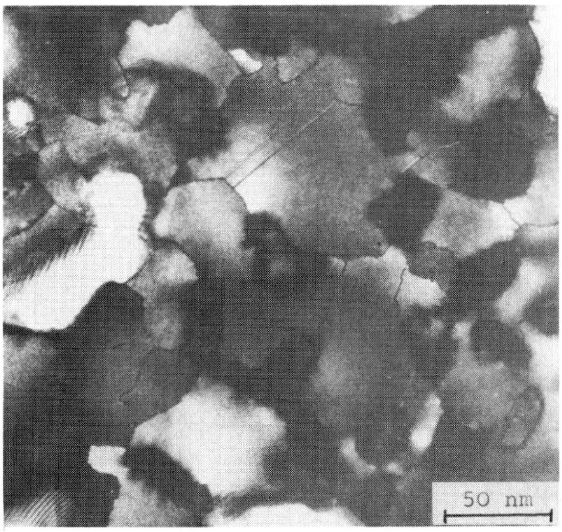

Fig. 1: Bright field image of the Nb_3Ge matrix surrounded by hexagonal Nb_5Ge_3 (dark); notice planar faults in Nb_3Ge grains.

white rings having the periodicity of the A15 lattice can be recognized. Comparing the projected A15 structure with the calculated image contrast in the lower part of Fig. 2, it is

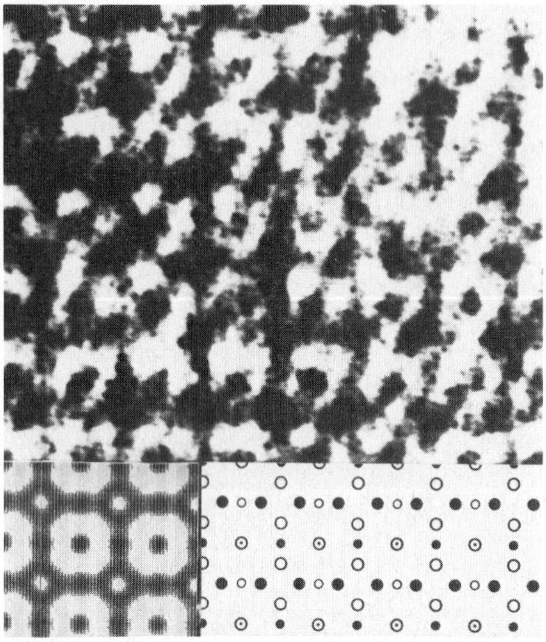

Fig. 2: Structure image of Nb_3Ge in the [001]-projection and scheme of the projected A15 structure. Circles: large = Nb, small = Ge, filled: z=0, open: z=1/2, dotted: z = ± 1/4. Calculated image contrast in lower left part (defocus: -700 Å, specimen thickness: 21 Å).

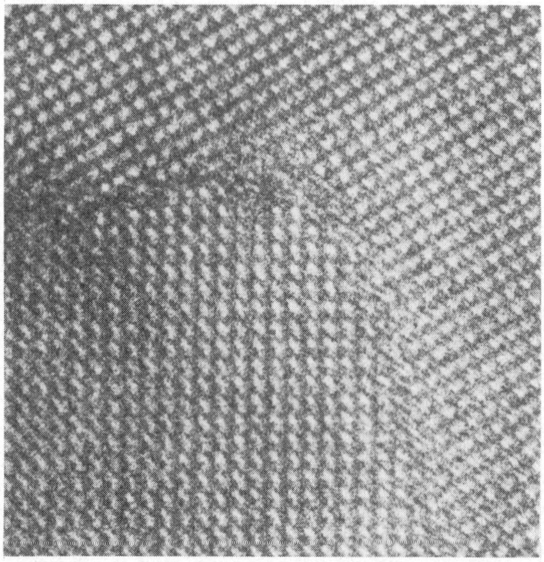

Fig. 3: Grain boundary between two A15 grains rotated by ∿ 29° around [001].

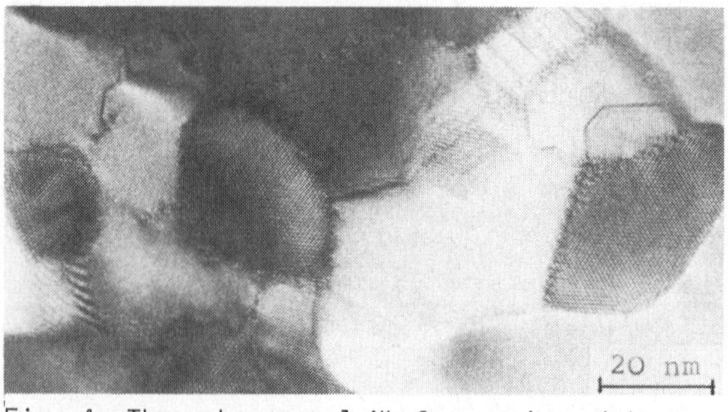

Fig. 4: Three hexagonal Nb_5Ge_3 grains with parallel orientation, separated by Nb_3Ge.

concluded that the black central dots correspond to Nb chains parallel to the [001]-direction. - A surprising orientation relation between adjacent A15 grains has been observed by SAD, which can be described by a rotation of ∿ 30° around a common [001]-axis. The Nb_5Ge_3 grains are oriented in parallel even though they are separated by Nb_3Ge. These SAD results are confirmed by the HREM images shown in Figs. 3 and 4. A curved plane grain boundary is reproduced in Fig. 3 with a rotation angle of 29.2° between two adjacent A15 grains. A statistical analysis of many similar images gives 28° as the most frequent rotation angle. In Fig. 4, three separate hexagonal Nb_5Ge_3 crystallites having a common orientation occur along the boundaries between A15 grains. - The following orientation relations have been established between the cubic A15 and the hexagonal ($D8_8$) grains:

$$(001)_{A15}//(0001)_{hex} \text{ and } [110]_{A15}//[1\bar{2}\bar{3}0]_{hex}$$

similar to what has been found for V_3Si (6). - A more complete analysis of textural relations and an atomic grain boundary model will be presented in a subsequent paper.

ACKNOWLEDGEMENT

We thank R. Wessicken for making excellent high resolution electron micrographs.

REFERENCES

(1) D. Yin, W. Schauer, F. Wüchner; IEEE Trans. Magn., MAG-19 (1983) 276.
(2) J.L. Jorda, R. Flükiger, J. Müller; J. of the Less-Com. Met. 62 (1978) 25.
(3) B. Krevet, W. Schauer, F. Wüchner; Appl. Phys. Lett. 36 (1980) 704.
(4) A.I. Braginski, G.W. Roland, A.T. Santhanam; IEEE Trans. Magn., MAG-15 (1979) 505.
(5) M. Konrad; KfK-Report 3244 (1981), unpubl.
(6) T. Ishimasa, Y. Fukano; Jap. J. Appl. Phys. 22 (1983) 6.

LT-17 (Contributed Papers)
U. Eckern, A. Schmid, W. Weber, H. Wühl (eds)

PRODUCTION OF A-15 COMPOUNDS BY CONDENSER DISCHARGE

Rui F.R.PEREIRA, Mãximo F.da SILVEIRA and Erich MEYER

Instituto de Física da Universidade Federal do Rio de Janeiro, Bl.A-4
Cidade Universitãria, 21944 - Rio de Janeiro, Brasil

Multiwire samples, consisting of two or more twisted wires of two different pure metals, are heated and melted by condenser discharge and quenched in a protective He atmosphere. The product is a large number of small spheres which represent the whole phase diagram (stable phases and metastable phase extensions), each sphere having typically a different composition, but being individually mostly homogeneous or having a second phase in microsegregation. Nb_3Al and Nb_3Au produced with this technique had T_c's up to 18.5 and 11.1K respectively. The described technique may be of interest not only for superconducting materials, but also for general stable and metastable phase diagram investigations.

1.INTRODUCTION

Contact-free melting, quenching and dendritic crystal growing of pure metals, with a technique which uses the capillary and pinch instability of wires molten by condenser discharge, has been reported before (1). The wire is melted by condenser discharge in a protective helium gas atmosphere. The liquid metal cylinder then becomes unstable because of the capillary and pinch forces and disintegrates into a "chain" of spherical samples. These liquid samples cool very rapidly by radiation and by contact with the He-gas (or liquid or superfluid He, if necessary) and solidify completely before touching any wall or the bottom of the apparatus. We modified this technique slightly, using two or more twisted wires of two different pure metals (e.g. Nb and Al or Nb and Au). In this way it is possible to obtain in a fraction of a second hundreds or different samples, all having statistically different compositions (because of the random loose mutual contact of the initial two wires), but all being individually homogeneous or containing a second phase in microsegregation. The statistical distribution of compositions can be changed by modifying the proportion of the two initial pure metal quantities e.g. by using more or thicker wires of material "A" and/or less or thinner wires of material "B". The large number and small size (0.1 - 0.7mm Ø) of the samples has the advantage, that the T_c's of many samples can be determined in a short time (~10min/sample) (2) and that entire samples can be analyzed e.g. in a Debye-Scherrer camera. Interesting samples (phases) can be selected by hardness or X-ray analysis and with some experience, just by the colour.

2.RESULTS

Table I shows the critical temperatures of some (as produced) Nb_3Al samples, before any annealing treatment, $T_c(b)$ (50% of the transition) and the width, $\Delta T_c(b)$ (10-90% of the transition). $T_c(a)$ and $\Delta T_c(a)$ are the corresponding values after a heat treatment in a protective argon atmosphere, as indicated in the last column. "a_o" is the lattice parameter, determined by the Nelson-Rilley extrapolation method (3) and "c(Al)" is the concentration of Al of the compound, obtained by using the experimental c(Al)-a_o relation of Flükiger et al. (4).

Nb-Au samples showed (as produced) exclusively A-1 (gold-rich) and A-2 (niobium-rich) structures. A-2 structure samples of favourable Au concentrations could be converted to the A-15 structure by annealing for 20-24 hours at 1000-1050C and showed T_c's up to 11.1K (5).

3.DISCUSSION AND CONCLUSIONS

Our samples showed lower critical temperatures than the highest reported in the literature, Nb_3Al: 19,0K (4) and Nb_3Au: 11.5K (6). However we believe that this is not due to a deficiency of our technique, but rather to incomplete annealing treatments. These heat treatments are in fact very delicate and time consuming (4) and work in this respect is in progress.

TABLE I
Critical temperatures of several Nb_3Al samples, before (b) and after (a) different annealing treatments, are shown. (w=week, d=day)

$T_c(b)$ (K)	$\Delta T_c(b)$ (K)	$T_c(a)$ (K)	$\Delta T_c(a)$ (K)	a_o (Å) (±0.002)	c(Al) (%)	annealing (K)
16.1	0.5	18.5	0.2	5.181	24.8	4w 750
16.0	0.4	18.3	0.3	5.182	24.4	1w 750 + 3w 700
15.8	0.7	17.6	0.4	5.183	24.1	1w 700
16.4	0.6	17.6	0.3	5.183	24.1	2w 650 + 1w 700
16.4	0.5	18.0	0.2	5.185	23.4	2d 700 + 4w 750

The described technique seems to be the simplest and fastest method for answering the question: "What stable and metastable phases and compositions can be obtained in a binary or multimetal system?"

ACKNOWLEDGEMENTS
The authors thank Profs. S.Moehlecke and A.R.Sweedler for useful discussions and the instituions CNPq, FINEP and CEPG for financial support.

REFERENCES
(1) E.Meyer and L.Rinderer, J.Crystal Growth 28 (1975) 199
(2) R.F.R.Pereira, E.Meyer and M.F.da Silveira, Rev.Sci.Instrum. 54 (1983) 899
(3) J.B.Nelson and D.P.Rilley, Proc. Phys.Soc. 57 (1945) 160
(4) R.Flükiger et al., Appl.Phys.Commun. 1 (1981) 9
(5) M.F.da Silveira, E.Meyer and R.F.R. Pereira, Rev.Bras. de Física 12 (1982) 867 (in portuguese)
(6) J.Muller et al., Proc. 13th Int. Conf.on Low Temp.Phys. Vol.3, eds. K.D.Timmerhaus et al. (Plenum Press N.Y. 1974) 446

LT-17 (Contributed Papers)
U. Eckern, A. Schmid, W. Weber, H. Wühl (eds)

NORMAL STATE ELECTRICAL RESISTIVITY OF THE A-15 SUPERCONDUCTOR, Ti_3Sb

V. SANKARANARAYANAN, G. RANGARAJAN, R. SRINIVASAN AND K.V.S. RAMA RAO

Department of Physics, Indian Institute of Technology, Madras-600 036, India

The normal state electrical resistivity of the A-15 superconductor, Ti_3Sb, as a function of temperature from 6 to 300 K is reported. The results are analysed on the basis of the generalised diffraction model which incorporates the Pippard-Ziman condition namely that the phonons with wavelength exceeding the mean free path of the electron are ineffective in scattering of electrons.

1. INTRODUCTION

In the case of high temperature superconductors such as A-15 and Chevrel phase compounds, the temperature variation of electrical resistivity exhibits a deviation from linearity and a tendency for saturation at high temperatures (1,2). In this paper the temperature variation of the normal state electrical resistivity of the A-15 superconductors, Ti_3Sb, from 6 to 300 K is reported.

2. RESULTS

The sample used in the present measurement was prepared by arc melting in the Institut fur Physikalische Chemie, Physikalische Chemie III, Technische Hochschule, Darmstadt. The details preparation and characterisation are reported in (3). The sample used in the present measurement was a square of side 0.5 cm and thickness about 0.2 cm. Electrical resistivity was measured using Montgomery's (4) four probe method. The details of the experimental arrangement are described elsewhere (5). The accuracies in the determination of electrical resistivity and temperature are respectively $\pm$ 0.5 μohm cm and 0.1 K respectively.

Figure 1 shows the variation of electrical resistivity of Ti_3Sb as a function of temperature. The value of the electrical resistivity just above the onset of superconducting transition is 26 μohm cm. This value is nearly twice that reported on anotehr sample by Ramakrishnan et al(6). The superconducting transition temperature of the present sample is 5.6 K. It is however to be mentioned that a range of T_c of 5.3 to 6.5 K has been reported by different workers (6,7,8).

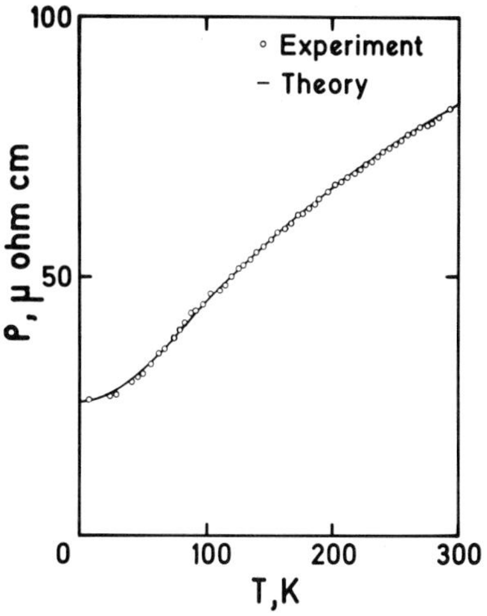

FIG.1

3. FIT TO COTE-MEISEL THEORY

Several models have been proposed to explain the temperature variation of resistivity of high temperature superconductors [Woodard and Cody (9), Morton et al (10), Cote and Meisel (11)]. Recently, Sankaranarayanan et al(12) found the Cote-Meisel model to give a better fit to the resistivity variation in a series of Chevrel phase compounds. The Cote-Meisel model, based on the diffraction model of Ziman, incorporates the Pippard-Ziman condition for the scattering of electrons by phonons. In this model one may write the expression for electrical resistivity as

$$\rho = (\rho_{ip}+\rho_o e^{-2W^*})/(\rho_{ip}/\rho^*+1) \qquad .. \quad (1)$$

Here, ρ_o is the resistivity just before the onset of the superconducting transition, ρ^* the saturation resistivity, $\exp(-2W^*)$ the averaged Debye-Waller factor and ρ_{ip} the ideal one phonon resistivity given by

$$\rho_{ip} = \beta'(T/\Theta_D)^2 \int_0^{\Theta_D/T} [x^2/(e^x-1)(1-e^{-x})]dx = \beta f(T) \qquad (2)$$

where $\beta = \beta'/\Theta_D^2$ is a constant and Θ_D is the Debye temperature. After substituting for ρ_{ip}, equation (1) may be rewritten as

$$(\rho-\rho_0)/\rho f(T) = \beta[(1/\rho)-(1/\rho^*)] \qquad .. (3)$$

The Debye temperature Θ_D was chosen as a parameter which has varied to give the best linear plot of $[(\rho-\rho_0)/\rho f(T)]$ against $(1/\rho)$. Figure 2 shows such a plot for the compound Ti_3Sb for Θ_D = 375 K.

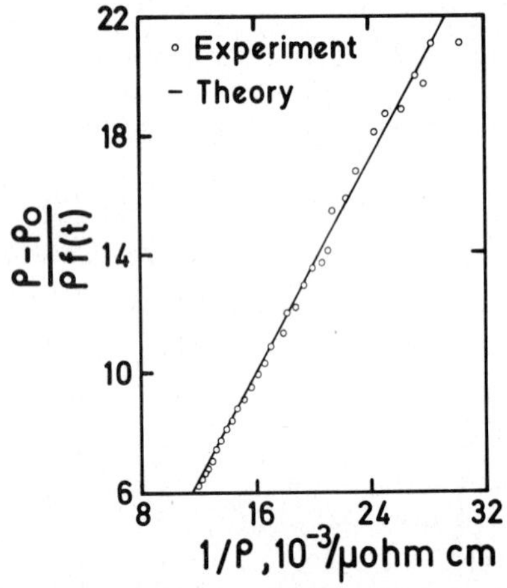

FIG. 2

From the slope and the intercept of the plot, the parameters β and ρ^* were obtained. The parameters obtained are shown in Table I and the continuous curve in Fig.1 is the theoretical fit. The RMS deviation of the experimental data points from the theoretical values is about 1%. Similar analysis of electrical resistivity data has been reported by Cote and Meisel (11) for the compounds Nb_3Sn and Nb_3Sb. It should be pointed out here that the parameter, β defined in the Cote-Meisel work is Debye temperature times the parameter, β used in the present work. The parameters obtained for Nb_3Sn and Nb_3Sb are also included in Table 1 for the sake of comparison. The value obtained for the parameter Θ_D agrees with the value of 360 K for V_3Sn (13) which is also an A-15 compound consisting of the neighbouring elements V and Sn.

TABLE I

	Ti_3Sb	Nb_3Sn & Nb_3Sb
ρ_0(μohm cm)	26.2	-
ρ^*(μohm cm)	206.3	200
β(10^{-3} μohm cm/K^2)	0.888	-
Θ_D(K)	375	-
$\beta\Theta_D$(μohm cm/K)	0.333	0.33

Thus the Cote-Meisel theory satisfactorily explains the variation of electrical resistivity with temperature in Ti_3Sb.

REFERENCES

(1) M. Milewits, S.J. Williamson and H. Taub, Phys. Rev. B 13 (1976) 5199.
(2) R.A. Martin, Thesis, University of California, Davis.
(3) K.V.S. Rama Rao, H. Sturm, B. Elschner and Alarich Weiss, Phys. Lett. 93A (1983) 492.
(4) H.C. Montgomery, J.Appl. Phys. 42 (1971) 2971.
(5) V. Sankaranarayanan, G. Rangarajan and R. Srinivasan, J. Phys. F: Met. Phys. 14 (1984) 691.
(6) S. Ramakrishnan and Girish Chandra, Proc. Indo-Soviet Conf. on Low Temp. Phys. held at the Indian Institute of Science, Bangalore, 1984 (to be published).
(7) B.T. Matthias, V.B. Compton and E. Corenzwit, J.Phys. Chem. Solids 19 (1961) 130.
(8) A. Junod, F. Heiniger, J. Muller and P. Spitzli, Helv. Phys. Acta, 43(1970) 59.
(9) D.W. Woodard and G.D. Cody, Phys. Rev. A 136 (1964) 166.
(10) N. Morton, B.W. James and G.H. Wostenholm, Cryogenics 18 (1978) 131.
(11) P.J. Cote and L.V. Meisel, Phys. Rev. Lett. 40 (1978) 1586.
(12) V. Sankaranarayanan, G. Rangarajan and R. Srinivasan, Proc. Indo-Soviet Conf. on Low Temp. Phys. held at the Indian Institute of Science, Bangalore, 1984 (to be published).
(13) F.Y. Fradin, G.S. Knapp, S.D. Bader, G. Cinader and C.W. Kimball, Proc. Second Rochester Conf. on Superconductivity in d- and f-band metals, Ed. D.H. Douglass, Plenum Press, New York and London (1976) 297.

LT-17 (Contributed Papers)
U. Eckern, A. Schmid, W. Weber, H. Wühl (eds)

EFFECTS OF CONCENTRATION AND ANNEALING ON THE SPECIFIC HEAT ANOMALY ASSOCIATED WITH SPIN-GLASS ORDERING IN CuMn*

G. E. Brodale, R. A. Fisher, W. E. Fogle,+ N. E. Phillips, and J. Van Curen

Department of Chemistry, University of California, Berkeley, California 94720, and Materials and Molecular Research Division, Lawrence Berkeley Laboratory, Berkeley, California 94720

The previously reported small specific heat anomaly in a 0.279 at.% CuMn sample has been shown to shift with T_g as the concentration is changed, providing additional support for the suggestion that it is a manifestation of the ordering processes that occur near T_g. The anomaly also depends on sample heat treatment, presumably reflecting changes in atomic order.

The specific heat, C, of an 0.279 at.% CuMn sample exhibits (1) a small anomaly near the spin-glass ordering temperature T_g. The existence of the anomaly is unambiguously established by structure in the first and second derivatives of C (or of C/T), but the data do not define the shape of the anomaly in detail. To study the field dependence of the anomaly it was somewhat arbitrarily defined relative to a background curve that had smooth first and second derivatives and reproduced the experimental data at both low and high temperatures (2). The same definition has been adopted to study the dependence of the anomaly on concentration, c, and on sample treatment, and those results are presented here.

The preparation of the annealed and unannealed samples is described in an accompanying paper (3) in which susceptibility measurements are reported.

To permit a comparison with scaling laws, the magnetic specific heat (total specific heat less that of pure Cu) for three annealed samples with different values of c is plotted as C/c vs T/c in Fig. 1. The 0.265 at.% sample was intended to duplicate as closely as possible the 0.279 at.% sample of reference (1) to provide a test of the sample-to-sample reproducibility of the anomaly. It was re-annealed for an additional 7 days with no significant effect on C. Data for annealed and unannealed samples are compared in Fig. 2, where the smooth background curves, defined as in reference(2), are presented by dotted curves.

The "anomalies", the difference between the background curves and the experimental points, are represented in Fig.3. For the three annealed samples, and for the 0.279 at.% sample, which is not included in the figures, the temperature of

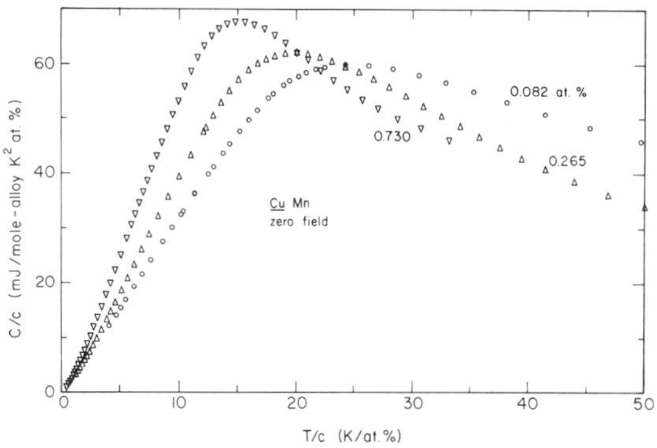

Figure 1. Scaling-law plot of the specific heat of three annealed CuMn samples.

the maximum in the anomaly, T_m, is strongly correlated with T_g. Values of the ratio T_m/T_g are given in Table I along with other parameters that characterize the anomaly. The entropy associated with the anomalies, ΔS, is reported as a percentage of the total spin entropy, Rln6. For the two samples that were studied in the unannealed form, T_m is lowered by about 30% and both the amplitude of the anomaly and ΔS are substantially reduced. The method of defining the anomaly tends to give it a width proportional to T_m, but the width of the structure in the second derivative itself shows that annealing does not appreciably effect the width of the anomaly relative to T_m.

The approximate proportionality of T_m to T_g provides additional evidence that the anomaly originally discovered in the 0.279 at.% sample

* This work was supported by the Director, Office of Energy Research, Office of Basic Energy Sciences, Materials Sciences Division of the U. S. Department of Energy under Contract No. W-7405-ENG-48.
+ Present address: National Bureau of Standards, B128, Bldg. 221, Washington D. C. 20234

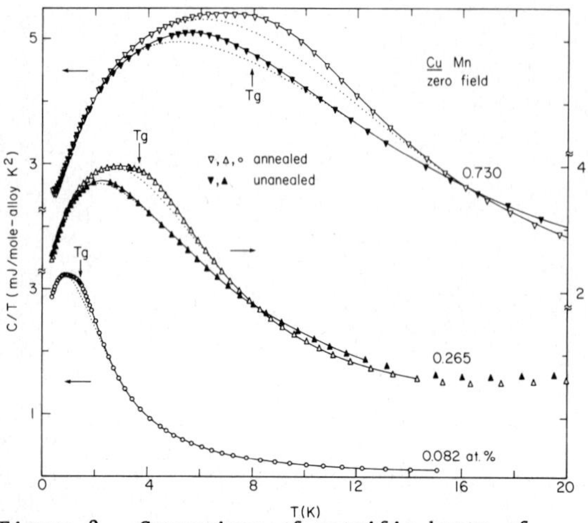

Figure 2. Comparison of specific heats of annealed and unannealed samples.

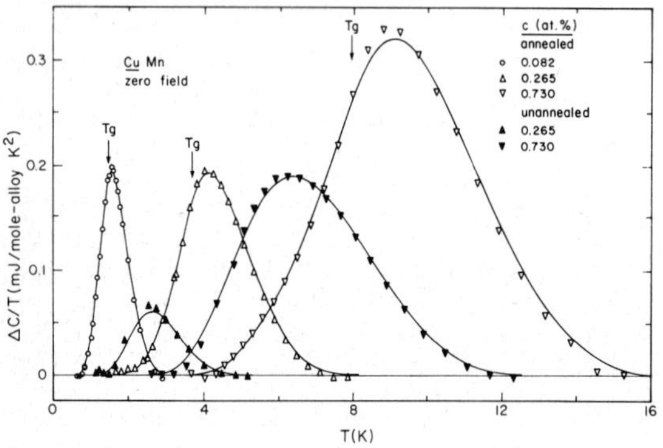

Figure 3. Specific heat anomalies in annealed and unannealed samples.

really is a manifestation of processes that occur at T_g. The sensitivity of the anomaly to heat treatment implies a sensitivity to the details of atomic order. It is obviously of considerable interest to understand that relation better than is possible on the basis of the available data on atomic order in this system. Two possible approaches to that end would be to study the anomaly in spin-glasses in which the atomic order is expected to be different, and to acquire more complete information on the relation between heat treatment and the resulting atomic order. In the meantime, it is significant that the heat treatment that gives the anomaly the greatest amplitude is also the one that produces the features in χ that are generally regarded as the signature of the spin-glass transition.

TABLE I. Characteristics of the specific heat anomalies.

c(at.%)	T_g(K)	T_m(K)	T_m/c (K/at.%)	T_m/T_g	$\frac{100\Delta S}{R\ln 6}$
Annealed samples					
0.082	1.45[a]	1.56	19.0	1.08	1.45
0.265	3.64	4.12	15.5	1.13	1.11
0.279	3.89	4.45	15.9	1.14	1.22
0.730	7.9	9.1	12.5	1.15	1.49
Unannealed samples					
0.265	---	2.64	10.0	---	0.24
0.730	---	6.4	8.8	---	0.73

[a] Estimated.

REFERENCES

(1) W. E. Fogle, J. D. Boyer, N. E. Phillips, and J. Van Curen, Phys. Rev. Lett. 47 (1981) 352.

(2) W. E. Fogle, J. D. Boyer, R. A. Fisher, and N. E. Phillips, Phys. Rev. Lett. 50 (1983) 1815.

(3) R. A. Fisher, E. W. Hornung, N. E. Phillips, and J. Van Curen, this volume.

LT-17 (Contributed Papers)
U. Eckern, A. Schmid, W. Weber, H. Wühl (eds)

MAGNETIC AND ELECTRICAL PROPERTIES OF Sn FILMS WITH 3at% Mn

U. HENGER and D. KORN

Fakultät für Physik, Universität Konstanz, Postfach 5560, D-7750 Konstanz, FRG

Sn+3at%Mn films are prepared by vapour condensation onto a liquid helium cooled substrate. The magnetic susceptibility and the electrical resistivity are measured in situ as a function of temperature after annealing to the temperatures Ta = 40 K, 80 K, 150 K. The electrical resistivity shows a maximum at low temperature. The slope above the temperature of the maximum decreases with annealing whereas the magnetic moment increases with annealing. This result is ascribed to the influence of lattice defects.

1. INTRODUCTION

When transition metals like Mn or Fe are dissolved in simple metals like Au or Cu the 3d levels of the impurities hybridize with the conduction band states of the host to form a virtual bound state (1). This means scattering of the conduction electrons into the 3d states. Localized magnetic moments are formed if the virtual bound states are not equally populated by spin up and spin down electrons. Best conditions for a localized moment are half filled d shells of the transition metals and a low Fermi energy of the host (2). The magnitude of the magnetic moment p_{eff} can be determined by measuring the magnetic susceptibility following the Curie Weiss law:

$$\chi = \frac{\mu_0 \cdot n \cdot p_{eff}^2 \cdot \mu_B^2}{3 \cdot k_B \cdot (T-\Theta)}$$

χ - mass susceptibility
μ_0 - vacuum permeability
n - number of magnetic atoms per mass unit
p_{eff} = $4 \cdot S(S+1)$
S - localized spin

The above mentioned dilute alloys show a resistance anomaly with a minimum at low temperatures. Kondo (3) described this effect by calculating the second order term in spin flip scattering. He found the temperature dependent part of the spin flip resistivity as

$$\rho_{spin} \propto c \cdot J \cdot S(S+1) \cdot \ln T$$

c - concentration of the localized spins
J - exchange interaction (J must be negative)

ρ_{spin} must be added to the temperature independent part of the magnetic resistivity and to contributions due to lattice defects and lattice vibrations. Further formulae for the spin dependent part of the resistivity are given by Fischer (4).

2. RESULTS AND DISCUSSION

The experimental procedure of the susceptibility measurement and the resistivity measurement is described elsewhere (5,6). Fig. 1 shows the inverse susceptibility χ^{-1} of Sn+3at%Mn. For all annealing stages χ^{-1} has a minimum at the same spin glass temperature T_f of

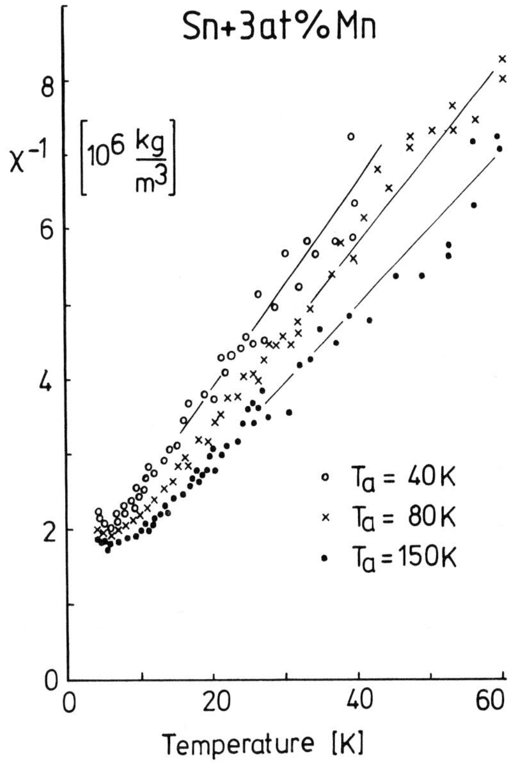

FIGURE 1

6 K . Above T_f the susceptibility is described by the Curie Weiss law. The slope of χ^{-1} decreases and hence the obtained p_{eff} increases with the annealing temperature T_a.

The resistivity ρ of Sn+3at%Mn (Fig. 2) shows for T_a = 40 K a maximum at 7.5 K and a minimum at 22 K. $-d\rho/dT$ is taken at 14 K. It only contains the temperature dependent parts of the resistivity due to the lattice vibrations and to the Mn spins. Assuming that the resistivity of the lattice vibrations is the same for all annealing temperatures, the variation of $-d\rho/dT$ with T_a means that the temperature dependent resistivity of the Mn spins changes with annealing.

p_{eff} and hence S(S+1) taken from Fig. 1 increases linearly with T_a (Fig. 3). Therefore

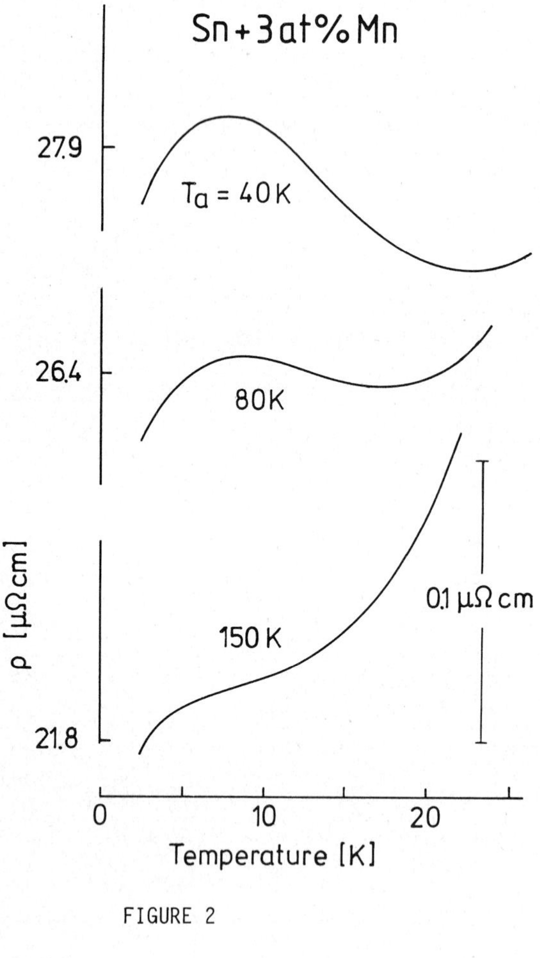

FIGURE 2

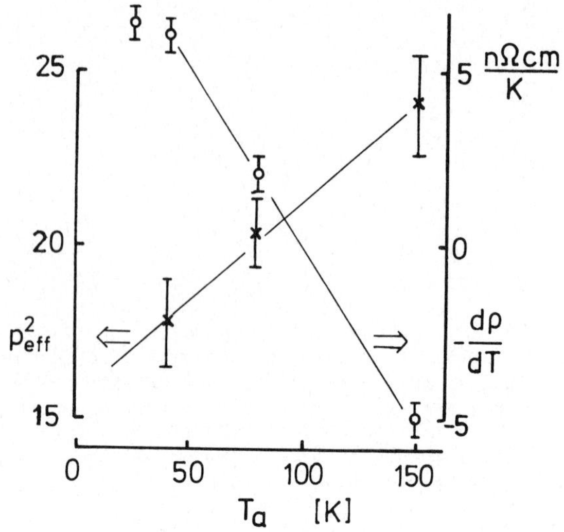

FIGURE 3

one expects that $-d\rho/dT$ also increases with T_a despite the high concentration. Fig. 3 however shows that $-d\rho/dT$ decreases linearly with T_a. This behaviour can be explained by the recovery of lattice defects. Lattice defects around the impurities strengthen the scattering of electrons into the d states. This causes a stronger resistance anomaly. On the other hand the stronger scattering decreases the localized magnetic moment of the impurity.

3. CONCLUSION

With regard to Sn+3at%Mn lattice defects contribute to the resistance anomaly. Also the magnitude of the localized moment depends on the lattice defects.

REFERENCES

(1) J. Friedel, J. Physique Radium 19 (1958) 573.
(2) P.W. Anderson, Phys. Rev. 124 (1961) 41.
(3) J. Kondo, Progr. Theo. Phys. 32 (1964) 37.
(4) K. Fischer, in: Landolt-Börnstein, Numerical Data and Relationships, Group III Vol. 15a, eds. K.H. Hellwege and J.L. Olsen (Springer; Berlin, Heidelberg, New York; 1982) pp. 289-291.
(5) G. Zibold and D. Korn, J. Phys. E 12 (1979) 490.
(6) D. Korn and K. Maurer, J. Physique 44 (1983) 490.

LT-17 (Contributed Papers)
U. Eckern, A. Schmid, W. Weber, H. Wühl (eds)
© Elsevier Science Publishers B.V., 1984

SPIN FREEZING TEMPERATURES IN DILUTE NOBLE METAL-Mn SPIN GLASSES: MEAN FREE PATH EFFECTS

J. R. THOMPSON, J. T. ELLIS and J. O. THOMSON

Department of Physics, University of Tennessee, Knoxville, Tennessee 37996-1200, USA

Spin interaction temperatures T_i in dilute noble metal-Mn alloys are related to the spin-spin interaction strength V_0. The influence of a reduced mean free path of the conduction electrons is examined; in $Cu_3Au(Mn)$, an exponential damping accounts satisfactorily for the attenuated interaction using transport values for the mfp.

1. INTRODUCTION

The existence of spin glasses is generally attributed to the presence of competing ferro- and antiferromagnetic interactions between magnetic species in a many body system. This competition produces "frustration" of the spins and can originate microscopically through many mechanisms, thereby providing spin glass behavior in insulating and semiconducting materials as well as conducting hosts. The long range, oscillatory RKKY interaction in metallic systems is the best known of these and the dominant (1) force in dilute noble metal based spin glasses.

In this paper, we discuss the spin interaction temperature T_i, obtained largely from dc magnetization measurements, in **dilute** noble metal-Mn alloys. The dependence of T_i on Mn concentration x is examined for Cu, Ag, Au, and Cu_3Au based alloys and related to the spin-spin interaction strength V_0, defined below. In Cu_3Au, we show that a reduction in electronic mean free path (mfp), which tends to decouple widely separated Mn spins (2), is accompanied by reduced spin interaction temperatures.

2. THEORY

We assume that the dominant interaction between magnetic ions with spins S_i and S_j has an isotropic RKKY form

$$H = \sum_{i,j}{}' V_0\, r_{ij}^{-3} \exp(-r_{ij}/\ell)[\cos 2k_F r_{ij}]\, \vec{S}_i \cdot \vec{S}_j \quad (1)$$

Here the separation between spins is r_{ij}, V_0 is the coupling strength, and the exponential damping term allows for mfp attenuation of the interaction in structurally disordered alloys. To proceed, we replace the exponential by an <u>average</u> factor $\exp(-\langle r\rangle/\ell)$, where $\langle r\rangle$ is the <u>mean</u> interimpurity separation and ℓ is the mfp. Since the r^{-3} factor is proportional to concentration x, it follows from scaling (3) that quantities such as T_i which depend on $\langle H\rangle$ should vary as

$$\langle H\rangle \sim k_B T_i = h\, V_0 x (N/V) \exp(-\langle r\rangle/\ell)\, S(S+1) \quad (2)$$

The factor "h", obtained in principle from a microscopic theory, is a dimensionless number of order unity which we determine experimentally.

At high temperature, the total dc magnetization of the dilute alloys follows a Curie law $M(T) = CH/T$. Near the spin freezing temperature, an M vs (1/T) curve has a knee, flattens, and approaches its zero temperature value M(0) with an $\alpha T^{3/2}$ dependence. We define(4) the spin interaction temperature as $T_i = CH/M(0)$, which is closely associated with T_f, the cusp in the ac susceptibility. From the coefficient α, we have obtained values(5) for V_0; see Table I.

3. EXPERIMENTAL RESULTS AND DISCUSSION

3.1 Scaling

SQUID-based magnetization studies of noble metal-Mn alloys with $x \leq 270$ at ppm have been performed in several laboratories. Reflecting weak spin-spin interactions, the T_i values are well below 1 K. In figure 1, we collect these data in a log-log plot of T_i vs $x \exp(-\langle r\rangle/\ell)$ for Cu(Mn) [Ref (4)], Ag(Mn) [Ref (6)], Au(Mn) [Ref (7)] and $Cu_3Au(Mn)$. We have assumed that $\ell^{-1}=0$ in the first three cases. In Cu_3Au the mean free path can be varied, which substantially affects the spin interaction temperature, and is discussed below.

TABLE I: Noble metal-Mn spin glass parameters

Alloy	Spin S	V_0[a]	n[b]
Cu(Mn)	1.67	14.2	0.887
Ag(Mn)	2.3	7.1	0.864
Au(Mn)	2.15	6.0	0.891
$Cu_3Au(Mn)$	1.6-1.9	11.4	0.888

[a]units of 10^{-37} erg·cm^3.
[b]n is exponent in $T_i \propto x^n$.

There are several features in Fig. 1 to be noted. First the data for Cu, Ag, and Au hosts lie on parallel lines, showing a common behavior in these spin glasses. As suggested by the scaling relation, Eq. 2, the data can be collapsed onto a "universal line", thereby giving a value for the factor h. By fitting Eq. 2 we obtain h = 0.50 ± 0.05. The error bars arise from both statistical uncertainty and a small, but systematic departure from the linear dependence on x in Eq. 2.

This systematic departure is noteworthy; although parallel, all lines have slopes which are slightly less than unity. If one assumes $T_i \propto x^n$, one finds an average slope n=0.88 with remarkable consistency between the Cu, Ag, and Au based alloys (Table I). Since the experimental work was performed in three separate laboratories with different chemical analysis techniques, etc., it appears that the weaker then linear dependence on concentration is real and not an experimental artifact. Why is T_i not linearly proportional to x? For more concentrated Cu(Mn) spin glasses, Walstedt (1) has utilized an exponential, self-damping term like that in Eq. 1, with a Mn cross section $\sigma = 3.1a^2$ (a = FCC lattice constant). The dependence of T_i on x is thereby reproduced, although the cross section used is substantially larger than the transport value of $0.54a^2$. This explanation appears to be untenable for the dilute alloys, where scaling should work best, since very large damping would be needed. We find that (σ/a^2) values of 32, 27, and 22, respectively, are required if we force the Cu, Ag, and Au(Mn) data in Fig. 1 to have unity slopes; such large cross sections seem unphysical. An alternate explanation is chemical clustering, but that too seems improbable in alloys where the Mn solubility is fairly large and the concentrations are quite low, $x \sim 10^{-5} - 10^{-4}$. At present, the origin of the deviation from ideal scaling is unclear.

3.2 Mean free path effects in $Cu_3Au(Mn)$

To investigate more directly the influence of diminished electronic mean free path on the spin-spin interaction, the order-disorder alloy Cu_3Au was used as a noble metal host (2). When structurally disordered the "FCC structure" and lattice constant are little changed, but the mfp is reduced by a factor of 3-10 relative to ordered Cu_3Au and becomes comparable with the average Mn-Mn separation <r>. Our experimental values of T_i in $Cu_3Au(Mn)$ are also plotted in Fig. 1, as a function of $x\exp(-\langle r\rangle/\ell)$. The mfp was calculated using the relation $\rho\ell = 0.85 \times 10^{-11} \Omega\cdot cm^2$, which is approximately the $\rho\ell$ product obtained from free electron theory for the electrical resistivity ρ. Large changes in T_i can result; for example, T_i decreased from 170 mK in ordered Cu_3Au to 97 mK when the same sample, containing 100 at ppm Mn, was disordered. It is evident from the figure that the behavior of $Cu_3Au(Mn)$ spin glass is very similar

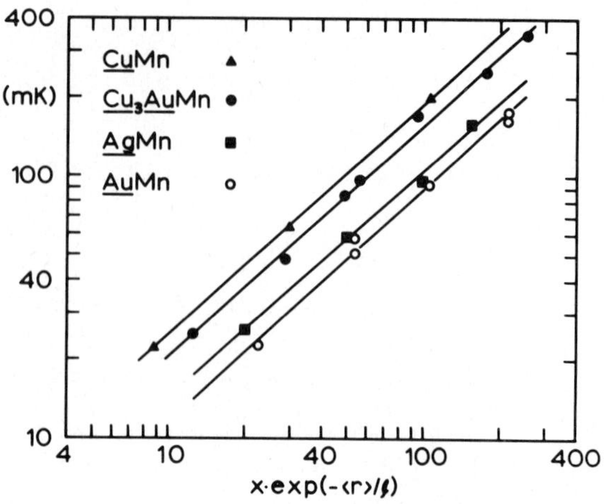

Fig. 1. The spin glass interaction temperature T_i vs. the product of Mn concentration x (at ppm) and an attenuation due to limited mean free path ℓ.

to that of the binary spin glasses, provided the exponential attenuation is included. The quality of this correction, which is not small, supports our assertion (2) that the observed decrease in T_i (and effective coupling) in $Cu_3Au(Mn)$ is due primarily to reduction of the mfp.

The Mn-Mn coupling strength V_0 also can be estimated, assuming $T_i \propto V_0$. Interpolating between the data for the Au and Cu based alloys in Fig. 1, one finds that the $Cu_3Au(Mn)$ data lie at 68% of their separation. The simple "lever rule" value is close, perhaps surprisingly so, to the 75 at % Cu composition in the host. This suggests that the Mn spins interact within a host comprised of "average atoms". From the T_i interpolation and values for V_0 in the binary alloys, we obtain $V_0 = 11.6 \times 10^{-37}$ erg·cm^3; this is within experimental uncertainty the value obtained from analysis of the low temperature variation of the dc magnetization.

REFERENCES

(1) R. E. Walstedt, Physica 110B, (1982) 1924).
(2) J. R. Thompson, J. T. Ellis, and J. O. Thomson, J. Appl. Phys., in print.
(3) J. Souletie and R. Tournier, J. Low Temp. Phys. 1 (1969) 95.
(4) E. C. Hirschkoff, O. G. Symko, and J. C. Wheatley, J. Low Temp. Phys. 5 (1971) 155.
(5) J. O. Thomson and J. R. Thompson, Physica 107B (1981) 637.
(6) J. C. Doran and O. G. Symko, AIP Conf. Proc. 18 (1971) 983 and Solid State Comm. 14 (1974) 719.
(7) J. R. Thompson and J. O. Thomson, Physica 107B (1981) 635.

LT-17 (Contributed Papers)
U. Eckern, A. Schmid, W. Weber, H. Wühl (eds)

SPIN-ORBIT COUPLING AND THE MAGNETORESISTANCE OF THE SPIN-GLASSES Au-8at%Mn AND Cu-4.6at%Mn

R.D. BARNARD

Department of Pure and Applied Physics, University of Salford, Salford, M5 4WT U.K.

Magnetoresistance measurements are reported on the spin-glasses Au-8at%Mn and Cu-4.6at%Mn as functions of both field and temperature. A sharp cusp is observed at the freezing temperature of 25K in Au-8at%Mn in a field of 130 gauss, but an almost undetectable coefficient was found in the case of Cu-4.6at%Mn. The dissimilar field dependences found in the two systems are discussed in terms of the presence of skew scattering in Au-Mn and its absence in Cu-Mn.

Previous measurements on the magnetoresistance of various spin-glasses (1) have not yielded any sharp anomaly at the freezing temperature T_f, and recent theoretical works on the transverse magnetoresistance (2) have so far assumed that spin-orbit effects are negligible. In this paper it is shown, contrary to the above observation, that the archetype spin-glass Au-8at%Mn possesses a very sharp anomaly at T_f provided the measurements are made in very low magnetic fields. We shall also argue that this occurrence appears to be closely associated with skew scattering from spin-orbit coupling, so far neglected in theoretical calculations of the magnetoresistance.

Au-8at%Mn is an example of a spin-glass in which the moment on the manganese atoms has an orbital contribution which is known to produce skew scattering from the spin-orbit coupling of the conduction electrons. This results in an enhanced Hall coefficient with a rounded peak at T_f when measured in moderately small magnetic fields (3). In figure 1 is shown our Hall results on Au-8at%Mn, (measured 20 days after a quench from 900°C) in fields as low as 10 and 2.5 gauss. It is evident that an extremely sharp peak occurs at 25K comparable with that observed in the A.C. susceptibility. On the other hand, our results on Cu-4.6at%Mn, (a system in which no orbital contribution to the moment obtains and hence no skew scattering) have shown no evidence of a peak at T_f and a value of the Hall coefficient nearly two orders of magnitude smaller than that of Au-8at%Mn.

The transverse magnetoresistance of Au-8at% Mn in a field of 130 gauss is shown in figure 2. This was the lowest field in which results to about 1% could be obtained and clearly a very sharp peak occurs in $-\Delta\rho/\rho_0$ at T_f. However, as

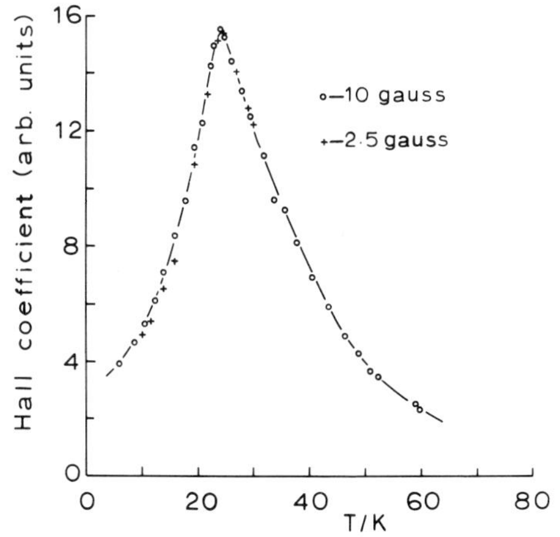

FIGURE 1
The Hall coefficient of Au-8at%Mn

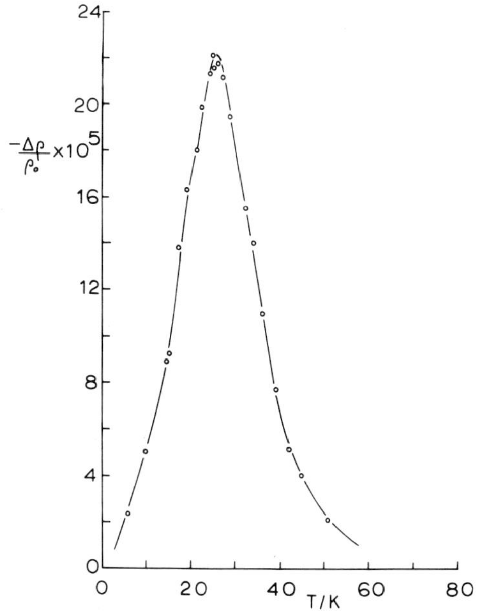

FIGURE 2
Magnetoresistance of Au-8at%Mn in 130 gauss.

with the Hall coefficient, the corresponding magnetoresistance of Cu-4.6at%Mn was so small ($-\Delta\rho/\rho_o < 10^{-6}$) that it was unplottable in figure 2.

Thus contrary to previously held views, the transverse magnetoresistance can now be included as another property of spin-glasses exhibiting a sharp anomaly at the freezing temperature. However, its appearance in Au-8at%Mn and its absence in Cu-4.6at%Mn strongly supports the view that the anomaly will only exist in those spin-glasses where asymmetrical skew scattering from spin-orbit coupling obtains.

To further clarify the role of spin-orbit coupling on the magnetoresistance, we have examined the field dependence of $-\Delta\rho/\rho_o$ in Au-8 at%Mn and Cu-4.6at%Mn. Figure 3 shows the

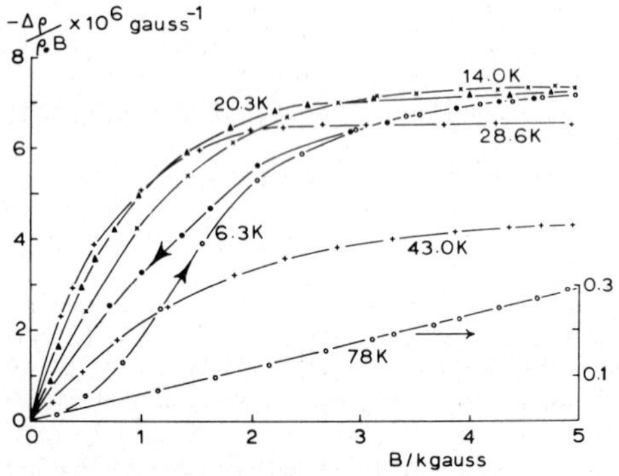

FIGURE 3
The field dependence of the magnetoresistance divided by B of Au-8at%Mn.

variation of $-\Delta\rho/\rho_0 B$ versus B for Au-8at%Mn at 6.3, 14.0, 20.3, 28.6, 43, and 78K. Conspicuous features of the curves are 1) rapidly rising coefficient of $-\Delta\rho/\rho_0$ at low fields, 2) a saturation of $-\Delta\rho/\rho_0 B$ for all temperatures where spin-glass effects are prominent, giving rise to

$$-\Delta\rho/\rho_o \propto B \quad ....(1)$$

at high fields, 3) for $T<T_f$, $-\Delta\rho/\rho_0 B$ saturates to the same value, -0.73×10^{-5} gauss^{-1}, and 4) hysteresis effects at 6.3K indicating a close connection between magnetoresistance and the magnetic state of the sample. Only at high temperatures, eg. 78K, does the expected B^2 dependence appear to be obeyed. (4).

In contrast to the above behaviour, in figure 4 is shown corresponding variations for the spin-only system Cu-4.6at%Mn. Here the essential variation is one of B^2 (4), although near T_f a variation of the form

$$-\Delta\rho/\rho_o = aB + bB^2 \quad .....(2)$$

is evident at high fields. The linear term may indicate the existence of a very small spin-orbit contribution in this system. But the

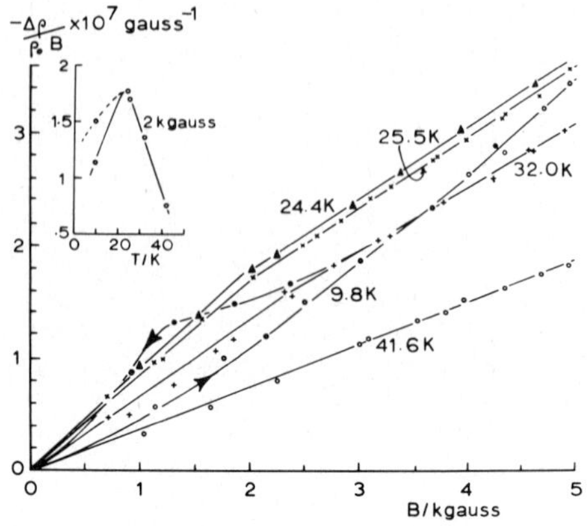

FIGURE 4
The field dependence of the magnetoresistance divided by B of Cu-4.6at%Mn.

essential fact is clear: the Au-8at%Mn sample with asymmetrical skew scattering possesses an entirely different field dependence from the spin-only system Cu-4.6at%Mn and a considerably enhanced coefficient at low fields. Furthermore, the magnetoresistance versus B variation in Au-8at%Mn is unchanged in form in the unfrozen paramagnetic state from that below T_f. Thus it appears that the characteristic shape of the $-\Delta\rho/\rho_0 B$ versus B variation, shown in figure 3, becomes established as the RKKY interaction progressively locks the manganese moments in cluster formation and the interactions which produce cluster freezing are not particularly important as far as $-\Delta\rho/\rho_o$ is concerned. Recognition of this feature should simplify analysis of these systems, where the role of spin-orbit scattering on the magnetoresistance clearly warrants further theoretical investigation.

ACKNOWLEDGEMENTS

This work is supported by a research grant from the SERC for which the author would like to record his thanks.

REFERENCES

(1) A.K. Nigam and A.K. Majumdar, Phys. Rev. B 27 (1983) 495.
(2) A. Mookerjee and D. Chowdhury, J. Phys. F. (Metal Phys.) 13 (1983) 365.
(3) S.P. McAlister and C.M. Hurd, J. Phys. F. (Metal Phys.) 8 (1978) 239.
(4) M.T. Béal-Monod and R.A. Weiner, Phys. Rev. 170 (1968) 552.

LT-17 (Contributed Papers)
U. Eckern, A. Schmid, W. Weber, H. Wühl (eds)

LOW TEMPERATURE BEHAVIOUR OF AMORPHOUS MAGNETIC SYSTEMS

C. BANSAL and V. SRINIVASAN

School of Physics, University of Hyderabad, Hyderabad 500 134, India

It is suggested that the magnetic behaviour of amorphous magnetic systems should be analysed by introducing a coupling between the crystalline and spin order parameters. The amorphous system will order in a ferromagnetic state if this coupling is weak and into a spin glass state if this coupling is strong.

1. INTRODUCTION

The nature of magnetic order and low temperature excitations in amorphous systems has been a subject of interest for a long time (1) and not so well understood. A useful starting point is to compare the magnetic properties of the non-crystalline system with a crystalline counterpart having the same constituents wherever possible or to compare very similar amorphous and crystalline systems. We give here a few typical examples from the numerous one's experimentally investigated to illustrate the observed effects.

Crystalline Mn Si is a ferromagnet with a Curie temperature of 30 K (2) whereas amorphous Mn Si shows spin glass behaviour with a transition temperature of 22 K (3). Here the removal of crystalline order results in a disordering of the magnetic long range order also and although the amorphous system has a high concentration of magnetic moment bearing atoms (50%) the magnetic order is into a spin-glass state. Similarly, crystalline $Y_{1-x}Fe_x$ alloys are ferromagnetic whereas amorphous alloys have a non-collinear spin structure (4) or spin glass behaviour (5). On the other hand the magnetic order in the $Fe_{1-x}B_x$ system essentially remains the same in amorphous and crystalline states (6) whereas the Curie temperature is lower for the amorphous alloys than the corresponding crystalline alloys (e.g. T_c = 878 K and 509 K respectively for crystalline and amorphous $Fe_{88}B_{12}$) (7). The alloy system $B_{80-x}Fe_xP_{20}$ studied for $13 < x < 44$ exhibits close resemblance to the crystalline $Au_{1-x}Fe_x$ system and the magnetic order goes over from a spin glass type to long range ferromagnetic order for $x > 26$ (8). On the other hand, amorphous $Zr_{40}Cu_{60-x}Fe_x$ and $Nb_{50}Ni_{50-x}Fe_x$ systems do not show spin glass behaviour even for low concentration of Fe.

Besides the type of magnetic order other properties that have been compared in literature are the spin wave excitations, critical exponents, magnetic anisotropy, average magnetic moment per atom, magnetostriction etc. (9) and differences and similarities in amorphous and crystalline behaviour pointed out. The number of systems and magnetic properties investigated is indeed very large now and an effort is required to understand them in general formalism.

In our earlier work we have described the crystalline and glass states as two different space-time dependent ground states of a system of interacting molecules where the translational invariance is broken spontaneously. The Lagragian for a system in the gaseous state is translationally invariant. When the temperature is lowered and it freezes into a solid the Lagrangian is no longer translationally invariant and this symmetry is broken spontaneously. Umezawa et al (10) have shown that the perfect crystal state or the crystal with a defect or dislocation are different time dependent ground states of the above Lagrangian. In general, the order parameter associated with the system:

$\langle\chi(x)\rangle = 0$ denotes the gaseous state

$\langle\chi(x)\rangle = \langle\chi(x+a)\rangle \neq 0$ (for the lattice periodicity a), gives the crystalline state and $\langle\chi(x)\rangle \neq \langle\chi(x+a)\rangle \neq 0$ describes a defect or a dislocation. If the density of defects in the system is large the long range order appears to be destroyed and this state is the glass state (11). The quasiparticles in this case were shown to be both two level excitations and phonons.

In a magnetic system described by a rotationally invariant spin Hamiltonian

$$H = - \sum_{i>j} J_{ij} \vec{S}_i \cdot \vec{S}_j$$

the ground state of the system wherein rotational invariance is broken spontaneously is the ferromagnetic or spin glass state (12). The order parameter for such a magnetically ordered system

$$\langle\psi\rangle \neq 0$$

It therefore seems a logical extension of these two microscopic models to describe the magnetic state of an amorphous system in terms of both order parameters $<\chi(x)>$ and $<\psi>$ and a coupling between the two order parameters. A physical basis of the origin of the coupling between the crystalline order and magnetic order is the magneto-crystalline anisotropy (13) which results due to the fact that orbital overlap of electrons in a solid depends on its structure and the spins are then influenced by the spin-orbit interaction of the orbital electrons.

The free energy of the system will now be given by a Landau-Ginsburg form for each separately together with an interaction between the two. The formalism need not be given as it proceeds along the same lines as adopted for the problem of coexistence of superconductivity and ferromagnetism.

A gas of molecules can therefore exist in the following states when condensed:

State	Order Parameters		Coupling
1. Crystalline order with long range magnetic order	$<\chi(x)>=<\chi(x+a)>\neq 0$	$<\psi>\neq 0$	Very weak
2. Amorphous with long range magnetic order	$<\chi(x)>\neq<\chi(x+a)>\neq 0$	$<\psi>\neq 0$	intermediate
3. Amorphous without long range magnetic order	$<\chi(x)>\neq\chi(x+a)>\neq 0$	$<\psi>=0$	strong

An experimental observation which supports our viewpoint is the existence of substantial magnetic anisotropy even in amorphous magnetic systems which suggests that the interplay of magnetic and crystalline order has a crucial role to play in the magnetic behaviour of amorphous materials.

REFERENCES

(1) A.I. Gubanov, Sov. Phys. Solid State 2, 468 (1960).

(2) J.H. Wernick, G.K. Werthein and R.G. Sherwood, Mat. Res. Bull. 7, 1431 (1972).

(3) J.J. Hauser, I.S.L. Hsu, G.W. Kammlott and J.V. Waszozak, Phys. Rev. B20, 3391 (1979).

(4) J.J. Ryne, J.H. Schilling and N.C. Koon, Phys. Rev. B10, 4672 (1974)

(5) J.M.D. Coey, D. Givord, A. Lienhard and J.P. Rebouillat, J. Phys. F. Metal Phys. 11, 2707 (1981).

(6) J.A. Mydosh and G.J. Nieuwenhuys in Ferromagnetic Materials ed. E.P. Wohlfastt (North Holland Publishing Co., Amsterdam) Vol.1, p.71 (1980).

(7) R. Hasegawa and R. Ray, J. Appl. Phys. 49, 4174 (1978).

(8) P.L. Maitrepierre, J. Appl. Phys. 40, 4826 (1969).

(9) H.S. Chen, Rep. Prog. Phys. 43, 353 (1980)

(10) M. Wadate, H. Matrumoto, Y. Takahashi and H. Umezawa, Phys. Lett.62A, 255 (1977); Fortschr. Phys. (1978).

(11) C. Bansal and V. Srinivasan, Phys. Lett. 75A, 420 (1980).

(12) C. Bansal and V. Srinivasan, Solid State Comm. 49, 455 (1984).

(13) See e.g. C. Kittel, Introduction to Solid State Physics (John Wiley and Sons, New York) p.490 IIIrd Edition.

LT-17 (Contributed Papers)
U. Eckern, A. Schmid, W. Weber, H. Wühl (eds)

NEW EXPERIMENTAL RESULTS FOR METALLIC SPIN GLASSES

Sheldon SCHULTZ

Department of Physics, B-019, University of California, San Diego, La Jolla, California 92093*

In conjunction with the collaborators cited, we report new experimental results for metallic spin glasses in the following areas. (1) The dependence of the spin glass transition temperature, T_g, on magnetic and non-magnetic impurities (Mr. David Vier), (2) New features in the magnetic field dependence of the parallel ($\chi_{\parallel}$) and perpendicular ($\chi_{\perp}$) ac susceptibilities (Mr. Dojun Youm), (3) The temperature dependence of the anisotropy in $AgMn_{4\%}Er_x$ spin glass alloys (Professor Saul Oseroff), and (4) Direct measurements of the triad modes in a spin glass (Dr. Eric Gullikson). The particular results in these experiments that suggest the need for new theoretical explanations will be discussed.

1. THE DEPENDENCE OF THE SPIN GLASS TRANSITION TEMPERATURE, T_g, ON MAGNETIC AND NON-MAGNETIC IMPURITIES (in collaboration with Mr. David Vier).

We have made a comprehensive study (over 200 samples) of the dependence of T_g on the addition of magnetic and non-magnetic impurities to the metallic spin glasses CuMn, AgMn, AuMn, AuCr, and AuFe. We determine T_g from the cusp in the dc susceptibility as measured in low field (<10G) with a superconducting quantum-interference detector (SQUID). We also measure the dc resistivity, ρ. We find the following.

1.1. The well-known deviations from a linear dependence of T_g on local moment concentrations above 0.5% is determined to be a function of the finite mean free path due to the self resistivity as proposed by Larsen (1), but a re-examination of the theoretical formulation used in reference (1) (and in subsequent work by others), reveals that it is in error (2), despite their apparent good fit to the data.

1.2. For several systems the dependence of T_g on ρ (due to an added non-magnetic impurity) is well represented by the form:

$$T_g(\rho) = T_g(\infty) + (T_g(0) - T_g(\infty))(1 - e^{-\rho/\rho_0})$$

1.3. T_g does not go to zero as $\rho \to \infty$ but instead there is a large residual spin glass transition temperature, $T(\infty)$.

1.4. $T_g(0)$ obtained from the data analysis is found to be linear in Mn concentration for AgMn, but not so for CuMn.

1.5. $T_g(\infty)$ obtained from the data analysis is found to be linear in Mn concentration for both AgMn and CuMn.

1.6. A study of the effects of impurities with small incremental resistivities and large spin-orbit scattering, suggest that in some cases there may be effects on T_g due to increased anisotropy.

1.7. The expected increase in T_g when other magnetic moments are added can be observed when the reduction due to their incremental resistivity is properly taken into account.

The need for a theoretical explanation of our evidence for a significant residual short range interaction (exemplified by the large $T_g(\infty)$) will be discussed.

2. NEW FEATURES IN THE MAGNETIC FIELD DEPENDENCE OF THE PARALLEL ($\chi_{\parallel}$) AND PERPENDICULAR ($\chi_{\perp}$) AC SUSCEPTIBILITIES (in collaboration with Mr. Dojun Youm).

We have made detailed measurements of the magnetic field dependence of $\chi_{\parallel}$ and $\chi_{\perp}$ for the spin glass alloys $Cu_{1-x}Mn_x$ (x = 2, 5, 8, and 10%, and $CuMn_5Ni_{0.5}$ and $CuMn_5Ge_5$ (all %), over the temperature range $T_g/10^5$ to 1.4 T_g. We find the following.

2.1. In contrast to the "rounding" of the $\chi_{\parallel}$ cusp due to increasing magnetic field characteristically reported in the literature, we find for dc fields, H_0, above 400 G, there is a distinct temperature independent region, or plateau.

2.2. For a given field, the plateau region terminates in a sharp cusp-like feature, with the line connecting such points given by $\Delta\chi/\chi_g = \alpha\Delta T/T_g$. The parameter α is close to 3 for all the pure alloys.

2.3. The $\chi_{\perp}$ data agree with values predicted from the integration of the field dependent $\chi_{\parallel}$ and also the dc $\chi \equiv M/H$, until a temperature, T_r below T_g where remanence, M_r, may be observed.

2.4. Below T_r the $\chi_{\perp}$ data may be analyzed to yield the temperature dependent anisotropy

---

*Work supported by National Science Foundation DMR-83-12450, and DMR-83-14655.

constant, K(T), utilizing the known values of H and M_r.

2.5. We find that the suppression of $\chi_\parallel$ with dc field can be represented by suitable scaling expressions both above and below T_g. The analysis of the data in terms of a temperature dependent cluster size, and the implication of the new features observed will be discussed, such as the relationship of the plateau cusp line to the deAlmeida-Thouless transition.

3. THE TEMPERATURE DEPENDENCE OF THE ANISOTROPY IN $AgMn_{4\%}Er_x$ SPIN GLASS ALLOYS (in collaboration with Professor Saul Oseroff).

From an observation of the ESR lineshift and measured magnetization, we have determined the behavior of the anisotropy constant, K, (3) for the spin glass $AgMn_{4\%}$ doped with Er from x = 0.5 to 1.2 at% over the temperature range $T_g/10$ to 2 T_g. We find that the low temperature slope of K vs T exhibits a strong and unexpected dependence on concentration of Er. The implication of these data for interpreting the relaxation times probed by the ESR measurements will be discussed.

4. DIRECT MEASUREMENTS OF THE TRIAD MODES IN A SPIN GLASS (in collaboration with Dr. Eric Gullikson).

Utilizing novel spectroscopic techniques, we have made direct measurements of the triad modes in the CuMn spin glass system. Comparison of these measurements with those made utilizing mode-mixing at level crossings (4) will be presented, and the implications of the data toward an understanding of relaxation times in the spin glass will be discussed.

REFERENCES

(1) U. Larsen, Sol. St. Comm. 22, 311 (1977).
(2) D. R. Fredkin and H. Shore, private communication.
(3) S. Schultz, E. M. Gullikson, D. R. Fredkin and M. Tovar, Phys. Rev. Lett. 45, #18, 1508 (1980).
(4) E. M. Gullikson, D. R. Fredkin, and S. Schultz, Phys. Rev. Lett. 50, #7, 537 (1983).

LT-17 (Contributed Papers)
U. Eckern, A. Schmid, W. Weber, H. Wühl (eds)

MAGNETIC ORDERING OF DILUTE PALLADIUM-IRON ALLOYS AT MILLIKELVIN TEMPERATURES

D.I.BRADLEY, A.M.GUÉNAULT, V.KEITH, C.J.KENNEDY, S.G.MUSSETT and G.R.PICKETT

Department of Physics, University of Lancaster, Lancaster LA1 4YB, United Kingdom

We have measured the magnetic susceptibility of two very dilute Pd-Fe alloys (220 and 145 ppm Fe) in the millikelvin range in order to investigate the magnetic ordering. Our results show broad, rounded maxima in the a.c. susceptibility indicating a transition from paramangetic to spin-glass behaviour. The peaks occur at temperatures similar to the thermopower cross-over temperatures measured on the same samples. In the paramagnetic region the results yield an effective moment of Fe in Pd of $(7.8 \pm 0.8)\mu_B$ in agreement with results of other workers.

1. INTRODUCTION

The magnetic ordering of dilute alloys of iron in palladium has been the subject of numerous theoretical and experimental investigations(1). Pure Pd shows a large enhancement of the susceptibility over the Pauli paramagnetic value indicating that it is very close to being ferromagnetic. The addition of 0.1% Fe to Pd results in ferromagnetic ordering and it has been predicted that at concentrations less than 0.06% of Fe palladium should show a paramagnetic to spin-glass transition (2).

We have previously studied the thermopower, S, of very dilute PdFe alloys (10 - 1000 ppm Fe) in the millikelvin range (3). The main feature of these experiments was the observation of a change of sign of S from being negative at high temperatures to positive at low temperatures, which is interpreted as a change from independent impurity behaviour to a magnetically ordered state. Following Foiles (4) one expects to be able to associate the temperature at which S is zero, T*, with the spin-glass ordering temperature. However, our work showed an unexpected variation of T* with Fe concentration c, T* being proportional to c to the power 1.9 to 2. In the present work we have extended our study to include the magnetic susceptibility of the same samples in order to observe the magnetic ordering process more directly.

2. EXPERIMENTAL DETAILS

The susceptibilities of two of the PdFe samples have so far been measured. As in the measurement of the thermopower the samples were cooled by a dilution refrigerator having a base temperature of 3 mK. The configuration of the sample attachment is shown in Fig.1. Thermal contact was made by a silver sinter placed in the dilute phase of the mixing chamber of the refrigerator. The sinter was connected via a high purity silver wire spot welded to the sample outside the mixing chamber. The thermometer, previously calibrated against CMN susceptibility, was a slice of 47Ω Speer carbon resistor glued to a copper plate spot welded to the sample via a second silver wire. A 10Ω heater of eureka wire was wound directly on the sample. One end of the sample was placed inside one end of an astatic pair of mutual inductance coils. External magnetic fields were shielded out by a niobium tube thermally anchored at the 20 mK plate of the refrigerator. The mutual inductance was measured with a SHE RLM bridge using a SQUID as the null detector.

Measurements were made at various frequencies in the 16-160 Hz range with excitation fields of order 10^{-2} gauss. The mutual inductance was measured as a function of temperature during the initial cool-down of the refrigerator and remeasured by applying heat to the sample after the refrigerator had reached its base temperature with identical results.

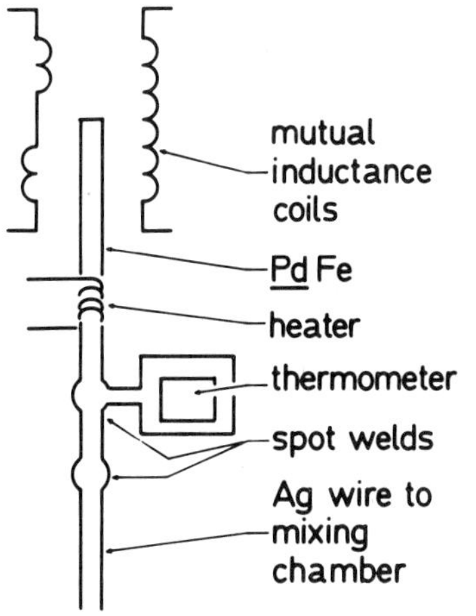

FIGURE 1
Schematic view of the experiment for measuring the susceptibility.

3. RESULTS AND DISCUSSION

A change in mutual inductance with temperature gives a measure of the change in susceptibility, χ, of the sample. Fig.2 shows the real part of the additional mutual inductance of the coils, the high temperature background M_0 having been subtracted. The same coil system was used for both samples hence the active length of each was the same although the cross sectional areas differed. The resistive part of the mutual inductance was also temperature dependent but the changes were very small being less than 1% of the variation of the real part.

Our results showed broad maxima in the ac susceptibility for both samples indicating that magnetic ordering not ferromagnetic in character occurs in the samples. The maximum suggests a paramagnetic to spin-glass transition is taking place. Usually in zero static field a much sharper cusp-like peak is observed defining a characteristic temperature, T_g, below which the spins associated with the impurities become frozen in random orientations (5). The sample containing 220 ppm Fe has a very broad maximum from 25 to 50 mK making it difficult to define a single characteristic temperature T_g. The 145 ppm sample has a well-defined peak at 18 mK.

We may note from Fig.2 that the temperatures of the peaks in χ are similar to the cross-over temperatures in the thermopower (3), as would be expected if these are two manifestations of the same magnetic transition. Furthermore there are similarities in the concentration dependence. If the results for the 220 ppm Fe sample are interpreted following Guy (6) as having a peak at 41 mK and an additional shoulder at 0.6 T_g then the results could support a c^2 dependence of T_g in agreement with the observed c^2 dependence of T* in the thermopower. However, there are problems. The value of T_g of 7 mK suggested by Webb et al. (7) for a Pd+ 1.7 ppm Fe sample seems very high in comparison with our results and one must conclude that the precise nature of the transition is not fully understood.

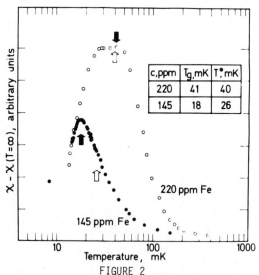

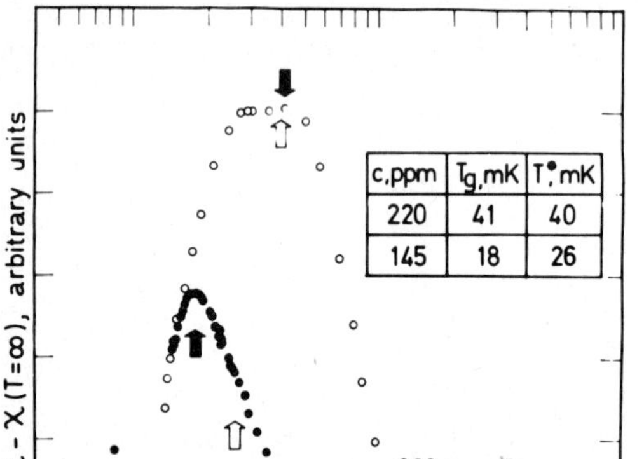

FIGURE 2

The relative ac susceptibility of two PdFe alloys at 79 Hz as a function of temperature. The peak to peak measuring field was 10^{-2} gauss. T_g is indicated by solid arrows and T* by hollow arrows.

In the paramagnetic region the situation is somewhat clearer. Neutron diffraction measurements (8) have shown that the giant polarization moment around an individual Fe ion in Pd extends out to about 10Å. Since for our 145 ppm and 220 ppm samples the mean impurity spacings are approximately 43 Å and 38 Å respectively any effects of polarization overlap are likely to be small. In this regime a limiting value of gJ has been suggested by Manuel and McDougald (9). Work by Webb et al. (7) on a very dilute 1.7 ppm alloy indicates this value to be $(7.6 + 0.7)\mu_B$. We have been able to fit our data in the paramagnetic region above T_g to a Curie-Weiss law using g=2 and g=3 after Manuel and McDougald. An average effective moment of $(7.8\pm0.8)\mu_B$ is obtained for the 145 ppm sample and a similar but less precise value for the 220 ppm sample. This is in agreement with the result of Webb et al. and confirms the prediction of a constant effective moment in the limit of low concentration.

ACKNOWLEDGEMENTS

We are grateful to the S.E.R.C. for support for this work. We thank N.S.Lawson and I.E.Miller for considerable contributions to this project and P.A.Schroeder for the loan of the samples.

REFERENCES

(1) G.J.Nieuwenhuys, Adv. Phys. 24, (1975) 515.
(2) G.J.Nieuwenhuys, Phys. Lett. 67A, (1978) 237.
(3) D.I.Bradley, A.M.Guénault, V.Keith, G.R.Pickett and W.P.Pratt, Jr., J.L.T.P. 45, (1981) 357.
(4) C.L.Foiles, Phys. Lett. 67A, (1978) 214.
(5) V.Cannella and J.A.Mydosh, Phys. Rev. B6, (1972) 4220; A.I.P. Conf. Proc. 10, (1973) 785.
(6) C.N.Guy, J. Phys. F: Metal Phys. 8, (1978) 1309.
(7) R.A. Webb, G.W.Crabtree and J.J. Vuillemin, Phys. Rev. Lett. 43, (1979) 796.
(8) G.G.Low and T.M.Holden, Proc. Phys. Soc. 89, (1966) 119.
(9) A.J.Manuel and M.McDougald, J. Phys. C: Solid Stat Phys. 3, (1970) 147.

LT-17 (Contributed Papers)
U. Eckern, A. Schmid, W. Weber, H. Wühl (eds)

ANOMALY OF THE ^{57}Fe HYPERFINE FIELD DISTRIBUTION AT THE REENTRANT SPIN-GLASS TRANSITION IN PdFeMn

G. CHANDRA*, M. MAURER, J.M. FRIEDT

Centre de Recherches Nucléaires B.P.20 67037 STRASBOURG Cedex (France)

We compare the temperature dependences of the hyperfine field distribution at ^{57}Fe in three Pd-Fe-Mn alloys displaying respectively reentrant spin-glass (RSG), giant moment ferromagnetic (F) and spin-glass (SG) transitions. A strong anomaly is observed at the reentrant transition in the RSG alloy whereas magnetic inhomogeneities induce broadenings in the vicinity of T_c (in F, RSG) or T_{SG} (SG) in all three samples.

1. INTRODUCTION

The understanding of local magnetic properties in SG or RSG and the nature of the RSG transition remain debated. In particular, the interpretation of hyperfine data in terms of spin canting and/or dynamical effects is still open. (1). We focus here on the temperature dependence of the hyperfine field distribution P(H) in zero external field. Among the many systems exhibiting the several SG, F, RSG phases as a consequence of competing exchange effects (e.g. AuFe, amorphous Fe alloys...), the PdFeMn alloys has been selected for a Mössbauer study because i) SG, F, and RSG states occur within a narrow range of concentrations (2) hence allowing a sensible comparison between the three RSG, F and SG phases. ii) impurity concentrations are well below solubility limits, thus minimizing possible segregation effects iii) in this cubic Pd host, isomershift distribution and quadrupole interactions are negligibly small, making the Mössbauer analysis reasonably safe and easy (paramagnetic line width = 0.26 mm/s).

2. EXPERIMENTAL PROCEDURE

RSG ($PdFe_{0.007}Mn_{0.05}$), F($PdFe_{0.07}$) and SG($PdFe_{0.0035}Mn_{0.05}$) alloys were prepared by arc melting high purity metals (70 % ^{57}Fe enriched iron). After several melting, buttons were rolled (40 μm thick), annealed (1 day, 1000°C) and quenched.

Mössbauer spectra were taken using a ^{57}Co:Rh source. Magnetic spectra were computer fitted for the hyperfine field distribution using an histogram method. The average field $\bar{H}$ and the mean square deviation ΔH^2 were directly computed. Since absorber are thin (0.13 mg $^{57}Fe/cm^2$) and the six lines are practically resolved, the relative intensity I of the lines 2,5 is determined directly, yielding the moments polarization with respect to the γ-ray propagation axis.

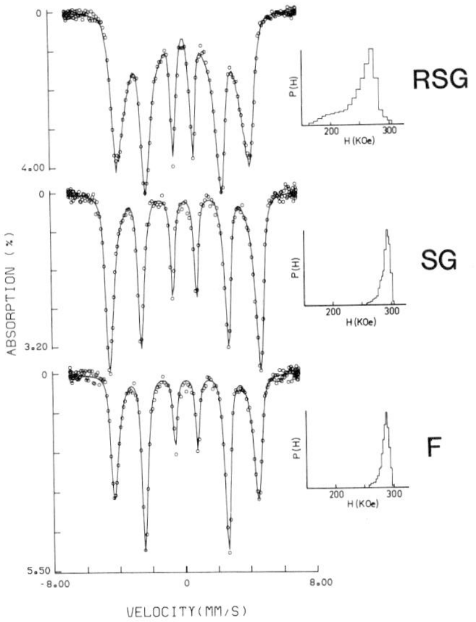

Fig.1 : ^{57}Fe Mössbauer spectra at $t \simeq 0.35$. Full line represent best fits with a hyperfine field distribution P(H) as represented in insert (histogram).

3. RESULTS AND DISCUSSION

AC susceptibility measurements (on the same samples) confirm the existence of a double transition in RSG (T_{RSG} = 4 K, T_c = 13.5 K) and of single transitions in SG (T_{SG} = 6.8 K) and F (T_c = 20.4 K). These transitions are sharp

*On leave of absence from Tata Institute of Fundamental Research, Bombay, India.

(except at T_{RSG}) proving macroscopic homogeneity.

Analysis of ^{57}Fe Mössbauer spectra (Fig.1) allows to draw the following primary conclusions :

i) the orientation of magnetic moments is isotropic (I = 2) in both SG and RSG below T_{RSG} whereas a small in-plane polarization develops above T_{RSG} (I = 2.3). Polarization is even stronger in F (I = 3.2), approaching full in-plane orientation (I = 4). These observations are consistent with data reported in external field (3).

ii) the hyperfine field distributions P(H) at saturation (1.5 K) are narrow. The saturation $\overline{H}$ are similar in the three RSG, F and SG alloys (Fig.2a). This demonstrates that P(H) hardly depends on the actual magnetic order (F or SG), likely due to weak hyperfine field contributions from neighbours. Hence, the ^{57}Fe hyperfine field mostly arises from the local atomic moment itself.

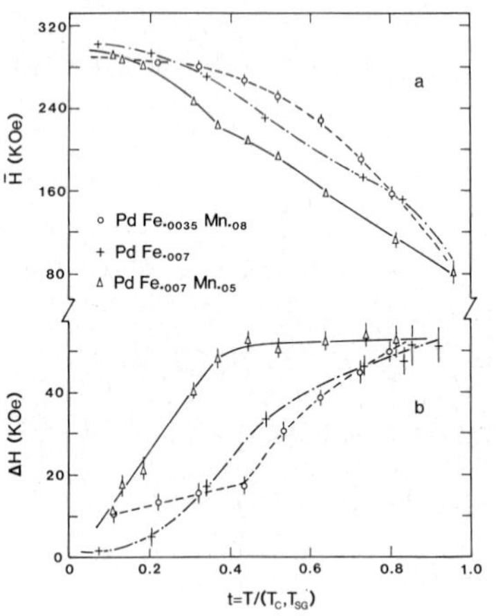

Fig.2 : Plot of (a) the average hyperfine field $\overline{H}$ and (b) ΔH (= $MSD^{1/2}$ of P(H)) versus reduced temperature t.

iii) $\overline{H}(t)$ where t is the reduced temperature : $t = T/(T_c, T_{SG})$ does not follow a Brillouin function (Fig.2a). Keeping apart the giant moment ferromagnet F, the difference between SG and RSG is striking.

iv) the field distribution broadens at $t \rightarrow 1$ in all alloys. A small distribution of the molecular field (i.e. of T_c or T_{SG}) can account for this trend. However, in RSG, a drastic broadening takes place at T_{RSG}, well below T_c. From the above remark (ii), a change of magnetic structure (spin glass to ferro) cannot explain either the flattening of $\overline{H}(t)$ (Fig.2a) or the broadening of P(H)(t) (Fig.2b). In order to reconcile the single-ion nature of H in PdFeMn with the picture of transverse component freezing at T_{RSG}, dynamical effect have been tentatively considered. Fitting spectra by either uniaxial or spherical relaxation mechanisms failed. However, both of these mechanisms may represent poor descriptions for fluctuations of the transverse component alone. More elaborate relaxation models need to be worked out. Additional experiments in an external field are in progress for the three RSG, SG and F sample with the aim to decide between static (within the Mössbauer time window $\simeq 10^{-8}$s), single - ion model and superparamagnetic - like behaviour.

1. See e.g. Proc. of HEIDELBERG COLLOQUIUM ON SPIN GLASS, published in LECTURE NOTES IN PHYSICS 192 (SPRINGER 1983) ed. van HEMMEN and MORGENSTERN

2. B.H. VERBEEK, G.J. NIEUWENHUYS, H. STOCKER and J.A. MYDOSH, Phys. Rev. Lett. 40 (1978) 586

3. R.D. TAYLOR and J.O. WILLIS, J. Magn. Magn. Mater. 15 - 18 (1980) 623

LT-17 (Contributed Papers)
U. Eckern, A. Schmid, W. Weber, H. Wühl (eds)

PAC STUDY OF MAGNETIC BEHAVIOUR IN $Au_{74}Fe_{26}$

J. VAN CAUTEREN and M. ROTS

Instituut voor Kern- en Stralingsfysika, Katholieke Universiteit Leuven, B-3030 Leuven, Belgium

Perturbed angular correlation experiments are reported studying the presumed coexistence of spin-glass ordering and ferromagnetism beyond the percolation limit. We found in $Au_{74}Fe_{26}$ the onset of a magnetic hyperfine field at T=230K, followed by a remarkable increase of growth rate of the hyperfine field at T=50K. The similarity between this anomalous temperature behaviour and reentrant magnetism is discussed.

1. INTRODUCTION

The nature of ferromagnetism in AuFe-alloys has been discussed in recent years, but still no definite conclusion can be drawn from existing data. In the concentration range above the percolation limit (15.7 at%Fe) different authors (1,2) suggest a reentrant magnetic behaviour i.e. upon cooling the alloy undergoes a paramagnetic to ferromagnetic transition followed at a lower temperature by a transition into a spin-glass like state. Whether the ferromagnetism is long range or confined to magnetic clusters remains an open question (3,4). Recent perturbed angular correlation (PAC) experiments (5) have shown that a $Au_{82}Fe_{18}$ alloy remains paramagnetic down to 50-60 K. It is interesting to note that this temperature matches with the one where an increase in the magnetic Bragg intensity was found (6) for the 17 at%Fe as well as the 19 at%Fe alloy. Moreover also at T=50K, a sudden increase in hyperfine field (7) and canting angle (8) with decreasing temperature was observed in Mössbauer spectra.

In order to discuss the magnetic states of these concentrated alloys the randomness of the solid-solution is crucial. The highest degree of disorder in a 15 at%Fe sample was obtained (9) by quenching from about 550K instead of quenching from high temperature. In a recent discussion of the AuFe Mössbauer data Violet and Borg (10) concluded that virtually all experimental work was done on samples that where not even approximately random solid solutions. Two different compositional phases were invoked to explain the existence of two hyperfine fields. If this interpretation is correct much of the reentrant magnetic behaviour in AuFe may be absent. In this paper we present the first results dealing with an extension of our previous PAC work to higher Fe-concentration.

2. EXPERIMENTAL

2.1 Sample preparation

The AuFe alloy was prepared by melting the high purity constituents in an induction furnace under vacuum (10^{-5} Torr). After rolling to a foil thickness of approximately 50 μm, the ^{111}In activity from a carrier-free solution was deposited and dried. The foil was then remelted under vacuum and kept in molten state a few minutes. A second rolling to foil thickness of nearly 30 μm was followed by thermal annealing for 10 hours at 900°C under continuous flow of dried argon. The heat treatment was terminated by quenching directly from the oven into water. With a second part of the original alloy we proceeded in the same way, but replaced the final quench by cooling the foil slowly in the oven. Microprobe analysis determined the concentration of our sample as $Au_{74}Fe_{26}$.

2.2 Outline of the method

The applied technique is perturbed angular correlation (PAC) using the radioactive probe ^{111}In, decaying by a γ-γ radiation cascade. The basic principle of measurement is the observation of the time evolution of the nuclear spin alignment produced simply by observing the first radiation in a fixed direction. The angular distribution of the second radiation is then spatially anisotropic relative to the emission direction of the first radiation. The angular correlation is measured as a function of delay time between the cascade radiations and equals nearly 40% at t=0. The interaction of nuclear moments with electromagnetic fields from the environment, introduces a characteristic time modulation in the anisotropy from which the interaction strength and the hyperfine fields may be deduced.

In the present context we will observe, at room temperature, an alloy in its paramagnetic state. Due to the random distribution of Fe impurities over the cubic gold lattice, an electric quadrupole interaction, with a broad distribution, at the probe site will be observed. Upon cooling we expect the onset of magnetic ordering and a non-zero magnetic hyperfine field will evolve, superimposed on the quadrupole interaction response.

2.3 Results

The measurements were performed, without external magnetic field, in a variable temperature cryostat in the range 4.2K to 300K. Time spectra of the correlation anisotropy are similar to those published earlier (5). Data analysis followed the procedure outlined in an earlier paper (11). First the room temperature spectrum is least-squares fitted using a function describing a pure electric quadrupole interaction. The derived interaction parameters (strength and distribution) are then kept fixed in an expression describing a collinear magnetic and electric interaction. In our data the onset of magnetic ordering upon cooling is detected when the fit function representing the room temperature data is no longer satisfactory. Over the whole temperature range the data are analysed with two free parameters: the magnetic interaction strength and its distribution. Therefore the magnitude of the mean magnetic hyperfine field (hff) is derived as a function of temperature as shown in fig. 1. It is obvious that upon cooling the growth in the hff determines two critical temperatures i.e. at 230K and 50K. The same measurements were also performed on an identical sample, which was slowly cooled from 900°C instead of quenched. The results are essentially reproduced with exactly the same critical temperatures and hff values.

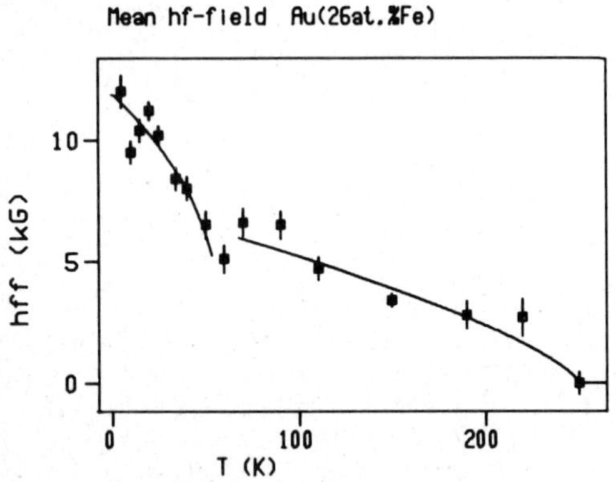

FIGURE 1
Temperature dependence of the mean hyperfine field at ^{111}Cd probe

3. CONCLUSION

The different heat treatment was given to the samples especially to produce differences in atomic clustering. No changes were observed showing that precipitation, if any, occurs much faster than the normal quench rate realised. This observation agrees with the experience (9) that the quenched-in state depends only on the quenching rate when annealed at high enough temperature. The x-ray determination (12) of short range order gives qualitatively the same result for the 14.4 at% concentration independent on quench temperature.

The remarkable result, as yet not understood, of the present experiments, however, concerns the origin of the sudden increase in hff at 50K. At the same temperature we observed (5) the onset of hff in a $Au_{82}Fe_{18}$ alloy. This temperature is also characteristic in the Mössbauer data (7,8) as well as neutron scattering work (6) on 17 or 19 at%Fe. Moreover our preliminary results on samples with 20 to 26 at%Fe, indicate the same value for the lower critical temperature independent on concentration.

The question arises whether the anomaly in hff at T=50K may be interpreted as the anticipated reentrant behaviour. The upper critical temperature from PAC is substantially smaller than the Curie temperature found in Mössbauer work (13) on the alloy concentration under study. Moreover the apparent concentration independence of the lower critical temperature is not expected if this were a transition to the spin-glass state.

ACKNOWLEDGEMENTS

It is a pleasure to express our gratitude to Prof. P.A. Beck for stimulating and clarifying comments. This work was sponsored by the Belgian "Interuniversitair Instituut voor Kernwetenschappen".

REFERENCES

(1) B.H. Verbeek and J.A. Mydosh, J. Phys. F8 (1978) L104.
(2) S. Crane and H. Claus, Phys. Rev. Lett. 46 (1981) 1693.
(3) P.A. Beck, Phys. Rev. B23 (1983) 2290.
(4) A.P. Murani, Phys. Rev. B28 (1983) 432.
(5) M. Rots, L. Hermans and J. Van Cauteren, Sol. Stat. Comm. 49 (1984) 131.
(6) A.P. Murani, Sol. Stat. Comm. 34 (1983) 705
(7) J. Lauer and W. Keune, Phys. Rev. Lett. 48 (1982) 1850.
(8) F. Varret, A. Hamzic and I.A. Campbell, Phys. Rev. B26 (1982) 5285.
(9) S. Crane and H. Claus, Sol. Stat. Comm. 35 (1980) 461.
(10) C.E. Violet and R.J. Borg, Phys. Rev. Lett. 51 (1983) 1073.
(11) M. Rots, L. Hermans and J. Van Cauteren, J. Appl. Phys (1984) in press.
(12) E. Dartyge, H. Bouchiat and P. Monod, Phys. Rev. B25 (1982) 6995.
(13) U. Gonser, R.W. Grant, C.S. Meechan, A.H. Muir and H. Widersich, J. Appl. Phys. 36 (1965) 2124

LT-17 (Contributed Papers)
U. Eckern, A. Schmid, W. Weber, H. Wühl (eds)

TRANSPORT PROPERTIES STUDIES OF THE FERRO- TO SPIN GLASS TRANSITION IN AMORPHOUS Fe-Ni BASED ALLOYS

R.H. CROOK, E.D. DAHLBERG

Department of Physics, University of Minnesota, Minneapolis, MN 55455, and

K.V. Rao

E & I T Sector Laboratories, 3M, St. Paul, MN 55133

Studies of the magnetic ordering in the amorphous alloys $Fe_{10}Ni_{65}B_{15}Si_{10}$ and $Fe_{15}Ni_{65}P_{16}B_6Al_3$ have been made using magnetoresistance as a probe. Both of these alloys order ferromagnetically and at a lower temperature become spin glasses. The data suggest the breakdown of ferromagnetic order occurs on a length scale of less than 1 nm.

1. INTRODUCTION

In recent years the appearance of a spin-glass like behavior at low temperatures in random alloys with competing magnetic exchange interactions, after a ferromagnetic transition at some higher temperature has been a subject of considerable study both experimentally and from a theoretical point of view. Considerable impetus to such studies was due to a generalized magnetic phase diagram calculated by Sherrington and Kirkpatrick (1) for an Ising spin system with infinite range interactions having a gaussian distribution with mean J_o, and a standard deviation J. They found paramagnetic to ferromagnetic transitions for values $(J_o/J)>1.2$, a paramagnetic to spin glass phase of the type described by Edwards and Anderson (2) for $(J_o/J)<1$, and for $1<(J_o/J)<1.2$ the loss of long range ferromagnetic order to a spin glass phase. The nature of the transition to this low temperature 'phase' from the ferromagnetic phase, and a microscopic description of this regime has been a topic of intense current research activities. More recent theoretical and experimental developments suggest that a microscopic description of this 're-entrant' spin-glass regime is by no means unique. One picture of this regime considers the role of the onset of strong random anisotropy either by the onset of strong transverse magnetization suggested by Gabay and Toulouse (3) by a transition of the coupled longitudinal and transverse magnetization analogous to the Almeida-Thouless (4) transition of spin glass in applied fields, or merely by the spin-freezing phenomena. Thus experimentally the focus has been in identifying the role of randomness and anisotropy which suggests the possible existence of only a spin-glass phase or a co-existence of ferromagnetic and spin-glass character in this regime of interest.

The experimental probes used thus far to study this phenomenon have been primarily large wavelength measurements such as susceptibility and magnetization. Magnetoresistance is a probe of short length scales, and has the added advantages of altering the local magnetic environment in a systematic fashion as a function of temperature. Magnetoresistance has been used by previous workers to investigate local anisotropy in crystalline disordered alloys (5). Because of the information one gains on the local environment with magnetoresistance we have made measurements on two amorphous soft magnetic alloys, $Fe_{10}Ni_{65}B_{15}Si_{10}$ and $Fe_{15}Ni_{65}P_{16}B_6Al_3$, in which the existence of the ferromagnetic transition followed by a spin glass transition at lower temperatures (1,6) has been demonstrated using other experimental techniques.

2. EXPERIMENTAL DETAILS AND RESULTS

The amorphous alloy ribbons used in this study were prepared by melt spinning and are made from the same batch of samples used in earlier studies (1,6). For our magnetoresistance studies samples typically of dimensions 1 mm wide, 15 μm thick, and about 2.8 cm long were used. The four terminal magnetoresistance measurements were made in a variable temperature cryostat over the temperture range of 4.2-280 K in applied fields up to 9 Tesla both parallel and transverse to the current in the plane of the sample.

Figure 1a shows the change in resistance $\{R(8T)-R(0)\}/R(0)$, for the $Fe_{10}Ni_{65}B_{15}Si_{10}$ sample in an applied field of 8 Tesla as a function of temperature, and figure 1b is a similar plot for the $Fe_{15}Ni_{65}P_{16}B_6Al_3$ sample. In both of these figures one should note the

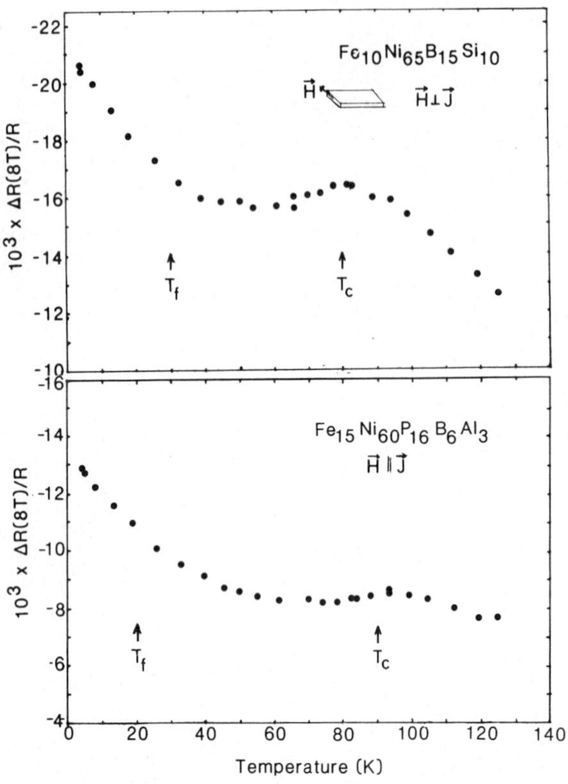

FIGURE 1
Change in resistance in an applied field of 8 Tesla, {R(8T)-R(0)}/R(0), as a function of temperature for a) $Fe_{10}Ni_{65}B_{15}Si_{10}$ and b) $Fe_{15}Ni_{65}P_{16}B_6Al_3$.

maxima in the effect of the applied fields at the Curie temperature followed by a decrease and then an increase once again as the temperature approaches the spin glass temperature. Within our error of measurements, the results of the measurements with the current parallel to the applied field are identical to those with the current perpendicular to the applied field.

3. DISCUSSION OF RESULTS

Previous work (7) on disordered magnets has considered the effects of spin correlations on the transport properties. These workers have found that transport measurements provide information on the spin correlations for short distances (on the order of 1 nm) in amorphous alloys. An analysis of the zero field resistance as function of temperature is difficult in this type of model since there may be another temperature dependent scattering contribution to the resistivity of non-magnetic origin (8). The magnetoresistance, however, allows one to measure the change in the spin correlations at fixed temperature and thus probe the local correlations more directly.

As mentioned previously, the resistance, R, is proportional to:

$$R \propto \sum_{i,j} \langle S_i \cdot S_j \rangle_\lambda$$

where the sum is restricted to spin i and j separated by a distance less than the electronic mean free path λ. In a ferromagnetic material one would expect the magnitude of the magnetoresistance to be a maximum at the Curie temperature where the susceptibility diverges. This is consistent with our observations.

As exhibited in fig 1 both the alloys have a maxima in the magnetoresistance at roughly their respective Curie temperatures. At lower temperatures where the ferromagnetic state is destroyed and the spin glass state sets in, the magnetoresistance again increases in both alloys. If the destruction of the long range ordered or ferromagnetic state occured on length scales much longer than the mean free path, the magnetoresistance would be unaffected by the change in state from ferro to spin glass. However, by interpreting the data in terms of the coherent exchange scattering model of Asomoza et al (8), the breakdown of the long range magnetic order is in fact on spatial scales of less than 1 nm.

ACKNOWLEDGEMENTS
We thank H.S. Chen and T. Egami for making available the amorphous ribbon samples. This work was funded in part by the Graduate School of the University of Minnesota. One of the authors, E.D.D., acknowledges receipt of an A.P. Sloan Fellowship.

REFERENCES
(1) D. Sherrington and S. Kirkpatrick, Phys. Rev. Letters 35, 1792 (1975)
(2) S.F. Edwards and P.W. Anderson, J. Phys F5, 965 (1975)
(3) M. Gabay, and G. Toulouse, Phys. Rev. Letters 47, 201 (1981)
(4) J.R.L. deAlmeida and D.J. Thouless, J. Phys. A11, 983 (1978)
(5) S. Senoussi, J. Phys (Paris) 42, L35 (1981)
(6) T. Kudo, T. Egami, and K.V. Rao, J. Appl. Phys. 53(3), 2214 (1982)
(7) S.R. Nagel, Phys. Rev. B16, 1694 (1977)
(8) R. Asomoza, A. Fert, I.A. Campbell, and R. Mayer, J. Phys F7, L327 (1977); A.K. Bhattacharjee and B. Coqblin J. Phy. F8, L221 (1978); R. Asomoza, I.A. Campbell, A. Fert, A. Lienard, and J.P.Rebouillat, J. Phys. F9, 349 (1979)

LT-17 (Contributed Papers)
U. Eckern, A. Schmid, W. Weber, H. Wühl (eds)

MAGNETIC PROPERTIES OF GADOLINIUM DOPED LEAD TELLURIDE

F.T. HEDGCOCK* AND P.C. SULLIVAN**

Department of Physics, McGill University, Montreal, Quebec, Canada. H3A 2T8

M. BARTKOWSKI

Department of Physics, Concordia University, Montreal, Quebec, Canada.

The a.c. susceptibility of $Pb_{1-x}Gd_xTe$ has been measured from 40K down to 1.1K for samples with Gd concentrations x of 1% and 5%. Deviations from the simple Curie law are observed below 4.2K.

1. INTRODUCTION

The group of compounds described as semi-magnetic semiconductors which have been investigated most extensively have been chalcogenides containing transition metal impurities and in particular the Mn^{++} ion which carries a bare spin of 5/2 (1). In most of these chalcogenides the charge carrier density is sufficiently low that direct magnetic interactions are the only possible interactions available to explain the occurance of magnetic freezing in these materials. For magnetic freezing to occur requires, therefore, quite large transition metal ion concentrations. An unwanted consequence is that at these large concentrations the band structure of the host is severely altered and in some cases the charge carrier concentration as well. The use of rare earths as the magnetic species could well be an advantage in these cases and in fact preliminary measurements of Savage et al (2) would indicate this to be so. In this communication we wish to report measurements of the temperature and concentration dependence of the low field a.c. susceptibility of 3 lead telluride alloys containing the rare earth gadolinium.

2. EXPERIMENTAL

The alloys were prepared by melting PbTe ingots supplied by Global Thermoelectric Power Systems with Gd_2Te_3 powder. Both Xray diffraction and microprobing indicate a single phase however homogeneity tests on various samples from the same melt indicated large concentration gradients. For this reason room temperature susceptibility measurements were used to adjust the concentration to correspond to a Gd ion spin of 7/2 (2). The three samples studied with listings of their nominal concentration, room temperature susceptibility and adjusted concentration are shown in Table 1. The room temperature susceptibilities were measured on a conventional Curie method servo balance. All of the samples studied showed a field independent magnetization up to fields of 15 kilo gauss. The absolute values of the susceptibilities were determined by calibrating the balance with high purity aluminum (0.607×10^{-6} emu/gm) and germanium (0.104×10^{-6} emu/gm). The value of susceptibility of the PbTe host was found by this method to be -0.341×10^{-6} emu/gm. The low temperature susceptibilities were determined using an ac field of a few gauss at a frequency of 87 Hz. The low temperature bridge was calibrated with $CuSO_4 5H_2O$.

3. EXPERIMENTAL RESULTS AND DISCUSSION

Shown in Figure 1 is a graph of $(\chi_{alloy} - \chi_{host})^{-1}$ vs temperature for samples 2 & 3. Deviations from a simple Curie plot are seen to occur for the more highly concentrated sample at temperatures below 4.2° K. As determined from the linear portion a value of 8.03 for Peff is found for sample 2 when the adjusted concentration as listed in Table 1 is used. The high temperature slope for the more concentrated sample yields a value of 7.80 which

*Research made possible by financial assistance of the Natural Sciences and Engineering Research Council of Canada.

**PCS would like to acknowledge financial support as a postdoctoral fellow through the Natural Sciences and Engineering Research Council of Canada.

Table 1. Concentration, effective magneton numbers, Curie-Weiss constant and values of $\sqrt{\Delta\chi T}$ for the three samples studied.

Sample	Nominal Conc.	Adjusted Conc.	χ_{rt} emu/g	P_{eff} μ_B	θ K	$\sqrt{\Delta\chi T}_{4.2}$ (emu K/g) 1/2	$\sqrt{\Delta\chi T}$ (emu K/g)$^{1/2}$
1	0.5	1.10	5.33x10^{-7}	8.12	-0.4	.01642	.01624
2	2.0	1.06	5.05x10^{-7}	8.03	-0.2	.01592	.01580
3	5.0	5.77	2.49x10^{-6}	7.80	-0.4	.02822	.02820

agrees with the value obtained from the slope of the graph of $SnAgGdT_e$ as reported by Savage. We have checked the constancy of the moment by evaluating $\sqrt{\Delta\chi T}$ at 4.2° K and at room temperature. These values, listed in Table 1, can be seen to be constant within experimental error. Below 4.2° K, however, the value of the moment decreases. Also included in Table 1 are the values for the Curie-Weiss constant, θ, as determined by extrapolating the Curie line. The sign and magnitude of θ is in agreement with qualitative statements made in previous work (2) and we see that whatever interactions are present they must be extremely week.

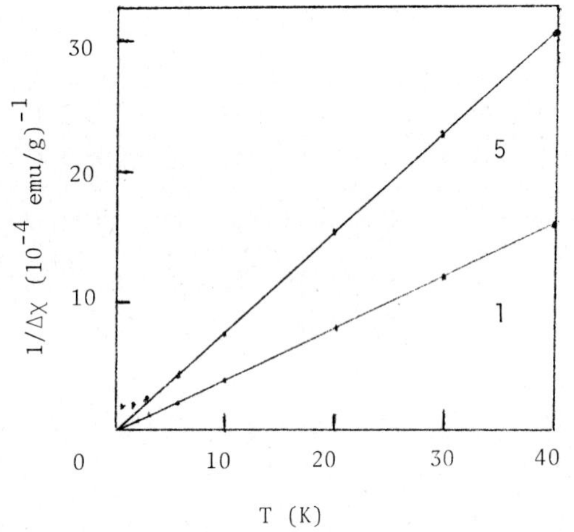

FIGURE 1
The reciprocal susceptibility $\Delta\chi^{-1} = (\chi_{alloy} - \chi_{host})^{-1}$ as a function of temperature.

In order to expand the low temperature scale the local moment gadolinium susceptibility $\Delta\chi$ is shown in Figure 2 as a function of T^{-1}. Deviations from the simple Curie law are obviously concentration dependent and for the nominal 5% sample show quite large deviations at temperatures below 4.2°K. These deviations which represent a decrease in the paramagnetism as the temperature is lowered could be interpreted as the onset of a spin glass transition occuring in these compounds. Novak et al (3) have observed similar behavior in the semimagnetic semiconductor Cd-Mn-Te in the milli degree Kelvin range. More recently McAlister et al (4) have reported very striking effects in the spin glass transition in $Zn_{1-x}Mn_xTe$ compounds where characteristic spin glass cusps were observed in the d.c. susceptibility when the Mn concentration is in excess of 30%. We plan to extend our magnetic studies to more concentrated samples and to look for field cooling effects in the d.c. susceptibility. In the meantime we think it of considerable interest that preliminary evidence indicates possible spin glass behavior in a rare earth semimagnetic material.

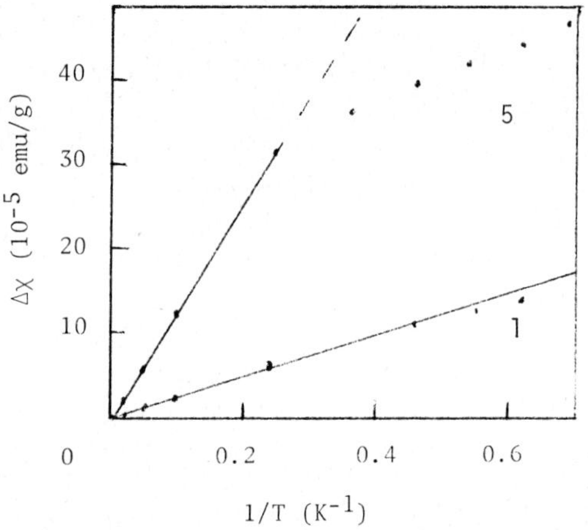

FIGURE 2
Gadolinium susceptibility $\Delta\chi$ as a function of reciprocal temperature.

(1) T. Hamasaki, Solid State Communications 32 (1979) 1069; J.L. Tholence, A. Benoit, A. Mauger, M. Escorne, and R. Triboulet, Solid State Communications 49 (1984) 417.
(2) H.T. Savage and J.J. Rhyne, AIP Conf. Proc. 5 (1971) 879.
(3) M. A. Novak, S. Oseroff, and O.G. Symko, Physica 107 B&C (1981) 313.
(4) S.P. McAlister, J.K. Furdyna, and W. Giriat, Phys. Rev. B 29, (1984) 1310.

LT-17 (Contributed Papers)
U. Eckern, A. Schmid, W. Weber, H. Wühl (eds)

CALORIMETRIC INVESTIGATION OF THE Y-Er MAGNETIC PHASE DIAGRAM

E.BONJOUR, R.CALEMCZUK, R.CAUDRON[*], H.SAFA[*], P.GANDIT[+] and P.MONOD[**]

Centre d'Etudes Nucléaires de Grenoble SBT/LCP 85 X 38041 Grenoble
[*]Direction des Matériaux, ONERA 29 Ave Leclerc 92320 Chatillon
[**]Physique des Solides, Université Paris-Sud 91405 Orsay
+CRTBT CNRS Ave des Martyrs 166 X 38042 Grenoble

Measurements of Y-Er alloys at 1%, 3%, 8%, 17%, 35%, 63% and 95% atomic Er show a sharp break in the specific heat at the same temperature where a maximum of susceptibility occurs. This fact confirms the magnetic phase diagram of Y-Er and raises the question of the specific heat of an Ising spin glass which should represent this system at low Er concentration.

We present a series of measurements of the specific heat of the system Y-Er throughout the whole concentration range of Er in Y. The purpose of this investigation is twofold : first to determine calorimetrically the boundaries of the phase diagram of this system which was already studied by neutron scattering (1) and by magnetic susceptibility (2), as shown on figure (1a); second we want to compare the properties of the low Er concentration range (a few %) to those of the higher concentration range where antiferromagnetic order exists (1). Indeed as is apparent from magnetic measurements (3),Y-Er becomes a spin-glass at low concentration. However the presence of a strong hexagonal crystalline field restricts the Er ion at low temperature to a Γ_7 doublet (4) with the Er moment parallel to the hexagonal c axis. Thus the Y-Er system at low Er concentration is a good candidate for being an S = 1/2 Ising spin-glass system (3).

Y-Er alloys form solid solutions in all proportions. The alloy samples were melted at least twice in a semi-levitation rf oven under Ar atmosphere and cooled rapidly, producing a highly polycrystalline material at nominal Er concentration 3%, 4%, 8%, 17%, 35%, 63%, and 95%. A small cylinder about 0.7g was cut out each alloy and used for *both* specific heat (5) *and* susceptibility measurements.

Fig.2 shows the specific heat results in Joule/K mole Er of the alloys (after subtraction of the Y measurement in order to eliminate most of the phonon part). As indicated by arrows the temperature of the main specific heat anomaly coincides with the susceptibility anomaly measured at low fields on the same samples. In particular it is noteworthy that a clear break in the specific heat is detected at the magnetic transition for the two lowest concentrations (1% and 3%) where an Ising spin glass state is likely to prevail(7). In the range 8%-17% the broad shoulder peaked at 25°K is due to the Er single ion crystal field Schottky contribution (4). In order to make a more quantitative comparison we have plotted on Fig.1b the discontinuity of the specific heat at the transition per mole Er at T_c : it is about constant at high concentration and we see that although this quantity is less well defined at low concentration there is a seven fold decrease around 20% Er. This drastic reduction coincides with the effective spin reduction from 15/2 to 1/2. However mean field theory only accounts for a 5/3 reduction in the specific heat jump (6) from $S = \infty$ to $S = 1/2$. We are led to conclude that this property may offer means to distinguish the A.F. regime ($c > 17\%$) from a possible Ising spin glass regime ($c < 8\%$) : this conclusion should be borne out by direct neutrons investigation.

Acknowledgements are due to H.Bouchiat, J. Chaussy, B.Elschner and E.F.Wasserman for help, samples and discussions.

REFERENCES

(1) H.R.Child, W.C.Koehler, E.O.Wollan and J.W. Cable, Phys.Rev.138, A1 1655 (1965).
(2) J.Pelzl et al, Phys.Lett.62 A, 117 (1977).
(3) A.Fert et al, Phys.Rev.B26, 5300 (1982). H.Albrecht et al, Phys.Rev.Lett.48,819(1982)
(4) G.Keller and J.M.Dixon, J.Phys.F6, 819(1976)
(5) R.Lagnier, J.Pierre and M.J.Mortimer, Cryogenics 17, 341 (1977).
(6) D.C.Mattis, "Theory of Magnetism",Harper and Row N.Y. (1965).
(7) D.Sherrington and S.Kirkpatrick, Phys.Rev. Lett.32, 1792 (1975) these authors predict a discontinuity in the slope of C(T) at T_c for mean field theory.

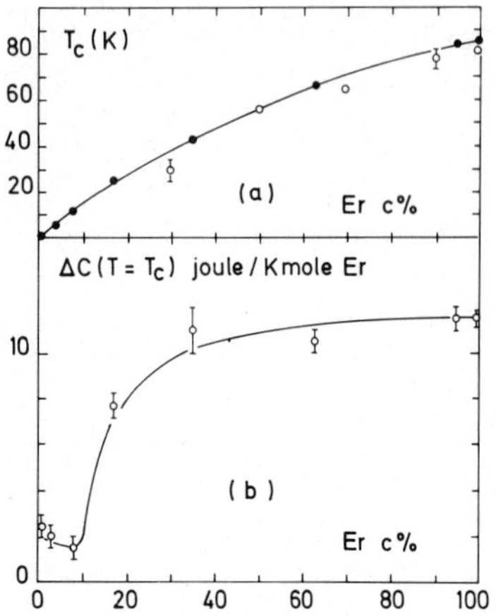

FIGURE 1
(a)Simplified magnetic phase diagram of the Y-Er system: closed symbols magnetic susceptibility(this work and ref(2)); open symbols neutron diffraction ref(1).
(b)Discontinuity of the specific heat per mole Er at the transition temperature versus Er concentration. For 1% to 17% we have plotted the peak height of the specific heat after removing the Schottky crystal field contribution.

FIGURE 2

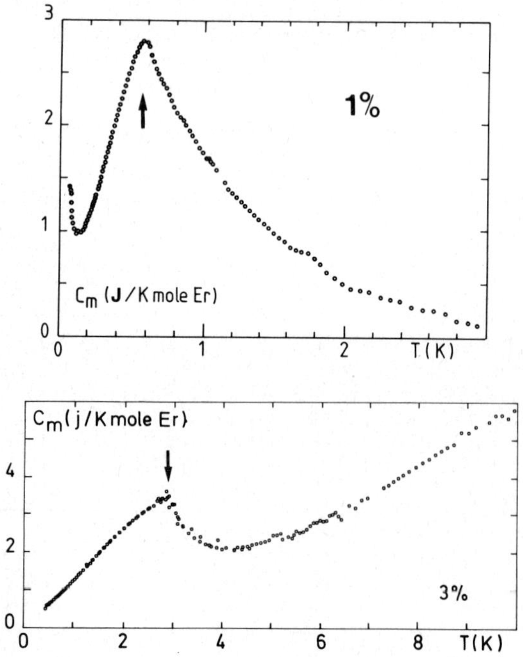

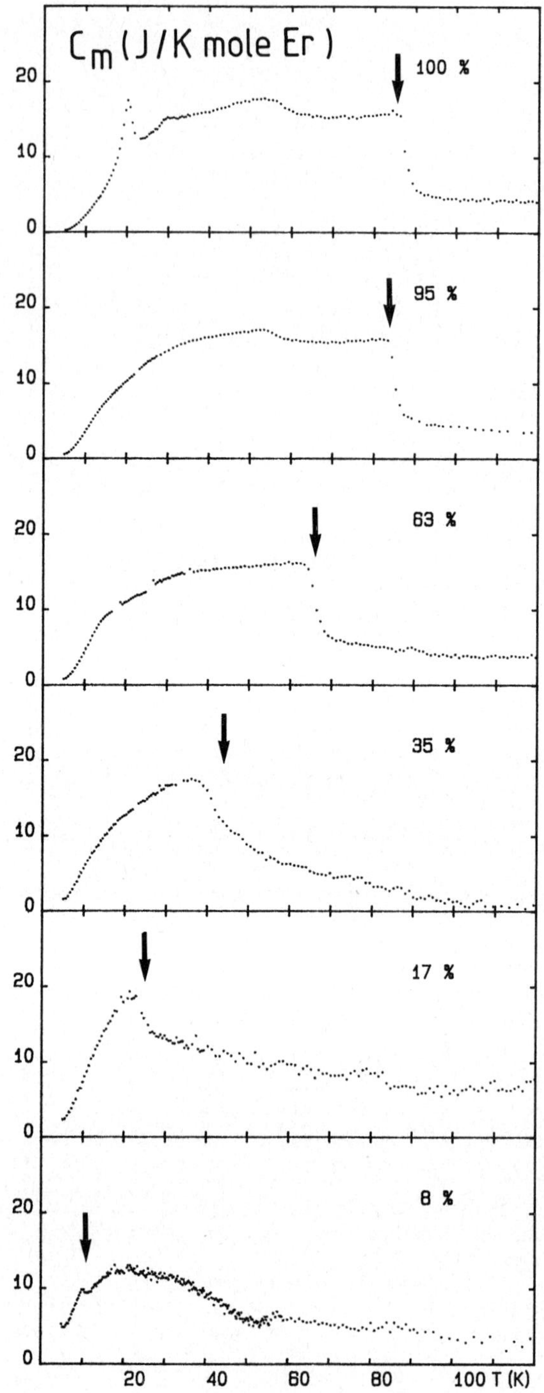

FIGURE 2
Magnetic specific heat of Y-Er alloys(per mole Er) - The arrows indicate the temperature of the susceptibility anomaly reported on Fig.1a.

LT-17 (Contributed Papers)
U. Eckern, A. Schmid, W. Weber, H. Wühl (eds)

NON-MEAN FIELD GENERALIZATION OF THE GABAY-TOULOUSE LINE

S.E. BARNES

Physics Department, University of Miami, Coral Gables, Fl 33124

A.P. MALOZEMOFF

IBM Thomas J. Watson Research Center, Box 218, Yorktown Heights, New York 10598

B. BARBARA

Labortoire Louis Neel, CNRS, F-38042 Grenoble, Cedex, France

Described is a scaling theory of spin-glasses which has both longitudinal (q_ℓ) and transverse (q_t) order parameters. The form of the term coupling these parameters is deduced by requiring the longitudinal but not the transverse transition to be suppressed in a field. If the large field susceptibility corresponds to $\chi \sim h^{2/\delta}$ then the transverse freezing, or Gabay-Toulouse, temperature is found to obey a scaling law $t_T \sim h^{4/\delta}$. For mean-field δ=2 to give $t_T \sim h^2$, in agreement with Gabay and Toulouse.

1. INTRODUCTION

It is quite remarkable that while the mean-field theory of the spin-glass transition yields correctly the famous 2/3 power for the de Almeida-Thouless line (1) corresponding to the onset or irreversibility it is wrong e.g. by factors of two in its prediction of δ the scaling exponent for the susceptibility in large fields $\chi \sim h^{2/\delta}$. The present authors have recently suggested (2) that the experimental agreement with the 2/3 power is accidental. We have correlated this reversibility cross-over for $T < T_G$, the glass temperature, with the $T > T_G$ cross-over from analytic to non-analytic field dependences of the susceptibility. These lines correspond to $t = A_\pm H^{2/\phi_\pm}$ respectively with both cross-over exponents being close to 3 and the amplitudes $A_\pm$ differing only by a factor of two. Of course the mean-field prediction for ϕ_- (the de Almeida-Thouless line) is $\phi_- = 3$, also in agreement with experiment, however the general cross-over exponent is 2 so mean-field would predict $\phi_- = 3$ but $\phi_+ = 2$ which does not agree with experiment.

For realistic, Heisenberg, systems there is also a mean-field prediction for the exponent associated with the Gabay-Toulouse (3) transverse freezing line i.e.

$$t_T \sim -h^2 \;;$$

the exponent is 2.

This contribution deals with the obvious next question namely, what is the non mean-field prediction for this exponent? It turns out this exponent is related to the susceptibility exponent δ defined above. The result is

$$t_T \sim \pm\, h^{4/\delta}.$$

The shift in the transverse ordering temperature might be in either direction and since $\delta \sim 4$-5 for CuMn the non-mean-field prediction differs significantly from the mean-field prediction $\delta = 2$.

2. CALCULATION

Our approach is inspired by that of Suzuki (4). We first write a Landau type free energy in order to first reproduce the mean-field results:

$$F = F_u + am^2 + bm^4 - cq_\ell^2 - dq_\ell^3 - 2q_t^2 - 2dq_t^3 + 2fq_\ell^2\, g_t^2 + eqm_\ell^2 - hm$$

In the usual way one writes $c = c_0(T-T_G) \equiv c_0 t$ and differentiate to obtain equations for the total magnetisation m and the transverse q_t and longitudinal q_ℓ order parameters. The novel feature of this free energy is the coupling term $2fq_\ell^2 q_t^2$. This form correctly reproduces the Gabay-Toulouse h^2 shift and provided one gives the coefficients their mean-field values also reproduces the de Almeida-Thouless 2/3 power.

The non-mean-field theory results from replacing the above by a scaling form of the same structure:

$$F_s = (a+eq_\ell)m^2 + bm^4 - hm - cq_\ell^2\, g^\pm(q_\ell|t|^{-\beta} - 2cq_t\, g^\pm(q_t|t|^{-\beta}) + f(q_\ell, q_t).$$

The $g^{\pm}(x)$ are scaling functions for $t \gtrless 0$ and $f(q_\ell, q_t)$ is a generalized coupling term. We define one exponent via $c = c_0 t^\gamma$ and will relate the other β to ϕ below. The quantity $h = g\mu_B H/kT_G$ is the reduced field.

The form of the coupling term is deduced from physical input suggested by mean-field theory. It is assumed that a finite field, while suppressing the longitudinal transition, simply shifts the transverse one. The latter implies the last two terms in F_s can be combined in the form;

$$-2c_0 \; |t'|^\gamma q_t^2 \; g^{\pm}(q_t |t'|^{-\beta})$$

where now $t' = t - \Delta T_G(q_\ell)/T_G^0$ corresponds to transition shifted by $\Delta T_G(q_\ell)$. At this point as is usual in scaling theories, we require $g^{\pm}(x)$ to be analytic for small x. For small enough fields h and close, but not too close, to the transition temperature both q_ℓ and q_t can be made arbitary small while still $t \gg \Delta T_G / T_G^0$ and we can expand in q_t to obtain a leading coupling between the order parameters of the form:

$$t^{\gamma-1} \Delta T_G(q_\ell) q_t^2$$

Finally we note that the axis of quantisation is arbitarly chosen for cubic systems so F_s must be symmetric in q_ℓ and q_t which directly implies

$$\Delta T_G \propto q_\ell^2$$

The last step is to relate q_ℓ to the susceptibility exponent. Minimizing F_s with respect to m gives a Fischer like formula:

$$\chi \cong (1/2a)\;(1 - (e/a)\; q_\ell)$$

so the singular part of χ and q_ℓ have the same exponent i.e. $q_\ell \sim h^{2/\delta}$ whence our result:

$$\Delta T_G \sim h^{4/\delta}$$

The only related experimental work corresponds to such high fields that it almost certainly lies outside the critical region (5).

ACKNOWLEDGEMENTS

Work by one of the authors (SEB) is supported in part by NSF - Solid State Theory Grant No. DMR81-20827

REFERENCES

(1) J.R.L. de Almeida and D.J. Thouless J.Phys. A11(1978), 983.

(2) A.P. Malozemoff, S.E. Barnes and B. Barbara Phys. Rev. Lett. 51 (1983), 1704.

(3) G. Toulouse and M. Gabay, J. Phys. (Paris), Lett. 42 (1981) L103

(4) M. Suzuki, Prog. Theo. Phys. 58 (1977), 1151.

(5) W.E. Fogle, J.D. Boyer, R.A. Fisher and N.E. Phillips, Phys. Rev. Lett. 50, (1983) 1815.

LT-17 (Contributed Papers)
U. Eckern, A. Schmid, W. Weber, H. Wühl (eds)

SPIN-GLASS PHASE IN THE LOW CONCENTRATION RANGE OF THE SEMICONDUCTOR SYSTEM : $Sn_{1-x}Mn_xTe$

M. GODINHO, J.L. THOLENCE, A. MAUGER+, M. ESCORNE+ and A. KATTY+

Centre de Recherches sur les Très Basses Températures, C.N.R.S., BP 166 X, 38042 Grenoble-Cédex, France
+ Laboratoire de Physique des Solides, 1 Place A. Briand, 92190 Meudon, France.

It is well known that $Sn_{1-x}Mn_xTe$ undergoes a para to ferromagnetic phase transition for x > 5 at.%, upon cooling. We have extended magnetic measurements down to lower temperatures (T > . 1 K) and lower concentrations x > 0.8 at.% and found a spin glass phase for x < 3 at.% as well as a so-called "reentrant" spin-glass phase in the ferromagnetic regime (x > 3 at.%).

1. INTRODUCTION

A spin glass like freezing has been observed in several magnetic semiconductors containing Mn impurities. This freezing of the localized spin S = 5/2 of the Mn^{2+} ions is observed only above some critical Mn concentration $x_c \sim 20$ at.% in $Cd_{1-x}Mn_xTe$ (1,2,3) or $Zn_{1-x}Mn_xTe$ (4), while x_c almost vanishes in $Hg_{1-x}Mn_xTe$ (5) and in $Pb_{1-x}Mn_xTe$ (6). In all these systems, the interaction between the Mn ions is antiferromagnetic and the spin glass behaviour is attributed to frustration inherent to the fcc sublattice partly occupied by Mn. On the contrary, $Sn_{1-x}Mn_xTe$ is ferromagnetic for x > 4 at.% (7) due to the Ruderman-Kittel-Kasuya-Yoshida (RKKY) interaction mediated via free carriers (holes) generated by Sn vacancies in concentration equal to few 10^{20} cm^{-3}. However, a change in the magnetic properties is expected at lower Mn concentrations, because the long range ferromagnetic order cannot be maintained when the range of the interactions becomes smaller than the average distance between the Mn^{2+} ions. In this paper, we report magnetic properties of $Sn_{1-x}Mn_xTe$ in the range 0.8 < x < 5.8 at.%, at low temperatures T > 0.1 K reached by adiabatic demagnetization.

2. EXPERIMENTS

All the samples have been grown by the Bridgman technique, and the Mn concentration in the samples has been checked by microprobe analysis. We have measured in the same time the magnetization of the sample in a d.c. field ranging from 0 to 2500 Oe and the a.c. susceptibility in an a.c. field of 1 Oe.

The most concentrated samples (x ≧ 3.5 at.%) exhibit a ferromagnetic ordering at a Curie temperature T_C characterized by an abrupt change in the d.c. and a.c. susceptibilities which reach the inverse of the demagnetizing form factor. When the temperature is further decreased, the real part of the a.c. magnetic susceptibility, χ' decreases monotonically, while its imaginary part χ'' goes through a maximum. These are common features of "reentrant spin-glasse" observed in some ferromagnets. The temperature of the maximum of χ'' (3.2 K, 2.3 K and 2 K for x = 3.5, 4.5 and 5.8 at.% respectively), corresponds to the inflexion point of $\chi'(T)$ and marks the reentrant phase which may differ from one system to the other since it has been attributed to the freezing of canted ferromagnetism in AuFe (8).

In more dilute samples (x < 3 at.%), $\chi'(T)$ goes through a single maximum at a temperature T_f. When the measuring frequency is ν = 112 Hz, we find T_f = 0.52 K and 2.05 K for x = 0.8 and 2.2 at.% respectively. This maximum is shifted towards higher temperatures when ν is increased, as shown in fig. 1a. The temperature dependence of χ'' is illustrated in fig. 1b at two frequencies for the same sample x = 2.2 at.%. Like in other spin glasses (9), χ'' is small and does not depend strongly on ν at very low temperature ; then χ'' goes through a maximum at a temperature lower than T_f before it vanishes at higher temperatures. For x = 2.2 at.% a small bump is also observed around 2.8 K in the $\chi''(T)$ curve. This may be due to some inhomogeneities in this sample which is close to the ferromagnetic regime. We have analyzed the frequency dependence of T_f for these dilute samples in terms of a Fulcher law (10) :

$$\tau_m = \tau_\circ \exp E/k_B(T_f-T_\circ)$$

where $\tau_m = 1/\nu$, and k_B is the Boltzman constant. $\tau_\circ$ is chosen equal to 10^{-13} s. A quantitative agreement with the experimental data is found with $T_\circ$ = 0.43 K and E/k_B = 3.15 K for x = 0.08 at.%, while $T_\circ$ = 1.6 K and E/k_B = 3.15 K for x = 2.2 at.%, in the whole range of frequencies investigated (ν ≦ 101.2 kHz). These values indicate a rather large frequency dependence of T_f, characterized by the relative shift of T_f per frequency decade : $R=\Delta T_f/(T_f \Delta \log_{10}\nu)$ = 22×10^{-3} for x = 0.8 at.% and 27×10^{-3} for x =

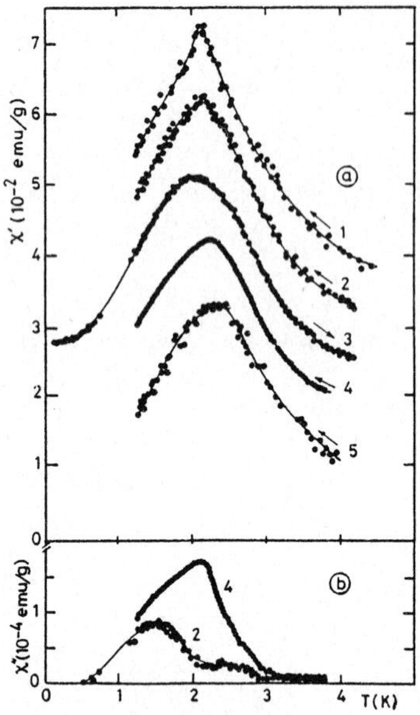

FIGURE 1
a) Real part of the a.c. susceptibility of $Sn_{0.978}Mn_{0.022}Te$ at different frequencies ν = 11.2 and 112 Hz, 1.12, 11,2 and 101.2 kHz for curves 1 to 5 respectively. The vertical scale is shown for the curve 5, the other curves being shifted by 0.075 emu/g. The relatively small T_f value of curve 3 might be attributed to the fact that this curve has been obtained by heating up instead of cooling down. b) Imaginary part χ'' at ν = 112 Hz and 11.2 kHz (curves 2 and 4 respectively).

2.2 at.%. In this low concentration range, R appears to be concentration independent, like in RKKY spin glasses (10). This is a clear distinction with the $Eu_xSr_{1-x}S$ system for which the relative frequency dependence of T_f decreases for increasing concentrations, starting from an Arrhenius law characteristic of the blocking of clusters ($T_o \cong 0$ (10)). However, the ratio R is larger than the values observed in canonical RKKY spin glasses (10) : 6×10^{-3} in Cu:Mn and Ag:Mn, $\sim 10^{-2}$ in Au:Fe, and is roughly equal to the value observed in the insulating system $Eu_{0.25}Sr_{0.75}S$ (11). Therefore the magnetic freezing of $Sn_{1-x}Mn_xTe$ presents some of the salient features of RKKY spin glasses, but also some features of insulating cluster glass systems. This points out the interest in further investigations on this material in which the RKKY magnetic interactions are mediated via a free carrier density much smaller than in metals.

3. CONCLUSION

We conclude that $Sn_{1-x}Mn_xTe$ forms a spin-glass phase at low Mn concentrations. A tentative magnetic phase diagram is shown in figure 2. In particular, a long range ferromagnetic order does not take place below a critical Mn concentration $\sim$ 3 at.%, contrary to previous assertions (12) deduced from experiments performed at too high magnetic fields.

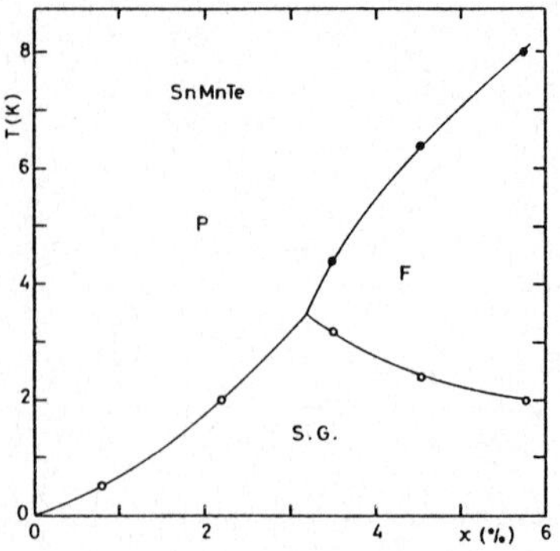

FIGURE 2
Magnetic phase diagram of $Sn_{1-x}Mn_xTe$ in the dilute Mn concentration limit. SG stands for spin-glass, F for ferromagnetic and P for paramagnetic configurations.

REFERENCES

(1) R.R. Galazka, S. Nagata and P.H. Keesom, B22 (1980) 3344.
(2) M. Escorne, A. Mauger, R. Triboulet and J.L. Tholence, Physica 107B (1981) 309.
(3) S.B. eroff, Phys. Rev. B25 (1982) 6584.
(4) S.P. McAlister, J.K. Furdyma and W. Giriat, Phys. Rev. B29 (1984) 1310.
(5) N.B. Brandt, V.V. Moshchalkov, A.O. Orlov, I. Skrbek, I.M. Tsidilkovskii, S.M. Chudinov, Zh. Eksp. Teor. Fiz. 84 (1983) 1050.
(6) J.L. Tholence, A. Benoit, A. Mauger, M. Escorne and R. Triboulet, Solid State Commun. 49 (1984) 417.
(7) M. Escorne and A. Mauger, Solid State Commun. 31 (1979) 893.
(8) See for instance I.A. Campbell, S. Senoussi, F. Varret, J. Teillet and A. Hamzić, Phys. Rev. Lett. 50 (1983) 1615.
(9) M.B. Salamon and J.L. Tholence, J. Magn. Magn. Mat. 31-34 (1983) 1375.
(10) J.L. Tholence, Solid State Commun. 35 (1980) 35 ; J. Appl. Phys. 50 (1979) 7310 ; Physica 108B, (1981) 1287.
(11) H. Maletta and W. Felsh, Phys. Rev. B 20 (1979) 1245.
(12) J. Cohen, A. Globa, P. Mollard, H. Rodot and M. Rodot, J. Phys. Colloque 29 (1968) C4-142.

LT-17 (Contributed Papers)
U. Eckern, A. Schmid, W. Weber, H. Wühl (eds)

EVIDENCE FOR A SPIN GLASS SCALING FUNCTION IN THE DISORDERED INSULATOR $CsNiFeF_6$

C.PAPPA, J.HAMMANN and C.JACOBONI*

DPh/SPSRM, CEN-SACLAY, 91191 Gif-sur-Yvette Cedex, France
*ERA,CNRS,N°609, Université du Maine, 72107 Le Mans Cedex, France

Magnetization data on $CsNiFeF_6$ have been analysed in the vicinity of T_g. A critical behavior of the parameter q defined by $M/H=C(1-q)/T$ has been found. The scaling relation $q=\tau^{\beta}F\{h^2/\tau^{\beta+\gamma}\}$ with $h=(H_a+J_oM)/T$ could be fitted above as well as below T_g.

Spin glass theories leading to a thermodynamic phase transition at the freezing temperature T_g, imply a critical behavior of the non linear part of the magnetization in the vicinity of T_g. A scaling relation has been proposed,which directly derives from that of a ferromagnet:the magnetization M is replaced by the non linear susceptibility χ_s and the magnetic field h by its square h^2[1]

$$\chi_s = \tau^{\beta}F(h^2/\tau^{\beta+\gamma}) \qquad (1)$$

where $\tau=|T-T_g|/T_g$ is the reduced temperature, and F(x) is a double valued function depending on whether $T>T_g$ or $T<T_g$

Many magnetization data have already been analysed in order to check these theoretical results. The divergence of the non linear susceptibility has been reported for several systems above T_g, with the corresponding values of the critical exponents [1]. Experimental scaling functions have been derived for $T>T_g$, but their extension below T_g has not yet been successful. In this paper we report on an analysis of magnetization data which show the validity of the scaling relation above as well as below T_g in the case of the disordered frustrated insulator $CsNiFeF_6$.

MAGNETIC PROPERTIES OF THE COMPOUND

The ac susceptibility in $CsNiFeF_6$ increases monotonously with decreasing temperature, departing from a high temperature Curie Weiss behavior below 150K. It is only around 5K that a sharp cusp appears. Below that temperature spontaneous local fields are observed by Mössbauer experiments, but no magnetic Bragg reflections could be detected by neutron diffraction.

Results of field cooled and zero field cooled magnetization measurements point to spin glass properties. A field-temperature phase diagram has been derived and analysed in terms of the mean field model for spin glasses [2]. The main conclusions which are of concern here are the following :

- the behavior of $CsNiFeF_6$ must be discussed in terms of a large value of J_o
- two characteristic temperatures T_m and Tr_1 have been defined. T_m is the temperature where the maxima of the field cooled curves occur and Tr_1 the temperature where irreversibilities first appear. $T_m(H=0)=4.75\pm.05K$ and $Tr_1(H=0)=5.35\pm.05K$.

From the discussion in ref[2] it was suggested that Tr_1 should be the critical temperature T_g.

DETERMINATION OF THE FREEZING TEMPERATURE

A new set of field cooled magnetization measurements has been performed in the temperature range of 4.2K to 6K with applied fields of up to 80G. The results have been corrected for residual field and demagnetizing effects.

The experimental data have been plotted on M/H versus H^2 curves. Their extrapolation to H=0 gives the values of the linear susceptibility $\chi_o(T)$. This susceptibility has a well marked maximum at 4.77K (in agreement with $T_m(0)$ previously mentioned) and it varies almost by two orders of magnitude between 6K and 4.77K. The temperature dependence of $1/\chi_o$ is linear above 5.35K and leads to an effective Curie Weiss temperature of θ=5.23K. This value of θ along with the large maximum of χ_o shows that $CsNiFeF_6$ has important effective ferromagnetic interactions. In the following discussion, the comparison with spin glass theories will thus be made by assuming a mean value J_o of the interactions equal to θ.

The large value of J_o makes it difficult to study the divergence of the non linear terms of the magnetization M above T_g. This is best seen in the expansion of the mean field solution for M, where the enhancement effect of the terms $1/(T-J_o)$ appears quite clearly :

$$T>T_g \qquad M = \frac{H}{T-J_o} - \frac{a_3}{T^2-T_g^2}\left(\frac{H}{T-J_o}\right)^3 \frac{T}{T-J_o} + \ldots (2)$$

Since $\chi_o(T)\propto 1/(T-J_o)$ for T>5.35K,relation (2) can be written as a function of $x=H^2\chi_o^3T$. Its generalization to non mean field theories would then be :

$$\frac{M}{\chi_o H} = 1 - \frac{A_3}{(T-Tg)^{\gamma}}\,x + \ldots \qquad (3)$$

In Fig.1, the experimental data have been plotted as M/χ_oH versus $x=H^2\chi_o^3T$. A large temperature dependence is observed from 5.92K down to 5.33K

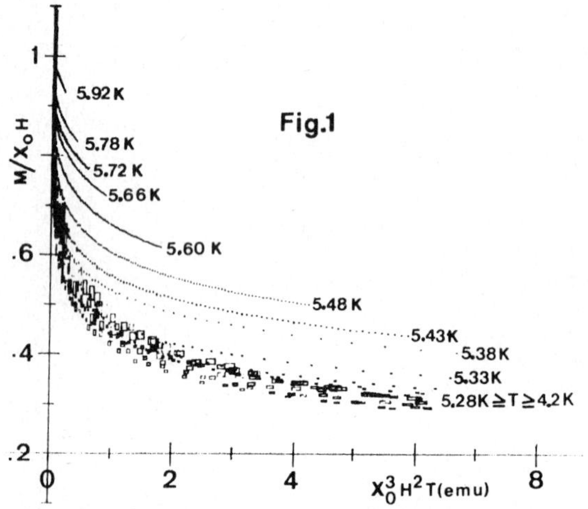

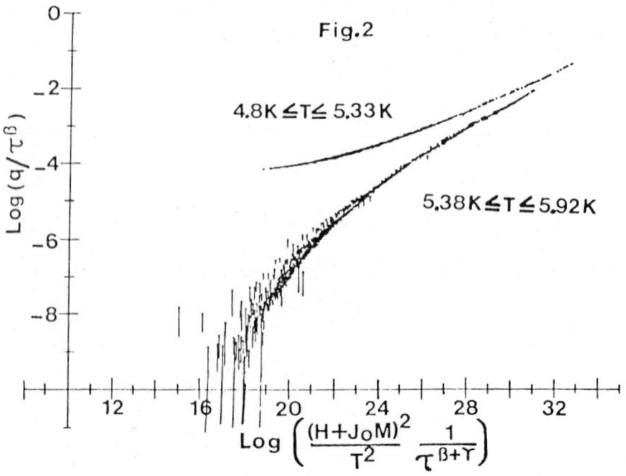

indicating that this is the temperature range where relation (3) should apply. The curves corresponding to T<5.35K almost coincide. The result points to the value of 5.35K for the freezing temperature T_g. This value is equal to the previously reported value of Tr_1 corresponding to the onset of irreversibilities in the system.

Analysing the curves of Fig.1, we have tried to define the temperature dependence of the first terms of expansion (3). But since x becomes too large in the vicinity of T_g (due to the increase of χ_0) it has been impossible to separate the different contributions. The first term could however be determined as the initial slope of the isotherms. It varies by two orders of magnitude between 5.92K and 5.43K and leads to the value of the exponent $\gamma=3\pm.5$.

TEMPERATURE DEPENDENCE OF THE ORDER PARAMETER

We use the expression $M=CH(1-q)/T$ which to first order in H relates the order parameter q to M [3] (below T_g,q is different from the Edwards Anderson parameter). This formula assumes $J_0=0$, if H stands for the applied field. In order to account for a non zero value of J_0 i.e. a Curie Weiss behavior $C/(T-\theta)$ for the linear susceptibility above T_g, it is convenient to take H as the mean local field : $H_1=H_a+J_0M$ ($J_0=\theta/C$).

The experimental values of M/H_1 have been extrapolated to $H_1=0$.C has been determined from the results obtained above 5.35K and the temperature dependence of q at zero field has then been derived from $q=1-T/C \lim_{H_1\to 0} M/H_1$.

q is found to behave as τ^{β} with $1.1\leq\beta\leq1.3$.

SCALING RELATION

From the previous discussion, it is now possible to gather the whole set of data on a single scaling function

$$q = \tau^{\beta} F\left\{\frac{(H_a+J_oM)^2}{T^2}\ \frac{1}{\tau^{\beta+\gamma}}\right\} \qquad (4)$$

where q is defined as $\frac{M}{H_a+J_oM} = \frac{C}{T}(1-q)$. (5)

For $T>T_g$,q/T is the non linear susceptibility.

For $T<T_g$,q/T is the difference between the total susceptibility and the extrapolation to this range of temperature of the linear part of the susceptibility as found above T_g.

The result is shown on Fig.2 for $\beta=1.3$, $\gamma=3$ and $T_g=5.35$K. The magnetization curves at many different temperatures are overlapping in exact continuation of each other. In the high temperature limit of the plot, the experimental errors on the very low field data are responsible for the observed scattering of the points.

In order to check the argumentation of [4],we have tried to scale our data to $T_g=0$K in the way suggested in ref[4]. Very large values of γ and Δ have been obtained : $\gamma=75\pm5$,$\Delta=50\pm5$. This result seems unphysical. Moreover, it was impossible to get a satisfactory overlap of the whole set of magnetization curves.

CONCLUSION

The excellent fit of the experimental data to the scaling relation (4) provides a strong argument in favor of the spin glass theories implying a thermodynamic phase transition. Two accomodations to the usual scaling relation had to be brought in to account for the temperature range $T<T_g$ and for a non zero value of J_0 :

- the parameter q to be scaled is determined by relation (5)
- the reduced parameter in the scaling function depends on the mean local field $H_1=H_a+J_oM$.

REFERENCES

[1] For a recent review see J.P.Renard,Congrès S.F.P.,Grenoble (Sept.83) Ed.de Phys.84
[2] C.Pappa,J.Hammann,C.Jacoboni,J.Phys.C:Solid State Phys. 17 (1984) 1303
[3] G.Parisi, Phys.Rev.Lett. vol.43,1754 (1979)
[4] K.Binder, Z.Phys.B, Condensed Matter 48,319 (1982).

LT-17 (Contributed Papers)
U. Eckern, A. Schmid, W. Weber, H. Wühl (eds)

ULTRASONIC MEASUREMENTS IN THE AMORPHOUS SPIN GLASS $(MnF_2)_{.65}$ $(NaPO_3)_{.15}$ $(BaF_2)_{.20}$

P. DOUSSINEAU★, A. LEVELUT★, M. MATECKI●, W. SCHÖN★, and W.D. WALLACE★☆

★*Laboratoire d'Ultrasons*†*, Université P. & M. Curie, Tour 13, 4 place Jussieu, 75230 Paris Cedex 05*

●*Laboratoire de Chimie Minérale D*†*, Université de Rennes -Beaulieu, 35042 Rennes Cedex, France.*

We have measured the temperature variation of the ultrasonic velocity and attenuation in the amorphous spin glass $(MnF_2)_{.65}(BaF_2)_{.20}(NaPO_3)_{.15}$ in a temperature range which includes the temperature T_f = 3.4 K . Our data between 0.1 K and 5 K can be accounted for by the standard two level system model. We find no definite anomaly at T_f.

1. INTRODUCTION

The temperature variation of the ultrasonic velocity in spin glasses has the appearance of possessing an anomaly associated with the temperature T_f at which the magnetic susceptibility has a cusp [1,2,3]. When the fractional velocity change $\delta v/v$ is plotted against temperature it appears that there is a decrease near T_f relative to what one might expect from the variation on either side of this temperature region. Such an anomaly would be expected if there was a phase transition at T_f. To date, a corresponding anomaly has not been observed in the ultrasonic attenuation.

If one assumes that the ultrasonic velocity is sensing a spin glass transition at T_f, one can attempt to substract the measured temperature variation of $\delta v/v$ from a smooth background variation to obtain the velocity change due solely to the transition. The difficulty lies in the fact that the apparent anomaly is rather broad in temperature and it is not a simple matter to determine the temperature variation of the background.

When a magnetic ion such as manganese is added to a glass and the results compared to a similar glass without the magnetic ion, there appears to be a large magnetic anomaly near T_f [2,3]. We show in this paper, however, that in fact there is very little, if any, anomaly near T_f in our samples in the usual sense of this term. Rather, we find that the same two level system (TLS) model which describes the ultrasonic behaviour in non-magnetic glasses describes quite well the temperature variation of $\delta v/v$ in our amorphous spin glass. The apparent large anomaly arises from the fact that the TLS parameters in our amorphous spin glass are significantly different from those for the non-magnetic glass.

2. THE TWO LEVEL SYSTEM MODEL

This model assumes amorphous materials contain entities with two energy levels which couple to phonons. Details of this model may be found in ref. [4] and we present only a summary here.

There is a *resonant* interaction with the two level systems which gives, for $\hbar\omega \ll k_B T$, a velocity variation

$$(\delta v/v)_{L,\tau} = C_{L,\tau} \ln T \qquad (1)$$

and a saturable attenuation not observed in this work. There is also a *relaxational* contribution of the TLS to the elastic constants which is more complex : at low temperatures in an insulating glass

$$(\delta v/v)_{L,\tau} = -(\pi^6/315)\, C_{L,\tau}\, K_3^2 T^6/\omega^2$$
$$\alpha_{L,\tau} = (\pi^4/96\, v_{L,\tau})\, C_{L,\tau}\, K_3\, T^3 \qquad (2)$$

while at high temperatures

$$(\delta v/v)_{L,\tau} = (C_{L,\tau}/2) \ln \omega - (3\, C_{L,\tau}/2) \ln T$$
$$\alpha_{L,\tau} = (\pi/2\, v_{L,\tau})\, C_{L,\tau}\, \omega \qquad (3)$$

In these expressions $\omega/2\pi$ is the ultrasonic frequency, the subscripts L, τ refer to longitudinal or transverse polarizations, α is the attenuation and $C_{L,\tau}$ and K_3 are constants for a given material. In Eqs 2 and 3 low temperatures means $\omega T_1^m >> 1$ and high temperatures means $\omega T_1^m \ll 1$, where T_1^m is the minimum relaxation time for a TLS of given energy. Measurements of $\delta v/v$ and α permit one to calculate C_L, C_T and K_3. The TLS spectral density, $\bar{P}$, is then given by $\bar{P} = \frac{4 k_B^3}{\pi \hbar^4} \frac{1}{K_3}\left(\frac{2 C_T}{v_T^3} + \frac{C_L}{v_L^3}\right)$ and the deformation potential by $\gamma_{L,\tau} = (\rho v_{L,\tau}^2 C_{L,\tau}/\bar{P})^{1/2}$ where ρ is the mass density.

In most amorphous insulators the values of $\bar{P}$, $\gamma_{L,\tau}$, K_3 have similar values, $\bar{P} \simeq 10^{32}$ erg^{-1} cm^{-3}, γ_L ($\simeq 1.5\, \gamma_T$) $\simeq$ 1 eV and $K_3 \simeq 10^9$ K^{-3} sec^{-1}.

☆Permanent address : Department of Physics, Oakland University, Rochester, Michigan 48063, U.S.A.
†Associated with the Centre National de la Recherche Scientifique.

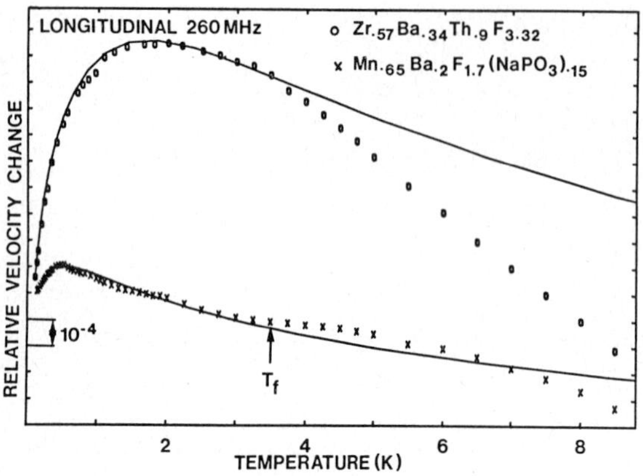

FIGURE 1
Relative velocity *vs* temperature on a linear scale. The X's are experimental points for the amorphous spin glass. The O's are data points for a non-magnetic fluoride glass, $(MnF_2)_{.65}$ $(NaPO_3)_{.15}(BaF_2)_{.20}$, measured in our laboratory. The curves are calculated by combining the resonant and relaxational contributions (Eqs 1,2,3).

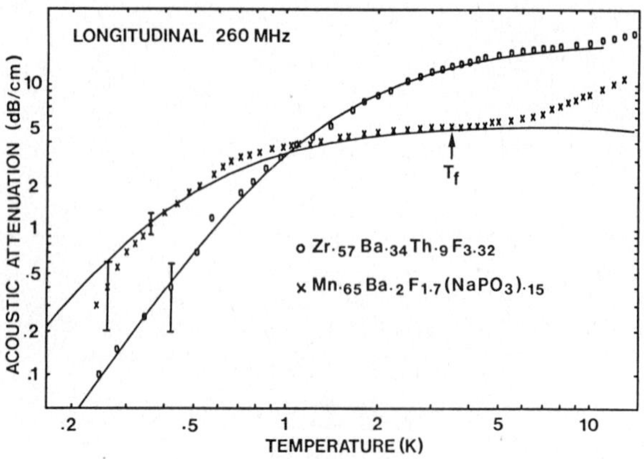

FIGURE 2
Attenuation as a function of temperature on a log-log scale. The data points are as described in figure 1. The curves are calculated from Eqs 2 and 3.

3. EXPERIMENTAL RESULTS

We have measured $\delta v/v$ and α *vs* T in $(MnF_2)_{.65}$ $(NaPO_3)_{.15}(BaF_2)_{.20}$ for both longitudinal and transverse waves at frequencies between 100 and 1000 MHz in two samples for $0.1 \text{ K} < T < 100 \text{ K}$. The magnetic susceptibility cusp occurs at $T_f = 3.4$ K in this material [5]. We observe a logarithmic variation of $\delta v/v$ (Eq. 1) and a T^3 variation of the attenuation (Eq. 2) in all measurements at low temperatures.

In this material, as was the case for a different fluoride glass with manganese reported earlier [3], we find that C_L and C_T are smaller by about a factor of 5 and K_3 is some 50 times larger than in non-magnetic fluoride glasses. We calculate the following values for the TLS parameters : $\bar{P} \simeq 1.6 \cdot 10^{30}$ erg^{-1} cm^{-3}, $\gamma_L = 4.6$ eV , $\gamma_T = 3.3$ eV . Note that spectral density $\bar{P}$ is smaller than usual for a glass. No anomaly is found in α near T_f, but the velocity variation has the appearance of a significant decrease in a temperature interval which includes T_f, if this data is compared to a non-magnetic fluoride glass, as can be seen in figure 1.

The magnitude of K_3 ($= 4.5 \cdot 10^{10}$ K^{-3} sec^{-1}) means (Eq. 2) that the *relaxational* contribution is much larger than usual for a glass. When the exact relaxational part of $\delta v/v$ is calculated (see ref. [4]) and added to the resonant part (Eq. 1), using our experimentally determined values for C_L, C_T and K_3 we find that the temperature variations in *both* $\delta v/v$ and α are substantially accounted for at frequencies from 100 to 1000 MHz [6]. A comparison between experiment and calculation for $\delta v/v$ is given in figure 1 and for α in figure 2 for longitudinal waves at 260 MHz. Note that for the most part, the experimental and calculated values are quite similar and that within the accuracy of our data, we see no anomaly near T_f.

4. CONCLUSION

In our amorphous spin glass we find that what appears to be a substantial decrease in $\delta v/v$ near T_f is actually the result of a large negative relaxational contribution from the TLS. We do not find any otherwise measurable anomaly in $\delta v/v$ or α. Although it is tempting to speculate, we cannot at this time say anything about the role of the manganese in determining the unusual values we find for the TLS parameters. Experiments are in progress which will hopefully shed light on this point. Finally, our results bring into question the validity, in general, of making extrapolations for use in obtaining a purely magnetic contribution to the ultrasonic velocity in amorphous spin glasses.

REFERENCES

[1] G.F. HAWKINS, R.L. THOMAS and A.M. de GRAAF, *J. Appl. Phys.* 50 (1979) 1709.
[2] M.J. LIN and R.L. THOMAS, in *Phonon Scattering in Condensed Matter*, Ed. H.J. Maris (Plenum Press, New York, 1980), p. 53.
[3] P. DOUSSINEAU, A. LEVELUT, M. MATECKI, and W.D. WALLACE (to be published)
[4] P. DOUSSINEAU, C. FRENOIS, R.G. LEISURE, A. LEVELUT and J.-Y. PRIEUR, *J. Physique* 41 (1980) 1193.
[5] C. DUPAS, J.P. RENARD and M. MATECKI (to be published)
[6] As is usually the case for non-magnetic glasses, a factor of about two must be added to Eq. 3 for α in order to match the temperature independent value measured experimentally.

LT-17 (Contributed Papers)
U. Eckern, A. Schmid, W. Weber, H. Wühl (eds)

ONSET OF HYSTERESIS IN $Al_2Mn_3Si_3O_{12}$ SPIN GLASS IN FINITE MAGNETIC FIELDS*

J.A. HAMIDA and S.J. WILLIAMSON

Department of Physics, New York University, New York, New York, 10003, U.S.A.

Measurements of the temperature dependence of the imaginary component of the magnetic susceptibility of $Al_2Mn_3Si_3O_{12}$ in weak magnetic fields reveal that the transition line marking the onset of hysteresis has negative curvature when the applied field approaches zero, as was observed previously for another insulating spin glass $Eu_{0.4}Sr_{0.6}S$.

1. INTRODUCTION

The onset of hysteresis in spin glasses in finite magnetic fields has been interpreted as a transverse spin freezing transition whose field near T_g was predicted (1,2) to have the form

$$H_c(T) = A(1 - T/T_g)^{\alpha} \tag{1}$$

where $\alpha = 3/2$. This "de Almeida-Thouless" line has been the focus of much interest, and the predicted field variation has been observed in several conducting spin glasses (3-7). Both T_g and the prefactor A exhibit a frequency dependence, as is expected from dynamical theories for spin freezing (8-11). Monte Carlo simulations by Young (12) for a spin glass model with short range interactions yield such lines as the loci for constant relaxation times. These calculations for a two dimensional Ising model predicted a departure from the form in Eq. 1 in the weak field limit, where α assumes a value less than unity. Such behavior has been observed in recent measurements on the insulating spin glass $Eu_{0.4}Sr_{0.6}S$, where the field range for α less than unity increases for higher measuring frequencies (13-15). The interpretation of these results, however, may be complicated by the fact that the high concentration of Eu in this system places it relatively near the crossover to ferromagnetism at 51% and furthermore produces a large susceptibility with accompanying demagnetization corrections.

The purpose of the present measurements is to explore the behavior of another insulating spin glass whose properties have been well characterized (16-19): manganese aluminosilicate with the chemical formula $Al_2Mn_3Si_3O_{12}$, having 15 at.% Mn. This has recently been the subject of extensive static and dynamic measurements by Beauvillain et al. (19), who give further references to prior studies. Their measurements of the temperature dependence of the imaginary component of the ac susceptibility χ'' at 133 Hz revealed an abrupt increase at the onset of hysteresis on cooling. Taking the inflection point as denoting T_c, the transition temperature in a given field, they reported that T_c decreases approximately linearly with field up to 180 Oe, in disagreement with Eq. 1.

2. EXPERIMENTAL METHOD

Measurements were conducted on the same cylindrical sample studied by Beauvillain et al. (19), with a diameter of 7 mm and length of 15 mm. The temperature dependence of the ac magnetic moment was measured using a SQUID magnetometer equipped with two coaxial copper solenoids, an outer one wound on a superconducting cylinder to provide a stabilized dc field and an inner one for the ac field. The ac field amplitude was less than 60 mOe. The ac component of the sample's magnetic moment was detected by the coaxial primary of a superconducting flux transformer whose secondary was magnetically coupled to a commercial SQUID system. The components χ' and χ'' of the longitudinal susceptibility were obtained by use of a lock-in amplifier whose reference phase was carefully controlled. Corrections for the sample's demagnetization factor are less than 3% and so were neglected in quoting field values.

3. RESULTS

Figure 1 illustrates the abrupt onset of χ'' for a frequency of 174 Hz. In zero field, the peak in χ' at T_g (arrow) does not coincide with the inflection point of χ''. Similar behavior is observed in the frequency range from 7 to 1500 Hz, with T_g increasing with frequency in agreement with previous studies. Application of a dc field causes a shift of the χ'' curves toward lower temperature, accompanied by a diminution of its peak. We define T_c by the inflection point in χ'' because it can be determined with reasonable precision. The results reported here would not be changed qualitatively had we instead chosen the temperature where χ'' reaches half the value of its zero field peak. Figure 2

* Supported in part by National Science Foundation grant DMR7917403.

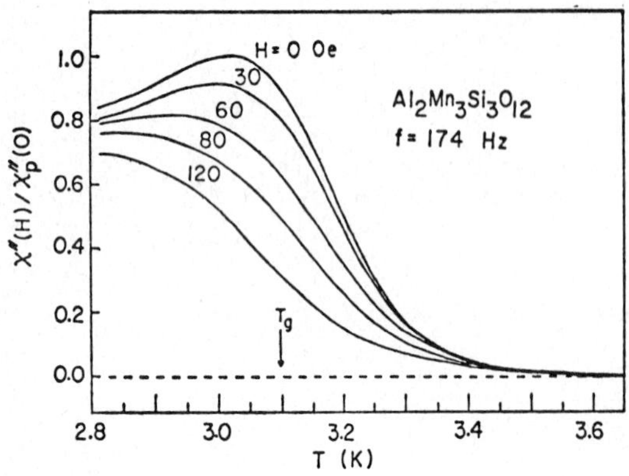

FIGURE 1
Temperature dependence of the imaginary component of the susceptibility at 174 Hz in various dc fields. The curves are normalized to the peak value of χ'' just below the onset in zero field.

shows how T_c shifts toward lower temperature increasing field. At high field the transition lines are consistent with a linear variation as reported in Ref. 19, although the field range available to us does not permit the exclusion of α being greater than unity in Eq. 1. However, our enhanced sensitivity permits an exploration in weak fields as well, and it can be seen that as H approaches zero these transition lines approach the temperature axis vertically. The data are consistent with $\alpha = 1/2$, indicating that the reduction of T_c is an analytic function

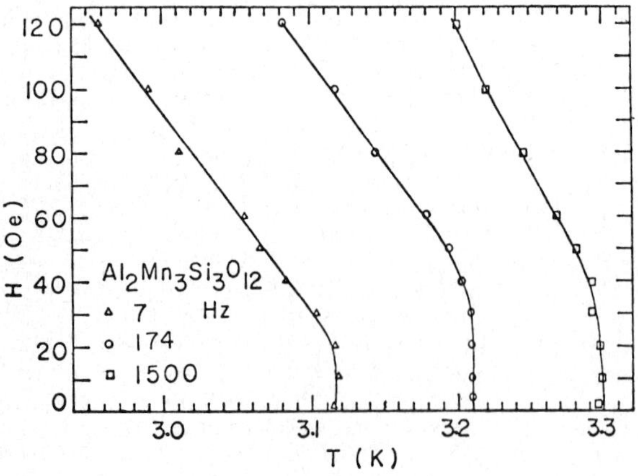

FIGURE 2
Transition lines marking the onset of hysteresis for three measuring frequencies. Lines are drawn as a guide to the eye.

of H. In addition, the field range for this approach is enhanced at higher frequencies. This is similar to the behavior seen in $Eu_{0.4}Sr_{0.6}S$ (13-15). Thus the crossover from linear behavior (or in the case of $Eu_{0.4}Sr_{0.6}S$ positive curvature) to negative curvature may be a common feature in insulating spin glasses with short range interactions, whereas there is yet to be a report on such behavior in conducting systems.

ACKNOWLEDGEMENTS
We thank K. Knorr and J.P. Renard for making available the sample used in these studies and C. Paulsen for discussions and technical help.

REFERENCES
(1) J.R.L. de Almeida and D.J. Thouless, J. Phys. A 11 (1978) 983.
(2) G.Thoulouse and M. Gabay, Phys. Rev. Lett. 47 (1981) 201.
(3) P. Monod and H. Bouchiat, J. Phys. (Paris) Lett. 43 (1982) L45.
(4) R.V. Chamberlin, M. Hardiman, L.A. Turkevich, and R. Orbach, Phys. Rev. B 25 (1982) 6720.
(5) M.B. Salamon and J.L. Tholence, J. Magn. Magn. Mat. 31-34 (1983) 1375.
(6) Y. Yeshurun and M. Sompolinsky, Phys. Rev. B 26 (1982) 1487.
(7) Y. Yeshurun, L.J.P. Ketelsen, and M.B. Salamon, Phys. Rev. B 26 (1982) 1491.
(8) H. Sompolinsky and A. Zippelius, Phys. Rev. Lett. 47 (1981) 359; Phys. Rev. B 25 (1982) 6860.
(9) W. Kinzel and K. Binder, Phys. Rev. Lett. 50 (1983) 1509.
(10) W. Kinzel and K. Binder, Phys. Rev. Lett. 29 (1984) 1300.
(11) K. Binder and A.P. Young, preprint.
(12) A.P. Young, Phys. Rev. Lett. 50 (1983) 1509.
(13) N. Bontemps, J. Rajchenbach, and R. Orbach, J. Phys. (Paris) 44 (1983) L47.
(14) J. Rajchenbach and N. Bontemps, J. Appl. Phys. 55 (1984) 1649.
(15) C. Paulsen, J.A. Hamida, S.J. Williamson, and H. Maletta, J. Appl. Phys. 55 (1984) 1652.
(16) R.A. Verhelst, R.W. Kline, A.M. de Graaf, and H.O. Hopper, Phys. Rev. B 11 (1975) 4427.
(17) H.R. Rechenberg and A.M. de Graaf, J. de Physique, Colloq. 39 (1978) C6-934.
(18) A.F.J. Morgownik, J.A. Mydosh, and L.E. Wenger, J. Appl. Phys. 53 (1982) 2211.
(19) P. Beauvillain, C. Dupas, J.P. Renard, and P. Veillet, Phys. Rev. B, in press.

LT-17 (Contributed Papers)
U. Eckern, A. Schmid, W. Weber, H. Wühl (eds)

SPIN GLASS-LIKE DYNAMICAL BEHAVIOR AROUND THE FREEZING TEMPERATURE OF KCl OH

H. GODFRIN, G. LAMARCHE*, J. GILCHRIST, D. THOULOUZE

Centre de Recherches sur les Très Basses Températures, C.N.R.S., B.P. 166X, 38042 Grenoble Cedex, France.

*Département de Physique, Université d'Ottawa, Ontario K1N 6N5, Canada.

The complex dielectric susceptibility $\varepsilon' - i\varepsilon''$ of KCl OH (concentration 800 ppm) has been measured as a function of temperature (4 mK - 4.2 K) and frequency (10 Hz - 100 kHz). Below T_F, the freezing temperature, ε' is linear in temperature, and ε'' becomes frequency independent, indicating a very wide range of relaxation times. The dynamical behavior of KCl OH is similar to that of spin glasses, when χ'' data are available.

1. INTRODUCTION

The properties of OH^- substitutional impurities in KCl crystals have been extensively studied (1). Dielectric susceptibility measurements (2-6) show that ε' has a maximum at a temperature T_F where k_BT_F is of the order of the average dipole-dipole interaction. At higher temperatures the behavior is paraelectric : below T_F the dipoles are "frozen". T_F varies linearly with the OH concentration and depends on the measuring frequency. A spin glass-like phase transition has been predicted by Fischer and Klein (7) in this system.

We report in this paper results of dielectric susceptibility measurements in a large temperature and frequency range.

2. EXPERIMENTAL DETAILS

The sample was made at the Cornell Material Science Center. Heat capacity and thermal conductivity were measured at Cornell and Grenoble (8). The OH concentration, determined by IR spectroscopy, is 800 ppm ($1.3\ 10^{19}\ cm^{-3}$).

The sample was cut to the dimensions $8 \times 4 \times 0.3\ mm^3$, with the smallest dimension parallel to a (100) direction. Silver electrodes were vapor deposited ; electrical contact was achieved by superconducting wires glued with silver epoxy. The sample was placed inside the mixing chamber of a dilution refrigerator.

The capacitance C and the loss angle δ were measured with a capacitance bridge and a dual phase lock-in amplifier, in the frequency range 10 Hz - 100 kHz.

The temperature was measured by a CMN mutual inductance, calibrated against the vapor pressure of ^4He and a superconducting fixed points device.

The complex susceptibility $\varepsilon = \varepsilon' - i\varepsilon''$ was deduced from the measured $C(T,f)$ and $\delta(T,f)$ by assuming a frequency and temperature independent susceptibility for the KCl matrix (5) : $\varepsilon_m = 4.5$.

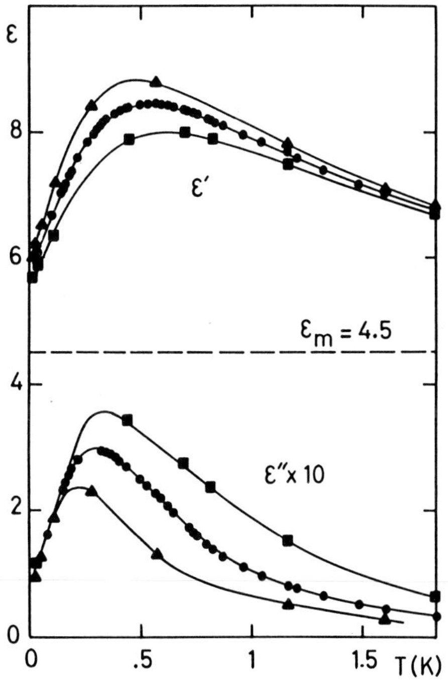

FIGURE 1
Dielectric constant $\varepsilon = \varepsilon' - i\varepsilon''$ of KCl OH (conc. 800 ppm) vs. temperature, at : Δ 39 Hz ; o 1 kHz ; ▪ 16 kHz.

3. RESULTS

Figure 1 shows part of the results. ε' has a maximum at a temperature $T_F(f)$; the frequency dependence can be described by the expression : $f = f_o \exp(-\Delta/T_F)$ with $\Delta = 13.5$ K and $f_o = 8\ 10^{13}$ Hz. Combining our results with those of ref. 2 we deduce that Δ increases linearly with the concentration.

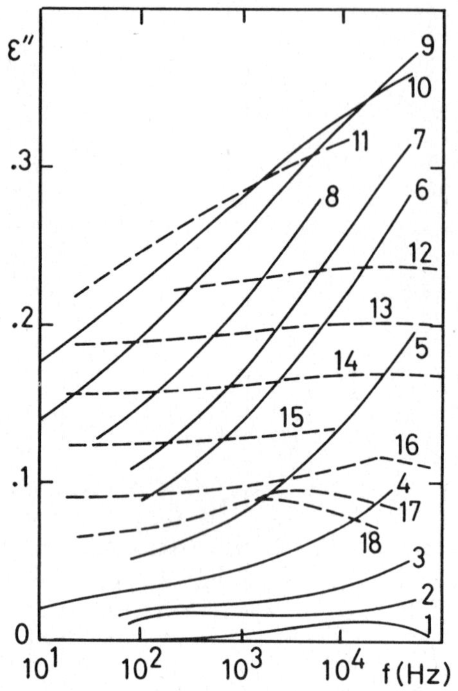

FIGURE 2
ε'' as a function of frequency for different temperatures :
1) 4.20 K ; 2) 2.81 K ; 3) 2.207 K ; 4) 1.602 K ; 5) 1.160 K ; 6) 0.814 K ; 7) 0.691 K ; 8) 0.569 K ; 9) 0.443 K ; 10) 0.347 K ; 11) 0.278 K ; 12) 0.155 K ; 13) 0.118 K ; 14) 0.089 K ; 15) 0.059 K ; 16) 0.028 K ; 17) 0.016 K ; 18) 0.008 K.
Note the difference between the high temperature (solid lines), and the low temperature behavior (dashed lines).

Above T_F the behavior is paraelectric and ε'' is small. Well below T_F (T <.2 K) ε' has a well defined linear variation with temperature (not adequately visible in figure 1). The slope increases as the frequency is reduced. The curves can be safely extrapolated to zero temperature ; ε'(T = 0) depends on the frequency and is larger than ε_m . The width of the maximum decreases with the frequency.

The maximum of ε'' occurs at a temperature $T_S < T_F$. Below T_S ε'' does not depend on frequency ; above T_F, a strong dependence is observed.

A detailed investigation of the frequency dependence of ε'' for different temperatures gives the results shown in figure 2. For temperatures between 2 and 4 K, a small maximum is observed ; it corresponds to thermally activated reorientation of nearest neighbour pairs, as observed on spin glasses (9) and in KCl OH (5).

Between .3 and 2 K, the evolution of ε''(f,T) corresponds to a distribution of relaxation times which become progressively longer as the temperature decreases.

Below .2 K ε'' is pratically frequency independent, indicating a dramatic increase in the distribution of the relaxation times. A similar behavior has been observed (10) in (Eu,Sr)S.

At the lowest temperatures (around 10 mK), ε'' decreases at high frequencies, showing possibly the upper limit of the distribution of relaxation frequencies at this temperature.

4. CONCLUSIONS

Accurate measurements of ε' and ε'' over a wide range of temperatures and frequencies provide a view of the evolution of the relaxation times with temperature. Our results on KCl OH are similar to those obtained in different spin-glasses (11). An extrapolation of the present data to zero frequency is not obvious, and the question of the existence of a phase transition in KCl OH is still open.

ACKNOWLEDGMENTS

We thank B. Castaing, J. Joffrin, J.C. Lasjaunias, K. Matho, J.J. Préjean, M. Saint-Paul, J. Souletie and J.L. Tholence for stimulating discussions.

REFERENCES

(1) V. Narayanamurti and R.O. Pohl, Rev. Mod. Phys. 42 (1970) 201.
(2) W. Känzig, H.R. Hart Jr. and S. Roberts, Phys. Rev. Lett. 13 (1964) 543.
(3) A.T. Fiory, Phys. Rev. B4 (1971) 614.
(4) U. Bosshard, R.W. Dreyfus and W. Känzig, Phys. Kondens. Materie 4 (1965) 254.
(5) R.C. Potter and A.C. Anderson, Phys. Rev. B24 (1981) 677.
(6) M. Saint-Paul, M. Mesa and R. Nava, Sol. State Com. 47 (1983) 183.
(7) B. Fischer and M.W. Klein, Phys. Rev. Lett. 37 (1976) 756 and 43 (1979) 289.
(8) J.C. Lasjaunias and H.v. Löhneysen, Sol. State Com. 40 (1981) 755.
J.J. De Yoreo, R.O. Pohl, J.C. Lasjaunias and H.v. Löhneysen, Sol. State. Com. 49 (1984) 7.
(9) J.L. Tholence, F. Holtzberg, H. Godfrin, H.v. Löhneysen and R. Tournier, J. Phys. (Paris) 39 (1978) C6-928.
(10) D. Hüser, L.E. Wenger, A.J. van Duyneveldt and J.A. Mydosh, Phys. Rev. B27 (1983) 3100.
(11) J.L. Tholence, EPS, Trends in Physics, Istambul (1981) Ed. I.A. Dorobantu, The Central Institute of Physics, Bucharest.

LT-17 (Contributed Papers)
U. Eckern, A. Schmid, W. Weber, H. Wühl (eds)

DYNAMICAL ASPECTS OF E.S.R. IN THE INSULATING SPIN-GLASS $Eu_{0.4}Sr_{0.6}S$

A.Deville, C.Blanchard, A.Landi

Lb.Electronique des milieux condensés, Aix-Marseille I, 13397 Marseille 13 (France)

The transient E.S.R. behaviour of the insulating spin-glass $Eu_{0.4}Sr_{0.6}S$ has been studied at 9.3 GHZ with the pulse saturation method, for $T \gtrsim T_g$. The r.f energy absorbed by the Zeeman reservoir is transmitted to the large exchange reservoir, vïa the dipolar coupling. This leads to a temporary shift of the E.S.R. line, which can be observed because of the presence of a phonon bottleneck.

I- INTRODUCTION

Little is presently known about the dynamical aspects of E.S.R. in spin-glasses. We present experimental results about the spin relaxation in $Eu_{0.4}\ Sr_{0.6}S$ (f.c.c. lattice ; d=4.21 Å between n.n Eu^{2+}; exchange integrals $J_1/k=0.2$ K (n.n), $J_2/k=-0.1$ K (n.n.n); $T_g=1.55$ K at $\omega=0$; at X band, absorption signal consisting of a single lorentzian line which broadens and shifts towards low fields below 10 K [1]).

II-EXPERIMENTAL CONDITIONS AND RESULTS

Measurements were done at ν=9.3 GHZ (spherical single crystal, $\emptyset$= 0.8 mm, in a cavity filled with liquid He; amplitude of the r.f linear magnetic field : $2H_1=4\sqrt{P}$ (H_1 gauss, P watt). In all experiments, the driving magnetic field was kept constant (unless otherwise stated , it was the resonance field at the temperature T of experiment); the spins were submitted to a pulse of intense r.f field (power P_{sat}, duration τ_{sat}) ; we then recorded the transient E.S.R signal using

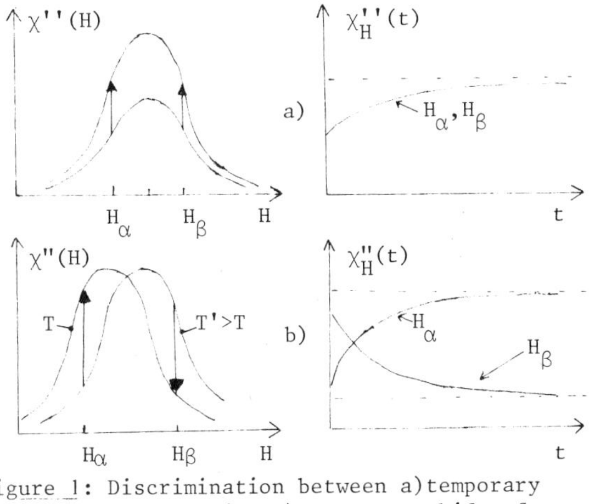

Figure 1: Discrimination between a) temporary decrease of absorption b) temporary shift of E.S.R line : in a) the signal increases with time for any value of the driving field H , contrary to b) (underloaded cavity) . (Principle) .

a weak r.f field. The first experiments, at 4.2 K, showed the signal to be energy-dependent: if P_{sat}=1mW then, for τ_{sat}=1ms the recovery was exponential ($\tau \sim 40$ ms) while for τ_{sat}=10 or 100 ms it fitted with $Ae^{-t/\tau'}+Be^{-t/\tau''}$ ($\tau' \sim 4$ ms, $\tau'' \sim 40$ ms; A/B 1.5 for 10 ms and 9 for 100 ms). At 1.4 k, for a strong energy the recovery time was $\sim$3 ms; we could not observe the longer time constant for a moderate energy.

Experiments at different values of the driving field (fig.1,2) proved that the recovery signal was mainly caused by a temporary shift of the E.S.R line, not by an intensity decrease of absorption as in paramagnets; an easy detection of this shift needed a sufficient saturating power. A similar result was obtained with a magnetron emitting the energy in a short time (1μs, 80W). Ordinary sample heating was ruled out (shift occuring within 1μs, heat capacity of the spins far greater than that of the phonons-cf.§III). We interpret this shift as a raising of the spin temperature by the r.f photons. We now discuss the origin of the longer time constant (from now on called τ_{obs} : $\tau_{obs} \sim 40$ ms) .

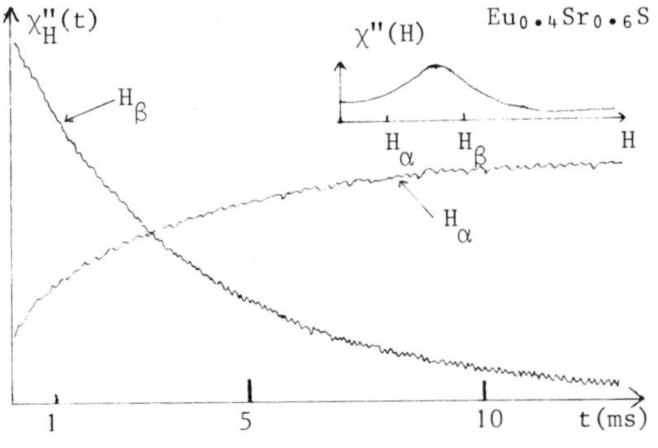

Figure 2: E.S.R recovery after a pulse with P_{sat}=80 W, τ_{sat}=1 μs ; (magnetron), driving field: α) $H_\alpha \sim 1300$ G; β) $H_\beta \sim 3300$ G; T=1.46 K) . Similar results were observed with P_{sat}=0.1W, τ_{sat}=1ms (klys.)

III-DISCUSSION

We used the Bloembergen-Wang reservoir model (fig.3) created for ferromagnets above T_c [2] : it should be a good starting point when $T \gtrsim T_g$, as it rests upon commutation relations [3] still valid in spin-glasses. The r.f field is coupled to the Zeeman reservoir,not to the exchange reservoir (S_x commutes with the exchange hamiltonian,not with the Zeeman one);these reservoirs are coupled by the dipolar coupling,and are in equilibrium at the end of our saturating pulse even if $\tau_{sat}=1\mu s$,as $\tau_{ZEx} \sim 10^{-10}$s (a value obtained from the high-temperature linewidth $\Delta H_{pp}=720$ G [1] together with Redfield theory [4] [5]: $Eu_{0.4}Sr_{0.6}S$ shows extreme narrowing

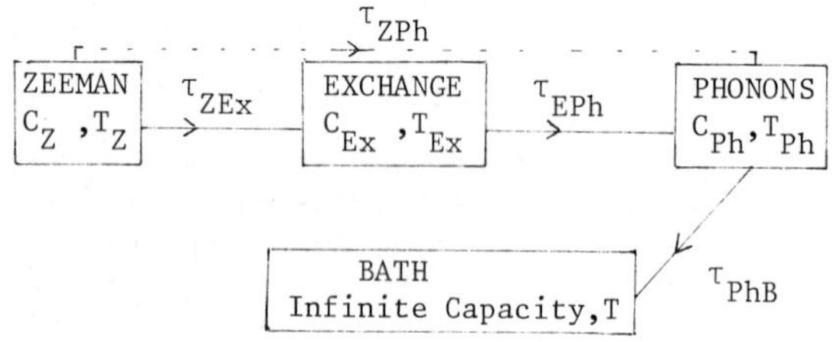

Figure 3:Bloembergen-Wang reservoir model; Heat Capacities:C_Z,C_{Ex},C_{Ph};Temperatures:T_Z,T_{Ex}, T_{Ph},T;Time constants: τ_{ZEx},τ_{EPh},τ_{PhB},τ_{ZPh} .

and $T_1=T_2$; T_1 must be identified with τ_{ZEx} ; (this should be reasonable down to T_g).As s=7/2 and $h\nu/kT \ll 1$,C_Z verifies $C_Z / N_o k \simeq \frac{21}{4}\left(\frac{h\nu}{kT}\right)^2 = \frac{1.04}{T^2}$ (N_o:number of spins); an approximation for C_{Ex} (Opechovski method [6], considering n.n only) gives

$$C_{Ex} /N_o k = \tfrac{2}{3}\left[J_1 s(s+1)\right]^2 cz/(kT)^2 \sim \frac{32}{T^2} \quad :$$

the capacity of the spins reduces to that of the exchange reservoir. If $T \ll T_D$ (Debye temperature $T_D \sim 300$ K) ,$C_{Ph}/N_o k = 1170(T/T_D)^3$ (acoustical phonons,Debye model): at 4.2K $C_{Ex}/C_{Ph} = 500$ (C_{Ph} reaches C_{Ex} near 16 K).

We calculated τ_{EPh} considering the J_1 modulation by the phonons -assuming $J_1 \propto e^{-\lambda x}$ (x: distance between n.n)- and neglecting the correlation between pairs in the perturbation [7] ; τ^{-1}_{EPh} is proportional to m_4 (direct process)or m_2 (Raman process), the 4th and 2nd moments of a spectral density; we calculated m_4 for a linear chain and arbitrary spin s, and multiplied it by $(0.4x11)^2$ (error made if s=1/2 : 4°/₀ [7]).For the direct process we found :

$$(\tau^{-1}_{EPh})_{dir} = \frac{16kT(d\lambda)^2 J_1^4}{9\pi\rho v_t^5 \hbar^4} s(s+1)\left[16s(s+1)-3\right]\left[c(z-1)\right]^2$$

(z:number of n.n ,c:Eu^{2+} concentration) ; taking $\lambda d=8$,$\rho=3800$ kg/m^3,$v_t=2000$ m/s ,then $(\tau^{-1}_{EPh})_{dir}=1.6\ 10^5 T$. We found that the Raman (exact calculation of m_2) and direct processes have the same efficiency at 30 K (the direct process is 10^5 times more efficient at 4.2 K, 30 times at 16 K). Thus at 4.2 K,$\tau_{EPh} \sim 1\ \mu s$; the exchange-phonon coupling by-passes the Zeeman-phonon coupling, met in paramagnets (in CaF_2:Eu, $\tau_{ZPh}=20$ ms at 4.2 K,[8];the τ_{ZPh} value should be comparable in SrS:Eu) .

Our experimental value ($\tau_{obs} \sim 40$ ms at 4.2 K) and the $C_{Ex}/C_{Ph} = 500$ ratio at 4.2 K suggest the presence of a phonon bottleneck [9]:after the thermalization of the Zeeman and Exchange reservoirs,the phonons shortly reach the spin temperature ($\tau_1 \sim \tau_{EPh} \cdot C_{Ph}/C_{Ex}$), then the spin-phonon system slowly reaches the bath temperature ($\tau_2 \sim \tau_{PhB} \cdot C_{Ex}/C_{Ph}$) ; the lifetime of the phonons is probably limited by collisions with spins (which is confirmed by thermal conductivity measurements [10]) rather than by inelastic scattering at the crystal boundaries,which would lead to $\tau_{obs}=(L/2v_t)\cdot(C_{Ex}/C_{Ph}) \sim 0.2$ ms (L:linear dimension of the crystal) .

This dynamical behaviour for $T \gtrsim T_g$, mainly associated with the high capacity of the exchange reservoir , and with the shift of the E.S.R line when the spin temperature approaches T_g , should be found in other insulating spin-glasses .

We are grateful to Dr. H.Maletta for supplying us with the samples.

REFERENCES

[1] A.Deville, C.Arzoumanian, B.Gaillard, C.Blanchard, J.P.Jamet, H.Maletta, J.Physique 42 (1981)1641.

[2] N.Bloembergen, S.Wang, Phys.Rev.93 (1954)72

[3] J.H.Van Vleck, Nuovo Cimento, Sup.Vol.VI , Ser.X, 3(1957)1081 .

[4] C.P.Slichter, Principles of magnetic resonance, Harper & Row, New-York (1964).

[5] A.Abragam, Les principes du magnétisme nucléaire, P.U.F. (1961).

[6] A.Herpin, Théorie du magnétisme, P.U.F.(1968)

[7] R.B.Griffiths, Phys.Rev. 124 (1961)1023.

[8] C.Y.Huang, Phys.Rev. 139,1A(1965)A241.

[9] A.Abragam, B.Bleaney, Résonance paramagnétique des ions de transition, P.U.F. (1971).

[10] C.Arzoumanian, A.M.de Goer, B.Salce, J.Physique Lettres 44 (1983)L39.

LT-17 (Contributed Papers)
U. Eckern, A. Schmid, W. Weber, H. Wühl (eds)

SPECULAR AND DIFFUSIVE SCATTERING OF HIGH FREQUENCY PHONONS AT SAPPHIRE SURFACES RELATED TO SURFACE TREATMENT

S. BURGER, W. EISENMENGER, K. LASSMANN

Physikalisches Institut, Universität Stuttgart
Federal Republic of Germany

The amount of specular and diffuse scattering of high frequency phonons at nonideal Al2O3 surfaces has been determined by comparison of reflection experiments with detailed Monte-Carlo calculations and has been found to have a characteristic frequency dependence related to the final polishing step.

Diffuse scattering of high frequency (>90 GHz) phonons at surfaces of solids may be effective either due to surface roughness because of their short wavelength or due to quantum interaction with surface states because of their high energies. Phonon scattering is therefore a promising means to probe nonideal surfaces.

By taking advantage of phonon focusing for a certain geometry of oblique backscattering Marx et. al. (1) have shown for silicon that the scattering of > 280 GHz-phonons is mainly diffusive irrespective of the surface treatment applied (mechanical, chemical or ion bombardment polish). In contrast a large contribution of specular reflection has been found for Al2O3 surfaces (2,3,4).

In this paper we show that in the case of Al2O3 surfaces the type of scattering depends significantly on the applied polishing procedure and on frequency. The experimental set-up is very similar to (1): The scattering surface is a (001) plane, the specular path lies in the C-X-plane and the distance (7 mm) between generator (constantan heater) and detector (Al- or Sn-tunnel junctions) has been chosen such that the expected reflection signal as computed by Monte-Carlo-simulations allows to discriminate between the specular and diffuse contributions. We have calculated not only the arrival times but also the amplitudes for specular reflection. The diffuse part was calculated as described in (1). The calculated specular and diffuse scattering signals are shown in Fig.1 a and b taking into account pulse length (100 ns), time constant (50 ns) and generator (.5 mm^2) and detector (.3 mm^2) areas. The specular part consists of essentially 3 peaks containing all nine possibilities of pure and mode converted reflections, whereas from the diffuse scattering

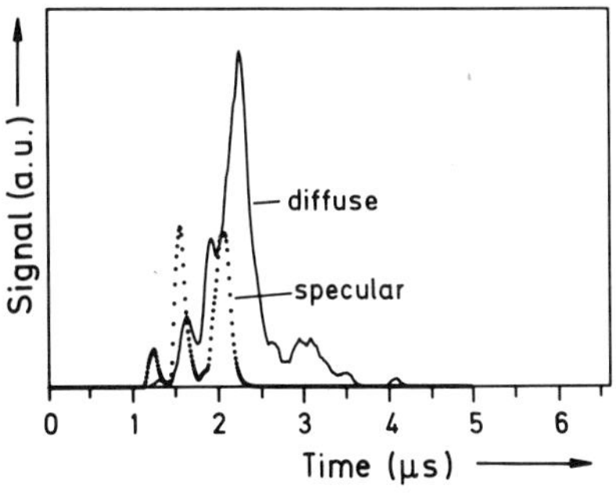

Fig. 1 Calculated detector signal for specular and diffuse reflection.

there is some peaked contribution within the time range of specular reflection and a large and sharp peak at a later time. Fits to the experimental curves have been accomplished by the weighted superposition of both constituents (Fig. 2 a and b).

Depending on polishing treatment two extreme cases have been found: (i) strong diffuse scattering not depending on frequency, if the final polishing agent was an alkaline suspension of colloidal silica (Fig. 3a) and (ii) mainly specular reflection at lower frequencies with increasing diffuse contribution at higher frequencies, if a suspension of .05 μm alumina in water or in paraffin oil was used (Fig. 3b). In the latter case even multiple specular reflections at the upper and lower Al2O3 surfaces can be resolved as peaks in the signal tail. The phonon frequencies appearing in the signal can be varied by changing the heater power or the detector threshold ($2\Delta_{Al}(T)$, $2\Delta_{Sn}(T)$).

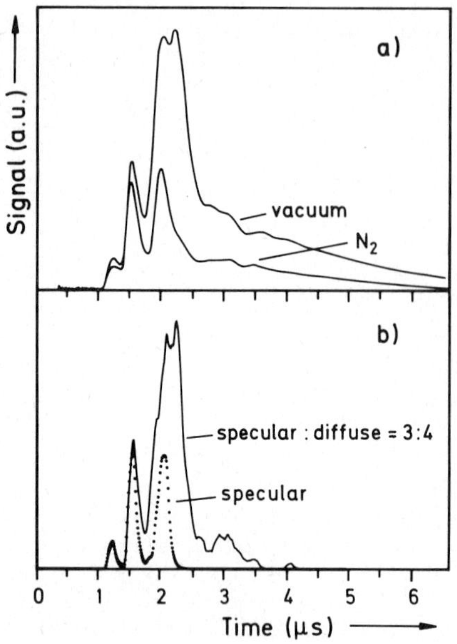

Fig. 2a Detector signal with scattering surface in vacuum or covered with solid nitrogen (heater power: .8 W/mm^2, Sn-detector).
Fig. 2b Calculated detector signals corresponding to the experimental situations in Fig. 2a.

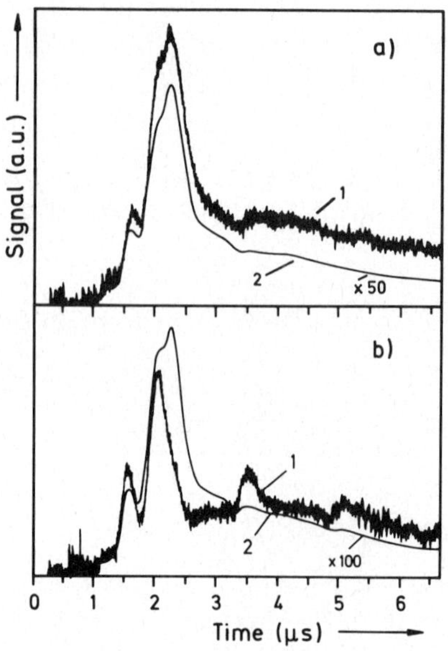

Fig. 3a and b Detector signals for different heater powers (1: 9 mW/mm^2, 2: .8 W/mm^2) and for different polishing treatments (a: case (i), b: case (ii)).

Covering the surface with liquid helium or solid nitrogen reduces only the diffuse contribution (Fig. 2a) (by the same amount in agreement with earlier experiments (1)). At the extreme of low frequencies (Al-detector and small heater power) case (ii) shows specular reflection with very small reduction by coverage (Fig. 4).

A twin geometry as described in (1) allowed direct comparison of optically smooth and rough surfaces in the same experiment. It turned out that the final polishing agent is more important than the apparent smoothness or roughness of the surface. In both cases (i) and (ii) the frequency dependence of the rough surface is the same as for the smooth surface. Of course in case (ii) the diffuse scattering of the rough surface is larger than for the smooth surface and there are no multiple specular reflections.

Repeated measurements showed that the results are reproducibly related to the final polishing step. Different cleaning procedures after this polishing (acetone or $K_2Cr_2O_7$ in H_2SO_4 in an ultrasonic bath) did not influence the signals.

We gratefully acknowledge the skilled technical assistance in polishing the crystals by Miss G. Untereiner, help in computer programming by W. Burger and M. Walter, and financial support by the Deutsche Forschungsgemeinschaft.

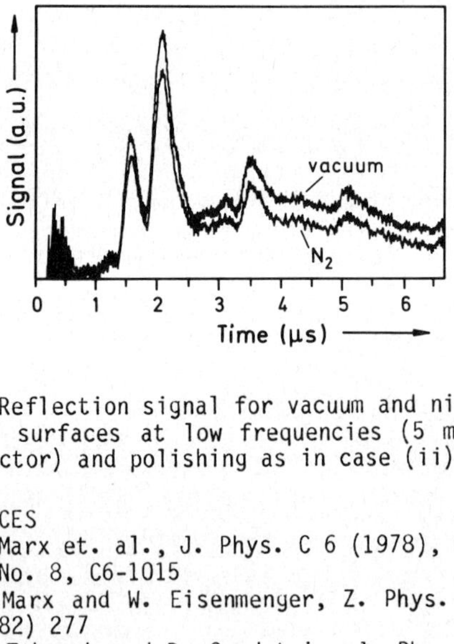

Fig. 4 Reflection signal for vacuum and nitrogen covered surfaces at low frequencies (5 mW/mm^2, Al-detector) and polishing as in case (ii).

REFERENCES

(1) D. Marx et. al., J. Phys. C 6 (1978), Suppl. to No. 8, C6-1015
D. Marx and W. Eisenmenger, Z. Phys. B 48 (1982) 277
(2) P. Taborek and D. Goodstein, J. Phys. C 12 (1979) 4737
(3) P. Taborek and D. Goodstein, Phys. Rev. B 22 (1980) 1550
(4) G. A. Northrop, Imaging of Specularly Reflected Phonons from a Crystal Boundary, in: Phonon Scattering in Condensed Matter, eds. W. Eisenmenger et. al., (Springer Series in Solid-State Sciences 51, 1984)

LT-17 (Contributed Papers)
U. Eckern, A. Schmid, W. Weber, H. Wühl (eds)

AMPLIFICATION OF SURFACE-MODE PHONONS IN n-TYPE InSb FILMS

C. C. WU

Institute of Electronics, National Chiao Tung University, Hsinchu, Taiwan, China

Amplification characteristics of surface-mode phonons in n-type InSb films have been investigated quantum mechanically in the GHz frequency region. Numerical results show that the frequency dependence of the amplification coefficient appears some oscillations which depend upon the mode of surface phonons. Moreover, the amplification coefficient depends also on the electronic screening effect in solids.

1. INTRODUCTION

In piezoelectric semiconductors such as n-type InSb, the interaction of surface phonons with conduction electrons is dominated by the deformation-potential and piezoelectric fields. The effect of nonparabolicity on the amplification of surface phonons becomes considerably important for the Rayleigh wave due to the nonlinear nature of energy bands in semiconductors (1). In an elastic medium with a stress-free plane boundary, acoustic waves can be propagated along the boundary of an elastic half-space and these waves are reflected on the boundary (2). Therefore there exist four different kinds of elastic waves other than the Rayleigh wave. These are referred to as the P-SV (pressure wave-shear wave with vertical polarization) mode, the SV-P (shear wave-pressure wave with vertical polarization) mode, the SH (shear wave with horizontal polarization) mode, and the TR (total reflection) mode. The TR mode can be understood as the special case of the SV-P mode and the electronic states of a semiconductor in the vicinity of its surface are not so simple, thus the TR mode wave will not be discussed simultaneously in here with three other mode waves. We use the quantum mechanical treatment in the GHz region such that $q\ell < 1$, where q is the wave number of surface phonons and ℓ is the mean free path of electrons. The energy band of semiconductors is assumed to be nonparabolic.

2. THEORY

A thin layer with the thickness d of a piezoelectric semiconductor is grown epitaxially on an insulating substrate with the same elastic properties as the semiconductor layer (3). As shown in Fig. 1, the Cartesian coordinates are fixed so that the material occupies the half-space $z > 0$ and has the stress-free surface parallel to the x-y plane. The potential along the z axis is assumed a square well which has infinitely high potential barriers at $z = 0$ and $z = d$. The field operator $\psi(\vec{r})$ of electrons in the second quantized form can be taken as (4)

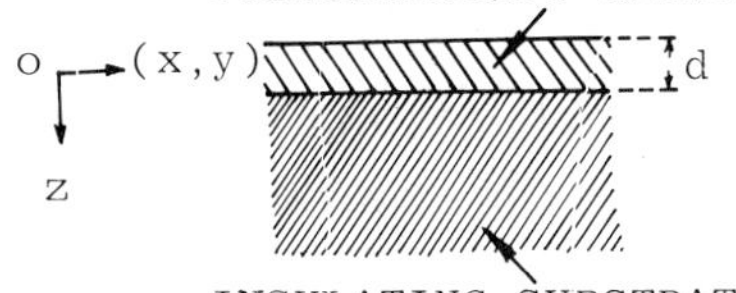

FIGURE 1

$$\psi(\vec{r}) = \left(\frac{2}{dS}\right)^{\frac{1}{2}} \sum_{n,\vec{k}} b_{\vec{k}n} \exp(i\vec{k}\cdot\vec{x}) \sin\left(\frac{n\pi z}{d}\right), \tag{1}$$

where $\vec{r} = (\vec{x},z) = (x,y,z)$, $\vec{k} = (k_x,k_y)$, S is the surface area of the film, $b_{\vec{k}n}$ is the operator of electrons satisfying the commutative relation of Fermi type. The energy of conduction electrons $E_{\vec{k}n}$ for the nonparabolic band structure is given by the relation (1)

$$E_{\vec{k}n}\left(1 + \frac{E_{\vec{k}n}}{E_g}\right) = \frac{\hbar^2\vec{k}^2}{2m^*} + \frac{\pi^2\hbar^2 n^2}{2m^* d^2}, \quad n = 1,2,3,\ldots. \tag{2}$$

where E_g is the energy gap between the conduction and valence bands, and m^* is the effective mass of electrons. The quantization of the elastic-wave field $\vec{u}(\vec{r},t)$ can be expanded in terms of the coefficient a_J as (2)

$$\vec{u}(\vec{r},t) = \sum_J \left(\frac{\hbar}{2\rho\omega_J S}\right)^{\frac{1}{2}} [a_J u_J(\vec{r}) \exp(-i\omega_J t) + \text{(Hermitian conjugate)}] , \tag{3}$$

where ρ is the mass density of solids, $J = (\vec{q},c,m)$ is a suitable set of quantum numbers, $\vec{q} = (q_x,q_y)$ is the wave vector of a surface phonon, c is the phase velocity defined by $\omega_J = c|\vec{q}| = cq$, and m specifies the propagation mode of phonons. a_J is the operator of surface-mode phonons obeying the commutative relation of Bose type. Using the Green's function method with the Born approximation (4), the amplification coefficient α_i of surface-mode phonons for the i-type electronic screening effect induced by the acoustic vibration of longitudinal or transverse waves in the piezoelectric coupling can be obtained as

$$\alpha_i^J = \left(\frac{x}{\hbar^3}\right)\left(\frac{2m^{*3}}{k_BT}\right)^{\frac{1}{2}} \sum_{s=1}^{\infty} (-1)^{s+1}\sqrt{s}\, \exp[sP(q)/k_BT]$$

$$\times\left\{ \sum_{nn'}{}' \exp[-sQ(q,n,n')/k_BT] \left[\frac{C^2}{[\varepsilon_\ell(q)]^2}|\Delta^J_{n'n}|^2 + \frac{16\pi^2 e^2 \beta_p^2}{\varepsilon_0^2[\varepsilon_i(q)]^2} |\phi^J_{n'n}|^2\right]\right\}, \qquad (4)$$

where Σ' indicates a summation over n and n' with $n \neq n'$, x is the drift parameter, $P(q) = \frac{1}{2}E_g + \mu - (m^*/2q^2\hbar^2)[\hbar\omega_J x - (q^2\hbar^2/2m^*)]$, $Q(q,n,n') = \frac{1}{2}E_g a_n + (m^*/2q^2\hbar^2)\{E_g[\hbar\omega_J x - (q^2\hbar^2/2m^*)](a_n - a_{n'}) + \frac{1}{4}E_g^2 (a_n - a_{n'})^2\}$, ($\mu$ is the chemical potential), and $\varepsilon_i(q) = 1 + (4\pi n_0 e^2 c_i^2/\varepsilon_0 c^2 k_B T q^2)$, ($n_0$ is the electron concentration). $i = \ell$ denotes the electronic screening effect induced by the acoustic vibrations of longitudinal waves with the longitudinal sound velocity c_ℓ, and i = t denotes that of transverse waves with the transverse sound velocity c_t (5). $\Delta^J_{n'n}$ and $\phi^J_{n'n}$ are the integral functions of the interaction between electrons and surface-mode phonons from Eqs. (1) and (3).

3. NUMERICAL RESULTS

We present numerical results as shown in Fig. 2. The relevant values of physical parameters are taken to be: d = 1 μm, $\beta_p = 1.8\times10^4$ esu/cm , $m^* = 0.013m_0$, ρ = 5.8 g/cm , $\varepsilon_0 = 18$, $c_\ell = 3.76\times 10^5$ cm/sec, $c_t = 1.61\times10^5$ cm/sec, C = 4.5 eV, E_g = 0.2 eV, T = 19.7 K, and $n_0 = 1.75\times10^{14}$ cm^{-3}.

For the P-SV mode, the phase velocity c is equal to $(\sqrt{6}/2)c_\ell$ and the wavevector $\vec{q}$ is parallel to [110] axis. The frequency dependence of amplification coefficient at x = 10 (corresponding to the applied electric field E = 8.51 V/cm) is shown as the curve A of Fig. 2. It shows that the amplification coefficient increases with frequency up to around ν = 15 GHz and then drops off rapidly with increasing frequency. Some oscillations with frequency can be observed in the high-frequency region. For the SV-P mode, c is equal to $\sqrt{6}c_t$ and $\vec{q}$ is also parallel to [110] axis. The frequency dependence of the amplification coefficient at x = 10 (E = 7.29 V/cm) is shown as the curve B of Fig. 2. The amplification coefficient increases with frequency up to around ν = 12 GHz and then decreases rapidly with increasing frequency. However, this decreasing will become slower after ν = 50 GHz. No oscillations can be observed except that the inflection point appears in the intermediate-frequency region. For the SH mode, c is equal to $\sqrt{2}c_t$ and $\vec{q}$ is taken in the [010] direction. The frequency dependence of the amplification coefficient at x = 10 (E = 4.21 V/cm) is shown as the curve C of Fig. 2. We can see that the amplification coefficient increases with frequency in the lower-frequency region and then decreases rapidly. From our numerical results presented here, it shows that the amplification coefficient due to the electronic screening of the transverse waves is larger than that of the longitudinal waves in the low-frequency region. However, when the frequency is coming into the high-frequency region, the screening effect will be vanished.

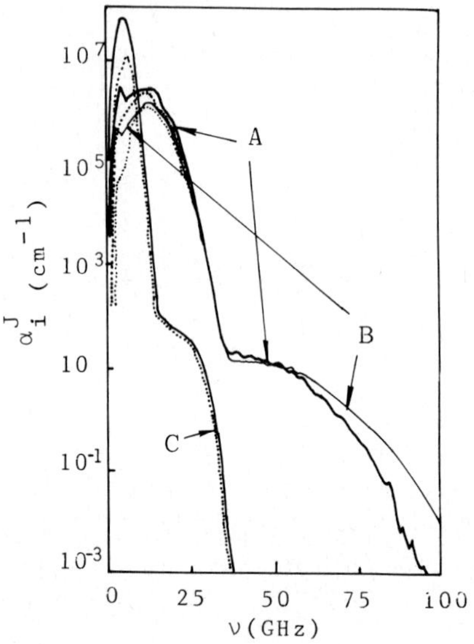

FIGURE 2
Amplification coefficient of surface-mode phonons versus frequency. Solid curve:the screening effect of the transverse waves; broken curve: the screening effect of the longitudinal waves.

REFERENCES

(1) C. C. Wu and J. Tsai, J. Low Temp. Phys. 51 (1983) 453.
(2) H. Ezawa, Ann. Phys. (N.Y.) 67 (1971) 438.
(3) S. M. Sze, Physics of Semiconductor Devices (Wiley-Interscience, New York, 1981).
(4) S. Tamura and T. Sakuma, Phys. Rev. B 16 (1977) 3638.
(5) F. García-Moliner and F. Flores, Introduction to the Theory of Solid Surface (Cambridge, 1979).

LT-17 (Contributed Papers)
U. Eckern, A. Schmid, W. Weber, H. Wühl (eds)

THE PHONON SPECTRA EMITTED BY SUPERCONDUCTING Pb AND PbBi TUNNEL JUNCTIONS

Paul BERBERICH and Helmut KINDER

Physik-Department E 10, Technische Universität München, D-8046 Garching, FRG

The phonon spectra emitted by superconducting Pb and PbBi tunnel junctions in the single particle tunneling regime were studied with a stress tuned Si : B spectrometer. An unexpected large contribution of phonons with an energy below 2 Δ was found. The results are compared with solutions of the nonlinear coupled kinetic equations for quasiparticles and phonons.

Single particle tunneling between two superconductors is known to result in a quasiparticle spectrum sharply peaked at eV - Δ and at Δ . The quasiparticles relax to the gap edge and then recombine. Therefore the emitted phonon spectrum consists of a broad band of relaxation phonons with a sharp high frequency cutoff at eV - 2Δ and of 2 Δ -phonons (1,2). These spectral features have been widely used for monochromatic phonon generation (3). For such experiments Pb and PbBi junctions provide a wide frequency range of up to 800 GHz because of their large gaps. However, there are two factors which may influence the emitted phonon spectrum of these junctions: (i) The large wave vector of the generated phonons ($q \leq 0.5\ q_{max}$) favors phonon decay processes. (ii) The junctions are easily driven into the overinjection regime due to the high phonon reabsorption rate by quasiparticles. In this paper we present the emission spectra of such junctions measured with a stress tuned Si:B phonon spectrometer (4,5).

This spectrometer consists of a Si crystal of dimensions 2.5x4x15 mm containing $5 \times 10^{13}/cm^3$ boron acceptors. The generator junction (1x1 mm, 300 nm) was evaporated on one 4x15 side and an Al detector junction on the opposite side. The sample was immersed in liquid He at 1 K. The "differential" spectra were recorded by superimposing small pulses (t_p = 200 ns) to the dc bias voltage. FT phonons were detected in $(1\bar{1}0)$ direction. When applying uniaxial (111) stress the acceptor ground state splits into a doublet the splitting being proportional to the applied stress. Phonons are resonantly scattered at this two level system. The spectrum is thus obtained by sweeping the line position. Inelastic phonon scattering by the acceptors is neglible because of the low concentration used. When the Pb junction was biased in the ac Josephson regime monochromatic phonons with the energy 2eV were emitted (6). This way the energy scale and the resolution of the spectrometer were determined.

Fig.1 shows the differential phonon emission spectra from a Pb junction (R_∞ = 6 mΩ). In trace (a) the junction was biased at the gap voltage V = 2.80 mV. The peak at 2.8 meV corresponds to 2 Δ -phonons. The width (20%) is larger than the spectrometer resolution (10%). There is also an unexpected large contribution of low energy phonons. In trace (b) the 2 Δ -phonon peak is shifted upwards to eV = 3.12 meV because of direct recombination of quasiparticles (3). The low energy background apparently increases because the differential spectra are recorded. The low energy cutoff at 0.3meV is caused by the Al detector used. At higher bias voltages (traces c-h) the relaxation of quasiparticles is getting faster than direct recombination. Consequently, the peak of the recombination phonons shifts back and "bremsstrahlung" phonon peaks appear at eV-2 Δ (2). The width of these phonon peaks is caused by the gap anisotropy and the spectrometer resolution. Reabsorption of the "bremsstrahlung" phonons sets in at V = 4 Δ/e (trace h).

The phonon emission spectra of PbBi alloy junctions are similar to that of Pb. Biased at the gap voltage V = 3.13 mV the low frequency background is a factor of 4 larger than in trace (a). The "bremsstrahlung" phonons at higher bias voltages are sharper than in traces (c) to (h) because of the smaller gap anisotropy of PbBi. Above about 2 meV their intensity decreases in a similar way as for Pb. In traces (i) to (k) we have depicted some of these spectra because of an interesting detail. Note there are peaks at (eV - 2 Δ)/2 -see arrows- indicating a subharmonic decay of the relaxation phonons.

Chang and Scalapino (7) have studied the properties of a thin superconducting film strongly driven into a spatially uniform steady state by quasiparticle injection. They solved the resulting nonlinear kinetic equations for quasiparticle and phonon distributions numerically. The steady state gap parameter was adjusted via the

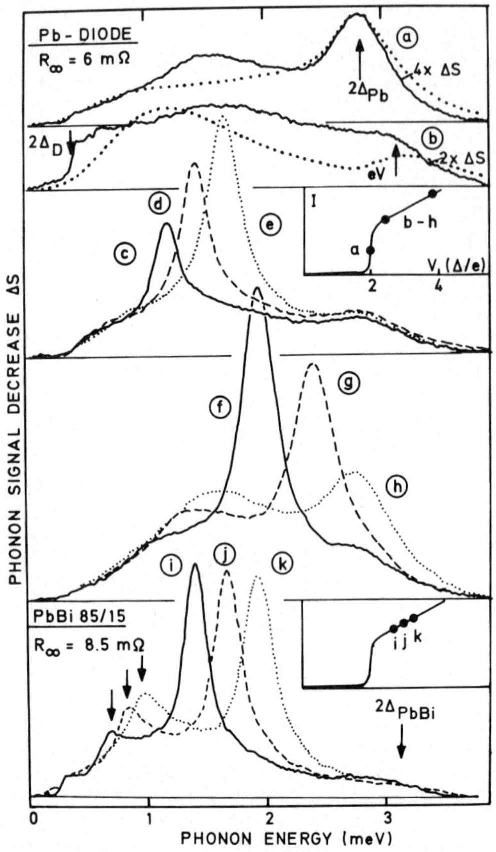

FIGURE 1
Differential phonon emission spectra. Modulation 8 mA. Bias voltages in mV: a 2.80, b 3.12, c-e 3.89, 4.14, 4.39, f-h 4.64, 5.14, 5.67, i-k 4.53, 4.78, 5.03. The dotted lines in a and b are the calculated spectra for s = 400.

BCS gap equation. In similar calculations we used for Pb the material parameters from (7) and varied only the trapping parameter $s = \tau_{es}/\tau_B$ where τ_{es} is the phonon escape time and τ_B is the pair breaking lifetime. The spectral emitted power $p(\Omega) = \Omega^3(n(\Omega) - n(\Omega,T))/\tau_{es}$ calculated for a bias point $eV = 2.07\ \Delta_0$ is shown in Fig.2. The small gap reduction Δ/Δ_0 was not corrected for. The calculations clearly show a broadening of the 2Δ-phonons and a background of low energy phonons increasing with s. The basic processes involved are: (i) An initial bandwidth of the quasiparticles is multiplied by many recombination and reabsorption events; (ii) Quasiparticles at Δ are up-scattered to 3Δ by absorbing 2Δ - phonons. We find the decrease of

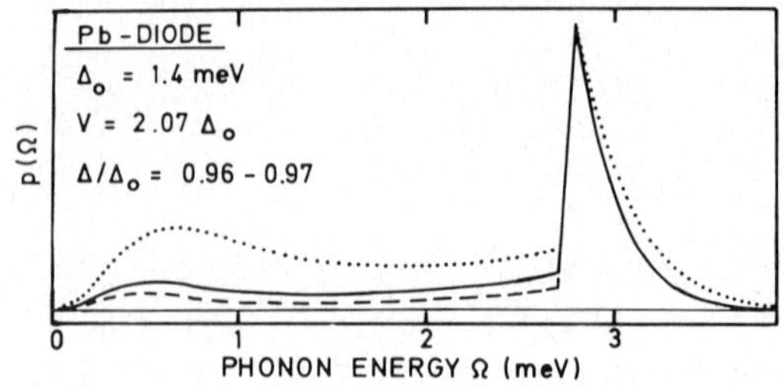

FIGURE 2
The calculated spectral power of emitted phonons for the Pb junction at $eV = 2.07\ \Delta$. Dashed/solid/dotted lines: s = 100/200/400.

the phonon signal ΔS by folding $p(\Omega)$ with a spectrometer weight function (5). Spectra calculated this way are depicted in Figs. 1a and b as dotted lines. At higher bias voltages -not shown- the calculated variation of the relaxation phonon peaks is similar in form as in Figs. 1 (c) to (h) but weaker.

In summary, the main spectral features are reproduced but the quantitative agreement is poor. The trapping parameter of 400 used is 4 times larger than determined from the transient response of laser irradiated films (8). Therefore other scattering processes may be effective as indicated by the PbBi results. The relaxation of electrons by two phonon emission (9) or phonon decay via lattice anharmonicity (10,11) are too weak and lead to a stronger frequency dependence than experimentally observed. Inelastic phonon scattering by impurity states might be strong enough but there is no experimental evidence for such modes in the metal films studied.

REFERENCES

(1) W. Eisenmenger and A.H. Dayem, Phys. Rev. Lett. 18 (1967) 125.
(2) H. Kinder, Phys. Rev. Lett. 28 (1972) 1564
(3) W. Eisenmenger, in: Physical Acoustics, Vol 12, eds. W. P. Mason and R. N. Thurston (Academic Press, New York, 1976) p.79
(4) R.C. Dynes, V. Narayanamurti and M. Chin, Phys. Rev. Lett. 26 (1971) 181.
(5) T. Fjeldly, T. Ishiguro and C. Elbaum, Phys. Rev. B7 (1973) 1392.
(6) P. Berberich, R. Buemann and H. Kinder, Phys. Rev. Lett. 49 (1982) 1500.
(7) J. J. Chang and D. J. Scalapino, J. Low Temp. Phys. 31 (1978) 1.
(8) C. C. Chi. M. M. T. Loy and D. C. Cronemeyer, Phys. Rev. B23 (1981) 124.
(9) L. Tewordt, Phys. Rev. 127 (1962) 371.
(10) R. Orbach and L. A. Vredevoe, Physics 1 (1964) 91.
(11) P. G. Klemens, J. Appl. Phys. 38 (1967) 4573

LT-17 (Contributed Papers)
U. Eckern, A. Schmid, W. Weber, H. Wühl (eds)

LASER INDUCED NONEQUILIBRIUM SUPERCONDUCTIVITY - A SPATIALLY RESOLVING PHONON DETECTOR

H. Schreyer, W. Dietsche, and H. Kinder

Physik - Department E10, TU München, D 8046 Garching, W. Germany

A focused laser beam was raster-scanned across a superconducting tunnel junction. The laser-light irradiation caused the quasiparticles to be locally out of equilibrium. In these areas the sensitivity for phonons was found to be different from the remainder of the junction. It is demonstrated that such a device can be used as a spatially resolving phonon detector.

Over the past years the physics of high-frequency phonons has progressed rapidly. However a whole class of interesting problems could not be tackled because of the lack of a practical spatially resolving phonon detector. To overcome this shortcoming Northrop and Wolfe used a movable heater, a laser-heated spot, to obtain their phonon-focusing images (1). This method is restricted to heat pulses, however. Experiments requiring monochromatic phonons are not possible with this method. Angular emission characteristics of fixed phonon sources are also impossible to study.

Recent attempts to realize a spatially resolving detector included placing several tunnel junction in series (2) and using the fountain pressure of a superfluid Helium film (3). These methods, however did not find wide-spread applications.

In this Paper we introduce a novel method which requires just a conventional superconducting tunnel junction as sensitive element. The spatial distribution of impinging phonons can be read out by scanning a laser beam across the junction.

The experimental set-up is sketched as insert in Fig. 1. We used as sample a (110)-oriented Si crystal of 2 mm thickness, and 15 mm diameter. On the bottom side of the sample, a Ag film of 50μm x 100μm area was placed serving as heat-pulse generator. On the top side a rhomb-shaped Al tunnel junction of 24 mm^2 area was prepared. The whole sample was immersed in liquid Helium at 1.1 K. An Ar laser beam in cw mode was focused (o.1 mm diameter) onto the Al-junction from outside the cryostat. Using two computer-controlled galvo-driven mirrors the laser beam could be raster-scanned across the whole junction.

The principle of operation is indicated in Fig. 1. It shows schematically the part of the IV-characteristic of the tunnel junction where single-particle tunneling sets on. With the Ar laser on, the characteristic is expected to change to the dotted line. The laser light produces quasiparticles at an high rate in the focal area. Under this nonequilibrium condition the energy gap decreases locally (4). This in turn leads to a locally increased current density if the junction is biased just below $2\Delta/e$. The situation can be visualized as a large unperturbed tunnel junction wired in parallel with a small one having a reduced gap.

For phonon detection we operated along the dotted load line. Phonons were generated by applying a sine wave to the heater and the signal was measured with a Lock-In. The data were recorded by the computer and displayed on a

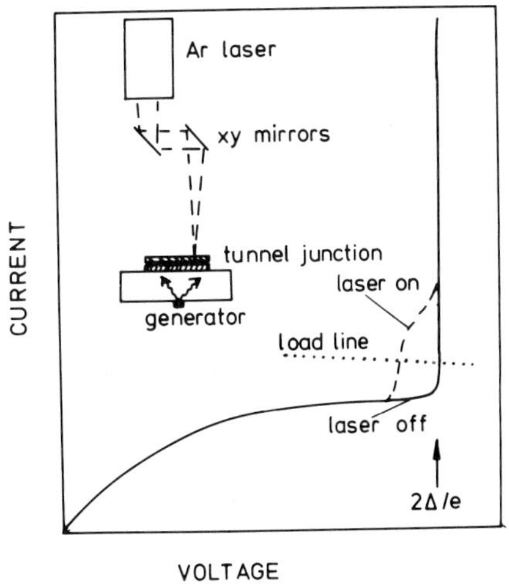

FIGURE 1
Schematic of an IV-characteristic of a tunnel junction used for phonon detection. The effect of the "laser on" (dashed line) is exaggerated for clarity. Operating point is the intersection of the load line and the "laser on" trace.

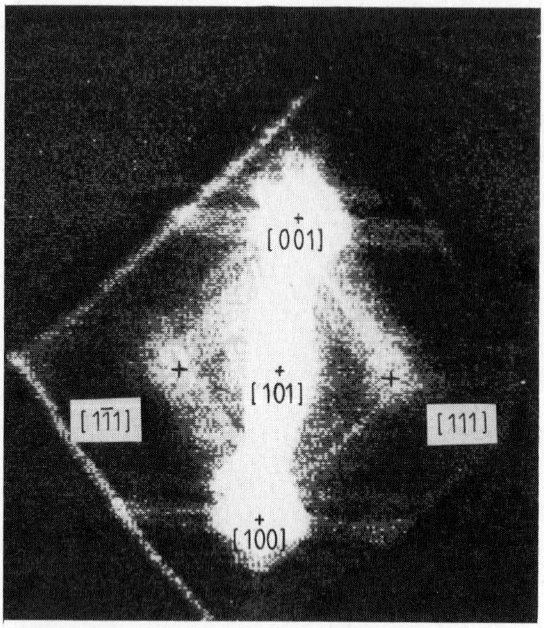

FIGURE 2
Photograph of the TV picture which displays the distribution of the phonon signal. The rhomb is the overlap area of the tunnel junction. Inside the rhomb the phonon focusing pattern is visible.

TV monitor (5). Fig. 2 shows the result. The grey tone of any point corresponds to the phonon intensity of the respective xy position of the laser beam on the crystal. The outline of the junction is easily recognized by its rhomb-like shape. Some symmetry directions of the crystal are marked.

Strikingly, structures can be seen in the phonon signal within the junction area. These are obviously correlated with the crystal symmetries. In fact the picture is very similar to the phonon focusing image obtained by Dietsche et al. (6) in Ge for the same crystal symmetry. Therefore we conclude that we observed the phonon focusing properties of Si by using this new spatially resolving detector.

Its basic detection mechanism is the further gap reduction caused by impinging phonons in the laser-excited area where the gap is already reduced. The voltage signal which corresponds to these phonons is not shorted by the remainder of the junction because that has a large dynamic resistance. On the other hand, the unexcited area gives rise to a background signal. This signal, however, is reduced because it is shorted by the excited area which has a small dynamic resistance. Furthermore the background signal is constant over the junction area. For the data of Fig. 2 an offset of three times full scale was used to suppress it.

In conclusion we have demonstrated that a spatially resolving phonon detector can be realized by using a strikingly simple technique. The scanning of the laser beam across the junction is very similar to the operation of a vidicon TV tube where the elctron beam is scanned across the light-sensitive element.

REFERENCES

(1) G. A. Northrop and J. P. Wolfe, Phys. Rev. Lett 43, 1424 (1979).
(2) H. Kinder, J. Weber, and W. Dietsche, in: Phonon Scattering in Condensed Matter, ed. H. J. Maris (Plenum, 1980) pp 173 - 180.
(3) W. Eisenmenger, in: Phonon Scattering in Condensed Matter, ed. H. J. Maris (Plenum, 1980) pp 303 - 308.
(4) For review see several articles in: Nonequilibrium Superconductivity, Phonons, and Kapitza Boundaries, ed. K. E. Gray (Plenum, 1981).
(5) The data processing was modeled along the lines developed by J. P. Wolfe and coworkers, see e. g. Ref. (1).
(6) W. Dietsche, G. A. Northrop, and J. P. Wolfe, Phys. Rev. Lett 47, 660 (1981).

LT-17 (Contributed Papers)
U. Eckern, A. Schmid, W. Weber, H. Wühl (eds)

TIME-RESOLVED SPECIFIC HEAT MEASUREMENTS ON HIGH-PURITY SINGLE CRYSTALS BETWEEN 50 mK AND 1 K

Wolfgang KNAAK and Michael MEIßNER

Institut für Festkörperphysik, Technische Universität Berlin, D-1000 Berlin 12, F.R.G.

We report on heat pulse experiments on dielectric crystals in the temperature range 50 mK to 1 K, where we measure the temperature-vs.-time profiles on a time scale 1 µsec to 1 sec. Using a relaxation method and thin film techniques for our sample arrangement we were able to measure the time-dependence of the specific heat of 5 high-purity single crystals: Si, Ge, KBr, NaF and SiO_2. At temperatures above ~400 mK the correct DEBYE specific heat C_D is measured for times (t > 10 µsec) shortly after the decay of the incoming ballistic phonon signal (t ≈ 3 µsec), whereas at lower temperatures internal thermalization processes (surface scattering, spurious impurities) produce time-dependent specific heat values $C_t \gtrless C_D$.

1. INTRODUCTION

In recent experiments on amorphous materials a logarithmic time dependence of the specific heat has been reported on a time scale ~5 µsec to ~10 msec (1)-(3). In these experiments the experimental basis of short-time thermometry has been tested by measuring the DEBYE specific heat C_D of crystalline Ge (1), SiO_2 (2) and KBr (3), showing a time-dependent contribution $\Delta C(t)$ to C_D below ~0.5 K. In this paper we present a more detailed study on time-resolved specific heat measurements on 5 high-purity single crystals (Si, Ge, KBr, NaF, SiO_2), demonstrating 1) that thermometry on a µsec-time scale is possible and 2) that the internal thermalization process in a crystal is influenced by surface and volume effects. As a consequence from the latter it can be argued that time-resolved specific heat measurements can give a sensitive test for the characterization of impurities in crystals.

2. EXPERIMENTAL

We used a relaxation method (inset of fig. 1) with an addenda (20 µg Au-film heater, 20 µg C-film thermometer, 0.5 mg silver paint) heat capacity of $\sim 2\cdot 10^{-9}$ J/K at 100 mK. The carbon bolometer, made by rubbing pencil material onto a roughned spot of the crystal surface, proofed to be fast enough to detect ballistic phonons down to 100 mK. In fig. 1 the subsequent temperature profiles for a Si-crystal are shown: at 900 mK an ideal DEBYE-behaviour is evident as the sample is thermalized on a µsec-time scale and only the sample-to-bath relaxation is observed. At 400 mK

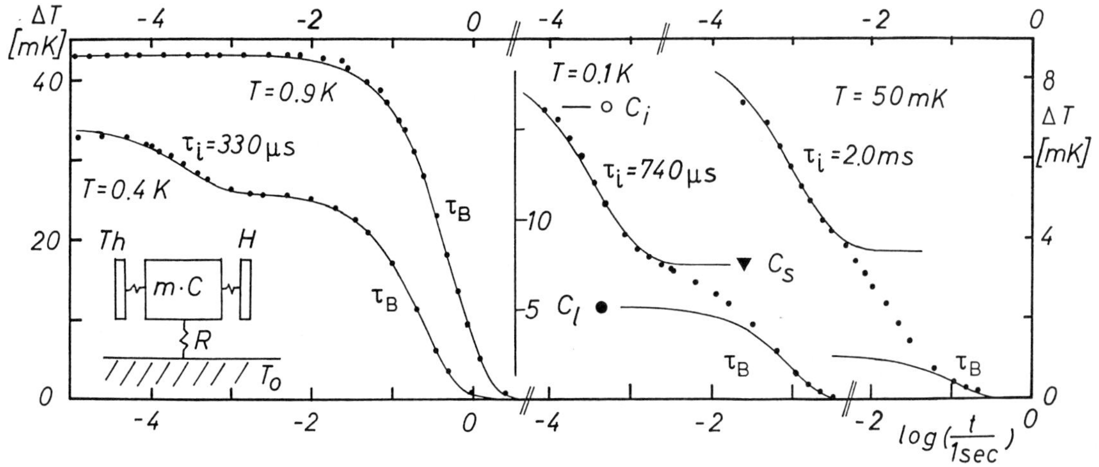

FIGURE 1
Temperature profiles ΔT vs. log(t) for the Si-crystal resulting from a defined heat pulse energy at four different base temperatures T. Solid lines are exponential fits to the sample-to-bath relaxation (τ_B) and to the internal thermalization (τ_i). Definitions for time-dependent specific heat values C_l (●), C_s (▼) and C_i (O) are shown as an example for T = 100 mK. The inset shows the sample set up.

an overshoot shows up at short times which decays exponentially as fitted by the solid lines. At 100 mK and 50mK the temperature profile looks more complicated. By fitting the overshoot profile at times t<1ms, only, we define from the initial temperature rise a specific heat value C_i and from the intermediate plateau a short-time value C_S. The long time specific heat is as usually evaluated from a log(ΔT)-vs.-t plot by extrapolating the sample-to-bath relaxation to t=0.

3. RESULTS AND DISCUSSION

As seen for Si, Ge and KBr in fig. 2 the long time specific heat C_l varies as T^3, except for the Si-crystal below 100 mK. This deviation can be explained by the addenda heat capacity because the Si-sample had the lowest heat capacity measured. However, the short time specific heat C_S follows a T^3-law down to 50 mK with a C/T^3-value very close to that calculated from the elastic constants. This shows that our sample arrangement and thermometry is correct.

In contrast, the initial specific heat values C_i are clearly below the elastic limit for all samples measured. Thus, we argue that the phonons were not able to reach thermal equilibrium at times as long as ~1 msec (at 100 mK). One can assume that the phonons emitted by the heater are reflected mainly elastically at the crystal walls

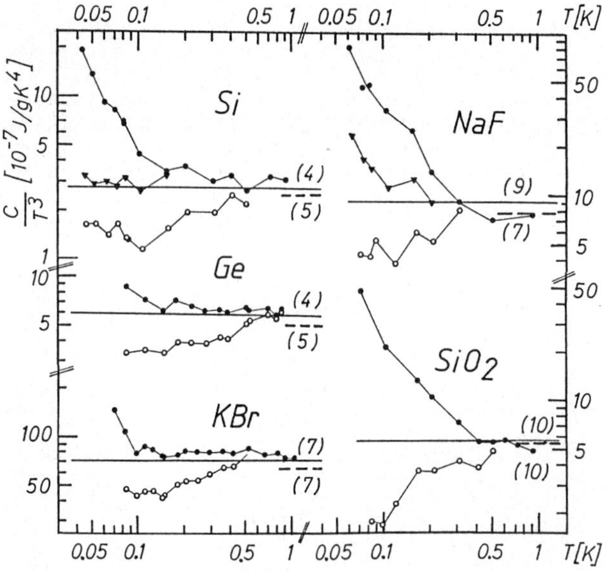

FIGURE 2

Specific heat C/T^3 vs. T on long (●) and 2 short (▼,○) time scales (as defined in fig.1) for Si (16g, zone refined and VPE-purified, B:~$5\cdot10^{12}cm^{-3}$ O,C:~$1\cdot10^{15}cm^{-3}$), Ge (55g, as used in (6)), KBr (9.3 g, as used in (3)), NaF (9.4 g, as used in (8)) and SiO_2 (44g, as used in (10)).References indicate: —— extrapolated caloric data C_D(0K); – – – extrapolated elastic data C_D(0K).

and the bolometer signal is determined by the energy flux of phonons incident to the bolometer area. Such a reverberation signal has been reported (11), where the time constant was found to be mainly dependent on phonon losses at the boundaries to the heater/detector films.

The deviation of the long time specific heat of NaF and SiO_2 from the T^3-law is too large to be due to addenda heat capacities. Although the NaF-sample was the one used by McNelly et al. (8) for the demonstration of second sound and much effort was taken to eliminate OH^--contamination, this additional heat capacity can be ascribed by a SCHOTTKY-anomaly due to residual 30 ppb OH^--concentration (9). As in NaF the coupling to the OH^--tunnelling systems is on a sub-μsec-time scale (8), the short-time specific heat data C_S below ~200 mK show only partial decoupling of the OH^--contribution to C_l. The additional long time specific heat for SiO_2 can be ascribed to excess modes as natural quartz is known to contain metallic impurities (10). Assuming an uniform distribution of low energy excitations, a concentration of only ~$3\cdot10^{15}cm^{-3}$ (0.3 ppm) is sufficient to account for the observed specific heat.

We summarize, that time-resolved specific heat measurements at low temperatures are a very sensitive test for residual impurities in perfect crystals. Furthermore, the relaxation time of ballistic phonons to achieve thermal equilibrium increases rapidly below 400mK and becomes $\tau_i \simeq$ 2ms at 50 mK.

ACKNOWLEDGEMENTS

We like to thank Prof. R.O. Pohl for supplying the Ge-, KBr-, NaF- and SiO_2-samples used in former investigations and for many valuable suggestions and discussions. Prof. R. Helbig was very kind to supply the Si-sample and gave us information on its purity.

REFERENCES

(1) M.T. Loponen et al. Phys. Rev.B 25 (1982) 1161.
(2) M. Meißner and K. Spitzmann, Phys. Rev. Lett. 46 (1981) 265; W. Knaak and M. Meißner in Proc. Conf. "Phonon Scattering in Condensed Matter", ed. by W. Eisenmenger, Springer-Verlag, Berlin (1984) in press.
(3) J.J. DeYoreo et al. Phys. Rev. Lett. 51 (1983) 1050.
(4) P.H. Keesom and G. Seidel, Phys. Rev. 113 (1959) 33.
(5) P. Flubacher et al. Phil. Mag. 4 (1959) 273.
(6) R.B. Stephens, Phys. Rev.B 8 (1973) 2896.
(7) J.T. Lewis et al. Phys. Rev. 161 (1967) 887.
(8) T.F. McNelly et al. Phys. Rev. Lett. 24 (1970) 100.
(9) J.H. Harrison et al. J. Chem. Phys. Solids 29 (1968) 557.
(10) R.C. Zeller and R.O. Pohl, Phys. Rev.B 4 (1971) 2029.
(11) H.J. Trumpp and W. Eisenmenger, Z. Physik B 28 (1977) 159.

LT-17 (Contributed Papers)
U. Eckern, A. Schmid, W. Weber, H. Wühl (eds)
© Elsevier Science Publishers B.V., 1984

POINT DEFECT RELAXATION IN $LaAl_2$

Michael SEEGEL, Siegfried EWERT, and Herbert SCHMIDT, Dieter LENZ

II. Phys. Institut, and Institut für Allg. Metallkunde, RWTH Aachen, 5100 Aachen, West Germany*

Ultrasonic attenuation and sound velocity measurements were used to study anelastic relaxation in $LaAl_2$ caused by mixed dumbbell rotation ($\nu_0 = 10^{13}$Hz; H = 35 meV).

1. INTRODUCTION

Measurements of ultrasonic attenuation α can be used to obtain important information on the LT electronic properties e.g. of superconductors. In the case of $LaAl_2$ (1) and dilute alloys based on this compound (2) the information has been found to be masked by anelastic α and sound velocity v effects (3,4) occurring at T<10K. In addition to these effects a pronounced anelastic process has been observed at 30-40K by α measurements(1). To clarify the possibility of correlation between the effects in both T ranges the high T process has been studied in more detail and by additional v measurements.

2. EXPERIMENTAL

The sound attenuation α and the sound velocity v have been measured by the pulse echo technique between RT and 4.2K at frequencies f between 5 and 250 MHz. The samples were cut (diamond wheel) from a 6 mm Ø, [100] oriented Czochralski crystal (5). They were lapped to flatness and parallelity of their endfaces of better 0.5 µm/cm. Nonaq Stopcock grease was used to bond the quartz transducers to the samples.

3. RESULTS AND DISCUSSION

Fig.1a,b shows the v, α results at 5 MHz: v(T) exhibits a sharp decrease (from pure elastic $v_0(T)$ behaviour (6)) which occurs in the narrow range of the pronounced attenuation peak ($T_p \simeq 32$K, half width $\Delta T_{1/2}=6$K). It is important to note that at temperatures $>T_p+\Delta T_{1/2}$ the anelastic $\Delta v = v_0 - v$ contribution decreases proportional to 1/T.

The observed α and Δv behaviour can be fully rationalized in terms of a single classical Debye relaxation process for which the damping Q^{-1} (= $3.66\times10^{-2}\alpha/f$ with α[dB/µs], f[MHz]) and the modulus defect $\Delta M/M$ (=$2\Delta v/v_0$) are given by

$$Q^{-1} = \frac{\Delta_0}{T}\,\frac{\omega\tau}{1+\omega^2\tau^2} \qquad (1)$$

$$\frac{\Delta M}{M} = \frac{\Delta_0}{T}\,\frac{1}{1+\omega^2\tau^2} \qquad (2)$$

Here Δ_0/T is the relaxation strength, $\omega = 2\pi f$ with f = frequency of measurement, τ the relaxation time. For a thermally activated process τ is given by

$$1/\tau \equiv \nu = \nu_0 \exp(-H/kT) \qquad (3)$$

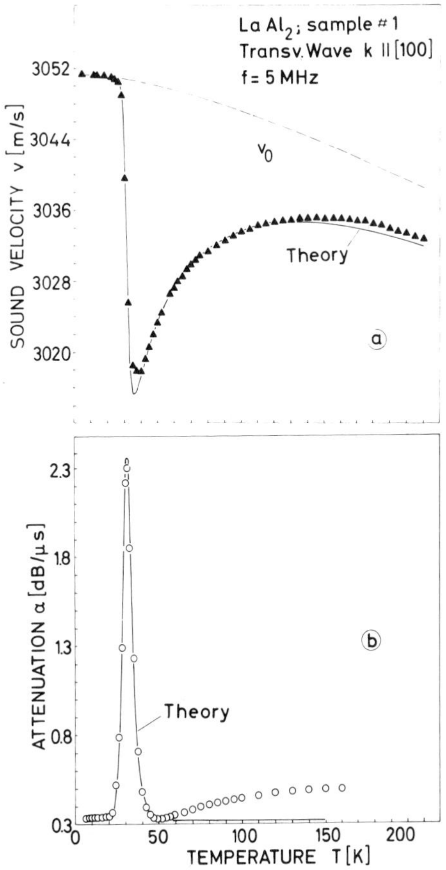

FIGURE 1
v(T) and α(T) compared with theory, equs.(1)-(3)

* Work financed by the Deutsche Forschungsgemeinschaft (SFB 125 Aachen-Jülich-Köln).

In order to derive the attempt frequency ν_0 and the activation energy H we plot in Fig.2 the measured data from Fig.1 (solid triangles) as $(\Delta M/M)\omega/Q^{-1}$ versus 1/T. This Arrhenius plot yields $\nu_0=(1.1 \pm 0.3)10^{13}$ Hz and H = (35 ± 2) meV.

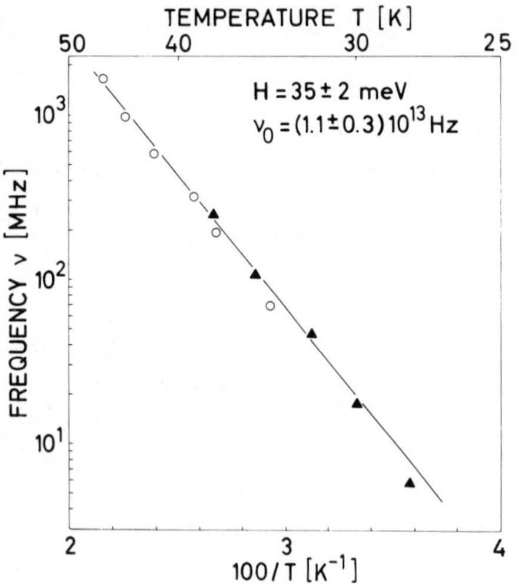

FIGURE 2
Temperature dependence of relaxation rate

In Fig.2 additional $\nu(T)$ data (open circles) are shown which are derived from longitudinal wave attenuation measurements between 10 and 250 MHz on a sample from the same crystal but with a deviation of 7 degrees from [100]. In this case the $\nu(T)$ data are derived from the shift of the Q^{-1} maximum to higher temperatures with increasing frequency. Using this values and equ.(3) the Debye equs.(1) and (2) then yield the theoretical $\alpha(T)$ and v(T) curves in Fig.1 which match the experimental data very closely. This proves that the underlaying process is of single nature and leads to the conclusion that the relaxation involves a well defined lattice defect. The ν_0 value is in agreement with a point defect mechanism (e.g. of Snoek-type). In addition to the observed $\nu(T)$ behaviour (Fig.2) also the 1/T behaviour of the relaxation strength proves that the process is thermally activated and not governed by quantum mechanical tunneling.

For further identification of the atomic origin of the process its lattice symmetry properties (7) were studied. For the transversal wave along [100] the relaxation occurs in the stiffness constant s_{44}. In contrast we have found that the longitudinal wave along [100] exhibits no relaxation phenomena at all. This means that $s_{11}-s_{12}$ is non-relaxing. Thus the defect is of trigonal symmetry (7). In the fcc $LaAl_2$ crystal (Fd3m space group) sites of such lattice symmetry can be found along the [111] directions with the exception of the sites of higher symmetry, i.e. the La sites and the centers of the Al tetrahedra. It is conceivable that a lattice defect with such crystallographic structure either originates from nonexact stöchiometry of the compound or is caused by impurities (foreign atoms FA). The low value of the activation energy is comparable in magnitude with activation energies for interstitial processes in normal fcc metals (e.g. Cu or Al) and thus directly points to a dumbbell interstitial in the present case. Possible configurations are <La-La> or <Al-Al> or <La-Al> or <La-FA> or <Al-FA> pairs on normal lattice sites with the pair axis along [111]. Under appropriate external stress then the energy levels of the 4 possible [111] orientations split and subsequent repartition of the dumbbell orientation distribution (by rotation of the defect) takes place with results in anelasticity (of Snoek-type). Energetically we favour the <La-FA> pair with a small FA (oxygen) inside the La tetrahedron with the 4 corner atoms forming a cage for the defect. This picture implies that only rotation but no long range diffusion occurs. Similar processes have been observed in normal metals with <interstitial-FA> pairs (e.g. <I-Fe> in Al with $H \simeq 15$ meV (8)).

The present interpretation is in variance with Herrmann and Bömmel (1) who explain their $\alpha(T)$ peak by dislocation relaxation (Bordoni effect). However, the Bordoni effect never shows sharp symmetry properties or single relaxation times which characterize the present process. Furthermore, since many atoms are involved, the attempt frequencies for dislocation processes are expected to be considerably (at least by an order of magnitude) lower than the observed ν_0 value. The present experiments (including additional results on doped $LaAl_2$) yield no direct evidence for the presumed correlation with the anelastic process occurring at T<10K (4) since different symmetry properties and no proportionality of the two relaxation strengths are obtained.

REFERENCES

(1) G. Herrmann and H.E. Bömmel, Appl. Physics 10 (1976) 81.
(2) D. Lenz and S. Ewert, Z.Physik B37 (1980) 47.
(3) G. Federle, K. Dransfeld, H.E. Bömmel, P. Rödhammer, Solid State Commun.34 (1980) 379.
(4) S. Ewert, A. Hof, D. Lenz, H. Schmidt, J. de Physique C5 (1981) 695.
(5) M. Beyss, J.M. Welter, T. Kaiser, J. Cryst. Growth 50 (1980) 419.
(6) R.J. Schiltz and J.F. Smith, J. Appl. Phys. 45 (1974) 4681.
(7) A.S. Nowick and W.R. Heller, Adv. in Physics 14 (1965) 101.
(8) G. Kollers, H. Jacques, L.E. Rehn, D.H. Robrock, J.de Physique C5 (1981) 729.

LT-17 (Contributed Papers)
U. Eckern, A. Schmid, W. Weber, H. Wühl (eds)

PHONON ECHOES FROM ACCEPTORS IN SILICON

B. GOLDING and W. H. HAEMMERLE

AT&T Bell Laboratories, Murray Hill, N.J. 07974, USA

Coherent microwave acoustic excitation of the inhomogeneously broadened ground state of the neutral acceptor In in crystalline Si has resulted in the first observation of phonon echoes below 0.1 K. The echoes allow a quantitative determination of the acceptor dephasing times, radiative lifetimes, and acoustic deformation potentials. In addition, the temperature-dependent sound velocity provides information on the inhomogeneous linewidth of the ground state in highly perfect crystals.

1. INTRODUCTION

The four-fold degenerate ground state of neutral acceptors in crystalline Si and Ge is broadened as a result of internal static strains and electric fields. The extent of the inhomogeneous splitting is related to the perfection of the host lattice, which is generally high in dislocation-free single crystals. Although these lattice distortions have a relatively small effect on the far infrared spectra they can play an important role in the low temperature, low frequency response (1-6). We describe the application of coherent microwave acoustic techniques to the study of shallow acceptors B and In in Si. Phonon echoes, the strain analog of spin echoes, are excited at temperatures below 0.1 K. Analysis of the echoes provides quantitative information on the acceptor-lattice relaxation time T_1, the acceptor dephasing time T_2, and the effective deformation potential coupling D_{eff} . In addition, sound velocity shifts can provide information on the distribution of ground-state energy splittings (7). This acoustic investigation complements recent coherent dielectric studies on the same Si crystals (8). The work reported below was carried out without the application of external static electric or magnetic fields.

2. RESULTS

The Si:In crystal studied in the present experiments was a cylinder, 4 mm by 14 mm in diameter, with polished parallel faces perpendicular to <111>. Longitudinal ultrasonic waves were excited by applying coherent microwave pulses to a ZnO thin film transducer sputtered onto one of the faces. The dislocation-free crystal, grown at AT&T Bell Laboratories, was In-doped to a level of $9\times10^{15}cm^{-3}$. Phonon echoes were excited at 1.54 GHz by propagating two equal 50 ns pulses through the crystal. Fig. 1 shows six pairs of reflections detected at the transducer surface. The small negative-going signal is an artifact due to receiver ringing. The phonon echoes, indicated by the arrows, are extremely weak and cannot be

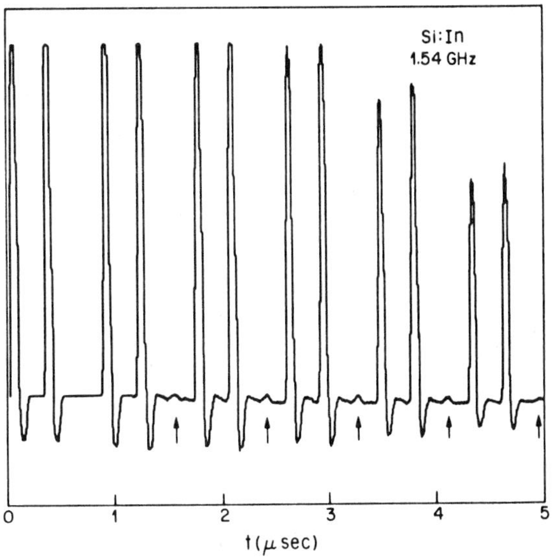

FIGURE 1
Acoustic reflections and phonon echoes (arrows) generated by two 50 ns 1.54 GHz longitudinal acoustic pulses. The sample is <111> oriented Si:In ($9\times10^{15}cm^{-3}$) at 30 mK.

observed until at least two passes of the wave through the sample. We believe that the small coherent signal originates in the low density of states at 1.5 GHz. It has proven impossible also to observe intensity-dependent absorption at this frequency in this sample i.e. $\alpha L << 1$.

Information on the distribution of energy splittings or density of states can be derived from sound velocities. Fig. 2 shows the relative variation in the 1.54 GHz <111> longitudinal sound velocity in Si:In from 30 mK to 12 K. The behavior of $\Delta v/v$ is similar to that reported earlier(7) in acceptor-doped Ge: a minimum at very low temperature, a logarithmic T-dependent region between 0.1 and 1.0 with positive slope, and a weaker T-dependence at higher temperatures. The region above 0.1 K is interpreted as follows: the logarithmic slope $d(\log v)/d(\log T)$ is given by $N(E=2k_BT)D_{eff}^2/\rho v^2$ where D_{eff} is the effective deformation coupling for this mode and ρ is the mass density. This result derives from the dispersive component of the resonant acceptor-phonon interaction (8,9). D_{eff} can be obtained from the two-pulse echo excitation maximum or from the decay of the three-pulse echo i.e., analysis of T_1 in terms of a one-phonon decay. This procedure gives $D_{eff} \approx 2.5$ eV. The decrease in the slope of $\Delta v/v$ above 1 K is interpreted as a decrease in the acceptor density of states above a few K (above 50 GHz). This interpretation is supported by integration of $d(\log v)/d(\log T)$, which is equivalent to integrating N(E) over energy. We find that this yields $10^{16}cm^{-3}$ for n_a , the acceptor concentration, in good agreement with the nominal In doping level. The general features of $\Delta v/v$ in Si:In are similar to glasses below 1 K(9) with the important exception that no significant relaxational contribution is observed above 1 K.

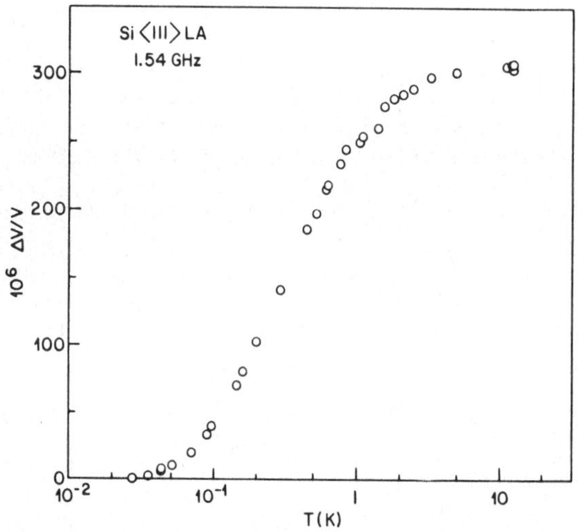

FIGURE 2
Relative change in the 1.54 GHz <111> longitudinal sound velocity in Si:In. The slope above 0.1 K is proportional to the acceptor density of states $N(E=2k_BT)$.

3. CONCLUSION

In conclusion, we note that the distribution of energy splittings appears to be highly broadened, with a width of about 0.5 meV. Since the absorption is extremely low at 1.5 GHz it appears that a gap at low energies ($\lesssim 2$ GHz) must exist. These conclusions are consistent with previous dielectric and electric echo results on similar crystals (8) but differ quantitatively from other acoustic investigations (6).

REFERENCES

1. K. Suzuki, N. Mikoshiba, Phys. Rev. B3 (1971) 2550.
2. T. Fjeldy, T. Ishiguro, C. Elbaum, Phys. Rev. B7 (1973) 1932.
3. Hp. Schad, K. Lassmann, Phys. Letters. 56A (1976) 409.
4. H. Tokumoto, T. Ishiguro, Phys. Rev. B15 (1977) 2099.
5. H. Tokumoto, T. Ishiguro, K. Kajimura, J. Phys. C 13 (1980) 4061.
6. H. Zeile, K. Lassmann, Phys. Stat. Sol. 111 (1982) 555.
7. B. Golding, J. E. Graebner, W. H. Haemmerle in: Proc. of the 14th International Conference on the Physics of Semiconductors, edited by B.L.H.Wilson (The Institute of Physics, London, 1978) 411.
8. B. Golding, Bull. A. P. S. 27 (1982) 232.
9. B. Golding, J. E. Graebner, A. B. Kane, Phys. Rev. Letters. 37 (1976) 1248.

LT-17 (Contributed Papers)
U. Eckern, A. Schmid, W. Weber, H. Wühl (eds)

STRUCTURE SCATTERING OF PHONONS IN A SEMI-CRYSTALLINE SOLID AT VERY LOW TEMPERATURES.

D.M. FINLAYSON and P.J. MASON

Department of Physics, University of St. Andrews, St. Andrews, Scotland

Abstract. The two-level tunnelling theory of Anderson et al (1) and Phillips (2), widely used and apparently successful for truly amorphous materials, is unable to explain the thermal conductivity of semi-crystalline material. Good agreement with experiment is obtained from the structure scattering theory of Morgan and Smith (3) using correlation lengths related to the known parameters of the solid.

1. INTRODUCTION

The thermal properties of amorphous solids below 2 K is characterised by a linear term in the specific heat and a T^2 dependence of the thermal conductivity. The two-level tunnelling states model of Anderson et al (1) and Phillips (2) gives a fairly coherent and plausible explanation of both specific heat and thermal conductivity in amorphous solids. The basic idea is that certain atoms or groups of atoms have available two mutually accessible potential minima differing in energy by a small amount Δ. The spread of asymmetries and barrier heights between potential wells gives rise to a constant density of states. Anderson et al have shown that such a system will give rise to a linear specific heat and to a T^2 dependence of the thermal conductivity.

An alternative explanation of the thermal conductivity of disordered solids at low temperatures first suggested by Klemens (4) has been developed by Morgan and Smith (3). This structure scattering theory is based on the idea of diffraction of lattice waves by regions of different density in the material.

We have been investigating the thermal properties of the semi-crystalline polymer, polyethylene. This is believed to contain lamellar regions of crystalline material typically 100 Å thick interspersed with regions of amorphous material. In an unextruded sample the lamellae join end to end to form ribbon like structures which grow out from nucleating centres in 3 dimensions to form spherulites with a typical diameter of 10 μm. These lamellae are randomly orientated giving rise to an isotropic structure. In an extruded sample, the lamellae are orientated with cylindrical symmetry at a fixed angle relative to the extrusion direction, giving rise to an anisotropic structure. The distance travelled by a phonon within a lamellae depends on the orientation of the lamellae relative to the direction of phonon propagation which is altered by extrusion. Hence semi-crystalline polyethylene is an ideal substance for the study of the relative merits of the two-level and structure scattering theories.

2. EXPERIMENTAL

The specific heat of 1 unextruded and 3 extruded samples of polyethylene was measured over the temperature range 0.15 to 1.5 K using a heat pulse technique. It was found that the specific heat (fig.1) varied as $C = aT + bT^3$, was the same for all samples and hence that extrusion had no effect on the specific heat. Accepting the Anderson et al two-level explanation of the low temperature specific heat then implies that extrusion has no influence on the number and distribution of the two-level systems.

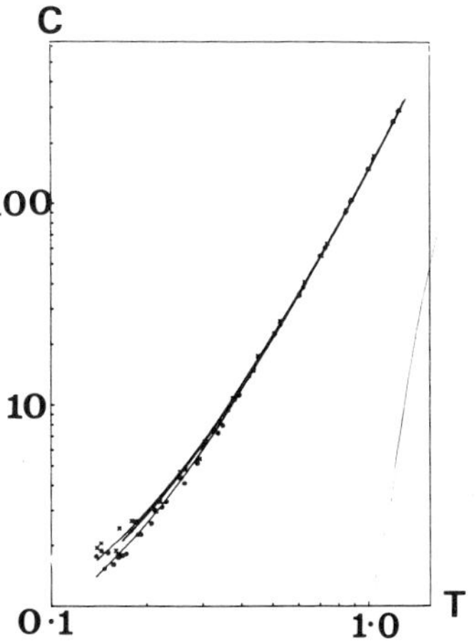

Figure 1.

Specific heat C (erg $g^{-1}K^{-1}$) versus temperature for 4 samples of polyethylene, one unextruded and three extruded.

The thermal conductivity of one unextruded and two extruded samples was measured over the temperature range 0.08 to 1.8 K using a steady state method. It was found (fig.2) that in sharp contrast to the purely amorphous case, the thermal conductivity showed an approximately linear dependence on temperature over a substantial range of temperature and that this range was different for the extruded and unextruded samples.

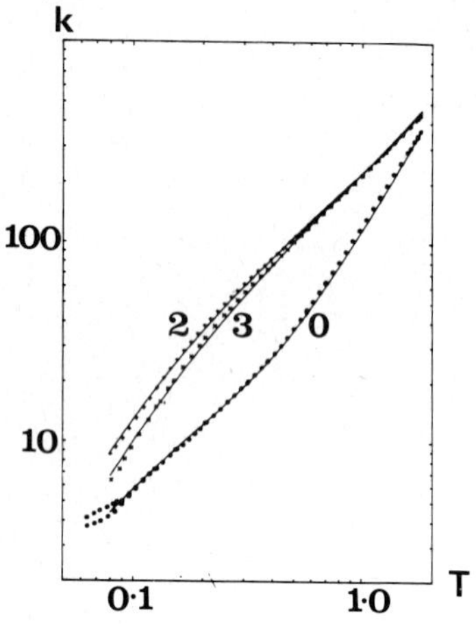

Figure 2.

Thermal conductivity k (erg $s^{-1}cm^{-1}K^{-1}$) versus temperature; sample 0 unextruded, samples 2 and 3 extruded. The points represent experimental values; the lines represent theoretical calculations.

3. DISCUSSION

Since we have seen from specific heat measurements that extrusion does not change the two-level system it is clear that a two-level theory cannot account even qualitatively for the observations. Instead we have turned to the structure scattering theory of Morgan and Smith (3).

To obtain the phonon mean free path, correlation lengths a_s and a_c are introduced. The value of the correlation length a_s (5 Å) is related to the dimensions of the structural units present within the amorphous regions of the semi-crystalline polymer; the value of the correlation length a_c (70 - 300 Å) is related to the dimensions and orientation of the crystalline lamellae.

To account for the thermal conductivity at the lowest temperatures, Morgan and Smith introduced a long correlation length a_L (1000 - 3000 Å). The experimental evidence for structure of these dimensions is scanty and so we have preferred to use a term associated with phonon scattering by tunnelling states. The existence of the latter, though not understood on an atomic scale, is supported by a variety of experimental evidence from work on amorphous solids and it is reasonable to suppose that at low enough temperatures, all solids will appear to be amorphous to very long wavelength phonons.

On this basis, relaxation curves were obtained relating phonon mean free path to wavelength and from these the thermal conductivity was calculated using parameters derived from a curve-fitting procedure. From fig.2 it will be seen that the calculated curves fit the experimental points very well over the whole temperature range. Indeed they can be extended up to 100 K to fit the experimental points of Burgess and Greig (5). The parameters derived from this curve fitting procedure fit in well with the known parameters of this polymer.

4. CONCLUSION

Our general conclusion is that structure scattering is essential for any explanation of the conductivity of semi-crystalline materials. It can moreover, account for the thermal conductivity of a wide range of amorphous and semi-crystalline materials over a range of temperature from 0.1 to 100 K and so from the point of view of thermal conductivity the two-level theory is unnecessary. However, structure scattering makes no attempt to explain the low temperature anomalies in specific heat and for this tunnelling state theory is required.

ACKNOWLEDGEMENTS

We are most grateful to the polymer group of Leeds University for the supply of samples and to the Science and Engineering Research Council of the U.K. for financial assistance.

REFERENCES

(1) P.W. Anderson, B.I. Halperin and C.M. Varma, Phil. Mag. 25 (1972) 1.
(2) W.A. Phillips, J. Low Temp. Phys. 7 (1972) 351.
(3) G.J. Morgan and D. Smith, J. Phys. C 7 (1974) 649.
(4) P.G. Klemens, Proc. Roy. Soc. Lond, A208 (1951) 108.
(5) S. Burgess and D. Greig, J. Phys. C 8 (1975) 1637.

LT-17 (Contributed Papers)
U. Eckern, A. Schmid, W. Weber, H. Wühl (eds)

THE LATTICE THERMAL CONDUCTIVITY OF POTASSIUM MEASURED BY THE CORBINO METHOD

R.J.M. VAN VUCHT, P.A. SCHROEDER*, H. VAN KEMPEN and P. WYDER

Research Institute for Materials, University of Nijmegen,
Tournooiveld, 6525 ED Nijmegen, The Netherlands
*permanent address: Dept. of physics, Michigan State University, East Lansing, Michigan, U.S.A.

The low temperature lattice thermal conductivity λ^g of potassium has been measured by various authors using different techniques. The results they have obtained vary over more than one order of magnitude. In the case of high field methods, where the lattice thermal conductivity is assumed to stay unaffected, the interpretation of the results has caused important difficulties. We have remeasured λ^g by the Corbino method, a high field method in which the major problems are circumvented. Our results confirm those of Stinson et al.

1. INTRODUCTION

The low temperature lattice thermal conductivity of pure simple metals is expected to be limited by the phonon-electron interaction, which would give it a T^2-dependence. This behaviour has indeed been found e.g. in aluminium (1) and indium (2). A simplified calculation based on the same idea predicted a similar behaviour for the 'simplest' metal potassium (3) (curve a in fig. 1). However, experiments intended to verify this prediction have lead to rather confusing results.

Two methods are popular for separating the lattice thermal conductivity λ^g from the electronic contribution λ^e which is normally $>>\lambda^g$: alloying and high magnetic fields. In potassium, the alloying method has been employed by two groups (4,5) giving results relatively close to Ekin's calculation (curves b and c in fig. 1). The results appeared to depend on the degree of alloying, however. The high field method has been applied by Tausch et al. (6) and by Stinson et al. (7). They come to strikingly different conclusions: Tausch et al. regard curve d in fig. 1 as the true result for λ^g, but a different interpretation of their data leads to curve e. Stinson et al. find curves f, indicating an important sample dependence.

As we shall show below, the difficulties are primarily caused by the fact that one wants to measure the thermal conductivity tensor $\boldsymbol{\lambda}$ which is simply the sum of λ^g and $\boldsymbol{\lambda}^e$, but the latter two groups measured the thermal resistivity $\mathbf{W} = \boldsymbol{\lambda}^{-1}$. Since $\boldsymbol{\lambda}$ has off-diagonal components in a magnetic field, measurements of both W_{xx} and W_{xy} are required to obtain $\boldsymbol{\lambda}$ and hence λ^g as the extrapolation to $B=\infty$. The Corbino method employs a circular geometry (fig. 2) in which the boundary conditions are such that λ_{xx} is measured directly.

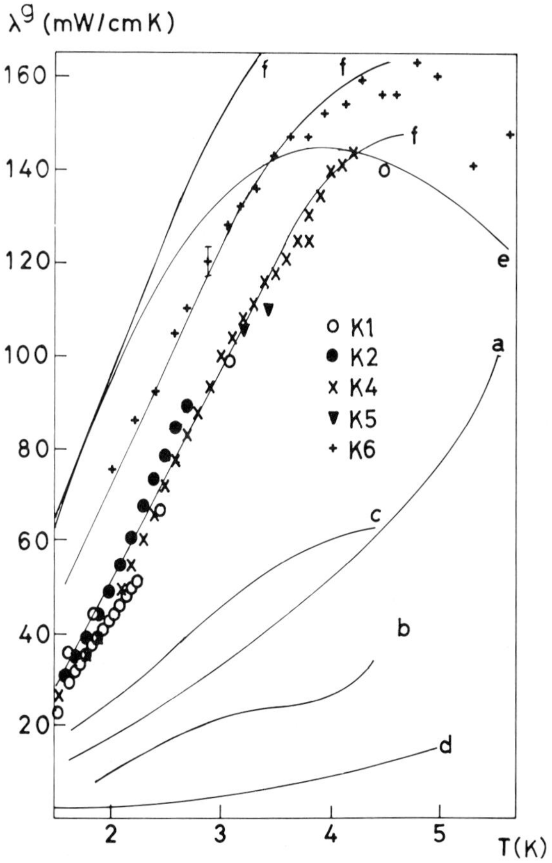

Fig. 1: λ^g as a function of T for our samples (symbols) and other theories and experiments (curves); the labels are explained in the text.

2. THERMAL TRANSPORT IN HIGH FIELDS

With $B = B\hat{z}$ such that $\omega_c\tau>>1$, the thermal resistivity tensor $\mathbf{W}$ in the absence of thermal conduction by the lattice has the components

$W^e_{xx} = W^e_{yy} = W^e_0(1+\alpha B)$ (linear thermal magnetoresistance), and $W^e_{xy} = -W^e_{yx} = A_{RL}B$ (A_{RL}, the Righi-Leduc coefficient, is the thermal equivalent of the Hall constant). Here W^e_0 is assumed to be equal to W^e at B=0, where

$$W^e_0 = \frac{\rho_0}{L_0 T} + bT^2 \qquad (1)$$

The first term arises from electron-impurity and lattice defect scattering, and the second term from electron-phonon scattering. A_{RL} is close to its free-electron value $(L_0Tne)^{-1}$. The tensor $\boldsymbol{\lambda}$ is found by inverting $\mathbf{W}$ and adding $\lambda^g I$. In the high field limit we obtain:

$$\lambda_{xx} = \lambda_{yy} \simeq W^e_0(1+\alpha B)(A_{RL}B)^{-2} + \lambda^g \qquad (2)$$

$$\lambda_{xy} = -\lambda_{yx} \simeq -(A_{RL}B)^{-1} \qquad (3)$$

The thermal resistivity $\mathbf{W}$ in the presence of λ^g can be found by inverting $\boldsymbol{\lambda}$ again. Both W_{xx} and W_{xy} appear to be affected by λ^g: it produces a term $\lambda^g(A_{RL}B)^2$ in W_{xx}, and a decrease of W_{xy} at high fields. Eq. (2) shows that λ^g is easily found by measuring λ_{xx} and plotting it as a function of $(1+\alpha B)B^{-2}$. In the Corbino geometry λ^g is indeed measured directly, but in the traditional rectangular geometry W_{xx} and W_{xy} are obtained (fig. 2). These are both needed to determine λ^g unless $\mathbf{W}^e$ is well known. A simultaneous measurement of both components was used by Stinson et al. (7), who finally obtained the results which we found more directly.

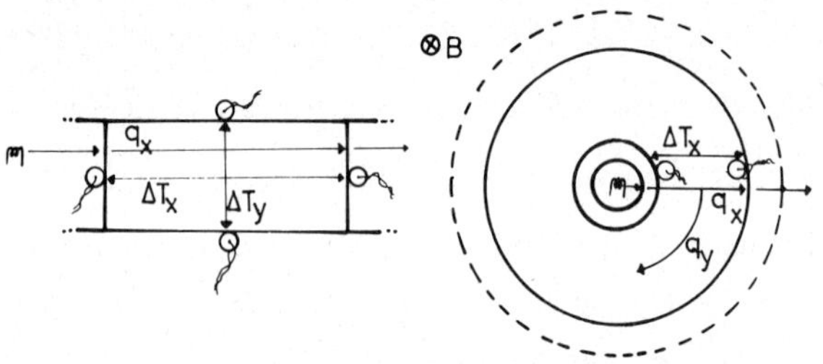

Fig. 2: Left: conventional geometry: the transverse heat current is zero. From $\nabla T = \mathbf{W}q$ follows that measurements of ∇T_x and ∇T_y will render W_{xx} and W_{xy}.
Right: Corbino geometry. For symmetry, the transverse T gradient is zero. Since $q = \boldsymbol{\lambda}\nabla T$, measuring ∇T_x gives λ_{xx}.

3. RESULTS

Measurements were made between 1.6 and 5.5 K, at fields between 1 and 9 T. Data taken at each temperature were fitted to:

$$\lambda_{xx} = \lambda^g + \lambda_2 \frac{1+\alpha B}{B^2} \qquad (4)$$

where $\lambda_2 = W^e_0(A_{RL})^{-2}$ (see eq. (2)). The fits were optimised by assuming for each sample an individual, temperature independent α.

The λ^g we obtained from the fits are found in fig. 1. The most important error source is always the uncertainty in the geometrical factor; therefore the temperature dependence of the data is quite reliable, but the absolute magnitude is less certain, except for sample K6. In particular, the apparent striking coincidence among samples K1, K2 and K4 as well as one of the samples of Stinson et al. is fortuitous. On the other hand, the data are consistent with the latter as to their temperature dependence and order of magnitude. The highest temperature points of sample K6 show that a turnover of λ^g at about 5 K is confirmed. This also follows from the data of Tausch et al. if they had interpreted their results along the line which they discredited.

The λ_2 data are directly coupled to W^e_0 (eq. (4)). Since $A_{RL} \propto T$, and $W^e_0(T)$ is given by eq. (1), a plot of $\lambda_2 T^{-1}$ as a function of T^3 should show straight lines with a slope related to b and T=0 values related to the residual resistivity. This is indeed found. The residual resistivities we find from these curves are in the range 2500-4000, i.e. in agreement with electrical measurements on the same material. The e-p coefficient b agrees with values measured in zero field by Newrock and Maxfield (8).

4. CONCLUSIONS

The unexpected magnitude of λ^g of K as first published by Stinson et al. is confirmed here. The explanation for this may be sought in a special group of phonons with momentum close to symmetry directions which have a strongly reduced p-e interaction. The remarkable sample dependence of λ^g may then be due to crystallite boundaries where these phonons experience a mismatch. The results from the alloying method appear to be obscured by strong phonon-impurity scattering.

Part of this work has been supported by the Stichting voor Fundamenteel onderzoek der Materie (FOM) with financial support from the Nederlandse Organisatie voor Zuiver Wetenschappelijk Onderzoek (ZWO).

REFERENCES

(1) H.N. de Lang, H. van Kempen and P. Wyder, J.Phys.F 8 (1978) L39
(2) H. van Kempen, H.N. de Lang, J.S. Lass and P. Wyder, Phys.Lett. 42 (1972) A277
(3) J.W. Ekin, Phys.Rev. B 6 (1972) 371
(4) M.A. Archibald, J.E. Dunick and M.H. Jericho, Phys.Rev. 153 (1967) 786
(5) T. Amundsen and J.A.M. Salter, Phys.Rev. B 23 (1981) 931
(6) P.J. Tausch and R.S. Newrock, Phys.Rev. B 16 (1977) 5381
(7) M.R. Stinson, R. Fletcher and C.R. Leavens, Phys.Rev. B 20 (1979) 3970
(8) R.S. Newrock and B.W. Maxfield, Phys.Rev. B 7 (1973) 1283

LT-17 (Contributed Papers)
U. Eckern, A. Schmid, W. Weber, H. Wühl (eds)

DYNAMICS OF LIBRATIONAL MOTION AND PHASE TRANSITIONS IN N_2-TYPE CRYSTALS

T.N. ANTSYGINA, V.A. SLUSAREV, Yu.A. FREIMAN and A.I. ERENBURG

Institute for Low Temperature Physics and Engineering, UkrSSR Academy of Sciences, Kharkov, USSR

The dynamics of librational motion in N -type crystals (α-N_2, α -CO, N_2O, CO_2) is treated by taking into account both anharmonic and correlation effects. The method used is similar to the Tyablikov's method in the theory of magnetism. The main thermodynamic characteristics of the librational subsystem are calculated: the order parameter, rms librational angle, librational mode frequencies and corresponding Grüneisen parameters, librational heat capacity, internal and free energies. The librational isotope effects for α - $^{14}N_2$ and α - $^{15}N_2$ is considered. An explanation of the anomaly isotope effects in heat capacity is proposed. A theory of the phase transition into orientationally disordered state is developed.

Diatomic crystals N_2 and CO, and triatomic crystals N_2O and CO_2 (so-called N_2-type crystals) are typical representatives of the most simple molecular crystals. At low temperatures these crystals are orientationally ordered, the molecular centers of inertia are located in fcc lattice sites, and the axes are directed along the four spatial diagonals of the cube (space group Pa3). At temperatures T=35.61 K and 61.55 K N_2 and CO crystals, respectively, undergo a phase transition from the orientationally ordered α-phase into the β -phase, having hcp lattice (space group P6$_3$ / mmc) and characterized by a high degree of orientational disorder. In N_2O and CO_2 crystals the noncentral interaction is so appreciable, that orientational disordering occurs only as a result of crystal melting. The main difficulty which appears in theoretical studies of libration motion in the molecular crystals is due to considerable anharmonicity of librational oscillations. Anharmonic effects are important even at temperatures much lower than the orientational disordering temperatures of these crystals.

In this paper we developed the theory of librational motion in N_2-type crystals taking into account both anharmonic and correlation effects. The method used is similar to the Tyablikov's method in the theory of magnetism. Adequancy of the suggested approach was verified by comparison of the analytical results with the results of computer experiments for the same model system.

To describe the properties of the crystals we have used the modified Kohin's potential where as a new variable parameter the distance between force centers has been introduced for the repulsive part of intermolecular interaction instead of the internuclear distance. This potential posseses a convenient and simple analytical form and permits to describe virtually all properties of the librational subsystem at equilibrium vapour pressures.

The main thermodynamic characteristics of the librational subsystem in the orientationally ordered phase have been calculated: the order parameter, rms librational angle, librational mode frequencies and corresponding Grüneisen parameters, librational heat capacity, internal and free energies. There is a good agreement between the calculated and observed results. The relative contributions to these characteristics of anharmonic and correlation effects have been analyzed. The most sensitive to the anharmonic effects is the librational heat capacity (Figure 1). If one neglects the anharmonic effects, they cannot be compensated by a choice of the form and parameters of the intermolecular interaction potential. At the same time, the librational mode frequencies at T=0 are defined primarily by the anisotropic interaction potential but their temperature

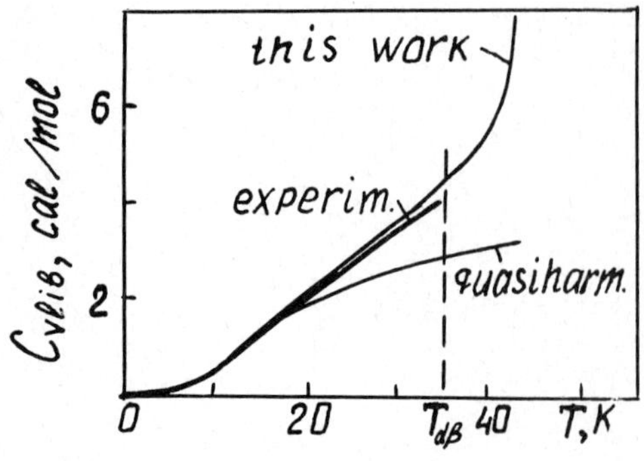

FIGURE 1

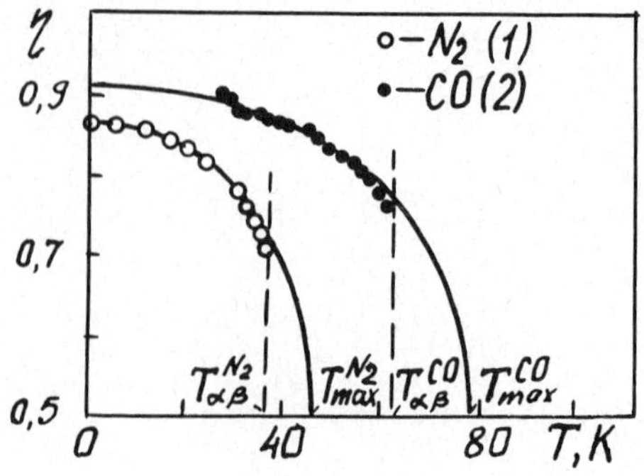

FIGURE 2

dependences are defined essentially by anharmonic and correlation effects. An appreciable softening of all librational modes is realized even for rigid lattice at constant volume. An additional softening appears as a result of taking into account the V(T) dependence. Correctness of description of the latter effects is dependent on to what extent the anisotropic interaction potential used reproduces the correct value of the quadrupole-quadrupole to repulsive force ratio which defines the Grüneisen parameters of the librational modes. The chosen value I.7 for the mean square weighted value of the Grüneisen parameters of the librational modes permits to obtain correct temperature dependences of the librational mode frequencies for all the substances of the group.

A theory of the phase transition for the system of interacting rotators into the orientationally disordered state is given. This phase transition is shown to be a strongly pronounced first order transition with a great jump in the order parameter at the transition point (Figure 2). The temperatures of absolute instability of the orientationally ordered phase (T_{MAX}) (overheating temperature) as well as the temperatures of equilibrium transition into orientationally disordered state were calculated. For N_2 and CO these latter temperatures are close to the respective α-β transition points, but for N_2O and CO_2 they exceed substantionally the melting points in compliance with the fact that in those crystals the orientationally ordered state persists up to the melting points.

The developed theory of orientational disordering in the rigid lattice permitted us to describe the observed phase transition implying that in the nature of the α-β transition the main role belongs to the processes in the librational subsystem and the translational-librational interaction is of minor importance, and furthermore, the fcc-hcp transition is a secondary phenomenon in the α-β transition.

REFERENCES

(1) T.A. Scott, Phys.Reports 27C (1976) 89.
(2) F. Li et al., J.Chem.Phys. 74 (1981) 3120.

LT-17 (Contributed Papers)
U. Eckern, A. Schmid, W. Weber, H. Wühl (eds)

A SCALED MODEL FOR THE LOW ENERGY VIBRATIONAL MODES OF A SINTER

C.J.LAMBERT

Department of Physics, University of Lancaster, Lancaster LA1 4YB, UK.

A scaled model is proposed for the low lying vibrational modes of a sintered metallic powder. Such states provide a possible contribution to the metal/liquid He Kapitza conductance anomaly found experimentally below 20 mK. The density of states and degree of localisation of the modes are calculated. The results support the "shaking box" viewpoint that the active states in the 10-20 mK region are formed from the collective motion of small localised groups of powder particles.

1. INTRODUCTION

Recent interest in the low energy vibrational modes of sintered metallic powders has arisen from the failure of acoustic mismatch theories (1,2) to describe the Kapitza conductance κ of solid/liquid ^{3}He and solid/^{3}He in ^{4}He boundaries below 20 mK. Such theories yield for the variation of κ with temperature, $\kappa \sim T^3$ and with modifications arising from finite lifetime effects and surface modes (3,4) are in agreement with experiment between 20 mK and 100 mK. However, below 20 mK anomalously high conductances are found (5).

The present paper deals with a possible mechanism for this discrepancy (5,6), arising from the low energy vibrational modes of a sinter. A dominant bulk phonon wavelength of the order of an average particle diameter at 20 mK, suggests that below this temperature vibrational modes can only be formed by the collective motion of a number of sinter particles. Rutherford et al (8) have adopted the viewpoint that these excitations are highly localised; formed by the collective motion of small groups of sinter particles. They coined the phrase "shaking boxes" and showed that with an assumed form for the density of states (D.O.S.) the correct temperature dependence of κ is obtained. The purpose of this note is to suggest a simple microscopic model for the shaking box modes and from this to compute the D.O.S. and degree of localisation of these states.

2. THE MODEL

Consider the form taken by the dynamical matrix H of a sinter. Within the harmonic approximation one may write $H = \sum_P H_P + \sum_{PP' (P \neq P')} H_{PP'}$,

where $H_P = \sum_{i_P, j_P} H_{i_P j_P}$ is the dynamical matrix for atoms contained within an isolated sinter particle P and $H_{PP'} = \sum_{i_P j_P} H_{i_P j_{P'}}$ is the interaction between atoms on the surface of P and those of P'. The matrix H contains information about the high energy as well as the low lying modes of the sinter. The former correspond to short wavelength excitations and can be eliminated by employing a scale transformation such as that used in the decimation approach to the Anderson transition in disordered systems (9,10). For example consider the situation in which there is one degree of freedom per atom and a total of N atoms in the sinter. The effect of removing atom $k_{P''}$ from particle P" is to transform the N x N matrix equation $H\,\psi(\omega) = \omega^2\,\psi(\omega)$ into the (N - 1) x (N - 1) equation $H^{(1)}\,\psi^{(1)}(\omega) = \omega^2\,\psi^{(1)}(\omega)$. $\psi^{(1)}(\omega)$ is the projection of $\psi(\omega)$ onto the subspace in which $k_{P''}$ is omitted and

$$H^{(1)}_{i_P j_{P'}} = H_{i_P j_{P'}} + H_{i_P k_{P''}}\, H_{k_{P''} j_{P'}} / (\omega^2 - H_{k_{P''} k_{P''}}) \quad (1)$$

In principle, one can envisage repeating this procedure until only a single atom per sinter particle remains. In this manner a scaled dynamical matrix $\tilde{H}$ is generated, whose elements $\tilde{H}_{PP'}$ describe particle-particle interactions. In practice the force constants needed to set up the initial matrix are not known. Furthermore, typical numbers of $\sim 10^9$ atoms/particle prevent the scale transformation from being carried out exactly. In order to circumvent these problems, we consider a simple parameterization of the scaled matrix $\tilde{H}$, which is expected to be valid in the small ω limit.

The model is based on the observation that within the harmonic approximation, the diagonal elements of the original matrix H satisfy $H_{i_P j_P} = -\sum_{P' (P' \neq P)} \sum_{j_{P'}} H_{i_P j_{P'}}$. This guarantees the existence of an $\omega = 0$ mode. In the limit $\omega \to 0$ it can be shown from (1) that the above equation is also satisfied by the elements of the *scaled* matrix $\tilde{H}$; ie as $\omega \to 0$, $\tilde{H}_{PP} = -\sum_{P' \neq P} \tilde{H}_{PP'}$.

Writing the off-diagonal elements in the form $\tilde{H}_{PP'} = -\exp[-\alpha_{PP'} |\underline{P} - \underline{P}'|]$, where $|\underline{P} - \underline{P}'|$ is the distance between P and P', we adopt the simplifying assumption that $\alpha_{PP'} = \alpha[1 + \chi_{PP'}]$. For the energy range of interest α is assumed to be independent of ω and $\chi_{PP'}$ is a random variable taken from a uniform distribution over the interval $-\Delta/2$ to $+\Delta/2$. Thus the proposed model contains two free parameters; α characterizes the average inverse range of particle-particle interactions and Δ simulates the disorder, which inevitably is introduced by the sintering process.

3. RESULTS

For a cubic lattice of $\tilde{N} = 216$ particles with periodic boundary conditions, the D.O.S. $\rho(\omega)$ has been computed numerically. Figure 1 shows the results for a range of Δ, with $\alpha = 1$ and 10 respectively. For ease of comparison the ω axes have been re-scaled by dividing the elements of $\tilde{H}$ by the average value of the diagonal elements $(\sum_P \tilde{H}_{PP})/\tilde{N}$. In the limit $\alpha \to 0$, $\Delta \to 0$ it is readily shown that $\tilde{H}$ possesses a single eigenvalue at $\omega = 0$ and a $(\tilde{N} - 1)$-fold degenerate eigenvalue at $\omega = 1$. Figure 1 shows that for $\Delta = 1$ the effect of disorder is to produce a spread of these states centred on $\omega = 1$. In the limit $\alpha \to \infty$, $\Delta \to 0$, $\tilde{H}$ reduces to a simple nearest neighbour "tight binding Hamiltonian", whose eigenvalues lie within the band $0 \leqslant \omega \leqslant \sqrt{2}$. Figure 1 shows that for large α increasing the disorder moves the upper band edge to higher frequencies, producing a tail in the D.O.S.

Besides the D.O.S. one is also interested in the degree of localisation of the low lying modes. A localised wavefunction $\psi(\omega)$ peaked at particle P', will decay away from this position with an envelope $\sim \exp[-\beta(\omega)\,|\underline{P} - \underline{P}'|]$. The inverse localisation length $\beta(\omega)$ is shown superposed on the D.O.S. plots. The states in the high frequency tails of the large α, $\Delta = 0.5$ and 1.0 results are highly localised and it is natural to identify these as the "shaking box" modes of Rutherford et al (8). These results support the viewpoint that the low lying states responsible for the Kapitza conductance anomaly are localised over only a small number of sinter particles. However, the D.O.S. is not simply a constant down to $\omega = 0$. Instead the D.O.S. in the tails increases slowly with decreasing ω until a peak is attained.

A more sophisticated model of a sinter, in which the assumption of a frequency independent α is relaxed, may modify the precise form of the D.O.S. in the tails. However, it seems unlikely that the peaks will disappear completely. If such peaks occur in practice and their positions can be varied by tuning the physical parameters associated with α and Δ, then this suggests an interesting possibility. The heat exchange within a sinter could be enhanced by choosing the peak to lie within the temperature range of interest.

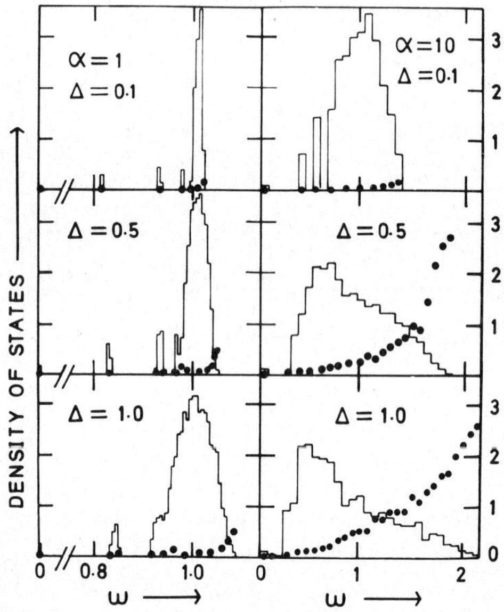

FIGURE 1
Numerical results for the density of states and inverse localisation length β as a function of frequency ω. The D.O.S. scale is arbitrary, while the β scale has units of (particle spacing)$^{-1}$. Of particular interest are the $\alpha = 10$, $\Delta = 0.5$ and $\Delta = 1.0$ results, which possess a tail of highly localised states for $\omega \gtrsim 1$.

ACKNOWLEDGEMENTS

The author thanks A.M.Guénault, P.M.Lee and G.R.Pickett for useful discussions.

REFERENCES

(1) I.L.Bekarevitch and F.M.Khalatnikov, Sov. Phys.JETP 12 1187 (1961).
(2) W.A.Little, Am.J.Phys. 37 334 (1959).
(3) H.Haug and K.Weiss, Phys.Lett. 40A 19 (1972).
(4) R.E.Peterson and A.C.Anderson, Phys.Lett. 40A 317 (1972).
(5) For a recent review see J.P.Harrison, JLTP 37 467 (1979).
(6) J.P.Harrison and D.B.McColl, J.Phys.C 10 L 297 (1977).
(7) B.Frisken et al, J.Phys. (Paris) 42, C6-858 (1981).
(8) A.R.Rutherford, J.P.Harrison and M.J.Scott (1984) preprint.
(9) H.Aoki, J.Phys. C13 (1980) 3369.
(10) C.J.Lambert and D.Weire, Phys.Stat.Sol.(b) 101 (1980) 591.

LT-17 (Contributed Papers)
U. Eckern, A. Schmid, W. Weber, H. Wühl (eds)

VIBRATIONAL MODES OF SINTERED METAL POWDER

J.H. PAGE, J.P. HARRISON and M. MALIEPAARD

Physics Dept., Queen's University, Kingston, Ontario, Canada, K7L 3N6

The results of an ultrasonic study of sintered metal powder support the postulate that the vibrational modes of sintered heat exchangers, excited in the millikelvin temperature range, are localised modes.

INTRODUCTION

Recently Rutherford et al (1) developed a model for the Kapitza resistance between liquid ^{3}He or liquid ^{3}He-^{4}He mixtures and sintered metal powders in the millikelvin temperature range. It was based upon a postulated spectrum of the vibrational modes of a sintered metal powder. This spectrum is shown schematically in figure (1) for a 0.5 μm sinter. Above 5 GHz there are the usual bulk metal Debye phonons. At 5 GHz the phonon wavelength is ∿ 0.5 μm and therefore lower frequency (or longer wavelength) phonons are cut off (2). At very long wavelengths the sinter will appear as a continuous effective medium. Earlier work has demonstrated that low frequency sound does propagate through sinter with about one quarter the bulk metal sound velocity and that this velocity agrees well with that expected from static measurements of the elastic modulii (3,4). In this early work the sinter was modelled as a Debye solid with a Debye frequency determined by the size of the metal particles; this was about 1 GHz corresponding to a Debye temperature of about 20 mK. Subsequently electron microscope pictures of sinter made it clear that there is a great deal of disorder in a sinter with inhomogeneities extending up to about 20 d where d is the powder diameter. We postulated that effective medium phonons with $\lambda \lesssim 20$ d would be strongly attenuated and that the vibrational spectrum would be better described by inhomogeneity or localised modes (1). These would extend from 50 MHz to 5 GHz for a 0.5 μm powder size sinter. In analogy with a glass, which has disorder on an atomic scale, we further postulated a constant density of states for these modes.

Although there has been no direct evidence of these inhomogeneity modes the measured linear heat capacities of packed insulating powders (2) agree fairly well with those calculated with the model; measured Kapitza resistances between ^{3}He and sinter also agree fairly well, given the difficulty of separating the Kapitza resistance from other series thermal resistances (1). This present experiment is an indirect test for these modes; localised modes are not propagating modes and hence sound in that frequency range should not propagate.

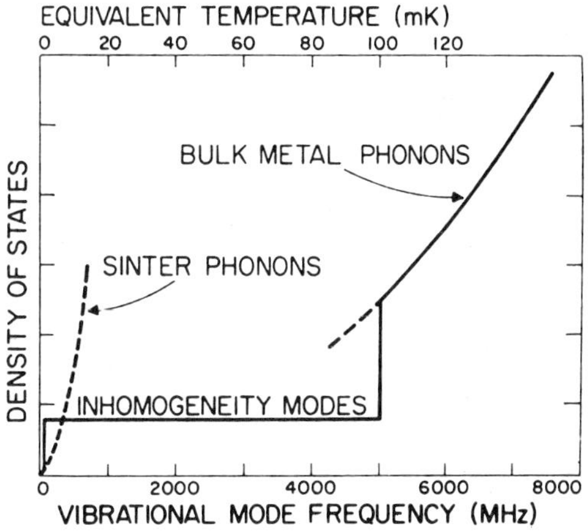

FIGURE 1
Postulated vibrational modes of a 0.5 μm sinter.

2. EXPERIMENT

To optimise the ultrasonic transmission through the sinter, the ultrasonic experiments were performed in the frequency range 1-20 MHz and the spectrum shown in figure (1) was moved into this range by working with 10 μm and 300 μm powder size sinters. The sinter samples were dry polished and placed between two quartz delay rods to which either longitudinal or shear wave ceramic transducers were bonded. Figure 2 shows oscilloscope traces for shear waves at different frequencies propagating in a 10 μm powder sinter of 49% packing factor; at each frequency the traces show both the ultrasonic input pulse reflected from the quartz/sinter interface and the transmitted signal which has passed through the sinter. Where possible, calibrated attenuators, as specified in the figure caption, were used to adjust the reflected and transmitted pulse heights to the same level on the oscilloscope. At the lowest

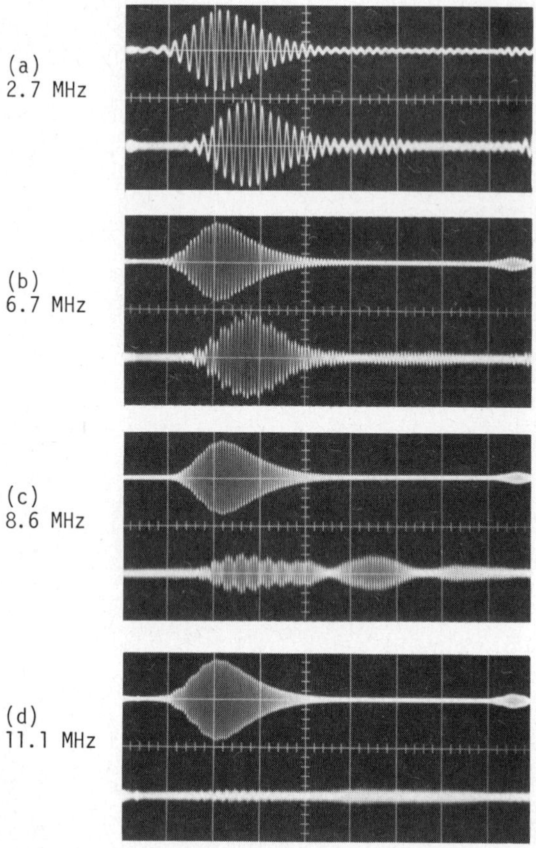

FIGURE 2
Reflected and transmitted signals (upper and lower traces) in a 10 μm powder sinter for the frequencies indicated. The attenuator settings for the reflected and transmitted signals were, respectively: (a) 96 dB, 63 dB; (b) 93 dB, 30 dB; (c) and (d) 93 dB, 8 dB. The oscilloscope time base was 2 μs per large division, the sample thickness was 0.6 mm and the ultrasonic velocity was 670 ms^{-1}.

frequency, the transmitted signal matches the input signal and the attenuation is relatively low (Figure 2(a)). Figures 2(b)-(d) show that as the frequency is increased, the transmitted signal is progressively distorted and attenuated. In Figure (3) these results are collected together as a plot of transmitted signal amplitude (relative to the reflected pulse) versus λ/d; the figure shows the dramatic decrease in wave propagation with decreasing wavelength as λ/d approaches $\sim$ 10, where the apparent attenuation length corresponds to about 2 powder particle diameters.

For the 300 μm particle sinters, it was necessary to reduce the sample thickness to a few powder diameters in order to observe transmitted signals in our frequency range; in this case, the inhomogeneities are largely suppressed, although disorder on the scale of the particle diameter remains. In a sample $\sim$ 2d thick, as the frequency is increased the transmitted signal remains distorted and small as $\lambda/d \sim 1$ is approached at which point the cross-over to bulk metal phonons occurs.

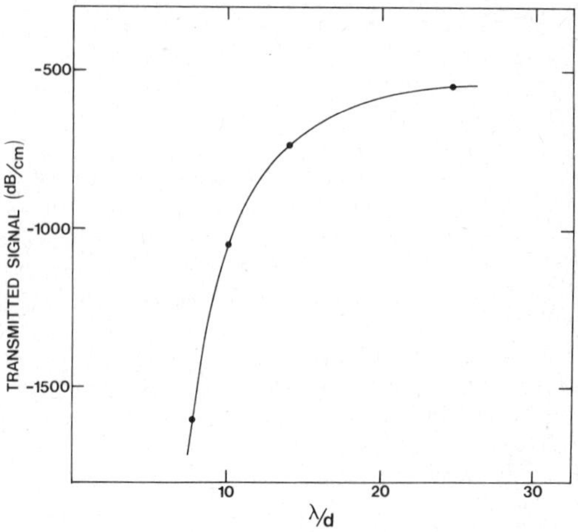

FIGURE 3
The relative transmitted signal as a function of λ/d.

3. DISCUSSION

The results have demonstrated the absence of propagating sound waves in sintered metal powder when $1 \lesssim \lambda \lesssim 20d$. The bandwidth between these two limits in fact extends over about two orders of magnitude in frequency because the low frequency propagating waves have a lower velocity than the high frequency propagating waves. This is strong support for the postulate that there is a band of frequencies where the vibrational modes are localised modes of one or a few powder particles.

ACKNOWLEDGEMENTS
This work has been supported by NSERC and ARC. JPH acknowledges a Killam Research Fellowship and MM an NSERC Summer Award.

REFERENCES
(1) A.R. Rutherford, J.P. Harrison and M.J. Stott, J.Low Temp.Phys. 55,157 (1984).
(2) R.H. Tait, Thesis, Cornell University (1975); R.O. Pohl in Topics in Current Physics, Vol.24, W.A. Phillips, ed. (Springer, 1981).
(3) B. Frisken et al, J. Phys. (Paris) 42 (Suppl. 12), C6-858 (1981).
(4) R.J. Robertson, F. Guillon and J.P.Harrison, Canad. J. Phys. 61, 164 (1983).

LT-17 (Contributed Papers)
U. Eckern, A. Schmid, W. Weber, H. Wühl (eds)

WELL CHARACTERIZED SINTERED MATERIAL FORMED FROM SUBMICRON CU POWDER FOR LOW TEMPERATURE HEAT EXCHANGER

K. ROGACKI, M. KUBOTA, E. G. SYSKAKIS, R. M. MUELLER, F. POBELL*

(Institut für Festkörperforschung der KFA-Jülich, D-5170 Jülich, F.R.Germany)

At temperatures below 10 mK heat transfer between sintered material and liquid He may be very much enhanced through some soft mode vibrations. In order to test such an idea we have been studying the process of sintering using "300 Å Japanese Cu powder" while varying conditions like: pretreatment, sintering temperature (20°C $\leqslant T \leqslant$ 240°C), H_2 or vacuum, mechanical pressure on the powder etc. We can now characterize the sinter well. A surface area of 2.9 m^2/g was achieved while keeping the electrical conductivity high.

1. INTRODUCTION

It is currently a common practice to use metallic sintered fine particles for the thermal conctact between liquid He and metals at low temperatures, however, the real mechanism of the heat transfer is still far from final understanding (1). Acoustic mismatch theory seems apply for well annealed clean bulk metallic surfaces below 100 mK where the average phonon wavelength becomes longer than the surface roughness of the metal, but surface imperfections play an important role in increasing the heat transfer between liquid He and the metals.

Quite different behaviour of the boundary resistance has been observed for sintered materials. Instead of the expected T^{-3} law sometimes one obtains T^{-2} or T^{-1} depencence below 10 mK. Nakayama pointed out that at such tmeperature heat transfer between sintered material and liquid He may be very much enhanced through some soft mode vibrations (2). He has also pointed out that the coupling between ^{3}He-^{4}He mixture quasiparticles and the soft mode phonons of the sinter should depend on the pore size of the sintered material (3).

In order to test such an idea, the sintering process of ultra fine Cu powder has been studied under various conditions and the sinter samples were characterized by: (a) softness, (b) size of particles, bridges and pores. Although work along similar lines has been attempted by other groups (4,5) we present here, for the first time, a systematic study of the sintering process for high density Cu powder formed from particles smaller then 1000 Å.

Cu was chosen mainly because it is known that the sintering process at room temperature is depressed by the existence of the oxide layer (6) while other noble metal submicron particles sinter at room temperature and change properties with time (7).

2. EXPERIMENTAL METHOD AND RESULTS

In order to control the sintering process the samples were prepared in a jig offering the possibility to adjust such parameters as: temperature in vicinity of the sample (T), mechanical pressure acting on the powder (P), hydrogen gas pressure inside of the jig (PH_2) and sintering time (t). In order to get a good description of the sample properties we have measured : specific surface area (S_W) by the BET adsorption isotherm technique, packing factor (α), specific electrical resistivity at room temperature (ρ_R) and liquid nitrogen temperature (ρLN_2), electrical resistance ratio R_R/RLN_2 and R_R/R_{LHe}, hardness (HV) by Vickers-method and the mechanical contact between sinter and Cu foil. The sintered samples were also tested by transmission and scanning electron microscopy.

The samples were made from so called "300 Å Japanese Cu powder" (8) which was badly oxidized upon receipt. The specific surface area for this "black powder" was in the range 4.4 m^2/g to 7.6 m^2/g, depending on the batch of powder. This coresponds to the calculated diameters of spherical grains from 1500 Å to 880 Å. The results were marvelously confirmed by the microphotos, which showed the average diameter of the deoxidized and sintered, therefore somewhat smaller grains, was between 800 Å and 1000 Å.

The first samples were made directly from the black powder, which was not reduced in hydrogen before placing into a form of 20 mm diameter and 0.7 to 7 mm thickness. Considerable inhomogeneity was seen in both hardness and in the color of cross-sections when the samples were broken.

In order to improve homogeneity of the samples, a presintering process with optimal conditions was introduced. In a quartz tube, with the volume 1.7 l, 10 g powder was heated to 120°C (± 2°C), for 100 minutes, with

* Experimental Physics V, University of Bayreuth, D-8580 Bayreuth, F.R. Germany.

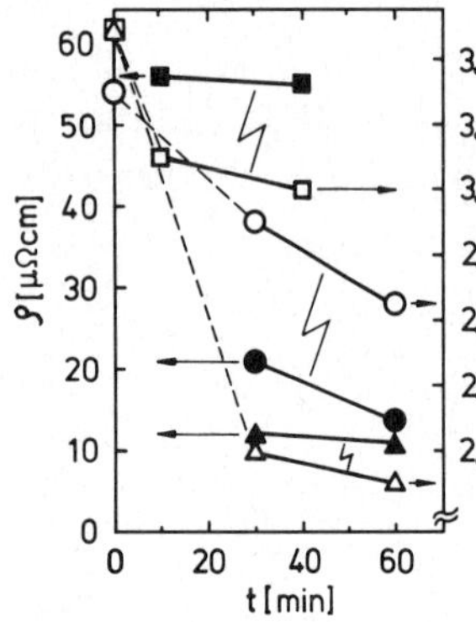

FIGURE 1

Influence of the sintering time on the electrical resistivity (closed symbols) and the specific surface area (open symbols) for samples made as follows:
(A) ■□: pressing in vacuum at T = 20°C, $P = 12.5\ kN/cm^2$ and $p = 10^{-5}$ Torr
(B) ●○: pressing in vacuum at T = 20°C, $P = 9.5\ kN/cm^2$, t = 10 min and $p = 10^{-5}$ Torr then heating at T = 100°C, $P = 0.6\ kN/cm^2$ and $PH_2 = 760$ Torr
(C) ▲△: heating at T = 130°C, P = $1.6\ kN/cm^2$ and $PH_2 = 760$ Torr.

a hydrogen atmosphere at 1 atm. During this process, from the "black powder" with the surface area $S_W^b \simeq 4.4\ m^2/g$, a brown-red powder with a high homogeneity and $S_W^o = 3.3\ m^2/g$ was obtained. This powder, or similar with a little larger $S_W^o = 3.5\ m^2/g$, was the starting material for all the next samples.

Various combinations of parameters were tried during the sintering: $20\ ^oC \leq T \leq 130\ ^oC$, $10\ min \leq t \leq 60\ min$, $0.6\ kN/cm^2 \leq P \leq 12.5\ kN/cm^2$, 10^{-5} Torr $\leq PH_2 \leq 760$ Torr. More than 40 samples were made, which yielded the various reproducible properties in the following ranges: $1.8\ m^2/g \leq S_W \leq 3.3\ m^2/g$, $44\ \% \leq \alpha \leq 53\%$, $9\ \mu\Omega cm \leq \rho_R \leq 80\ \mu\Omega cm$, $1.7 \leq R_R/R_{LN2} \leq 2.8$, $2.0 \leq R_R/R_{LHe} \leq 3.7$, $18\ kp/mm^2 \leq HV\ (0.2\ kp) \leq 48\ kp/mm^2$.

On the basis of the results, influence of the various parameters on the sinter properties was determined. For example, in Fig. 1 the influence of the sintering time on the resistivity ρ_R and surface area S_W for samples prepared in three different ways at temperatures (A) 20°C, (B) 100°C and (C) 130°C is shown (see explanation under figure). Dashed lines connect the surface area S_W^o for starting material (presintered powder) with S_W for sinters. The samples B were first pressed at room temperature for 10 minutes and under pressure $P = 9.5\ kN/cm^2$, then sintered at 100°C with reduced pressure and hydrogen atmosphere.

For samples prepared in 20°C and 130°C (A and B) parameters ρ and S_W are only slightly dependent on time t, indicating a slow sintering process. Samples A have large S_W and simultanously relatively high ρ_R in spite of high mechanical pressure P used. Resistivity ρ_R of samples C is significantly lower but on the other hand S_W is 40 % smaller than that of presintered powder despite the low applied pressure. A compromise between ρ_R and S_W had to be made for sample B which had been initially pressed to the required packing factor followed by sintering. Here a strong dependence of ρ and S_W as a function of t is observed.

3. SUMMARY

We have studied the sintering process and finished sintered material for submicron Cu powder having a true diameter 800 - 1000 Å (nominal 300 Å). By performing a presintering process we could produce homogenous powder and achieve reproducible results. This presintering process changed the characteristic surface area of the powder only 25 % from $4.4\ m^2/g$ to $3.4\ m^2/g$. There are a number of parameters which influence the sintering process. There is a characteristic temperature at which the sintering process starts, for example ca. 100°C for the presintered powder with $S_W^o \simeq 3.3\ m^2/g$. A typical sample shows $S_W = 2.8\ m^2/g$, $\rho_R = 14\ \mu\Omega cm$ and $R_R/R_{LHe} = 3.0$.

REFERENCES

(1) J.P. Harrison, J. Low Temp. Phys. 37, 467 (1979)
(2) N. Nishiguchi and T. Nakayama, Sol. St. Comm. 45, 877 (1983)
(3) T. Nakayama, private communication
(4) M. Krusius, D.N. Paulson and J.C. Wheatley, Cryogenics 18, 649 (1978)
(5) R.J. Robertson, F. Guillon and J.P. Harrison, Can. J. Phys. 61, 164 (1983)
(6) K. Rogacki, et al. to be published
(7) S. Iwama and K. Hayakawa, Jap. J. Appl. Phys. 20, 335 (1981)
(8) Vacuum Metallurgical Co., LTD. Nihon Typewriter Building, 11-2, 1-chome, Kyobashi, Chuo-ku Tokyo 104, Japan.

LT-17 (Contributed Papers)
U. Eckern, A. Schmid, W. Weber, H. Wühl (eds)

ANGULAR DEPENDENCE OF PHONON FLUX TRANSMISSION FROM He II INTO SOLID AT 0.1 K*

E. E. Nothdurft and K. Luszczynski

Department of Physics, Washington University, St. Louis, Missouri 63130, U.S.A.

Measurements have been made at SVP and at 20 atm of the angular dependence of phonon flux transmitted from He II into aluminum at 0.1 K. The detecting surface was that of a 100-nm layer of aluminum forming an Al-AlO$_x$-Al detector which was vacuum deposited on a glass surface. The detector could be used either as a superconducting bolometer, sensing the entire phonon spectrum, or as a superconducting tunnel junction, sensitive mainly to high frequency phonons. For both detection modes the signal strength shows a narrow pressure independent peak centered on the normal angle of incidence and a broad base which depends on pressure and the detector mode.

Angular dependence of the transmission of thermal phonons from He II into solid can provide further insight into the properties of the He II-solid interface that is not disturbed by large energy flux, normally present at the source in heat pulse experiments. The results presented here, obtained during a series of experiments on phonon propagation in He II at 0.1 K and pressures up to 24 atm, are the first reported measurements on the angular dependence of phonon flux transmitted from He II into a solid with a nominally rough surface. A review of some of the related work on the propagation and transmission of phonons in He II at low temperatures has been given by Wyatt (1).

The basic experimental technique used in the present experiments, described in detail elsewhere (2), consists of generating a pulse of thermal phonons at a small source (0.85 mm x 0.85 mm) on one side of a relatively large volume (115 cm^3) of He II and detecting it 1 to 2 cm away with a small (0.85 mm x 0.85 mm) detector. To a good approximation the experimental arrangement makes it possible to observe ballistic propagation of phonons from a point source in an unbounded region; time resolution of signals is used to eliminate wall reflections. The orientation of the detector can be changed during the experiment so that the angle of incidence α can be varied from 0 to 90°. The detector consisted of an Al-AlO$_x$-Al sandwich vacuum deposited on a clean but otherwise untreated glass slide surface. The solid surface subjected to the He II phonon flux is that of the first aluminum layer, which is about 100 nm thick. The detector, placed 17 mm from the source, can be used either as a superconducting bolometer or as a superconducting tunnel junction. The two detector modes are obtained by changing only the external biasing arrangements so that both modes have the same He II-solid (aluminum) interface. The bolometer senses the entire spectrum of the incident phonon flux, but the tunnel junction should detect only phonons with k values above the threshold value k_J determined by the gap of the superconductor used and the phonon velocity in He II; here k_J changes from 0.22 Å^{-1} at SVP to 0.15 Å^{-1} at 20 atm.

The measured signal strength, defined here as the slope of the leading edge of the detected signal (2), is plotted as a function of α in Fig. 1. For purposes of comparison with calculated results the experimental data in Fig. 1 were used to derive the effective transmission coefficient τ, defined as the ratio of the measured and predicted signal strengths; τ characterizes the transmission of the net phonon flux across the interface, rather than the behavior of individual phonons. In calculating the predicted signal strength, which does not include any angular dependence of phonon transmission across the boundary, two geometrical factors are taken into account: the "cosα factor" associated with the effective area of the detector normal to the incident beam and the "finite size factor" arising from the distribution of flight times (distances) between the source and the edges of the detector; the second factor also includes the effects of spectral distribution and dispersion in the propagating phonon beam (2).

The angular dependence of the effective transmission coefficient τ, as measured by the bolometer and the tunnel junction, is shown in Fig. 2. In each case $\tau(\alpha)$ is normalized to $\alpha = 0$. For small angles, i.e., $\alpha < 20°$, both detectors show very similar angular dependence,

*This work was supported in part by National Science Foundation-Low Temperature Physics-Grant DMR 80-10818.

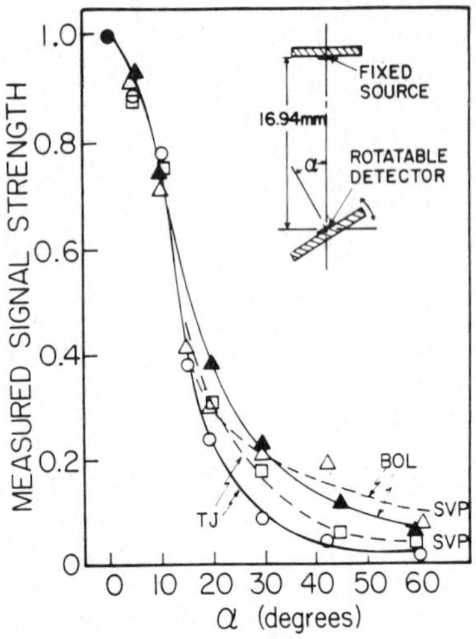

FIGURE 1
Angular dependence of the measured signal strength normalized to $\alpha = 0$ (solid circle). Solid and open triangles—bolometer at SVP and 20 atm; squares and circles—tunnel junction at SVP and 20 atm. The lines are added as visual guides only. Inset shows the relative positions of the source and the detector. The input pulse at the heater is 3-μs long with power density of about 5 mW/mm^2 corresponding to an effective source temperature of about 1.85 K (2,3).

insensitive to pressure, with a narrow central peak whose half-width at half-height is about 15°. For larger values of α the angular dependence of τ is more complicated and varies with pressure and depends on the detector mode. A rather remarkable feature of these results is that the peaks are so narrow, not unlike those observed for freshly cleaved crystal surfaces (1).

The present results for $\tau(\alpha)$ obtained for a nominally rough surface suggest that sharply peaked transmission of phonons from He II into solid is not limited to atomically flat (cleaved) surfaces. One of the implications is that He II-solid surfaces are apparently much smoother than expected, at least with regard to transmission of thermal phonons.

The results obtained from detailed comparisons between calculated and observed signal strengths for the two detector modes as a function of pressure will be presented elsewhere

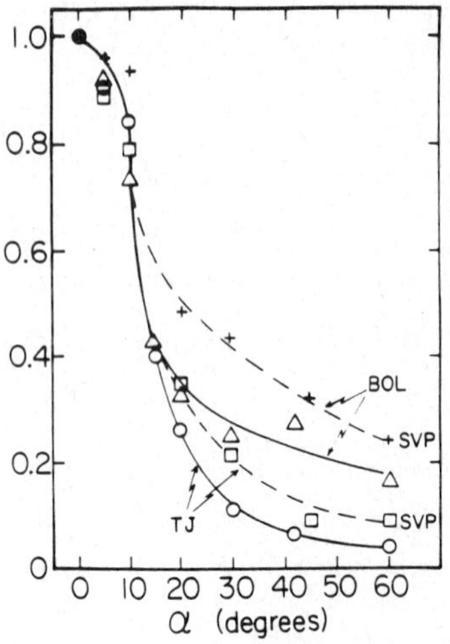

FIGURE 2
Angular dependence of the effective transmission coefficient τ normalized to $\alpha = 0$ (solid circle). Triangles and crosses—bolometer at SVP and 20 atm; squares and circles—tunnel junction at SVP and 20 atm. The lines are added as visual guides only.

REFERENCES

(1) A. F. G. Wyatt, in Nonequilibrium Superconductivity, Phonons, and the Kapitza Boundaries, K. E. Gray (ed.), (Plenum Press, New York, 1981), p. 31.
(2) E. E. Nothdurft, Ph.D. Thesis (Washington University, 1980), unpublished: Dissertation Abstract 41 (1981) 3079.
(3) E. E. Nothdurft and K. Luszczynski, J. de Physique C6-39 (1978) 252.

LT-17 (Contributed Papers)
U. Eckern, A. Schmid, W. Weber, H. Wühl (eds)

KAPITZA RESISTANCE OF SINGLE CRYSTAL $Gd_3Ga_5O_{12}$

Xu Yun-hui, Zheng Jia-qi and Guan Wei-yan

Institute of Physics, Academia Sinica, Beijing, China.

The thermal boundary resistance between liquid HeII and $Gd_3Ga_5O_{12}$ as a function of the temperature has been measured by dc method in the temperature range of 1.39 to 2.02K. The temperature dependence of R_k of GGG with a clean surface is $R_k=27.6\times T^{-1.92}$ K cm²/W.

The paramagnetic salt godolinium gallium garnet $Gd_3Ga_5O_{12}$ (GGG) is an active magnetic element in the magnetic refrigerators with a high efficiency and a high cooling power[1,2]. The thermodynamic properties (specific heat, entropy, thermal conductivity and diffusivity) of GGG between 0.05K and 25K have been studied[3,4]. The Kapitza resistance is one of the important characteristics of materials suitable for magnetic refrigeration. However, up to now there is no information reported about this parameter. The Kapitza resistance between HeII and single crystal GGG is measured by dc method as a function of the temperature in range of 1.39 to 2.02K in this paper.

The specimen of single crystal GGG in the form of a short cylinder with (111) plane ends polished specularly and with diameter 17.6mm, height 11mm is glued to the stainless steel tube by Stycast 1266. The heater of manganin wire is mounted on the upper end of the specimen in the vacuum can. T_1, T_2 and T_3 are three calibrated A-B carbon resister thermometers. The experimental apparatus is shown in the insert of Fig.1.

During the measurements the temperature of HeII bath is kept a constant to an accuracy of $\pm5\times10^{-5}$K by adjusting vapor pressure of helium and by using an electronic thermostabilizer. The vacuum in the can is kept better than 5×10^{-6} Torr.

A steady heat flow is supplied by the heater and the temperature difference ΔT across the interface is determined by using the sensitivity $\frac{dR}{dT}$ of the thermometers. The measuring errors of T and ΔT are less than $\pm3.5\times10^{-3}$K and $\pm1\times10^{-4}$K respectively.

The experimental results show that when ΔT is small enough as below 25mK, the ΔT is proportional to the heat flow per

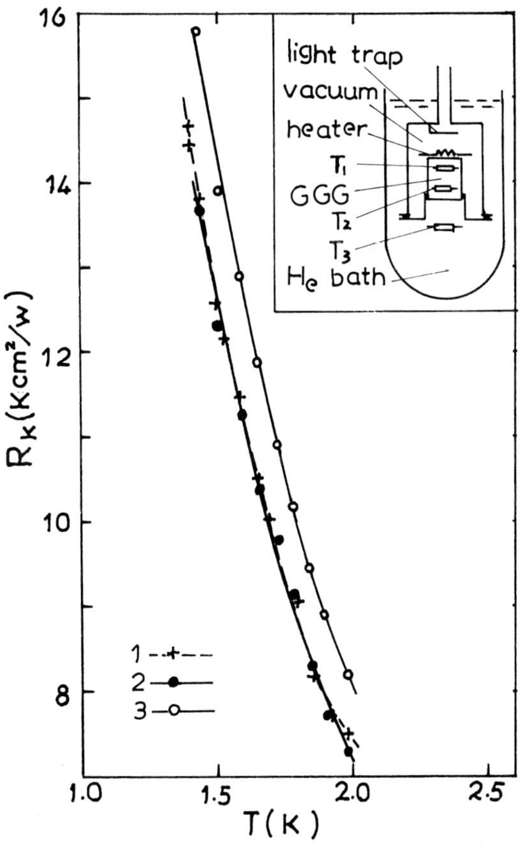

FIGURE 1
Kapitza resistance R_k between HeII and Single Crystal GGG with (111) surface vs. Temperature.

Run	Temperature dependence of R_k (Kcm^2/w)	Experimental details
1	$27.6T^{-1.92}$	Cleaned the surface with alcohol; precooled in He gas of 1atm. at 77K for 14hrs.
2	$27.6T^{-1.93}$	Same operations as run 1; pumped out He gas; added pure O_2 gas to 10 Torr for 1.7min. and to 30 Torr for 5min. at 77K.
3	$32.2T^{-1.99}$	Cleaned the surface with alcohol; precooled in the mixture gases of air(220Torr) and He gas (540 Torr)for 15 hrs.

Table 1. Measured Kapitza Resistance Data and experimental details.

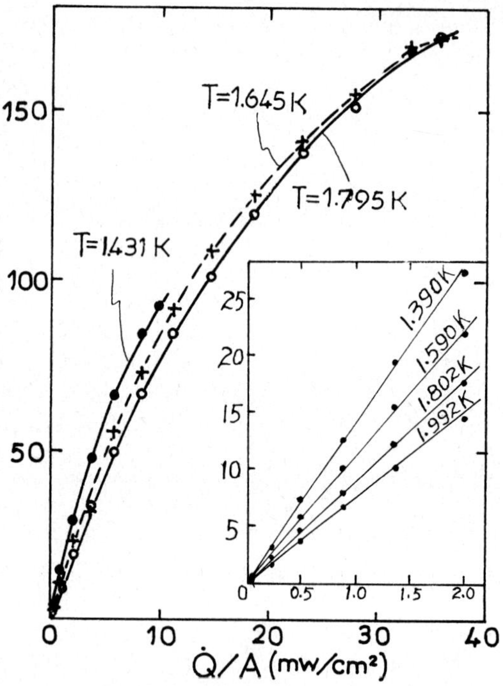

FIGURE 2
Temperature difference ΔT on the interface vs. heat flow density.

unit area of the interface, $\dot{Q}/A$.

The values of Kapitza resistance $R_k=\Delta T/(\dot{Q}/A)$ ($\Delta T \ll T$) can be obtained from the slopes of R_k-T curves. The temperature dependence of R_k is plotted in Fig.1. For the case of the clean surface R_k is found to be $27.6T^{-1.92}K.cm^2/w$ (see Table 1). It has been noted that the results are influenced by experimental conditions, but they are reproducible with a high accuracy in the same run.

The temperature dependences of R_k for three different runs are listed in Table 1. The deviations from straightlines are observed when the heat flow densities $\dot{Q}/A$ are greater than $2.5mW/cm^2$

The Kapitza resistance of single crystal GGG calculated with the acoustic mismatch theory[5] is $1193.6T^{-3}Kcm^2/W$ which is large by a factor of 20-50 of our experimental results that are reasonably close to the published data on paramagnetic salts CeES and PrES[6].

REFERENCES.

[1] C.Delpuech, R. Beranger, G. Bon Mardion G. Claudet and A. A. Lacaze, Cryogenics, 21(1981)579.
[2] A.F. Lacaze, R. Beranger, G. Bon Mardion, G. Claudet, C. Delpuech, A.A. Lacaze and J. Verdier, Proc. CEC 81, San Diego (1981).
[3] B. Daudin, R. Lagnier and B. Salce, J. Magnetism and Magnetic Materials 27 (1982) 315.
[4] J. A. Barclay and W. A. Steyert, Cryogenics, 22 (1982) 73.
[5] I.M. Khalatnikov, Zh. Eksp. Teor. Fiz. 22 (1952)687.
[6] H. Glattli, Can.J. Phys. 46(1968) 103.

LT-17 (Contributed Papers)
U. Eckern, A. Schmid, W. Weber, H. Wühl (eds)

OBSERVATION OF FLUXON AND ANTIFLUXON COLLISION IN A JOSEPHSON TRANSMISSION LINE

Azusa MATSUDA and Tsuyoshi KAWAKAMI

Musashino Electrical Communication Laboratory, Nippon Telegraph and Telephone Public Corporation, Tokyo 180 Japan

The behavior of fluxon-antifluxon collision is experimentally investigated, using a direct measurement system on a fluxon waveform. The experiment shows the increase in propagation delay time due to a collision. This result, which is contradictory to the free soliton theory, can be qualitatively explained as the effect of dissipation.

1. INTRODUCTION

The fluxon propagation properties of a Josephson transmission line (JTL) have been obtained experimentally using direct fluxons observation methods by a signal processing(1-3) and by a Josephson sampling technique(4). However, very little experimental information has been obtained concerning fluxon interaction properties. This paper describes the results of an fluxon-antifluxon collision experiment using the direct measurement system based on a signal processing technique.

In the case of unperturbed fluxon and antifluxon collision, phase shift is analytically obtained and is negative. Therefore, the propagation delay time for a fluxon will decrease when a collision occurs. However, if actual JTLs are considered, several perturbation terms exist and substantially affect the results. For example, under the existence of a loss term with a certain magnitude, it is theoretically predicted that fluxon and antifluxon collision produces a breather soliton which ultimately dies out due to dissipation (5). An analysis of fluxon collision behavior provides information on perturbation properties as well as on basic soliton behavior.

2. EXPERIMENT

The JTL is a Nb/Nb-oxide/Pb tunnel junction with a 50μm width and a 1cm length. It was constructed by conventional photolithographic techniques. The Nb oxide tunnel layer was made by RF plasma discharge. Critical current density for the measured JTL was 9.8 A/cm^2. Calculated Josephson penetration depth and plasma period were 143μm and 17ps, respectively. The input pulse was fed to the terminated end of the JTL through a 1dB step attenuator. The other end remained open. The voltage pulse, which appeared at this end, was detected by the signal processing system.

Under these termination conditions, when two pulses, P_1 and P_2, are fed into the input end in sequence, the first fluxon generated by P_1 is reflected as an antifluxon at the open end and collides with the second fluxon generated by P_2. The collision effect is observed through the delay change in the second fluxon.

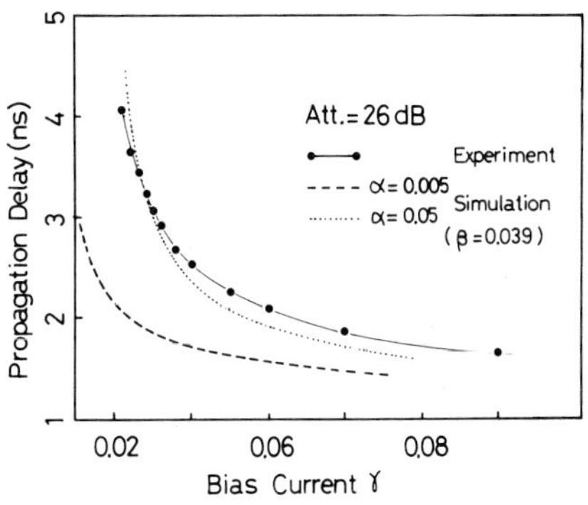

Fig.1 Propagation delay time bias current dependence.

3. RESULTS

In Fig.1, propagation delay time for a fluxon is plotted as a function of normalized bias current $\gamma(=I_B/I_{Jmax})$ for a single pulse input. This JTL shows a somewhat long propagation time, compared with the theoretical time shown by the dashed line in the figure. This propagation time is calculated using the parameters adopted in Ref.(2). The collision experiment gives further information about this point.

The results of the collision experiment are shown in Fig.2. In this figure, propagation times for the second fluxon are plotted as a function of P_1 attenuation(=Att_1, 0dB=820mV in 50Ω system) with varying pulse separation times Td. γ is fixed at 0.036. The arrows indicate single and double fluxon thresholds for P_1 and denote As and Ad, respectively. When Att_1 >As,

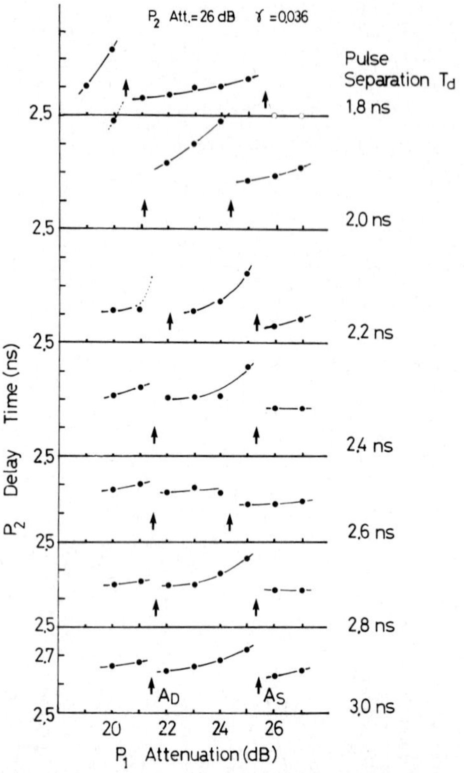

Fig.2 Experimental results of second fluxon delay time as a function of P_1 attenuation.

no fluxon is propagated. When $Ad<Att_1 \leqq As$, a single fluxon is propagated. Delay times for the first fluxon were approximately 2.9ns for $Att_1=As$ and 2.2ns for $Att_1=Ad$. When $Att_1 \leqq Ad$, two fluxons propagate at a time for a single input pulse. Figure 2 clearly shows the effects of a collision as a sudden increase in delay times. Especially noteworthy is that at Td=2ns and 2.2ns, the second fluxon is destroyed by sequential collisions with 2 antifluxons at double fluxon threshold. These increases are commonly observed at all thresholds and accompany subsequent decreases. The decreases are caused by the change in antifluxons velocity and suggest the velocity sensitive nature of the interaction. It is thought that these features, which contradict the free soliton theory, show the specific interaction properties of fluxons in this JTL, although the variations due to the change in the colliding position are also observed. Other JTLs, having similar critical current densities, also showed features similar to those explained here.

4. DISCUSSION

Numerical analyses were performed to explain

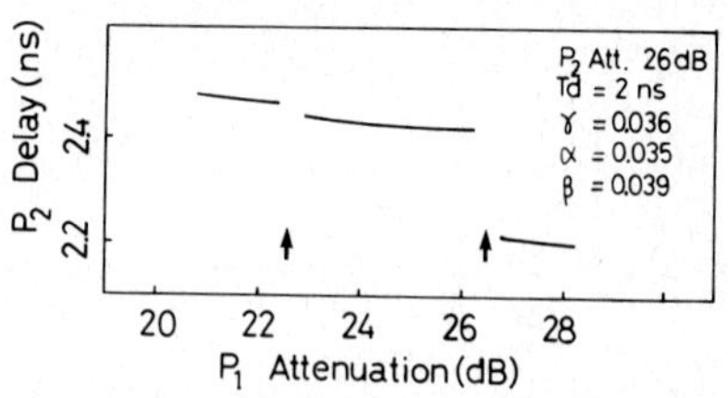

Fig.3 Simulated result of second fluxon delay time as a function of P_1 attenuation.

the present experimental results. The modified S-G equation,

$$\phi_{xx}-\phi_{tt}=\sin\phi+\alpha\phi_t-\beta\phi_{xxt}-\gamma,$$

is assumed for the JTL. Here, α is a quasiparticle loss parameter and β is an rf loss parameter. α is adjusted to reproduce the bias current dependence of delay time. β is fixed at 0.039. If $\alpha=0.05$, the theoretical calculation agrees quite well with the experimental result, as shown by a dotted line in Fig.1. However, in this case, no incident fluxon is ever reflected at the open end. In Fig.3, the input attenuation dependence of delay time is plotted, assuming $\alpha=0.035$. This calculation shows the delay time increases due to a collision, although they do not accompany the subsequent decreases. It was found, from analyzing the trajectories of the colliding fluxons, that this increase is caused by the long range interaction which attracts a fluxon and an antifluxon after the collision. Although details have not been clarified, the present experimental results can possibly be explained by such an interaction.

5. CONCLUSION

Fluxon-antifluxon collision behavior was experimentally investigated. It is observed that propagation delay substantially increases when a collision occurs. These characteristics are qualitaively explained by the effect of the perturbation term which describes dissipation.

ACKNOWLEDGEMENT

The authors are grateful to N.Kuroyanagi and H.Okamoto for their continuous support as well as to J.Nitta for his helpful discussions.

REFERENCES

(1) A.Matsuda and S.Uehara, Appl.Phys.Lett. 41 (1982)770
(2) A.Matsuda and T.Kawakami, Phys.Rev.Lett. 51 (1983)694
(3) J.Nitta, A.Matsuda and T.Kawakami, J.Appl.Phys. in print.
(4) S.Sakai et.al, Jpn.J.Appl.Phys. 22 (1983)L479
(5) D.W.McLaughlin and A.C.Scott, Phys.Rev. A18 (1979)1652

LT-17 (Contributed Papers)
U. Eckern, A. Schmid, W. Weber, H. Wühl (eds)

FLUXON DYNAMICS IN THE ABSENCE OF BOUNDARY COLLISIONS

B. Dueholm*, A. Davidson, C.C. Tsuei, M.J. Brady, K.H. Brown, A.C. Callegari, M.M. Chen, J.H. Greiner, H.C. Jones, K.K. Kim, A.W. Kleinsasser, H.A. Notarys, G. Proto, R.H. Wang, T. Yogi

IBM Watson Research Laboratory, Yorktown Heights, NY 10598

We report the first experimental results on fluxon motion in long annular Josephson tunnel junctions. Though common in theories, the annular geometry is novel in real structures. The reason to use it is to eliminate fluxon-boundary collision effects. We find that the zero-field steps are in good agreement with simulations of the perturbed sine-Gordon equation, and differ substantially from the perturbation theory result at high fluxon density and at low velocity. We attribute the difference to fluxon-antifluxon collisions. Odd numbered Fiske steps never appeared, even when attempting to trap flux.

1. INTRODUCTION

Experiments on fluxon propagation in Josephson junctions have been largely confined to junctions having a rectangular geometry (1-5). There have been some studies of cylindrical geometries (6,7), but without direct measurement of fluxon induced voltage steps. In this paper we report the first observation of fluxon induced voltage steps in long annular junctions. The annular geometry, unlike the cylindrical geometry, is compatible with advanced thin film processing. But like the cylindrical geometry it eliminates the effects of fluxon-boundary collisions, thereby allowing for a more direct comparison with the perturbation theory (1). Fluxon-antifluxon collisions, however, still play a role and hence deviations are to be expected.

2. LAYOUT

Fig. 1 schematically shows the layout of our annulus. The junction, placed above a superconducting groundplane, was made on the edge of the base electrode, and was 0.3 microns wide and 300 microns in diameter, with a nominal current density of about 4000 amperes per cm^2. These junctions were produced by the Josephson development line at IBM using essentially their standard process. The chips were mounted so that the junctions were aligned with superconducting loops a few microns above the chip surface, in order to efficiently apply a magnetic field. The mounted chips were placed on a temperature controlled stage in a vacuum can inside a magnetically shielded dewar within a screened room.

*Permanent Address: Physics Laboratory I, The Technical University of Denmark, DK-2800 Lyngby, Denmark.

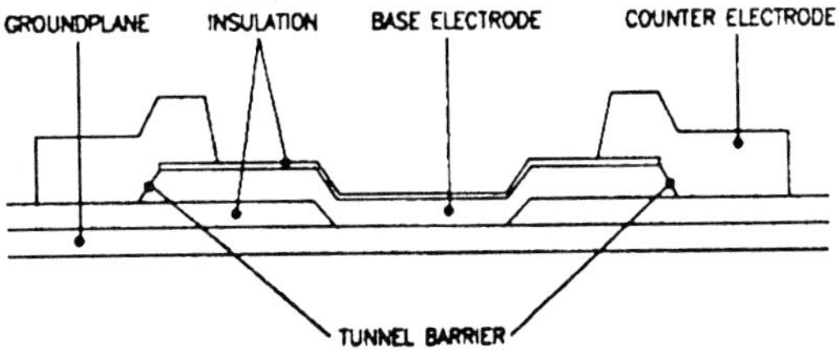

Fig. 1. Cross sectional view of the annular junction structure. The base electrode connects to a groundplane to help remove angular dependence in the current distribution. Current from the counter electrode is collected some distance away by a resistor network also designed to minimize variation in the current distribution.

3. ZERO FIELD STEPS

To see steps with no applied field the temperature had to be raised to above roughly 6K. The bias current was then varied by hand to trace out each step. The signal from the junction was amplified, digitized, and averaged in a micro computer. The known junction parameters are consistent with a plasma frequency of 65 GHz, a Josephson penetration depth of 33 microns, a McCumber parameter β_c of 300, and critical current of 4.3 milliamps at the measurement temperature.

The lowest voltage step with the largest current range that we have observed is shown in Fig. 2. The voltage is that expected from a single fluxon and a single antifluxon in counter rotation. Also shown are the results of the perturbation theory and those of a recently published numerical simulation (8). The theories have been fitted to yield the

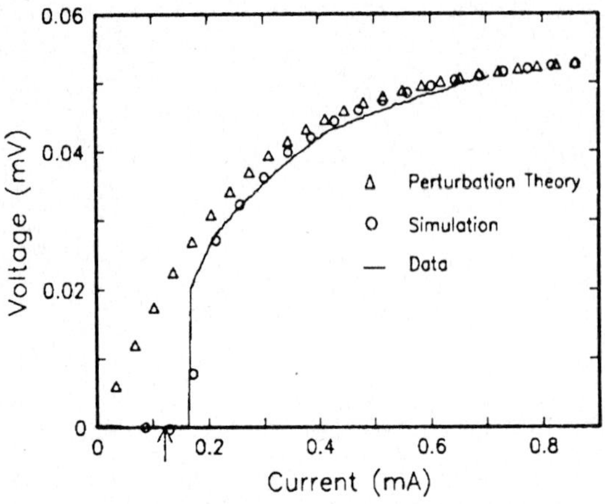

Fig. 2. Voltage as a function of current for the experiment (solid line) perturbation theory (triangles) and numerical simulation (circles) for a counter-rotating fluxon-antifluxon pair. The parameters used in the theory and simulation were $\beta_c = 300$, Ic = 4.3 milliamps, $\omega_p = 2\pi$ x 65GHz, and $\lambda_j = 33$ microns. These parameters are consistent with other independent measurements of junction parameters. The arrow on the current axis indicates the analytical annihilation threshold.

observed asymptotic value (corresponding to the velocity of light in the junction). The perturbation theory result for collisionless fluxon propagation is

$$V_n = \frac{\phi_0 \omega_p n}{l} \frac{1}{\sqrt{1+(4\alpha/\pi\eta)^2}} \tag{1}$$

Here n is the number of moving fluxons, l is the length of the junction normalized to the Josephson penetration depth, $\alpha = 1/\sqrt{\beta_c}$, and η is the bias current normalized to the critical current. ϕ_0 is the superconducting flux quantum, and ω_p is the zero bias plasma frequency of the junction.

As can be seen there is quite good agreement between the simulation and the experimental curve. For the perturbation theory (Eq. 1) the agreement is good on the upper portion of the step, while large deviations occur for small bias currents. We believe that this is because the perturbation theory does not take into account the fluxon-antifluxon collision process, which tends to slow down the fluxons thereby decreasing the average voltage. As the bias current is increased the fluxons are Lorentz contracted and the fluxon-antifluxon interaction becomes less important, resulting in better agreement with perturbation theory. As the current is lowered a threshold is reached below which the fluxon and antifluxon annihilate , and the junction resets to zero voltage. This feature, nicely reproduced by the simulation, is obviously not included in Eq. 1. The theory, however, may be expanded (1) to yield this annihilation threshold and the result is shown in Fig. 2. As predicted in Ref. 8 this threshold is lower than that observed experimentally.

4. FISKE STEPS

In the case of an applied magnetic field the situation in the annular junction is expected to be different from the usual rectangular geometry (9). Since the counterelectrode is a superconducting ring, flux in the junction can only be introduced as fluxon-antifluxon pairs. The result is that odd numbered Fiske steps are predicted to be absent, and none were observed, even with field substantial enough to cause large modulation of both the even numbered steps and the Josephson current. If, however, the junction with magnetic field applied is cooled from a temperature above Tc of the top electrode (Pb alloy) but below Tc of the base (Nb), then flux may be trapped in the junction. We would then expect the supercurrent and the even numbered steps to be completely suppressed, and for the odd numbered Fiske steps to appear. Moreover, we expected the first Fiske step to be in agreement with the perturbation theory result since there would be neither fluxon collisions, nor boundary collisions.

We have tried to observe these unusual Fiske steps in several ways: by using the temperature cycle described above with fields up to 10 gauss (0.1 gauss should have been sufficient); by rapid cooling to make thermally induced trapped flux; and by dropping all precautions normally used to prevent trapped flux. For reasons not yet known, none of these methods have worked. It is as if there is some mechanism that expels flux from the ring as it cools through its superconducting transition.

ACKNOWLEDGMENTS

We gratefully acknowledge many ideas and suggestions by D.J. Frank that were helpful in this work. One of us (Dueholm) is grateful for the hospitality of IBM Yorktown Heights, during the preparation of this work.

REFERENCES

(1) D.W. McLaughlin, A.C. Scott, Phys. Rev. A18, 1652 (1978).
(2) R.D. Parmentier, Solitons in Action (Academic, New York, 1978)
(3) P.M. Marcus and Y. Imry, Solid State Commun. 33, 345 (1980).
(4) N.F. Pedersen, Proceedings of the NATO ASI, Advances in Superconductivity, Erice (Plenum, New York, 1982).
(5) M. Buettiker and R. Landauer, "Transport and Fluctuations in Linear Arrays of Multistable Systems," in Nonlinear Phenomena at Phase Transitions and Instabilities, edited by T. Riste (Plenum, New York, 1982).
(6) M. Bhushan and M.D. Sherrill, Physica 108B+C, 735 (1981).
(7) M. Kuwada, Y. Onodera, and Y. Sawada, Phys. Rev. B, 27, 5486 (1983).
(8) A. Davidson and N.F. Pedersen, Appl. Phys. Lett. 44 465 (1984).
(9) B. Dueholm, E. Joergensen, O.A. Levring, J. Mygind, N.F. Pedersen, M.R. Samuelson, O.H. Olsen, M. Cirillo, Physica 108B+C, 1303 (1981).

LT-17 (Contributed Papers)
U. Eckern, A. Schmid, W. Weber, H. Wühl (eds)

FLUXON-PAIR STATE IN JOSEPHSON JUNCTION

Masanori SUGAHARA

Faculty of Engineering, Yokohama National University, Tokiwadai 156, Hodogaya, Yokohama 240, Japan

The pairing state of fluxons with opposite polarity in a Josephson junction gets into the lowest energy state with negative value of nucleation energy when it is in motion with velocity c_J, the light velocity in the junction. Threshold external fields necessary to destroy the bound state are calculated and compared with experimental value observed in a hollow cylindrical Josephson junction.

1. INTRODUCTION

The fluxon pair state in 2 dimensional superconductor is attracting attention in connection with Kosterlitz-Thouless (KT) phase transition. (1) In this report we treat properties of fluxon pairs in Josephson junctions in which topological fluxon orientation has only 1 degree of freedom (1DF). Examples of 1DF junction are i) a long junction whose junction length L is much larger than the width h, and ii) a junction with cylindrical structure whose diameter of the circumference (length L) is comparable or smaller than the junction height h.

It is known in a junction the penetration depth λ_J, flux intrusion field B_{c1J} and energy ε_J of fluxon with length h are respectively given by $\lambda_J=(2ej_c\mu_0 \times\Lambda/\hbar)^{-1/2}$ with $\Lambda=2\lambda+d$, $B_{c1J}=2\Phi_0/\pi^2\Lambda\lambda_J$ and $\varepsilon_J=2h\Phi_0{}^2/\pi^2\mu_0\Lambda\lambda_J$, (2) where e is the electron charge, j_c is the threshold junction current density, μ_0 is the permeability of vacuum, λ is the penetration depth of junction electrodes, d is the thickness of junction tunnel barrier, and Φ_0 is the flux quantum. Since λ_J is so large that sometimes it is comparable to junction size, we may discuss the pair dissociation temperature T_{KTJ} after KT theory (3) and get

$$k_BT_{KTJ}=2\Phi_0{}^2/\pi^2\mu_0(\Lambda\lambda_J/h)\ell n(L/\Lambda_J),$$

where k_B is the Boltzmann constant and Λ_J is the effective core size of a fluxon. Using practical values for quantities, we find that T_{KTJ} is much higher than T_c for any superconductor. Therefore we can safely neglect the pair dissociation by thermal effect in 1DF junctions. In the following we study electromagnetic properties of the pairing state of fluxons in 1DF condition.

2. NUCLEATION ENERGY OF FLUXON PAIR

Suppose a 1DF junction taking a coordinate frame with x axis along the junction length L, y axis perpendicular to tunnel barrier, and z axis along the fluxon length h. If a fluxon is at x=0, the distribution of local magnetic flux density is $b(x)=(\Phi_0/\pi\Lambda\lambda_J)\mathrm{sech}(x/\lambda_J)$. The magnetic interaction energy between the fluxon and the other one put at $x=\ell$ is $\pm\Phi_0hb(\ell)/\mu_0=M_AM_B\beta(\ell)$, where M_A and M_B are the magnetic moments of fluxons with value $\pm\Phi_0h/\mu_0=\pm\mu_f$, $\beta(\ell)=(\mu_0/\pi\lambda_J\Lambda h)\times\mathrm{sech}(\ell/\lambda_J)$.

If the fluxon at the origin is in motion with velocity v_L in the x direction, the induced voltage between two electrodes is $V(x)=v_Lb(x)\Lambda$, and the charge induced on electrode surfaces is $\pm Q=\pm h\int\varepsilon[V(x)/d]dx=\pm h\varepsilon v_L\Phi_0/d$, where ε is dielectric constant of the tunnel barrier. The electrostatic energy between two fluxons moving with velocity v_L keeping their interval ℓ is $\pm QV=P_AP_B\times\gamma(\ell)$, where P_A and P_B are the dipole moments of fluxons with value $\pm Qd=\pm h\varepsilon\times v_L\Phi_0$ and $\gamma(\ell)=(\pi\varepsilon h\lambda_Jd)^{-1}\mathrm{sech}(\ell/\lambda_J)$. The nucleation energy of a pair in external electric field E_e and magnetic field B_e is found to be

$$U_{PJ}(\ell,v_L)=\beta(\ell)M_AM_B-B_e(M_A+M_B)$$
$$+\gamma(\ell)P_AP_B-E_e(P_A+P_B)+2\varepsilon_J. \qquad [1]$$

In order to include the effect of the "Lorentz contraction" (4) we have only to replace $\lambda_J\rightarrow\lambda_J(1-v_L{}^2/c_J{}^2)^{1/2}$.

For a pair of opposite polarity fluxons we must make $M_A+M_B=P_A+P_B=0$, and have

$$U_{PJ\uparrow\downarrow}(\ell,v_L)=-\mu_f{}^2\beta(\ell)-(\varepsilon hv_L\Phi_0)^2\gamma(\ell)+2\varepsilon_J.$$

The maximum velocity of a fluxon in a junction may correspond to the maximum group velocity which is equal to the light velocity $c_J=(d/\Lambda\varepsilon\mu_0)^{1/2}$. Since $U_{PJ\uparrow\downarrow}$ is decreasing function on v_L^2, we have the minimum of the energy when $|v_L|=c_J$ getting

$$U_{PJ\uparrow\downarrow}(\ell,c_J)=-\mu_f^2\beta(\ell)-p_f^2\gamma(\ell)+2\varepsilon_J$$
$$=(4h\Phi_0^2/\pi^2\mu_0\Lambda\lambda_J)[1-(\pi/2)\operatorname{sech}(\ell/\lambda_J)]$$

with $p_f=\varepsilon c_J h\Phi_0$. It must be noted that $U_{PJ\uparrow\downarrow}(\ell,c_J)$ can be negative for $\ell<1.02\lambda_J$. According to Kulik we must replace $\lambda_J\to\lambda_J(1-v_L^2/c_J^2)^{1/2}$ for a fluxon in motion. However, it seems unphysical in real junctions that, in the limit $v_L\to c_J$, the fluxon size in the x direction vanishes, or that the self energy of the moving fluxon diverges. Therfore we put λ_J as it is in the equation above in taking the limit.

The nucleation energy of a pair of two fluxons with equal polarity moving with velocity $\pm c_J$ and keeping interfluxon distance ℓ is also obtained from Eq. [1];

$$U_{PJ\uparrow\uparrow}(\ell,c_J)=(4h\Phi_0^2/\pi^2\mu_0\Lambda\lambda_J)[1+(\pi/2)$$
$$\times\operatorname{sech}(\ell/\lambda_J)]-(2\mu_f/\Lambda)(\Lambda B_e\pm dE_e/c_J).$$

3. THE DESTRUCTION OF THE BOUND STATE BY EXTERNAL FIELDS

The transition between the two pair states $(\uparrow\downarrow)$ and $(\uparrow\uparrow)$ may take place when $U_{PJ\uparrow\downarrow}=U_{PJ\uparrow\uparrow}$. In the threshold of transition the following condition is found for the external fields.

$$\Lambda B_e\pm c_J^{-1}dE_e=(2\Phi_0/\pi\lambda_J)\operatorname{sech}(\ell/\lambda_J) \qquad [2]$$

It must be noted that phase velocity v_{ph} in a long junction with junction phase difference $\theta(x,t)$ is given by (2)

$$v_{ph}=(d\theta/dt)/(d\theta/dx)=E_{in}d/B_{in}\Lambda,$$

where E_{in} and B_{in} are the fields induced by the motion of phase singularity. On the other hand, E_e and B_e appearing in Eq. [2] is the externally applied fields.

It is known that the basic relations on fluxons in motion in a long junction is symmetrical for the replacement $x_J\rightleftarrows c_Jt$, $B\Lambda\rightleftarrows Ed$ and $\theta\rightleftarrows\pi+\theta$, where x_J is the position coordinate along L, B and E are fields in the junction. (4) Then instead of Eq. (2) we must use the following threshold condition.

$$\pm\Lambda B_e\pm c_J^{-1}dE_e=(2\Phi_0/\pi\lambda_J)\operatorname{sech}(\ell/\lambda_J) \qquad [3]$$

Since the pair $(\uparrow\downarrow)$ moving at the light velocity has the lowest nucleation energy with negative value when $\ell<\lambda_J$, the following properties can be expected on the bound state. i) Once a pair $(\uparrow\downarrow)$ is formed, it can not be destroyed by recombination or by thermal effect if external fields are within the threshold given by Eq. [3]. ii) No energy dissipation follows in the pair motion.

4. COMPARISON WITH EXPERIMENT

Equation [3] shows that the threshold magnetic field at $E_e=0$ is $\simeq\Phi_0/\Lambda\lambda_J$, which is several times larger than B_{c1J}. This means that intruding free fluxons will smear the effect due to the destruction of bound pairs by external fields in a long junction. On the other hand, a cylindrical junction with hollow has some merits in the observation of the effect. i) The intrusion of free fluxons are restricted owing to the shielding effect of outer electrode. ii) Fluxon pairs $(\uparrow\downarrow)$ can selectively be introduced in the junction region by way of flux trapping in the hollow of the structure. The trapped flux induces shielding current on the surfaces of two electrodes. With the help of thermal fluctuation the Lorentz force formed between the trapped flux and shielding current works to make fluxon pairs expending the trapped flux.

Kuwada et al. made an observation on the relation between voltage V=Ed and applied magnetic field H_z on hollow cylindrical junctions. (5) When trapping field $H_{z,t}$ is 1.357 Oe they found structures in V-H_z relation as shown in Fig.1.

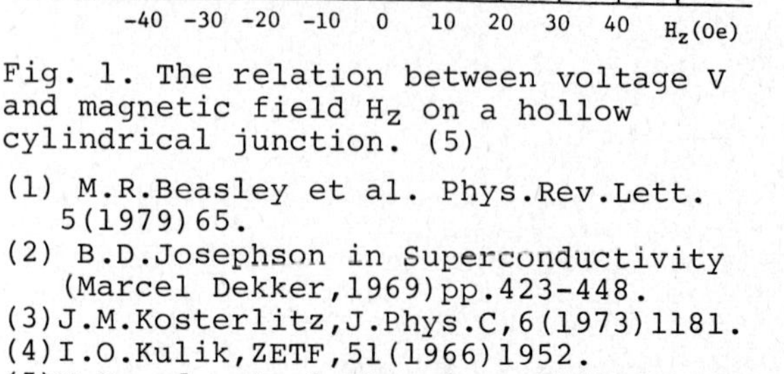

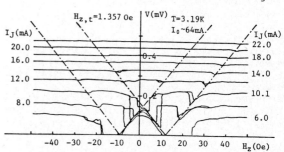

Fig. 1. The relation between voltage V and magnetic field H_z on a hollow cylindrical junction. (5)

(1) M.R.Beasley et al. Phys.Rev.Lett. 5(1979)65.
(2) B.D.Josephson in Superconductivity (Marcel Dekker,1969)pp.423-448.
(3) J.M.Kosterlitz,J.Phys.C,6(1973)1181.
(4) I.O.Kulik,ZETF,51(1966)1952.
(5) M.Kuwada et al.Phys.Rev.B27(1983)5486.

LT-17 (Contributed Papers)
U. Eckern, A. Schmid, W. Weber, H. Wühl (eds)

NONLINEAR PROPAGATION OF FLUXONS IN LONG, CROSS-TYPE JOSEPHSON JUNCTIONS

Roman SOBOLEWSKI, Piotr GIERŁOWSKI and Stanisław J. LEWANDOWSKI

Instytut Fizyki Polskiej Akademii Nauk, Al. Lotników 32/46, 02-668 Warszawa, Poland

Dynamics of fluxon motion in long, cross-type Josephson junctions has been investigated by measuring the flux-flow voltage across the junction. For fixed external parameters the resonant propagation was possible only in one direction of the junction. Propagation of fluxons was found to be highly nonlinear. The velocity of the flux flow was a nonlinear and multiple valued function of the external magnetic field. The limiting velocities of the flux motion are identified with phase velocities of corresponding transverse electromagnetic modes of the junction.

1. INTRODUCTION

Studies of the dynamics of magnetic flux propagation in long Josephson junctions have recently attracted much attention, partly because of possible computer applications (1). Physically the problem is similar to the resistive behavior of type-II superconductors (2). Most of the theoretical and experimental work has been restricted to the in-line geometry of the junction.

In this paper we show that in cross-line geometry the dynamics of the vortex motion is qualitatively different than in overlap junctions. The resonant vortex propagation is possible in only one direction of the junction and leads to significant asymmetry in the I-V characteristics. The flux-flow voltage lines are nonlinear even in the small voltage regime and several voltage lines exist simultaneously for each value of the external magnetic field, H_e. Each line represents a flux flow whose velocity is limited in the high field regime by the phase velocity of some electromagnetic mode of the junction. This latter phenomenon manifests itself as multiple-resonances of traveling waves, first observed by Jung (3).

2. EXPERIMENTAL PROCEDURES

Experimental data were obtained for Sn-Insulator-Sn Josephson tunnel junctions shown schematically in Fig. 1.

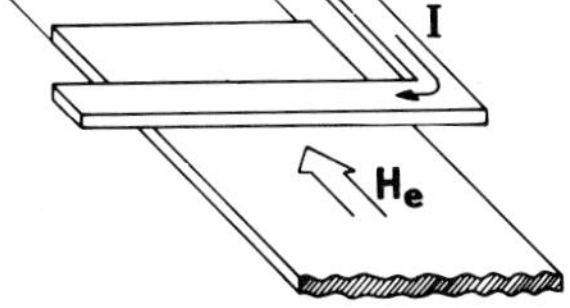

FIGURE 1
Geometry of a cross-type Josephson junction. An external magnetic field is applied in the direction of the arrow.

Junction width was equal to 0.06 mm. The length was equal either to 0.6 mm or to 0.8 mm. The junctions were fabricated essentially in the same manner as in Ref. 3. Some technological changes were introduced, however, in order to obtain very thin and uniform insulating layer (4). With H_e applied parallel to the long edge, the junctions exhibited high Q, multiple- resonances of standing waves and the critical current could be completely suppressed by H_e. For measurements we applied a similar technique to that reported by Nakajima <u>et al</u>. (5) and recorded the junction-voltage <u>vs the</u> external magnetic field (V-H characteristics) at different values of the bias current. The data were taken for the entire range of the bias currents up to voltages corresponding to the energy gap. We have also recorded I-V characteristics for fixed values of H_e.

3. EXPERIMENTAL RESULTS AND DISCUSSION

V-H characteristics presented in Fig. 2 show that with H_e increasing above H_{c1} the flux-flow voltage line splits into two different lines, which are nonlinear functions (a logarithmic curve represents the best fit) of the external magnetic field. The observed phenomena could be probably fully explained only by suitable computer simulations. Some of the complexities of the problem can be seen from the analysis carried out by Olsen and Samuelsen (6). The authors considered only the simplest possible model - the reflection of a single fluxon propagating in a Josephson line cavity influenced by an external magnetic field. Even in this case one is involved with an interaction of at least three solitons, one of them static and corresponding to Zharkov's solution (7), and the other two propagating in opposite directions. Our experimental conditions approach the limit of high H_e and high fluxon velocities (region V in Fig. 8 in Ref. 6) where, to quote the authors, "some highly nonlinear phenomena take place" resulting in the

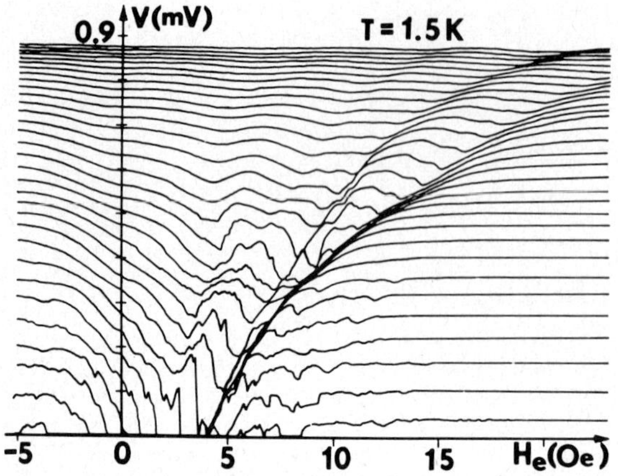

FIGURE 2
V-H characteristics of a 0.06 x 0.6 mm^2 junction. The curves are plotted for constant bias currents at 5 mA intervals up to 160 mA.

conversion of the incident fluxon into breather waves or its fission into a higher number of fluxons (6). Thus, it appears that we should describe our observations in terms of propagation of bundles of fluxons rather than in terms of single fluxon propagation.

Olsen and Samuelsen suggest that in the discussed region the mechanism responsible for the Fiske steps is likely to appear. The family of V-H characteristics presented in Fig. 3 seems to support this statement. The unusual dip observed in the current range 40 - 48 mA manifests itself on the I-V curve as a sharp current spike resembling a Fiske step, but with magnetic field dependent position. For higher bias currents this current spike evolves into a traveling-wave resonance.

In high voltage region the flux flow lines are continued as indentations in the V-H characteristics, which correspond on the I-V characteristics to traveling-wave resonance peaks (c.f. Ref. 5). Several (up to 4) such traveling wave resonance branches can be distinguished. In this region the number of fluxons participating in the flow is increased to such a degree that the flow can be considered as a propagation of current waves. The maximum phase velocities of these are equal to the electromagnetic wave propagation velocity - the usual condition of the traveling-wave resonance (8). The separate branches of traveling-wave resonances would correspond in this picture to different transverse modes of electromagnetic propagation. Note that in this high frequency region nonlinearities follow from dispersion and changing penetration depth.

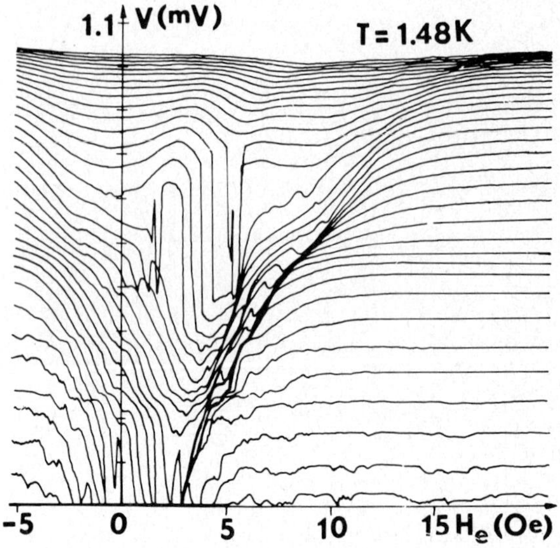

FIGURE 3
V-H characteristics of a 0.06 x 0.8 mm^2 junction The curves are plotted for constant bias currents at 2 mA intervals up to 70 mA.

Another characteristic feature of the presented V-H curves is the strong asymmetry with respect to H_e. We ascribe it to our cross-line geometry in which one of the junction edges carries more current than the other and the junction boundary conditions are asymmetric (9). Observe that such boundary conditions may even further complicate the fluxon propagation.

ACKNOWLEDGEMENTS
The authors are indebted to Dr. J. Zagrodziński for many very valuable discussions.

REFERENCES
(1) T.V. Rajeevakumar, IEEE Trans. Magn. MAG-17 (1981) 591.
(2) I.O. Kulik, Zh. Eksp. Teor. Fiz. 51 (1966) 1952 [Sov. Phys. JETP 24 (1967) 1307].
(3) G. Jung, J. Appl. Phys. 53 (1982) 576.
(4) P. Gierłowski, S.J. Lewandowski, V. Miliaiev, A.V. Shirkov and R. Sobolewski, to be published.
(5) K. Nakajima, H. Ichimura and Y. Onodera, J. Appl. Phys. 49 (1978) 4881.
(6) O.H. Olsen and M.R. Samuelsen, J. Appl. Phys. 52 (1981) 6247.
(7) G.F. Zharkov, Zh. Eksp. Teor. Fiz. 71 (1976) 1951 [Sov. Phys. JETP 44 (1976) 1023].
(8) R.E. Eck, D.J. Scalapino and B.N. Taylor, Phys. Rev. Lett. 13 (1964) 15.
(9) A. Barone, W.J. Johnson and R. Vaglio, J. Appl. Phys. 46 (1975) 3628.

LT-17 (Contributed Papers)
U. Eckern, A. Schmid, W. Weber, H. Wühl (eds)

LOCALIZATION OF FLUXONS ON INHOMOGENEITIES IN LONG JOSEPHSON JUNCTIONS

Yu.S. GALPERN, A.N. VYSTAVKIN

Institute of Radio Engineering and Electronics, USSR
Academy of Sciences, Marx av. 18, GSP-3, Moscow 103907, USSR

Some possibilities of observing effects of fluxon bound states localized on microinhomogeneities in Josephson transmission lines are discussed. Simple methods for generating such states are proposed, and direct ways of observing them by using a laser scanning technique are considered. A possibility of inducing transitions between the bound states near bifurcation points in extertal magnetic fields is pointed out.

In Josephson transmission lines (JTL), i.e. in Josephson junctions with $l \gg l_j \gg w$ (Fig. 1), where l_j is the Josephson length, there exist non-trivial effects of interaction of fluxons with inhomogeneities inside the junction. The most interesting effects

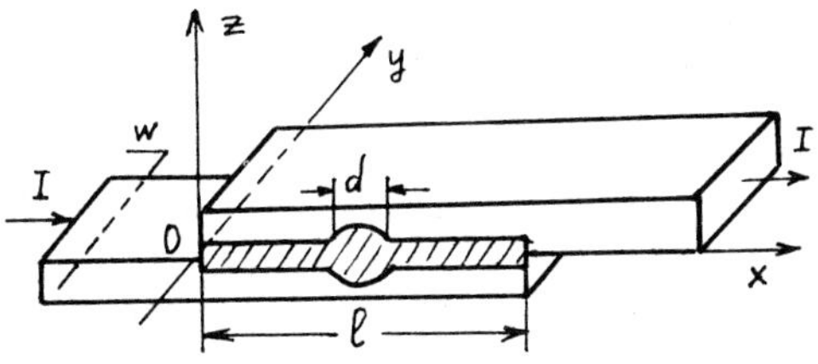

Fig. 1

occur if the fluxons interact with small ($d < l_j$) inhomogeneities attracting the fluxons, we call them "microresistances" (MR), in contrast to repulsive "microshorts" (MS). In JTL with MR the magnetic flux can form stable static bound states (BS) localized on inhomogeneities[1-3]. Any bound state can be characterized by its total flux $F_B \equiv nF_0$ (F_0 is a quantum of the magnetic flux, n is ingeneral non-integer), total energy E_B, and fundamental frequency, f_B, of its pulsations in response to small external perturbations. The number of the stable BS depends on the number of inhomogeneities, their dimensions, d, their positions, and on the external magnetic field h. If the external parameters p (such as h, positions of MR, etc.) are changed adiabatically, any stable BS deforms smoothly as long as $f_B(p)$ is positive. If, for some critical values $p = p_c$, the fundamental frequency $f_B(p)$ vanishes, the number of the bound states changes abruptly. This is called a bifurcation phenomenon. At the bifurcation point $p = p_c$ the state of JTL can be rapidly switched by a slow change in p. These results have been obtained in.[1-3] The usual soliton perturbation theory applied earliar for treating the interactions of fluxons with microshorts[4-6] is inapplicable to the case of MR due to strong deformations of the fluxons. In view of this, a special bifurcational perturbation theory has been developed in[3] providing a description of the bound states near the bifurcation points.

Here we consider the problem of the experimental observation of bound states. The simplest realization of the MR is provided by a local thickening of the dielectric layer in the JTL as shown in Fig. 1. Near such an inhomogeneity the Josephson current $j(x)$ can be approximated by $j(x) = j_c \mathrm{th}^2 [2(x-x_0)/d]$. If $d \ll l_j$, the simpler approximation is possible: $j(x) = j_c(1 - d\,\delta(x-x_c))$.[1-3] The numerical calculations of the bound states and their bifurcations show that both approximations are practically equivalent even for $d < l_j/2$ as long as the inhomogeneities are not too close to each other and to the edges of the JTL. To create a bound state on the MR, one may form (e.g. by a pulse of the external current I) a fluxon-like pulse near the edge of the junction. If the flux and the energy of this pulse are greater than F_B and E_B respectively, it will move to the MR eventually forming the bound state.[8] The relaxation time of this process strongly depends on the dissipative effects related to the normal tunnelling, on the parameters of the bound state, and on the initial pulse. It can be calculated only in numerical experiments.[8]

A simplest direct way of observing the bound states is provided by a technique of scanning a soft focused laser beam along the JTL.[7] If the BS is rigid enough (i.e. its fundamental frequency f_B is comparable to the Josephson frequency f_j), the perturbation caused by the laser beam will not significantly change the initial current distribution and one can compare its form with the one theoretically predicted. By using an additional, stronger laser beam a MR with controllable position can be created. This gives an opportunity to observe evolution of BS upon a slow change of the position of the MR. The scanning technique can also be used as a tool for diagnostics of JTL with inhomogeneities. Each JTL has an individual "portrait" of the bound states, and having a rich enough collection of experimental current distributions one can, in principle, solve the inverse problem - to find the parameters characterizing inhomogeneities.

Finally, the most interesting problem is to observe the phenomena occurring near bifurcation points. Considering for simplicity a semi-infinite junction with an MR at $x = x_o$ one can study a dependence of the BS on the magnetic field h_o at the edge $x = o$. The dependence of the spectrum of the bound states on h_o is shown in Fig. 2.

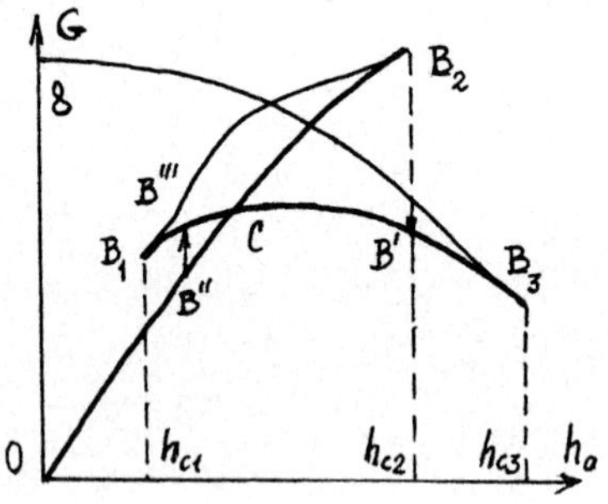

Fig. 2

We here use, instead of the energy E, the generalized energy $G = E + 2\pi h_o (1 - F_B/F_o)$ which includes the boundary energy. The branch OB_2 corresponds to stable BS having small F_B. (Fig.3a). The branch B_1B_3 represents stable BS with large F_B, this is simply a fluxon deformed by the external field h_o and by interaction with the MR (Fig. 3b)

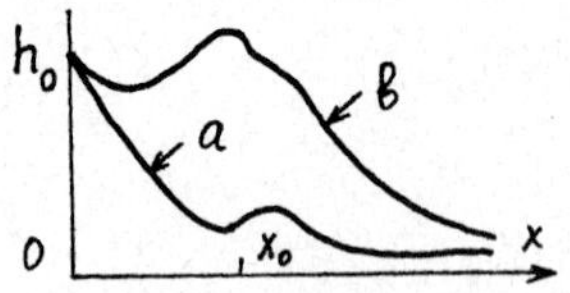

Fig. 3

The branches B_1B_2 and O^1B_3 correspond to unstable states with imaginary f_B, their lifetime is $|f_B|^{-1}$. The critical values h_{c1}, h_{c2}, and h_{c3} correspond to bifurcation points. Suppose the initial value of h_o is 0 and the junction is in the ground state. If h_o is slowly increasing from 0 to h_{c2} the enegy G is slowly growing from 0 to G_{c2} (on the curve OB_2). For $h_o \gg h_{c2}$ the junction rapidly changes its state, and after dissipating the redundant energy $G_{c2} - G^1$ it will eventually reach the state B' on the curve B_1B_3, which now is the ground state. Another way to reach the curve B_1B_3 is depicted by the arrow $B''B'''$ in Fig. 2. It requires supplying the junction with some amount of energy. Existence of the two easily controllable stable states (bistability) is of a significant importance for applications. Finally, note that the number of stable BS is rapidly growing with the number of MR.

We are indebted to Dr. A.T. Filippov for invaluable collaboration.

REFERENCES

(1) Yu.S. Galpern, A.T. Filippov, JETP Lett. 35 (1982) 580.
(2) A.T. Filippov, Yu.S. Galpern, Solid State Comm. 48 (1983) 665.
(3) Yu.S. Galpern, A.T. Filippov, JETP 36 (1984) 1527.
(4) R.D. Parmentier, in "Solitons in Action", Acad. Press, N.Y. (1978).
(5) T.A. Fulton et al., Proc. IEEE 61 (1973) 28
(6) M.B. Fogel et al., Phys. Rev. B 15 (1977), 1578.
(7) M.Scheuermann et al. Phys. Rev. Lett. 50 (1983) 74.
(8) G.S. Kazacha, S.I. Serdyukova, A.T. Filippov, JINR communication P-11-76, Dubna (1984)

LT-17 (Contributed Papers)
U. Eckern, A. Schmid, W. Weber, H. Wühl (eds)

ON THE MAGNETIC FIELD DEPENDENCE OF THE SUPERCURRENT IN A JOSEPHSON JUNCTION OF INTERMEDIATE SIZE

J. Bindslev Hansen

Physics Laboratory I, H.C. Ørsted Institute, University of Copenhagen, DK-2100 Copenhagen, Denmark

G.F. Eriksen, J. Mygind, M.R. Samuelsen

Physics Laboratory I, The Technical University of Denmark, DK-2800 Lyngby, Denmark

S.A. Vasenko

Department of Physics, Moscow State University, 117234 Moscow, USSR

The maximum supercurrent through a long and narrow Josephson Nb-Nboxide-Pb tunnel junction is measured as a function of applied magnetic field. The observed rounding of the central (zero-vortex) lobe is shown to be in agreement with numerical calculations based on the sine-Gordon equation. As a consequence the critical current density and the Josephson penetration depth can be determined in the overlap geometry from the $I_c(B)$ pattern at all normalized lengths.

Numerical calculations of the maximum supercurrent through long ($L >> \lambda_J$) and narrow ($W << \lambda_J$) Josephson tunnel junctions in an applied magnetic field are used to determine the junction parameters. They are carried out for an overlap junction with a bias current distribution as the one in a long thin superconducting strip, and they are based on the modified sine-Gordon equation which is known to describe the static properties of the junction. Lengths are normalized to the Josephson penetration depth $\lambda_J = (h/4\pi\mu_o eJd)^{\frac{1}{2}}$, where J is the maximum Josephson current density and d the effective magnetic thickness. The magnetic field is applied in the plane of the junction perpendicular to the long direction.

For short junctions ($L << \lambda_J$) this curve has the well-known (Fraunhofer) diffraction pattern shape: $|\sin z/z|$, $z = \pi\phi/\phi_o$, where ϕ is the applied magnetic flux and $\phi_o = h/2e$ is the flux quantum. The $I_c(B)$ curves computed (1) using the realistic non-uniform bias current distribution exhibit a "rounding" of the peak of the central lobe around B=0. The depression of $I_{co} = I_c(0)$ was found to increase with increasing length. For ℓ = 5, 10, and 20 the relative reduction $\Delta I_{co}/JWL$ was predicted to be 4%, 24%, and 47%, respectively. In Fig. 1 this is shown together with the results obtained for I_c^*, where I_c^* is the intercept on the I_c-axis when the linear part of the $I_c(B)$ curve is extrapolated back to B=0.

The $I_c(B)$ curves representing the threshold for the zero-vortex state defines a magnetic induction B_{c1} through $I_c(B) = 0$. In Fig. 2 we have plotted B_{c1} normalized to $\mu_o\lambda_J J$ versus ℓ (2).

Two Nb-Nboxide-Pb tunnel junctions with relatively large excess currents were fabricated less than 2 mm apart in the same process on a glass substrate. The Nb film (3500Å) was dc-magnetron sputtered, whereas the Pb film (4500Å) was thermally evaporated. The oxide was grown in a rf-glow discharge. The longer junction had an overlap geometry: W×L=25.5×396.0 μm^2.The shorter junction (22.0×18.5 μm^2) was intended as a reference junction. The sample was mounted in a temperature-stabilized MIC box inside a vacuum can that was immersed in liquid helium. Careful electrical and magnetical shielding protected the sample from external noise sources. A long calibrated solenoid could produce a magnetic field in the plane of the oxide barrier. For each junction at a fixed temperature the dc IV curve was recorded and the $I_c(B)$ curve was traced out point by point using a storage

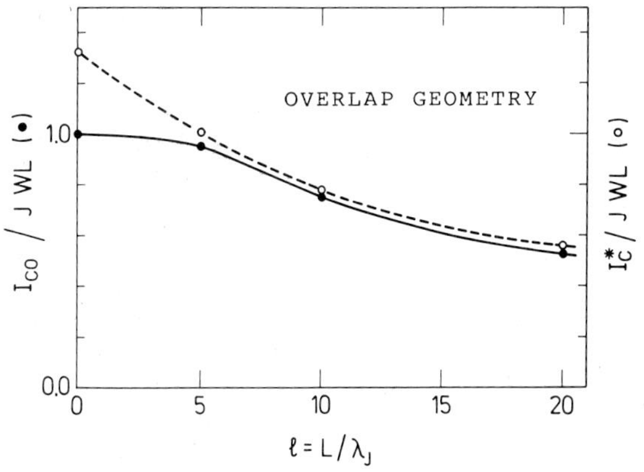

FIGURE 1

The normalized maximum supercurrent I_{co}/JWL and I_c^*/JWL versus the normalized length $\ell = L/\lambda_J$. Numerically except for ℓ = 0.

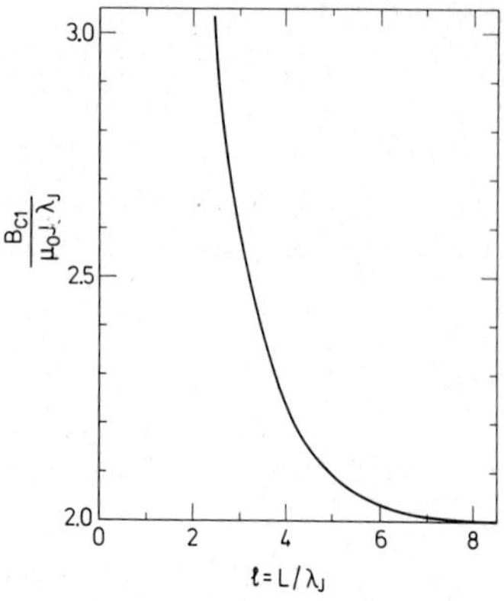

FIGURE 2

The normalized magnetic field $B_{c1}/\mu_0 J\lambda_J$ (=2/k) versus the normalized length $\ell = L/\lambda_J$ (=2kK(k)).

oscilloscope. For the short junction ($\ell \sim 0.2$) the $I_c(B)$ curve closely reproduced the theoretical Fraunhofer pattern. For the long junction part of the $I_c(B)$ curves at two different temperatures are shown in Fig. 3. The rounding of the peak of the central (zero-fluxon) lobe is clearly seen. The experimental values I_{co}, I_c^*, B_{c1}, L, W and the Figs. 1 and 2 determine λ_J and J completely. The measured and the calculated values at four temperatures are collected in the Table. The magnitude and temperature dependence of the measured reduction agree with the calculated values. For the short junction at 4.2 K we found d = 136 nm in agreement with the value found by other workers for similar (short) Nb-Pb junctions, whereas for the long junction, the value of d is about 30% lower. Both junctions exhibited a slight ageing effect (3). When stored at room temperature over a period of one month the critical current decreased by 4% and 1.4% for the short and the long junction, respectively. The IV curves were typical for Nb-Nboxide-Pb junctions (3). Comparing the two junctions we found that their quasiparticle curves scaled perfectly (within 1%) although their pair currents after scaling deviated by $\sim$10%, the long junction having the smaller value of I_{co} (scaled). The critical current density of the short junction was, however, 30% lower than that of the long junction implying that the short junction had a reduced effective area when compared with the long junction. We think this is due to edge effects (3), the short junction having a higher edge/length ratio than the long one.

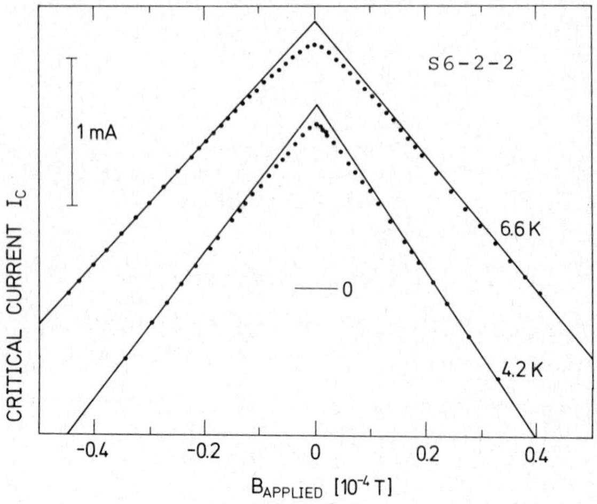

FIGURE 3

Central lobe of the measured $I_c(B)$ curves for an intermediate length ($\ell \sim 5$) junction at 6.6 K and 4.2 K. Note: the curves are reproduced with a current off-set of 0.72 and 3.82 mA, respectively.

CONCLUSIONS

It is shown that the Josephson critical current density J and penetration length λ_J can be determined from the magnetic interference pattern in overlap geometry tunnel junctions independent of the normalized length. This method is of special importance in junctions with edge defects, excess currents, and strong coupling effects.

T [K]	2.71	4.23	5.59	6.58
I_{co} [mA]	5.50	4.93	3.83	2.38
B_{c1} [10^{-4}T]	1.027	0.939	0.784	0.579
$\Delta I_{co}/I_c^*$ (meas.) %	2.6±0.8	2.7±0.3	4.9±0.5	7.8±0.4
$\Delta I_{co}/I_c^*$ (cal.) %	3.6±1.5	4±2	4±2	6±2
J [A/cm^2]	58.9	52.4	39.9	24.5
λ_J [μm]	68.0	69.7	75.4	87.9
d [nm]	96.3	103	116	139
ℓ	5.82	5.68	5.25	4.51

REFERENCES

(1) M.R. Samuelsen and S.A. Vasenko, this vol.
(2) C.S. Owen and D.J. Scalapino, Phys.Rev. 164 (1967) 538.
(3) A. Matsuda, T. Inamura, and H. Yoshiko, J. Appl. Phys. 51 (1980) 4310.

LT-17 (Contributed Papers)
U. Eckern, A. Schmid, W. Weber, H. Wühl (eds)
© Elsevier Science Publishers B.V., 1984

I-V CHARACTERISTICS OF TWO-DIMENSIONAL JOSEPHSON TUNNEL JUNCTIONS CONTAINING MANY VORTICES

I.P. NEVIRKOVETS, E.M. RUDENKO

Institute of Metal Physics, Ukrainian Academy of Sciences, Kiev, USSR

I-V characteristics of long asymmetrically biased tunnel junctions Pb-I-Pb and Sn-I-Sn have been measured. In the case of very low-resistance junctions with a large number of vortices the strong asymmetry in I-V curves has been found for different directions of the magnetic field. In one of the field directions there appears almost vertical resistive branch, but it is absent for the opposite field direction.

The processes of vortex motion in distributed Josephson junctions are of both theoretical and practical interest due to the possibility of their application in superconducting electronics. Functioning of devices similar to that described in Ref. (I) is based on I-V characteristics of long Josephson junctions. Resistive branches may appear on the I-V curves of such junctions as nearly normal current steps due to the motion of vortices when they attain the velocity $\bar{c}$ of electromagnetic waves in the junction. The voltage position V of resistive branch may be varied with magnetic field induced by the control electrode. For the quasi-one-dimensional junctions used in (I) and in the case of a symmetric current application the I-V curves were symmetric for two opposite directions of the magnetic field H (or, the same, for the opposite directions of a tunnel current with the same direction of the magnetic field).

It has been found that this is not the case with the asymmetric current application where the I-V curve shape may differ essentially for different directions of the magnetic field (2). Asymmetric current application is most easily implemented for two-dimensional cross-like tunnel junctions (TJ) where one of dimensions, $L > \lambda_J$ (see the inset in Fig.I). We have studied the TJ Sn-I-Sn and Pb-I-Pb of such a shape by varying the magnitude of L and w in the range $L, w \geqslant \lambda_J$ to $L \gg \lambda_J$. The magnetic field was applied normally to L dimension in two opposite directions.

Fig.I shows I-V curves of the TJ Pb-I-Pb at T=4.2 K. The TJ had the tunnel resistance $R_t \sim 10^{-5}\ \Omega\cdot cm^2$. The Josephson penetration depth was λ_J=0.003 mm. The junction L dimension was not much more than λ_J. As one can see, in this case the I-V curves containing resistive branches, display a weak asymmetry in the magnetic field. This asymmetry seems to be due to the effect of the magnetic field induced by its own cur-

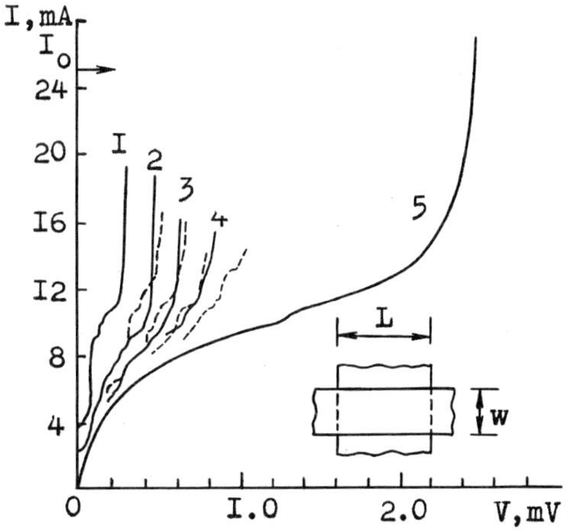

FIGURE I
I-V curves of a Pb-I-Pb TJ at T=4.2 K. Solid lines are characteristics for one of the magnetic field directions, normal to L. Curves I through 5 corresponds respectively to magnetic fields of 4.8, 6.4, 8.0, 9.6, 32 G. Dashed lines refer to magnetic fields for the cases I-4 but in opposite direction.

rent. The situation changes essentially when the barrier thickness is decreased so that the junction comprises a very large number of vortices. Actually, the vortex characteristic size λ_J diminishes with the j_o growth (j_o is the critical current density of dc Josephson current): $\lambda_J=(\Phi_o/2\mu_o j_o d)^{I/2}$ where Φ_o is the magnetic flux quantum; $d=t+2\lambda$; λ is the London penetration depth. We hawe found that in this case the asymmetry becomes rather significant and is especially well observed on the I-V curves of the TJ Sn-I-Sn. Qualitatively the same picture is also observed for very low-resistance TJ Pb-I-Pb. These junctions, however, have significant leakage currents at small barrier thickness, this perhaps being the cause of negligible magnitude of a dc component in Josephson current for such junctions at $V\neq0$. Contrary to Pb junctions, the TJ on tin base allows an easy decrease in R_t to $IO^{-7}\,\Omega\cdot cm^2$ with barrier homogeneity remaining rather good.

Fig.2 shows I-V curves for TJ Sn-I-Sn with the tunnel resistance $R_t\sim IO^{-6}\,\Omega\cdot cm^2$ at T=2.I K. The junctions dimensions were such that $L\gg\lambda_J$ and $w\geqslant\lambda_J$. The

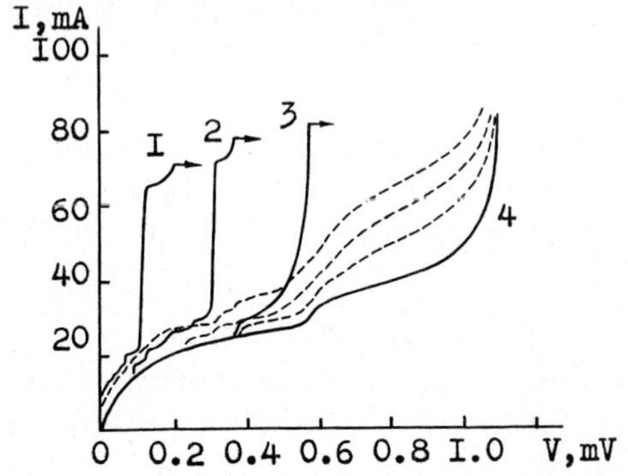

FIGURE 2
I-V curves of Sn-I-Sn TJ at T=I.8 K. The almost vertical resistive branch can be seen for one of the field directions. Curves I-4 corresponds respectively to magnetic fields of 4.8, 6.6, 9.0, 30 G. Dashed lines refer to the cases I-3 but in the opposite field direction.

external magnetic field was applied normally to L dimension. As one can see, in one of the magnetic field directions (opposite to the corresponding component of the magnetic field induced by tunnel current the almost vertical resistive branch in the I-V curve is observed which is absent when the field is applied in the opposite direction.

We conclude that the I-V curves strongly asymmetric in the magnetic field appear for the junctions with $L\gg\lambda_J$, $w\geqslant\lambda_J$ and at current concentration on the junction edge typical for distributed junctions with asymmetric current application (3). On the junction edge every new vortex which enters the TJ under the external field, is subjected to Lorentz force from the tunnel current. If the direction of this force is such that it helps the vortex entrance then the vortex will not be able to accelerate efficiently because of repulsive action of numerous vortices which are already present in the TJ. But for the opposite direction of Lorentz force it will be efficiently accelerated now experiencing no resistance. When the vortex attain the velosity $v=\bar{c}$, the current peak appears at $V=(td/\epsilon)^{I/2}H$ (4), where ϵ is the dielectric constant of the oxide layer. But estimation has given the value $v=\bar{c}/2$, perhaps due to the Lorentz contraction (2,5).

The voltage position of the resistive branch in I-V curve (Fig.2) may be controlled by a change in the current passed through the narrow film which is equivalent to application of external magnetic field. Such a system functions similarly to that described in (I).

The authors are thankful to K.K. Likharev for a helpful discussion.

(I) T.V. Rajeevakumar, Appl. Phys. Lett. 39 (I98I) 439.
(2) I.P. Nevirkovets, E.M. Rudenko, Fiz. Nizk. Temp. IO (I984) 56.
(3) A. Barone, F. Esposito, K.K. Likharev et al., J. Appl. Phys. 53 (I982) 6802.
(4) R.E. Eck, D.J. Scalapino, B.N. Taylor, Phys. Rev. Lett. I3 (I964) I5.
(5) T.V. Rajeevakumar, J.X. Przybysz, J.T. Chen, Phys. Rev. B 2I (I980) 5432.

LT-17 (Contributed Papers)
U. Eckern, A. Schmid, W. Weber, H. Wühl (eds)

MAXIMUM CURRENT AMPLITUDE OF ZERO-FIELD SINGULARITIES IN LONG JOSEPHSON JUNCTIONS

Matteo CIRILLO, Sandro PACE, and Bonaventura SAVO

Dipartimento di Fisica, Università di Salerno, 84100 Salerno, Italy *

Fluxon reflections at the edges of long Josephson junctions could be a determining physical phenomenon limiting the maximum current amplitude of zero-field singularities. Using a mechanical analog, phase oscillations generated by 2π-kink reflections are studied as a function of the bias current and of the length of the junction. Critical conditions for stable fluxon oscillations are found.

1. INTRODUCTION

Zero-field singularities (ZFS) appear in the D.C. current-voltage (I-V) characteristics of Josephson tunnel junctions long with respect to the Josephson penetration depth (λ_j). ZFS show the maximum current amplitude in zero external magnetic field and are explained in terms of stable fluxon oscillations. The fluxons travel with a velocity $\upsilon \leq c$ (the electromagnetic wave velocity in the junction), which corresponds to resonant voltages V_n, where n is both the order number of ZFS and the number of fluxons in the junction. ZFS have been studied by numerical computations (1), perturbation theory (2), approximate analytical solutions (3), and mechanical analog simulations (4). In spite of these numerous different approaches some important features have not been explained. In particular the maximum current amplitude I_n is still an unsolved problem. In niobium-lead junctions I_1 is typically about 0.5 - 0.8 times the maximum Josephson current I_j. Numerical computations (1) give values of I_1 similar to the experimental ones, but they have not yet given informations on the processes determining the amplitude of the singularities. Perturbation theory (2) fits in a reasonable way the shape of the singularities, but this power balance analysis gives a divergent value of the current as the voltage approaches the resonant value. In this paper the instability conditions for periodic fluxon oscillations are briefly discussed. Experimental data relative to 2π-kink reflections on a mechanical analog are reported. These oscillations seem to be the determining phenomenon limiting the ZFS maximum current.

* Work partially supported by C.N.R. - G.N.S.M.

2. INSTABILITY CONDITIONS FOR FLUXON OSCILLATIONS

In this paper we restrict our analysis to one-dimensional overlap junctions long in the direction perpendicular to the bias current. This junctions are described by the sine-Gordon equation with bias and losses. In the absence of external magnetic field a reasonable ansatz for the phase $\Phi(x,t)$ describing fluxon oscillations, is given by (3):

$$\Phi(x,t) = \arcsin\gamma + \Phi^*(x,t)$$

where $\Phi^*(x,t)$ describes the oscillating fluxon, and $\arcsin\gamma$ is the constant value of the phase in the portion of the junction in which the fluxon is not present. In long junctions $\arcsin\gamma$ is equal to the static solution. For quadratic losses (3) this approach gives the relation between γ and υ, so that the I-V characteristic of ZFS can be fitted (6). Finite values of I_1 are given, but the experimental values are lower than the theoretical ones. Experimental data (5) clearly show the existence of a threshold value γ^* of the bias current density, so that for $\gamma \geq \gamma^*$ fluxon oscillations are unstable. Moreover for long junctions γ^* is independent of the length. All that makes reasonable the hypothesis that the instability problem does not arise inside the junction but at the edges, where reflections occur. This hypothesis is completely confirmed by experimental data taken on a mechanical analog. The details of the analog are described in reference (4): in the standard version it is composed of 24 elestically coupled units and it has a length $L = 7\lambda_j$. Each unit is biased by an external torque supplied by compressed air. The torque, which corresponds to the bias current density in the junction, is measured by the pressure value P. Increasing the bias, transitions

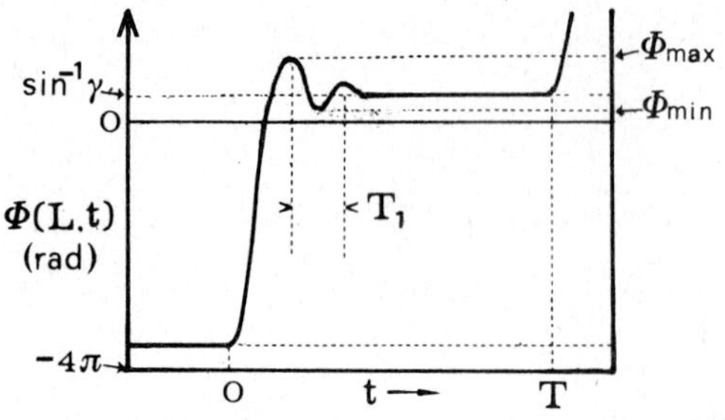

FIGURE 1
Qualitative behavior of the phase as a function of the time in one edge (x=L)of the analog.

from ZFS to the zero-fluxon state occur after a reflection. For bias lower than the critical one, it is easy to observe large oscillations arising at the edges of the mechanical analog after a reflection (fig.1). Oscillations are localized in a neighborhood of the edge and are damped by the losses. The amplitude A of the first oscillation increases with the fluxon velocity. A diverges as the bias approaches the critical value P*, that is for P > P* a continuous increasing of the phase (i.e. the transition to the zero fluxon state) occurs.

3. RESULTS AND COMMENTS

The figs. 2a,b show in detail the data of the oscillations taken on the mechanical analog consisting respectively of N = 7, 24 coupled units. As a function of the external pressure P, Φ_{max} , Φ_{min} , arcsin γ_o are reported. As indicated in fig.1, Φ_{max} is the maximum value of the phase at one edge of the analog after a reflection, Φ_{min} is the minimum value of the phase in the first oscillation, and arcsin γ_o is the static equilibrium value. Let us consider first the case N = 24. For P $\leq$ 1 atm oscillations are completely neglegible: after a reflection the phase reaches the value Φ = arcsin γ which is equal to the static value. For 1 atm < P < 2.2 atm Φ_{max} increases with P, Φ_{min} is not monotonic decreasing. Since the period T_1 and the damping time of the oscillations are smaller than the period T = 2L/v, the interactions between the oscillations and the fluxon are neglegible. An analogous behavior can be observed for N=10,12,16,20. For N $\geq$ 10 the critical value of the bias is independent of N; the instability occurs for P*$\simeq$ 2.1 - 2.2 atm corresponding to a value of $\Phi \simeq 31°$-$33°$. For N < 8 the behavior is in some way different. As shown in fig. 2b for N=7 we have $\Phi_{max} \geq \Phi_{min}$ > arcsin γ : after a reflec-

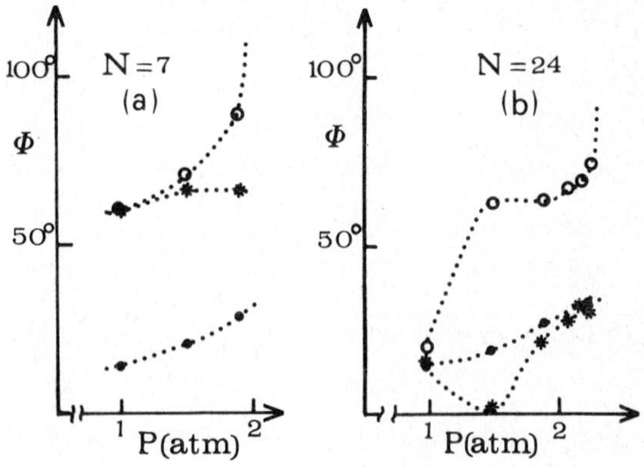

FIGURE 2
Pressure dependence of Φ_{max} (o), Φ_{min} (*), arcsin γ_o(•) for different length (number of sections) of the mechanical analog.

tion the phase reaches a quasi static equilibrium value greater than the constant static value. For N < 7 the period T has a value similar to T_1. In this conditions the interaction of the oscillating fluxon with the edge oscillations can play some determinant role in the fluxon dynamics. In any case Φ_{max} diverges as P approaches a critical value P* but, for N < 8, P* is a decreasing function of N: for N = 7 we have P* = 2.0 atm (Φ = 29°).

In conclusion the data taken on the mechanical analog of a long Josephson junction clearly show the processes determining the maximum height of the first ZFS. Large phase oscillations are generated at the edges of the junctions by the reflections of the traveling fluxons. These oscillations make unstable the periodic fluxon propagation when the bias reaches a critical value.

REFERENCES

(1) P.S. Lomdahl, O.H. Soerensen, and P.L. Christiansen,Phys. Rev. B25(1982) 5737.
(2) N.F. Pedersen, and D. Welner, Report 253 Lyngby, Denmark (1983); to be published on Phys. Rev. B
(3) S. Pace, B. Savo, and R. Vaglio, Phys. Lett. A, 82A (1981) 362.
(4) M. Cirillo, R.D. Parmentier, and B. Savo, Physica 3D (1981) 565.
(5) A.M. Cucolo et al., Physica 107 B+C(1981)547.
(6) M. Cirillo, S. Pace, and B. Savo, Shape of zero-field singularities in long Josephson junctions, internal report, Dipartimento di Fisica, Università di Salerno, Italy (1984).

LT-17 (Contributed Papers)
U. Eckern, A. Schmid, W. Weber, H. Wühl (eds)

PERTURBED FLUXON STATE AND CURRENT STEP IN JOSEPHSON JUNCTION

Kazuyuki WATANABE and Chikara ISHII

Department of Physics, Science University of Tokyo, Shinjuku-ku, Tokyo 162, Japan

The properties of zero field steps in a relatively long Josephson junction are studied on the basis of fluxon oscillation picture. Modified fluxon solution of barrier equation in the presence of magnetic field as well as the resistive current is obtained. This is used to calculate the dependences of step height of zero field step on the magnetic field and the quality factor, yielding the results consistent with experiments and prior theories.

The resonant step structures such as the Fiske steps(FS)[1),2)] and zero field steps(ZFS)[3)] have been receiving a continuous interest both theoretically and experimentally. Two lines of approach have been proposed for the theoretical interpretation of these steps. The first is the cavity mode picture[4)-6)] which is able to explain most of the observed features of the resonant steps in a junction with length shorter than the Josephson penetration length λ_J. On the other hand, it has been suggested that ZFS in a longer junction come out as a result of periodic propagation of soliton like nonlinear waves[7),8)] (to be called fluxons). Recently a half fluxon wave[9),10)], a combined sequence of fluxon and plasmon like motion in a period, has been found in the numerical simulations for the nonlinear waves in a long junction in the presence of the magnetic field. This is expected to provide the origin of the FS in this system. It should be noted, however, that the detailed properties of the resonant steps (their dependence on the applied magnetic field or on the temperature) have not been examined on the basis of the fluxon oscillation picture.

With this motivation, we try here to investigate how the characteristics of ZFS in a relatively long junction in a magnetic field may be derived in the framework of the fluxon oscillation model. To do this, we start from the barrier equation (perturbed sine-Gordon equation),

$$\phi_{xx}-\phi_{tt}-\sin\phi=\alpha\phi_t-\gamma\ , \qquad (1)$$

for the flux ϕ ($\phi_t=\partial\phi/\partial t$, etc.). Here we are considering a junction of length L, extending x direction, and space, time, and flux variables are scaled by corresponding characteristic quantities: Josephson penetration length and plasma frequency, and flux quantum Φ_0. Further, γ is a bias current scaled by maximum Josephson current, and the reduced dissipation coefficient α is related to the junction quality factor Z_1 by $\alpha=L/\pi Z_1$[6)]. The eq.(1) should be solved under the boundary conditions

$$\phi_x(0,t)=\phi_x(L,t)=\frac{2\pi}{L}X\ , \qquad (2)$$

in the presence of the applied magnetic field. It is well known that the unperturbed barrier equation (eq.(1) with $\alpha=\gamma=0$) with open end boundary conditions (eq.(2) with $X=0$) has the analytical fluxon solution $\phi^f(x,t)$[11)]. Making use of this solution, we try to construct the solution of eq.(1) in the form

$$\phi(x,t)=\phi^f(x,t)+\phi^b(x)+\tilde{\phi}(x,t)\ , \qquad (3)$$

where $\phi^b(x)$ is introduced to satisfy the boundary conditions (2), and it represents a pair of virtual static fluxons outside the junction[12)].

$$\phi^b(x)=4\tan^{-1}[\exp(x+\xi)]-2\pi+4\tan^{-1}[\exp(x-L-\xi)]. \qquad (4)$$

Here the parameter ξ is a function of the applied field, and determined self-consistently from eq.(2). The function $\tilde{\phi}(x,t)$ appears as the result of the presence of perturbations and/or the applied field. Since this obeys the open end boundary conditions, we expand it in the form

$$\tilde{\phi}(x,t)=\sum_n\phi_n(t)\cos(k_nx),\quad k_n=\frac{n\pi}{L}\ . \qquad (5)$$

The eq.(1) is rewritten as an equation for $\phi_n(t)$

$$\phi_{n,tt}+\alpha\phi_{n,t}+k_n^2\phi_n=S_n(t)\ ,$$
$$S_n(t)=-2\langle\cos(k_nx)[\sin(\phi^f+\phi^b+\tilde{\phi})-\sin\phi^f-\phi^b_{xx}+\alpha\phi^f_t-\gamma]\rangle, \qquad (6)$$

where the symbol $\langle\cdots\rangle$ stands for a space average. Hereafter we neglect $\tilde{\phi}$ in the source term $S_n(t)$, assuming small perturbations and weak magnetic field. Then the source term $S_n(t)$ becomes a periodic function of time with fundamental frequency $\omega=\pi V_r/L$ (V_r is the average fluxon velocity[13]) and we can assume the following form of ϕ_n

$$\phi_n(t)=\frac{1}{2}a_n^0+\sum_m[a_n^m\cos(\omega_m t)-b_n^m\sin(\omega_m t)] \quad . \qquad (7)$$

Making use of the solution $\phi(x,t)$ obtained in this way, we can calculate the I-V characteristics as follows: Demanding the balance between the average input power $\gamma\langle\langle\phi_t\rangle\rangle$ and power dissipation $\alpha\langle\langle\phi_t^2\rangle\rangle$, we have a relation; $\gamma=\alpha\langle\langle\phi_t^2\rangle\rangle/\langle\langle\phi_t\rangle\rangle$. Here, $\langle\langle\phi_t\rangle\rangle$ is the reduced voltage and calculated to be $\Phi_0\omega_J V_r/L$. Substituting the explicit form of $\phi(x,t)$ in eq.(3) into the power balance condition[11),13)], and subtracting the Ohmic part $\gamma_o=2\pi\alpha V_r/L$, we obtain the expression of the supercurrent step height γ_s of the form

$$\gamma_s=\frac{\alpha L}{\pi V_r}[\langle\langle\phi_t^{f2}\rangle\rangle-2(\frac{\pi V_r}{L})^2+\sum_{n,m}\{\frac{\omega_m^2}{8}(a_n^{m2}+b_n^{m2}) -2\omega_m b_n^m\langle\langle\cos(\omega_m t)\cos(k_n x)\phi_t^f\rangle\rangle\}] . \qquad (8)$$

The prediction of this results is illustrated in Fig.1 to show how the height of first ZFS behaves as a function of quality factor Z_1. The parameters, L=2,V_r=0.99,X=0 are used. This is interpreted as representing the effective temperature dependence of the step height, since an increase in Z_1 means a decrease in the temperature. The

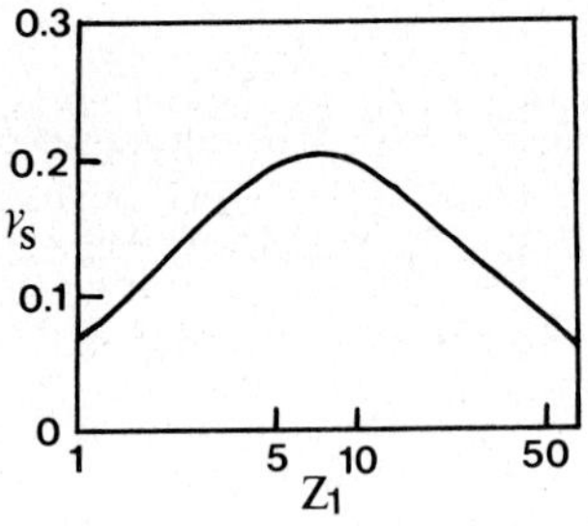

Fig.1: γ_s-Z_1 characteristics

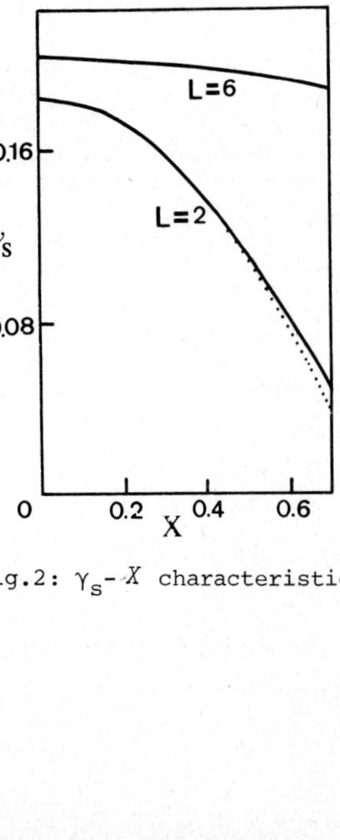

Fig.2: γ_s-X characteristics

main features of this γ_s-Z_1 characteristics are consistent with the observed one and also with the prediction of other theories[4),5)] based on the cavity mode picture. (In our previous treatment[13)] in which the effect of ϕ was completely neglected, the step height γ_s was a monotonically decreasing function of Z_1.) The dependence on the applied field X is shown in Fig.2 in case of reduced length L=2 and 6. The parameters,V_r=0.99,α=0.55 are used. The dotted line is from the approximation X^2. It is clearly seen that the step height decreases with increase in the field strength,as was observed[2)] and predicted in the cavity mode theory[6)]. The relatively weak effect of the field in case of longer junction with L=6 is naturally interpreted as the result of less frequent collisions with boundaries.

It should be commented that the present scheme of calculation may be applied to the analysis of the FS, if we replace ϕ^f in eq.(4) with proper expression (a candidate is $\phi^f-\omega t$) to construct perturbed half-fluxon state, suggested in the numerical works[9),10)]. This line of study is now in progress.

In conclusion,the modification of fluxon state arizing from the dissipation and change in the boundary conditions due to magnetic field is essential to understand the detailed properties of the nonlinear resonant structures in the current-voltage characteristics of a relatively long Josephson junction.

References

1) Y.S.Gou and R.I.Gayley, Phys.Rev.B10(1974) 4584.
2) G.Paterno and J.Nordman, J.Appl.Phys.49(1978) 2456.
3) J.T.Chen and D.N.Langenberg, in *Low Temperature Physics LT13*, ed.K.D.Timmerhaus, et al. (Plenum, New York, 1974), p.289.
4) K.Takanaka, Solid State Commun.29(1979)443.
5) Y.S.Gou and C.S.Chung, J.Low Temp.Phys.37 (1979)367.
6) K.Enpuku, K.Yoshida, and F.Irie, J.Appl.Phys. 52(1981)344.
7) T.A.Fulton and R.C.Dynes, Solid State Commun. 12(1973)57.
8) P.S.Lomdahl, O.H.Soerensen, and P.L. Christiansen, Phys.Rev.B25 (1982) 5737.
9) B.Dueholm, E.Joergensen,O.A.Levring, J.Mygind, N.F.Pedersen, M.R.Samuelsen, O.H.Olsen and M.Cirillo, Physica 108B(1981)1303.
10) S.N.Erne, A.Ferrigno and R.D.Parmentier, Phys.Rev.B27(1983)5440.
11) G.Costabile, R.D.Parmentier, B.Savo, D.W.Mclaughlin and A.C.Scott, Appl.Phys.Lett. 32(1978)587.
12) O.H.Olsen and M.R.Samuelsen, J.Appl.Phys.52 (1981)6247.
13) K.Watanabe and C.Ishii, Prog.Theor.Phys.66 (1981)749.

LT-17 (Contributed Papers)
U. Eckern, A. Schmid, W. Weber, H. Wühl (eds)

JOSEPHSON JUNCTION DYNAMICS BY MULTIMODE THEORY AND NUMERICAL SIMULATION

M.P. SOERENSEN and R.D. PARMENTIER*

Laboratory of Applied Mathematical Physics, The Technical University of Denmark, DK-2800 Lyngby, Denmark

A perturbed sine-Gordon model of an overlap-geometry Josephson tunnel junction of normalized length 2 is solved by multimode theory and by direct numerical simulation. The two approaches are in good agreement as regards the magnetic field dependence of both the height of the first zero-field step/second Fiske step and the frequency content and power levels of emitted microwave radiation.

1. INTRODUCTION

Enpuku et al. (1) and Chang et al. (2) have developed multimode extensions of the classic Kulik theory (3) of Fiske steps in the current-voltage characteristics of one-dimensional Josephson tunnel junctions. As noted by Chang et al. (2), such multimode theories put Fiske steps (FS) and zero-field steps (ZFS) on an equal theoretical footing. Enpuku et al. (1) moreover showed that an appropriate choice of parameter values in such a theory allows a reasonable fitting to experimental data on the magnetic field dependence of the step heights, at least for junctions having $L/\lambda_J \gtrsim 1$ (L = physical length; λ_J = Josephson penetration length).

Soerensen et. al. (4) have recently reported a comparison between experimental and numerical simulation results on the magnetic field dependence of the microwave radiation emitted by junctions of normalized length $L/\lambda_J \sim 2$, current-biased on ZFS1/FS2. The comparison is qualitatively and semi-quantitatively quite satisfactory. In this work Soerensen et al. (4) also demonstrated an excellent quantitative agreement between the multimode theory and the numerical simulation result as regards the height of ZFS1/FS2 in magnetic field.

In the present note we extend the comparison between multimode theory and direct simulation to the frequency content and the power levels of the emitted microwave radiation. The agreement that emerges here is also good.

2. MATHEMATICAL MODEL AND CALCULATION TECHNIQUES

The mathematical model studied is the perturbed sine-Gordon equation

$$\phi_{xx} - \phi_{tt} - \sin\phi = \alpha\phi_t - \beta\phi_{xxt} - \gamma, \quad (1a)$$

$$\phi_x(0,t) = \phi_x(l,t) = \eta. \quad (1b)$$

A derivation of this equation, details of the normalizations, and a description of the implicit finite-difference method used in the numerical simulation may be found in Ref.(5). Details of the computation procedures are given in Ref. (4). The following parameter values have been employed; shunt loss, $\alpha = 0.05$; surface resistance loss, $\beta = 0.02$; length $l = L/\lambda_J = 2$; bias current, $0 \lesssim \gamma \lesssim 1$; magnetic field, $0 \lesssim \eta \lesssim 6$.

The multimode theory assumes as an approximate solution of Eqs.(1) the ansatz

$$\phi(x,t) = \eta x + \omega t + \sum_{n=1}^{N} \phi_n(t)\cos(n\pi x/l). \quad (2)$$

Insertion of Eq.(2) into Eq.(1a) yields a set of N coupled second-order nonlinear ordinary differential equations for the functions ϕ_n (note that Eq.(1b) is satisfied automatically). An approximate solution by Krylov-Bogoliubov equivalent linearization (6) yields the identification $\phi_n(t) = A_n\cos(\omega_n t + \theta_n)$, in which $\omega_n \equiv n\pi/l$, with 2N coupled nonlinear functional equations for the constants A_n and θ_n. In the present study we have arbitrarily fixed N = 5. Moreover, we have considerd only the resonant case, i.e., in Eq.(2) we have set $\omega = \omega_2 = 2\pi/l$, which implies that the multimode theory is used to calculate the peak value of ZFS1/FS2.

3. RESULTS

fig. 1 shows the magnetic field dependence of the height of ZFS1/FS2, as reported in Ref.(4).

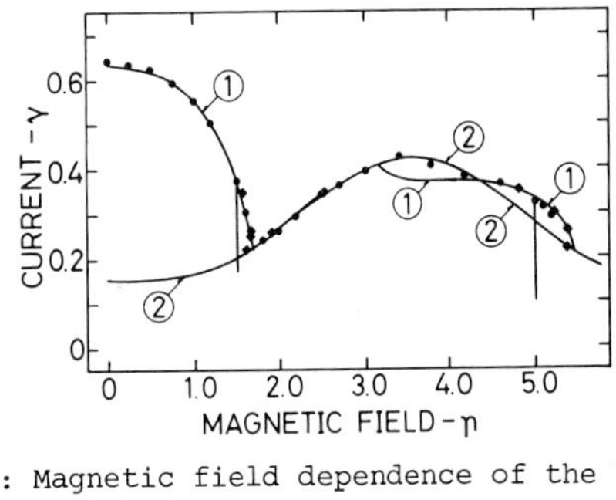

Fig. 1: Magnetic field dependence of the height of ZFS1/FS2.

*Permanent address: Dipartimento di Fisica dell' Universitá, I-84100 Salerno, Italy.

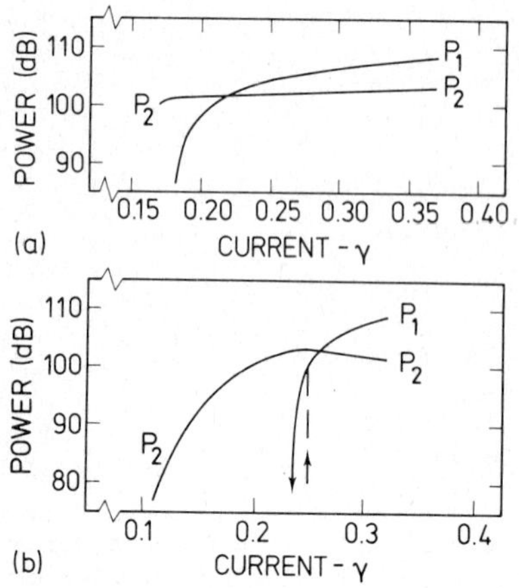

Fig. 2: Current dependence of power levels at (a) $\eta = 1.5$; (b) $\eta = 5.0$.

The circles and diamonds are the numerical simulation results. Circles have been calculated by increasing γ at constant η and diamonds by increasing or decreasing η at constant γ. The solid curve is the result of the multimode theory. The current in this theory is composed of an ohmic component plus a supercurrent component. Defining voltage as ϕ_t, the resonance frequency $\omega = \omega_2$ corresponds, from Eq.(2), to an average voltage $\langle\phi_t\rangle = \omega_2 = \pi$. Accordingly, the ohmic component is $\alpha\langle\phi_t\rangle = 0.157$, independent of η. The supercurrent component is the space- and time-average of $\sin\phi$, with ϕ given by Eq.(2), in which the A_n and the θ_n have been determined as described above. We underline that no parameter fitting is used in this procedure.

For $0 \leq \eta \leq 1.7$ and for $3.1 \leq \eta \leq 5.5$ the multimode theory predicts two different multimode configurations. The curves labelled 1 in Fig. 1 correspond to configurations in which the dominant component is at $\omega = \omega_1$, whereas $\omega = \omega_2$ is the dominant component for the curve labelled 2. As can be seen, the simulation points almost always correspond to the higher of the two curves.

Fig. 2 shows the power levels P_1 and P_2 at ω_1 and ω_2 obtained by numerical simulation along the two vertical solid lines indicated in Fig. 1. These have been calculated by applying a Fast Fourier Transform to the voltage waveform at the left (x=0) end of the junction. The levels have been normalized arbitrarily to $\phi_t^2 = 1.0 \times 10^{-10}$.

Both Fig. 2a and Fig. 2b show a crossover from an ω_2-dominant to an ω_1-dominant configuration with increasing γ. This point is at $\gamma = 0.22$ in Fig. 2a and at $\gamma = 0.26$ in Fig. 2b. In Fig. 1 the solid vertical lines cross the curve labelled 2 at $\gamma = 0.20$ for $\eta = 1.5$ and at $\gamma = 0.27$ for $\eta = 5.0$. Thus, the lower multimode curve in these regions gives a reasonable estimate of the numerically simulated power crossover points. However, we note that fine details, such as the hysteresis shown in the P_1 curve in Fig. 2b, are not reproduced by the multimode theory.

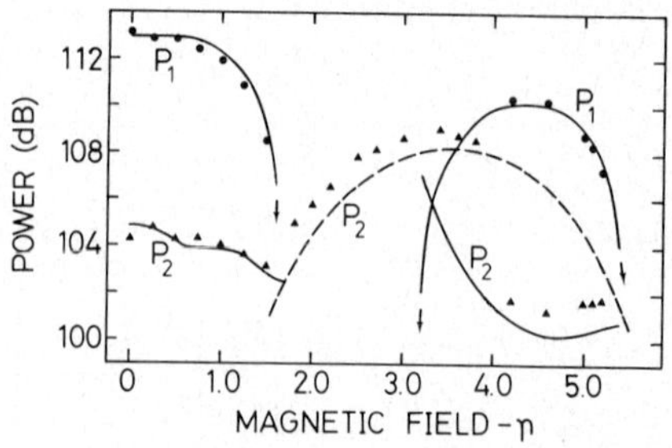

Fig. 3: Magnetic field dependence of power levels at the peak of ZFS1/FS2.

Fig. 3 shows a comparison of the P_1 and P_2 levels at the peak of ZFS1/FS2 in magnetic field as obtained by direct simulation (circles are P_1; triangles are P_2) with those obtained from the multimode theory (solid curves are P_1 and P_2 for the ω_1-dominant configuration; the dashed curve is P_2 for the ω_2-dominant configuration). The simulation points have been calculated as described above. To calculate power levels from the multimode theory we note, from Eq.(2), that the voltage at x = 0 is

$$\phi_t(0,t) = \omega - \sum_{n=1}^{N} \omega_n A_n \sin(\omega_n t + \theta_n). \qquad (3)$$

Accordingly, we define the power at $\omega = \omega_n$ as $P_n(\text{dB}) = 20 \log_{10}(\omega_n A_n/\sqrt{2}) + 100$. (the additive factor of 100 dB gives the same normalization as that used in the simulation). As can be seen the agreement between the two results is good.

4. CONCLUSIONS

Our results suggest that, at least for junctions with L/λ_J not too large, the multimode theory is a useful alternative to direct numerical simulation inasmuch as it gives reasonably reliable predictions of experimentally measurable quantities at a lower computational cost.

REFERENCES

(1) K. Enpuku, K. Yoshida and F. Irie, J. Appl. Phys. 52 (1981) 344.
(2) J.-J. Chang, J.T. Chen and M.R. Scheuerman, Phys. Rev. B25 (1982) 151.
(3) I.O. Kulik, Zh. Tekn. Fiz. 37 (1967) 157 [Sov. Phys.-Tech. Phys. 12 (1967) 111].
(4) M.P. Soerensen, R.D. Parmentier, P.L. Christiansen, O. Skovgaard, B. Dueholm, E. Joergensen, V.P. Koshelets, O.A. Levring, R. Monaco, J. Mygind, N.F. Pedersen and M.R. Samuelsen, Phys. Rev. B (in print)
(5) P.S. Lomdahl, O.H. Soerensen and P.L. Christiansen, Phys. Rev. B25 (1982) 5737.
(6) See, e.g., N. Minorsky, Nonlinear Oscillations (Van Nostrand, New York, 1962), Chap. 14.

LT-17 (Contributed Papers)
U. Eckern, A. Schmid, W. Weber, H. Wühl (eds)
© Elsevier Science Publishers B.V., 1984

A NEW METHOD TO DETERMINE JOSEPHSON JUNCTION PARAMETERS

B. Kofoed, J. Mygind, N. F. Pedersen, M. R. Samuelsen, D. Welner, and C. A. D. Winther

Physics Laboratory I, The Technical University of Denmark, DK-2800 Lyngby, Denmark

A new way to determine Josephson junction parameters is presented. The method is based on the properties of fluxons in long overlap junctions. As an example results for two niobium-lead junctions are given.

The determination of Josephson junction parameters in tunnel junctions is of fundamental interest not only for the understanding of circuit performance, but also as a characterization of junction quality. Here we demonstrate how the shunt damping parameter α ($\alpha = 1/\sqrt{\beta_c}$, where β_c is the McCumber parameter) together with the surface impedance damping parameter, β, may be determined by a simple dc experimental technique. The method is applicable for a wide range of temperatures and current densities. It may replace more ordinary methods that either have a more restricted range of applicability or are difficult experimentally. (As an example α is typically determined by a microwave plasma resonance experiment).

Our method employs the properties of fluxon propagation in long overlap Josephson junctions. Experimentally the dc-curves of the zero field steps are measured, and α and β are obtained by a simple fit to the theoretical IV curve to be discussed below.

For a junction with normalized length and width such that $\ell = L/\lambda_J >> 1$ and $w = W/\lambda_J << 1$ the fluxon propagation with uniform current distribution is described by the sine-Gordon equation[1].

$$\beta\phi_{xxt} + \phi_{xx} - \phi_{tt} - \alpha\phi_t - \sin\phi = \eta \qquad (1)$$

with boundary conditions $\phi_x(0) = \phi_x(\ell) = 0$. Here x is measured in units of the Josephson penetration depth λ_J and t in units of the inverse maximum plasma frequency, ω_o^{-1}. η is the dc-bias current, I_{dc}, normalized to the maximum zero voltage current, $I_o = J\cdot W\cdot L$. $\alpha = g(\hbar/2eJC)^{\frac{1}{2}}$ where g, J, and C are the shunt conductance, the current density, and the capacitance per unit area, respectively. β is the ratio of the imaginary and real part of the film surface impedance.

Equation 1 has solutions corresponding to a fluxon (anti-fluxon) propagating back and forth with velocity u (normalized to the Swihart velocity $\bar{c} = \lambda_J\cdot\omega_o$).[1] Within perturbation theory u is determined as a balance between the energy input represented by η and the losses represented by α and β. The result may be written[1]

$$\eta = \frac{4u}{\pi\sqrt{1-u^2}}\left[\alpha + \frac{\beta}{3(1-u^2)}\right] \qquad (2)$$

The propagating fluxon gives rise to the so-called zero field step (ZFS) with an IV-curve given by

$$I = I_o\cdot\eta, \qquad V = \left(\frac{h}{2e}\frac{\bar{c}}{L}\right)u \qquad (3)$$

where η and u are related through Eq. 2.

In the fit the measured quantities are I(V) of the ZFS, L, and I_o. If a ground plane is not used I_o may be depressed due to current spikes at the ends. The value of I_o may then be estimated from the current increase at the gap voltage or from measurements on a small junction. The quantities obtained by a least squares iteration (on a desk-computer) are α, β, and $\bar{c}$. The goodness-of-fit measure which is the estimated standard deviation of the measured points relative to the points calculated by Eq. 2 is always found to be equal to or less than the standard deviation of the measured points. For the very low current values there is a systematic deviation to be discussed later. The confidence limits on α, β, and $\bar{c}$ obtained by a Monte Carlo experiment are better than 10%, 5%, and 1%. Further the fitted values of $\bar{c}$ agree with independent measurements.

In our investigations we have concentrated on Nb-Nboxide-Pb junctions. Such junctions are of technological importance and are known to suffer from fabrication problems (e.g.leakage current) which may affect α and β. A temperature dependence of α and β is expected even in ideal junctions since the losses depend on the number of excited quasiparticles. In ideal junctions at a given temperature we expect the shunt losses (α) to scale with $\sqrt{J}$ and the series losses (β) to be independent of J. For nearly ideal high current density junctions ($\sim$500 A/cm^2) it was found experimentally[2] that α-losses dominate, whereas for low-current density junctions (1-20 A/cm^2) β-losses were most important[3]. In nonideal junctions the excess current may increase both α and β.

Our results for two non-ideal Nb-Nboxide-Pb

junctions with intermediate current densities are shown in Fig. 1. Junction I had $J \sim 24$ A/cm^2, $L \times W = 418 \times 44$ μm^2, $\ell \sim 4$ and some excess

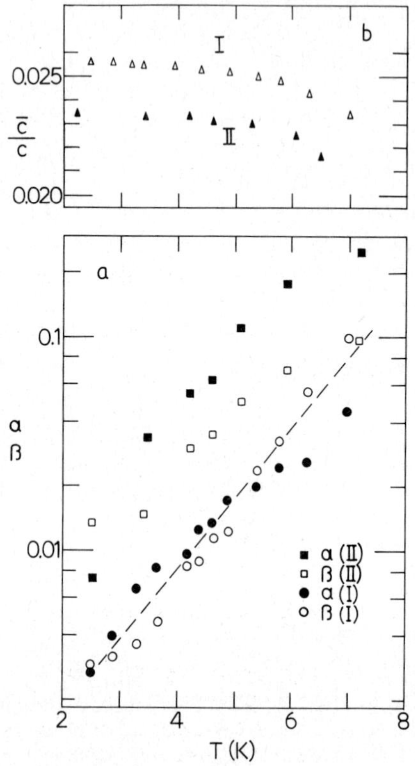

FIGURE 1
a) α and β for junctions I and II. b) $\bar{c}/c$ for junctions I and II.

current below the gap. Junction II had $J \sim 185$ A/cm^2, $L \times W = 17 \times 372$ μm^2, $\ell \sim 12$, and larger excess current than junction I. For junction I we find α and β to be comparable. For junction II we find that the α-losses dominate the β-losses in agreement with the previous discussion. When comparing junction II with junction I we find that α-losses have increased although somewhat more than expected[4]. Also the β-losses have increased. Both these effects could be due to a larger damage of the Nb-surface of junction II during fabrication (larger excess current)[5]. The dashed curve in Fig. 1 shows results for β for a low current density junction[3]. The agreement with junction I is very good.

We found that a satisfactory fit between the experiment and the results of the perturbation calculation Eqs. 2, 3 could be obtained. However, for the smallest values of η a systematic deviation (which does not affect the obtained values of α and β appreciably) was always observed. The voltages were measured too small due to a collision delay not included in Eqs.2,3.

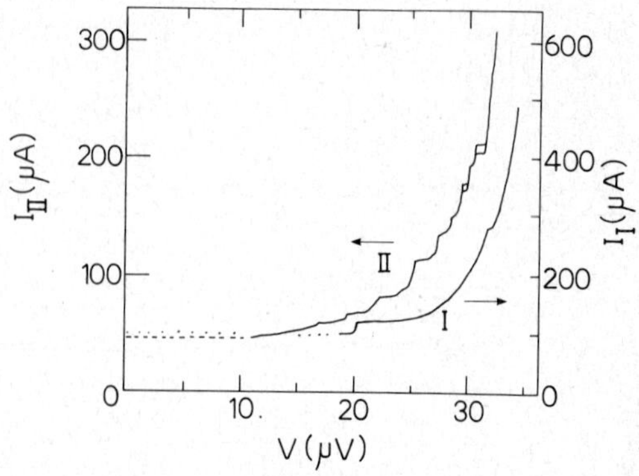

FIGURE 2
The lowest part ($\zeta \lesssim 0.05$) of the n = 1 zero field step for junctions I and II.

In addition, most pronounced at low temperatures, a fine structure as shown in Fig. 2 was observed. In[6] this fine structure was identified as fractional cavity modes determined by $\bar{c}/L$. We find that the number and appearance of these steps depend on the current density and not on $\bar{c}/L$ (which is almost identical for junctions I and II), as demonstrated in Fig. 2. We believe that the structure is due to fluxon interaction with $\kappa \simeq 0$ plasma oscillations, i.e. it should appear at voltages given by $n\,\omega_s = n \cdot 2eV/2h = \omega_o$.

REFERENCES

(1) D.W. Mc Laughlin and A.C. Scott, Phys. Rev. A18, 1672 (1978).
(2) N.F. Pedersen and D. Welner, Phys. Rev. B, (in press).
(3) A.M. Cucolo, S. Pace, R. Vaglio, V. Laquaniti, and G. Marullo, IEEE trans.mag. MAG-17, 812 (1981).
(4) For a quantitative comparison it should be noted that for junction II the critical current was depressed by 35%. This is not included in Fig. 1. Correcting for that reduces α and β for junction II with 35% but leaves the general conclusions unchanged.
(5) A. Matsuda, T. Inamura, and H. Yoshiko, J. Appl. Phys. 51, 4310 (1980).
(6) M. R. Scheuermann, T.V. Rajeevakumar, J.-J. Chang, and J.T. Chen, Physica B107, 543 (1981).

AUTHOR INDEX TO PARTS I AND II – CONTRIBUTED PAPERS